W9-AYX-402

ENCYCLOPEDIA OF MATHEMATICS AND ITS APPLICATIONS

Volterra Integral and Functional Equations

G. GRIPENBERG,
University of Helsinki

S.-O. LONDEN & O. STAFFANS
Helsinki University of Technology

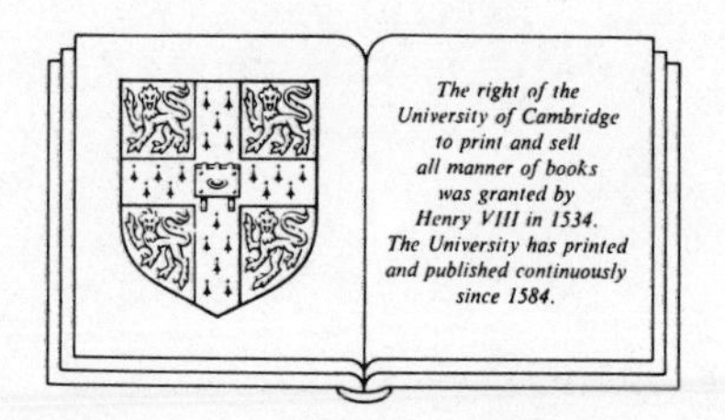
The right of the
University of Cambridge
to print and sell
all manner of books
was granted by
Henry VIII in 1534.
The University has printed
and published continuously
since 1584.

CAMBRIDGE UNIVERSITY PRESS

Cambridge

New York Port Chester

Melbourne Sydney

Published by the Press Syndicate of the University of Cambridge
The Pitt Building, Trumpington Street, Cambridge CB2 1RP
40 West 20th Street, New York, NY 10011, USA
10 Stamford Road, Oakleigh, Melbourne 3166, Australia

First published 1990

Printed in Great Britain at the University Press, Cambridge

British Library cataloguing in publication data available

Library of Congress cataloguing in publication data available

ISBN 0 521 37289 5

ENCYCLOPEDIA OF MATHEMATICS AND ITS APPLICATIONS

EDITED BY G.-C. ROTA

Volume 34

Volterra Integral and Functional Equations

ENCYCLOPEDIA OF MATHEMATICS AND ITS APPLICATIONS

1 Luis A. Santalo Integral Geometric Probability
2 George E. Andrews The Theory of Partitions
3 Robert J. McEliece The Theory of Information and Coding: A Mathematical Framework for Communication
4 Willard Miller, Jr. Symmetry and Separation of Variables
5 David Ruelle Thermodynamic Formalism: the Mathematical Structures of Classical Equilibrium Statistical Mechanics
6 Henryk Minc Permanents
7 Fred S. Roberts Measurement Theory with Applications to Decisionmaking, Utility, and the Social Services
8 L. C. Biedenharn and J. D. Louck Angular Momentum in Quantum Physics: Theory and Application
9 L. C. Biedenharn and J. D. Louck The Racah-Wigner Algebra in Quantum Theory
10 W. Dollard and Charles N. Friedman Product Integration with Application to Differential Equations
11 William B. Jones and W. J. Thron Continued Fractions: Analytic Theory and Applications
12 Nathaniel F. G. Martin and James W. England Mathematical Theory of Entropy
13 George A. Baker, Jr. and Peter R. Graves-Morris Padé Approximants, Part I: Basic Theory
14 George A. Baker, Jr. and Peter R. Graves-Morris Padé Approximants, Part II: Extensions and Applications
15 E. C. Beltrametti and G. Cassinelli The Logic of Quantum Mechanics
16 G. D. James and A. Kerber The Representation Theory of the Symmetric Group
17 M. Lothaire Combinatorics on Words
18 H. O. Fattorini The Cauchy Problem
19 G. G. Lorentz, K. Jetter, and S. D. Riemenschneider Birkhoff Interpolation
20 Rudolf Lidl and Harald Niederreiter Finite Fields
21 William T. Tutte Graph Theory
22 Julio R. Bastida Field Extensions and Galois Theory
23 John R. Cannon The One-Dimensional Heat Equation
24 S. Wagon The Banach-Tarski Paradox
25 A. Salomaa Computation and Automata
26 N. White (ed) Theory of Matroids
27 N. Bingham, C. Goldie & J. L. Teugels Regular Variation
28 P. Petrushev, & P. Popov Rational Approximation of Real Variables
29 N. White (ed) Combinatorial Geometries
30 M. Pohst and H. Zassenhaus Algorithmic Algebraic Number Theory
31 J. Aczel & J. D. Hombres Functional Equations containing Several Variables
32 M. Kuczma, B. Chozewski & R. Ger Iterative Functional Equations
34 G. Gripenberg, S.-O. Londen and O. Staffans Volterra Integral and Functional Equations
35 G. Gasper & M. Rahman Basic Hypergeometric Series

CONTENTS

Chapter 9
Linear Nonconvolution Equations

Chapter 10
Linear Nonconvolution Equations with Measure Kernels

Part II: General Nonlinear Theory

Chapter 11
Perturbed Linear Equations

Chapter 12
Existence of Solutions of Nonlinear Equations

Part III: Frequency Domain and Monotonicity Techniques

Chapter 20
Monotonicity Methods

PREFACE

During the past 25 years the theories of Volterra integral equations, Volterra integrodifferential equations, and functional differential equations have undergone rapid developments. What began as a few scattered papers on specific equations, and on particular applied problems, has grown to branches of applied analysis of considerable size, having rich structures of their own. The growth has been strongly promoted by the large number of applications that these theories have found in physics, engineering, and biology. Our understanding of those comparatively simple, usually linear, problems that were present from the beginning has increased significantly. In addition, knowledge has been obtained about more general and about more complex equations. We observe, for example, that the main part of the asymptotic theory for nonlinear Volterra integral and integrodifferential equations is of fairly recent origin. The same observation can be made about equations involving parameters or additional variables, and more generally about equations in infinite-dimensional spaces.

It is very common to make a fairly sharp distinction between Volterra integral and integrodifferential equations on the one hand, and functional differential equations on the other hand. In many respects this distinction is artificial, being due more to the different backgrounds of the researchers than to any inherent differences. In the sequel we refer to these groups of equations as 'Volterra integral and functional equations', or 'Volterra equations' for short, and many of the results that we give apply equally well to functional differential equations and to Volterra integral or integrodifferential equations. However, our general attitude to these equations resembles the attitude that has been prevailing for the last few years among people working in Volterra integral and integrodifferential equations. In particular, we pay no attention to questions that are specific for finite delay problems.

Although our knowledge of Volterra equations is swiftly growing, the point may well be made that the study of Volterra equations in finite-dimensional spaces has reached a certain maturity, which both motivates and makes possible a coherent presentation. This book testifies to the fact that we share this opinion. Certainly, we do not believe that Volterra equations in $\mathbf{R}^n$ or $\mathbf{C}^n$ constitute a closed chapter. It suffices to consider the scarcity of results on periodic solutions of nonlinear equations, or the so far unsatisfactory resolvent theory for nonintegrable kernels, to realize that the field is wide open for progress. But we do believe that the available research results make up a certain whole which may profitably be presented in the form of a book. In addition, this will make articles scattered throughout the literature readily available to the working mathematician.

The class of ordinary differential equations is subsumed under the class of Volterra equations. However, the importance of this fact should not be exaggerated. The present text makes it clear that the theory of Volterra equations exhibits a rich variety of features not present in the theory of ordinary differential equations. The converse statement also holds: some of the more specific results available for ordinary differential equations cannot possibly be extended to Volterra equations.

Chapter 1 gives an introduction to Volterra equations and a general overview of the contents. The remaining part of the book comprises 19 chapters and is divided into three parts. In Part I (Chapters 2–10) we consider the linear theory, and in Part II (Chapters 11–15) we deal with quasilinear equations and existence problems for general nonlinear equations, and give some general asymptotic results. Part III (Chapters 16–20) is devoted to frequency domain methods in the study of nonlinear equations.

Obviously, to some extent the contents and the general approach to the subject reflect the research interests of the authors. This is particularly true of Chapters 15–20. In fact, a presentation of the results in Part III provided the initial impetus for the writing of this book. It was soon realized, however, that a treatment of the linear theory, together with the quasilinear and existence chapters of Part II, was equally important. The project thus grew into an exposé of both linear and nonlinear Volterra equations in finite-dimensional spaces. Within these limits, we have attempted to incorporate as much as possible of what we feel is essential to the understanding of Volterra equations. Where limitation of space has prohibited the inclusion of the complete proof, we have tried to give at least an outline. Additional results are frequently touched upon in the comment sections at the end of each chapter.

The major part of the text is made up of results that have so far at most been available in research papers. In particular, this is true for Chapters 4–10, 12, and 15–20. A number of the results given here have not appeared in print before, and many previously existing results have been modified and improved.

At an early stage it was decided not to include a number of worthy topics in Volterra equations: equations in infinite-dimensional spaces, stochastic equations, geometric and degree theory, attractors, stable and unstable manifolds, eigenfunction expansions, bifurcation theory, control and optimization problems, oscillation results, and numerical methods. Very little is said about neutral equations. The primary motivation for these omissions is simply a question of volume. There is more than enough in the basic theory of Volterra equations in finite-dimensional spaces to fill a good-sized book. A particular motivation was that many of these fields, for example abstract Volterra equations, are still in a state of flux, with new basic results continuously emerging.

With a few exceptions, the entire text analyses n-dimensional rather than scalar equations. This, of course, gives the results a greater generality and a wider applicability, and it facilitates generalizations to infinite-dimensional spaces. In Parts I and II this setting is quite natural, and it does not overly complicate the proofs. The n-dimensional results in Chapters 16–19 have appeared earlier in the literature almost exclusively in a scalar setting. It does, however, appear natural and consistent to present the nonlinear asymptotic theory in the same general n-dimensional setting as we use in the earlier chapters. The theory of scalar kernels of positive type can be extended to matrix-valued kernels without too much additional complexity, and this is done in Chapter 16. In some of the theorems in Chapters 17–19 one has to make the significant additional assumption that the nonlinearity is the gradient of a scalar-valued function; when $n = 1$ this assumption is automatically satisfied.

A substantial part of the text concerns equations with measure-valued kernels. In particular, this is true for Chapters 3, 4, 10, 16, and for parts of Chapters 17–19. Our results may therefore be applied to delay and differential-delay equations. The approach that we take to functional differential equations differs, at least in appearance, from the common one. The main difference is that we stress resolvent theory, and avoid semigroup techniques. In Chapter 8 a semigroup theory is presented briefly, but the topic is not pursued further. One consequence of our approach is that there is no need for a separate treatment of functional differential equations with finite delay. Hence, we do not discuss typical finite delay questions, such as eigenfunction expansions, solutions that vanish after a finite time, and backward continuation of solutions. Instead, we attach primary importance to the integrability properties of the kernels and the resolvents. Also, we put the emphasis on a study of the Laplace transform of the kernel instead of on a study of the spectrum of the generator of the semigroup (which, however, in reality amounts to the same thing).

Every effort has been made to achieve a reasonably self-contained presentation. We have been forced to assume some basic knowledge of analysis, and do occasionally fall back on results in Rudin [2] and [3], Hille and Phillips [1], and Hewitt and Stromberg [1]. Even so, a significant part of the text has been devoted to analysis results which are difficult to find in the literature, but which are needed for the development of the theory. In particular, we present a number of theorems on vector measures, on convolutions, and on Laplace transforms. These results have been separated from the main body of the text, and several chapters contain appendices describing results of a technical nature.

Exercises and some comments on further references, historical developments, etc., are included at the end of each chapter. We warn the reader that the exercises cover a wide range of difficulty. Some of the problems are easy applications of the text, others require a much deeper understanding of the topic.

One of our intentions when writing this book was to stimulate further research. Consequently, we have tried to make the text reflect as much as possible of the present state of knowledge in the area of the theory of Volterra equations that we cover. In the case where a result included in the text is not the best available, we have tried to incorporate a statement to this effect, together with further references in the corresponding comment section.

We sustain a hope that the text will bring about more interaction between research workers in the two fields of functional differential equations on the one hand and Volterra integral and integrodifferential equations on the other hand. We believe that both areas, which have been unduly separated from each other, would profit from such an interaction. In addition, we hope that an even larger number of concrete applications would find their way into the present theory of Volterra equations. Without the feedback from and the incitement supplied by such applications in physics, engineering, and biology, the future of the field would look much bleaker.

This text was composed and typeset at the Institute of Mathematics of the Helsinki University of Technology. It is a pleasure for us to acknowledge the excellent working conditions and facilities at our Institute that made this voluminous venture possible. At various stages of our efforts the Academy of Finland relieved some of us from teaching and administrative duties. Needless to say, we appreciate this support. Visits to Madison, Blacksburg, Graz, and Carnegie-Mellon also contributed to the final product.

Over the years John A. Nohel has actively encouraged us in our endeavour. We express our gratitude for his continuous and enthusiastic support. Our colleagues at Virginia Tech, Kenneth B. Hannsgen and Robert L. Wheeler, have given us much time and advice. In particular we wish to thank them for their fruitful criticism of earlier versions of the manuscript. Our thanks also go to Ismo Sedig for comments on Chapter 15.

We are very pleased to see our work appear in the series 'Encyclopedia of Mathematics' of the Cambridge University Press and the pleasant cooperation with David Tranah is gratefully acknowledged.

Our wives have undauntedly endured several years of book-writing with subsequent neglect of family duties and delights. All things considered, without their persistent support and teasing comments this book would never have been finished.

Gustaf Gripenberg
Stig–Olof Londen
Olof Staffans

References

E. Hewitt and K. Stromberg

1. *Real and Abstract Analysis*, Springer-Verlag, Berlin, 1965.

E. Hille and R. S. Phillips

1. *Functional Analysis and Semi-Groups*, Amer. Math. Soc., Providence, R. I., 1957.

W. Rudin

2. *Functional Analysis*, McGraw–Hill, New York, 1973.
3. *Real and Complex Analysis,* 3rd *ed.*, McGraw–Hill, New York, 1986.

LIST OF SYMBOLS

Basic Sets and Spaces

$\mathbf{R}$	$(-\infty, \infty)$.
$\mathbf{R}^+$	$[0, \infty)$.
$\mathbf{R}^-$	$(-\infty, 0]$.
$\mathbf{C}$	The complex plane.
$\mathbf{E}$	Either $\mathbf{R}$ or $\mathbf{C}$.
$\mathbf{Z}$	The set of integers.
$\mathbf{N}$	The set of nonnegative integers.
$\mathbf{R}^n$	The set of n-dimensional column vectors with real entries; various norms and inner products.
$\mathbf{C}^n$	The set of n-dimensional column vectors with complex entries; various norms and inner products.
$\mathbf{R}^{n\times m}$	The set of $(n \times m)$-dimensional matrices with real entries; a norm adapted to the corresponding norms in $\mathbf{R}^m$ and $\mathbf{R}^n$.
$\mathbf{C}^{n\times m}$	The set of $(n \times m)$-dimensional matrices with complex entries; a norm adapted to the corresponding norms in $\mathbf{C}^m$ and $\mathbf{C}^n$.
$\mathcal{T}_S$	The real line $\mathbf{R}$ where the points $t+mS$, $m \in \mathbf{Z}$, are identified.
$\emptyset$	The empty set.

Function Spaces

$V(J;Q)$	Functions of type V ($= L^p$, BC etc.) with domain J and range contained in Q ($= \mathbf{R}^n$, $\mathbf{C}^{n\times n}$ etc.).
$V(J;\eta;Q)$	Functions of type V ($= L^p$, BC etc.) with domain J, weight function η, and range contained in Q.
$V_{\text{loc}}(J;Q)$	Functions of type V with domain J that belong to $V(K;Q)$ for every compact subset K of J. If $V(K;Q)$ is a Banach space, then $V_{\text{loc}}(J;Q)$ is given the topology induced by the

metric $d(\psi,\phi)=\sum_{j=1}^{\infty}\{2^{-j}\|\psi-\phi\|_{V(K_j)}/(\|\psi-\phi\|_{V(K_j)}+1)\}$ where $\bigcup_{j=1}^{\infty}K_j=J$.

$V(\mathcal{T}_S;Q)$ — S-periodic functions whose restrictions to each finite interval are of type V with range in Q; $\|\phi\|_{V(\mathcal{T}_S)}=\|\phi\|_{V([0,S))}$.

$U \dotplus V$ — The space of functions of the type $\{\, f+g \mid f\in U,\ g\in V\,\}$, where U and V are some function spaces with $U\cap V=\{0\}$.

AC — Absolutely continuous functions.

AP — Almost-periodic functions; sup-norm.

B^{∞} — Borel measurable, bounded functions; sup-norm.

B_0^{∞} — Functions in B^{∞} that tend to zero at infinity.

B_ℓ^{∞} — Functions in B^{∞} with limits at infinity (not necessarily the same limit at $-\infty$ and $+\infty$ if the domain is $\mathbf{R}$).

$B_{\mathrm{loc}}^{\infty}$ — Functions locally in B^{∞}, i.e., their restrictions to compact subsets of the domain belong to B^{∞}; see $V_{\mathrm{loc}}(J,Q)$ above.

$BBV(J)$ — Functions in $BV_{\mathrm{loc}}(J)$, where $J=\mathbf{R}$ or $\mathbf{R}^+$, satisfying $\|\phi\|_{BBV(J)}=\sup_{[s,s+1]\subset J}\|\phi\|_{BV([s,s+1])}<\infty$.

$BBV_0(J)$ — Functions in $BBV(J)$ such that $\|\phi\|_{BV([s,s+1])}$ tends to zero at infinity. Here $J=\mathbf{R}$ or $\mathbf{R}^+$.

BC — Bounded continuous functions; sup-norm.

BC_0 — Functions in BC that tend to zero at infinity.

BC_ℓ — Functions in BC with limits at infinity (not necessarily the same limit at $-\infty$ and $+\infty$ if the domain is $\mathbf{R}$).

$BL^p(J)$, $1\le p\le\infty$: Functions in $L^p_{\mathrm{loc}}(J)$ with a finite norm $\|\phi\|_{BL^p(J)}=\sup_{(s,s+1)\subset J}\|\phi\|_{L^p(s,s+1)}<\infty$. Here $J=\mathbf{R}$ or $\mathbf{R}^+$.

$BL_0^p(J)$, $1\le p\le\infty$: Functions in $BL^p(J)$ such that $\|\phi\|_{L^p(s,s+1)}$ tends to zero at infinity. Here $J=\mathbf{R}$ or $\mathbf{R}^+$.

BUC — Bounded uniformly continuous functions; sup-norm.

BUC^1 — Bounded uniformly continuous functions with bounded uniformly continuous derivative.

BV — Functions of bounded variation; the norm is the sum of the total variation and the sup-norm.

BV_{loc} — Functions locally of bounded variation, i.e., their restrictions to compact subsets of the domain are of bounded variation, see $V_{\mathrm{loc}}(J,Q)$ above.

C — Continuous functions; sup-norm on each compact set. The same space as BC_{loc}.

C_c — Continuous functions with compact support.

C_c^{∞} — Infinitely many times differentiable functions with compact support.

L^p, $1 \le p < \infty$: Measurable functions with finite norm $\left\{\int |\phi(t)|^p \, dt\right\}^{1/p}$.

L^∞ Measurable functions with finite norm $\operatorname{ess\,sup}|\phi(t)|$.

L_0^∞ Functions in L^∞ with an essential limit 0 at infinity.

L_ℓ^∞ Functions in L^∞ with essential limits at infinity (not necessarily the same limit at $-\infty$ and $+\infty$ if the domain is $\mathbf{R}$).

L_{loc}^p, $1 \le p \le \infty$: Functions locally in L^p, i.e., their restrictions to compact subsets of the domain belong to L^p; see $V_{\text{loc}}(J, Q)$ above.

$W^{1,p}$, $1 \le p \le \infty$: Locally absolutely continuous functions that together with their derivative belong to L^p; norm $\|\phi\|_{W^{1,p}(J)} = \|\phi\|_{L^p(J)} + \|\phi'\|_{L^p(J)}$.

$\tilde{L}^1$, $\hat{L}^1$, $\widehat{M}$ Functions that are Fourier or Laplace transforms of L^1-functions or finite measures; see Sections **6**.3 and **7**.3.

$\mathcal{D}$ Infinitely many times differentiable functions with compact support.

$\mathcal{D}'$ Distributions; the dual of $\mathcal{D}$.

$\mathcal{S}$ Infinitely many times differentiable functions that decay rapidly at infinity.

$\mathcal{S}'$ Tempered distributions; the dual of $\mathcal{S}$.

$\mathcal{V}$ Volterra kernels of various types; see Chapters **9** and **10**.

$\mathcal{F}$ Fredholm kernels of various types; see Chapter **9**.

$\mathcal{K}$ This letter stands for one of the letters $\mathcal{F}$ or $\mathcal{V}$.

$[V; J]$ Kernels of type V on J.

Measure Spaces

$M(J; Q)$ Finite measures on J with range contained in Q and total variation norm; see Section **3**.2.

$M(J; \eta; Q)$ Measures on J with range contained in Q and total variation norm weighted by the function η; see Section **4**.3.

$M_{\text{loc}}(J; Q)$ Set functions on J that belong to $M(K; Q)$ for every compact subset K of J.

PT Kernels of positive type; see Chapter **16**.

PT_{aco} Kernels of anti-coercive type; see Section **16**.5.

$\mathcal{M}$ Nonconvolution measure kernels of various types; see Chapter **10**.

Transforms

$\hat{\mu}$ Laplace–Stieltjes transform of the measure μ, $\hat{\mu}(z) = \int_{\mathbf{R}^+} \mathrm{e}^{-zs}\mu(\mathrm{d}s)$ if μ is measure on $\mathbf{R}^+$, and $\hat{\mu}(z) = \int_{\mathbf{R}} \mathrm{e}^{-zs}\mu(\mathrm{d}s)$ if μ is a measure on $\mathbf{R}$.

$\tilde{\mu}$ Fourier–Stieltjes transform of the measure μ, $\tilde{\mu}(\omega) = \hat{\mu}(\mathrm{i}\omega)$.

$\hat{a}$ Laplace transform of the function a, $\hat{a}(z) = \int_0^\infty \mathrm{e}^{-zs}a(s)\,\mathrm{d}s$ if a is defined on $\mathbf{R}^+$, and $\hat{a}(z) = \int_{-\infty}^\infty \mathrm{e}^{-zs}a(s)\,\mathrm{d}s$ if a is defined on $\mathbf{R}$.

$\tilde{a}$ Fourier transform of the function a, $\tilde{a}(\omega) = \hat{a}(\mathrm{i}\omega)$.

Operators and various other Symbols

$\langle u, v\rangle$ Inner product of u and v.

$|v|$ Absolute value of v if v is a scalar; norm of v if v is a vector or a matrix.

$|\mu|(E)$ Total variation measure, $\sup\sum_{j=1}^N|\mu(E_j)|$, where $\{E_j\}_{j=1}^N$ is a partition of E.

$|v|_+$ $\max\{0, v\}$. Here v is real.

$|v|_-$ $\max\{0, -v\}$. Here v is real.

$\|\phi\|_V$ Norm of ϕ in the space V. Here ϕ is a function or a measure.

$\|\phi\|_{\sup(J)}$ Supremum of $|\phi|$ on J.

$\|\phi\|_{\mathrm{var}(J)}$ Total variation of ϕ on J.

$|||k|||_{L^p(J)}$ $\sup_{\substack{\|g\|_{L^q(J)}\le 1\\ \|f\|_{L^p(J)}\le 1}} \int_J\int_J|g(t)k(t,s)f(s)|\,\mathrm{d}s\,\mathrm{d}t$, $1/p + 1/q = 1$; see Section **9**.2.

$|||k|||_{L^{p,q}(J)}$ $\sup_{\substack{\|f\|_{L^p(J)}\le 1\\ \|g\|_{L^{q'}(J)}\le 1}} \int_J\int_J|g(t)k(t,s)f(s)|\,\mathrm{d}s\,\mathrm{d}t$, $1/q + 1/q' = 1$; see Section **9**.2.

$|||k|||_{B^\infty(J)}$ $\sup_{t\in J}\int_J|k(t,s)|\,\mathrm{d}s$; see Section **9**.5.

$|||\kappa|||_{B^\infty(J)}$ $\sup_{t\in J}|\kappa|(t, J)$; see Section **10**.2.

$|||\kappa|||_{[B^\infty;L^1](J)}$: $\int_J|\kappa|(t,J)\,\mathrm{d}t$; see Section **10**.3.

$|||k|||_{[M;B^\infty](J)}$: $\sup_{t,s\in J}|k(t,s)|$; see Section **10**.3.

$\Re$ The real part, or the symmetric part of a matrix.

$\Im$ The imaginary part, or the anti-symmetric part of a matrix.

$\sigma(\varphi)$ The spectrum of the function φ; see Section **15**.4.

$\Gamma(\varphi)$ The limit set of the function φ; see Section **15**.2.

τ_h Translation operator: $(\tau_h\varphi)(t) = \varphi(t+h)$.

$*$	Convolution (see Sections **2**.2, **3**.2 and **4**.1).
$\star$	The convolution-like product defined in Sections **9**.2, **10**.2 and **10**.3.
$\succeq$	$A \succeq 0$ if $\langle y, Ay \rangle \geq 0$ for all $y \in \mathbf{C}^n$, where $\langle \cdot, \cdot \rangle$ is some given inner product; $u \succeq v$ if $u - v \in K$ where K is a given cone.
$\succ$	$A \succ 0$ if $\langle y, Ay \rangle > 0$ for all $y \in \mathbf{C}^n \setminus \{0\}$; $u \succ v$ if $u - v$ belongs to the interior of a given cone K.
$\overline{z}$	Complex conjugate of z.
$\overline{\Omega}$	Closure of the set Ω.
$\alpha\{\Omega\}$	In Chapter **12**, the measure of noncompactness of the set Ω.
$\operatorname{conv}\{\Omega\}$	Convex hull of the set Ω.
$\operatorname{diam}\{\Omega\}$	Diameter of the set Ω; $\operatorname{diam}\{\Omega\} = \sup\{ \|x - y\| \mid x, y \in \Omega \}$.
A^*	The adjoint of the matrix A with respect to some given inner product.
$\det[A]$	Determinant of the matrix A.
$\operatorname{diag}[\alpha_1, \ldots, \alpha_n]$:	Diagonal matrix with entries $\alpha_1, \ldots, \alpha_n$.
$\ker[A]$	Kernel (nullspace) of the matrix A.
$\operatorname{range}[A]$	Range of the matrix A.
$\operatorname{span}\{\Omega\}$	Linear span of the set Ω of vectors.
ln	Logarithm with base e.
lim	Limit.
lim sup	Superior limit.
lim inf	Inferior limit.
inf	Infimum.
sup	Supremum.
ess inf	Essential infimum.
ess lim	Essential limit.
ess sup	Essential supremum.
$[\nu, \phi]$	The value of the distribution ν evaluated at the test function ϕ.
$[\![\mu, \phi]\!]$	$\Re \int_{\mathbf{R}} \langle \phi(t), (\mu * \phi)(t) \rangle \, dt$; see Section **16**.2.
δ_t	Unit point mass at t, $\delta = \delta_0$.
μ_a	Absolutely continuous part of the measure μ.
μ_d	Discrete part of the measure μ.
μ_s	Singular part of the measure μ.
χ_E	Characteristic function of the set E.
I	Identity matrix.

$Q(\phi, \mu, T)$	$\int_0^T \langle \phi(t), \int_{[0,t]} \mu(\mathrm{d}s)\phi(t-s) \rangle \, \mathrm{d}t$, used in Chapters **17**–**19**.
$T_{\max}$	In Chapter **12**, the upper limit for the interval of existence of a noncontinuable solution.
x_M	In Section **13**.4, maximal solution.
x_m	In Section **13**.4, minimal solution.

1

Introduction and Overview

We give a brief introduction to Volterra equations, review some simple mathematical models involving such equations, and give an overview of the contents of the book.

1. Introduction

Let $T(\xi, t)$ be the temperature in a semi-infinite ($\xi \geq 0$) one-dimensional bar. Assume that the bar loses energy at a rate proportional to $T(0, t)$ at the point $\xi = 0$, and that an external source generates heat proportional to the function $q(t)$ at this end of the bar. If the bar is insulated at all other parts, and has temperature zero at time zero, then T is a solution of the following initial-boundary value problem (where some proportionality constants have been set equal to 1):

$$\left.\begin{aligned} T_t(\xi, t) &= T_{\xi\xi}(\xi, t), && \xi > 0, \quad t > 0, \\ T_\xi(0, t) &= \alpha T(0, t) - q(t), && \qquad\quad t > 0, \\ T(\xi, 0) &= 0, && \xi \geq 0, \\ \lim_{\xi \to \infty} T(\xi, t) &= 0, && \qquad\quad t \geq 0. \end{aligned}\right\} \tag{1.1}$$

Using Laplace transforms it is not hard to show that T satisfies the equation

$$T(\xi, t) = \frac{1}{\sqrt{\pi}} \int_0^t (t-s)^{-\frac{1}{2}} \mathrm{e}^{-\xi^2/(4(t-s))} \big(q(s) - \alpha T(0, s)\big) \,\mathrm{d}s, \quad t > 0, \quad \xi \geq 0.$$

Consequently, the function T can be obtained once $T(0, t)$ is known. Taking $\xi = 0$ above, we get the convolution Volterra integral equation

$$x(t) + \int_0^t k(t-s) x(s) \,\mathrm{d}s = f(t), \quad t \geq 0, \tag{1.2}$$

where $x(t) = T(0, t)$, $k(t) = \frac{\alpha}{\sqrt{\pi}} t^{-\frac{1}{2}}$, and $f(t) = \frac{1}{\sqrt{\pi}} \int_0^t (t-s)^{-\frac{1}{2}} q(s) \,\mathrm{d}s$.

Equation (1.2) is a typical example of the equations to be studied in this book. Of course, one might just as well write the equation in the form

$$x(t) = \int_0^t h(t-s)x(s)\,\mathrm{d}s + f(t), \quad t \geq 0,$$

where $h(t) = -k(t)$, but we shall usually write the integral on the left-hand side. (Observe that, in this example, the forcing function f is of the special form $\int_0^t k(t-s)\phi(s)\,\mathrm{d}s$. In some of our results we shall take advantage of this fact.) In (1.2), the unknown function $x(t)$ appears both by itself and under the integral sign multiplied by the kernel k. This fact qualifies this equation as a Volterra equation of the second kind.

As we will see later, the existence of a solution of an equation of the second kind usually presents less of a problem. A more difficult and challenging task is to determine how the solution behaves as $t \to \infty$.

As another example, let us briefly consider a control system with negative, delayed feedback. The simplest model for such a system is the differential-delay equation

$$\begin{aligned} x'(t) &= -ax(t-\tau), \quad && t \geq 0; \\ x(t) &= \varphi(t), && t \in [-\tau, 0], \end{aligned} \tag{1.3}$$

where $\tau > 0$. One possible approach to solve this equation is to use semigroup theory, but in this book we will mostly take this equation as an example of the integrodifferential equation

$$x'(t) + \int_{[0,t]} \mu(\mathrm{d}s)x(t-s) = f(t), \quad t \geq 0; \quad x(0) = x_0. \tag{1.4}$$

To get (1.3) from (1.4), let μ be a point mass with weight a at the point τ, i.e., $\mu(E) = a$ if $\tau \in E$; otherwise $\mu(E) = 0$. Furthermore, define f by $f(t) = -a\chi_{[0,\tau)}(t)\varphi(t-\tau)$, and let $x_0 = \varphi(0)$.

A common way of writing a linear functional differential-delay equation is

$$\begin{aligned} x'(t) &= \int_{(-\infty,0]} \eta(\mathrm{d}s)x(t+s) + F(t), \quad && t \geq 0; \\ x(t) &= \varphi(t), && t \leq 0, \end{aligned}$$

where the initial function φ is prescribed. However, by defining $f(t) = F(t) + \int_{(-\infty,-t)} \eta(\mathrm{d}s)\varphi(t+s)$ and writing $\mu(E) = -\eta(-E)$, this equation may as well be reduced to (1.4).

We shall frequently use the fact that (1.4) can be put into the form (1.2) by integration.

The two preceding examples share one property: neither of them can be reduced to a finite-dimensional ordinary differential equation. Another common feature is that, in both cases, the feedback is time invariant, i.e., the influence of the past values $x(s)$ on $x(t)$ (or on $x'(t)$) depends only on the difference $t-s$, and not on the specific values of t and s. In some cases this assumption is too restrictive; more realistic models often call for time-dependent feedback. The equation (1.2) should then be replaced by

$$x(t) + \int_0^t k(t,s)x(s)\,\mathrm{d}s = f(t), \quad t \geq 0. \tag{1.5}$$

A similar change is required in (1.4). Equation (1.5) is a nonconvolution Volterra equation of the second kind.

In the equations above the feedback is assumed to be linear; normally this is an acceptable first approximation. However, in a more detailed analysis one must often take nonlinearities into account. For example, the state at time t may depend on $x(s)$ for $s < t$ through a nonlinear function g (and through the feedback kernel k). If both the nonlinearity and the feedback are taken to be time invariant, then we end up with the nonlinear integral equation

$$x(t) + \int_0^t k(t-s)g(x(s))\,\mathrm{d}s = f(t), \quad t \geq 0,$$

as the description of our system. To complicate things further, we may have a nonlinear, time-dependent feedback, we may have time-dependent nonlinearities, etc. Note that one need not always have the nonlinearity inside the integral. If, for example, one takes $y(t) = g(x(t))$ above, then y satisfies the equation

$$y(t) = g\left(f(t) - \int_0^t k(t-s)y(s)\,\mathrm{d}s\right), \quad t \geq 0.$$

A feature common to all of the integral or integrodifferential equations above is the fact that the supports of the kernels k and μ are such that the independent variable t appears as the upper limit of integration. This is, of course, only another way of stating that the present depends on the past, but not on the future.

Equations involving integrals where the integration does not extend past the independent variable take their name from Vito Volterra, who was one of the first to investigate equations of this type. (For a short biography of Volterra, 1860–1940, see Volterra [4]. This reference also contains a complete list of Volterra's publications.) The class of equations which one gets by dropping this requirement on the upper limit of integration, i.e., equations of the form

$$x(t) + \int_0^b k(t,s)x(s)\,\mathrm{d}s = f(t), \quad 0 \leq t \leq b,$$

are called integral equations of Fredholm type. Thus, formally a Volterra equation is a special case of a Fredholm equation. However, much of the classical theory of Fredholm equations reduces to mere trivialities when applied to Volterra equations. On the other hand, Volterra equations exhibit a variety of phenomena unknown to Fredholm theory. In addition, it is generally true that Fredholm equations describe boundary value problems and are related to operator theory, whereas Volterra equations describe initial value problems and are related to dynamical systems. Hence there is every reason to consider these fields separately; the fact that integrals appear in both types of equations should not make one overlook the basic differences. This book concentrates on the Volterra theory, but those results that are common to Volterra and Fredholm equations are formulated in such generality that they can be applied to Fredholm equations as well. This is, in particular, true for equations that are required to hold on the whole real line $(-\infty, \infty)$.

The overwhelming majority of the integral equations we analyse are of the second kind. Only in Chapter 5 do we treat equations of the first kind in more detail. These are equations where the unknown function only appears under the integral sign. Thus,

$$\int_0^t k(t-s)x(s)\,\mathrm{d}s = f(t), \quad t > 0,$$

is a convolution Volterra equation of the first kind. Analogously,

$$\int_0^t k(t,s)x(s)\,\mathrm{d}s = f(t), \quad t > 0,$$

is a nonconvolution Volterra equation of the first kind. Trivially, if the kernel k and the nonhomogeneous term f are sufficiently smooth and k satisfies $0 < |k(0)| < \infty$, then the former equation may be reduced to an equation of the second kind by differentiation. A similar comment applies to the nonconvolution equation.

2. Some Examples of Volterra Equations

Volterra equations appear in a variety of applied problems. Below we give a few examples.

2.1 Example

A model for cosmic ray transport was presented in Klimas and Sandri [1] and subjected to further investigation in Hanson *et al.* [1] and [2]. It is shown that the transport of charged particles in a turbulent plasma, e.g., the cosmic rays in the interplanetary solar wind, is described by the coupled equations

$$\begin{aligned} &\frac{\partial I}{\partial \tau} + \alpha \nabla \cdot \Phi = 0, \\ &\frac{\partial \Phi}{\partial \tau} + \epsilon \Omega \cdot \Phi = -(\epsilon\eta)^2 \int_0^\tau \mathcal{K}(\epsilon, \lambda) \cdot \Phi(\tau - \lambda)\,\mathrm{d}\lambda - \tfrac{1}{3}\alpha \nabla I. \qquad (2.1) \end{aligned}$$

Here I is the omnidirectional intensity, Φ is the flux, and the parameters $(\epsilon\eta)^2$ and α are small. The term $\frac{1}{3}\alpha\nabla I$ is a source term due to gradients in the density. The tensor kernel $\mathcal{K}(\epsilon, \lambda)$ is computed from the power spectrum associated with the random magnetic field, and decays as $\frac{1}{\lambda}$ for λ large. The fact that the kernel is nonintegrable over $\mathbf{R}^+$ causes the adiabatic approximation to be invalid. (See Klimas and Sandri [1] for details.)

For (2.1), the scalar approximation

$$\frac{\partial}{\partial t}F(t,\alpha) + \alpha \int_0^t k(t-s)F(s,\alpha)\,ds = -\sigma(t), \quad t > 0; \quad F(0,\alpha) = 1, \tag{2.2}$$

was presented in Hanson *et al.* [2]. The kernel $k(t)$ is assumed to satisfy $k(t) = (1+t)^{-1}$; α plays the role of $(\epsilon\eta)^2$ in (2.1). The nonhomogeneous term $\sigma(t)$ represents the density gradient. The goal is to acquire information on the behaviour of the flux $F(t,\alpha)$, in particular as $t \to \infty$ and $\alpha \downarrow 0$.

In Hanson *et al.* [1], a detailed analysis of (2.2) is carried out using Laplace transforms. Several results on how $F(t,\alpha)$ behaves for small α are obtained. See also Hanson *et al.* [2] for numerical work, and Ling [1] for some qualitative extensions. Here we observe that, by Theorem **3**.3.3, one has $F(t,\alpha) = r_\alpha(t) - (r_\alpha * \sigma)(t)$ where r_α is the differential resolvent of $\alpha/(1+t)$. Since the kernel is completely monotone, one may apply Theorem **5**.4.1. This leads to the conclusion that for every $\alpha > 0$ the function r_α is both bounded and integrable on $\mathbf{R}^+$. Consequently, if σ is either bounded or integrable on $\mathbf{R}^+$, then one has $\sup_{t\in\mathbf{R}^+}|F(t,\alpha)| < \infty$. If we, in addition, employ Exercise **6**.13, and assume that σ is integrable, then we conclude that this bound is uniform with respect to α. Moreover, one can show that

$$r_\alpha(t) = \left(\alpha t \ln^2(t)\right)^{-1} + O\left(\left(t\ln^3(t)\right)^{-1}\right),$$

as $t \to \infty$. See Wong and Wong [1], Theorem 4. Further results on how $F(t,\alpha)$ and r_α depend on α may be obtained by consulting, e.g., Hannsgen [11] and [15]. See also Jordan and Wheeler [6], p. 106.

2.2 Example

Population dynamics constitutes a major source of Volterra equations, the classical reference being Volterra [5]. Let us briefly sketch one example from this field.

Consider a population having an age distribution $y(t,a)$, $t \geq 0$, $a \geq 0$. Thus $\int_A y(t,a)\,da$ is the number of individuals with age in the set A at time t. Assume that the process of aging and dying is modelled by the balance law

$$\frac{\partial y(t,a)}{\partial t} + \frac{\partial y(t,a)}{\partial a} = -m(a)y(t,a). \tag{2.3}$$

The nonnegative function m denotes the age-dependent death rate. Suppose that the birth process satisfies

$$y(t,0) = \int_{\mathbf{R}^+} y(t,a)b(a)\,da, \tag{2.4}$$

where b is the age-dependent fertility. Finally assume that the initial age distribution $y(0,a) = \phi(a)$ is known.

If one solves (2.3) by the method of characteristics and makes use of the initial condition, then one obtains

$$y(t,a) = \begin{cases} \phi(a-t)\exp\left(-\int_0^t m(s+a-t)\,\mathrm{d}s\right), & 0 \le t < a, \\ y(t-a,0)\exp\left(-\int_0^a m(s)\,\mathrm{d}s\right), & t \ge a. \end{cases} \tag{2.5}$$

Inserting the expression for $y(t,a)$ given by (2.5) into (2.4), we obtain the linear Volterra equation

$$y(t,0) + \int_0^t k(t-s)y(s,0)\,\mathrm{d}s = f(t), \quad t \ge 0, \tag{2.6}$$

where

$$k(t) = -b(t)\exp\left(-\int_0^t m(s)\,\mathrm{d}s\right),$$

$$f(t) = \int_{\mathbf{R}^+} \phi(s)\exp\left(-\int_0^t m(s+\sigma)\,\mathrm{d}\sigma\right)b(t+s)\,\mathrm{d}s.$$

Equation (2.6) is the classical renewal equation; see Feller [1].

A more realistic model is obtained if one includes density-dependence as well. Then

$$\begin{aligned} \frac{\partial y}{\partial t} + \frac{\partial y}{\partial a} &= -m(a, N(t))y(t,a), \\ y(t,0) &= \int_{\mathbf{R}^+} y(t,a)b(a,N(t))\,\mathrm{d}a, \\ y(0,a) &= \phi(a), \\ N(t) &= \int_{\mathbf{R}^+} y(t,a)\,\mathrm{d}a. \end{aligned}$$

This model was analysed in Gurtin and MacCamy [1] and has been extensively developed. For notes on this development, and for a comprehensive account of age-dependent population dynamics in general (including a wealth of references), see Webb [1]. A detailed account of mathematical models for physiologically structured populations can be found in Metz and Diekmann [1].

For a general treatment of age-dependent processes and renewal theory, see Chapter IV of Athreya and Ney [1].

2.3 Example

The spread of an epidemic in a population of fixed size can be mathematically described as follows (Diekmann [2]). Assume that the population consists of infected $(I(t))$, and susceptible $(S(t))$ individuals with $I(t)+S(t)=P$. Further assume that transitions from the infected state to the susceptible state cannot occur. Then $I'(t)$ is the rate at which susceptible individuals become infected. We take the infected population $I(t)$ to be structured according to the time since infection; i.e., $i(t,s)\,\mathrm{d}s$ denotes the number of individuals that were infected between $t-s$ and $t-s+\mathrm{d}s$. Then

$$I(t) = \int_{\mathbf{R}^+} i(t,s)\,\mathrm{d}s, \quad t \geq 0,$$

and $\frac{\partial i}{\partial t} + \frac{\partial i}{\partial s} = 0$. Hence,

$$i(t,s) = i(t-s,0), \quad t > s,$$
$$i(t,s) = i(0,s-t), \quad t < s.$$

Moreover,

$$\frac{\mathrm{d}S}{\mathrm{d}t} = -\frac{\mathrm{d}I}{\mathrm{d}t} = -i(t,0), \quad t > 0.$$

Suppose that there exists a nonnegative function A such that the infectivity B (the rate at which susceptibles become infected) is given by

$$B(t) = \int_{\mathbf{R}^+} i(t,s)A(s)\,\mathrm{d}s.$$

Then

$$\frac{\mathrm{d}S}{\mathrm{d}t} = -S(t)\int_{\mathbf{R}^+} i(t,s)A(s)\,\mathrm{d}s, \quad t > 0,$$

with $S(0) = S_0 > 0$ given. Divide by $S(t)$ and integrate over $(0,t)$ to obtain

$$\ln\left(\frac{S(t)}{S_0}\right) = \int_0^t \big(S(t-s) - S_0\big)A(s)\,\mathrm{d}s - f(t),$$

where

$$f(t) = \int_0^t \int_{\mathbf{R}^+} i(0,\tau)A(s+\tau)\,\mathrm{d}\tau\,\mathrm{d}s.$$

If we let $x(t) = -\ln(S(t)/S_0)$, then we end up with the nonlinear Volterra equation

$$x(t) = S_0 \int_0^t A(t-s)g(x(s))\,\mathrm{d}s + f(t), \quad t \geq 0, \tag{2.7}$$

where $g(x) = 1 - \mathrm{e}^{-x}$.

The classical paper on this model is Kermack and McKendrick [1]. An asymptotic analysis of the solutions of (2.7) was done in Diekmann [1], in Diekmann and Kaper [1], and in Gripenberg [27]. Space-dependence was incorporated in Diekmann [2]. For models including susceptible, infected, and removed individuals, see, e.g., Stech and Williams [1]. In Gripenberg [20], the spread of an infectious disease that does not induce permanent immunity is studied.

For a survey of epidemics in homogeneous populations, see Hethcote, Stech, and Van den Driessche [1]. Further work on mathematical epidemiology is to be found in Capasso, Grosso, and Paveri-Fontana [1], pp. 106–189 and in Hethcote, Lewis, and Van den Driessche [1].

2.4 Example
Volterra equations have frequently been used to describe control systems with feedback loops. We illustrate by the single closed-loop, time invariant, and linear system shown below:

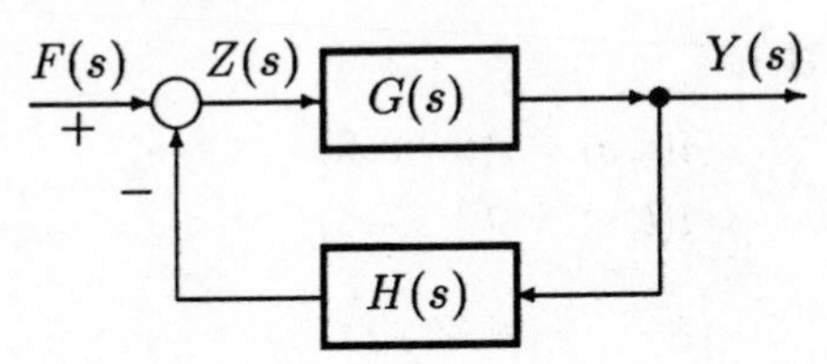

Here $F(s)$ and $Y(s)$ are, respectively, the transforms of the input $f(t)$ and the output $y(t)$. The system transfer function $G(s)$ is defined as the ratio of the transform of the output $Y(s)$ to the transform of the input $Z(s)$. The feedback transfer function is denoted by $H(s)$. Consequently,

$$Y(s) = G(s)\Big[F(s) - Y(s)H(s)\Big],$$

or

$$Y(s) = \frac{G(s)}{1 + G(s)H(s)}F(s).$$

Let g and h be the inverse transforms of G and H, and write $k = g * h$, $f_1 = g * f$. We then have, in the time domain,

$$y(t) + \int_0^t k(t-s)y(s)\,\mathrm{d}s = f_1(t), \quad t \ge 0.$$

If the feedback is built up by a nonlinear term N with no memory and a linear time invariant part with memory, then the resulting equation is

$$y(t) + \int_0^t k(t-s)N\big(y(s)\big)\,\mathrm{d}s = f_1(t), \quad t \ge 0.$$

In case the system is time varying, we are led to a nonconvolution equation. Of course, a stable system is the foremost goal. In particular one wishes to have input–output stability in an L^p-sense, frequently for $p = \infty$ or $p = 2$.

The literature in the field is huge; see MacFarlane [1] for a review and reprints of important papers. For more recent work, see, e.g., Mousa, Miller, and Michel [1] and the references mentioned therein.

2.5 Example
The point model equations for a nuclear reactor with linear feedback may be written as

$$\left.\begin{aligned}
\tfrac{\mathrm{d}}{\mathrm{d}t}p(t) &= \frac{k(t)-\beta}{l}p(t) + \sum_{i=1}^{m}\lambda_i c_i(t),\\
\tfrac{\mathrm{d}}{\mathrm{d}t}c_i(t) &= \frac{\beta_i}{l}p(t) - \lambda_i c_i(t), \quad i = 1,\ldots,m,\\
k(t) &= -\int_0^t K(t-s)\big(p(s)-p_0\big)\,\mathrm{d}s + f(t), \quad t \ge 0.
\end{aligned}\right\} \tag{2.8}$$

Above p denotes the neutron density, k is the reactivity, β_i, $i = 1, \ldots, m$, are the delayed neutron fractions ($\beta = \sum_{i=1}^{m} \beta_i$), l is the neutron average lifetime, and λ_i are the decay constants of the delayed neutrons. The functions c_i represent the delayed neutron densities, K is the reactivity feedback kernel, p_0 denotes the equilibrium power level, and f is the external control. It is obviously of primary interest to have stability of the equilibrium solution $p = p_0$, $c_i = \beta_i p_0/(l\lambda_i)$.

A large number of results on stability criteria for (2.8) exist. See Londen [1], [3], and [4], Podowski [1] and [2], and the references mentioned in these articles. General treatises on nuclear reactor dynamics are, for example, Ash [1] and Akcasu, Lellouche, and Shotkin [1].

Observe that if the delayed neutrons are neglected then (2.8) reduces to

$$\frac{\left(\frac{\mathrm{d}}{\mathrm{d}t}p(t)\right)}{p(t)} + \frac{1}{l}\int_0^t K(t-s)\bigl(p(s) - p_0\bigr)\,\mathrm{d}s = \frac{f(t)}{l}.$$

Define $x(t) = \ln(p(t)/p_0)$, $a(t) = p_0 K(t)/l$, and $h(t) = f(t)/l$. Then

$$x'(t) + \int_0^t a(t-s)(\mathrm{e}^{x(s)} - 1)\,\mathrm{d}s = h(t).$$

We end up with a nonlinear Volterra integrodifferential equation having a nonlinearity which is *a priori* bounded from below.

More complicated models may be constructed, for example, with interacting cores or including space-dependence.

2.6 Example

The theory of superfluidity has motivated several authors to study the following problem.

Let $(x, y, z) \in \mathbf{R}^3$ and let two infinite plates be located at $x = 0$ and $x = L$, respectively. Assume that the region between the plates contains liquid helium. Suppose that the boundary plates are given sinusoidal oscillations in the y-direction. This will create a one-dimensional flow in the liquid. Denote the velocity profile at the point (t, x, y, z) by $u(t, x)$. Then u will be governed by the partial differential equation

$$u_t = u_{xx}, \quad t > 0, \quad 0 < x < L,$$

subject to the conditions

$$\begin{aligned} u(0,x) &= u_0(x), \quad 0 < x < L, \\ u_x(t,0) &= c_1\bigl(u(t,0) - c_2\sin(k_1 t)\bigr)^3, \quad t > 0, \\ u_x(t,L) &= -c_1\bigl(u(t,L) - c_2\sin(k_2 t)\bigr)^3, \quad t > 0, \end{aligned}$$

where c_1, c_2, k_1, k_2 are positive constants, and u_0 is the initial velocity profile. This problem may be converted to the system of nonlinear Volterra equations

$$x_1(t) + \int_0^t a_1(t-s) g_1\big(s, x_1(s)\big)\,ds + \int_0^t a_2(t-s) g_2\big(s, x_2(s)\big)\,ds = f_1(t),$$

$$x_2(t) + \int_0^t a_2(t-s) g_1\big(s, x_1(s)\big)\,ds + \int_0^t a_1(t-s) g_2\big(s, x_2(s)\big)\,ds = f_2(t),$$

where $x_1(t) = u(t,0)$, $x_2(t) = u(t,L)$, and

$$g_i(t,u) = c_1\big(u - c_2 \sin(k_i t)\big)^3, \quad i = 1, 2.$$

Clearly, these functions g_i are periodic in t. The kernels a_1 and a_2 are defined by

$$a_1(t) = \frac{1}{L}\Big(1 + 2\sum_{n=1}^{\infty} \exp(-(n\pi/L)^2 t)\Big),$$

and

$$a_2(t) = \frac{1}{L}\Big(1 + 2\sum_{n=1}^{\infty} (-1)^n \exp(-(n\pi/L)^2 t)\Big).$$

The nonhomogeneous terms $f_1(t)$ and $f_2(t)$ depend on the initial function u_0. These two terms are asymptotically almost-periodic.

For this problem, one can show that solutions exist for all $t \geq 0$ and are asymptotically almost-periodic. See Miller [7], [8], and [12], and Gripenberg [21], for this result and extensions.

In Levinson [1], the corresponding semi-infinite problem is considered, i.e., the equation

$$u_t = u_{xx}, \quad x > 0, \quad t > 0,$$

subject to

$$u(x,0) = 0, \quad x > 0; \quad u_x(0,t) = \Phi\big(u(0,t) - f(t)\big), \quad t > 0.$$

Here Φ is monotone increasing, locally Lipschitz and satisfies $\Phi(0) = 0$ and $f(t)$ is periodic. The problem is converted to

$$u(0,t) + \frac{1}{\sqrt{\pi}} \int_0^t (t-s)^{-\frac{1}{2}} \Phi\big(u(0,s) - f(s)\big)\,ds = 0, \quad t \geq 0,$$

and conditions are given under which $\lim_{t\to\infty}(u(0,t) - \varphi(t)) = 0$, where φ is periodic. Note that from $u(0,t)$ on may easily compute $u(x,t)$.

2.7 Example

Volterra equations are common in mathematical viscoelasticity pertaining to materials with memory. These materials are characterized by constitutive relations which are functionals of the past history of the material. In this context, we examine only a simple example and refer the reader to Renardy, Hrusa, and Nohel [1] for an in-depth overview of the field. (In particular, see Section IV 4 of this monograph.) For a brief, easily accessible account of initial value, integrodifferential equations in viscoelasticity, see Hrusa, Nohel, and Renardy [1].

Specifically, let us show how a partial integrodifferential problem can be reduced to a parameter-dependent scalar Volterra equation. For this purpose, consider

$$\left.\begin{aligned} u_{tt}(x,t) &= \sigma_x(x,t) + F(x,t), && 0 < x < 1, && t > 0,\\ u(x,t) &= U_0(x,t), && 0 \le x \le 1 && -\infty < t \le 0,\\ u_t(x,0+) &= U_1(x), && 0 \le x \le 1, && \\ u(0,t) &= u(1,t) = 0, && && t \ge 0, \end{aligned}\right\} \tag{2.9}$$

with the constitutive law

$$\sigma(x,t) = \int_{\mathbf{R}^+} A(s)\, \frac{\partial}{\partial t}\, u_x(x, t-s)\, \mathrm{d}s.$$

Here U_0, U_1, F, and A are real-valued given functions. It is assumed that A is nonnegative, nonincreasing and convex on $(0,\infty)$, and integrable on $(0,1)$. Physically, $u(x,t)$ is the displacement at time t of the point x on a one-dimensional rod located at $0 \le x \le 1$. The partial derivative u_x is then the strain and σ is the stress which is expressed as a convolution integral of the strain. The weight function A is the linear stress relaxation modulus and F denotes the external force. See, e.g., Hannsgen and Wheeler [5].

If the constitutive law is used to replace σ in (2.9), then one obtains, after integrating with respect to t,

$$\begin{aligned} u_t(x,t) &= \int_0^t A(t-s) u_{xx}(x,s)\, \mathrm{d}s + f(x,t), \quad t \ge 0,\\ u(x,0) &= u_0(x) = U_0(x,0), \quad u(0,t) = u(1,t) = 0, \end{aligned} \tag{2.10}$$

where f includes F, the initial velocity U_1, and the strain history. Let $Lu = u''$ for $u \in \{\varphi'' \in L^2(0,1) \mid \varphi(0) = \varphi(1) = 0\}$. Then (2.10) can be written in the abstract form

$$U'(t) = \int_0^t A(t-s) LU(s)\, \mathrm{d}s + G(t), \quad t \ge 0; \quad U(0) = u_0. \tag{2.11}$$

One can show that (2.11) has the solution

$$U(t) = R(t) u_0 + \int_0^t R(t-s) G(s)\, \mathrm{d}s,$$

where the resolvent R is the solution of (2.11) with $G \equiv 0$; see Carr and Hannsgen [1] and [2]. The self-adjoint operator L has the eigenvalues $\lambda_n = -(n\pi)^2$ with the corresponding normalized eigenfunctions $\varphi_n(x) = \sqrt{2}\sin(n\pi x)$. Separation of variables yields, for $y \in L^2(0,1)$,

$$R(t)y = \sum_{n=1}^{\infty} r_n(t) \left\{ \int_0^1 y(x)\varphi_n(x)\, \mathrm{d}x \right\} \varphi_n,$$

where the scalar functions r_n are the solutions of

$$r_n'(t) + |\lambda_n| \int_0^t A(t-s) r_n(s)\, \mathrm{d}s = 0; \quad r_n(0) = 1.$$

Once we have the functions r_n we get R and hence also $u(x,t)$.

2.8 Example
The flutter problem for an aerofoil section in a two-dimensional flow has been considered by Burns, Herdman, and Turi [2] and Burns, Cliff, and Herdman [1]. (See these articles for a detailed derivation of the model; here we only briefly mention the structure of the resulting equations.)

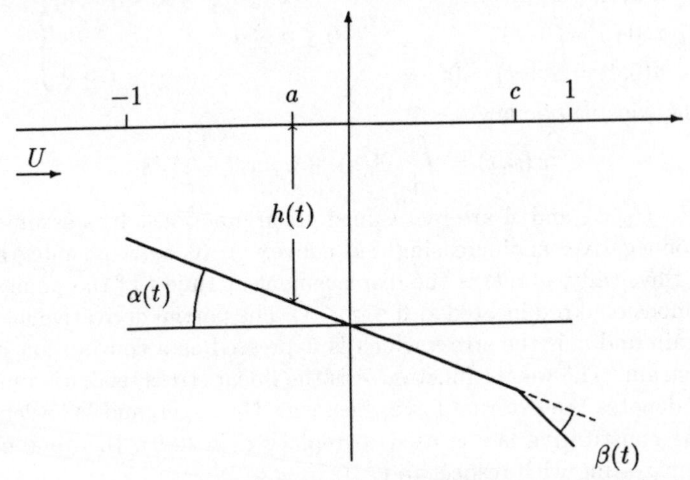

Consider the aerofoil section above, and assume that the aerofoil is placed in an incompressible, inviscid two-dimensional flow with undisturbed stream velocity U. Let $u(t)$ be a control torque on the flap. Define the vector $x(\cdot) \in \mathbf{R}^8$ by

$$x(t) = \big(h'(t), \alpha'(t), \beta'(t), h(t), \alpha(t), \beta(t), \Gamma_q(t), \Gamma_q'(t)\big)^T,$$

where $\Gamma_q : \mathbf{R} \mapsto \mathbf{R}$ is the total aerofoil circulation. The equations of motion may then be written

$$\frac{\mathrm{d}}{\mathrm{d}t}\Big(A_0 x(t) + \int_{-\infty}^{0} A_1(s)x(t+s)\,\mathrm{d}s\Big)$$
$$= B_0 x(t) + \int_{-\infty}^{0} B_1(s)x(t+s)\,\mathrm{d}s + \tilde{G}u(t), \quad (2.12)$$

where A_0, A_1, B_0, B_1 and $\tilde{G}$ are certain 8×8 matrices.

The matrices A_i, B_i have a special structure; in particular, A_0 is singular. Therefore, (2.12) may be viewed as a mixture of a Volterra equation of the first kind and a Volterra equation of the second kind. See Burns, Herdman, and Turi [1] and Burns *et al.* [1] for work on this problem.

2.9 Example
Consider a one-dimensional homogeneous bar of unit length, let $u(t, x)$ denote the temperature of the bar at the point $x \in [0, 1]$ and at time $t \geq 0$, and assume that the temperature at the end-points is kept at zero.

Let the internal energy ϵ and the heat flux q be functionals of the past history of u and the gradient of u, respectively, that is, assume that the material has memory. A reasonable mathematical model is to take

$$\epsilon(t,x) = b_0 u(t,x) + \int_0^t \beta(t-s)u(s,x)\,\mathrm{d}s,$$

$$q(t,x) = -c_0\sigma\big(u_x(t,x)\big) + \int_0^t \gamma(t-s)\sigma\big(u_x(s,x)\big)\,\mathrm{d}s,$$

for $t \geq 0$ and $0 \leq x \leq 1$. Here b_0, c_0 are positive constants, and $\beta, \gamma : \mathbf{R}^+ \mapsto \mathbf{R}$. The nonlinear function σ is monotone increasing, sufficiently smooth, and satisfies $\sigma(0) = 0$. Denote the externally supplied heat by $h(t,x)$. The balance equation is then

$$\epsilon_t = -\frac{\partial q}{\partial x} + h$$

which may be reformulated as

$$b_0 u(t,x) + \int_0^t \beta(t-s)u(s,x)\,\mathrm{d}s = \int_0^t c(t-s)\big(\sigma(u_x(s,x))\big)_x\,\mathrm{d}s + G(t,x). \quad (2.13)$$

The kernel c is defined by $c(t) = c_0 - \int_0^t \gamma(s)\,\mathrm{d}s$; the function G is determined by the external heat supply and the history prior to $t = 0$.

For details and for references to the underlying physical theory, we refer the reader to Nohel [3].

Owing to the space-dependence of u, the analysis of (2.13) does *per se* fall outside the scope of the present work. Nevertheless, since the kernels β and γ are usually taken to be of positive type, the analysis does use much of the theory developed in Chapters 16–20. In particular, strong positivity and the theory of completely positive kernels are useful in this context. See Nohel [3], Sections 3.2, 4.2, and 4.3.

3. Summary of Chapters 2–20

The text is divided into three parts, devoted respectively to linear theory, general nonlinear theory, and asymptotic theory for nonlinear equations.

The first part on linear theory may further be subdivided according to the distinction convolution/nonconvolution. Chapters 2–8 deal with convolution equations and therefore rely heavily on transform theory. Chapters 9 and 10 analyse nonconvolution equations.

The general nonlinear part, Chapters 11–15, addresses questions like existence of solutions, continuous dependence, perturbations, Lyapunov methods, and general asymptotics.

In the final part, Chapters 16–20, we discuss frequency domain and monotonicity methods. Most of the equations investigated in this part are of convolution type, and for the development of their asymptotics the theory of kernels of positive type plays a decisive role.

At the end of each chapter there are exercises and some comments on references, historical developments, etc. For a wealth of further comments on earlier work, especially on the Soviet literature, we refer the reader to Corduneanu [10].

A brief survey of the contents of each chapter follows.

PART I

3.2 Chapter 2: Linear Convolution Integral Equations

We begin our study of Volterra equations by considering the linear convolution equation

$$x(t) + \int_0^t k(t-s)x(s)\,\mathrm{d}s = f(t), \quad t \in \mathbf{R}^+, \tag{\textbf{2}.1.1}$$

where k is an $n \times n$ matrix with complex entries, and x and f are complex-valued n-vectors. We use the notation $(k * x)(t) \stackrel{\text{def}}{=} \int_0^t k(t-s)x(s)\,\mathrm{d}s$. By successive iteration, we formally derive the *variation of constants formula* $x = f - r * f$ where the *resolvent* r should satisfy the equations $r = k - k*r = k - r*k$. Before progressing further we list some basic properties of convolutions and of Laplace and Fourier transforms of functions (the proofs are deferred to an appendix). Then, in Section **2**.3, we prove the existence of a unique solution r of the resolvent equations, and deduce the variation of constants formula.

Section **2**.4 is devoted to the Paley–Wiener theorem. This fundamental result says that if $k \in L^1(\mathbf{R}^+; \mathbf{C}^{n\times n})$, then $r \in L^1(\mathbf{R}^+; \mathbf{C}^{n\times n})$ if and only if ($\hat{}$ denotes the Fourier transform)

$$\det[I + \hat{k}(z)] \neq 0, \quad \Re z \geq 0.$$

The proof is executed in a series of lemmas in Section **2**.5. In addition to kernels supported on $\mathbf{R}^+$ (Volterra kernels) we consider kernels supported on $\mathbf{R}$ (Fredholm kernels). Thus, some of our results apply to the equation

$$x(t) + \int_{-\infty}^{\infty} k(t-s)x(s)\,\mathrm{d}s = f(t), \quad t \in \mathbf{R}. \tag{\textbf{2}.4.3}$$

To realize the importance of the condition $r \in L^1(\mathbf{R}^+; \mathbf{C}^{n\times n})$, observe that, if $f \in L^p(\mathbf{R}^+; \mathbf{C}^n)$ and $r \in L^1(\mathbf{R}^+; \mathbf{C}^{n\times n})$, then $r * f \in L^p(\mathbf{R}^+; \mathbf{C}^n)$. Consequently, the solution $x = f - r * f$ of (**2**.1.1) belongs to the same space as f. Theorems **2**.4.5–**2**.4.7 give further results along the same line.

An analysis of $f \mapsto r * f$ as a map of $L^2(\mathbf{R}^+; \mathbf{C}^n)$ into itself, together with Nyquist's theorem on the number of zeros of $\det[I + \hat{k}(z)]$ in $\Re z > 0$ conclude the chapter.

3.3 Chapter 3: Linear Integrodifferential Convolution Equations

In the third chapter we analyse the integrodifferential equations

$$x'(t) + \int_{[0,t]} \mu(\mathrm{d}s)x(t-s) = f(t), \quad t \in \mathbf{R}^+; \quad x(0) = x_0, \qquad (\mathbf{3}.1.2)$$

and

$$x'(t) + \int_{\mathbf{R}} \mu(\mathrm{d}s)x(t-s) = f(t), \quad t \in \mathbf{R}. \qquad (\mathbf{3}.1.6)$$

The kernel μ in (**3**.1.2) is a locally bounded matrix-valued Borel measure supported on $\mathbf{R}^+$ (a Volterra kernel), whereas μ in (**3**.1.6) is a bounded matrix-valued measure on $\mathbf{R}$ (a Fredholm kernel). We denote these classes of measures by $M_{\mathrm{loc}}(\mathbf{R}^+; \mathbf{C}^{n\times n})$ and $M(\mathbf{R}; \mathbf{C}^{n\times n})$ respectively. The functions x and f take their values in $\mathbf{C}^n$. Convolution is again denoted by $*$, that is, $(\mu * x)(t) = \int_{\mathbf{R}} \mu(\mathrm{d}s)x(t-s)$. In case μ and x are supported on $\mathbf{R}^+$, then $(\mu * x)(t) = \int_{[0,t]} \mu(\mathrm{d}s)x(t-s)$. Note that if μ consists of point masses only (a case which we do not exclude), then (**3**.1.2) reduces to a differential-delay equation.

Linear functional differential equations are included in the formulation (**3**.1.2), but no special consideration is given to equations with finite delay.

Following an expositional introduction, we study the convolution of a measure and a function, and consider Laplace transforms of locally finite measures. Section **3**.3 contains the main results of the chapter. We first deduce the existence of the *differential resolvent* of a measure $\mu \in M_{\mathrm{loc}}(\mathbf{R}^+; \mathbf{C}^{n\times n})$, that is, we prove the existence of a unique locally absolutely continuous function r defined on $\mathbf{R}^+$ satisfying $r(0) = I$ and the equations

$$r'(t) + (\mu * r)(t) = r' + (r * \mu)(t) = 0, \quad t \in \mathbf{R}^+.$$

With the aid of the differential resolvent we can express the solution of (**3**.1.2) in the *variation of constants* form

$$x(t) = r(t)x_0 + (r * f)(t), \quad t \in \mathbf{R}^+.$$

The second part of Section **3**.3 gives the Paley–Wiener theorems applicable to (**3**.1.2) and (**3**.1.6). Assume that μ in (**3**.1.2) satisfies $\mu \in M(\mathbf{R}^+; \mathbf{C}^{n\times n})$. Then the resolvent r of μ satisfies $r \in L^1(\mathbf{R}^+; \mathbf{C}^{n\times n})$ if and only if $\det[zI + \hat{\mu}(z)] \neq 0$ for $\Re z \geq 0$. Similarly, if μ in (**3**.1.6) satisfies $\mu \in M(\mathbf{R}; \mathbf{C}^{n\times n})$, then μ has a resolvent $r \in L^1(\mathbf{R}; \mathbf{C}^{n\times n})$ if and only if $\det[zI + \hat{\mu}(z)] \neq 0$ for $\Re z = 0$. If the latter resolvent (required to satisfy $r' + \mu * r = r' + r * \mu = 0$ on $\mathbf{R}$ and $r(0+) - r(0-) = I$) exists, then we can express the solution of (**3**.1.6) in the slightly different *variation of constants* form

$$x = r * f,$$

provided that f belongs to some suitable space of functions. Theorems **3**.3.9–**3**.3.10 demonstrate the significance of (the existence of) integrable resolvents. We conclude the section by giving some results on periodic solutions and on L^2-solutions.

At the end of Chapter 3 there are five appendices, i.e., Sections **3**.4–**3**.8, containing supplementary results, needed in this as well as in later chapters. Section **3**.4 formulates different versions of Fubini's theorem, and Section **3**.5 discusses the total variation of a vector measure. In Section **3**.6 we establish some properties of the convolution of a measure and a function. Derivatives of these convolutions are analysed in Section **3**.7. Finally, in Section **3**.8 we give a few results on Laplace transforms of measures.

3.4 Chapter 4: Equations in Weighted Spaces

In this chapter we analyse the convolution integral equations

$$x(t) + \int_{[0,t]} \mu(\mathrm{d}s)x(t-s) = f(t), \quad t \in \mathbf{R}^+, \qquad (\mathbf{4}.1.1)$$

and

$$x(t) + \int_{\mathbf{R}} \mu(\mathrm{d}s)x(t-s) = f(t), \quad t \in \mathbf{R}, \qquad (\mathbf{4}.1.2)$$

with *measure-valued* kernels. Furthermore, we give some additional results on (**3**.1.2) and (**3**.1.6). In (**4**.1.1) we assume that $\mu \in M_{\mathrm{loc}}(\mathbf{R}^+; \mathbf{C}^{n\times n})$ (a Volterra kernel), and in (**4**.1.2) that $\mu \in M(\mathbf{R}; \mathbf{C}^{n\times n})$ (a Fredholm kernel). If μ consists of point-masses only, then the equation becomes a difference equation.

The resolvent ρ of μ is required to satisfy $\rho + \mu * \rho = \rho + \rho * \mu = \mu$; hence ρ (if it exists) is now a measure. This fact forces us to define the convolution of two measures, and to analyse the properties of this type of convolution (the proof is deferred to an appendix). This being done, we show in Theorem **4**.1.5 that a measure $\mu \in M_{\mathrm{loc}}(\mathbf{R}^+; \mathbf{C}^{n\times n})$ has a unique resolvent $\rho \in M_{\mathrm{loc}}(\mathbf{R}^+; \mathbf{C}^{n\times n})$ if and only if $\det[I + \mu(\{0\})] \neq 0$ (where $\mu(\{0\})$ denotes a possible point mass of μ at zero; observe that the integral in (**4**.1.1) contains a hidden term $\mu(\{0\})x(t)$, so that the actual coefficient of $x(t)$ is $I+\mu(\{0\})$). With this measure resolvent ρ at our disposal, we can write the solution of (**4**.1.1) in the variation of constants form $x = f - \rho * f$.

The remaining part of Chapter 4 is formulated in the framework of *weighted spaces*. Sections **4**.2 and **4**.3 introduce the reader to weight functions, and define weighted function spaces and the weighted measure spaces $M(\mathbf{R}^+; \varphi)$ and $M(\mathbf{R}; \varphi)$. Here the weight function $\varphi : \mathbf{R} \to \mathbf{R}^+$ is positive, locally bounded, submultiplicative (that is, $\varphi(t+s) \le \varphi(t)\varphi(s)$), and satisfies $\varphi(0) = 1$. For example, the space $M(\mathbf{R}^+; \varphi; \mathbf{C}^{n\times n})$ consists of those matrices of locally finite Borel measures that satisfy $\int_{\mathbf{R}^+} |\mu|(\mathrm{d}s)\varphi(s) < \infty$.

The key question of Section **4**.4 is the following. Suppose that $\mu \in M(\mathbf{R}^+; \varphi; \mathbf{C}^{n\times n})$. When does it follow that the resolvent ρ belongs to the same space? A crucial role in the development of the answers (i.e., Theorems **4**.4.3, **4**.4.4, and their corollaries) is played by Theorem **4**.4.2, due to Gel′fand, concerning the invertibility of a scalar measure. Note that in this result a size restriction must be imposed on the singular part of μ. Once we know that $\rho \in M(\mathbf{R}^+; \varphi; \mathbf{C}^{n\times n})$, or $\rho \in M(\mathbf{R}; \varphi; \mathbf{C}^{n\times n})$, then various results about the solution (Theorems **4**.4.11 and **4**.4.12) can be deduced.

The integrodifferential equations (**3**.1.2) and (**3**.1.6) can be analysed in weighted spaces too. The results are actually sharper than those for the integral equation in the sense that the conditions obtained are both necessary and sufficient; see Theorems **4**.4.13 and **4**.4.14.

Equations of so-called neutral type are discussed among the exercises of this chapter.

3.5 Chapter 5: Completely Monotone Kernels

Scalar convolution kernels that are locally integrable and *completely monotone* on $(0,\infty)$, i.e., kernels satisfying $(-1)^j k^{(j)}(t) \geq 0$ for $t > 0$ and $j = 0, 1, 2, 3, \ldots$, have been the object of numerous studies. In Chapter 5 we present some of the results obtained and apply them to Volterra equations. We extend the concept of complete monotonicity to matrices by requiring $\langle v, k(\cdot)v\rangle$ to be completely monotone for all $v \in \mathbf{C}^n$ (where $\langle \cdot, \cdot \rangle$ is some inner product).

Section **5**.2 contains basic definitions of matrix algebra needed both in Chapter 5 and in Part III. The well-known theorem by Bernstein on the representation of a completely monotone function as the Laplace transform of a nonnegative measure is given. Subsequently, a characterization of the Laplace transform $F(z)$ of a completely monotone function is presented. We show that such a transform can be analytically extended to $\mathbf{C} \setminus \mathbf{R}^-$, is real on the real axis, satisfies $\Im F(z) \leq 0$ for $\Im z > 0$, and $\lim_{x\to\infty} F(x) = 0$. Conversely, a function $F(z)$ having these four properties is shown to be the transform of a completely monotone function. This result is extremely useful in applications.

As a consequence of Bernstein's theorem we establish, in Theorem **5**.2.8, that the logarithm of a completely monotone function is convex.

It is well known that the resolvent of a scalar completely monotone function is completely monotone. We prove this fact for the matrix case in Theorem **5**.3.1.

In Section **5**.4 we analyse the differential resolvent r of a completely monotone, locally integrable function k. It is easy to show that r cannot be completely monotone on $(0,\infty)$. However, $-r$ is the sum of a completely monotone function and an exponentially decaying one. The proof of this fact contains some elaborate analysis.

Section **5**.5 briefly considers linear *Volterra equations of the first kind*, i.e., equations of the type

$$\int_0^t k(t-s)x(s)\,\mathrm{d}s = f(t), \quad t \in \mathbf{R}^+. \tag{5.1.1}$$

If r satisfies $(r * k)(t) = (k * r)(t) = I$ for $t > 0$, then r is a *resolvent of the first kind* of k, and the solution of (**5**.1.1) is $x(t) = r(t)f(0) + (r * f')(t)$. As an application of the characterization of the Laplace transform of a completely monotone function, we obtain a result on the existence of a resolvent of the first kind. Furthermore, the case where the (scalar) kernel k is nonnegative and nonincreasing is studied in Theorem **5**.5.5.

3.6 Chapter 6: Nonintegrable Kernels with Integrable Resolvents

The Paley–Wiener theorems in Chapters 2–4 which yield the existence of an integrable resolvent all require the original kernel to be integrable. In general, if the conclusion is to be retained, then we cannot remove this requirement without imposing some extra conditions. Although some answers, as to what extra conditions suffice, do exist, the problem is still to some extent open. Chapter 6 treats this question in some detail.

In Section **6**.2 we show that, if in the scalar case the kernel k is locally integrable, nonnegative, nonincreasing, and convex, then $\|r\|_{L^1(\mathbf{R}^+)} < \infty$. In fact, we record an explicit upper bound for $\|r\|_{L^1(\mathbf{R}^+)}$, independent of k.

One may combine the Paley–Wiener theorems on $\mathbf{R}^+$ with this result on the size of the resolvent of a scalar convex kernel, to obtain additional results on the integrability of resolvents. To do this, it is necessary to study functions that are locally equal to Laplace transforms. This is done in Section **6**.3. The actual combination is presented in Section **6**.4, together with some further extensions of the Paley–Wiener results. Observe that two of the extensions, i.e., Theorems **6**.4.2 and **6**.4.4, relate to systems.

In the appendix in Section **6**.5, we state and prove an inequality due to Hardy and Littlewood that is needed in the proof of the fact that the resolvent of a convex kernel is integrable.

3.7 Chapter 7: Unbounded and Unstable Solutions

Section **7**.2 is devoted to the cases where the conditions $\det[zI + \hat{\mu}(z)] \neq 0$, $\Re z \geq 0$ or $\det[I + \hat{\mu}(z)] \neq 0$, $\Re z \geq 0$ are violated in the open right half plane $\Re z > 0$. In these cases, the differential resolvent or resolvent of μ includes exponentially growing terms. Theorems **7**.2.3 and **7**.2.5 describe the behaviour of solutions of equations containing such a kernel μ. We remark that these results are formulated in terms of weighted spaces.

The case where $\det[zI + \hat{\mu}(z)]$ or $\det[I + \hat{\mu}(z)]$ has zeros on the critical line is studied in Section **7**.3. The results are similar to those obtained in Section **7**.2, but stronger assumptions have to be made. The analysis is much more intricate, and many details are omitted. As an application, some consequences pertaining to the renewal equation are given.

3.8 Chapter 8: Volterra Equations as Semigroups

In Chapter 8 we briefly analyse Volterra equations from a *semigroup* point of view. In Section **8**.2 we define the *initial function semigroup* T and the *forcing function semigroup* S, and compute their infinitesimal generators.

Theorem **8**.2.8 ascertains that the adjoint semigroup of T is the transposed forcing function semigroup.

It is shown in Section **8**.3 that by considering *extended semigroups* one may remove some of the limitations that appear in the context of initial function and forcing function semigroups.

This semigroup approach is not used elsewhere in the book.

3.9 Chapter 9: Linear Nonconvolution Equations

Here we present the basic theory on Volterra equations of the form

$$x(t) + \int_0^t k(t,s)x(s)\,\mathrm{d}s = f(t), \quad t \in \mathbf{R}^+. \qquad (\mathbf{9}.1.1)$$

Much of the analysis is performed in the setting of various *Banach algebras*.

In Section **9**.2 we first define *kernels of type* L^p on some interval J. These are the matrix-valued kernels k for which the mapping $f \mapsto \int_J |k(t,s)| f(s)\,\mathrm{d}s$ takes $L^p(J)$ continuously into itself. If multiplication is defined by

$$(k_1 \star k_2)(t,s) = \int_J k_1(t,u)k_2(u,s)\,\mathrm{d}u,$$

then this class of kernels constitutes a Banach algebra. The subalgebra of Volterra kernels consists of those kernels for which $k(t,s) = 0$ when $s > t$. The convenient Proposition **9**.2.7 gives conditions (sufficient when $1 < p < \infty$, necessary and sufficient when $p = 1$ or $p = \infty$) for k to be a kernel of type L^p.

Section **9**.3 is devoted to the existence of a resolvent of a kernel of type L^p. The resolvent r of k is required to be of the same type as k, and to satisfy $r + k \star r = r + r \star k = k$ on J^2. If k has a resolvent, then the solution of equation (**9**.1.1) can be written as $x = f - r \star f$.

Theorem **9**.3.13 is a key result on the existence of a resolvent of a non-convolution Volterra kernel. Here it is shown that the global problem of finding a resolvent of k on J can be reduced to the local problem of finding a resolvent of k on J_i, where $J = \bigcup_{i=1}^m J_i$. It follows from Banach-algebra arguments that k has a unique resolvent on J_i if $|||k|||_{L^p(J_i)} < 1$. Corollary **9**.3.18 gives the classical representation of the resolvent as a sum of iterated kernels.

In practice we often have $J = \mathbf{R}^+$. For this case, it is convenient to have local versions of the theorems giving the existence of a resolvent. Such versions are found in Section **9**.4.

Kernels of so-called bounded and continuous types on J are considered in Section **9**.5. Kernels of *bounded type* map $B^\infty(J)$ into itself, and kernels of *continuous type* map bounded subsets of $L^\infty(J)$ into equicontinuous sets of functions on J. Section **9**.6 analyses certain subalgebras of kernels of bounded type, in particular, kernels of *bounded periodic type*, and kernels of *asymptotically periodic type*.

In the applications of fixed point theorems one frequently requires the kernel k, or the resolvent r, to generate a *compact* mapping. In Section **9**.7 we formulate sufficient conditions for this to happen in the case where k is of type L^p, with $1 \le p \le \infty$.

For equations with nonconvolution kernels, the global theory is much more complicated than for convolution equations. Only in the scalar case has it been possible to obtain interesting results on the asymptotic behaviour of solutions of (**9**.1.1). These results (in Sections **9**.8 and **9**.9) are based on monotonicity or sign properties of the kernel k. Section **9**.8 starts out by presenting a generalization of the well-known Gronwall inequality and continues with results exposing the advantage of having a nonnegative resolvent.

The appendix in Section **9**.10 contains some lemmas needed earlier in the characterization of kernels of type L^p on J.

3.10 Chapter 10: Linear Nonconvolution Equations with Measure Kernels

In this chapter we investigate the equations

$$x(t) + \int_{[0,t]} \kappa(t, \mathrm{d}s)x(s) = f(t), \quad t \in \mathbf{R}^+, \tag{\textbf{10}.1.5}$$

and

$$x'(t) + \int_{[0,t]} \kappa(t, \mathrm{d}s)x(s) = f(t), \quad t \in \mathbf{R}^+; \quad x(0) = x_0. \tag{\textbf{10}.1.6}$$

The kernel κ is now assumed to be a t-dependent matrix of complex measures. Delay equations of the type

$$x(t) + a(t)x\big(t - g(t)\big) = f(t),$$
$$x'(t) + a(t)x\big(t - g(t)\big) = f(t),$$

are therefore included.

We begin Section **10**.2 by defining Volterra *measure kernels of type* B^∞ on an interval $J \subset \mathbf{R}$. Specifically, κ is of type B^∞ on J if, for each Borel set $E \subset J$, the function $\kappa(\cdot, E)$ is Borel measurable, if κ is of Volterra type, i.e., if for each $t \in J$, $\kappa(t, \cdot)$ is supported on $J \cap (-\infty, t]$, and if $\sup_{t\in J}|\kappa|(t, J) < \infty$. The set of these kernels can be made a Banach algebra by defining the norm of κ to be $\sup_{t\in J}|\kappa|(t, J)$, and defining multiplication by $(\kappa_1 \star \kappa_2)(t, E) = \int_J \kappa_1(t, \mathrm{d}s)\kappa_2(s, E)$.

The resolvent ρ of a kernel κ of type B^∞ is required to be of the same type as κ, and to satisfy $\rho + \kappa \star \rho = \rho + \rho \star \kappa = \kappa$. If κ has a resolvent, then the variation of constants formula $x = f - \int_J \rho(t, \mathrm{d}s)f(s)$ for the solution of (**10**.1.5) is valid. Theorem **10**.2.8 and Corollary **10**.2.9 show that if J can be divided into subintervals J_i such that the norm of κ on each J_i is < 1, then κ has a resolvent on J. Some local versions of these claims and a perturbation result bring the section to an end.

Section **10**.3 analyses (**10**.1.6), and begins by defining a class of measures that generate maps of $B^\infty(J)$ into $L^1(J)$. To quote the main requirements, κ is said to be a Volterra *measure kernel of type* $[B^\infty; L^1]$ on J if the measure $\kappa(t, \mathrm{d}s)$ is, for each $t \in J$, supported on $J \cap (-\infty, t]$, and if $\int_J |\kappa|(t, J)\,\mathrm{d}t < \infty$. With $\int_J |\kappa|(t, J)\,\mathrm{d}t$ as norm, these measures form a Banach space.

To supplement this space of kernels it is appropriate to define *kernels of type* $[M; B^\infty]$ mapping M (hence L^1) into B^∞. These are simply the functions $k \in B^\infty(J^2; \mathbf{C}^{n\times n})$ that vanish for $s > t$. In Theorems **10**.3.4 and **10**.3.5, we prove various results on the $\star$-product of the three different spaces of kernels that have been introduced.

The differential resolvent r of κ is required to satisfy $r(t,t) = I$, $r(t,s) = 0$, $s > t$, and

$$\frac{\partial}{\partial t} r(t,s) + \int_{[s,t]} \kappa(t, \mathrm{d}u) r(u,s) = 0,$$

$$\frac{\partial}{\partial s} r(t,s) - \int_{[s,t]} r(t,u)\kappa(u, \mathrm{d}s)\,\mathrm{d}u = 0$$

(where the second relation should be interpreted as a measure equation). As Theorem **10**.3.7 exhibits, if κ is locally of type $[B^\infty; L^1]$ on $\mathbf{R}^+$, then it has a unique differential resolvent r which is locally of type $[M; B^\infty]$ on $\mathbf{R}^+$. The solution of (**10**.1.6) can then be written as $x(t) = r(t,0)x_0 + \int_0^t r(t,s) f(s)\,\mathrm{d}s$.

Section **10**.4 contains some global results for (**10**.1.6). It is worth pointing out that all available results of this kind are of perturbation type. In particular, they are based on the fact that if $\kappa = \kappa_1 + \kappa_2$ (both locally of type $[B^\infty; L^1]$ on $\mathbf{R}^+$), and if r_1 and r are the differential resolvents of respectively κ_1 and κ, then $r + r_1 \star \kappa_2 \star r = r + r \star \kappa_2 \star r_1 = r_1$. Various examples are given where r_1 originates in an ordinary differential equation, or in a Volterra equation of convolution type. The section is concluded with a brief discussion of equation (**10**.1.6) with $\mathbf{R}^+$ replaced by $\mathbf{R}$.

PART II

3.11 Chapter 11: Perturbed Linear Equations

In this chapter we launch the study of nonlinear equations by analysing perturbations of linear equations. Specifically, we consider the equations

$$x(t) + \int_0^t k(t,s)x(s)\,\mathrm{d}s = G(x)(t), \quad t \in \mathbf{R}^+, \tag{\textbf{11}.1.1}$$

and

$$x'(t) + \int_{[0,t]} \kappa(t, \mathrm{d}s)x(s) = G(x)(t), \quad t \in \mathbf{R}^+; \quad x(0) = x_0, \tag{\textbf{11}.1.2}$$

where we assume that G can be written as the sum of a small x-dependent term and a term independent of x. Using the variation of constants formula, we rewrite equations (**11**.1.1) and (**11**.1.2) as

$$x(t) = G(x)(t) - \big(r_1 \star G(x)\big)(t),$$
$$x(t) = r_2(t)x_0 + \big(r_2 \star G(x)\big)(t),$$

respectively. Here r_1 is the resolvent of k, and r_2 is the differential resolvent of κ. The product $\star$ includes the normal convolution as a special case.

In Section **11**.2, we formulate two fixed point theorems which are applied in subsequent sections. Section **11**.3 treats the convolution case of (**11**.1.1) (i.e., $k(t,s)$ is replaced by $k(t-s)$), with $k \in L^1(\mathbf{R}^+;\mathbf{R}^{n\times n})$, $G(x)(t) = \int_0^t h(t-s,x(s))\,\mathrm{d}s$, and $h(t,y) = o(|y|)$ as $y \to 0$. Under the key assumption that $\det[I + \hat{k}(z)] \neq 0$ for $\Re z \geq 0$, we obtain asymptotic results on x, including results on the rate of convergence to zero. An analogous result is given for the corresponding equation on $\mathbf{R}$. In Section **11**.4 we analyse the convolution case of (**11**.1.2) (then $\kappa(t,\mathrm{d}s)x(s)$ is replaced by $\mu(\mathrm{d}s)x(t-s)$), without giving G any more explicit form. Section **11**.5 deals with perturbations of ordinary differential equations.

In the last Section **11**.6, we work in the framework of $L^2(\mathbf{R}^+;\mathbf{R}^n)$. Conditions are imposed on the kernel which guarantee that the resolvent maps $L^2(\mathbf{R}^+;\mathbf{R}^n)$ into itself. Under certain hypotheses on the nonlinearities we obtain $x \in L^2(\mathbf{R}^+;\mathbf{R}^n)$. Several additional perturbation theorems are included in the exercises of Chapter 11, in particular, some theorems on nonconvolution equations.

3.12 Chapter 12: Existence of Solutions of Nonlinear Equations

In Part I, where we developed the linear theory, the existence of a unique solution followed once the existence of a unique resolvent was established. In Chapter 11, the existence of a solution is a consequence of various fixed point theorems (except for Theorems **11**.3.3 and **11**.4.3 where it is assumed). However, in the remaining part of the text the existence of a solution is frequently part of the hypothesis; the main point of interest being the asymptotic behaviour of the solution(s). It is therefore appropriate at this point to give some existence and uniqueness results, general enough to cover all equations analysed later.

After a simple introductory result, Theorem **12**.1.1, on the existence of a continuous solution of the equation

$$x(t) = f(t) + \int_0^t a(t,s)h\big(s,x(s)\big)\,\mathrm{d}s, \quad t \geq 0,$$

we state the fixed point Theorem **12**.1.3 which we use for the proofs of our existence results. This theorem, based on the measure of noncompactness, is reasonably general and includes both Schauder's fixed point theorem and the contraction mapping principle.

Theorem **12**.2.1 is our main result on the existence of continuous solutions. It is formulated for the quite general equation

$$x(t) = G(x)(t), \quad t \geq 0, \tag{12.1.6}$$

where G maps a suitably restricted class of continuous functions defined on $[0,T]$ into $C([0,T];\mathbf{R}^n)$. In addition, G is required to be causal and to satisfy a certain combined contraction and equicontinuity hypothesis. The conclusion contains statements about existence and about the behaviour of maximally defined solutions. The proof of Theorem **12**.2.1 uses some elaborate analysis. Corollary **12**.2.2 gives an easy uniqueness result. Theorems **12**.2.4 and **12**.2.5 consider the question whether the solutions can be forced to stay in some *a priori* defined set. This problem is of importance for comparison studies. Section **12**.2 is concluded by several existence results for

$$x(t) = f(t) + \int_0^t g\big(t,s,x(s)\big)\,\mathrm{d}s, \quad t \geq 0, \tag{\textbf{12}.1.7}$$

where g satisfies the Carathéodory conditions.

Section **12**.3 is devoted to the functional differential equation

$$\tfrac{\mathrm{d}}{\mathrm{d}t}x(t) = H(x)(t), \quad t \geq 0; \quad x(0) = x_0, \tag{\textbf{12}.3.1}$$

where the causal operator H maps a suitably restricted class of continuous functions into integrable functions. By integrating (**12**.3.1) and then applying Theorem **12**.2.1 we obtain an existence theorem for (**12**.3.1).

We begin Section **12**.4 by proving an existence and uniqueness result for L^p-solutions of (**12**.1.6). Here G is assumed to be locally Lipschitzian in the L^p-norm. An analogous result on B^∞-solutions follows. Subsequently, we pursue the investigation of (**12**.1.7), obtaining a theorem on the existence of L^p-solutions. The proof uses compactness in L^1. The section is concluded by two results on (**12**.1.7) in a B^∞- and L^∞-setting, respectively.

Section **12**.5 is devoted to (**12**.1.7) in the case where, for each (t,s), the function $y \mapsto g(t,s,y)$ is locally bounded but not necessarily continuous. The concept of a solution must now be enlarged. A result on the existence of solutions in this enlarged sense is obtained.

Section **12**.6 contains two examples. One illustrates the differences between the assumptions of Theorem **12**.2.1 and those of Theorem **12**.2.8; the other indicates the necessity of one of the assumptions in Theorem **12**.2.1.

3.13 Chapter 13: Continuous Dependence, Differentiability, and Uniqueness

Chapter 13 on the *continuous dependence* on parameters, and on the *differentiability and uniqueness* of solutions, is closely related to the previous chapter. First, we consider the simple equation

$$x(\lambda,t) = f(\lambda,t) + \int_0^t a(\lambda,t,s)h\big(\lambda,s,x(\lambda,s)\big)\,\mathrm{d}s, \quad t > 0. \tag{\textbf{13}.1.6$_\lambda$}$$

In Theorem **13**.2.1 we study the parameter-dependence of the solutions of

$$x(\lambda,t) = G\big(\lambda,x(\lambda,\cdot)\big)(t), \quad t > 0, \tag{\textbf{13}.1.1$_\lambda$}$$

under the same hypotheses on G as those that were made in Theorem **12**.2.1. Here uniqueness is not assumed. Theorem **13**.2.4, related to

Theorem **12**.4.1, deals with the parameter-dependence of L^p-solutions of $(\mathbf{13}.1.1)_\lambda$. In this case, a Lipschitz condition assures uniqueness.

We analyse the differentiability of the solutions of $(\mathbf{13}.1.6)_\lambda$ and $(\mathbf{13}.1.1)_\lambda$ with respect to the parameter λ in Section **13**.3. The key result of the section is Theorem **13**.3.1, where it is shown that under the given hypotheses the solution $x(\lambda,\cdot)$ of $(\mathbf{13}.1.1)_\lambda$ has, in a certain sense, a right-hand derivative with respect to λ at $\lambda = 0$. Theorem **13**.1.2 is a corollary pertaining to $(\mathbf{13}.1.6)_\lambda$.

In Section **13**.4, *maximal and minimal solutions* are discussed, and a number of comparison results are given.

The last two sections deal with uniqueness of solutions, in particular with uniqueness of solutions of the equation

$$x(t) = f(t) + \int_0^t g\big(t, s, x(s)\big)\,\mathrm{d}s, \quad t > 0.$$

A comparison approach together with a monotonicity assumption gives us the uniqueness Theorem **13**.5.1. In Theorem **13**.5.4 we formulate some results spelling out hypotheses under which the key condition of Theorem **13**.5.1 is satisfied.

3.14 Chapter 14: Lyapunov Techniques

In Chapter 14 we study Lyapunov techniques. We stress the main ideas involved, and give some general principles and a few simple examples. As is well known, the actual construction of Lyapunov functions for more complicated equations is frequently an insurmountable task.

After some introductory comments, we discuss, in Section **14**.2, the scalar differential-delay equation

$$\begin{cases} x'(t) + a(t)g(x(t)) + b(t)g\big(x(t-1)\big) = f(t), & t \geq 0; \\ x(t) = \varphi(t), & -1 \leq t \leq 0. \end{cases}$$

Through the analysis of this equation we are led to two *boundedness principles.* In Proposition **14**.2.9, we apply Lyapunov techniques to the scalar nonconvolution integrodifferential equation

$$x'(t) + \int_{[0,t]} \kappa(t, \mathrm{d}s)g(x(s)) = f(t), \quad t \in \mathbf{R}^+; \quad x(0) = x_0,$$

and obtain a boundedness result.

In Section **14**.3, we examine the existence of limits using Lyapunov methods. Again we emphasize main ideas. We formulate two *limit principles*, and apply them to a differential-delay equation and to a nonlinear Volterra equation. The appendix in Section **14**.4 contains some technical lemmas.

3.15 Chapter 15: General Asymptotics

Some of the tools needed for the study of the asymptotics of various nonlinear equations are investigated in Chapter 15. In particular, we study *limit sets, limit equations*, and some results in *spectral analysis*. Using these tools and some additional analysis we obtain a large number of asymptotic results in Chapters 17–20.

We begin Section **15**.2 by defining the positive and negative limit sets $\Gamma_{\pm}(\varphi)$ of a bounded uniformly continuous function $\varphi : \mathbf{R} \to \mathbf{C}^n$ as

$$\Gamma_{\pm}(\varphi) = \{\, \psi : \mathbf{R} \to \mathbf{C}^n \mid \varphi(t+t_k) \to \psi(t) \textit{ uniformly on compact subsets of } \mathbf{R} \textit{ for some sequence } t_k \to \pm\infty \,\},$$

and state some elementary facts about these limit sets. Autonomous operators, the concept of *narrow convergence*, and limit equations are introduced. For the simple equation

$$x(t) + \int_0^t a(t-s)g(x(s))\,\mathrm{d}s = f(t), \quad t \in \mathbf{R}^+, \tag{15.1.1}$$

where $a \in L^1(\mathbf{R}^+; \mathbf{R}^{n\times n})$ and f vanishes at ∞, the limit equation is

$$y(t) + \int_{-\infty}^t a(t-s)g(y(s))\,\mathrm{d}s = 0, \quad t \in \mathbf{R}. \tag{15.1.2}$$

If x satisfies (**15**.1.1) and $y \in \Gamma_+(x)$, then y satisfies (**15**.1.2). Some further results in this direction are obtained in Theorems **15**.2.9 and **15**.2.11. Limit equations associated with periodic equations constitute our final topic in this section.

In Section **15**.3, we first give some statements on the limit sets of asymptotically slowly varying functions. We then define finitely generated limit sets and prove that every limit set of this category contains a periodic function.

Section **15**.4 characterizes bounded functions in terms of their spectral properties. We begin by introducing the *spectrum* $\sigma(\mu)$ of a finite measure μ (this is the support of the Fourier transform of μ), and continue with a definition of the spectrum of a bounded continuous function φ (this spectrum is defined as $\sigma(\varphi) = \bigcap_{a\in N(\varphi)}\{\, \omega \mid \tilde{a}(\omega) = 0 \,\}$ where $N(\varphi) = \{\, a \in L^1(\mathbf{R}; \mathbf{C}) \mid a * \varphi \equiv 0 \,\}$). In Theorem **15**.4.6 we establish some of the properties of these spectra; in particular, we show that $\sigma(\mu * \varphi) \subset \sigma(\mu) \cap \sigma(\varphi)$. Subsequently, a version of Wiener's Tauberian theorem is presented, together with some ramifications.

An important question that frequently arises in asymptotic studies is the following: under what conditions does the inclusion $\sigma(\varphi) \subset \{\, \omega \in \mathbf{R} \mid \tilde{\mu}(\omega) = 0 \,\}$ imply that $\mu * \varphi \equiv 0$? Some results related to this question are given in Theorem **15**.4.12, in the following corollaries, and in Theorem **15**.4.15. Theorem **15**.4.17 characterizes those bounded continuous functions which have (at most) countable spectra.

In Section **15**.5, *directional spectra* are defined and some of their properties are proved. Observe that this section is meaningful only when $n > 1$.

Using the *asymptotic (directional) spectrum* we then obtain some asymptotic results. An application to the renewal equation is worked out in Section **15**.7. The chapter ends with the proof of a certain Lyapunov-type result which is needed later.

PART III

3.16 Chapter 16: Convolution Kernels of Positive Type

In this part we first develop, in Chapter 16, the theory of convolution kernels of *positive type*. These results are of utmost importance for most of the asymptotic theory derived in later chapters.

A locally finite matrix-valued Borel measure μ supported on $\mathbf{R}^+$ is said to be of positive type if for every $\varphi \in L^2(\mathbf{R}; \mathbf{C}^n)$ with compact support one has

$$[\![\mu, \varphi]\!] \stackrel{\text{def}}{=} \Re \int_{\mathbf{R}} \Big\langle \varphi(t), \int_{\mathbf{R}^+} \mu(\mathrm{d}s)\varphi(t-s) \Big\rangle \,\mathrm{d}t \geq 0.$$

As Theorem **16**.2.5 points out, measures of positive type define tempered distributions. Moreover, μ is of positive type if and only if the real part of the Laplace transform of μ is nonnegative. The equivalence statements are formulated in Theorem **16**.2.4 (where an extra size assumption is made on μ) and in the more general Theorem **16**.2.6. The section ends with Bochner's Theorem **16**.2.7, which characterizes the class of continuous functions of positive type, and with a result on the existence of a weak limit at infinity of a continuous function of positive type.

We give some examples of measures of positive type in Section **16**.3. A classic example in the scalar case is $\mu(\mathrm{d}s) = a(s)\,\mathrm{d}s$ with a locally integrable, positive, nonincreasing, and convex. Propositions **16**.3.5–**16**.3.7 give some useful results describing how positivity is preserved under multiplication and convolution. The last theorem of this section relates the positivity of a function to the negativity of its second distributional derivative.

In Sections **16**.4 and **16**.5, certain subclasses of the measures of positive type are investigated in more detail. We first define measures of *strong positive type*: a measure μ is, roughly speaking, of this type if the real part of its Fourier transform is bounded from below by $\epsilon(1+\omega^2)^{-1}$ for some $\epsilon > 0$. Hypotheses are given under which an absolutely continuous measure $\mu(\mathrm{d}s) = a(s)\,\mathrm{d}s$ is of strong positive type. Next, we define measures of *strict positive type*: a measure μ is said to be of this type if there exists a function $a \in L^1(\mathbf{R}^+; \mathbf{C}^{n\times n})$ satisfying $\langle y, \Re\tilde{a}(\omega)y\rangle > 0$ for all $y \in \mathbf{C}^n \setminus \{0\}$ and all $\omega \in \mathbf{R}$, such that $\nu(\mathrm{d}t) = \mu(\mathrm{d}t) - a(t)\,\mathrm{d}t$ is of positive type. We prove that μ is of strict positive type if and only if for all $\tau \in \mathbf{R}$

$$\inf_{|y|=1} \liminf_{\substack{z\to i\tau \\ \Re z>0}} \big\langle y, \Re\hat{\mu}(z)y\big\rangle > 0.$$

Proposition **16**.4.6 maps out the hypotheses under which a nonnegative, nonincreasing, and convex function is of strict positive type.

Section **16**.5 is devoted to measures of positive type that satisfy a particular inverse coercivity inequality. Specifically, a locally finite matrix-valued measure μ is of *anti-coercive type*, with coercivity constant $q > 0$, if for every $\varphi \in L^2(\mathbf{R}; \mathbf{C}^n)$ with compact support we have $\mu * \varphi \in L^2(\mathbf{R}; \mathbf{C}^n)$ and

$$\|\mu * \varphi\|^2_{L^2(\mathbf{R})} \le q[\![\mu, \varphi]\!].$$

One immediately sees that if a measure is of anti-coercive type then the operator $\varphi \mapsto \mu * \varphi$ maps $L^2(\mathbf{R}; \mathbf{C}^n)$ continuously into itself.

In Theorem **16**.5.6, we characterize measures of anti-coercive type through some equivalence relations. In particular, note that anti-coercivity is equivalent (in the scalar case) to $|\hat{\mu}(z)|^2 \le q\Re\hat{\mu}(z)$ for $\Re z > 0$. Moreover, we record the important fact that if a kernel of positive type has a resolvent then this resolvent is of anti-coercive type. Some examples of anti-coercive measures are given in Propositions **16**.5.7 and **16**.5.8.

In Section **16**.6 we analyse bounds of the form

$$\left| \int_{\mathbf{R}} \langle b(t), \varphi(t) \rangle \, \mathrm{d}t \right|^2 \le q[\![\mu, \varphi]\!],$$

where μ is of positive type, $b \in L^2_{\mathrm{loc}}(\mathbf{R}; \mathbf{C}^n)$, and $\varphi \in L^2(\mathbf{R}; \mathbf{C}^n)$ with compact support. The estimates obtained play a significant role in some of the proofs included in the remaining chapters.

In the first appendix of Chapter 16 we define *semi-inner products*, prove a lemma on these products, and apply it to positive matrices, obtaining some useful inequalities. The second appendix contains some elementary properties of distributions, in particular of *tempered distributions* and of their Fourier and Laplace transforms.

3.17 Chapter 17: Frequency Domain Methods: Basic Results

The analysis concentrates on the two equations

$$x(t) + \int_0^t a(s) g\big(x(t-s)\big) \, \mathrm{d}s = f(t), \quad t \in \mathbf{R}^+, \qquad (\mathbf{17}.1.1)$$

and

$$x'(t) + \int_{[0,t]} \mu(\mathrm{d}s) g\big(x(t-s)\big) = f(t), \quad t \in \mathbf{R}^+; \quad x(0) = x_0, \qquad (\mathbf{17}.1.2)$$

where g is a continuous map of $\mathbf{R}^n$ into itself. Occasionally, we allow g to depend explicitly on t. (Chapter 20 contains results on somewhat more complicated equations.)

The methods that we employ in the analysis of (**17**.1.1) and (**17**.1.2) are based on the theory of kernels of positive type. Despite the fact that the equations are nonlinear, we make frequent use of transform techniques.

Before discussing Chapter 17 in detail, let us briefly mention the distinctive features of each of Chapters 17–19. Chapter 17 analyses (**17**.1.1) and (**17**.1.2) under certain size assumptions on the nonhomogeneous term. For example, to obtain asymptotic statements on the solution x of (**17**.1.2) it is typically assumed that $f \in L^1(\mathbf{R}^+;\mathbf{R}^n)$. Chapter 18 takes a somewhat different approach, interpolating between the differentiated and the undifferentiated version of the particular equation under discussion. Finally, in Chapter 19, we allow nonhomogeneous terms f which vanish at infinity, but do not necessarily satisfy integrability assumptions such as $f \in L^p(\mathbf{R}^+;\mathbf{R}^n)$ for $p < \infty$.

Variants of a particular technique are present in several of the proofs of Chapters 17–19. We sketch this technique in the introductory section of Chapter 17. Section **17**.2 investigates boundedness of solutions of (**17**.1.2). Most of these results are formulated in the context of $\mathbf{R}^n$, and require g to be the gradient of a function $G \in C^1(\mathbf{R}^n;\mathbf{R})$. Observe that the conclusion of the various theorems invariably includes

$$\sup_{T>0} Q(\mu, g\circ x, T) < \infty, \tag{3.1}$$

where the quadratic form Q is defined by

$$Q(\mu, g\circ x, T) = \int_0^T \Big\langle g(x(t)), \int_{[0,t]} \mu(\mathrm{d}s) g\big(x(t-s)\big)\Big\rangle\,\mathrm{d}t. \tag{3.2}$$

In Section **17**.3 we start out from the assumption that x is a bounded solution of (**17**.1.2) which satisfies (3.1) and then gain a more detailed knowledge about the asymptotic behaviour of x. Lemma **17**.3.4 and its consequence Theorem **17**.3.5 are important results. They state that, if μ is of strict positive type, $f \in L^1_{\mathrm{loc}}(\mathbf{R}^+;\mathbf{R}^n)$, g is continuous, and if $x \in BUC(\mathbf{R}^+;\mathbf{R}^n)$ is a locally absolutely continuous solution of (**17**.1.2) on $\mathbf{R}^+$ satisfying (3.1), then $\lim_{t\to\infty} g(x(t)) = 0$.

The case where μ is not strictly positive is considered in Theorems **17**.3.7 and **17**.3.8. It is shown that under the same assumptions as in Theorem **17**.3.5, but with strict positivity replaced by mere positivity, one may conclude that the asymptotic spectrum of $g\circ x$ is contained in the set where the real part of the Fourier transform of μ vanishes.

Section **17**.4 gives hypotheses under which a solution x of (**17**.1.2) satisfies $x \in L^2(\mathbf{R}^+;\mathbf{R}^n)$. In Section **17**.5 we deduce various asymptotic facts concerning solutions of (**17**.1.1).

3.18 Chapter 18: Frequency Domain Methods: Additional Results

We continue our frequency domain analysis by presenting some interpolation results. First, in Section **18**.2, we interpolate between the integrodifferential and the corresponding integral equation. In Section **18**.3 the equation is remoulded by partial integration; in the subsequent section some further methods of recasting the equations are given. Primarily, these methods involve taking the convolution of both sides by a suitable function so that the theory of measures of positive type can be applied. These ideas are applied to integrodifferential equations in Section **18**.5.

The miscellaneous assumptions (in terms of Laplace transforms) on the kernel a in

$$x(t) + \int_0^t a(s)g\big(x(t-s)\big)\,\mathrm{d}s = f(t), \quad t \in \mathbf{R}^+, \tag{18.1.1}$$

and the kernel μ in the integrodifferential equation

$$x'(t) + \int_{[0,t]} \mu(\mathrm{d}s)g\big(x(t-s)\big) = f(t), \quad t \in \mathbf{R}^+; \quad x(0) = x_0, \tag{18.1.2}$$

are summarized in the introductory section.

For the historical records, we note that the ideas of this chapter were among the first to be exploited in the analysis of nonlinear Volterra equations.

3.19 Chapter 19: Combined Lyapunov and Frequency Domain Methods

Here we consider the equation

$$x'(t) + \int_{[0,t]} \mu(\mathrm{d}s)g\big(x(t-s)\big) = f(t), \quad t \in \mathbf{R}^+; \quad x(0) = x_0, \tag{19.1.1}$$

under the weakest possible assumptions on the forcing function f. As a consequence, we have to make stronger assumptions on the kernel and the nonlinearity: in Theorem **19**.2.1 we require that $\int_{\mathbf{R}^+} t|\mu|(\mathrm{d}t) < \infty$, and in Theorem **19**.3.1 we require that g be locally Lipschitz-continuous. We assume throughout that there exists a bounded solution. The principal conclusion is that x is asymptotically slowly varying.

In Section **19**.4 we use resolvent theory to deal with the case where the kernel μ is not necessarily a finite measure.

3.20 Chapter 20: Monotonicity Methods

Chapter 20 assembles miscellaneous results on the qualitative behaviour of solutions of integral equations. The proofs are based on various kinds of monotonicity techniques.

Section **20**.2 is devoted to a brief presentation of *nonconvolution kernels of positive type* under certain convexity assumptions.

The scalar equation

$$x(t) + \int_0^t k(t,s)g\big(s, x(s)\big)\,\mathrm{d}s = f(t), \quad t \in \mathbf{R}^+, \tag{3.3}$$

is studied in Section **20**.3 under the assumption that the kernel k is log-convex, i.e., that it satisfies

$$k(v,s)k(t,u) \le k(t,s)k(v,u), \quad \text{for almost all } s \le u \le v \le t.$$

In Section **20**.4 we define and study kernels of *anti-accretive* (with respect to some L^p-space) and *totally invariant* and *totally positive types.* We give some characterizations of these concepts, and prove results on integral equations with kernels of these categories. In particular, Theorems **20**.4.16

and **20**.4.17 describe kernels which are, respectively, of L^∞-anti-accretive type and of totally positive type.

In our final section, we study nonlinear convolution equations of the form

$$x(t) + \int_0^t k\big(t-s, x(s)\big)\,\mathrm{d}s = f(t), \quad t \in \mathbf{R}^+, \tag{20.5.1}$$

and the corresponding integrodifferential equations

$$x'(t) + \int_0^t k\big(t-s, x(s)\big)\,\mathrm{d}s = f(t), \quad t \in \mathbf{R}^+; \quad x(0) = x_0. \tag{20.5.2}$$

Note that if, for example, one adds several equations of the form (**17**.1.2) then one gets an equation of this kind. We prove some asymptotic results on the solutions of these equations using monotonicity assumptions on k in both of its variables.

4. Exercises

1. Rewrite the equation

$$x(t) = \int_S^t h\big(t, s, x(s)\big)\,\mathrm{d}s + g(t), \quad t \ge S,$$

in the form

$$y(t) = \int_0^t g\big(t, s, y(s)\big)\,\mathrm{d}s, \quad t \ge 0.$$

2. Show that the temperature $T(0,t)$ in (1.1) satisfies equation (1.2).
3. Rewrite the equation

$$x(t) = h\left(\int_0^t k_1\big(t, s, x(s)\big)\,\mathrm{d}s, \ldots, \int_0^t k_m\big(t, s, x(s)\big)\,\mathrm{d}s\right) + f(t), \quad t \ge 0,$$

in the form

$$y(t) = \int_0^t g\big(t, s, y(s)\big)\,\mathrm{d}s, \quad t \ge 0.$$

4. Derive (2.6) from (2.3) and (2.4).
5. Rewrite the equation

$$x(t) = \int_t^T h\big(t, s, x(s)\big)\,\mathrm{d}s + g(t), \quad t \le T < \infty,$$

in the form

$$y(t) = \int_0^t g\big(t, s, y(s)\big)\,\mathrm{d}s, \quad t \ge 0.$$

6. Rewrite the equation

$$\begin{aligned} x'(t) &= -\sum_{i=1}^{\infty} a_i x(t-\tau_i), & t \in \mathbf{R}^+, \\ x(t) &= \varphi(t), & t \in \mathbf{R}^-, \end{aligned}$$

where each $\tau_i \geq 0$, in the form (1.4). What conditions have to be imposed on the coefficients a_i for the resulting measure μ to satisfy $\mu \in M(\mathbf{R}^+)$?

7. Rewrite the equation

$$x(t) = \int_S^{\alpha(t)} h\bigl(t, s, x(\beta(s))\bigr)\, ds + g(t), \quad t \geq S,$$

where $\beta(S) = S$, β is strictly increasing and continuously differentiable, and $\beta(\alpha(t)) \leq t$, in the form

$$y(t) = \int_0^t g\bigl(t, s, y(s)\bigr)\, ds, \quad t \geq 0.$$

References

Z. Akcasu, G. S. Lellouche, and L. S. Shotkin

1. *Mathematical Methods in Nuclear Reactor Dynamics*, Academic Press, New York, 1971.

M. Ash

1. *Nuclear Reactor Kinetics*, McGraw–Hill, New York, 1965.

K. B. Athreya and P. E. Ney

1. *Branching Processes*, Springer-Verlag, Berlin, 1972.

J. A. Burns, E. M. Cliff, and T. L. Herdman

1. A state-space model for an aeroelastic system, in *Proc. of the* 22nd *IEEE Conference on Decision and Control 1983, San Antonio, Texas*, pp. 1074–1077.

J. A. Burns, E. M. Cliff, T. L. Herdman, and J. Turi (Burns *et al.*)

1. On integral transforms appearing in the derivation of the equations of an aeroelastic system, in *Nonlinear Analysis and Applications*, V. Lakshmikantham, ed., Lecture Notes in Pure and Applied Mathematics **109**, Marcel Dekker, New York, 1987, pp. 89–98.

J. A. Burns, T. L. Herdman, and J. Turi

1. Well-posedness of functional differential equations with nonatomic D operators, in *Trends in the Theory and Practice of Non-Linear Analysis*, V. Lakshmikantham, ed., Proc. 6th Int. Conf., Arlington, Math. Stud. **110**, Elsevier, 1985, pp. 71–77.
2. Neutral functional integro-differential equations with weakly singular kernels, *J. Math. Anal. Appl.*, to appear 1989.

V. Capasso, E. Grosso, and S. L. Paveri-Fontana

1. *Mathematics in Biology and Medicine*, V. Capasso, E. Grosso and S. L. Paveri-Fontana, eds., Proceedings, Bari 1983, Lecture Notes in Biomathematics **57**, Springer-Verlag, Berlin, 1985.

R. W. Carr and K. B. Hannsgen

1. A nonhomogeneous integrodifferential equation in Hilbert space, *SIAM J. Math. Anal.* **10** (1979), pp. 961–984.
2. Resolvent formulas for a Volterra equation in Hilbert space, *SIAM J. Math. Anal.* **13** (1982), pp. 459–483.

C. Corduneanu

10. *Integral Equations and Applications*, Cambridge University Press, Cambridge, to appear 1990.

O. Diekmann

1. Limiting behaviour in an epidemic model, *Nonlinear Anal.* **1** (1977), pp. 459–470.
2. Thresholds and travelling waves for the geographical spread of infection, *J. Math. Biol.* **6** (1978), pp. 109–130.

O. Diekmann and H. G. Kaper

1. On the bounded solutions of a nonlinear convolution equation, *Nonlinear Anal.* **2** (1978), pp. 721–737.

W. Feller

1. On the integral equation of renewal theory, *Ann. Math. Statist.* **12** (1941), pp. 243–267.

G. Gripenberg

20. On some epidemic models, *Quart. Appl. Math.* **39** (1981), pp. 317–327.
21. On the convergence of solutions of Volterra equations to almost-periodic functions, *Quart. Appl. Math.* **39** (1981), pp. 363–373.
27. An estimate for the solution of a Volterra equation describing an epidemic, *Nonlinear Anal.* **7** (1983), pp. 161–165.

M. E. Gurtin and R. C. MacCamy

1. Nonlinear age-dependent population dynamics, *Arch. Ratl. Mech. Anal.* **54** (1974), pp. 281–300.

K. B. Hannsgen

11. Uniform L^1 behavior for an integrodifferential equation with parameter, *SIAM J. Math. Anal.* **8** (1977), pp. 626–639.
15. A uniform approximation for an integrodifferential equation with parameter, *J. Integral Equations* **2** (1980), pp. 117–131.

K. B. Hannsgen and R. L. Wheeler

5. Time delays and boundary feedback stabilization in one-dimensional viscoelasticity, in *Proc. Conf. Control of Distributed Parameter Systems*, Vorau, 1986.

F. B. Hanson, A. Klimas, G. V. Ramanathan, and G. Sandri (Hanson *et al.*)

1. Uniformly valid asymptotic solution to a Volterra equation on an infinite interval, *J. Math. Phys.* **14** (1973), pp. 1592–1600.
2. Analysis of a model for transport of charged particles in a random magnetic field, *J. Math. Anal. Appl.* **44** (1973), pp. 786–798.

H. W. Hethcote, M. A. Lewis, and P. Van den Driessche

1. An epidemiological model with a delay and a nonlinear incidence rate, *J. Math. Biol.* **27** (1989), pp. 49–64.

H. W. Hethcote, H. W. Stech, and P. Van den Driessche

1. Periodicity and stability in epidemic models: a survey, in *Differential equations and applications in ecology, epidemics, and population problems*, K. L. Cooke, ed., Proc. Conf., Claremont/Calif. 1981, Academic Press, New York, 1981, pp. 65–82.

W. J. Hrusa, J. A. Nohel, and M. Renardy

1. Initial value problems in viscoelasticity, *Appl. Mech. Rev.* **41** (1988), pp. 371–378.

G. S. Jordan and R. L. Wheeler

6. Rates of decay of resolvents of Volterra equations with certain non-integrable kernels, *J. Integral Equations* **2** (1980), pp. 103–110.

W. O. Kermack and A. G. McKendrick

1. A contribution to the mathematical theory of epidemics, *Proc. Roy. Soc. London Ser. A* **115** (1927), pp. 700–721.

A. J. Klimas and G. Sandri

1. Foundation of the theory of cosmic-ray transport in random magnetic fields, *Astrophys. J.* **169** (1971), pp. 41–56.

N. Levinson

1. A nonlinear Volterra equation arising in the theory of superfluidity, *J. Math. Anal. Appl.* **1** (1960), pp. 1–11.

R. Ling

1. Asymptotic behavior of integral and integrodifferential equations, *J. Math. Phys.* **18** (1977), pp. 1574–1576.

S-O. Londen

1. On the asymptotic behavior of the solution of a nonlinear integrodifferential equation, *SIAM J. Math. Anal.* **2** (1971), pp. 356–367.
3. Stability analysis of nonlinear point reactor kinetics, in *Adv. Nuclear Sci. Tech.* **6**, 1972, pp. 45–63.
4. On some nonlinear Volterra integrodifferential equations, *J. Differential Equations* **11** (1972), pp. 169–179.

A. G. J. MacFarlane

1. *Frequency-Response Methods in Control Systems*, A. G. J. MacFarlane, ed., Selected Reprint Series, IEEE Press, New York, 1979.

J. A. J. Metz and O. Diekmann

1. *The Dynamics of Physiologically Structured Populations*, J. A. J. Metz and O. Diekmann, eds., Lecture Notes in Biomathematics **68**, Springer-Verlag, Berlin, 1986.

R. K. Miller

7. Almost-periodic behavior of solutions of a nonlinear Volterra system, *Quart. Appl. Math.* **28** (1971), pp. 553–570.
8. Asymptotically almost periodic solutions of a nonlinear Volterra system, *SIAM J. Math. Anal.* **2** (1971), pp. 435–444.
12. A system of Volterra integral equations arising in the theory of superfluidity, *An. Ştiinţ. "Al. I. Cuza" Iaşi Secţ. I a Mat (N.S.)* **19** (1973), pp. 349–364.

M. S. Mousa, R. K. Miller, and A. N. Michel

1. Stability analysis of hybrid composite dynamical systems: descriptions involving operators and differential equations, *IEEE Trans. Automat. Control* **31** (1986), pp. 216–226.

J. A. Nohel

3. Nonlinear Volterra equations for heat flow in materials with memory, in *Integral and Functional Differential Equations*, T. L. Herdman, S. M. Rankin III, and H. W. Stech, eds., Lecture Notes in Pure and Applied Mathematics **67**, Marcel Dekker, New York, 1981, pp. 3–82.

M. Z. Podowski

1. Nonlinear stability analysis for a class of differential-integral systems arising from nuclear reactor dynamics, *IEEE Trans. Automat. Control* **31** (1986), pp. 98–107.
2. A study of nuclear reactor models with nonlinear reactivity feedbacks: stability criteria and power overshoot evaluation, *IEEE Trans. Automat. Control* **31** (1986), pp. 108–115.

M. Renardy, W. J. Hrusa, and J. A. Nohel

1. *Mathematical Problems in Viscoelasticity*, Longman Scientific & Technical and J. Wiley, Essex and New York, 1987.

H. Stech and M. Williams

1. Stability in a class of cyclic epidemic models with delay, *J. Math. Biol.* **11** (1981), pp. 95–103.

V. Volterra

4. *Theory of Functionals and of Integral and Integro-Differential Equations*, Dover Publications, New York, 1959.
5. *Leçons sur la Théorie Mathématique de la Lutte pour la Vie*, Gauthier–Villars, Paris, 1931.

G. F. Webb

1. *Theory of Nonlinear Age-Dependent Population Dynamics*, Marcel Dekker, New York, 1985.

J. S. W. Wong and R. Wong

1. Asymptotic solutions of linear Volterra integral equations with singular kernels, *Trans. Amer. Math. Soc.* **189** (1974), pp. 185–200.

2

Linear Convolution Integral Equations

We analyse the equation $x + k * x = f$ ($*$ denotes convolution). We define the resolvent kernel r, and derive the variation of constants formula $x = f - r * f$ for the solution. In addition, we prove the Paley–Wiener theorem, which gives a necessary and sufficient condition for the resolvent to be globally integrable. Some consequences of the Paley–Wiener theorem for the asymptotic behaviour of x are formulated. A short section on convolutions and on Laplace transforms is included.

1. Introduction

This chapter is devoted to a study of linear Volterra integral equations of the type

$$x(t) + \int_0^t k(t-s)x(s)\,\mathrm{d}s = f(t), \quad t \in \mathbf{R}^+. \tag{1.1}$$

Later on we shall consider systems of such equations, but for the time being it suffices to let x, k and f be scalar functions defined on $\mathbf{R}^+$. The functions k and f are known, and we want to find out whether (1.1) has a solution x, and what properties such a solution x may have. We say that k is the *kernel* and that f is the *forcing function*, and we assume that these functions are (at least) locally integrable. Under appropriate additional assumptions on the kernel and on the forcing function, we prove that (1.1) has a unique solution x which behaves roughly in the same way that f does. For example, we give conditions implying that x is bounded on $\mathbf{R}^+$ iff f is bounded on $\mathbf{R}^+$, that x tends to a limit at infinity iff f tends to a limit at infinity, and that x is asymptotically periodic iff f is asymptotically periodic.

First let us show, without going into all the technical details, that similarly to what is true in the theory of linear ordinary differential equations with constant coefficients, the solution of (1.1) can be expressed by the *variation of constants formula*

$$x(t) = f(t) - \int_0^t r(t-s) f(s)\,\mathrm{d}s, \quad t \in \mathbf{R}^+. \tag{1.2}$$

Here r is the analogue of the fundamental solution of an ordinary differential equation and it is called the *resolvent* of the kernel k.

How does one prove the validity of the formula (1.2)? The idea is very simple, and it consists of setting up an iteration to solve (1.1). We define x_m for $m \geq 0$ by

$$\begin{aligned} x_0(t) &= f(t), \\ x_m(t) &= f(t) - \int_0^t k(t-s) x_{m-1}(s)\,\mathrm{d}s, \quad m \geq 1. \end{aligned} \tag{1.3}$$

This set of equations can be solved recursively for x_m, and one gets an expression for x_m which is an alternating sum of $m+1$ terms, and where each term contains one more integral than the preceding. To be able to deal with these cumbersome expressions we introduce a convenient notation for iterated integrals.

Let us define the *convolution* $a * b$ of two functions a and b, given on $\mathbf{R}^+$, by

$$(a * b)(t) = \int_0^t a(t-s) b(s)\,\mathrm{d}s, \quad t \in \mathbf{R}^+.$$

Then the second equation in (1.3) becomes

$$x_m = f - k * x_{m-1},$$

and the solutions x_m of (1.3) are given by

$$x_m = f - k * f + k * (k * f) - \cdots + (-1)^m \underbrace{(k * (k * \cdots * (k}_{m \text{ times}} * f) \ldots)),$$

where the last term contains a total of m convolution factors k and one factor f. We will prove below that if a, b, and c are three locally integrable functions defined on $\mathbf{R}^+$ then $(a * (b * c))(t) = ((a * b) * c)(t)$. In other words, the convolution operator is associative. When we use this fact in the preceding expression for x_m, we obtain

$$x_m = f - r_m * f,$$

where

$$r_m = k - k * k + k * k * k - \cdots + (-1)^{m+1} (\underbrace{k * k * \cdots * k}_{m \text{ times}}).$$

If we let k^{*j}, $j \geq 1$, denote the $(j-1)$-fold convolution of k with itself ($k^{*1} = k$), then

$$r_m = \sum_{j=1}^{m} (-1)^{j-1} k^{*j}.$$

We hope that, as $m \to \infty$, r_m converges to a limit r equal to

$$r = \sum_{j=1}^{\infty} (-1)^{j-1} k^{*j}, \tag{1.4}$$

and that x_m tends to the solution x of (1.1). If this is so, then passing to the limit in the equation $x_m = f - r_m * f$ we get $x = f - r * f$, which is (1.2).

The function r does not only serve as the resolvent of k in the sense that all solutions can be expressed with the aid of formula (1.2), but r itself satisfies a slightly modified version of (1.1). This can be seen as follows. If we factor out $k*$ to the left (or $*k$ to the right) on the right-hand side of (1.4), then we arrive at

$$r = k - k * r = k - r * k,$$

and we conclude that r is the solution of the two *resolvent equations*

$$r(t) + \int_0^t k(t-s) r(s)\, \mathrm{d}s = k(t), \quad t \in \mathbf{R}^+, \tag{1.5}$$

and

$$r(t) + \int_0^t r(t-s) k(s)\, \mathrm{d}s = k(t), \quad t \in \mathbf{R}^+. \tag{1.6}$$

In the scalar case it is trivially true that $k*r = r*k$, so these two equations are the same, and they are the particular case of (1.1) where f has been replaced by k. When we deal with systems of equations, then k and r are square matrices, and for matrix-valued functions it is not in general true that $a * b = b * a$. However, the two preceding equations show that even in the system case the kernel k and the resolvent r satisfy $k * r = r * k$. In this case the two equations above are not quite of the form (1.1), because r and k are square matrices, whereas x and f are column vectors.

The preceding iteration scheme is straightforward and gives good results as far as local existence of solutions of (1.1) is concerned. For global results the method has to be substantially refined.

One interesting fact is that it is not important how one constructs a solution r of (1.5) and (1.6). The essential thing is that the solution r of (1.5) and (1.6), if it exists, is unique, and that the existence of a solution r of (1.5) and (1.6) implies that for every appropriately chosen f the equation (1.1) has a unique solution x. Moreover, this solution is given by (1.2). One can see that this is true as follows. Let x satisfy (1.1), and convolve (1.1) by r to get $r * x + r * k * x = r * f$. By (1.6), this implies $k * x = r * f$. Consequently, $x = f - k * x = f - r * f$, and so x is given by (1.2). Conversely, let y be defined by $y = f - r * f$, convolve this relation by k, and use (1.5) to get $k * y = r * f$. Thus $y = f - r * f = f - k * y$, and y satisfies (1.1). (The same argument, with f replaced by k and x replaced by r, can be used to show that the solution r of (1.5) and (1.6) is unique.)

After this sketchy argument we proceed to a more rigorous treatment. First we establish, in Section 3, local existence of the resolvent of k, and then we continue with global, i.e., asymptotic, considerations in Section 4. Before that, however, we briefly summarize some elementary facts about convolutions and their Laplace transforms.

2. Convolutions and Laplace Transforms

In this section we give a brief description of the convolution of two functions, and of the Laplace and Fourier transforms of functions which are locally or globally integrable. All the proofs are deferred to Section 7. Later on, in Chapters 3, 4 and 16, we shall discuss convolutions where one or both factors are measures or distributions. A brief presentation of the Laplace and Fourier transforms of measures is given is Chapter 3, and Laplace and Fourier transforms of distributions are discussed in Chapter 16.

All of the results in this section are valid both for functions with scalar values in $\mathbf{R}$ or $\mathbf{C}$, and for functions with $(m \times n)$-matrix values, i.e., with values in $\mathbf{R}^{m\times n}$ or $\mathbf{C}^{m\times n}$. Of course, in all the convolution results we have to require that the dimensions of the matrices should be compatible with respect to matrix multiplication.

2.1 Definition

The convolution $a * b$ of two functions a and b defined on $\mathbf{R}$ is the function

$$(a * b)(t) = \int_{-\infty}^{\infty} a(t-s)b(s)\,\mathrm{d}s, \tag{2.1}$$

defined for all $t \in \mathbf{R}$ for which the integral exists. The convolution $a * b$ of two functions a and b defined on an interval $[0, T)$ is the function

$$(a * b)(t) = \int_{0}^{t} a(t-s)b(s)\,\mathrm{d}s, \tag{2.2}$$

which is defined for all those $t \in [0, T)$ for which the integral exists.

In the above $0 < T \leq \infty$, and the integrals are interpreted in the sense of Lebesgue; in particular, the integrals are required to converge absolutely. Observe that the convolution on $[0, T)$ can be regarded as a special case of the convolution on $\mathbf{R}$ in the sense that we get (2.2) from (2.1) by defining $a(t) = b(t) = 0$ for $t \notin [0, T)$, applying the convolution on $\mathbf{R}$, and then restricting $a * b$ to $[0, T)$. The case $[0, \infty)$ is especially important.

Occasionally we use a mixture of the two definitions above. For example, if a is defined on $\mathbf{R}^+$ and b is defined on $\mathbf{R}$, then we integrate from $-\infty$ to t.

In the next theorem, and in Theorems 4.5 and 4.6, we let $U \oplus V$ stand for the space of functions of the type $\{\, f + g \mid f \in U,\ g \in V \,\}$, where U and V are some function spaces, whose intersection contains nothing but the zero function. Most of the time both U and V will be closed subspaces of some other space W, and we let the norm in $U \oplus V$ be the same as the norm in W. (If U and V are not closed subspaces of some common space W, then we let the norm of $u+v \in U \oplus V$ be, e.g., $\|u+v\|_{U\oplus V} = \|u\|_U + \|v\|_V$.) In particular, $C(\mathcal{T}_S) \oplus BC_0(\mathbf{R}^+)$ and $AP(\mathbf{R}^+) \oplus BC_0(\mathbf{R}^+)$ can be identified, respectively, with the spaces of continuous, asymptotically periodic, and continuous, asymptotically almost-periodic functions, both with the sup-norm. Similarly, since $L^p(\mathcal{T}_S)$ is the space of S-periodic L^p-functions, and $BL^p_0(\mathbf{R}^+)$ is the space of locally L^p-integrable functions that tend to zero at infinity in an L^p-sense, we find that $L^p(\mathcal{T}_S) \oplus BL^p_0(\mathbf{R}^+)$ is the space of functions that are asymptotically L^p-periodic, endowed with the norm of BL^p.

2.2 Theorem

Let J be $\mathbf{R}$, $\mathbf{R}^+$, or an interval with left end-point 0 and right end-point T, where $0 < T < \infty$.

(i) *Let $a \in L^1(J)$ and $b \in L^p(J)$ for some $p \in [1,\infty]$. Then $a*b \in L^p(J)$, and $\|a*b\|_{L^p(J)} \le \|a\|_{L^1(J)}\|b\|_{L^p(J)}$. If $p=\infty$, then $a*b \in BUC(J)$, and $\|a*b\|_{\sup(J)} \le \|a\|_{L^1(J)}\|b\|_{L^\infty(J)}$. If $J = \mathbf{R}$ or $\mathbf{R}^+$, and if $b \in L^\infty_\ell(J)$ then $a*b \in BC_\ell(J)$, and if $b \in L^\infty_0(J)$ then $a*b \in BC_0(J)$. In particular, if b belongs to one of the spaces $B^\infty(J)$, $BC(J)$, or $BUC(J)$, or if $J = \mathbf{R}^+$ or $\mathbf{R}$ and b belongs to one of the spaces $L^\infty_\ell(J)$, $L^\infty_0(J)$, $B^\infty_\ell(J)$, $B^\infty_0(J)$, $BC_\ell(J)$, or $BC_0(J)$, then $a*b$ belongs to the same space as b.*

(ii) *Let $J = \mathbf{R}$ or $\mathbf{R}^+$, $a \in L^1(J)$, and $b \in BL^p(J)$ for some $p \in [1,\infty]$. Then $a*b \in BL^p(J)$, and $\|a*b\|_{BL^p(J)} \le \|a\|_{L^1(J)}\|b\|_{BL^p(J)}$. If $b \in BL^p_0(J)$, then $a*b \in BL^p_0(J)$.*

(iii) *Let $a \in L^1(J)$ and $b \in BV(J)$. Then $a*b \in BV(J)$, and $\|a*b\|_{BV(J)} \le \|a\|_{L^1(J)}\|b\|_{BV(J)}$.*

(iv) *Let $J = \mathbf{R}$ or $\mathbf{R}^+$, $a \in L^1(J)$, and $b \in BBV(J)$. Then $a*b \in BBV(J)$, and $\|a*b\|_{BBV(J)} \le \|a\|_{L^1(J)}\|b\|_{BBV(J)}$. If $b \in BBV_0(J)$, then $a*b \in BBV_0(J)$.*

(v) *Let $a \in L^1(\mathbf{R})$. If $S>0$, and b belongs to one of the spaces $L^p(\mathcal{T}_S)$, $p \in [1,\infty]$, $B^\infty(\mathcal{T}_S)$, $C(\mathcal{T}_S)$, $BV(\mathcal{T}_S)$, or $AP(\mathbf{R})$, then $a*b$ belongs to the same space as b.*

(vi) *Let $a \in L^1(J)$. If $J = \mathbf{R}$ or $\mathbf{R}^+$, $S>0$, and b belongs to one of the spaces $L^p(\mathcal{T}_S) \oplus BL^p_0(J)$, $p \in [1,\infty]$, $B^\infty(\mathcal{T}_S) \oplus B^\infty_0(J)$, $C(\mathcal{T}_S) \oplus BC_0(J)$, $BV(\mathcal{T}_S) \oplus BBV_0(J)$, or $AP(J) \oplus BC_0(J)$, then $a*b$ belongs to the same space as b.*

(vii) *Let $J = \mathbf{R}$ or $\mathbf{R}^+$, and let $1<p,\,q<\infty$, $1/p+1/q=1$, $a \in L^p(J)$ and $b \in L^q(J)$. Then $a*b \in BC_0(J)$ and $\|a*b\|_{\sup(J)} \le \|a\|_{L^p(J)}\|b\|_{L^q(J)}$.*

(viii) *If $a \in L^p(J)$, $b \in L^1(J)$ and $c \in L^q(J)$, for some p and q in $[1,\infty]$ with $1/p+1/q=1$, then $((a*b)*c)(t) = (a*(b*c))(t)$, $((b*c)*a)(t) = (b*(c*a))(t)$ and $((c*a)*b)(t) = (c*(a*b))(t)$ for all $t \in J$.*

(ix) *If a, $b \in L^1(J)$, and $c \in BL^1(J)$, then $(a*b)*c = a*(b*c)$, $(b*c)*a = b*(c*a)$, and $(c*a)*b = c*(a*b)$.*

(x) *If a or b is scalar-valued, then $a * b = b * a$.*

The analogous results that one gets by replacing $a * b$ in (i)–(vi) by $b * a$ are true as well.

The proof of Theorem 2.2 is given in Section 7.

2.3 Corollary

(i) *Let $a \in L^1_{\mathrm{loc}}(\mathbf{R}^+)$, and $b \in L^p_{\mathrm{loc}}(\mathbf{R}^+)$ for some $p \in [1, \infty]$. Then $a * b \in L^p_{\mathrm{loc}}(\mathbf{R}^+)$. Moreover, if $c \in L^1_{\mathrm{loc}}(\mathbf{R}^+)$, then $(a*b)*c = a*(b*c)$ and $c * (a * b) = (c * a) * b$.*

(ii) *Let $a \in L^1_{\mathrm{loc}}(\mathbf{R}^+)$, and $b \in BV_{\mathrm{loc}}(\mathbf{R}^+)$. Then $a * b \in BV_{\mathrm{loc}}(\mathbf{R}^+)$.*

(iii) *Let $a \in L^p_{\mathrm{loc}}(\mathbf{R}^+)$ and $b \in L^q_{\mathrm{loc}}(\mathbf{R}^+)$, with $1 \le p, q \le \infty$ and $1/p + 1/q = 1$. Then $a * b \in C(\mathbf{R}^+)$ and $(a * b)(0) = 0$. Moreover, if $c \in L^1_{\mathrm{loc}}(\mathbf{R}^+)$, then $((c*a)*b)(t) = (c*(a*b))(t)$ and $((b*c)*a)(t) = (b * (c * a))(t)$ for all $t \ge 0$.*

This follows directly from Theorem 2.2 because the restriction of $a * b$ to a finite interval $[0, T)$ depends only on the restrictions of a and b to $[0, T)$.

There is a partial converse to Theorem 2.2(i).

2.4 Theorem

Let J be $\mathbf{R}$, or $\mathbf{R}^+$, or an interval with left end-point 0 and right end-point T, where $0 < T < \infty$, and let a be measurable on J. Then each of the following conditions implies that $a \in L^1(J)$:

(i) *$a * b \in L^\infty(J)$ for every $b \in L^\infty(J)$;*

(ii) *$a * b \in L^\infty(J)$ for every $b \in BC_0(J)$;*

(iii) *$a * b \in L^1(J)$ for every $b \in L^1(J)$.*

The proof is given in Section 7.

On each finite interval $[0, T]$, the convolution operator $b \mapsto a*b$ is compact in $L^p(0, T)$, for all $p \in [1, \infty]$.

2.5 Theorem

Let $0 < T < \infty$, $a \in L^1(0, T)$, and let B be a bounded subset of $L^p(0, T)$, for some $p \in [1, \infty]$. Then the set $\{\, a * b \mid b \in B \,\}$ is totally bounded in $L^p(0, T)$.

The proof of this, too, is given in Section 7. (A linear operator is compact iff it maps bounded sets into totally bounded sets, i.e., into sets that can be covered by finitely many balls of radius ϵ, for every $\epsilon > 0$.)

Next, we discuss Laplace and Fourier transforms.

2.6 Definition

The Laplace transform $\hat{a}$ of a function a, defined on $\mathbf{R}^+$, is the function

$$\hat{a}(z) = \int_0^\infty \mathrm{e}^{-zt} a(t)\,\mathrm{d}t,$$

defined for those $z \in \mathbf{C}$ for which the integral exists as a (absolutely convergent) Lebesgue integral. The (bilateral) Laplace transform of a function a, defined on $\mathbf{R}$, is the function

$$\hat{a}(z) = \int_{-\infty}^\infty \mathrm{e}^{-zt} a(t)\,\mathrm{d}t,$$

which is defined for those $z \in \mathbf{C}$ for which the integral exists.

Observe that in both cases we have $\hat{a}(z) = (e_z * a)(0)$, where $e_z(t) = \mathrm{e}^{zt}$ for $t \in \mathbf{R}$. (In the former case, define $a(t) = 0$ for $t < 0$.) Clearly, if a is matrix-valued, then so is $\hat{a}$.

In the above we have required the Laplace transforms to converge in an absolute sense. In the literature it is quite common to use instead a weaker, Riemann-type convergence condition in the definition of Laplace transforms, but we will not do so here.

In harmonic analysis the *Fourier transform* is used more frequently than the bilateral Laplace transform. They differ from each other only by a rotation in the complex plane; the Fourier transform $\tilde{a}$ of a being defined by $\tilde{a}(z) = \hat{a}(\mathrm{i}z)$ for $z \in \mathbf{C}$. Most of the time when we use Fourier transforms we let the argument z be real.

2.7 Theorem

(i) *If $a \in L^1(\mathbf{R})$, then $\hat{a}(z)$ is defined and continuous on the line $\Re z = 0$, $\hat{a}(\mathrm{i}\omega) \to 0$ as $|\omega| \to \infty$, and $\hat{a}(\mathrm{i}\omega) = 0$ for all $\omega \in \mathbf{R}$ if and only if $a(t) = 0$ a.e.*

(ii) *If $a \in L^1_{\mathrm{loc}}(\mathbf{R})$, and $\hat{a}(z_0)$ is defined for some $z_0 \in \mathbf{C}$, then $\hat{a}(z)$ is defined on the vertical line $\Re z = \Re z_0$.*

(iii) *If $a \in L^1_{\mathrm{loc}}(\mathbf{R})$ and $b \in L^1_{\mathrm{loc}}(\mathbf{R})$, then $\widehat{(a * b)}(z) = \hat{a}(z)\hat{b}(z)$ for all those $z \in \mathbf{C}$ for which both $\hat{a}(z)$ and $\hat{b}(z)$ are defined.*

The statement in (i) that $\hat{a}(\mathrm{i}\omega) \to 0$ as $|\omega| \to \infty$ is commonly called the Riemann–Lebesgue lemma.

2.8 Theorem

(i) *If $a \in L^1(\mathbf{R}^+)$, then $\hat{a}(z)$ is defined and continuous in the closed half plane $\Re z \geq 0$, it is analytic in $\Re z > 0$, and $\hat{a}(z) \to 0$ as $|z| \to \infty$ in the half plane $\Re z \geq 0$. Moreover, $\hat{a}(z) = 0$ for all z in the half plane $\Re z \geq 0$ if and only if $a(t) = 0$ a.e.*

(ii) *If $a \in L^1_{\mathrm{loc}}(\mathbf{R}^+)$, and $\hat{a}(z_0)$ is defined for some $z_0 \in \mathbf{C}$, then $\hat{a}(z)$ is defined in the closed half plane $\Re z \geq \Re z_0$.*

(iii) *If $a \in L^1_{\mathrm{loc}}(\mathbf{R}^+)$ and $b \in L^1_{\mathrm{loc}}(\mathbf{R}^+)$, then $\widehat{(a * b)}(z) = \hat{a}(z)\hat{b}(z)$ for all those $z \in \mathbf{C}$ for which both $\hat{a}(z)$ and $\hat{b}(z)$ are defined.*

Observe that Theorem 2.7 uses bilateral Laplace transforms and Theorem 2.8 standard Laplace transforms. The proofs of these two theorems are given in Section 7.

The following properties of Laplace transforms are easy to verify, so their proofs are left to the reader.

2.9 Theorem
Let a and b be two locally integrable functions defined on $\mathbf{R}$ or $\mathbf{R}^+$.

(i) *If c is a complex constant, $a_c(t) = \mathrm{e}^{-ct}a(t)$, and $b_c(t) = \mathrm{e}^{-ct}b(t)$, then $(a_c * b_c)(t) = \mathrm{e}^{-ct}(a * b)(t)$, and $\widehat{a_c}(z) = \hat{a}(z + c)$.*

(ii) *If $a_c(t) = a(t/c)$ for some $c > 0$, then $\widehat{a_c}(z) = c\hat{a}(cz)$.*

(iii) *If a is defined on $\mathbf{R}$, c is real, and $a_c(t) = a(t+c)$, then $\widehat{a_c}(z) = \mathrm{e}^{cz}\hat{a}(z)$.*

3. Local Existence and Uniqueness

Let us return to the Volterra integral equation (1.1). With the convolution notation, this equation can be rewritten in the form

$$x(t) + (k * x)(t) = f(t), \quad t \in \mathbf{R}^+. \tag{3.1}$$

If we suppose that $k \in L^1_{\mathrm{loc}}(\mathbf{R}^+; \mathbf{C}^{n\times n})$ and that $x \in L^p_{\mathrm{loc}}(\mathbf{R}^+; \mathbf{C}^n)$ for some $p \in [1, \infty]$, then, by Corollary 2.3(i), f belongs to $L^p_{\mathrm{loc}}(\mathbf{R}^+; \mathbf{C}^n)$. The converse question is the interesting one: if $k \in L^1_{\mathrm{loc}}(\mathbf{R}^+; \mathbf{C}^{n\times n})$ and $f \in L^p_{\mathrm{loc}}(\mathbf{R}^+; \mathbf{C}^n)$, then does it follow that (3.1) has a solution $x \in L^p_{\mathrm{loc}}(\mathbf{R}^+; \mathbf{C}^n)$? As we shall see below, the answer to this question is affirmative.

Before we continue, let us agree on what we mean by a solution of (3.1). When all the terms in (3.1) are defined everywhere (e.g., if they are continuous), then we call x a solution of (3.1) on an interval $[0, T]$ or $[0, T)$ if (3.1) holds for all t in this interval. In the L^p-case, where one or several of the terms need not be everywhere defined, we call x a solution of (3.1) on an interval $[0, T)$ if (3.1) holds for almost all $t \in [0, T)$.

To prove that (3.1) has a solution in $L^p_{\mathrm{loc}}(\mathbf{R}^+; \mathbf{C}^n)$ whenever $k \in L^1_{\mathrm{loc}}(\mathbf{R}^+; \mathbf{C}^{n\times n})$ and $f \in L^p_{\mathrm{loc}}(\mathbf{R}^+; \mathbf{C}^n)$, we first show that (3.1) has a 'fundamental' solution r. We give the space $L^1_{\mathrm{loc}}(\mathbf{R}^+; \mathbf{C}^{n\times n})$ the topology of convergence in the L^1-norm on each compact subset of $\mathbf{R}^+$.

3.1 Theorem
Let $k \in L^1_{\mathrm{loc}}(\mathbf{R}^+; \mathbf{C}^{n\times n})$. Then there is a solution $r \in L^1_{\mathrm{loc}}(\mathbf{R}^+; \mathbf{C}^{n\times n})$ of the two equations $r + k * r = r + r * k = k$. This solution r is unique and depends continuously on k in the topology of $L^1_{\mathrm{loc}}(\mathbf{R}^+; \mathbf{C}^{n\times n})$.

It will be obvious from the proof given below that if $k \in L^1_{\mathrm{loc}}(\mathbf{R}^+; \mathbf{R}^{n\times n})$, then $r \in L^1_{\mathrm{loc}}(\mathbf{R}^+; \mathbf{R}^{n\times n})$.

Proof of Theorem 3.1 First, let us show that a solution $r \in L^1([0,T];\mathbf{C}^{n\times n})$ of the equations $r + k * r = r + r * k = k$ on an interval $[0,T]$ has to be unique. Suppose that $q \in L^1([0,T];\mathbf{C}^{n\times n})$ satisfies $q + k * q = k$, and that $r \in L^1([0,T];\mathbf{C}^{n\times n})$ satisfies $r + r * k = k$. Then

$$r = k - r * k = k - r * (q + k * q) = k - (r + r * k) * q = k - k * q = q.$$

Thus we have uniqueness.

Next, let us observe that it suffices to prove that for each $T \in (0,\infty)$ there is a function $r_T \in L^1([0,T];\mathbf{C}^{n\times n})$ satisfying

$$r_T + k * r_T = r_T + r_T * k = k$$

on $[0,T]$. If this is true, then by the uniqueness proved above, for each $j \in N$, the restriction of r_{j+1} to $[0,j)$ must be equal to r_j a.e. We get a unique solution $r \in L^1_{\text{loc}}(\mathbf{R}^+;\mathbf{C}^{n\times n})$ of $r + k * r = r + r * k = k$ on $\mathbf{R}^+$ by defining $r(t) = r_{j+1}(t)$ for $t \in [j, j+1)$.

In the special case where $\int_0^T |k(t)|\,\mathrm{d}t < 1$, we can use the iteration method outlined in Section 1 to get the function r_T. Define

$$r_m = \sum_{j=1}^{m} (-1)^{j-1} k^{*j},$$

where k^{*j} is the $(j-1)$-fold convolution of k by itself. By Theorem 2.2(i),

$$\|k^{*j}\|_{L^1(0,T)} \leq \|k\|^j_{L^1(0,T)}.$$

This means that $\{r_m\}_{m=1}^{\infty}$ is a Cauchy sequence in $L^1([0,T];\mathbf{C}^{n\times n})$. Therefore, since $L^1([0,T];\mathbf{C}^{n\times n})$ is complete, there exists a function $r_T \in L^1([0,T];\mathbf{C}^{n\times n})$ to which r_m tends in $L^1([0,T];\mathbf{C}^{n\times n})$ as $m \to \infty$. Each r_m, $m \geq 2$, satisfies

$$r_m + k * r_{m-1} = r_m + r_{m-1} * k = k.$$

By Theorem 2.2(i), $k * r_{m-1} \to k * r$ and $r_{m-1} * k \to r * k$ in $L^1([0,T];\mathbf{C}^{n\times n})$; hence r must satisfy $r + k * r = r + r * k = k$.

To finish the existence proof, it suffices to show that we may always, without loss of generality, take $\int_0^T |k(t)|\,\mathrm{d}t < 1$. By Lebesgue's dominated convergence theorem, the function $\mathrm{e}^{-\sigma t}k(t)$ tends to 0 in $L^1([0,T];\mathbf{C}^{n\times n})$ as $\sigma \to \infty$, and so, by choosing σ large enough, we may assume that the function $a(t) = \mathrm{e}^{-\sigma t}k(t)$, $t \in [0,T]$, satisfies $\|a\|_{L^1(0,T)} < 1$. Suppose that the equations $q + a * q = q + q * a = a$ have a solution q in $L^1([0,T];\mathbf{C}^{n\times n})$. Define

$$r(t) = \mathrm{e}^{\sigma t} q(t), \quad t \in [0,T].$$

Then $r \in L^1([0,T];\mathbf{C}^{n\times n})$, and it follows from the definition of the convolution that r satisfies $r + k * r = r + r * k = k$. Thus we have found a solution $r \in L^1([0,T];\mathbf{C}^{n\times n})$ of the original equation.

We leave the proof of the continuous dependence to the reader. (This claim follows immediately from Lemma **9**.3.11.) □

3.2 Definition
The function r in Theorem 3.1 is called the resolvent of k.

In the course of the proof of Theorem 3.1 we make the following two observations that will be useful later.

3.3 Lemma
If $r \in L^1_{\text{loc}}(\mathbf{R}^+; \mathbf{C}^{n\times n})$ is the resolvent of $k \in L^1_{\text{loc}}(\mathbf{R}^+; \mathbf{C}^{n\times n})$, and σ is a complex constant, then the function $e^{\sigma t} r(t)$ is the resolvent of $e^{\sigma t} k(t)$.

3.4 Lemma
If the function $e^{-\sigma t} k(t)$ belongs to $L^1(\mathbf{R}^+; \mathbf{C}^{n\times n})$ for some $\sigma \in \mathbf{R}$, then there is a constant c such that the function $e^{-ct} r(t)$ belongs to $L^1(\mathbf{R}^+; \mathbf{C}^{n\times n})$, where r is the resolvent of k.

In Corollary 4.2 we will get a much sharper version of Lemma 3.4 (the conclusion of Lemma 3.4 is true for every $c \geq \sigma$ such that $\det[I + \hat{k}(z)] \neq 0$ in the half plane $\Re z \geq c$).

Observe that if k is $\mathbf{R}^{n\times n}$-valued then it follows from the preceding construction that r is $\mathbf{R}^{n\times n}$-valued, too.

With the resolvent at our disposal we can solve (3.1).

3.5 Theorem
Let $k \in L^1_{\text{loc}}(\mathbf{R}^+; \mathbf{C}^{n\times n})$. Then, for every $f \in L^1_{\text{loc}}(\mathbf{R}^+; \mathbf{C}^n)$, there is a unique solution $x \in L^1_{\text{loc}}(\mathbf{R}^+; \mathbf{C}^n)$ of (3.1). This solution is given by the variation of constants formula

$$x(t) = f(t) - (r * f)(t), \quad t \in \mathbf{R}^+, \tag{3.2}$$

where r is the resolvent of k. If V is one of the spaces $L^p_{\text{loc}}(\mathbf{R}^+; \mathbf{C}^n)$, with $1 \leq p \leq \infty$, $B^\infty_{\text{loc}}(\mathbf{R}^+; \mathbf{C}^n)$, $C(\mathbf{R}^+; \mathbf{C}^n)$, or $BV_{\text{loc}}(\mathbf{R}^+; \mathbf{C}^n)$, and if $f \in V$, then $x \in V$.

Proof Let $f \in L^1_{\text{loc}}(\mathbf{R}^+; \mathbf{C}^n)$ and define x by (3.2). Then $x \in L^1_{\text{loc}}(\mathbf{R}^+; \mathbf{C}^n)$, and

$$x + k * x = x + k * (f - r * f) = x + (k - k * r) * f = x + r * f = f,$$

so x satisfies (3.1).

Conversely, let x be an arbitrary solution of (3.1) in $L^1_{\text{loc}}(\mathbf{R}^+; \mathbf{C}^n)$. Then $f \in L^1_{\text{loc}}(\mathbf{R}^+; \mathbf{C}^n)$, and

$$x = f - k * x = f - (r + r * k) * x = f - r * (x + k * x) = f - r * f,$$

and so x is given by (3.2).

It follows from Corollary 2.3 that x belongs to the same space as f. □

To get an explicit solution of the resolvent equation one can sometimes use Laplace transforms. Suppose that for some sufficiently large σ the function $k(t)e^{-\sigma t}$ belongs to $L^1(\mathbf{R}^+; \mathbf{C}^{n\times n})$. Then $\hat{k}(z)$ exists for $\Re z \geq \sigma$, and, by Lemma 3.4, $\hat{r}(z)$ exists in some other half plane $\Re z \geq c$. Using

Theorem 2.8 we conclude that the resolvent equation may be transformed, and that

$$\hat{r}(z) = \hat{k}(z)[I + \hat{k}(z)]^{-1}, \quad \Re z \geq \max\{\sigma, c\}.$$

An inversion of the Laplace transform (which occasionally can be done in a closed form) gives r.

4. Asymptotic Behaviour and the Paley–Wiener Theorems

Suppose that we know something about the asymptotic behaviour of the forcing function f in (3.1). For example, f could be bounded or asymptotically periodic, or f might satisfy $f \in L^p(\mathbf{R}^+; \mathbf{C}^n)$ or $f \in BL^p(\mathbf{R}^+; \mathbf{C}^n)$ for some $p \in [1, \infty]$. What conditions do we need on the kernel k in order to conclude that the solution x of (3.1) has the same asymptotic behaviour as f?

If $f \in L^1(\mathbf{R}^+; \mathbf{C}^n)$, or $f \in L^\infty(\mathbf{R}^+; \mathbf{C}^n)$, then a necessary and sufficient condition exists, namely $r \in L^1(\mathbf{R}^+; \mathbf{C}^{n\times n})$. This is true because, by Theorems 2.2(i) and 2.4, the convolution operator $r*$ maps $L^1(\mathbf{R}^+; \mathbf{C}^n)$ into $L^1(\mathbf{R}^+; \mathbf{C}^n)$ iff $r \in L^1(\mathbf{R}^+; \mathbf{C}^{n\times n})$, and similarly, it maps $L^\infty(\mathbf{R}^+; \mathbf{C}^n)$ into $L^\infty(\mathbf{R}^+; \mathbf{C}^n)$ iff $r \in L^1(\mathbf{R}^+; \mathbf{C}^{n\times n})$. In the remaining cases mentioned above, we see, by Theorem 2.2, that the condition $r \in L^1(\mathbf{R}^+; \mathbf{C}^{n\times n})$ is sufficient to imply that x behaves asymptotically in the same way that f does. If k is integrable, then the integrability of r turns out to be a necessary condition as well.

When is it true that $r \in L^1(\mathbf{R}^+; \mathbf{C}^{n\times n})$? As we shall see below, if $k \in L^1(\mathbf{R}^+; \mathbf{C}^{n\times n})$, then one can give a complete answer to this question in terms of a Laplace transform condition. If $k \notin L^1(\mathbf{R}^+; \mathbf{C}^{n\times n})$, then it is possible to find some necessary conditions, but in this case the problem of finding sharp sufficient conditions is very much open. (We return to this problem in Chapter 6.)

The case where $k \in L^1(\mathbf{R}^+; \mathbf{C}^{n\times n})$ is settled by the following theorem.

4.1 The Half Line Paley–Wiener Theorem
Let $k \in L^1(\mathbf{R}^+; \mathbf{C}^{n\times n})$. Then the resolvent r of k satisfies

$$r \in L^1(\mathbf{R}^+; \mathbf{C}^{n\times n})$$

if and only if

$$\det[I + \hat{k}(z)] \neq 0, \quad \Re z \geq 0.$$

It is easy to see that the Laplace transform condition is necessary. For, if $r \in L^1(\mathbf{R}^+; \mathbf{C}^{n\times n})$ and $r + k * r = k$, then we can transform the equation to get

$$\hat{r}(z) + \hat{k}(z)\hat{r}(z) = \hat{k}(z), \quad \Re z \geq 0$$

(cf. Theorem 2.8). Equivalently,

$$[I + \hat{k}(z)]\,[I - \hat{r}(z)] = I, \quad \Re z \geq 0. \tag{4.1}$$

Clearly this implies that $\det[I + \hat{k}(z)] \neq 0$, $\Re z \geq 0$.

We return later to the more difficult sufficiency part of the proof of Theorem 4.1.

An extension of Theorem 4.1 is given in Section **7**.2, where it is assumed that $\det[I+\hat{k}(z)]$ has no zeros on the imaginary axis, but does have zeros in the open half plane $\Re z > 0$. The case where we have zeros on the imaginary axis is treated in Section **7**.3.

Theorem 4.1 implies the following sharper version of Lemma 3.4.

4.2 Corollary

If the function $\mathrm{e}^{-\sigma t}k(t)$ *belongs to* $L^1(\mathbf{R}^+; \mathbf{C}^{n\times n})$ *for some* $\sigma \in \mathbf{R}$*, then the function* $\mathrm{e}^{-ct}r(t)$*, where* r *is the resolvent of* k*, belongs to* $L^1(\mathbf{R}^+; \mathbf{C}^{n\times n})$ *for each constant* $c \geq \sigma$ *such that* $\det[I + \hat{k}(z)] \neq 0$ *for* $\Re z \geq c$.

In the above we assumed that k and r are defined on $\mathbf{R}^+$, and we defined the convolutions $k * r$ and $r * k$ accordingly. In some instances it is useful to study the more general problem where $k \in L^1(\mathbf{R}; \mathbf{C}^{n\times n})$ and $r \in L^1(\mathbf{R}; \mathbf{C}^{n\times n})$, and where k and r satisfy the whole line versions of the resolvent equations, i.e.,

$$r(t) + \int_{-\infty}^{\infty} k(t-s)r(s)\,\mathrm{d}s = r(t) + \int_{-\infty}^{\infty} r(t-s)k(s)\,\mathrm{d}s = k(t), \quad t \in \mathbf{R},$$

or, with our notation for convolutions,

$$r(t) + (k * r)(t) = r(t) + (r * k)(t) = k(t), \quad t \in \mathbf{R}. \tag{4.2}$$

This problem comes up naturally in the course of the proof of Theorem 4.1, and it is of interest in its own right.

The analogue of Theorem 4.1 for equation (4.2) can be stated as follows.

4.3 The Whole Line Paley–Wiener Theorem

Let $k \in L^1(\mathbf{R}; \mathbf{C}^{n\times n})$*. Then there is a function*

$$r \in L^1(\mathbf{R}; \mathbf{C}^{n\times n})$$

satisfying the two equations $r + k * r = r + r * k = k$*, if and only if*

$$\det[I + \hat{k}(z)] \neq 0, \quad \Re z = 0.$$

This function r *is unique.*

4.4 Definition
The function r in Theorem 4.3 is called the whole line resolvent of k.

Observe that Theorem 4.3 uses bilateral Laplace transforms as opposed to the standard Laplace transform used in Theorem 4.1. Again, the necessity of the Laplace transform condition is easily proved.

The proof of the sufficiency of the conditions on the determinants in Theorems 4.1 and 4.3 is rather long and somewhat technical. Therefore, let us postpone it for another moment, and show instead how Theorems 4.1 and 4.3 can be applied, respectively, to (3.1), and to the whole line equation

$$x(t) + \int_{-\infty}^{\infty} k(t-s)x(s)\,\mathrm{d}s = f(t), \quad t \in \mathbf{R}.$$

With the convolution notation this equation becomes

$$x(t) + (k * x)(t) = f(t), \quad t \in \mathbf{R}. \tag{4.3}$$

4.5 Theorem
Let $k \in L^1(\mathbf{R}^+; \mathbf{C}^{n\times n})$, and let V be one of the following spaces:

(1) $L^p(\mathbf{R}^+; \mathbf{C}^n)$, $p \in [1,\infty]$, $L^\infty_\ell(\mathbf{R}^+; \mathbf{C}^n)$, *or* $L^\infty_0(\mathbf{R}^+; \mathbf{C}^n)$;
(2) $B^\infty(\mathbf{R}^+; \mathbf{C}^n)$, $B^\infty_\ell(\mathbf{R}^+; \mathbf{C}^n)$, *or* $B^\infty_0(\mathbf{R}^+; \mathbf{C}^n)$;
(3) $BC(\mathbf{R}^+; \mathbf{C}^n)$, $BUC(\mathbf{R}^+; \mathbf{C}^n)$, $BC_\ell(\mathbf{R}^+; \mathbf{C}^n)$, *or* $BC_0(\mathbf{R}^+; \mathbf{C}^n)$;
(4) $BL^p(\mathbf{R}^+; \mathbf{C}^n)$, *or* $BL^p_0(\mathbf{R}^+; \mathbf{C}^n)$, $p \in [1,\infty]$;
(5) $BV(\mathbf{R}^+; \mathbf{C}^n)$;
(6) $BBV(\mathbf{R}^+; \mathbf{C}^n)$, *or* $BBV_0(\mathbf{R}^+; \mathbf{C}^n)$;
(7) $L^p(\mathcal{T}_S; \mathbf{C}^n) \oplus BL^p_0(\mathbf{R}^+; \mathbf{C}^n)$, $p \in [1,\infty]$, $B^\infty(\mathcal{T}_S; \mathbf{C}^n) \oplus B^\infty_0(\mathbf{R}^+; \mathbf{C}^n)$, $C(\mathcal{T}_S; \mathbf{C}^n) \oplus BC_0(\mathbf{R}^+; \mathbf{C}^n)$, *or* $BV(\mathcal{T}_S; \mathbf{C}^n) \oplus BBV_0(\mathbf{R}^+; \mathbf{C}^n)$, *where* $S > 0$;
(8) $AP(\mathbf{R}^+; \mathbf{C}^n) \oplus BC_0(\mathbf{R}^+; \mathbf{C}^n)$.

Then the following four conditions are equivalent:

(i) *the resolvent r of k belongs to $L^1(\mathbf{R}^+; \mathbf{C}^{n\times n})$;*
(ii) $\det[I + \hat{k}(z)] \neq 0$ *for all $z \in \mathbf{C}$ with $\Re z \geq 0$;*
(iii) *for every $f \in V$, the solution x of the equation $x + k * x = f$ on $\mathbf{R}^+$ belongs to V;*
(iv) *for every $f \in V$, the solution x of the equation $x + k * x = f$ on $\mathbf{R}^+$ belongs to V, and x depends continuously on f (in the norm of V).*

That (i) and (ii) are equivalent is exactly the conclusion of the Half Line Paley–Wiener Theorem 4.1. Obviously, (iv) implies (iii). To see that (i) implies (iv) one uses the variation of constants formula (3.2) and Theorem 2.2. The rest of the proof is given in Section 5.

A part of Theorem 4.5 is extended in Section **7**.2 to the case where $\det[I + \hat{k}(z)]$ has zeros in the open right half plane, and in Section **7**.3 to the case where there are zeros on the imaginary axis.

When we consider equations on the entire real line, the results are practically the same as those for the half line. (There is one exceptional case that does not appear in Theorems 4.5 and 4.6, namely the case where f is periodic. See Theorem 4.7 below.)

4.6 Theorem
Let $k \in L^1(\mathbf{R}; \mathbf{C}^{n\times n})$, and let V be one of the spaces (1)–(8) in Theorem 4.5, with $\mathbf{R}^+$ replaced by $\mathbf{R}$, or
(9) $AP(\mathbf{R}; \mathbf{C}^n)$.
Then the following four conditions are equivalent:
(i) *k has a whole line resolvent, which belongs to $L^1(\mathbf{R}; \mathbf{C}^{n\times n})$;*
(ii) *$\det[I + \hat{k}(z)] \neq 0$ for all $z \in \mathbf{C}$ with $\Re z = 0$;*
(iii) *for every $f \in V$, the equation $x + k * x = f$ has a unique solution x in V;*
(iv) *for every $f \in V$, the equation $x + k * x = f$ has a unique solution x in V, and x depends continuously on f (in the norm of V).*
When these conditions hold, then the solution x in (iii) and (iv) is given by $x = f - r * f$, where r is the whole line resolvent of k.

Observe that in (iii) and (iv) the solution is required to be *unique in* V; it does not preclude the possibility that the equation may have a solution that does not belong to any of the spaces V listed above.

The equivalence above of (i) and (ii) is the conclusion of the Whole Line Paley–Wiener Theorem 4.3. To prove the variation of constants formula $x = f - r * f$, one simply repeats the algebraic manipulations in the proof of Theorem 3.5. This formula, together with Theorem 2.2, gives the implication (i) $\Rightarrow$(iv). Clearly, (iv) implies (iii). The rest of the proof is given in Section 5.

Finally, let us consider the case of periodic functions.

4.7 Theorem
Let $k \in L^1(\mathbf{R}; \mathbf{C}^{n\times n})$, $S > 0$, and let V be one of the following spaces:
(1) $L^p(\mathcal{T}_S; \mathbf{C}^n)$, $1 \le p \le \infty$;
(2) $B^\infty(\mathcal{T}_S; \mathbf{C}^n)$;
(3) $C(\mathcal{T}_S; \mathbf{C}^n)$;
(4) $BV(\mathcal{T}_S; \mathbf{C}^n)$.
Then the following four conditions are equivalent:
(i) *there exists a function $r_S \in L^1(\mathcal{T}_S; \mathbf{C}^{n\times n})$, satisfying*

$$r_S(t) + \int_0^S k_S(t-s) r_S(s)\,\mathrm{d}s = r_S(t) + \int_0^S r_S(t-s) k_S(s)\,\mathrm{d}s = k_S(t), \quad t \in \mathbf{R},$$

where $k_S \in L^1(\mathcal{T}_S; \mathbf{C}^{n\times n})$ is defined by $k_S(t) = \sum_{j=-\infty}^{\infty} k(t + jS)$, $t \in \mathbf{R}$;
(ii) *$\det\big[I + \hat{k}\,(\mathrm{i}\,2\pi m/S)\big] \neq 0$ for all $m \in \mathbf{Z}$;*
(iii) *for every $f \in V$, the equation $x + k * x = f$ has a unique solution x in V;*

(iv) *for every $f \in V$, the equation $x + k * x = f$ has a unique solution x in V, and x depends continuously on f (in the norm of V).*

When these conditions hold, then the function r_S in (i) *is unique, and, for all $f \in V$, the solution x in* (iii) *and* (iv) *is given by*

$$x(t) = f(t) - \int_0^S r_S(t-s)f(s)\,\mathrm{d}s, \quad t \in \mathbf{R}.$$

Again, (iii) and (iv) do not exclude the possibility that the equation may have solutions that are not S-periodic.

The proof of this, too, is given in Section 5.

4.8 Definition

The function r_S in Theorem 4.7(i) *is called the S-periodic resolvent of k.*

5. Proofs of the Paley–Wiener Theorems

First, we demonstrate how one deduces Theorem 4.1 from Theorem 4.3.

Proof of Theorem 4.1, assuming Theorem 4.3 The necessity of the Laplace transform condition has already been established. Conversely, suppose that $\det[I + \hat{k}(z)] \neq 0$, for $\Re z \geq 0$. Let $k(t) = 0$ for $t < 0$. By Theorem 4.3, there is a function $r \in L^1(\mathbf{R}; \mathbf{C}^{n \times n})$ satisfying (4.2). The only thing that we have to show is that this function vanishes a.e. on $\mathbf{R}^-$. This we do as follows. Define

$$r_1(t) = \begin{cases} r(t), & t < 0, \\ 0, & t \geq 0, \end{cases} \qquad r_2(t) = \begin{cases} 0, & t < 0, \\ r(t), & t \geq 0. \end{cases}$$

Then, for $\Re z = 0$,

$$\hat{r}(z) = \hat{k}(z)\big[I + \hat{k}(z)\big]^{-1} = \widehat{r_1}(z) + \widehat{r_2}(z).$$

Rewrite this as

$$\widehat{r_1}(z) = \hat{k}(z)\big[I + \hat{k}(z)\big]^{-1} - \widehat{r_2}(z), \quad \Re z = 0.$$

Since $\det[I + \hat{k}(z)] \neq 0$ for $\Re z \geq 0$, both $\hat{k}(z)\big[I + \hat{k}(z)\big]^{-1}$ and $\widehat{r_2}(z)$ are continuous in $\Re z \geq 0$, and analytic in $\Re z > 0$ (cf. Theorem 2.8). Similarly, because we can interpret $\widehat{r_1}(-z)$ as an ordinary Laplace transform of the function $r(-t)$, we see that $\widehat{r_1}(z)$ is continuous in $\Re z \leq 0$, and analytic in $\Re z < 0$. It follows from, e.g., Rudin [3], Theorem 16.8, (which essentially is a corollary to Morera's theorem) that the function $\varphi(z)$ defined by

$$\varphi(z) = \begin{cases} \widehat{r_1}(z), & \Re z \leq 0, \\ \hat{k}(z)\big[I + \hat{k}(z)\big]^{-1} - \widehat{r_2}(z), & \Re z > 0, \end{cases}$$

is entire. By Theorem 2.8(i), $|\widehat{r_2}(z)| \to 0$ and $|\hat{k}(z)| \to 0$ as $|z| \to \infty$ in $\Re z \geq 0$, and this implies that $\varphi(z)$ is bounded in $\Re z \geq 0$. Similarly, $\widehat{r_1}(z)$

is bounded in $\Re z \le 0$, so φ is bounded in the whole plane. It follows from Liouville's theorem that φ has to be a constant. Since $\varphi(\mathrm{i}\omega) \to 0$ as $|\omega| \to \infty$, this constant is zero. Thus, in particular, $\widehat{r_1}(z) = 0$ for $\Re z \le 0$. By the uniqueness theorem for Laplace transforms, $r_1(t) = 0$ for almost all $t < 0$. Consequently, $r(t) = 0$ for almost all $t < 0$. □

The proof of Theorem 4.1 given above can be interpreted as a proof of the following lemma.

5.1 The Paley–Wiener Lemma
Let $a \in L^1(\mathbf{R}; \mathbf{C}^{n\times n})$, let φ be a function that is bounded and continuous in the closed half plane $\Re z \ge 0$ and analytic in the open half plane $\Re z > 0$, and suppose that $\hat{a}(\mathrm{i}\omega) = \varphi(\mathrm{i}\omega)$ for all $\omega \in \mathbf{R}$. Then $a(t) = 0$ for almost all $t < 0$, and $\hat{a}(z) = \varphi(z)$ for all $z \in \mathbf{C}$ with $\Re z \ge 0$.

Next, we turn to the proof of Theorem 4.3. We begin with some lemmas. In these lemmas it is more convenient to use the Fourier transform than to use the Laplace transform. Recall that (for real arguments) the Fourier transform $\tilde{a}$ of $a \in L^1(\mathbf{R}; \mathbf{C}^{n\times n})$ is defined by

$$\tilde{a}(\omega) = \hat{a}(\mathrm{i}\omega) = \int_{\mathbf{R}} \mathrm{e}^{-\mathrm{i}\omega t} a(t)\,\mathrm{d}t, \quad \omega \in \mathbf{R}.$$

Our first lemma is a reformulation of the fairly obvious fact that if the L^1-norm of the kernel is less than 1 then a resolvent exists.

5.2 Lemma
Let a and $b \in L^1(\mathbf{R}; \mathbf{C}^{n\times n})$, and suppose that $\|b\|_{L^1(\mathbf{R})} < 1$. Then there exists a function $c \in L^1(\mathbf{R}; \mathbf{C}^{n\times n})$ such that $\tilde{c}(\omega) = \tilde{a}(\omega)\big[I + \tilde{b}(\omega)\big]^{-1}$ for all $\omega \in \mathbf{R}$.

Proof Define $c_m = a + \sum_{j=1}^{m}(-1)^j a * b^{*j}$, where the notation is the same as in the proof of Theorem 3.1. In the same way as in that proof we conclude that c_m tends to a limit function $c \in L^1(\mathbf{R}; \mathbf{C}^{n\times n})$ satisfying $c + c * b = a$. Transforming this equation we get $\tilde{c}(\omega) = \tilde{a}(\omega)\big[I + \tilde{b}(\omega)\big]^{-1}$ for $\omega \in \mathbf{R}$. □

In general, because of the size condition on b, we cannot apply this result directly, with a and b both replaced by k. Instead we 'localize' the problem. More specifically, we show that in a neighbourhood of each point of the real axis, (including the point at infinity), the Fourier transform of an integrable function is equal to a constant plus the Fourier transform of an integrable function with small L^1-norm. Using this fact, we are able to convert the original problem into a finite number of problems, where we can employ the previous lemma. Finally, the solutions of the auxiliary problems are patched together to give us the solution of the original problem.

In the localization procedure mentioned above, we need a scalar function η whose transform $\tilde{\eta}$ is constant in a neighbourhood of zero. An easy way to get such a function is to start with the *Fejér kernel*

$$\zeta(t) = \frac{1}{\pi t^2}(1 - \cos t), \tag{5.1}$$

whose Fourier transform is

$$\tilde{\zeta}(\omega) = \begin{cases} 1 - |\omega|, & |\omega| \le 1, \\ 0, & |\omega| > 1. \end{cases}$$

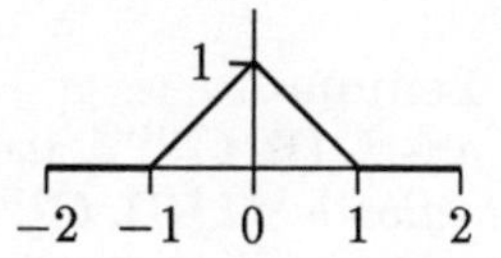

If we define $\eta(t) = 4\zeta(2t) - \zeta(t)$, then

$$\eta(t) = \frac{1}{\pi t^2}(\cos t - \cos 2t), \tag{5.2}$$

and

$$\tilde{\eta}(\omega) = \begin{cases} 1, & |\omega| \le 1, \\ 2 - |\omega|, & 1 < |\omega| \le 2, \\ 0, & |\omega| > 2. \end{cases}$$

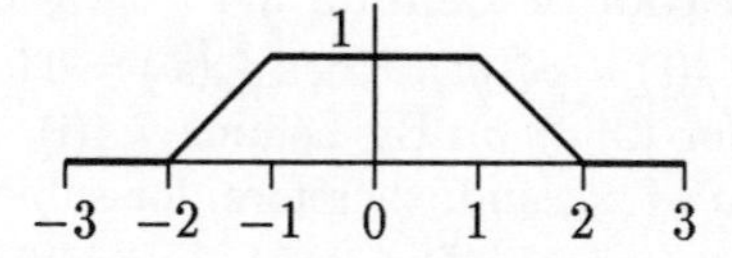

5.3 Lemma
Let $a \in L^1(\mathbf{R}; \mathbf{C}^{n\times n})$, $\epsilon > 0$, and $\omega_0 \in \mathbf{R}$. Then there exist a number $\delta > 0$, independent of ω_0, and a function $b \in L^1(\mathbf{R}; \mathbf{C}^{n\times n})$ with $\|b\|_{L^1(\mathbf{R})} \le \epsilon$, satisfying $\tilde{a}(\omega) = \tilde{a}(\omega_0) + \tilde{b}(\omega)$ for $|\omega - \omega_0| \le \delta$.

Proof Let η be the function in (5.2), and, for each $\delta > 0$, define η_δ by $\eta_\delta(t) = \delta\eta(\delta t)$ for $t \in \mathbf{R}$. Then $\widetilde{\eta_\delta}(\omega) = \tilde{\eta}(\omega/\delta) = 1$ for $|\omega| \le \delta$. If we define $\widetilde{b_\delta}$ by

$$\widetilde{b_\delta}(\omega) = \widetilde{\eta_\delta}(\omega - \omega_0)\big(\tilde{a}(\omega) - \tilde{a}(\omega_0)\big),$$

then $\tilde{a}(\omega) = \tilde{a}(\omega_0) + \widetilde{b_\delta}(\omega)$ for $|\omega - \omega_0| \le \delta$. The function $\widetilde{b_\delta}$ is the Fourier transform of

$$\begin{aligned} b_\delta(t) &= \int_{\mathbf{R}} \eta_\delta(t-s)\mathrm{e}^{\mathrm{i}\omega_0(t-s)}a(s)\,\mathrm{d}s - \eta_\delta(t)\mathrm{e}^{\mathrm{i}\omega_0 t}\int_{\mathbf{R}} \mathrm{e}^{-\mathrm{i}\omega_0 s}a(s)\,\mathrm{d}s \\ &= \int_{\mathbf{R}} \big(\eta_\delta(t-s) - \eta_\delta(t)\big)\mathrm{e}^{\mathrm{i}\omega_0(t-s)}a(s)\,\mathrm{d}s, \end{aligned}$$

and, therefore,

$$\begin{aligned} \int_{\mathbf{R}} |b_\delta(t)|\,\mathrm{d}t &\le \int_{\mathbf{R}}\int_{\mathbf{R}} \big|\eta_\delta(t-s) - \eta_\delta(t)\big||a(s)|\,\mathrm{d}s\,\mathrm{d}t \\ &= \int_{\mathbf{R}} |a(s)| \int_{\mathbf{R}} \big|\eta(t - \delta s) - \eta(t)\big|\,\mathrm{d}t\,\mathrm{d}s. \end{aligned}$$

Observe that this estimate is independent of ω_0. The function

$$s \mapsto |a(s)| \int_{\mathbf{R}} |\eta(t-\delta s) - \eta(t)|\,\mathrm{d}t$$

is bounded by $2|a(s)|\|\eta\|_{L^1(\mathbf{R})}$, and it follows from Lemma 7.3(iii) that for each fixed s it tends to zero as δ tends to zero. Therefore, Lebesgue's dominated convergence theorem implies $\|b_\delta\|_{L^1(\mathbf{R})} \to 0$ as $\delta \to 0$. This shows that by choosing δ small enough we can make the L^1-norm of b_δ less than ϵ. □

5.4 Lemma
Let $a \in L^1(\mathbf{R}; \mathbf{C}^{n\times n})$ and $\epsilon > 0$. Then there exist a number $M > 0$ and a function $b \in L^1(\mathbf{R}; \mathbf{C}^{n\times n})$, with $\|b\|_{L^1(\mathbf{R})} \le \epsilon$, satisfying $\tilde{a}(\omega) = \tilde{b}(\omega)$ for $|\omega| \ge M$.

In particular, the Riemann–Lebesgue lemma, i.e., the fact that $\tilde{a}(\omega) \to 0$ as $|\omega| \to \infty$, follows from Lemma 5.4, because trivially $|\tilde{b}(\omega)| \le \epsilon$.

Proof of Lemma 5.4 Let ζ be the function in (5.1), and define ζ_ρ by $\zeta_\rho(t) = \rho\zeta(\rho t)$. Then $\tilde{\zeta_\rho}(\omega) = 0$ for $|\omega| \ge \rho$, and so $\tilde{a}(\omega) = \tilde{a}(\omega) - \tilde{\zeta_\rho}(\omega)\tilde{a}(\omega)$ for $|\omega| \ge \rho$. By Lemma 7.4(i), $\zeta_\rho * a$ converges to a in $L^1(\mathbf{R}; \mathbf{C}^{n\times n})$ as $\rho \to \infty$, and, therefore, for sufficiently large ρ, $b = a - \zeta_\rho * a$ has a small norm in $L^1(\mathbf{R}; \mathbf{C}^{n\times n})$. In addition, $\tilde{b}(\omega) = \tilde{a}(\omega)$ for $|\omega| \ge \rho$, so b is of the required type. □

5.5 Lemma
Let $a \in L^1(\mathbf{R}; \mathbf{C}^{n\times n})$. Then there exist a constant $M > 0$ and a function $r \in L^1(\mathbf{R}; \mathbf{C}^{n\times n})$, such that $\tilde{r}(\omega) = \tilde{a}(\omega)\big[I + \tilde{a}(\omega)\big]^{-1}$ for $|\omega| \ge M$. Moreover, for each $\omega_0 \in \mathbf{R}$ such that $\det[I + \tilde{a}(\omega_0)] \ne 0$, there exist a $\delta > 0$ and a function $r \in L^1(\mathbf{R}; \mathbf{C}^{n\times n})$, such that $\tilde{r}(\omega) = \tilde{a}(\omega)\big[I + \tilde{a}(\omega)\big]^{-1}$ for $|\omega - \omega_0| \le \delta$. If $\det[I + \tilde{a}(\omega)] \ne 0$ for all $\omega \in \mathbf{R}$, then the constant δ can be chosen independently of ω_0.

Proof By Lemma 5.4, there exist a constant M and a function $b \in L^1(\mathbf{R}; \mathbf{C}^{n\times n})$, with $\|b\|_{L^1(\mathbf{R})} < 1$, satisfying $\tilde{b}(\omega) = \tilde{a}(\omega)$ for $|\omega| \ge M$. It follows from Lemma 5.2 that there exists a function $r \in L^1(\mathbf{R}; \mathbf{C}^{n\times n})$ such that $\tilde{r}(\omega) = \tilde{a}(\omega)[I + \tilde{b}(\omega)]^{-1}$ for all $\omega \in \mathbf{R}$. Since $\tilde{b}(\omega) = \tilde{a}(\omega)$ when $|\omega| \ge M$, we have $\tilde{r}(\omega) = \tilde{a}(\omega)[I + \tilde{a}(\omega)]^{-1}$ for these values of ω. This proves the first claim.

To prove the remaining two claims, we define $A = [I + \tilde{a}(\omega_0)]^{-1}$, and use Lemma 5.3 to find a constant δ and a function b that satisfy $\tilde{b}(\omega) = \tilde{a}(\omega) - \tilde{a}(\omega_0)$ for $|\omega - \omega_0| \le \delta$ and $\|b\|_{L^1(\mathbf{R})} < 1/|A|$. Note that Lemma 5.3 implies that δ can be chosen independently of ω_0 when $\det[I + \tilde{a}(\omega)] \ne 0$ for all $\omega \in \mathbf{R}$. Since this is the case, we may require b to satisfy

$$\|b\|_{L^1(\mathbf{R})} < \frac{1}{\sup_{\omega\in\mathbf{R}}\big|\big[I + \tilde{a}(\omega)\big]^{-1}\big|}.$$

Also observe that, for $|\omega - \omega_0| \le \delta$,

$$I + \tilde{a}(\omega) = I + \tilde{a}(\omega_0) + \tilde{b}(\omega) = A^{-1} + \tilde{b}(\omega) = [I + \tilde{b}(\omega)A]A^{-1},$$

which implies

$$\tilde{a}(\omega)[I+\tilde{a}(\omega)]^{-1} = \tilde{a}(\omega)A[I+\tilde{b}(\omega)A]^{-1}.$$

Now we can again invoke Lemma 5.2 (with a and b replaced by aA and bA, respectively) and obtain the desired conclusion. □

Proof of Theorem 4.3 We commented on the necessity of the Laplace transform condition immediately after the statement of Theorem 4.1, and the same comments apply here (with the half plane $\Re z \geq 0$ replaced by the imaginary axis $\Re z = 0$). Uniqueness is proved exactly in the same way as in Theorem 3.1. Therefore, let us suppose that

$$\det[I+\tilde{k}(\omega)] \neq 0 \text{ for } \omega \in \mathbf{R}, \tag{5.3}$$

and let us show that there exists a function $r \in L^1(\mathbf{R};\mathbf{C}^{n\times n})$ that satisfies $r + k * r = r + r * k = k$.

By the first part of Lemma 5.5, there exist $r_\infty \in L^1(\mathbf{R};\mathbf{C}^{n\times n})$ and $M > 0$, such that

$$\widetilde{r_\infty}(\omega) = \tilde{k}(\omega)\big[I+\tilde{k}(\omega)\big]^{-1}, \quad |\omega| \geq M. \tag{5.4}$$

Without loss of generality, let M be an integer. By the continuity of $\tilde{k}(\omega)$, by (5.3), and by the second part of Lemma 5.5, there exists an integer $m > 0$ such that, for every $j = 0, \pm 1, \pm 2, \ldots, \pm mM$, there is a function $r_j \in L^1(\mathbf{R};\mathbf{C}^{n\times n})$ satisfying

$$\widetilde{r_j}(\omega) = \tilde{k}(\omega)\big[I+\tilde{k}(\omega)\big]^{-1}, \quad |\omega - j/m| \leq 1/m.$$

Define $\psi_j(t) = (1/m)\mathrm{e}^{-\mathrm{i}jt/m}\zeta(t/m)$, where $\zeta(t) = (\pi t^2)^{-1}(1-\cos t)$ is the Fejér kernel. Then

$$\widetilde{\psi_j}(\omega) = \begin{cases} 1 - |m\omega - j|, & |\omega - j/m| \leq 1/m, \\ 0, & |\omega - j/m| > 1/m. \end{cases}$$

$\frac{j-1}{m}$ $\frac{j}{m}$ $\frac{j+1}{m}$

Finally, let

$$r = \sum_{j=-mM}^{mM} \psi_j * (r_j - r_\infty) + r_\infty.$$

It is obvious that the function r is integrable. Our choice of the functions ψ_j guarantees that $1 - \sum_{j=-mM}^{mM}\widetilde{\psi_j}(\omega) = 0$ when $|\omega| \leq M$, and hence it follows from (5.4) that

$$\Big(1 - \sum_{j=-mM}^{mM}\widetilde{\psi_j}(\omega)\Big)\widetilde{r_\infty}(\omega) = \Big(1 - \sum_{j=-mM}^{mM}\widetilde{\psi_j}(\omega)\Big)\tilde{k}(\omega)\big[I+\tilde{k}(\omega)\big]^{-1}, \quad \omega \in \mathbf{R}.$$

On the other hand, we have from our choice of the functions r_j, and from the fact that $\widetilde{\psi_j}(\omega) = 0$ when $|\omega - j/m| > 1/m$, that

$$\widetilde{\psi_j}(\omega)\widetilde{r_j}(\omega) = \widetilde{\psi_j}(\omega)\tilde{k}(\omega)\big[I+\tilde{k}(\omega)\big]^{-1}, \quad \omega \in \mathbf{R}, \quad j = 0, \pm 1, \pm 2, \ldots, \pm mM.$$

Adding all these identities, and recalling the definition of the function r, we obtain $\tilde{r}(\omega) = \tilde{k}(\omega)\big[I + \tilde{k}(\omega)\big]^{-1}$ for all $\omega \in \mathbf{R}$, or, equivalently,

$$\tilde{r}(\omega) + \tilde{r}(\omega)\tilde{k}(\omega) = \tilde{k}(\omega).$$

By the uniqueness theorem for Fourier transforms, this implies $r + r * k = k$. On the other hand, for each $\omega \in \mathbf{R}$,

$$\tilde{k}(\omega)\big[I + \tilde{k}(\omega)\big]^{-1} = \big[I + \tilde{k}(\omega)\big]^{-1}\tilde{k}(\omega),$$

and so $\tilde{r}(\omega) + \tilde{k}(\omega)\tilde{r}(\omega) = \tilde{k}(\omega)$ for $\omega \in \mathbf{R}$. This implies that $r + k * r = k$. □

Proof of Theorem 4.5 The comments made immediately after Theorem 4.5 showed that (i) and (ii) are equivalent, that (i) implies (iv), and that (iv) implies (iii).

To prove that (iii) implies (iv) one uses the open mapping theorem. The operator that maps x into f according to the formula $f = x + k * x$ is continuous (this follows from Theorem 2.2). By Theorem 3.5, it is one-to-one. Thus, if this mapping is onto, then its inverse must be continuous (see Rudin [2], Corollary 2.12). Hence (iii) implies (iv).

In the course of the proof we shall need one more condition that is equivalent to (i)–(iv), namely

(v) *the function* $\mathbf{e} * r$ *belongs to* $L^1(\mathbf{R}^+; \mathbf{C}^{n\times n})$.

Here r is the resolvent of k, and $\mathbf{e}(t) = \mathrm{e}^{-t}$ for $t \geq 0$. Clearly, by Lemma 2.2(i), (i) implies (v). Conversely, if (v) holds, then it follows from Theorem 2.8 and the equation defining r that

$$\widehat{(\mathbf{e} * r)}(z)[I + \hat{k}(z)] = \hat{\mathbf{e}}(z)\hat{k}(z), \quad \Re z \geq 0,$$

or, equivalently,

$$\big[\widehat{(\mathbf{e} * r)}(z) - \hat{\mathbf{e}}(z)I\big][I + \hat{k}(z)] = -\hat{\mathbf{e}}(z)I = -\frac{1}{z+1}I, \quad \Re z \geq 0.$$

Therefore, we must have $\det[I + \hat{k}(z)] \neq 0$ for $\Re z \geq 0$. Thus (v) implies (ii), and we conclude that (i), (ii), and (v) are equivalent.

(Observe that the function $\mathbf{e}$ in (v) could have been replaced by any $a \in L^1(\mathbf{R}^+; \mathbf{C}^{n\times n})$ satisfying $\det[\hat{a}(z)] \neq 0$ for $\Re z \geq 0$.)

To complete the proof, we must show that (iv), or (iii), implies one of the conditions (i), (ii), or (v). The choice of (i), (ii), or (v), will depend on V.

When $V = L^1(\mathbf{R}^+; \mathbf{C}^n)$, or $BC_0(\mathbf{R}^+; \mathbf{C}^n) \subset V \subset L^\infty(\mathbf{R}^+; \mathbf{C}^n)$, it follows from Theorem 2.4, together with the variation of constants formula, that (iii) implies (i).

Next, let us take $V = L^p(\mathbf{R}^+; \mathbf{C}^n)$, with $1 < p < \infty$ (this approach works for $p = 1$ as well). Suppose that (ii) does not hold, i.e., suppose that there exists a point z_0 with $\Re z_0 \geq 0$, such that $\det[I + \hat{k}(z_0)] = 0$. Then there exists a nonzero vector $v \in \mathbf{C}^n$ satisfying $v + \hat{k}(z_0)v = 0$. For each $T > 0$, define f_T by $f_T(t) = -\chi_{[0,T)}(t) \int_t^\infty \mathrm{e}^{z_0(t-s)} k(s) v \, \mathrm{d}s$. Then, on the interval $[0,T)$, the solution x_T of the equation $x_T + k * x_T = f_T$ is equal to $\mathrm{e}^{z_0 t} v$. If $\Re z_0 > 0$, then we choose $T = \infty$. In this case, f_∞ is the convolution of an L^p- and an L^1-function, so $f_\infty \in L^p(\mathbf{R}^+; \mathbf{C}^n)$. On the other hand, $x_\infty \notin L^p(\mathbf{R}^+; \mathbf{C}^n)$. This contradicts (iii). For the case $\Re z_0 = 0$, observe that $\|x_T\|_{L^p(\mathbf{R}^+)} \geq \|x_T\|_{L^p(0,T)} = T^{1/p}|v|$, whereas $|f_T(t)| \leq \chi_{[0,T)}(t)|v| \int_t^\infty |k(s)| \, \mathrm{d}s$, hence $\|f_T\|_{L^p(\mathbf{R}^+)} = o(T^{1/p})$ as $T \to \infty$. Thus the mapping $f \mapsto f - r * f$ cannot be continuous from $L^p(\mathbf{R}^+; \mathbf{C}^n)$ into itself, and this contradicts (iv).

Suppose that (iii) is true with $V = BV(\mathbf{R}^+; \mathbf{C}^n)$. Let $f \in L^1(\mathbf{R}^+; \mathbf{C}^n)$ be arbitrary, and define $g(t) = \int_0^t f(s) \, \mathrm{d}s$ for $t \geq 0$. Then $g \in BV(\mathbf{R}^+; \mathbf{C}^n)$; hence by (iii), the solution y of the equation $y + k * y = g$ satisfies $y \in BV(\mathbf{R}^+; \mathbf{C}^n)$. However, it is easy to show that $y(t) = \int_0^t x(s) \, \mathrm{d}s$ for $t \geq 0$, where x is the solution of the equation $x + k * x = f$ (use, e.g., Corollary **3**.7.3(i)). Since $y \in BV(\mathbf{R}^+; \mathbf{C}^n)$, we conclude that $x \in L^1(\mathbf{R}^+; \mathbf{C}^n)$. We find that for every $f \in L^1(\mathbf{R}^+; \mathbf{C}^n)$ the solution x of $x + k * x = f$ belongs to $L^1(\mathbf{R}^+; \mathbf{C}^n)$. Therefore, by the L^1-result that we proved above, (i) must be true. Thus, (iii) implies (i) when $V = BV(\mathbf{R}^+; \mathbf{C}^n)$.

When V is one of the spaces $BBV(\mathbf{R}^+; \mathbf{C}^n)$, $BBV_0(\mathbf{R}^+; \mathbf{C}^n)$, or $BV(\mathcal{T}_S; \mathbf{C}^n) \simeq BBV_0(\mathbf{R}^+; \mathbf{C}^n)$, we show that (iii) implies (v). Suppose that (iii) is true for one of these spaces. Let $f \in BC_0(\mathbf{R}^+; \mathbf{C}^n)$ be arbitrary. Then $\mathbf{e} * f \in V \cap BC(\mathbf{R}^+; \mathbf{C}^n)$. Thus, by (iii), the solution y of the equation $y + k * y = \mathbf{e} * f$ belongs to V. In addition, it is continuous, hence it belongs to $BC(\mathbf{R}^+; \mathbf{C}^n)$. But $y = \mathbf{e} * f - r * \mathbf{e} * f$, hence $(r * \mathbf{e}) * f \in BC(\mathbf{R}^+; \mathbf{C}^n)$. Now one can apply Theorem 2.4(ii) to get (v).

In all the remaining cases, we have $BC_0(\mathbf{R}^+; \mathbf{C}^n) \subset V$, but $V \not\subset L^\infty(\mathbf{R}^+; \mathbf{C}^n)$. On the other hand, we do have $V \subset BL^1(\mathbf{R}^+; \mathbf{C}^n)$. Thus to complete the proof it suffices to show that the following condition implies (v).

(vi) *For every $f \in BC_0(\mathbf{R}^+; \mathbf{C}^n)$, the solution x of the equation $x + k * x = f$ on $\mathbf{R}^+$ belongs to $BL^1(\mathbf{R}^+; \mathbf{C}^n)$.*

This we do as follows. The convolution operator $x \mapsto \mathbf{e} * x$ maps $BL^1(\mathbf{R}^+; \mathbf{C}^n)$ into $BC(\mathbf{R}^+; \mathbf{C}^n)$; see Exercise 1. Therefore, if (vi) holds, then for every $f \in BC_0(\mathbf{R}^+; \mathbf{C}^n)$ the solution $x = f - r * f$ of the equation $x + k * x = f$ satisfies $\mathbf{e} * x \in BC(\mathbf{R}^+; \mathbf{C}^n)$. In particular, $(\mathbf{e} * r) * f = \mathbf{e} * f - \mathbf{e} * x \in BC(\mathbf{R}^+; \mathbf{C}^n)$. This, together with Theorem 2.4(ii), implies (v), and completes the proof of Theorem 4.5. □

The main difference between the proof of Theorem 4.5 given above and that of Theorem 4.6, to be given next, is that below we can use a 'localization' technique.

Proof of Theorem 4.6 The comments made immediately after Theorem 4.6 showed that (i) and (ii) are equivalent, that (i) implies (iv), and that

(iv) implies (iii). To show that (iii) $\Rightarrow$(iv) one argues in the same way as in the proof of Theorem 4.5, i.e., one uses the open mapping theorem.

It remains to prove that (iv) implies (ii). Assume that (ii) does not hold, i.e., that $\det[I+\hat{k}(i\omega_0)]=0$ for some $\omega_0\in\mathbf{R}$, or equivalently that $\det[I+\tilde{k}(\omega_0)]=0$ for some $\omega_0\in\mathbf{R}$. Then there exists a nonzero vector $v\in\mathbf{C}^n$ such that $v+\tilde{k}(\omega_0)v=0$. This implies that $x(t)=e^{i\omega_0 t}v$ is a solution, not identically zero, of the equation $x+k*x=0$. Hence, if the space V is large enough to contain this function, then the solution of the equation $x+k*x=f$ is not unique in V. Thus we conclude that (iii) implies (ii) when V is one of the spaces $L^\infty(\mathbf{R};\mathbf{C}^n)$, $B^\infty(\mathbf{R};\mathbf{C}^n)$, $BC(\mathbf{R};\mathbf{C}^n)$, $BUC(\mathbf{R};\mathbf{C}^n)$, $BL^p(\mathbf{R};\mathbf{C}^n)$, $1\le p\le\infty$, $BBV(\mathbf{R};\mathbf{C}^n)$, or $AP(\mathbf{R};\mathbf{C}^n)$.

In the remaining cases we have to use a slightly more complicated argument (this argument applies to the spaces listed above as well). We still suppose that (ii) does not hold, and in addition we assume (iv) to be valid, and look for a contradiction. By (iv), the mapping $K(f)=x$, where x is the unique solution of the equation $x+k*x=f$, is continuous, i.e., it has a finite norm M. Of course, K^{-1} is the mapping that takes x into f. Clearly, the mapping K^n which we get by iterating K n times is continuous, and its norm is at most M^n. Since $\det[I+\tilde{k}(\omega_0)]=0$, we can split $\mathbf{C}^n$ into the null space Z of $[I+\tilde{k}(\omega_0)]^n$ (which is the same as the generalized eigenspace of $I+\tilde{k}(\omega_0)$ corresponding to the eigenvalue zero) and the range Y of $[I+\tilde{k}(\omega_0)]^n$ (which is the sum of the generalized eigenspaces of $I+\tilde{k}(\omega_0)$ corresponding to the nonzero eigenvalues). Let P be the projection (we shall identify P with an $n\times n$ matrix) whose null space is Y and range Z, and let w be an eigenvector corresponding to the zero eigenvalue of $I+\tilde{k}(\omega_0)$. Then $w\ne 0$, $Pw=w$, and $[I+\tilde{k}(\omega_0)]^nP=0$.

Let $x(t)=e^{i\omega_0 t}\rho\zeta(\rho t)w$, where ζ is the Fejér kernel defined in (5.1), and $\rho\in(0,1)$ is some small number that we will fix later. Then $Px=x$, and $\tilde{x}(\omega)=\tilde{\zeta}((\omega-\omega_0)/\rho)w$ vanishes for $|\omega-\omega_0|\ge\rho$. Note that $x\in V$ for all the spaces V under consideration, and that $\|x\|_V>0$. Define $f=K^{-n}(x)$. Then, by Theorem 2.2, $f\in V$, and $\|f\|_V>0$. Moreover, by Theorem 2.7, f is uniquely determined by the fact that

$$\tilde{f}(\omega)=\big[I+\tilde{k}(\omega)\big]^n\tilde{x}(\omega),\quad \omega\in\mathbf{R}.$$

Let a be the L^1-function whose Fourier transform is given by

$$\tilde{a}(\omega)=\tilde{\eta}(\omega-\omega_0)\big[I+\tilde{k}(\omega)\big]^nP,\quad \omega\in\mathbf{R},$$

where η is the function defined in (5.2). Then $\tilde{a}(\omega)=[I+\tilde{k}(\omega)]^nP$ for $|\omega-\omega_0|\le 1$. In particular, $\tilde{a}(\omega_0)=0$. Since $P\tilde{x}=\tilde{x}$, and since $\tilde{x}(\omega)$ vanishes for $|\omega-\omega_0|\ge 1$ (recall that we take $\rho<1$), we find that f has the same Fourier transform as the function $a*x$. Therefore, $f=a*x$. Now use Lemma 5.3 to conclude that there exist a number $\delta>0$ and a function $b\in L^1(\mathbf{R};\mathbf{C}^{n\times n})$ with $\|b\|_{L^1(\mathbf{R})}\le\frac{1}{2}M^{-n}$, such that $\tilde{a}(\omega)=\tilde{b}(\omega)$ for $|\omega-\omega_0|\le\delta$. Choose $\rho<\delta$. Then, by an argument similar to the one used above, we see that $f = b * x$. Hence, it follows from Theorem 2.2 that

$$\|f\|_V \leq \|b\|_{L^1(\mathbf{R})}\|x\|_V \leq \frac{1}{2M^n} M^n \|f\|_V = \frac{1}{2}\|f\|_V.$$

This contradiction shows that with all different choices of V, (iv) implies (ii). □

Proof of Theorem 4.7 As usual, the equivalence of (iii) and (iv) follows from the open mapping theorem.

Suppose that (ii) does not hold, i.e., suppose that there exists an integer m such that $\det[I + \hat{k}(\mathrm{i}2\pi m/S)] = 0$, or equivalently that $\det[I + \tilde{k}(2\pi m/S)] = 0$. Then there exists a nonzero vector $v \in \mathbf{C}^n$ such that $v + \tilde{k}(2\pi m/S)v = 0$. This implies that the function x defined by $x(t) = \mathrm{e}^{\mathrm{i}2\pi mt/S}v$ is a solution of the equation $x + k * x = 0$. This function belongs to V, and hence the solution of the equation $x + k * x = f$ is not unique in V, as required by (iii). Thus (iii) implies (ii).

Next, we want to show that (ii) implies (i). Let us start with some comments on (i), and, in particular, on the function k_S. Since $L^1([0,S];\mathbf{C}^{n\times n})$ is complete, the sum defining k_S converges in the L^1-sense on the interval $[0,S]$. Moreover, k_S is clearly S-periodic. Let κ_m be the m-th Fourier coefficient of k_S:

$$\kappa_m = \int_0^S \mathrm{e}^{-\mathrm{i}2\pi mt/S} k_S(t)\,\mathrm{d}t, \quad m \in \mathbf{Z}.$$

(In the sequel, when we use the words 'm-th Fourier coefficient', we mean an integral of the type computed above.) Then $\kappa_m = \tilde{k}(2\pi m/S)$ for all $m \in \mathbf{Z}$.

It is well known (and easy to prove) that one gets the Fourier coefficients of the periodic convolutions in (i) by multiplying the Fourier coefficients of the factors (see, e.g., Zygmund [1], Theorem (1.5), p. 36). This means that the Fourier coefficients

$$\rho_m = \int_0^S \mathrm{e}^{-\mathrm{i}\,2\pi mt/S} r_S(t)\,\mathrm{d}t, \quad m \in \mathbf{Z},$$

of the function r_S in (i) must satisfy

$$[I + \kappa_m]\rho_m = \rho_m[I + \kappa_m] = \kappa_m, \quad m \in \mathbf{Z},$$

or, equivalently,

$$\rho_m = [I + \kappa_m]^{-1}\kappa_m = \left[I + \tilde{k}(2\pi m/S)\right]^{-1}\tilde{k}(2\pi m/S), \quad m \in \mathbf{Z}. \tag{5.5}$$

(In particular, (i) implies (ii).) Conversely, if we can find a function $r_S \in L^1(\mathcal{T}_S;\mathbf{C}^{n\times n})$ whose Fourier coefficients ρ_m satisfy (5.5), then, because of the fact that the Fourier coefficients of a function determine this function uniquely (see Zygmund [1], Theorem (6.2), p. 12), r_S must satisfy the two equations in (i). Thus, to prove that (ii) implies (i), it suffices to show that, when (ii) holds, it is possible to find a function $r_S \in L^1(\mathcal{T}_S;\mathbf{C}^{n\times n})$ whose Fourier coefficients satisfy (5.5).

To prove that there is such a function we argue as follows. As a consequence of Lemma 5.4, we can find an integer $M > 0$, and a function $b \in L^1(\mathbf{R}; \mathbf{C}^{n\times n})$, with $\|b\|_{L^1(\mathbf{R})} \leq \frac{1}{2}$, such that $\tilde{k}(\omega) = \tilde{b}(\omega)$ for $|\omega| \geq 2\pi(M+1)/S$. Define the function b_S by

$$b_S(t) = \sum_{j=-\infty}^{\infty} b(t + jS), \quad t \in \mathbf{R}.$$

Then b_S belongs to $L^1(\mathcal{T}_S; \mathbf{C}^{n\times n})$, $\|b_S\|_{L^1(\mathcal{T}_S)} \leq \frac{1}{2}$, and the m-th Fourier coefficient of b_S is $\tilde{b}(2\pi m/S)$. By using the contraction mapping principle, one finds that the equation

$$q(t) + \int_0^S b_S(t-s)q(s)\,\mathrm{d}s = b_S(t), \quad t \in \mathbf{R},$$

has a unique solution $q \in L^1(\mathcal{T}_S; \mathbf{C}^{n\times n})$. The m-th Fourier coefficient γ_m of this function will be $[I + \tilde{b}(2\pi m/S)]^{-1}\tilde{b}(2\pi m/S)$. In particular, for $|m| \geq M+1$, $\gamma_m = [I + \tilde{k}(2\pi m/S)]^{-1}\tilde{k}(2\pi m/S)$. For $|m| \leq M$, define α_m by

$$\alpha_m = \left[I + \tilde{k}(2\pi m/S)\right]^{-1}\tilde{k}(2\pi m/S) - \gamma_m,$$

and define r_S by

$$r_S(t) = \frac{1}{S}\sum_{m=-M}^{M} \mathrm{e}^{\mathrm{i}\,2\pi m t/S}\alpha_m + q(t), \quad t \in \mathbf{R}.$$

We leave it to the reader to check that this function has the correct Fourier coefficients, as specified in (5.5).

To complete the proof, we must show that (i) implies (iv). Let $f \in V$, and define x by

$$x(t) = f(t) - \int_0^S r_S(t-s)f(s)\,\mathrm{d}s, \quad t \in \mathbf{R}. \tag{5.6}$$

Then, by periodicity, $x(t) = f(t) - \int_{t-S}^{t} r_S(t-s)f(s)\,\mathrm{d}s$ for $t \in \mathbf{R}$. Thus, by Theorem 2.2, $x \in V$, and x depends continuously on f. Let $\varphi = x + k * x$. Since x is periodic, the function φ can be expressed in the alternative form

$$\varphi(t) = x(t) + \int_0^S k_S(t-s)x(s)\,\mathrm{d}s.$$

This, together with (5.5), (5.6), and the fact that the Fourier coefficients of a convolution of the type that we have in (5.6) is the product of the Fourier coefficients of the factors, implies that f has the same Fourier coefficients as φ. Since the Fourier coefficients of a periodic function determine the function uniquely, we find that $\varphi = f$. Consequently, x is a solution of the equation $x + k * x = f$.

The proof of the uniqueness of x is quite similar, and is left to the reader. □

6. Further Results

First, we want to study the following question: what can be said about the size of the L^1-norm of the resolvent in the case where the Paley–Wiener theorem is applicable? We shall formulate an answer to this question for the whole line resolvent, and only remark that one gets the corresponding result for the resolvent on $\mathbf{R}^+$ by extending k as zero to the negative axis.

Recall that $\lceil s \rceil$ denotes the smallest integer $\geq s$.

6.1 Theorem
Let $k \in L^1(\mathbf{R}; \mathbf{C}^{n\times n})$ satisfy $\det[I + \hat{k}(z)] \neq 0$ for $\Re z = 0$. Define the number q by

$$q = \sup_{\omega\in\mathbf{R}} \left|[I + \hat{k}(\mathrm{i}\omega)]^{-1}\right|,$$

and choose positive numbers T and δ satisfying

$$\int_{|s|\geq T} |k(t)|\,\mathrm{d}t \leq \frac{1}{12q}, \qquad \sup_{0<s<\delta} \int_{-\infty}^{\infty} |k(t) - k(t-s)|\,\mathrm{d}t \leq \frac{1}{4}.$$

Then the whole line resolvent r of k satisfies

$$\|r\|_{L^1(\mathbf{R})} \leq \Big(8\big\lceil 6qT\|k\|_{L^1(\mathbf{R})}\big\rceil \big\lceil 8\|k\|_{L^1(\mathbf{R})}/\delta\big\rceil + 6\Big) q\|k\|_{L^1(\mathbf{R})}.$$

Proof The idea of the proof is simply to go through Lemmas 5.2–5.5, to compute explicit bounds on the L^1-norms of all the functions one gets, and then to use this information in the last stage of the proof of Theorem 4.3. The following relations are needed (the proofs are left to the reader)

$$\|\zeta_\rho\|_{L^1(\mathbf{R})} = 1, \quad \int_{|t|\geq\tau} |\zeta(t)|\,\mathrm{d}t < \frac{1}{\tau},$$

$$\|\eta_\rho\|_{L^1(\mathbf{R})} = \frac{2\sqrt{3}}{\pi} + \frac{1}{3}, \quad \operatorname{var}(\eta; \mathbf{R}) < \frac{3}{2}.$$

We choose the integers m and M used in the proof of Theorem 4.3 to be $m = \lceil 6qT\|k\|_{L^1(\mathbf{R})}\rceil$ and $M = \lceil 8\|k\|_{L^1(\mathbf{R})}/\delta\rceil$. From our choice of T and m we see that $\|r_j\|_{L^1(\mathbf{R})} \leq 2q\|k\|_{L^1(\mathbf{R})}$, and from our choice of δ and M it follows that $\|r_\infty\|_{L^1(\mathbf{R})} \leq 2\|k\|_{L^1(\mathbf{R})}$, where r_j and r_∞ are the functions appearing in the proof of Theorem 4.3. Now it is straightforward to complete the proof (we leave the details to the reader). □

We noted in Theorem 2.4 that the condition $r \in L^1(\mathbf{R}^+)$ is necessary and sufficient for the convolution operator $f \mapsto r * f$ to map L^1 into L^1 and L^∞ into L^∞. For the spaces $L^p(\mathbf{R}^+)$, with $1 < p < \infty$, this condition, although sufficient, is no longer necessary, unless r is the resolvent of an integrable function (see Theorem 4.6). The problem of finding weaker conditions that imply that a convolution operator maps $L^p(\mathbf{R}^+)$ into itself in the intermediate case $1 < p < \infty$ has been extensively studied. No simple necessary and sufficient conditions are known, except when $p = 2$.

6.2 Theorem
Let $k \in L^1_{\mathrm{loc}}(\mathbf{R}^+; \mathbf{C}^{n\times n})$ satisfy $\int_0^\infty \mathrm{e}^{-\sigma t}|k(t)|\,\mathrm{d}t < \infty$ for every $\sigma > 0$, and let r be the resolvent of k. Then the convolution operator $f \mapsto r * f$ maps $L^2(\mathbf{R}^+; \mathbf{C}^n)$ continuously into itself if and only if

$$\sup_{\Re z>0} \big|\hat{k}(z)[I + \hat{k}(z)]^{-1}\big| < \infty.$$

Proof The assumptions and claims above do not depend on the norm that we use in $\mathbf{C}^n$, so we may, without loss of generality, let the norm be the standard Euclidean norm. This has the advantage of allowing us to relate the norm of the operator $f \mapsto r*f$ to the constant $\sup_{\Re z>0}\big|\hat{k}(z)[I+\hat{k}(z)]^{-1}\big|$.

First, let us assume that $\hat{k}(z)[I + \hat{k}(z)]^{-1}$ is bounded in $\Re z > 0$. Take $f \in L^2(\mathbf{R}^+; \mathbf{C}^n)$ and define f_σ by $f_\sigma(t) = \mathrm{e}^{-\sigma t} f(t)$ for $t \in \mathbf{R}^+$ and $\sigma > 0$. Then the Fourier transform of f_σ is the function $\omega \mapsto \hat{f}(\sigma + \mathrm{i}\omega)$. The L^2-norm of f_σ is majorized by the L^2-norm of f, and therefore the Plancherel theorem (see Theorem **16**.8.2; this theorem requires that one use a norm induced by an inner product, such as the Euclidean norm) implies that

$$\sup_{\sigma>0} \frac{1}{2\pi} \int_{-\infty}^{\infty} |\hat{f}(\sigma + \mathrm{i}\omega)|^2 \mathrm{d}\omega = \sup_{\sigma>0} \|f_\sigma\|^2_{L^2(\mathbf{R}^+)} = \|f\|^2_{L^2(\mathbf{R}^+)}.$$

If $\sup_{\Re z>0}|\hat{k}(z)[I + \hat{k}(z)]^{-1}| = \sup_{\Re z>0}|\hat{r}(z)| = M$, then

$$\sup_{\sigma>0} \frac{1}{2\pi} \int_{-\infty}^{\infty} |\hat{r}(\sigma + \mathrm{i}\omega)\hat{f}(\sigma + \mathrm{i}\omega)|^2 \mathrm{d}\omega \le M^2 \|f\|^2_{L^2(\mathbf{R}^+)}.$$

Using the Plancherel theorem once more, one gets

$$\sup_{\sigma>0} \|r_\sigma * f_\sigma\|_{L^2(\mathbf{R}^+)} \le M \|f\|_{L^2(\mathbf{R}^+)},$$

where $r_\sigma(t) = \mathrm{e}^{-\sigma t} r(t)$ and $(r_\sigma * f_\sigma)(t) = \mathrm{e}^{-\sigma t}(r * f)(t)$, $t \in \mathbf{R}^+$. Letting $\sigma \downarrow 0$ we obtain $\|r * f\|_{L^2(\mathbf{R}^+)} \le M\|f\|_{L^2(\mathbf{R}^+)}$.

Next, assume that the operator $f \mapsto r*f$ maps $L^2(\mathbf{R}^+; \mathbf{C}^n)$ continuously into itself. Let M be the norm of this mapping. Choose $\sigma > 0$ to be so large that $\det[I + \hat{k}(z)] \ne 0$ in the half plane $\Re z \ge \sigma$; it follows from Theorem 2.8(i) that this can always be done. By Lemma 3.3, and by the Half Line Paley–Wiener Theorem 4.1, we have $\int_0^\infty \mathrm{e}^{-\sigma t}|r(t)|\,\mathrm{d}t < \infty$. This implies that $(r * \varphi)(t) = \int_{-\infty}^t r(t - s)\varphi(s)\,\mathrm{d}s$ is well defined for functions $\varphi \in L^2(\mathbf{R}; \mathbf{C})$ satisfying $\varphi(t) = O(\mathrm{e}^{\sigma t})$ as $t \to -\infty$. Now use the fact that convolution commutes with translation to show that for such φ one has $r * \varphi \in L^2(\mathbf{R}; \mathbf{C}^n)$ and $\|r * \varphi\|_{L^2(\mathbf{R})} \le M\|\varphi\|_{L^2(\mathbf{R})}$, where M is the norm of the convolution operator $f \mapsto r * f$ as a mapping from $L^2(\mathbf{R}^+; \mathbf{C}^n)$ into itself. If one applies this result to the function

$$\varphi(t) = \begin{cases} \mathrm{e}^{zt}, & t \le 0, \\ 0, & t > 0, \end{cases}$$

with $\Re z \ge \sigma$, then one finds that $|\hat{r}(z)|$ satisfies

$$|\hat{r}(z)| = \big|\hat{k}(z)[I + \hat{k}(z)]^{-1}\big| \le M, \quad \Re z \ge \sigma. \tag{6.1}$$

This bound is independent of σ, as long as $\det[I + \hat{k}(z)]$ does not have any zeros in the half plane $\Re z \geq \sigma$. Therefore, we can decrease the value of σ, and still have (6.1) valid, until either $\sigma = 0$, or σ equals the real part of one of the zeros of $\det[I + \hat{k}(z)]$ in the open half plane $\Re z > 0$. To complete the proof, it suffices to show that the second of the two alternatives is impossible. Suppose the contrary, i.e., that $\det[I + \hat{k}(z_0)] = 0$, and that $\Re z_0 = \sigma > 0$. Then, by continuity, $\det[I + \hat{k}(z)] \to 0$ as $z \to z_0$. But this is impossible, because for $\Re z > \sigma$ it follows from (4.1) that $\det[I + \hat{k}(z)] \det[I - \hat{r}(z)] = 1$, and, by (6.1), $\det[I - \hat{r}(z)]$ is bounded as z approaches z_0 from the right. This contradiction shows that the second of the two alternatives mentioned above cannot occur, and the proof is complete. □

It follows from Theorem 4.5 that the problem of whether a given function $k \in L^1(\mathbf{R}^+; \mathbf{C}^{n\times n})$ satisfies $\det[I + \hat{k}(z)] \neq 0$ for all z with $\Re z \geq 0$ is extremely important. Sometimes this question can be settled with the following criterion.

6.3 Nyquist's Theorem
Let $k \in L^1(\mathbf{R}^+; \mathbf{C}^{n\times n})$ *satisfy* $\det[I + \hat{k}(\mathrm{i}\omega)] \neq 0$ *for all* $\omega \in \mathbf{R}$. *Then the number of zeros of* $\det[I + \hat{k}(z)]$ *in* $\Re z > 0$, *counted according to their multiplicities, equals the index of the curve* $\det[I + \hat{k}(\mathrm{i}\omega)]$, $\infty \geq \omega \geq -\infty$ *(that is, it is the same as the number of times that this curve circles anticlockwise around the origin as* ω *goes from* $+\infty$ *to* $-\infty$*).*

Proof By Theorem 2.8, $|\hat{k}(z)| \to 0$ as $|z| \to \infty$ in the half plane $\Re z \geq 0$. Therefore, the set of zeros of $\det[I + \hat{k}(z)]$ in $\Re z \geq 0$ is bounded. It is also finite, because an infinite number of zeros would imply that $\det[I+\hat{k}(z)]$ has at least one zero on the imaginary axis, contrary to the original assumption (if the set of zeros of an analytic function is infinite, then this set always has a cluster point at the boundary of the domain of analyticity). Let us denote the total number of zeros of $\det[I + \hat{k}(z)]$ in $\Re z \geq 0$, counted according to their multiplicities, by N. Choosing σ small enough and R large enough, we can make all these zeros lie in a semidisc whose boundary consists of the straight line $\Gamma_1(\omega) = \sigma + \mathrm{i}\omega$, $R \geq \omega \geq -R$, and the semicircle $\Gamma_2(\varphi) = \sigma + R\mathrm{e}^{\mathrm{i}\varphi}$, $-\pi/2 \leq \varphi \leq \pi/2$. By Rudin [3], Theorem 10.43(a), the index of the closed curve $\det[I + \hat{k}(z)](\Gamma_1 \cup \Gamma_2)$ with respect to the origin equals N. If we let $R \to \infty$, then, by Theorem 2.8, $\det[I + \hat{k}(z)] \to 1$ for all $z \in \Gamma_2$. This means that also the index of the curve $\Gamma_\sigma(\omega) = \det[I + \hat{\mu}(\sigma + \mathrm{i}\omega)]$, $\infty \geq \omega \geq -\infty$, with respect to the origin is N. As $\sigma \downarrow 0$, $\Gamma_\sigma(\omega) \to \Gamma(\omega) = \det[I + \hat{k}(\mathrm{i}\omega)]$, uniformly in ω. Therefore the index of Γ, with respect to the origin, is N as well. □

6.4 Example

Let us consider the following application of Nyquist's theorem. Take

$$k(t) = \begin{cases} -1, & 0 \le t < 1, \\ 1, & 1 \le t < 2, \\ 0, & \text{otherwise.} \end{cases}$$

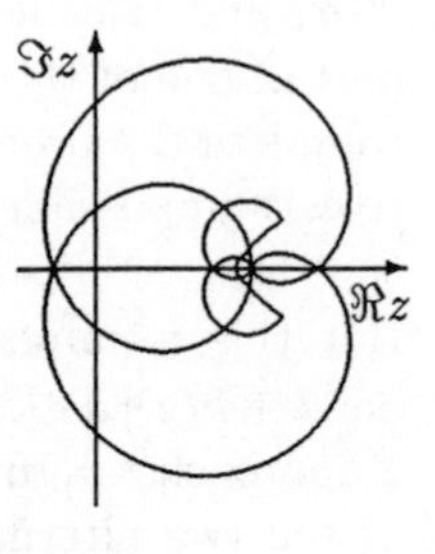

The Laplace transform of k is $\hat{k}(z) = -\frac{2}{z}(1 - \mathrm{e}^{-z}) + \frac{1}{z}(1 - \mathrm{e}^{-2z})$ and if we choose $\sigma = 0$ and $R = 10$ in the proof above then we see that the curve $(1 + \hat{k}(z))(\Gamma_1 \cup \Gamma_2)$ circles twice around the origin. Thus there are two zeros of $1 + \hat{k}(z)$ in $\Re z \ge 0$.

6.5 Corollary

Let $k \in L^1(\mathbf{R}^+; \mathbf{C}^{n\times n})$. Then $\det[I + \hat{k}(z)]$ has no zeros in the closed half plane $\Re z \ge 0$ if and only if the curve $\det[I + \hat{k}(\mathrm{i}\omega)]$, $\infty \ge \omega \ge -\infty$, neither passes through the origin, nor encircles the origin.

The following is another result that sometimes turns out to be useful. Under the assumption that the kernel k has a finite first moment, it relates two different properties of the resolvent to each other, i.e., convergence to zero at infinity, and integrability.

6.6 Theorem

Let $k \in L^1(\mathbf{R}^+; \mathbf{C}^{n\times n}) \cap BC_0(\mathbf{R}^+; \mathbf{C}^{n\times n})$ have a finite first moment, i.e., let $\int_0^\infty t|k(t)|\,\mathrm{d}t < \infty$. Then the resolvent r of k belongs to $L^1(\mathbf{R}^+; \mathbf{C}^{n\times n})$ if and only if $r(t) \to 0$ as $t \to \infty$.

Proof If $r \in L^1(\mathbf{R}^+; \mathbf{C}^{n\times n})$, then it follows from either one of the two resolvent equations that $r(t) \to 0$ as $t \to \infty$.

Conversely, suppose that $r \notin L^1(\mathbf{R}^+; \mathbf{C}^{n\times n})$. Then it follows from the Half Line Paley–Wiener Theorem 4.1 that there exist $z_0 \in \mathbf{C}$ with $\Re z_0 \ge 0$ and a nonzero $v \in \mathbf{C}^n$, such that

$$[I + \hat{k}(z_0)]v = 0.$$

Define $x(t) = \mathrm{e}^{z_0 t}v$ for $t \ge 0$. Then $x(t) \not\to 0$ as $t \to \infty$. Define

$$f(t) = x(t) + (k * x)(t), \quad t \in \mathbf{R}^+.$$

Then

$$f(t) = \mathrm{e}^{z_0 t}v + \int_0^t \mathrm{e}^{z_0(t-s)}k(s)v\,\mathrm{d}s = -\int_t^\infty \mathrm{e}^{-z_0(s-t)}k(s)v\,\mathrm{d}s,$$

so

$$|f(t)| \le \int_t^\infty |k(s)|\,\mathrm{d}s.$$

This, together with the moment condition on k, implies that $f \in L^1(\mathbf{R}^+; \mathbf{C}^n) \cap BC_0(\mathbf{R}^+; \mathbf{C}^n)$. If it were true that $r(t) \to 0$ as $t \to \infty$, then the variation of constants formula (3.2), together with Theorem 2.2(i), would imply that $x(t) \to 0$ as $t \to \infty$. This contradicts our earlier observation that $x(t) \not\to 0$ as $t \to \infty$. □

7. Appendix: Proofs of some Auxiliary Results

Below we follow the same convention as in Section 2, and do not explicitly write out the spaces in which the values of the functions and measures lie (i.e., $\mathbf{R}$, $\mathbf{C}$, $\mathbf{R}^{m\times n}$, or $\mathbf{C}^{m\times n}$). Of course, the dimensions should be compatible with respect to matrix addition and multiplication.

In the proof of Theorem 2.2 (as well as in some other places) we need the following notation and results.

7.1 Definition
If a is an arbitrary function defined on $\mathbf{R}$ and $h \in \mathbf{R}$, then we define $\tau_h a$ to be the translate $\tau_h a(t) = a(t+h)$, $t \in \mathbf{R}$.

7.2 Lemma
If a and b belong to $L^1_{\mathrm{loc}}(\mathbf{R})$ and t and $h \in \mathbf{R}$, then $(\tau_h(a * b))(t) = ((\tau_h a) * b)(t) = (a * (\tau_h b))(t)$ whenever one of these expressions exists.

The proof is obvious.

7.3 Lemma

(i) *If $a \in BUC(\mathbf{R})$, then $\tau_h a \to a$ uniformly on $\mathbf{R}$ as $h \to 0$.*
(ii) *If $a \in BC(\mathbf{R})$, then $\tau_h a \to a$ uniformly on compact sets as $h \to 0$.*
(iii) *If $a \in L^p(\mathbf{R})$ for some $p \in [1,\infty)$, then $\tau_h a \to a$ in $L^p(\mathbf{R})$ as $h \to 0$.*
(iv) *If $a \in BL^p_0(\mathbf{R})$ for some $p \in [1,\infty)$, then $\tau_h a \to a$ in $BL^p_0(\mathbf{R})$ as $h \to 0$.*
(v) *If $a \in BL^p(\mathbf{R})$ for some $p \in [1,\infty)$, then $\tau_h a \to a$ in $L^p_{\mathrm{loc}}(\mathbf{R})$ as $h \to 0$.*

Proof It is obvious that (i) and (ii) are true. This implies, in particular, that if $a \in C_c(\mathbf{R})$, that is, a is a continuous function with compact support, then $\tau_h a \to a$ in $L^p(\mathbf{R})$ as $h \to 0$ for every $p \in [1,\infty]$. The space $C_c(\mathbf{R})$ is dense in $L^p(\mathbf{R})$ and in $BL^p_0(\mathbf{R})$ when $p \in [1,\infty)$, and therefore, (iii) and (iv) are true. That (v) holds follows directly from (iii). □

Proof of Theorem 2.2 We prove only the case $J = \mathbf{R}$, and leave the proof of the other cases to the reader (define all the functions to be zero outside J).

(i) First, let us suppose that $p \in [1,\infty)$. By Jensen's inequality (see Rudin [3], Theorem 3.3, where one takes $\mu(\mathrm{d}s) = (|a(s)|/\|a\|_{L^p})\,\mathrm{d}s)$, for each $t \in \mathbf{R}$,

$$|(a*b)(t)|^p \le \left\{ \int_{\mathbf{R}} |a(s)||b(t-s)|\,\mathrm{d}s \right\}^p \le \|a\|_{L^1(\mathbf{R})}^{p-1} \int_{\mathbf{R}} |a(s)||b(t-s)|^p \mathrm{d}s. \tag{7.1}$$

Integrating over $\mathbf{R}$, and using Fubini's theorem (for a proof of the fact that the function $(s,t) \mapsto |a(s)||b(t-s)|$ is measurable in $\mathbf{R}^2$, see Hewitt and Stromberg [1], pp. 396-397, or, alternatively, assume that b is Borel measurable and use Rudin [3], p. 171) we get

$$\|a*b\|_{L^p(\mathbf{R})}^p \le \|a\|_{L^1(\mathbf{R})}^{p-1} \int_{\mathbf{R}} |a(s)| \int_{\mathbf{R}} |b(t-s)|^p \mathrm{d}t\,\mathrm{d}s = \|a\|_{L^1(\mathbf{R})}^p \|b\|_{L^p(\mathbf{R})}^p,$$

hence $\|a*b\|_{L^p(\mathbf{R})} \le \|a\|_{L^1(\mathbf{R})}\|b\|_{L^p(\mathbf{R})}$. The measurability of $a*b$ follows from Fubini's theorem (for each $T>0$ the function $(s,t)\mapsto a(s)b(t-s)$ is integrable on $\mathbf{R}\times[-T,T]$), and, therefore, $a*b \in L^p(\mathbf{R})$.

In the case where $p=\infty$ one gets the inequality $|(a*b)(t)| \le \|a\|_{L^1(\mathbf{R})}\|b\|_{L^\infty(\mathbf{R})}$ in a trivial way. The uniform continuity of $a*b$ follows from Lemmas 7.2 and 7.3, because, for each $h \in \mathbf{R}$,

$$\sup_{t\in\mathbf{R}} |(a*b)(t+h) - (a*b)(t)| = \sup_{t\in\mathbf{R}} |((\tau_h a - a)*b)(t)| \le \|\tau_h a - a\|_{L^1(\mathbf{R})}\|b\|_{L^\infty(\mathbf{R})},$$

and $\|\tau_h a - a\|_{L^1(\mathbf{R})} \to 0$ as $h \downarrow 0$.

Let $b \in L^\infty(\mathbf{R})$ satisfy $\lim_{t\to\infty} \operatorname{ess\,sup}_{s\ge t}|b(s)| = 0$. Then

$$|(a*b)(t)| \le \int_{-\infty}^{t/2} |a(t-s)||b(s)|\,\mathrm{d}s + \int_{t/2}^{\infty} |a(t-s)||b(s)|\,\mathrm{d}s \le \|b\|_{L^\infty(\mathbf{R})} \int_{t/2}^{\infty} |a(s)|\,\mathrm{d}s + \|a\|_{L^1(\mathbf{R})} \operatorname*{ess\,sup}_{s\ge t/2}|b(s)|.$$

Therefore, $(a*b)(t) \to 0$ as $t\to\infty$. A similar computation shows that $(a*b)(t) \to 0$ as $t \to -\infty$, provided $\lim_{t\to-\infty} \operatorname{ess\,sup}_{s\le t}|b(s)| = 0$. If $b \in L^\infty_\ell(\mathbf{R})$, then we can apply the same argument to the functions $b - b(\infty)$ and $b - b(-\infty)$, to conclude that the two limits $\lim_{t\to\pm\infty}(a*b)(t)$ exist.

(ii) The proof of (ii) is very similar to the proof of (i), and is left to the reader.

(iii) This claim, together with (iv), is proved in Section **3**.6 (they are special cases of (v) and (vi) in Theorem **3**.6.1).

(v) If $b \in C(\mathcal{T}_S)$, then b is bounded, and $b(t+S) = b(t)$ for all $t \in \mathbf{R}$. In this case,

$$(a*b)(t+S) = \int_{\mathbf{R}} a(t+S-s)b(s)\,\mathrm{d}s = \int_{\mathbf{R}} a(t-s)b(s+S)\,\mathrm{d}s = \int_{\mathbf{R}} a(t-s)b(s)\,\mathrm{d}s = (a*b)(t),$$

and so $a*b \in C(\mathcal{T}_S)$. The same argument shows that $a*b \in C(\mathcal{T}_S)$ whenever $b \in L^\infty(\mathcal{T}_S)$, and that $a*b \in BV(\mathcal{T}_S)$ whenever $b \in BV(\mathcal{T}_S)$.

The proof of the claim that $a*b \in L^p(\mathcal{T}_S)$ whenever $a \in L^1(\mathbf{R})$ and $b \in L^p(\mathcal{T}_S)$ is left to the reader.

Almost-periodicity of b means that for each $\epsilon > 0$ there exists a $T > 0$ such that every subinterval of $\mathbf{R}$ of length T contains at least one point h such that $\|\tau_h b - b\|_{\sup(\mathbf{R})} \le \epsilon$. Now

$$\begin{aligned}\|\tau_h(a*b) - (a*b)\|_{\sup(\mathbf{R})} &= \|a*(\tau_h b - b)\|_{\sup(\mathbf{R})}\\ &\le \|a\|_{L^1(\mathbf{R})}\|\tau_h b - b\|_{\sup(\mathbf{R})} \le \epsilon\|a\|_{L^1(\mathbf{R})},\end{aligned}$$

and, therefore, $a*b$ has the same property as b, i.e., it is almost-periodic.

(The same proof, with sup-norms replaced by BL^1-norms, shows that if $a \in L^1(\mathbf{R})$ and if b is almost-periodic in the sense of Stephanov, then $a*b$ is almost-periodic in the sense of Stephanov.)

(vi) The asymptotically periodic and asymptotically almost-periodic cases can be reduced to the preceding ones, because if b is such a function then $b = c + d$, where $c(t) \to 0$ as $t \to \infty$ (either pointwise, or in the BL^p- or BBV-sense), and d is periodic or almost-periodic. Now $a*b = a*c + a*d$, with $(a*c)(t) \to 0$ as $t \to \infty$, and $a*d$ is periodic or almost-periodic. Thus $a*b$ is of the same type as b.

(vii) The norm inequality follows directly from Hölder's inequality, and the continuity is proved in the same way as in (i). To show that $(a*b)(t) \to 0$ as $t \to \infty$ one argues as in the proof of the $L_0^\infty(\mathbf{R})$-claim in (i), again using Hölder's inequality.

(viii) It follows from Fubini's theorem and Hölder's inequality that for each $t \in \mathbf{R}$ the function $(u,s) \mapsto |a(t-s)|\,|b(s-u)|\,|c(u)|$ is integrable:

$$\begin{aligned}&\int_{\mathbf{R}}\int_{\mathbf{R}} |a(t-s)|\,|b(s-u)|\,|c(u)|\,\mathrm{d}u\,\mathrm{d}s\\ &\quad = \int_{\mathbf{R}} |b(u)| \int_{\mathbf{R}} |a(t-s)|\,|c(s-u)|\,\mathrm{d}s\,\mathrm{d}u \le \|b\|_{L^1(\mathbf{R})}\|a\|_{L^p(\mathbf{R})}\|c\|_{L^q(\mathbf{R})}.\end{aligned}$$

This permits a use of Fubini's theorem without the absolute values, and a straightforward computation gives $((a*b)*c)(t) = (a*(b*c))(t)$ for all $t \in \mathbf{R}$. The proofs of the facts that $((b*c)*a)(t) = (b*(c*a))(t)$ and $((c*a)*b)(t) = (c*(a*b))(t)$ for all $t \in \mathbf{R}$ are very similar, and are left to the reader.

(ix) It follows from (ii) that all the listed convolutions are defined a.e. The same is true if we replace a, b and c by their absolute values, and, therefore, we can use Fubini's theorem to get equality a.e.

(x) This is obvious (change the order of the factors, and also change the integration variable). □

Proof of Theorem 2.4 Before discussing the different cases, let us observe that all the cases imply that a is locally integrable on J. This is because we require the integrals in Definition 2.1 to converge absolutely and therefore it follows that, if we let b be a continuous function with compact support in J, then we find that, for some appropriately chosen $\epsilon > 0$ and almost all t, $\int_0^\epsilon |a(t-s)|\,\mathrm{d}s < \infty$. Clearly, this implies local integrability of a. (The same argument shows that (i) is trivially true when $J = \mathbf{R}$; simply choose $b \equiv 1$.)

Now, let us consider the different cases. Claim (i) is a special case of Lemma **9**.10.2, combined with Proposition **9**.2.7(i). Claim (ii) is a special case of Lemma **9**.10.3, combined with Proposition **9**.2.7(i). Finally, claim (iii) is a special case of Lemma **9**.10.4, combined with Proposition **9**.2.7(ii). □

Proof of Theorem 2.5 If $a \in C([0,T])$ and $a(0) = a(T) = 0$, and if we define $a(t) = 0$ for $t \notin [0,T]$, then $\tau_h a \to a$ uniformly as $h \to 0$. From this, one concludes easily that the set $\{\, a * b \mid b \in B \,\}$ is equicontinuous. In addition, it is pointwise bounded. By the Arzelà–Ascoli theorem (see Theorem 7.5) this set is totally bounded in $C([0,T])$; hence in $L^p(0,T)$ as well.

In the general case we can take a sequence $\{a_m\}$ of functions of the type described above, converging to a in $L^1(0,T)$. Since each set $\{\, a_m * b \mid b \in B \,\}$ is totally bounded, the set $\{\, a * b \mid b \in B \,\}$ must be totally bounded. □

Proof of Theorem 2.7 The assertion (i) follows from Rudin [3], Theorems 9.6 and 9.12. Clearly, (ii) is a direct consequence of Definition 2.6. Property (iii) follows from Rudin [3], Theorem 9.2(c), combined with Theorem 2.9(i). □

Proof of Theorem 2.8 Most of the claim follows directly from the fact that one can apply Theorem 2.7, provided one first defines $a(t) = b(t) = 0$ for $t < 0$. Only the assertion about the analyticity of $\hat{a}(z)$ in $\Re z > 0$, and the claim that $\hat{a}(z) \to 0$ as $|z| \to \infty$ in $\Re z \geq 0$, require a separate argument.

To prove the analyticity it suffices to observe that one can differentiate under the integral sign, see the computation in Section **16**.8, or alternatively compare Rudin [3], Theorem 9.2(f), where the same computation is performed employing a real argument instead of a complex argument; the complex case is only slightly more difficult than the real case.

If $z = \sigma + \mathrm{i}\omega$, then

$$|\hat{a}(z)| \leq \int_{\mathbf{R}} \mathrm{e}^{-\sigma t} |a(t)|\,\mathrm{d}t,$$

and, therefore, by Lebesgue's dominated convergence theorem, $\hat{a}(\sigma + \mathrm{i}\omega) \to 0$ as $\sigma \to \infty$, uniformly in ω. It follows from the Riemann–Lebesgue lemma (see the comment following Lemma 5.4 above) that for each fixed σ the function $\omega \mapsto \hat{a}(\sigma + \mathrm{i}\omega)$ belongs to $BC_0(\mathbf{R})$. The mapping that takes σ into the function $t \mapsto \mathrm{e}^{-\sigma t} a(t)$ is continuous from $\mathbf{R}^+$ into $L^1(\mathbf{R}^+)$, and this implies that the mapping that takes σ into the function $\omega \mapsto \hat{a}(\sigma + \mathrm{i}\omega)$

is continuous from $\mathbf{R}^+$ into $BC_0(\mathbf{R})$. This, together with the fact that the function $\omega \mapsto \hat{a}(\sigma + \mathrm{i}\omega)$ tends to 0 in $BC_0(\mathbf{R})$ as $\sigma \to \infty$, implies that the set of functions $\{\, \omega \mapsto \hat{a}(\sigma + \mathrm{i}\omega) \mid \sigma \in \mathbf{R}^+ \}$ is compact in $BC_0(\mathbf{R})$. Every compact set in $BC_0(\mathbf{R})$ tends uniformly to zero at infinity, and so we may conclude that $\hat{a}(\sigma + \mathrm{i}\omega) \to 0$ as $\omega \to \infty$, uniformly in σ. Together with the fact that $\hat{a}(\sigma + \mathrm{i}\omega) \to 0$ as $\sigma \to \infty$, this implies that $\hat{a}(z) \to 0$ as $|z| \to \infty$ in the closed right half plane. □

In the proof of Lemma 5.4 we needed a result on *approximate identities.* A similar but more general lemma will be needed later, and, consequently, we prove a more general lemma than we needed above.

7.4 Lemma

Let $a \in L^1(\mathbf{R})$, denote $\int_{\mathbf{R}} a(s)\,\mathrm{d}s$ by A, and for each $\epsilon > 0$ define a_ϵ by $a_\epsilon(t) = a(t/\epsilon)/\epsilon$.

(i) *If $b \in L^p(\mathbf{R})$ for some $p \in [1, \infty)$, then $a_\epsilon * b \to Ab$ in $L^p(\mathbf{R})$ as $\epsilon \downarrow 0$.*
(ii) *If $b \in BUC(\mathbf{R})$, then $a_\epsilon * b \to Ab$ uniformly on $\mathbf{R}$ as $\epsilon \downarrow 0$.*
(iii) *If $b \in BC(\mathbf{R})$, then $a_\epsilon * b \to Ab$ uniformly on compact sets as $\epsilon \downarrow 0$.*
(iv) *If $b \in BL^p_0(\mathbf{R})$ for some $p \in [1, \infty)$, then $a_\epsilon * b \to Ab$ in $BL^p_0(\mathbf{R})$ as $\epsilon \downarrow 0$.*
(v) *If $b \in BL^p(\mathbf{R})$ for some $p \in [1, \infty)$, the $a_\epsilon * b \to Ab$ in $L^p_{\mathrm{loc}}(\mathbf{R})$ as $\epsilon \downarrow 0$.*
(vi) *Let $b \in BL^1(\mathbf{R})$, and suppose that the function c defined by $c(t) = \operatorname{ess\,sup}_{|s| \ge t} |a(s)|$ is integrable. Then $(a_\epsilon * b)(t) \to Ab(t)$ for every t in the Lebesgue set of b. In particular, $a_\epsilon * b \to Ab$ almost everywhere, and $(a_\epsilon * b)(t) \to Ab(t)$ at all points of continuity of b.*

Proof (i) We have $\int_{\mathbf{R}} a_\epsilon(t)\,\mathrm{d}t = A$ for all $\epsilon > 0$, and, therefore,

$$\begin{aligned}
\big|(a_\epsilon * b)(t) - Ab(t)\big| &= \left| \int_{\mathbf{R}} a_\epsilon(s)\big(b(t-s) - b(t)\big)\,\mathrm{d}s \right| \\
&\le \frac{1}{\epsilon} \int_{\mathbf{R}} |a(s/\epsilon)|\big|b(t-s) - b(t)\big|\,\mathrm{d}s \\
&\le \int_{\mathbf{R}} |a(s)|\big|b(t-\epsilon s) - b(t)\big|\,\mathrm{d}s.
\end{aligned}$$

Using Jensen's inequality and Fubini's theorem as we did in the proof of Theorem 2.2(i), we get

$$\|a_\epsilon * b - Ab\|^p_{L^p(\mathbf{R})} \le \|a\|^{p-1}_{L^1(\mathbf{R})} \int_{\mathbf{R}} |a(s)|\, \|\tau_{-\epsilon s} b - b\|_{L^p(\mathbf{R})}\,\mathrm{d}s.$$

The function $|a(s)|\, \|\tau_{-\epsilon s} b - b\|_{L^p(\mathbf{R})}$ is bounded by the integrable function $2|a(s)|\, \|b\|_{L^p(\mathbf{R})}$, and, by Lemma 7.3, it tends to zero as $\epsilon \downarrow 0$ for each fixed s. Hence, by Lebesgue's dominated convergence theorem, $|a(s)|\, \|\tau_{-\epsilon s} b - b\|_{L^p(\mathbf{R})}$ tends to zero in $L^1(\mathbf{R})$ as $\epsilon \downarrow 0$, and this implies the conclusion of (i).

(ii)–(iii) These proofs are left to the reader (start out as above, and split the integral into $\int_{\mathbf{R}} = \int_{-\infty}^{-T} + \int_{-T}^{T} + \int_{T}^{\infty}$ for some large T).

(iv) This follows immediately from (i), from Theorem 2.2(ii), and from the fact that $L^p(\mathbf{R})$ is dense in $BL^p(\mathbf{R})$.

(v) The proof is similar to the proof of (iii), and is omitted.

(vi) This claim is proved in Stein and Weiss [1], Theorem 1.25, p. 13, with BL^1 replaced by L^p, and, with a very small modification, the same proof shows that (vi) is true (in the estimate for I_2 in Stein and Weiss [1], p. 15, one uses the fact that the function c satisfies $\sum_{k=1}^{\infty} \sup_{t\in[k,k+1]} c(t) < \infty$). □

Above, in the proof of Theorem 2.5, we needed the well-known Arzelà–Ascoli theorem that can be formulated, e.g., as follows.

7.5 The Arzelà–Ascoli Theorem
Let J be a compact subinterval of $\mathbf{R}$, and let A be a subset of $C(J)$. Then the following four conditions are equivalent:

(i) *A is pointwise bounded and equicontinuous, i.e., for every $t \in J$, $\sup_{f\in A}|f(t)| < \infty$, and $\lim_{s\to t, s\in J} \sup_{f\in A}|f(s) - f(t)| = 0$;*

(ii) *A is totally bounded;*

(iii) *the closure of A is compact;*

(iv) *every sequence $\{a_m\} \subset A$ contains a subsequence that converges uniformly to some function in $C(J)$.*

Proof That (i) implies (ii) is proved, e.g., in Rudin [2], Theorem A5, p. 369, and the converse is more or less trivial. The conditions (ii), (iii) and (iv) are equivalent in every complete metric space (cf. Rudin [2], Theorem A4, p. 369). □

Theorem 7.5, combined with a diagonalization argument, gives us the following modified version of the Arzelà–Ascoli theorem.

7.6 The Arzelà–Ascoli Theorem
Let J be a subinterval of $\mathbf{R}$, and let A be a pointwise bounded, equicontinuous subset of $C(J)$. Then every sequence $\{a_m\} \subset A$ contains a subsequence that converges uniformly on compact subsets of J.

8. Exercises

1. Let $\mathbf{e}$ be the function $\mathbf{e}(t) = \mathrm{e}^{-t}$ for $t \geq 0$, and $\mathbf{e}(t) = 0$ for $t < 0$, and let $J = \mathbf{R}$ or $\mathbf{R}^+$. Show that the convolution operator $f \mapsto \mathbf{e} * f$ maps the following spaces V continuously into the corresponding spaces W (in the following list, V is the space to the left of the arrow, and W the space to the right of the arrow)

(i) $BL^p(J) \to BC(J)$, $1 \le p \le \infty$.
(ii) $BL^p_0(J) \to BC_0(J)$, $1 \le p \le \infty$.
(iii) $L^p(\mathcal{T}_s) \to C(\mathcal{T}_S)$, $1 \le p \le \infty$.
(iv) $L^p(\mathcal{T}_s) \oplus BL^p_0(J) \to C(\mathcal{T}_S) \oplus BC_0(J)$, $1 \le p \le \infty$.
(v) $BC_0(J) \to BBV_0(J)$.

2. Add the following spaces V and W to the list of spaces in Exercise 1:
(vi) $BL^p(J) \to BUC(J)$, $1 < p \le \infty$.
Give an example of a function $f \in BL^1(\mathbf{R})$ for which $\mathbf{e}*f \notin BUC(\mathbf{R})$.
3. Let $J = \mathbf{R}$ or $\mathbf{R}^+$, and $p \in [1, \infty]$. Suppose that $f \in BL^p(J)$ is locally absolutely continuous, and that $f' \in BL^p(J)$. Does this imply that f is uniformly continuous? (The answer depends on p; cf. Exercise 2.)
4. Let $0 < T \le \infty$, let a be measurable on $[0, T)$, and suppose that $a * b \in BV([0, T))$ for every $b \in BV([0, T))$. Prove that $a \in L^1(0, T)$.
5. Let $k(t) = at^m$ for $t \ge 0$, where $a \ge 0$, and m is a nonnegative integer. Compute the Laplace transform of the resolvent r of k. Is it true that $r \in L^1(\mathbf{R}^+; \mathbf{R})$? Can you get an explicit expression for r?
6. Let $k \in L^1(\mathbf{R}^+; \mathbf{C})$ satisfy one of the following two sets of conditions:
(i) $\Re\hat{k}(z) > -1$ *for* $\Re z \ge 0$;
(ii) $\Re\hat{k}(\sigma) > -1$ *for all real* $\sigma \ge 0$, *and* $\Im\hat{k}(z) \ne 0$ *for all* z *with* $\Re z \ge 0$ *and* $\Im z \ne 0$.
Show that in either case, the resolvent of k is integrable.
7. Let $k \in L^1(\mathbf{R}; \mathbf{C})$ satisfy one of the following two sets of conditions:
(i) $\Re\hat{k}(i\omega) > -1$ *for* $\omega \in \mathbf{R}$;
(ii) $\Re\hat{k}(0) \ne -1$, *and* $\Im\hat{k}(i\omega) \ne 0$ *for nonzero* $\omega \in \mathbf{R}$.
Show that in either case, k has a whole line resolvent.
8. Let $k \in L^1(\mathbf{R}; \mathbf{C})$ satisfy $k(-t) = -\overline{k(t)}$ for almost all $t \in \mathbf{R}$. Show that k has a whole line resolvent.
9. Let $k(t) = e^{-at}\cos(bt)$ for $t \ge 0$, where a and b are nonnegative numbers. Find the resolvent r of k. Is it true that $r \in L^1(\mathbf{R}^+; \mathbf{R})$?
10. Let $k(t) = 1$ for $t \in [0, 1]$, and $k(t) = 0$ for $t > 1$. Find the resolvent r of k. Is it true that $r \in L^1(\mathbf{R}^+; \mathbf{R})$?
11. Let $k(t) = 1 - t$ for $t \in [0, 1]$, and $k(t) = 0$ for $t > 1$. Can you find the resolvent r of k? Is it true that $r \in L^1(\mathbf{R}^+; \mathbf{R})$?
12. Let $k(t) = e^{at}\cos(bt)$ for $t \le 0$, with $a > 0$ and $b \ge 0$, and $k(t) = 0$ for $t > 0$. Find the whole line resolvent of k. Hint: If you replace t by $-t$, then you get the kernel in Exercise 9.
13. The bilateral Laplace transform of the resolvent in Exercise 12 is a rational function $p(z)$, which has at least one pole in the right half plane. Can you find a kernel $k \in L^1_{\mathrm{loc}}(\mathbf{R}^+; \mathbf{R})$, whose resolvent r satisfies $\hat{r}(z) = p(z)$ for sufficiently large $\Re z$?
14. Define $k(t) = -2e^{-t}$ for $t \ge 0$, and $k(t) = 0$ for $t < 0$. This kernel has both a resolvent in $L^1_{\mathrm{loc}}(\mathbf{R}^+; \mathbf{R})$ and a whole line resolvent in $L^1(\mathbf{R}; \mathbf{R})$. Find them both.
15. Let $k(t) = ae^{-b|t|}$ for $t \in \mathbf{R}$, with $a > 0$ and $b > 0$. Try to find the

whole line resolvent of k.

16. Prove that if $k \in L^1(\mathbf{R}^+;\mathbf{C}^{n\times n})$ satisfies $\det\left[I+\hat{k}(z)\right] \neq 0$ for $\Re z \geq 0$, then the mapping $k \mapsto r$ is continuous in the L^1-norm in a neighbourhood of k. Also prove the analogous result for the whole line resolvent.
17. Let $k \in L^1(\mathbf{R};\mathbf{C}^{n\times n})$ have a whole line resolvent r, and let $S > 0$. Show that k has an S-periodic resolvent r_S, and that this resolvent is given by $r_S = \sum_{j=-\infty}^{\infty} r(t+jS)$, $t \in \mathbf{R}$.
18. Let $k \in L^1(\mathbf{R};\mathbf{C}^{n\times n})$ have an S-periodic resolvent r_S, and let $T = S/k$ for some positive integer k. Show that k has a T-periodic resolvent r_T. How do you compute r_T from r_S?
19. Let $k_S \in L^1(\mathcal{T}_S;\mathbf{C}^{n\times n})$. Show that the following two conditions are equivalent:

    (i) *there exists a function* $r_S \in L^1(\mathcal{T}_S;\mathbf{C}^{n\times n})$ *satisfying*
    $$\begin{aligned} r_S(t) + \int_0^S k_S(t-s)r_S(s)\,\mathrm{d}s &= r_S(t) + \int_0^S r_S(t-s)k_S(s)\,\mathrm{d}s \\ &= k_S(t), \quad t \in \mathbf{R}; \end{aligned}$$

    (ii) $\det[I+\hat{\kappa}_m] \neq 0$ *for all* $m \in \mathbf{Z}$, *where* $\kappa_m = \int_0^S \mathrm{e}^{-\mathrm{i}\,2\pi m t/S} k_S(t)\,\mathrm{d}t$ *is the m-th Fourier coefficient of* k_S.

    What are the analogues of (iii) and (iv) in Theorem 4.7? Hint: Define $k(t) = k_S(t)$ for $t \in [0,S)$, and $k(t) = 0$ for $t \notin [0,S)$. Then apply Theorem 4.7.
20. Prove the following result.

    8.1 Theorem
    Let $k \in L^1(\mathbf{R};\mathbf{C}^{n\times n})$, $S > 0$, and let $f \in V$, where V is one of the spaces in (1)–(4) of Theorem 4.7. Then the equation $x + k * x = f$ has a solution x in V if and only if, for all $m \in \mathbf{Z}$, the mth Fourier coefficient of f belongs to the range of $\left(I + \hat{k}(\mathrm{i}\,2\pi m/S)\right)$.
21. Let $k \in L^1(\mathbf{R};\mathbf{C}^{n\times n})$ have an S-periodic resolvent. Show that there is some $\epsilon > 0$ such that, for all T satisfying $|T - S| \leq \epsilon$, k has a T-periodic resolvent. (In other words, the set of $S \in \mathbf{R}$ for which k has an S-periodic resolvent is open.)
22. Let $k \in C^1(\mathbf{R}^+;\mathbf{R})$ with $k(0) = 1$, and let r be the resolvent of k. Can it happen that $k = r'$?
23. Prove the following result.

    8.2 Theorem
    Let $k \in L^1_{\mathrm{loc}}(\mathbf{R}^+;\mathbf{C}^{n\times n})$ satisfy $\int_{\mathbf{R}^+} \mathrm{e}^{-\sigma t}|k(t)|\,\mathrm{d}t < \infty$ for all $\sigma > 0$ and $\sup_{\Re z>0}\left(|\hat{k}(z)| + \left|[I+\hat{k}(z)]^{-1}\right|\right) < \infty$. Then the resolvent r of k is integrable on $\mathbf{R}^+$ if and only if k is integrable on $\mathbf{R}^+$.
24. For each space V considered in Theorems 4.6 and 4.7, let V^m be the corresponding space of functions ϕ whose derivatives up to order

$m-1$ are locally absolutely continuous, and that satisfy $\phi^{(j)} \in V$ for $1 \le j \le m$. Show that one can replace V in (iii) and (iv) of the two theorems by V^m.

25. Let $k \in L^1(\mathbf{R}^+; \mathbf{C}^{n\times n})$. Show that the following two conditions are equivalent to each other, and that they are equivalent to (i)–(iv) in Theorem 4.5 when V is one of the spaces in (1)–(8):
    (v) *for some $a \in L^1(\mathbf{R}^+; \mathbf{C}^{n\times n})$ satisfying $\det[\hat{a}(z)] \neq 0$ for $\Re z \ge 0$, and for every $f \in V$, the solution x of the equation $x + k*x = f$ on $\mathbf{R}^+$ satisfies $a * x \in V$;*
    (vi) *for some $a \in L^1(\mathbf{R}^+; \mathbf{C}^{n\times n})$ satisfying $\det[\hat{a}(z)] \neq 0$ for $\Re z \ge 0$, and for every $f \in V$, the solution x of the equation $x + k*x = f$ on $\mathbf{R}^+$ satisfies $a * x \in V$, and $a * x$ depends continuously on f (in the norm of V).*

    Hint: Imitate the proof of Theorem 4.5.
26. Let $k \in L^1(\mathbf{R}; \mathbf{C}^{n\times n})$. Show that the following condition is equivalent to (i)–(iv) in Theorem 4.7 when V is one of the spaces in (1)–(4).
    (v) *For every $f \in V$, the equation $x + k*x = f$ has a solution in V.*
27. Let $p \in (1,\infty)$, $1/p + 1/q = 1$, $k \in L^1(\mathbf{R}^+; \mathbf{C}^{n\times n}) \cap L^p(\mathbf{R}^+; \mathbf{C}^{n\times n})$, and suppose that $\int_0^\infty \left(\int_t^\infty |k(s)|\, ds\right)^q dt < \infty$. Show that the resolvent r of k belongs to $L^1(\mathbf{R}^+; \mathbf{C}^{n\times n})$ if and only if it belongs to $L^p(\mathbf{R}^+; \mathbf{C}^{n\times n})$.
28. Let $J = \mathbf{R}$ or $\mathbf{R}^+$, $1 \le p \le \infty$, and let $L^{p,1}(J)$ be the set of functions $\phi \in L^p_{\mathrm{loc}}(J)$ that satisfy $\|\phi\|_{L^{p,1}(J)} \stackrel{\mathrm{def}}{=} \sum_{j=-\infty}^{\infty} \|\phi\|_{L^p(j,j+1)} < \infty$. Add the following claims to Theorem 2.2.
    (xi) *Let $J = \mathbf{R}$ or $\mathbf{R}^+$, $a \in L^{p,1}(J)$, and $b \in BL^q(J)$, for some p and q in $[1,\infty]$ with $1/p + 1/q = 1$. Then $a * b \in BC(J)$, and $\|a*b\|_{\sup(J)} \le \|a\|_{L^{p,1}(J)} \|b\|_{BL^q(J)}$. If $b \in BL^q_0(J)$, then $a * b \in BC_0(J)$.*
    (xii) *If $a \in L^{p,1}(J)$, $b \in L^1(J)$ and $c \in BL^q(J)$, for some p and q in $[1,\infty]$ with $1/p + 1/q = 1$, then $((a*b)*c)(t) = (a*(b*c))(t)$, $((b*c)*a)(t) = (b*(c*a))(t)$ and $((c*a)*b)(t) = (c*(a*b))(t)$ for all $t \in J$.*
29. Let $J = \mathbf{R}$ or $\mathbf{R}^+$, $1 \le p \le \infty$, $1 \le q < \infty$, and let $L^{p,q}(J)$ be the set of functions $\phi \in L^p_{\mathrm{loc}}(J)$ that satisfy $\|\phi\|_{L^{p,q}(J)} < \infty$ where

$$\|\phi\|_{L^{p,q}(J)} \stackrel{\mathrm{def}}{=} \left\{ \sum_{j=-\infty}^{\infty} \|\phi\|^q_{L^p(j,j+1)} \right\}^{1/q} < \infty.$$

Analogously, let $L^{p,\infty}$ be the set of functions $\phi \in L^p_{\mathrm{loc}}(J)$ that satisfy $\|\phi\|_{L^{p,\infty}(J)} \stackrel{\mathrm{def}}{=} \sup_{j\in\mathbf{Z}} \|\phi\|_{L^p(j,j+1)} < \infty$. (Clearly, $L^{p,\infty}$ is the same space as BL^p, but with a different norm.) Add the following claims to Theorem 2.2.

(xiii) *Let $J = \mathbf{R}$ or $\mathbf{R}^+$, $a \in L^1(J)$, and $b \in L^{p,q}(J)$ for some p, $q \in [1,\infty]$. Then $a * b \in L^{p,q}(J)$, and there is some constant*

$K \ge 1$ such that $\|a*b\|_{L^{p,q}(J)} \le K\|a\|_{L^1(J)}\|b\|_{L^{p,q}(J)}$.

(xiv) *Let $J = \mathbf{R}$ or $\mathbf{R}^+$, and let $1 < p, q < \infty$, $1/p + 1/p' = 1$, $1/q + 1/q' = 1$, $a \in L^{p,q}(J)$ and $b \in L^{p',q'}(J)$. Then $a*b \in BC(J)$, and there is a constant $K \ge 1$ such that $\|a*b\|_{\sup(J)} \le K\|a\|_{L^{p,q}(J)}\|b\|_{L^{p',q'}(J)}$.*

(xv) *If $a \in L^{p,q}(J)$, $b \in L^1(J)$ and $c \in L^{p',q'}(J)$, for some p, q, p', and q' in $[1,\infty]$ with $1/p + 1/p' = 1$, $1/q + 1/q' = 1$, then $((a*b)*c)(t) = (a*(b*c))(t)$, $((b*c)*a)(t) = (b*(c*a))(t)$ and $((c*a)*b)(t) = (c*(a*b))(t)$ for all $t \in J$.*

30. Show that the spaces $V = L^{p,q}(J)$, $1 \le p, q \le \infty$, can be added to the spaces V listed in Theorems 4.5 and 4.6 (cf. Exercise 29).
31. Show that if $k \in L^1(\mathbf{R}^+; \mathbf{C}^n)$ satisfies $\det[I + \hat{k}(z)] \ne 0$ for $\Re z \ge 0$, then one cannot, in general, find a bound on the L^1-norm of the resolvent r that depends only on $\sup_{\Re z \ge 0}\big(|\hat{k}(z)| + \big|[I + \hat{k}(z)]^{-1}\big|\big)$. Hint: Cf. Theorem 8.2 and use the function $\mathrm{e}^{-\epsilon t} t^{-1/2} J_1(t^{1/2})$.
32. Complete the proof of Theorem 6.1.
33. Complete the proof of Lemma 7.4.
34. Plot the curve $1 + \hat{k}(\mathrm{i}\omega)$, $\infty \ge \omega \ge -\infty$, and check whether the resolvent of k is integrable in the following cases:
    (i) $k(t) = 4\mathrm{e}^{-t/2} - 3\mathrm{e}^{-t}$;
    (ii) $k(t) = 6\mathrm{e}^{-t/2} - 5\mathrm{e}^{-t}$;
    (iii) $k(t) = \begin{cases} 2, & 0 \le t < 1, \\ 1, & 1 \le t < 2, \\ 0, & 2 \le t < \infty. \end{cases}$
    (iv) $k(t) = \begin{cases} 1, & 0 \le t < 1, \\ 3, & 1 \le t < 2, \\ 0, & 2 \le t < \infty. \end{cases}$

9. Comments

Section 1:
The Dover edition of Volterra [4] contains both a biography of Vito Volterra, written by E. T. Whittaker, and a list of Volterra's publications. For some of the first papers on integral equations of Volterra type, see Volterra [1], [2], and [3]. The variation of constants formula (1.2) (for the nonconvolution case) is found already in Volterra [1].

Section 2:
For Laplace transforms of functions, one may consult Doetsch [1] and Widder [2]. The literature on Fourier transforms is abundant. See, e.g., Stein and Weiss [1], or Katznelson [1].

The basic results on L^p-spaces, B^∞-spaces, and functions of bounded variation, are found in most books on real analysis, e.g., in Rudin [3], and in Hewitt and Stromberg [1]. The importance of the spaces BL^p, etc., in asymptotic analysis has been widely recognized. Wiener [1] calls a function of type $BL^2(\mathbf{R})$ *nearly bounded.* Our notations BL^p, BL^p_0, BBV, and BBV_0 are not standard. For example, Massera and Schäffer [1] denote BL^1 by M and BL^1_0 by M_0, but we reserve the letter M for use in connection with measures.

Almost-periodic functions are discussed, e.g., in Amerio and Prouse [1], Besicovitch [1], Bohr [1], Corduneanu [5], and Fink [1].

Section 3:
Theorems 3.1 and 3.5 are classical, see e.g., Miller [10], p. 13, or Fenyő and Stolle [1], Part 3, p. 255. Put in a different perspective, the content of these theorems is that $I + k*$ can be inverted, and that the inverse has nice properties. The proof of Theorem 3.1 shows that the inverse can be expressed as a Neumann series. Observe that $r*$ does *not* (cf. semigroup theory) define the inverse of $I+k*$. Instead, formally, $\left[I+k*\right]^{-1} = I - r*$.

Sections 4 and 5:
Theorems 4.1 and 4.3 and their various extensions are fundamental to the theory of linear Volterra equations. They are proved in Paley and Wiener [1], p. 60, by essentially the same argument as here. Paley and Wiener state only Theorem 4.1 explicitly, but Theorem 4.3 is an obvious consequence of their proof. Observe that the label 'Paley–Wiener theorem' is frequently used to denote something entirely different from Theorems 4.1 and 4.3, i.e., to denote certain results on Fourier transforms of L^2-functions; see, e.g., Rudin [3], Theorems 19.2 and 19.3.

Note that the whole line resolvent is sometimes called the *Green's function* of equation (4.3).

In Chapter 3 we formulate Paley–Wiener results for the integrodifferential equation $x' + k * x = f$. In Chapter 6 we revisit the Paley–Wiener theorems in connection with analytic mappings of Fourier transforms. In Chapter 4 we will return to the integrability of the resolvent in a more general setting (weighted spaces and measure-valued kernels), employing Banach algebra techniques.

The parts of Theorems 4.5 and 4.6 where $V = L^1$ and $V = L^\infty$ are classical. Invariance results for Volterra equations in fractional order Sobolev spaces are discussed in Kappel and Kunisch [1].

Section 6:
If k is scalar, nonnegative, nonincreasing, and convex on $\mathbf{R}^+$, then the resolvent r of k satisfies $\|r\|_{L^1(\mathbf{R}^+)} \le 20$ (see Theorem **6**.2.1). If k is scalar, nonnegative, and nonincreasing, then $\|r\|_{L^1(\mathbf{R}^+)} = O(\|k\|^5_{L^1(\mathbf{R}^+)})$ as $\|k\|_{L^1(\mathbf{R}^+)} \to \infty$, see Gripenberg [11].

Nyquist's Theorem 6.3—part of the folklore in control theory—was originally published in Nyquist [1]. For a reprint of Nyquist [1], and for extensions and further developments, see MacFarlane [1].

Section 8:
The spaces $L^{p,q}$ are discussed in Holland [1].

References

L. Amerio and G. Prouse
1. *Almost-Periodic Functions and Functional Equations*, Van Nostrand Reinhold, New York, 1971.

A. S. Besicovitch
1. *Almost Periodic Functions*, Cambridge University Press, Cambridge, 1932.

H. Bohr
1. *Almost Periodic Functions*, Chelsea, New York, 1947.

C. Corduneanu
5. *Almost Periodic Functions*, Interscience, New York, 1968.

G. Doetsch
1. *Handbuch der Laplace-Transformation. I–III*, Birkhäuser, Basel, 1950–1956.

I. Fenyő and H. W. Stolle
1. *Theorie und Praxis der linearen Integralgleichungen, 1–4*, Birkhäuser, Basel, 1982–1984.

A. M. Fink
1. *Almost Periodic Differential Equations*, Springer-Verlag, Berlin, 1974.

G. Gripenberg
11. On the resolvents of Volterra equations with nonincreasing kernels, *J. Math. Anal. Appl.* **76** (1980), pp. 134–145.

E. Hewitt and K. Stromberg
1. *Real and Abstract Analysis*, Springer-Verlag, Berlin, 1965.

F. Holland
1. Harmonic analysis on amalgams of L^p and l^q, *J. London Math. Soc.(2)* **10** (1975), pp. 295–305.

F. Kappel and K. Kunish
1. Invariance results for delay and Volterra equations in fractional order Sobolev spaces, *Trans. Amer. Math. Soc.* **304** (1987), pp. 1–51.

Y. Katznelson
1. *An Introduction to Harmonic Analysis*, Dover Publications, New York, 1976.

A. G. J. MacFarlane
1. *Frequency-Response Methods in Control Systems*, A. G. J. MacFarlane, ed., Selected Reprint Series, IEEE Press, New York, 1979.

J. L. Massera and J. J. Schäffer
1. *Linear Differential Equations and Function Spaces*, Academic Press, New York, 1966.

R. K. Miller

10. *Nonlinear Volterra Integral Equations*, W. A. Benjamin, Menlo Park, Calif., 1971.

H. Nyquist

1. Regeneration theory, *Bell System Tech. J.* **11** (1932), pp. 126–147.

R. E. A. C. Paley and N. Wiener

1. *Fourier Transforms in the Complex Domain*, Amer. Math. Soc., Providence, R. I., 1934.

W. Rudin

2. *Functional Analysis*, McGraw–Hill, New York, 1973.
3. *Real and Complex Analysis,* 3rd *ed.*, McGraw–Hill, New York, 1986.

E. M. Stein and G. Weiss

1. *Introduction to Fourier Analysis on Euclidean Spaces*, Princeton University Press, Princeton, 1971.

V. Volterra

1. Sulla inversione degli integrali definiti, *R. C. Accad. Lincei (5)* **5** (1896), pp. 177–185.
2. Sulla inversione degli integrali multipli, *R. C. Accad. Lincei (5)* **5** (1896), pp. 289–300.
3. Sull' inversione degli integrali definiti, *Atti Accad. Torino* **31** (1896), pp. 311–323, 400–408, 537–567, 693–708.
4. *Theory of Functionals and of Integral and Integro-Differential Equations*, Dover Publications, New York, 1959.

D. V. Widder

2. *The Laplace Transform*, Princeton University Press, Princeton, 1946.

N. Wiener

1. On the representation of functions by trigonometrical integrals, *Math. Z.* **24** (1925), pp. 575–616.

A. Zygmund

1. *Trigonometric Series, I–II,* 2nd *ed.*, Cambridge University Press, Cambridge, 1979.

3

Linear Integrodifferential Convolution Equations

We study the integrodifferential equation $x' + \mu * x = f$, $x(0) = x_0$. We define the differential resolvent kernel r, and derive the variation of constants formula $x = rx_0 + r * f$ for the solution. We give necessary and sufficient conditions under which r is integrable. Some technical results on measures, convolutions, and Laplace transforms are presented.

1. Introduction

A differentiation of the Volterra integral equation

$$x(t) + \int_0^t k(t-s)x(s)\,\mathrm{d}s = f(t), \quad t \in \mathbf{R}^+ \tag{1.1}$$

(i.e., the same equation as (**2**.1.1)), yields

$$x'(t) + k(0)x(t) + \int_0^t k'(t-s)x(s)\,\mathrm{d}s = f'(t), \quad t \in \mathbf{R}^+; \quad x(0) = f(0).$$

This is an example of the Volterra integrodifferential equation

$$x'(t) + \int_{[0,t]} \mu(\mathrm{d}s)x(t-s) = f(t), \quad t \in \mathbf{R}^+; \quad x(0) = x_0, \tag{1.2}$$

to be discussed in this chapter. Here μ is a locally finite Borel measure. Another equation of the same type is the differential-delay equation

$$x'(t) + \sum_{0 \le t_k \le t} a_k x(t - t_k) = f(t), \quad t \in \mathbf{R}^+; \quad x(0) = x_0,$$

obtained from (1.2) by taking $\mu = \sum_k a_k \delta_{t_k}$, where δ_{t_k} is the unit point mass at t_k.

The functional differential equation

$$x'(t) = \int_{(-\infty,0]} \eta(\mathrm{d}s)x(t+s) + F(t), \quad t \geq 0,$$

$$x(t) = \varphi(t), \quad t \leq 0,$$

can be written in the form (1.2) if one takes $\mu(E) = -\eta(-E)$ for every Borel set $E \subset \mathbf{R}^+$, $x_0 = \varphi(0)$, and $f(t) = F(t) + \int_{(-\infty,-t)} \eta(\mathrm{d}s)\varphi(s)$.

Proceeding formally, in much the same way as in Section **2**.1, one can show that the solution x of (1.2) ought to be given by a variation of constants formula, similar to the corresponding formula for an ordinary differential equation. The fastest way to get the appropriate expression is to use Laplace transforms. Taking transforms of both sides of (1.2), one gets

$$z\hat{x}(z) - x_0 + \hat{\mu}(z)\hat{x}(z) = \hat{f}(z),$$

and hence

$$\hat{x}(z) = \left[zI + \hat{\mu}(z)\right]^{-1}(x_0 + \hat{f}(z)).$$

(Above, $\hat{\mu}(z) = \int_{\mathbf{R}^+} \mu(\mathrm{d}s)\mathrm{e}^{-zs}$. As we show in Sections 2, 6, 7, and 8, convolutions where one of the factors is a locally finite measure, and the Laplace transforms of such convolutions and measures, have almost the same properties as we encountered in Chapter 2 for convolutions and Laplace transforms of locally integrable functions.)

Let us define the convolution $\mu * a$ of a measure μ and a function a (both given on $\mathbf{R}^+$) as the function

$$(\mu * a)(t) = \int_{[0,t]} \mu(\mathrm{d}s)a(t-s).$$

For more details, see Definition 2.1 and Theorem 6.1.

Assume that there exists a locally integrable function r whose Laplace transform is $\hat{r}(z) = \left[zI + \hat{\mu}(z)\right]^{-1}$. Then x is given by the *variation of constants formula*

$$x(t) = r(t)x_0 + (r * f)(t), \quad t \in \mathbf{R}^+. \tag{1.3}$$

Here the *differential resolvent* r corresponds to the fundamental solution of an ordinary differential equation, and many writers do in fact refer to r as the fundamental solution of the integrodifferential equation (1.2). There are good reasons for the name 'fundamental solution', because r satisfies a modified version of (1.2). To see this, one need only observe that

$$\hat{r}(z)[zI + \hat{\mu}(z)] = [zI + \hat{\mu}(z)]\hat{r}(z) = I.$$

Hence r is a solution of the two equations

$$r'(t) + (\mu * r)(t) = r'(t) + (r * \mu)(t) = 0, \quad t \in \mathbf{R}^+; \quad r(0) = I. \tag{1.4}$$

The discussion above was inspired by the equation that one gets by differentiating the half line Volterra integral equation (1.1). If instead one differentiates the whole line integral equation

$$x(t) + \int_{-\infty}^{\infty} k(t-s)x(s)\,\mathrm{d}s = f(t), \quad t \in \mathbf{R} \tag{1.5}$$

(this is the same equation as (**2**.4.3)), then the result is an equation of the type

$$x'(t) + (\mu * x)(t) = f(t), \quad t \in \mathbf{R}. \tag{1.6}$$

Here it is not appropriate to add an initial condition; instead one adds a size condition on x at plus and minus infinity, to guarantee that the convolution exists.

The solution of (1.6) can be expressed by the slightly different variation of constants formula

$$x(t) = (r * f)(t), \quad t \in \mathbf{R}, \tag{1.7}$$

where the *whole line differential resolvent* r is locally absolutely continuous, except for a jump discontinuity at zero, and satisfies

$$r'(t)+(\mu*r)(t) = r'(t)+(r*\mu)(t) = 0, \quad t \in \mathbf{R}; \quad r(0+)-r(0-) = I. \tag{1.8}$$

Here we have interpreted r' as the pointwise derivative of r, and the equation will in general hold only for almost all t. If instead one wants to use distribution derivatives, then the right-hand side of (1.8) should be replaced by δI, where δ is the (scalar) unit point mass at zero, to account for the fact that r has a jump discontinuity at the origin.

Before discussing (1.2) and (1.6) in more detail, we give a short description of measures, of convolutions in which one of the factors is a measure and the other factor is a function, and of Laplace transforms of measures (a more extensive treatment can be found in Sections 6 and 8).

2. Measures, Convolutions and Laplace Transforms

The measures that we use in the sequel are matrices of scalar measures, where each scalar measure is defined as a countably additive mapping from a collection of measurable subsets of $\mathbf{R}$ into $\mathbf{C}$, or as a linear continuous mapping from $BC_0(\mathbf{R})$ into $\mathbf{C}$, or as the distribution derivative of a function of bounded variation from $\mathbf{R}$ into $\mathbf{C}$. The collection of measurable subsets will always contain all the Borel sets, and it will be complete in the sense assigned to this word in measure theory. It follows from the Riesz representation theorem, and from standard measure theory, that these three different characterizations give the same class of measures.

We denote the Banach space of matrix-valued complete (bounded) Borel measures on an interval $J \subset \mathbf{R}$ by $M(J; \mathbf{C}^{n\times n})$, and the corresponding space of scalar measures by $M(J; \mathbf{C})$. The norms in $M(J; \mathbf{C}^{n\times n})$ and $M(J; \mathbf{C})$ are $\|\mu\|_{M(J)} = |\mu|(J)$, where $|\mu|$ is the total variation of μ (see Section 5). The support of μ is the complement of the largest open set E such that $|\mu|(E) = 0$. If it is clear from the context in what space the values of the measure lie, or if the result is true with all different choices of spaces, then we simply write $M(J)$. In particular, $M(\mathbf{R})$ is the space of measures on $\mathbf{R}$, and $M(\mathbf{R}^+)$ is the space of measures on $\mathbf{R}^+$. In general, we extend measures on $\mathbf{R}^+$ to measures on $\mathbf{R}$ by defining them to be zero on $(-\infty, 0)$ (i.e., we define $\mu(E) = \mu(E \cap \mathbf{R}^+)$ for every Borel set $E \subset \mathbf{R}$), and with this interpretation $M(\mathbf{R}^+) \subset M(\mathbf{R})$. The notations $M_{\text{loc}}(\mathbf{R})$ and $M_{\text{loc}}(\mathbf{R}^+)$ are used for *local measures*, i.e., set functions that locally behave like bounded measures (for each $T > 0$ the function that maps the Borel set E into $\mu(E \cap [-T, T])$ is a bounded measure). These measures are also called *Radon measures.*

It is a consequence of the Radon–Nikodym theorem that every local measure μ can be split into three parts, the *absolutely continuous part* (i.e., absolutely continuous with respect to Lebesgue measure), the *discrete part* which consists of a sum of point masses, and the *singular part.* We shall frequently use a differential type notation, i.e., we let the notation $\mu(\mathrm{d}s)$ represent the measure μ, and regard $\mathrm{d}s$ as a dummy argument. With this notation, we can write an arbitrary measure μ in the form

$$\mu(\mathrm{d}s) = b(s)\,\mathrm{d}s + \sum_{k=1}^{\infty} a_k \delta_{t_k}(\mathrm{d}s) + \mu_{\mathrm{s}}(\mathrm{d}s), \tag{2.1}$$

where b represents the absolutely continuous part of μ (with respect to Lebesgue measure $\mathrm{d}s$), δ_{t_k} is the (scalar) unit point mass at the point t_k, the coefficients a_k are constants (whose dimension is the same as the dimension of μ), the sum is the discrete part of μ, and μ_{s} is the singular part of μ.

2.1 Definition

The convolution $\mu * a$ of a measure $\mu \in M(\mathbf{R})$ and a Lebesgue measurable function a, defined on $\mathbf{R}$, is the function

$$(\mu * a)(t) = \int_{\mathbf{R}} \mu(\mathrm{d}s) a(t-s),$$

defined for those t for which the function $s \to a(t-s)$ is $|\mu|$-integrable. The convolution $\mu * a$ of a measure $\mu \in M([0,T])$ and a Lebesgue measurable function a, defined on $[0,T]$, is the function

$$(\mu * a)(t) = \int_{[0,t]} \mu(\mathrm{d}s) a(t-s),$$

defined for those t for which the function $s \to a(t-s)$ is $|\mu|$-integrable. In both cases the convolution $a * \mu$ is defined analogously, with $\mu(\mathrm{d}s)a(t-s)$ replaced by $a(t-s)\mu(\mathrm{d}s)$.

The values of μ and a should be compatible, so that the (matrix) product $\mu(\mathrm{d}s)a(t-s)$ (or $a(t-s)\mu(\mathrm{d}s)$) is defined. The convolution integral can be computed componentwise, so if, e.g., $\mu \in M(\mathbf{R}; \mathbf{C}^{n\times n})$ and a takes column vector values in $\mathbf{C}^n$ then $\mu * a$ takes column vector values in $\mathbf{C}^n$, with components $(\mu * a)_i = \sum_{j=1}^{n} \mu_{ij} * a_j$. In Theorem 6.1, it is shown that if, for example, $\mu \in M(\mathbf{R})$ and $a \in BL^1(\mathbf{R})$ then $\mu * a$ is defined almost everywhere. If $\mu \in M(\mathbf{R})$ and $a \in B^\infty(\mathbf{R})$, then $\mu * a$ is defined everywhere.

According to Definition 2.1, one has $(\delta_s * a)(t) = a(t - s)$ (the notation is the same as in (2.1)). In other words, translation of a function can be represented as a convolution of this function by a measure. Furthermore, we observe that the equation one gets by differentiating (1.1), can be written

$$x' + \mu * x = f',$$

where $\mu(\mathrm{d}s) = k(0)\delta_0(\mathrm{d}s) + k'(s)\,\mathrm{d}s$.

The reader who is not familiar with convolutions where one of the factors is a measure should consult Section 6. The results listed there show that these convolutions behave in almost exactly the same way as convolutions where the measure has been replaced by an L^1-function. We use the results of Section 6 extensively in the proofs given below.

To handle the asymptotics of the solutions of our integrodifferential equations, we introduce Laplace transforms of measures.

2.2 Definition

The Laplace transform $\hat{\mu}(z)$ of a locally finite measure μ on $\mathbf{R}^+$ is the function

$$\hat{\mu}(z) = \int_{[0,\infty)} \mathrm{e}^{-zt}\mu(\mathrm{d}t),$$

defined for those $z \in \mathbf{C}$ for which this integral converges absolutely. The (bilateral) Laplace transform of a locally finite measure μ on $\mathbf{R}$ is the function

$$\hat{\mu}(z) = \int_{\mathbf{R}} \mathrm{e}^{-zt}\mu(\mathrm{d}t),$$

which is defined for those $z \in \mathbf{C}$ for which the integral converges absolutely.

Laplace transforms of measures behave in many respects in the same way as Laplace transforms of L^1-functions (see Section 8).

3. The Integrodifferential Equation

Let us return to the two equations

$$x'(t) + (\mu * x)(t) = f(t), \quad t \in \mathbf{R}^+; \quad x(0) = x_0, \tag{3.1}$$

and

$$x'(t) + (\mu * x)(t) = f(t), \quad t \in \mathbf{R}, \tag{3.2}$$

mentioned in the introduction. First, we study the local behaviour of (3.1), prove the existence of the differential resolvent, and establish the variation of constants formula (1.3). After that we discuss a global theory for (3.2).

In the same way as in Chapter 2, let us make a general agreement on what we mean when we say that x is a solution of (3.1). In most cases x cannot satisfy (3.1) everywhere, because even if f is continuous the convolution $\mu * x$ will have jump discontinuities at the points of $(0, \infty)$ where μ has point masses (if μ has no point masses in $(0, \infty)$, and if f is continuous, then $\mu * x$ is continuous). At these points x' is not defined, because the left-hand and the right-hand derivatives of x are different. Owing to this fact we adopt a weaker notion, and call x a solution of (3.1) on an interval $[0, T)$ whenever x is locally absolutely continuous, $x(0) = x_0$, and $x'(t) + (\mu * x)(t) = f(t)$ for almost all $t \in [0, T)$. Similar comments apply to the whole line integrodifferential equation (3.2), and to the resolvent equations (1.4) and (1.8).

3.1 Theorem
Let $\mu \in M_{\mathrm{loc}}(\mathbf{R}^+; \mathbf{C}^{n\times n})$. Then there is a unique locally absolutely continuous function r on $\mathbf{R}^+$ with $r(0) = I$ that satisfies the equations $r'(t) + (\mu * r)(t) = r'(t) + (r * \mu)(t) = 0$ for almost all $t \in \mathbf{R}^+$. Furthermore, the derivative r' is equal a.e. to a function which is locally of bounded variation.

It will be obvious from the proof of this theorem that if μ is $\mathbf{R}^{n\times n}$-valued, then so is r.

3.2 Definition
The function r in Theorem 3.1 is called the differential resolvent of μ.

3.3 Theorem
Let $\mu \in M_{\mathrm{loc}}(\mathbf{R}^+; \mathbf{C}^{n\times n})$. Then, for every $f \in L^1_{\mathrm{loc}}(\mathbf{R}^+; \mathbf{C}^n)$, the equation (3.1) has a unique locally absolutely continuous solution x. This solution is given by the variation of constants formula $x(t) = r(t)x_0 + (r * f)(t)$ for $t \in \mathbf{R}^+$, where r is the differential resolvent of μ. If $f \in L^p_{\mathrm{loc}}(\mathbf{R}^+; \mathbf{C}^n)$ for some $p \in (1, \infty]$, then $x' \in L^p_{\mathrm{loc}}(\mathbf{R}^+; \mathbf{C}^n)$, and if $f \in BV_{\mathrm{loc}}(\mathbf{R}^+; \mathbf{C}^n)$ then x' is equal a.e. to a function in $BV_{\mathrm{loc}}(\mathbf{R}^+; \mathbf{C}^n)$.

Proof of Theorem 3.1 Define $k(t) = \mu([0, t))$. Then k is locally of bounded variation. Let q be the resolvent of k, that is, the solution of $q + k * q = q + q * k = k$. By Theorem **2**.3.1, and by Corollary **2**.2.3(ii), $q \in BV_{\mathrm{loc}}(\mathbf{R}^+; \mathbf{C}^{n\times n})$. Define $r(t) = I - \int_0^t q(s)\, ds$. Then $r(0) = I$, and by differentiating $k * r$ and $r * k$ in two different ways (see Corollaries 7.3(ii) and 7.4(i)) we get

$$\mu * r = (k * r)' = k + k * r',$$

and

$$r * \mu = (r * k)' = k + r' * k.$$

But $r' = -q$, so we have

$$\mu * r = k - k * q = q = -r',$$

and

$$r * \mu = k - q * k = q = -r'.$$

Hence r satisfies (1.4).

To prove uniqueness, let r and s be two locally absolutely continuous functions satisfying $r(0) = s(0) = I$ and $r' + \mu * r = s' + s * \mu = 0$. Then, by Corollary 7.4,

$$r + s' * r = (s * r)' = s + s * r',$$

or, equivalently,

$$r - (s * \mu) * r = s - s * (\mu * r).$$

Since the convolution operator is associative (see Corollary 6.2(v)), this implies that $r = s$. □

Proof of Theorem 3.3 Let $f \in L^1_{\mathrm{loc}}(\mathbf{R}^+; \mathbf{C}^n)$, and define x by $x = rx_0 + r * f$. Then $x(0) = x_0$. By Corollary 7.4(ii), x is locally absolutely continuous and

$$\begin{aligned} x' &= r'x_0 + f + r' * f \\ &= -(\mu * r)x_0 - (\mu * r) * f + f \\ &= -\mu * (rx_0 + r * f) + f \\ &= -\mu * x + f. \end{aligned}$$

This shows that x is a solution of (3.1).

Conversely, let x be an arbitrary solution of (3.1). Then x is locally absolutely continuous, and so $x + r' * x = (r * x)' = rx_0 + r * x'$. Hence

$$\begin{aligned} x &= rx_0 + r * x' - r' * x \\ &= rx_0 + r * (-\mu * x + f) + (r * \mu) * x \\ &= rx_0 + r * f, \end{aligned}$$

and x satisfies (1.3). □

As the following theorem shows, the differential resolvent depends continuously on the kernel.

3.4 Theorem

Let $\mu_j \in M_{\mathrm{loc}}(\mathbf{R}^+; \mathbf{C}^{n\times n})$ and $\mu \in M_{\mathrm{loc}}(\mathbf{R}^+; \mathbf{C}^{n\times n})$, and suppose that $\mu_j([0,\cdot)) \to \mu([0,\cdot))$ in $L^1_{\mathrm{loc}}(\mathbf{R}^+; \mathbf{C}^{n\times n})$ as $j \to \infty$. Let r_j and r be the differential resolvents of μ_j and μ, respectively. Then $r_j \to r$ uniformly on compact subsets of $\mathbf{R}^+$, and $r'_j \to r'$ in $L^1_{\mathrm{loc}}(\mathbf{R}^+; \mathbf{C}^{n\times n})$ as $j \to \infty$.

The proof is left to the reader (combine the argument used in the proof of Theorem 3.1 with Theorem **2**.3.1).

Above we have addressed the question of existence and uniqueness of solutions of (3.1). The next question that presents itself is: can one say something about the asymptotic behaviour of the solution? In particular, is it true that x behaves asymptotically roughly in the same way as f

does? In analogy to what is true for the integral equation (1.1), to get an affirmative answer to this question we need an estimate on the asymptotic size of r. It should not be surprising that the appropriate condition is $r \in L^1(\mathbf{R}^+; \mathbf{C}^{n\times n})$.

In the same way as in Chapter 2, it is possible to give necessary and sufficient conditions, in terms of Laplace transforms, for r to belong to $L^1(\mathbf{R}^+; \mathbf{C}^{n\times n})$ in the case where μ is bounded, i.e., $\mu \in M(\mathbf{R}^+; \mathbf{C}^{n\times n})$.

3.5 Theorem

Let $\mu \in M(\mathbf{R}^+; \mathbf{C}^{n\times n})$. Then the differential resolvent r of μ satisfies

$$r \in L^1(\mathbf{R}^+; \mathbf{C}^{n\times n})$$

if and only if

$$\det[zI + \hat{\mu}(z)] \neq 0, \quad \Re z \geq 0.$$

Moreover, if $r \in L^1(\mathbf{R}^+; \mathbf{C}^{n\times n})$, then $r' \in L^1(\mathbf{R}^+; \mathbf{C}^{n\times n})$, and r' is equal a.e. to a function of bounded variation.

The demonstration of how Theorem 3.5 follows from Theorem **2**.4.1 is entirely analogous to the proof of Theorem 3.7 given below (one throughout replaces convolutions on $\mathbf{R}$ by convolutions on $\mathbf{R}^+$). Alternatively, one can prove Theorem 3.5 by combining Theorem 3.7 with Lemma **2**.5.1. We leave this proof to the reader.

In analogy to Corollary **2**.4.2 we have the following result.

3.6 Corollary

If the measure $e^{-\sigma t}\mu(dt)$ belongs to $M(\mathbf{R}^+; \mathbf{C}^{n\times n})$ for some $\sigma \in \mathbf{R}$, then the function $e^{-ct}r(t)$, where r is the differential resolvent of μ, belongs to $L^1(\mathbf{R}^+; \mathbf{C}^{n\times n})$ for each constant $c \geq \sigma$ such that $\det[zI + \hat{\mu}(z)] \neq 0$ for $\Re z \geq c$.

We proceed to formulate a similar result for equation (3.2). Observe that there is no local theory for (3.2), only a global one (in this respect the situation is the same as in (1.5)).

3.7 Theorem

Let $\mu \in M(\mathbf{R}; \mathbf{C}^{n\times n})$. Then

> *the equations $r' + \mu * r = r' + r * \mu = 0$ have a solution $r \in L^1(\mathbf{R}; \mathbf{C}^{n\times n})$, which is locally absolutely continuous, except for a jump discontinuity at zero with size $r(0+) - r(0-) = I$,*

if and only if

$$\det[zI + \hat{\mu}(z)] \neq 0, \quad \Re z = 0.$$

This solution r is unique in $L^1(\mathbf{R}; \mathbf{C}^{n\times n})$, and it satisfies $r' \in L^1(\mathbf{R}; \mathbf{C}^{n\times n})$. Moreover, r' is equal a.e. to a function of bounded variation.

Above we interpret r' as the ordinary (not distribution) derivative of r. If one wants to interpret the two equations satisfied by r' in the distribution sense, then they should be written in the form $r' + \mu * r = r' + r * \mu = \delta I$, where δ is the (scalar) unit point mass at zero.

3.8 Definition

The function r in Theorem 3.7 is called the whole line differential resolvent of μ.

Proof of Theorem 3.7 Suppose that there exists a function r with the given properties. We claim that

$$\hat{r}(z) = \big[zI + \hat{\mu}(z)\big]^{-1}, \quad \Re z = 0. \tag{3.3}$$

Clearly, this implies that

$$\det[zI + \hat{\mu}(z)] \neq 0, \quad \Re z = 0.$$

To prove (3.3), observe first that the equation $r' = -\mu * r$, together with $r \in L^1(\mathbf{R}; \mathbf{C}^{n\times n})$ and $\mu \in M(\mathbf{R}; \mathbf{C}^{n\times n})$, implies that $r' \in L^1(\mathbf{R}; \mathbf{C}^{n\times n})$; see Theorem 6.1(i). Consequently $r(t) \to 0$, as $|t| \to \infty$. Either use Theorem 8.4(i), or integrate by parts, to show that, for $\Re z = 0$,

$$\widehat{r'}(z) = \left(\int_{-\infty}^{0} + \int_{0}^{\infty}\right) \mathrm{e}^{-zt} r'(t)\,\mathrm{d}t = z\hat{r}(z) - I,$$

and take bilateral Laplace transforms in the equation $r' + \mu * r = 0$ to get

$$z\hat{r}(z) + \hat{\mu}(z)\hat{r}(z) = I, \quad \Re z = 0.$$

This is equivalent to (3.3).

Conversely, suppose that

$$\det[zI + \hat{\mu}(z)] \neq 0, \quad \Re z = 0.$$

We wish to find a function $r \in L^1(\mathbf{R}; \mathbf{C}^{n\times n})$ that satisfies (3.3). Define $\mathbf{e}(t) = \mathrm{e}^{-t}$ for $t \geq 0$, and $\mathbf{e}(t) = 0$ for $t < 0$. Then $\hat{\mathbf{e}}(z) = (1+z)^{-1}$ for $\Re z = 0$, and

$$\begin{aligned}\big[zI + \hat{\mu}(z)\big]^{-1} &= [(1+z)I + (\hat{\mu}(z) - I)]^{-1} \\ &= \hat{\mathbf{e}}(z)[I + \hat{\mathbf{e}}(z)(\hat{\mu}(z) - I)]^{-1} \\ &= \hat{\mathbf{e}}(z)[I + \hat{d}(z)]^{-1},\end{aligned}$$

where $\hat{d}(z)$ is the transform of $d = -\mathbf{e}I + \mathbf{e} * \mu$. Note that $d \in L^1(\mathbf{R}; \mathbf{C}^{n\times n})$, and that

$$I + \hat{d}(z) = \hat{\mathbf{e}}(z)[zI + \hat{\mu}(z)],$$

which means that $\det[I + \hat{d}(z)] \neq 0$ for $\Re z = 0$. Accordingly, by Theorem 2.4.3, the function d has a whole line resolvent q. This function q satisfies $\hat{q}(z) = \hat{d}(z)[I + \hat{d}(z)]^{-1}$ for $\Re z = 0$, hence

$$I - \hat{q}(z) = [I + \hat{d}(z)]^{-1}, \quad \Re z = 0.$$

We end up with the equation

$$[zI + \hat{\mu}(z)]^{-1} = \hat{\mathbf{e}}(z)[I - \hat{q}(z)], \quad \Re z = 0.$$

Thus, to get a function $r \in L^1(\mathbf{R}; \mathbf{C}^{n\times n})$ that satisfies (3.3) it suffices to define $r = \mathbf{e}I - \mathbf{e} * q$. Note that d is of bounded variation, hence so is q, and r is locally absolutely continuous, except for a jump discontinuity at zero with $r(0+) - r(0-) = I$.

To prove that $r' + \mu * r = r' + r * \mu = 0$, we observe that (3.3) implies

$$\hat{r}(z)[zI + \hat{\mu}(z)] = [zI + \hat{\mu}(z)]\hat{r}(z) = I, \quad \Re z = 0.$$

Thus, by the uniqueness of the Laplace transform, it suffices to show that $z\hat{r}(z) - I = \widehat{r'}(z)$ for $\Re z = 0$. It is a simple consequence of Theorem 8.4(ii) that this is true, but it can also be proved in the following elementary way: we defined r to be $r = \mathbf{e}I - \mathbf{e} * q$, and therefore, $r' = -\mathbf{e}I - q + \mathbf{e} * q$, and for $\Re z = 0$,

$$\hat{q}(z) = \frac{\hat{\mathbf{e}}(z)I - \hat{r}(z)}{\hat{\mathbf{e}}(z)} = I - \frac{\hat{r}(z)}{\hat{\mathbf{e}}(z)} = I - (1+z)\hat{r}(z),$$

and

$$\begin{aligned}\widehat{r'}(z) &= (\hat{\mathbf{e}}(z) - 1)\hat{q}(z) - \hat{\mathbf{e}}(z)I \\ &= \Big(\frac{1}{z+1} - 1\Big)\hat{q}(z) - \frac{1}{z+1}I \\ &= z\hat{r}(z) - I. \ \square\end{aligned}$$

The analogue of Theorem **2**.4.5 for equation (3.1) reads as follows.

3.9 Theorem

Let $\mu \in M(\mathbf{R}^+; \mathbf{C}^{n\times n})$ and let V be one of the following spaces:

(1) $L^p(\mathbf{R}^+; \mathbf{C}^n)$, $p \in [1, \infty]$, $L^\infty_\ell(\mathbf{R}^+; \mathbf{C}^n)$, *or* $L^\infty_0(\mathbf{R}^+; \mathbf{C}^n)$;

(2) $B^\infty(\mathbf{R}^+; \mathbf{C}^n)$, $B^\infty_\ell(\mathbf{R}^+; \mathbf{C}^n)$, *or* $B^\infty_0(\mathbf{R}^+; \mathbf{C}^n)$;

(3) $BC(\mathbf{R}^+; \mathbf{C}^n)$, $BUC(\mathbf{R}^+; \mathbf{C}^n)$, $BC_\ell(\mathbf{R}^+; \mathbf{C}^n)$, *or* $BC_0(\mathbf{R}^+; \mathbf{C}^n)$;

(4) $BL^p(\mathbf{R}^+; \mathbf{C}^n)$, *or* $BL^p_0(\mathbf{R}^+; \mathbf{C}^n)$, $p \in [1, \infty]$;

(5) $BV(\mathbf{R}^+; \mathbf{C}^n)$;

(6) $BBV(\mathbf{R}^+; \mathbf{C}^n)$, *or* $BBV_0(\mathbf{R}^+; \mathbf{C}^n)$;

(7) $L^p(\mathcal{T}_S; \mathbf{C}^n) \oplus BL^p_0(\mathbf{R}^+; \mathbf{C}^n)$, $p \in [1, \infty]$, $B^\infty(\mathcal{T}_S; \mathbf{C}^n) \oplus B^\infty_0(\mathbf{R}^+; \mathbf{C}^n)$, $C(\mathcal{T}_S; \mathbf{C}^n) \oplus BC_0(\mathbf{R}^+; \mathbf{C}^n)$, *or* $BV(\mathcal{T}_S; \mathbf{C}^n) \oplus BBV_0(\mathbf{R}^+; \mathbf{C}^n)$, *where* $S > 0$;

(8) $AP(\mathbf{R}^+; \mathbf{C}^n) \oplus BC_0(\mathbf{R}^+; \mathbf{C}^n)$.

Then the following four conditions are equivalent:

(i) *the differential resolvent r of μ belongs to $L^1(\mathbf{R}^+; \mathbf{C}^{n\times n})$;*

(ii) $\det[zI + \hat{\mu}(z)] \neq 0$ *for all $z \in \mathbf{C}$ with $\Re z \geq 0$;*

(iii) *for every $f \in V$, the solution x of the equation $x' + \mu * x = f$ on $\mathbf{R}^+$ belongs to V;*

(iv) *for every $f \in V$, the solution x of the equation $x' + \mu * x = f$ on $\mathbf{R}^+$ belongs to V, and x depends continuously on f (in the norm of V).*

When these conditions hold, the following two additional claims are valid: if V contains some discontinuous function, and if $f \in V$, then x' is equal a.e. to a function in V; if f is continuous and μ has no point masses in $(0, \infty)$ or $f(0) = 0$, then $x' \in V$.

Proof That (i) and (ii) are equivalent is nothing but Theorem 3.5. Trivially, (iv) implies (iii).

Let (i) hold. Then, by Theorem 3.5, one has $r' \in L^1(\mathbf{R}^+; \mathbf{C}^{n\times n})$, and so $r \in V$ for any choice of V. According to the variation of constants formula, $x(t) = r(t)x_0 + (r * f)(t)$ for $t \in \mathbf{R}^+$, and therefore, by Theorem **2**.2.2, (i) implies (iv).

Next, we show that (iii) implies (ii). Consider the particular case where $x(0) = x_0 = 0$. Let f be an arbitrary function in V. Then, by Theorem 6.1, $x' = f - \mu * x$ belongs to V. Let $u = x + x'$, and define the function d by $d = -\mathbf{e}I + \mathbf{e} * \mu$ (where $\mathbf{e}(t) = \mathrm{e}^{-t}$, $t > 0$). Then u satisfies the equation $u + d * u = f$. Since $d \in L^1(\mathbf{R}^+; \mathbf{C}^{n\times n})$, we can apply Theorem **2**.4.5, and conclude that $\det[I + \hat{d}(z)] \neq 0$ for $\Re z \geq 0$. But $I + \hat{d}(z) = \hat{\mathbf{e}}(z)(zI + \hat{\mu}(z))$, and thus we get (ii).

The claim concerning x' follows from (3.1) combined with Theorem 6.1. □

We proceed to the asymptotic analysis of equation (3.2).

3.10 Theorem

Let $\mu \in M(\mathbf{R}; \mathbf{C}^{n\times n})$ and let V be one of the spaces (1)–(8) in Theorem 3.9 with $\mathbf{R}^+$ replaced by $\mathbf{R}$, or

(9) *$AP(\mathbf{R}; \mathbf{C}^n)$.*

Then the following four conditions are equivalent:

(i) *μ has a whole line differential resolvent that belongs to $L^1(\mathbf{R}; \mathbf{C}^{n\times n})$;*

(ii) *$\det[zI + \hat{\mu}(z)] \neq 0$ for all $z \in \mathbf{C}$ with $\Re z = 0$;*

(iii) *for every $f \in V$, the equation $x' + \mu * x = f$ has a unique, locally absolutely continuous solution x in V;*

(iv) *for every $f \in V$, the equation $x' + \mu * x = f$ has a unique, locally absolutely continuous solution x in V, and x depends continuously on f (in the norm of V).*

When these conditions hold, the solution x in (iii) and (iv) is given by $x = r * f$, where r is the whole line differential resolvent of μ. Moreover, x' is equal a.e. to a function in V, and if, in addition, f is continuous then $x' \in V$.

As in Theorem **2**.4.6, the uniqueness requirements in (iii) and (iv) specifically refer to *uniqueness in V*, i.e., the equation may have other solutions that do not belong to V. Of course, this is in sharp contrast to Theorem 3.9. In addition, observe that there is no need to restrict the discrete part of μ in order to get a continuously differentiable solution in the case where f is continuous.

Proof of Theorem 3.10 That (i) and (ii) are equivalent is nothing but Theorem 3.7. Trivially, (iv) implies (iii).

Let (i) hold. A differentiation of the expression

$$x(t) = (r * f)(t) = \int_{-\infty}^{t} r(t-s)f(s)\,\mathrm{d}s + \int_{t}^{\infty} r(t-s)f(s)\,\mathrm{d}s, \quad t \in \mathbf{R},$$

together with the facts that $r' + \mu * r = 0$ and $r(0+) - r(0-) = I$, shows that the function x defined in this way satisfies (3.2) (see Theorem 7.1(ii)). Similarly, if x is an arbitrary solution of (3.2) and x is in the same space as f, then (again, use Theorem 7.1)

$$x + r' * x = (r * x)' = r * x',$$

and so (use Theorem 6.1(x)),

$$\begin{aligned} x &= r * x' - r' * x \\ &= r * (f - \mu * x) + (r * \mu) * x \\ &= r * f. \end{aligned}$$

This proves uniqueness. By Theorem **2**.2.2, $x = r * f$ belongs to the same space as f, and x depends continuously on f in the norm of V. Moreover, $r' \in L^1(\mathbf{R}; \mathbf{C}^{n\times n})$, and so Theorem **2**.2.2 says that $f + r' * f$ and f are elements of the same space. This proves (iv) as well as the claim about x', because $x' = f + r' * f$ a.e.

To prove that (iii) implies (ii) we proceed in the same manner as in the corresponding part of the proof of Theorem 3.9, invoking Theorem **2**.4.6. □

As in Chapter 2, for the case of periodic functions we have a slightly different result.

3.11 Theorem

Let $\mu \in M(\mathbf{R}; \mathbf{C}^{n\times n})$, $S > 0$, and let V be one of the following spaces:

(1) $L^p(\mathcal{T}_S; \mathbf{C}^n)$, $1 \le p \le \infty$;

(2) $B^\infty(\mathcal{T}_S; \mathbf{C}^n)$;

(3) $C(\mathcal{T}_S; \mathbf{C}^n)$;

(4) $BV(\mathcal{T}_S; \mathbf{C}^n)$.

Then the following four conditions are equivalent:

(i) *there exists a function $r_S \in BV(\mathcal{T}_S; \mathbf{C}^{n\times n})$, that is locally absolutely continuous, except for jump discontinuities at the points jS of size $r(jS+) - r(jS-) = I$, $j \in \mathbf{Z}$, and that satisfies*

$$r'(t) + \int_{[0,S)} \mu_S(\mathrm{d}s) r(t-s) = r'(t) + \int_{[0,S)} r(t-s)\mu_S(\mathrm{d}s) = 0, \quad t \in \mathbf{R},$$

where $\mu_S \in M(\mathcal{T}_S; \mathbf{C}^{n\times n})$ is defined by $\mu_S(E) = \sum_{j=-\infty}^{\infty} \mu(j + E)$ for all bounded Borel sets E;

(ii) $\det[(\mathrm{i}\,2\pi m/S)I + \hat{\mu}(\mathrm{i}\,2\pi m/S)] \neq 0$ *for all $m \in \mathbf{Z}$;*

(iii) *for every $f \in V$, the equation $x' + \mu * x = f$ has a unique locally absolutely continuous solution x in V;*

(iv) *for every $f \in V$, the equation $x' + \mu * x = f$ has a unique locally absolutely continuous solution x in V, and x depends continuously on f (in the norm of V).*

When these conditions hold, the function r_S in (i) *is unique, and the solution x in* (iii) *and* (iv) *is given by*

$$x(t) = \int_0^S r_S(t-s)f(s)\,\mathrm{d}s, \quad t \in \mathbf{R}.$$

Moreover, x' is equal a.e. to a function in V, and if, in addition, f is continuous then $x' \in V$.

3.12 Definition

The function r_S in Theorem 3.11(i) *is called the S-periodic differential resolvent of k.*

Proof of Theorem 3.11 The equivalence of (iii) and (iv) follows from the open mapping theorem. To see that (iii) implies (ii), one argues as in the corresponding part of the proof of Theorem **2**.4.7.

The proof of the fact that (i) and (ii) are equivalent is analogous to the proof of Theorem 3.7, and the proof of the fact that (i) implies (iv) is very similar to the proof of Theorem 3.10. We omit these proofs. (Define d in the same way as in the proof of Theorem 3.7, let q_S be the S-periodic resolvent of d, and define $r_S(t) = \sum_{j=-\infty}^{\infty} \mathbf{e}(t+jS) - (\mathbf{e} * q_S)(t)$. Also consult the proof of Theorem **2**.4.7, given in Section **2**.5.) □

By Theorem 3.3, the solution of (3.1) is given by $x = r(t)x_0 + (r * f)(t)$ for $t \in \mathbf{R}^+$. Above we gave conditions that imply $r \in L^1(\mathbf{R}^+;\mathbf{C}^{n\times n})$, and hence that x and f belong to the same space. Sometimes it is difficult, or even impossible, to prove that $r \in L^1(\mathbf{R}^+;\mathbf{C}^{n\times n})$. In particular, this may be true when μ is unbounded, i.e., when $\mu \notin M(\mathbf{R}^+;\mathbf{C}^{n\times n})$. In this case one may have to settle for the condition $r \in L^p(\mathbf{R}^+;\mathbf{C}^{n\times n})$ for some $p > 1$. In general, the question of finding sufficient conditions that imply $r \in L^p(\mathbf{R}^+;\mathbf{C}^{n\times n})$ is far from solved. However, for the case $p = 2$ a complete answer exists.

3.13 Theorem

Let $\mu \in M_{\mathrm{loc}}(\mathbf{R}^+;\mathbf{C}^{n\times n})$ satisfy $\int_{\mathbf{R}} \mathrm{e}^{-\sigma t}|\mu|(\mathrm{d}t) < \infty$ for all $\sigma > 0$. Then the differential resolvent r of μ belongs to $L^2(\mathbf{R}^+;\mathbf{C}^{n\times n})$ if and only if $\det[zI + \hat{\mu}(z)] \neq 0$ *for* $\Re z > 0$ *and*

$$\sup_{\sigma>0} \int_{\mathbf{R}} \left| [(\sigma + \mathrm{i}\omega)I + \hat{\mu}(\sigma + \mathrm{i}\omega)]^{-1} \right|^2 \mathrm{d}\omega < \infty.$$

A closely related result is the following analogue of Theorem **2**.6.2.

3.14 Theorem
Let $\mu \in M_{\text{loc}}(\mathbf{R}^+; \mathbf{C}^{n\times n})$ satisfy $\int_0^\infty e^{-\sigma t}|\mu|(dt) < \infty$ for every $\sigma > 0$. Then the convolution operator $f \mapsto r * f$ induced by the differential resolvent r of μ maps $L^2(\mathbf{R}^+; \mathbf{C}^n)$ continuously into itself if and only if the function $z \mapsto \left|[zI + \hat{\mu}(z)]^{-1}\right|$ is bounded in $\Re z > 0$.

The proofs of the two previous theorems are left to the reader (cf. the proof of Theorem **2**.6.2).

Nyquist's theorem can be adapted to the integrodifferential equation, e.g., as follows.

3.15 Theorem
Let $\mu \in M(\mathbf{R}^+; \mathbf{C}^{n\times n})$ satisfy $\det[i\omega I + \hat{\mu}(i\omega)] \neq 0$ for all $\omega \in \mathbf{R}$. Let $R > \|\mu\|_{M(\mathbf{R}^+)}$, and let Γ_R be the closed curve $\{\, i\omega \mid R \geq \omega \geq -R \,\} \cup \{\, Re^{i\varphi} \mid -\pi/2 \leq \varphi \leq \pi/2 \,\}$, oriented in the anticlockwise direction. Then the number of zeros of $\det[zI + \hat{\mu}(z)]$ in $\Re z > 0$, counted according to their multiplicities, equals the index of the curve $\det[zI + \hat{\mu}(z)]$, $z \in \Gamma_R$ (that is, it is the same as the number of times that this curve circles anticlockwise around the origin).

This proof is left to the reader as well.

3.16 Corollary
Let $\mu \in M(\mathbf{R}^+; \mathbf{C}^{n\times n})$, let $R > \|\mu\|_{M(\mathbf{R}^+)}$, and let Γ_R be the closed curve $\{\, i\omega \mid R \geq \omega \geq -R \,\} \cup \{\, Re^{i\varphi} \mid -\pi/2 \leq \varphi \leq \pi/2 \,\}$. Then $\det[zI + \hat{\mu}(z)]$ has no zeros in the closed half plane $\Re z \geq 0$ if and only if the curve $\det[zI + \hat{\mu}(z)]$, $z \in \Gamma_R$, neither passes through the origin, nor encircles the origin.

The following result parallels Theorem **2**.6.6.

3.17 Theorem
Let $\mu \in M(\mathbf{R}^+; \mathbf{C}^{n\times n})$ have a finite first moment, i.e., let $\int_{\mathbf{R}^+} t|\mu|(dt) < \infty$. Then the differential resolvent r of μ belongs to $L^1(\mathbf{R}^+; \mathbf{C}^{n\times n})$ if and only if $r(t) \to 0$ as $t \to \infty$.

The proof is left to the reader (see Exercise 18).

It is well known that, if r has a Fourier transform that belongs to $L^q(\mathbf{R}; \mathbf{C}^{n\times n})$ for some $q \in [1, 2]$, then $r \in L^p(\mathbf{R}^+; \mathbf{C}^{n\times n})$, where $1/p + 1/q = 1$; see, e.g., Titchmarsh [1], Theorem 74, p. 96. In the case where $p \in (1, 2)$, one may apply a different result, described in Titchmarsh [1], Theorem 84, p. 115. According to that result, if the Fourier transform of r is smooth in a certain sense then $r \in L^p(\mathbf{R}^+; \mathbf{C}^{n\times n})$.

4. Appendix: Fubini's Theorem

All of the results in this appendix and in later appendices of this chapter are valid both for functions and measures with scalar values in $\mathbf{R}$ or $\mathbf{C}$, and for functions and measures with $m \times n$-matrix values, i.e., with values in $\mathbf{R}^{m\times n}$ or $\mathbf{C}^{m\times n}$. Of course, we have to require that the dimensions of the matrices should be compatible with respect to matrix addition and multiplication. The easiest way to prove the results that involve matrices is to reduce them to the scalar case (one can do all the computations componentwise).

In the proofs of the convolution results given in Section **2**.7, we used Lebesgue measure on $\mathbf{R}^2$, and these proofs were heavily based on Fubini's theorem for Lebesgue integrable functions. In Section 6 below, we prove similar results for the convolution of a function with a measure. The proofs given there are very similar to those in Section **2**.7. However, when one works with some other measure than Lebesgue measure on $\mathbf{R}$, it may not always be obvious what measure one should use on $\mathbf{R}^2$. If μ is a measure on, e.g., $\mathbf{R}$ and a is a function on $\mathbf{R}$, and we want to prove something about $\mu * a$, then in all the L^p-cases (the cases where $a \in L^p_{\mathrm{loc}}(\mathbf{R})$ for some $p \in [1, \infty]$) one should work with the completion of the cross product of μ and the Lebesgue measure rather than with this cross product itself (the Lebesgue measure in $\mathbf{R}^2$ is not the product of the Lebesgue measure in $\mathbf{R}$ by itself, but is instead the completion of this product). In particular, this means that in the L^p-cases one should work with the completed version of Fubini's theorem, which in the positive, scalar case can be stated as follows.

4.1 Fubini's Theorem for Positive Functions and Measures

Let $(X, \mathcal{M}, \mu)$ and $(Y, \mathcal{N}, \nu)$ be complete, σ-finite measure spaces with μ and ν (scalar-valued and) positive, and let $\overline{\mathcal{M} \times \mathcal{N}}$ be the completion of $\mathcal{M} \times \mathcal{N}$, relative to the measure $\mu \times \nu$. If f is a nonnegative, extended real-valued $\overline{\mathcal{M} \times \mathcal{N}}$-measurable function on $X \times Y$, then

(i) *the function $x \mapsto f(x, y)$ is $\mathcal{M}$-measurable for ν-almost all $y \in Y$,*
(ii) *the function $y \mapsto f(x, y)$ is $\mathcal{N}$-measurable for μ-almost all $x \in X$,*
(iii) *the function $y \mapsto \int_X f(x, y)\mu(\mathrm{d}x)$ is $\mathcal{N}$-measurable,*
(iv) *the function $x \mapsto \int_Y f(x, y)\nu(\mathrm{d}y)$ is $\mathcal{M}$-measurable,*

and

$$\begin{aligned}\int_{X\times Y} f(x,y)\overline{(\mu\times\nu)}(\mathrm{d}x\,\mathrm{d}y) &= \int_Y\Big(\int_X f(x,y)\mu(\mathrm{d}x)\Big)\nu(\mathrm{d}y)\\ &= \int_X\Big(\int_Y f(x,y)\nu(\mathrm{d}y)\Big)\mu(\mathrm{d}x).\end{aligned}$$

This is Hewitt and Stromberg [1], Theorem (21.16), and it is contained in Rudin [3], Theorem 8.12.

Whenever we apply Fubini's theorem to functions or measures that are not (scalar-valued and) positive, we use the following variant of this theorem.

4.2 Fubini's Theorem

Let $(X, \mathcal{M}, \mu)$ and $(Y, \mathcal{N}, \nu)$ be complete σ-finite measure spaces, and let $\overline{\mathcal{M} \times \mathcal{N}}$ be the completion of $\mathcal{M} \times \mathcal{N}$, relative to the measure $|\mu| \times |\nu|$. Let f be a $\overline{\mathcal{M} \times \mathcal{N}}$-measurable function on $X \times Y$ and suppose that one of the three integrals

$$\int_{X \times Y} |f(x, y)|(|\mu| \times |\nu|)(\mathrm{d}x\, \mathrm{d}y),$$

$$\int_Y \left(\int_X |f(x, y)||\mu|(\mathrm{d}x) \right) |\nu|(\mathrm{d}y),$$

and

$$\int_X \left(\int_Y |f(x, y)||\nu|(\mathrm{d}y) \right) |\mu|(\mathrm{d}x)$$

is finite (if one of these integrals is finite, then so are the others). Then

(i) *the function $x \mapsto f(x, y)$ is in $L^1(X; \mathcal{M}; \mu)$ for ν-almost all $y \in Y$,*
(ii) *the function $y \mapsto f(x, y)$ is in $L^1(Y; \mathcal{N}; \nu)$ for μ-almost all $x \in X$,*
(iii) *the function $y \mapsto \int_X f(x, y)\mu(\mathrm{d}x)$ is in $L^1(Y; \mathcal{N}; \nu)$,*
(iv) *the function $x \mapsto \int_Y f(x, y)\nu(\mathrm{d}y)$ is in $L^1(X; \mathcal{M}; \mu)$,*

and

$$\begin{aligned}\int_{X \times Y} f(x, y)\overline{(\mu \times \nu)}(\mathrm{d}x\, \mathrm{d}y) &= \int_Y \left(\int_X f(x, y)\mu(\mathrm{d}x) \right) \nu(\mathrm{d}y) \\ &= \int_X \left(\int_Y f(x, y)\nu(\mathrm{d}y) \right) \mu(\mathrm{d}x).\end{aligned}$$

This theorem is a more or less immediate consequence of Theorem 4.1 (cf. Hewitt and Stromberg [1], Note (21.17), or Rudin [3], Theorem 8.12).

In the case where we have a convolution of a measure μ and a function a, where a is continuous and bounded, or of bounded variation, or, more generally, a is Borel measurable and bounded, then it is possible to define $\mu * a$ everywhere rather than almost everywhere. To show this, one may use the following result (which is part of the collection of theorems and lemmas that share the name 'Fubini's theorem').

4.3 Theorem

Let $(X, \mathcal{M}, \mu)$ and $(Y, \mathcal{N}, \nu)$ be locally finite measure spaces, let f be a $\mathcal{M} \times \mathcal{N}$-measurable function on $X \times Y$, and suppose that the function $y \mapsto \int_X |f(x, y)||\mu|(\mathrm{d}x)$ is locally bounded. Then the function $y \mapsto \int_X f(x, y)\mu(\mathrm{d}x)$ is locally bounded and $\mathcal{N}$-measurable.

Above, when we say that $(X, \mathcal{M}, \mu)$ and $(Y, \mathcal{N}, \nu)$ are locally finite measure spaces, we mean that these spaces are locally compact and σ-compact (in particular, they have a topology), and that μ and ν are Radon measures. That is, μ and ν are complex-valued set functions that, restricted to compact sets, are complex Borel measures.

Theorem 4.3 is a fairly direct consequence of the usual noncompleted version of Theorem 4.1, see Hewitt and Stromberg [1], Theorem (21.13), or Rudin [3], Theorem 8.8. Observe in particular that $\mathcal{N}$ need not be complete (when we apply this theorem, we let $\mathcal{N}$ be the noncomplete σ-algebra of all the Borel sets).

5. Appendix: Total Variation of a Matrix Measure

In our equations we use Borel measures that are matrix-valued. Most of the theory of these measures can be deduced directly from the corresponding theory for scalar measures. One simply applies the scalar theory to each component separately. There is one exception to this general rule: the computation of the total variation of a matrix-valued measure must be computed directly for the matrix case, because its value depends on the norm that one uses in $\mathbf{C}^{m\times n}$. For this reason, we briefly discuss the total variation of matrix-valued measures below.

One defines a matrix-valued or matrix function of bounded variation in the same way as in the scalar case.

5.1 Definition

The total variation of a $\mathbf{C}^{m\times n}$-valued function a over an interval $J \subset \mathbf{R}$ is given by

$$\|a\|_{\mathrm{var}(J)} = \sup \sum_{j=1}^{N} |a(t_j) - a(t_{j-1})|,$$

where the supremum is taken over all N and over all sets of points $t_j \in J$ such that $t_i < t_j$ for $i < j$.

Similarly, the standard definition of the total variation of a scalar measure has an obvious extension to the matrix-valued case.

5.2 Definition

The total variation $|\mu|(E)$ of the measure $\mu \in M(\mathbf{R};\mathbf{C}^{m\times n})$ on a set E is given by

$$|\mu|(E) = \sup \sum_{j=1}^{N} |\mu(E_j)|,$$

where the supremum is taken over all N and over all partitions $\{E_j\}$ of E.

Here we call $\{E_j\}_{j=1}^{N}$ a partition of E if $\bigcup_{j=1}^{N} E_j = E$ and $E_i \cap E_j = \emptyset$ for all $i \neq j$.

5.3 Theorem
Let $\mu \in M(\mathbf{R}; \mathbf{C}^{m\times n})$. The set function that to each Borel set E assigns the total variation $|\mu|(E)$ of μ on E is a positive measure.

The proof of this theorem is an exact copy of the proof of the scalar case, see Rudin [3], Theorem 6.2.

It is well known that in the scalar case, if μ is a finite measure on $\mathbf{R}$, and if a is defined by $a(t) = \mu((-\infty, t))$ for all $t \in \mathbf{R}$, then a is of bounded variation, and the total variations of a and μ, computed as in Definitions 5.1 and 5.2, respectively, are equal. That the same claim is true for matrix-valued measures follows from Theorem 5.7 below.

For scalar measures, the total variation can be computed in a different way.

5.4 Theorem
Let $\mu \in M(\mathbf{R}; \mathbf{C})$. Then $|\mu|(\mathbf{R}) = \sup\left|\int_{\mathbf{R}} \mu(\mathrm{d}s)\varphi(s)\right|$, where the supremum is taken over all functions $\varphi \in BC_0(\mathbf{R}; \mathbf{C})$ satisfying $|\varphi(t)| \le 1$ for all $t \in \mathbf{R}$.

This is a consequence of the Riesz representation theorem (see Rudin [3], Theorem 6.19).

The following counterexample shows that the preceding result is not valid for $n > 1$.

5.5 Example
Let $n = 2$, and define

$$A_0 = \begin{pmatrix} 1 & 0 \\ 0 & 0 \end{pmatrix}, \quad A_1 = \begin{pmatrix} 0 & 0 \\ 0 & 1 \end{pmatrix},$$

let δ_0 and δ_1 be scalar unit point masses at 0 and 1, and define $\mu = A_0\delta_0 + A_1\delta_1$. Let the norm in $\mathbf{C}^2$ be the Euclidean norm $|(x, y)^T| = (x^2 + y^2)^{1/2}$, and let the norm in $\mathbf{C}^{2\times 2}$ be the corresponding operator norm. Then $|\mu|(\mathbf{R}) = 2$, and

$$\sup\left|\int_{\mathbf{R}} \mu(\mathrm{d}s)\varphi(s)\right| = \sqrt{2},$$

where the supremum is taken over all functions $\varphi \in BC_0(\mathbf{R}; \mathbf{C}^2)$ satisfying $|\varphi(t)| \le 1$ for all $t \in \mathbf{R}$.

Proof We have $|A_0| = |A_1| = 1$, and therefore $|\mu| = \delta_0 + \delta_1$, and $|\mu|(\mathbf{R}) = 2$. If $\varphi \in BC_0(\mathbf{R}; \mathbf{C}^2)$, $\varphi = (\psi, \xi)^T$, then

$$\int_{\mathbf{R}} \mu(\mathrm{d}s)\varphi(s) = \begin{pmatrix} \psi(0) \\ \xi(1) \end{pmatrix},$$

and $|\int_{\mathbf{R}} \mu(\mathrm{d}s)\varphi(s)| = \left(|\psi(0)|^2 + |\xi(1)|^2\right)^{1/2}$. To maximize this expression, it suffices to choose $|\psi(0)| = |\xi(1)| = 1$ (which implies $\xi(0) = \psi(1) = 0$). Doing so, one gets

$$\max\left|\int_{\mathbf{R}} \mu(\mathrm{d}s)\varphi(s)\right| = \sqrt{2},$$

as claimed. □

Note that the counterexample above depends strongly on the norm one uses in $\mathbf{C}^2$. If, instead, one had used the l^1-norm $|(x,y)^T| = |x| + |y|$, then the two numbers would have been the same.

The following result is needed for the norm inequalities in, e.g., Theorems 6.1 and 4.3.5.

5.6 Theorem
Let $\mu \in M(\mathbf{R}; \mathbf{C}^{m\times n})$, and let φ be a $\mathbf{C}^{n\times p}$-valued $|\mu|$-integrable function on $\mathbf{R}$. Then

$$\left| \int_{\mathbf{R}} \mu(\mathrm{d}s)\varphi(s) \right| \le \int_{\mathbf{R}} |\mu|(\mathrm{d}s)|\varphi(s)|. \tag{5.1}$$

Proof By the definition of the total variation measure, if χ_E is the characteristic function of a Borel set E, then

$$\left| \int_{\mathbf{R}} \mu(\mathrm{d}s)\chi_E(s) \right| = |\mu(E)| \le |\mu|(E) = \int_{\mathbf{R}} |\mu|(\mathrm{d}s)\chi_E(s).$$

Therefore, (5.1) is true for characteristic functions. This, together with the triangle inequality, gives (5.1) for simple functions (i.e., for functions of the form $\varphi = \sum_{j=1}^N A_j\chi_{E_j}$, where the sets E_j are pairwise disjoint and their union is $\mathbf{R}$, and the constants A_j are matrices, vectors or scalars):

$$\begin{aligned}
\left| \int_{\mathbf{R}} \mu(\mathrm{d}s)\varphi(s) \right| &\le \sum_{j=1}^N \left| \int_{\mathbf{R}} \mu(\mathrm{d}s)A_j\chi_{E_j}(s)\,\mathrm{d}s \right| = \sum_{j=1}^N |\mu(E_j)A_j| \\
&\le \sum_{j=1}^N |\mu|(E_j)|A_j| = \sum_{j=1}^N \int_{E_j} |\mu|(\mathrm{d}s)|A_j| \\
&= \sum_{j=1}^N \int_{E_j} |\mu|(\mathrm{d}s)|\varphi(s)| = \int_{\mathbf{R}} |\mu|(\mathrm{d}s)|\varphi(s)|.
\end{aligned}$$

If φ is an arbitrary $|\mu|$-integrable function, then we can find a sequence of simple functions φ_j converging to φ in $L^1(|\mu|)$. Since both sides of (5.1) depend continuously on φ in the norm of $L^1(|\mu|)$, the limit function φ must satisfy (5.1) as well. □

In Chapter 10 we shall need the following alternative way of computing the total variation.

5.7 Theorem
Let $\mu \in M(\mathbf{R}; \mathbf{C}^{m\times n})$. For every Borel set E one has

$$|\mu|(E) = \lim_{m\to\infty} \sum_{k=-2^{2m}}^{2^{2m}} \left| \mu(E \cap [k2^{-m}, (k+1)2^{-m})) \right|.$$

Proof To simplify the notations, let us denote $[k2^{-m}, (k+1)2^{-m})$ by $E_{k,m}$, for all $m > 0$ and all $k \in \mathbf{Z}$.

Clearly, $|\mu|(E) \geq \sum_{k=-2^{2m}}^{2^{2m}} |\mu(E \cap E_{k,m})|$, since $\{E \cap E_{k,m}\}_{k=-\infty}^{\infty}$ is a partition of E. The sum is a monotone function of m which is bounded from above, so it tends to a finite limit as $m \to \infty$. This limit must satisfy $\lim_{m\to\infty} \sum_{k=-2^{2m}}^{2^{2m}} |\mu(E \cap E_{k,m})| \leq |\mu|(E)$.

As a first step in the proof of the converse inequality we show that for all finite numbers $\alpha < \beta$ we have

$$|\mu([\alpha, \beta))| \leq F(\beta) - F(\alpha), \tag{5.2}$$

where

$$F(t) = \lim_{m\to\infty} \sum_{k=-2^{2m}}^{2^{2m}} |\mu((-\infty, t) \cap E_{k,m})|. \tag{5.3}$$

If α and β both are multiples of 2^{-m} for some $m > 0$, then

$$\begin{aligned}
F(\beta) &= F(\alpha) + \lim_{m\to\infty} \sum_{k=-2^{2m}}^{2^{2m}} |\mu([\alpha, \beta) \cap E_{k,m})| \\
&\geq F(\alpha) + \lim_{m\to\infty} \left| \sum_{k=-2^{2m}}^{2^{2m}} \mu([\alpha, \beta) \cap E_{k,m}) \right| \\
&= F(\alpha) + |\mu([\alpha, \beta))|,
\end{aligned}$$

so in this case (5.2) is true. The general case can be deduced from this case together with the fact that both sides of (5.2) are left-continuous functions of α and β. Thus, the claim (5.2) is true.

Let λ be the nonnegative measure induced by the nondecreasing function F in (5.3), cf. Rudin [3], Exercise 13, p. 157. Then, by (5.2), the inequality

$$|\mu(E)| \leq \lambda(E)$$

holds if E is an interval of the form $[\alpha, \beta)$. Since every open set in $\mathbf{R}$ is a countable disjoint union of such intervals, (5.2) holds for every open set, hence for every Borel set. Definition 5.2 now implies that $|\mu| \leq \lambda$. In particular,

$$|\mu|(\mathbf{R}) \leq \lambda(\mathbf{R}) = F(\infty) = \lim_{m\to\infty} \sum_{k=-2^{2m}}^{2^{2m}} |\mu(E_{k,m})|.$$

The preceding argument applies to any measure μ. If we apply it to the measure μ_E defined by $\mu_E(F) = \mu(E \cap F)$ for all Borel sets F, then we get the desired

$$|\mu|(E) \leq \lim_{m\to\infty} \sum_{k=-2^{2m}}^{2^{2m}} |\mu(E \cap E_{k,m})|. \ \square$$

We conclude this section by giving two lemmas that are needed in the proof of Theorem **16**.6.2. The first gives us information on the values of

f if we know its averages; the second concerns polar decomposition in the matrix case.

5.8 Lemma
Let $\mu \in M(W; \mathbf{R}^+)$ where $W \subset \mathbf{R}$, let f map W into $\mathbf{C}^{n\times n}$ and assume that $\int_W \mu(\mathrm{d}s)|f|(s) < \infty$. Let $S \subset \mathbf{C}^{n\times n}$ be closed and suppose that

$$\frac{1}{\mu(E)}\int_E \mu(\mathrm{d}s) f(s) \in S$$

for all Borel sets $E \subset W$ with $\mu(E) > 0$. Then $f(t) \in S$ for almost all $t \in W$.

The proof is the same as in the scalar case; see, e.g., Rudin [3], Theorem 1.40.

5.9 Lemma
Let $\mu \in M(\mathbf{R}; \mathbf{C}^{n\times n})$. Then there exists a μ-measurable function h with $|h(s)| = 1$ such that $\mu(\mathrm{d}s) = h(s)|\mu|(\mathrm{d}s)$.

To prove this it suffices to observe that the proof of Rudin [3], Theorem 6.12, can be adapted to the matrix case, with Theorem 1.40 in Rudin [3] replaced by Lemma 5.8.

6. Appendix: the Convolution of a Measure and a Function

Convolutions of two functions were considered in some detail in Sections **2**.2 and **2**.7. In Section 2, we defined the convolution of a measure and a function. As we have mentioned repeatedly, the properties of these two types of convolutions do not differ too much. Below we make this statement more precise. Since the proofs of many of the results given below are, to a high degree, analogous to the corresponding proofs given in Section **2**.7, we leave some of them to the reader.

Our first theorem is an extension of Theorem **2**.2.2.

6.1 Theorem
Let J be $\mathbf{R}$, $\mathbf{R}^+$, or an interval with left end-point 0 and right end-point T, where $0 < T < \infty$, and let $\mu \in M(J)$.

(i) *If $a \in L^p(J)$ for some $p \in [1, \infty]$, then $\mu * a \in L^p(J)$ and $\|\mu * a\|_{L^p(J)} \le \|\mu\|_{M(J)}\|a\|_{L^p(J)}$. If $J = \mathbf{R}$ or $\mathbf{R}^+$, and if $a \in L^\infty_\ell(J)$ then $\mu * a \in L^\infty_\ell(J)$, and if $a \in L^\infty_0(J)$ then $\mu * a \in L^\infty_0(J)$.*

(ii) *If $a \in B^\infty(J)$, then $\mu * a \in B^\infty(J)$ and $\|\mu * a\|_{\sup(J)} \le \|\mu\|_{M(J)}\|a\|_{\sup(J)}$. If $J = \mathbf{R}$ or $\mathbf{R}^+$, and if $a \in B^\infty_\ell(J)$ then $\mu * a \in B^\infty_\ell(J)$, and if $a \in B^\infty_0(J)$ then $\mu * a \in B^\infty_0(J)$.*

(iii) *If* $J = \mathbf{R}$, *and if* a *belongs to one of the spaces* $BC(J)$, $BUC(J)$, $BC_\ell(J)$ *or* $BC_0(J)$, *then* $\mu * a$ *belongs to the same space as* a. *The same is true if* $J = \mathbf{R}^+$, *provided* μ *has no point masses in* $(0, \infty)$, *or* $a(0) = 0$. *If* $J = [0, T)$, *if* $a \in BC(J)$ *or* $a \in BUC(J)$, *and if* μ *has no point masses in* $(0, T)$, *or* $a(0) = 0$, *then* $\mu * a$ *belongs to the same space as* a.

(iv) *If* $J = \mathbf{R}$ *or* $\mathbf{R}^+$, *and if* $a \in BL^p(J)$ *for some* $p \in [1, \infty]$, *then* $\mu * a \in BL^p(J)$, *and* $\|\mu * a\|_{BL^p(J)} \le \|\mu\|_{M(J)} \|a\|_{BL^p(J)}$. *If* $a \in BL^p_0(J)$, *then* $\mu * a \in BL^p_0(J)$.

(v) *If* $a \in BV(J)$, *then* $\mu * a \in BV(J)$, *and* $\|\mu * a\|_{BV(J)} \le \|\mu\|_{M(J)} \|a\|_{BV(J)}$.

(vi) *If* $a \in BBV(J)$, *then* $\mu * a \in BBV(J)$, *and* $\|\mu * a\|_{BBV(J)} \le \|\mu\|_{M(J)} \|a\|_{BBV(J)}$. *If* $a \in BBV_0(J)$, *then* $\mu * a \in BBV_0(J)$.

(vii) *If* $J = \mathbf{R}$, $S > 0$, *and* a *belongs to one of the spaces* $L^p(\mathcal{T}_S)$, $p \in [1, \infty]$, $B^\infty(\mathcal{T}_S)$, $C(\mathcal{T}_S)$, $BV(\mathcal{T}_S)$, *or* $AP(\mathbf{R})$, *then* $\mu * a$ *belongs to the same space as* a.

(viii) *If* $J = \mathbf{R}$ *or* $\mathbf{R}^+$, $S > 0$, *and* a *belongs to one of the spaces* $L^p(\mathcal{T}_S) \oplus BL^p_0(J)$, $p \in [1, \infty]$, $B^\infty(\mathcal{T}_S) \oplus B^\infty_0(J)$, *or* $BV(\mathcal{T}_S) \oplus BBV_0(J)$, *then* $\mu * a$ *belongs to the same space as* a. *If* $J = \mathbf{R}$, $S > 0$, *and* a *belongs to* $C(\mathcal{T}_S) \oplus BC_0(J)$ *or* $AP(J) \oplus BC_0(J)$, *then* $\mu * a$ *belongs to the same space as* a. *The last statement is true if* $J = \mathbf{R}^+$ *as well, provided* μ *has no point masses in* $(0, \infty)$, *or* $a(0) = 0$.

(ix) *If* $a \in L^p(J)$ *and* $b \in L^q(J)$ *for some* p *and* q *in* $[1, \infty]$ *with* $1/p + 1/q = 1$, *then* $((\mu * a) * b)(t) = (\mu * (a * b))(t)$, $((a * \mu) * b)(t) = (a * (\mu * b))(t)$, *and* $((a * b) * \mu)(t) = (a * (b * \mu))(t)$ *for all* $t \in J$.

(x) *If* $a \in L^1(J)$, *and* $b \in BL^1(J)$, *then* $(\mu * a) * b = \mu * (a * b)$, $(a * b) * \mu = a * (b * \mu)$, *and* $(b * \mu) * a = b * (\mu * a)$.

(xi) *If* μ *or* a *is scalar-valued, then* $\mu * a = a * \mu$.

Proof Most of the proof of Theorem 6.1 is completely analogous to the proof of Theorem **2**.2.2 given in Section **2**.7, and may consequently be omitted (see the comments in Section 4 on the use of Fubini's theorem, and Theorem 5.6). We shall prove only the statements concerning $B^\infty(\mathbf{R})$, $B^\infty_0(\mathbf{R})$, $BV(\mathbf{R})$, and $BBV(\mathbf{R})$. Note that it suffices to examine the case $J = \mathbf{R}$, because all the other cases can be reduced to this case, provided one defines μ and a to be zero outside J.

If $a \in B^\infty(\mathbf{R})$, i.e., a is Borel measurable and bounded, then, by Theorem 5.6, $\mu * a$ is bounded. That $\mu * a$ is Borel measurable follows from Theorem 4.3.

If $a \in B^\infty_0(\mathbf{R})$, i.e., $a \in B^\infty(\mathbf{R})$ and $a(t) \to 0$ as $|t| \to \infty$, then a computation, analogous to the one made at the end of (i) in the proof of Theorem **2**.2.2 (see Section **2**.7), shows that $(\mu * a)(t) \to 0$ as $|t| \to \infty$.

Let $a \in BV(\mathbf{R})$. Recall that, by definition, $\|a\|_{BV(\mathbf{R})} = \|a\|_{\sup(\mathbf{R})} + \|a\|_{\mathrm{var}(\mathbf{R})}$. Thus, by (ii), to prove (v) is suffices to show that $\|\mu * a\|_{\mathrm{var}(\mathbf{R})} \le \|\mu\|_{M(\mathbf{R})} \|a\|_{\mathrm{var}(\mathbf{R})}$. This is done as follows: if

$-\infty < t_0 < t_1 < \cdots < t_N < \infty$, then

$$\begin{aligned}\sum_{j=1}^{N}|(\mu * a)(t_j) - (\mu * a)(t_{j-1})| &= \sum_{j=1}^{N}\left|\int_{\mathbf{R}} \mu(\mathrm{d}s)(a(t_j - s) - a(t_{j-1} - s))\right| \\ &\le \sum_{j=1}^{N}\int_{\mathbf{R}} |\mu|(\mathrm{d}s)|a(t_j - s) - a(t_{j-1} - s)| \\ &= \int_{\mathbf{R}} |\mu|(\mathrm{d}s)\sum_{j=1}^{N}|a(t_j - s) - a(t_{j-1} - s)| \\ &\le \int_{\mathbf{R}} |\mu|(\mathrm{d}s)\|a\|_{\mathrm{var}(\mathbf{R})} = \|\mu\|_{M(\mathbf{R})}\|a\|_{\mathrm{var}(\mathbf{R})}.\end{aligned}$$

Clearly, this proves that $\mu * a \in BV(\mathbf{R})$, and that $\|\mu * a\|_{\mathrm{var}(\mathbf{R})} \le \|\mu\|_{M(\mathbf{R})}\|a\|_{\mathrm{var}(\mathbf{R})}$.

If $a \in BBV(\mathbf{R})$, then one makes the same computation, with the restriction that $t_N - t_0 < 1$, and obtains the inequality

$$\|\mu * a\|_{\mathrm{var}([t,t+1])} \le \int_{\mathbf{R}} |\mu|(\mathrm{d}s)\|a\|_{\mathrm{var}([t-s,t-s+1])}, \quad t \in \mathbf{R}.$$

This inequality combined with (ii) gives the desired conclusion. If $a \in BBV_0(\mathbf{R})$, then, by Lebesgue's dominated convergence theorem, $\|\mu * a\|_{\mathrm{var}([t,t+1))} \to 0$ as $|t| \to \infty$. □

6.2 Corollary

Let $\mu \in M_{\mathrm{loc}}(\mathbf{R}^+)$.

(i) *If $a \in L^p_{\mathrm{loc}}(\mathbf{R}^+)$ for some $p \in [1, \infty]$, then $\mu * a \in L^p_{\mathrm{loc}}(\mathbf{R}^+)$.*

(ii) *If $a \in B^\infty_{\mathrm{loc}}(\mathbf{R}^+)$, then $\mu * a \in B^\infty_{\mathrm{loc}}(\mathbf{R}^+)$.*

(iii) *If $a \in C(\mathbf{R}^+)$, and μ has no point masses in $(0, \infty)$, or $a(0) = 0$, then $\mu * a \in C(\mathbf{R}^+)$.*

(iv) *If $a \in BV_{\mathrm{loc}}(\mathbf{R}^+)$, then $\mu * a \in BV_{\mathrm{loc}}(\mathbf{R}^+)$.*

(v) *Let $a \in L^p_{\mathrm{loc}}(\mathbf{R}^+)$ and $b \in L^q_{\mathrm{loc}}(\mathbf{R}^+)$, with $1 \le p, q \le \infty$ and $1/p + 1/q = 1$. Then $((\mu * a) * b)(t) = (\mu * (a * b))(t)$, $((a * \mu) * b)(t) = (a * (\mu * b))(t)$ and $((a * b) * \mu)(t) = (a * (b * \mu))(t)$ for all $t \ge 0$.*

(vi) *If $a, b \in L^1_{\mathrm{loc}}(J)$, then $(\mu * a) * b = \mu * (a * b)$, $(a * b) * \mu = a * (b * \mu)$, and $(b * \mu) * a = b * (\mu * a)$.*

This is an immediate consequence of Theorem 6.1.

7. Appendix: Derivatives of Convolutions

The theorems in this appendix describe derivatives of convolutions. They show, essentially, that differentiation commutes with convolution. Again, the results are valid both for scalar-valued and for matrix-valued functions and measures.

7.1 Theorem

(i) *Let $\mu \in M(\mathbf{R})$, and let $a \in BC(\mathbf{R}) \cap AC_{\mathrm{loc}}(\mathbf{R})$, and let $a' \in BL^1(\mathbf{R})$. Then $\mu * a \in AC_{\mathrm{loc}}(\mathbf{R})$ and $(\mu * a)' = \mu * a'$. In particular, $(\mu * a)' \in BL^1(\mathbf{R})$.*

(ii) *Let $a \in L^1(\mathbf{R}) \cap BV(\mathbf{R})$ and let $b \in BL^1(\mathbf{R})$. Then $(a * b)$ is locally absolutely continuous and $(a * b)' = \mu * b$ where μ is the measure induced by the distribution derivative of a, i.e., the measure that satisfies $\mu((-\infty, t]) = a(t)$ at all points of continuity of a. In particular, $(a * b)' \in BL^1(\mathbf{R})$.*

Proof (i) We know from Theorem 6.1(iv) that if $a' \in BL^1(\mathbf{R})$ then $\mu * a' \in BL^1(\mathbf{R})$. Therefore, to prove (i), it suffices to show that, for all $t \in \mathbf{R}$,

$$(\mu * a)(t) = (\mu * a)(0) + \int_0^t (\mu * a')(s)\, \mathrm{d}s.$$

If $t > 0$ (the case $t < 0$ is entirely similar), we define $\eta_t = \chi_{[-t,0)}$ to be the (scalar) characteristic function of the interval $[-t, 0)$. Clearly, as $a' \in BL^1(\mathbf{R})$, $\eta_t * a' \in BC(\mathbf{R})$. This means that one can use Fubini's theorem in the same way as in the proofs of Theorem **2**.2.2(x) and Theorem 6.1(ix) to show that

$$((\eta_t * \mu) * a')(0) = (\mu * (\eta_t * a'))(0) = (\eta_t * (\mu * a'))(0) \tag{7.1}$$

(cf. Exercise **2**.28(xii)). Thus,

$$\begin{aligned}
\int_0^t (\mu * a')(s)\, \mathrm{d}s &= (\eta_t * (\mu * a'))(0) \\
&= (\mu * (\eta_t * a'))(0) \\
&= \int_{\mathbf{R}} \mu(\mathrm{d}s) \int_{-s}^{t-s} a'(u)\, \mathrm{d}u \\
&= \int_{\mathbf{R}} \mu(\mathrm{d}s) a(t-s) - \int_{\mathbf{R}} \mu(\mathrm{d}s) a(-s) \\
&= (\mu * a)(t) - (\mu * a)(0).
\end{aligned}$$

This proves (i).

(ii) By Theorem 6.1(iv), $\mu * b \in BL^1(\mathbf{R})$. To prove (ii), it suffices to show that, for all $t \in \mathbf{R}$, $(a * b)(t) = (a * b)(0) + \int_0^t (\mu * b)(s)\, \mathrm{d}s$. Define η_t in the same way as in the proof of (i) (without loss of generality, we take $t > 0$ and suppose that a is right continuous), and use (7.1) with a' replaced by

b, to get

$$\begin{aligned}\int_0^t (\mu * b)(s)\,\mathrm{d}s &= (\eta_t * (\mu * b))(0)\\ &= ((\eta_t * \mu) * b)(0)\\ &= \int_{\mathbf{R}}\int_{(-s,t-s]} \mu(\mathrm{d}u)b(s)\,\mathrm{d}s\\ &= \int_{\mathbf{R}} (a(t-s) - a(-s))b(s)\,\mathrm{d}s\\ &= (a * b)(t) - (a * b)(0).\end{aligned}$$

This completes the proof. □

Concerning Theorem 7.1, observe the obvious fact that, under the same assumptions on μ, a, and b, we have, in case (i), $(a * \mu)' = a' * \mu$ and, in case (ii), $(b * a)' = b * \mu$.

7.2 Corollary

(i) *Let $a \in L^1(\mathbf{R})$, let $1 \le p \le \infty$, and let $b \in L^p(\mathbf{R})$ be locally absolutely continuous and satisfy $b' \in L^p(\mathbf{R})$. Then $a * b$ is locally absolutely continuous and $(a * b)' = a * b'$. In particular, $(a * b)' \in L^p(\mathbf{R})$, and if $p = \infty$ then $(a * b)'$ is continuous.*

(ii) *Let $a \in L^1(\mathbf{R})$ be locally absolutely continuous with $a' \in L^1(\mathbf{R})$, let $1 \le p \le \infty$, and let $b \in L^p(\mathbf{R})$. Then $a * b$ is locally absolutely continuous and $(a * b)' = a' * b$. In particular, $(a * b)' \in L^p(\mathbf{R})$, and if $p = \infty$ then $(a * b)'$ is continuous.*

This follows from Theorem 7.1 combined with Theorem **2**.2.2(i).

Our next corollary is an immediate consequence of Theorem 7.1 combined with Corollary 6.2.

7.3 Corollary

(i) *Let $\mu \in M_{\mathrm{loc}}(\mathbf{R}^+)$, and let a be locally absolutely continuous on $\mathbf{R}^+$, and satisfy $a(0) = 0$ and $a' \in L^p_{\mathrm{loc}}(\mathbf{R}^+)$ for some $p \in [1, \infty]$. Then $\mu * a$ is locally absolutely continuous on $\mathbf{R}^+$, and $(\mu * a)' = \mu * a'$. In particular, $(\mu * a)' \in L^p_{\mathrm{loc}}(\mathbf{R}^+)$.*

(ii) *Let $a \in BV_{\mathrm{loc}}(\mathbf{R}^+)$ and $b \in L^p_{\mathrm{loc}}(\mathbf{R}^+)$ for some $p \in [1, \infty]$. Then $(a*b)$ is locally absolutely continuous on $\mathbf{R}^+$, and $(a * b)' = \mu * b$ where μ is the measure induced by the distribution derivative of a, i.e., the measure that satisfies $\mu([0, t]) = a(t)$ at all points of continuity of a. In particular, $(a * b)' \in L^p_{\mathrm{loc}}(\mathbf{R}^+)$.*

In the sequel we shall refer to the measure μ in part (ii) of Corollary 7.3 as the distribution derivative of a. Observe that if a is locally absolutely continuous on $\mathbf{R}^+$ then $\mu(\mathrm{d}s) = a(0)\delta(\mathrm{d}s) + a'(s)\,\mathrm{d}s$, where δ is the (scalar) unit point mass at zero.

7.4 Corollary

(i) *Let* $a \in L^1_{\text{loc}}(\mathbf{R}^+)$, *let* $1 \le p \le \infty$, *and let* $b \in L^p_{\text{loc}}(\mathbf{R}^+)$ *be locally absolutely continuous on* $\mathbf{R}^+$ *and satisfy* $b' \in L^p_{\text{loc}}(\mathbf{R}^+)$. *Then* $a * b$ *is locally absolutely continuous and* $(a * b)'(t) = a(t)b(0) + (a * b')(t)$ *for almost all* t. *In particular, if* $a \in L^p_{\text{loc}}(\mathbf{R}^+)$ *or* $b(0) = 0$, *then* $(a * b)' \in L^p_{\text{loc}}(\mathbf{R}^+)$, *and if* a *is continuous, or* $b(0) = 0$ *and* $p = \infty$, *then* $(a * b)'$ *is continuous.*

(ii) *Let* a *be locally absolutely continuous on* $\mathbf{R}^+$, *let* $1 \le p \le \infty$, *and let* $b \in L^p_{\text{loc}}(\mathbf{R}^+)$. *Then* $a * b$ *is locally absolutely continuous and* $(a * b)'(t) = a(0)b(t) + (a' * b)(t)$. *In particular,* $(a * b)' \in L^p_{\text{loc}}(\mathbf{R}^+)$, *and if* b *is continuous then so is* $(a * b)'$.

Both part (i) and part (ii) of this corollary can be deduced from Corollary 7.3(ii) combined with Corollary **2**.2.3.

Observe that if both a and b are locally absolutely continuous then Corollary 7.4 gives the formula

$$a(t)b(0) + (a * b')(t) = a(0)b(t) + (a' * b)(t),$$

which, of course, is an equivalent way of writing the standard formula for integration by parts.

8. Appendix: Laplace Transforms of Measures and Convolutions

This appendix discusses Laplace transforms of measures, and Laplace transforms of derivatives of measures or functions. Again, the results are valid both for scalar-valued and for matrix-valued functions and measures. See Section 2 for the definitions of Laplace transforms of measures.

8.1 Theorem

(i) *If* $\mu \in M(\mathbf{R})$, *then* $\hat{\mu}(z)$ *is defined, bounded and uniformly continuous on the line* $\Re z = 0$, *and* $\mu = 0$ *if and only if* $\hat{\mu}(\mathrm{i}\omega) = 0$ *for all* $\omega \in \mathbf{R}$.

(ii) *If* $\mu \in M_{\text{loc}}(\mathbf{R})$, *and* $\hat{\mu}(z_0)$ *is defined for some* $z_0 \in \mathbf{C}$, *then* $\hat{\mu}(z)$ *is defined on the entire vertical line* $\Re z = \Re z_0$.

(iii) *If* $\mu \in M_{\text{loc}}(\mathbf{R})$ *and* $a \in L^1_{\text{loc}}(\mathbf{R})$, *then* $\widehat{(\mu * a)}(z) = \hat{\mu}(z)\hat{a}(z)$ *for all those* $z \in \mathbf{C}$ *for which both* $\hat{\mu}(z)$ *and* $\hat{a}(z)$ *are defined.*

Proof (i) It follows directly from Definition 2.2 that $\hat{\mu}(\mathrm{i}\omega)$ is defined for all $\omega \in \mathbf{R}$ when $\mu \in M(\mathbf{R})$, and that $\hat{\mu}(\mathrm{i}\omega)$ is bounded. The uniform continuity of $\hat{\mu}$ on the line $\Re z = 0$ follows from the fact that, for all $\omega \in \mathbf{R}$ and $\xi \in \mathbf{R}$,

$$|\hat{\mu}(\mathrm{i}\xi) - \hat{\mu}(\mathrm{i}\omega)| = \left|\int_{\mathbf{R}} (\mathrm{e}^{-\mathrm{i}\xi t} - \mathrm{e}^{-\mathrm{i}\omega t})\mu(\mathrm{d}t)\right| \le \int_{\mathbf{R}} |\mathrm{e}^{-\mathrm{i}(\omega-\xi)t} - 1||\mu|(\mathrm{d}t),$$

and, by Lebesgue's dominated convergence theorem, the last integral tends to zero as $|\xi - \omega| \to 0$.

If $\mu = 0$, then $\hat{\mu} = 0$. If $\hat{\mu}(\mathrm{i}\omega) = 0$ for all ω, then $\int_{\mathbf{R}} \mu(\mathrm{d}t) f(t) = 0$ for all $f \in BC_0(\mathbf{R})$ (see, e.g., Katznelson [1], p. 132).

(ii) If $\hat{\mu}(\sigma + \mathrm{i}\omega)$ is defined for some complex $\sigma + \mathrm{i}\omega$, then the measure $\mathrm{e}^{-\sigma t}\mu(\mathrm{d}t)$ belongs to $M(\mathbf{R})$, and $\hat{\mu}(\sigma + \mathrm{i}\xi)$ is defined for all $\xi \in \mathbf{R}$.

(iii) If $\mu \in M(\mathbf{R})$, $a \in L^1(\mathbf{R})$ and $\omega \in \mathbf{R}$, then we can use Fubini's theorem to get

$$\begin{aligned}\widehat{(\mu * a)}(\mathrm{i}\omega) &= \int_{\mathbf{R}} \mathrm{e}^{-\mathrm{i}\omega t} \int_{\mathbf{R}} \mu(\mathrm{d}s) a(t - s)\, \mathrm{d}t \\ &= \int_{\mathbf{R}} \int_{\mathbf{R}} \mathrm{e}^{-\mathrm{i}\omega s} \mu(\mathrm{d}s) \mathrm{e}^{-\mathrm{i}\omega t} a(t)\, \mathrm{d}t \\ &= \hat{\mu}(\mathrm{i}\omega)\hat{a}(\mathrm{i}\omega).\end{aligned}$$

The general case follows from this special case, if one uses the fact that, for all $\sigma \in \mathbf{R}$, we have $\hat{\mu}(\sigma + \mathrm{i}\omega) = \widehat{\mu_\sigma}(\mathrm{i}\omega)$ and $\hat{a}(\sigma + \mathrm{i}\omega) = \widehat{a_\sigma}(\mathrm{i}\omega)$, where $\mu_\sigma(\mathrm{d}t) = \mathrm{e}^{-\sigma t}\mu(\mathrm{d}t)$ and $a_\sigma(t) = \mathrm{e}^{-\sigma t}a(t)$. □

8.2 Theorem

(i) *If $\mu \in M(\mathbf{R}^+)$, then $\hat{\mu}(z)$ is defined, bounded and uniformly continuous in the closed half plane $\Re z \geq 0$, and analytic in the open half plane $\Re z > 0$. Moreover, $\mu = 0$ if and only if $\hat{\mu}(z) = 0$ for all z in the closed half plane $\Re z \geq 0$.*

(ii) *If $\mu \in M_{\mathrm{loc}}(\mathbf{R}^+)$, and $\hat{\mu}(z_0)$ is defined for some $z_0 \in \mathbf{C}$, then $\hat{\mu}(z)$ is defined in the closed half plane $\Re z \geq \Re z_0$.*

(iii) *If $\mu \in M_{\mathrm{loc}}(\mathbf{R}^+)$ and $a \in L^1_{\mathrm{loc}}(\mathbf{R}^+)$, then $\widehat{(\mu * a)}(z) = \hat{\mu}(z)\hat{a}(z)$ for all those $z \in \mathbf{C}$ for which both $\hat{\mu}(z)$ and $\hat{a}(z)$ are defined.*

The proof of Theorem 8.2 is left to the reader.

The next result is obvious.

8.3 Theorem

Let J be either $\mathbf{R}$ or $\mathbf{R}^+$, and let $\mu \in M_{\mathrm{loc}}(J)$ and $b \in L^1_{\mathrm{loc}}(J)$.

(i) *If c is a complex constant, and $\mu_c(\mathrm{d}t) = \mathrm{e}^{-ct}\mu(\mathrm{d}t)$, $b_c(t) = \mathrm{e}^{-ct}b(t)$, then $(\mu_c * b_c)(t) = \mathrm{e}^{-ct}(\mu * b)(t)$, and $\widehat{\mu_c}(z) = \hat{\mu}(z + c)$.*

(ii) *If $\mu_c(E) = \mu(E/c)$ for some $c > 0$ (and for every Borel set E), then $\widehat{\mu_c}(z) = c\hat{\mu}(cz)$.*

(iii) *If $J = \mathbf{R}$, c is real, and $\mu_c(E) = \mu(E + c)$ (for every Borel set E), then $\widehat{\mu_c}(z) = \mathrm{e}^{cz}\hat{\mu}(z)$.*

Our final results concern Laplace transforms of derivatives.

8.4 Theorem

(i) *Let* $a \in BV_{\mathrm{loc}}(\mathbf{R})$ *and let* μ *be the distribution derivative of* a. *Then* $\hat{\mu}(z) = z\hat{a}(z)$ *for all those* $z \in \mathbf{C}$ *for which both* $\hat{a}(z)$ *and* $\hat{\mu}(z)$ *are defined. In particular, if* $\hat{a}(z_0)$ *and* $\hat{\mu}(z_0)$ *are defined for some* $z_0 \in \mathbf{C}$, *then* $\hat{\mu}(z) = z\hat{a}(z)$ *for all* $z \in \mathbf{C}$ *with* $\Re z = \Re z_0$.

(ii) *Let* $\mu, \nu \in M(\mathbf{R})$ *and let* $z\hat{\mu}(z) = \hat{\nu}(z)$ *for all* $z \in \mathbf{C}$ *with* $\Re z = 0$. *Then* $\mu(\mathrm{d}s) = a(s)\,\mathrm{d}s$ *for some function* $a \in L^1(\mathbf{R}) \cap BV(\mathbf{R})$, *and* ν *is the distribution derivative of* a.

Proof (i) If $a \in L^1(\mathbf{R}) \cap BV(\mathbf{R})$ and $\omega \in \mathbf{R}$, and if we define $\mathbf{e}_\omega(t) = \mathrm{e}^{\mathrm{i}\omega t}$ for $t \in \mathbf{R}$, then we get $(a * \mathbf{e}_\omega)(t) = \mathrm{e}^{\mathrm{i}\omega t}\hat{a}(\mathrm{i}\omega)$ and $(\mu * \mathbf{e}_\omega)(t) = \mathrm{e}^{\mathrm{i}\omega t}\hat{\mu}(\mathrm{i}\omega)$. Differentiating $(a * \mathbf{e}_\omega)$, and using Theorem 7.1(ii), we get

$$\mathrm{i}\omega \mathrm{e}^{\mathrm{i}\omega t}\hat{a}(\mathrm{i}\omega) = \frac{\mathrm{d}}{\mathrm{d}t}\left(\mathrm{e}^{\mathrm{i}\omega t}\hat{a}(\mathrm{i}\omega)\right) = (a * \mathbf{e}_\omega)'(t) = (\mu * \mathbf{e}_\omega)(t) = \mathrm{e}^{\mathrm{i}\omega t}\hat{\mu}(\mathrm{i}\omega).$$

Therefore, the identity $\hat{\mu}(z) = z\hat{a}(z)$ is satisfied for $\Re z = 0$. The general case follows from this special case and from Theorem 8.3(i).

(ii) The easiest way to prove (ii) is to use distribution Fourier transform theory (see Section **16**.8). If we identify μ with a distribution b, then the distribution Fourier transform of b can be identified with the function $\hat{\mu}(\mathrm{i}\omega)$, $\omega \in \mathbf{R}$. This implies that the Fourier transform of the distribution derivative of b can be identified with the function $\mathrm{i}\omega\hat{\mu}(\mathrm{i}\omega)$, $\omega \in \mathbf{R}$. This is also the transform of ν, and consequently, by the uniqueness theorem for distribution Fourier transforms, $b' = \nu$ in the distribution sense. If the measure μ has a distribution derivative which belongs to $M(\mathbf{R})$, then necessarily $\mu(\mathrm{d}s) = a(s)\,\mathrm{d}s$ for some function $a \in L^1(\mathbf{R}) \cap BV(\mathbf{R})$, and this proves the claim. □

8.5 Corollary

(i) *Let* $a \in L^1(\mathbf{R})$ *be locally absolutely continuous with* $a' \in L^1(\mathbf{R})$. *Then* $\widehat{a'}(z) = z\hat{a}(z)$ *for* $\Re z = 0$.

(ii) *Let* $a \in L^1(\mathbf{R})$, *and suppose that* $z\hat{a}(z) = \hat{b}(z)$ *for some function* $b \in L^1(\mathbf{R})$ *and for* $\Re z = 0$. *Then* a *is locally absolutely continuous, and* $a'(t) = b(t)$ a.e.

This follows directly from Theorem 8.4.

8.6 Theorem

(i) *Let* $a \in BV_{\mathrm{loc}}(\mathbf{R}^+)$, *and let* μ *be the distribution derivative of* a. *Then* $\hat{\mu}(z) = z\hat{a}(z)$ *for all those* $z \in \mathbf{C}$ *for which* $\hat{a}(z)$ *and* $\hat{\mu}(z)$ *are defined. In particular, if* $\hat{a}(z_0)$ *and* $\hat{\mu}(z_0)$ *are defined for some* $z_0 \in \mathbf{C}$, *then* $\hat{\mu}(z) = z\hat{a}(z)$ *for all* $z \in \mathbf{C}$ *with* $\Re z \geq \Re z_0$.

(ii) *Let* μ, $\nu \in M_{\mathrm{loc}}(\mathbf{R}^+)$, *let* $\hat{\nu}(z) = z\hat{\mu}(z)$ *on some vertical line* $\Re z = \sigma$. *Then* $\mu(\mathrm{d}s) = a(s)\,\mathrm{d}s$ *for some function* $a \in BV_{\mathrm{loc}}(\mathbf{R}^+)$, *and* ν *is the distribution derivative of* a.

The claims in Theorem 8.6 can be reduced to the corresponding claims in Theorem 8.4. We leave these reductions to the reader.

8.7 Corollary

(i) *Let a be defined and locally absolutely continuous on $\mathbf{R}^+$ and suppose that $\hat{a}(z_0)$ and $\widehat{a'}(z_0)$ are defined at some point $z_0 \in \mathbf{C}$. Then $\widehat{a'}(z) = z\hat{a}(z) - a(0)$ for all z with $\Re z \geq \Re z_0$.*

(ii) *Let $a \in L^1_{\text{loc}}(\mathbf{R}^+)$ and $b \in L^1_{\text{loc}}(\mathbf{R}^+)$, $c \in \mathbf{C}^{n\times n}$, and suppose that $\hat{b}(z) = z\hat{a}(z) - c$ in some half plane $\Re z > \sigma$. Then a is locally absolutely continuous, $a(0) = c$, and $a'(t) = b(t)$ a.e. on $\mathbf{R}^+$.*

For the proof it suffices to apply Theorem 8.6 (recall that we have defined the distribution derivative of a locally absolutely continuous function a defined on $\mathbf{R}^+$ in such a way that it contains a point mass of size $a(0)$ at zero).

9. Exercises

1. Let $\mu \in M(\mathbf{R}^+; \mathbf{C})$ satisfy one of the following two sets of conditions:
   (i) $\Re\hat{\mu}(z) > -\Re z$ for $\Re z \geq 0$;
   (ii) $\Re\hat{\mu}(\sigma) > -\sigma$ for real $\sigma \geq 0$, $\mu(\mathbf{R}^+) \neq 0$, and $(\Im\hat{\mu}(z))/(\Im z) > -1$ for $\Re z \geq 0$ and $\Im z \neq 0$.
   Show that in either case, the differential resolvent of μ is integrable.
2. Let $\mu \in M(\mathbf{R}; \mathbf{C})$ satisfy one of the following two sets of conditions:
   (i) $\Re\hat{\mu}(i\omega) > 0$ for $\omega \in \mathbf{R}$;
   (ii) $\mu(\mathbf{R}) \neq 0$, and $\Im\hat{\mu}(i\omega)/\omega > -1$ for nonzero $\omega \in \mathbf{R}$.
   Show that in either case, μ has a whole line differential resolvent.
3. Let $\mu \in M(\mathbf{R}; \mathbf{C})$ satisfy $\mu(-E) = \overline{\mu(E)}$ for all Borel sets E, and suppose that $\mu(\mathbf{R}) \neq 0$. Show that μ has a whole line differential resolvent.
4. Let $\mu(\mathrm{d}s) = (a_1\mathrm{e}^{-b_1 s} + a_2\mathrm{e}^{-b_2 s})\,\mathrm{d}s$, $s \geq 0$, with a_1, a_2, b_1, $b_2 > 0$. Show that the differential resolvent of μ is integrable.
5. Let $\mu(\mathrm{d}s) = a\mathrm{e}^{-b|s|}\,\mathrm{d}s$, $s \in \mathbf{R}$, with $a, b > 0$. Show that μ has a whole line differential resolvent. Compute this resolvent when $a = 3$ and $b = 1$. Can you prove that the kernel $\mu(\mathrm{d}s) = (a_1\mathrm{e}^{-b_1|s|} + a_2\mathrm{e}^{-b_2|s|})\,\mathrm{d}s$, with a_1, a_2, b_1, $b_2 > 0$, has a whole line differential resolvent?
6. Let $\mu(\mathrm{d}s) = \left(\sum_{n=1}^{\infty} a_n\mathrm{e}^{-b_n s}\right)\mathrm{d}s$, $s \geq 0$, with $a_n, b_n > 0$ for all n. Assume that $\sum_{n=1}^{\infty} a_n/b_n < \infty$. Does the differential resolvent r of μ satisfy $r \in L^1(\mathbf{R}^+; \mathbf{R})$?
7. Show how Theorem 3.5 follows from Theorem 3.7.
8. Prove or disprove the following claim. If the (half line) differential resolvent r of $\mu \in M(\mathbf{R}^+; \mathbf{C}^{n\times n})$ satisfies $r \in L^1(\mathbf{R}^+; \mathbf{C}^{n\times n})$, then r,

$r' \in L^p(\mathbf{R}^+; \mathbf{C}^{n\times n})$ for all $p \in [1,\infty]$. Does the analogous statement hold for the whole line differential resolvent of $\mu \in M(\mathbf{R}; \mathbf{C}^{n\times n})$?

9. Consider the scalar equation

$$x'(t) + ax(t) + \int_0^t b(t-s)x(s)\,\mathrm{d}s = 0, \quad t \geq 0; \quad x(0) = x_0.$$

Suppose that $\int_{\mathbf{R}^+} t|b(t)|\,\mathrm{d}t < \infty$, and that $a + \int_{\mathbf{R}^+} b(t)\,\mathrm{d}t > 0$. Show that the resolvent r belongs to $L^1(\mathbf{R}^+; \mathbf{C})$. Hint: Write the equation in the form $x'(t) + Ax(t) + \frac{\mathrm{d}}{\mathrm{d}t}\int_0^t B(t-s)x(s)\,\mathrm{d}s = 0$, where $A = a + \int_{\mathbf{R}^+} b(t)\,\mathrm{d}t$ and $B(t) = -\int_t^\infty b(s)\,\mathrm{d}s$, and observe that $|z+A| \geq |z|$ when $\Re z \geq 0$. (See Brauer [4].)

10. (Hard) Consider the scalar equation

$$x''(t) + a_0 x'(t) + a_1 x(t) + \int_0^t b(t-s)x(s)\,\mathrm{d}s = 0, \quad t \geq 0,$$

with initial conditions $x(0) = x_0$, $x'(0) = x_1$. Assume that a_0 and a_1 are nonnegative constants, and that $b \geq 0$ is nonincreasing and satisfies $b \not\equiv 0$, $b \in L^1(\mathbf{R}^+; \mathbf{R})$. Can you say anything about the behaviour of $x(t)$ as $t \to \infty$?

11. Consider the scalar, retarded delay equation

$$x'(t) + ax(t) + bx(t-T) = 0, \quad t \geq 0, \quad T > 0, \tag{9.1}$$

obtained from (3.1) by taking $\mu(\mathrm{d}s) = a\delta(\mathrm{d}s) + b\delta_T(\mathrm{d}s)$.

   (i) Show that all the roots of the characteristic equation $z + \hat{\mu}(z) = 0$ lie to the left of some vertical line in the complex plane.

   (ii) (Hard) Show that all the roots of the characteristic equation have strictly negative real parts if and only if

$$-aT < 1, \qquad -aT < bT < \frac{u_1}{\sin(u_1)},$$

   where u_1 is the root of $u = -aT\tan(u)$ in $0 \leq u < \pi$, with $u_1 = \frac{\pi}{2}$ if $aT = 0$. (See Hayes [1], Bellman and Cooke [1], p. 444, or Hale [5], p. 108.)

   (iii) Show that the characteristic equation has, in general, infinitely many roots and obtain their asymptotic location. (See Bellman and Cooke [1], Chs. 4 and 12.)

12. Discuss conditions on a, b and T under which $\mu(\mathrm{d}s) = a\delta(\mathrm{d}s) + b\delta_T(\mathrm{d}s)$ has a whole line differential resolvent r satisfying $r \in L^1(\mathbf{R}; \mathbf{R})$. Consider both the case $T \geq 0$ and the case $T < 0$. Hint: See Exercise 11. The case $T < 0$ can be reduced to the case $T > 0$ by a reflection of the real axis.

13. Do the details of the proof of Theorem 3.11.

14. Consider the scalar higher-order delay equation ($n < \infty$, $m \leq \infty$),

$$\sum_{j=0}^{n}\sum_{k=0}^{m} a_{jk} u^{(j)}(t - T_k) = f(t), \quad t \geq 0.$$

Give conditions on a_{jk} and T_k under which this equation can be written in the form (3.1) with $\mu \in M(\mathbf{R}^+; \mathbf{C})$.

15. Prove Theorems 3.13 and 3.14.
16. Let $k(t) = 1$ for $t \in [0,1]$, and $k(t) = 0$ for $t > 1$. Try to find the differential resolvent r of k. Is it true that $r \in L^1(\mathbf{R}^+; \mathbf{R})$?
17. Let $k(t) = c(1-t)$ for $t \in [0,1]$, and $k(t) = 0$ for $t > 1$, where c is a positive constant. Is it true that the differential resolvent r of k satisfies $r \in L^1(\mathbf{R}^+; \mathbf{R})$? Are you capable of getting an explicit expression for r? (See Hannsgen [1].)
18. Prove Theorem 3.17. Hint: See the proof of Theorem **2**.6.6.
19. Does the scalar equation
$$x'(t) + \int_{-\infty}^{\infty} b(t-s)x(s)\,\mathrm{d}s = \sin(\alpha t), \quad t \in \mathbf{R},$$
have a periodic solution if
 (i) $b(t) = c$, $|t| \le 1$, $b(t) = 0$, $|t| > 1$,
 (ii) $b(t) = c(1-t)$, $|t| \le 1$, $b(t) = 0$, $|t| > 1$?
In both cases, $c > 0$, $\alpha \in \mathbf{R}$. Can you extend these results to the more general case where
 (iii) b is real and even, and $\int_{\mathbf{R}} b(t)\,\mathrm{d}t \neq 0$?
20. Do the following exercises in Chapter 2, throughout replacing resolvents by differential resolvents (define $\mu(\mathrm{d}s) = k(s)\,\mathrm{d}s$, and throughout replace $[I + \hat{k}(z)]$ by $[zI + \hat{k}(z)]$):
 (i) Exercise **2**.5;
 (ii) Exercise **2**.16. Hint: $\left[z + \hat{\mu}_1(z)\right]^{-1} - \left[z + \hat{\mu}_2(z)\right]^{-1} = \left[z + \hat{\mu}_1(z)\right]^{-1}\left[\hat{\mu}_2(z) - \hat{\mu}_1(z)\right]\left[z + \hat{\mu}_2(z)\right]^{-1}$;
 (iii) Exercise **2**.17;
 (iv) Exercise **2**.18;
 (v) Exercise **2**.19;
 (vi) Exercise **2**.20;
 (vii) Exercise **2**.21;
 (viii) Exercise **2**.24;
 (ix) (Hard) Exercise **2**.30.
Try to replace the kernel $k \in L^1(\mathbf{R}^+; \mathbf{C}^{n\times n})$ by a measure $\mu \in M(\mathbf{R}^+; \mathbf{C}^{n\times n})$ in as many of the preceding results as possible.
21. Let the assumptions of Theorem 3.11 hold. Show that the following condition is equivalent to (i)–(iv) in Theorem 3.11.
 (v) *For every $f \in V$, the equation $x' + \mu * x = f$ has a locally absolutely continuous solution x in V.*
22. Let $\mu(\mathrm{d}t) = k(t)\,\mathrm{d}t$, where k is one of the following kernels:
 (i) $k(t) = 3\mathrm{e}^{-t} - \mathrm{e}^{-t/2}$ for $t \in \mathbf{R}^+$;
 (ii) $k(t) = 2\mathrm{e}^{-t/2} - 3\mathrm{e}^{-t}$ for $t \in \mathbf{R}^+$;
 (iii) $k(t) = \begin{cases} 2, & 0 \le t < 1, \\ 1, & 1 \le t < 2, \\ 0, & 2 \le t < \infty; \end{cases}$

$$\text{(iv)}\ k(t) = \begin{cases} 1, & 0 \le t < 1, \\ 2, & 1 \le t < 2, \\ 0, & 2 \le t < \infty. \end{cases}$$

For each k, choose some $T > \|k\|_{L^1(\mathbf{R}^+)}$, plot the curve mentioned in Corollary 3.16, and determine, whether the differential resolvent of k is integrable on $\mathbf{R}^+$.

23. Let μ be one of the following measures (δ_j is a unit point mass at j):
    (i) $\mu = 2\delta_0 - \delta_1 + 2\delta_2$;
    (ii) $\mu = \delta_0 + 2\delta_1 - 2\delta_2$;
    (iii) $\mu = \delta_0 - 2\delta_1 + \delta_2$.
    In each case, choose some $T > \|\mu\|_{M(\mathbf{R}^+)}$, plot the curve mentioned in Corollary 3.16, and determine, whether the differential resolvent of μ is integrable on $\mathbf{R}^+$.
24. Let $\mu \in M_{\text{loc}}(\mathbf{R}^+; \mathbf{C}^{n\times n})$ and f, $g \in L^1_{\text{loc}}(\mathbf{R}^+; \mathbf{C}^n)$. Show that the equation
    $$(x+g)'(t) + (\mu * x)(t) = f(t), \quad t \in \mathbf{R}^+; \quad (x+g)(0) = y_0,$$
    has a unique solution x in the sense that there exists a unique function $x \in L^1_{\text{loc}}(\mathbf{R}^+; \mathbf{C}^n)$ such that $x + g$ is locally absolutely continuous, $(x+g)(0) = y_0$, and such that the equation holds for almost all t. Hint: x is given a.e. by the variation of constants formula
    $$x(t) = r(t)y_0 - g(t) + (r * f)(t) - (r' * g)(t),$$
    where r is the differential resolvent of μ.

10. Comments

Sections 1 and 3:

An early paper where the variation of constants formula (for solutions of (3.1)) can be found is Grossman and Miller [1]. In addition, this reference contains a version of Theorem 3.13 and a formulation of the equivalence of (iii) and (iv) in Theorem 3.9.

The linear autonomous retarded functional differential equation

$$x'(t) + \int_{[0,r]} \mu(\mathrm{d}s)x(t-s) = 0; \quad x(t) = \phi(t), \quad -\infty < -r \le t \le 0, \tag{10.1}$$

is obviously a particular case of (3.1) (extend μ to a measure on $\mathbf{R}^+$ by defining $\mu(E) = \mu(E \cap [0,r])$ for each Borel set E, and include the contribution of ϕ in f). A classical reference to the study of (10.1) (above all, to the case where μ has only point masses) is Bellman and Cooke [1], where a wealth of examples and historical references can be found. For an extensive treatment of (10.1), see Hale [5], Chs. 6 and 7, where the variation of constants formula and the resolvent (the fundamental matrix) are given. In Hale [5], Ch. 7, the solution operator is viewed as a semigroup, and eigenfunction expansions are obtained. In our approach, the question of whether the delay is finite or not, i.e., whether μ has compact support or not, is irrelevant and consequently we do not discuss questions that are typical for the case where r is finite. For some decomposition results and for Volterra equations in the framework of semigroup theory, see Chapters 7 and 8, respectively.

The case where the measure μ in (3.1) is of the form $\mu(\mathrm{d}s) = A\delta(\mathrm{d}s) + B(s)\,\mathrm{d}s$ is considered in Miller [9]. It is shown that the condition $r \in L^1(\mathbf{R}^+)$ is equivalent to the uniform asymptotic stability of the zero solution of $x' + \mu * x = 0$. The equivalence of the transform condition $zI + A + \hat{B}(z) \neq 0$, $\Re z \geq 0$, and the uniform asymptotic stability of the zero solution are shown under the hypothesis that $\int_{\mathbf{R}^+}(1+t)|B(t)|\,\mathrm{d}t < \infty$. This condition is weakened to $B \in L^1(\mathbf{R}^+)$ in Miller [11] and in Grossman and Miller [2] by two different methods. In the former reference, the method of proof is basically what we employ in the proof of Theorem 3.7. The more cumbersome method of proof in the latter reference uses uniform asymptotic stability.

The case of Theorem 3.5 where the measure μ has no singular part is established in Corduneanu [8], Lemma 1, p. 107, with a proof similar to ours.

For equation (3.1) in weighted spaces, see Chapter 4.

Theorem 3.9 raises the question of how to determine whether or not the equation $\det[zI + \hat{\mu}(z)] = 0$ has any roots in $\Re z \geq 0$. This question has been dealt with, e.g., by Jordan [1]. For the location of the roots of characteristic polynomials one may consult Bellman and Cooke [1], or Hale [5], Appendix, pp. 337–341.

For early work on (3.1), see Halanay [2], Ch. 4.

Note that the whole line differential resolvent is sometimes called the *Green's function* of equation (3.2).

Levin and Shea [1] analyse the structure of solutions of (3.2) (and of solutions of $\mu * x = f$) that *a priori* are assumed to satisfy $x \in L^\infty(\mathbf{R})$. They also consider equations on $\mathbf{R}^+$. For more detailed comments on this reference, see Section **15**.11.

Langenhop [1] contains some results on periodic solutions of Volterra integrodifferential equations.

Sections 2 and 6–8:

For the fundamentals of measure theory, see, e.g., Rudin [3], Ch. 6, or Hewitt and Stromberg [1], Chs. III and V. For convolutions of measures and functions, and for their transforms, see Hewitt and Ross [1], Chs. V and VI, Katznelson [1], Ch. VI, or Schwartz [1], Chs. VI and VII. Laplace transforms can be studied in Doetsch [1] and Widder [2].

Section 4:

Fubini's theorem can be found in Rudin [3], Ch. 8, and in Hewitt and Stromberg [1], Ch. VI.

References

R. Bellman and K. L. Cooke

1. *Differential-Difference Equations*, Academic Press, New York, 1963.

F. Brauer

4. Asymptotic stability of a class of integro-differential equations, *J. Differential Equations* **28** (1978), pp. 180–188.

C. Corduneanu

8. Some differential equations with delay, in *Proceedings of Equadiff III, 1972*, M. Ráb and J. Vosmanský, eds., Czechoslovak Conference on Differential Equations and their Applications, J. E. Purkyně University, Brno, 1973, pp. 105–114.

G. Doetsch

1. *Handbuch der Laplace-Transformation. I–III*, Birkhäuser, Basel, 1950–1956.

S. I. Grossman and R. K. Miller

1. Perturbation theory for Volterra integrodifferential systems, *J. Differential Equations* **8** (1970), pp. 457–474.
2. Nonlinear Volterra integrodifferential systems with L^1-kernels, *J. Differential Equations* **13** (1973), pp. 551–566.

A. Halanay

2. *Differential Equations, Stability, Oscillations, Time Lags*, Academic Press, New York, 1966.

J. K. Hale

5. *Theory of Functional Differential Equations*, Springer-Verlag, Berlin, 1977.

K. B. Hannsgen

1. Indirect Abelian theorems and a linear Volterra equation, *Trans. Amer. Math. Soc.* **142** (1969), pp. 539–555.

N. D. Hayes

1. Roots of the transcendental equation associated with a certain difference-differential equation, *J. London Math. Soc.* **25** (1950), pp. 226–232.

E. Hewitt and K. A. Ross

1. *Abstract Harmonic Analysis, 1–2*, Springer-Verlag, Berlin, 1963–1970.

E. Hewitt and K. Stromberg

1. *Real and Abstract Analysis*, Springer-Verlag, Berlin, 1965.

G. S. Jordan

1. Asymptotic stability of a class of integrodifferential systems, *J. Differential Equations* **31** (1979), pp. 359–365.

Y. Katznelson

1. *An Introduction to Harmonic Analysis*, Dover Publications, New York, 1976.

C. E. Langenhop

1. Periodic and almost periodic solutions of Volterra integral differential equations with infinite memory, *J. Differential Equations* **58** (1985), pp. 391–403.

J. J. Levin and D. F. Shea

1. On the asymptotic behavior of the bounded solutions of some integral equations, I, II, III, *J. Math. Anal. Appl.* **37** (1972), pp. 42–82, 288–326, 537–575.

R. K. Miller

9. Asymptotic stability properties of linear Volterra integrodifferential equations, *J. Differential Equations* **10** (1971), pp. 485–506.

11. Asymptotic stability and perturbations for linear Volterra integrodifferential systems, in *Delay and Functional Differential Equations and their Applications*, Academic Press, New York, 1972, pp. 257–268.

W. Rudin

3. *Real and Complex Analysis,* 3rd *ed.*, McGraw–Hill, New York, 1986.

L. Schwartz

1. *Théorie des Distributions, nouv. éd.*, Hermann, Paris, 1966.

E. C. Titchmarsh

1. *Introduction to the Theory of Fourier Integrals,* 2nd *ed.*, Oxford University Press, London, 1948.

D. V. Widder

2. *The Laplace Transform*, Princeton University Press, Princeton, 1946.

4

Equations in Weighted Spaces

We investigate linear convolution integral and integrodifferential equations with measure-valued kernels in weighted spaces. A result due to Gel'fand is used to determine under what conditions the resolvent (now a measure) belongs to a weighted space.

1. Introduction

In this chapter we study the convolution equations

$$x(t) + \int_{[0,t]} \mu(\mathrm{d}s)x(t-s) = f(t), \quad t \in \mathbf{R}^+, \tag{1.1}$$

and

$$x(t) + \int_{\mathbf{R}} \mu(\mathrm{d}s)x(t-s) = f(t), \quad t \in \mathbf{R}, \tag{1.2}$$

where (at least) $\mu \in M_{\mathrm{loc}}(\mathbf{R}; \mathbf{C}^{n \times n})$, and where the functions x and f take their values in $\mathbf{C}^n$. Both local results, and asymptotic results that involve specific growth or decay rates, are given. In addition, we perform a more detailed asymptotic analysis of the solutions of the two integrodifferential equations

$$x'(t) + \int_{[0,t]} \mu(\mathrm{d}s)x(t-s) = f(t), \quad t \in \mathbf{R}^+; \quad x(0) = x_0, \tag{1.3}$$

and

$$x'(t) + \int_{\mathbf{R}} \mu(\mathrm{d}s)x(t-s) = f(t), \quad t \in \mathbf{R}, \tag{1.4}$$

considered in Chapter 3.

We begin with a local result (Theorem 1.5) on the resolvent ρ (now a measure) of the measure μ in (1.1). Formally, this result is almost identical to the corresponding result, Theorem **2**.3.1, for the case where $\mu(\mathrm{d}s) = k(s)\,\mathrm{d}s$, with $k \in L^1_{\mathrm{loc}}(\mathbf{R}^+)$. However, as a consequence of the fact that the locally integrable functions k and r have been replaced by locally finite measures μ and ρ, respectively, we need the concept of the convolution of two measures. Out of several equivalent definitions, we prefer to employ the following.

1.1 Definition

The convolution $\mu * \nu$ of two measures $\mu, \nu \in M(\mathbf{R})$ is the completion of the measure that to each Borel set $E \subset \mathbf{R}$ assigns the value

$$(\mu * \nu)(E) = \int_{\mathbf{R}} \mu(\mathrm{d}t)\nu(E - t),$$

where $E - t$ is the set $\{ s - t \mid s \in E \}$. The convolution $\mu * \nu$ of two measures defined on $[0, T)$ is obtained in the same way, once μ and ν have been defined on every Borel set $E \subset (-\infty, T)$ through $\mu(E) = \mu(E \cap \mathbf{R}^+)$ and $\nu(E) = \nu(E \cap \mathbf{R}^+)$.

If χ_E is the characteristic function of the set E, then $s \mapsto \chi_E(s + t)$ is the characteristic function of $E - t$, and $\nu(E - t) = \int_{\mathbf{R}} \chi_E(s + t)\nu(\mathrm{d}s)$. In particular, this combined with Theorem **3**.6.1(ii) shows that the function $t \mapsto \nu(E - t)$ belongs to $B^\infty(\mathbf{R})$, hence the definition of $(\mu * \nu)(E)$ above makes sense. One can also write $(\mu*\nu)(E)$ in the form (use Fubini's theorem to get the last line)

$$\begin{aligned}(\mu * \nu)(E) &= \int_{\mathbf{R}} \mu(\mathrm{d}t)\nu(E - t)\\ &= \int_{\mathbf{R}} \int_{\mathbf{R}} \chi_E(s + t)\mu(\mathrm{d}t)\nu(\mathrm{d}s)\\ &= \int_{\mathbf{R}} \mu(E - s)\nu(\mathrm{d}s).\end{aligned} \tag{1.5}$$

Still another way to write this is to use convolution notation. If we define $-E$ to be the reflected set $-E = \{ s \in \mathbf{R} \mid -s \in E \}$, then (recall Definition **3**.2.1)

$$\begin{aligned}\nu(E - t) &= \int_{\mathbf{R}} \chi_E(s + t)\nu(\mathrm{d}s) = \int_{\mathbf{R}} \chi_{-E}(-s - t)\nu(\mathrm{d}s)\\ &= (\nu * \chi_{-E})(-t).\end{aligned}$$

Similarly, $\mu(E - t) = (\mu * \chi_{-E})(-t)$, and (1.5) can be written in the form

$$(\mu * \nu)(E) = (\mu * (\nu * \chi_{-E}))(0) = ((\mu * \chi_{-E}) * \nu)(0). \tag{1.6}$$

If the two measures μ and ν have no singular part, say

$$\begin{aligned}\mu(\mathrm{d}s) &= \sum_{k=1}^{\infty} p_k \delta_{t_k}(\mathrm{d}s) + a(s)\,\mathrm{d}s,\\ \nu(\mathrm{d}s) &= \sum_{l=1}^{\infty} q_l \delta_{t_l}(\mathrm{d}s) + b(s)\,\mathrm{d}s\end{aligned}$$

(here we use the same conventions as in (**3**.2.1)), then $\mu * \nu$ has no singular part, the discrete part of $\mu * \nu$ is the sum

$$\sum_{k=1}^{\infty}\sum_{l=1}^{\infty} p_k q_l \delta_{(t_k+t_l)},$$

and the absolutely continuous part of $\mu * \nu$ is the L^1-function

$$\sum_{k=1}^{\infty} p_k b(t-t_k) + \sum_{l=1}^{\infty} a(t-t_l)q_l + (a*b)(t)$$

(multiplied by $\mathrm{d}t$). If $\mu \in M(\mathbf{R})$ is arbitrary, and $\nu \in M(\mathbf{R})$ is absolutely continuous, i.e., $\nu(\mathrm{d}s) = a(s)\,\mathrm{d}s$ for some function $a \in L^1(\mathbf{R})$, then $\mu * \nu$ is absolutely continuous and $(\mu * \nu)(\mathrm{d}s) = (\mu * a)(s)\,\mathrm{d}s$.

The following theorem shows that $\mu * \nu$ is well defined, and extends some of the claims in Theorem **3**.6.1.

1.2 Theorem
Let J be $\mathbf{R}$, $\mathbf{R}^+$, or $[0,T)$ where $0 < T < \infty$, and let μ, ν and ρ belong to $M(J)$. Then
(i) $\mu * \nu \in M(J)$, *and* $\|\mu * \nu\|_{M(J)} \le \|\mu\|_{M(J)}\|\nu\|_{M(J)}$,
(ii) *if* $f \in B^\infty(J)$, *then* $((\mu*\nu)*f)(t) = (\mu*(\nu*f))(t)$ *and* $((\mu*f)*\nu)(t) = (\mu*(f*\nu))(t)$ *for all* $t \in J$,
(iii) *if* $f \in BL^1(J)$, *then* $(\mu*\nu)*f = \mu*(\nu*f)$ *and* $(\mu*f)*\nu = \mu*(f*\nu)$,
(iv) $(\mu * \nu) * \rho = \mu * (\nu * \rho)$,
(v) *if* μ *or* ν *is scalar-valued, then* $\mu * \nu = \nu * \mu$.

Theorem 1.2 is proved in Section 5.

Combining Theorem 1.2(i) and (iv) with the fact that $M(J)$ is a Banach space, we get the following result.

1.3 Corollary
$M(J)$ is a Banach algebra with product $*$, when $J = \mathbf{R}$, $\mathbf{R}^+$, or $[0,T)$, where $0 < T < \infty$.

It is obvious that many of the statements in Theorem 1.2 hold locally on $\mathbf{R}^+$.

1.4 Corollary
Let μ, ν, and ρ belong to $M_{\mathrm{loc}}(\mathbf{R}^+)$. Then
(i) $\mu * \nu \in M_{\mathrm{loc}}(\mathbf{R}^+)$,
(ii) *if* $f \in B^\infty_{\mathrm{loc}}(\mathbf{R}^+)$, *then* $((\mu * \nu) * f)(t) = (\mu * (\nu * f))(t)$ *and* $((\mu * f) * \nu)(t) = (\mu * (f * \nu))(t)$ *for all* $t \in \mathbf{R}^+$,
(iii) *if* $f \in L^1_{\mathrm{loc}}(\mathbf{R}^+)$, *then* $(\mu*\nu)*f = \mu*(\nu*f)$ *and* $(\mu*f)*\nu = \mu*(f*\nu)$,
(iv) $(\mu * \nu) * \rho = \mu * (\nu * \rho)$.

With these results at our disposal, we can get the following theorem (cf. Theorem **2**.3.1) on the existence of the resolvent ρ of a measure μ supported on $\mathbf{R}^+$.

1.5 Theorem
Suppose that $\mu \in M_{\text{loc}}(\mathbf{R}^+; \mathbf{C}^{n\times n})$. Then there exists a unique measure $\rho \in M_{\text{loc}}(\mathbf{R}^+; \mathbf{C}^{n\times n})$ satisfying $\rho + \mu * \rho = \rho + \rho * \mu = \mu$ if and only if $\det[I + \mu(\{0\})] \neq 0$. This measure, when it exists, depends continuously on μ in the topology of $M_{\text{loc}}(\mathbf{R}^+; \mathbf{C}^{n\times n})$.

Proof Suppose that there exists a measure ρ such that $\rho + \mu * \rho = \mu$. This implies that $(\delta + \mu) * (\delta - \rho) = \delta$, hence $[I + \mu(\{0\})][I - \rho(\{0\})] = I$. It follows that $\det[I + \mu(\{0\})] \neq 0$.

Next we assume that $\det[I + \mu(\{0\})] \neq 0$. Although we now work with measures, the proof closely follows that of Theorem **2**.3.1. Therefore, we sketch it only briefly. First, one observes that it suffices to prove that for each positive T there exists a unique $\rho_T \in M([0,T])$ such that

$$\rho_T + \mu * \rho_T = \rho_T + \rho_T * \mu = \mu. \tag{1.7}$$

Uniqueness of ρ_T follows as in the proof of Theorem **2**.3.1. Define $A = \mu(\{0\})$. To show existence, one claims that, without loss of generality, one may take $\left|[I + A]^{-1}\right| |\mu|((0,T]) < 1$. To substantiate this claim, define $\lambda_m(\mathrm{d}s) = \mathrm{e}^{-ms}\mu(\mathrm{d}s)$, note that for sufficiently large m one has $\left|[I + A]^{-1}\right| |\lambda_m|((0,T]) < 1$, and observe that, if ν satisfies

$$\nu + \nu * \lambda_m = \nu + \lambda_m * \nu = \lambda_m, \tag{1.8}$$

then $\rho_T(\mathrm{d}s) = \mathrm{e}^{ms}\nu(\mathrm{d}s)$ satisfies (1.7).

Consider the equation $\rho + \mu * \rho = \rho + \rho * \mu = \mu$ on $[0,T]$, assuming $\left|[I + A]^{-1}\right| |\mu|((0,T]) < 1$. Then

$$\begin{aligned} \rho &= [I + A]^{-1}\mu - [I + A]^{-1}(\mu - A\delta) * \rho \\ &= \mu[I + A]^{-1} - \rho * (\mu - A\delta)[I + A]^{-1}. \end{aligned} \tag{1.9}$$

Now use either of these two equations as an iteration formula, and recall that the regular complex Borel measures on $[0,T]$ form a Banach space with norm $\|\mu\|_{M([0,T])} = |\mu|([0,T])$.

We leave the proof of the continuous dependence to the reader. □

1.6 Definition
The measure ρ in Theorem 1.5 is called the resolvent of μ.

Not surprisingly, our next result concerns the solutions of (1.1).

1.7 Theorem
Let $\mu \in M_{\text{loc}}(\mathbf{R}^+; \mathbf{C}^{n\times n})$ with $\det[I + \mu(\{0\})] \neq 0$. Then, for every $f \in L^1_{\text{loc}}(\mathbf{R}^+; \mathbf{C}^n)$, there is a unique solution $x \in L^1_{\text{loc}}(\mathbf{R}^+; \mathbf{C}^n)$ of (1.1). This solution is given by

$$x(t) = f(t) - \int_{[0,t]} \rho(\mathrm{d}s) f(t-s), \quad t \in \mathbf{R}^+, \tag{1.10}$$

where ρ is the resolvent of μ. If f belongs to $L^p_{\text{loc}}(\mathbf{R}^+; \mathbf{C}^n)$ for some $p > 1$, or to $B^\infty_{\text{loc}}(\mathbf{R}^+; \mathbf{C}^n)$ or $BV_{\text{loc}}(\mathbf{R}^+; \mathbf{C}^n)$, then x belongs to the same space *as f. If $f \in C(\mathbf{R}^+; \mathbf{C}^n)$ and μ has no point masses in $(0, \infty)$ or $f(0) = 0$, then $x \in C(\mathbf{R}^+; \mathbf{C}^n)$.*

The proof is analogous to that of Theorem **2**.3.5. In order to deduce that x belongs to the same space as f, one uses Corollary **3**.6.2. Note that if μ_1, $\mu_2 \in M_{\mathrm{loc}}(\mathbf{R}^+)$, and if μ_1 has no discrete part, then neither has $\mu_1 * \mu_2$. Therefore, if μ has no point masses in $(0, \infty)$, then (1.9) implies that ρ cannot have any point masses in $(0, \infty)$.

2. Introduction to Weighted Spaces

To prove statements concerning the asymptotic behaviour of the solutions of (1.1), and to be able to solve (1.2), we have to impose additional conditions on μ. For the purpose of obtaining more specific results, it is appropriate to work in weighted function spaces. Under suitable assumptions, this approach will both give us the information that x vanishes at infinity, and tell us something about the rate of convergence of $x(t)$ to zero as $t \to \infty$. The key idea is simply to incorporate a given convergence rate in the definition of a function space, and to give conditions that imply that if f belongs to a particular weighted function space then x belongs to the same weighted space.

We begin with some elementary facts concerning weighted function spaces. Following this presentation, we give asymptotic results of Paley–Wiener type for (1.1)–(1.4). The reader who prefers to work in the usual nonweighted spaces should take $\eta(t) \equiv \rho(t) \equiv 1$ below.

2.1 Definition

A function η defined on an interval $J \subset \mathbf{R}$ is called a weight function on J if η is positive, Borel measurable, locally bounded, and locally bounded away from zero on J.

Observe that a weight function on J need be neither bounded from above nor bounded away from zero at an end-point of J that does not belong to J.

Let η be a weight function. We define the weighted L^p-spaces $L^p(J; \eta)$, $1 \le p \le \infty$, to be the set of functions f on J that satisfy $\eta f \in L^p(J)$, with norm

$$\|f\|_{L^p(J;\eta)} = \|\eta f\|_{L^p(J)} = \begin{cases} \left(\int_J (\eta(t)|f(t)|)^p \mathrm{d}t\right)^{1/p}, & 1 \le p < \infty, \\ \operatorname{ess\,sup}_{t \in J} |\eta(t) f(t)|, & p = \infty. \end{cases}$$

For $J = \mathbf{R}$ or $\mathbf{R}^+$, the space $BL^p(J; \eta)$, $1 \le p \le \infty$, is defined to be the set of functions f satisfying $\eta f \in BL^p(J)$, with $\|f\|_{BL^p(J;\eta)} = \|\eta f\|_{BL^p(J)}$.

The space $BL_0^p(J;\eta)$, $1 \le p \le \infty$, consists of those functions f for which $\eta f \in BL_0^p(J)$.

Similarly, we define the weighted Borel space $B^\infty(J;\eta)$ to be the set of Borel functions f on J that satisfy $\eta f \in B^\infty(J)$, with norm $\|f\|_{B^\infty(J;\eta)} = \|f\|_{\sup(J;\eta)} = \|\eta f\|_{B^\infty(J)} = \sup_{t\in J}|\eta(t)f(t)|$. We let $BC(J;\eta)$ be the subspace of $B^\infty(J;\eta)$ consisting of continuous functions, and define $BC_0(\mathbf{R}^+;\eta)$ ($BC_0(\mathbf{R};\eta)$) as the set of functions in $BC(\mathbf{R}^+;\eta)$ ($BC(\mathbf{R};\eta)$) that satisfy $\eta(t)f(t) \to 0$ as $t \to \infty$ ($|t| \to \infty$). Analogously, the space $B_0^\infty(\mathbf{R}^+;\eta)$ ($B_0^\infty(\mathbf{R};\eta)$) consists of those locally bounded Borel functions f for which $\eta f \to 0$ as $t \to \infty$ ($|t| \to \infty$).

Weighted spaces of bounded uniformly continuous functions are defined as follows. For $J = \mathbf{R}$ we let $BUC(\mathbf{R};\eta)$ be the subset of functions f in $BC(\mathbf{R};\eta)$ that satisfy $\lim_{h\to 0}\|\tau_h f - f\|_{\sup(\mathbf{R};\eta)} = 0$, where τ_h is the translation operator $\tau_h f(t) = f(t+h)$. Similarly, we define $BUC(\mathbf{R}^+;\eta)$ to be the space of functions $f \in BC(\mathbf{R}^+;\eta)$ that satisfy $\lim_{h\to 0}\sup_{t\ge |h|}\eta(t)|f(t+h) - f(t)| = 0$.

Finally, we let $BV(J;\eta)$ consist of those $f \in BV_{\mathrm{loc}}(J)$ that satisfy $\|f\|_{BV(J;\eta)} = \|f\|_{\sup(J;\eta)} + \int_J \eta(t)|\nu_f|(\mathrm{d}t) < \infty$, where $\nu_f \in M_{\mathrm{loc}}(J)$ is the locally finite measure generated by the variation of f (i.e., the distribution derivative of f). The space $BBV(J;\eta)$ consists of those $f \in BV_{\mathrm{loc}}(J)$ that satisfy $\|f\|_{BBV(J;\eta)} = \sup_{[s,s+1]\subset J}\|f\|_{BV([s,s+1];\eta)} < \infty$ and $BBV_0(J;\eta)$ is the subspace of $BBV(J;\eta)$ containing the functions for which $\|f\|_{BV([s,s+1];\eta)} \to 0$ as s tends to infinity.

All the weighted spaces defined above are Banach spaces. If we want to emphasize in what space the values of f lie, we write $f \in L^p(J;\eta;\mathbf{R})$, or $f \in L^p(J;\eta;\mathbf{R}^n)$, etc.

Weighted measure spaces are introduced in Section 3.

Some results are quite easy to generalize from nonweighted spaces (where $\eta(t) \equiv 1$) to certain weighted spaces, simply because f belongs to one of these weighted spaces if and only if ηf belongs to the corresponding nonweighted space (this is not the case for all spaces). For example, the integral operator

$$f \mapsto (Kf)(t) = \int_S^t k(t,s)f(s)\,\mathrm{d}s$$

maps $L^p((S,T);\eta)$ continuously into itself if and only if the operator

$$(K_\eta g)(t) \stackrel{\mathrm{def}}{=} \int_S^t \eta(t)k(t,s)\eta^{-1}(s)g(s)\,\mathrm{d}s$$

maps $L^p(S,T)$ continuously into itself. If one applies this to the nonconvolution equation

$$x(t) + \int_0^t k(t,s)x(s)\,\mathrm{d}s = f(t), \quad t \ge 0,$$

discussed in Chapter 9, then one gets the following result.

2.2 Lemma
*Let η be a weight function on (S,T), and let k be a kernel satisfying the hypothesis of Corollary **9**.4.3. Then the resolvent r of k maps $L^p((S,T);\eta)$ continuously into itself if and only if the resolvent of the kernel $\eta(t)k(t,s)/\eta(s)$ maps $L^p(S,T)$ continuously into itself.*

This is true because the resolvent of the kernel $\eta(t)k(t,s)/\eta(s)$ is $\eta(t)r(t,s)/\eta(s)$, where r is the resolvent of $k(t,s)$. (To see this, multiply the resolvent equation

$$k(t,s) - r(t,s) = \int_s^t k(t,u)r(u,s)\,\mathrm{d}u = \int_s^t r(t,u)k(u,s)\,\mathrm{d}u$$

by $\eta(t)$, and divide by $\eta(s)$.)

In the convolution case, this transformation is less successful because if we start with a convolution kernel $k(t-s)$, then $\eta(t)k(t-s)/\eta(s)$ is no longer a convolution kernel (unless η is an exponential). Therefore, we have to impose additional conditions on η.

3. Regular Weight Functions

Let η be a weight function defined on J, where J stands for $\mathbf{R}^+$ or $\mathbf{R}$. To obtain results on convolutions we have to impose some regularity condition on η. Out of several possible, slightly different, sets of conditions, we prefer to use the following. We define

$$\varphi(t) = \sup_{s\in J,\, s+t\in J} \frac{\eta(s+t)}{\eta(s)}, \quad t \in \mathbf{R}, \tag{3.1}$$

and require φ to be locally bounded. Note that φ is defined on $\mathbf{R}$, even if $J = \mathbf{R}^+$.

The following lemma shows how the local boundedness of φ regulates the behaviour of η.

3.1 Lemma
Let η be a weight function on J (equal to $\mathbf{R}$ or $\mathbf{R}^+$), and suppose that φ, defined in (3.1), is locally bounded. Then $\varphi(0) = 1$, and

$$\eta(s+t) \le \eta(s)\varphi(t), \quad s, s+t \in J, \tag{3.2}$$

$$\varphi(s+t) \le \varphi(s)\varphi(t), \quad s,t \in \mathbf{R}, \tag{3.3}$$

$$1/\varphi(-t) \le \eta(t)/\eta(0) \le \varphi(t), \quad t \in J. \tag{3.4}$$

Moreover, if we denote $\limsup_{s\to 0}\varphi(s)$ by K, then

$$K^{-1}\limsup_{s\to t}\eta(s) \le \eta(t) \le K \liminf_{s\to t}\eta(s), \quad s,t \in J, \tag{3.5}$$

$$K^{-1}\limsup_{s\to t}\varphi(s) \le \varphi(t) \le K \liminf_{s\to t}\varphi(s), \quad s,t \in \mathbf{R}. \tag{3.6}$$

In particular, if φ is continuous at zero, then η is continuous on J, and φ is continuous on $\mathbf{R}$.

Proof Clearly, $\varphi(0) = 1$ and $\varphi(t) > 0$ for all $t \in \mathbf{R}$. By (3.1), the inequality (3.2) is true.

If $J = \mathbf{R}$, then one gets (3.3) from the simple computation

$$\begin{aligned}\varphi(s+t) &= \sup_{u\in J, u+s+t\in J} \frac{\eta(u+s+t)}{\eta(u)} \\ &= \sup_{u+t\in J, u\in J, u+s+t\in J} \frac{\eta(u+s+t)}{\eta(u+t)} \frac{\eta(u+t)}{\eta(u)} \\ &\le \varphi(s)\varphi(t).\end{aligned}$$

The same computation is valid when $J = \mathbf{R}^+$ as well, because we can (without loss of generality) let $t \ge s$, and observe that then $u + t \ge 0$ if $u \ge 0$ and $u + s + t \ge 0$.

That (3.4) is valid follows directly from (3.2) (take $t = -s$ to get the lower bound and $s = 0$ to get the upper bound). The same is true as regards (3.5), because for all $t \in J$ and $s + t \in J$, we have $\eta(t) \le \eta(t + s)\varphi(-s)$ and $\eta(t) \ge \eta(t + s)/\varphi(s)$. To prove (3.6), one argues in exactly the same way, replacing (3.2) with (3.3). □

As the next definition shows, the function φ in (3.1) is a particular example of a submultiplicative function.

3.2 Definition

A positive function φ defined on $\mathbf{R}$ is called submultiplicative, if $\varphi(0) = 1$, and

$$\varphi(s+t) \le \varphi(s)\varphi(t),$$

for all s, $t \in \mathbf{R}$. We say that a weight function η is dominated by a submultiplicative function φ if

$$\eta(s+t) \le \eta(s)\varphi(t),$$

for all $s \in J$ and all t satisfying $t + s \in J$, where J is the domain of η. A weight function η is regular if $\eta(0) = 1$, and if η is dominated by some locally bounded submultiplicative function φ.

Obviously the smallest possible submultiplicative function that dominates a weight function η, is the function φ in (3.1). Observe that a submultiplicative weight function φ is a regular weight function, and that it is dominated by itself.

(We do not require a submultiplicative function to be measurable, since we shall encounter nonmeasurable submultiplicative functions in the proof of Theorem 4.2. It follows from (3.4), with η replaced by φ, that a locally bounded submultiplicative function is locally bounded away from zero. Most of the time, our submultiplicative functions will be required to be weight functions, i.e., Borel measurable, locally bounded, and locally bounded away from zero.)

The final part of the proof of Lemma 3.1 may be used to yield the following result.

3.3 Corollary

Let η be a regular weight function dominated by a submultiplicative weight function φ. Then the inequalities (3.2)–(3.6) hold.

Some frequently occurring submultiplicative weight functions φ are

$$\left.\begin{aligned} \varphi_1(t) &= (1+|t|)^{\delta_1}, & \delta_1 &\geq 0, \\ \varphi_2(t) &= \big(1+\ln(1+|t|)\big)^{\delta_2}\varphi_1(t), & \delta_2 &\geq 0, \\ \varphi_3(t) &= \mathrm{e}^{\alpha t}\varphi_2(t), & \alpha &\in \mathbf{R}. \end{aligned}\right\} \tag{3.7}$$

To get an example of a weight function η dominated by a submultiplicative function φ (for which φ and η are not identical), take for $\delta > 0$

$$\eta(t) = \begin{cases} (1+|t|)^{-\delta}, & t < 0, \\ 1, & t \geq 0, \end{cases} \tag{3.8}$$

and

$$\varphi(t) = \begin{cases} 1, & t < 0, \\ (1+t)^{\delta}, & t \geq 0. \end{cases} \tag{3.9}$$

Let φ be a weight function (which in most cases is submultiplicative). We define the weighted measure spaces $M(\mathbf{R}^+;\varphi)$ and $M(\mathbf{R};\varphi)$ analogously to the weighted function spaces. Thus, a locally finite Borel measure μ belongs to $M(\mathbf{R}^+;\varphi)$ or $M(\mathbf{R};\varphi)$ if and only if the measure $\varphi(t)\mu(\mathrm{d}t)$ belongs to $M(\mathbf{R}^+)$ or $M(\mathbf{R})$. The norm of μ in the weighted space is the norm of $\varphi(t)\mu(\mathrm{d}t)$ in the nonweighted space.

3.4 Theorem

Let J be $\mathbf{R}$ or $\mathbf{R}^+$, and let φ be a submultiplicative weight function. Then $M(J;\varphi)$ is a Banach algebra with product $*$. In particular,

$$\|\mu * \nu\|_{M(J;\varphi)} \leq \|\mu\|_{M(J;\varphi)}\|\nu\|_{M(J;\varphi)}. \tag{3.10}$$

The proof is given in Section 5.

In analogy to Theorem **3**.6.1, we have the following result on the weighted norm of $\mu * f$ when $\mu \in M(\mathbf{R};\varphi)$.

3.5 Theorem

Let J be $\mathbf{R}$ or $\mathbf{R}^+$, let η be a regular weight function on J, dominated by a submultiplicative weight function φ, and let $\mu \in M(J;\varphi)$.

(i) *If $a \in L^p(J;\eta)$ for some $p \in [1,\infty]$, then $\mu * a \in L^p(J;\eta)$ and $\|\mu * a\|_{L^p(J;\eta)} \leq \|\mu\|_{M(J;\eta)}\|a\|_{L^p(J;\eta)}$. If $a \in L^\infty_0(J;\eta)$, then $\mu * a \in L^\infty_0(J;\eta)$.*

(ii) *If $a \in B^\infty(J;\eta)$, then $\mu * a \in B^\infty(J;\eta)$ and $\|\mu * a\|_{\sup(J;\eta)} \leq \|\mu\|_{M(J;\eta)}\|a\|_{\sup(J;\eta)}$. If $a \in B^\infty_0(J;\eta)$, then $\mu * a \in B^\infty_0(J;\eta)$.*

(iii) *If $J = \mathbf{R}$, and if a belongs to one of the spaces $BC(J;\eta)$, $BUC(J;\eta)$, or $BC_0(J;\eta)$, then $\mu * a$ belongs to the same space as a. The same is true if $J = \mathbf{R}^+$, provided μ has no point masses in $(0,\infty)$, or $a(0) = 0$.*

(iv) *If $a \in BL^p(J;\eta)$ for some $p \in [1,\infty]$, then $\mu * a \in BL^p(J;\eta)$, and $\|\mu * a\|_{BL^p(J;\eta)} \le \|\mu\|_{M(J;\eta)}\|a\|_{BL^p(J;\eta)}$. If $a \in BL^p_0(J;\eta)$, then $\mu * a \in BL^p_0(J;\eta)$.*

(v) *If $a \in BV(J;\eta)$, then $\mu * a \in BV(J;\eta)$, and $\|\mu * a\|_{BV(J;\eta)} \le \|\mu\|_{M(J;\eta)}\|a\|_{BV(J;\eta)}$.*

(vi) *If $a \in BBV(J;\eta)$, then $\mu * a \in BBV(J;\eta)$, and $\|\mu * a\|_{BBV(J;\eta)} \le \|\mu\|_{M(J;\eta)}\|a\|_{BBV(J;\eta)}$. If $a \in BBV_0(J;\eta)$, then $\mu * a \in BBV_0(J;\eta)$.*

This proof, too, is given in Section 5.

If Theorem 3.5 is to be applied to the solutions of the convolution equations discussed in Section 1, then we are led to the question: when is it true that the resolvent of μ belongs to the weighted space $M(\mathbf{R}^+;\varphi)$ or to $M(\mathbf{R};\varphi)$? In the case where the resolvent can be represented by a function, the corresponding question is: under what assumptions does this function belong to $L^1(\mathbf{R}^+;\varphi)$ or to $L^1(\mathbf{R};\varphi)$? These questions are answered in the next section.

4. Gel′fand's Theorem

Let us first prove a lemma which says that every locally bounded submultiplicative function has a growth rate or decay rate at infinity which is of exponential type.

4.1 Lemma

Let φ be a locally bounded submultiplicative function. Then the two limits

$$\alpha_\varphi \stackrel{\text{def}}{=} -\lim_{t\to-\infty}\frac{\ln(\varphi(t))}{t}, \quad \omega_\varphi \stackrel{\text{def}}{=} -\lim_{t\to\infty}\frac{\ln(\varphi(t))}{t}, \tag{4.1}$$

exist, and $-\infty < \omega_\varphi \le \alpha_\varphi < \infty$. Moreover,

$$\varphi(t) \ge \max\{e^{-\alpha_\varphi t}, e^{-\omega_\varphi t}\}, \quad t \in \mathbf{R}. \tag{4.2}$$

Proof To prove that the limits in (4.1) exist, let us define the two numbers α_φ and ω_φ in a different way, namely as

$$\alpha_\varphi = -\sup_{t<0}\frac{\ln(\varphi(t))}{t}, \quad \omega_\varphi = -\inf_{t>0}\frac{\ln(\varphi(t))}{t}. \tag{4.3}$$

For every t, $s \in \mathbf{R}$ and every $m = 0, 1, 2, \ldots.$, we have

$$\varphi(t) \le \varphi(s)^m\varphi(t - ms).$$

Trivially, $\omega_\varphi > -\infty$ (with ω_φ defined as in (4.3)). Let $\gamma > -\omega_\varphi$, and choose some $s > 0$ such that $s^{-1}\ln(\varphi(s)) < \gamma$. Let $ms \le t < (m+1)s$, where $m = 0, 1, 2, \ldots$. Then

$$\begin{aligned} -\omega_\varphi \le \frac{\ln(\varphi(t))}{t} &\le \frac{ms}{t}\frac{\ln(\varphi(s))}{s} + \frac{\ln(\varphi(t-ms))}{t} \\ &< \frac{ms\gamma}{t} + \frac{\ln(\varphi(t-ms))}{t}. \end{aligned}$$

Letting $t \to \infty$ we get

$$\limsup_{t\to\infty} \frac{\ln(\varphi(t))}{t} \le \gamma,$$

and this shows that $\lim_{t\to\infty} \frac{1}{t}\ln(\varphi(t)) = -\omega_\varphi$. A completely similar argument gives

$$\lim_{t\to-\infty} \frac{\ln(\varphi(t))}{t} = -\alpha_\varphi.$$

The inequality $\omega_\varphi \le \alpha_\varphi$ follows from the fact that $\varphi(-t)\varphi(t) \ge 1$, which implies

$$\frac{\ln(\varphi(-t))}{-t} \le \frac{\ln(\varphi(t))}{t}, \quad t > 0.$$

We leave it to the reader to check that (4.3), together with the inequality $\omega_\varphi \le \alpha_\varphi$, implies (4.2). □

The next few results will make it obvious that the numbers α_φ and ω_φ are of crucial importance when one wishes to invert a measure in $M(\mathbf{R}^+;\varphi)$ or in $M(\mathbf{R};\varphi)$. Note that if $\nu \in M(\mathbf{R}^+;\varphi)$ then $\hat{\nu}(z)$ is well defined for $\Re z \ge \omega_\varphi$ (see (4.2)). Analogously, if $\nu \in M(\mathbf{R};\varphi)$, then $\hat{\nu}(z)$ is well defined for $\omega_\varphi \le \Re z \le \alpha_\varphi$.

The following (scalar) result is basic. Here the subindices d and s denote, respectively, the discrete part and the singular part of the measure in question. The unit point mass at zero is denoted by δ.

4.2 Theorem

Let φ be a submultiplicative weight function on $\mathbf{R}$ and let $\mu \in M(\mathbf{R};\varphi;\mathbf{C})$. Suppose that

$$\hat{\mu}(z) \ne 0, \quad \omega_\varphi \le \Re z \le \alpha_\varphi, \tag{4.4}$$

and

$$\inf_{\omega_\varphi \le \Re z \le \alpha_\varphi} |\hat{\mu}_d(z)| > \|\mu_s\|_{M(\mathbf{R};\varphi)}. \tag{4.5}$$

Then there exists a measure $\nu \in M(\mathbf{R};\varphi;\mathbf{C})$ such that $\mu * \nu = \nu * \mu = \delta$.

Proof The space $M(\mathbf{R};\varphi;\mathbf{C})$ is a commutative Banach algebra, with convolution as multiplication, and with δ as unit. By a well-known result, see, e.g., Rudin [2], Theorem 11.5, there exists a measure $\nu \in M(\mathbf{R};\varphi;\mathbf{C})$ such that $\mu * \nu = \nu * \mu = \delta$, if and only if $h(\mu) \ne 0$ for every complex homomorphism h on $M(\mathbf{R};\varphi;\mathbf{C})$.

Let h be an arbitrary complex homomorphism on $M(\mathbf{R};\varphi;\mathbf{C})$. We proceed to prove that $h(\mu) \neq 0$. Define the function $\chi : \mathbf{R} \to \mathbf{C}$ by

$$\chi(t) = h(\delta_t), \quad t \in \mathbf{R}, \tag{4.6}$$

where δ_t is the unit point mass at t. Since h is a homomorphism, and $\delta_t * \delta_s = \delta_{t+s}$, we find that χ satisfies

$$\chi(t+s) = \chi(t)\chi(s), \quad t, s \in \mathbf{R}, \quad \chi(0) = 1. \tag{4.7}$$

Moreover, $|\chi(t)| \leq \|\delta_t\|_{M(\mathbf{R};\varphi)} = \varphi(t)$ for $t \in \mathbf{R}$, so χ is locally bounded. This implies that χ is locally bounded away from zero (take $s = -t$ in (4.7)). Thus, both $|\chi|$ and $1/|\chi|$ are submultiplicative and locally bounded. Consequently, one can use Lemma 4.1 to show that there exists some γ, $\omega_\varphi \leq \gamma \leq \alpha_\varphi$, such that

$$|\chi(t)| = \mathrm{e}^{-\gamma t}, \quad t \in \mathbf{R}. \tag{4.8}$$

First we consider the so-called absolutely continuous case where there exists a function $f \in L^1(\mathbf{R};\varphi;\mathbf{C})$ such that $h(f) \neq 0$ (i.e., there is a measure μ with no discrete part and no singular part, and $h(\mu) \neq 0$; we identify μ with its absolutely continuous part f). Without loss of generality, we may assume that f is a C^∞-function with compact support, because this class of functions is dense in L^1. Take any such f. Define the translation operator τ_t by $(\tau_t f)(s) = f(t+s)$. Then $\tau_{-t} f = \delta_t * f$, and hence

$$h(\tau_{-t} f) = h(\delta_t * f) = h(\delta_t)h(f) = \chi(t)h(f).$$

Thus,

$$\chi(t) = \frac{h(\tau_{-t} f)}{h(f)}, \quad t \in \mathbf{R}.$$

Due to our choice of the function f, the mapping $t \mapsto \tau_{-t} f$ is continuously differentiable from $\mathbf{R}$ into $L^1(\mathbf{R};\varphi;\mathbf{C})$ (in fact, C^∞), and this, together with the preceding formula, implies that χ is continuously differentiable. Now differentiate (4.7) to conclude that χ must be an exponential, say $\chi(t) = \mathrm{e}^{-zt}$, for some $z \in \mathbf{C}$ and all $t \in \mathbf{R}$. Because of (4.8), we must have $\Re z = \gamma$; in particular, $\omega_\varphi \leq \Re z \leq \alpha_\varphi$.

We claim that in the preceding case $h(\mu) = \hat{\mu}(z)$, i.e.,

$$h(\mu) = \int_{\mathbf{R}} \mathrm{e}^{-zs} \mu(\mathrm{d}s). \tag{4.9}$$

Let us first show that this is true if μ is absolutely continuous, i.e., $\mu(\mathrm{d}s) = g(s)\,\mathrm{d}s$ for some $g \in L^1(\mathbf{R};\varphi;\mathbf{C})$. As the dual space of $L^1(\mathbf{R};\varphi;\mathbf{C})$ is $L^\infty(\mathbf{R};1/\varphi;\mathbf{C})$, and since the restriction of h to $L^1(\mathbf{R};\varphi;\mathbf{C})$ is an element of this dual space, we must have

$$h(g) = \int_{\mathbf{R}} \xi(s)g(s)\,\mathrm{d}s, \tag{4.10}$$

for some $\xi \in L^\infty(\mathbf{R};1/\varphi;\mathbf{C})$, and all $g \in L^1(\mathbf{R};\varphi;\mathbf{C})$. In particular, this is true for all C^∞-functions g with compact support. Recall that $h(\tau_{-t} g) =$

$\chi(t)h(g) = \mathrm{e}^{-zt}h(g)$, replace g in (4.10) by $\tau_{-t}g$, and make a change of variable to get

$$\int_{\mathbf{R}} \xi(t+s)g(s)\,\mathrm{d}s = \mathrm{e}^{-zt}\int_{R} \xi(s)g(s)\,\mathrm{d}s.$$

Multiplying both sides by e^{zt}, and replacing $\mathrm{e}^{-zs}g(s)$ by $g(s)$, we observe that, in the distribution sense, $\mathrm{e}^{zt}\xi(t)$ is a constant. Thus, there exists a constant K such that $\xi(t) = K\mathrm{e}^{-zt}$ for almost all t. We conclude that $h(g) = K\hat{g}(z)$ for all $g \in L^1(\mathbf{R};\varphi;\mathbf{C})$. Clearly $K \neq 0$, because $h(f) \neq 0$, where f is the function that we used in the proof of the fact that $\chi(t) = \mathrm{e}^{-zt}$.

Now it is easy to prove that (4.9) holds. Simply use the result that we just obtained to compute

$$Kh(\mu)\hat{f}(z) = h(\mu)h(f) = h(\mu * f) = K(\widehat{\mu * f})(z) = K\hat{\mu}(z)\hat{f}(z),$$

and divide by $K\hat{f}(z)$.

The conclusion of the preceding discussion is that, in the absolutely continuous case, $h(\mu) = \hat{\mu}(z)$ for some z with $\omega_\varphi \le \Re z \le \alpha_\varphi$. Hence, it follows from (4.4) that in this case $h(\mu) \neq 0$.

Next, consider the case where $h(f) = 0$ for every $f \in L^1(\mathbf{R};\varphi;\mathbf{C})$ (i.e., $h(\mu) = 0$ whenever $\mu = \mu_a$; here μ_a is the absolutely continuous part of μ). Then, in particular, $h(\mu_a) = 0$. Thus, by linearity, $h(\mu) = h(\mu_d) + h(\mu_s)$ (where μ_d and μ_s are the discrete and the singular part of μ, respectively). Suppose we can prove that

$$h(\mu_d) \in \overline{\{\, \hat{\mu}_d(z) \mid \omega_\varphi \le \Re z \le \alpha_\varphi \,\}}. \tag{4.11}$$

Then, by (4.5) and the fact that $|h(\mu_s)| \le \|\mu_s\|_{M(\mathbf{R};\varphi)}$,

$$\begin{aligned}|h(\mu)| &= |h(\mu_d) + h(\mu_s)| \ge |h(\mu_d)| - \|\mu_s\|_{M(\mathbf{R};\varphi)}\\ &\ge \inf_{\omega_\varphi \le \Re z \le \alpha_\varphi} |\hat{\mu}_d(z)| - \|\mu_s\|_{M(\mathbf{R};\varphi)} > 0.\end{aligned}$$

We conclude that if we can establish (4.11) then the entire proof is complete.

Let us make the immediate observation that it suffices to prove (4.11) in the case where μ has no absolutely continuous part and no singular part (these parts do not appear in (4.11)). Moreover, it suffices to consider the case where μ is supported on a finite set, say

$$\mu = \sum_{j=1}^{k} \alpha_j \delta_{s_j}, \tag{4.12}$$

because every discrete measure can be approximated by a measure of this type in the norm of $M(\mathbf{R};\varphi;\mathbf{C})$.

We claim that the function χ defined in (4.6) has the following property:

for every finite set $\{\, s_1, s_2, \ldots, s_k \,\}$ *and every* $\epsilon > 0$*, there is a* $\xi \in \mathbf{R}$ *such that* $|\chi(t) - \mathrm{e}^{-(\gamma + \mathrm{i}\xi)t}| \le \epsilon$ *for* $t = s_1, \ldots, s_k$. (4.13)

Here γ is the constant in (4.8). Suppose that this is true. By (4.6), (4.12), and the linearity of h,

$$h(\mu) = \sum_{j=1}^{k} \alpha_j \chi(s_j).$$

On the other hand,

$$\hat{\mu}(\gamma + \mathrm{i}\xi) = \sum_{j=1}^{k} \alpha_j \mathrm{e}^{-(\gamma+\mathrm{i}\xi)s_j}.$$

Thus, if ξ has been chosen as in (4.13), then

$$|h(\mu) - \hat{\mu}(\gamma + \mathrm{i}\xi)| \leq \epsilon \sum_{j=1}^{k} |\alpha_j|.$$

As ϵ can be made arbitrarily small, this proves (4.11). Thus, to complete the proof it suffices to show that (4.13) holds.

Let $t_1, \ldots, t_n$ be a rationally independent basis for $s_1, \ldots, s_k$. Without loss of generality, choose the numbers t_m in such a way that

$$s_j = \sum_{m=1}^{n} p_{j,m} t_m, \tag{4.14}$$

where each $p_{j,m}$ is an integer. By (4.8), we may set $\chi(t_m) = \mathrm{e}^{-(\gamma+\mathrm{i}\xi_m)t_m}$ for some real numbers ξ_m. By (4.14) and (4.7),

$$\begin{aligned} \chi(s_j) = \chi\Big(\sum_{m=1}^{n} p_{j,m} t_m\Big) &= \prod_{m=1}^{n} \chi(t_m)^{p_{j,m}} \\ &= \mathrm{e}^{-\gamma s_j - \mathrm{i}\sum_{m=1}^{n} p_{j,m}\xi_m t_m}. \end{aligned}$$

According to Kronecker's theorem (see Hille and Phillips [1], Theorem 4.17.5, p. 146), given $\eta > 0$, there exists a $\xi \in \mathbf{R}$ and integers $q_1, \ldots, q_n$ such that $|\xi t_m - \xi_m t_m - 2\pi q_m| \leq \eta$ for $m = 1, \ldots, n$ (in other words, modulo 2π, $|(\xi - \xi_m)t_m| \leq \eta$). But this means that

$$\begin{aligned} \big|\chi(s_j) - \mathrm{e}^{-(\gamma+\mathrm{i}\xi)s_j}\big| &= \mathrm{e}^{-\gamma s_j}\Big|\mathrm{e}^{-\mathrm{i}\sum_{m=1}^{n} p_{j,m}(\xi_m - \xi)t_m} - 1\Big| \\ &\leq \mathrm{e}^{-\gamma s_j} \sum_{m=1}^{n} |p_{j,m}|\eta. \end{aligned}$$

Hence, choosing $\eta = \epsilon/C$, where $C = \max_{1\leq j\leq k} \mathrm{e}^{-\gamma s_j} \sum_{m=1}^{n} |p_{j,m}|$, we get (4.13). □

For $\nu \in M(\mathbf{R}; \varphi; \mathbf{C}^{n\times n})$ we let $\det[\nu]$ be defined as the scalar measure that one obtains after performing the standard operations by which the determinant of a matrix is computed, but with multiplications replaced by convolutions. Clearly, $\det[\nu] \in M(\mathbf{R}; \varphi; \mathbf{C})$. Analogously, from $\nu \in M(\mathbf{R}; \varphi; \mathbf{C}^{n\times n})$ we obtain $\mathrm{adj}[\nu] \in M(\mathbf{R}; \varphi; \mathbf{C}^{n\times n})$, by doing the usual operations by which the adjoint of a matrix is computed, but with multiplication replaced by convolution. Routine calculations show that $\det[\hat{\nu}(z)]$

(here, of course, one uses ordinary multiplication when computing the determinant) is the transform of $\det[\nu]$, and that $\operatorname{adj}[\hat{\nu}(z)]$ is the transform of $\operatorname{adj}[\nu]$.

Making use of Theorem 4.2, we can obtain results of Paley–Wiener type in weighted spaces (observe that we give only sufficient, and not necessary, conditions; this is contrary to what we did in Chapters 2 and 3).

4.3 The Half Line Gel'fand Theorem
Let φ be a submultiplicative weight function, and let $\mu \in M(\mathbf{R}^+;\varphi;\mathbf{C}^{n\times n})$ satisfy

$$\det[I+\hat{\mu}(z)] \neq 0, \quad \Re z \geq \omega_\varphi, \tag{4.15}$$

and

$$\inf_{\Re z \geq \omega_\varphi} \left|\det[I+\widehat{\mu_d}(z)]\right| > \left\|\left(\det[\delta I+\mu]\right)_s\right\|_{M(\mathbf{R}^+;\varphi)}. \tag{4.16}$$

Then the resolvent ρ of μ belongs to $M(\mathbf{R}^+;\varphi;\mathbf{C}^{n\times n})$.

Observe that (4.16) implies that $\det[I+\mu(\{0\})] \neq 0$ (because $\mu(\{0\}) = \lim_{x\to+\infty}\widehat{\mu_d}(x)$) and hence the existence of a unique resolvent $\rho \in M_{\text{loc}}(\mathbf{R}^+;\mathbf{C}^{n\times n})$ is guaranteed; see Theorem 1.5. Also observe that one may ignore the absolutely continuous part in the computation of $(\det[\delta I+\mu])_s$, since $(\det[\delta I+\mu])_s = (\det[\delta I+\mu_d+\mu_s])_s$.

4.4 The Whole Line Gel'fand Theorem
Let φ be a submultiplicative weight function, and let $\mu \in M(\mathbf{R};\varphi;\mathbf{C}^{n\times n})$ satisfy

$$\det[I+\hat{\mu}(z)] \neq 0, \quad \omega_\varphi \leq \Re z \leq \alpha_\varphi, \tag{4.17}$$

and

$$\inf_{\omega_\varphi \leq \Re z \leq \alpha_\varphi} \left|\det[I+\widehat{\mu_d}(z)]\right| > \|(\det[\delta I+\mu])_s\|_{M(\mathbf{R};\varphi)}. \tag{4.18}$$

Then there exists a solution $\rho \in M(\mathbf{R};\varphi;\mathbf{C}^{n\times n})$ of the equations $\rho+\mu*\rho = \rho+\rho*\mu = \mu$. This solution is unique in $M(\mathbf{R};\varphi;\mathbf{C}^{n\times n})$.

Theorem 4.3 follows from Theorem 4.4 in the same fashion as Theorem **2**.4.1 follows from Theorem **2**.4.3. We omit the details.

Proof of Theorem 4.4 Recall that for $\nu \in M(\mathbf{R};\varphi;\mathbf{C}^{n\times n})$ the inverse of the matrix $\hat{\nu}(z)$ (assuming $\det\hat{\nu} \neq 0$) is given by

$$[\hat{\nu}(z)]^{-1} = \frac{1}{\det[\hat{\nu}(z)]}\operatorname{adj}[\hat{\nu}(z)]. \tag{4.19}$$

Since the Laplace transform of $(\det[\delta I+\mu])_d$ is equal to the Laplace transform of $\det[\delta I+\mu_d]$ which is equal to $\det[I+\widehat{\mu_d}(z)]$ it follows from (4.17) and (4.18) that Theorem 4.2 may be applied to yield the existence of $\psi \in M(\mathbf{R};\varphi;\mathbf{C})$ such that

$$\hat{\psi}(z) = \left(\det\left[I+\hat{\mu}(z)\right]\right)^{-1}, \quad \omega_\varphi \leq \Re z \leq \alpha_\varphi.$$

Consequently, by (4.19),

$$[I + \hat{\mu}(z)]^{-1} = \hat{\psi}(z)\,\mathrm{adj}[I + \hat{\mu}(z)], \quad \omega_\varphi \le \Re z \le \alpha_\varphi.$$

Define $\rho = \delta I - \psi * \mathrm{adj}[\delta I + \mu]$. Then $\rho \in M(\mathbf{R}; \varphi; \mathbf{C}^{n\times n})$ and

$$\hat{\rho}(z) = I - \hat{\psi}(z)\,\mathrm{adj}[I + \hat{\mu}(z)] = I - [I + \hat{\mu}(z)]^{-1},$$

from which we conclude that $\rho + \mu * \rho = \rho + \rho * \mu = \mu$.

The argument that proves uniqueness can be copied, e.g., from the proof of Theorem **2**.3.1 (replace k and r by μ and ρ, respectively), or one can use Lemma **9**.3.3. □

4.5 Definition

The measure ρ in Theorem 4.4 is called the whole line measure resolvent of μ corresponding to the submultiplicative weight function φ.

The reader should be aware of the fact that a measure μ may have many different whole line resolvents, corresponding to different submultiplicative weight functions φ. See Exercises 10 and 11.

4.6 Remark

For the conclusions of Theorems 4.2 and 4.4 to hold, some condition on the singular part is necessary. This is shown by the fact that there exists a measure $\mu \in M(\mathbf{R}; \mathbf{C})$, with compact support, satisfying $\inf_{\Re z=0}|1 + \hat{\mu}(z)| \ge 1$, such that no measure $\rho \in M(\mathbf{R}; \mathbf{C})$ exists for which $\rho + \rho * \mu = \mu$. The following (brief outline of a) proof of this fact is adapted from Gel′fand, Raikov, and Shilov [1], Theorem 3, p. 191.

Let P be a perfect set in $\mathbf{R}$ with measure zero, whose points are linearly independent (with respect to the integers), and let $(k)P$ have measure zero for all $k \ge 1$, where $(k)P = \{\sum_{j=1}^{k} p_j \mid p_j \in P, \quad j = 1, \dots, k\}$. Let $\mathcal{F}$ be the collection of all countable unions of sets of the form $(k)F - t$, where F is a countable union of closed subsets of P, k is a positive integer, and t is a real number.

Let $M_{\mathcal{F}^c}$ be the set of all measures $\psi \in M(\mathbf{R}; \mathbf{C})$ satisfying $\psi(A) = 0$ for all $A \in \mathcal{F}$, and let $M_{\mathcal{F}}$ denote the set of all measures $\psi \in M(\mathbf{R}; \mathbf{C})$ whose support is contained in some set in $\mathcal{F}$. It is possible to show that $M_{\mathcal{F}^c}$ is an ideal, and that $M_{\mathcal{F}}$ is a subalgebra of $M(\mathbf{R}; \mathbf{C})$. Furthermore, one can show that each measure $\psi \in M(\mathbf{R}; \mathbf{C})$ can be represented as $\psi = \psi_{\mathcal{F}} + \psi_{\mathcal{F}^c}$ where $\psi_{\mathcal{F}} \in M_{\mathcal{F}}$ and $\psi_{\mathcal{F}^c} \in M_{\mathcal{F}^c}$.

Next, let ν be a continuous nonzero measure with compact support in P. Then there must exist a number ω such that $\int_R \mathrm{e}^{-\mathrm{i}\omega t}\nu(\mathrm{d}t) = \zeta \ne 0$. Define the measure μ by $\mu(E) = -\zeta^{-1}\nu(E) + \overline{\zeta}^{-1}\overline{\nu(-E)}$. It follows immediately that $\Re\hat{\mu}(z) = 0$ for $\Re z = 0$, and hence $\inf_{\Re z=0}|1 + \hat{\mu}(z)| \ge 1$. On the other hand, we can define a homomorphism h on $M(\mathbf{R}; \mathbf{C})$ by $h(\psi) = \int_{\mathbf{R}} \mathrm{e}^{-\mathrm{i}\omega t}\psi_{\mathcal{F}}(\mathrm{d}t)$. If we let ν^* be the measure defined by $\nu^*(E) = \overline{\nu(-E)}$, then one can prove that $\nu^*_{\mathcal{F}} = 0$. Since it follows from the definition that

$h(\nu) = \zeta$, we have $h(\delta + \mu) = 0$ and, therefore, this measure cannot be invertible in $M(\mathbf{R}; \mathbf{C})$.

If μ has no singular part, then the Gel′fand theorems above become somewhat simpler.

4.7 Corollary

Let φ be a submultiplicative weight function, and assume that the measure $\mu \in M(\mathbf{R}^+; \varphi; \mathbf{C}^{n\times n})$ has no singular part. Then the resolvent ρ of μ satisfies

$$\rho \in M(\mathbf{R}^+; \varphi; \mathbf{C}^{n\times n})$$

if and only if

$$\inf_{\Re z \geq \omega_\varphi} \left|\det[I + \hat{\mu}(z)]\right| > 0. \tag{4.20}$$

This corollary can be deduced from the following result.

4.8 Corollary

Let φ be a submultiplicative weight function, and assume that the measure $\mu \in M(\mathbf{R}; \varphi; \mathbf{C}^{n\times n})$ has no singular part. Then there is a measure

$$\rho \in M(\mathbf{R}; \varphi; \mathbf{C}^{n\times n})$$

satisfying the two equations $\rho + \mu * \rho = \rho + \rho * \mu = \mu$ if and only if

$$\inf_{\omega_\varphi \leq \Re z \leq \alpha_\varphi} \left|\det[I + \hat{\mu}(z)]\right| > 0. \tag{4.21}$$

This measure ρ is unique in $M(\mathbf{R}; \varphi; \mathbf{C}^{n\times n})$.

Proof Suppose that $\rho \in M(\mathbf{R}; \varphi; \mathbf{C}^{n\times n})$ satisfies $\rho + \mu * \rho = \rho + \rho * \mu = \mu$. Then

$$[I + \hat{\mu}(z)][I - \hat{\rho}(z)] = I, \quad \omega_\varphi \leq \Re z \leq \alpha_\varphi,$$

and (4.21) certainly holds.

To prove the converse, note that, since μ has no singular part,

$$\det[I + \hat{\mu}(z)] = \det[I + \hat{\mu}_a(z) + \hat{\mu}_d(z)].$$

By the Riemann–Lebesgue lemma (see the proof of Theorem **2**.2.8(i) given in Section **2**.7), $\hat{\mu}_a(z) \to 0$ as $|\Im z| \to \infty$, uniformly for $\omega_\varphi \leq \Re z \leq \alpha_\varphi$, and therefore (4.21) implies that there exists $K < \infty$ such that

$$\inf_{\omega_\varphi \leq \Re z \leq \alpha_\varphi, |z| \geq K} |\det[I + \hat{\mu}_d(z)]| > 0.$$

The Fourier transform of a finite, discrete measure is almost-periodic (it can be uniformly approximated by a finite sum of periodic functions), and, consequently, $\hat{\mu}_d$ is almost-periodic along any line parallel to the imaginary axis. Hence the relation above must hold without the condition $|z| \geq K$. Now apply Theorem 4.4. □

Let us next formulate some easy consequences of Theorems 4.3 and 4.4 for the solutions of (1.1) and (1.2). We begin by considering the case $\eta \equiv \varphi \equiv 1$. Then $\alpha_\varphi = \omega_\varphi = 0$.

4.9 Theorem

Let $\mu \in M(\mathbf{R}^+; \mathbf{C}^{n\times n})$, and assume that

$$\det[I + \hat{\mu}(z)] \neq 0, \quad \Re z \geq 0, \tag{4.22}$$

and

$$\inf_{\Re z \geq 0} \left|\det[I + \hat{\mu}_d(z)]\right| > \|(\det[\delta I + \mu])_s\|_{M(\mathbf{R}^+)}. \tag{4.23}$$

Let $f \in V$, where V is one of the following spaces:

(1) $L^p(\mathbf{R}^+; \mathbf{C}^n)$, $p \in [1, \infty]$, $L^\infty_\ell(\mathbf{R}^+; \mathbf{C}^n)$, *or* $L^\infty_0(\mathbf{R}^+; \mathbf{C}^n)$;
(2) $B^\infty(\mathbf{R}^+; \mathbf{C}^n)$, $B^\infty_\ell(\mathbf{R}^+; \mathbf{C}^n)$, *or* $B^\infty_0(\mathbf{R}^+; \mathbf{C}^n)$;
(3) $BL^p(\mathbf{R}^+; \mathbf{C}^n)$, *or* $BL^p_0(\mathbf{R}^+; \mathbf{C}^n)$, $p \in [1, \infty]$;
(4) $BV(\mathbf{R}^+; \mathbf{C}^n)$;
(5) $BBV(\mathbf{R}^+; \mathbf{C}^n)$, *or* $BBV_0(\mathbf{R}^+; \mathbf{C}^n)$;
(6) $L^p(\mathcal{T}_S; \mathbf{C}^n) \oplus BL^p_0(\mathbf{R}^+; \mathbf{C}^n)$, $p \in [1, \infty]$, $B^\infty(\mathcal{T}_S; \mathbf{C}^n) \oplus B^\infty_0(\mathbf{R}^+; \mathbf{C}^n)$, *or* $BV(\mathcal{T}_S; \mathbf{C}^n) \oplus BBV_0(\mathbf{R}^+; \mathbf{C}^n)$, *where* $S > 0$;
(7) $AP(\mathbf{R}^+; \mathbf{C}^n) \oplus B^\infty_0(\mathbf{R}^+; \mathbf{C}^n)$.

Then the solution x of (1.1) belongs to V.

For the proof, combine Theorems **3**.6.1 (take $J = \mathbf{R}^+$) with Theorems 1.7 and 4.3.

Observe that if μ has no singular part then (4.22) and (4.23) can be combined into one single condition, namely

$$\inf_{\Re z \geq 0} \left|\det[I + \hat{\mu}(z)]\right| > 0$$

(cf. Corollary 4.7). Similar observations apply to Theorems 4.10–4.12.

Concerning solutions of (1.2) we have the following result.

4.10 Theorem

Let $\mu \in M(\mathbf{R}; \mathbf{C}^{n\times n})$ satisfy

$$\det[I + \hat{\mu}(z)] \neq 0, \quad \Re z = 0,$$

and

$$\inf_{\Re z = 0} \left|\det[I + \widehat{\mu_d}]\right| > \|(\det[\delta I + \mu])_s\|_{M(\mathbf{R})}.$$

Let $f \in V$, where V is one of the spaces (1)–(7) in Theorem 4.9 with $\mathbf{R}^+$ replaced by $\mathbf{R}$, or one of the following spaces:

(8) $BC(\mathbf{R}; \mathbf{C}^n)$, $BUC(\mathbf{R}; \mathbf{C}^n)$, $BC_\ell(\mathbf{R}; \mathbf{C}^n)$, *or* $BC_0(\mathbf{R}; \mathbf{C}^n)$;
(9) $AP(\mathbf{R}; \mathbf{C}^n)$;
(10) $C(\mathcal{T}_S; \mathbf{C}^n) \oplus BC_0(\mathbf{R}; \mathbf{C}^n)$, *where* $S > 0$;
(11) $AP(\mathbf{R}; \mathbf{C}^n) \oplus BC_0(\mathbf{R}; \mathbf{C}^n)$.

Then (1.2) has a unique solution $x \in V$, given by

$$x = f - \rho * f, \tag{4.24}$$

where ρ is the whole line resolvent of μ.

For the proof one uses Theorem **3**.6.1 (with $J = \mathbf{R}$) and Theorem 4.4. The corresponding weighted results can be stated as follows.

4.11 Theorem
Let η be a regular weight function on $\mathbf{R}^+$, dominated by the submultiplicative weight function φ. Let $\mu \in M(\mathbf{R}^+;\varphi;\mathbf{C}^{n\times n})$ and suppose that

$$\det[I + \hat{\mu}(z)] \neq 0, \quad \Re z \geq \omega_\varphi,$$

and

$$\inf_{\Re z \geq \omega_\varphi} \left|\det[I + \widehat{\mu_d}(z)]\right| > \left\|\left(\det[\delta I + \mu]\right)_s\right\|_{M(\mathbf{R}^+;\varphi)}.$$

Let $f \in V$, where V is one of the following spaces:
(1) $L^p(\mathbf{R}^+;\eta;\mathbf{C}^n)$, $p \in [1,\infty]$, *or* $L_0^\infty(\mathbf{R}^+;\eta;\mathbf{C}^n)$;
(2) $B^\infty(\mathbf{R}^+;\eta;\mathbf{C}^n)$, *or* $B_0^\infty(\mathbf{R}^+;\eta;\mathbf{C}^n)$;
(3) $BL^p(\mathbf{R}^+;\eta;\mathbf{C}^n)$, *or* $BL_0^p(\mathbf{R}^+;\eta;\mathbf{C}^n)$, $p \in [1,\infty]$;
(4) $BV(\mathbf{R}^+;\eta;\mathbf{C}^n)$;
(5) $BBV(\mathbf{R}^+;\eta;\mathbf{C}^n)$, *or* $BBV_0(\mathbf{R}^+;\eta;\mathbf{C}^n)$.
Then the solution x of (1.1) belongs to V.

Proof Recall Theorems 1.7, 3.5, and 4.3. □

4.12 Theorem
Let η be a regular weight function on $\mathbf{R}$, dominated by the submultiplicative weight function φ. Let $\mu \in M(\mathbf{R};\varphi;\mathbf{C}^{n\times n})$ satisfy

$$\det[I + \hat{\mu}(z)] \neq 0, \quad \omega_\varphi \leq \Re z \leq \alpha_\varphi,$$

and

$$\inf_{\omega_\varphi \leq \Re z \leq \alpha_\varphi} \left|\det[I + \widehat{\mu_d}(z)]\right| > \left\|\left(\det[\delta I + \mu]\right)_s\right\|_{M(\mathbf{R};\varphi)}.$$

Let $f \in V$, where V is one of the spaces (1)–(5) *in Theorem 4.11 with $\mathbf{R}^+$ replaced by $\mathbf{R}$, or*
(6) $BC(\mathbf{R};\eta;\mathbf{C}^n)$, $BUC(\mathbf{R};\eta;\mathbf{C}^n)$, *or* $BC_0(\mathbf{R};\eta;\mathbf{C}^n)$.
Then (1.2) has a unique solution $x \in V$, given by (4.24).

Proof Use Theorems 3.5 and 4.4. □

We proceed to the differential resolvent of μ, and to the integrodifferential equations (1.3) and (1.4).

Recall that the existence of the differential resolvent r of $\mu \in M_{\text{loc}}(\mathbf{R}^+)$ (i.e., a locally absolutely continuous function r defined on $\mathbf{R}^+$, satisfying $r(0) = I$ and $r' + r * \mu = r' + \mu * r = 0$) was established in Theorem **3**.3.1. In addition, in Chapter 3 we obtained Paley–Wiener type results for the size of r, in the case where $\mu \in M(\mathbf{R}^+;\varphi)$, with $\varphi(t) \equiv 1$. We now consider arbitrary weight functions.

4.13 Theorem
Let φ be a submultiplicative weight function, and let $\mu \in M(\mathbf{R}^+;\varphi;\mathbf{C}^{n\times n})$. Then the differential resolvent r of μ satisfies

$$r \in L^1(\mathbf{R}^+;\varphi;\mathbf{C}^{n\times n})$$

if and only if

$$\det[zI+\hat{\mu}(z)] \neq 0, \quad \Re z \geq \omega_\varphi.$$

Moreover, if $r \in L^1(\mathbf{R}^+;\varphi;\mathbf{C}^{n\times n})$, then $r' \in L^1(\mathbf{R}^+;\varphi;\mathbf{C}^{n\times n})$, and r' is equal a.e. to a function in $BV(\mathbf{R}^+;\varphi;\mathbf{C}^{n\times n})$.

The whole line case where $\mu \in M(\mathbf{R};\varphi;\mathbf{C}^{n\times n})$, with $\varphi \equiv 1$, was considered in Theorem **3**.3.7. For arbitrary φ the corresponding result is the following.

4.14 Theorem
Let φ be a submultiplicative weight function, and let $\mu \in M(\mathbf{R};\varphi;\mathbf{C}^{n\times n})$. Then

> *the equations $r' + \mu * r = r' + r * \mu = 0$ have a solution $r \in L^1(\mathbf{R};\varphi;\mathbf{C}^{n\times n})$, which is locally absolutely continuous, except for a jump discontinuity at zero with size $r(0+) - r(0-) = I$,*

if and only if

$$\det[zI+\hat{\mu}(z)] \neq 0, \quad \omega_\varphi \leq \Re z \leq \alpha_\varphi. \tag{4.25}$$

This solution r is unique in $L^1(\mathbf{R};\varphi;\mathbf{C}^{n\times n})$, and $r' \in L^1(\mathbf{R};\varphi;\mathbf{C}^{n\times n})$. Moreover, r' is equal a.e. to a function in $BV(\mathbf{R};\varphi;\mathbf{C}^{n\times n})$.

Again note that it suffices to prove Theorem 4.14.

4.15 Definition
The function r in Theorem 4.14 is called the whole line differential resolvent of μ corresponding to the submultiplicative weight function φ.

We ask the reader to bear in mind that a measure μ may have many whole line differential resolvents, corresponding to different functions φ. See Exercise 10.

Proof of Theorem 4.14 To demonstrate the necessity of (4.25), one assumes that there exists a function r with the given properties. Working as in the proof of Theorem **3**.3.7, one obtains

$$\hat{r}(z) = [zI+\hat{\mu}(z)]^{-1}, \quad \omega_\varphi \leq \Re z \leq \alpha_\varphi.$$

To prove sufficiency, suppose that

$$\det[zI+\hat{\mu}(z)] \neq 0, \quad \omega_\varphi \leq \Re z \leq \alpha_\varphi.$$

Define $\mathbf{e}(t) = \mathrm{e}^{-(1-\omega_\varphi)t}$ for $t \in \mathbf{R}^+$ and $\mathbf{e}(t) = 0$ for $t \in \mathbf{R}^-$. Then $\mathbf{e} \in L^1(\mathbf{R};\varphi;\mathbf{C})$ and $\hat{\mathbf{e}}(z) = [1-\omega_\varphi+z]^{-1}$ for $\Re z \geq \omega_\varphi$. As in the proof of Theorem **3**.3.7, one finds that

$$[zI+\hat{\mu}(z)]^{-1} = \hat{\mathbf{e}}(z)[I+\hat{d}(z)]^{-1},$$

where $d = (-1+\omega_\varphi)\mathbf{e}I + \mathbf{e} * \mu$. The remainder of the proof parallels the proof of Theorem 3.3.7. □

Applying Theorems 4.13 and 4.14 to equations (1.3) and (1.4), we get the following theorems.

4.16 Theorem
Let η be a regular weight function on $\mathbf{R}^+$, dominated by the submultiplicative weight function φ. Let $\mu \in M(\mathbf{R}^+;\varphi;\mathbf{C}^{n\times n})$ and suppose that

$$\det[zI + \hat{\mu}(z)] \neq 0, \quad \Re z \geq \omega_\varphi.$$

Let $f \in V$, where V is one of the following spaces:
(1) $L^p(\mathbf{R}^+;\eta;\mathbf{C}^n)$, $p \in [1,\infty]$, *or* $L_0^\infty(\mathbf{R}^+;\eta;\mathbf{C}^n)$;
(2) $B^\infty(\mathbf{R}^+;\eta;\mathbf{C}^n)$, *or* $B_0^\infty(\mathbf{R}^+;\eta;\mathbf{C}^n)$;
(3) $BC(\mathbf{R}^+;\eta;\mathbf{C}^n)$, $BUC(\mathbf{R}^+;\eta;\mathbf{C}^n)$, *or* $BC_0(\mathbf{R}^+;\eta;\mathbf{C}^n)$;
(4) $BL^p(\mathbf{R}^+;\eta;\mathbf{C}^n)$, *or* $BL_0^p(\mathbf{R}^+;\eta;\mathbf{C}^n)$, $p \in [1,\infty]$;
(5) $BV(\mathbf{R}^+;\eta;\mathbf{C}^n)$;
(6) $BBV(\mathbf{R}^+;\eta;\mathbf{C}^n)$, *or* $BBV_0(\mathbf{R}^+;\eta;\mathbf{C}^n)$.
Then (1.3) has a unique locally absolutely continuous solution $x \in V$, given by $x(t) = r(t)x_0 + (r*f)(t)$, $t \in \mathbf{R}^+$. If V contains some discontinuous function, and if $f \in V$ then x' is equal a.e. to a function in V. If f is continuous and μ has no point masses in $(0,\infty)$ or $f(0)=0$, then $x' \in V$.

4.17 Theorem
Let η be a regular weight function on $\mathbf{R}$, dominated by the submultiplicative weight function φ. Let $\mu \in M(\mathbf{R};\varphi;\mathbf{C}^{n\times n})$ and suppose that

$$\det[zI + \hat{\mu}(z)] \neq 0, \quad \omega_\varphi \leq \Re z \leq \alpha_\varphi.$$

Let $f \in V$, where V is one of the spaces (1)–(6) in Theorem 4.16 with $\mathbf{R}^+$ replaced by $\mathbf{R}$. Then (1.4) has a unique locally absolutely continuous solution $x \in V$, given by $x = r * f$, where $r \in L^1(\mathbf{R};\varphi;\mathbf{C}^{n\times n})$ is the whole line differential resolvent of μ. Moreover, x' is equal a.e. to a function in V, and if, in addition, f is continuous then $x' \in V$.

The proofs of these two theorems are left to the reader.

5. Appendix: Proofs of some Convolution Theorems

Proof of Theorem 1.2 Without loss of generality, let $J = \mathbf{R}$.

(i) The fact that $\mu * \nu$ is a measure, i.e., a countably additive set function, follows directly from Lebesgue's dominated convergence theorem and from

(1.5), because if $E = \bigcup_{i=1}^{\infty} E_i$, with $E_i \cap E_j = \emptyset$ for $i \neq j$, then

$$\begin{aligned}(\mu * \nu)(E) &= \int_{\mathbf{R}} \int_{\mathbf{R}} \chi_E(s+t)\mu(\mathrm{d}t)\nu(\mathrm{d}s) \\ &= \lim_{j\to\infty} \sum_{i=1}^{j} \int_{\mathbf{R}} \int_{\mathbf{R}} \chi_{E_i}(s+t)\mu(\mathrm{d}t)\nu(\mathrm{d}s) \\ &= \lim_{j\to\infty} \sum_{i=1}^{j} (\mu * \nu)(E_i).\end{aligned}$$

Moreover, for each finite partition $\{E_j\}_{j=1}^N$ of $\mathbf{R}$,

$$\begin{aligned}\sum_{j=1}^{N} \bigl|(\mu * \nu)(E_j)\bigr| &= \sum_{j=1}^{N} \left| \int_{\mathbf{R}} \mu(\mathrm{d}t)\nu(E_j - t) \right| \\ &\le \sum_{j=1}^{N} \int_{\mathbf{R}} |\mu|(\mathrm{d}t) |\nu(E_j - t)| \\ &= \int_{\mathbf{R}} |\mu|(\mathrm{d}t) \sum_{j=1}^{N} |\nu(E_j - t)| \\ &\le \int_{\mathbf{R}} |\mu|(\mathrm{d}t) \|\nu\|_{M(\mathbf{R})} = \|\mu\|_{M(\mathbf{R})} \|\nu\|_{M(\mathbf{R})},\end{aligned}$$

and this proves the claim about the norm, i.e., the total variation, of $\mu * \nu$.

(ii) Let E be a Borel set, and let $t \in \mathbf{R}$. Define F to be the set $F = t - E$, and use (1.6) to get

$$((\mu * \nu) * \chi_{-F})(0) = (\mu * (\nu * \chi_{-F}))(0) = ((\mu * \chi_{-F}) * \nu)(0).$$

Here

$$\begin{aligned}(\nu * \chi_{-F})(s) &= \int_{\mathbf{R}} \nu(\mathrm{d}u)\chi_{-F}(s-u) = \int_{\mathbf{R}} \nu(\mathrm{d}u)\chi_{E-t}(s-u) \\ &= \int_{\mathbf{R}} \nu(\mathrm{d}u)\chi_E(s+t-u) = (\nu * \chi_E)(s+t).\end{aligned}$$

By using the translation operator τ_t, we can write this result in the form $\nu * \chi_{-F} = \tau_t(\nu * \chi_E)$. Similar identities hold with ν replaced by μ and by $\mu * \nu$. Hence, because translation commutes with convolution, we get

$$((\mu * \nu) * \chi_E)(t) = (\mu * (\nu * \chi_E))(t) = ((\mu * \chi_E) * \nu)(t).$$

In other words, if a is a characteristic function, then $((\mu * \nu) * a)(t) = (\mu * (\nu * a))(t) = ((\mu * a) * \nu)(t)$, for all $t \in \mathbf{R}$. Every simple function is a linear combination of characteristic functions (with scalar, vector or matrix coefficients), and, therefore, the two associative properties mentioned in (ii) are valid for simple functions. From this fact one immediately concludes that the same is true for arbitrary B^∞-functions, because these functions can be uniformly approximated by simple functions, and Lebesgue's dominated convergence theorem applies.

(iii) By Theorem 3.6.1(iv) and Theorem 1.2(i), the listed convolutions are defined (as absolutely convergent Lebesgue integrals) a.e. To show that $((\mu * f) * \nu)(t) = (\mu * (f * \nu))(t)$ for almost all t, one simply uses Fubini's theorem. If $f \in BL_0^1(\mathbf{R})$, then it follows from (ii), and from the fact that the set of B^∞-functions with compact support is dense in $BL_0^1(\mathbf{R})$, that $((\mu * \nu) * f)(t) = (\mu * (\nu * f))(t)$ for almost all t. If $f \notin BL_0^1(\mathbf{R})$, then one uses the following approximation method. Suppose that μ and ν are both supported on a finite interval $[-m, m]$. Then $\mu * \nu$ is supported on $[-2m, 2m]$, and, for t in a finite interval $[-T, T]$, the values of both $((\mu * \nu) * f)(t)$ and $(\mu * (\nu * f))(t)$ depend only on the values of f in $[-T - 2m, T + 2m]$. This means that one may redefine f outside this interval to make it a function in $BL_0^1(\mathbf{R})$, so by the preceding argument $((\mu * \nu) * f)(t) = (\mu * (\nu * f))(t)$ for almost all $t \in [-T, T]$. This is true for all T, so we find that, in the case where μ and ν have compact support, $((\mu*\nu)*f)(t) = (\mu*(\nu*f))(t)$ for almost all t. Since the set of measures with compact support is dense in $M(\mathbf{R})$, the same must be true for arbitrary μ and ν.

(iv) By (1.6) and Theorem 1.2(ii),

$$\begin{aligned}((\mu * \nu) * \rho)(E) &= ((\mu * \nu) * (\rho * \chi_{-E}))(0) = (\mu * (\nu * (\rho * \chi_{-E})))(0)\\ &= (\mu * ((\nu * \rho) * \chi_{-E}))(0) = (\mu * (\nu * \rho))(E).\end{aligned}$$

This proves that $(\mu * \nu) * \rho = \mu * (\nu * \rho)$.

(v) This is an immediate consequence of (1.5). □

Proof of Theorem 3.4 Because of (4.2), we may without loss of generality suppose that $\varphi(t) \geq 1$ for all t (observe that the mapping $\mu(\mathrm{d}t) \mapsto \mathrm{e}^{-\omega_\varphi t}\mu(\mathrm{d}t)$ commutes with convolution).

Since we assume that $\varphi(t) \geq 1$, most of Theorem 3.4 follows directly from Corollary 1.3. The only new ingredient is the inequality (3.10). To prove this inequality, it suffices to use Fubini's theorem in the computation

$$\begin{aligned}\int_J \varphi(t)|(\mu * \nu)|(\mathrm{d}t) &\leq \int_J \varphi(t)(|\mu| * |\nu|)(\mathrm{d}t)\\ &= \int_J\int_J \varphi(t+s)|\mu|(\mathrm{d}t)|\nu|(\mathrm{d}s)\\ &\leq \int_J\int_J \varphi(t)|\mu|(\mathrm{d}t)\varphi(s)|\nu|(\mathrm{d}s)\\ &= \|\mu\|_{M(J;\varphi)}\|\nu\|_{M(J;\varphi)}.\ \square\end{aligned}$$

In the proof of Theorem 3.5 we need the following auxiliary result.

5.1 Theorem

Let J be $\mathbf{R}$ or $\mathbf{R}^+$, let η be a regular weight function on J, dominated by a submultiplicative weight function φ, let $\mu \in M(J;\varphi)$, and let $\nu \in M_{\mathrm{loc}}(J)$ satisfy $\sup_{\tau\in J}\int_{[\tau,\tau+1]}\eta(t)|\nu|(\mathrm{d}t) < \infty$. Then $\mu * \nu \in M_{\mathrm{loc}}(J)$, and $\sup_{\tau\in J}\int_{[\tau,\tau+1]}\eta(t)|(\mu * \nu)|(\mathrm{d}t) < \infty$.

Proof Without loss of generality, take $J = \mathbf{R}$.

First we show that $\mu * \nu$ is a measure in $M_{\mathrm{loc}}(\mathbf{R})$. The proof is exactly the same as that of Theorem 1.2(i), but in order to invoke the dominated convergence theorem we have to assume that E is a bounded set, observe that $\inf_{\sigma \in E} \eta(\sigma) > 0$, and, in addition, observe that for all s and t with $s + t \in E$,

$$\inf_{\sigma \in E} \eta(\sigma) \le \eta(t+s) \le \varphi(t)\eta(s);$$

hence

$$\begin{aligned} &\int_{\mathbf{R}} \int_{\mathbf{R}} \chi_E(s+t)|\mu|(\mathrm{d}t)|\nu|(\mathrm{d}s) \\ &\qquad \le \frac{1}{\inf_{\sigma \in E} \eta(\sigma)} \int_{\mathbf{R}} \varphi(t)|\mu|(\mathrm{d}t) \int_{\mathbf{R}} \chi_E(s+t)\eta(s)|\nu|(\mathrm{d}s) < \infty. \end{aligned}$$

To prove that

$$\sup_{\tau \in \mathbf{R}} \int_{[\tau,\tau+1]} \eta(t)|(\mu * \nu)|(\mathrm{d}t) \le \|\mu\|_{M(\mathbf{R};\varphi)} \sup_{\tau \in \mathbf{R}} \int_{[\tau,\tau+1]} \eta(t)|\nu|(\mathrm{d}t) < \infty,$$

one uses a similar computation, namely

$$\begin{aligned} \int_{[\tau,\tau+1]} \eta(t)|(\mu * \nu)|(\mathrm{d}t) &\le \int_{[\tau,\tau+1]} \eta(t)(|\mu| * |\nu|)(\mathrm{d}t) \\ &= \int_{\mathbf{R}} \int_{\mathbf{R}} \chi_{[\tau,\tau+1]}(s+t)\eta(s+t)|\mu|(\mathrm{d}t)|\nu|(\mathrm{d}s) \\ &\le \int_{\mathbf{R}} \varphi(t)|\mu|(\mathrm{d}t) \int_{\mathbf{R}} \chi_{[\tau,\tau+1]}(s+t)\eta(s)|\nu|(\mathrm{d}s) \\ &\le \|\mu\|_{M(\mathbf{R};\varphi)} \sup_{\tau \in \mathbf{R}} \int_{[\tau,\tau+1]} \eta(t)|\nu|(\mathrm{d}t). \ \square \end{aligned}$$

Proof of Theorem 3.5 Let us first prove (i), (ii), and (iv). Without loss of generality, take $J = \mathbf{R}$. By (3.2), we have

$$\begin{aligned} \eta(t)\big|(\mu * a)(t)\big| = \eta(t)\left|\int_{\mathbf{R}} \mu(\mathrm{d}s)a(t-s)\right| &\le \eta(t) \int_{\mathbf{R}} |\mu|(\mathrm{d}s)\big|a(t-s)\big| \\ &\le \int_{\mathbf{R}} \varphi(s)|\mu|(\mathrm{d}s)\eta(t-s)\big|a(t-s)\big|, \end{aligned}$$

and Theorem **3**.6.1 applied to $\varphi\mu$ and ηa gives the desired claims.

(iii) If $a \in BC(\mathbf{R};\eta)$ then, for each $t \in \mathbf{R}$ and $h \in \mathbf{R}$,

$$\begin{aligned} &\eta(t)\big|(\mu * a)(t+h) - (\mu * a)(t)\big| \\ &\qquad \le \int_{\mathbf{R}} \varphi(s)|\mu|(\mathrm{d}s)\eta(t-s)\big|a(t+h-s) - a(t-s)\big|, \end{aligned} \tag{5.1}$$

and it follows from this inequality, and from Lebesgue's dominated convergence theorem, that $(\mu * a)(t+h) \to (\mu * a)(t)$ as $h \to 0$. Thus we see that $\mu * a \in BC(\mathbf{R};\eta)$. The same inequality shows also that $\|\tau_h(\mu * a) - \mu * a\|_{\sup(\mathbf{R};\eta)} \le \|\mu\|_{M(\mathbf{R};\varphi)}\|\tau_h a - a\|_{\sup(\mathbf{R};\eta)}$, hence

$\mu * a \in BUC(\mathbf{R};\eta)$ whenever $a \in BUC(\mathbf{R};\eta)$. Since $BC_0 = B_0^\infty \cap BC$, the claim about BC_0 is true as well. The case where $J = \mathbf{R}^+$ is proved in a similar way.

If $a \in BV(\mathbf{R};\eta)$ or $BBV(\mathbf{R};\eta)$, then we let ν be the distribution derivative of a, and observe that $\mu * \nu$ is the distribution derivative of $\mu * a$. Thus (v) and (vi) follow from (ii), combined with Theorems 3.4 and 5.1, respectively.

The proof for the case where $V = BBV_0(\mathbf{R};\eta)$ is left to the reader. □

6. Exercises

1. Let c be a (scalar) bounded, nonnegative, and nonincreasing function on $\mathbf{R}^+$ and let γ be the distribution derivative of c. Show that γ has a resolvent $\rho \in M(\mathbf{R}^+;\mathbf{R})$. Hint: $\delta + \gamma$ has a point mass at zero that dominates the rest of γ.
2. Consider the equation $x + \mu * x = f$ on $\mathbf{R}^+$, where $\mu \in M_{\text{loc}}(\mathbf{R}^+;\mathbf{C}^{n\times n})$ satisfies $\det[I + \mu(\{0\})] \neq 0$. Let u be the (matrix) solution of the equation $u(t) + (\mu * u)(t) = I$, $t \geq 0$. Determine the relation between u and the resolvent ρ of μ, and express x in terms of u.
3. Let the scalar measure μ be $\mu(\mathrm{d}t) = \frac{1}{2}\delta_{t_k}$, for some $t_k \geq 0$. Compute the resolvent of μ.
4. Let $\mu = A\delta_0 + \lambda$, where A is a matrix and the measure λ has no point mass at zero. Assume that $\det[I+A] \neq 0$, and let ν be the resolvent of $[I+A]^{-1}\lambda$. Show that the resolvent of μ is $A[I+A]^{-1}\delta_0 + \nu[I+A]^{-1}$. Can you express the resolvent of μ in terms of the resolvent of $\lambda[I+A]^{-1}$?
5. Show that the functions given in (3.7) are submultiplicative, and determine α_φ, ω_φ (as defined in (4.3)) for these functions.
6. Show that the function η given in (3.8) is dominated by the given function φ. Try to construct another pair η, φ for which η and φ are not identical.
7. Let φ be a submultiplicative weight function on $\mathbf{R}$, and define $\eta_1(t) = 1/\varphi(-t)$ and $\eta_2(t) = 1/\varphi(t)$ for $t \in \mathbf{R}$. Show that η_1 and η_2 are regular weight functions on $\mathbf{R}$, and determine the smallest possible dominating submultiplicative functions.
8. Let η be a regular weight function on J, with $J = \mathbf{R}$ or $J = \mathbf{R}^+$. Show that $W^{1,1}(J;\eta) \subset BC_0(J;\eta) \subset BUC(J;\eta)$. (These inclusions are needed in the proof of Theorem 4.16.)
9. Let $\nu \in M(\mathbf{R};\varphi;\mathbf{C}^{n\times n})$. Convince yourself that $\det[\hat{\nu}(z)]$ is the transform of $\det[\nu]$, and that $\operatorname{adj}[\hat{\nu}(z)]$ is the transform of $\operatorname{adj}[\nu]$.
10. Define $k(t) = -\mathrm{e}^{-t}$ for $t \geq 0$, and $k(t) = 0$ for $t \leq 0$.
    (i) What is the (half line) resolvent of k?

(ii) How many different whole line resolvents does k have (corresponding to different weight functions)?
(iii) How many different whole line differential resolvents does k have (corresponding to different weight functions)?

11. Construct a kernel $k \in L^1(\mathbf{R})$ that has three different whole line resolvents (corresponding to different weight functions).
12. Prove the following lemma, and show that Theorem 4.3 follows from Theorem 4.4.

6.1 Lemma
Let $\mu \in M(\mathbf{R}; \mathbf{C}^{n\times n})$, let φ be a function that is bounded and continuous in the closed half plane $\Re z \geq 0$ and analytic in the open half plane $\Re z > 0$, and suppose that $\hat{\mu}(\mathrm{i}\omega) = \varphi(\mathrm{i}\omega)$ for all $\omega \in \mathbf{R}$. Then μ is supported on $\mathbf{R}^+$, and $\hat{\mu}(z) = \varphi(z)$ for all $z \in \mathbf{C}$ with $\Re z \geq 0$.

13. Prove Theorems 4.16 and 4.17.
14. Let $T \in (0, \infty)$, and let $J = [0, T]$.
    (i) Show that a measure $\mu \in M(J, \mathbf{C}^{n\times n})$ is invertible in the Banach algebra $M(J, \mathbf{C}^{n\times n})$ if and only if $\mu(\{0\})$ is invertible.
    (ii) Show that the maximal ideal space of $M(J, \mathbf{C})$ consists of one single point, namely the complex homomorphism $\mu \mapsto \mu(\{0\})$.
15. Prove the following theorem (see Gel'fand [1]):

6.2 Theorem
Let φ be a submultiplicative weight function on $\mathbf{R}$ and let $M(\mathbf{Z}; \varphi; \mathbf{C})$ be the set of all discrete measures in $M(\mathbf{R}; \varphi; \mathbf{C})$ supported on $\mathbf{Z}$. Then $M(\mathbf{Z}; \varphi; \mathbf{C})$ is a Banach algebra, and every complex homomorphism h on $M(\mathbf{Z}; \varphi; \mathbf{C})$ is given by

$$h(\mu) = \hat{\mu}(z),$$

for some $z \in \mathbf{C}$ with $\omega_\varphi \leq \Re z \leq \alpha_\varphi$ and $0 \leq \Im z < 2\pi$.

16. If $k \in L^1(\mathbf{R}; \mathbf{C}^{n\times n})$ and $\det[I + \hat{k}(z)] \neq 0$ for $\Re z = 0$, then we know that the resolvent r of k belongs to $L^1(\mathbf{R}; \mathbf{C}^{n\times n})$. Show that it is impossible to find a bound on $\|r\|_{L^1(\mathbf{R})}$ that depends on $\|k\|_{L^1(\mathbf{R})}$ and $\sup_{\Re z=0}|[I + \hat{k}(z)]^{-1}|$ only. Hint: Use the result from Gel'fand, Raikov, and Shilov [1], Theorem 3, p. 191, that says that there exists a measure $\mu \in M(\mathbf{R}; \mathbf{C})$ satisfying $\Re\hat{\mu}(z) = 0$ for $\Re z = 0$, but there is no measure $\rho \in M(\mathbf{R}; \mathbf{C})$ satisfying $\rho + \rho * \mu = \mu$.
17. Let μ, $\nu \in M_{\mathrm{loc}}(\mathbf{R}^+; \mathbf{C}^{n\times n})$, and f, $g \in L^1_{\mathrm{loc}}(\mathbf{R}^+; \mathbf{C}^n)$, and suppose that $\det[I + \mu(\{0\})] \neq 0$. Show that the equation

$$\begin{aligned} x(t) + (\mu * x)(t) + g(t) &= y(t), \quad t \in \mathbf{R}^+, \\ y'(t) + (\nu * x)(t) &= f(t), \quad t \in \mathbf{R}^+, \\ y(0) &= y_0, \end{aligned}$$

has a unique pair of solutions (x, y) in the sense that there exist unique functions $x \in L^1_{\mathrm{loc}}(\mathbf{R}^+; \mathbf{C}^n)$ and $y \in AC_{\mathrm{loc}}(\mathbf{R}^+; \mathbf{C}^n)$ such that the first equation holds for all $t \in \mathbf{R}^+$, and the second equation holds for almost all $t \in \mathbf{R}^+$. Hint: x and y are given a.e. by the variation of constants formula

$$x(t) = r(t)y_0 - g(t) + (r * f)(t) - (r' * g)(t),$$
$$y(t) = \big(r(t) + (\mu * r)(t)\big)y_0 + \big((r + \mu * r) * f\big)(t) + (\nu * r * g)(t),$$

where r is the solution of the equations

$$(r + r * \mu)' + r * \nu = (r + \mu * r)' + \nu * r = 0, \quad t \in \mathbf{R}^+; \quad r(0) = I.$$

(This equation is of so-called *neutral* type.)

7. Comments

Section 1:

For the convolution of two measures see, e.g., Hewitt and Ross [1], Secs. 19–20, Rudin [2], p. 219, or Rudin [3], p. 175.

Measure resolvents are more or less briefly considered in several sources, e.g. in Staffans [25], p. 1038.

Sections 2 and 3:

The literature on weighted spaces is abundant. Weighted L^1-spaces on $\mathbf{R}$ are analysed in Gel′fand [1] and Gel′fand, Raikov, and Shilov [1], pp. 113–116. For $M(\mathbf{R}^+;\varphi)$, see Hille and Phillips [1], pp. 141–150. Additional, partly more technical, references are the following: Coleman and Mizel [1] and [2], Hale [2], Hale and Kato [1], Kappel and Schappacher [2], Leitman and Mizel [1], Lima [1], Naito [2] and [3], Staffans [16] (where the major part of Lemma 1.2 can be found), [18], and [25].

Section 4:

A nonweighted version of Theorem 4.2 was proved in the thirties by Wiener and Pitt [1], and a few years later Lemma 4.1 and Theorem 4.2 were proved in Gel′fand [1]. Our proofs of Lemma 4.1 and Theorem 4.2 are similar to those given in Hille and Phillips [1], p. 244 and pp. 144–150. An $\mathbf{R}^+$-version of Theorem 4.2 is given in Hille and Phillips [1], Theorem 4.18.5, p. 149, and a version of Corollary 4.7 is given in Hille and Phillips [1], Theorem 4.18.6, p. 150. These two results use slightly weaker assumptions on φ than we do, since they require φ to be defined and submultiplicative only on $\mathbf{R}^+$.

In the case where μ has compact support, (1.1) and its characteristic equation (with $\varphi \equiv 1$) are identical to the difference equation and its characteristic equation employed in the study of autonomous neutral functional differential equations; see Hale [5], Ch. 12, Secs. 3–4, and Hale and Meyer [1], p. 36. Thus, (1.1) may be called a difference equation with infinite delay.

Theorem 4.13 was obtained in Shea and Wainger [1], p. 318.

For some theorems that yield $r \in L^1(\mathbf{R}^+;\varphi)$ when the kernel k does not belong to $L^1(\mathbf{R}^+;\varphi)$ and $1+\hat{k}(z) \neq 0$ for $\Re z \geq 0$, see Gripenberg [6], Jordan and Wheeler [6] and Jordan, Staffans, and Wheeler [1], Sec. 6.

References

B. D. Coleman and V. J. Mizel

1. Norms and semi-groups in the theory of fading memory, *Arch. Ratl. Mech. Anal.* **23** (1966), pp. 87–123.
2. On the general theory of fading memory, *Arch. Ratl. Mech. Anal.* **29** (1968), pp. 18–31.

I. M. Gel′fand

1. Über absolut konvergente trigonometrische Reihen und Integrale, *Mat. Sb.* **9** (1941), pp. 51–66.

I. M. Gel′fand, D. A. Raikov, and G. E. Shilov

1. *Commutative Normed Rings*, Chelsea, New York, 1964.

G. Gripenberg

6. On rapidly decaying resolvents of Volterra equations, *J. Integral Equations* **1** (1979), pp. 241–247.

J. K. Hale

2. Dynamical systems and stability, *J. Math. Anal. Appl.* **26** (1969), pp. 39–59.
5. *Theory of Functional Differential Equations*, Springer-Verlag, Berlin, 1977.

J. K. Hale and J. Kato

1. Phase space for retarded equations with infinite delay, *Funkcial. Ekvac.* **21** (1978), pp. 11–41.

J. K. Hale and K. R. Meyer

1. *A Class of Functional Equations of Neutral Type*, Mem. Amer. Math. Soc. **76**, Amer. Math. Soc., Providence, R. I., 1967.

E. Hewitt and K. A. Ross

1. *Abstract Harmonic Analysis, 1–2*, Springer-Verlag, Berlin, 1963–1970.

E. Hille and R. S. Phillips

1. *Functional Analysis and Semi-Groups*, Amer. Math. Soc., Providence, R. I., 1957.

G. S. Jordan, O. J. Staffans, and R. L. Wheeler

1. Local analyticity in weighted L^1-spaces and applications to stability problems for Volterra equations, *Trans. Amer. Math. Soc.* **274** (1982), pp. 749–782.

G. S. Jordan and R. L. Wheeler

6. Rates of decay of resolvents of Volterra equations with certain non-integrable kernels, *J. Integral Equations* **2** (1980), pp. 103–110.

F. Kappel and W. Schappacher

2. Some considerations to the fundamental theory of infinite delay equations, *J. Differential Equations* **37** (1980), pp. 141–183.

M. J. Leitman and V. J. Mizel

1. On fading memory spaces and hereditary integral equations, *Arch. Ratl. Mech. Anal.* **55** (1974), pp. 18–51.

D. F. Lima

1. *Hopf Bifurcation in Equations with Infinite Delay*, Ph. D. Thesis, Brown University, Providence, R. I., 1977.

T. Naito

2. On autonomous linear functional differential equations with infinite retardations, *J. Differential Equations* **21** (1976), pp. 297–315.
3. On linear autonomous retarded equations with an abstract phase space for infinite delay, *J. Differential Equations* **33** (1979), pp. 74–91.

W. Rudin

2. *Functional Analysis*, McGraw–Hill, New York, 1973.
3. *Real and Complex Analysis,* 3rd *ed.*, McGraw–Hill, New York, 1986.

D. F. Shea and S. Wainger

1. Variants of the Wiener–Lévy theorem, with applications to stability problems for some Volterra integral equations, *American J. Math.* **97** (1975), pp. 312–343.

O. J. Staffans

16. On a neutral functional differential equation in a fading memory space, *J. Differential Equations* **50** (1983), pp. 183–217.
18. The null space and the range of a convolution operator in a fading memory space, *Trans. Amer. Math. Soc.* **281** (1984), pp. 361–388.
25. Extended initial and forcing function semigroups generated by a functional equation, *SIAM J. Math. Anal.* **16** (1985), pp. 1034–1048.

N. Wiener and H. R. Pitt

1. On absolutely convergent Fourier–Stieltjes transforms, *Duke Math. J.* **4** (1938), pp. 420–436.

5

Completely Monotone Kernels

We study kernels that are completely monotone, and show that the resolvent of a completely monotone kernel is completely monotone. A related result is true for differential resolvents. We also discuss Volterra equations of the first kind, in particular equations of the first kind with completely monotone kernels.

1. Introduction

In this chapter we study the concept of complete monotonicity, and show how this concept can be used in the theory of Volterra integral equations. The principal result is that the resolvent of a completely monotone kernel is completely monotone. A related result is true for the differential resolvent of a completely monotone kernel. This differential resolvent is no longer completely monotone, but it can be written as the difference of an exponentially decaying function and a completely monotone function.

The concept of complete monotonicity enables us to solve certain Volterra equations of the first kind, i.e., equations of the type

$$\int_0^t k(t-s)x(s)\,\mathrm{d}s = f(t), \quad t \in \mathbf{R}^+. \tag{1.1}$$

If k and f are locally absolutely continuous and $k(0)$ is invertible, then we can simply differentiate (1.1) to get

$$x(t) + \left[k(0)\right]^{-1} \int_0^t k'(t-s)x(s)\,\mathrm{d}s = \left[k(0)\right]^{-1} f'(t), \quad t \in \mathbf{R}^+,$$

and this is a Volterra equation of the second kind, i.e., the equation that we studied in Chapter 2. Therefore we do not need a new theory for this case. However, if e.g. $|k(t)| \to \infty$ as $t \to 0$, then this approach can no longer be pursued. In this situation it is, instead, sometimes possible to make use of the notion of a resolvent of the first kind of the kernel, and to get a new variation of constants formula. In particular, if the kernel is completely monotone, or if it is real-valued, nonnegative, and nonincreasing, then this method will work.

In Section 2 we develop the basic theory of completely monotone kernels. This theory is applied in Section 3 to Volterra integral equations of the second kind, and in Section 4 we study Volterra integrodifferential equations. Finally, in Section 5 we discuss Volterra equations of the first kind.

2. Basic Properties and Definitions

In the scalar case one defines a completely monotone function to be a nonnegative function with infinitely many derivatives that alternate in sign. In other words, a is completely monotone on an interval $I \subset \mathbf{R}$ if

$$(-1)^j a^{(j)}(t) \geq 0, \quad t \in I, \quad j = 0, 1, 2, \ldots.$$

In the matrix case one can get different notions of complete monotonicity by using different concepts of positivity. One possibility is to call a matrix positive if all its elements are positive, as one frequently does, e.g., in probabilistic arguments. However, this does not lead to a satisfactory result here. Instead, we define a matrix to be positive if it is positive semi-definite.

In the sequel, we let $\langle \cdot, \cdot \rangle$ denote some inner product on $\mathbf{C}^n$, and we let the norm in $\mathbf{C}^n$ be the norm induced by this inner product.

2.1 Definition

Let A be an $n \times n$ matrix. We say that A is positive and write $A \succeq 0$ iff $\langle v, Av \rangle \geq 0$ for all vectors $v \in \mathbf{C}^n$, and we say that A is strictly positive and write $A \succ 0$ iff $\langle v, Av \rangle > 0$ for all nonzero $v \in \mathbf{C}^n$. If $-A \succeq 0$ or $-A \succ 0$, then we say that A is negative or strictly negative, respectively, and write $A \preceq 0$ or $A \prec 0$.

The adjoint of A is denoted by A^* and it is defined by the formula $\langle u, Av \rangle = \langle A^* u, v \rangle$ for all u and $v \in \mathbf{C}^n$. If $A = A^*$ then A is self-adjoint, and if $A = -A^*$ then A is anti-symmetric. The matrix $\Re A = \frac{1}{2}(A + A^*)$ is called the real part of A, and the matrix $\Im A = \frac{1}{2\mathrm{i}}(A - A^*)$ the imaginary part of A.

A measure $\alpha \in M_{\mathrm{loc}}(\mathbf{R}; \mathbf{C}^{n \times n})$ is said to be positive if $\alpha(E) \succeq 0$ for every bounded Borel set $E \subset \mathbf{R}$.

Here some cautionary words are in order. In general, neither $\Re A$ nor $\Im A$ need be real in the sense that they belong to $\mathbf{R}^{n \times n}$, although they are self-adjoint. Frequently $\Re A$ is called the symmetric or self-adjoint part of A, and $\mathrm{i}\Im A$ the anti-symmetric part of A. The names 'real part' and 'imaginary part', which we use for the quantities $\Re A$ and $\Im A$, are motivated by the fact that they have properties similar to those of the real and imaginary parts of a complex number, and that, in the scalar case, they coincide with these.

The two most important properties of $\Re A$ and $\Im A$ (besides their self-adjointness) are the facts that $A = \Re A + \mathrm{i}\Im A$, and that for all $v \in \mathbf{C}^n$ we have $\Re\langle v, Av\rangle = \langle v, (\Re A)v\rangle$ and $\Im\langle v, Av\rangle = -\langle v, (\Im A)v\rangle$. Observe that $A = \Re A$ iff A is self-adjoint, and that $A = \mathrm{i}\Im A$ iff A is anti-symmetric.

It is not true that $Av = 0$ implies $(\Re A)v = (\Im A)v = 0$ (cf. Exercise 10). However, the following result is true.

2.2 Lemma

Let $A \in \mathbf{C}^{n\times n}$, and suppose that at least one of the conditions

(i) $\Re A \succeq 0$,
(ii) $\Re A \preceq 0$,
(iii) $\Im A \succeq 0$,
(iv) $\Im A \preceq 0$

holds. Then, for each $v \in \mathbf{C}^n$, $Av = 0$ iff $(\Re A)v = (\Im A)v = 0$. In particular, if one of (i)–(iv) *holds in a strict sense (i.e., with $\succeq$ replaced by $\succ$, or with $\preceq$ replaced by $\prec$), then A is invertible.*

Proof Clearly, if $(\Re A)v = (\Im A)v = 0$, then $Av = (\Re A + \mathrm{i}\Im A)v = 0$.

Conversely, let $Av = 0$. Then $\langle v, Av\rangle = 0$, hence $\langle v, (\Re A)v\rangle = \Re\langle v, Av\rangle = 0$, and $\langle v, (\Im A)v\rangle = -\Im\langle v, Av\rangle = 0$. If, e.g., $\Re A \succeq 0$, or $\Re A \preceq 0$, then this implies $(\Re A)v = 0$ (see (**16**.2.2)), hence $(\Im A)v = \frac{1}{\mathrm{i}}(A - \Re A)v = 0$. The other cases are proved analogously. □

Note that if $\langle v, Av\rangle$ is real for all $v \in \mathbf{C}^n$ then A is self-adjoint. For a proof, see Section **16**.2.

When we use the preceding definition of positivity, the notion of complete monotonicity is extended to matrix-valued functions as follows.

2.3 Definition

Let $J \subset \mathbf{R}$ be an interval. A matrix-valued function $a : J \to \mathbf{C}^{n\times n}$ is completely monotone on J if a is infinitely many times differentiable, and

$$(-1)^j a^{(j)}(t) \succeq 0, \quad t \in J, \quad j = 0, 1, 2, \ldots.$$

The assumptions that a is infinitely many times differentiable and that the derivatives of a alternate in sign may be replaced by appropriate difference inequalities. It turns out that these difference inequalities imply that a is C^∞, in fact analytic (the last claim follows from Theorem 2.5 below).

Usually one takes the interval J to be $\mathbf{R}^+$ or $(0, \infty)$.

The following result is an easy consequence of Definition 2.3.

2.4 Lemma
A function $a : J \to \mathbf{C}^{n\times n}$ is completely monotone if and only if for each $v \in \mathbf{C}^n$ the function $\langle v, a(\cdot)v\rangle$ is completely monotone.

Proof If a is completely monotone then by Definition 2.3 the function $\langle v, a(\cdot)v\rangle$ is completely monotone for each $v \in \mathbf{C}^n$.

Conversely, let $\langle v, a(\cdot)v\rangle$ be completely monotone for each $v \in \mathbf{C}^n$. For $j = 1, \dots, n$, let u_j and v_j be vectors in $\mathbf{C}^n$ such that $\langle Au_j, v_k\rangle$ equals the element A_{jk} for every matrix $A \in \mathbf{C}^{n\times n}$. Then each element $a_{jk}(\cdot)$ of the matrix function $a(\cdot)$ is the complex conjugate of $\langle v_k, a(\cdot)u_j\rangle$. Since

$$\begin{aligned} 4\langle v_k, a(\cdot)u_j\rangle = {} & \langle (v_k + u_j), a(\cdot)(v_k + u_j)\rangle \\ & - \langle (v_k - u_j), a(\cdot)(v_k - u_j)\rangle \\ & + \mathrm{i}\langle (v_k + \mathrm{i}u_j), a(\cdot)(v_k + \mathrm{i}u_j)\rangle \\ & - \mathrm{i}\langle (v_k - \mathrm{i}u_j), a(\cdot)(v_k - \mathrm{i}u_j)\rangle, \end{aligned} \tag{2.1}$$

it follows that $a_{jk}(t)$ is a linear combination of four completely monotone functions. Consequently $a_{jk}(t)$ and, therefore, $a(t)$ are infinitely many times differentiable and

$$\langle v, (-1)^j a^{(j)}(t)v\rangle = (-1)^j \Big(\frac{\mathrm{d}}{\mathrm{d}t}\Big)^j \langle v, a(t)v\rangle \geq 0,$$

for $t \in J$, $j = 0, 1, 2, \dots$, and an arbitrary $v \in \mathbf{C}^n$. We conclude that a is completely monotone. □

It is well known that every scalar function which is completely monotone on $(0, \infty)$ is the Laplace transform of a positive, locally finite measure. The same is true in the matrix case.

2.5 Bernstein's Theorem
A function $a : (0, \infty) \to \mathbf{C}^{n\times n}$ is completely monotone on $(0, \infty)$ if and only if a is the restriction to the positive real axis of the Laplace transform of a positive measure $\alpha \in M_{\mathrm{loc}}(\mathbf{R}^+; \mathbf{C}^{n\times n})$ satisfying $\int_{\mathbf{R}^+} \mathrm{e}^{-\sigma t}|\alpha|(\mathrm{d}\sigma) < \infty$ for every $t > 0$, that is,

$$a(t) = \int_{\mathbf{R}^+} \mathrm{e}^{-\sigma t}\alpha(\mathrm{d}\sigma), \quad t > 0. \tag{2.2}$$

Moreover, $a \in L^1_{\mathrm{loc}}(\mathbf{R}^+; \mathbf{C}^{n\times n})$ if and only if $\int_{\mathbf{R}^+}(1+\sigma)^{-1}|\alpha|(\mathrm{d}\sigma) < \infty$, and $a \in L^1(\mathbf{R}^+; \mathbf{C}^{n\times n})$ if and only if $\alpha(\{0\}) = 0$ and $\int_{(0,\infty)} \sigma^{-1}|\alpha|(\mathrm{d}\sigma) < \infty$.

Recall that α is positive if and only if $\alpha(E) \succeq 0$ for every bounded Borel set $E \subset \mathbf{R}^+$.

The proof given below is not self-contained, since it only reduces the matrix case to the scalar case, which can be found in Widder [2].

Proof of Theorem 2.5 If a has an integral representation of the form given in (2.2), then a is positive and analytic, and by differentiating under the integral sign one shows that the derivatives of a alternate in sign.

Conversely, let us suppose that a is completely monotone. As (2.1) shows, we may write an arbitrary element a_{jk} of a as a linear combination of four completely monotone scalar functions. Therefore, to prove the representation formula in the matrix case it suffices to prove it in the scalar case. For this proof, see Widder [2], p. 160.

Once we have the matrix representation formula, we see that

$$\langle v, a(t)v \rangle = \int_{\mathbf{R}^+} \mathrm{e}^{-\sigma t} \langle v, \alpha(\mathrm{d}\sigma)v \rangle, \quad t > 0, \quad v \in \mathbf{C}^n,$$

and then it follows from the scalar case that the measure $\langle v, \alpha(\cdot)v \rangle$ is positive. Since v is arbitrary, it follows that $\alpha(E) \succeq 0$ for every Borel set $E \subset \mathbf{R}^+$.

Next we show that a is locally integrable on $\mathbf{R}^+$ if and only if $\int_{\mathbf{R}^+} (1 + \sigma)^{-1} |\alpha|(\mathrm{d}\sigma) < \infty$. As above, we find that it suffices to prove this in the scalar case. When a is scalar-valued we use Bernstein's representation formula and Fubini's theorem to get

$$\int_0^1 (a(t) - a(\infty))\, \mathrm{d}t = \int_{(0,\infty)} \frac{1 - \mathrm{e}^{-\sigma}}{\sigma} \alpha(\mathrm{d}\sigma).$$

Because of the monotonicity assumption on a, we have $a \in L^1_{\mathrm{loc}}(\mathbf{R}^+; \mathbf{R})$ if and only if $\int_0^1 (a(t) - a(\infty))\, \mathrm{d}t < \infty$, and, on the other hand, the right-hand side above is finite if and only if $\int_{\mathbf{R}^+} (1 + \sigma)^{-1} \alpha(\mathrm{d}\sigma) < \infty$.

By using a similar argument, one shows that $a \in L^1(\mathbf{R}^+; \mathbf{C}^{n \times n})$ if and only if $\alpha(\{0\}) = 0$ and $\int_{(0,\infty)} \sigma^{-1} |\alpha|(\mathrm{d}\sigma) < \infty$. □

The following result characterizes the Laplace transform of a completely monotone, locally integrable function.

2.6 Theorem
The Laplace transform $F(z) = \hat{a}(z)$ of a function $a : (0, \infty) \mapsto \mathbf{C}^{n \times n}$ that is locally integrable on $\mathbf{R}^+$ and completely monotone on $(0, \infty)$ has the following properties:

(i) *F has an analytic extension to the region $\mathbf{C} \setminus \mathbf{R}^-$;*
(ii) *$F(x) = F^*(x)$ for $x \in (0, \infty)$;*
(iii) *$\lim_{x \to +\infty} F(x) = 0$;*
(iv) *$\Im F(z) \preceq 0$ for $\Im z > 0$;*
(v) *$\Im\{zF(z)\} \succeq 0$ for $\Im z > 0$, and $F(x) \succeq 0$ for $x \in (0, \infty)$.*

Moreover, $a \in L^1(\mathbf{R}^+; \mathbf{C}^{n \times n})$ if and only if

(vi) *$\limsup_{x \downarrow 0} |F(x)| < \infty$.*

Conversely, every function F that satisfies (i)–(iii), *together with* (iv) *or* (v), *is the Laplace transform of a function $a : (0, \infty) \mapsto \mathbf{C}^{n \times n}$, which is locally integrable on $\mathbf{R}^+$ and completely monotone on $(0, \infty)$.*

Proof It follows (after some calculations) from Lemma 2.4, (2.1), and the relation $\Im\langle v, F(z)v \rangle = -\langle v, \Im F(z)v \rangle$, that we may, without loss of generality, take $n = 1$.

If F is the Laplace transform of a completely monotone, locally integrable scalar function a, then we can use the Bernstein Theorem 2.5, Fubini's theorem, and the fact that $\int_{\mathbf{R}^+} e^{-(z+\sigma)t}\,dt = \frac{1}{z+\sigma}$, to write $F(z)$ for $\Re z > 0$ in the form

$$F(z) = \int_{\mathbf{R}^+} \frac{\alpha(d\sigma)}{z+\sigma}. \tag{2.3}$$

(Thus F is the Stieltjes transform of α; see Widder [2], Ch. VIII.) Use this formula to extend F analytically to all of $\mathbf{C}\setminus\mathbf{R}^-$. It is then easy to verify the properties (ii)–(v). To get the last claim (vi) we recall the last part of the Bernstein Theorem 2.5.

To prove the converse statement, it suffices to show that F has a representation of the form (2.3) for some positive measure α satisfying $\int_{\mathbf{R}^+}(1+\sigma)^{-1}\alpha(d\sigma) < \infty$, because then we can reverse the computation above to show that F is the Laplace transform of the Laplace transform of α. Theorem 2.5 yields that the Laplace transform of α is completely monotone on $(0,\infty)$. Hence the conclusion follows.

Let us suppose that F is a function satisfying conditions (i)–(iii) and (iv). The function $u(z) = (z-i)/(z+i)$ maps the upper half plane $\Im z > 0$ onto the unit disk $\{\,|u| < 1\,\}$. Its inverse function $z(u) = i(1+u)/(1-u)$ maps the unit disk onto the upper half plane, and therefore, if we define

$$G(u) = iF(z(u)) = iF\left(\frac{i(1+u)}{1-u}\right),$$

then, by (i) and (iv), $\Re G$ is a positive harmonic function in the unit disk. This implies (see, e.g., Rudin [3], Theorem 11.30) that $\Re G$ can be written as the Poisson integral

$$\Re G(re^{i\theta}) = \frac{1}{2\pi}\int_{[0,2\pi)} P_r(\theta-\phi)\mu(d\phi)$$

of a positive measure μ, where the Poisson kernel $P_r(\theta-\phi)$ is given by

$$P_r(\theta-\phi) = \Re\left\{\frac{e^{i\phi}+re^{i\theta}}{e^{i\phi}-re^{i\theta}}\right\}.$$

The function $z \mapsto (z-i)/(z+i)$ maps the positive real axis onto the semicircle $\{\, e^{i\phi} \mid \pi \le \phi < 2\pi\,\}$, and, therefore, it follows from (ii) that $\lim_{r\uparrow 1} G(re^{i\theta}) = 0$ for $\pi < \theta < 2\pi$. Consequently, by Rudin [3], Theorem 11.24 and Exercise 19, p. 251, μ is supported on $[0,\pi]$.

Take the formula above for $\Re G$ and replace the Poisson kernel $P_r(\theta-\phi)$ by $(e^{i\phi}+re^{i\theta})/(e^{i\phi}-re^{i\theta})$. Then we get an analytic function that has the same real part as the function G. If two analytic functions have the same real parts, then their difference is a constant, and therefore there is a constant, which we choose to denote by iC, such that

$$G(u) = \frac{1}{2\pi}\int_{[0,\pi]} \frac{e^{i\phi}+u}{e^{i\phi}-u}\mu(d\phi) + iC, \quad |u| < 1.$$

Let us return to the function $F(z) = -\mathrm{i}G(u(z)) = -\mathrm{i}G\big((z-\mathrm{i})/(z+\mathrm{i})\big)$. We make the substitution $u = (z-\mathrm{i})/(z+\mathrm{i})$ in the integral above, and make a change of variable $\mathrm{e}^{\mathrm{i}\phi} = (t-\mathrm{i})/(t+\mathrm{i})$, or equivalently $t = \mathrm{i}(1+\mathrm{e}^{\mathrm{i}\phi})/(1-\mathrm{e}^{\mathrm{i}\phi}) = -\cot(\phi/2)$. If we let β be the positive measure satisfying $\beta((-\infty, -\cot(\phi/2)]) = \frac{1}{2\pi}\mu((0,\phi])$ for $\phi \in (0,\pi]$, then we get

$$F(z) = \int_{\mathbf{R}^-} \frac{1+tz}{z-t}\beta(\mathrm{d}t) + Az + C, \quad z \in \mathbf{C}\setminus\mathbf{R}^-, \tag{2.4}$$

where $A = -\frac{1}{2\pi}\mu(\{0\})$ (originally we get the identity only for $\Im z > 0$, but since both sides are analytic on $\mathbf{C}\setminus\mathbf{R}^-$, they must be equal in this larger region). If we divide (2.4) by z, let $z \to +\infty$, and use assumption (iii), we get $A = 0$. To obtain a size estimate on $\int_{\mathbf{R}^-}|t|\beta(\mathrm{d}t)$, we let $z \to +\infty$ in (2.4) (without dividing by z), and find from Fatou's lemma and (iii) that

$$\int_{\mathbf{R}^-}|t|\beta(\mathrm{d}t) \le \liminf_{x\to\infty}\int_{\mathbf{R}^-} -\frac{1+tx}{x-t}\beta(\mathrm{d}t) = C < \infty.$$

Rewrite (2.4) as

$$F(z) = \int_{\mathbf{R}^-}\frac{1+t^2}{z-t}\beta(\mathrm{d}t) + \int_{\mathbf{R}^-} t\beta(\mathrm{d}t) + C.$$

If we once more let $z \to +\infty$, and use (iii), Lebesgue's dominated convergence theorem, and the fact that $\int_{\mathbf{R}^-}(1+|t|)\beta(\mathrm{d}t) < \infty$, we find that $\int_{\mathbf{R}^-} t\beta(\mathrm{d}t) + C = 0$. Thus

$$F(z) = \int_{\mathbf{R}^-}\frac{1+t^2}{z-t}\beta(\mathrm{d}t). \tag{2.5}$$

Define the positive, locally finite measure α by $\alpha([0,\sigma]) = \int_{[-\sigma,0]}(1+t^2)\beta(\mathrm{d}t)$. Then it follows from the estimate above on $\int_{\mathbf{R}^-}|t|\beta(\mathrm{d}t)$ that $\int_{\mathbf{R}^+}(1+\sigma)^{-1}\alpha(\mathrm{d}\sigma) < \infty$. Moreover, (2.5) becomes (2.3). As we already observed above, this implies that F is the Laplace transform of a completely monotone, locally integrable function (i.e., of the Laplace transform of α).

Finally, let us assume that F satisfies conditions (i)–(iii) and (v). By an argument almost identical to the one used above, one sees that there exists a positive measure β supported on $\mathbf{R}^-$, and constants A and C such that

$$zF(z) = -\int_{\mathbf{R}^-}\frac{1+tz}{z-t}\beta(\mathrm{d}t) + Az + C, \quad z \in \mathbf{C}\setminus\mathbf{R}^-. \tag{2.6}$$

If we divide (2.6) by z, let $z \to +\infty$, and use assumption (iii), we get $A = 0$. By Fatou's lemma and the second part of assumption (v) (take z real and let $z \downarrow 0$),

$$\beta(\{0\}) = 0, \qquad \int_{(-\infty,0)}\frac{\beta(\mathrm{d}t)}{|t|} \le C.$$

Rewrite (2.6) in the form

$$F(z) = -\int_{(-\infty,0)}\frac{1}{z-t}\Big(t+\frac{1}{t}\Big)\beta(\mathrm{d}t) + \frac{1}{z}\Big\{\int_{(-\infty,0)}\frac{\beta(\mathrm{d}t)}{t} + C\Big\}. \tag{2.7}$$

Define the positive, locally finite measure α on $\mathbf{R}^+$ by

$$\alpha(\{0\}) = \int_{(-\infty,0)} \frac{\beta(\mathrm{d}t)}{t} + C, \quad \alpha((0,\sigma]) = -\int_{[-\sigma,0)} \Big(t + \frac{1}{t}\Big)\beta(\mathrm{d}t).$$

Then $\int_{\mathbf{R}^+}(1+\sigma)^{-1}\alpha(\mathrm{d}\sigma) < \infty$, and from (2.7) we again get (2.3). As before, this implies that F is the Laplace transform of a completely monotone, locally integrable function. □

Inspecting the proof of Theorem 2.6 one finds that the following alternative result is true as well.

2.7 Corollary

Let F satisfy (i)–(iii) *below, together with* (iv) *or* (v)*:*

(i) *F has an analytic extension to the region $\mathbf{C} \setminus \mathbf{R}^-$;*

(ii) *$F(x) \succeq 0$ for $x \in (0,\infty)$;*

(iii) *$\limsup_{x\to+\infty}|F(x)| < \infty$;*

(iv) *$\Im F(z) \preceq 0$ for $\Im z > 0$;*

(v) *$\Im\{zF(z)\} \succeq 0$ for $\Im z > 0$.*

Then $A \stackrel{\text{def}}{=} \lim_{x\to\infty} F(x)$ exists, $A \succeq 0$, and $F(\cdot) - A$ is the Laplace transform of a function $a : (0,\infty) \mapsto \mathbf{C}^{n\times n}$, which is locally integrable on $\mathbf{R}^+$ and completely monotone on $(0,\infty)$.

It is clear from Definition 2.3 that every scalar completely monotone function is convex. In fact even more is true, namely its logarithm is convex.

2.8 Theorem

Let a be a scalar, completely monotone function on $(0,\infty)$, satisfying $a(t) > 0$ for some $t > 0$. Then $a(t) > 0$ for all $t > 0$, and $\ln(a)$ is convex on $(0,\infty)$.

Proof Let t, $s > 0$ and $q \in (0,1)$. Using the Bernstein Theorem 2.5 and Hölder's inequality, we get

$$\begin{aligned} a\big(qt + (1-q)s\big) &= \int_{\mathbf{R}^+} \mathrm{e}^{-\sigma q t}\mathrm{e}^{-\sigma(1-q)s}\alpha(\mathrm{d}\sigma) \\ &\le \Big(\int_{\mathbf{R}^+} \mathrm{e}^{-\sigma t}\alpha(\mathrm{d}\sigma)\Big)^q \Big(\int_{\mathbf{R}^+} \mathrm{e}^{-\sigma s}\alpha(\mathrm{d}\sigma)\Big)^{1-q} \\ &= a(t)^q a(s)^{1-q}. \end{aligned}$$

This inequality shows that, either $a \equiv 0$ (which is excluded by the hypothesis), or $a(t) > 0$ for all $t > 0$. Taking logarithms of both sides we get the claim. □

3. Volterra Equations with Completely Monotone Kernels

We proceed to analyse what implications the preceding section has concerning the behaviour of the resolvent of a completely monotone function. Recall that the resolvent r of k is the solution of

$$k(t) = r(t) + (r * k)(t) = r(t) + (k * r)(t), \quad t \in \mathbf{R}^+.$$

The principal result is the following.

3.1 Theorem

Let $k \in L^1_{\mathrm{loc}}(\mathbf{R}^+; \mathbf{C}^{n\times n})$, and let r be the resolvent of k. Then k is completely monotone on $(0,\infty)$ if and only if r is completely monotone on $(0,\infty)$, $r \in L^1(\mathbf{R}^+; \mathbf{C}^{n\times n})$, and $\int_{\mathbf{R}^+} r(t)\,\mathrm{d}t \preceq I$. Moreover, $k \in L^1(\mathbf{R}^+; \mathbf{C}^{n\times n})$ if and only if $\int_{\mathbf{R}^+} r(t)\,\mathrm{d}t \prec I$.

Proof Suppose that k is completely monotone. By Lemma **2**.3.4, there is a constant $c \geq 0$ such that, for all z with $\Re z \geq c$, the Laplace transform $\hat{r}(z)$ exists, and

$$\hat{r}(z) = I - \left[I + \hat{k}(z)\right]^{-1}. \tag{3.1}$$

We know that $\hat{k}(z)$ satisfies conditions (i)–(iv) (as well as (v)) in Theorem 2.6, and we want to show that the same conditions hold for the function $\hat{r}(z)$.

To show that $\hat{r}$ satisfies (i), we have to check that $\det[I + \hat{k}(z)] \neq 0$ for $z \in \mathbf{C} \setminus \mathbf{R}^-$. When $\Im z = 0$, this follows from Lemma 2.2 and from the fact that, for real x, we have $I + \hat{k}(x) \succeq I$ (because $k(t) \succeq 0$ for $t > 0$, hence $\hat{k}(x) \succeq 0$ for $x > 0$). Suppose that $\det[I + \hat{k}(z_0)] = 0$ for some z_0 with $\Im z_0 > 0$ (the case $\Im z_0 < 0$ is treated analogously). Then there is some nonzero $v \in \mathbf{C}^n$ such that $(I + \hat{k}(z_0))v = 0$, hence $\langle v, \Re\{I + \hat{k}(z_0)\}v\rangle = 0$ and $\langle v, \Im\{I + \hat{k}(z_0)\}v\rangle = 0$. By condition (iv) in Theorem 2.6, the function $\langle v, \Im\{I + \hat{k}(z)\}v\rangle = \langle v, \Im\{\hat{k}(z)\}v\rangle$ is a nonpositive harmonic function that has a zero, so, by the maximum principle for harmonic functions, $\langle v, \Im\{\hat{k}(z)\}v\rangle$ must vanish identically. This again implies that $\langle v, \Re\{I + \hat{k}(z)\}v\rangle$ must be a constant. This constant function is zero at z_0, so we conclude that $\langle v, (I + \hat{k}(z))v\rangle \equiv 0$, which is not true (certainly not for z positive real). Therefore, we conclude that $\det[I+\hat{k}(z)] \neq 0$ for all $z \in \mathbf{C} \setminus \mathbf{R}^-$, and we have proved that $\hat{r}(z)$ satisfies (i) in Theorem 2.6.

It is obvious that $\hat{r}(z)$ satisfies conditions (ii) and (iii) in Theorem 2.6.

Let us check that (iv) holds as well. By (3.1), for every $v \in \mathbf{C}^n$ we have

$$\begin{aligned}\langle v, \Im\{\hat{r}(z)\}v\rangle &= \Big\langle v, \frac{1}{2\mathrm{i}}\Big(\big[I+\hat{k}^*(z)\big]^{-1} - \big[I+\hat{k}(z)\big]^{-1}\Big)v\Big\rangle \\ &= \Big\langle v, \big[I+\hat{k}^*(z)\big]^{-1}\Im\{\hat{k}(z)\}\big[I+\hat{k}(z)\big]^{-1}v\Big\rangle \\ &= \Big\langle \big[I+\hat{k}(z)\big]^{-1}v, \Im\{\hat{k}(z)\}\big[I+\hat{k}(z)\big]^{-1}v\Big\rangle.\end{aligned}$$

It follows from condition (iv) in Theorem 2.6 that the last line in this chain is nonpositive for $\Im z > 0$.

Since $\hat{r}(z)$ satisfies (i)–(iv) of Theorem 2.6, we conclude that r is completely monotone.

That r is integrable follows from Theorem 2.6(vi), (3.1), and the fact that for all $x > 0$, $\hat{k}(x) \succeq 0$, hence $I + \hat{k}(x) \succeq I$ and $0 \preceq \hat{r}(x) \prec I$. Moreover, letting $x \downarrow 0$ we conclude that $\int_{\mathbf{R}^+} r(t)\,\mathrm{d}t = \hat{r}(0) \preceq I$. If k is integrable, then $\int_{\mathbf{R}^+} r(t)\,\mathrm{d}t = I - [I + \hat{k}(0)]^{-1} \prec I$.

We leave the proof of the converse part to the reader. It is very similar to the proof above. Observe that (3.1) can be rewritten in the form

$$\hat{k}(z) = [I - \hat{r}(z)]^{-1} - I,$$

and that the condition $\int_{\mathbf{R}^+} r(t)\,\mathrm{d}t \preceq I$ implies that $\hat{r}(x) \prec I$ for $x > 0$. □

4. Volterra Integrodifferential Equations with Completely Monotone Kernels

In this section we consider the Volterra integrodifferential equation

$$x'(t) + (k * x)(t) = f(t), \quad t \in \mathbf{R}^+; \quad x(0) = x_0. \tag{4.1}$$

As we recall from Chapter 3, the solution of this equation is

$$x(t) = r(t)x_0 + (r * f)(t), \quad t \in \mathbf{R}^+, \tag{4.2}$$

where the differential resolvent r satisfies

$$r'(t) + (k * r)(t) = r'(t) + (r * k)(t) = 0, \quad t \in \mathbf{R}^+; \quad r(0) = I. \tag{4.3}$$

Suppose that k is completely monotone. To see that r is not completely monotone it suffices to observe that $r(0) = I$ and $r'(0) = 0$ and, unless r is a constant (which is true iff $k = 0$), this implies that r cannot be completely monotone. We do, however, have the following result.

4.1 Theorem
Assume that $k \in L^1_{\mathrm{loc}}(\mathbf{R}^+;\mathbf{C}^{n\times n})$ is completely monotone on $(0,\infty)$ and satisfies $k'(t) \prec 0$ for some $t > 0$, and let r be the differential resolvent of k. Then there exist positive numbers K and ϵ and a positive measure $\mu \in M([0,\epsilon];\mathbf{C}^{n\times n})$ such that

$$\left| r(t) + \int_{[0,\epsilon]} \mathrm{e}^{-\sigma t}\mu(\mathrm{d}\sigma) \right| \leq K\mathrm{e}^{-\epsilon t}, \quad t \in \mathbf{R}^+.$$

Consequently, $-r$ is the sum of a completely monotone function and a function that decays exponentially to zero as $t \to \infty$. Moreover, $r \in L^1(\mathbf{R}^+;\mathbf{C}^{n\times n})$.

The claim that r is integrable is a simple consequence of the main claim together with Theorem 2.6(vi) and (4.3), so we leave the proof of this claim to the reader as an exercise. We remark that the completely monotone part in the decomposition may be zero (see Exercise 21). However, the exponentially decaying part is always nonzero, since $r(0) = I \succ 0$.

Proof of Theorem 4.1 (preliminaries) It follows from Bernstein's Theorem 2.5, and from our assumptions on k, that there exists a locally finite, positive Borel measure α such that

$$k(t) = \int_{\mathbf{R}^+} \mathrm{e}^{-\sigma t}\alpha(\mathrm{d}\sigma), \quad t > 0.$$

The Laplace transform of k can be expressed as the Stieltjes transform

$$\hat{k}(z) = \int_{\mathbf{R}^+} \frac{\alpha(\mathrm{d}\sigma)}{z+\sigma}, \quad z \in \mathbf{C}\setminus\mathbf{R}^-. \tag{4.4}$$

For all $z \in \mathbf{C}\setminus\mathbf{R}^-$, let

$$D(z) \stackrel{\mathrm{def}}{=} zI + \hat{k}(z).$$

If we let r denote the differential resolvent of k, then $\hat{r}(z) = [D(z)]^{-1}$.

It follows from the Laplace inversion formula that one can write r in the form

$$r(t) = \lim_{T\to\infty} \frac{1}{2\pi\mathrm{i}} \int_{c-\mathrm{i}T}^{c+\mathrm{i}T} \mathrm{e}^{zt}[D(z)]^{-1}\,\mathrm{d}z, \tag{4.5}$$

where c is some sufficiently large number, and the integral is a line integral in the complex plane. The fact that this relation holds a.e. on $\mathbf{R}^+$ is a consequence of results in Widder [2], p. 81. It follows from Hannsgen [1], p. 548, that it holds for every t.

The idea behind the proof of Theorem 4.1 is to shift the path of integration into the left half plane so that one is able to factor out an exponentially decaying factor $\mathrm{e}^{-\epsilon t}$ from the integral. Unfortunately, a simple translation of the path of integration to the left is not possible, because, as one can see from (4.4), the function $\hat{k}$, and therefore also D, will in general have singularities on the negative real axis $\mathbf{R}^-$. Instead, one has to use a more complicated method, and distort the path of integration close to the origin.

Before we can carry out the argument sketched above, we have to check two things: we must show that $D(z)$ is invertible in some half plane $\Re z \geq -\epsilon$, where $\epsilon > 0$, as long as one stays away from $\mathbf{R}^-$, and that $[D(z)]^{-1}$ behaves well as one approaches the negative real axis.

Write $z = x + \mathrm{i}y$, and split D into its real and imaginary parts to get (recall that $\Re\alpha(E) = \alpha(E)$ for every Borel set E, since α is positive)

$$\begin{aligned}\Re D(x+\mathrm{i}y) &= xI + \int_{\mathbf{R}^+} \frac{x+\sigma}{(x+\sigma)^2+y^2}\alpha(\mathrm{d}\sigma),\\ \Im D(x+\mathrm{i}y) &= yI - \int_{\mathbf{R}^+} \frac{y}{(x+\sigma)^2+y^2}\alpha(\mathrm{d}\sigma), \quad x+\mathrm{i}y \in \mathbf{C}\setminus\mathbf{R}^-. \end{aligned} \tag{4.6}$$

We claim that $\det[D(z)] \neq 0$ for $\Re z \geq 0$, $z \neq 0$. To see this, observe that the first of the relations above implies that in this region, $\Re D(z) \succ 0$ (since (2.2) together with the assumption that k is nonconstant implies that α cannot vanish on $(0,\infty)$), and use Lemma 2.2.

Next, let us study the behaviour of D on $\mathbf{R}^-$. We claim that

$D(x) \stackrel{\text{def}}{=} \lim_{y\downarrow 0} D(x+\mathrm{i}y)$ exists and is invertible for almost all $x \in \mathbf{R}^-$. (4.7)

To prove (4.7), we first study an auxiliary function A, defined by $A(z) = \hat{k}(z) - \mathrm{i}I$. We observe that, by (iv) in Theorem 2.6, $\Im A(z) \preceq -I$ for $\Im z > 0$. In particular, $A(z)$ is invertible and $|[A(z)]^{-1}| \leq 1$ for z in $\Pi^+ \stackrel{\text{def}}{=} \{\, z \in \mathbf{C} \mid \Im z > 0 \,\}$ (because $|v||A(z)v| \geq |\langle v, A(z)v\rangle| \geq |\Im\langle v, A(z)v\rangle| \geq |v|^2$; hence $|A(z)v| \geq |v|$ for every $v \in \mathbf{C}^n$). Let us denote $[A(z)]^{-1}$ by $B(z)$. Then each component of B belongs to the Hardy space $H^\infty(\Pi^+)$, and hence $B(x) \stackrel{\text{def}}{=} \lim_{y\downarrow 0} B(x+\mathrm{i}y)$ exists for almost all $x \in \mathbf{R}$ (see, e.g., Rudin [3], Theorem 11.32). Also note that $\det[B(z)] \in H^\infty(\Pi^+)$, and therefore $\lim_{y\downarrow 0}\det[B(x+\mathrm{i}y)]$ exists and is nonzero for almost all $x \in \mathbf{R}$ (see Rudin [3], Theorem 17.18). Clearly this limit equals $\det[B(x)]$. We conclude that $\lim_{y\downarrow 0} A(x+\mathrm{i}y) = \lim_{y\downarrow 0}[B(x+\mathrm{i}y)]^{-1}$ exists and equals $[B(x)]^{-1}$ for almost all $x \in \mathbf{R}$.

Now it is easy to show that the limit in (4.7) exists for almost all $x \in \mathbf{R}^-$: simply observe that

$$D(z) = (z+\mathrm{i})\,A(z)\Big(B(z) + \frac{1}{z+\mathrm{i}}I\Big), \quad z \in \Pi^+,$$

and use the fact that the limits of A and B exist. To see that the limit $D(x)$ is invertible for almost all $x \in \mathbf{R}^-$, write D^{-1} in the form

$$[D(z)]^{-1} = \frac{1}{z+\mathrm{i}}\Big[B(z) + \frac{1}{z+\mathrm{i}}I\Big]^{-1} B(z),$$

and note that $B(x)$ is well defined for almost all $x \in \mathbf{R}$. Furthermore, observe that $\det\big[B(z) + \frac{1}{z+\mathrm{i}}I\big] \in H^\infty(\Pi^+)$, hence $\det\big[B(x) + \frac{1}{x+\mathrm{i}}I\big]$ exists and is nonzero for almost all $x \in \mathbf{R}$. Consequently, $D(x)$ is invertible for almost all $x \in \mathbf{R}$, as claimed.

Before continuing the proof of Theorem 4.1, let us prove the following key estimate on $D(z)$ in a neighbourhood of an interval $[-\eta, 0]$.

4.2 Lemma
Let k be as in Theorem 4.1. There exists a positive constant η such that for all $z = x + iy$ with $-\eta \le x \le 0$, $0 < |y| \le \eta$, and for all $v \in \mathbf{C}^n$, either

$$\frac{1}{y}\langle v, \Im\{D(z)\}v\rangle \le -|v|^2, \tag{4.8}$$

or

$$\langle v, \Re\{D(z)\}v\rangle \ge \eta |v|^2. \tag{4.9}$$

In particular, $D(z)$ is invertible for these values of z and $\left|[D(z)]^{-1}\right| \le \frac{1}{|y|}$.

Proof of Lemma 4.2 First, let us show how the statement about the invertibility of $D(z)$ follows from the first claim. To do this, we observe that

$$|v|^2 \left|D(z)v\right|^2 \ge \left|\langle v, D(z)v\rangle\right|^2 = \langle v, \Re\{D(z)\}v\rangle^2 + \langle v, \Im\{D(z)\}v\rangle^2.$$

Thus, if (4.8) or (4.9) holds we have

$$|D(z)v| \ge |y|\,|v|,$$

and we get the second statement.

Clearly, it suffices to prove the claims (4.8) and (4.9) in the scalar case, and throughout the rest of the proof of the lemma we shall assume that the dimension $n = 1$, and take $v = 1$.

First, let us consider the case where $\int_{(0,\infty)} \sigma^{-2}\alpha(d\sigma) = \infty$ (by Theorem 2.5 and by the fact that k is locally integrable on $\mathbf{R}^+$, this means that $\int_{(0,1)} \sigma^{-2}\alpha(d\sigma) = \infty$). Choose η so small that

$$\int_{(\eta,\infty)} \frac{\alpha(d\sigma)}{\sigma^2 + \eta^2} \ge 2.$$

Then, if $-\eta \le x \le 0$ and $0 < |y| \le \eta$ (note that $|x + \sigma| \le \sigma$ when $\sigma \ge -x$),

$$\int_{\mathbf{R}^+} \frac{\alpha(d\sigma)}{(x+\sigma)^2 + y^2} \ge \int_{(\eta,\infty)} \frac{\alpha(d\sigma)}{(x+\sigma)^2 + y^2} \ge \int_{(\eta,\infty)} \frac{\alpha(d\sigma)}{\sigma^2 + \eta^2} \ge 2,$$

and (4.8) follows from the second half of (4.6).

Consider next the case where $c \overset{\text{def}}{=} \int_{(0,\infty)} \sigma^{-2}\alpha(d\sigma) < \infty$. By choosing γ small enough we have

$$\int_{(\gamma,\infty)} \frac{\sigma\alpha(d\sigma)}{\gamma(\sigma^2 + \gamma^2)} \ge \int_{(\gamma,\infty)} \frac{\alpha(d\sigma)}{\sigma^2 + \gamma^2} \ge \frac{c}{2}.$$

Let $0 \le \eta \le \gamma$, and suppose that, for some $z = x + iy$ with $-\eta \le x \le 0$ and $0 < |y| \le \eta$, the inequality (4.8) does not hold (we shall fix η in a moment). Then it follows from the second half of (4.6) that

$$\int_{\mathbf{R}^+} \frac{\alpha(d\sigma)}{(x+\sigma)^2 + y^2} \le 2,$$

which combined with the first half of (4.6) gives (keep in mind that $x \le 0$)

$$\begin{aligned}\Re D(z) &\ge \int_{\mathbf{R}^+} \frac{\sigma\alpha(\mathrm{d}\sigma)}{(x+\sigma)^2+y^2} - 3|x| \ge \int_{(\gamma,\infty)} \frac{\sigma\alpha(\mathrm{d}\sigma)}{(x+\sigma)^2+y^2} - 3|x| \\ &\ge \int_{(\gamma,\infty)} \frac{\sigma\alpha(\mathrm{d}\sigma)}{\sigma^2+\eta^2} - 3|x| \ge \frac{c\gamma}{2} - 3\eta.\end{aligned}$$

Thus, if we choose η to be so small that $c\gamma/2 - 3\eta \ge \eta$, then the inequality (4.9) holds, and the proof of Lemma 4.2 is complete. □

Proof of Theorem 4.1 (continued) Rewrite (4.4) in the form

$$\hat{k}(z) = \int_{\mathbf{R}^+} \frac{(1+\sigma)\ \alpha(\mathrm{d}\sigma)}{(z+\sigma)\ (1+\sigma)}, \quad z \in \mathbf{C} \setminus \mathbf{R}^-,$$

and recall that $\int_{\mathbf{R}^+} (1+\sigma)^{-1}|\alpha|(\mathrm{d}\sigma) < \infty$. From this equation we deduce, using the dominated convergence theorem, that

$$\begin{aligned}&\hat{k}(x+\mathrm{i}y) \to 0, \text{ as } |y| \to \infty, \\ &\qquad \textit{uniformly in every half plane } x \ge \lambda > -\infty.\end{aligned} \tag{4.10}$$

Take any $\eta > 0$ given by Lemma 4.2, and choose a number $\eta_0 \ge \eta$ large enough that $D(x+\mathrm{i}y)$ is invertible when $-\eta \le x < \infty$ and $|y| \ge \eta_0$. Since $\det[D(z)]$ is nonconstant and analytic for $z \in \mathbf{C} \setminus \mathbf{R}^-$, it can have at most a finite number of zeros in the compact set $\{\, z = x+\mathrm{i}y \mid -\eta \le x \le 0, \eta \le |y| \le \eta_0 \,\}$. Because none of the zeros lies on the imaginary axis, it is possible (by choosing $\epsilon \le \eta$ sufficiently small) to find a region $\{\, z = x+\mathrm{i}y \mid x \ge 0, x^2+y^2 \ge \eta^2 \,\} \cup \{\, z = x+\mathrm{i}y \mid -\epsilon \le x \le 0, |y| \ge \eta \,\}$ in which $D(z)$ is invertible. It is easy to see that in this region the inverse is bounded. By using (4.7) and Lemma 4.2, and, if necessary, by decreasing the value of ϵ, we find that there exists a positive number ϵ such that

$$\det[D(z)] \ne 0, \quad z \in \mathbf{C} \setminus \mathbf{R}^-, \quad \Re z \ge -\epsilon, \tag{4.11}$$

and

$$D(-\epsilon) = \lim_{y \downarrow 0} D(-\epsilon + \mathrm{i}y) \textit{ exists and is invertible.} \tag{4.12}$$

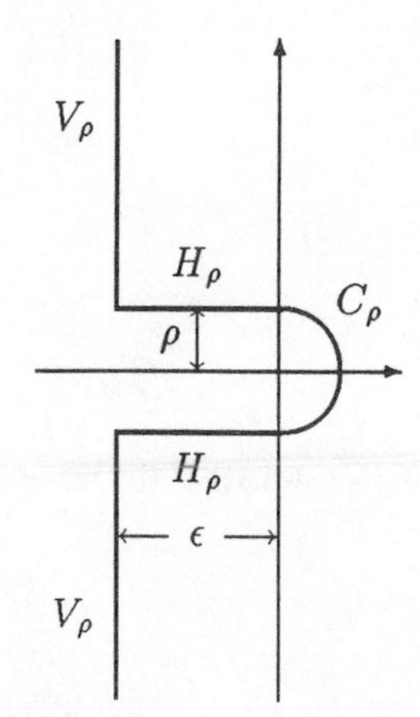

Let us recall the Laplace inversion formula

$$r(t) = \lim_{T\to\infty} \frac{1}{2\pi\mathrm{i}} \int_{c-\mathrm{i}T}^{c+\mathrm{i}T} \mathrm{e}^{zt}[D(z)]^{-1}\,\mathrm{d}z. \tag{4.5}$$

Using (4.10), (4.11) and Cauchy's theorem we can push the path of integration in (4.5) to the left, and obtain for every number $\rho \in (0, \eta)$,

$$r(t) = C_\rho(t) + V_\rho(t) + H_\rho(t), \quad t > 0, \tag{4.13}$$

where

$$C_\rho(t) = \frac{\rho}{2\pi} \int_{-\frac{\pi}{2}}^{\frac{\pi}{2}} \mathrm{e}^{t\rho\mathrm{e}^{\mathrm{i}\theta}} [D(\rho\mathrm{e}^{\mathrm{i}\theta})]^{-1} \mathrm{e}^{\mathrm{i}\theta}\,\mathrm{d}\theta$$

is the integral over a semicircle in the right half

plane with radius ρ,

$$V_\rho(t) = \lim_{T\to\infty} \frac{1}{2\pi}\left(\int_{-T}^{-\rho} + \int_{\rho}^{T}\right) \mathrm{e}^{(-\epsilon+\mathrm{i}y)t}[D(-\epsilon+\mathrm{i}y)]^{-1}\,\mathrm{d}y$$

is the integral over two segments of the line $\Re z = -\epsilon$, and

$$H_\rho(t) = \frac{1}{2\pi\mathrm{i}}\int_{-\epsilon}^{0} \left(\mathrm{e}^{(x-\mathrm{i}\rho)t}[D(x-\mathrm{i}\rho)]^{-1} - \mathrm{e}^{(x+\mathrm{i}\rho)t}[D(x+\mathrm{i}\rho)]^{-1}\right)\mathrm{d}x$$

arises from the integrals over the two intervals $[-\epsilon, 0] \times \{\pm\mathrm{i}\rho\}$.

Our purpose is to let ρ tend to zero and to show that $C_\rho \to 0$, that V_ρ is bounded by an exponentially decaying function as $\rho \downarrow 0$, and that $-H_\rho$ turns into a completely monotone function.

The claim that $C_\rho(t) \to 0$ as $\rho \downarrow 0$ is quite easy to prove. If $k \in L^1(\mathbf{R}^+; \mathbf{C}^{n\times n})$, then $\hat{k}(z) \to \hat{k}(0)$ as $z \to 0$, $\Re z \geq 0$. Now $\hat{k}(0) \succ 0$ (because $k(t) \succ 0$ for $t > 0$), and therefore $[D(z)]^{-1}$ is bounded as $z \to 0$, $\Re z \geq 0$. Clearly, this implies that

$$\lim_{\rho\downarrow 0} C_\rho(t) = 0, \quad t > 0. \tag{4.14}$$

If $k \notin L^1(\mathbf{R}^+; \mathbf{C}^{n\times n})$, then (see Exercise 18) we can write k as the sum of two completely monotone functions k_1 and k_2 where k_1 is integrable and satisfies $\hat{k}_1(0) \succ 0$. This, together with the first half of (4.6), implies that we can bound $\Re D(z)$ from below by $\Re D(z) \succeq K_1 I$ for some positive scalar K_1 in some neighbourhood of zero in $\Re z \geq 0$. Thus $[D(z)]^{-1}$ is uniformly bounded as $z \to 0$, $\Re z \geq 0$, and we again get (4.14).

Next, we consider $V_\rho(t)$. Since α is self-adjoint we have $D(\overline{z}) = D^*(z)$, and so

$$V_\rho(t) = \frac{\mathrm{e}^{-\epsilon t}}{\pi} \lim_{T\to\infty} \int_\rho^T \Re\left\{\mathrm{e}^{\mathrm{i}yt}[D(-\epsilon+\mathrm{i}y)]^{-1}\right\}\mathrm{d}y.$$

We split the integral in two parts over $[\rho, 1]$ and $[1, T]$, respectively. Using (4.11) and (4.12) we see immediately that there exists a constant K_2 (independent of ρ) such that

$$\left|\frac{1}{\pi}\int_\rho^1 \Re\left\{\mathrm{e}^{\mathrm{i}yt}[D(-\epsilon+\mathrm{i}y)]^{-1}\right\}\mathrm{d}y\right| \leq K_2. \tag{4.15}$$

In the second integral we integrate by parts to obtain

$$\begin{aligned}
&\int_1^T \Re\left\{\mathrm{e}^{\mathrm{i}yt}[D(-\epsilon+\mathrm{i}y)]^{-1}\right\}\mathrm{d}y \\
&\quad = \Re\left\{\frac{1}{it}\left(\mathrm{e}^{\mathrm{i}tT}[D(-\epsilon+\mathrm{i}T)]^{-1} - \mathrm{e}^{\mathrm{i}t}[D(-\epsilon+\mathrm{i})]^{-1}\right)\right. \\
&\qquad \left. + \frac{1}{t}\int_1^T \mathrm{e}^{\mathrm{i}yt}[D(-\epsilon+\mathrm{i}y)]^{-1}D'(-\epsilon+\mathrm{i}y)[D(-\epsilon+\mathrm{i}y)]^{-1}\,\mathrm{d}y\right\}.
\end{aligned}$$

To estimate D', we differentiate (4.4) to get

$$\hat{k}'(z) = -\int_{\mathbf{R}^+} \frac{\alpha(\mathrm{d}\sigma)}{(z+\sigma)^2},$$

and the dominated convergence theorem, together with Theorem 2.5 and the assumption $k \in L^1_{\text{loc}}(\mathbf{R}^+;\mathbf{C}^{n\times n})$, implies that $\hat{k}'(\epsilon+\mathrm{i}y) \to 0$ as $|y|\to\infty$. It is easy to estimate directly from (4.4) that $\hat{k}(\epsilon+\mathrm{i}y)/\mathrm{i}y \to 0$ as $|y|\to\infty$. This together with (4.15) shows that there exists a constant K such that

$$\left|V_\rho(t)\right| \le K\mathrm{e}^{-\epsilon t}, \quad t\ge 1, \quad \rho>0. \tag{4.16}$$

The term $H_\rho(t)$ remains to be analysed. (As we mentioned above, this is the term that will give us the completely monotone part of $-r$.) Once more we use the fact that $D(\overline{z}) = D^*(z)$, and rewrite $H_\rho(t)$ as

$$\begin{aligned} H_\rho(t) &= -\frac{1}{\pi}\int_{-\epsilon}^{0} \mathrm{e}^{xt}\cos(\rho t)\,\Im\big\{[D(x+\mathrm{i}\rho)]^{-1}\big\}\,\mathrm{d}x \\ &\quad - \frac{1}{\pi}\int_{-\epsilon}^{0} \mathrm{e}^{xt}\sin(\rho t)\,\Re\big\{[D(x+\mathrm{i}\rho)]^{-1}\big\}\,\mathrm{d}x. \end{aligned}$$

For the second integrand we have, by Lemma 4.2, the uniform bound

$$\left|\mathrm{e}^{xt}\sin(\rho t)\,\Re\big\{[D(x+\mathrm{i}\rho)]^{-1}\big\}\right| \le \frac{\rho t}{\rho} = t.$$

In addition, by (4.7),

$$\lim_{\rho\downarrow 0}\big(\mathrm{e}^{xt}\sin(\rho t)\,\Re\big\{[D(x+\mathrm{i}\rho)]^{-1}\big\}\big) = 0 \quad \text{for almost all } x\in[-\epsilon,0].$$

From the dominated convergence theorem we conclude that

$$\lim_{\rho\downarrow 0}\frac{1}{\pi}\int_{-\epsilon}^{0} \mathrm{e}^{xt}\sin(\rho t)\,\Re\big\{[D(x+\mathrm{i}\rho)]^{-1}\big\}\,\mathrm{d}x = 0.$$

Thus,

$$H_\rho(t) = -\frac{1}{\pi}\int_{-\epsilon}^{0} \mathrm{e}^{xt}\cos(\rho t)\,\Im\big\{[D(x+\mathrm{i}\rho)]^{-1}\big\}\,\mathrm{d}x. \tag{4.17}$$

For a further analysis of $H_\rho(t)$, we make the crucial observation that it follows from (4.13), (4.14), (4.16) and (4.17) that

$$\limsup_{\rho\downarrow 0}\left|\int_{-\epsilon}^{0} \mathrm{e}^{xt}\cos(\rho t)\,\Im\big\{[D(x+\mathrm{i}\rho)]^{-1}\big\}\,\mathrm{d}x\right| < \infty, \quad t\ge 1. \tag{4.18}$$

It is not true that $\Im\{[D(x+\mathrm{i}\rho)]^{-1}\}\succeq 0$, but it is almost true in the sense that, as we shall see in a moment, for all $-\eta\le x\le 0$ and $0<|y|\le\eta$,

$$\Im\big\{[D(x+\mathrm{i}y)]^{-1}\big\} \succeq -\frac{y}{\eta^2}I, \tag{4.19}$$

where η is the number in Lemma 4.2. This together with (4.18) implies that (fix, e.g., $t=1$, and argue componentwise in the same way as we did in the proof of Lemma 2.4)

$$\limsup_{\rho\downarrow 0}\int_{-\epsilon}^{0} \big|\Im\big\{[D(x+\mathrm{i}\rho)]^{-1}\big\}\big|\,\mathrm{d}x < \infty.$$

The unit ball in $M([0,\epsilon];\mathbf{C}^{n\times n})$ is sequentially weak* compact, and therefore, we can find some sequence $\{\rho_k\}$ tending to zero and a nonnegative measure $\mu\in M([0,\epsilon];\mathbf{C}^{n\times n})$ such that the sequence of functions $\Im\{[D(x+\mathrm{i}\rho_k)]^{-1}\}$ converges in the weak* sense to μ. In particular, this means that

$$\lim_{\rho_k\downarrow 0} H_{\rho_k}(t) = -\int_{[0,\epsilon]} \mathrm{e}^{-\sigma t}\mu(\mathrm{d}\sigma), \quad t>0. \tag{4.20}$$

By combining this with (4.13), (4.14) and (4.16) we conclude that

$$\left| r(t) + \int_{[0,\epsilon]} \mathrm{e}^{-\sigma t}\mu(\mathrm{d}\sigma)\right| \le K\mathrm{e}^{-\epsilon t}, \quad t\ge 1$$

(the restriction $t\ge 1$ is inherited from (4.16)). But r is continuous and the measure μ is finite, so the restriction that t should not be less than 1 can be removed, at the expense of enlarging K.

To complete the proof, we still have to verify (4.19). To do this, we have to show that, for all $v\in\mathbf{C}^n$ and all $z=x+\mathrm{i}y$ with $-\eta\le x\le 0$ and $0<|y|\le\eta$,

$$\left\langle v, \Im\{[D(z)]^{-1}\}v\right\rangle \ge -\frac{y}{\eta^2}|v|^2. \tag{4.21}$$

Fix x, y and v as above, and define

$$w = [D(z)]^{-1}v.$$

Then, by the second part of (4.6),

$$\langle v, \Im\{[D(z)]^{-1}\}v\rangle = -\langle w, \Im\{D(z)\}w\rangle \ge -y|w|^2. \tag{4.22}$$

If $\langle w, \Im\{D(z)\}w\rangle \le 0$, then (4.21) clearly holds. If not, then by Lemma 4.2,

$$\langle w, \Re\{D(z)\}w\rangle \ge \eta|w|^2.$$

This implies (see the first lines in the proof of Lemma 4.2) that

$$|w|\le \frac{1}{\eta}|v|.$$

Substituting this into (4.22) we get (4.21). □

5. Volterra Equations of the First Kind

In this section we study the equation

$$\int_0^t k(t-s)x(s)\,\mathrm{d}s = f(t), \quad t\in\mathbf{R}^+, \tag{1.1}$$

which is a Volterra equation of the first kind. In the introduction we pointed out that no new theory is needed for (1.1) if k and f are locally absolutely continuous and $k(0)$ is invertible. Furthermore, if $k(0)=0$ but k' and f'

are locally absolutely continuous and $k'(0)$ is invertible, then we obtain a Volterra equation of the second kind by differentiating (1.1) twice. If $k(0)$ is neither invertible, nor completely vanishing, then one gets a mixture of a Volterra integral and a Volterra integrodifferential equation. It is possible to develop a theory for such equations along the lines of Chapters 2 and 3, but it is also possible to rewrite an equation of this type as a Volterra equation of the second kind.

This procedure for the reduction of a Volterra equation of the first kind to a Volterra equation of the second kind is, in general, applicable whenever k and f are smooth enough to allow the evaluation of sufficiently many derivatives. However, there are many cases where this procedure is not applicable. Below, we consider some of these cases.

In order to see what the 'resolvent' should look like in this case, let us review what we have done earlier. For the equation $x + k * x = f$ we defined r to be the solution of $r + r * k = k$, and noted that $x = (\delta - r) * (x + k * x) = (\delta - r) * f = f - r * f$. In this calculation the crucial point is that $(\delta - r) * (\delta + k) = \delta$. In (1.1) it is clear that one cannot find a measure ρ such that $\rho * k = \delta$. However, one may be able to find a measure (frequently an absolutely continuous one) such that

$$(k * \rho)(t) = (\rho * k)(t) = I, \quad t \in \mathbf{R}^+. \tag{5.1}$$

If ρ is a solution of these equations, then it follows from (1.1) that $\int_0^t x(s)\,\mathrm{d}s = (\rho * k * x)(t) = (\rho * f)(t)$, and differentiating this equation with respect to t we get a new variation of constants formula, namely

$$x(t) = \tfrac{\mathrm{d}}{\mathrm{d}t}(\rho * f)(t), \quad t \in \mathbf{R}^+. \tag{5.2}$$

We refer to equations (5.1) as the resolvent equations of the first kind, and to (5.2) as the variation of constants formula of the first kind.

Note that, in general, we allow ρ to be a measure. One could, for reasons of symmetry, let the kernel k be a measure as well, but we shall not pursue this approach.

One can give a different motivation for equations (5.1) based on a singular perturbation analysis. Let us look at the Volterra integral equation

$$\epsilon x(t) + \int_0^t k(t-s)x(s)\,\mathrm{d}s = f(t), \quad t \in \mathbf{R}^+, \tag{5.3}$$

where ϵ is a small positive constant. Let q_ϵ be the resolvent of $\frac{1}{\epsilon}k$. Then q_ϵ satisfies

$$\epsilon q_\epsilon(t) + (q_\epsilon * k)(t) = \epsilon q_\epsilon(t) + (k * q_\epsilon)(t) = k(t), \quad t \in \mathbf{R}^+,$$

and the solution of (5.3) is given by

$$x_\epsilon(t) = \frac{1}{\epsilon}\big(f(t) - (q_\epsilon * f)(t)\big), \quad t \in \mathbf{R}^+.$$

Define $r_\epsilon(t) = \frac{1}{\epsilon}\big(I - \int_0^t q_\epsilon(s)\,\mathrm{d}s\big)$ for $t \in \mathbf{R}^+$. Then it is easy to show that r_ϵ satisfies

$$\epsilon r_\epsilon(t) + (r_\epsilon * k)(t) = \epsilon r_\epsilon(t) + (k * r_\epsilon)(t) = I, \quad t \in \mathbf{R}^+,$$

and that x_ϵ can be expressed in the form

$$x_\epsilon(t) = \tfrac{\mathrm{d}}{\mathrm{d}t}(r_\epsilon * f)(t), \quad t \in \mathbf{R}^+.$$

If we let $\epsilon \downarrow 0$, and assume that r_ϵ tends to a measure ρ and that x_ϵ tends to x, then, formally, we get (5.1) and (5.2).

5.1 Definition

Let $k \in L^1_{\mathrm{loc}}(\mathbf{R}^+; \mathbf{C}^{n\times n})$ and $\rho \in M_{\mathrm{loc}}(\mathbf{R}^+; \mathbf{C}^{n\times n})$ satisfy (5.1). Then ρ is said to be a (measure) resolvent of the first kind of k. If $k \in L^1_{\mathrm{loc}}(\mathbf{R}^+; \mathbf{C}^{n\times n})$ and $r \in L^1_{\mathrm{loc}}(\mathbf{R}^+; \mathbf{C}^{n\times n})$ satisfy (5.1) (with ρ replaced by r), then r is said to be a (function) resolvent of the first kind of k.

Note the obvious fact that, if $r \in L^1_{\mathrm{loc}}$ is a function resolvent of the first kind of $k \in L^1_{\mathrm{loc}}$, then k is a function resolvent of the first kind of r.

Usually we refer to both ρ and r above simply as 'a resolvent of the first kind of k', and specify separately if the resolvent is a measure or a function. Clearly, the class of function resolvents of the first kind can be identified with the class of absolutely continuous measure resolvents of the first kind: if r is a function resolvent of the first kind of k, then the measure $\rho(\mathrm{d}s) = r(s)\,\mathrm{d}s$ is an absolutely continuous measure resolvent of the first kind of k; conversely, if ρ is an absolutely continuous measure resolvent of k, then ρ is of the form $\rho(\mathrm{d}s) = r(s)\,\mathrm{d}s$ for some $r \in L^1_{\mathrm{loc}}$, and r is a function resolvent of k.

5.2 Theorem

A kernel $k \in L^1_{\mathrm{loc}}(\mathbf{R}^+; \mathbf{C}^{n\times n})$ can have at most one resolvent $\rho \in M_{\mathrm{loc}}(\mathbf{R}^+; \mathbf{C}^{n\times n})$ of the first kind. (In particular, if it has a function resolvent r, the only measure resolvent is $\rho(\mathrm{d}s) = r(s)\,\mathrm{d}s$.) Moreover, the restriction of the resolvent ρ to some interval $[0, T]$ depends on the values of k on that interval only.

The proof of this uniqueness result is left to the reader, and so is the proof of the following variation of constants formula.

5.3 Theorem

Let $k \in L^1_{\mathrm{loc}}(\mathbf{R}^+; \mathbf{C}^{n\times n})$ have a resolvent $\rho \in M_{\mathrm{loc}}(\mathbf{R}^+; \mathbf{C}^{n\times n})$ of the first kind. Then, for every $f \in L^1_{\mathrm{loc}}(\mathbf{R}^+; \mathbf{C}^n)$ such that $\rho * f \in AC_{\mathrm{loc}}(\mathbf{R}^+; \mathbf{C}^n)$ with $(\rho * f)(0) = 0$ (in particular, these conditions are satisfied if $f \in AC_{\mathrm{loc}}(\mathbf{R}^+; \mathbf{C}^n)$ and $f(0) = 0$), there is a unique solution $x \in L^1_{\mathrm{loc}}(\mathbf{R}^+; \mathbf{C}^n)$ of (1.1) given by $x(t) = \frac{\mathrm{d}}{\mathrm{d}t}(\rho * f)(t)$.

The classical example of a Volterra equation of the first kind is the scalar Abel equation

$$\int_0^t \frac{x(s)}{(t-s)^{1/2}}\,\mathrm{d}s = f(t), \quad t > 0,$$

which was discussed in Abel [1]. In this case, the resolvent of the first kind is the function $r(t) = \pi^{-1}t^{-1/2}$ (cf. Lemma **13**.6.3), and so

$$x(t) = \frac{f(0)}{\pi t^{1/2}} + \pi^{-1}\int_0^t \frac{f'(s)}{(t-s)^{1/2}}\,\mathrm{d}s.$$

The issue regarding what conditions on the kernel k imply the existence of a resolvent of the first kind is far from solved in the general case. We shall only give two positive results: a completely monotone kernel has a resolvent (Theorem 5.4) consisting of a point mass at zero and a completely monotone function, and a nonincreasing scalar kernel has a measure resolvent (Theorem 5.5). (See also Exercise 25.)

5.4 Theorem

Let $k \in L^1_{\mathrm{loc}}(\mathbf{R}^+; \mathbf{C}^{n\times n})$ be completely monotone on $(0,\infty)$, and suppose that $k(t) \succ 0$ for some $t > 0$. Then k has a resolvent of the first kind. This resolvent is the sum of a point mass at zero and a completely monotone function. The point mass at zero is invertible iff $\limsup_{t\downarrow 0}|k|(t) < \infty$, and it is absent iff $\lim_{t\downarrow 0}\langle v, k(t)v\rangle = \infty$ for all nonzero vectors $v \in \mathbf{C}^n$.

Below we sketch the proof of this theorem, omitting some of the details. Observe that the condition that $k(t) \succ 0$ for some $t > 0$ is necessary for the existence of a resolvent of the first kind (if $\langle v, k(t)v\rangle = 0$ for some $t > 0$, then $k(t)v \equiv 0$, and one cannot possibly have $\rho * k = I$).

Proof of Theorem 5.4 Using arguments much like those in the proof of Theorem 3.1, we find that $\det[z\hat{k}(z)] \neq 0$ for $z \in \mathbf{C} \setminus \mathbf{R}^-$ and conclude that $F(z) \stackrel{\mathrm{def}}{=} [z\hat{k}(z)]^{-1}$ is analytic in this region.

We claim that $\limsup_{x\to+\infty}|F(x)| < \infty$. To see this, observe that, by (2.3), with F replaced by $\hat{k}$,

$$[F(x)]^{-1} = x\hat{k}(x) = \int_{\mathbf{R}^+} \frac{x}{x+\sigma}\,\alpha(\mathrm{d}\sigma) \tag{5.4}$$

hence $[F(x)]^{-1}$ is a monotone nondecreasing, strictly positive function on $\mathbf{R}^+$.

Note that $\Im\hat{k}(z) \preceq 0$ for $\Im z > 0$ gives $\Im\{zF(z)\} \succeq 0$ for $\Im z > 0$, and that $\hat{k}(x) \succ 0$ for $x > 0$ gives $F(x) \succ 0$ for $x > 0$. Thus, by Corollary 2.7, F is the sum of a positive constant A and the transform of a completely monotone, locally integrable function r, and $\lim_{x\to+\infty} F(x) = A$. We find that the measure $A\delta(\mathrm{d}t) + r(t)\,\mathrm{d}t$ is a resolvent of the first kind of k.

The proofs of the claims which relate A to the behaviour of k at zero rely on the fact that, by (2.2) and (5.4), for all $v \in \mathbf{C}^n$,

$$\begin{aligned}\lim_{t\downarrow 0}\langle v, k(t)v\rangle &= \int_{\mathbf{R}^+}\langle v, \alpha(\mathrm{d}\sigma)v\rangle \\ &= \lim_{x\to+\infty}\int_{\mathbf{R}^+}\frac{x}{x+\sigma}\langle v, \alpha(\mathrm{d}\sigma)v\rangle = \lim_{x\to+\infty}\langle v, [F(x)]^{-1}v\rangle.\end{aligned} \tag{5.5}$$

The easy proofs of the facts that A is invertible iff $\limsup_{t\downarrow 0}|k|(t) < \infty$, and that $\lim_{t\downarrow 0}\langle v, k(t)v\rangle = \infty$ for all nonzero $v \in \mathbf{C}^n$ if $A = 0$ are left to the reader (see Exercise 3). The proof of the fact that $A = 0$ if $\lim_{t\downarrow 0}\langle v, k(t)v\rangle = \infty$ for all nonzero $v \in \mathbf{C}^n$ is slightly more complicated. We claim that this condition implies that

$$\lim_{x\to+\infty} \inf_{|v|=1} \int_{\mathbf{R}^+} \frac{x}{x+\sigma}\langle v, \alpha(\mathrm{d}\sigma)v\rangle = \infty, \tag{5.6}$$

i.e., we claim that $\langle v, [F(x)]^{-1}v\rangle$ tends to infinity, uniformly for $|v| = 1$, as $x \to +\infty$. If not, then there are a finite M, and sequences x_j, and v_j, with $x_j \to \infty$ as $j \to \infty$, such that

$$\lim_{j\to\infty} \int_{\mathbf{R}^+} \frac{x_j}{x_j+\sigma}\langle v_j, \alpha(\mathrm{d}\sigma)v_j\rangle = M.$$

Without loss of generality, suppose that $v_j \to v$ as $j \to \infty$, for some $v \in \mathbf{C}^n$ with $|v| = 1$ (pass to a subsequence). Then, by Lebesgue's dominated convergence theorem, for each finite T we have

$$\int_{[0,T]} \langle v, \alpha(\mathrm{d}\sigma)v\rangle = \lim_{j\to\infty} \int_{[0,T]} \frac{x_j}{x_j+\sigma}\langle v_j, \alpha(\mathrm{d}\sigma)v_j\rangle \le M.$$

This being true for all T, we conclude that $\limsup_{x\to+\infty}\langle v, [F(x)]^{-1}v\rangle \le M$, contradicting the assumption that $\lim_{x\to+\infty}\langle v, [F(x)]^{-1}v\rangle = \infty$ for all nonzero $v \in \mathbf{C}^n$. Thus, (5.6) holds.

Once (5.6) is known to hold, it is not difficult to show that $A = 0$. We leave this to the reader. □

In the scalar case we have the following more general result on the existence of the resolvent of the first kind of a nonnegative and nonincreasing kernel. (It is mainly this result that motivates us to allow the resolvent of a function to be a measure.)

5.5 Theorem
Let $k \in L^1_{\mathrm{loc}}(\mathbf{R}^+;\mathbf{R})$ be nonnegative and nonincreasing on $(0,\infty)$, and not identically zero. Then k has a positive (measure) resolvent ρ of the first kind. This resolvent has no discrete part iff $\lim_{t\downarrow 0} k(t) = +\infty$.

We leave it as an exercise to show that, even when $\lim_{t\downarrow 0} k(t) = +\infty$, the resolvent above need not be (induced by) a locally integrable function. In Corollary 5.6 we give conditions on the kernel k that imply that the resolvent of the first kind is a locally integrable function r.

Proof of Theorem 5.5 Let κ be the (measure) derivative of $-k$, i.e., κ is the locally finite measure on $(0,\infty)$ satisfying $\kappa((t,\infty)) = k(t)$ for $t > 0$ (without loss of generality we take k to be right-continuous). Then κ is nonnegative, and $\int_{(1,\infty)} \kappa(\mathrm{d}t)$ is finite. The monotonicity of k implies that $0 \le tk(t) \le \int_0^t k(s)\,\mathrm{d}s$, and, therefore, $tk(t) \to 0$ as $t \downarrow 0$. Moreover,

$\int_{(0,1]} t\kappa(\mathrm{d}t) = \int_0^1 k(t)\,\mathrm{d}t - k(1)$ is finite. Integrating by parts, and using these facts, we get

$$x\hat{k}(x) = k(\infty) + \int_{(0,\infty)} (1 - \mathrm{e}^{-xt})\kappa(\mathrm{d}t), \quad x > 0.$$

Define $\varphi(x) = x\hat{k}(x)$ for $x > 0$. Then $\varphi(x) > 0$ and $(-1)^j\varphi^{(j)}(x) \le 0$ for $x > 0$ and $j = 1, 2, \ldots$, i.e., φ is a Bernstein function (see Berg and Forst [1], p. 61). Since φ is not identically 0, it follows that $1/\varphi$ is completely monotone; see Berg and Forst [1], p. 66. Thus by Bernstein's Theorem 2.5, there exists a positive measure $\rho \in M_{\mathrm{loc}}(\mathbf{R}^+; \mathbf{R})$, satisfying $\int_{\mathbf{R}^+} \mathrm{e}^{-\epsilon t}|\rho|(\mathrm{d}t) < \infty$ for all $\epsilon > 0$, such that $\hat{k}(x)\hat{\rho}(x) = \hat{\rho}(x)\hat{k}(x) = 1/x$ for $x > 0$. From the uniqueness of Laplace transforms we get (5.1).

If $\lim_{t\downarrow 0} k(t) = k(0) < \infty$, then it is clear that $\rho(\{0\}) = 1/k(0)$ and ρ has a discrete part. Conversely, if ρ has a discrete part, i.e., if ρ has a point mass somewhere, then there exist constants $t_0 \ge 0$ and $c_0 > 0$ such that

$$1 = \int_{[0,t]} k(t-s)\rho(\mathrm{d}s) \ge c_0 k(t - t_0), \quad t \ge t_0.$$

But this is impossible if $\lim_{t\downarrow 0} k(t) = +\infty$. □

5.6 Corollary

Let $k \in L^1_{\mathrm{loc}}(\mathbf{R}^+; \mathbf{R})$ be nonnegative and nonincreasing on $(0,\infty)$, with $\lim_{t\downarrow 0} k(t) = \infty$. In addition, let k be locally absolutely continuous on $(0,\infty)$. Then the resolvent of the first kind of k is (induced by) a locally integrable function.

Proof To show that ρ is absolutely continuous with respect to Lebesgue measure, note that $\hat{\rho}(z) = 1/(z\hat{k}(z))$, and differentiate this relation to obtain

$$\frac{\mathrm{d}\hat{\rho}(z)}{\mathrm{d}z} = -\left(\hat{\rho}(z)\right)^2 \frac{\mathrm{d}}{\mathrm{d}z}\left(z\hat{k}(z)\right), \quad \Re z > 0.$$

Now $\mathrm{e}^{-t}tk'(t) \in L^1(\mathbf{R}^+; \mathbf{R})$ and $\frac{\mathrm{d}}{\mathrm{d}z}(z\hat{k}(z))$ is the transform of $tk'(t)$. We recall from the proof of Theorem 5.5 that $\int_{\mathbf{R}^+} \mathrm{e}^{-t}|\rho|(\mathrm{d}t) < \infty$. Thus $\frac{\mathrm{d}}{\mathrm{d}z}\hat{\rho}(z)$ is the transform of a function ϕ satisfying $\int_{\mathbf{R}^+} \mathrm{e}^{-t}|\phi(t)|\,\mathrm{d}t < \infty$. But $\frac{\mathrm{d}}{\mathrm{d}z}\hat{\rho}(z)$ is the transform of the measure $t\rho(\mathrm{d}t)$. Since ρ has no point mass at the origin, we conclude that ρ is absolutely continuous. □

We end this chapter with an easy perturbation result.

5.7 Theorem

Let $k \in L^1_{\mathrm{loc}}(\mathbf{R}^+; \mathbf{C}^{n\times n})$ have a resolvent of the first kind. Let $\alpha \in M_{\mathrm{loc}}(\mathbf{R}^+; \mathbf{C}^{n\times n})$ satisfy $\det[I + \alpha(\{0\})] \ne 0$. Then the kernel $k + k * \alpha$ has a resolvent of the first kind.

Proof Let $\rho_0 \in M_{\mathrm{loc}}(\mathbf{R}^+; \mathbf{C}^{n\times n})$ be the resolvent of the first kind of k. Let β be the measure resolvent of α (Theorem **4**.1.5 guarantees the existence of β), and let $\rho = \rho_0 - \beta * \rho_0$. Some straightforward calculations, where one uses the facts that $(\delta + \alpha) * (\delta - \beta) = (\delta - \beta) * (\delta + \alpha) = \delta$ and that $k * \rho_0 = \rho_0 * k = I$, show that ρ is the resolvent of the first kind of $k + k * \alpha$. □

6. Exercises

1. Show that $\|A\| \stackrel{\text{def}}{=} \sup_{|v|=1} |\langle v, Av \rangle|$ defines a norm on $\mathbf{C}^{n \times n}$, and that $\|A\| \le |A|$ for every $A \in \mathbf{C}^{n \times n}$. Is this norm compatible with matrix multiplication, i.e., is it true for all matrices A and $B \in \mathbf{C}^{n \times n}$ that $\|AB\| \le \|A\| \|B\|$? Hint: You can find the answer to the second question by studying 2×2 matrices with three elements out of four equal to zero.
2. Let $A \in \mathbf{C}^{n \times n}$, and suppose that $\alpha \stackrel{\text{def}}{=} \inf_{|v|=1} \langle v, Av \rangle > 0$. Show that A is invertible, and that $|A^{-1}| \le \frac{1}{\alpha}$.
3. Let $A \in \mathbf{C}^{n \times n}$ satisfy $A \succ 0$. Prove that
$$\inf_{|v|=1} \langle v, Av \rangle = \left\{ \sup_{|w|=1} \langle w, A^{-1} w \rangle \right\}^{-1}.$$
4. Show that every inner product on $\mathbf{C}^n$ is of the form $\langle \cdot, B \cdot \rangle$ where $\langle \cdot, \cdot \rangle$ denotes some fixed inner product and B is a matrix that is strictly positive with respect to this inner product. Hint: See Rudin [2], Theorem 12.8.
5. Let $\langle \cdot, \cdot \rangle$ be a given inner product on $\mathbf{C}^n$, let $\succeq$ and $\succ$ refer to this inner product, and let $A \in \mathbf{C}^{n \times n}$. Show that the following conditions are equivalent.
   (i) A is positive with respect to *some* inner product in $\mathbf{C}^n$;
   (ii) $BA \succeq 0$ for some $B \succ 0$;
   (iii) $AB \succeq 0$ for some $B \succ 0$;
   (iv) $B^{-1}AB \succeq 0$ for some $B \succ 0$.

   Also prove the corresponding result where the positivity has been replaced by strict positivity. Hint for (iv): Every strictly positive matrix has a strictly positive square root.
6. Let $\langle \cdot, \cdot \rangle$ be some inner product on $\mathbf{C}^n$. Show that there exist vectors u_j and $v_j \in \mathbf{C}^n$, $j = 1, \ldots, n$, such that $\langle Au_j, v_k \rangle = A_{jk}$ for every matrix $A \in \mathbf{C}^{n \times n}$ with elements A_{jk}. Hint: See Exercise 4.
7.
   (i) Let A and $B \in \mathbf{C}^{n \times n}$ be invertible. Show that
   $$A^{-1} - B^{-1} = A^{-1}(B - A)B^{-1} = B^{-1}(B - A)A^{-1}.$$
   (ii) Let A be a continuously differentiable $\mathbf{C}^{n \times n}$-valued function. Show that A^{-1} is continuously differentiable whenever it exists,

and that

$$\frac{\mathrm{d}}{\mathrm{d}t}[A(t)]^{-1} = -[A(t)]^{-1}A'(t)[A(t)]^{-1}$$

(this result was needed in the proof of Theorem 4.1).

(iii) In addition to (ii), suppose that $A \succ 0$, and $A' \preceq 0$. Show that $\frac{\mathrm{d}}{\mathrm{d}t}[A(t)]^{-1} \succeq 0$.

(iv) In addition to (ii), suppose $A \succ 0$ is two times continuously differentiable, and that $A'' \preceq 0$. Show that the second derivative of A^{-1} is positive.

8. Let $A \in \mathbf{C}^{n\times n}$ be invertible. Show that
   (i) $\Re\{A^{-1}\} = [A^{-1}]^*\Re\{A\}A^{-1} = A^{-1}\Re\{A\}[A^{-1}]^*$,
   (ii) $\Im\{A^{-1}\} = -[A^{-1}]^*\Im\{A\}A^{-1} = -A^{-1}\Im\{A\}[A^{-1}]^*$.
9. Let $A \in \mathbf{C}^{n\times n}$ be invertible. Show that
   (i) $\Re A \succeq 0 \Rightarrow \Re\{A^{-1}\} \succeq 0$,
   (ii) $\Re A \succ 0 \Rightarrow \Re\{A^{-1}\} \succ 0$,
   (iii) $\Im A \succeq 0 \Rightarrow \Im\{A^{-1}\} \preceq 0$,
   (iv) $\Im A \succ 0 \Rightarrow \Im\{A^{-1}\} \prec 0$.
10. Let
$$A = \begin{pmatrix} 0 & 1 \\ 0 & 0 \end{pmatrix}.$$
Find a vector $v \in \mathbf{C}^2$ such that $Av = 0$, but $(\Re A)v \neq 0 \neq (\Im A)v$.
11. Let A, B be positive matrices such that $AB = BA$. Show that AB is positive.
12. Find positive matrices A and B such that AB is not positive.
13. Give an example of a matrix for which the claim $\Re A \succeq 0$ is true for some inner product, but not for all inner products.
14. Show that e^{-At} is completely monotone iff $A \succeq 0$. Hint: Cf. Exercise 11.
15. Add the following claim to Theorem 2.6. The Laplace transform of a completely monotone function $a \in L^1_{\mathrm{loc}}(\mathbf{R}^+; \mathbf{C}^{n\times n})$ satisfies $\Re\hat{a}(z) \succeq 0$ for $\Re z > 0$. (Thus, it is of positive type; see Chapter 16.)
16. Let $a \in L^1_{\mathrm{loc}}(\mathbf{R}^+; \mathbf{R})$ be completely monotone on $\mathbf{R}^+$ and let m be a positive integer. Show that there exists a completely monotone function $b \in L^1_{\mathrm{loc}}(\mathbf{R}^+; \mathbf{R})$ so that $b^{*m} = a$.
17. Suppose a is completely monotone on an interval $(0, T)$ where $T < \infty$. Is it always possible to extend a as a completely monotone function to $\mathbf{R}^+$?
18. Show that, if a is completely monotone and strictly positive on $(0, \infty)$, and if a is not integrable over $\mathbf{R}^+$, then a can be written as the sum of two completely monotone functions a_1 and a_2, where a_1 is integrable over $\mathbf{R}^+$ and satisfies $\hat{a}_1(0) \succ 0$.
19. Let $F(z) = \sum_{j=1}^{N} a_j(z + b_j)^{-p_j}$, where N is finite, each $a_j \neq 0$ is real, each $b_j \geq 0$ is real, and each $p_j > 0$ is an integer. Moreover, assume that no two terms in the sum can be combined into one, i.e.,

assume that if $b_j = b_k$ then $p_j \neq p_k$. Show that F is the transform of a completely monotone, locally integrable function if and only if each $p_j = 1$ and each $a_j > 0$.

20. Prove the converse part of Theorem 3.1 and the integrability claim in Theorem 4.1.
21. Let k be the scalar kernel $k(t) = ae^{-bt}$, where a and b are strictly positive constants. Show that for this kernel the completely monotone part of the differential resolvent in the decomposition in Theorem 4.1 may be taken identically equal to 0. Determine the largest possible value of ϵ as a function of a and b.
22. Let $k \in L^1_{\mathrm{loc}}(\mathbf{R}^+; \mathbf{C}^{n\times n})$ be completely monotone, and let $A \succ 0$. Show that the equations $Aq(t) + (k * q)(t) = q(t)A + (q * k)(t) = I$ for $t \geq 0$ have a unique completely monotone solution q. In what way is q related to the solution of the equation $Ax + k * x = f$? Can you extend this to the case where $A \succeq 0$, but A is not invertible?
23. Give an example of two functions a and b such that $a \in C((0,\infty); \mathbf{R}) \cap L^1(\mathbf{R}^+; \mathbf{R})$ and $a(t) \to 0$ as $t \to \infty$ and $b \in C(\mathbf{R}^+; \mathbf{R}) \cap L^1(\mathbf{R}^+; \mathbf{R})$, but $(a * b)(t) \not\to 0$ as $t \to \infty$.
24. Solve
$$\int_0^t \frac{x(s)}{(t-s)^\alpha}\, ds = f(t),$$
for $t > 0$ with $\alpha \in (0,1)$ fixed.
25. Suppose that $k \in BV_{\mathrm{loc}}(\mathbf{R}^+; \mathbf{C}^{n\times n})$ is right-continuous. Show that k has a resolvent of the first kind iff $k(0)$ is invertible. Moreover, show that the resolvent of k is the sum of a point mass at zero and an absolutely continuous measure iff $k \in AC_{\mathrm{loc}}(\mathbf{R}^+; \mathbf{C}^{n\times n})$. Hint: Compute the (ordinary) measure resolvent of the distribution derivative of $k - I$.
26. Suppose that $k \in L^1_{\mathrm{loc}}(\mathbf{R}^+; \mathbf{C}^{n\times n})$ satisfies $k(t) \to 0$ as $t \to \infty$. Show that k cannot have a (function) resolvent r of the first kind satisfying $r \in L^1(\mathbf{R}^+; \mathbf{C}^{n\times n})$.
27. Prove Theorem 5.2.
28. Prove Theorem 5.3.
29. Show that the completely monotone part r of the resolvent given by Theorem 5.4 belongs to $L^1(\mathbf{R}^+; \mathbf{C}^{n\times n})$ if and only if $k(\infty) \succ 0$, and that $r(\infty) \succ 0$ if and only if $k \in L^1(\mathbf{R}^+; \mathbf{C}^{n\times n})$.
30. Prove the following result.

6.1 Theorem
Let $k = k_1 + k_2$, where $k_1 \in L^1_{\mathrm{loc}}(\mathbf{R}^+; \mathbf{R})$ is nonnegative and nonincreasing on $(0,\infty)$ and satisfies $\lim_{t\downarrow 0} k_1(t) = \infty$ and where $k_2 \in BV_{\mathrm{loc}}(\mathbf{R}^+; \mathbf{R})$. Then k has a resolvent of the first kind.

31. Give an example of a nonincreasing, nonnegative locally integrable function k with $\lim_{t\downarrow 0} k(t) = +\infty$, such that the resolvent of the

first kind of k is not an absolutely continuous measure. Hint: See Gripenberg [12].

32. Construct a matrix function F with $F(x) \succ 0$ for $x > 0$ such that $\langle v, [F(x)]^{-1} v\rangle \to \infty$ as $x \to \infty$ for all nonzero $v \in \mathbf{C}^n$, but $F(x) \not\to 0$ as $x \to \infty$ (compare this to the situation in the proof of Theorem 5.4).

7. Comments

Section 2:

Theorem 2.5 was first obtained by Bernstein [1]. See also Widder [1] and [2], p. 144. In Bochner [2], complete monotonicity is used to discuss the Laplace operator on a torus. A more general version of Theorem 2.5 is given in Bochner [3].

The scalar version of Theorem 2.6 can be found in Gripenberg [12]. It is closely related to Shohat and Tamarkin [1], Lemma 2.1, where the essential part of the scalar argument is given.

Theorem 2.8 is the same result as is found in Miller [4], Lemma 2, p. 326.

Section 3:

In Reuter [1] it was proved, apparently for the first time, that the resolvent of a completely monotone scalar function is completely monotone. The same result can be found in Miller [4], with a reference to Friedman [1] for the proof. This proof is more complicated than the one given here, and it does not rely on Theorem 2.6. For the scalar case the converse part of Theorem 3.1 was proved in Gripenberg [2], Theorem 3.

Friedman [1] has raised the question about intermediate results, i.e., he asks for conditions that imply $(-1)^j r^{(j)}(t) \geq 0$ for $t > 0$ and $j = 1, 2, \ldots, m$. This question has not yet been answered, but it seems that log-convexity plays an important role. As we shall see in Section **9**.8, if k is log-convex and nonincreasing, then the resolvent of k is nonnegative. Gripenberg [2] has proved that if, in addition, $-k'$ is log-convex then r is nonincreasing.

Section 4:

Theorem 4.1 and its proof are taken from Hannsgen and Wheeler [1]. They also discuss the relationship between the measure α corresponding to k and the measure μ mentioned in the conclusion of Theorem 4.1. Some further results in this direction can be found in Hannsgen and Wheeler [4]. In particular, it is shown that the numbers M and ϵ of Theorem 4.1 can be chosen uniformly over certain classes of kernels. For an application of Theorem 4.1, see Hannsgen and Wheeler [3], Section 4. Some earlier results on the differential resolvent of a completely monotone function are given in Hannsgen [3].

Section 5:

The classical reference to Volterra equations of the first kind is Abel [1]. Most of the results in this section are taken from Gripenberg [12], where additional results can be found. For example, if k in Corollary 5.6 is log-convex, then the resolvent r of first kind is nonincreasing, and if $-k'$ is log-convex, then r is convex. A result closely related to Theorem 5.5 is given in Theorem **20**.4.16 and Corollary **20**.4.18. The proof of Theorem 5.5 given in Gripenberg [12] is much more complicated than the one given here, suggested to us by Ph. Clément and J. Prüss.

Various modifications of the Abel equation have been studied by, e.g., Atkinson [1]. Vessella [1] gives estimates of the solutions of generalized Abel equations. For an extensive review of the Abel equation and of its applications, see Gorenflo and Vessella [1]. See also Reynolds [1].

References

N. H. Abel

1. Résolution d'un problème de mécanique, in *Oeuvres* **1**, Christiania, 1881, pp. 97–101.

K. E. Atkinson

1. An existence theorem for Abel integral equations, *SIAM J. Math. Anal.* **5** (1974), pp. 729–736.

C. Berg and G. Forst

1. *Potential Theory on Locally Compact Abelian Groups*, Springer-Verlag, Berlin, 1975.

S. Bernstein

1. Sur les fonctions absolument monotones, *Acta Math.* **51** (1928), pp. 1–66.

S. Bochner

2. Completely monotone functions of the Laplace operator for torus and sphere, *Duke Math. J.* **3** (1937), pp. 488–502.
3. Completely monotone functions in partially ordered spaces, *Duke Math. J.* **9** (1942), pp. 519–526.

A. Friedman

1. On integral equations of Volterra type, *J. Analyse Math.* **11** (1963), pp. 381–413.

R. Gorenflo and S. Vessella

1. *Basic theory and some applications of Abel integral equations*, Preprint 237/1986, Fachbereich Mathematik, Freie Universität, Berlin.

G. Gripenberg

2. On positive, nonincreasing resolvents of Volterra equations, *J. Differential Equations* **30** (1978), pp. 380–390.
12. On Volterra equations of the first kind, *Integral Equations Operator Theory* **3/4** (1980), pp. 473–488.

K. B. Hannsgen

1. Indirect Abelian theorems and a linear Volterra equation, *Trans. Amer. Math. Soc.* **142** (1969), pp. 539–555.
3. A Volterra equation with completely monotonic convolution kernel, *J. Math. Anal. Appl.* **31** (1970), pp. 459–471.

K. B. Hannsgen and R. L. Wheeler

1. Complete monotonicity and resolvents of Volterra integrodifferential equations, *SIAM J. Math. Anal.* **13** (1982), pp. 962–969.
3. A singular limit problem for an integrodifferential equation, *J. Integral Equations* **5** (1983), pp. 199–209.
4. Uniform L^1-behavior in classes of integrodifferential equations with completely monotone kernels, *SIAM J. Math. Anal.* **15** (1984), pp. 579–594.

R. K. Miller

4. On Volterra integral equations with nonnegative integrable resolvents, *J. Math. Anal. Appl.* **22** (1968), pp. 319–340.

G. E. H. Reuter

1. Über eine Volterrasche Integralgleichung mit totalmonotonem Kern, *Arch. Math.* **7** (1956), pp. 59–66.

D. W. Reynolds

1. On linear singular Volterra integral equations of the second kind, *J. Math. Anal. Appl.* **103** (1984), pp. 230–262.

W. Rudin

2. *Functional Analysis*, McGraw–Hill, New York, 1973.
3. *Real and Complex Analysis*, 3rd *ed.*, McGraw–Hill, New York, 1986.

J. A. Shohat and J. D. Tamarkin

1. *The Problem of Moments*, Amer. Math. Soc., Providence, R. I., 1943.

S. Vessella

1. Stability results for Abel equations, *J. Integral Equations* **9** (1985), pp. 125–134.

D. V. Widder

1. Necessary and sufficient conditions for the representation of a function as a Laplace integral, *Trans. Amer. Math. Soc.* **33** (1931), pp. 851–892.
2. *The Laplace Transform*, Princeton University Press, Princeton, 1946.

6

Nonintegrable Kernels with Integrable Resolvents

First we prove that the resolvent of a real, nonnegative, nonincreasing, and convex function is integrable. Then we extend this result to a more general class of kernels by considering functions that are locally equal to Fourier transforms of integrable functions.

1. Introduction

We have previously, in Chapter 2, stressed the fact that the asymptotic behaviour of the resolvent r is of crucial importance for the asymptotic behaviour of the solution of

$$x(t) + \int_0^t k(t-s)x(s)\,\mathrm{d}s = f(t), \quad t \in \mathbf{R}^+. \tag{1.1}$$

In particular, it was shown that if $r \in L^1(\mathbf{R}^+)$ then the behaviour of f at infinity is inherited by x, since $x = f - r * f$. Analogous considerations apply to the integrodifferential equation $x' + \mu * x = f$; see Chapter 3. We even observed (see Theorem **2**.2.4) that a necessary condition for the convolution operator $f \mapsto r * f$ to map $L^\infty(\mathbf{R}^+)$ (or $L^1(\mathbf{R}^+)$) into itself is that $r \in L^1(\mathbf{R}^+)$. Moreover, the applicability of perturbation techniques frequently depends on a knowledge of the size of the resolvent (see Chapter 11).

Thus it becomes very important to answer the question: what conditions on the kernel k imply $r \in L^1(\mathbf{R}^+)$? In the case where $k \in L^1(\mathbf{R}^+)$, we obtained a complete answer in Theorem **2**.4.1: a necessary and sufficient condition is that $I + \hat{k}(z)$ be invertible for all z with $\Re z \geq 0$. In the case of the integrodifferential equation $x' + \mu * x = f$, with $\mu \in M(\mathbf{R}^+)$, the condition is that $zI + \hat{\mu}(z)$ should be invertible for the same values of z, see Theorem **3**.3.5. For the integrability properties of the resolvent in weighted spaces, see Chapter 4. A large number of applications do, however, exhibit kernels that are not integrable at infinity, but behave asymptotically as, e.g., $t^{-\alpha}$ for some $\alpha \in (0, 1)$. In several cases one can apply the results from Chapter 5 where it is assumed that the kernel is completely monotone. But there are clearly good reasons to search for other hypotheses on a nonintegrable kernel k under which $r \in L^1(\mathbf{R}^+)$.

Rather few explicit results in this direction are currently known, and in fact this question constitutes one of the major open problems in the theory of linear Volterra equations. A Paley–Wiener type condition is clearly necessary, but it is not sufficient when the kernel is not integrable. This is seen from the fact that there exist kernels of positive type (for such kernels, see Definition **16**.2.1 and Theorem **16**.2.4) whose resolvents are not integrable; one such example is given in Exercise 4. Frequently one is able to compute an expression for $\hat{r}(z)$, valid in the half plane $\Re z > 0$, or even an expression for the boundary function $\tilde{r}(\omega)$; the difficulty then consists in showing that $\tilde{r}(\omega)$ is the Fourier transform of an integrable function. Necessary conditions are obvious: $\tilde{r}(\omega)$ should be bounded, continuous and vanish at infinity.

One of the few results that can be succinctly stated, and proved without unreasonable effort, is Theorem 2.1 below, due to Shea and Wainger [1]. By this result, a locally integrable, nonnegative, nonincreasing, and convex kernel k has an integrable resolvent. In addition, it is true that the L^1-norm of the resolvent is bounded independently of k.

The result of Shea and Wainger can be extended to give interesting results for the integrodifferential equation as well. The basic idea is the observation that, in order to prove that $\tilde{r}(\omega)$ is the transform of an integrable function, it suffices to show that this is true locally, i.e., in the neighbourhood of each point (including infinity). It turns out that this is often much easier to do.

2. The Shea–Wainger Theorem

In this section we prove that the resolvent of a nonnegative, nonincreasing, and convex function is integrable. The proof combines certain estimates on the Fourier transforms of such kernels with a classic result (Theorem 5.1) on Fourier transforms of functions in a Hardy space over a half plane. In our case, this result says that $\|r\|_{L^1(\mathbf{R}^+)} \le \frac{1}{2}\|\tilde{r}'\|_{L^1(R)}$. In principle, this method of proof can be applied to other classes of kernels as well, but in general it is very difficult to obtain the necessary estimates.

2.1 The Shea–Wainger Theorem

Let $k \in L^1_{\mathrm{loc}}(\mathbf{R}^+; \mathbf{R})$ be nonnegative, nonincreasing, and convex on $(0, \infty)$. Then the resolvent r of k is integrable, and $\|r\|_{L^1(\mathbf{R}^+)} \le 20$.

To prove this theorem, we first prove a lemma that gives the necessary estimates on the Laplace transform of a nonnegative, nonincreasing, and convex function, and then we use these estimates, together with Theorem 5.1, to complete the proof.

2.2 Lemma
Let $k \in L^1_{\text{loc}}(\mathbf{R}^+; \mathbf{R})$ be nonnegative, nonincreasing, and convex on $(0, \infty)$. Then the Laplace transform $\hat{k}(z)$ of k exists in the open half plane $\Re z > 0$, it satisfies $\Re \hat{k}(z) \geq 0$ and $\Im z \, \Im \hat{k}(z) \leq 0$ in this half plane, and it can be extended to a continuously differentiable function on $\{ \Re z \geq 0 \mid z \neq 0 \}$. If we denote this extended function by $\hat{k}$ as well, then, for each real $\omega \neq 0$,

$$\frac{1}{\sqrt{2}} \int_0^{1/|\omega|} k(t)\,\mathrm{d}t \leq |\hat{k}(\mathrm{i}\omega)| \leq 2 \int_0^{1/|\omega|} k(t)\,\mathrm{d}t \tag{2.1}$$

and

$$|\hat{k}'(\mathrm{i}\omega)| \leq 5 \int_0^{1/|\omega|} t k(t)\,\mathrm{d}t. \tag{2.2}$$

Proof Trivially, if $k(t) \equiv k(\infty) \overset{\text{def}}{=} \lim_{t\to\infty} k(t)$, then all the assertions are true. Therefore, let $k(t) \not\equiv k(\infty)$.

For each $t > 0$, we let $k'(t)$ denote the right-hand derivative of k at t. Then k' is right-continuous, nonpositive, nondecreasing, and locally of bounded variation on $(0, \infty)$. If $0 < s < t < \infty$, then we can write $k'(t)$ in the form

$$k'(t) = k'(s) + \int_{(s,t]} \gamma(\mathrm{d}u),$$

where γ is a nonnegative, locally finite measure on $\mathbf{R}^+$.

In the sequel, we perform several integrations by parts. To be able to carry out these integrations we need to know that $tk(t) \to 0$ and $t^2 k'(t) \to 0$ as $t \downarrow 0$, that $tk'(t) \to 0$ as $t \to \infty$, and that the integrals $\int_0^1 t|k'(t)|\,\mathrm{d}t$, $\int_1^\infty |k'(t)|\,\mathrm{d}t$, $\int_{(0,1]} t^2\gamma(\mathrm{d}t)$ and $\int_{(1,\infty)} t\gamma(\mathrm{d}t)$ are finite. Let us prove these facts. The monotonicity of k implies that $0 \leq tk(t) \leq \int_0^t k(s)\,\mathrm{d}s$, and therefore $t\,k(t) \to 0$ as $t \downarrow 0$. Because k' is nonpositive, and since $\int_0^1 t\,k'(t)\,\mathrm{d}t = k(1) - \int_0^1 k(t)\,\mathrm{d}t$ and $\int_1^\infty k'(t)\,\mathrm{d}t = k(\infty) - k(1)$, we see that both $\int_0^1 t|k'(t)|\,\mathrm{d}t$ and $\int_1^\infty |k'(t)|\,\mathrm{d}t$ are finite. Using the monotonicity of k', we get $t^2|k'(t)| \leq 2\int_0^t s|k'(s)|\,\mathrm{d}s$ and

$$\frac{t}{2}|k'(t)| \leq \int_{t/2}^t |k'(s)|\,\mathrm{d}s = k(t/2) - k(t).$$

Therefore, $t^2 k'(t) \to 0$ as $t \downarrow 0$ and $t\,k'(t) \to 0$ as $t \to \infty$. Finally, the finiteness of $\int_{(0,1]} t^2\gamma(\mathrm{d}t)$ and $\int_{(1,\infty)} t\,\gamma(\mathrm{d}t)$ follow from the facts that $\int_{(0,1]} t^2\gamma(\mathrm{d}t) = k'(1) - 2\int_0^1 t\,k'(t)\,\mathrm{d}t$ and $\int_{(1,\infty)} t\,\gamma(\mathrm{d}t) = -k'(1) - \int_1^\infty k'(s)\,\mathrm{d}s$.

When $\Re z > 0$ we can integrate by parts twice, to get

$$\begin{aligned}\hat{k}(z) &= \frac{k(\infty)}{z} + \int_0^\infty \mathrm{e}^{-zt}\big(k(t) - k(\infty)\big)\,\mathrm{d}t \\ &= \frac{k(\infty)}{z} - \frac{1}{z}\int_0^\infty (1 - \mathrm{e}^{-zt})k'(t)\,\mathrm{d}t \\ &= \frac{k(\infty)}{z} - \frac{1}{z^2}\int_{(0,\infty)} (1 - zt - \mathrm{e}^{-zt})\gamma(\mathrm{d}t) \end{aligned} \tag{2.3}$$

(observe that $1-\mathrm{e}^{-zt} = O(t)$ and $1-zt-\mathrm{e}^{-zt} = O(t^2)$ as $t \downarrow 0$). Using this formula, we can extend $\hat{k}$ to a continuous function on $\{\, \Re z \geq 0 \mid z \neq 0 \,\}$. To show that the extended function is continuously differentiable, one can simply differentiate under the integral sign to get

$$\begin{aligned}\hat{k}'(z) &= -\frac{1}{z^2}k(\infty) \\ &\quad + \frac{1}{z^3}\int_{(0,\infty)} \big(2(1 - zt - \mathrm{e}^{-zt}) + zt - zt\mathrm{e}^{-zt}\big)\gamma(\mathrm{d}t), \\ &\Re z \geq 0, \quad z \neq 0. \end{aligned} \tag{2.4}$$

If we substitute $z = \mathrm{i}\omega$ into (2.3) with ω real and nonzero, then we end up with

$$\Re\hat{k}(\mathrm{i}\omega) = \frac{1}{\omega^2}\int_{(0,\infty)} \big(1 - \cos(\omega t)\big)\gamma(\mathrm{d}t) \tag{2.5}$$

and

$$\Im\hat{k}(\mathrm{i}\omega) = -\frac{1}{\omega}k(\infty) - \frac{1}{\omega^2}\int_{(0,\infty)} \big(\omega t - \sin(\omega t)\big)\gamma(\mathrm{d}t). \tag{2.6}$$

Clearly, this together with the fact that γ is nonnegative implies that $\Re\hat{k}(\mathrm{i}\omega) \geq 0$, and that $\omega\Im\hat{k}(\mathrm{i}\omega) \leq 0$ for all real nonzero ω. To show that the same sign conditions are true at points $z = \sigma + \mathrm{i}\omega$, with $\sigma > 0$, one can apply the same argument, but with $k(t)$ replaced by $\mathrm{e}^{-\sigma t}k(t)$. This proves our claim that $\Re\hat{k}(z) \geq 0$ and $\Im z\,\Im\hat{k}(z) \leq 0$ in $\{\, \Re z \geq 0 \mid z \neq 0 \,\}$.

To prove the second half of (2.1) one can use (2.3) to get, for all nonzero real ω,

$$\big|\hat{k}(\mathrm{i}\omega)\big| = \big|\hat{k}(\mathrm{i}|\omega|)\big| \leq \frac{1}{|\omega|}k(\infty) + \frac{1}{|\omega|}\int_0^\infty \big|F(|\omega|t)\big|\,\big|k'(t)\big|\,\mathrm{d}t, \tag{2.7}$$

where

$$\big|F(s)\big| = \big|1 - \mathrm{e}^{\mathrm{i}s}\big| = \big|\mathrm{e}^{-\mathrm{i}s/2} - \mathrm{e}^{\mathrm{i}s/2}\big| = 2\big|\sin(s/2)\big|.$$

On the other hand, integrating by parts one has

$$\begin{aligned}\int_0^{1/|\omega|} k(t)\,\mathrm{d}t &= \frac{1}{|\omega|}k(\infty) - \int_0^{1/|\omega|} tk'(t)\,\mathrm{d}t - \frac{1}{|\omega|}\int_{1/|\omega|}^\infty k'(t)\,\mathrm{d}t \\ &= \frac{1}{|\omega|}k(\infty) + \frac{1}{|\omega|}\int_0^\infty f(|\omega|t)\big|k'(t)\big|\,\mathrm{d}t, \end{aligned} \tag{2.8}$$

where

$$f(s) = \begin{cases} s, & 0 \le s < 1, \\ 1, & 1 \le s < \infty. \end{cases}$$

One can easily show that $|F(s)| \le 2f(s)$ for all $s \ge 0$, and this inequality, together with (2.7) and (2.8), implies the second half of (2.1).

To prove the first part of (2.1), one uses (2.5) and (2.6) to get

$$\begin{aligned}\sqrt{2}|\hat{k}(\mathrm{i}\omega)| &= \sqrt{2}|\hat{k}(\mathrm{i}|\omega|)| \ge \Re\hat{k}(\mathrm{i}|\omega|) - \Im\hat{k}(\mathrm{i}|\omega|) \\ &= \frac{1}{|\omega|}k(\infty) + \frac{1}{\omega^2}\int_{(0,\infty)} G(|\omega|t)\gamma(\mathrm{d}t),\end{aligned}$$

where

$$G(s) = s + 1 - \sin(s) - \cos(s), \quad s \in \mathbf{R}^+.$$

Integrating by parts once more in (2.8), we can also show that

$$\int_0^{1/|\omega|} k(t)\,\mathrm{d}t = \frac{1}{|\omega|}k(\infty) + \frac{1}{\omega^2}\int_0^\infty g(|\omega|t)\gamma(\mathrm{d}t),$$

where $g(s) = \int_0^s f(t)\,\mathrm{d}t$ is the function

$$g(s) = \begin{cases} s^2/2, & 0 \le s < 1, \\ s - 1/2, & 1 \le s < \infty. \end{cases}$$

For $0 \le s < 1$ one can expand $G(s)$ into a Taylor series to obtain

$$G(s) \ge \frac{s^2}{2} + \frac{s^3}{6} - \frac{s^4}{24} - \frac{s^5}{120} \ge \frac{s^2}{2},$$

and for $s \ge 1$ one can use the fact that $\sin(s) + \cos(s) = \sqrt{2}\cos(s - \pi/4)$, to get

$$G(s) \ge s + 1 - \sqrt{2} \ge s - 1/2.$$

Therefore, for all s, $G(s) \ge g(s)$, and this clearly implies the first part of (2.1).

To prove (2.2) one invokes (2.4), obtaining, for all real $\omega \ne 0$,

$$|\hat{k}'(\mathrm{i}\omega)| = |\hat{k}'(\mathrm{i}|\omega|)| \le \frac{1}{\omega^2}k(\infty) + \frac{1}{|\omega|^3}\int_{(0,\infty)} |H(|\omega|t)|\gamma(\mathrm{d}t),$$

where

$$H(s) = 2(1 - \mathrm{e}^{-\mathrm{i}s}) - \mathrm{i}s(1 + \mathrm{e}^{-\mathrm{i}s}), \quad s \in \mathbf{R}^+.$$

Factoring out $\mathrm{e}^{-\mathrm{i}s/2}$ from H, one finds that H can be written in the form

$$H(s) = -4\mathrm{i}\mathrm{e}^{-\mathrm{i}s/2}\Big(\tfrac{s}{2}\cos\left(\tfrac{s}{2}\right) - \sin\left(\tfrac{s}{2}\right)\Big), \quad s \in \mathbf{R}^+.$$

Hence

$$|H(s)| = 4\Big|\tfrac{s}{2}\cos\left(\tfrac{s}{2}\right) - \sin\left(\tfrac{s}{2}\right)\Big|, \quad s \in \mathbf{R}^+.$$

On the other hand, integrating by parts one can show that

$$\begin{aligned}
\int_0^{1/|\omega|} tk(t)\,\mathrm{d}t &= \frac{1}{2\omega^2}k(\infty) - \int_0^{1/|\omega|} \frac{t^2}{2}k'(t)\,\mathrm{d}t - \frac{1}{\omega^2}\int_{1/|\omega|}^{\infty} \frac{1}{2}k'(t)\,\mathrm{d}t \\
&= \frac{1}{2\omega^2}k(\infty) + \int_{(0,1/|\omega|]} \frac{t^3}{6}\gamma(\mathrm{d}t) \\
&\quad + \frac{1}{|\omega|^3}\int_{(1/|\omega|,\infty)} \left(\frac{|\omega|t}{2} - \frac{1}{3}\right)\gamma(\mathrm{d}t) \\
&= \frac{1}{2\omega^2}k(\infty) + \frac{1}{\omega^3}\int_{(0,\infty)} h(\omega t)\gamma(\mathrm{d}t),
\end{aligned}$$

where

$$h(s) = \begin{cases} s^3/6, & 0 \le s < 1, \\ s/2 - 1/3, & 1 \le s < \infty. \end{cases}$$

Therefore, to complete the proof of (2.2) it suffices to show that

$$|H(s)| \le 5h(s), \quad s \in \mathbf{R}^+,$$

or equivalently, that

$$\sup_{s>0} m(s) \le 5,$$

where $m(s) = |H(2s)|/h(2s)$ is given by

$$m(s) = \begin{cases} 3s^{-3}|s\cos(s) - \sin(s)|, & 0 < s < 1/2, \\ 12(3s-1)^{-1}|s\cos(s) - \sin(s)|, & 1/2 \le s < \infty. \end{cases}$$

This function satisfies $\lim_{s\to 0} m(s) = 1$ and $\limsup_{s\to\infty} m(s) = 4$, and it has a global maximum which lies well below 5 (it is approximately 4.85, and appears near $s = 2.67$). Therefore (2.2) is satisfied. □

Examining the proof of Lemma 2.2, we observe that it in fact implies the following result.

2.3 Lemma

(i) *Let $k \in BV_{\mathrm{loc}}((0,\infty);\mathbf{C})$ be right-continuous, denote the distribution derivative of k by κ, and suppose that $\int_{(0,1)} t|\kappa|(\mathrm{d}t) < \infty$, and that $\int_{(1,\infty)}|\kappa|(\mathrm{d}t) < \infty$. Then the Laplace transform $\hat{k}(z)$ of k converges in the open half plane $\Re z > 0$, and it can be extended to a continuous function in $\{\,\Re z \ge 0 \mid z \ne 0\,\}$. If we denote the extended function by $\hat{k}$ as well, then, for each real nonzero ω,*

$$\left|\hat{k}(\mathrm{i}\omega)\right| \le \frac{1}{|\omega|}|k(\infty)| + 2\int_0^{1/|\omega|} |\kappa|((t,\infty))\,\mathrm{d}t.$$

(ii) *Let $k \in L^1_{\mathrm{loc}}(\mathbf{R}^+;\mathbf{R})$ be nonnegative and nonincreasing on $(0,\infty)$. Then the conclusion of (i) holds. In addition, $\Im z\,\Im\hat{k}(z) \le 0$ when $\Re z \ge 0$ and $z \ne 0$, and*

$$\left|\hat{k}(\mathrm{i}\omega)\right| \le 2\int_0^{1/|\omega|} k(t)\,\mathrm{d}t - \frac{1}{|\omega|}k(\infty).$$

(iii) *Let $k \in AC_{\mathrm{loc}}((0,\infty);\mathbf{C})$ have a right-derivative k' which is locally of bounded variation on $(0,\infty)$. Denote the distribution derivative of k' by γ, and suppose that $\int_{(0,1)} t^2|\gamma|(\mathrm{d}t) < \infty$ and $\int_{(1,\infty)} t|\gamma|(\mathrm{d}t) < \infty$. Then the conclusion of* (i) *holds. In addition, $\hat{k}$ is continuously differentiable in $\{\,\Re z \geq 0 \mid z \neq 0\,\}$, and*

$$|\hat{k}'(\mathrm{i}\omega)| \leq \frac{1}{\omega^2}|k(\infty)| + 5\int_0^{1/|\omega|} tM(t)\,\mathrm{d}t,$$

where $M(t) = \int_t^\infty |\gamma|((s,\infty))\,\mathrm{d}s$.

Now it is time to return to the proof of Theorem 2.1.

Proof of Theorem 2.1 First, let us show that we may assume that $k \in L^1(\mathbf{R}^+;\mathbf{R})$. This is true, because by Lemma **2**.3.3, if r is the resolvent of k, then, for each $\sigma > 0$, the function $r_\sigma(t) = \mathrm{e}^{-\sigma t}r(t)$ is the resolvent of $k_\sigma(t) = \mathrm{e}^{-\sigma t}k(t)$. The function k_σ is nonnegative, nonincreasing, and convex, and it belongs to $L^1(\mathbf{R}^+;\mathbf{R})$. Therefore, if Theorem 2.1 is true for this class of functions, we are able to conclude that $r_\sigma \in L^1(\mathbf{R}^+;\mathbf{R})$, and that $\int_0^\infty \mathrm{e}^{-\sigma t}|r(t)|\,\mathrm{d}t \leq 20$. Letting $\sigma \downarrow 0$ we get the general conclusion. (We make no claim whatsoever that the number 20 is the best possible; it can in fact be decreased.) We may, of course, also assume that $k \not\equiv 0$.

In the case where $k \in L^1(\mathbf{R}^+;\mathbf{R})$ we shall use Theorem 5.1 to show that $\|r\|_{L^1(\mathbf{R}^+)} \leq 20$. Let us verify that the assumptions of Theorem 5.1 hold. By Lemma 2.2, we have $\Re[1+\hat{k}(z)] \geq 1$ for $\Re z \geq 0$, in particular, $1+\hat{k}(z) \neq 0$ for $\Re z \geq 0$. Therefore, the Paley–Wiener Theorem **2**.4.1 implies that $r \in L^1(\mathbf{R}^+;\mathbf{R})$, hence $\hat{r}$ is continuous in $\{\,z \mid \Re z \geq 0\,\}$. Since

$$\hat{r}(z) = \frac{\hat{k}(z)}{1+\hat{k}(z)}, \quad \Re z \geq 0, \tag{2.9}$$

it follows from Lemma 2.2 that $\hat{r}$ is continuously differentiable in $\{\,\Re z \geq 0 \mid z \neq 0\,\}$. Thus, if we can show that

$$\int_{-\infty}^{\infty} |\hat{r}'(\mathrm{i}\omega)|\,\mathrm{d}\omega \leq 40, \tag{2.10}$$

then Theorem 5.1 implies that $\|r\|_{L^1(\mathbf{R})} \leq 20$, as desired.

Clearly, (2.10) is equivalent to

$$\int_{-\infty}^{\infty} \frac{|\hat{k}'(\mathrm{i}\omega|}{|1+\hat{k}(\mathrm{i}\omega)|^2}\,\mathrm{d}\omega \leq 40. \tag{2.11}$$

Fix some $\delta > 0$. It follows from Lemma 2.2 (since $\Re\hat{k}(\mathrm{i}\omega) \geq 0$ implies

$\left|1 + \hat{k}(\mathrm{i}\omega)\right| \geq 1$) that

$$\begin{aligned}\int_{|\omega|\geq 1/\delta} \frac{|\hat{k}'(\mathrm{i}\omega)|}{|1+\hat{k}(\mathrm{i}\omega)|^2}\,\mathrm{d}\omega &\leq 10 \int_{1/\delta}^{\infty}\int_0^{1/\omega} tk(t)\,\mathrm{d}t\,\mathrm{d}\omega \\ &= 10\int_0^{\delta} s^{-2}\int_0^{s} tk(t)\,\mathrm{d}t\,\mathrm{d}s \\ &= 10\int_0^{\delta} tk(t)\int_t^{\delta} s^{-2}\,\mathrm{d}s\,\mathrm{d}t \\ &= 10\int_0^{\delta}\left(1-\frac{t}{\delta}\right)k(t)\,\mathrm{d}t \leq 10\int_0^{\delta} k(t)\,\mathrm{d}t.\end{aligned}$$

Because $\Re\hat{k}(\mathrm{i}\omega) \geq 0$ we have $\left|1+\hat{k}(\mathrm{i}\omega)\right| \geq |\hat{k}(\mathrm{i}\omega)|$, and therefore it follows from Lemma 2.2 that

$$\begin{aligned}\int_{-1/\delta}^{1/\delta} \frac{|\hat{k}'(\mathrm{i}\omega)|}{|1+\hat{k}(\mathrm{i}\omega)|^2}\,\mathrm{d}\omega &\leq 20\int_0^{1/\delta}\frac{\int_0^{1/\omega} tk(t)\,\mathrm{d}t}{\left(\int_0^{1/\omega} k(t)\,\mathrm{d}t\right)^2}\,\mathrm{d}\omega \\ &= 20\int_{\delta}^{\infty}\frac{1}{s^2}\Psi(s)\,\mathrm{d}s,\end{aligned}$$

where

$$\Psi(s) = \frac{\int_0^s tk(t)\,\mathrm{d}t}{\left(\int_0^s k(t)\,\mathrm{d}t\right)^2}.$$

Since $k \in L^1(\mathbf{R}^+;\mathbf{R})$ is nonincreasing, we have

$$\frac{t}{2}k(t) \leq \int_{t/2}^{t} k(s)\,\mathrm{d}s \leq \int_{t/2}^{\infty} k(s)\,\mathrm{d}s,$$

and therefore $tk(t) \to 0$ as $t \to \infty$. This implies that $s^{-1}\int_0^s tk(t)\,\mathrm{d}t \to 0$ as $s \to \infty$; hence $s^{-1}\Psi(s) \to 0$ as $s \to \infty$ as well. We can integrate by parts to get

$$\int_{\delta}^{\infty}\frac{1}{s^2}\Psi(s)\,\mathrm{d}s = \frac{1}{\delta}\Psi(\delta) + \int_{\delta}^{\infty}\frac{1}{s}\Psi'(s)\,\mathrm{d}s.$$

But

$$\frac{1}{\delta}\Psi(\delta) \leq \frac{1}{\int_0^{\delta} k(t)\,\mathrm{d}t}$$

and

$$\Psi'(s) \leq \frac{sk(s)}{\left(\int_0^s k(t)\,\mathrm{d}t\right)^2}.$$

Therefore,

$$\begin{aligned}\int_{\delta}^{\infty}\frac{1}{s^2}\Psi(s)\,\mathrm{d}s &\leq \frac{1}{\int_0^{\delta} k(t)\,\mathrm{d}t} + \int_{\delta}^{\infty}\frac{k(s)}{\left(\int_0^s k(t)\,\mathrm{d}t\right)^2}\,\mathrm{d}s \\ &= \frac{2}{\int_0^{\delta} k(t)\,\mathrm{d}t} - \frac{1}{\int_0^{\infty} k(t)\,\mathrm{d}t}.\end{aligned}$$

Combining the preceding estimates, we get

$$\int_{-\infty}^{\infty} \frac{|\hat{k}'(\mathrm{i}\omega)|}{|1+\hat{k}(\mathrm{i}\omega)|^2}\,\mathrm{d}\omega \le 10\int_0^{\delta} k(t)\,\mathrm{d}t + \frac{40}{\int_0^{\delta} k(t)\,\mathrm{d}t} - \frac{20}{\int_0^{\infty} k(t)\,\mathrm{d}t}.$$

If $\|k\|_{L^1(\mathbf{R}^+)} > 2$, then choose δ so that $\int_0^{\delta} k(t)\,\mathrm{d}t = 2$ to obtain

$$\int_{-\infty}^{\infty} \frac{|\hat{k}'(\mathrm{i}\omega)|}{|1+\hat{k}(\mathrm{i}\omega)|^2}\,\mathrm{d}\omega \le 40.$$

If $2-\sqrt{2} \le \|k\|_{L^1(\mathbf{R}^+)} \le 2$, then let $\delta \to \infty$ to get

$$\int_{-\infty}^{\infty} \frac{|\hat{k}'(\mathrm{i}\omega)|}{|1+\hat{k}(\mathrm{i}\omega)|^2}\,\mathrm{d}\omega \le 10\int_0^{\infty} k(t)\,\mathrm{d}t + \frac{20}{\int_0^{\infty} k(t)\,\mathrm{d}t} \le 40.$$

Thus, if $\|k\|_{L^1(\mathbf{R}^+)} \ge 2-\sqrt{2}$, then $\int_0^{\infty} |\hat{k}'(\mathrm{i}\omega)||1+\hat{k}(\mathrm{i}\omega)|^{-2}\,\mathrm{d}\omega < 40$, hence by Theorem 5.1, $\|r\|_{L^1(\mathbf{R}^+)} < 20$. If $\|k\|_{L^1(\mathbf{R}^+)} \le 2-\sqrt{2}$, then recall that

$$\|r\|_{L^1(\mathbf{R}^+)} \le \frac{\|k\|_{L^1(\mathbf{R}^+)}}{1-\|k\|_{L^1(\mathbf{R}^+)}} \le \sqrt{2}.$$

Therefore, we have $\|r\|_{L^1(\mathbf{R}^+)} \le 20$ independently of the magnitude of $\|k\|_{L^1(\mathbf{R}^+)}$, and the proof of Theorem 2.1 is complete. □

3. Analytic Mappings of Fourier Transforms

In the scalar case it is possible to combine the Half Line Paley–Wiener Theorem **2**.4.1 and the Shea–Wainger Theorem into one result, which contains both these theorems. It is also possible to obtain an analogous result for the integrodifferential equation. To do this, one has to realize that one of the basic ideas in the proof of the Paley–Wiener Theorem **2**.4.1, is the observation that it suffices to have the transform of r *locally* (on the imaginary axis) equal to the transform of an integrable function.

We begin by giving a definition. Recall that we denote the Fourier transform by $\tilde{}$, and that $\tilde{a}(\omega) = \hat{a}(\mathrm{i}\omega)$.

3.1 Definition

A (matrix-valued, vector-valued or scalar-valued) function ψ belongs locally to $\tilde{L}^1$ at a point $\omega_0 \in \mathbf{R}$, if there exist a function $a \in L^1(\mathbf{R})$ and a constant $\epsilon > 0$ such that $\psi(\omega) = \tilde{a}(\omega)$ for $|\omega - \omega_0| \le \epsilon$ (in particular, ψ is defined at least for these values of ω). The function ψ belongs locally to $\tilde{L}^1$ at infinity, if there exist a function $a \in L^1(\mathbf{R})$, and a constant $M \ge 0$, such that $\psi(\omega) = \tilde{a}(\omega)$ for $|\omega| \ge M$. If ψ belongs locally to $\tilde{L}^1$ at every $\omega_0 \in \mathbf{R}$ and also at infinity, then we say that ψ belongs locally to $\tilde{L}^1$ on $\overline{\mathbf{R}}$.

Here $\overline{\mathbf{R}}$ represents $\overline{\mathbf{R}} = \mathbf{R} \cup \{\infty\}$ (the compactified real line, where $-\infty$ has been identified with ∞).

Before giving examples of functions that belong locally to $\tilde{L}^1$, we state and prove a theorem that essentially was established in the proof of Theorem **2**.4.3.

3.2 Theorem
Every function that belongs locally to $\tilde{L}^1$ on $\overline{\mathbf{R}}$ is the Fourier transform of a function in $L^1(\mathbf{R})$.

Proof Let ψ be such a function. Since ψ belongs locally to $\tilde{L}^1$ at infinity, we can find an integer M, and a function $a_\infty \in L^1(\mathbf{R})$, such that $\psi(\omega) = \tilde{a}_\infty(\omega)$ for $|\omega| \geq M$. Moreover, for every $\xi \in [-M, M]$ we can find an open interval U_ξ containing ξ, and a function $a_\xi \in L^1(\mathbf{R})$, such that $\psi(\omega) = \tilde{a}_\xi(\omega)$ for $\omega \in U_\xi$. The interval $[-M, M]$ is compact, and therefore there is an integer m such that each interval $[(j-1)/m, (j+1)/m]$, $j = 0, \pm 1, \ldots, \pm mM$, is contained in one of the intervals U_ξ. That is, for each of these values of j we find a function $a_j \in L^1(\mathbf{R})$ such that $\psi(\omega) = \tilde{a}_j(\omega)$ when $|\omega - j/m| \leq 1/m$. Now we can continue in the same way as we did in the proof of Theorem **2**.4.3. □

If k is nonnegative, nonincreasing, and convex, then $\hat{k}(i\omega)$ does not belong locally to $\tilde{L}^1$ at zero; except, of course, if k is integrable. However, unless $k \equiv 0$, it is true that $1/\hat{k}(i\omega)$ belongs locally to $\tilde{L}^1$ at zero (this will be proved below). In order to be able to use this observation in the most efficient manner, we give some basic results involving functions locally equal to Fourier transforms of L^1-functions.

3.3 Theorem

(i) *A function ψ that is analytic at a point $\omega_0 \in \mathbf{R}$ belongs locally to $\tilde{L}^1$ at ω_0.*

(ii) *If $\mu \in M(\mathbf{R})$, then $\tilde{\mu}$ belongs locally to $\tilde{L}^1$ at ω_0 for every $\omega_0 \in \mathbf{R}$.*

(iii) *Let $\psi_1, \psi_2, \ldots, \psi_m$ be scalar functions that belong locally to $\tilde{L}^1$ at a point $\omega_0 \in \mathbf{R}$, and let ϕ be a function of $m+1$ variables that is analytic at the point $(\omega_0, \psi_1(\omega_0), \psi_2(\omega_0), \ldots, \psi_m(\omega_0))$. Then the function $\omega \mapsto \phi(\omega, \psi_1(\omega), \psi_2(\omega), \ldots, \psi_m(\omega))$ belongs locally to $\tilde{L}^1$ at ω_0.*

Before proving Theorem 3.3, we formulate the corresponding result for the point at infinity.

3.4 Theorem

(i) *If ψ is analytic at zero and $\psi(0) = 0$, then the function $\omega \mapsto \psi(1/\omega)$ belongs locally to $\tilde{L}^1$ at infinity. In particular, the function $\omega \mapsto 1/\omega$ belongs locally to $\tilde{L}^1$ at infinity.*

(ii) *If ψ belongs locally to $\tilde{L}^1$ at infinity and $\mu \in M(\mathbf{R})$, then $\psi\tilde{\mu}$ belongs locally to $\tilde{L}^1$ at infinity. In particular, the function $\omega \mapsto \omega^{-1}\tilde{\mu}(\omega)$ belongs locally to $\tilde{L}^1$ at infinity.*

(iii) *Let $\psi_1, \psi_2, \ldots, \psi_m$ be scalar functions that belong locally to $\tilde{L}^1$ at infinity, and let ϕ be a function of $m+1$ variables that is analytic at the point $(0,0,\ldots,0)$, and satisfies $\phi(0,0,\ldots,0) = 0$. Then the function $\omega \mapsto \phi\big(\omega^{-1}, \psi_1(\omega), \psi_2(\omega), \ldots, \psi_m(\omega)\big)$ belongs locally to $\tilde{L}^1$ at infinity.*

Clearly, the major part of Lemma **2**.5.5 is a consequence of Theorems 3.3(iii) and 3.4(iii)

The proofs of these two theorems are very similar. Below we prove Theorem 3.3, and leave the proof of Theorem 3.4 as an exercise.

If we combine (ii) and (iii) in Theorem 3.3 with (ii) in Theorem 3.4 and use results in Section **16**.8, concerning the Fourier transform of the derivative of a distribution, then we conclude that the distribution Fourier transform of every function of bounded variation belongs locally to $\tilde{L}^1$ at all points in $\overline{\mathbf{R}}$, except perhaps at zero.

Proof of Theorem 3.3 (i) The function $\tilde{\mathbf{e}}(\omega) = 1/(1+\mathrm{i}\omega)$ is the Fourier transform of the L^1-function $\mathbf{e}$ defined by $\mathbf{e}(t) = \mathrm{e}^{-t}$ for $t \geq 0$, $\mathbf{e}(t) = 0$ for $t < 0$, and therefore it belongs locally to $\tilde{L}^1$ at every $\omega_0 \in \mathbf{R}$ (and also at infinity). Since $\omega = \mathrm{i} + (\mathrm{i}\tilde{\mathbf{e}}(\omega))^{-1}$, we can express the function ψ in (i) in the form $\psi(\omega) = \psi_0(\mathbf{e}(\omega))$, where $\psi_0(z) = \psi(\mathrm{i} + (\mathrm{i}z)^{-1})$, and this means that (i) is a special case of (iii) (where ψ does not depend explicitly on ω), which will be proved below.

(ii) Define ξ to be the function $\xi(t) = (|\omega_0|+1)\eta((|\omega_0|+1)t)$, where η is the function $\pi^{-1}t^{-2}(\cos(t) - \cos(2t))$ (see (**2**.5.2)), which satisfies $\tilde{\eta}(\omega) = 1$ for $|\omega| \leq 1$. Then $\tilde{\xi}(\omega) = 1$ for $|\omega| \leq |\omega_0| + 1$. In particular, $\tilde{\mu}(\omega) = \tilde{\xi}(\omega)\tilde{\mu}(\omega)$ for $|\omega| \leq |\omega_0|+1$. In addition, $\tilde{\xi}\tilde{\mu}$ is the Fourier transform of the integrable function $\xi * \mu$, and, therefore, $\tilde{\mu}$ belongs locally to $\tilde{L}^1$ at ω_0.

(iii) First, consider the case where ϕ does not depend explicitly on ω. Analyticity of ϕ at the point $(\psi_1(\omega_0), \psi_2(\omega_0), \ldots, \psi_m(\omega_0))$ means that, in a neighbourhood of this point, ϕ can be expressed as an absolutely convergent power series. More precisely, there exist constants $\epsilon_1, \epsilon_2, \ldots, \epsilon_m$ such that for $|z_j - \psi_j(\omega_0)| \leq \epsilon_j$, $1 \leq j \leq m$, we can write $\phi(z_1, z_2, \ldots, z_m)$ in the form

$$\phi(z_1, z_2, \ldots, z_m) = \sum_{j_1=0}^{\infty}\sum_{j_2=0}^{\infty}\cdots\sum_{j_m=0}^{\infty}\Big\{\alpha_{j_1,j_2,\ldots,j_m}(z_1 - \psi_1(\omega_0))^{j_1} \times (z_2 - \psi_2(\omega_0))^{j_2}\cdots(z_m - \psi_m(\omega_0))^{j_m}\Big\},$$

and this series converges absolutely. Because each ψ_j belongs locally to $\tilde{L}^1$ at ω_0, it follows from Lemma **2**.5.3 that, for each j, we can find a function $a_j \in L^1(\mathbf{R})$ with $\|a_j\|_{L^1(\mathbf{R})} \leq \epsilon_j$, such that $\widetilde{a_j}(\omega) = \psi_j(\omega) - \psi_j(\omega_0)$ in some neighbourhood U_j of ω_0. Define

$$b = \sum_{j_1=0}^{\infty}\sum_{j_2=0}^{\infty}\cdots\sum_{j_m=0}^{\infty}\Big\{\alpha_{j_1,j_2,\ldots,j_m}a_1^{*j_1} * a_2^{*j_2} * \cdots * a_m^{*j_m}\Big\},$$

where $a_k^{*0} = \delta$, $a_k^{*1} = a_k$, and a_k^{*j} is the $(j-1)$-fold convolution of a_k with itself for $j \geq 2$. Then $\alpha_{0,0,\ldots,0} = \phi(\psi_1(\omega_0), \psi_2(\omega_0), \ldots, \psi_m(\omega_0))$ is the size of the point mass of b at zero, and the remainder of the series that defines b converges in $L^1(\mathbf{R})$ to an L^1-function. In particular, $b \in M(\mathbf{R})$. We have

$$\phi(\psi_1(\omega), \psi_2(\omega), \ldots, \psi_m(\omega)) = \tilde{b}(\omega),$$

for $\omega \in \bigcap_{j=1}^m U_j$. This, together with (ii), shows that the composite function $\phi(\psi_1, \psi_2, \ldots, \psi_m)$ belongs locally to $\tilde{L}^1$ at ω_0.

Observe that the case considered above, combined with the comments that we made earlier on (i), completes the proof of (i).

Finally, consider (iii) in the case where $\phi = \phi(\omega, \psi_1(\omega), \ldots, \psi_m(\omega))$. Define $\psi_0(\omega) = \omega$ for $\omega \in \mathbf{R}$. By (i), ψ_0 belongs locally to $\tilde{L}^1$ at every $\omega \in \mathbf{R}$. Hence,

$$\phi(\psi_0(\omega), \psi_1(\omega), \ldots, \psi_m(\omega)) = \phi(\omega, \psi_1(\omega), \ldots, \psi_m(\omega))$$

belongs locally to $\tilde{L}^1$ at ω_0. □

Let us end this section with an additional comment on the proof of Theorem **2**.4.3. Some easy modifications in the proof of Lemma **2**.5.5 give us the following result.

3.5 Lemma

(i) *Let ψ belong locally to $\tilde{L}^1(\mathbf{R}; \mathbf{C}^{n\times n})$ at $\omega_0 \in \mathbf{R}$, and suppose that $\det[\psi(\omega_0)] \neq 0$. Then there exist a $\delta > 0$ and a function $a \in L^1(\mathbf{R}; \mathbf{C}^{n\times n})$, such that $\tilde{a}(\omega) = [\psi(\omega)]^{-1}$ for $|\omega - \omega_0| \leq \delta$.*

(ii) *Let ψ_1 and ψ_2 belong locally to $\tilde{L}^1(\mathbf{R}; \mathbf{C}^{n\times n})$ at infinity. Then there exists a constant $M > 0$ and a function $a \in L^1(\mathbf{R}; \mathbf{C}^{n\times n})$ such that $\tilde{a}(\omega) = \psi_1(\omega)[I + \psi_2(\omega)]^{-1}$ for $|\omega| \geq M$.*

An immediate consequence of Lemma 3.5 is the following.

3.6 Corollary

(i) *Let ψ belong locally to $\tilde{L}^1(\mathbf{R}; \mathbf{C}^{n\times n})$ at $\omega_0 \in \mathbf{R}$, and suppose that $\det[\psi(\omega_0)] \neq 0$. Then $[\psi(\omega)]^{-1}$ belongs locally to $\tilde{L}^1(\mathbf{R}; \mathbf{C}^{n\times n})$ at ω_0.*

(ii) *Let ψ_1 and ψ_2 belong locally to $\tilde{L}^1(\mathbf{R}; \mathbf{C}^{n\times n})$ at infinity. Then $\psi_1(\omega)[I + \psi_2(\omega)]^{-1}$ belongs locally to $\tilde{L}^1(\mathbf{R}; \mathbf{C}^{n\times n})$ at infinity.*

4. Extensions of the Paley–Wiener Theorems

We mentioned earlier that in the scalar case the Paley–Wiener Theorem **2**.4.1 and Theorem 2.1 may be combined into a single result. Let us formulate and prove this statement.

4.1 Theorem

Let $k = a + b$, where $a \in L^1_{\text{loc}}(\mathbf{R}^+;\mathbf{R})$ is nonnegative, nonincreasing, and convex on $(0,\infty)$, and $b \in L^1(\mathbf{R}^+;\mathbf{C})$. Define $\hat{a}$ on the imaginary axis by $\hat{a}(\mathrm{i}\omega) = \lim_{\sigma\downarrow 0}\hat{a}(\sigma + \mathrm{i}\omega)$ for $\omega \in \mathbf{R}$ (if $a \notin L^1(\mathbf{R}^+;\mathbf{R})$, then define $\hat{a}(0) = \infty$). Then the resolvent r of k satisfies

$$r \in L^1(\mathbf{R}^+;\mathbf{C})$$

if and only if

$$1 + \hat{a}(z) + \hat{b}(z) \neq 0 \text{ in the half plane } \Re z \geq 0.$$

Proof If the resolvent r of k satisfies $r \in L^1(\mathbf{R}^+;\mathbf{C})$, then obviously $1 + \hat{k}(z) \neq 0$, for $\Re z \geq 0$ (because $\big(1 + \hat{k}(z)\big)\big(1 - \hat{r}(z)\big) = 1$ for nonzero z with $\Re z \geq 0$, and $\hat{k}(0) = \infty$ if $k \notin L^1(\mathbf{R}^+;\mathbf{C})$).

Conversely, let $1 + \hat{k}(z) \neq 0$ for $\Re z \geq 0$. We have to show that the resolvent of k belongs to $L^1(\mathbf{R}^+;\mathbf{C})$. Obviously, we may assume that a is not integrable, because otherwise we apply Theorem **2**.4.1.

Without loss of generality, suppose that $a(0) < \infty$, hence $a' \in L^1(\mathbf{R}^+;\mathbf{C})$ (if necessary, transfer one part of a into b). We have $\hat{k}(z) = \hat{a}(z) + \hat{b}(z) = \frac{1}{z}\big(a(0) + \widehat{a'}(z)\big) + \hat{b}(z)$ for $\Re z > 0$, and thus

$$\tilde{k}(\omega) = \frac{1}{\mathrm{i}\omega}\big(a(0) + \widetilde{a'}(\omega)\big) + \tilde{b}(\omega), \quad \omega \neq 0.$$

By definition, $\widetilde{a'}$ and $\tilde{b}$ belong locally to $\tilde{L}^1$ on $\overline{\mathbf{R}}$. Apply Theorem 3.3(iii) and Theorem 3.4(iii) with $\phi(\omega,\psi_1,\psi_2)$ or $\phi(\omega^{-1},\psi_1,\psi_2)$, respectively, given by $1/(\mathrm{i}\omega)\big(a(0)+\psi_1\big)+\psi_2$, $\psi_1 = \widetilde{a'}$, and $\psi_2 = \tilde{b}$, to conclude that $\tilde{k}$ belongs locally to $\tilde{L}^1$ on $\overline{\mathbf{R}}\setminus\{0\}$. This, together with the assumption $1 + \tilde{k} \neq 0$ and Theorems 3.3(iii) and 3.4(iii) (with $\phi(\psi_1) = \psi_1/(1+\psi_1)$, and $\psi_1 = \tilde{k}$), implies that $\tilde{k}/(1+\tilde{k})$ belongs locally to $\tilde{L}^1$ on $\overline{\mathbf{R}}\setminus\{0\}$.

The point $\omega = 0$ remains to be studied. By Theorem 2.1, the resolvent q of a belongs to $L^1(\mathbf{R}^+;\mathbf{R})$. Since$(1-\tilde{q})(1+\tilde{a}) = 1$, and since $\tilde{a}(0) = \infty$, we must have $\tilde{q}(0) = 1$. In a neighbourhood of zero we can write $1/\tilde{a}$ in the form $1/\tilde{a} = (1-\tilde{q})/\tilde{q}$. Now use Theorem 3.3(iii) (with $\phi(\psi_1) = (1-\psi_1)/\psi_1$, and $\psi_1 = \tilde{q}$ to show that $1/\tilde{a}$ belongs locally to $\tilde{L}^1$ at zero. Close to zero we have

$$\frac{\tilde{k}(\omega)}{1+\tilde{k}(\omega)} = \frac{\tilde{a}(\omega)+\tilde{b}(\omega)}{1+\tilde{a}(\omega)+\tilde{b}(\omega)} = \frac{1+\tilde{b}(\omega)/\tilde{a}(\omega)}{1+(1+\tilde{b}(\omega))/\tilde{a}(\omega)}.$$

Use Theorem 3.3(iii) with $\phi(\psi_1,\psi_2) = \big(1+\psi_1\psi_2\big)/\big(1+(1+\psi_1)\psi_2\big)$, $\psi_1 = \tilde{b}$, and $\psi_2 = 1/\tilde{a}$, to conclude that the function $\tilde{k}/(1+\tilde{k})$ belongs locally to $\tilde{L}^1$ at zero.

By Theorem 3.2, there is a function $r \in L^1(\mathbf{R};\mathbf{C})$ such that $\tilde{r}(\omega) = \tilde{k}(\omega)/(1+\tilde{k}(\omega))$ for all $\omega \in \mathbf{R}$. It follows from Lemma **2**.5.1 that r vanishes a.e. on $\mathbf{R}^-$, and that $\hat{r}(z) = \hat{k}(z)/(1+\hat{k}(z))$ for all $z \in \mathbf{C}$ with $\Re z \geq 0$. Therefore, by the uniqueness theorem for Laplace transforms (combine

Lemma **2**.3.4 with Theorems **2**.2.8(i) and **2**.2.9(i)), r is the resolvent of k. This proves that the resolvent of k belongs to $L^1(\mathbf{R}^+;\mathbf{C})$. □

Below we give three additional extensions of the Paley–Wiener theorem, of which we prove only the last one. The reader should be able to do the omitted proofs, i.e., Exercise 9.

The first result concerns systems of integral equations.

4.2 Theorem
Let $k = a + b$, where $a \in BV(\mathbf{R}^+;\mathbf{C}^{n\times n})$ satisfies $\det[a(\infty)] \neq 0$, and $b \in L^1(\mathbf{R}^+;\mathbf{C}^{n\times n})$. Define $\hat{a}$ on the imaginary axis by $\hat{a}(\mathrm{i}\omega) = \lim_{\sigma\downarrow 0}\hat{a}(\sigma+\mathrm{i}\omega)$ for $\omega \neq 0$. Then the resolvent r of k satisfies

$$r \in L^1(\mathbf{R}^+;\mathbf{C}^{n\times n})$$

if and only if

$$\det[I + \hat{a}(z) + \hat{b}(z)] \neq 0 \textit{ for all nonzero } z \in \mathbf{C} \textit{ with } \Re z \geq 0.$$

The next result is the analogue of Theorem 4.1 for differential resolvents.

4.3 Theorem
Let $\mu(\mathrm{d}s) = \nu(\mathrm{d}s) + a(s)\,\mathrm{d}s$, where $\nu \in M(\mathbf{R}^+;\mathbf{C})$, and $a \in L^1_{\mathrm{loc}}(\mathbf{R}^+;\mathbf{R})$ is nonnegative, nonincreasing, and convex on $(0,\infty)$. Define $\hat{a}$ on the imaginary axis by $\hat{a}(\mathrm{i}\omega) = \lim_{\sigma\downarrow 0}\hat{a}(\sigma + \mathrm{i}\omega)$ for $\omega \neq 0$ (if $a \notin L^1(\mathbf{R}^+;\mathbf{R})$, then define $\hat{a}(0) = \infty$). Then the differential resolvent r of μ satisfies

$$r \in L^1(\mathbf{R}^+;\mathbf{C})$$

if and only if

$$z + \hat{a}(z) + \hat{\nu}(z) \neq 0 \textit{ in the half plane } \Re z \geq 0.$$

Finally, we give the analogue of Theorem 4.2 for differential resolvents.

4.4 Theorem
Let $\mu(\mathrm{d}s) = \nu(\mathrm{d}s) + a(s)\,\mathrm{d}s$, where $\nu \in M(\mathbf{R}^+;\mathbf{C}^{n\times n})$ and where $a \in BV(\mathbf{R}^+;\mathbf{C}^{n\times n})$ satisfies $\det[a(\infty)] \neq 0$. Define $\hat{a}$ on the imaginary axis by $\hat{a}(\mathrm{i}\omega) = \lim_{\sigma\downarrow 0}\hat{a}(\sigma + \mathrm{i}\omega)$ for $\omega \neq 0$. Then the differential resolvent r of μ satisfies

$$r \in L^1(\mathbf{R}^+;\mathbf{C}^{n\times n})$$

if and only if

$$\det[zI + \hat{a}(z) + \hat{\nu}(z)] \neq 0 \textit{ for all nonzero } z \in \mathbf{C} \textit{ with } \Re z \geq 0.$$

Proof Let $\alpha \in M(\mathbf{R}^+;\mathbf{C}^{n\times n})$ be the distribution derivative of the function a (including the point mass of size $a(0)$ at zero). By Theorem **3**.8.6(i) (recall that we defined $\hat{a}(\omega)$ by $\hat{a}(\omega) = \lim_{\sigma\downarrow 0}\hat{a}(\sigma + \mathrm{i}\omega)$) we have $\tilde{a}(\omega) = (\mathrm{i}\omega)^{-1}\tilde{\alpha}(\omega)$ for $\omega \neq 0$. Define $\tilde{r}$ by

$$\tilde{r}(\omega) = \mathrm{i}\omega\left[(\mathrm{i}\omega)^2 I + \tilde{\alpha}(\omega) + \mathrm{i}\omega\tilde{\nu}(\omega)\right]^{-1}, \quad \omega \in \mathbf{R}.$$

From Theorem 3.3, the assumption $\det[a(\infty)] \neq 0$ (which is equivalent to $\det[\tilde{\alpha}(0)] \neq 0$), the fact that $\det[\mathrm{i}\omega I + (\mathrm{i}\omega)^{-1}\tilde{\alpha}(\omega) + \tilde{\nu}(\omega)] \neq 0$ for $\omega \neq 0$, and from Corollary 3.6(i), we conclude that $\tilde{r}(\omega)$ belongs locally to $\tilde{L}^1$ for every $\omega \in \mathbf{R}$.

It remains to show that $\tilde{r}$ belongs locally to $\tilde{L}^1$ at infinity. For $\omega \neq 0$ we have

$$\tilde{r}(\omega) = (\mathrm{i}\omega)^{-1}\big[I + \tilde{\phi}(\omega)\big]^{-1},$$

where $\tilde{\phi}(\omega) = (\mathrm{i}\omega)^{-2}\tilde{\alpha}(\omega) + (\mathrm{i}\omega)^{-1}\tilde{\nu}(\omega)$ belongs locally to $\tilde{L}^1$ at infinity, according to Theorem 3.4. By Corollary 3.6(ii), the same statement is true for $\tilde{r}(\omega)$.

It is now a consequence of Theorem 3.2 that $\tilde{r}(\omega)$ is the Fourier transform of a function $r \in L^1(\mathbf{R}; \mathbf{C}^{n\times n})$. To show that r is the differential resolvent of μ is suffices to apply Lemma **2**.5.1 and Corollary **3**.3.6, and to use the uniqueness theorem for Laplace transforms. □

5. Appendix: the Hardy–Littlewood Inequality

5.1 Theorem
Let $r \in L^1_{\mathrm{loc}}(\mathbf{R}^+; \mathbf{C})$ have a Laplace transform $\hat{r}(z)$ that is defined (as an absolutely convergent integral) in the open half plane $\Re z > 0$. Moreover, suppose that $\hat{r}$ is bounded and has a bounded continuous extension to the closed half plane $\Re z \geq 0$, and that the boundary function $\tilde{r}(\omega) = \lim_{\sigma\downarrow 0}\hat{r}(\sigma + \mathrm{i}\omega)$ is locally absolutely continuous and satisfies $\tilde{r}' \in L^1(\mathbf{R}; \mathbf{C})$. Then $r \in L^1(\mathbf{R}^+; \mathbf{C})$, and $\|r\|_{L^1(\mathbf{R}^+)} \leq \frac{1}{2}\|\tilde{r}'\|_{L^1(\mathbf{R})}$.

Note that the proof of this theorem is very easy if r is nonnegative: Then we have $\tilde{r}(0) = \|r\|_{L^1(\mathbf{R}^+)}$, and because $\lim_{|\omega|\to\infty}\tilde{r}(\omega) = 0$ it follows that the total variation of $\tilde{r}$ on $\mathbf{R}$ is at least $2\|r\|_{L^1(\mathbf{R}^+)}$. Thus, if $\tilde{r}$ is locally absolutely continuous, one has $\|\tilde{r}'\|_{L^1(\mathbf{R})} \geq 2\|r\|_{L^1(\mathbf{R}^+)}$.

Proof of Theorem 5.1 It follows from Koosis [1], p. 143, that $\hat{r}(\sigma + \mathrm{i}\omega) = (P_\sigma * \tilde{r})(\omega)$ for all $\sigma > 0$ and all $\omega \in \mathbf{R}$, where $P_\sigma(\omega) = \sigma^{-1}P(\omega/\sigma)$ for $\sigma > 0$, and P is the Poisson kernel $P(x) = 1/(\pi(1+x^2))$ (alternatively, one could derive this result by first mapping the right half plane onto the unit disc, then applying Rudin [2], Theorem 11.30(b), and finally mapping the result back onto the right half plane; cf. the proof of Theorem **5**.2.6). This kernel satisfies $\int_{\mathbf{R}} P(x)\,\mathrm{d}x = 1$. By Corollary **3**.7.2(i) combined with Theorem **2**.2.2(i), the function $\omega \mapsto \hat{r}(\sigma + \mathrm{i}\omega)$ is, for each $\sigma > 0$, locally absolutely continuous and satisfies $\int_{\mathbf{R}}|\frac{\partial}{\partial\omega}\hat{r}(\sigma + \mathrm{i}\omega)|\,\mathrm{d}\omega \leq \|\tilde{r}'\|_{L^1(\mathbf{R})}$. We know that $\hat{r}$ is analytic in the set $\{\, z \in \mathbf{C} \mid \Re z > 0 \,\}$, hence $\frac{\partial}{\partial\omega}\hat{r}(\sigma + \mathrm{i}\omega) = \mathrm{i}\hat{r}'(\sigma + \mathrm{i}\omega)$. By definition, this means that $\hat{r}'$ belongs to the Hardy space H^1 over this half space.

By the assumption, the function $t \mapsto t\mathrm{e}^{-\sigma t}r(t)$ is integrable for each $\sigma > 0$. Its Fourier transform, which is the function $\omega \mapsto \hat{r}'(\sigma + \mathrm{i}\omega)$ is also integrable. Therefore, we can use the Fourier inversion theorem (see Rudin [2], Theorem 9.11) to get, for all $\sigma > 0$ and almost all $t \geq 0$,

$$t\mathrm{e}^{-\sigma t}r(t) = \frac{1}{2\pi}\int_{\mathbf{R}} \mathrm{e}^{\mathrm{i}\omega t}\hat{r}'(\sigma + \mathrm{i}\omega)\,\mathrm{d}\omega.$$

As $\sigma \downarrow 0$, the function $\omega \mapsto \hat{r}'(\sigma + \mathrm{i}\omega)$ tends to $-\mathrm{i}\tilde{r}'$ in $L^1(\mathbf{R};\mathbf{C})$ (see Lemma **2**.7.4(i)), and $t\mathrm{e}^{\sigma t}r(t) \to tr(t)$. Therefore, for almost all $t \geq 0$,

$$tr(t) = \frac{1}{\mathrm{i}\,2\pi}\int_{\mathbf{R}} \mathrm{e}^{\mathrm{i}\omega t}\tilde{r}'(\omega)\,\mathrm{d}\omega.$$

By the half plane analogue of the Hardy–Littlewood inequality (see Hille and Tamarkin [1], Theorems 3.1 and 4.2),

$$\int_{\mathbf{R}^+} |r(t)|\,\mathrm{d}t \leq \frac{1}{2}\int_{\mathbf{R}} |\tilde{r}'(\omega)|\,\mathrm{d}\omega,$$

and the proof is complete. □

6. Exercises

1. Show that if a is positive and nonincreasing on $\mathbf{R}^+$, with $0 < a(\infty) \leq a(0) < \infty$, then its resolvent r is integrable.
2. Prove Lemma 2.3.
3. Improve the constant 20 in Theorem 2.1. How far can you go? Show that this constant must be greater than one.
4. Show that $k(t) = t^{-1/2}J_1(t^{1/2})$, $t > 0$, is of positive type and does not have an integrable resolvent. Here J_1 is the Bessel function of first order. (Kernels of positive type are defined in Chapter 16.) Hint: The Laplace transform of this kernel is $2\big(1 - \mathrm{e}^{-1/(4z)}\big)$.
5. Prove Theorem 3.4.
6. Show that if $a \in BV(\mathbf{R})$ then the distribution Fourier transform of a belongs locally to $\tilde{L}^1$ at all nonzero points ω and at infinity. (See the comment after Theorem 3.4.)
7. Let u be a finite sum of finite order distribution derivatives of functions in $L^1(\mathbf{R})$. Show that the distribution Fourier transform of u belongs locally to $\tilde{L}^1$ at all points $\omega \in \mathbf{R}$. (Distributions of this type are sometimes referred to as integrable distributions.)
8. Let φ belong locally to $\tilde{L}^1(\mathbf{R};\mathbf{C})$ at $\omega_0 \in \mathbf{R}$, and suppose that $\varphi(\omega_0) \neq 0$. Show that $\overline{\varphi}$, $|\varphi|$, and $\sqrt{|\varphi|}$ belong locally to $\tilde{L}^1$ at ω_0.
9. Prove Theorems 4.2 and 4.3.
10. Suppose that the Laplace transform conditions in Theorems 4.1–4.4 are satisfied, so that the resolvent r, or the differential resolvent r, is

integrable. Compute $\int_{\mathbf{R}^+} r(t)\,dt$ in all cases. Also compute the same integral for the resolvent in Theorem 2.1.

11. Consider the scalar, higher-order equation

$$\sum_{k=1}^{n} \alpha_k x^{(k)}(t) + \int_0^t a(t-s)x(s)\,ds = f(t), \quad t \geq 0,$$

where α_k, $k = 1, \ldots, n$, are constants, and $\alpha_n \neq 0$. Assume that $a \in L^1_{\mathrm{loc}}(\mathbf{R}^+; \mathbf{R})$ is nonnegative, nonincreasing, and convex on $(0, \infty)$, and that

$$\sum_{k=1}^{n} \alpha_k z^k + \hat{a}(z) \neq 0, \quad \Re z \geq 0.$$

Rewrite this equation as a first order system of integrodifferential equations, and show that the resolvent of this system is integrable. In particular, if $f \in L^p(\mathbf{R}^+)$ for some $p \in [1, \infty]$, then $x \in L^p(\mathbf{R}^+)$. (Shea and Wainger [1], p. 316).

12. Let $a \in L^1_{\mathrm{loc}}(\mathbf{R}^+; \mathbf{R})$ be nonnegative, nonincreasing, and convex for $t > 0$, and not identically zero. Determine necessary and sufficient conditions on a (given for a itself, not for the Laplace transform of a) under which the differential resolvent r of a satisfies $r \in L^1(\mathbf{R}^+; \mathbf{R})$. Moreover, compute the value of $\int_0^\infty r(s)\,ds$. Hint: See (2.5) and (2.6). Show that $\Re\hat{a}(z) > 0$ for $\Re z > 0$, and that $\Re\hat{a}(i\omega) = 0$ for some nonzero $\omega \in \mathbf{R}$ iff γ is supported on the set $\{k2\pi/\omega\}_{k=1}^{\infty}$, i.e., a is piecewise linear, and changes its slope only at points belonging to the set above. See Hannsgen [1].

13. Show that if $a \in L^1_{\mathrm{loc}}(\mathbf{R}^+; \mathbf{R})$ is nonnegative, nonincreasing, and convex on $(0, \infty)$, then its differential resolvent r satisfies $|r(t)| \leq 1$ for $t \geq 0$. Hint: Show that the function $r^2(t) + a(t)\big(\int_0^t r(s)\,ds\big)^2 - \int_0^t \big(\int_{t-\tau}^t r(s)\,ds\big)^2 a'(\tau)\,d\tau$ is nonincreasing. See Hannsgen [2] or [4]. (This result is a special case of Corollary **17**.2.3.)

14. Let $a \in L^1_{\mathrm{loc}}(\mathbf{R}^+)$ be nonnegative, nonincreasing, and convex on $(0, \infty)$. Let $\lambda_0 > 0$, let r_λ be the differential resolvent of λa for $\lambda > 0$, and suppose that $r_{\lambda_0} \in L^1(\mathbf{R}^+; \mathbf{R})$ (cf. Exercise 12). Show that, for some constant c,

$$\left| \int_0^t r_\lambda(s)\,ds \right| \leq c \int_0^\infty |r_{\lambda_0}(s)|\,ds$$

for $\lambda_0 \leq \lambda < \infty$ and $0 \leq t < \infty$. How small can you make c? Hint: Show that

$$\int_0^t r_\lambda(t)\,dt = \int_0^t r_{\lambda_0}(t)\,dt + \frac{\lambda - \lambda_0}{\lambda} \int_0^t (r_\lambda(t-s) - 1) r_{\lambda_0}(s)\,ds,$$

and use Exercise 13. See Hannsgen [5].

15. Prove the following result.

6.1 Theorem
Let k be a kernel such that $e^{-\sigma t}k(t) \in L^1(\mathbf{R}^+;\mathbf{C})$ for all $\sigma > 0$ and assume that the resolvent r of k is integrable on $\mathbf{R}^+$. Define $\tilde{k}(\omega) = \lim_{\sigma\downarrow 0} \hat{k}(\sigma + i\omega)$. Then $\tilde{k}$ belongs locally to $\tilde{L}^1(\mathbf{R};\mathbf{C})$ at infinity and at all points for which $\tilde{r}(\omega) \neq 1$, and $1/\tilde{k}$ belongs to $\tilde{L}^1(\mathbf{R};\mathbf{C})$ at all points where $\tilde{r}(\omega) = 1$.

16. Prove the following result (See Jordan, Staffans, and Wheeler [1], Proposition 7.3).

6.2 Theorem
Let $k \in L^1_{loc}(\mathbf{R}^+;\mathbf{C})$, and let r be the resolvent of k. Prove that $r \in L^1(\mathbf{R}^+;\mathbf{C})$ if and only if the following four conditions hold:
(i) *$\int_0^\infty e^{-\sigma t}|k(t)|\,dt < \infty$ for some sufficiently large σ;*
(ii) *$\hat{k}$ has a meromorphic extension to $\Re z \geq 0$ in the sense that there exist functions φ and ψ, analytic in $\Re z > 0$ and continuous in $\Re z \geq 0$, and without common zeros, such that $\hat{k}(z) = \varphi(z)/\psi(z)$ for $\Re z \geq \sigma$;*
(iii) *the extended function $\hat{k}$ (i.e., $\hat{k}(z) = \varphi(z)/\psi(z)$, unless $\psi(z) = 0$, in which case $\hat{k}(z) = \infty$) satisfies $1 + \hat{k}(z) \neq 0$ for $\Re z \geq 0$, and $\hat{k}(z)/\big(1 + \hat{k}(z)\big)$ is bounded in $\Re z \geq 0$;*
(iv) *if $\tilde{k}(\omega) = \hat{k}(i\omega)$, then, at all points $\omega \in \mathbf{R}$, $\tilde{k}$ or $1/\tilde{k}$ belongs locally to $\tilde{L}^1(\mathbf{R};\mathbf{C})$, and at infinity, $\tilde{k}$ belongs locally to $\tilde{L}^1(\mathbf{R};\mathbf{C})$.*

17. Prove the following result.

6.3 Lemma
Let $\mu \in M(\mathbf{R}^+)$ and $\nu \in M(\mathbf{R}^+)$, with $\tilde{\mu}(0) = 0$ and $\tilde{\nu}(0) = 0$, and suppose that $\int_{\mathbf{R}^+} |\mu|((t,\infty))|\nu|((t,\infty))\,dt < \infty$. Then $\frac{1}{\omega}(\tilde{\mu}(\omega)\tilde{\nu}(\omega))$ belongs locally to $\tilde{L}^1(\mathbf{R})$, and $\lim_{\omega\to 0} \frac{1}{\omega}(\tilde{\mu}(\omega)\tilde{\nu}(\omega)) = 0$.

Hint: Show that, for each $T > 0$,

$$\begin{aligned}
\int_0^T \left| \int_{[0,t]} \mu([0,t-s])\nu(ds) \right| dt
&\leq \int_{\mathbf{R}^+} |\mu|((t,\infty))|\nu|((t,\infty))\,dt \\
&\quad + \int_{[0,T]} \int_s^T |\mu((t-s,t])|\,dt\,|\nu|(ds) \\
&= \int_{\mathbf{R}^+} |\mu|((t,\infty))|\nu|((t,\infty))\,dt \\
&\quad + \int_{[0,T]} \left(\int_0^{T-s} - \int_s^T \right) |\mu((t,\infty))|\,dt\,|\nu|(ds) \\
&\leq 2\int_{\mathbf{R}^+} |\mu|((t,\infty))|\nu|((t,\infty))\,dt.
\end{aligned}$$

Moreover, show that

$$\int_0^T \int_{[0,t]} \mu([0,t-s])\nu(\mathrm{d}s)\,\mathrm{d}t = \int_0^T \mu((T-s,\infty))\nu((s,\infty))\,\mathrm{d}s$$
$$\leq \int_0^T |\mu|((T-s,\infty))|\nu|((s,\infty))\,\mathrm{d}s,$$

and that the last term tends to zero as $T \to \infty$ (it is dominated by $|\mu|((s,\infty))|\nu|((s,\infty))$ for $s \in [0, T/2]$).

18. Prove the following result.

6.4 Theorem
Let $k = a + b$ where
(i) $\mathrm{e}^{-\sigma t}a(t) \in L^1(\mathbf{R}^+;\mathbf{C}^{n\times n})$ *for $\sigma > 0$, and the resolvent r_a of a satisfies $r_a \in L^1(\mathbf{R}^+;\mathbf{C}^{n\times n})$ and $\int_{\mathbf{R}^+} r_a(t)\,\mathrm{d}t = I$,*
(ii) $b \in BV(\mathbf{R}^+;\mathbf{C}^{n\times n})$ *and $b(\infty) = 0$,*
(iii) $\det[I + \hat{a}(z) + \hat{b}(z)] \neq 0$ *for all nonzero $z \in \mathbf{C}$ with $\Re z \geq 0$,*
(iv) $\int_0^\infty \|b\|_{\mathrm{var}((t,\infty))} \int_t^\infty |r_a(s)|\,\mathrm{d}s\,\mathrm{d}t < \infty$.
Then the resolvent r of k satisfies $r \in L^1(\mathbf{R}^+;\mathbf{C}^{n\times n})$.

Hint: Apply Lemma 6.3 with $\mu([0,t]) = b(t)$ and $\nu([0,t]) = I - \int_0^t r_a(s)\,\mathrm{d}s$. (Gripenberg [17].)

19. Let $a \in L^1_{\mathrm{loc}}(\mathbf{R}^+)$ be nonnegative, nonincreasing, and convex on $(0,\infty)$. Moreover, assume that $-a'$ is convex on $(0,\infty)$. Show that, for some $c > 0$ (independent of a),

$$|\tilde{a}''(\omega)| \leq c \int_0^{1/|\omega|} t^2 a(t)\,\mathrm{d}t, \quad \omega \neq 0.$$

See Carr and Hannsgen [1], Lemma 5.1.

20. Consider the kernel $k(t) = t^{-\alpha}\cos(t)$, $t > 0$, where the constant $\alpha \in [0,1)$. Show that the resolvent r of k is integrable.

21. Let $k(t) = a(t) + a_\infty$, where $a \in L^1(\mathbf{R}^+;\mathbf{C}^{n\times n})$ and a_∞ is a constant matrix. Discuss conditions under which the resolvent r of k satisfies $r \in L^1(\mathbf{R}^+;\mathbf{C}^{n\times n})$. Are your conditions necessary as well?

7. Comments

Section 2:

In Reuter [1] it was shown that the resolvent of a locally integrable, completely monotone (therefore convex) function is integrable. This result was extended in Miller [4] to the case where $a \in L^1_{\mathrm{loc}}(\mathbf{R}^+)$, a is strictly positive, and $\ln(a(t))$ is convex for $t > 0$. In fact, under these conditions, $0 \leq r(t) \leq a(t)$, and $\int_{\mathbf{R}^+} r(t)\,\mathrm{d}t \leq 1$ (see Corollary **9**.8.8). Theorem 2.1 first appeared in Shea and Wainger [1], but without the uniform bound on the L^1-norm of the resolvent. The first uniform bound was given in Gripenberg [8].

An easy example of a kernel k whose Laplace transform satisfies the Paley–Wiener condition $1 + \hat{k}(z) \neq 0$ in $\Re z \geq 0$ but that does not have an integrable resolvent is given in Exercise 4. A more subtle example is found in Shea and Wainger [1], Section 5. There the transform $\hat{k}$ is in addition continuous in $\Re z \geq 0$, and $|\hat{k}(z)| \leq \big(1 + |\Im z|\big)^{-1}$ for $\Re z \geq 0$.

The fact that the convexity assumption cannot be removed from the assumptions of Theorem 2.1 was demonstrated in Gripenberg [4], where a nonnegative, nonincreasing function $a \in L^1_{\mathrm{loc}}(\mathbf{R}^+)$ was constructed such that the corresponding resolvent r is not in $L^1(\mathbf{R}^+)$. For a related result, see Ibragimov [1]. For nonnegative, nonincreasing, and integrable kernels it is possible to give an explicit bound on the L^1-norm of the resolvent in terms of the norm of the kernel, see Gripenberg [11].

Sections 3–4:

Theorem 4.1 was established in Jordan and Wheeler [3]. The idea of applying the concept of local analyticity, in one form or another, has also been used in Gripenberg [17], and in Jordan, Staffans, and Wheeler [1].

Another result, given in Londen [14], can be summarized as follows. Let $a \in L^1_{\mathrm{loc}}(\mathbf{R}^+)$, $a \notin L^1(\mathbf{R}^+)$, $a(t) = b(t)c(t)$, $0 < t < \infty$, where $b(t)$ is completely monotone on $\mathbf{R}^+$, and $c(t)$ is of positive type. Then $b(t) = \int_{\mathbf{R}^+} \mathrm{e}^{-st}\mu(\mathrm{d}s)$, $c(t) = \lim_{T\to\infty} \int_{[0,T]} (1 - s/T)\cos(st)\nu(\mathrm{d}s)$, where μ and ν are positive, locally bounded measures. Extend ν to $\mathbf{R}^-$ as an odd measure (i.e., $\nu(-E) = -\nu(E)$ for each bounded Borel set E). Then, for $\Re z$ sufficiently large, $\hat{a}(z) = \frac{1}{2}\int_{\mathbf{R}} \hat{b}(z - ix)\nu(\mathrm{d}x)$. Using this singular convolution one can give conditions on μ and ν which imply that the resolvent r of a satisfies $r \in L^1(\mathbf{R}^+)$. Explicit bounds on $\|r\|_{L^1(\mathbf{R}^+)}$ can also be obtained in this case.

One cannot have a uniform bound on the L^1-norm of the differential resolvent, if the kernel satisfies the condition

(i) *$a \in L^1_{\mathrm{loc}}(\mathbf{R}^+;\mathbf{R})$ and a is nonnegative, nonincreasing, and convex on* $(0,\infty)$.

This follows from the fact that there are piecewise linear functions that satisfy condition (i), but that do not satisfy the Paley–Wiener condition $z + \hat{a}(z) \neq 0$ for $\Re z \geq 0$. See Exercise 12. However, one does have the bounds given in Exercises 13 and 14. See also Hannsgen [5].

A large number of papers have been written on how the differential resolvent r_λ of the kernel λa, i.e., the solution of the equation

$$r'_\lambda(t) + \lambda \int_0^t r_\lambda(t-s)a(s)\,\mathrm{d}s = 0, \quad t \geq 0, \quad r_\lambda(0) = 1,$$

depends on the positive parameter λ. In Carr and Hannsgen [1] it is shown (among other things) that, if the kernel a satisfies condition (i) above as well as the assumptions

(ii) $\Re\tilde{a}(\omega) > 0$ *for* $\omega \in \mathbf{R}$,
(iii) $\limsup_{\omega\to\infty} |\Im\tilde{a}(\omega)| \,/\, |\omega\Re\tilde{a}(\omega)| < \infty$,
and
(iv) $\int_1^\infty \frac{1}{t}a(t)\,\mathrm{d}t < \infty$,
then for every fixed $\Lambda > 0$ we have

$$\lim_{t\to\infty} \sup_{\Lambda\le\lambda<\infty} |r_\lambda(t)| = 0 \quad \textit{and} \quad \int_0^\infty \sup_{\lambda_0\le\lambda<\infty} |r_\lambda(t)|\,\mathrm{d}t < \infty.$$

In Hannsgen [11] it is shown that assumption (iii) is crucial for the conclusion above. For an example of a kernel where this condition fails, see Nohel and Shea [2]. Further results about this kind of parameter-dependence can be found in Carr and Hannsgen [2], Hannsgen [4], [6]–[10], [12], and [15], Hannsgen and Wheeler [2], and Noren [1] and [3].

A different type of parameter-dependence of the differential resolvent has been considered in Hannsgen and Wheeler [3]. Let r_c be the solution of the equation

$$r_c'(t) + \int_0^t r_c(t-s)\big(a(s)+c\big)\,\mathrm{d}s = 0, \quad t > 0, \quad r_c(0) = 1,$$

where $c \in [0,1]$ is a constant, and where the kernel a satisfies (i) and (ii). Then $\sup_{0\le c\le 1}\int_0^\infty |r_c(t)|\,\mathrm{d}t < \infty$. Recall (see Exercise 12) that only certain piecewise linear kernels are excluded by the condition $\Re\hat{a}(i\omega) > 0$. For results giving $\int_0^\infty \sup_{0\le c\le 1}|r_c(t)|\,\mathrm{d}t < \infty$, see Hannsgen and Wheeler [3] and Noren [2].

For asymptotic representations of the differential resolvents of kernels behaving asymptotically as t^γ ($\gamma < 0$), see Wong and Wong [1].

Section 5:
The Hardy–Littlewood inequality was first obtained in Hardy and Littlewood [1]. The original version of Theorem 5.1 (in particular, the half plane analogue of the Hardy–Littlewood inequality) appears in Hille and Tamarkin [1]. In addition, they show that the constant $\frac{1}{2}$ is the best possible.

References

R. W. Carr and K. B. Hannsgen
1. A nonhomogeneous integrodifferential equation in Hilbert space, *SIAM J. Math. Anal.* **10** (1979), pp. 961–984.
2. Resolvent formulas for a Volterra equation in Hilbert space, *SIAM J. Math. Anal.* **13** (1982), pp. 459–483.

G. Gripenberg

4. A Volterra equation with nonintegrable resolvent, *Proc. Amer. Math. Soc.* **73** (1979), pp. 57–60.
8. On the asymptotic behavior of resolvents of Volterra equations, *SIAM J. Math. Anal.* **11** (1980), pp. 654–662.
11. On the resolvents of Volterra equations with nonincreasing kernels, *J. Math. Anal. Appl.* **76** (1980), pp. 134–145.
17. Integrability of resolvents of systems of Volterra equations, *SIAM J. Math. Anal.* **12** (1981), pp. 585–594.

K. B. Hannsgen

1. Indirect Abelian theorems and a linear Volterra equation, *Trans. Amer. Math. Soc.* **142** (1969), pp. 539–555.
2. On a nonlinear Volterra equation, *Michigan Math. J.* **16** (1969), pp. 365–376.
4. A Volterra equation with parameter, *SIAM J. Math. Anal.* **4** (1973), pp. 22–30.
5. Note on a family of Volterra equations, *Proc. Amer. Math. Soc.* **46** (1974), pp. 239–243.
6. A Volterra equation in Hilbert space, *SIAM J. Math. Anal.* **5** (1974), pp. 412–416.
7. A linear Volterra equation in Hilbert space, *SIAM J. Math. Anal.* **5** (1974), pp. 927–940.
8. Uniform boundedness in a class of Volterra equations, *SIAM J. Math. Anal.* **6** (1975), pp. 689–697.
9. Continuous parameter dependence in a class of Volterra equations, *SIAM J. Math. Anal.* **7** (1976), pp. 45–58.
10. The resolvent kernel of an integrodifferential equation in Hilbert space, *SIAM J. Math. Anal.* **7** (1976), pp. 481–490.
11. Uniform L^1 behavior for an integrodifferential equation with parameter, *SIAM J. Math. Anal.* **8** (1977), pp. 626–639.
12. An L^1 remainder theorem for an integrodifferential equation with asymptotically periodic solution, *Proc. Amer. Math. Soc.* **73** (1979), pp. 331–337.
15. A uniform approximation for an integrodifferential equation with parameter, *J. Integral Equations* **2** (1980), pp. 117–131.

K. B. Hannsgen and R. L. Wheeler

2. Behavior of the solution of a Volterra equation as a parameter tends to infinity, *J. Integral Equations* **7** (1984), pp. 229–237.
3. A singular limit problem for an integrodifferential equation, *J. Integral Equations* **5** (1983), pp. 199–209.

G. H. Hardy and J. E. Littlewood

1. Some new properties of Fourier constants, *Math. Ann.* **97** (1926), pp. 159–209.

E. Hille and J. D. Tamarkin

1. On the absolute integrability of Fourier transforms, *Fund. Math.* **25** (1935), pp. 329–352.

I. A. Ibragimov

1. A remark on the ergodic theorem for Markov chains, *Theor. Probability Appl.* **20** (1975), pp. 174–176.

G. S. Jordan, O. J. Staffans, and R. L. Wheeler

1. Local analyticity in weighted L^1-spaces and applications to stability problems for Volterra equations, *Trans. Amer. Math. Soc.* **274** (1982), pp. 749–782.

G. S. Jordan and R. L. Wheeler

3. A generalization of the Wiener–Lévy theorem applicable to some Volterra equations, *Proc. Amer. Math. Soc.* **57** (1976), pp. 109–114.

P. Koosis

1. *Introduction to H_p Spaces*, Cambridge University Press, Lecture Note Series 40, Cambridge, 1980.

S-O. Londen

14. On some nonintegrable Volterra kernels with integrable resolvents including some applications to Riesz potentials, *J. Integral Equations* **10** (1985), pp. 241–289.

R. K. Miller

4. On Volterra integral equations with nonnegative integrable resolvents, *J. Math. Anal. Appl.* **22** (1968), pp. 319–340.

J. A. Nohel and D. F. Shea

2. Frequency domain methods for Volterra equations, *Adv. in Math.* **22** (1976), pp. 278–304.

R. Noren

1. A singular limit problem for a linear Volterra equation, *Quart. Appl. Math.* **46** (1988), pp. 169–179.
2. A singular limit problem for a Volterra equation, *SIAM J. Math. Anal.* **19** (1988), pp. 1103–1107.
3. Uniform L^1 behavior for the solution of a Volterra equation with a parameter, *SIAM J. Math. Anal.* **19** (1988), pp. 270–286.

G. E. H. Reuter

1. Über eine Volterrasche Integralgleichung mit totalmonotonem Kern, *Arch. Math.* **7** (1956), pp. 59–66.

W. Rudin

2. *Functional Analysis*, McGraw–Hill, New York, 1973.

D. F. Shea and S. Wainger

1. Variants of the Wiener–Lévy theorem, with applications to stability problems for some Volterra integral equations, *American J. Math.* **97** (1975), pp. 312–343.

J. S. W. Wong and R. Wong

1. Asymptotic solutions of linear Volterra integral equations with singular kernels, *Trans. Amer. Math. Soc.* **189** (1974), pp. 185–200.

7

Unbounded and Unstable Solutions

We study linear Volterra equations of convolution type where the characteristic equation has zeros in the right half plane. The unbounded solutions are analysed, and from this analysis we obtain conditions for the existence of bounded but unstable solutions.

1. Introduction

In Chapter 3 we studied the equation

$$x'(t) + (\mu * x)(t) = f(t), \quad t \in \mathbf{R}^+; \quad x(0) = x_0. \tag{1.1}$$

If $\mu \in M(\mathbf{R}^+)$ and $\det[zI + \hat{\mu}(z)] = 0$ for some $z \in \mathbf{C}$ with $\Re z \geq 0$, then by Theorem **3**.3.9, there exist an initial value x_0 and a bounded function f for which the corresponding solution x of (1.1) is unbounded. Below we discuss two problems, both of which are related to this situation.

- How does an unbounded solution x of (1.1) behave asymptotically?
- Under what conditions are solutions of (1.1) bounded on $\mathbf{R}^+$, in spite of the fact that $\det[zI + \hat{\mu}(z)]$ has zeros in $\Re z \geq 0$?

The answers to these two questions depend on how the zeros of $\det[zI + \hat{\mu}(z)]$ are located. In the 'critical' case, where $\det[zI + \hat{\mu}(z)]$ has zeros on the boundary of the domain of $\hat{\mu}$, i.e., typically on the imaginary axis, the situation is quite complicated, but in the 'noncritical' case where $\det[zI + \hat{\mu}(z)]$ has zeros in the open half plane $\Re z > 0$, but no zeros on the imaginary axis, it is fairly easy to give a complete answer. Below, in Section 2 we consider the noncritical case, and in Section 3 we briefly discuss the critical case.

The same questions are relevant for the equation

$$x(t) + \int_0^t k(t-s)x(s)\,\mathrm{d}s = f(t), \quad t \in \mathbf{R}^+, \tag{1.2}$$

studied in Chapter 2, although the function $zI + \hat{\mu}(z)$ is, of course, to be replaced by $I + \hat{k}(z)$. Furthermore, as we shall see below, we may just as well consider the more general equation

$$x(t) + \int_{[0,t]} \mu(\mathrm{d}s)x(t-s) = f(t), \quad t \in \mathbf{R}^+, \tag{1.3}$$

previously studied in Chapter 4.

2. Characteristic Exponents in the Open Right Half Plane

Suppose that $\mu \in M(\mathbf{R}^+; \mathbf{C}^{n\times n})$ and that $\det[zI + \hat{\mu}(z)] = 0$ for some $z \in \mathbf{C}$ with $\Re z > 0$, but that $\det[zI + \hat{\mu}(z)] \neq 0$ for $\Re z = 0$. We claim that this implies that the total number of zeros of $\det[zI + \hat{\mu}(z)]$ in $\Re z > 0$ must be finite. This is because the existence of an infinite number of zeros of $\det[zI + \hat{\mu}(z)]$ would imply that there exists a sequence of zeros of $\det[zI + \hat{\mu}(z)]$ which either converges to a finite zero z_0 of $\det[zI + \hat{\mu}(z)]$, or tends to infinity. The latter case is impossible, because $\left|\det[zI + \hat{\mu}(z)]\right| \to \infty$ as $|z| \to \infty$. The former case is also impossible, because z_0 cannot satisfy $\Re z_0 > 0$ (since $\det[zI + \hat{\mu}(z)]$ is analytic for $\Re z > 0$ and since the zeros of an analytic function are isolated), and because we explicitly assumed that $\det[zI + \hat{\mu}(z)]$ has no zeros with $\Re z = 0$.

In the preceding case it is quite easy to prove that the differential resolvent of μ is the sum of an exponential polynomial and a remainder in $L^1(\mathbf{R}^+; \mathbf{C}^{n\times n})$ (throughout, we use the name 'exponential polynomial' for a function that is a sum of products of exponentials and ordinary polynomials). Moreover, we can use the weighted spaces introduced in Chapter 4 without much extra effort.

2.1 Theorem

Let φ be a submultiplicative weight function defined on $\mathbf{R}$ for which $\alpha_\varphi = \omega_\varphi$, and let $\mu \in M(\mathbf{R}^+; \varphi; \mathbf{C}^{n\times n})$ satisfy $\det[zI + \hat{\mu}(z)] \neq 0$ for all $z \in \mathbf{C}$ with $\Re z = \omega_\varphi$. Then the differential resolvent r of μ is of the form

$$r(t) = \sum_{\ell=1}^{m} \sum_{j=0}^{p_\ell - 1} K_{\ell,j}\, t^j e^{z_\ell t} + q(t), \quad t \in \mathbf{R}^+, \tag{2.1}$$

where each exponent z_ℓ is a scalar with $\Re z_\ell > \omega_\varphi$, each $K_{\ell,j}$ is a matrix, and $q \in L^1(\mathbf{R}^+; \varphi; \mathbf{C}^{n\times n})$ is locally absolutely continuous, and satisfies $q' \in L^1(\mathbf{R}^+; \varphi; \mathbf{C}^{n\times n})$.

More specifically, the exponents z_ℓ are the zeros of $\det[zI + \hat{\mu}(z)]$ in $\Re z > \omega_\varphi$, and the coefficient matrices $K_{\ell,j}$ are determined by the principal *part of* $[zI+\hat{\mu}(z)]^{-1}$ *at* z_ℓ *in the sense that, in some neighbourhood of* z_ℓ,

$$[zI+\hat{\mu}(z)]^{-1}=\sum_{j=0}^{p_\ell-1}\frac{j!\,K_{\ell,j}}{(z-z_\ell)^{(j+1)}}+\hat{q}_\ell(z),$$

where the remainder $\hat{q}_\ell(z)$ *is analytic at* z_ℓ. *Moreover, if we extend* μ *to* $\mathbf{R}$ *by defining* $\mu(E)=\mu(E\cap\mathbf{R}^+)$ *for every Borel set* $E\subset\mathbf{R}$, *and if we extend* q *to* $\mathbf{R}$ *by defining*

$$q(t)=-\sum_{\ell=1}^{m}\sum_{j=0}^{p_\ell-1}K_{\ell,j}\,t^j\mathrm{e}^{z_\ell t},\quad t\in(-\infty,0),\tag{2.2}$$

then q *is the whole line differential resolvent of* μ *corresponding to the weight function* φ *(with* $q(0)$ *defined to be equal to* $q(0+)$*).*

Sometimes the words 'singular part expansion' are used instead of 'principal part'. Observe that definition (2.2) is consistent with (2.1) in the sense that if we define $r(t)$ to be zero for $t<0$, then

$$r(t)=\sum_{\ell=1}^{m}\sum_{j=0}^{p_\ell-1}K_{\ell,j}\,t^j\mathrm{e}^{z_\ell t}+q(t),\quad t\in\mathbf{R}.\tag{2.3}$$

The preceding result (together with Theorems 2.3–2.5 below) motivates the following definition.

2.2 Definition
The equations $\det[zI+\hat{\mu}(z)]=0$ *and* $\det[I+\hat{\mu}(z)]=0$ *are called the characteristic equations of (1.1) and (1.3), respectively. The roots of these equations are called characteristic exponents of (1.1) and (1.3).*

Proof of Theorem 2.1 Let q be the whole line differential resolvent (see Theorem 4.4.4) of the extended measure μ, and define $r(t)=0$ for $t<0$. Then the restriction of q to $\mathbf{R}^+$ has the stated growth and smoothness properties. Thus, to complete the proof, it suffices to show that the difference $r-q$ is equal to the double sum in (2.3) for all $t\in\mathbf{R}$.

To prove (2.3), we use the inversion formula for Laplace transforms given, e.g., in Doetsch [1], Vol. I, Satz 1, p. 210. If we fix $\sigma>0$ so large that all the characteristic exponents in $\Re z\geq\omega_\varphi$ lie in the strip $\omega_\varphi<\Re z<\sigma$, then, by Theorem 4.4.13, $\int_{\mathbf{R}^+}\mathrm{e}^{-\sigma t}|r(t)|\,\mathrm{d}t<\infty$. Define $r(t)=0$ for $t<0$. Then we can use the inversion formula to get, for all $t\neq0$ (both r and q are locally absolutely continuous except at zero),

$$r(t)=\lim_{T\to\infty}\frac{1}{2\pi\mathrm{i}}\int_{\sigma-\mathrm{i}T}^{\sigma+\mathrm{i}T}\mathrm{e}^{zt}\big[zI+\hat{\mu}(z)\big]^{-1}\,\mathrm{d}z,$$

$$q(t)=\lim_{T\to\infty}\frac{1}{2\pi\mathrm{i}}\int_{\omega_\varphi-\mathrm{i}T}^{\omega_\varphi+\mathrm{i}T}\mathrm{e}^{zt}\big[zI+\hat{\mu}(z)\big]^{-1}\,\mathrm{d}z.$$

If we define $q(0) = q(0+)$, then for all $t \in \mathbf{R}$

$$r(t) - q(t) = \lim_{T\to\infty} \frac{1}{2\pi\mathrm{i}} \left\{ \int_{\sigma-\mathrm{i}T}^{\sigma+\mathrm{i}T} - \int_{\omega_\varphi-\mathrm{i}T}^{\omega_\varphi+\mathrm{i}T} \right\} \mathrm{e}^{zt} \big[zI + \hat{\mu}(z)\big]^{-1} \,\mathrm{d}z.$$

The function $\big[zI + \hat{\mu}(z)\big]^{-1}$ tends to zero as $|\Im z| \to \infty$, uniformly in the strip $\omega_\varphi \le \Re z \le \sigma$. Therefore, if we define Γ_T to be the closed curve in the complex plane that is composed of four straight lines, and joins the points $\omega_\varphi - \mathrm{i}T$, $\sigma - \mathrm{i}T$, $\sigma + \mathrm{i}T$, and $\omega_\varphi + \mathrm{i}T$ in the anticlockwise direction, then

$$r(t) - q(t) = \lim_{T\to\infty} \frac{1}{2\pi\mathrm{i}} \int_{\Gamma_T} \mathrm{e}^{zt} \big[zI + \hat{\mu}(z)\big]^{-1} \,\mathrm{d}z, \quad t \in \mathbf{R}.$$

By the residue theorem, the integral on the right-hand side is the sum of the residues of the integrand inside Γ_T. Expanding e^{zt} and $\big[zI + \hat{\mu}(z)\big]^{-1}$ into power series around each characteristic exponent, one immediately gets the desired formula. □

Theorem 2.1 implies the following result.

2.3 Theorem

Let φ be a submultiplicative weight function defined on $\mathbf{R}$ for which $\alpha_\varphi = \omega_\varphi$. Let $\mu \in M(\mathbf{R}^+; \varphi; \mathbf{C}^{n\times n})$, and suppose that $\det[zI + \hat{\mu}(z)] \ne 0$ for all $z \in \mathbf{C}$ with $\Re z = \omega_\varphi$. Let $f \in V$, where V is one of the spaces

(1) $L^p(\mathbf{R}^+; \varphi; \mathbf{C}^n)$, $p \in [1, \infty]$, *or* $L^\infty_0(\mathbf{R}^+; \varphi; \mathbf{C}^n)$,
(2) $B^\infty(\mathbf{R}^+; \varphi; \mathbf{C}^n)$, *or* $B^\infty_0(\mathbf{R}^+; \varphi; \mathbf{C}^n)$,
(3) $BC(\mathbf{R}^+; \varphi; \mathbf{C}^n)$, $BUC(\mathbf{R}^+; \varphi; \mathbf{C}^n)$, *or* $BC_0(\mathbf{R}^+; \varphi; \mathbf{C}^n)$,
(4) $BL^p(\mathbf{R}^+; \varphi; \mathbf{C}^n)$, *or* $BL^p_0(\mathbf{R}^+; \varphi; \mathbf{C}^n)$, $p \in [1, \infty]$,
(5) $BV(\mathbf{R}^+; \varphi; \mathbf{C}^n)$,
(6) $BBV(\mathbf{R}^+; \varphi; \mathbf{C}^n)$, *or* $BBV_0(\mathbf{R}^+; \varphi; \mathbf{C}^n)$,

or, for $\omega_\varphi = 0$, one of the spaces

(7) $L^p(\mathcal{T}_S; \mathbf{C}^n) \oplus BL^p_0(\mathbf{R}^+; \varphi; \mathbf{C}^n)$ *where* $p \in [1, \infty]$, $B^\infty(\mathcal{T}_S; \mathbf{C}^n) \oplus B^\infty_0(\mathbf{R}^+; \varphi; \mathbf{C}^n)$, $C(\mathcal{T}_S; \mathbf{C}^n) \oplus BC_0(\mathbf{R}^+; \varphi; \mathbf{C}^n)$, *or* $BV(\mathcal{T}_S; \mathbf{C}^n) \oplus BBV_0(\mathbf{R}^+; \varphi; \mathbf{C}^n)$, *where* $S > 0$,
(8) $AP(\mathbf{R}; \mathbf{C}^n) \oplus BC_0(\mathbf{R}^+; \varphi; \mathbf{C}^n)$.

Then the solution x of (1.1) is of the form

$$x(t) = \sum_{\ell=1}^{m} \sum_{j=0}^{p_\ell-1} \alpha_{\ell,j}\, t^j \mathrm{e}^{z_\ell t} + y(t), \quad t \in \mathbf{R}^+, \tag{2.4}$$

where the exponents z_ℓ are the zeros of $\det[zI + \hat{\mu}(z)]$ in $\Re z > \omega_\varphi$, each $\alpha_{\ell,j}$ is a vector, and $y \in V$.

More specifically, the coefficients $\alpha_{\ell,j}$ satisfy

$$\alpha_{\ell,j} = K_{\ell,j} x_0 + \sum_{i=0}^{p_\ell-j-1} \binom{j+i}{j} K_{\ell,j+i} \hat{f}^{(i)}(z_\ell), \tag{2.5}$$

where $K_{\ell,j}$ are the coefficients in (2.1), and $y(t) = qx_0 + \int_0^\infty q(t-s)f(s)\,\mathrm{d}s$, where q is the function in Theorem 2.1, extended to $\mathbf{R}$ as in (2.2). In *particular, $x \in V$ if and only if*

$$K_{\ell,j}x_0 + \sum_{i=0}^{p_\ell - j - 1} \binom{j+i}{j} K_{\ell,j+i}\hat{f}^{(i)}(z_\ell) = 0,$$
$$1 \le \ell \le m, \quad 0 \le j \le p_\ell - 1. \quad (2.6)$$

Note that if there are zeros z_ℓ of $\det[zI + \hat{\mu}(z)]$ in $\Re z > \omega_\varphi$ and (2.6) is satisfied so that the solution belongs to V, then it is *unstable* in the sense that arbitrarily small perturbations of the function f give solutions that do not belong to V.

Proof of Theorem 2.3 By Theorem **3**.3.3, the solution x of (1.1) is given by the variation of constants formula $x(t) = r(t)x_0 + \int_0^t r(t-s)f(s)\,\mathrm{d}s$ for $t \ge 0$. If we define $r(t) = 0$ for $t < 0$, and decompose r as in the L^1-remainder Theorem 2.1, then we get

$$x(t) = \sum_{\ell=1}^{m}\sum_{j=0}^{p_\ell - 1}\Big(t^j \mathrm{e}^{z_\ell t}K_{\ell,j}x_0 + \int_{\mathbf{R}^+} (t-s)^j \mathrm{e}^{z_\ell (t-s)} K_{\ell,j} f(s)\,\mathrm{d}s \Big)$$
$$+ q(t)x_0 + \int_{\mathbf{R}^+} q(t-s)f(s)\,\mathrm{d}s, \quad t \in \mathbf{R}^+.$$

Using the binomial expansion for $(t-s)^j$, the fact that $\int_{\mathbf{R}^+} s^i \mathrm{e}^{-z_\ell s} f(s)\,\mathrm{d}s = (-1)^i \hat{f}^{(i)}(z_\ell)$, and a change of summation variable, we get the expression given in Theorem 2.3. That $x \in V$ under the additional condition follows from Theorems **3**.6.1 and **4**.3.5. □

Analogous results are true for the Volterra equation (1.3) (or (1.2)). Note that some additional condition (here (2.8)) is needed in order to guarantee that there are only finitely many characteristic exponents z_ℓ satisfying $\Re z_\ell > \omega_\varphi$. This assumption holds automatically, however, in the case where the kernel is an integrable function instead of a measure. For simplicity, as in Corollary **4**.4.7, we assume that μ has no singular part.

2.4 Theorem

Let φ be a submultiplicative weight function defined on $\mathbf{R}$ for which $\alpha_\varphi = \omega_\varphi$. Let $\mu \in M(\mathbf{R}^+; \varphi; \mathbf{C}^{n\times n})$ have no singular part and suppose that

$$\inf_{\Re z = \omega_\varphi} \big|\det[I + \hat{\mu}(z)]\big| > 0, \quad (2.7)$$

and

$$\liminf_{\substack{|z|\to\infty \\ \Re z \ge \omega_\varphi}} \big|\det[I + \hat{\mu}(z)]\big| > 0. \quad (2.8)$$

Then the resolvent ρ of μ is of the form

$$\rho(\mathrm{d}t) = \sum_{\ell=1}^{m}\sum_{j=0}^{p_\ell - 1} K_{\ell,j}\, t^j \mathrm{e}^{z_\ell t}\,\mathrm{d}t + \xi(\mathrm{d}t), \quad t \in \mathbf{R}^+, \quad (2.9)$$

where each exponent z_ℓ is a scalar with $\Re z_\ell > \omega_\varphi$, each $K_{\ell,j}$ is a matrix, and $\xi \in M(\mathbf{R}^+;\varphi;\mathbf{C}^{n\times n})$.

More specifically, the exponents z_ℓ are the zeros of $\det[I+\hat\mu(z)]$ in $\Re z > \omega_\varphi$, and the coefficient matrices $K_{\ell,j}$ are determined by the principal part of $\hat\mu(z)[I+\hat\mu(z)]^{-1}$ at z_ℓ in the sense that, in some neighbourhood of z_ℓ,

$$\hat\mu(z)\,[I+\hat\mu(z)]^{-1} = \sum_{j=0}^{p_\ell - 1} \frac{j!\,K_{\ell,j}}{(z-z_\ell)^{(j+1)}} + \hat\xi_\ell(z),$$

where the remainder $\hat\xi_\ell(z)$ is analytic at z_ℓ. Moreover, if we extend μ to $\mathbf{R}$ by defining $\mu(E) = \mu(E\cap\mathbf{R}^+)$ for every Borel set $E\subset\mathbf{R}$, and if we extend ξ to $\mathbf{R}$ by defining

$$\xi(\mathrm{d}t) = -\sum_{\ell=1}^{m}\sum_{j=0}^{p_\ell-1} K_{\ell,j}\, t^j \mathrm{e}^{z_\ell t}\,\mathrm{d}t, \quad t\in(-\infty,0), \tag{2.10}$$

then ξ is the whole line resolvent of μ corresponding to the weight function φ.

Again, observe that (2.9) can be extended into

$$\rho(\mathrm{d}t) = \sum_{\ell=1}^{m}\sum_{j=0}^{p_\ell-1} K_{\ell,j}\, t^j \mathrm{e}^{z_\ell t}\,\mathrm{d}t + \xi(\mathrm{d}t), \quad t\in\mathbf{R}. \tag{2.11}$$

Proof of Theorem 2.4 By (2.7) and (2.8) there are at most a finite number of characteristic exponents z_ℓ for which $\Re z_\ell > \omega_\varphi$. Therefore, by Corollary 4.4.7, the resolvent ρ of μ satisfies $\int_{\mathbf{R}^+}|\rho|(\mathrm{d}t)\mathrm{e}^{-\sigma t} < \infty$, provided $\sigma > \omega_\varphi$ is sufficiently large. By Corollary 4.4.8, the (extended) measure μ has a unique whole line resolvent $\nu\in M(\mathbf{R};\varphi;\mathbf{C}^{n\times n})$.

Let φ be any nonnegative C^∞-function vanishing outside the interval $[-1,1]$, and satisfying $\int_{\mathbf{R}}\varphi(x)\,\mathrm{d}x = 1$. Define $\varphi_n(t) = n\varphi(nt)$, $\rho_n = \varphi_n * \rho$, and $\nu_n = \varphi_n * \nu$ for $n = 1,2,\ldots$. Then one has, after arguing as in the proof of Theorem 2.1,

$$\rho_n(t) - \nu_n(t) = \lim_{T\to\infty}\frac{1}{2\pi\mathrm{i}}\left(\int_{\sigma-\mathrm{i}T}^{\sigma+\mathrm{i}T} - \int_{\omega_\varphi-\mathrm{i}T}^{\omega_\varphi+\mathrm{i}T}\right) \mathrm{e}^{zt}\widehat{\varphi_n}(z)\hat\mu(z)[I+\hat\mu(z)]^{-1}\,\mathrm{d}z, \quad t\in\mathbf{R}.$$

For each n, the integrand tends to zero as $|\Im z|\to\infty$, uniformly in the strip $\omega_\varphi \le \Re z \le \sigma$. Therefore, the residue theorem may be applied to get

$$\rho_n(t) - \nu_n(t) = \sum_{\ell=1}^{m}\sum_{j=0}^{p_\ell-1} K_{\ell,j,n} t^j \mathrm{e}^{z_\ell t}, \quad t\in\mathbf{R}.$$

For any continuous function f with compact support, we have

$$\lim_{n\to\infty}\int_{\mathbf{R}}\rho_n(t)f(t)\,\mathrm{d}t = \int_{R}\rho(\mathrm{d}t)f(t).$$

Since analogous relations hold for ν_n, and for the sum, we obtain (2.9). □

This theorem has the following consequence for the behaviour of unbounded solutions of (1.3).

2.5 Theorem

Let φ be a submultiplicative weight function defined on $\mathbf{R}$, for which $\alpha_\varphi = \omega_\varphi$. Let $\mu \in M(\mathbf{R}^+;\varphi;\mathbf{C}^{n\times n})$ satisfy the conditions of Theorem 2.4, and let $f \in V$, where V is one of the spaces

(1) $L^p(\mathbf{R}^+;\varphi;\mathbf{C}^n)$, $p \in [1,\infty]$, *or* $L_0^\infty(\mathbf{R}^+;\varphi;\mathbf{C}^n)$,
(2) $B^\infty(\mathbf{R}^+;\varphi;\mathbf{C}^n)$, *or* $B_0^\infty(\mathbf{R}^+;\varphi;\mathbf{C}^n)$,
(3) $BL^p(\mathbf{R}^+;\varphi;\mathbf{C}^n)$, *or* $BL_0^p(\mathbf{R}^+;\varphi;\mathbf{C}^n)$, $p \in [1,\infty]$,
(4) $BV(\mathbf{R}^+;\varphi;\mathbf{C}^n)$,
(5) $BBV(\mathbf{R}^+;\varphi;\mathbf{C}^n)$, *or* $BBV_0(\mathbf{R}^+;\varphi;\mathbf{C}^n)$,

or, for $\omega_\varphi = 0$, one of the spaces

(6) $L^p(\mathcal{T}_S;\mathbf{C}^n) \oplus BL_0^p(\mathbf{R}^+;\varphi;\mathbf{C}^n)$ *where* $p \in [1,\infty]$, $B^\infty(\mathcal{T}_S;\mathbf{C}^n) \oplus B_0^\infty(\mathbf{R}^+;\varphi;\mathbf{C}^n)$, *or* $BV(\mathcal{T}_S;\mathbf{C}^n) \oplus BBV_0(\mathbf{R}^+;\varphi;\mathbf{C}^n)$, *where* $S > 0$,
(7) $AP(\mathbf{R}^+;\mathbf{C}^n) \oplus B_0^\infty(\mathbf{R}^+;\varphi;\mathbf{C}^n)$.

Then the solution x of (1.3) is of the form

$$x(t) = \sum_{\ell=1}^{m}\sum_{j=0}^{p_\ell-1} \alpha_{\ell,j}\, t^j e^{z_\ell t} + y(t), \quad t \in \mathbf{R}^+, \tag{2.12}$$

where each exponent z_ℓ is a scalar with $\Re z_\ell > \omega_\varphi$, each $\alpha_{\ell,j}$ is a vector, and $y \in V$.

More specifically, the exponents z_ℓ are the zeros of $\det[I + \hat{\mu}(z)]$ in $\Re z > \omega_\varphi$, and the coefficients $\alpha_{\ell,j}$ satisfy

$$\alpha_{\ell,j} = -\sum_{i=0}^{p_\ell-j-1} \binom{j+i}{j} K_{\ell,j+i}\hat{f}^{(i)}(z_\ell), \tag{2.13}$$

where $K_{\ell,j}$ are the coefficients in (2.9), and $y(t) = f(t) - \int_{-\infty}^{t} \xi(\mathrm{d}s) f(t-s)$, where ξ is the measure in Theorem 2.4, extended to $\mathbf{R}$ as in (2.10). In particular, $x \in V$ if and only if

$$\sum_{i=0}^{p_\ell-j-1} \binom{j+i}{j} K_{\ell,j+i}\hat{f}^{(i)}(z_\ell) = 0, \quad 1 \le \ell \le m, \quad 0 \le j \le p_\ell - 1. \tag{2.14}$$

Proof Combine Theorem 2.4 with Theorems **3**.6.1 and **4**.3.5. □

In this case too, we observe that the solutions that belong to V owing to the fact that (2.14) holds, are unstable.

3. Characteristic Exponents on the Critical Line

We proceed to a few brief comments on the case where some of the characteristic exponents lie on the critical line $\Re z = \omega_\varphi$. In this case, if we desire results analogous to those above, we have to make some extra assumptions. One obvious hypothesis is to assume explicitly that there are only finitely many characteristic exponents z_ℓ satisfying $\Re z_\ell \geq \omega_\varphi$. In the previous theorems, this assumption is a consequence of the other conditions. Another new requirement is that the transform $\hat{\mu}$ have certain smoothness properties at the characteristic points. To see that some assumption of this kind is necessary, let us look at a simple scalar example, and assume that the resolvent r of a function $k \in L^1(\mathbf{R}^+;\mathbf{C})$ is of the form

$$r(t) = at^{p-1}e^{z_0 t} + q(t), \quad t \in \mathbf{R}^+, \tag{3.1}$$

where $a \in \mathbf{C}$ is nonzero, $p \geq 1$ is an integer, $\Re z_0 = 0$, and $q \in L^1(\mathbf{R}^+;\mathbf{C})$. Let $\lambda = a(p-1)!$. Then

$$\frac{\hat{k}(z)}{1+\hat{k}(z)} = \lambda(z-z_0)^{-p} + \hat{q}(z), \quad \Re z \geq 0, \quad z \neq z_0,$$

and consequently (use the continuity of $\hat{k}$ at the point z_0),

$$\hat{k}(z) = -1 - \lambda^{-1}(z-z_0)^p + (z-z_0)^{2p}\hat{g}(z), \quad \Re z \geq 0,$$

where

$$\hat{g}(z) = \lambda^{-2}\frac{\hat{q}(z)-1}{1+\lambda^{-1}(z-z_0)^p(\hat{q}(z)-1)}.$$

Let $z = i\omega$, $z_0 = i\omega_0$. By Theorem **6**.3.3(i), the function $(i\omega - i\omega_0)^p$ belongs locally to $\tilde{L}^1$ at ω_0. Since $\hat{q}(z)$ belongs locally to $\tilde{L}^1$ everywhere on $\mathbf{R}$, and because the denominator of $\hat{g}(z)$ does not vanish for z sufficiently close to z_0, we conclude from Theorem **6**.3.3(iii) that $\tilde{g}(\omega)$ belongs locally to $\tilde{L}^1$ at ω_0. Therefore, there is some $h \in L^1(\mathbf{R};\mathbf{C})$ and some $\epsilon > 0$ such that

$$\tilde{k}(\omega) = -1 - \lambda^{-1}(i\omega - i\omega_0)^p + (i\omega - i\omega_0)^{2p}\tilde{h}(\omega), \quad |\omega - \omega_0| < \epsilon. \tag{3.2}$$

Thus, in order for (3.1) to be valid, $\tilde{k}(\omega)$ must be of the form (3.2), i.e., for ω close to ω_0, $1+\tilde{k}$ must be the sum of an analytic function of z, and a function which is the product of the factor $(\omega-\omega_0)^{2p}$ and the Fourier transform of an L^1-function.

The preceding discussion motivates us to make some definitions. First, we extend the notion introduced in Chapter 6 of a function belonging locally to $\tilde{L}^1$ to include the effect of weight functions. (In the new context, Definition **6**.3.1 is not appropriate. Instead, we have to use the characterizations in Theorems **6**.3.3 and **6**.3.4.)

3.1 Definition
Let φ be a submultiplicative weight function on $\mathbf{R}$*. A (matrix-valued, vector-valued, or scalar-valued) function ϕ is said to belong locally to $\hat{L}^1(\varphi)$ at a point $z_0 \in \mathbf{C}$, $\omega_\varphi \le \Re z_0 \le \alpha_\varphi$, if there exist (scalar) measures $\mu_1, \ldots, \mu_k$ in $M(\mathbf{R};\varphi)$ and a function $\psi(z, \xi_1, \ldots, \xi_k)$, analytic at $(z_0, \widehat{\mu_1}(z_0), \ldots, \widehat{\mu_k}(z_0))$, such that, for some $\epsilon > 0$, $\phi(z) = \psi(z, \widehat{\mu_1}(z), \ldots, \widehat{\mu_k}(z))$, when $|z - z_0| < \epsilon$ and $\omega_\varphi \le \Re z \le \alpha_\varphi$.*

The function ϕ is said to belong locally to $\hat{L}^1(\varphi)$ at infinity if there exist (scalar) functions $a_1, \ldots, a_m$ in $L^1(\mathbf{R};\varphi)$, measures $\mu_1, \ldots, \mu_k$ in $M(\mathbf{R};\varphi)$, and a function $\psi(z, \eta_1, \ldots, \eta_m, \xi_1, \ldots, \xi_k)$ analytic at $(0, 0, \ldots, 0)$, such that, for some $\epsilon > 0$, $\phi(z) = \psi(\frac{1}{z}, \widehat{a_1}(z), \ldots, \widehat{a_m}(z), \frac{1}{z}\widehat{\mu_1}(z), \ldots, \frac{1}{z}\widehat{\mu_k}(z))$, when $|z| > \epsilon$ and $\omega_\varphi \le \Re z \le \alpha_\varphi$.

If ϕ belongs locally to $\hat{L}^1(\varphi)$ at every point z_0 with $\omega_\varphi \le \Re z_0 \le \alpha_\varphi$ and at infinity, then we say that ϕ belongs locally to $\hat{L}^1(\varphi)$.

The following theorem motivates the use of the words 'belongs locally to $\hat{L}^1(\varphi)$' in Definition 3.1. For the proof, see Gel′fand, Raikov, and Shilov [1], Theorem 1, p. 82, and Jordan, Staffans, and Wheeler [1], Prop. 2.3, p. 755.

3.2 Theorem
Let φ be a submultiplicative weight function on $\mathbf{R}$*, and let ϕ belong locally to $\hat{L}^1(\varphi)$ and satisfy $\phi(\infty) = 0$. Then there exists a unique function $a \in L^1(\mathbf{R};\varphi)$ such that $\hat{\phi}(z) = \hat{a}(z)$ for $\omega_\varphi \le \Re z \le \alpha_\varphi$.*

Since some of our resolvents are measures instead of functions, we need the following extension of Definition 3.1.

3.3 Definition
Let φ be a submultiplicative weight function on $\mathbf{R}$*. A (matrix-valued, vector-valued, or scalar-valued) function ϕ is said to belong locally to $\widehat{M}(\varphi)$ at infinity if there exist (scalar) measures $\mu_1, \ldots, \mu_k$ in $M(\mathbf{R};\varphi)$, and a function $\psi(z, \xi_1, \ldots, \xi_k)$ analytic on $(0, U_1, \ldots, U_k)$, where U_i is an open set containing the closure of the set*

$$\left\{ w \in \mathbf{C} \mid |w - \widehat{(\mu_i)_d}(z)| \le \|(\mu_i)_s\|_{M(\mathbf{R};\varphi)} \text{ for some } z,\ \omega_\varphi \le \Re z \le \alpha_\varphi \right\},$$

such that, for some $\epsilon > 0$, $\phi(z) = \psi(\frac{1}{z}, \widehat{\mu_1}(z), \ldots, \widehat{\mu_k}(z))$, when $|z| > \epsilon$ and $\omega_\varphi \le \Re z \le \alpha_\varphi$.

If ϕ belongs locally to $\hat{L}^1(\varphi)$ at every point z_0 with $\omega_\varphi \le \Re z_0 \le \alpha_\varphi$, and ϕ belongs locally to $\widehat{M}(\varphi)$ at infinity, then we say that ϕ belongs locally to $\widehat{M}(\varphi)$.

Corresponding to Theorem 3.2 we have

3.4 Theorem
Let φ be a submultiplicative weight function on $\mathbf{R}$, and let ϕ belong locally to $M(\varphi)$. Then there exists a unique measure $\mu \in M(\mathbf{R};\varphi)$ such that $\hat{\mu}(z) = \phi(z)$ for $\omega_\varphi \leq \Re z \leq \alpha_\varphi$.

For the proof, see Gel'fand, Raikov, and Shilov [1], p. 63, and Jordan, Staffans, and Wheeler [1], p. 779.

For our current purposes we need to define what is meant by the order of smoothness and by the order of a zero.

3.5 Definition
Let φ be a submultiplicative weight function on $\mathbf{R}$ and let $p \geq 1$ be an integer. A function ψ is $\hat{L}^1(\varphi)$-smooth of order p at a point z_0, $\omega_\varphi \leq \Re z_0 \leq \alpha_\varphi$, if it is of the form $\phi(z) = w(z) + (z - z_0)^p\zeta(z)$ in a neighbourhood of z_0, where w is analytic at z_0, and ζ belongs locally to $\hat{L}^1(\varphi)$ at z_0.

3.6 Definition
Let φ be a submultiplicative weight function on $\mathbf{R}$ and let $p \geq 1$ be an integer. A scalar function ϕ has a $\hat{L}^1(\varphi)$-zero of order p at z_0, $\omega_\varphi \leq \Re z_0 \leq \alpha_\varphi$, if ϕ is of the form $\phi(z) = (z - z_0)^p\zeta(z)$ in a neighbourhood of z_0, where ζ belongs locally to $\hat{L}^1(\varphi)$ at z_0 and $\zeta(z_0) \neq 0$.

Thus, in (3.2), $1 + \hat{k}$ is $\hat{L}^1(\varphi)$-smooth of order $2p$ and has a $\hat{L}^1(\varphi)$-zero of order p at z_0, with respect to the weight function $\varphi \equiv 1$.

As the following theorem shows, the situation described above is typical.

3.7 Theorem
Let φ be a submultiplicative weight function on $\mathbf{R}$, and let $\mu \in M(\mathbf{R}^+;\varphi;\mathbf{C}^{n\times n})$ satisfy $\det[zI + \hat{\mu}(z)] \neq 0$ for all z with $\Re z \geq \omega_\varphi$, except when z belongs to a finite set $Z = \{z_1, \ldots, z_m\}$. Suppose that, for each $\ell = 1, \ldots, m$, the function $\det[zI + \hat{\mu}(z)]$ has a $\hat{L}^1(\varphi)$-zero at z_ℓ, of order p_ℓ, and is $\hat{L}^1(\varphi)$-smooth of order $2p_\ell$ at z_ℓ. Then the differential resolvent r of μ is of the form

$$r(t) = \sum_{\ell=1}^{m}\sum_{j=0}^{p_\ell - 1} K_{\ell,j}t^j e^{z_\ell t} + q(t), \quad t \in \mathbf{R}^+, \tag{3.3}$$

where $q \in L^1(\mathbf{R}^+;\varphi;\mathbf{C}^{n\times n})$ and each $K_{\ell,j}$ is a constant matrix.

For the proof, see Jordan, Staffans, and Wheeler [1], p. 758. It is evident that Theorem 3.2 plays a key role in this proof.

An analogous result may be formulated for the resolvent of $\mu \in M(\mathbf{R}^+;\varphi;\mathbf{C}^{n\times n})$ in the case where there are finitely many characteristic exponents in $\Re z \geq \omega_\varphi$. The proof is based on Theorem 3.4.

4. The Renewal Equation

We conclude this chapter by considering the well-known renewal theorem. The renewal equation on the positive half line is

$$x(t) - \int_{[0,t]} \beta(\mathrm{d}s)x(t-s) = f(t), \quad t \in \mathbf{R}^+. \tag{4.1}$$

The crucial assumption is that β is a real-valued positive measure with total mass 1. Therefore, $z = 0$ will be a characteristic exponent. Here we make two simplifying assumptions, namely that β has a nonzero absolutely continuous component, and a finite first moment. These assumptions will be relaxed in Chapter 15 (at the expense of a weaker conclusion).

4.1 Theorem
*Let $\beta \in M(\mathbf{R}^+;\mathbf{R})$ be positive with $\beta(\mathbf{R}^+) = 1$ and $\int_{\mathbf{R}^+} t\beta(\mathrm{d}t) < \infty$. Furthermore, assume that β has a nontrivial absolutely continuous component. Let $-\mu$ be the resolvent of $-\beta$ (which exists, according to Theorem **4**.1.5). Then μ is a positive measure, and μ is of the form*

$$\mu(\mathrm{d}t) = \nu(\mathrm{d}t) + \nu([0,t])\,\mathrm{d}t, \tag{4.2}$$

where $\nu \in M(\mathbf{R}^+;\mathbf{R})$, and $\nu(\mathbf{R}^+) = \left(\int_{\mathbf{R}^+} t\beta(\mathrm{d}t)\right)^{-1}$. If, in addition, $\int_{\mathbf{R}^+} t^2\beta(\mathrm{d}t) < \infty$, then μ is of the form

$$\mu(\mathrm{d}t) = \gamma(\mathrm{d}t) + \frac{\mathrm{d}t}{\int_{\mathbf{R}^+} s\beta(\mathrm{d}s)}, \tag{4.3}$$

where $\gamma \in M(\mathbf{R}^+;\mathbf{R})$.

Proof It is straightforward to show that μ is positive (write μ in the form $\mu = \sum_{j=1}^{\infty} \beta^{*j}$). Recall that

$$\hat{\mu}(z) = \frac{\hat{\beta}(z)}{1-\hat{\beta}(z)} = \left(1 + \frac{1}{z}\right)\frac{z}{z+1}\frac{\hat{\beta}(z)}{1-\hat{\beta}(z)}.$$

Observe that the moment condition on β implies that the function $b(t) \stackrel{\text{def}}{=} 1 - \beta([0,t])$ belongs to $L^1(\mathbf{R}^+;\mathbf{R})$. The Laplace transform of this function is $\hat{b}(z) = (1-\hat{\beta}(z))/z$ for $\Re z \geq 0$, $z \neq 0$, and $\hat{b}(0) \neq 0$. Therefore the function

$$\phi(z) \stackrel{\text{def}}{=} \frac{z}{z+1}\frac{\hat{\beta}(z)}{1-\hat{\beta}(z)}$$

belongs locally to $\hat{L}^1(\varphi)$, with $\varphi \equiv 1$, at all points on the imaginary axis $\Re z = 0$. Moreover, owing to our assumption that β has a nontrivial absolutely continuous component, ϕ belongs locally to $\widehat{M}(\varphi)$ at infinity. From Lemma **4**.6.1 and Theorem 3.4, we conclude that there exists a measure $\nu \in M(\mathbf{R}^+;\mathbf{R})$ such that $\hat{\nu}(z) = \phi(z)$. Clearly, (4.2) follows from the fact that $\hat{\mu}(z) = \left(1 + \frac{1}{z}\right)\hat{\nu}(z)$. To prove the claim about $\nu(\mathbf{R}^+)$ it suffices to observe that $\nu(\mathbf{R}^+) = -\phi(0) = -1/\hat{b}(0)$, and that $\hat{b}(0) = \int_{\mathbf{R}^+} t\beta(\mathrm{d}t)$.

If $\int_{\mathbf{R}^+} t^2\beta(\mathrm{d}t) < \infty$, then the same argument, with $\varphi \equiv 1$ replaced by $\varphi(t) = (1+|t|)$, shows that $\int_{\mathbf{R}^+} t|\nu|(\mathrm{d}t) < \infty$. Therefore, we get the second representation (4.3) by defining $\gamma(\mathrm{d}t) = \nu(\mathrm{d}t) - \nu((t,\infty))\,\mathrm{d}t$. □

5. Exercises

1. Let $\alpha \in \mathbf{C}$, and define
$$a(t) = \begin{cases} \mathrm{e}^{\alpha t}, & t \geq 0, \\ 0, & t < 0, \end{cases} \qquad b(t) = \begin{cases} 0, & t \geq 0, \\ -\mathrm{e}^{\alpha t}, & t < 0. \end{cases}$$
Compute the bilateral Laplace transforms of a and b. How does the answer relate to the theory presented in Sections 2 and 3?
2. Let the kernel k in (1.2) be of the form $k = \sum_{j=1}^{q} A_j \mathrm{e}^{B_j t}$, where A_j and B_j, $1 \leq j \leq q$, are $n \times n$-matrices. Show that the resolvent r of k can be written in the form
$$r(t) = \sum_{\ell=1}^{m} \sum_{j=0}^{p_\ell - 1} K_{\ell,j} t^j \mathrm{e}^{z_\ell t}, \quad t \in \mathbf{R}^+,$$
where each z_ℓ is a characteristic exponent. In this sum there is no restriction of the type $\Re z_\ell \geq 0$.
3. Prove the analogous result for the integrodifferential equation (1.3), with $\mu(\mathrm{d}s)$ replaced by $k(s)\,\mathrm{d}s$, where k is of the same type as in the preceding exercise.
4. Extend the results in the preceding two exercises to the case where $\hat{r}(z)$ is an arbitrary quotient of two matrix polynomials, satisfying
   (i) $\hat{r}(z) \to 0$ as $|z| \to \infty$ in Exercise 2,
   (ii) $\hat{r}(z)$ is bounded for large $|z|$ in Exercise 3.
5. Write the solution of
$$x'(t) + ax(t) + bx(t-T) = 0, \quad t \geq 0, \quad T > 0,$$
as an infinite series.
6. Formulate and prove a result analogous to Theorem 3.7 for the resolvent ρ of a measure $\mu \in M(\mathbf{R}^+;\varphi)$.
7. Write down explicit functions $k \in L^1(\mathbf{R}^+;\mathbf{C})$ that have resolvents r of the form given by (3.1) with $p = 1$ and $p = 2$.
8. What happens if one replaces the condition $\beta(\mathbf{R}^+) = 1$ in Theorem 4.1 by either $\beta(\mathbf{R}^+) > 1$ or $\beta(\mathbf{R}^+) < 1$?
9. Give an example of a nonnegative kernel $\beta \in M(\mathbf{R}^+;\mathbf{R})$ with total mass 1 and a finite first moment, for which the conclusion of Theorem 4.1 does not hold. Can you choose β in such a way that zero is the only characteristic exponent of (4.1)?

6. Comments

Section 2:

In the simplest possible case, the case of the ordinary differential equation

$$x'(t) = Ax(t), \quad t \in \mathbf{R}^+, \tag{6.1}$$

where A is an $n \times n$ matrix, the characteristic equation is $\det[zI - A] = 0$. This means that the characteristic exponents are the eigenvalues of the matrix A. The decomposition of the solutions of (6.1) into components with different exponential growth rate is a central theme in the theory of linear ordinary differential equations. See, e.g., Hale [3], Section III.4.

The next step towards more generality is to replace (6.1) by the finite delay functional differential equation

$$x'(t) + \int_{[0,r]} \mu(\mathrm{d}s)x(t-s) = 0; \quad x(t) = \phi(t), \quad -\infty < -r \le t \le 0. \tag{6.2}$$

In this equation the characteristic function $\det[zI + \hat{\mu}(z)]$ is entire, and the situation is quite different from the case encountered in Sections 2–3 where an infinite delay is not excluded. (Recall that in the infinite delay case there is usually a 'critical line' to the left of which the characteristic function is not defined). In particular, in the finite delay case, the equation always has at least one critical exponent, and in most cases it has infinitely many. This immediately raises the question about the possibility of writing a general solution of (6.2) as an infinite sum of characteristic solutions, and this question occupies a central position in the theory of functional differential equations with finite delay. See, e.g., Hale [5], Chapter 7, Salamon [1], Section 2.4 and Chapter 3, and Lunel [2].

In most of the early work on functional differential equations with finite delay, the decomposition mentioned above has been achieved with a different method from the one we use: the equation is shown to generate a semigroup, and an eigenvalue expansion is performed for the generator of the semigroup. This is the approach used in, e.g., Hale [5], Chapters 7 and 12. However, our Laplace transform formula has been known for a long time. Already Krasovskiĭ [1] identifies the eigenvalues of the generator of the semigroup with the characteristic exponents. Further results are found, e.g., in Levinger [1], Kappel and Wimmer [1], Kappel [1] and [2], and Lunel [2].

A Volterra equation with infinite delay, and with characteristic values in the open right half plane, has been studied in Miller and Nohel [1].

Russell [1], Theorem 2, has developed a theory similar to the one in Theorems 2.4 and 2.5 for a difference equation in a Banach space.

Section 3:

The case where the characteristic equation $\det[zI + \hat{\mu}(z)] = \det[zI + A + \hat{B}(z)] = 0$ has roots in the closed right half plane is analysed in Miller [14]. See also Jordan and Wheeler [4], [5], and [7], Krisztin and Terjéki [1],

Miller and Nohel [1], Jordan, Staffans, and Wheeler [1]–[3], and Staffans [16], [18], and [28].

The smoothness assumption in Theorem 3.7 can be relaxed somewhat: the number p_ℓ can be replaced by the order of the pole of the characteristic function, and this order is, in general, smaller than the order of the zero of the determinant. See Jordan, Staffans, and Wheeler [2] and [3].

Section 4:
For results on renewal theory, see, e.g., Bingham, Goldie, and Teugels [1], p. 359, or Feller [2], Ch. XI, and references therein. Systems of renewal equations are treated in Miller [15] and Wagner [2]. An earlier result for a special case of systems is to be found in Tsalyuk [1]. An asymptotic result on a nonlinear renewal equation is given in Ney [1].

References

N. H. Bingham, C. M. Goldie, and J. L. Teugels
1. *Regular Variation*, Cambridge University Press, Cambridge, 1987.

G. Doetsch
1. *Handbuch der Laplace-Transformation. I–III*, Birkhäuser, Basel, 1950–1956.

W. Feller
2. *An Introduction to Probability Theory and its Applications, II*, 2nd *ed.*, J. Wiley, New York, 1971.

I. M. Gel'fand, D. A. Raikov, and G. E. Shilov
1. *Commutative Normed Rings*, Chelsea, New York, 1964.

J. K. Hale
3. *Ordinary Differential Equations*, J. Wiley, New York, 1969.
5. *Theory of Functional Differential Equations*, Springer-Verlag, Berlin, 1977.

G. S. Jordan, O. J. Staffans, and R. L. Wheeler
1. Local analyticity in weighted L^1-spaces and applications to stability problems for Volterra equations, *Trans. Amer. Math. Soc.* **274** (1982), pp. 749–782.
2. Convolution operators in a fading memory space: the critical case, *SIAM J. Math. Anal.* **18** (1987), pp. 366–386.
3. Subspaces of stable and unstable solutions of a functional differential equation in a fading memory space: the critical case, *SIAM J. Math. Anal.* **18** (1987), pp. 1323–1340.

G. S. Jordan and R. L. Wheeler

4. Asymptotic behavior of unbounded solutions of linear Volterra integral equations, *J. Math. Anal. Appl.* **55** (1976), pp. 596–615.
5. Structure of resolvents of Volterra integral and integrodifferential systems, *SIAM J. Math. Anal.* **11** (1980), pp. 119–132.
7. Weighted L^1-remainder theorems for resolvents of Volterra equations, *SIAM J. Math. Anal.* **11** (1980), pp. 885–900.

F. Kappel

1. *Laplace-transform methods and linear autonomous functional-differential equations*, Berichte der Mathematisch-statistischen Sektion im Forschungszentrum Graz, Report nr. 64, 1976.
2. *Linear autonomous functional differential equations in the state space C*, Technical Report 34, Technische Universität Graz, Universität Graz, 1984.

F. Kappel and H. K. Wimmer

1. An elementary divisor theory for autonomous linear functional differential equations, *J. Differential Equations* **21** (1976), pp. 134–147.

N. N. Krasovskiĭ

1. *Stability of Motion*, Stanford University Press, Stanford, California, 1963 [translation of *Nekotorye zadachi teoriĭ ustoĭchivosti dvizheniya*, Gos. Fiz. Mat. Lit., Moscow, 1959.]

T. Krisztin and J. Terjéki

1. On the rate of convergence of solutions of linear Volterra equations, *Boll. Un. Mat. Ital. B (7)* **2** (1988), pp. 427–444.

B. W. Levinger

1. A folk theorem in functional differential equations, *J. Differential Equations* **4** (1968), pp. 612–619.

S. M. V. Lunel

2. *Exponential Type Calculus for Linear Delay Equations*, Ph. D. Thesis, Centrum voor Wiskunde en Informatica, Amsterdam, 1988.

R. K. Miller

14. Structure of solutions of unstable linear Volterra integrodifferential equations, *J. Differential Equations* **15** (1974), pp. 129–157.
15. A system of renewal equations, *SIAM J. Appl. Math.* **29** (1975), pp. 20–34.

R. K. Miller and J. A. Nohel

1. A stable manifold theorem for a system of Volterra integro-differential equations, *SIAM J. Math. Anal.* **6** (1975), pp. 506–522.

P. Ney

1. The asymptotic behavior of a Volterra-renewal equation, *Trans. Amer. Math. Soc.* **228** (1977), pp. 147–155.

D. L. Russell

1. A Floquet decomposition for Volterra equations with periodic kernel and a transform approach to linear recursion equations, *J. Differential Equations* **68** (1987), pp. 41–71.

D. Salamon

1. *Control and Observation of Neutral Systems*, Pitman, London, 1984.

O. J. Staffans

16. On a neutral functional differential equation in a fading memory space, *J. Differential Equations* **50** (1983), pp. 183–217.
18. The null space and the range of a convolution operator in a fading memory space, *Trans. Amer. Math. Soc.* **281** (1984), pp. 361–388.
28. On the stable and unstable subspaces of a critical functional differential equation, *J. Integral Equations Appl.*, to appear 1989.

Z. B. Tsalyuk

1. Asymptotic properties of solutions of the regeneration equation, *Differential Equations* **6** (1970), pp. 852–854.

E. Wagner

2. Zur Asymptotik der Lösungen linearer Volterrascher Integralgleichungssysteme zweiter Art vom Faltungstyp mit nichtnegativen Kernen, *Math. Nachr.* **90** (1979), pp. 173–187.

8

Volterra Equations as Semigroups

We consider the semigroup approach to Volterra equations. The initial function and the forcing function semigroups are introduced, and the relations between them are investigated. In addition, a combined initial and forcing function semigroup is studied.

1. Introduction

The convolution equations

$$x(t) + \int_{\mathbf{R}^+} \mu(\mathrm{d}s)x(t-s) = f(t), \quad t \in \mathbf{R}^+, \tag{1.1}$$

and

$$x'(t) + \int_{\mathbf{R}^+} \mu(\mathrm{d}s)x(t-s) = f(t), \quad t \in \mathbf{R}^+, \tag{1.2}$$

may be studied within the framework of semigroup theory. With both (1.1) and (1.2) one associates the initial condition (note that the integrations in (1.1) and (1.2) extend over $\mathbf{R}^+$ and not over $[0,t]$)

$$x(t) = \psi(t), \quad t \in \mathbf{R}^-. \tag{1.3}$$

The finite delay case, where μ has its support in a bounded interval $[0,\sigma]$ with $\sigma < \infty$, is contained in the above formalism. In this case ψ needs to be given only on $[-\sigma, 0)$.

The outline of this chapter is as follows. First, in Section 2, we consider equation (1.1) in a quite simple setting: we take $\mu(\mathrm{d}s) = k(s)\,\mathrm{d}s$ with $k \in L^1(\mathbf{R}^+)$, let the spaces on which the semigroups act be L^2, and do not bring in any weight functions. Even in this simple case we shall be able to distinguish some important characteristic features. We present two different semigroups generated by (1.1), compute their infinitesimal generators, and show that one is the (transpose of the) adjoint of the other.

In Section 3 we analyse briefly the extended semigroups generated by (1.1). We employ weight functions, let $\mu \in M(\mathbf{R}^+;\varphi)$, and consider L^p-spaces with arbitrary $p \in (1,\infty)$. Note, however, that the basic results remain much the same.

For technical reasons we concentrate on (1.1). For references to semigroup work on (1.2), and for the cases $p = 1$ and $p = \infty$, see Section 5.

2. The Initial and Forcing Function Semigroups

Let $\mathcal{B}$ be a Banach space. We say that T is a strongly continuous (linear) semigroup on $\mathcal{B}$ if, for each $t, s \geq 0$, $T(t)$ is a linear continuous operator on $\mathcal{B}$, $T(t)T(s) = T(t+s)$, $T(0) = I$, and $T(t)x \to x$ as $t \downarrow 0$ for each $x \in \mathcal{B}$. The infinitesimal generator A of the semigroup T is defined by $Ax = \lim_{t\downarrow 0} \frac{1}{t}(T(t)x - x)$ for all x in the domain

$$\mathcal{D}(A) \stackrel{\text{def}}{=} \left\{ x \in \mathcal{B} \mid \lim_{t\downarrow 0} \tfrac{1}{t}(T(t)x - x) \text{ exists} \right\}.$$

Let us consider (1.1) with an absolutely continuous measure as kernel, and initial conditions (1.3), i.e., let us consider the initial value problem

$$\begin{aligned} x(t) + \int_0^\sigma k(s)x(t-s)\,\mathrm{d}s &= f(t), \qquad & t \in \mathbf{R}^+, \\ x(t) &= \psi(t), & -\sigma < t < 0, \end{aligned} \tag{2.1}$$

where $0 < \sigma \leq \infty$, $k \in L^1((0,\sigma);\mathbf{C}^{n\times n})$, $\psi \in L^2((-\sigma,0);\mathbf{C}^n)$ and $f \in L^2_{\mathrm{loc}}(\mathbf{R}^+;\mathbf{C}^n)$. Let $r \in L^1_{\mathrm{loc}}(\mathbf{R}^+;\mathbf{C}^{n\times n})$ be the resolvent of k. Then, by Theorem **2**.3.5, the solution of (2.1) is given by

$$x(t) = f(t) + (C\psi)(t) - (r * f)(t) - (r * C\psi)(t),$$

where the *initial function correction* $C\psi$ is defined through

$$(C\psi)(t) = \begin{cases} -\int_t^\sigma k(s)\psi(t-s)\,\mathrm{d}s, & 0 \leq t < \sigma, \\ 0, & t \geq \sigma. \end{cases} \tag{2.2}$$

If one takes $f = 0$ in (2.1), then

$$\begin{aligned} x(t) + \int_0^\sigma k(s)x(t-s)\,\mathrm{d}s &= 0, \qquad & t \in \mathbf{R}^+, \\ x(t) &= \psi(t), & -\sigma < t < 0. \end{aligned} \tag{2.3}$$

This equation is translation invariant in the following sense. Fix $h \geq 0$, and define

$$\begin{aligned} x_h(t) &= x(t+h), \qquad & t \geq -\sigma, \\ \psi_h(t) &= x(t+h), & -\sigma < t < 0. \end{aligned}$$

Thus x_h is x translated to the left by h, and ψ_h is the restriction of x_h to $[-\sigma, 0)$. Since translation commutes with convolution, we see that the translated function x_h satisfies (2.3) with ψ replaced by ψ_h, i.e.,

$$\begin{aligned} x_h(t) + \int_0^\sigma k(s)x_h(t-s)\,ds &= 0, \qquad & t \in \mathbf{R}^+, \\ x_h(t) &= \psi_h(t), & -\sigma < t < 0. \end{aligned} \tag{2.4}$$

As the next theorem shows, the mappings $\psi \mapsto \psi_h$ produce a strongly continuous semigroup.

2.1 Theorem
Let $0 < \sigma \leq \infty$, and let $k \in L^1((0,\sigma); \mathbf{C}^{n\times n})$. For each $\psi \in L^2((-\sigma,0); \mathbf{C}^n)$, let x be the solution of (2.3), and define $(T(h)\psi)(t) = x(t+h)$ for $-\sigma < t < 0$ and $h \geq 0$. Then T is a strongly continuous semigroup on $L^2((-\sigma,0); \mathbf{C}^n)$.

Proof Trivially, $T(0) = I$. One easily shows that, for each fixed h, $T(h)$ is a linear continuous map of $L^2((-\sigma,0); \mathbf{C}^n)$ into itself, and that $\lim_{h\downarrow 0} T(h)\psi = \psi$ for $\psi \in L^2((-\sigma,0); \mathbf{C}^n)$. The uniqueness of the solution of (2.4) implies that $T(h_1 + h_2) = T(h_1)T(h_2)$ for $h_1, h_2 \geq 0$. Thus $T(h)$, $h \geq 0$, is a strongly continuous semigroup. □

2.2 Definition
The semigroup T described in Theorem 2.1 is called the initial function semigroup determined by k on $(0,\sigma)$. The semigroup determined by k^T on $(0,\sigma)$ is denoted by T_T, and is called the transposed initial function semigroup determined by k on $(0,\sigma)$.

The preceding semigroups are well defined both in the case where σ is finite, and in the case where $\sigma = \infty$.

Above we took $f = 0$ in (2.1). If we let, instead, $\psi = 0$, and define $k(t) = 0$ for $t > \sigma$ if $\sigma < \infty$, then we get the familiar Volterra equation

$$x(t) + \int_0^t k(s)x(t-s)\,ds = f(t), \quad t \in \mathbf{R}^+. \tag{2.5}$$

It is possible to obtain a semigroup from this equation as well. The precise formulation is as follows.

2.3 Theorem
Let $0 < \sigma \leq \infty$, and let $k \in L^1((0,\sigma); \mathbf{C}^{n\times n})$. For each $f \in L^2((0,\sigma); \mathbf{C}^n)$, let x be the solution of (2.5) (where k and f have been extended as zero to (σ,∞) if $\sigma < \infty$), and for each $h > 0$ let $S(h)$ be the operator that maps f to the function $(S(h)f)(t) = f(t+h) - \int_t^{t+h} k(s)x(t-s+h)\,ds$, $t \in [0,\sigma)$. Then S is a strongly continuous semigroup on $L^2((0,\sigma); \mathbf{C}^n)$.

Proof Fix $h \geq 0$ and define

$$\begin{aligned} x_h(t) &= x(t+h), \quad t \geq -h, \\ f_h(t) &= f(t+h), \quad t \in \mathbf{R}^+. \end{aligned}$$

Then x_h satisfies

$$x_h(t) + \int_0^{t+h} k(s)x_h(t-s)\,\mathrm{d}s = f_h(t), \quad t \in \mathbf{R}^+.$$

By moving one part of the integral over to the other side, we get an equation of the same type as (2.5), namely

$$x_h(t)+\int_0^t k(s)x_h(t-s)\,\mathrm{d}s = f_h(t)-\int_t^{t+h} k(s)x_h(t-s)\,\mathrm{d}s, \quad t \in \mathbf{R}^+. \tag{2.6}$$

We defined $S(h)$ to be the operator that maps f into the right-hand side of (2.6) (restricted to $[0,\sigma)$, if $\sigma < \infty$). It is obvious from the definition of S (and from the fact that x depends continuously on f) that $S(h)$ is continuous from $L^2(\mathbf{R}^+;\mathbf{C}^n)$ into itself for each fixed h, that $S(0) = I$, and that $\lim_{h\downarrow 0} S(h)f = f$ for $f \in L^2(\mathbf{R}^+;\mathbf{C}^n)$. To get the semigroup property $S(h_1+h_2) = S(h_1)S(h_2)$ for $h_1, h_2 \geq 0$, it suffices to observe that, as (2.6) shows, the solution of (2.5) with f replaced by $S(h_2)f$ is x_{h_2}, hence

$$\begin{aligned}
\big(S(h_1)S(h_2)f\big)(t) &= \big(S(h_2)f\big)(t+h_1) - \int_t^{t+h_1} k(s)x_{h_2}(t-s+h_1)\,\mathrm{d}s \\
&= f(t+h_1+h_2) - \int_{t+h_1}^{t+h_1+h_2} k(s)x_{h_2}(t+h_1-s)\,\mathrm{d}s \\
&\quad - \int_t^{t+h_1} k(s)x_{h_2}(t-s+h_1)\,\mathrm{d}s \\
&= f(t+h_1+h_2) - \int_t^{t+h_1+h_2} k(s)x(t-s+h_1+h_2)\,\mathrm{d}s \\
&= \big(S(h_1+h_2)f\big)(t).
\end{aligned}$$

Consequently, $S(h)$ is a strongly continuous semigroup on $L^2((0,\sigma);\mathbf{C}^n)$. □

2.4 Definition

The semigroup S described in Theorem 2.3 is called the forcing function semigroup determined by k on $(0,\sigma)$. The forcing function semigroup determined by k^T is denoted by S_T, and it is called the transposed forcing function semigroup determined by k on $(0,\sigma)$.

Let us next compute the infinitesimal generators of the two semigroups we have defined. For simplicity, we take $\sigma < \infty$. (The infinite delay case is included in the extended semigroups considered in the next section.) We begin with the initial function semigroup.

2.5 Theorem

Let $0 < \sigma < \infty$, let $k \in L^1((0,\sigma);\mathbf{C}^{n\times n})$, and let A be the infinitesimal generator of the initial function semigroup T determined by k. Then

$$\mathcal{D}(A) = \left\{ \psi \in W^{1,2}([-\sigma,0];\mathbf{C}^n) \;\middle|\; \psi(0) + \int_0^\sigma k(s)\psi(-s)\,\mathrm{d}s = 0 \right\},$$

and

$$A\psi = \psi', \quad \psi \in \mathcal{D}(A).$$

Proof Let $\psi \in \mathcal{D}(A)$. Then, by a standard property of semigroups, $T(s)\psi \in \mathcal{D}(A)$ for all $s \geq 0$, and

$$\lim_{h\downarrow 0} \frac{T(s+h)\psi - T(s)\psi}{h} = AT(s)\psi$$

(see, e.g., Pazy [1], Theorem 2.4(c), p. 5). By the definition of T, this means that, for all $s \geq 0$, the function $t \mapsto \frac{1}{h}(x(s+t+h) - x(s+t))$ tends to the function $AT(s)$ in $L^2((-\sigma,0);\mathbf{C}^n)$ as $h \downarrow 0$. Therefore, $x \in W^{1,2}_{\text{loc}}([-\sigma,\infty);\mathbf{C}^n)$, and $(AT(s))(t) = x'(t+s)$ a.e. (see, e.g., Hille and Phillips [1], p. 535). Taking $s = 0$ we get $A\psi = \psi'$, as claimed. Moreover, we observe that x is continuous on $[-\sigma,\infty)$. In particular, x is continuous at zero. This, combined with (2.3), gives the extra necessary condition $\psi(0) + \int_0^\sigma k(s)\psi(-s)\,\mathrm{d}s = 0$.

Conversely, it is easy to show that, if the solution x of (2.3) belongs to $W^{1,2}_{\text{loc}}([-\sigma,\infty);\mathbf{C}^n)$, then the initial function ψ in (2.3) belongs to $\mathcal{D}(A)$. Thus, to complete the proof it suffices to show that the conditions $\psi \in W^{1,2}([-\sigma,0];\mathbf{C}^n)$ and $\psi(0)+\int_0^\sigma k(s)\psi(-s)\,\mathrm{d}s = 0$ together imply $x \in W^{1,2}_{\text{loc}}([-\sigma,\infty);\mathbf{C}^n)$. This one may do, e.g., as follows. Define

$$\xi(t) = \begin{cases} \psi(t), & -\sigma \leq t < 0, \\ \psi(0), & t \geq 0, \end{cases}$$

and let $y = x - \xi$. Then y vanishes on $[\sigma, 0)$, and

$$y(t) + \int_0^t k(s)y(t-s)\,\mathrm{d}s = g(t), \quad t \in \mathbf{R}^+,$$

where $g(t) = -\psi(0) - \int_0^\sigma k(s)\xi(t-s)\,\mathrm{d}s$. Observe that the condition $\psi \in W^{1,2}([-\sigma,0];\mathbf{C}^n)$ implies $\xi \in W^{1,2}_{\text{loc}}([-\sigma,\infty);\mathbf{C}^n)$, hence $g \in W^{1,2}_{\text{loc}}(\mathbf{R}^+;\mathbf{C}^n)$. By the variation of constants formula, $y = g - r * g$. This implies that y is locally absolutely continuous, and that $y'(t) = g'(t) - r(t)g(0) - (r*g')(t)$ a.e. If $\psi(0)+\int_0^\sigma k(s)\psi(-s)\,\mathrm{d}s = 0$, then $g(0) = 0$, and we conclude that $y \in W^{1,2}_{\text{loc}}(\mathbf{R}^+;\mathbf{C}^n)$, hence $x \in W^{1,2}_{\text{loc}}(\mathbf{R}^+;\mathbf{C}^n)$, as claimed. □

Next we compute the generator of the forcing function semigroup.

2.6 Theorem

Let $0 < \sigma < \infty$, let $k \in L^1((0,\sigma);\mathbf{C}^{n\times n})$ and let B be the infinitesimal generator of the forcing function semigroup S determined by k on $(0,\sigma)$. Then

$$\mathcal{D}(B) = \big\{\, f \in W^{1,1}([0,\sigma];\mathbf{C}^n) \;\big|\; f(\sigma) = 0, \text{ and } f'(\cdot) - k(\cdot)f(0) \in L^2((0,\sigma);\mathbf{C}^n) \,\big\},$$

and

$$(Bf)(\cdot) = f'(\cdot) - k(\cdot)f(0), \quad f \in \mathcal{D}(B).$$

Proof Assume that $f \in W^{1,1}([0,\sigma];\mathbf{C}^n)$, $f'(\cdot)-k(\cdot)f(0) \in L^2((0,\sigma);\mathbf{C}^n)$, and that $f(\sigma) = 0$. Extend f and k to $\mathbf{R}^+$, defining $f(t) = k(t) = 0$ for $t > \sigma$. Then $f \in W^{1,1}(\mathbf{R}^+;\mathbf{C}^n)$ and $f'(\cdot)-k(\cdot)f(0) \in L^2(\mathbf{R}^+;\mathbf{C}^n)$. Solve x from (2.5), using the variation of constants formula, and differentiate the result, to observe that $x \in W^{1,1}_{\text{loc}}(\mathbf{R}^+;\mathbf{C}^n)$ (and that $x' = f' - rf(0) - r * f'$). As $x \in W^{1,1}_{\text{loc}}(\mathbf{R}^+;\mathbf{C}^n)$, one can differentiate (2.5) to get (observe that $x(0) = f(0)$)

$$x'(t) + \int_0^t k(s)x'(t-s)\,\mathrm{d}s = f'(t) - k(t)f(0). \tag{2.7}$$

By Theorem **2**.3.5, this, together with the assumption on f', implies that $x' \in L^2_{\text{loc}}(\mathbf{R}^+;\mathbf{C}^n)$. In other words, $x \in W^{1,2}_{\text{loc}}(\mathbf{R}^+;\mathbf{C}^n)$.

For $h \geq 0$, we can write $S(h)f$ in the equivalent form (use (2.6))

$$(S(h)f)(t) = x_h(t) + \int_0^t k(s)x_h(t-s)\,\mathrm{d}s, \quad 0 < t < \sigma. \tag{2.8}$$

This, together with the fact that $x \in W^{1,2}_{\text{loc}}(\mathbf{R}^+;\mathbf{C}^n)$, implies that $S(h)f$ is differentiable in $L^2((0,\sigma);\mathbf{C}^n)$ (i.e., $f \in \mathcal{D}(B)$), and that

$$(Bf)(t) = x'(t) + \int_0^t k(s)x'(t-s)\,\mathrm{d}s, \quad 0 < t < \sigma.$$

By (2.7), this can be written in the form given in the statement of the theorem, namely $(Bf)(t) = f'(t) - k(t)f(0)$, and we have proved one half of the theorem.

Before we prove the converse statement, let us make one simple observation. Let $f \in L^2((0,\sigma);\mathbf{C}^n)$. Then (2.8) holds. By (2.5), the right-hand side of (2.8) is zero for $t > \sigma$ (recall that $k(s) = 0$ for $s > \sigma$), so if we define $(S(h)f)(t)$ to be zero for $t > \sigma$ then (2.8) holds for (almost) all values of $t \in \mathbf{R}^+$. Moreover, $S(h)f$ is differentiable with respect to h in $L^2((0,\sigma);\mathbf{C}^n)$ if and only if the extended function is differentiable in $L^2(\mathbf{R}^+;\mathbf{C}^n)$.

Now suppose that $f \in \mathcal{D}(B)$. Then the extended function $S(h)f$ is differentiable with respect to h in $L^2(\mathbf{R}^+;\mathbf{C}^n)$. If we use the variation of constants formula to solve (2.8) for x_h, then we find that x_h is differentiable with respect to h in $L^2_{\text{loc}}(\mathbf{R}^+;\mathbf{C}^n)$. This means that $x \in W^{1,2}_{\text{loc}}(\mathbf{R}^+;\mathbf{C}^n)$. Now use (2.5) and (2.7) to conclude that $f \in W^{1,1}_{\text{loc}}(\mathbf{R}^+;\mathbf{C}^n)$, and that $f' - kf(0) \in L^2_{\text{loc}}(\mathbf{R}^+;\mathbf{C}^n)$. To get the final conclusion $f(\sigma) = 0$ it suffices to observe that f is continuous and vanishes on (σ,∞). □

The following theorem tells us somewhat more about the generators of T and S. Recall that the spectrum $\sigma(A)$ of A is the set of $\lambda \in \mathbf{C}$ for which $\lambda I - A$ is not invertible. The resolvents of A are the operators $[\lambda I - A]^{-1}$ where λ does not belong to the spectrum.

2.7 Theorem
Let $0 < \sigma < \infty$, and let $k \in L^1((0,\sigma);\mathbf{C}^{n\times n})$. The spectra of the generators of the initial function and the forcing function semigroups determined by k on $(0,\sigma)$ are pure point spectra, and they are given by $\{\,\lambda \mid \det[I+\hat{k}(\lambda)] = 0\,\}$. The resolvents of these generators are compact.

We leave the proof as an exercise.

It is an interesting fact that the nature of the spectra changes drastically if we take $\sigma = \infty$. In particular, the spectra contain the entire half plane $\{\,\lambda \mid \Re\lambda \le 0\,\}$. See Exercise 2.

One important relation between S and T is given in the following result. Again, take $\sigma < \infty$. Let T^* be the *adjoint semigroup* that one gets from T by defining $T^*(t) = (T(t))^*$ for all $t \ge 0$; here $(T(t))^*$ is the adjoint of the operator $T(t)$, operating on the dual Banach space. Since T is strongly continuous and L^2 is reflexive, it follows that T^* is strongly continuous, see Pazy [1], Corollary 10.6, p. 41. To analyse the connection between T and T^* it is convenient to regard $L^2((0,\sigma);\mathbf{C}^n)$ and $L^2((-\sigma,0);\mathbf{C}^n)$ as dual spaces, with the duality mapping given by $\langle f,\psi\rangle = \int_0^\sigma f(s)^T\psi(-s)\,ds$ for $f \in L^2((0,\sigma);\mathbf{C}^n)$ and $\psi \in L^2((-\sigma,0);\mathbf{C}^n)$. (Note that we do not define an inner product, hence there are no complex conjugates in the definition above.)

2.8 Theorem
Let $0 < \sigma < \infty$, and let $k \in L^1((0,\sigma);\mathbf{C}^{n\times n})$. The adjoint semigroup of T is S_T and the adjoint semigroup of S is T_T.

Proof We show that $T^* = S_T$. To do this, it suffices to show that the adjoint of the generator A of T is the generator B_T of S_T, see Pazy [1], Theorem 2.6, p. 6, and Corollary 10.6, p. 41. Let $\psi \in \mathcal{D}(A)$ and $f \in \mathcal{D}(B_T)$. Then

$$\begin{aligned}
\langle f, A\psi\rangle &= \int_0^\sigma f(t)^T\psi'(-t)\,dt = \int_0^\sigma f'(t)^T\psi(-t)\,dt + f(0)^T\psi(0)\\
&= \int_0^\sigma \big(f'(t)^T - f(0)^T k(t)\big)\psi(-t)\,dt\\
&= \int_0^\sigma \big(f'(t) - k(t)^T f(0)\big)^T\psi(-t)\,dt\\
&= \langle B_T f, \psi\rangle.
\end{aligned}$$

Thus $\mathcal{D}(B_T) \subset \mathcal{D}(A^*)$, and A^* restricted to $\mathcal{D}(B_T)$ equals B_T. Combining this with the fact that $\mathcal{D}(B_T)$ is dense and B_T is closed, one gets $A^* = B_T$.

We leave the proof of the second part as an exercise. □

Two semigroups T_1 and T_2 are called *equivalent* if there exists a one-to-one bicontinuous mapping G of the state space of T_1 onto the state space of T_2, and $T_1 = G^{-1}T_2G$, hence $T_2 = GT_1G^{-1}$. Clearly, two equivalent semigroups may be considered as two different representations of the same semigroup.

One may ask whether the semigroups S and T are equivalent. As the next theorem and a simple counterexample show, the answer is affirmative if $\sigma < \infty$, but negative if $\sigma = \infty$. (As usual, we define the translation operator τ_h by $(\tau_h \psi)(t) = \psi(t+h)$.)

2.9 Theorem

Suppose that $0 < \sigma < \infty$, and that $k \in L^1((0,\sigma); \mathbf{C}^{n\times n})$. Extend f as zero to (σ,∞). Let Gf be the restriction to $(0,\sigma)$ of the solution of (2.5). Then G is a one-to-one, bicontinuous map of $L^2((0,\sigma); \mathbf{C}^n)$ onto itself. The inverse of G is

$$(G^{-1}\psi)(t) = \psi(t) + \int_0^t k(t-s)\psi(s)\,\mathrm{d}s, \quad 0 \le t \le \sigma,$$

and

$$S(h) = G^{-1}\tau_{-\sigma}T(h)\tau_\sigma G, \quad T(h) = \tau_\sigma G S(h) G^{-1}\tau_{-\sigma}, \quad h \ge 0.$$

Proof It is convenient to formulate the proof in terms of the shifted initial function semigroup $\tilde{T}$, defined on $L^2((0,\sigma); \mathbf{C}^n)$, and constructed from the equation

$$\begin{aligned} x(t) + \int_0^\sigma k(s)x(t-s)\,\mathrm{d}s &= 0, \quad \sigma \le t < \infty, \\ x(t) &= \psi(t), \quad 0 < t < \sigma, \end{aligned} \tag{2.9}$$

in the same way as T was constructed from (2.3). Obviously, $\tilde{T} = \tau_{-\sigma}T\tau_\sigma$. Moreover, if $\psi = Gf$, then the solutions of (2.5) and (2.9) coincide for $t \ge 0$. Combining these two observations one gets the assertion. □

To see that the answer is negative if $\sigma = \infty$, one can take the simple example $k = 0$ (we urge the reader to construct other examples). Then $(T(h)\psi)(t) = 0$ for $-h \le t < 0$, $(T(h)\psi)(t) = \psi(t+h)$ for $-\infty < t < -h$, and $(S(h)f)(t) = f(t+h)$ for $0 \le t < \infty$. Now observe that $T(h)\psi$ does not tend to zero as t tends to infinity for any nonzero $\psi \in L^2(\mathbf{R}^-; \mathbf{C}^n)$, but that $S(h)f \to 0$ as $h \to \infty$ for every $f \in L^2(\mathbf{R}^+; \mathbf{C}^n)$. This, of course, excludes the possibility that S and T could be equivalent.

Undoubtedly it is somewhat artificial to require either $\psi = 0$ or $f = 0$. It should be possible to allow both the initial and the forcing function to be nonzero and yet obtain a strongly continuous semigroup from (1.1), with initial condition (1.3). In fact this is possible, and it may be accomplished with the use of so-called extended initial function and forcing function semigroups. These semigroups, moreover, have the advantage of being equivalent even if $\sigma = \infty$. We present these extended semigroups briefly in the next section.

3. Extended Semigroups

Let η be a regular weight function on $\mathbf{R}$ dominated by a submultiplicative weight function φ. Recall that the translation operator τ_h is defined by $(\tau_h y)(t) = y(t+h)$ for $t, h \in \mathbf{R}$. For fixed h and $p \in [1, \infty)$, this operator is a continuous linear operator mapping $L^p(\mathbf{R}; \eta; \mathbf{C}^n)$ into itself, with norm $\|\tau_h\| \leq \varphi(-h)$. Moreover, this mapping is strongly continuous in h, i.e., for each $f \in L^p(\mathbf{R}; \eta; \mathbf{C}^n)$ the mapping $h \mapsto \tau_h f$ is continuous from $\mathbf{R}$ into $L^p(\mathbf{R}; \eta; \mathbf{C}^n)$. (In other words, $h \mapsto \tau_h$ is a strongly continuous group, the so-called *translation group* in $L^p(\mathbf{R}; \eta; \mathbf{C}^n)$.)

Suppose that the kernel μ in (1.1) satisfies $\mu \in M(\mathbf{R}^+; \varphi; \mathbf{C}^{n\times n})$, and that $\det[I + \mu(\{0\})] \neq 0$. Let ρ be the resolvent of μ. Then, by Theorem **4**.1.7, the solution of (1.1), with initial condition (1.3), can be written as $x = \kappa * (f + C\psi)$, where $\kappa = \delta - \rho$ and where, analogously to (2.2), the initial function correction $C\psi$ is defined as

$$(C\psi)(t) = -\int_{(t,\infty)} \mu(\mathrm{d}s)\psi(t-s), \quad t \geq 0, \tag{3.1}$$

and as zero for $t < 0$. We let f and ψ be zero on $\mathbf{R}^-$ and $\mathbf{R}^+$, respectively. Then

$$x(t) = (\psi + \kappa * (f + C\psi))(t), \quad t \in \mathbf{R}.$$

The easy proof of the following theorem is left to the reader as an exercise.

3.1 Theorem

Let η be a regular weight function, dominated by the submultiplicative weight function φ. Let $p \in (1, \infty)$ and define $\mathcal{I} = L^p(\mathbf{R}^-; \eta; \mathbf{C}^n)$, $\mathcal{F} = L^p(\mathbf{R}^+; \eta; \mathbf{C}^n)$. Suppose that $\mu \in M(\mathbf{R}^+; \varphi; \mathbf{C}^{n\times n})$ satisfies $\det[I + \mu(\{0\})] \neq 0$. For each $(\psi, f) \in \mathcal{I} \times \mathcal{F}$, let $x(\psi, f)$ be the solution of (1.1), with initial condition (1.3), and let $T_e(h)(\psi, f) \stackrel{\text{def}}{=} (x_h(\psi, f), f_h)$ where $h \geq 0$, $x_h(\psi, f)$ is the restriction of $\tau_h x(\psi, f)$ to $\mathbf{R}^-$, and f_h is the restriction of $\tau_h f$ to $\mathbf{R}^+$. Then T_e is a strongly continuous semigroup on $\mathcal{I} \times \mathcal{F}$.

3.2 Definition

The semigroup T_e, described in Theorem 3.1, is called the extended initial function semigroup determined by μ.

Note that by taking $f = 0$, and by considering only the first component of $T_e(h)(\psi, f)$, we are back to the original initial function semigroup.

Let us next define the *extended forcing function semigroup*.

3.3 Theorem
Let $p, \mathcal{I}, \mathcal{F}$ and μ be as in Theorem 3.1. For each $(\psi, f) \in \mathcal{I} \times \mathcal{F}$, let

$$x(\psi, f)(t) = \psi(t), \quad t < 0, \tag{3.2}$$

and for $t \geq 0$ let $x(\psi, f)(t)$ be the solution of the equation

$$x(t) + \int_{[0,t]} \mu(\mathrm{d}s)x(t-s) = f(t), \quad t \geq 0. \tag{3.3}$$

Define $S_e(h)(\psi, f) = (x_h(\psi, f), f_h + C(x_h(\psi, f) - \psi_h))$, where $x_h(\psi, f)$ is the restriction of $\tau_h x(\psi, f)$ to $\mathbf{R}^-$, f_h is the restriction of $\tau_h f$ to $\mathbf{R}^+$, and

$$C(x_h(\psi, \tau) - \psi_h)(t) = -\int_{(t,t+h]} \mu(\mathrm{d}s)x(t+h-s), \quad t \geq 0.$$

Then S_e is a strongly continuous semigroup on $\mathcal{I} \times \mathcal{F}$.

The proof is not difficult and is left to the reader. Observe that the restriction of $x(\psi, f)$ to $\mathbf{R}^+$ does not depend on ψ. This means that, independently of ψ, the second component of $S_e(h)(\psi, f)$ is the forcing function semigroup that we considered in the previous section. (The only role that the first component has is to record the past values of $x(\psi, f)$, without affecting the future behaviour of the semigroup.)

3.4 Definition
The semigroup S_e described in Theorem 3.3 is called the extended forcing function semigroup determined by μ.

The adjoint semigroup T_e^* of T_e is of type S_e. The exact statement (given below) is somewhat technical. Without loss of generality, we may assume that $\mu(\{0\}) = 0$ (otherwise multiply (1.1) by $[I + \mu(\{0\})]^{-1}$). For the proof, and for a more detailed discussion, see Staffans [25].

3.5 Theorem
Let η be a regular weight function, dominated by the submultiplicative weight function φ. Let $p \in (1, \infty)$, $p^{-1} + q^{-1} = 1$, and define η^* by $\eta^*(t) = \eta(-t)^{-1}$, $t \in \mathbf{R}$. Take $\mathcal{I} = L^p(\mathbf{R}^-; \eta; \mathbf{C}^n)$, and $\mathcal{F} = L^p(\mathbf{R}^+; \eta; \mathbf{C}^n)$, and identify the duals of $\mathcal{I}$ and $\mathcal{F}$ with $\mathcal{I}^* = L^q(\mathbf{R}^+; \eta^*; \mathbf{C}^n)$ and $\mathcal{F}^* = L^q(\mathbf{R}^-; \eta^*; \mathbf{C}^n)$, using the duality mapping $\langle f, \psi \rangle = \int_{\mathbf{R}^+} f(s)^T \psi(-s)\,\mathrm{d}s$. Suppose that $\mu \in M(\mathbf{R}^+; \varphi; \mathbf{C}^{n\times n})$ has no point mass at zero. Then the (strongly continuous) adjoint semigroup T_e^* of T_e on $\mathcal{F}^* \times \mathcal{I}^*$ can be described as follows. Let x^* be the solution of

$$\begin{aligned} x^*(t) &= f^*(t), && t < 0, \\ x^*(t) + \int_{[0,t]} \mu^T(\mathrm{d}s)x^*(t-s) &= \psi^*(t), && t \geq 0, \end{aligned} \tag{3.4}$$

let x_h^* be the restriction of $\tau_h x^*$ to $\mathbf{R}^-$, and let ψ_h^* be the restriction of $\tau_h \psi^*$ to $\mathbf{R}^+$. Then

$$T_e^*(h)(f^*, \psi^*) = (x_h^*, \psi_h^* + C^*(x_h^* - f_h^*)),$$

where

$$C^*(x_h^* - f_h^*)(t) = -\int_{(t,t+h]} \mu^T(\mathrm{d}s)x^*(t+h-s), \quad t \geq 0.$$

Comparing Theorem 3.5 with Theorem 3.3 one finds that the adjoint semigroup T_e^* is a semigroup of the type described in Theorem 3.3: simply replace the notations that refer to adjoints by the notation used in (1.1) and (1.3). More specifically, replace η^* by η, the transpose μ^T of μ by μ, x^* by x, f^* by ψ, ψ^* by f, and C^* by C.

Our next purpose is to show that the extended initial function semigroup and the extended forcing function semigroup are equivalent. Take p, $\mathcal{I}$, $\mathcal{F}$ as in Theorem 3.1. Define the continuous map Q from $\mathcal{I} \times \mathcal{F}$ into itself by $Q(\psi, f) = (0, -C\psi)$, with C defined as in (3.1), and let $D = I + Q$, thus $D(\psi, f) = (\psi, f - C\psi)$. Because $Q^2 = 0$, it follows that $D^{-1} = I - Q$.

3.6 Theorem
Let T_e and S_e be as in Theorems 3.1 and 3.3. Then $S_e(t) = D^{-1}T_e(t)D$, and $T_e(t) = DS_e(t)D^{-1}$, where $D(\psi, f) = (\psi, f - C\psi)$ and C is the initial function correction $(C\psi)(t) = -\int_{(t,\infty)} \mu(\mathrm{d}s)\psi(t-s)$.

Proof Let $(\psi, f) \in \mathcal{I} \times \mathcal{F}$, let x be the solution of (3.3), with initial condition (3.2). Observe that we may write (3.3) in the form of (1.1) by subtracting $C\psi$ from both sides to get

$$x(t) + (\mu * x)(t) = f(t) - (C\psi)(t), \quad t \in \mathbf{R}^+.$$

Define $\psi(t) = 0$ for $t \geq 0$, and let ψ_t be the restriction to $\mathbf{R}^-$ of $\tau_t\psi$. Then

$$T_e(t)D(\psi, f) = (x_t, f_t - C\psi_t).$$

On the other hand, by definition,

$$S_e(t)(\psi, f) = (x_t, f_t + Cx_t - C\psi_t),$$

Thus

$$\begin{aligned} T_e(t)D(\psi, f) &= (x_t, f_t - C\psi_t) = (x_t, f_t - C\psi_t + Cx_t) - (0, Cx_t) \\ &= S_e(t)(\psi, f) + QS_e(t)(\psi, f) = DS_e(t)(\psi, f). \ \square \end{aligned}$$

One can compute the generator A_e of T_e by an argument which is almost identical to the argument used in the proof of Theorem 2.5. The result given below says that $(\psi, f) \in \mathcal{D}(A_e)$ iff $(\psi, f) \in \mathcal{I} \times \mathcal{F}$, $f' \in \mathcal{F}$, x is locally absolutely continuous, and the functions $\tau_h x'$ restricted to $\mathbf{R}^-$ are in $\mathcal{I}$. Moreover, $A_e(\psi, f) = (\psi', f')$ for $(\psi, f) \in \mathcal{D}(A_e)$.

3.7 Theorem
Let the assumptions of Theorem 3.1 hold, and let A_e be the infinitesimal generator of T_e. Then

$$\mathcal{D}(A_e) = \{ (\psi, f) \in \mathcal{I} \times \mathcal{F} \mid (\psi', f') \in \mathcal{I} \times \mathcal{F}, \psi(0) + (\mu * \psi)(0) = f(0) \},$$

and $A_e(\psi, f) = (\psi', f')$ for $(\psi, f) \in \mathcal{D}(A_e)$.

Now that we know the generator of T_e, we may use Theorem 3.6 to compute the generator of S_e: clearly, since $S_e(t) = D^{-1}T_e(t)D$ for all $t \geq 0$, we must have $B_e = D^{-1}A_eD$, with $\mathcal{D}(B) = D^{-1}\mathcal{D}(A_e)$.

3.8 Theorem
Let the assumptions of Theorem 3.3 hold, with $\mu(\{0\}) = 0$, and let B_e be the infinitesimal generator of S_e. Then

$$B_e(\psi, f) = \left(\psi', \tfrac{\mathrm{d}}{\mathrm{d}\cdot}\big(f(\cdot) - \mu([0, \cdot])f(0)\big)\right)$$

for $(\psi, f) \in \mathcal{D}(B_e)$, where

$$\mathcal{D}(B_e) = \{ (\psi, f) \in \mathcal{I} \times \mathcal{F} \mid B_e(\psi, f) \in \mathcal{I} \times \mathcal{F} \text{ and } \psi(0) = f(0) \}.$$

Note that the requirement $B_e(\psi, f) \in \mathcal{I} \times \mathcal{F}$ in the description of $\mathcal{D}(B_e)$ means that the functions ψ and $t \mapsto f(t) - \mu([0, t])f(0)$ should be locally absolutely continuous, and have derivatives in $\mathcal{I}$ and $\mathcal{F}$, respectively. (In general, the function $t \mapsto f(t) - \mu([0, t])f(0)$ will not itself belong to $\mathcal{F}$.)

Proof of Theorem 3.8 Let $(\psi, f) \in \mathcal{D}(B_e)$. Then, as $\mathcal{D}(A_e) = D\mathcal{D}(B_e)$, we have $D(\psi, f) = (\psi, f - C\psi) \in \mathcal{D}(A_e)$. This, combined with Theorem 3.7, implies that ψ and $f - C\psi$ are locally absolutely continuous, that $(\psi', (f - C\psi)') \in \mathcal{I} \times \mathcal{F}$, and that $\psi(0) + (\mu * \psi)(0) = (f - C\psi)(0)$. But $(C\psi)(0) = -(\mu * \psi)(0)$, and so $\psi(0) = f(0)$. Moreover, as $B_e = D^{-1}A_eD$, it follows from Theorem 3.7 that

$$\begin{aligned} B_e(\psi, f) &= D^{-1}A_eD(\psi, f) = D^{-1}(\psi', (f - C\psi)') \\ &= (I - Q)\big(\psi', (f - C\psi)'\big) \\ &= \big(\psi', (f - C\psi)' + C\psi'\big). \end{aligned}$$

Define $y(t) = 0$ for $t < 0$, and $y(t) = \psi(0)$ for $t \geq 0$. In addition, let $\psi(t) \equiv 0$ for $t \geq 0$. Then, for $t \geq 0$, $(C\psi)(t) = -(\mu * \psi)(t)$, and, for these values of t,

$$(C\psi')(t) = -\big(\mu * (\psi + y)'\big)(t) = -\big(\mu * (\psi + y)\big)'(t)$$

(note that $(\psi + y)'(t) = 0$ for $t > 0$). Consequently, for $t \geq 0$,

$$\begin{aligned} &(f - C\psi)'(t) + C\psi'(t) \\ &\qquad = (f + \mu * \psi)'(t) - (\mu * (\psi + y))'(t) = (f - \mu * y)'(t). \end{aligned}$$

Now use the fact that $(\mu * y)(t) = \mu([0, t])\psi(0) = \mu([0, t])f(0)$ to get the desired expression for $B_e(\psi, f)$.

To arrive at the characterization of $\mathcal{D}(B_e)$, note that, for $t \geq 0$,

$$(f - C\psi)'(t) = (f - \mu * y)'(t) - C\psi'(t),$$

hence $(f - C\psi)' \in \mathcal{F}$ if and only if $(f - \mu * y)' \in \mathcal{F}$. □

4. Exercises

1. Prove Theorem 2.7.
2. Let η be a regular weight function on $\mathbf{R}$ dominated by a submultiplicative function φ for which $\omega_\varphi = \alpha_\varphi$, let $\mu \in M(\mathbf{R}^+; \varphi; \mathbf{C}^{n\times n})$ have no point mass at zero. Prove that the spectra of the semigroups T_e and S_e contain the half plane $\Re z \le \omega_\varphi$. Prove the same result for the semigroups T and S in the case of an infinite delay. Hint: See Burns and Herdman [2], pp. 734–736.
3. Prove the second half of Theorem 2.8.
4. Prove Theorem 3.1.
5. How do the formulas in Theorems 3.5 and 3.8 change if you allow μ to have a point mass at zero, with $\det[I + \mu(\{0\})] \neq 0$?
6. Let η be a regular weight function on $\mathbf{R}$ dominated by a submultiplicative function φ for which $\omega_\varphi = \alpha_\varphi$, let $\mu \in M(\mathbf{R}^+; \varphi; \mathbf{C}^{n\times n})$ have no point mass at zero, and let $\Re z_\ell > \alpha_\varphi$. Show that the following conditions are equivalent:
   (i) *z_ℓ is a characteristic exponent, i.e., it is a zero of the characteristic equation;*
   (ii) *z_ℓ is an eigenvalue of the generator of the initial function semigroup T that one gets by restricting T_e to $\mathcal{I}$;*
   (iii) *z_ℓ is an eigenvalue of the generator of the forcing function semigroup S that one gets by ignoring the first component of S_e;*
   (iv) *z_ℓ is an eigenvalue of the generator of the extended initial function semigroup T_e;*
   (v) *z_ℓ is an eigenvalue of the generator of the extended forcing function semigroup S_e.*
7. Show that the eigenfunctions corresponding to the eigenvalue z_ℓ with $\Re z_\ell > \alpha_\varphi$ in the preceding exercise can be described as follows.
   (i) *A function ψ is an eigenfunction of the generator of the initial function semigroup T if and only if*

$$\psi(t) = \sum_{j=0}^{p_\ell - 1} \alpha_{\ell,j}\, t^j e^{z_\ell t}, \quad t \in \mathbf{R}^-,$$

   where the coefficients $\alpha_{\ell,j}$ satisfy

$$\alpha_{\ell,j} = \sum_{i=0}^{p_\ell - j - 1} \binom{j+i}{j} K_{\ell,j+i} v_{\ell,j},$$

   *where $K_{\ell,j}$ are the coefficients in (**7**.2.9), and $v_{\ell,j}$ are some arbitrarily chosen vectors.*

(ii) *A function f is an eigenfunction of the generator of the forcing function semigroup S if and only if it is of the type $C\psi$, where C stands for the initial function correction, and ψ is an eigenfunction of the generator of the initial function semigroup T.*

(iii) *A pair of functions (ψ, f) is an eigenfunction of the generator of the extended initial function semigroup T_e if and only if ψ is an eigenfunction of the generator of the initial function semigroup T and $f \equiv 0$.*

(iv) *A pair of functions (ψ, f) is an eigenfunction of the generator of the extended forcing function semigroup S_e if and only if ψ is an eigenfunction of the generator of the initial function semigroup T, and $f = C\psi$, where C represents the initial function correction.*

8. Extend the results of this chapter to semigroups generated by the integrodifferential equation (1.2).

5. Comments

Section 2:

The first semigroup approach to delay equations resulted in an initial function semigroup for a finite delay integrodifferential equation of the type (1.2); see Krasovskiĭ [1], § 29, and Hale [1]. There the state space was the space $C([-\sigma, 0])$ of continuous functions, as opposed to our L^2-space. The same initial function semigroup prevailed for a long time; for various results with continuous initial functions, see Barbu and Grossman [1], Borisovič and Turbabin [1], Hale [5], and Kappel [2]; for L^p-results, see Bernier and Manitius [1], Burns and Herdman [1], Delfour [1] and [2], Delfour and Manitius [1], Hale [4], Manitius [1], Naito [1] and [2], Stech [1], and Vinter [1]; and for results in other spaces, see Hale and Kato [1], Kappel and Schappacher [2], and Naito [3]. Initial function semigroups for our equation (1.1) are found in Diekmann [3], van Gils [1], Hale [5], Henry [2], and Salamon [1].

The forcing function semigroup is of a younger origin than the initial function semigroup. Interestingly enough, it emerged first in connection with our equation (1.1) in Miller and Sell [4], and was later adapted to the integrodifferential equation (1.2) in Miller [13]. In addition, forcing function semigroups are found in, e.g., Burns and Herdman [2], Diekmann [3], Diekmann and van Gils [1], van Gils [1], and Stech [1]. As we have seen above, the forcing function semigroup is the adjoint of the initial function semigroup, and in this form it is present in several other papers too, but usually without an explicit description of the type given above.

In our setting, the adjoint of an initial function semigroup is a forcing function semigroup, and the adjoint of a forcing function semigroup is an initial function semigroup. In much of the earlier literature the situation is different: the adjoint of an initial function semigroup is of initial function type. This is due to the fact that a different duality mapping has been in common use; one imbeds $\mathcal{I}$ into $\mathcal{F}$ by using the forcing function correction C, observing that a function $\psi \in \mathcal{I}$ induces an element of the dual space of $\mathcal{I}$ through the linear functional $\langle \psi, \phi \rangle = \int_0^\sigma (C\psi)^T(s)\phi(-s)\,\mathrm{d}s$. For a detailed treatment of this approach, see Hale [5], where our adjoint is labelled 'true adjoint'. The first explicit duality characterization of the type described here is found in Burns and Herdman [1]; a more implicit one is given already in Henry [1].

For some current studies on semigroups in the nonreflexive space of continuous functions, and in its dual and bidual spaces, see Clément, Diekmann, *et al.* [1]–[5] and Clément, Heijmans, *et al.* [1], Section 3.4.

Section 3:

The extended initial function semigroup seems to appear for the first time in Staffans [22]. For further results on extended semigroups, and results on the cases $p = 1$ and $p = \infty$, see Staffans [25]. Corresponding results for the integrodifferential equation are given in Staffans [26].

As we mentioned earlier, the translation operator generates a strongly continuous group in $L^p(\mathbf{R}; \eta; \mathbf{C}^n)$. This is due to our definition of regularity, given in Chapter 4. It is possible to define regularity in a different and weaker way, and in this approach, the translation operator generates only a strongly continuous semigroup in $L^p(\mathbf{R}; \eta; \mathbf{C}^n)$. This was the view taken when weighted L^p-spaces were first introduced in the study of delay equations; see Coleman and Mizel [1]–[3].

The results of this chapter are largely based on Staffans [21], where further comments may be found.

References

V. Barbu and S. I. Grossman

1. Asymptotic behavior of linear integrodifferential systems, *Trans. Amer. Math. Soc.* **173** (1972), pp. 277–288.

C. Bernier and A. Manitius

1. On semigroups in $\mathbf{R}^n \times L^p$ corresponding to differential equations with delays, *Canad. J. Math.* **30** (1978), pp. 897–914.

J. G. Borisovič and A. S. Turbabin

1. On the Cauchy problem for linear non-homogeneous differential equations with retarded argument, *Soviet Math. Doklady* **10** (1969), pp. 401–405.

J. A. Burns and T. L. Herdman

1. Adjoint semigroup theory for a Volterra integrodifferential system, *Bull. Amer. Math. Soc.* **81** (1975), pp. 1099–1102.
2. Adjoint semigroup theory for a class of functional differential equations, *SIAM J. Math. Anal.* **7** (1976), pp. 729–745.

Ph. Clément, O. Diekmann, M. Gyllenberg, H. J. A. M. Heijmans, and H. R. Thieme (Clément, Diekmann, *et al.*)

1. Perturbation theory for dual semigroups I. The sun-reflexive case, *Math. Ann.* **277** (1987), pp. 709–725.
2. Perturbation theory for dual semigroups II. Time-dependent perturbations in the sun-reflexive case, *Proc. Roy. Soc. Edinburgh Ser A.* **109** (1988), pp. 145–172.
3. Perturbation theory for dual semigroups III. Nonlinear Lipschitz continuous perturbations in the sun-reflexive case, in *Volterra Integrodifferential Equations in Banach Spaces and Applications*, G. da Prato and M. Iannelli, eds., Proc., Trento 1987, Research Notes in Mathematics, Pitman, London, to appear 1989.
4. Perturbation theory for dual semigroups IV. The intertwining formula and the canonical pairing, in *Trends in Semigroup Theory and Applications*, Ph. Clément, S. Invernizzi, E. Mitidieri, and I. I. Vrabie, eds., Proc., Trieste, 1987, Marcel Dekker, New York, 1989, pp. 95–116.
5. A Hille-Yosida theorem for a class of weakly* continuous semigroups, *Semigroup Forum* **38** (1989), pp. 157–178.

Ph. Clément, H. J. A. M. Heijmans, S. Angenent, C. J. van Duijn, and B. de Pagter (Clément, Heijmans, *et al.*)

1. *One-Parameter Semigroups*, North-Holland, Amsterdam, 1987.

B. D. Coleman and V. J. Mizel

1. Norms and semi-groups in the theory of fading memory, *Arch. Ratl. Mech. Anal.* **23** (1966), pp. 87–123.
2. On the general theory of fading memory, *Arch. Ratl. Mech. Anal.* **29** (1968), pp. 18–31.
3. On the stability of solutions of functional-differential equations, *Arch. Ratl. Mech. Anal.* **30** (1968), pp. 173–196.

M. C. Delfour

1. The largest class of hereditary systems defining a C_0 semigroup on the product space, *Canad. J. Math.* **32** (1980), pp. 969–978.
2. Status of the state space theory of linear hereditary differential systems with delays in state and control variables, in *Analysis and Optimization of Systems*, A. Bensoussan and J. L. Lions, eds., Proc. of the 4th Internat. Conf. on Analysis and Optimization of Systems, Lecture Notes in Control and Information Sciences **28**, Springer-Verlag, Berlin, 1980, pp. 83–96.

M. C. Delfour and A. Manitius

1. The structural operator F and its role in the theory of retarded systems I–II, *J. Math. Anal. Appl.* **73–74** (1980), pp. 466–490,359–381.

O. Diekmann

3. *Volterra integral equations and semigroups of operators*, Report TW 197, Stichting Mathematisch Centrum, Amsterdam, 1980.

O. Diekmann and S. A. van Gils

1. Invariant manifolds of Volterra integral equations of convolution type, *J. Differential Equations* **54** (1984), pp. 139–180.

S. A. van Gils

1. *Linear Volterra convolution equations: semigroups, small solutions and convergence of projection operators*, Report TW 248, Stichting Mathematisch Centrum, Amsterdam 1983.

J. K. Hale

1. Linear functional differential equations with constant coefficients, in *Contributions to Differential Equations* **2**, a serial issued under the auspices of RIAS and the University of Maryland, J. Wiley, New York, 1963, pp. 291–319.
4. Functional differential equations with infinite delays, *J. Math. Anal. Appl.* **48** (1974), pp. 276–283.
5. *Theory of Functional Differential Equations*, Springer-Verlag, Berlin, 1977.

J. K. Hale and J. Kato

1. Phase space for retarded equations with infinite delay, *Funkcial. Ekvac.* **21** (1978), pp. 11–41.

D. Henry

1. The adjoint of a linear functional differential equation and boundary value problems, *J. Differential Equations* **9** (1971), pp. 55–66.
2. Linear autonomous neutral functional differential equations, *J. Differential Equations* **15** (1974), pp. 106–128.

E. Hille and R. S. Phillips

1. *Functional Analysis and Semi-Groups*, Amer. Math. Soc., Providence, R. I., 1957.

F. Kappel

2. *Linear autonomous functional differential equations in the state space* C, Technical Report 34, Technische Universität Graz, Universität Graz, 1984.

F. Kappel and W. Schappacher

2. Some considerations to the fundamental theory of infinite delay equations, *J. Differential Equations* **37** (1980), pp. 141–183.

N. N. Krasovskiĭ

1. *Stability of Motion*, Stanford University Press, Stanford, California, 1963 [translation of *Nekotorye zadachi teoriĭ ustoĭchivosti dvizheniya*, Gos. Fiz. Mat. Lit., Moscow, 1959.]

A. Manitius

1. Completeness and F-completeness of eigenfunctions associated with retarded functional differential equations, *J. Differential Equations* **35** (1980), pp. 1–29.

R. K. Miller

13. Linear Volterra integrodifferential equations as semigroups, *Funkcial. Ekvac.* **17** (1974), pp. 39–55.

R. K. Miller and G. R. Sell

4. *Volterra Integral Equations and Topological Dynamics*, Mem. Amer. Math. Soc. **102**, Amer. Math. Soc., Providence, R. I., 1970.

T. Naito

1. Adjoint equations of autonomous linear functional differential equations with infinite retardations, *Tôhoku Math. J. (2)* **28** (1976), pp. 135–143.
2. On autonomous linear functional differential equations with infinite retardations, *J. Differential Equations* **21** (1976), pp. 297–315.
3. On linear autonomous retarded equations with an abstract phase space for infinite delay, *J. Differential Equations* **33** (1979), pp. 74–91.

A. Pazy

1. *Semigroups of Linear Operators and Applications to Partial Differential Equations*, Springer-Verlag, Berlin, 1983.

D. Salamon

1. *Control and Observation of Neutral Systems*, Pitman, London, 1984.

O. J. Staffans

21. Semigroups generated by a convolution equation, in *Infinite-Dimensional Systems.*, F. Kappel and W. Schappacher, eds., Proc. of the Conference on Operator Semigroups and Applications held in Retzhof, Austria, June 5–11, 1983, Lecture Notes in Mathematics **1076**, Springer-Verlag, Berlin, 1984, pp. 209–226.
22. Some well-posed functional equations which generate semigroups, *J. Differential Equations* **58** (1985), pp. 157–191.
25. Extended initial and forcing function semigroups generated by a functional equation, *SIAM J. Math. Anal.* **16** (1985), pp. 1034–1048.
26. Semigroups generated by a neutral functional differential equation, *SIAM J. Math. Anal.* **17** (1986), pp. 46–57.

H. W. Stech

1. On the adjoint theory for autonomous linear functional differential equations with unbounded delays, *J. Differential Equation* **27** (1978), pp. 421–443.

R. B. Vinter

1. On the evolution of the state of linear differential delay equations in M^2: properties of the generator, *J. Inst. Math. Appl.* **21** (1978), pp. 13–23.

9

Linear Nonconvolution Equations

We study nonconvolution kernels and the corresponding nonconvolution Volterra equations $x(t) + \int_0^t k(t,s)x(s)\,ds = f(t)$, $t \geq 0$, in the framework of various Banach algebras. The existence of resolvents is established, and we analyse a number of different types of kernels. Some asymptotic results are also given.

1. Introduction

The Volterra equations studied so far (integral equations, integrodifferential equations, equations with functions or with measures as kernels) have all been of convolution type. This fact greatly facilitated the analysis, and made possible the use of various transform techniques. Here, and in the next chapter, the study is extended to nonconvolution (i.e., not necessarily convolution) equations.

The present chapter is devoted to the basic theory of linear Volterra equations (of the second kind) of the form

$$x(t) + \int_0^t k(t,s)x(s)\,ds = f(t), \quad t \in \mathbf{R}^+. \tag{1.1}$$

Integrodifferential equations with kernels that are not of convolution type will be discussed in Chapter 10. There we shall also analyse (1.1) with k replaced by a measure κ. The *kernel* k in (1.1) is a known function defined for $0 \leq s \leq t < \infty$, with values in the space $\mathbf{C}^{n\times n}$ of n by n matrices. The *forcing function* f is defined for $t \geq 0$, and it takes (column vector) values in $\mathbf{C}^n$. We study (1.1) in various Banach space settings, and, depending on the Banach space in question, we require a solution x on $[0,T)$ to satisfy (1.1) for either all, or almost all, $t \in [0,T)$. Obviously, we get the earlier convolution case analysed in Chapter 2 by taking k to be a function of the difference $t - s$ alone.

As in the convolution case, we want to give results saying that, if f belongs to some specific Banach space of functions on $[0, T)$, then (1.1) has exactly one solution x on $[0, T)$ belonging to the same Banach space as f. For T finite we shall obtain quite satisfactory results, but in the case where T is infinite only some rather restricted results are known. In particular, there do not (yet) exist any nonconvolution versions of the Paley–Wiener theorems.

One can rewrite (1.1) in such a way that it resembles the convolution equation discussed in Chapter 2. If we denote the function $t \mapsto \int_0^t k(t,s)x(s)\,ds$ in (1.1) by $k \star x$, then (1.1) becomes

$$x(t) + (k \star x)(t) = f(t), \quad t \in \mathbf{R}^+. \tag{1.2}$$

Arguing formally in the same way as in the convolution case, one finds that the solution of (1.2) should be given by the *variation of constants formula* (cf. Theorem **2**.3.5),

$$x(t) = f(t) - (r \star f)(t), \quad t \in \mathbf{R}^+. \tag{1.3}$$

Here the *resolvent* r is the solution of the two equations (cf. (**2**.1.5) and (**2**.1.6))

$$k(t,s) - r(t,s) = (r \star k)(t,s) = (k \star r)(t,s), \quad 0 \le s \le t < \infty, \tag{1.4}$$

where $(r\star k)(t,s) = \int_s^t r(t,u)k(u,s)\,du$ and $(k\star r)(t,s) = \int_s^t k(t,u)r(u,s)\,du$. As we shall see below, this formal reasoning can be made precise.

Essentially, the theory of (1.2) is simply the theory of how the inverse of the operator $x \mapsto x + k \star x$ on the left-hand side of (1.2) behaves. This inverse can be expressed in terms of the resolvent of k, and so most of the theory of (1.1) has, in one way or another, something to do with this resolvent.

In Sections 2–4 we work in L^p-spaces over an arbitrary interval. We begin the analysis in Section 2 by defining kernels of type L^p, and by formulating some of the basic properties of this class. The key result is Theorem 2.4, which says that, for each p, the kernels of type L^p constitute a Banach algebra.

In Section 3 we consider resolvents of type L^p, and, in particular, we give conditions under which a Volterra kernel of type L^p has a resolvent in the same algebra. The following section treats kernels that are only locally in L^p.

Some further types of kernels are introduced in Sections 5 and 6. We analyse kernels that are of continuous type, of bounded type, of periodic type, etc. Again we stress the fact that the classes of kernels that we define constitute Banach algebras.

In Section 7 we analyse a compactness problem: under what assumptions do kernels of type L^p generate compact mappings from L^p into itself?

Sections 8 and 9 deal with asymptotic results for Volterra equations with nonconvolution kernels. Certain sign and monotonicity properties make up the essential assumptions on the kernels.

In Section 2, and in the first half of Section 3, we do not require the kernels to be of Volterra type (i.e., $k(t,s) = 0$ for $s > t$). The results of Sections 6 and 7, and partially those of Section 5, are also independent of whether the kernels are of Volterra type or not.

2. Kernels of Type L^p

We begin our study of (1.1) with the case where $f \in L^p(0,T)$, $1 \le p \le \infty$, and $0 < T \le \infty$. Our goal is to find a solution x belonging to the same space. The case $p = \infty$ is the most important one, because it is closely related to the existence of, e.g., bounded, or periodic, or asymptotically periodic solutions.

Although intervals of the type $(0,T)$ are the most common in applications, it is advantageous to consider a more general interval J. Correspondingly, we shall take the kernel k to be defined on the square J^2.

Before we can proceed we must define a suitable class of kernels to be used, and define the $\star$-product introduced above.

First, we define what we mean by a *Fredholm kernel* and by a *Volterra kernel.*

2.1 Definition
Let $J \subset \mathbf{R}$ be an interval. A measurable function $k : J^2 \to \mathbf{C}^{n\times n}$ is called a Fredholm kernel on J. A Fredholm kernel k that satisfies $k(t,s) = 0$ when $s > t$ and $(t,s) \in J^2$ is called a Volterra kernel.

To be able to solve integral equations we must know much more about the kernel in question. For this reason, we introduce a number of different types of kernels.

2.2 Definition
Let $J \subset \mathbf{R}$ be an interval, and let p and $q \in [1,\infty]$ satisfy $1/p + 1/q = 1$. A function $k : J^2 \to \mathbf{C}^{n\times n}$ is called a kernel of type $[L^p; J; \mathbf{C}^{n\times n}]$ (or of type L^p on J) if k is measurable and $|||k|||_{L^p(J)} < \infty$, where

$$|||k|||_{L^p(J)} \overset{\text{def}}{=} \sup_{\substack{\|g\|_{L^q(J)}\le 1 \\ \|f\|_{L^p(J)}\le 1}} \int_J \int_J |g(t)k(t,s)f(s)| \, \mathrm{d}s \, \mathrm{d}t, \tag{2.1}$$

and the supremum is taken over (scalar) functions g and f in $L^q(J;\mathbf{C})$ and $L^p(J;\mathbf{C})$, respectively.

The class of all Fredholm kernels of type $[L^p; J; \mathbf{C}^{n\times n}]$ is denoted by $\mathcal{F}(L^p; J; \mathbf{C}^{n\times n})$, and the class of all Volterra kernels of type $[L^p; J; \mathbf{C}^{n\times n}]$ is denoted by $\mathcal{V}(L^p; J; \mathbf{C}^{n\times n})$.

The symbol $\mathcal{K}$ is to be interpreted as either $\mathcal{F}$ or $\mathcal{V}$.

Of course, the interpretation of the letter $\mathcal{K}$ should be the same throughout a particular theorem and its proof.

If it is obvious from the context (or if it is irrelevant) in what space the values of k lie, then we abbreviate $\mathcal{F}(L^p; J; \mathbf{C}^{n\times n})$ and $\mathcal{V}(L^p; J; \mathbf{C}^{n\times n})$ by $\mathcal{F}(L^p; J)$ and $\mathcal{V}(L^p; J)$.

2.3 Definition

Let $J \subset \mathbf{R}$, let a and b be measurable functions on J^2, and let f be a measurable function on J. Then, for all those t and s in J for which the functions $u \mapsto a(t,u)b(u,s)$ and $u \mapsto a(t,u)f(u)$ are integrable over J, we define the $\star$-products $(a\star b)(t,s)$ and $(a\star f)(t)$ to be

$$(a\star b)(t,s) = \int_J a(t,u)b(u,s)\,\mathrm{d}u,$$

$$(a\star f)(t) = \int_J a(t,u)f(u)\,\mathrm{d}u.$$

By Lemma 10.1, formula (2.1) can also be written as

$$|||k|||_{L^p(J)} = \sup_{\|f\|_{L^p(J)}\le 1} \big\| |k| \star |f| \big\|_{L^p(J)}. \tag{2.2}$$

In particular, this means that $k \in \mathcal{F}(L^p; J; \mathbf{C}^{n\times n})$ if and only if k is measurable and the operator $f \mapsto |k| \star f$ is continuous from $L^p(J; \mathbf{C})$ into itself. Furthermore, the norm of this operator is $|||k|||_{L^p(J)}$. If we let $\|k\|_{L^p(J)}$ denote the norm of the mapping $f \mapsto k\star f$ from $L^p(J)$ into itself, then $\|k\|_{L^p(J)} \le |||k|||_{L^p(J)}$. Strict inequality is possible, because the class of kernels $k(t,s)$ for which $k\star f$ is continuous from $L^p(J; \mathbf{C}^n)$ into itself (instead of $|k|\star f$ from $L^p(J;\mathbf{C})$ into itself) is larger than $\mathcal{F}(L^p; J; \mathbf{C}^{n\times n})$. For an example, see Section 12.

2.4 Theorem

The class $\mathcal{K}(L^p; J; \mathbf{C}^{n\times n})$ of kernels of type L^p on an interval J forms a Banach algebra with product $\star$ and norm $|||\cdot|||_{L^p(J)}$ if one identifies kernels that are equal a.e. on J^2. Moreover, $L^p(J; \mathbf{C}^n)$ is a left Banach module over $\mathcal{K}(L^p; J; \mathbf{C}^{n\times n})$.

In other words, we claim that $\mathcal{K}(L^p; J; \mathbf{C}^{n\times n})$ is a Banach space, and that, for all $a, b, c \in \mathcal{K}(L^p; J; \mathbf{C}^{n\times n})$, all $f, g \in L^p(J; \mathbf{C}^n)$, and all $\alpha \in \mathbf{C}$, we have $a\star b \in \mathcal{K}(L^p; J; \mathbf{C}^{n\times n})$, $a\star f \in L^p(J; \mathbf{C}^n)$, $(a\star b)\star c = a\star(b\star c)$, $a\star(b+c) = a\star b + a\star c$, $(a+b)\star c = a\star c + b\star c$, $(a\star b)\star f = a\star(b\star f)$, $(a+b)\star f = a\star f + b\star f$, $a\star(f+g) = a\star f + a\star g$, $\alpha(a\star b) = (\alpha a)\star b = a\star(\alpha b)$, $\alpha(a\star f) = (\alpha a)\star f = a\star(\alpha f)$, and

$$|||a\star b|||_{L^p(J)} \le |||a|||_{L^p(J)} |||b|||_{L^p(J)}, \tag{2.3}$$

$$\|a\star f\|_{L^p(J)} \le |||a|||_{L^p(J)} \|f\|_{L^p(J)}. \tag{2.4}$$

(We do not require a Banach algebra to have a unit.) For a general discussion of modules, see Marcus [1], p. 17. (In connection with the module $L^p(J;\mathbf{C}^n)$, the word *left* is used to specify that the kernel $k \in \mathcal{K}(L^p;J;\mathbf{C}^{n\times n})$ must appear to the left of $f \in L^p(J;\mathbf{C}^n)$ in the product $k \star f$.)

Proof of Theorem 2.4 It is quite easy to see that the sums and $\star$-products of Volterra kernels are Volterra kernels (provided they are well defined as Fredholm kernels). Therefore, we may assume that $\mathcal{K} = \mathcal{F}$ below.

The distributive claims concerning the product $\star$, i.e., the claims that $a\star(b+c) = a\star b + a\star c$, $(a+b)\star c = a\star c + b\star c$, $(a+b)\star f = a\star f + b\star f$, and $a\star(f+g) = a\star f + a\star g$ for all $a,b,c \in \mathcal{K}(L^p;J;\mathbf{C}^{n\times n})$ and all $f,g \in L^p(J;\mathbf{C}^n)$, are obvious, once one has shown that $\star$ is well defined. The same applies to the assertions concerning multiplication by α. It is also easy to show that $\mathcal{K}(L^p;J;\mathbf{C}^{n\times n})$ is a normed linear space.

We claim that $\mathcal{K}(L^p;J;\mathbf{C}^{n\times n})$ is complete, hence a Banach space. To prove this, let us first show that convergence in $\mathcal{K}(L^p;J;\mathbf{C}^{n\times n})$ implies convergence in $L^1(K^2;\mathbf{C}^{n\times n})$ for every bounded subinterval K of J. If K is a bounded subinterval of J with length $m(K)$, then the characteristic function χ_K of K belongs to $L^q(J;\mathbf{R}) \cap L^p(J;\mathbf{R})$, with $\|\chi_K\|_{L^q(J)} = m(K)^{1/q}$ and $\|\chi_K\|_{L^p(J)} = m(K)^{1/p}$. Therefore, for every $k \in \mathcal{K}(L^p;J;\mathbf{C}^{n\times n})$,

$$\int_K\int_K |k(t,s)|\,\mathrm{d}s\,\mathrm{d}t = \int_J\int_J |\chi_K(t)k(t,s)\chi_K(s)|\,\mathrm{d}s\,\mathrm{d}t \le m(K)|||k|||_{L^p(J)}.$$

In other words, the norm of k in $L^1(K^2;\mathbf{C}^{n\times n})$ is majorized by a constant times $|||k|||_{L^p(J)}$. Thus, convergence in $\mathcal{K}(L^p;J;\mathbf{C}^{n\times n})$ implies convergence in $L^1(K^2;\mathbf{C}^{n\times n})$, as claimed.

Let $\{k_i\}_{i=1}^\infty$ be a Cauchy sequence in $\mathcal{K}(L^p;J;\mathbf{C}^{n\times n})$. Then, for each bounded subinterval $K \subset J$, $\{k_i\}_{i=1}^\infty$ is a Cauchy sequence in $L^1(K^2;\mathbf{C}^{n\times n})$. Therefore, there is a subsequence $\{k_{\ell_j}\}_{j=1}^\infty$ which converges a.e. on J^2 to a measurable function k. By Fatou's lemma, we have, for each $i \ge 1$, $g \in L^q(J;\mathbf{C})$, and each $f \in L^p(J;\mathbf{C})$,

$$\begin{aligned}\int_J\int_J &|g(t)||k_i(t,s) - k(t,s)||f(s)|\,\mathrm{d}s\,\mathrm{d}t\\ &= \int_J\int_J \lim_{j\to\infty} |g(t)||k_i(t,s) - k_{\ell_j}(t,s)||f(s)|\,\mathrm{d}s\,\mathrm{d}t\\ &\le \liminf_{j\to\infty}\int_J\int_J |g(t)||k_i(t,s) - k_{\ell_j}(t,s)||f(s)|\,\mathrm{d}s\,\mathrm{d}t.\end{aligned}$$

Taking the supremum over all g and all f with norm at most 1, we get

$$|||k_i - k|||_{L^p(J)} \le \liminf_{j\to\infty} |||k_i - k_{\ell_j}|||_{L^p(J)}.$$

This, together with the fact that $\{k_i\}_{i=1}^\infty$ is a Cauchy sequence, implies that $k_i \to k$ in $\mathcal{K}(L^p;J;\mathbf{C}^{n\times n})$ as $i \to \infty$, proving that $\mathcal{K}(L^p;J;\mathbf{C}^{n\times n})$ is complete.

Next, let us show that the operator $f \mapsto a \star f$ maps $L^p(J;\mathbf{C}^n)$ into itself, and that (2.4) holds. Let $g \in L^q(J;\mathbf{C})$, $a \in \mathcal{K}(L^p;J;\mathbf{C}^{n\times n})$, and

$f \in L^p(J;\mathbf{C}^n)$. Then the function $(t,s) \mapsto g(t)a(t,s)f(s)$ is integrable on J^2. Fubini's theorem applies, and, together with (2.1), it tells us that the function $t \mapsto g(t)(a \star f)(t)$ is measurable, and satisfies

$$\int_J |g(t)(a \star f)(t)| \,\mathrm{d}t \le \|g\|_{L^q(J)} |||a|||_{L^p(J)} \|f\|_{L^p(J)}.$$

This being true for all $g \in L^q(J;\mathbf{C}^n)$, we can use Lemma 10.1 to conclude that $a \star f \in L^p(J;\mathbf{C}^n)$, and that (2.4) holds.

Let b be another kernel in $\mathcal{K}(L^p;J;\mathbf{C}^{n\times n})$. Then, for all g, a and f as above,

$$\begin{aligned}\int_J\int_J\int_J &|g(t)||a(t,u)||b(u,s)||f(s)| \,\mathrm{d}s\,\mathrm{d}u\,\mathrm{d}t \\ &\le \|g\|_{L^q(J)} |||a|||_{L^p(J)} \big\| |b| \star |f| \big\|_{L^p(J)} \\ &\le \|g\|_{L^q(J)} |||a|||_{L^p(J)} |||b|||_{L^p(J)} \|f\|_{L^p(J)},\end{aligned}$$

where the last step is a consequence of (2.4). In particular, $(t,u,s) \mapsto g(t)a(t,u)b(u,s)f(s)$ is integrable on J^3. Fubini's theorem applies, and it tells us, e.g., that $(t,s) \mapsto g(t)(a \star b)(t,s)f(s)$ is measurable, and that, for almost all t, $g(t)((a\star b)\star f)(t) = g(t)(a\star(b\star f))(t)$. Both g and f could be chosen arbitrarily, and therefore $a\star b$ is measurable, and $(a\star b)\star f = a\star(b\star f)$. Moreover, it follows directly from the preceding computation that (2.3) holds.

There is only one more thing to prove, namely that if c is a third kernel in $\mathcal{K}(L^p;J;\mathbf{C}^{n\times n})$ then $(a\star b)\star c = a\star(b\star c)$. To prove this, it suffices to argue as above to show that the function $(t,v,u,s) \mapsto g(t)a(t,v)b(v,u)c(u,s)f(s)$ is integrable on J^4, and then to apply Fubini's theorem. □

For completeness, we include a definition of kernels of type $[L^p;L^q]$ (which map L^p into L^q). These are further commented on in Section 12. See also Exercises 2 and 3, and Lemma 11.3.

2.5 Definition

Let $J \subset \mathbf{R}$ be an interval, let $p,q \in [1,\infty]$ and let q, q' satisfy $1/q + 1/q' = 1$. A function $k : J^2 \mapsto \mathbf{C}^{n\times n}$ is called a (Fredholm) kernel of type $[L^p;L^q;J;\mathbf{C}^{n\times n}]$ (or of type $[L^p;L^q]$ on J) if k is measurable and $|||k|||_{L^{p,q}(J)} < \infty$, where

$$|||k|||_{L^{p,q}(J)} = \sup_{\substack{\|f\|_{L^p(J)}\le 1\\ \|g\|_{L^{q'}(J)}\le 1}} \int_J\int_J |g(t)k(t,s)f(s)| \,\mathrm{d}s\,\mathrm{d}t,$$

and the supremum is taken over (scalar) functions g and f in $L^{q'}(J;\mathbf{C})$ and $L^p(J;\mathbf{C})$, respectively.

It is a noteworthy fact that a kernel which is of type L^p for two different values of p, is of type L^p for all intermediate values of p.

2.6 Theorem
If k is a kernel of both type L^p and L^r on J, with $1 \le p < r \le \infty$, then k is of type L^s for every $s \in (p,r)$, and

$$|||k|||_{L^s(J)} \le |||k|||^t_{L^p(J)} |||k|||^{1-t}_{L^r(J)},$$

where $t \in (0,1)$ is defined through the relation $1/s = t/p + (1-t)/r$.

This follows immediately from the Riesz interpolation theorem (see Stein and Weiss [1], p. 179) applied to the operator $f \mapsto |k| \star f$ (cf. the discussion following Definition 2.3).

It can be difficult to check directly from Definition 2.2 whether $|||k|||_{L^p(J)}$ is finite or not. The following sets of conditions under which k is of type L^p on J are much easier to apply than (2.1). Note, however, that these conditions are not necessary when $1 < p < \infty$.

2.7 Proposition
Let $k : J^2 \to \mathbf{C}^{n\times n}$ be measurable. Then the following conclusions hold.

(i) *k is of type L^∞ on J if and only if $\operatorname{ess\,sup}_{t\in J} \int_J |k(t,s)|\,\mathrm{d}s < \infty$. Moreover,*

$$|||k|||_{L^\infty(J)} = \operatorname*{ess\,sup}_{t\in J} \int_J |k(t,s)|\,\mathrm{d}s.$$

(ii) *k is of type L^1 on J if and only if $\operatorname{ess\,sup}_{s\in J} \int_J |k(t,s)|\,\mathrm{d}t < \infty$. Moreover,*

$$|||k|||_{L^1(J)} = \operatorname*{ess\,sup}_{s\in J} \int_J |k(t,s)|\,\mathrm{d}t.$$

(iii) *If $1 < p < \infty$, $1/p + 1/q = 1$, and*

$$\int_J \Big(\int_J |k(t,s)|^q\,\mathrm{d}s \Big)^{p/q} \mathrm{d}t < \infty, \text{ or } \int_J \Big(\int_J |k(t,s)|^p\,\mathrm{d}t \Big)^{q/p} \mathrm{d}s < \infty,$$

then k is of type L^p on J, and

$$|||k|||_{L^p(J)} \le \min \Bigg\{ \bigg(\int_J \Big(\int_J |k(t,s)|^q\,\mathrm{d}s \Big)^{p/q} \mathrm{d}t \bigg)^{1/p}, \bigg(\int_J \Big(\int_J |k(t,s)|^p\,\mathrm{d}t \Big)^{q/p} \mathrm{d}s \bigg)^{1/q} \Bigg\}.$$

Proof When $p = \infty$, the function $f(s) \equiv 1$ maximizes the integral in (2.1) over the set of permitted functions f. This means that in this case (2.1) is equivalent to

$$|||k|||_{L^\infty(J)} = \sup_{\|g\|_{L^1(J)}\le 1} \int_J |g(t)| \int_J |k(t,s)|\,\mathrm{d}s\,\mathrm{d}t,$$

and (i) follows from Lemma 10.1. The proof of (ii) is completely similar: first take $g \equiv 1$ in (2.1), then use Fubini's theorem, and finally maximize over f.

To prove the first half of (iii), one uses Hölder's inequality twice to get

$$\begin{aligned}\int_J\left(\int_J |g(t)k(t,s)f(s)|\,\mathrm{d}s\right)\mathrm{d}t \\ &\leq \left(\int_J |g(t)|\left(\int_J |k(t,s)|^q\,\mathrm{d}s\right)^{1/q}\mathrm{d}t\right)\|f\|_{L^p(J)} \\ &\leq \|g\|_{L^q(J)}\left(\int_J\left(\int_J |k(t,s)|^q\,\mathrm{d}s\right)^{p/q}\mathrm{d}t\right)^{1/p}\|f\|_{L^p(J)}.\end{aligned}$$

An analogous computation proves the second half of (iii). □

Observe, in particular, that it follows from (iii) that, if $1 < p < \infty$, $p^{-1} + q^{-1} = 1$, and $k(t,s) = a(t)b(s)$ with $a \in L^p(J)$ and $b \in L^q(J)$, then k is of type L^p on J. By Definition 2.2 and Lemma 10.1, we have, in fact, $|||k|||_{L^p(J)} = \|a\|_{L^p(J)}\|b\|_{L^q(J)}$.

To get examples where $|||k|||_{L^p(J)} < \infty$, and yet

$$\min\left\{\left(\int_J\left(\int_J |k(t,s)|^q\,\mathrm{d}s\right)^{p/q}\mathrm{d}t\right)^{1/p}, \left(\int_J\left(\int_J |k(t,s)|^p\,\mathrm{d}t\right)^{q/p}\mathrm{d}s\right)^{1/q}\right\} = \infty,$$

take $p = q = 2$, $J = \mathbf{R}$, $k(t,s) = a(t-s)$ with $a \in L^1(\mathbf{R}) \setminus L^2(\mathbf{R})$, and recall Theorem **2**.2.2(i).

If a kernel k on an interval J satisfies $|k(t,s)| \leq a(t-s)$ for some $a \in L^1(\mathbf{R})$ and almost all t and s, then k is of type L^p for all p, $1 \leq p \leq \infty$ (this follows from Theorem 2.6 and Proposition 2.7(i) and (ii)).

3. Resolvents of Type L^p

In this section we develop and justify the formal arguments presented in Section 1 concerning the existence and uniqueness of the resolvent.

3.1 Definition

Let $J \subset \mathbf{R}$ be an interval, let $1 \leq p \leq \infty$, and let k be a Fredholm kernel of type L^p on J. A Fredholm kernel r of type L^p on J is called a Fredholm resolvent (of type L^p) of k on J if $r + k \star r = r + r \star k = k$ on J^2. If both k and r are Volterra kernels, then r is called a Volterra resolvent (of type L^p) of k. If k is a Volterra kernel and r is a Volterra resolvent of k, or if k is not a Volterra kernel and r is a Fredholm resolvent of k, then r is simply called a resolvent of k.

It is possible for a Volterra kernel of type L^p to have a Fredholm resolvent of type L^p, but no Volterra resolvent of type L^p. To see that this can happen it suffices to consider the convolution case, with $a \in L^1(\mathbf{R}^+; \mathbf{C}^{n\times n})$ and $\det[I+\hat{a}(z)] \neq 0$ for $\Re z = 0$, but $\det[I+\hat{a}(z_0)] = 0$ for some z_0 with $\Re z_0 > 0$. It follows from the Paley–Wiener Theorems **2**.4.1 and **2**.4.3 that a has a whole line (i.e., Fredholm) resolvent in $L^1(\mathbf{R}; \mathbf{C}^{n\times n})$, but no half line (i.e., Volterra) resolvent in $L^1(\mathbf{R}^+; \mathbf{C}^{n\times n})$. Of course, a does have a Volterra resolvent in $L^1_{\text{loc}}(\mathbf{R}^+; \mathbf{C}^{n\times n})$. For simplicity, we require the resolvent of a Volterra kernel to be a Volterra kernel. With this requirement, Definition 3.1 can be seen to be a special case of the following, more general definition.

3.2 Definition

Let $\mathcal{A}$ be an associative algebra with product $\star$. If $k \in \mathcal{A}$ and $r \in \mathcal{A}$, and $r + k \star r = k + r \star k = k$, then r is called a resolvent of k.

Observe that, if k is a Volterra kernel that has a (Volterra) resolvent r on some interval J, then r is also a resolvent of k on every subinterval of J. This is *not* true for Fredholm kernels.

Note also that, if k is an element of a Banach algebra $\mathcal{A}$ with unit ε, then one usually defines the resolvent of k in a different way, i.e., the resolvent of k is the inverse of $\lambda\varepsilon - k$, defined for all those $\lambda \in \mathbf{C}$ for which this inverse exists in $\mathcal{A}$.

Concerning the uniqueness of the resolvent we have the following statement.

3.3 Lemma

Let $\mathcal{A}$ be an associative algebra with product $\star$. Let q, k, $r \in \mathcal{A}$, and let $r + k \star r = q + q \star k = k$. Then $r = q$.

An immediate consequence is that, if k is of type L^p on J, then k can have at most one resolvent of type L^p on J.

Proof of Lemma 3.3 We have $q\star(k-r) = q\star k\star r = (k-q)\star r$. Cancelling out $q \star r$ one gets $q \star k = k \star r$. Thus, $k - q = q \star k = k \star r = k - r$, and so $r = q$. □

The general algebraic interpretation of our standard variation of constants formula is the following.

3.4 Lemma

Let $\mathcal{A}$ be an associative algebra with product $\star$, and let F be a left module over $\mathcal{A}$. If $k \in \mathcal{A}$ has a resolvent $r \in \mathcal{A}$, and if $f \in F$, then the equation

$$x + k \star x = f \tag{3.1}$$

has a unique solution $x \in F$. This solution is given by the variation of constants formula

$$x = f - r \star f. \tag{3.2}$$

Proof Define x by (3.2). Then $x \in F$, and $k \star x = k \star f - k \star r \star f = k \star f - (k - r) \star f = r \star f$. This means that $x = f - k \star x$, i.e., x satisfies (3.1).

Conversely, if $x \in F$ is a solution of (3.1), then $r \star f = r \star x + r \star k \star x = r \star x + (k - r) \star x = k \star x$. This means that $x = f - k \star x = f - r \star f$, i.e., x is given by (3.2). □

The same argument can be used to prove the following modification of Lemma 3.4.

3.5 Lemma

Let $\mathcal{A}$ be an associative algebra with product $\star$, let $k \in \mathcal{A}$ have a resolvent $r \in \mathcal{A}$, and let $a \in \mathcal{A}$. Then the equations

$$x + k \star x = a$$

and

$$y + y \star k = a$$

have unique solutions $x \in \mathcal{A}$ and $y \in \mathcal{A}$. These solution are given by the variation of constants formulas

$$x = a - r \star a,$$

and

$$y = a - a \star r.$$

Combining Lemma 3.4 with Theorem 2.4, we get the following result.

3.6 Theorem

If k is a kernel of type L^p on J that has a resolvent of type L^p on J, and if $f \in L^p(J; \mathbf{C}^n)$, then the equation

$$x(t) + \int_J k(t,s)x(s)\,\mathrm{d}s = f(t), \quad t \in J, \tag{3.3}$$

has a unique solution x in $L^p(J; \mathbf{C}^n)$. This solution is given by the variation of constants formula

$$x(t) = f(t) - \int_J r(t,s)f(s)\,\mathrm{d}s, \quad t \in J. \tag{3.4}$$

Of course, equations (3.3) and (3.4) are equalities in $L^p(J)$, and therefore these equations need only hold for almost all $t \in J$.

One can get a (rather weak) result on the existence of a resolvent of type L^p on J by applying a general Banach algebra result.

3.7 Lemma

Let $\mathcal{A}$ be a Banach algebra with norm $\|\cdot\|$. Let $k \in \mathcal{A}$ be of the form $k = k_1 + k_2$, where $k_1 \in \mathcal{A}$ and $k_2 \in \mathcal{A}$, and suppose that k_1 has a resolvent r_1 in $\mathcal{A}$. If $\|k_2\| < \left(1 + \|r_1\|\right)^{-1}$; then k has a resolvent $r \in \mathcal{A}$.

Proof Denote the product in $\mathcal{A}$ by $\star$, and consider the equation

$$r + (k_2 - r_1 \star k_2) \star r = k - r_1 \star k. \tag{3.5}$$

The fact that this equation has a solution r in $\mathcal{A}$ follows from the contraction mapping principle, and the fact that $\|k_2 - r_1 \star k_2\| \le \|k_2\|(1 + \|r_1\|) < 1$.

To show that the kernel r that we get satisfies the original equation $r + k \star r = k$ one multiplies (3.5) by k_1 from the left, and uses the fact that $r_1 + k_1 \star r_1 = k_1$, to obtain $k_1 \star r + k_1 \star k_2 \star r - (k_1 - r_1) \star k_2 \star r = k_1 \star k - (k_1 - r_1) \star k$. This simplifies to $k_1 \star r = r_1 \star k - r_1 \star k_2 \star r$, which substituted into (3.5) gives $r + (k_1 + k_2) \star r = k$. In a similar way it is proved that the equation $r + r \star (k_1 + k_2) = k$ has a solution. It follows from Lemma 3.3 that the solutions of the two equations are the same, and hence k has a resolvent of type L^p on J. □

In particular, taking k_1 to be zero, we get

3.8 Lemma

Let $\mathcal{A}$ be a Banach algebra. Then every $k \in \mathcal{A}$ with norm $|||k||| < 1$ has a resolvent in $\mathcal{A}$.

When we apply the previous two lemmas to the Banach algebras $\mathcal{F}(L^p; J; \mathbf{C}^{n\times n})$ and $\mathcal{V}(L^p; J; \mathbf{C}^{n\times n})$, we get the following two results.

3.9 Theorem

Let $J \subset \mathbf{R}$ be an interval. Let k be of the form $k = k_1 + k_2$, where $k_1, k_2 \in \mathcal{K}(L^p; J; \mathbf{C}^{n\times n})$, and suppose that k_1 has a resolvent r_1 of type L^p on J. If $|||k_2|||_{L^p(J)} < \left(1 + |||r_1|||_{L^p(J)}\right)^{-1}$, then k has a resolvent $r \in \mathcal{K}(L^p; J; \mathbf{C}^{n\times n})$.

3.10 Corollary

If k is a kernel of type L^p on an interval J with $|||k|||_{L^p(J)} < 1$, then k has a resolvent of type L^p on J.

Another immediate corollary to Lemma 3.7 is the following continuous dependence result.

3.11 Lemma

Let $\mathcal{A}$ be a Banach algebra with product $\star$, and let F be a left module over $\mathcal{A}$. If $k \in \mathcal{A}$ has a resolvent $r \in \mathcal{A}$, and if $k_m \to k$ in $\mathcal{A}$ as $m \to \infty$, then k_m has a resolvent $r_m \in \mathcal{A}$ for all sufficiently large m, and $r_m \to r$ in $\mathcal{A}$ as $m \to \infty$. If, moreover, $f \in F$ and $f_m \to f$ in F, then (3.3), with k and f replaced by k_m and f_m, has a solution $x_m \in F$ for all sufficiently large m, and $x_m \to x$ in F as $m \to \infty$.

Proof That r_m exists follows from Lemma 3.7 with k_1, k, r_1 and r replaced by k, k_m, r and r_m, respectively. To see that $r_m \to r$ as $m \to \infty$, first note that (3.5) (with the same replacements as above) gives, for large m,

$$\|r_m\| \leq \frac{\|k_m - r \star k_m\|}{1 - \|k_m - k\|(1 + \|r\|)};$$

hence $\limsup_{m\to\infty} \|r_m\| < \infty$. Then observe that the difference $q_m \stackrel{\text{def}}{=} r_m - r$ satisfies (define $a_m = k_m - k$)

$$q_m + k \star q_m = a_m - a_m \star r_m.$$

By the variation of constants formula,

$$q_m = a_m - a_m \star r_m - r \star a_m + r \star a_m \star r_m,$$

and the right-hand side tends to zero as $m \to \infty$. □

In particular, applying Lemma 3.11 to the algebras $\mathcal{F}(L^p; J; \mathbf{C}^{n\times n})$ and $\mathcal{V}(L^p; J; \mathbf{C}^{n\times n})$, we obtain the following result.

3.12 Corollary

If k is a kernel of type L^p on J that has a resolvent r of type L^p on J, and if $k_m \to k$ in $\mathcal{K}(L^p; J; \mathbf{C}^{n\times n})$ as $m \to \infty$, then, for all sufficiently large m, it is true that k_m has a resolvent r_m of type L^p on J, and that $r_m \to r$ in $\mathcal{K}(L^p; J; \mathbf{C}^{n\times n})$ as $m \to \infty$. If, moreover, $f \in L^p(J; \mathbf{C}^n)$ and $f_m \to f$ in $L^p(J; \mathbf{C}^n)$, then, for all sufficiently large m, it is true that (3.3), with k and f replaced by k_m and f_m, has a solution $x_m \in L^p(J; \mathbf{C}^n)$, and that $x_m \to x$ in $L^p(J; \mathbf{C}^n)$ as $m \to \infty$.

In our next theorem we make crucial use, for the first time, of a Volterra assumption on the kernel. We prove that, if k is a Volterra kernel that has a resolvent of type L^p on small intervals, then k has a resolvent on the whole interval J.

3.13 Theorem

Let k be a Volterra kernel of type L^p for some p, $1 \leq p \leq \infty$, on an interval $J \subset \mathbf{R}$, and suppose that J can be divided into finitely many subintervals J_i in such a way that k has a resolvent of type L^p on each subinterval J_i. Then k has a resolvent of type L^p on J.

Proof It suffices to consider the case where J is divided into two subintervals J_1 and J_2, since one can use induction to get the general case.

Let $-\infty \leq S < U < T \leq \infty$, let $J = (S,T)$, $J_1 = (S,U)$, and $J_2 = (U,T)$. Let r_i be the resolvent of k on J_i, $i = 1, 2$. Extend r_i to J^2 by defining $r_i(t,s) = 0$ for $(t,s) \in J^2 \setminus J_i^2$, and define k_i by $k_i(t,s) = k(t,s)$

for $(t,s) \in J_i^2$, $k_i(t,s) = 0$ for $(t,s) \in J^2 \setminus J_i^2$, $i = 1, 2$:

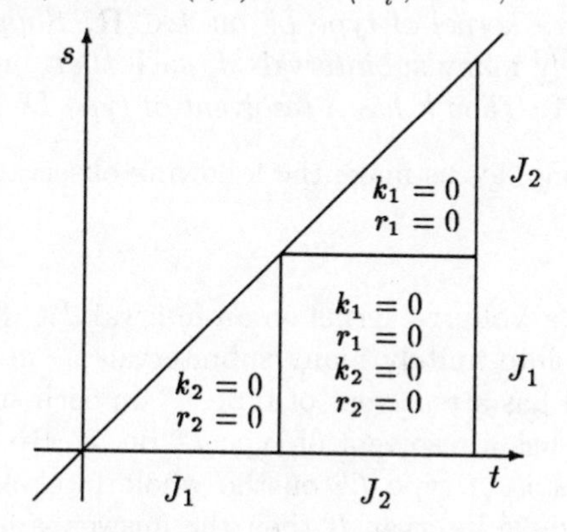

Then r_i is the resolvent of k_i on all of J, $i = 1, 2$ (and not only on J_i). Define r on J^2 by

$$r = k - k \star r_1 - r_2 \star k + r_2 \star k \star r_1. \tag{3.6}$$

Then r is of type L^p on J, $r(t,s) = 0$ for $s > t$, and since $r_1(t,s)$ vanishes for $s \in J_2$ and $r_2(t,s)$ vanishes for $t \in J_1$ we have $r = k - k \star r_1 = r_1$ on J_1^2 and $r = k - r_2 \star k = r_2$ on J_2^2. If we multiply (3.6) by k_2 on the left, and use the fact that r_2 is the resolvent of k_2, then we find that

$$\begin{aligned} k_2 \star r &= k_2 \star k - k_2 \star k \star r_1 - (k_2 - r_2) \star k + (k_2 - r_2) \star k \star r_1 \\ &= r_2 \star k - r_2 \star k \star r_1. \end{aligned}$$

Consequently (3.6) implies

$$r = k - k \star r_1 - k_2 \star r.$$

It is also true that

$$(k \star r_1)(t,s) = \chi_{J_1}(s) \int_{J_1} k(t,u) r_1(u,s)\, du = \int_{J_1} k(t,u) r(u,s)\, du$$

and that

$$(k_2 \star r)(t,s) = \chi_{J_2}(t) \int_{J_2} k_2(t,u) r(u,s)\, du = \int_{J_2} k(t,u) r(u,s)\, du,$$

where χ_{J_1} and χ_{J_2} are the characteristic functions of the intervals J_1 and J_2. Therefore, $k \star r_1 + k_2 \star r = k \star r$, and we have proved that $r = k - k \star r$. To show that r satisfies $r = k - r \star k$, one multiplies (3.6) by k_1 on the right to get $r = k - r_2 \star k - r \star k_1$, which in the same way as above can be seen to be equivalent to $r = k - r \star k$. □

We have the following immediate consequence of Theorem 3.13 and Lemma 3.8.

3.14 Corollary

Let k be a Volterra kernel of type L^p on $J \subset \mathbf{R}$. Suppose that J can be divided into finitely many subintervals J_i such that, on each J_i, it is true that $|||k|||_{L^p(J_i)} < 1$. Then k has a resolvent of type L^p on J.

Before proceeding, let us make the following observation.

3.15 Remark

Suppose that k is a Volterra kernel on an interval $J \subset \mathbf{R}$, and assume that J can be divided into finitely many subintervals J_i in such a way that k is of type L^p, and has a resolvent of type L^p on each subinterval J_i. Does this imply that k has a resolvent of type L^p on J? By Corollary 3.14, the answer is yes, if k is of type L^p on the whole interval J. If k is not of type L^p on the whole interval J, then the answer is no, as the following counterexample shows. Let $J = [0,1]$, and suppose that $k(t,s) = 0$ for all $t \le \frac{1}{2}$, and for all $s > \frac{1}{2}$. Moreover, suppose that k is not of type L^p on the whole interval J, but that k has a resolvent r of type L^p on J. We claim that this leads to a contradiction. Take $J_1 = [0,1/2]$ and $J_2 = (1/2,1]$. Clearly, k has the resolvent $r_1 \equiv 0$ on J_1, and the resolvent $r_2 \equiv 0$ on J_2. Because of the uniqueness of a resolvent of type L^p, $r(t,s)$ must vanish a.e. on the sets $0 \le s \le t \le \frac{1}{2}$ and $\frac{1}{2} < s \le t \le 1$. This implies that $(r \star k)(t,s) = (k \star r)(t,s) = 0$ for almost all $0 \le s \le t \le 1$. Therefore, by the resolvent equations, $r = k$. This contradicts the assumption that r is of type L^p on J, whereas k is not of type L^p on J.

If $1 < p < \infty$, then one may use Proposition 2.7(iii) to check whether the condition $|||k|||_{L^p(J_i)} < 1$ is satisfied. The resulting conclusion is of some importance.

3.16 Corollary

Let $1 < p < \infty$, and let $J \subset \mathbf{R}$ be an interval. Every Volterra kernel k that satisfies one of the two conditions in Proposition 2.7(iii) has a resolvent of type L^p on J.

This is true, because J can be divided into finitely many parts J_i in such a way that, when the outer integrals in Proposition 2.7(iii) are taken over J_i rather than over J, the result is strictly less than 1.

The Banach algebra $\mathcal{K}(L^p; J; \mathbf{C}^{n\times n})$ does not have a unit, although it is possible to attach an artificial unit; see Rudin [2], p. 228. Let us denote this artificial unit by ε. In an algebra $\mathcal{A}$ with unit ε one defines the spectrum of an element $k \in \mathcal{A}$ to be the set $\lambda \in \mathbf{C}$ for which $\lambda\varepsilon - k$ does not have an inverse in $\mathcal{A}$. In our case, where the unit has been artificially added, if k belongs to the original algebra $\mathcal{A}$, then $\lambda\varepsilon - k$ is invertible in the extended algebra iff $\lambda \ne 0$, and $\varepsilon - \frac{1}{\lambda}k$ is invertible. It is not difficult to see that the inverse must be of the form $\varepsilon - r$ for some $r \in \mathcal{A}$. Let us denote the multiplication in $\mathcal{A}$ by $\star$. Then $(\varepsilon - \frac{1}{\lambda}k) \star (\varepsilon - r) = (\varepsilon - r) \star (\varepsilon - \frac{1}{\lambda}k) = \varepsilon$, hence $r - \frac{1}{\lambda}k \star r = r - r \star \frac{1}{\lambda}k = -\frac{1}{\lambda}k$, i.e., r is the resolvent of $-\frac{1}{\lambda}k$. This observation motivates us to make the following definition.

3.17 Definition
Let $\mathcal{A}$ be an associative algebra without unit, and let $k \in \mathcal{A}$. A number $\lambda \in \mathbf{C}$ belongs to the spectrum $\sigma(k)$ of k (with respect to $\mathcal{A}$) if either $\lambda = 0$, or $-\frac{1}{\lambda}k$ does not have a resolvent. The spectral radius of k is the number $\sup\{\,|\lambda| \mid \lambda \in \sigma(k)\,\}$.

Our last corollary shows that the classical representation of the resolvent r as a power series (i.e., as a sum of iterated kernels) is an immediate consequence of Corollary 3.14 (whenever k satisfies the conditions of this corollary).

3.18 Corollary

(i) *Let k be as in Corollary 3.14. If we multiply k by an arbitrary complex constant λ satisfying $|\lambda| \le 1$, then the new kernel has a resolvent of type L^p. Thus, the spectral radius of k in $\mathcal{V}(L^p; J; \mathbf{C}^{n\times n})$ is less than 1.*

(ii) *Let k be as in Corollary 3.16. Then the spectral radius of k in $\mathcal{V}(L^p; J; \mathbf{C}^{n\times n})$ is zero, i.e., k is quasinilpotent.*

In both cases, r is the limit as $m \to \infty$ in $\mathcal{V}(L^p; J; \mathbf{C}^{n\times n})$ of the sums $r_m = \sum_{j=1}^{m}(-1)^{j-1}k^{\star j}$, where $k^{\star j}$ is the $(j-1)$-fold $\star$-product of k by itself.

To prove the final statement one uses the fact that the spectral radius in a Banach algebra can be obtained from the formula $\rho(k) = \lim_{n\to\infty}\|k^{\star n}\|^{1/n}$ (see Rudin [2], p. 235). Hence, there exist $\epsilon > 0$ and N such that $\|k^{\star n}\| \le 1 - \epsilon$ for $n \ge N$, and the power series converges.

Other classes of quasinilpotent kernels may be constructed; we leave the task to the reader (Exercise 11).

Let us end this section with an application of Theorem 3.13. It is a perturbation result, where we apply the Half Line Paley–Wiener Theorem to the main part of the kernel.

3.19 Theorem
Let $J = [S, \infty)$, where $-\infty < S < \infty$, and let $1 \le p \le \infty$. Suppose that k is a Volterra kernel of the form $k(t,s) = a(t-s) + b(t,s)$ for $s, t \in J$, $t \ge s$, where

(i) *$a \in L^1(\mathbf{R}^+; \mathbf{C}^{n\times n})$, and $\det\big[I + \hat{a}(z)\big] \ne 0$ for $\Re z \ge 0$,*

and

(ii) *b is a Volterra kernel of type L^p on J that has a resolvent of type L^p on every compact subinterval of J and satisfies $\lim_{T\to\infty} |||b|||_{L^p(T,\infty)} = 0$.*

Then k has a resolvent of type L^p on J.

(In particular, taking $a = 0$, one concludes that b has a resolvent of type L^p on J.)

Proof of Theorem 3.19 It follows from the Paley–Wiener Theorem 2.4.1 that a has a convolution resolvent $q \in L^1(\mathbf{R}^+; \mathbf{C}^{n\times n})$. If we define $r_1(t,s) = q(t-s)$ for $t \geq s$ and $r_1(t,s) = 0$ for $t < s$, then, by Proposition 2.7(i) and (ii), r_1 is both of type L^1 and L^∞ on J. Theorem 2.6 implies that r_1 is of type L^p on J.

Define $c(t,s) = a(t-s)$ for $t \geq s$ and $c(t,s) = 0$ for $t < s$. Then $k = c+b$, and r_1 is the resolvent of c on every interval $K \subset J$. Using the asymptotic assumption on b, and Theorem 3.9, we find that k has a resolvent of type L^p on an interval $[T, \infty)$ for some sufficiently large T.

It follows from Proposition 2.7(i) and (ii) that the norms $|||c|||_{L^1(s,s+h)}$ and $|||c|||_{L^\infty(s,s+h)}$ tend to zero as $h \to 0$, uniformly for $s \in J$. This, together with Theorem 2.6, implies that $|||c|||_{L^p(s,s+h)} \to 0$ as $h \to 0$, uniformly for $s \in J$. Therefore, there exist intervals J_i, $i = 1, \ldots, N$, with $\cup_{i=1}^N J_i = (S,T)$, such that $|||c|||_{L^p(J_i)} < \left(1 + |||r_2|||_{L^p(J_i)}\right)^{-1}$, $i = 1, \ldots, N$, where r_2 is the resolvent of b. By Theorem 3.9, k has a resolvent of type L^p on every J_i, $i = 1, \ldots, N$.

According to Theorem 3.13, k has a resolvent of type L^p on J. □

4. Volterra Kernels of Type L^p_{loc}

In the Volterra case we can formulate local versions of many of the preceding theorems.

4.1 Definition

Let $J \subset \mathbf{R}$ be an interval, and let $1 \leq p \leq \infty$. A function $k : J^2 \to \mathbf{C}^{n\times n}$ is called a (Volterra) kernel of type $[L^p_{\text{loc}}; J; \mathbf{C}^{n\times n}]$ if for every compact subinterval K of J the restriction of k to K^2 is a Volterra kernel of type $[L^p; K; \mathbf{C}^{n\times n}]$. The class of all (Volterra) kernels of type $[L^p_{\text{loc}}; J; \mathbf{C}^{n\times n}]$ is denoted by $\mathcal{V}(L^p_{\text{loc}}; J; \mathbf{C}^{n\times n})$.

It is easy to see that $\mathcal{V}(L^p_{\text{loc}}; J; \mathbf{C}^{n\times n})$ is an associative algebra, since the restriction of $a \star b$ to $K \times K$, where $K \subset J$, depends only on the values of a and b on $K \times K$. (The corresponding statement is not true for Fredholm kernels, and this makes it impossible to develop a local Fredholm theory.) As before, we define $r \in \mathcal{V}(L^p_{\text{loc}}; J; \mathbf{C}^{n\times n})$ to be the resolvent of $k \in \mathcal{V}(L^p_{\text{loc}}; J; \mathbf{C}^{n\times n})$ if $r + k \star r = r + r \star k = k$. We see from Lemma 3.3 that if k is of type L^p_{loc} then k can have at most one resolvent of type L^p_{loc}.

As usual, if we have a resolvent, then we can solve our equation (cf. Lemma 3.4).

4.2 Theorem

Let J be an interval of the type $[S,T)$, where $-\infty < S < T \leq \infty$. If k is a kernel of type L^p_{loc} on J which has a resolvent of type L^p_{loc} on J, and if $f \in L^p_{\mathrm{loc}}(J;\mathbf{C}^n)$, then the function x defined in (3.4) is the unique solution of (3.3) in $L^p_{\mathrm{loc}}(J;\mathbf{C}^n)$.

A local version of Corollary 3.12 may easily be formulated, and one can introduce kernels of type $[L^p_{\mathrm{loc}}; L^q_{\mathrm{loc}}; J; \mathbf{C}^{n\times n}]$. We omit the exact statements.

The local version of Corollary 3.14, combined with the final statement in Corollary 3.18, is of great practical significance, and reads as follows.

4.3 Corollary

Let k be a kernel of type L^p_{loc} on $J \subset \mathbf{R}$. Suppose that each compact subinterval K of J can be divided into finitely many subintervals K_i, such that, on each K_i, it is true that $|||k|||_{L^p(K_i)} < 1$. Then k has a resolvent of type L^p_{loc} on J. Moreover, r is the limit as $m \to \infty$ in $\mathcal{V}(L^p_{\mathrm{loc}}; J; \mathbf{C}^{n\times n})$ of the sums $r_m = \sum_{j=1}^m (-1)^{j-1} k^{\star j}$, where $k^{\star j}$ is the $(j-1)$-fold $\star$-product of k by itself.

Explicitly, the last claim is that $\lim_{m\to\infty} |||r - r_m|||_{L^p(K)} = 0$ for every compact subinterval K of J.

In the local results above, we have worked throughout on compact subintervals of J. When J is an interval of the form $J = (-\infty, T)$, with $-\infty < T \leq \infty$, then one can equally well work on subintervals of the type $(-\infty, U]$, where $-\infty < U < T$, and results analogous to Theorem 4.2 and Corollary 4.3 are true in this setting.

5. Kernels of Bounded and Continuous Types

In the preceding sections we considered (1.1) in different L^p-spaces over an arbitrary interval J. Another important task (especially from the point of view of applications), is to analyse this equation in the setting of continuous functions, and in the setting of bounded Borel measurable functions on J. This will be done in this, and in the next section.

First we define kernels of bounded type. Compared with kernels of type L^∞, the only difference is that we require a certain amount of Borel measurability, and that we use supremum in the norm, instead of essential supremum.

5.1 Definition
Let $J \subset \mathbf{R}$ be an interval. A function $k : J^2 \mapsto \mathbf{C}^{n\times n}$ is called a kernel of type $[B^\infty; J; \mathbf{C}^{n\times n}]$ (or of bounded type on J) if k is (Lebesgue) measurable and satisfies the following conditions: for every $t \in J$ the function $s \mapsto k(t,s)$ is integrable, for every $f \in L^\infty(J;\mathbf{C}^n)$ the function $t \mapsto \int_J k(t,s)f(s)\,\mathrm{d}s$ belongs to $B^\infty(J;\mathbf{C}^n)$ (i.e., it is bounded and Borel measurable), and $|||k|||_{B^\infty(J)} < \infty$, where

$$|||k|||_{B^\infty(J)} \stackrel{\mathrm{def}}{=} \sup_{t\in J} \int_J |k(t,s)|\,\mathrm{d}s.$$

We denote the class of Fredholm kernels of bounded type on J by $\mathcal{F}(B^\infty; J; \mathbf{C}^{n\times n})$, and the class of Volterra kernels of bounded type on J by $\mathcal{V}(B^\infty; J; \mathbf{C}^{n\times n})$. The class of kernels that belong to $\mathcal{V}(B^\infty; K; \mathbf{C}^{n\times n})$ for every compact interval $K \subset J$ is denoted by $\mathcal{V}(B^\infty_{\mathrm{loc}}; J; \mathbf{C}^{n\times n})$.

Obviously, every kernel of bounded type induces a kernel of type L^∞. Conversely, given any kernel of type L^∞, one can always find a kernel (in fact, many different kernels) of bounded type that is equal to the given kernel in the L^∞-sense (e.g., one can redefine the given kernel on a set of measure zero so that it becomes a Borel measurable function).

Next, we define kernels of continuous type.

5.2 Definition
Let $J \subset \mathbf{R}$ be an interval. A function $k : J^2 \mapsto \mathbf{C}^{n\times n}$ is called a kernel of type $[C; J; \mathbf{C}^{n\times n}]$ (continuous type on J), $[BC; J; \mathbf{C}^{n\times n}]$ (bounded continuous type on J), or $[BUC; J; \mathbf{C}^{n\times n}]$ (bounded uniformly continuous type on J), if k is measurable, if for every $t \in J$ the function $s \mapsto k(t,s)$ is integrable, and if the function $t \mapsto (s \mapsto k(t,s))$ is, respectively, a continuous, a bounded and continuous, or a bounded and uniformly continuous function, from J into $L^1(J;\mathbf{C}^{n\times n})$.

We denote the classes of Fredholm kernels of continuous, bounded continuous, and bounded uniformly continuous types on J by $\mathcal{F}(C; J; \mathbf{C}^{n\times n})$, $\mathcal{F}(BC; J; \mathbf{C}^{n\times n})$ and $\mathcal{F}(BUC; J; \mathbf{C}^{n\times n})$, respectively, and the corresponding classes of Volterra kernels by $\mathcal{V}(C; J; \mathbf{C}^{n\times n})$, $\mathcal{V}(BC; J; \mathbf{C}^{n\times n})$ and $\mathcal{V}(BUC; J; \mathbf{C}^{n\times n})$.

We use the norm $|||\cdot|||_{B^\infty(J)}$ (or $|||\cdot|||_{L^\infty(J)}$) in the classes $\mathcal{K}(BC; J; \mathbf{C}^{n\times n})$ and $\mathcal{K}(BUC; J; \mathbf{C}^{n\times n})$. Unless J is compact, the class $\mathcal{K}(C; J; \mathbf{C}^{n\times n})$ is not a Banach space, but one can easily define a metric on this space.

It is clear that one could, in addition, define kernels of uniformly (but not necessarily bounded) continuous type.

Again, we let the letter $\mathcal{K}$ represent either $\mathcal{F}$ or $\mathcal{V}$ if a result is true for both classes of kernels. Explicitly, the continuity assumption in Definition 5.2 means that, for each $t \in J$,

$$\lim_{h\to 0} \int_J |k(t+h,s) - k(t,s)|\,\mathrm{d}s = 0.$$

At all interior points of J this is a two-sided limit, and at an end-point of J that belongs to J it is a one-sided limit. In the uniform case, the approach to the limit should be uniform in t. Clearly, every kernel of bounded continuous type is a kernel of bounded type. Observe that, if J is closed and bounded (i.e., compact), then every kernel of continuous type on J must be of bounded uniformly continuous type on J.

Every (Volterra type) convolution kernel in $L^1_{\mathrm{loc}}(\mathbf{R}^+;\mathbf{C}^{n\times n})$ is of continuous type on every interval which is bounded to the left, and every (Fredholm type) convolution kernel in $L^1(\mathbf{R};\mathbf{C}^{n\times n})$ is of bounded uniformly continuous type on $\mathbf{R}$.

The following theorem gives the basic properties of our new classes of kernels.

5.3 Theorem

Let $J \subset \mathbf{R}$ be an interval.

(i) *If a is a kernel of one of the types $[B^\infty;J;\mathbf{C}^{n\times n}]$, $[C;J;\mathbf{C}^{n\times n}]$, $[BC;J;\mathbf{C}^{n\times n}]$, or $[BUC;J;\mathbf{C}^{n\times n}]$, and b is a kernel of type L^∞ on J, then $a \star b$ is a kernel of the same type as a.*

(ii) *$\mathcal{K}(B^\infty;J;\mathbf{C}^{n\times n})$, $\mathcal{K}(BC;J;\mathbf{C}^{n\times n})$, and $\mathcal{K}(BUC;J;\mathbf{C}^{n\times n})$ are Banach algebras, and $\mathcal{K}(BC;J;\mathbf{C}^{n\times n})$, and $\mathcal{K}(BUC;J;\mathbf{C}^{n\times n})$ are closed subalgebras of $\mathcal{K}(B^\infty;J;\mathbf{C}^{n\times n})$.*

(iii) *If k is a kernel of continuous (bounded continuous) [bounded uniformly continuous] type on J, then the operator $f \mapsto k \star f$ maps bounded sets in $L^\infty(J;\mathbf{C}^n)$ into equicontinuous (bounded and equicontinuous) [bounded and uniformly equicontinuous] sets of functions on J. In particular, if J is bounded and k is of bounded uniformly continuous type on J, then $f \mapsto k \star f$ is compact from $L^\infty(J;\mathbf{C}^n)$ into $BUC(J;\mathbf{C}^n)$.*

It follows from (i) and (ii) that $\mathcal{K}(BC;J;\mathbf{C}^{n\times n})$ and $\mathcal{K}(BUC;J;\mathbf{C}^{n\times n})$ are closed left ideals in $\mathcal{K}(B^\infty;J;\mathbf{C}^{n\times n})$.

Proof of Theorem 5.3 (i) If a is a kernel of bounded type or of continuous type, and b is a kernel of type $[L^\infty;J;\mathbf{C}^{n\times n}]$, then one sees from Proposition 2.7 that, for all $t \in J$,

$$\int_J |a(t,u)| \int_J |b(u,s)|\,\mathrm{d}s\,\mathrm{d}u < \infty.$$

Therefore, by Fubini's theorem, for all $t \in J$, $(a \star b)(t,s)$ is well defined for almost all $s \in J$. Moreover, in the case where a is a kernel of bounded type, using Fubini's theorem one can easily show that $a \star b$ has all the properties required from a kernel of type $[B^\infty;J;\mathbf{C}^{n\times n}]$, so $a \star b \in \mathcal{K}(B^\infty;J;\mathbf{C}^{n\times n})$. In particular, one finds that

$$|\!|\!| a \star b |\!|\!|_{B^\infty(J)} \le |\!|\!| a |\!|\!|_{B^\infty(J)} |\!|\!| b |\!|\!|_{L^\infty(J)}. \tag{5.1}$$

In the continuous cases, one observes that, whenever both t and $t+h$ belong to J,

$$\begin{aligned}\int_J \big|(a\star b)(t+h,s)-(a\star b)(t,s)\big|\,\mathrm{d}s &\le \int_J\int_J \big|a(t+h,u)-a(t,u)\big|\big|b(u,s)\big|\,\mathrm{d}u\,\mathrm{d}s\\ &\le |||b|||_{L^\infty(J)}\int_J \big|a(t+h,u)-a(t,u)\big|\,\mathrm{d}u.\end{aligned}$$

Therefore, in these cases, the function $t\mapsto(s\mapsto(a\star b)(t,s))$ has the same continuity properties from J into L^1 as the function $t\mapsto(s\mapsto a(t,s))$. This proves (i).

(ii) It is easy to show that the three different classes of kernels are normed spaces, and by (i) they are algebras. The norm in the three spaces is the same, and the standard norm inequality clearly follows from (5.1). Thus to prove (ii) it suffices to show that $\mathcal{K}(B^\infty;J;\mathbf{C}^{n\times n})$ is complete, and that $\mathcal{K}(BC;J;\mathbf{C}^{n\times n})$ and $\mathcal{K}(BUC;J;\mathbf{C}^{n\times n})$ are closed subspaces of $\mathcal{K}(B^\infty;J;\mathbf{C}^{n\times n})$.

To show that $\mathcal{K}(B^\infty;J;\mathbf{C}^{n\times n})$ is complete, we note that if $\{k_m\}$ is a Cauchy sequence in this space then the sequence $\{k_m(t,\cdot)\}$ is, for each $t\in J$, a Cauchy sequence in $L^1(J;\mathbf{C}^{n\times n})$. Consequently, there exists a function k such that $\lim_{m\to\infty}\int_J|k_m(t,s)-k(t,s)|\,\mathrm{d}s=0$ for each $t\in J$. Now it is straightforward to show that k is a kernel of type B^∞, and that $|||k_m-k|||_{B^\infty(J)}\to 0$ as $m\to\infty$.

To realize that $\mathcal{K}(BC;J;\mathbf{C}^{n\times n})$ and $\mathcal{K}(BUC;J;\mathbf{C}^{n\times n})$ are closed subspaces of $\mathcal{K}(B^\infty;J;\mathbf{C}^{n\times n})$, it suffices to observe that the limit of a uniformly convergent sequence of (uniformly) continuous functions is (uniformly) continuous.

(iii) If k is a kernel of bounded continuous type on J and $f\in L^\infty(J;\mathbf{C}^n)$, and both t and $t+h$ belong to J, then

$$|(k\star f)(t+h)-(k\star f)(t)|\le \|f\|_{L^\infty(J)}\int_J |k(t+h,s)-k(t,s)|\,\mathrm{d}s.$$

Now apply the Arzelà–Ascoli Theorem **2**.7.5. □

As an immediate consequence we have the following theorem.

5.4 Theorem

If k is a kernel of bounded type (bounded continuous type) [bounded uniformly continuous type] on an interval $J\subset\mathbf{R}$, and k has a resolvent r of type L^∞ on J, then r is of the same type as k (after redefinition on a set of measure zero). Moreover, for every bounded and Borel measurable (bounded and continuous) [bounded and uniformly continuous] function f on J, there is a unique bounded and Borel measurable (bounded and continuous) [bounded and uniformly continuous] solution of (3.3). This solution is given by (3.4).

Proof If we let r_1 denote the (almost everywhere defined) resolvent of k of type L^∞, and define r by $r = k - k \star r_1$, then, by Theorem 5.3(i), r is of the same type as k, and, in the L^∞-sense, r defines the same kernel as r_1. In particular, $r + k \star r = k$. Thus, in order to show that r is the resolvent of k in the algebra $\mathcal{K}(B^\infty; J; \mathbf{C}^{n\times n})$, it only remains to show that $r \star k = k \star r$ (in the sense of equality in $\mathcal{K}(B^\infty; J; \mathbf{C}^{n\times n})$). To do this, we observe on the one hand that, if we take the $\star$-product of the equation $r + k \star r = k$ and k, we get

$$r \star k = k \star k - k \star r \star k.$$

On the other hand, taking the $\star$-product of k and the equation $r_1 + r_1 \star k = k$, we get

$$k \star r_1 + k \star r_1 \star k = k \star k,$$

or equivalently,

$$k \star r = k \star k - k \star r \star k.$$

This shows that $r \star k = k \star r$, and completes the proof of the fact that r is the resolvent of k in $\mathcal{K}(B^\infty; J; \mathbf{C}^{n\times n})$.

The existence and uniqueness of the solution of (3.3) follow from Lemma 3.4. □

In the remainder of this section we consider Volterra kernels only. For these kernels one has the following important result.

5.5 Theorem

(i) *Every Volterra kernel of continuous type on some interval J of the type $J = [S, T)$, with $-\infty < S < T \le \infty$, has a resolvent of continuous type on J.*

(ii) *Every Volterra kernel of bounded uniformly continuous type on some bounded interval $J \subset \mathbf{R}$ has a resolvent of bounded uniformly continuous type on J.*

Proof (i) Because of the Volterra nature of the kernel and of the resolvent, it suffices to show that, on each compact subinterval $K \subset J$, it is true that k has a resolvent of continuous type on K. However, this follows directly from (ii) (every kernel of continuous type on a compact interval is necessarily of bounded uniformly continuous type). Thus, it suffices to prove (ii).

(ii) By the uniform continuity of the function $t \mapsto (s \mapsto k(t,s))$ from J into $L^1(J; \mathbf{C}^{n\times n})$, we can find a constant $h > 0$ such that if both v and t belong to J, with $0 \le t - v \le h$, then

$$\int_v^t |k(t,s)|\,ds = \int_v^t |k(t,s) - k(v,s)|\,ds \le \int_J |k(t,s) - k(v,s)|\,ds \le \tfrac{1}{2}.$$

Now apply Corollary 3.14 with $p = \infty$ to get the existence of a resolvent of type L^∞, and then use Theorem 5.4. □

One has the following result (cf. Theorem 4.2) on the existence of continuous solutions to the Volterra equation (3.3).

5.6 Corollary
Let J be an interval of the type $[S,T)$, where $-\infty < S < T \le \infty$. Let k be a Volterra kernel of continuous type on J, and let $f \in C(J; \mathbf{C}^n)$. Then (3.3) has a unique continuous solution on J. This solution is given by (3.4).

So far we have only considered kernels that are continuous in t with respect to the L^1-norm in s. Of course, one can also study the question of pointwise continuity of the resolvent on the triangle

$$\Delta_2(J) = \{\, (t,s) \in J^2 \mid s \le t \,\},$$

and, in particular, whether pointwise continuity of the kernel implies pointwise continuity of the resolvent. As might be expected, this is in fact true.

5.7 Theorem
Let $J \subset \mathbf{R}$ be an interval. Let $k : \Delta_2(J) \to \mathbf{C}^{n\times n}$ be continuous. Then the Volterra kernel k has a resolvent which is a continuous function from $\Delta_2(J)$ into $\mathbf{C}^{n\times n}$.

Proof It suffices to demonstrate that r is continuous on $\Delta_2(K)$ for every compact subinterval K of J. First observe that k is of continuous type on K (if we let $k(t,s) = 0$ for $s > t$), hence, by Theorem 5.5(i), k has a resolvent r of continuous type on K. Thus, we only have to prove that r is continuous on $\Delta_2(J)$. Now, k is bounded and uniformly continuous on $\Delta_2(K)$, and since $r = k - r \star k$ on $\Delta_2(K)$, we find that r is continuous on $\Delta_2(K)$ if $r \star k$ is. We know that r is of bounded uniformly continuous type on K. For all $s,t,u,v \in K$,

$$\begin{aligned}
&|(r\star k)(v,u) - (r\star k)(t,s)| \\
&\quad\le \int_K |r(v,\sigma)k(\sigma,u) - r(t,\sigma)k(\sigma,s)|\,\mathrm{d}\sigma \\
&\quad\le \int_K |r(v,\sigma) - r(t,\sigma)||k(\sigma,u)|\,\mathrm{d}\sigma + \int_K |r(t,\sigma)||k(\sigma,u) - k(\sigma,s)|\,\mathrm{d}\sigma \\
&\quad\le \sup_{\tau\in K}|k(\tau,u)| \int_K |r(v,\sigma) - r(t,\sigma)|\,\mathrm{d}\sigma \\
&\qquad + \sup_{\tau\in K}|k(\tau,u) - k(\tau,s)| \int_K |r(t,\sigma)|\,\mathrm{d}\sigma.
\end{aligned}$$

The right-hand side tends to zero as $v \to t$ and $u \to s$, and therefore $r \star k$ is continuous on $\Delta_2(K)$. □

6. Some Special Classes of Kernels

In this section we shall look at various special classes of kernels. These classes of kernels are closed subalgebras of $\mathcal{K}(B^\infty; J; \mathbf{C}^{n\times n})$, and they map various subspaces of $B^\infty(J; \mathbf{C}^n)$ back into themselves. As before, $\mathcal{K}$ stands for either $\mathcal{F}$ or $\mathcal{V}$. Throughout this section we suppose that the right end-point of the interval J is $+\infty$ but even in the case where $J = \mathbf{R}$ we will not consider limits at $-\infty$, that is, functions in $B_0^\infty(\mathbf{R})$ and $B_\ell^\infty(\mathbf{R})$ are not required to have limits at $-\infty$.

First, let us look at a class of kernels that map $B_0^\infty(J; \mathbf{C}^n)$ back into itself.

6.1 Definition

Let $J \subset \mathbf{R}$ be an interval whose right end-point is $+\infty$. A function $k : J^2 \mapsto \mathbf{C}^{n\times n}$ is called a kernel of type $[B_0^\infty; J; \mathbf{C}^{n\times n}]$ if k is a kernel of type $[B^\infty; J; \mathbf{C}^{n\times n}]$, and if, in addition, $\lim_{t\to\infty} \int_{s\le S, s\in J} |k(t,s)|\,\mathrm{d}s = 0$ for each $S \in J$.

We denote the class of kernels of type B_0^∞ on J by $\mathcal{K}(B_0^\infty; J; \mathbf{C}^{n\times n})$. Furthermore, we write $\mathcal{K}(BC_0; J; \mathbf{C}^{n\times n}) = \mathcal{K}(B_0^\infty; J; \mathbf{C}^{n\times n}) \cap \mathcal{K}(C; J; \mathbf{C}^{n\times n})$.

Clearly, every (Volterra type) convolution kernel in $L^1(\mathbf{R}^+; \mathbf{C}^{n\times n})$, and every (Fredholm type) convolution kernel in $L^1(\mathbf{R}; \mathbf{C}^{n\times n})$, is of type BC_0. Moreover, if a Borel measurable kernel k on an interval J with right end-point $+\infty$ satisfies $|k(t,s)| \le a(t-s)$ for some $a \in L^1(\mathbf{R})$, all t, and almost all s, then k is of type B_0^∞ on J.

The class $\mathcal{K}(B_0^\infty; J; \mathbf{C}^{n\times n})$ is an algebra.

6.2 Theorem

The class $\mathcal{K}(B_0^\infty; J; \mathbf{C}^{n\times n})$ of kernels of type B_0^∞ on an interval J is a closed subalgebra of the Banach algebra $\mathcal{K}(B^\infty; J; \mathbf{C}^{n\times n})$ of kernels of bounded type on J, and $B_0^\infty(J; \mathbf{C}^n)$ is a left Banach module over $\mathcal{K}(B_0^\infty; J; \mathbf{C}^{n\times n})$.

Proof Without loss of generality, let $J = R$ (define all functions to be zero outside of their original domain).

Clearly, $\mathcal{K}(B_0^\infty; J; \mathbf{C}^{n\times n})$ is closed under addition and scalar multiplication, so it is a linear subspace of $\mathcal{K}(B^\infty; J; \mathbf{C}^{n\times n})$. This means that there are essentially only three things to show, namely that this subspace is closed, that $a \star b \in \mathcal{K}(B_0^\infty; J; \mathbf{C}^{n\times n})$ if both a and b belong to $\mathcal{K}(B_0^\infty; J; \mathbf{C}^{n\times n})$, and that $a \star f \in B_0^\infty(J; \mathbf{C}^n)$ if $a \in \mathcal{K}(B_0^\infty; J; \mathbf{C}^{n\times n})$ and $f \in B_0^\infty(J; \mathbf{C}^n)$.

First we show that $\mathcal{K}(B_0^\infty; J; \mathbf{C}^{n\times n})$ is closed in $\mathcal{K}(B^\infty; J; \mathbf{C}^{n\times n})$. Let $a_n \in \mathcal{K}(B_0^\infty; J; \mathbf{C}^{n\times n})$, $n \ge 1$, and suppose that $a_n \to a$ in $\mathcal{K}(B^\infty; J; \mathbf{C}^{n\times n})$ as $n \to \infty$. Fix $S \in J$. Given $\epsilon > 0$, there exist an integer $n \ge 1$ such that $\int_{s\le S} |a_n(t,s) - a(t,s)|\,\mathrm{d}s \le \int_J |a_n(t,s) - a(t,s)|\,\mathrm{d}s \le \epsilon$ for $t \in J$, and a constant $T \in J$ such that $\int_{s\le S} |a_n(t,s)|\,\mathrm{d}s \le \epsilon$ for $t \ge T$. Thus,

$\int_{s\le S}|a(t,s)|\,\mathrm{d}s \le 2\epsilon$, $t \ge T$, and we conclude that $a \in \mathcal{K}(B_0^\infty; J; \mathbf{C}^{n\times n})$. This proves that $\mathcal{K}(B_0^\infty; J; \mathbf{C}^{n\times n})$ is closed in $\mathcal{K}(B^\infty; J; \mathbf{C}^{n\times n})$.

Next, let us show that if $a \in \mathcal{K}(B_0^\infty; J; \mathbf{C}^{n\times n})$ and $f \in B_0^\infty(J; \mathbf{C}^n)$, then $a \star f \in B_0^\infty(J; \mathbf{C}^n)$. Given $\epsilon > 0$, we can find constants S and T in J such that $|f(s)| \le \epsilon$ for $s \ge S$, $T \ge S$, and such that $\int_{s\le S}|a(t,s)|\,\mathrm{d}s \le \epsilon$ for all $t \ge T$. Then, for $t \ge T$,

$$\begin{aligned}
|(a\star f)(t)| &= \left|\int_J a(t,s)f(s)\,\mathrm{d}s\right| \\
&\le \int_{s\le S}|a(t,s)||f(s)|\,\mathrm{d}s + \int_{s\ge S}|a(t,s)||f(s)|\,\mathrm{d}s \\
&\le \epsilon\|f\|_{B^\infty(J)} + \epsilon|||a|||_{B^\infty(J)}.
\end{aligned}$$

Since ϵ is arbitrary, we conclude that $a \star f \in B_0^\infty(J; \mathbf{C}^n)$.

It remains to show that $a \star b \in \mathcal{K}(B_0^\infty; J; \mathbf{C}^{n\times n})$ if both a and b belong to $\mathcal{K}(B_0^\infty; J; \mathbf{C}^{n\times n})$. Fix $S \in J$. Given $\epsilon > 0$ we can choose V so large that $\int_{s\le S}|b(v,s)|\,\mathrm{d}s \le \epsilon$ for all $v \ge V$, and then T so large that $\int_{v\le V}|a(t,v)|\,\mathrm{d}v \le \epsilon$ for all $t \ge T$. Then, for $t \ge T$,

$$\begin{aligned}
\int_{s\le S}|(a\star b)(t,s)|\,\mathrm{d}s &\le \int_{s\le S}\int_J |a(t,v)||b(v,s)|\,\mathrm{d}v\,\mathrm{d}s \\
&\le \int_{v\le V}|a(t,v)|\int_{s\le S}|b(v,s)|\,\mathrm{d}s\,\mathrm{d}v \\
&\quad + \int_{v\ge V}|a(t,v)|\int_{s\le S}|b(v,s)|\,\mathrm{d}s\,\mathrm{d}v \\
&\le \epsilon|||b|||_{B^\infty(J)} + \epsilon|||a|||_{B^\infty(J)},
\end{aligned}$$

and we conclude that $a \star b \in \mathcal{K}(B_0^\infty; J; \mathbf{C}^{n\times n})$. □

Because of the algebraic nature of $\mathcal{K}(B_0^\infty; J; \mathbf{C}^{n\times n})$, the interesting question is, of course, what assumptions are needed on $k \in \mathcal{K}(B_0^\infty; J; \mathbf{C}^{n\times n})$ for k to have a resolvent $r \in \mathcal{K}(B_0^\infty; J; \mathbf{C}^{n\times n})$. An immediate consequence of Lemma 3.8 and Theorem 6.2 is the following simplistic result.

6.3 Theorem

Let J be an interval whose right end-point is $+\infty$. Let $k \in \mathcal{K}(B_0^\infty; J; \mathbf{C}^{n\times n})$ with $|||k|||_{B^\infty(J)} < 1$. Then k has a resolvent $r \in \mathcal{K}(B_0^\infty; J; \mathbf{C}^{n\times n})$.

To get another example of a kernel of type $[B_0^\infty; \mathbf{R}^+; \mathbf{C}^{n\times n}]$ whose resolvent is of the same type, one may combine Theorem 8.6 with inequality (8.6). Some further results in this direction are given in Theorems 7.5 and 11.1.

An important subclass of B_0^∞ can be defined as follows.

6.4 Definition
Let $J \subset \mathbf{R}$ be an interval whose right end-point is $+\infty$. A function $k : J^2 \mapsto \mathbf{C}^{n\times n}$ is called a kernel of type $[B_\ell^\infty; J; \mathbf{C}^{n\times n}]$ (or of type B_ℓ^∞ on J) if k is a kernel of type B_0^∞ on J, and if, in addition, $\lim_{t\to\infty} \int_J k(t,s)\,\mathrm{d}s$ exists. We denote the class of kernels of type B_ℓ^∞ on J by $\mathcal{K}(B_\ell^\infty; J; \mathbf{C}^{n\times n})$.

Every (Volterra type) convolution kernel in $L^1(\mathbf{R}^+; \mathbf{C}^{n\times n})$, and every (Fredholm type) convolution kernel in $L^1(\mathbf{R}; \mathbf{C}^{n\times n})$, is of type BC_ℓ.

Some properties of $\mathcal{K}(B_\ell^\infty; J; \mathbf{C}^{n\times n})$ are given in the following theorem.

6.5 Theorem
The class $\mathcal{K}(B_\ell^\infty; J; \mathbf{C}^{n\times n})$ of kernels of type B_ℓ^∞ on an interval J is a closed subalgebra of the Banach algebra $\mathcal{K}(B^\infty; J; \mathbf{C}^{n\times n})$ of kernels of bounded type on J, and $B_\ell^\infty(J; \mathbf{C}^n)$ is a left Banach module over $\mathcal{K}(B_\ell^\infty; J; \mathbf{C}^{n\times n})$.

If a and b both belong to $\mathcal{K}(B_\ell^\infty; J; \mathbf{C}^{n\times n})$ and $f \in B_\ell^\infty(J; \mathbf{C}^n)$, and $\int_J a(t,s)\,\mathrm{d}s \to A$, $\int_J b(t,s)\,\mathrm{d}s \to B$ and $f(t) \to F$ as $t \to \infty$, then $\int_J (a \star b)(t,s)\,\mathrm{d}s \to AB$ and $(a \star f)(t) \to AF$ as $t \to \infty$.

Proof Without loss of generality, let $J = R$.

That $\mathcal{K}(B_\ell^\infty; J; \mathbf{C}^{n\times n})$ is a closed subspace of $\mathcal{K}(B^\infty; J; \mathbf{C}^{n\times n})$ can be proved with an argument similar to the corresponding argument in the proof of Theorem 6.2.

Let $f(t) \to F$ and $\int_J a(t,s)\,\mathrm{d}s \to A$ as $t \to \infty$. Then, given $\epsilon > 0$, we can find $S \in J$ such that $|f(s) - F| \le \epsilon$ for all $s \ge S$, and $T \ge S$ such that $\int_{s\le S} |a(t,s)|\,\mathrm{d}s \le \epsilon$ and $|\int_J a(t,s)\,\mathrm{d}s - A| \le \epsilon$ for all $t \ge T$. Then, for $t \ge T$,

$$\begin{aligned}
|(a \star f)(t) - AF| &\le \int_J |a(t,s)||f(s) - F|\,\mathrm{d}s + \left|\int_J a(t,s)\,\mathrm{d}s - A\right||F| \\
&\le \int_{s\le S} |a(t,s)||f(s) - F|\,\mathrm{d}s \\
&\quad + \int_{s\ge S} |a(t,s)||f(s) - F|\,\mathrm{d}s + \epsilon|F| \\
&\le 3\epsilon\|f\|_{B^\infty(J)} + \epsilon|\!|\!|a|\!|\!|_{B^\infty(J)},
\end{aligned}$$

and we conclude that $(a \star f)(t) \to AF$.

It remains to show that $\int_J (a \star b)(t,s)\,\mathrm{d}s \to AB$ as $t \to \infty$ if both a and b belong to $\mathcal{K}(B_\ell^\infty; J; \mathbf{C}^{n\times n})$, and $\int_J a(t,s)\,\mathrm{d}s \to A$ and $\int_J b(t,s)\,\mathrm{d}s \to B$ as $t \to \infty$. Given $\epsilon > 0$ we can find $V \in J$ such that $\left|\int_J b(v,s)\,\mathrm{d}s - B\right| \le \epsilon$ if $v \ge V$, and $T \ge V$ such that $\int_{v\le V} |a(t,v)|\,\mathrm{d}v \le \epsilon$ if $t \ge T$. Then, for each $t \ge T$,

$$\begin{aligned}
&\left|\int_J (a \star b)(t,s)\,\mathrm{d}s - AB\right| \\
&\qquad = \left|\int_J a(t,v) \int_J b(v,s)\,\mathrm{d}s\,\mathrm{d}v - AB\right|
\end{aligned}$$

$$\begin{aligned}
&\leq \int_J |a(t,v)| \left| \int_J b(v,s)\,\mathrm{d}s - B \right| \mathrm{d}v + \left| \int_J a(t,v)\,\mathrm{d}v - A \right| |B| \\
&\leq \int_{v\leq V} |a(t,v)| \left| \int_J b(v,s)\,\mathrm{d}s - B \right| \mathrm{d}v \\
&\quad + \int_{v\geq V} |a(t,v)| \left| \int_J b(v,s)\,\mathrm{d}s - B \right| \mathrm{d}v + \epsilon |B| \\
&\leq 3\epsilon |||b|||_{B^\infty(J)} + \epsilon |||a|||_{B^\infty(J)}
\end{aligned}$$

and we conclude that $\int_J (a \star b)(t,s)\,\mathrm{d}s \to AB$ as $t \to \infty$. □

Sometimes one needs kernels that map the space $L^\infty(J)$ into $B_0^\infty(J)$ (in particular, they map $B^\infty(J)$ into $B_0^\infty(J)$).

6.6 Definition

Let $J \subset \mathbf{R}$ be an interval whose right end-point is $+\infty$. A function $k : J^2 \mapsto \mathbf{C}^{n\times n}$ is called a kernel of type $[L^\infty; B_0^\infty; J; \mathbf{C}^{n\times n}]$ if k is a kernel of type B^∞ on J, and if, in addition, $\lim_{t\to\infty} \int_J |k(t,s)|\,\mathrm{d}s = 0$. We denote the class of kernels of type $[L^\infty; B_0^\infty]$ on J by $\mathcal{K}(L^\infty; B_0^\infty; J; \mathbf{C}^{n\times n})$.

The proof of the following result is left to the reader.

6.7 Theorem

Let $J \subset \mathbf{R}$ be an interval.

(i) *If a is a kernel of type $[L^\infty; B_0^\infty; J; \mathbf{C}^{n\times n}]$, and b is a kernel of type L^∞ on J, then $a \star b$ is a kernel of type $[L^\infty; B_0^\infty; J; \mathbf{C}^{n\times n}]$.*

(ii) *If a is a kernel of type $[B_0^\infty; J; \mathbf{C}^{n\times n}]$ and b is a kernel of type $[L^\infty; B_0^\infty; J; \mathbf{C}^{n\times n}]$, then $a \star b$ is a kernel of type $[L^\infty; B_0^\infty; J; \mathbf{C}^{n\times n}]$.*

(iii) *$\mathcal{K}(L^\infty; B_0^\infty; J; \mathbf{C}^{n\times n})$ is a closed subalgebra of the Banach algebra $\mathcal{K}(B_0^\infty; J; \mathbf{C}^{n\times n})$.*

Next, we briefly consider periodic kernels.

6.8 Definition

Let $S > 0$. A function $k : \mathbf{R}^2 \to \mathbf{C}^{n\times n}$ is called a kernel of type $[B^\infty; \mathcal{T}_S; \mathbf{C}^{n\times n}]$ (or of bounded periodic type with period S) if k is a kernel of bounded type on $\mathbf{R}$, and if, in addition, for all $t \in \mathbf{R}$ and almost all s, $k(t+S, s+S) = k(t,s)$. If, moreover, k is a kernel of continuous type, then we call k a kernel of type $[C; \mathcal{T}_S; \mathbf{C}^{n\times n}]$ (or of continuous periodic type with period S).

Kernels of type $[B^\infty; \mathcal{T}_S; \mathbf{C}^{n\times n}]$ are denoted by $\mathcal{K}(B^\infty; \mathcal{T}_S; \mathbf{C}^{n\times n})$, and kernels of type $[C; \mathcal{T}_S; \mathbf{C}^{n\times n}]$ are denoted by $\mathcal{K}(C; \mathcal{T}_S; \mathbf{C}^{n\times n})$.

In the case of the class $[C; \mathcal{T}_S]$, for example, the conditions on k are that the function $t \mapsto (s \mapsto k(t, t+s))$ be continuous and periodic from $\mathbf{R}$ into $L^1(\mathbf{R}; \mathbf{C}^{n\times n})$, with period S. Note that $C(\mathcal{T}_S; \mathbf{C}^{n\times n}) = BC(\mathcal{T}_S; \mathbf{C}^{n\times n}) = BUC(\mathcal{T}_S; \mathbf{C}^{n\times n})$ due to periodicity. For this reason, one could also denote $\mathcal{K}(C; \mathcal{T}_S; \mathbf{C}^{n\times n})$ by $\mathcal{K}(BC; \mathcal{T}_S; \mathbf{C}^{n\times n})$, or by $\mathcal{K}(BUC; \mathcal{T}_S; \mathbf{C}^{n\times n})$. Note that every (Volterra type) convolution kernel in $L^1(\mathbf{R}^+;\mathbf{C}^{n\times n})$, and every (Fredholm type) convolution kernel in $L^1(\mathbf{R};\mathbf{C}^{n\times n})$, is of continuous periodic type with period S, for every $S>0$.

6.9 Theorem
Let $S>0$. The classes $\mathcal{K}(B^\infty;\mathcal{T}_S;\mathbf{C}^{n\times n})$ and $\mathcal{K}(C;\mathcal{T}_S;\mathbf{C}^{n\times n})$ are closed subalgebras of $\mathcal{K}(B^\infty;\mathbf{R};\mathbf{C}^{n\times n})$. Moreover, $B^\infty(\mathcal{T}_S;\mathbf{C}^n)$ is a left Banach module over $\mathcal{K}(B^\infty;\mathcal{T}_S;\mathbf{C}^{n\times n})$, and $C(\mathcal{T}_S;\mathbf{C}^n)$ is a left Banach module over $\mathcal{K}(C;\mathcal{T}_S;\mathbf{C}^{n\times n})$.

Proof Let $\mathcal{B}$ stand for either B^∞ or C. It is obvious that $\mathcal{K}(\mathcal{B};\mathcal{T}_S;\mathbf{C}^{n\times n})$ is a closed subspace of $\mathcal{K}(B^\infty;\mathbf{R};\mathbf{C}^{n\times n})$. That it is a subalgebra follows from the fact that, if both a and b belong to $\mathcal{K}(\mathcal{B};\mathcal{T}_S;\mathbf{C}^{n\times n})$, then

$$\begin{aligned}(a\star b)(t+S,s+S)&=\int_{\mathbf{R}}a(t+S,v)b(v,s+S)\,dv\\&=\int_{\mathbf{R}}a(t+S,v+S)b(v+S,s+S)\,dv\\&=\int_{\mathbf{R}}a(t,v)b(v,s)\,dv=(a\star b)(t,s).\end{aligned}$$

A completely analogous computation shows that $a\star f\in\mathcal{B}(\mathcal{T}_S;\mathbf{C}^n)$ whenever $a\in\mathcal{K}(\mathcal{B};\mathcal{T}_S;\mathbf{C}^{n\times n})$ and $f\in L^\infty(\mathcal{T}_S;\mathbf{C}^n)$. □

Kernels of periodic type have the interesting property that, if they have a resolvent of bounded type, then this resolvent must be periodic.

6.10 Theorem
Let $S>0$. If k is a kernel of type $[B^\infty;\mathcal{T}_S;\mathbf{C}^{n\times n}]$ or $[C;\mathcal{T}_S;\mathbf{C}^{n\times n}]$, and k has a resolvent r of bounded type, then r is of type $[B^\infty;\mathcal{T}_S;\mathbf{C}^{n\times n}]$ or $[C;\mathcal{T}_S;\mathbf{C}^{n\times n}]$, respectively.

Recall (see Theorem 5.4) that, if k has a resolvent of type L^∞, then (after redefinition on a set of measure zero) this resolvent is of bounded type, and in the continuous case it is of continuous type.

Proof of Theorem 6.10 Because of the periodicity of k, we have

$$\begin{aligned}r(t+S,s+S)&=k(t+S,s+S)-\int_{\mathbf{R}}k(t+S,u)r(u,s+S)\,du\\&=k(t,s)-\int_{\mathbf{R}}k(t+S,u+S)r(u+S,s+S)\,du\\&=k(t,s)-\int_{\mathbf{R}}k(t,u)r(u+S,s+S)\,du,\end{aligned}$$

so the function $q(t,s)=r(t+S,s+S)$ satisfies $q+k\star q=k$. A similar computation shows that it satisfies $q+q\star k=k$. By the uniqueness of the resolvent, we must have $r=q$. This means that r is of periodic type with period S. □

Finally, let us pay some attention to asymptotically periodic kernels.

6.11 Definition

Let $S > 0$ and let $J \subset \mathbf{R}$ be an interval whose right end-point is $+\infty$. A kernel k of bounded type on J is called a kernel of type $[B^\infty(\mathcal{T}_S) ⌓ B_0^\infty(J); \mathbf{C}^{n\times n}]$ (or a kernel of bounded asymptotically periodic type with period S on J) if k can be written as the sum of two kernels, $k = k_1 + k_2$, where

(i) k_1 is the restriction to J^2 of a kernel, which is both of bounded periodic type with period S and of type $[B_0^\infty; \mathbf{R}; \mathbf{C}^{n\times n}]$,

and

(ii) k_2 is a kernel of type $[L^\infty; B_0^\infty; J; \mathbf{C}^{n\times n}]$.

If, in addition, both k_1 and k_2 are of continuous type, then we call k a kernel of type $[C(\mathcal{T}_S) ⌓ BC_0(J); \mathbf{C}^{n\times n}]$ (or a kernel of continuous asymptotically periodic type with period S on J).

These two classes of kernels are denoted by $\mathcal{K}(B^\infty(\mathcal{T}_S) ⌓ B_0^\infty(J); \mathbf{C}^{n\times n})$ and $\mathcal{K}(C(\mathcal{T}_S) ⌓ BC_0(J); \mathbf{C}^{n\times n})$, respectively.

The basic properties of the class of asymptotically periodic kernels are given in the next theorem.

6.12 Theorem

The two classes $\mathcal{K}(B^\infty(\mathcal{T}_S) ⌓ B_0^\infty(J); \mathbf{C}^{n\times n})$ and $\mathcal{K}(C(\mathcal{T}_S) ⌓ BC_0(J); \mathbf{C}^{n\times n})$ of asymptotically periodic kernels are closed subalgebras of the Banach algebra $\mathcal{K}(B_0^\infty; J; \mathbf{C}^{n\times n})$. Moreover, $B^\infty(\mathcal{T}_S; \mathbf{C}^n) ⌓ B_0^\infty(J; \mathbf{C}^n)$ is a left Banach module over $\mathcal{K}(B^\infty(\mathcal{T}_S) ⌓ B_0^\infty(J); \mathbf{C}^{n\times n})$, and $C(\mathcal{T}_S; \mathbf{C}^n) ⌓ BC_0(J; \mathbf{C}^n)$ is a left Banach module over $\mathcal{K}(C(\mathcal{T}_S) ⌓ BC_0(J); \mathbf{C}^{n\times n})$.

Proof Let $J = [S, \infty)$ with $S > -\infty$ (the easier case $J = \mathbf{R}$ is left to the reader). Obviously, $\mathcal{K}(B^\infty(\mathcal{T}_S) ⌓ B_0^\infty(J); \mathbf{C}^{n\times n})$ and $\mathcal{K}(C(\mathcal{T}_S) ⌓ BC_0(J); \mathbf{C}^{n\times n})$ are subspaces of $\mathcal{K}(B_0^\infty; J; \mathbf{C}^{n\times n})$, and it is not difficult to show that these subspaces are closed.

We claim that $a, b \in \mathcal{K}(B^\infty(\mathcal{T}_S) ⌓ B_0^\infty(J); \mathbf{C}^{n\times n})$ implies $a \star b \in \mathcal{K}(B^\infty(\mathcal{T}_S) ⌓ B_0^\infty(J); \mathbf{C}^{n\times n})$.

By definition, $a = a_1 + a_2$, $b = b_1 + b_2$, where a_1 and b_1 belong to $\mathcal{K}(B^\infty; \mathcal{T}_S; \mathbf{C}^{n\times n}) \cap \mathcal{K}(B_0^\infty; \mathbf{R}; \mathbf{C}^{n\times n})$, and a_2 and b_2 belong to $\mathcal{K}(L^\infty; B_0^\infty; J; \mathbf{C}^{n\times n})$. Write

$$a \star b = \int_{\mathbf{R}} a_1(t,u) b_1(u,s)\,\mathrm{d}u - \int_{\mathbf{R}\setminus J} a_1(t,u) b_1(u,s)\,\mathrm{d}u + a_2 \star b_1 + a_1 \star b_2 + a_2 \star b_2.$$

By Theorem 6.2 and Theorem 6.9, the function that maps $(t,s) \in \mathbf{R}^2$ to $\int_{\mathbf{R}} a_1(t,u) b_1(u,s)\,\mathrm{d}u$ belongs to $\mathcal{K}(B^\infty; \mathcal{T}_S; \mathbf{C}^{n\times n}) \cap \mathcal{K}(B_0^\infty; \mathbf{R}; \mathbf{C}^{n\times n})$.

Next consider

$$\int_J \left| \int_{\mathbf{R}\setminus J} a_1(t,u) b_1(u,s)\,\mathrm{d}u \right| \mathrm{d}s \le \int_{\mathbf{R}\setminus J} |a_1(t,u)| \int_J |b_1(u,s)|\,\mathrm{d}s\,\mathrm{d}u. \tag{6.1}$$

Since $a_1 \in \mathcal{K}(B_0^\infty; \mathbf{R}; \mathbf{C}^{n\times n})$ and $\int_J |b_1(u,s)|\,ds \le |||b_1|||_{B^\infty(\mathbf{R})}$, we see that

$$\int_J \left| \int_{\mathbf{R}\setminus J} a_1(t,u) b_1(u,s)\,du \right| ds \to 0 \text{ as } t \to \infty.$$

In other words, this term belongs to $\mathcal{K}(L^\infty; B_0^\infty; J; \mathbf{C}^{n\times n})$.

In the B^∞-case, it remains to show that $a_2 \star b_1 + a_1 \star b_2 + a_2 \star b_2$ belongs to $\mathcal{K}(L^\infty; B_0^\infty; J; \mathbf{C}^{n\times n})$. The demonstration of this, as well as the remaining details of the proof, is left to the reader (cf. Theorems 5.3 and 6.7). □

7. L^p-Kernels Defining Compact Mappings

Several applications, for example various perturbation problems where one wishes to apply the Schauder fixed point theorem, require the resolvent operator to be compact. Below we formulate conditions for this to hold for kernels $k \in \mathcal{F}(L^p; J; \mathbf{C}^{n\times n})$, where $p \in [1,\infty]$.

First, we remark that, if for some p both k and its resolvent r belong to $\mathcal{F}(L^p; J; \mathbf{C}^{n\times n})$, then k defines a compact mapping from $L^p(J; \mathbf{C}^{n\times n})$ into itself if and only if r does so. To see this, it suffices to recall that $r = k - r \star k$, and to observe that the compactness of k implies that $r \star k$ is compact, since $f \mapsto r \star f$ is a continuous map of $L^p(J; \mathbf{C}^{n\times n})$ into itself. Consequently, it is enough to consider the compactness of k.

The results will differ depending upon whether $p = 1$, $p = \infty$, or $p \in (1,\infty)$. We begin with an example showing that not all kernels of type L^1 generate compact mappings.

7.1 Remark

It is *not* true that every kernel k of type L^1 on an interval J generates a compact mapping of $L^1(J)$ into itself. To see this, take the following scalar example. Let $J = [0,1]$, let $\{ J_k \}_{k=1}^\infty$ be a sequence of disjoint intervals $\subset [0,1]$, and let χ_k be the characteristic function of J_k. Let $\{ \varphi_k \}_{k=1}^\infty$ be any noncompact set of functions in $L^1(0,1)$ satisfying $0 \le \varphi_k(t) \le 1$ for all k and for all $t \in [0,1]$. Define $k(t,s) = \sum_{k=1}^\infty \varphi_k(t)\chi_k(s)$. Then $\sup_{(t,s)\in J^2} |k(t,s)| \le 1$ and so, by Proposition 2.7(ii), k is of type L^1 on J. That k does not generate a compact mapping can be seen as follows. Define $f_k(s) = \chi_k(s)[m(J_k)]^{-1}$. Clearly, $\|f_k\|_{L^1(0,1)} = 1$, so the set $\{ f_k \}_{k=1}^\infty$ is bounded in L^1. But $\int_J k(t,s) f_k(s)\,ds = \varphi_k(t)$; hence the image under k of this set is not compact in L^1. Thus, k does not generate a compact mapping.

The second fact that we observe is that the cases $p = 1$ and $p = \infty$, and more generally all conjugate indices p and q, are related in the following sense.

7.2 Proposition
Let $J \subset \mathbf{R}$ be an interval, and let p, $q \in [1, \infty]$ satisfy $1/p + 1/q = 1$. Let $k \in \mathcal{F}(L^p; J; \mathbf{C}^{n\times n})$, and let h be defined by $h(t,s) = k(s,t)^T$, $(t,s) \in J$. Then $h \in \mathcal{F}(L^q; J; \mathbf{C}^{n\times n})$, and the mapping $f \mapsto k \star f$ is compact from $L^p(J; \mathbf{C}^n)$ into itself if and only if the mapping $f \mapsto h \star f$ is compact from $L^q(J; \mathbf{C}^n)$ into itself.

Proof Because an operator mapping one Banach space into another is compact if and only if its adjoint is compact, we have to show that the (Banach space) adjoint (restricted to L^1 if $p = \infty$) of the mapping $f \mapsto k \star f$ is given by $f \mapsto h \star f$. That this is true can be seen from the following calculation, based on Fubini's theorem:

$$\int_J g(t)^T (k \star f)(t)\, dt$$
$$= \int_J \int_J g(t)^T k(t,s) f(s)\, ds\, dt = \int_J (h \star g)(s)^T f(s)\, ds. \quad \square \qquad (7.1)$$

Combining Remark 7.1 and Proposition 7.2, we get an example of a kernel of type L^∞ that does not generate a compact mapping from L^∞ into itself.

As we showed in Theorem 5.3(iii), for a kernel k of bounded type to generate a compact mapping from L^∞ into B^∞ (in fact, into BUC), it suffices to have $\int_J |k(t,s)|\, ds$ uniformly continuous in t, and J bounded. This, combined with Proposition 7.2, gives us the following corollary.

7.3 Corollary
Let $J \subset \mathbf{R}$ be a bounded interval. If $k : J^2 \to \mathbf{C}^{n\times n}$ is measurable, if for every s the function $t \mapsto k(t,s)$ is integrable, and if the function $s \mapsto (t \mapsto k(t,s))$ is uniformly continuous from J into $L^1(J; \mathbf{C}^{n\times n})$, then the mapping $f \mapsto k \star f$ is compact from $L^1(J; \mathbf{C}^n)$ into itself.

The continuity condition above is sufficient for compactness, but not necessary. A necessary hypothesis for k to generate a compact map of $L^1(J; \mathbf{C}^n)$ into itself is

$$\lim_{m(D)\to 0} \operatorname*{ess\,sup}_{s\in J} \int_D |k(t,s)|\, dt = 0,$$

where $m(D)$ is the Lebesgue measure of D. This condition is in fact necessary and sufficient for k to be weakly compact (see Krasnosel'skiĭ *et al.* [1], p. 117). If it holds, then $k \star k$ is compact (see the comment in Krasnosel'skiĭ *et al.* [1], p. 118). One observes that if k is a Volterra kernel, if J is bounded, and if this condition holds, then, by Theorem 3.18(i), k has a resolvent r, given by $r = k + \sum_{j=2}^{\infty} (-1)^{j-1} k^{\star j}$. It follows that in this case $r - k = r \star k$ is compact.

We finally note that, if $m(J) < \infty$ and if k maps L^1 into L^1, then k (and therefore r) generates a compact mapping of $L^{1+\epsilon}(J)$ into $L^1(J)$ for every $\epsilon > 0$ (see Krasnosel'skiĭ *et al.* [1], Theorem 5.4, p. 83).

Next, we consider the case $p \in (1, \infty)$, and show that the kernels studied in Proposition 2.7(iii) generate compact mappings.

7.4 Theorem

Let $J \subset \mathbf{R}$ be an interval, and let $1 < p < \infty$, and $1/p + 1/q = 1$. Let $k : J^2 \to \mathbf{C}^{n\times n}$ be measurable, and suppose that

$$\int_J \left(\int_J |k(t,s)|^q \, \mathrm{d}s \right)^{p/q} \mathrm{d}t < \infty,$$

$$\text{or} \quad \int_J \left(\int_J |k(t,s)|^p \, \mathrm{d}t \right)^{q/p} \mathrm{d}s < \infty. \tag{7.2}$$

Then $f \mapsto k \star f$ is a compact mapping of $L^p(J; \mathbf{C}^n)$ into itself.

Proof First, suppose that $m(J) < \infty$, take intervals $A \subset J$, $B \subset J$ and consider the kernel $h(t,s) = \chi_A(t)\chi_B(s)$. (Recall that χ_S is the characteristic function of the set S.) Obviously, h generates a compact mapping of $L^p(J; \mathbf{C}^n)$ into itself. The same is true for any finite sum

$$h_j(t,s) = \sum_{i=1}^{j} \chi_{A_i}(t)\chi_{B_i}(s).$$

Take an arbitrary set $E \subset J^2$ and fix $\epsilon > 0$. Choose finitely many disjoint rectangles $A_i \times B_i$, $i = 1, \ldots, j$, such that the measure of the symmetric difference between E and $E_j \stackrel{\text{def}}{=} \bigcup_{i=1}^{j}(A_i \times B_i)$, i.e., the set $D = (E \setminus E_j) \cup (E_j \setminus E)$, is less than ϵ. If $p \geq q$, then

$$\begin{aligned}
\int_J \left(\int_J |\chi_E(t,s) - h_j(t,s)|^q \, \mathrm{d}s \right)^{p/q} \mathrm{d}t & \\
&= \int_J \left(\int_J \chi_D(t,s) \, \mathrm{d}s \right)^{p/q} \mathrm{d}t \\
&= \int_J \left(\int_J \chi_D(t,s) \, \mathrm{d}s \right) \left(\int_J \chi_D(t,s) \, \mathrm{d}s \right)^{p/q-1} \mathrm{d}t \\
&\leq m(J)^{p/q-1}\epsilon.
\end{aligned}$$

Consequently, $|||\chi_E(t,s) - h_j(t,s)|||_{L^p(J)} \leq m(J)^{1/q-1/p}\epsilon^{1/p}$ by Proposition 2.7(iii). If $p < q$, then we can make the same computation, but with p and q interchanged, so in this case $|||\chi_E(t,s) - h_j(t,s)|||_{L^p(J)} \leq m(J)^{1/p-1/q}\epsilon^{1/q}$. Each h_j generates a compact mapping, and so the same is true for χ_E. But then any finite sum

$$k_m(t,s) = \sum_{i=1}^{m} \alpha_i \chi_{E_i}(t,s),$$

where $E_i \subset J^2$ and $\alpha_i \in \mathbf{C}^{n\times n}$, generates a compact mapping from L^p into L^p as well. Now choose $k_m(t,s)$, $m = 1,2,\ldots$, each of the type depicted above, in such a way that

$$|k_m(t,s)| \le |k(t,s)|, \text{ a.e. on } J^2, \quad k_m(t,s) \to k(t,s), \quad \text{as } m \to \infty.$$

Assume that the first of the two conditions in (7.2) is satisfied (the second case is completely analogous). Then it follows from Lebesgue's dominated convergence theorem that, for almost all t, $\int_J |k_m(t,s) - k(t,s)|^q \, ds \to 0$ as $m \to \infty$. Moreover, the function $t \mapsto \left(\int_J |k_m(t,s) - k(t,s)|^q \, ds\right)^{p/q}$ is dominated by the L^1-function $t \mapsto 2^p\left(\int_J |k(t,s)|^q \, ds\right)^{p/q}$. Thus, applying first Lebesgue's dominated convergence theorem once more, and then Proposition 2.7(iii), we find that $|||k_m - k|||_{L^p(J)} \to 0$ for $m \to \infty$. Consequently, k induces a compact mapping.

Finally, let us remove the assumption $m(J) < \infty$. Take $J_i \subset J$ such that $m(J_i) < \infty$, $J_i \subset J_{i+1}$ and $\bigcup_{i=1}^{\infty} J_i = J$. Each kernel $\chi_{J_i^2} k$ generates a compact mapping, and the final argument in the preceding paragraph shows that $|||k - \chi_{J_i^2} k|||_{L^p(J)} \to 0$ as $i \to \infty$. One concludes that k generates a compact mapping of $L^p(J; \mathbf{C}^n)$ into itself. □

The preceding result gives only sufficient criteria for compactness. For a necessary and sufficient condition (which, however, is difficult to use), see the comments in Section 12.

It is a well-known fact that the spectrum of a compact operator is countable. This fact can be used to prove the following result.

7.5 Theorem

Let $J \subset \mathbf{R}$ be an interval, let k belong to some closed subalgebra $\mathcal{A}$ of $\mathcal{K}(B^\infty; J; \mathbf{C}^{n\times n})$, and suppose that the mapping $f \mapsto k \star f$ is compact in $B^\infty(J; \mathbf{C}^n)$. Then k has a resolvent in $\mathcal{A}$ if and only if it has a resolvent in $\mathcal{K}(B^\infty; J; \mathbf{C}^{n\times n})$, or equivalently, if and only if it has a resolvent in the Banach algebra of all bounded linear operators on $B^\infty(J; \mathbf{C}^n)$.

Proof Let us denote the Banach algebra of all bounded linear operators on $B^\infty(J; \mathbf{C}^n)$ by $\mathcal{B}$. It is a consequence of Lemma 10.1 (cf. the proof of Lemma 10.2) that $\mathcal{K}(B^\infty; J; \mathbf{C}^{n\times n})$ is closed in $\mathcal{B}$. Since the operator $f \mapsto k \star f$ is compact, its spectrum is countable, see, e.g., Rudin [2], Theorem 4.24, and the complement of the spectrum has no bounded component. By Rudin [2], Theorem 10.18, the spectra of k with respect to $\mathcal{A}$, $\mathcal{K}(B^\infty; J; \mathbf{C}^{n\times n})$, and $\mathcal{B}$ are the same. □

8. Volterra Kernels with Nonpositive or Nonnegative Resolvents

The theory of the preceding sections is quite satisfactory, as long as one is interested in the behaviour of the solutions of (1.1) on a finite interval. However, the results given there are of rather limited value when applied to an infinite interval. We do know, by Theorem 4.2 and Corollary 4.3, that if k is of Volterra type on $\mathbf{R}^+$ then, under quite weak assumptions, k has a unique resolvent r, and the solution of (1.1) is given by $f - r \star f$. However, up to now we know very little about the behaviour of x or r at infinity. For example, if we apply our results to the convolution case, then we are only able to conclude that if a convolution kernel k belongs to $L^1(\mathbf{R}^+; \mathbf{C}^{n\times n})$, and if the L^1-norm of k is less than 1, then k has a resolvent in L^1. In particular, we have no nonconvolution versions of the Paley–Wiener theorems (nor will such versions be given below).

Below we shall develop some results for the case of an infinite interval. They are based on sign or on monotonicity properties of the kernel, and they apply exclusively to scalar Volterra equations.

We begin with a simple result on the sign of r.

8.1 Proposition

Let k be a scalar nonpositive Volterra kernel of type L^p for some p, $1 \le p \le \infty$, on an interval $J \subset \mathbf{R}$, and suppose that each compact subinterval K of J can be divided into finitely many subintervals K_i such that, on each K_i, $|||k|||_{L^p(K_i)} < 1$. Then the resolvent r of k (which exists, according to Corollary 4.3) is nonpositive.

This follows directly from the fact that $r_m \to r$, where $r_m = \sum_{j=1}^{m}(-1)^{j-1}k^{\star j}$ (see Corollary 4.3).

The conclusion of Proposition 8.1 is used as an assumption in the following simple but useful comparison theorem.

8.2 A Generalized Gronwall Lemma

Let k be a scalar Volterra kernel of type L^p_{loc} for some p, $1 \le p \le \infty$, on an interval $J \subset \mathbf{R}$ of the type $[S, T)$, with $-\infty < S < T \le \infty$, and assume that $-k$ has a nonpositive resolvent r of the same type. Let x, $f \in L^p_{\text{loc}}(J)$, and suppose that $x(t) \le (k \star x)(t) + f(t)$ for almost all $t \in J$. Then $x(t) \le y(t)$ for almost all $t \in J$, where y is the solution of the comparison equation $y(t) = (k \star y)(t) + f(t)$.

Proof Obviously, we may write $x = k \star x + f - g$ for some $g \ge 0$. By Theorem 4.2, $x = f - g - r \star (f - g)$, and so

$$k \star x = k \star (f - r \star f) - k \star (g - r \star g) = -r \star f + r \star g.$$

Therefore,

$$x \le k \star x + f = f - r \star f + r \star g \le f - r \star f = y. \ \square$$

So far in this section we have discussed kernels whose resolvents are nonpositive. Now it is time to turn to the more subtle question of when a resolvent is nonnegative. Our first observation in this direction is that if a kernel k is nonnegative, and if the resolvent r of k is nonnegative as well, then it follows from the resolvent equation that $r \le k$. Consequently, we have the following statement.

8.3 Proposition
Let k be a scalar Volterra kernel of type L^p for some p, $1 \le p \le \infty$, on an interval $J \subset \mathbf{R}$, and suppose that k has a resolvent r of type L^p_{loc} on J. If both k and r are nonnegative, then r is of type L^p on J.

Applying this theorem to the convolution case we get

8.4 Corollary
Let $k \in L^1(\mathbf{R}^+;\mathbf{R})$ be nonnegative and have a nonnegative convolution resolvent r. Then $r \in L^1(\mathbf{R}^+;\mathbf{R})$.

This result can easily be improved. By integrating the resolvent equation over $\mathbf{R}^+$, and interchanging the order of integration, one finds that

$$\int_{\mathbf{R}^+} r(s)\,\mathrm{d}s = \frac{\int_{\mathbf{R}^+} k(s)\,\mathrm{d}s}{1 + \int_{\mathbf{R}^+} k(s)\,\mathrm{d}s}.$$

In particular, if r and k are nonnegative, then $\|r\|_{L^1(\mathbf{R}^+)} < 1$. As such, this computation is not valid for nonconvolution kernels. However, one can prove the following somewhat weaker result.

8.5 Theorem
Let $J \subset \mathbf{R}$ be an interval, and let k be a scalar nonnegative Volterra kernel that is of type $[L^p_{\mathrm{loc}}; J; \mathbf{R}]$ for some p, $1 \le p \le \infty$, and has a nonnegative resolvent r of type $[L^p_{\mathrm{loc}}; J; \mathbf{R}]$. Then

$$|||r|||_{L^\infty(J)} \le \operatorname*{ess\,sup}_{(t,u,s)\in\Delta_3(J)} \frac{k(t,s)}{k(u,s)}, \tag{8.1}$$

and

$$|||r|||_{L^1(J)} \le \operatorname*{ess\,sup}_{(t,u,s)\in\Delta_3(J)} \frac{k(t,s)}{k(t,u)}, \tag{8.2}$$

where

$$\Delta_3(J) = \{\, (t,u,s) \in J^3 \mid s \le u \le t \,\},$$

and $\alpha/0$ is interpreted as ∞ for every $\alpha > 0$, but $0/0$ is defined to be 0.

Proof The proofs of (8.1) and (8.2) are completely similar, so let us prove (8.1) only. Since r is nonnegative, it suffices to show that, for almost all

$t \in J$,

$$\int_S^t r(t,u)\,\mathrm{d}u \le c, \tag{8.3}$$

where S is the left end-point of J, and c is an arbitrary constant such that

$$k(t,s) \le ck(u,s) \tag{8.4}$$

for almost all $(t,u,s) \in \Delta_3(J)$ (if no such constant c exists, then there is nothing to prove). The nonnegativity of k and r, together with the resolvent equation

$$r(t,s) + \int_s^t r(t,u)k(u,s)\,\mathrm{d}u = k(t,s), \tag{8.5}$$

implies that

$$0 \le r(t,s) \le k(t,s) \tag{8.6}$$

for almost all $(t,s) \in J^2$ such that $s \le t$. Now, let $t \in J$ be such that (8.4) holds for almost all $(u,s) \in J^2$ satisfying $s \le u \le t$, and such that (8.5), hence also (8.6), holds for almost all $s \le t$. (Only on a set of measure zero may it happen that these conditions on t are not satisfied). Then, for almost all $s \le t$,

$$\begin{aligned} k(t,s)\int_s^t r(t,u)\,\mathrm{d}u &\le c\int_s^t r(t,u)k(u,s)\,\mathrm{d}u \\ &= c\big(k(t,s) - r(t,s)\big) \le ck(t,s). \end{aligned} \tag{8.7}$$

If we let $(S, v(t))$ be the maximal interval with left end-point S on which $s \mapsto k(t,s)$ vanishes a.e., i.e.,

$$v(t) = \sup\{\, T \mid k(t,s) = 0 \text{ for almost all } s \in (S,T) \,\},$$

then it follows from (8.6) that $u \mapsto r(t,u)$ vanishes a.e. on $(S, v(t))$, and from (8.7) that

$$\int_{v(t)}^t r(t,u)\,\mathrm{d}u \le c.$$

This implies (8.3), and completes the proof of Theorem (8.1). □

Both Proposition 8.3 and Theorem 8.5 require k to have a nonnegative resolvent. But when is the resolvent nonnegative? The following is one answer to this question.

8.6 Theorem

Let k be a nonnegative kernel that is locally of type L^p for some p, $1 \le p \le \infty$, on an interval J. In addition, suppose that, for almost all $(t,v,u,s) \in J^4$ satisfying $s \le u \le v \le t$, it is true that

$$k(v,s)k(t,u) \le k(t,s)k(v,u). \tag{8.8}$$

Then k has a nonnegative resolvent which is locally of type L^p_{loc} on J^2, and it is strictly positive a.e. on the set where k is strictly positive. If k and r are continuous, then $r(t,s) > 0$ at all points where $k(t,s) > 0$. Moreover, if k is of type L^p on J, then so is r, the spectral radius of r is less than 1, the spectrum of k is contained in the closed right half plane, and the spectrum of r is contained in the circle $\left|z - \frac{1}{2}\right| \le \frac{1}{2}$.

Observe that in this theorem we obtain the existence of a resolvent without making any assumptions of the type used in, e.g., Corollary 3.14.

Before proving Theorem 8.6, let us make some comments on (8.8). If k is continuous and strictly positive on $\Delta_2(J) = \{\,(t,s) \in J^2 \mid s \le t\,\}$, then (8.8) is satisfied if and only if $t \mapsto k(t,u)/k(t,s)$ is nonincreasing for all $t \ge u \ge s$, or equivalently, if and only if $s \mapsto k(t,s)/k(v,s)$ is nonincreasing for all $s \le v \le t$. If, in addition, k is twice continuously differentiable, then all these conditions are equivalent to (see Exercise 30)

$$\frac{\partial^2}{\partial s \partial t} \ln\bigl(k(t,s)\bigr) \le 0. \tag{8.9}$$

If k is a convolution kernel, i.e., $k(t,s) = a(t-s)$ for some function a, then (8.8) is satisfied if and only if the function $t \mapsto \ln(a(t))$ is convex.

Observe that (8.8) is preserved if we multiply k by a nonnegative function of the type $b(t)c(s)$, so it is easy to construct examples of kernels that satisfy (8.8) and vanish on a set of positive measure. For these kernels, (8.9) does not make sense.

Proof of Theorem 8.6 Without loss of generality, let J be compact. Let

$$\begin{aligned}
\Lambda &= \bigl\{\,\lambda \ge 0 \bigm| \lambda k \text{ has a resolvent } r_\lambda \in \mathcal{V}(L^p; J; \mathbf{R})\,\bigr\},\\
\Lambda_\star &= \bigl\{\,\lambda \in \Lambda \bigm| |||\,|r_\lambda|^{\star j}\,||| < 1 \text{ for some } j \ge 1\,\bigr\},\\
\Lambda_+ &= \bigl\{\,\lambda \in \Lambda \bigm| r_\lambda \ge 0 \text{ a.e. on } J^2\,\bigr\}.
\end{aligned}$$

The mapping $\lambda \to r_\lambda$ is continuous, and therefore the set $\Lambda_\star$ is open in Λ and Λ_+ is closed in Λ. It follows that if we want to prove that $\Lambda_+ = \Lambda$ it suffices to prove that $\Lambda_+ = \Lambda_\star$. Moreover, observe that if $\lambda \in \Lambda_+$, then $r_\lambda(t,s) \le \lambda k(t,s)$ and this implies that if $0 < |\mu| < 1/(|||k|||_{L^p(J)} + \lambda |||k|||^2_{L^p(J)})$ and $\mu + \lambda \ge 0$, then $\lambda + \mu \in \Lambda$ (cf. Theorem 3.9). This means that, if $\Lambda_+ = \Lambda$, then necessarily $\Lambda = \mathbf{R}^+$. Thus, if we can prove that $\Lambda_+ = \Lambda_\star$, then $\Lambda_+ = \Lambda_\star = \Lambda = \mathbf{R}^+$.

First we assume that $\lambda \in \Lambda_+$ and prove that $\lambda \in \Lambda_\star$. To simplify the notation, we assume that $\lambda = 1$ and write $r_1 = r$ (replace k by λk). Let $f \in L^p(J;\mathbf{R}^+)$ be nonnegative but otherwise arbitrary. Iterating the resolvent equation we get

$$0 \le \sum_{j=1}^{\infty} (r^{\star j} \star f) \le k \star f.$$

If we let

$$g = \sum_{j=1}^{\infty} (r^{\star j} \star f),$$

then $\|g\|_{L^p(J)} \le |||k|||_{L^p(J)} \|f\|_{L^p(J)}$. Repeating this argument with f replaced by g, we obtain

$$\left\| \sum_{j=1}^{\infty} r^{\star j} \star g \right\|_{L^p(J)} \le |||k|||_{L^p(J)}^2 \|f\|_{L^p(J)}.$$

On the other hand, by Fubini's theorem,

$$\sum_{j=1}^{\infty} (r^{\star j} \star g) = \sum_{j=2}^{\infty} (j-1) r^{\star j} \star f.$$

Owing to the nonnegativity of all the functions involved,

$$(j-1)\|r^{\star j} \star f\|_{L^p(J)} \le |||k|||_{L^p(J)}^2 \|f\|_{L^p(J)}, \quad j \ge 2.$$

For nonnegative kernels the norm $|||\cdot|||_{L^p(J)}$ is identical with the operator norm, and since f was arbitrary we get the desired conclusion that $1 \in \Lambda_\star$.

We proceed to show that if $\lambda \in \Lambda_\star$ then $\lambda \in \Lambda_+$. Again, we may, without loss of generality, assume that $\lambda = 1$. We claim that the equation

$$q(t,s) = k(t,s) - \int_s^t k(t,u)|q(u,s)|_+ \, du, \quad (t,s) \in J^2, \tag{8.10}$$

has a solution $q \in \mathcal{V}(L^p; J; \mathbf{R})$. To prove this, note that we can, with the aid of the resolvent, write this equation in the form

$$q(t,s) = r(t,s) - \int_s^t r(t,u)|q(u,s)|_- \, du, \quad (t,s) \in J^2.$$

Since we assumed that $1 \in \Lambda_\star$, it is possible to use an iteration procedure to solve this equation and obtain a solution of equation (8.10).

Next we claim that the solution q of (8.10) is nonnegative. To prove this, we first show that q is strictly positive a.e. on the set where k is strictly positive. First, we fix $s \in J$ in such a way that (8.8) holds for almost all $(t,v,u) \in J^3$ satisfying $s \le u \le v \le t$, and such that, for almost all $t \ge s$, $k(t,s)$ and $\int_s^t k(t,u)k(u,s)\,du$ are finite, and (8.10) holds. Then we pick $t \in J$, $t \ge s$, in such a way that (8.8) holds for almost all $(v,u) \in J^2$ satisfying $s \le u \le v \le t$, and such that $k(t,s)$ and $\int_s^t k(t,u)k(u,s)\,du$ are finite, and (8.10) holds. Observe that the set of points (s,t) that we have excluded in this way has measure zero, and that no points (s,t) have been excluded if k and r are continuous. Finally, choose $v \in (s,t)$ in such a way that (8.8) holds for almost all $u \in (s,v)$, and such that $k(v,s)$ and $\int_s^v k(v,u)k(u,s)\,du$ are finite, and (8.10) holds with (t,s) replaced by (v,s).

Then, by (8.8) and (8.10),

$$
\begin{aligned}
k(t,s)q(v,s)&-k(v,s)q(t,s)\\
&= -\int_s^v \Big(k(t,s)k(v,u) - k(v,s)k(t,u)\Big)|q(u,s)|_+\,\mathrm{d}u\\
&\quad + k(v,s)\int_v^t k(t,u)|q(u,s)|_+\,\mathrm{d}u\\
&\le k(v,s)\int_v^t k(t,u)|q(u,s)|_+\,\mathrm{d}u. \qquad (8.11)
\end{aligned}
$$

Suppose that $k(t,s) > 0$ but $q(t,s) \le 0$. Then it follows from the previous inequality that

$$
|q(v,s)|_+ \le \frac{k(v,s)}{k(t,s)}F(v), \qquad (8.12)
$$

where

$$
F(v) = \int_v^t k(t,u)|q(u,s)|_+\,\mathrm{d}u,
$$

is a nonnegative continuous function on J. Multiply (8.12) by $k(t,v)$, and integrate over (w,t), where $w \in J$, $w \le t$, to get

$$
0 \le F(w) \le k(t,s)^{-1}\int_w^t k(t,v)k(v,s)F(v)\,\mathrm{d}v.
$$

We picked (t,s) in such a way that $v \mapsto k(t,v)k(v,s)$ is integrable on J, and therefore it follows from Gronwall's inequality that $F(w) = 0$ for $w \le t$. In particular, taking w to be the left end-point of J, we get $\int_J k(t,u)|q(u,s)|_+ du = 0$, which substituted into (8.10) gives $q(t,s) = k(t,s)$. This contradicts the original assumption that $k(t,s) > 0$ and $q(t,s) \le 0$, and shows that k is strictly positive a.e. on the set where q is strictly positive.

It remains to show that q is nonnegative a.e. on the set where k vanishes. Choose s, t and v as above. Suppose that $k(t,s) = 0$ and $q(t,s) < 0$. Then, since (8.10) implies that $q(t,s) \le k(t,s)$, it follows from (8.11) that, for almost all $v \in J$, $v \le t$,

$$
k(v,s) \le |q(t,s)|^{-1}k(v,s)F(v),
$$

where now $F(v) = \int_v^t k(t,u)k(u,s)\,\mathrm{d}u$. Multiply the inequality above by $k(t,v)$, integrate over (w,t), and apply Gronwall's inequality to get $\int_J k(t,u)k(u,s)\,\mathrm{d}u = 0$. Since $|q(u,s)|_+ \le k(u,s)$ for almost all $u \in J$, this implies $\int_J k(t,u)|q(u,s)|_+ du = 0$, and hence, by (8.10), $q(t,s) = k(t,s) = 0$. This contradiction shows that q is nonnegative a.e. on the set where k vanishes.

Since q is nonnegative, equation (8.10) shows that $q + k \star q = k$. We assumed that k has a resolvent r and hence it follows from Lemma 3.3 that $r = q$. Thus $1 \in \Lambda_+$ and $\Lambda_+ = \Lambda_\star$. As we observed earlier, this implies that $\Lambda_+ = \Lambda_\star = \Lambda = \mathbf{R}^+$.

The fact that $1 \in \Lambda_\star$ implies that the resolvent r of k exists, and the spectral radius of r is less than 1, i.e., the spectrum of r is contained in the circle $|z| < 1$. Let us denote the artificial unit in $\mathcal{V}(L^p; J; \mathbf{C})$ by ε (cf. the discussion preceding Definition 3.17). Then $\varepsilon + k$ is invertible in the extended algebra, and so is $z\varepsilon - r$ whenever $|z| \geq 1$. Now observe that

$$\begin{aligned} z\varepsilon - r &= z\varepsilon - k \star (\varepsilon + k)^{-1} = (z\varepsilon + zk - k) \star (\varepsilon + k)^{-1} \\ &= (1 - z)\left(\tfrac{z}{1-z}\,\varepsilon - k\right) \star (\varepsilon + k)^{-1}. \end{aligned}$$

This implies that $\left(\frac{z}{1-z}\,\varepsilon - k\right)$ is invertible when $|z| > 1$, i.e., $\frac{z}{1-z}$ does not belong to the spectrum of k. The mapping $z \mapsto \frac{z}{1-z}$ takes the region $|z| > 1$ into the half plane $\Re z < -\frac{1}{2}$, and we conclude that this half plane does not belong to the spectrum of k. If we repeat the same argument with k replaced by λk and r replaced by r_λ, with $\lambda > 0$, we conclude that the half plane $\Re z < -\frac{1}{2\lambda}$ does not belong to the spectrum of k. Letting $\lambda \to \infty$, we find that the spectrum of k is contained in the closed right half plane.

We leave it to the reader to show that the fact that the spectrum of k belongs to the right half plane implies that the spectrum of r must belong to the circle $\left|z - \frac{1}{2}\right| \leq \frac{1}{2}$. □

Theorem 8.6 is closely related to the following result.

8.7 Theorem

Let J be an interval of the form $[S, T)$, where $-\infty < S < T \leq \infty$. Let k be a nonnegative kernel that is locally of type L^p for some p, $1 \leq p \leq \infty$, on J, and suppose that each compact subinterval K of J can be divided into finitely many subintervals K_i such that, on each K_i, $|||k|||_{L^p(K_i)} < 1$. Let f be a nonnegative function in $L^p_{\mathrm{loc}}(J; \mathbf{R})$. In addition, suppose that, for almost all $(t, v, u) \in J^3$ satisfying $u \leq v \leq t$, it is true that

$$f(v)k(t, u) \leq f(t)k(v, u). \tag{8.13}$$

Then the solution x of (3.3) is nonnegative a.e. on J, and is strictly positive a.e. on the set where f is strictly positive.

If k is of continuous type and f is continuous, and if, for every $t \in J$, it is true that (8.13) holds for almost all $(v, u) \in J^2$ satisfying $u \leq v \leq t$, then x is continuous and nonnegative, and $x(t) > 0$ at all points where $f(t) > 0$.

If k and f are continuous and strictly positive, then (8.13) is satisfied if and only if for every u the function $t \mapsto k(t, u)/f(t)$ is nonincreasing for all $t \geq u$. Again it is easy to construct an example of a pair (f, k) that satisfies (8.13), and for which f is discontinuous or vanishes on a set of positive measure (multiply both $f(t)$ and $k(t, u)$ by a nonnegative function $g(t)$, and let g be discontinuous or vanish on a set of positive measure).

The proof of Theorem 8.7 is almost exactly the same as that of Theorem 8.6, and is therefore omitted; (replace $k(t, s)$ by $f(t)$ and $k(v, s)$ by $f(v)$, replace q by a function that does not depend on s and use the fact that the spectral radius of k is less than 1 to solve the modified version of (8.10)).

Combining Corollary 8.4 with Theorem 8.6, we get the following convolution result.

8.8 Corollary

Let k be strictly positive and nonincreasing on $(0, \infty)$, with $\int_0^1 k(s)\,\mathrm{d}s < \infty$, and suppose that $\ln(k)$ is convex. Then the resolvent r of k is positive, and $\int_{\mathbf{R}^+} r(s)\,\mathrm{d}s \le 1$.

Compare this with Theorem **5**.3.1.

9. An Asymptotic Result

In the preceding section we obtained some bounds on the size of the resolvent r by assuming (or establishing) its nonnegativity. In this section, we deduce some asymptotic results for the resolvent of a nonnegative kernel that is nonincreasing in a certain sense. First, we briefly consider the convolution case in order to motivate the more general discussion.

We claim that if $k \in L^1(\mathbf{R}^+; \mathbf{R})$ is nonnegative and nonincreasing on $\mathbf{R}^+$ then the resolvent of k is integrable. To prove this claim, it suffices to show that the half line Paley–Wiener condition, i.e., the condition $1 + \hat{k}(z) \neq 0$ for all $z \in \mathbf{C}$ with $\Re z \ge 0$, is satisfied (this we do in the next paragraph), and to apply the Half Line Paley–Wiener Theorem **2**.4.1.

The nonnegativity of k implies that $\hat{k}(\sigma) = \int_{\mathbf{R}^+} \mathrm{e}^{-\sigma t} k(t)\,\mathrm{d}t \ge 0$ for every $\sigma \ge 0$ and, therefore, $1 + \hat{k}(z) \neq 0$ when $\Im z = 0$. In case $z = \mathrm{i}\omega$ for some $\omega > 0$, then one can integrate by parts, to show that

$$\Im \hat{k}(\mathrm{i}\omega) = \int_{\mathbf{R}^+} k(t) \sin(\omega t)\,\mathrm{d}t = \omega^{-1} \int_{\mathbf{R}^+} \big(1 - \cos(\omega t)\big)\kappa(\mathrm{d}t) \ge 0,$$

where $\kappa([t, \infty)) = k(t)$. If the last inequality is strict, then clearly $1 + \hat{k}(\mathrm{i}\omega) \neq 0$. If it is not strict, then k is a constant in each interval of the form $(2\pi n/\omega, 2\pi(n+1)/\omega)$, and, therefore, $\Re \hat{k}(\mathrm{i}\omega) = 0$. Thus, in this case, too, $1 + \hat{k}(\mathrm{i}\omega) \neq 0$. The case where $z = \mathrm{i}\omega$ for some $\omega < 0$ is treated analogously (in this case $\Im \hat{k}(\mathrm{i}\omega) \le 0$, with equality only when $\Re \hat{k}(\mathrm{i}\omega) = 0$). Finally, if $z = \sigma + \mathrm{i}\omega$, with $\sigma > 0$ and $\omega \in \mathbf{R}$, $\omega \neq 0$, then we can apply the preceding argument, with $k(t)$ replaced by $\mathrm{e}^{-\sigma t} k(t)$, to show that again $1 + \hat{k}(z) \neq 0$. This proves our claim that the Paley–Wiener condition is satisfied, and hence the resolvent of k is integrable.

By using entirely different arguments, one can prove an analogous result for the nonconvolution case.

9.1 Theorem
Let k be a scalar Volterra kernel on $\mathbf{R}$ (i.e., $k : \mathbf{R}^2 \to \mathbf{R}$ is measurable and $k(t,s) = 0$ when $s > t$), satisfying the following conditions:

(i) *k is nonnegative, and $t \mapsto k(t,s)$ is nonincreasing in t on $[s, \infty)$ for each $s \in \mathbf{R}$;*

(ii) *$\int_t^{t+\delta} k(s,s)\, \mathrm{d}s \to 0$ as $\delta \downarrow 0$, uniformly for $t \in \mathbf{R}$;*

(iii) *the function $t \mapsto \int_{-\infty}^t k(t,s)\, \mathrm{d}s$ is continuous on $\mathbf{R}$ and the two limits $\lim_{t\to\pm\infty} \int_{-\infty}^t k(t,s)\, \mathrm{d}s$ exist;*

(iv) *$\lim_{T\to\infty} \sup_{t\in\mathbf{R}} \int_{-\infty}^{t-T} k(t,s)\, \mathrm{d}s < 1$.*

Then k has a resolvent r of bounded uniformly continuous type on $\mathbf{R}$.

Recall that the conclusion is equivalent to

> $\sup_{t\in\mathbf{R}} \int_{\mathbf{R}} |r(t,s)|\, \mathrm{d}s < \infty$, *and* $\int_{\mathbf{R}} |r(t+h,s) - r(t,s)|\, \mathrm{d}s \to 0$ *as* $h \to 0$, *uniformly in* $t \in \mathbf{R}$.

Also note that $\mathbf{R}$ can be replaced by an arbitrary interval of the type $J = [S, \infty)$ (one defines $k(t,s)$ to be zero when $s \le S$ or $t \le S$, and applies the theorem above).

As one may see from a careful analysis of the proof of Theorem 9.1 given below, condition (iii) can be slightly weakened. It suffices to assume that the function $t \mapsto \int_{-\infty}^t k(t,s)\, \mathrm{d}s$ is bounded and continuous, and slowly varying at plus and minus infinity in the sense that, for each $T > 0$, $\lim_{t\to\pm\infty} \{\int_{-\infty}^{t+v} k(t+v,s)\, \mathrm{d}s - \int_{-\infty}^t k(t,s)\, \mathrm{d}s\} = 0$, uniformly for v in $[0,T]$.

Comparing Theorem 9.1 with the convolution result mentioned above, we find that all the hypotheses in Theorem 9.1 are satisfied by the convolution kernel, except for (ii), which is satisfied iff the convolution kernel is bounded at zero.

Proof of Theorem 9.1 Let us begin the proof by showing that k is of bounded uniformly continuous type on $\mathbf{R}$. That k is of type L^∞ follows from (i), (iii) and Proposition 2.7(i). To prove continuity, we take $t \ge v$, and use (i) to get

$$
\begin{aligned}
\int_{-\infty}^t |k(t,s) - k(v,s)|\, \mathrm{d}s &= \int_{-\infty}^v |k(t,s) - k(v,s)|\, \mathrm{d}s + \int_v^t k(t,s)\, \mathrm{d}s \\
&= \int_{-\infty}^v k(v,s)\, \mathrm{d}s - \int_{-\infty}^t k(t,s)\, \mathrm{d}s + 2\int_v^t k(t,s)\, \mathrm{d}s \\
&\le \int_{-\infty}^v k(v,s)\, \mathrm{d}s - \int_{-\infty}^t k(t,s)\, \mathrm{d}s + 2\int_v^t k(s,s)\, \mathrm{d}s.
\end{aligned}
$$

By (ii) and (iii), the right-hand side tends to zero as $t - v \to 0$, uniformly in t and v. Thus k is of bounded uniformly continuous type on $\mathbf{R}$.

Since k is of continuous type, we know that k has a resolvent of continuous type on $\mathbf{R}$ (Theorem 5.5(i)), and we also know that if r is of type L^∞ on $\mathbf{R}$ then it must in fact be of bounded uniformly continuous type on $\mathbf{R}$

(Theorem 5.4). By Proposition 2.7(i), r is of type L^∞ if and only if

$$\operatorname*{ess\,sup}_{t\in\mathbf{R}} \int_{-\infty}^{t} |r(t,s)|\,\mathrm{d}s < \infty.$$

Equivalently, using the second part of Lemma 10.1 (where one takes $p = 1$ and $q = \infty$), and the fact that r is of continuous type, we find that r is of type L^∞ if and only if there is a constant K such that, for all $f \in BUC(\mathbf{R};\mathbf{R})$ that vanish on some interval $(-\infty, S]$ (where S depends on f) and satisfy $\|f\|_{\sup(\mathbf{R})} \le 1$, one has

$$\left| \int_{-\infty}^{t} r(t,s) f(s)\,\mathrm{d}s \right| \le K, \quad t \in \mathbf{R}. \tag{9.1}$$

If we let x be the solution of the equation

$$x(t) + \int_{-\infty}^{t} k(t,s) x(s)\,\mathrm{d}s = f(t), \quad t \in \mathbf{R}, \tag{9.2}$$

that vanishes to the left of the support of f, then $x = f - r \star f$, and (9.1) is satisfied for some constant K if and only if x is bounded, uniformly for all f with $\|f\|_{\sup(\mathbf{R})} \le 1$. Therefore, to prove the theorem it suffices to find such a uniform bound.

The argument that shows that the solution x of (9.2) is bounded, uniformly in f, is based on a differentiation of the equation. For this, k and f must be sufficiently smooth. Let us for the moment assume this to be the case; we shall later remove this simplifying assumption. As we show at the end of the proof we may also assume, without loss of generality, that

$$\sup_{t\in\mathbf{R}} \int_{t}^{t+1} |f'(s)|\,\mathrm{d}s \le M < \infty, \tag{9.3}$$

where the constant M is independent of f as long as $\|f\|_{\sup(\mathbf{R})} \le 1$.

To get a uniform bound on x, it suffices to get a uniform bound on $x(t)$ at all those points t where $|x(t)| = \max\{\, |x(s)| \mid s \le t \,\}$. This set can be split into two parts, i.e., those points where $x(t) > 0$, and those points where $x(t) < 0$. One can pass from one of these sets to the other by replacing f by $-f$ and x by $-x$. Therefore, let us restrict our attention to the set

$$E = \{\, t \in \mathbf{R} \mid |x(s)| \le x(t) \text{ for } s \le t \,\}. \tag{9.4}$$

By (iv), we can find constants T and γ such that

$$\int_{-\infty}^{t-T} k(t,s)\,\mathrm{d}s \le \gamma < 1, \quad t \in \mathbf{R}. \tag{9.5}$$

The bound that we get on $x(t)$ for $t \in E$ will depend on whether x changes sign on $(t-T, t]$ or not. Take the easy case first. If $t \in E$ and $x(s) \ge 0$ for all $s \in [t-T,t]$, then by (1.1), (9.4), and (9.5)

$$\begin{aligned} x(t) &= -\int_{-\infty}^{t-T} k(t,s)x(s)\,\mathrm{d}s - \int_{t-T}^{t} k(t,s)x(s)\,\mathrm{d}s + f(t) \\ &\le -\int_{-\infty}^{t-T} k(t,s)x(s)\,\mathrm{d}s + f(t) \\ &\le \gamma x(t) + \|f\|_{\sup(\mathbf{R})}. \end{aligned}$$

Thus, in this case, $x(t) \le (1-\gamma)^{-1}\|f\|_{\sup(\mathbf{R})}$.

To handle the more difficult case, where $t \in E$ and $x(u) = 0$ for some $u \in [t-T,t]$, we differentiate (1.1) to obtain, for almost all $v \in \mathbf{R}$,

$$x'(v) + k(v,v)x(v) = -\int_{-\infty}^{v} \frac{\partial}{\partial v}k(v,s)x(s)\,\mathrm{d}s + f'(v),$$

and solve this equation, using the variation of constants formula for ordinary differential equations, to get

$$\begin{aligned} x(t) &= -\int_{u}^{t} \exp\left(-\int_{v}^{t} k(s,s)\,\mathrm{d}s\right)\int_{-\infty}^{v} \frac{\partial}{\partial v}k(v,s)x(s)\,\mathrm{d}s\,\mathrm{d}v \\ &\quad + \int_{u}^{t} \exp\left(-\int_{v}^{t} k(s,s)\,\mathrm{d}s\right) f'(v)\,\mathrm{d}v \\ &\le a(t)x(t) + MT, \end{aligned}$$

where (9.3) and (9.4) have been invoked, and

$$a(t) = \int_{t-T}^{t} \exp\left(-\int_{v}^{t} k(s,s)\,\mathrm{d}s\right)\int_{-\infty}^{v} \left|\frac{\partial}{\partial v}k(v,s)\right|\,\mathrm{d}s\,\mathrm{d}v. \tag{9.6}$$

If we could show that $a(t) \le \alpha$ for some constant $\alpha < 1$, then $x(t) \le (1-\alpha)^{-1}TM$, and the proof would be complete. Unfortunately, this inequality will not in general be valid. As we state below, it is only true that $a(t) \le \alpha < 1$ for $|t| \ge V$ for some sufficiently large constant V (independent of f), and consequently we have to do some more analysis.

Suppose that $a(t) \le \alpha < 1$ for $|t| \ge V$. Then the preceding argument gives a uniform bound on x in the interval $(-\infty,-V]$, and this implies (by the same argument as we gave above—see (9.1) and (9.2)) that r is of type L^∞ on $(-\infty,-V]$. Similarly, if f and x vanish on $(-\infty,V]$, then we get a uniform bound on x on $[V,\infty)$, and this implies that r is of type L^∞ on $[V,\infty)$. We already know that r is of type L^∞ on $(-V,V)$. Therefore, by Theorem 3.13, r is of type L^∞ on $\mathbf{R}$, and we have obtained the desired conclusion.

It remains to prove that a is bounded in the above fashion. For almost all v,

$$\frac{\mathrm{d}}{\mathrm{d}v}\left(-\int_{v}^{t} k(s,s)\,\mathrm{d}s - K(v)\right) = \int_{-\infty}^{v} \left|\frac{\partial}{\partial v}k(v,s)\right|\,\mathrm{d}s,$$

where

$$K(t) = \int_{-\infty}^{t} k(t,s)\,\mathrm{d}s, \quad t \in \mathbf{R}.$$

Clearly, a can be trivially estimated from above by

$$a(t) \le \Big(\max_{t-T\le v\le t} \exp(K(v))\Big)$$
$$\times \int_{t-T}^{t} \exp\Big(-\int_{v}^{t} k(s,s)\,\mathrm{d}s - K(v)\Big) \int_{-\infty}^{v} \Big|\frac{\partial}{\partial v} k(v,s)\Big| \,\mathrm{d}s\,\mathrm{d}v,$$

and the integration on the right-hand side of this estimate can easily be performed. After an integration one gets

$$\begin{aligned} a(t) &\le \Big(\max_{t-T\le v\le t} \exp(K(v))\Big)\Big/_{t-T}^{t} \exp\Big(-\int_{v}^{t} k(s,s)\,\mathrm{d}s - K(v)\Big) \\ &= \max_{t-T\le v\le t} \exp(K(v) - K(t)) \\ &\quad - \max_{t-T\le v\le t} \exp\big(K(v) - K(t-T)\big) \exp\Big(-\int_{t-T}^{t} k(s,s)\,\mathrm{d}s\Big). \end{aligned}$$

By the last part of (iii), we can force $\max_{t-T\le v\le t} \exp(K(v) - K(t))$ and $\max_{t-T\le v\le t} \exp(K(v) - K(t-T))$ to be arbitrary close to 1 by choosing $|t|$ to be sufficiently large, and by (ii) $\exp\big(-\int_{t-T}^{t} k(s,s)\,\mathrm{d}s\big)$ is bounded away from zero, uniformly in t. Therefore, it is possible to find constants $V \ge 0$ and $\alpha < 1$, such that the right-hand side is less than α for $|t| \ge V$. Thus, the proof is complete under our simplifying smoothness assumptions.

We now turn to the elimination of the auxiliary smoothness conditions. First, we approximate k by a differentiable kernel. For each $h \in (0,1]$, define

$$k_h(t,s) = \frac{1}{h} \int_{t}^{t+h} k(v,s)\,\mathrm{d}v. \tag{9.7}$$

We claim that k_h satisfies (i)–(iv) with k replaced by k_h, uniformly in h, and that $k_h \to k$ in $\mathcal{V}(BUC;\mathbf{R};\mathbf{R})$ as $h \downarrow 0$. The proofs of these claims are straightforward. Clearly, by (i) and (9.7), k_h is nonnegative, $t \to k_h(t,s)$ is nonincreasing in t on $[s,\infty)$ for each $s \ge 0$, and

$$k_h(t,s) \le k(t,s), \quad -\infty < s \le t < \infty.$$

In particular, (ii) and (iv) hold with k replaced by k_h, uniformly in h. That (iii) holds with k replaced by k_h, uniformly in h, follows from the fact that

$$\int_{-\infty}^{t} k_h(t,s)\,\mathrm{d}s = \frac{1}{h} \int_{t}^{t+h} \int_{-\infty}^{t} k(v,s)\,\mathrm{d}s\,\mathrm{d}v.$$

Thus, (i)–(iv) hold with k replaced by k_h, uniformly in h. Moreover, because of (i),

$$\int_{\mathbf{R}} |k_h(t,s) - k(t,s)|\,\mathrm{d}s \le \frac{1}{h} \int_{t}^{t+h} \int_{\mathbf{R}} |k(v,s) - k(t,s)|\,\mathrm{d}s\,\mathrm{d}v,$$

so the fact that k is of bounded uniformly continuous type on $\mathbf{R}$ implies that $k_h \to k$ in $\mathcal{V}(BUC;\mathbf{R};\mathbf{R})$ as $h \downarrow 0$. In other words, the claims made above about the kernels k_h are true. In particular, note that the constants T, γ, V, and α that we picked in the first part of the proof can be chosen independently of h.

Let us use the kernels k_h constructed above to show that one can, without loss of generality, let f in (9.2) not only belong to $BUC(\mathbf{R};\mathbf{R})$ and vanish on an interval $(-\infty, S]$, but also have a derivative that satisfies (9.3). Choose h so small that $|||k_h - k|||_{B^\infty(\mathbf{R})} \leq 1/2$. Then, by Corollary 3.10 and Theorem 5.4, the equation $y + (k - k_h) \star y = f$ has a solution y, which vanishes to the left of the support of f, and is bounded, uniformly in f for $\|f\|_{\sup(\mathbf{R})} \leq 1$. Define $z = y - x$. Then

x is bounded, uniformly in f, if and only if z is bounded and z is the solution of the equation $z + k \star z = k_h \star y$.

The new forcing function $k_h \star y$ belongs to $BUC(\mathbf{R};\mathbf{R})$ and vanishes to the left of the support of f, it is bounded, uniformly in f, it is locally absolutely continuous, and its derivative satisfies (9.3) with f replaced by $k_h \star y$, because, for almost all t (note that h is now fixed),

$$(k_h \star y)'(t) = k_h(t,t)y(t) + \frac{1}{h}\int_{-\infty}^{t} \big(k(t+h,s) - k(t,s)\big)y(s)\,\mathrm{d}s,$$

and, consequently,

$$\sup_{t\in\mathbf{R}} \int_t^{t+1} |(k_h \star y)'(s)|\,\mathrm{d}s \leq \left(\sup_{t\in\mathbf{R}} \int_t^{t+1} k(s,s)\,\mathrm{d}s + \frac{1}{h}|||k|||_{B^\infty(\mathbf{R})}\right) \|y\|_{\sup(\mathbf{R})}.$$

The task of showing that the solution x of the original equation (9.2) is uniformly bounded for $\|f\|_{\sup(\mathbf{R})} \leq 1$ is now seen to be equivalent to showing that for any fixed M the solution of $z + k \star z = f$ is uniformly bounded for f in the set

$$S_M = \Big\{ \in BUC(\mathbf{R};\mathbf{R}) \,\Big|\, \|f\|_{\sup(\mathbf{R})} \leq 1 \text{ and } \sup_{t\in\mathbf{R}} \int_t^{t+1} |f'(s)|\,\mathrm{d}s \leq M \Big\}.$$

Consider the equation $z_h + k_h \star z_h = f$, which approximates $z + k \star z = f$. By the local version of Corollary 3.12, we have $z_h \to z$ for any $t \in \mathbf{R}$ as $h \downarrow 0$. Consequently, if we can bound $\|z_h\|_{\sup(\mathbf{R})}$ uniformly for $h \in (0,1]$ and $f \in S_M$ (with M arbitrary but fixed), then z will be uniformly bounded for $f \in S_M$, and hence, by the previous paragraph, x will be uniformly bounded for $\|f\|_{\sup(\mathbf{R})} \leq 1$, and we are through.

Apply the first part of the proof to the approximating equation $z_h + k_h \star z_h = f$. Because k_h and f are now sufficiently smooth, this can be done. Then recall that the bound that we obtained on the solution depended only on T, α, V, and γ (provided M is fixed). As we observed earlier, these quantitities may all be chosen independently of h, and therefore we have a bound on $\|z_h\|_{\sup(\mathbf{R})}$, which is uniform in $h \in (0,1]$ and in $f \in S_M$.

This completes the proof of Theorem 9.1. □

Theorem 9.1 immediately implies the following corollary.

9.2 Corollary
Let k be a scalar Volterra kernel on $\mathbf{R}$ (i.e., $k : \mathbf{R}^2 \to \mathbf{R}$ is measurable and $k(t,s) = 0$ when $s > t$), satisfying the following conditions:
(i) *k is nonnegative, and $s \mapsto k(t,s)$ is nondecreasing in s on $(-\infty, t]$ for each $t \in \mathbf{R}$;*
(ii) *$\int_s^{s+\delta} k(t,t)\,\mathrm{d}t \to 0$ as $\delta \downarrow 0$, uniformly for $s \in \mathbf{R}$;*
(iii) *the function $s \mapsto \int_s^\infty k(t,s)\,\mathrm{d}t$ is continuous on $\mathbf{R}$ and the two limits $\lim_{s\to\pm\infty} \int_s^\infty k(t,s)\,\mathrm{d}t$ exist;*
(iv) *$\lim_{T\to\infty} \sup_{s\in\mathbf{R}} \int_{s+T}^\infty k(t,s)\,\mathrm{d}t < 1$.*
Then k has a resolvent r of type L^1 on $\mathbf{R}$.

Proof Define $m(t,s) = k(-s,-t)$ for $(t,s) \in \Delta_2(\mathbf{R})$, and apply Theorem 9.1 to get

$$\sup_{t\in\mathbf{R}} \int_{-\infty}^t |q(t,s)|\,\mathrm{d}s < \infty,$$

where q is the resolvent of m. Define $r(t,s) = q(-s,-t)$ for $(s,t) \in \mathbf{R}^2$. Then r is the resolvent of k, and

$$\sup_{s\in\mathbf{R}} \int_s^\infty |r(t,s)|\,\mathrm{d}t < \infty.$$

By Proposition 2.7(ii), r is of type L^1 on $\mathbf{R}$. □

10. Appendix: Some Admissibility Results

10.1 Lemma
Let J be an interval, $1 \le p \le \infty$, and $1/p + 1/q = 1$. Then, for every measurable scalar function f,

$$\sup_{\|g\|_{L^q(J)}\le 1} \int_J |g(t)f(t)|\,\mathrm{d}t = \begin{cases} \|f\|_{L^p(J)}, & \text{if } f \in L^p(J;\mathbf{C}), \\ \infty, & \text{if } f \notin L^p(J;\mathbf{C}). \end{cases}$$

Moreover, there is a countable set $\{g_n \mid n \in \mathbf{N}\}$ of continuous scalar functions g_n (independent of f) with compact support in J satisfying $\|g_n\|_{L^q(J)} \le 1$, such that

$$\sup_{n\in\mathbf{N}} \left| \int_J g_n(t)f(t)\,\mathrm{d}t \right| = \begin{cases} \|f\|_{L^p(J)}, & \text{if } f \in L^p(J;\mathbf{C}), \\ \infty, & \text{if } f \notin L^p(J;\mathbf{C}). \end{cases}$$

Proof For $1 < p \le \infty$ the first assertion follows from the fact that L^p is the dual of L^q. The second assertion follows since, for example, the set of L^q-functions with compact support is dense in $L^q(J)$, and since on each compact subinterval K of J the set of polynomials with rational coefficients is dense in $L^q(K)$. For $p = 1$, we choose g to be $g(t) = \overline{f(t)}/|f(t)|$, to get $\|g\|_{L^\infty(J)} = 1$ and $\int_J f(t)g(t)\,\mathrm{d}t = \|f\|_{L^1(J)}$, and recall that the set of L^∞-functions with compact support is weak*-dense in $L^\infty(J)$, and that the set of polynomials with rational coefficients is weak*-dense in $L^\infty(K)$ for compact subintervals K of J. □

10.2 Lemma

Let k be measurable on J^2, with values in $\mathbf{C}^{n\times n}$, and suppose that $k \star f \in L^\infty(J;\mathbf{C}^n)$ for every $f \in L^\infty(J;\mathbf{C}^n)$. Then k is of type L^∞ on J.

Proof It suffices to prove the lemma in the scalar case $n = 1$, because one can get the general case by applying the scalar case to each component separately.

First, let us show that $f \mapsto k \star f$ is continuous from $L^\infty(J;\mathbf{C})$ into itself. To prove this, it suffices to show that this operator is closed, and to apply the closed graph theorem. Let $f_n \to f$ in $L^\infty(J;\mathbf{C})$, and suppose, in addition, that $k \star f_n \to g$ in $L^\infty(J;\mathbf{C})$. The function $v(t) \equiv 1$ belongs to $L^\infty(J;\mathbf{C})$, so, by the hypothesis on k, $k \star v$ belongs to $L^\infty(J;\mathbf{C})$. In particular, $(k\star v)(t)$ is defined for almost all t, and this means, by Definition 2.3, that, for almost all t, the function $s \mapsto k(t,s)v(s)$, or equivalently $s \mapsto k(t,s)$, is integrable on J. By Lebesgue's dominated convergence theorem, for those t, $(k\star f_n)(t) \to (k\star f)(t)$. On the other hand, $k\star f_n \to g$ in $L^\infty(J;\mathbf{C})$, so $k \star f_n$ converges pointwise a.e. to g. Therefore, $k \star f = g$ a.e., and so $f \mapsto k \star f$ is closed, hence continuous.

It remains to show that if the operator $f \mapsto k \star f$ is continuous from $L^\infty(J;\mathbf{C})$ into itself, then k is of type L^∞ on J. To do this, we choose a countable set $F = \{\, g_n \mid n \in \mathbf{N}\,\}$ of continuous functions with compact support, satisfying $\|g_n\|_{\sup(\mathbf{R})} \le 1$, of the type described in Lemma 10.1 (with $p = 1$ and $q = \infty$). Since $f \mapsto k\star f$ is continuous, there is a constant K such that $\|k \star f\|_{L^\infty(J)} \le K$ for all $f \in L^\infty(J;\mathbf{C})$ with $\|f\|_{L^\infty(J)} \le 1$. In particular,

$$\begin{aligned} K \ge \sup_{n\in\mathbf{N}} \|k \star g_n\|_{L^\infty(J)} &= \sup_{n\in\mathbf{N}} \operatorname*{ess\,sup}_{t\in J} \left| \int_J k(t,s) g_n(s)\,\mathrm{d}s \right| \\ &= \operatorname*{ess\,sup}_{t\in J} \sup_{n\in\mathbf{N}} \left| \int_J k(t,s) g_n(s)\,\mathrm{d}s \right| \\ &= \operatorname*{ess\,sup}_{t\in J} \int_J |k(t,s)|\,\mathrm{d}s, \end{aligned} \qquad (10.1)$$

and, by Proposition 2.7(i), k is of type L^∞ on J. □

The observant reader will realize that the proof above in fact gives us something more than Lemma 10.2, namely, the following result.

10.3 Lemma
Let $J \subset \mathbf{R}$ be an interval with right end-point $+\infty$, and let k be measurable on J^2, with values in $\mathbf{C}^{n\times n}$. If $k\star f \in L^\infty(J;\mathbf{C}^n)$ for every $f \in BC_0(J;\mathbf{C}^n)$, then k is of type L^∞ on J.

By duality, from Lemma 10.2 we get an analogous result in L^1.

10.4 Lemma
Let $k : J^2 \to \mathbf{C}^{n\times n}$ be measurable, and suppose that $k \star f \in L^1(J;\mathbf{C}^n)$ for every $f \in L^1(J;\mathbf{C}^n)$. Then k is of type L^1 on J.

Proof Without loss of generality, we again consider the scalar case, i.e., we take $n = 1$.

First, let us use the closed graph theorem to show that the mapping $f \mapsto k \star f$ is continuous from $L^1(J;\mathbf{C})$ into itself. Let $f_j \to f$ in $L^1(J;\mathbf{C})$, and assume that $k \star f_j \to g$ in $L^1(J;\mathbf{C})$. Without loss of generality, assume that both sequences converge pointwise a.e., and that $\sum_{j=1}^\infty \|f_j - f_{j+1}\|_{L^1(J)} < \infty$ (replace the original sequence by a subsequence). Then

$$F \stackrel{\text{def}}{=} \sum_{j=1}^{\infty} |f_j - f_{j+1}| \in L^1(J;\mathbf{C}),$$

and hence $k\star F \in L^1(J;\mathbf{C})$. By Definition 2.3, this means that the function $s \mapsto k(t,s)F(s)$ belongs to $L^1(J;\mathbf{C})$ for almost all $t \in J$. Since $|f_j| \le |f| + F$, we may apply Lebesgue's dominated convergence theorem to show that $(k\star f_j)(t) \to (k\star f)(t)$ for almost all $t \in J$. We conclude that $g(t) = (k\star f)(t)$ a.e. on J, and the closed graph theorem now tells us that the mapping $f \mapsto k\star f$ is continuous from $L^1(J;\mathbf{C})$ into itself.

Let φ be some strictly positive function in $L^1(J;\mathbf{R})$. Then, since $k \star \varphi$ is defined a.e., we have $\int_J |k(t,s)\varphi(s)|\,\mathrm{d}s < \infty$ for almost all $t \in J$. Thus, if for each $i \in \mathbf{N}$ we define $E_i = \{\, t \in J \mid \int_J |k(t,s)\varphi(s)|\,\mathrm{d}s > i\varphi(t)\,\}$, then $m(E_i) \to 0$ as $i \to \infty$. Moreover, for each i, the function $(s,t) \mapsto \chi_{J\setminus E_i}(t)k(t,s)\varphi(s)$ belongs to $L^1(J^2;\mathbf{C})$. In particular, for almost all $s \in J$, it is true that $\int_J \bigl|\chi_{J\setminus E_i}(t)k(t,s)\varphi(s)\bigr|\,\mathrm{d}t < \infty$. Since φ is strictly positive, this means that $\int_J \bigl|\chi_{J\setminus E_i}(t)k(t,s)\bigr|\,\mathrm{d}t < \infty$ for almost all $s \in J$. Define $F_{i,j} = \{\, s \in J \mid \int_J \bigl|\chi_{J\setminus E_i}(t)k(t,s)\bigr|\,\mathrm{d}t > j\,\}$. Then $m(F_{i,j}) \to 0$ as $j \to \infty$.

Define $k_{i,j}(t,s) = \chi_{J\setminus E_i}(t)k(t,s)\chi_{J\setminus F_{i,j}}(s)$. Let α be the norm of the mapping $f \mapsto k \star f$, i.e., let

$$\alpha = \sup_{\|f\|_{L^1(J)}\le 1} \|k \star f\|_{L^1(J)}.$$

Then, for each $f \in L^1(J)$, we have

$$\begin{aligned}\|k_{i,j} \star f\|_{L^1(J)} &= \|k \star (\chi_{J\setminus F_{i,j}} f)\|_{L^1(J\setminus E_i)} \\ &\le \alpha \|\chi_{J\setminus F_{i,j}} f\|_{L^1(J)} \le \alpha \|f\|_{L^1(J)}.\end{aligned}$$

Thus, the mapping $f \mapsto k_{i,j} \star f$ is continuous from $L^1(J)$ into itself, and the norm of this mapping is at most α.

By Proposition 2.7(ii), and the definition of the sets $F_{i,j}$, the kernel $k_{i,j}$ is of type L^1 on J. Therefore, the adjoint kernel $h_{i,j}(t,s) \stackrel{\text{def}}{=} k_{i,j}(s,t)$ is of type L^∞ (see (7.1)). Also, by (10.1),

$$\begin{aligned}\operatorname*{ess\,sup}_{t\in J}\int_J |h_{i,j}(t,s)|\,\mathrm{d}s &= \sup_{\|g\|_{L^\infty(J)}\le 1} \|h_{i,j}\star g\|_{L^\infty(J)}\\ &= \sup_{\|f\|_{L^1(J)}\le 1} \|k_{i,j}\star f\|_{L^1(J)} \le \alpha.\end{aligned}$$

Consequently,

$$\operatorname*{ess\,sup}_{s\in J\setminus F_{i,j}}\int_{J\setminus E_i} |k(t,s)|\,\mathrm{d}t \le \alpha.$$

If we first let $j\to\infty$, and then $i\to\infty$, we get

$$\operatorname*{ess\,sup}_{s\in J}\int_J |k(t,s)|\,\mathrm{d}t \le \alpha,$$

and this, combined with Proposition 2.7(ii), completes the proof. □

11. Exercises

1. Let a be a measurable function defined on $\mathbf{R}$, and define $k(t,s) = a(t-s)$. Suppose that the restriction of k to J^2 is of type L^p, for some finite interval $J = (S,T)$, and some $p \in [1,\infty]$. Show that $a \in L^1(S-T, T-S)$.
2. Let a be a kernel of type $[L^p; L^q]$, b a kernel of type $[L^r; L^p]$, and f a function in L^r. Show that $a \star b$ is a kernel of type $[L^r; L^q]$, and that $b \star f \in L^p$.
3. Let $J \subset \mathbf{R}$ be an interval, let $p \in (1,\infty)$ and $1/p + 1/q = 1$, and let $k : J^2 \to \mathbf{C}^{n\times n}$ be measurable. Prove the following claims.
   (i) *k is of type $[L^p; L^\infty]$ on J if and only if*

   $$\operatorname*{ess\,sup}_{t\in J}\left(\int_J |k(t,s)|^q\,\mathrm{d}s\right)^{1/q} < \infty.$$

   Moreover,

   $$|||k|||_{L^{p,\infty}(J)} = \operatorname*{ess\,sup}_{t\in J}\left(\int_J |k(t,s)|^q\,\mathrm{d}s\right)^{1/q}.$$

   (ii) *k is of type $[L^1; L^p]$ on J if and only if*

   $$\operatorname*{ess\,sup}_{s\in J}\left(\int_J |k(t,s)|^p\,\mathrm{d}t\right)^{1/p} < \infty.$$

Moreover,

$$|||k|||_{L^{1,p}(J)} = \operatorname*{ess\,sup}_{s\in J}\left(\int_J |k(t,s)|^p\,dt\right)^{1/p}.$$

(iii) *k is of type $[L^1; L^\infty]$ on J if and only if $k \in L^\infty(J^2)$, and*

$$|||k|||_{L^{1,\infty}(J)} = \|k\|_{L^\infty(J^2)}.$$

4. (Cf. Proposition 2.7.) Let $p \in (1,\infty)$, and $1/p + 1/q = 1$. Show that
$$\left\{\int_J\left(\int_J |k(t,s)|^q\,ds\right)^{p/q} dt\right\}^{1/p} \ge \left\{\int_J\left(\int_J |k(t,s)|^p\,dt\right)^{q/p} ds\right\}^{1/q},$$
when $p \le q$, and that the inequality is reversed when $p \ge q$. What is the analogous result when $p = 1$ or $p = \infty$?
5. (Very easy) Let r be the resolvent of k. Find the resolvent of $-r$.
6. Let $p \in [1,\infty]$ and $1/p + 1/q = 1$. Let k be a Volterra kernel of type $[L^p; J; \mathbf{C}^{n\times n}]$ on some interval $J \subset \mathbf{R}$, and suppose that k has a resolvent of type L^p on J. Show that for each (row-vector-valued) $g \in L^q(J; \mathbf{C}^n)$, the equation
$$y(s) + \int_J y(t)k(t,s)\,dt = g(s), \quad s \in J,$$
has a unique (row-vector-valued) solution y in $L^q(J; \mathbf{C}^n)$. How is this solution computed? Also prove the analogous local result when J is an interval of the form $J = (S,T]$, with $-\infty \le S < T < \infty$, and k is a Volterra kernel of type $[L^p_{\text{loc}}; J; \mathbf{C}^{n\times n}]$. (The equation above is usually called the *adjoint* equation of (3.3).)
7. Let k be the Fredholm kernel $k(t,s) = a(t)b(s)$, where $a \in L^p(J)$ and $b \in L^q(J)$, with $p \in [1,\infty]$ and $1/p+1/q = 1$. Under what assumption does k have a resolvent of type L^p on J? Find this resolvent.
8. Try to modify the previous exercise assuming that $k(t,s) = a(t)b(s)$ for $t \ge s$, but $k(t,s) = 0$ for $t < s$. Does this change the situation at all?
9. Give an example of a Volterra kernel of type L^∞ on $[0,T]$, $T > 0$, that does not satisfy the assumptions of Corollary 3.14, but nevertheless has a resolvent of type L^∞. Hint: Try $k(t,s) = t^{-1}$, $0 < s \le t \le 1$.
10. Give a metric in the space $\mathcal{K}(C; J; \mathbf{C}^{n\times n})$.
11. Give an example of a scalar Volterra kernel k of bounded type on the interval $[0,1]$ that satisfies
$$\liminf_{\epsilon\downarrow 0} |||k|||_{B^\infty([\frac{1}{2}-\epsilon,\frac{1}{2}+\epsilon])} > 1,$$
but that nevertheless has a resolvent of bounded type on $[0,1]$. Can you construct a kernel of continuous type with the same properties? Hint: Cf. Exercise 9.
12. In Section 6 we throughout considered limits at $+\infty$. Can you prove analogous results concerning limits at $-\infty$?
13. Show that there are kernels of type B^∞ that cannot be approximated

(in the norm of $\mathcal{K}(B^\infty)$) by kernels of type BC.

14. Let J be an interval of the type $J = [T, \infty)$, for some $T \in \mathbf{R}$. Prove the analogue of Theorem 3.19 for the following classes of kernels:
   (i) $\mathcal{V}(B^\infty; J; \mathbf{C}^{n\times n})$;
   (ii) $\mathcal{V}(BC; J; \mathbf{C}^{n\times n})$;
   (iii) $\mathcal{V}(BUC; J; \mathbf{C}^{n\times n})$;
   (iv) $\mathcal{V}(B_0^\infty; J; \mathbf{C}^{n\times n})$;
   (v) $\mathcal{V}(BC_0; J; \mathbf{C}^{n\times n})$;
   (vi) $\mathcal{V}(B_\ell^\infty; J; \mathbf{C}^{n\times n})$;
   (vii) $\mathcal{V}(BC_\ell; J; \mathbf{C}^{n\times n})$;
   (viii) $\mathcal{V}(B^\infty(\mathcal{T}_S) \oplus B_0^\infty(J); \mathbf{C}^{n\times n})$;
   (ix) $\mathcal{V}(C(\mathcal{T}_S) \oplus BC_0(J); \mathbf{C}^{n\times n})$.
15. In Theorem 3.19, replace J by $\mathbf{R}$, and assume, in addition, that $\lim_{S\to-\infty} |||b|||_{L^p(-\infty,S)} = 0$. Prove that k has a resolvent of type L^p on $\mathbf{R}$.
16. Improve the result in Exercise 15 by allowing k to have one representation of the type $k(t,s) = a(t-s) + b(t,s)$ for $-\infty < s \le t \le 0$, and a different representation of the same type for $0 < s \le t < \infty$.
17. Redo Exercise 14 with J replaced by $\mathbf{R}$, and Theorem 3.19 replaced by Exercise 16.
18. Prove Theorem 6.7.
19. Show that the Banach algebras discussed in this chapter do not have units.
20. Prove the following theorem (see Theorem 7.5).

   11.1 Theorem
   Let $J \subset \mathbf{R}$ be an interval, and let k belong to some closed subalgebra $\mathcal{A}$ of $\mathcal{K}(B^\infty; J; \mathbf{C}^{n\times n})$. Let $\lambda \in C([0,1]; \mathbf{C})$ satisfy $\lambda(0) = 0$ and $\lambda(1) = 1$, and suppose that for all $s \in [0,1]$, $\lambda(s)k$ has a resolvent in the Banach algebra of all bounded linear operators on $B^\infty(J; \mathbf{C}^n)$. Then, for all $s \in [0,1]$, $\lambda(s)k$ has a resolvent in $\mathcal{A}$.

21. Prove the following result.

   11.2 Theorem
   Let $J \subset \mathbf{R}$ be an interval, and let $k \in \mathcal{V}(B^\infty; J; \mathbf{C}^{n\times n})$ satisfy the assumption of Corollary 4.3, with L^p replaced by B^∞. Then the spectrum of k in the Banach algebra $\mathcal{V}(B^\infty; J; \mathbf{C}^{n\times n})$ is the same as the spectrum of the operator $f \mapsto k \star f$ in the Banach algebra of all bounded linear operators on $B^\infty(J; \mathbf{C}^n)$.

22. Let $J = \mathbf{R}$ or $\mathbf{R}^+$, and let $k \in L^1(J; \mathbf{C}^{n\times n})$. Show that the spectrum of k in the Banach algebra $L^1(J; \mathbf{C}^{n\times n})$ (with the convolution product) is given by
   (i) $\{0\} \cup \left\{\bigcup_{\Re z = 0} \sigma(\hat{k}(z))\right\}$, if $J = \mathbf{R}$,
   (ii) $\{0\} \cup \left\{\bigcup_{\Re z \ge 0} \sigma(\hat{k}(z))\right\}$, if $J = \mathbf{R}^+$.

Above $\sigma(\hat{k}(z))$ is the spectrum of $\hat{k}(z)$ in the Banach algebra $\mathbf{C}^{n\times n}$. Moreover, show that in both cases, the spectrum of k is connected.

23. Let $J = \mathbf{R}$ or $\mathbf{R}^+$, let $k \in L^1(J;\mathbf{C}^{n\times n})$, and let V be one of the spaces in Theorem **2**.4.5 or **2**.4.6. Show that the spectrum of k in the Banach algebra $L^1(J;\mathbf{C}^{n\times n})$ is is the same as the spectrum of the operator $f \mapsto k * f$ in the Banach algebra of all bounded linear operators on V.
24. Let $k \in L^1(J;\mathbf{C}^{n\times n})$ be completely monotone. Show that the spectrum of k in the Banach algebra $L^1(J;\mathbf{C}^{n\times n})$ is contained in the circle $\left|z - \frac{\lambda}{2}\right| \le \frac{\lambda}{2}$, if λ is chosen so that $\int_{\mathbf{R}^+} k(t)\,\mathrm{d}t \preceq \lambda I$. Hint: If you slightly increase the value of λ, then $\frac{1}{\lambda}k$ is the resolvent of a completely monotone integrable function. See also Theorem 8.6.
25. Let $k(t) = \mathrm{e}^{-\epsilon t}$ for $t \ge 0$. Find the resolvent r of k, and compute the spectra of k and r in the Banach algebra $L^1(\mathbf{R}^+;\mathbf{C})$. What happens when $\epsilon \downarrow 0$. Compare your result to Theorem 8.6 and Exercise 24.
26. Define the following classes of kernels, and formulate the analogues of the appropriate results in Sections 5 and 6:
    (i) $\mathcal{K}(L_0^\infty;J;\mathbf{C}^{n\times n})$;
    (ii) $\mathcal{K}(L_\ell^\infty;J;\mathbf{C}^{n\times n})$;
    (iii) $\mathcal{K}(L^\infty;L_0^\infty;J;\mathbf{C}^{n\times n})$;
    (iv) $\mathcal{K}(L^\infty;\mathcal{T}_S;\mathbf{C}^{n\times n})$;
    (v) $\mathcal{K}(L^\infty(\mathcal{T}_S) \oplus L_0^\infty(J);\mathbf{C}^{n\times n})$.
27. Formulate and prove a generalization of Proposition 8.1 in $\mathbf{R}^n$. Hint: Consider a matrix to be nonnegative if all its elements are nonnegative.
28. Formulate and prove a version of Lemma 8.2 with $J = \mathbf{R}$ (and k and r of type L^p instead of type L^p_{loc}.)
29. Let k be a scalar Volterra kernel of type L^p_{loc} on some interval J, and suppose that, for each $\lambda \in [0,1]$, λk has a resolvent r_λ. Moreover, suppose that the resolvent r of k is nonnegative. Show that k is nonnegative. Hint: Try to find the smallest possible $\alpha \ge 0$ such that r_λ is positive for all $\lambda \in [\alpha, 1]$.
30. Assume that $k : J^2 \to (0,\infty)$ is twice continuously differentiable and positive. Show that $k(v,s)k(t,u) \le k(t,s)k(v,u)$ when $s \le u \le v \le t$ if and only if $\frac{\partial^2}{\partial s \partial t}\ln(k(t,s)) \le 0$.
31. Prove Lemma 10.3.
32. Let k be measurable on J^2, let p, $q \in [1,\infty]$, and suppose that $k \star f \in L^q$ whenever $f \in L^p$. Prove that the operator $f \mapsto k \star f$ is continuous from L^p into L^q. Hint: See the proof of Lemma 10.4.
33. Prove the following result.

11.3 Lemma

Let k be measurable on J^2, and let $p \in [1, \infty]$. Then the following claims hold.

(i) *k is of type $[L^p; L^\infty]$ on J if and only if $k \star f \in L^\infty(J)$ for every $f \in L^p(J)$.*

(ii) *k is of type $[L^1; L^p]$ on J if and only if $k \star f \in L^p(J)$ for every $f \in L^1(J)$.*

12. Comments

Section 2:

The theory of Fredholm kernels of type L^p, and, more generally, the theory of kernels of type $[L^p; L^q]$, is developed in Zaanen [1] in a more general Orlicz-space setting. In particular, see Zaanen [1], Ch. 9, Sec. 7 and Ch. 13, Sec. 3. Here references to earlier work can be found. In Krasnosel'skiĭ *et al.* [1], this class of integral operators is called regular (see Krasnosel'skiĭ *et al.* [1], Theorem 4.2, p. 68). In this connection, Dunford and Schwartz [1], Ch. VI, is also relevant. A brief, more recent treatment is Jörgens [1]. See also Fenyő and Stolle [1], part I, Sec. 4.5, p. 286. For L^2-theory in particular, see Smithies [1].

The following example from Krasnosel'skiĭ *et al.* [1], p. 73, shows that k may define a continuous map of $L^p(J)$ into itself although $|||k|||_{L^p(J)} = \infty$. Take $n = 1$, $p = 2$ and define, for $i, j = 1, 2, \ldots$,

$$k(t,s) = \frac{2^{\frac{i+j}{2}}}{i - j + 1/2}, \quad 2^{-i} \le t < 2^{-i+1}, \quad 2^{-j} \le s < 2^{-j+1}.$$

Then $f \mapsto k \star f$ is a continuous map of $L^2(0,1)$ into itself, but $|||k|||_{L^2(0,1)} = \infty$.

By Lemma 10.1, one has (cf. equation (2.2))

$$|||k|||_{L^{p,q}(J)} = \sup_{\|f\|_{L^p(J)} \le 1} \| \, |k| \star |f| \, \|_{L^q(J)},$$

and so k is of type $[L^p; L^q]$ if and only if k is measurable and $f \mapsto |k| \star f$ is continuous from $L^p(J; \mathbf{C})$ into $L^q(J; \mathbf{C})$. Moreover, one can show that the kernels of type $[L^p; L^q]$ form a Banach space with norm $|||\cdot|||_{L^{p,q}(J)}$. An example of a kernel of this type is $k(t,s) = |t-s|^{\alpha-1}$, with $\alpha \in (0,1)$. If $q^{-1} = p^{-1} - \alpha$, $p > 1$ and $q < \infty$, then k is a Fredholm kernel of type $[L^p; L^q]$ on $\mathbf{R}$. See, e.g., Stein [1], p. 119. It is also possible to prove, by using basically the same arguments as in the proof of Theorem 2.4, that if a is of type $[L^q; L^r]$ and b is of type $[L^p; L^q]$ then $a \star b$ is of type $[L^p; L^r]$ with $|||a \star b|||_{L^{p,r}(J)} \le |||a|||_{L^{q,r}(J)} |||b|||_{L^{p,q}(J)}$.

In the previous literature on integral equations, the term admissibility (which has its origins in Massera and Schäffer [1]) has been employed to denote the following. Suppose F_1, F_2 are two spaces of functions defined on J and taking values in $\mathbf{C}^n$. The pair (F_1, F_2) is said to be *admissible* with respect to the integral operator defined by the kernel k if $(k \star f) \in F_2$ for $f \in F_1$. See, e.g., Corduneanu [9], Ch. 2, p. 23, Gollwitzer [1], and Miller [10], Ch. V, p. 252. Using this concept, one part of Theorem 2.4 may be reformulated as follows. If k is of type $[L^p; J; \mathbf{C}^{n\times n}]$, then the pair $(L^p(J; \mathbf{C}^n); L^p(J; \mathbf{C}^n))$ is admissible with respect to k. Lemma 10.2 says that the admissibility of the integral operator defined by the kernel k, with respect to the pair $(L^\infty(J; \mathbf{C}^n); L^\infty(J; \mathbf{C}^n))$, implies that k is of type L^∞. Lemmas 10.3, 10.4, and 11.3 can be given similar formulations.

No reference has been made to the theory of Calderón–Zygmund singular integral operators since this topic falls outside the scope of the present work.

Section 3:

The existence of the resolvent of a Volterra kernel of type L^p in the case where $p \in (1, \infty)$, and both the numbers $\int_J (\int_J |k(t,s)|^q \, ds)^{p/q} \, dt$ and $\int_J (\int_J |k(t,s)|^p \, dt)^{q/p} \, ds$ are finite, is proved in Miller [10], Ch. 4. Pointwise continuous kernels are also considered in this reference. See also Fenyő and Stolle [1], Part 3, Ch. 13.

Kernels of type L^∞ are analysed in Gripenberg [10], where it is shown that if $\int_0^t |k(t,s)| \, ds \in L^\infty(0,T)$, and

$$\limsup_{h \downarrow 0} \left(\operatorname*{ess\,sup}_{t \in (0,T)} \int_{|t-h|_+}^t |k(t,s)| \, ds \right) < 1,$$

then k has a resolvent satisfying $\int_0^t |r(t,s)| \, ds \in L^\infty(0,T)$. The corresponding results for kernels of type L^1 are obtained in Gripenberg [10] and in Staffans [23]. One version of Theorem 3.13 is stated in Staffans [23], p. 23. Examples of kernels $a(t,s)$ with the structure $a(t,s) = a_1(t-s)a_2(s)$ having resolvents of types L^∞ and L^1 are given in Miller, Nohel, and Wong [1]. (Here a_1 is supposed to be log-convex; cf. Section 8.)

Scalar singular Volterra equations of the type

$$x(t) = \int_0^t k(t,s)h(s)x(s) \, ds + f(t), \quad t > 0,$$

where $\int_0^1 |h(s)| \, ds = \infty$ have been considered by Reynolds [1]. In particular, it is shown that, if h_+ and h_- (the positive and negative parts of h) satisfy $\int_0^1 h_+(s) \, ds = \infty$, $\int_0^1 h_-(s) \, ds < \infty$, and $f \equiv 0$, then there is a one-parameter family $\{x_\delta\}$ of solutions of the form

$$x_\delta(t) = \delta e(t) \exp\left(-\int_t^1 h(s) \, ds \right), \quad t > 0, \quad \delta \in \mathbf{R},$$

where e is an absolutely continuous function with $e(0) = 1$.

Sections 5–6:

Kernels of bounded continuous type and of type BC_0 on $\mathbf{R}^+$ have been frequently used, see, e.g., Strauss [1], Miller, Nohel, and Wong [1], Kaplan

[1], Jordan and Wheeler [1], and Gripenberg [3] and [10]. Theorem 6.3 is proved in Strauss [1], Proposition 4, p. 573.

Periodic nonconvolution kernels are briefly mentioned in Kaplan [1] and in Miller, Nohel, and Wong [1]. See also Russell [1].

Section 7:

For compactness results, see Jörgens [1], Ch. 4, Sec. 11, Krasnosel'skiĭ *et al.* [1], Ch. 2, Secs. 5 and 6, and also Fenyő and Stolle [1], part 1, p. 291.

Let $k \in \mathcal{F}(L^p; J; \mathbf{C}^{n\times n})$ for some $p \in (1,\infty)$. Suppose that $m(J) < \infty$, or that k can be approximated by kernels k_n having compact support in the sense that $|||k - k_n|||_{L^p(J)} \to 0$ for $n \to \infty$. Then a necessary and sufficient condition that k be compact as a map of $L^p(J; \mathbf{C}^n)$ into itself is that

$$\lim_{\substack{m(D_1)+m(D_2)\to 0\\ D_1,D_2\subset J}} \left(\sup_{\substack{\|f\|_{L^p(J)}\le 1\\ \|g\|_{L^q(J)}\le 1}} \int_{D_1}\int_{D_2} |g(t)k(t,s)f(s)|\,\mathrm{d}s\,\mathrm{d}t \right) = 0.$$

See Krasnosel'skiĭ *et al.* [1], Ch. 1, Sec. 3.

Sections 8–9:

An early version of Theorem 8.5 combined with Theorem 8.6 is given in Miller, Nohel, and Wong [1], Theorem 6. This version is extended in Gripenberg [3].

Some further results on the behaviour of resolvents of log-convex convolution kernels are given in Gripenberg [25]. A slightly weaker version of Theorem 9.1 is given in Gripenberg [10], Theorem 5.

References

C. Corduneanu

9. *Integral Equations and Stability of Feedback Systems*, Academic Press, New York, 1973.

N. Dunford and J. T. Schwartz

1. *Linear Operators, Part I: General Theory*, J. Wiley, New York, 1957.

I. Fenyő and H. W. Stolle

1. *Theorie und Praxis der linearen Integralgleichungen, 1–4*, Birkhäuser, Basel, 1982–1984.

H. E. Gollwitzer

1. Admissibility and integral operators, *Math. Systems Theory* **7** (1973), pp. 219–231.

G. Gripenberg

3. *On Volterra equations with nonconvolution kernels*, Report-HTKK-MAT-A118, Helsinki University of Technology, 1978.
10. On the resolvents of nonconvolution Volterra kernels, *Funkcial. Ekvac.* **23** (1980), pp. 83–95.
25. Asymptotic estimates for resolvents of Volterra equations, *J. Differential Equations* **46** (1982), pp. 230–243.

G. S. Jordan and R. L. Wheeler

1. On the asymptotic behavior of perturbed Volterra integral equations, *SIAM J. Math. Anal.* **5** (1974), pp. 273–277.

K. Jörgens

1. *Linear Integral Operators*, Pitman, London, 1982.

J. L. Kaplan

1. On the asymptotic behavior of Volterra integral equations, *SIAM J. Math. Anal.* **3** (1972), pp. 148–156.

M. A. Krasnosel'skiĭ, P. P. Zabreĭko, E. I. Pustyl'nik, and P. E. Sobolevskiĭ (Krasnosel'skiĭ *et al.*)

1. *Integral Operators in Spaces of Summable Functions*, Noordhoff, Leyden, 1976.

M. Marcus

1. *Introduction to Modern Algebra*, Marcel Dekker, New York, 1978.

J. L. Massera and J. J. Schäffer

1. *Linear Differential Equations and Function Spaces*, Academic Press, New York, 1966.

R. K. Miller

10. *Nonlinear Volterra Integral Equations*, W. A. Benjamin, Menlo Park, Calif., 1971.

R. K. Miller, J. A. Nohel, and J. S. W. Wong

1. Perturbations of Volterra integral equations, *J. Math. Anal. Appl.* **25** (1969), pp. 676–691.

D. W. Reynolds

1. On linear singular Volterra integral equations of the second kind, *J. Math. Anal. Appl.* **103** (1984), pp. 230–262.

W. Rudin

2. *Functional Analysis*, McGraw–Hill, New York, 1973.

D. L. Russell

1. A Floquet decomposition for Volterra equations with periodic kernel and a transform approach to linear recursion equations, *J. Differential Equations* **68** (1987), pp. 41–71.

F. Smithies

1. *Integral Equations*, Cambridge University Press, Cambridge, 1958.

O. J. Staffans

23. On a nonconvolution Volterra resolvent, *J. Math. Anal. Appl.* **108** (1985), pp. 15–30.

E. M. Stein

1. *Singular Integrals and Differentiability Properties of Functions*, Princeton University Press, Princeton, 1970.

E. M. Stein and G. Weiss

1. *Introduction to Fourier Analysis on Euclidean Spaces*, Princeton University Press, Princeton, 1971.

A. Strauss

1. On a perturbed Volterra integral equation, *J. Math. Anal. Appl.* **30** (1970), pp. 564–575.

A. C. Zaanen

1. *Linear Analysis*, North-Holland, Amsterdam, 1964.

10

Linear Nonconvolution Equations with Measure Kernels

We investigate nonconvolution measure-valued Volterra kernels. First we introduce the class of kernels of type B^∞, and then we show that under weak conditions, kernels of this type have resolvents of the same type. For the study of integrodifferential equations, measure kernels of type $[B^\infty; L^1]$ and kernels of type $[M; B^\infty]$ are used.

1. Introduction

In Chapter 2 we studied the convolution equation

$$x(t) + \int_0^t k(t-s)x(s)\,\mathrm{d}s = f(t), \quad t \in \mathbf{R}^+. \tag{1.1}$$

In connection with the analysis of (1.1) in weighted spaces (Chapter 4), we replaced the kernel k by a measure μ, obtaining the more general equation

$$x(t) + \int_{[0,t]} \mu(\mathrm{d}s)x(t-s) = f(t), \quad t \in \mathbf{R}^+. \tag{1.2}$$

In Chapter 9 we generalized (1.1) in a different direction. There we studied the nonconvolution equation

$$x(t) + \int_0^t k(t,s)x(s)\,\mathrm{d}s = f(t), \quad t \in \mathbf{R}^+. \tag{1.3}$$

The differentiated version of (1.2), i.e.,

$$x'(t) + \int_{[0,t]} \mu(\mathrm{d}s)x(t-s) = f(t), \quad t \in \mathbf{R}^+; \quad x(0) = x_0, \tag{1.4}$$

was studied in Chapter 3.

Obviously, both (1.2) and (1.3) are particular cases of the equation

$$x(t) + \int_{[0,t]} \kappa(t, \mathrm{d}s)x(s) = f(t), \quad t \in \mathbf{R}^+, \tag{1.5}$$

where for each t the kernel $\kappa(t, \cdot)$ is a matrix of measures. Analogously, one may view (1.4) as a particular case of the equation

$$x'(t) + \int_{[0,t]} \kappa(t, \mathrm{d}s)x(s) = f(t), \quad t \in \mathbf{R}^+; \quad x(0) = x_0. \tag{1.6}$$

In this chapter, we formulate a theory for nonconvolution equations with measure kernels. In addition to (1.5) and (1.6), we study the equations

$$x(t) + \int_{(-\infty,t]} \kappa(t, \mathrm{d}s)x(s) = f(t), \quad t \in \mathbf{R}, \tag{1.7}$$

and

$$x'(t) + \int_{(-\infty,t]} \kappa(t, \mathrm{d}s)x(s) = f(t), \quad t \in \mathbf{R}. \tag{1.8}$$

Since the kernel $\kappa(t, \cdot)$ is assumed to be a t-dependent matrix of measures, the equations we consider include, e.g., delay equations of the form

$$x(t) + a(t)x\big(t - r(t)\big) = f(t),$$

and

$$x'(t) + a(t)x\big(t - r(t)\big) = f(t),$$

in addition to (1.1)–(1.4).

It is very tempting to assume throughout that the solutions x of the equations (1.5)–(1.8) are continuous. For the two integrodifferential equations (1.6) and (1.8), this will indeed be the case, and for (1.7) this hypothesis does not lead to any significant loss of generality. However, if one requires the solution of (1.5) to be continuous, then one excludes an interesting convolution equation, namely (1.2) in the case where μ has point masses in $(0, \infty)$. To see this, note that, if f in (1.2) is continuous but $f(0) \neq 0$, then the solution x of (1.2) will, in general, exhibit jump discontinuities at the locations of the point masses of μ. For this reason, we must permit a solution of (1.5) to be discontinuous. On the other hand, if one allows x to be an arbitrary function in L^∞, then the integral $\int_{[0,t]} \kappa(t, \mathrm{d}s)x(s)$ need not be well defined. For example, if we choose $\kappa(t, \cdot)$ to be independent of t and to represent a unit point mass at a point t_0, then for $t \geq t_0$ we have, formally, $\int_{[0,t]} \kappa(t, \mathrm{d}s)x(s) = x(t_0)$. However, since the point evaluation operator is not well defined in L^∞, this is not satisfactory. This difficulty was encountered earlier in the convolution case, but there it could be overcome. Here it prevents us from ignoring sets of measure zero.

Consequently, we require a solution x of (1.5) to be Borel measurable (instead of Lebesgue measurable), and to be locally bounded (rather than locally essentially bounded). Analogously, we require (1.5) and (1.7) to hold everywhere rather than almost everywhere.

Our approach to (1.5)–(1.8) resembles our approach in Chapters 2–4 and 9. First we develop a resolvent theory for (1.5) and (1.7), and then we reduce (1.6) and (1.8) to (1.5) and (1.7) by integration.

2. Integral Equations with Measure Kernels of Type B^∞

Let $J \subset \mathbf{R}$ be an interval. Both (1.5) and (1.7) are special cases of the more general integral equation

$$x(t) + \int_J \kappa(t, \mathrm{d}s)x(s) = f(t), \quad t \in J, \tag{2.1}$$

where we assume that, for each t, the measure $\kappa(t, \cdot)$ is supported on $J \cap (-\infty, t]$, i.e., the kernel κ is of Volterra type. (For simplicity, we do not consider measure kernels of Fredholm type.)

We require the kernel κ in (2.1) to be of the following type.

2.1 Definition

Let $J \subset \mathbf{R}$ be an interval. We say that κ is a Volterra measure kernel of type $[B^\infty; J; \mathbf{C}^{n\times n}]$ (or of type B^∞ on J), if

(i) *for each $t \in J$ $\kappa(t, \cdot)$ is an $n \times n$ matrix of complex Borel measures on J with support in $J \cap (-\infty, t]$,*

(ii) *$\sup_{t\in J}|\kappa|(t, J) < \infty$, where $|\kappa|(t, \cdot)$ is the total variation of $\kappa(t, \cdot)$ (see Section* **3**.*5),*

(iii) *for each Borel set $E \subset J$ the function $t \mapsto \kappa(t, E)$ is Borel measurable.*

The class of all Volterra measure kernels of type $[B^\infty; J; \mathbf{C}^{n\times n}]$ is denoted by $\mathcal{M}(B^\infty; J; \mathbf{C}^{n\times n})$.

As always, if it is clear from the context in what space the range of the kernels lies, we abbreviate $\mathcal{M}(B^\infty; J; \mathbf{C}^{n\times n})$ by $\mathcal{M}(B^\infty; J)$. If the measures $\kappa(t, \cdot)$ are $n \times n$ matrices of real measures, then we say that κ is a measure kernel of type $[B^\infty; J; \mathbf{R}^{n\times n}]$.

Observe the close connection between Definition 2.1 and Definition **9**.5.1 (which motivates the use of the notation $[B^\infty; J; \mathbf{C}^{n\times n}]$ in both cases). In particular, note that, in the simple case where $\kappa(t, \mathrm{d}s) = k(t, s)\mu(\mathrm{d}s)$, with μ scalar and positive, condition (ii) requires that

$$\sup_{t\in J} \int_J |k(t, s)|\mu(\mathrm{d}s) < \infty,$$

i.e., $k(t, s)$ must be of bounded type on J with respect to the measure μ. We return to this connection in Theorem 2.3.

The measurability assumption on κ can be expressed in a different form. If κ is a kernel of the type described in Definition 2.1, then the distribution function $a(t, s) = \kappa(t, J \cap [s, \infty))$ is a bounded function on J^2 that is Borel measurable in t for each fixed $s \in J$, vanishes for $s > t$, and is of bounded variation and left-continuous in s for each fixed $t \in J$. In addition, the total variation of a in its second argument is bounded, uniformly in t. Conversely, every function a with these properties generates a Volterra measure kernel κ of type B^∞ whose distribution function is a. The only nontrivial part of the proof of this claim is the verification of the fact that if E is an arbitrary Borel set, and κ is the measure whose distribution function is a, i.e., $\kappa(t, J \cap [s, \infty)) = a(t,s)$, then the function $t \mapsto \kappa(t, E)$ is Borel measurable. This claim follows from the fact (not hard to show) that the collection $\mathcal{B}$ of all subsets E of J for which the function $t \mapsto \kappa(t, E)$ is Borel measurable is a σ-algebra that contains all subintervals of J of the form $[s, v)$ with $s < v$. Every such σ-algebra must also contain all open subsets of J, hence $\mathcal{B}$ contains the σ-algebra of all Borel subsets of J, and this proves the assertion.

An important property of the class of Volterra measure kernels of type B^∞ is that it is a Banach algebra with norm

$$|||\kappa|||_{B^\infty(J)} = \sup_{t \in J} |\kappa|(t, J).$$

As a first step towards a proof of this fact, we formulate an auxiliary result. In this theorem we do not specify explicitly in what space the values of our measures and functions lie (it is one of the spaces $\mathbf{C}^{m \times n}$, $\mathbf{C}^n$, and $\mathbf{C}$), since the results are true for all compatible combinations.

2.2 Theorem

Let $\kappa \in \mathcal{M}(B^\infty; J)$, $f \in B^\infty(J)$, and $\mu \in M(J)$. Then the following claims are true:

(i) *the total variation kernel $|\kappa|(t, \cdot)$ belongs to $\mathcal{M}(B^\infty; J; \mathbf{R})$;*

(ii) *the function $(\kappa \star f)(t) \stackrel{\text{def}}{=} \int_J \kappa(t, \mathrm{d}s) f(s)$, $t \in J$, belongs to $B^\infty(J)$, and $\|\kappa \star f\|_{B^\infty(J)} \le |||\kappa|||_{B^\infty(J)} \|f\|_{B^\infty(J)}$;*

(iii) *the function $\mu \star \kappa$ which to each Borel set E assigns the value $(\mu \star \kappa)(E) = \int_J \mu(\mathrm{d}s) \kappa(s, E)$ is a measure, and for each Borel set E, $|\mu \star \kappa|(E) \le \int_J |\mu|(\mathrm{d}s) |\kappa|(s, E) \le \|\mu\|_{M(J)} |||\kappa|||_{B^\infty(J)}$;*

(iv) *the function $\kappa \star f$ and the measure $\mu \star \kappa$ defined above satisfy $\int_J \mu(\mathrm{d}s)(\kappa \star f)(s) = \int_J (\mu \star \kappa)(\mathrm{d}s) f(s)$.*

Proof (i) Obviously, conditions (i) and (ii) of Definition 2.1 are satisfied. Thus it only remains to show that for every Borel set E the function $t \mapsto |\kappa|(t, E)$ is Borel measurable. This fact follows, however, from Theorem **3**.5.7, because

$$|\kappa|(t, E) = \lim_{m \to \infty} \sum_{k=-2^{2m}}^{2^{2m}} \left| \kappa\big(t, E \cap [k2^{-m}, (k+1)2^{-m})\big) \right|$$

is the pointwise limit of a sequence of Borel functions.

(ii) It follows from Definition 2.1 that, if f is the characteristic function of a Borel set $E \subset J$, then $k \star f$ is a bounded Borel function. This implies that, if f is a simple function, i.e., a function of the type $f = \sum_{i=1}^{m} a_i \chi_{E_i}$, where each a_i is a constant (scalar, vector or matrix), and χ_{E_i} is the characteristic function of the Borel set E_i, then $k \star f$ is a bounded Borel function. An arbitrary bounded Borel function f can be approximated by a uniformly convergent sequence f_i of simple functions. Using Lebesgue's dominated convergence theorem, we find that, for all $t \in J$,

$$(\kappa \star f)(t) = \int_J \kappa(t, \mathrm{d}s) f(s) = \lim_{i \to \infty} \int_J \kappa(t, \mathrm{d}s) f_i(s),$$

i.e., $\kappa \star f$ is the pointwise limit of a sequence of Borel measurable functions. This implies that $\kappa \star f$ is Borel measurable. It follows from Theorem **3**.5.6 that $\|\kappa \star f\|_{B^\infty(J)} \leq |||\kappa|||_{B^\infty(J)} \|f\|_{B^\infty(J)}$.

(iii) In order to prove that $\mu \star \kappa$ is a measure, we have to show that $\mu \star \kappa$ is countably additive. Let E be the union $E = \bigcup_{i=1}^{\infty} E_i$ of countably many disjoint Borel sets $E_i \subset J$. Define $F_m = \bigcup_{i=1}^{m} E_i$. For each t, $\kappa(t, \cdot)$ is a measure, and therefore $\kappa(t, F_m) \to \kappa(t, E)$ as $m \to \infty$. The sequence of functions $t \mapsto \kappa(t, F_m)$ is uniformly bounded by $|||\kappa|||_{B^\infty(J)}$, and therefore, by Lebesgue's dominated convergence theorem,

$$\begin{aligned}(\mu \star \kappa)(E) &= \int_J \mu(\mathrm{d}s) \lim_{m \to \infty} \kappa(s, F_m) \\ &= \lim_{m \to \infty} \sum_{i=1}^{m} \int_J \mu(\mathrm{d}s) \kappa(s, E_i) = \sum_{i=1}^{\infty} (\mu \star \kappa)(E_i).\end{aligned}$$

This shows that $\mu \star \kappa$ is countably additive, hence a measure.

To prove the claim about the total variation of $\mu \star \kappa$, we let E be the union $E = \bigcup_{i=1}^{m} E_i$ of finitely many disjoint Borel sets E_i. Then (use Theorem **3**.5.6 for the inequality)

$$\begin{aligned}\sum_{i=1}^{m} |(\mu \star \kappa)(E_i)| &= \sum_{i=1}^{m} \left| \int_J \mu(\mathrm{d}s) \kappa(s, E_i) \right| \leq \sum_{i=1}^{m} \int_J |\mu|(\mathrm{d}s) |\kappa|(s, E_i) \\ &= \int_J |\mu|(\mathrm{d}s) \sum_{i=1}^{m} |\kappa|(s, E_i) = \int_J |\mu|(\mathrm{d}s) |\kappa|(s, E).\end{aligned}$$

Since $|\mu \star \kappa|(E) = \sup \sum_{i=1}^{m} |(\mu \star \kappa)(E_i)|$, where the sup is taken over all such divisions of E, we get the desired inequality.

(iv) We defined $\mu \star \kappa$ in (iii) in such a way that the claim (iv) is true if f is the characteristic function of a Borel set. As in the proof of (ii), this implies that the same formula is valid if f is a simple function, and Lebesgue's dominated convergence theorem permits us to extend the formula to general Borel functions f. □

Define the $\star$-product $\alpha \star \beta$ of two kernels $\alpha, \beta \in \mathcal{M}(B^\infty; J; \mathbf{C}^{n\times n})$ by

$$(\alpha \star \beta)(t, E) = \int_J \alpha(t, \mathrm{d}s)\beta(s, E).$$

The next result shows that $\alpha\star\beta \in \mathcal{M}(B^\infty; J; \mathbf{C}^{n\times n})$, and that this product turns $\mathcal{M}(B^\infty; J; \mathbf{C}^{n\times n})$ into a Banach algebra. The space $\mathcal{V}(B^\infty; J; \mathbf{C}^{n\times n})$ has been introduced in Definition **9**.5.1.

2.3 Theorem

The set $\mathcal{M}(B^\infty; J; \mathbf{C}^{n\times n})$ of Volterra measure kernels of type B^∞ on an interval J forms a Banach algebra with multiplication $\star$ and norm $|||\cdot|||_{B^\infty(J)}$. The space $B^\infty(J; \mathbf{C}^n)$ is a left Banach module over $\mathcal{M}(B^\infty; J; \mathbf{C}^{n\times n})$, and $M(J; \mathbf{C}^{n\times n})$ is a right Banach module over $\mathcal{M}(B^\infty; J; \mathbf{C}^{n\times n})$.

The Banach algebra $\mathcal{V}(B^\infty; J; \mathbf{C}^{n\times n})$ of all Volterra kernels of bounded type on J forms a closed subalgebra of $\mathcal{M}(B^\infty; J; \mathbf{C}^{n\times n})$, if each function $k \in \mathcal{V}(B^\infty; J; \mathbf{C}^{n\times n})$ is identified with the corresponding measure $E \mapsto \int_E k(t, s)\,\mathrm{d}s$.

Proof First, we show that $\mathcal{M}(B^\infty; J; \mathbf{C}^{n\times n})$ is a Banach space. It is easy to see that it is a normed linear space, and to establish completeness one argues as follows. Let $\{\alpha_m\}_{m=1}^\infty$ be a Cauchy sequence in $\mathcal{M}(B^\infty; J; \mathbf{C}^{n\times n})$. Then, for each fixed t, the sequence of measures $\{\alpha_m(t, \cdot)\}_{m=1}^\infty$ is a Cauchy sequence in $M(J; \mathbf{C}^{n\times n})$, hence it converges to a measure $\alpha(t, \cdot)$ in $M(J; \mathbf{C}^{n\times n})$. This convergence is uniform in t, so $\alpha_m(t, \cdot)$ tends to $\alpha(t, \cdot)$ in the norm of $\mathcal{M}(B^\infty; J; \mathbf{C}^{n\times n})$. If E is a Borel set, then, for each $t \in J$, $\alpha_m(t, E)$ tends to $\alpha(t, E)$ as $m \to \infty$. Pointwise convergence preserves Borel measurability, and therefore the function $t \mapsto \alpha(t, E)$ is Borel measurable. This shows that $\alpha \in \mathcal{M}(B^\infty; J; \mathbf{C}^{n\times n})$.

To prove the remaining claims, one applies Theorem 2.2 as follows. Let α, $\beta \in \mathcal{M}(B^\infty; J; \mathbf{C}^{n\times n})$ and define $\alpha \star \beta$ as above. To see that for each fixed t the mapping $E \mapsto (\alpha \star \beta)(t, E)$ is a Borel measure supported on $J \cap (-\infty, t]$, with $\sup_{t\in J}|(\alpha\star\beta)|(t, J) < \infty$, it suffices to take $\mu(\cdot) = \alpha(t, \cdot)$ and $\kappa(s, E) = \beta(s, E)$ in Theorem 2.2(iii). To conclude that for each Borel set $E \subset J$ the function $t \mapsto (\alpha \star \beta)(t, E)$ is Borel measurable, one takes $\kappa(t, \cdot) = \alpha(t, \cdot)$, $f(s) = \beta(s, E)$ in (ii). Thus $\alpha\star\beta \in \mathcal{M}(B^\infty; J; \mathbf{C}^{n\times n})$. The final details of the proof are left to the reader. □

Applying Definition **9**.3.2 to the algebra $\mathcal{M}(B^\infty; J; \mathbf{C}^{n\times n})$, we get

2.4 Definition

Let κ be a Volterra measure kernel of type B^∞ on an interval J. We say that ρ is a measure resolvent of κ of type B^∞ on J, if ρ is a Volterra measure kernel of type B^∞ on J, and $\rho + \kappa \star \rho = \rho + \rho \star \kappa = \kappa$.

According to Lemma **9**.3.3, each $\kappa \in \mathcal{M}(B^\infty; J; \mathbf{C}^{n\times n})$ has at most one resolvent in $\mathcal{M}(B^\infty; J; \mathbf{C}^{n\times n})$.

It follows from the Volterra nature of the kernels that, if ρ is the resolvent of κ on an interval J, then ρ is the resolvent of κ on each subinterval of J as well. This makes it meaningful to consider kernels that are locally of type B^∞. We return to this question in Definition 2.11.

Recalling Lemma **9**.3.4, and invoking Theorem 2.3, we obtain:

2.5 Theorem

If κ is a Volterra measure kernel of type B^∞ on J that has a resolvent ρ of type B^∞ on J, and $f \in B^\infty(J; \mathbf{C}^n)$, then (2.1) has a unique solution x in $B^\infty(J; \mathbf{C}^n)$. This solution is given by the variation of constants formula

$$x(t) = f(t) - \int_J \rho(t, \mathrm{d}s) f(s), \quad t \in J. \tag{2.2}$$

The above result again points to the fact that the existence of the resolvent is the basic problem. The remaining part of this section is devoted to an analysis of this problem.

We get the two following crude results on the existence of a resolvent by applying Lemmas **9**.3.7 and **9**.3.8 to $\mathcal{M}(B^\infty; J; \mathbf{C}^{n\times n})$.

2.6 Theorem

Let $J \subset \mathbf{R}$ be an interval. Let κ be of the form $\kappa = \kappa_1 + \kappa_2$, where κ_1, $\kappa_2 \in \mathcal{M}(B^\infty; J; \mathbf{C}^{n\times n})$, and suppose that κ_1 has a measure resolvent ρ_1 of type B^∞ on J. If $|||\kappa_2|||_{B^\infty(J)} < (1 + |||\rho_1|||_{B^\infty(J)})^{-1}$, then κ has a measure resolvent $\rho \in \mathcal{M}(B^\infty; J; \mathbf{C}^{n\times n})$.

2.7 Corollary

If κ is a Volterra measure kernel of type $[B^\infty; J; \mathbf{C}^{n\times n}]$, and $|||\kappa|||_{B^\infty(J)} < 1$, then κ has a measure resolvent of type B^∞ on J.

Clearly, the hypothesis of Corollary 2.7 prevents κ from having large point masses at the line $s = t$. This assumption was not needed in our discussion of equation (1.2) in Chapter 4. Exercise 4 can be used to relax this smallness assumption.

We leave the application of Lemma **9**.3.11 to $\mathcal{M}(B^\infty; J; \mathbf{C}^{n\times n})$ to the reader.

Note that the results we have obtained so far are true for Fredholm measure kernels of type B^∞ on J as well.

It is again a pleasant fact (cf. Theorem **9**.3.13) that the Volterra nature of the measure kernel makes it possible to deduce the existence of the resolvent on $J = \bigcup_{i=1}^n J_i$ from local existence on J_i:

2.8 Theorem

Let κ be a Volterra measure kernel of type B^∞ on an interval $J \subset \mathbf{R}$, and suppose that J can be divided into finitely many subintervals J_i in such a way that κ has a resolvent of type B^∞ on each subinterval J_i. Then κ has a resolvent of type B^∞ on J.

Proof The proof is essentially the same as that of Theorem **9**.3.13. As in that proof, it suffices to consider the case where J is divided into two disjoint intervals J_1 and J_2, with J_2 located to the right of J_1.

Define $\kappa_i(t, \mathrm{d}s) = \chi_{J_i}(t)\kappa(t, \mathrm{d}s)\chi_{J_i}(s)$, $i = 1, 2$, where χ_{J_i} is the characteristic function of the interval J_i. Let ρ_i be the resolvent of κ on J_i, $i = 1, 2$, and extend ρ_i, $i = 1, 2$, to a measure kernel on J by defining $\rho_i(t, E) = \rho_i(t, E \cap J_i)$ if $t \in J_i$, and $\rho_i(t, E) = 0$ if $t \in J \setminus J_i$ for every Borel set $E \subset J$. Then ρ_i is the resolvent of κ_i on all of J. Define ρ to be

$$\rho = \kappa - \kappa \star \rho_1 - \rho_2 \star \kappa + \rho_2 \star \kappa \star \rho_1.$$

Then ρ is a Volterra measure kernel, and one can easily check that $\chi_{J_i}(t)\rho(t, \mathrm{d}s)\chi_{J_i}(s) = \rho_i(t, \mathrm{d}s)$. In the same way as in Chapter 9, one gets $\rho = \kappa - \kappa \star \rho_1 - \kappa_2 \star \rho$. For each $\phi \in B^\infty(J; \mathbf{C}^n)$, we have

$$(\kappa \star \rho_1 \star \phi)(t) = \int_{J_1} \kappa(t, \mathrm{d}u) \int_{J_1} \rho(u, \mathrm{d}s)\phi(s) = \int_{J_1} \kappa(t, \mathrm{d}u) \int_J \rho(u, \mathrm{d}s)\phi(s),$$

and

$$\begin{aligned}(\kappa_2 \star \rho \star \phi)(t) &= \chi_{J_2}(t) \int_{J_2} \kappa(t, \mathrm{d}u) \int_J \rho(u, \mathrm{d}s)\phi(s) \\ &= \int_{J_2} \kappa(t, \mathrm{d}u) \int_J \rho(u, \mathrm{d}s)\phi(s).\end{aligned}$$

Therefore, $\kappa \star \rho_1 + \kappa_2 \star \rho = \kappa \star \rho$, so ρ satisfies $\rho = \kappa - \kappa \star \rho$. Analogously one proves that $\rho = \kappa - \rho \star \kappa$. □

Corollary 2.7 and Theorem 2.8 obviously imply:

2.9 Corollary

Let κ be a Volterra measure kernel of type B^∞ on $J \subset \mathbf{R}$. Suppose that J can be divided into finitely many subintervals J_i such that, on each J_i, $|||\kappa|||_{B^\infty(J_i)} < 1$. Then κ has a resolvent of type B^∞ on J.

As opposed to the Banach algebras in Chapter 9, the algebra $\mathcal{M}(B^\infty; J; \mathbf{C}^{n \times n})$ does have a unit, namely the kernel that for each $t \in J$ has a unit point mass at t, i.e., $\varepsilon(t, E) = \chi_E(t)$ for all Borel sets $E \subset J$. As a consequence, one defines the spectrum of a member of this algebra in a slightly different way:

2.10 Definition

Let $\mathcal{A}$ be an associative algebra with unit ε, and let $\kappa \in \mathcal{A}$. The spectrum $\sigma(\kappa)$ of $\kappa \in \mathcal{A}$ (with respect to $\mathcal{A}$) is composed of those $\lambda \in \mathbf{C}$ for which $\lambda\varepsilon - \kappa$ is not invertible in $\mathcal{A}$. Equivalently, $0 \in \sigma(\kappa)$ iff κ is not invertible, and $\lambda \neq 0$ belongs to $\sigma(\kappa)$ iff $-\frac{1}{\lambda}\kappa$ does not have a resolvent. The spectral radius of κ is the number $\sup\{\,|\lambda| \mid \lambda \in \sigma(\kappa)\,\}$.

Observe that the spectral radius of κ in Corollary 2.9 is less than 1 (cf. Corollary **9**.3.18). In particular, the resolvent ρ of κ is the limit in $\mathcal{M}(B^\infty; J; \mathbf{C}^{n\times n})$, as $m \to \infty$, of the sums $\rho_m = \sum_{j=1}^m (-1)^{j-1}\kappa^{\star j}$, where $\kappa^{\star j}$ is the $(j-1)$-fold $\star$-product of κ by itself.

As in Chapter 9, the Volterra nature of the kernel makes it possible to give local versions of many of the preceding theorems. For this, we first define the class $\mathcal{M}(B^\infty_{\mathrm{loc}}; J; \mathbf{C}^{n\times n})$:

2.11 Definition

Let $J \subset \mathbf{R}$ be an interval. We say that κ is a Volterra measure kernel of type $[B^\infty_{\mathrm{loc}}; J; \mathbf{C}^{n\times n}]$, on J if, for every compact subinterval K of J, the restriction of κ to K^2 is a Volterra measure kernel of type $[B^\infty; K; \mathbf{C}^{n\times n}]$.

The class of all $\mathbf{C}^{n\times n}$-valued Volterra measure kernels of type B^∞_{loc} on J is denoted by $\mathcal{M}(B^\infty_{\mathrm{loc}}; J; \mathbf{C}^{n\times n})$.

As before, we define $\rho \in \mathcal{M}(B^\infty_{\mathrm{loc}}; J; \mathbf{C}^{n\times n})$ to be the resolvent of $\kappa \in \mathcal{M}(B^\infty_{\mathrm{loc}}; J; \mathbf{C}^{n\times n})$ if $\rho + \kappa\star\rho = \rho + \rho\star\kappa = \kappa$. By Lemma **9**.3.3, the resolvent of κ is unique.

As usual (cf. Lemma **9**.3.4), if the resolvent exists, then we can solve (2.1):

2.12 Theorem

Let J be an interval of the type $[S, T)$, where $-\infty < S < T \le \infty$. If κ is a Volterra measure kernel of type B^∞_{loc} on J which has a resolvent ρ of type B^∞_{loc} on J, and $f \in B^\infty_{\mathrm{loc}}(J; \mathbf{C}^n)$, then the function x defined in (2.2) is the unique solution in $B^\infty_{\mathrm{loc}}(J; \mathbf{C}^n)$ of (2.1).

Concerning the existence of the resolvent of $\kappa \in \mathcal{M}(B^\infty_{\mathrm{loc}}; J; \mathbf{C}^{n\times n})$, we have the following theorem.

2.13 Theorem

Let κ be a Volterra measure kernel of type B^∞_{loc} on $J \subset \mathbf{R}$. Suppose that each compact subinterval K of J can be divided into finitely many subintervals K_i such that, on each K_i, $|||\kappa|||_{B^\infty(K_i)} < 1$. Then κ has a resolvent of type B^∞_{loc} on J.

A B^∞_{loc}-version for measure kernels of Lemma **9**.3.11 may also be formulated. We omit the exact statement.

In the local results above, we have worked throughout on compact subintervals of J. When J is an interval of the form $J = (-\infty, T)$, where

$-\infty < T \le \infty$, then one can equally well work on subintervals of the type $(-\infty, U]$, where $-\infty < U < T$, and the results above are true also in this setting.

As in Chapter 9 (see Theorem **9**.3.19), one may prove a perturbation result, where the asymptotically dominating part of the kernel is of convolution type:

2.14 Theorem
Let $\kappa = \kappa_1 + \kappa_2$ be a Volterra measure kernel of type B^∞ on an interval $J = [S, \infty)$, with $-\infty < S < \infty$, and assume that κ_1 and κ_2 satisfy the following conditions:

(i) *$\kappa_1(t, E) = \int_{\mathbf{R}^+} \mu(\mathrm{d}s)\chi_E(t-s)$, for each Borel set $E \subset J$ and each $t \in J$, where $\mu \in M(\mathbf{R}^+; \mathbf{C}^{n \times n})$ has no singular part, no point mass at zero, and satisfies $\inf_{\Re z \ge 0} |\det[I + \hat{\mu}(z)]| > 0$;*

(ii) *κ_2 is a Volterra measure kernel of type $[B^\infty; J; \mathbf{C}^{n \times n}]$ that has a measure resolvent of the same type and, in addition, satisfies $\lim_{T \to \infty} |||\kappa_2|||_{B^\infty([T,\infty))} = 0$.*

Then κ has a resolvent of type B^∞ on J.

(By using Exercise 4, one can relax the assumption that μ must not have a point mass at zero.)

Proof of Theorem 2.14 Since μ has no point mass at the origin, the norm $|||\kappa_1|||_{B^\infty([t,t+h])}$ tends to zero as $h \to 0$, uniformly in $t \ge 0$. Therefore, it follows from Theorem 2.6 that every finite interval $[S, T]$ can be divided into finitely many subintervals J_i in such a way that on each interval J_i the kernel κ has a resolvent of type B^∞.

To handle the infinite interval $[T, \infty)$, we observe first that Corollary **4**.4.7 shows that μ has a convolution resolvent $\rho \in M(\mathbf{R}^+; \mathbf{C}^{n \times n})$. If we define $\rho_1(t, E) = \int_{\mathbf{R}^+} \rho(\mathrm{d}s)\chi_E(t-s)$, then ρ_1 is a Volterra measure kernel of type B^∞ on J, and ρ_1 is the resolvent of κ_1 on J, hence on $[T, \infty)$. Using the asymptotic assumption on κ_2 and Theorem 2.6, we find that κ has a resolvent of type B^∞ on an interval $[T, \infty)$, for some sufficiently large T.

By Theorem 2.8, κ has a resolvent of type B^∞ on J. □

The generalized Gronwall Lemma **9**.8.2 can be extended to equation (2.1) as follows (the proof remains the same).

2.15 A Generalized Gronwall Lemma
Let κ be a scalar Volterra measure kernel of type B^∞_{loc} on an interval $J \subset \mathbf{R}$ of the type $[S, T)$, with $-\infty < S < T \le \infty$, and assume that $-\kappa$ has a nonpositive resolvent ρ of the same type. Let $f \in B^\infty_{\mathrm{loc}}(J)$, and suppose that $x(t) \le (\kappa \star x)(t) + f(t)$ for all $t \in J$. Then $x(t) \le y(t)$ for all $t \in J$, where y is the solution of the comparison equation $y(t) = (\kappa \star y)(t) + f(t)$.

3. Nonconvolution Integrodifferential Equations: Local Theory

We proceed to the analysis of the two integrodifferential equations (1.6) and (1.8). In addition to measure kernels of type B^∞ (recall that by (ii) of Theorem 2.2 they map B^∞ into itself) we need two more classes of measure kernels. These map B^∞ into L^1 and L^1 into B^∞, respectively. Below we first define these classes, and in Theorems 3.2, 3.4, and 3.5 we present some of their properties.

3.1 Definition

Let $J \subset \mathbf{R}$ be an interval. We say that κ is a Volterra measure kernel of type $[B^\infty; L^1; J; \mathbf{C}^{n\times n}]$ (or of type $[B^\infty; L^1]$ on J), if

(i) *for almost all $t \in J$ $\kappa(t,\cdot)$ is an $n\times n$ matrix of complex Borel measures on J supported on $J \cap (-\infty, t]$,*

(ii) *the function $t \mapsto |\kappa|(t, J)$ belongs to $L^1(J; \mathbf{R})$,*

(iii) *for every Borel set E, the function $t \mapsto \kappa(t, E)$ is Lebesgue measurable.*

The class of all $\mathbf{C}^{n\times n}$-valued Volterra measure kernels of type $[B^\infty; L^1]$ on J is denoted by $\mathcal{M}(B^\infty; L^1; J; \mathbf{C}^{n\times n})$.

In this definition we allow J to be either a finite or an infinite interval, but in most of the applications we take J to be finite. To see why it in many cases is unrealistic to have J unbounded, consider the simple example $\kappa(t, \mathrm{d}s) = a(t, s)\,\mathrm{d}s$. Condition (ii) then becomes $\int_J\int_J |a(t, s)|\,\mathrm{d}t\,\mathrm{d}s < \infty$. If, for example, a is of convolution type, then this is satisfied only if J is finite (or $a \equiv 0$).

It should come as no surprise that the Volterra measure kernels of type $[B^\infty; L^1]$ on an interval J form a Banach space with norm

$$|||\kappa|||_{[B^\infty;L^1](J)} = \int_J |\kappa|(t, J)\,\mathrm{d}t,$$

that they map B^∞ into L^1, and, generally speaking, that they behave as one would expect them to do. In the following theorem (comparable to Theorem 2.2) we again ignore the ranges of the kernel and the functions, since the results are true for all compatible choices of ranges.

3.2 Theorem

Let $\kappa \in \mathcal{M}(B^\infty; L^1; J)$, and $f, g \in B^\infty(J)$. Then the following claims are true:

(i) *the total variation kernel $|\kappa|(t,\cdot)$ belongs to $\mathcal{M}(B^\infty; L^1; J; \mathbf{R})$;*

(ii) *the function $(\kappa \star f)(t) = \int_J \kappa(t, \mathrm{d}s) f(s)$, $t \in J$, belongs to $L^1(J)$, and $\|\kappa \star f\|_{L^1(J)} \le |||\kappa|||_{[B^\infty;L^1](J)} \|f\|_{B^\infty(J)}$;*

(iii) *the function $g \star \kappa$ that to each Borel set E assigns the value $(g \star \kappa)(E) = \int_J g(t)\kappa(t, E)\,\mathrm{d}t$ is a measure, and, for each Borel set E,*

$|g \star \kappa|(E) \le \int_J |g(t)||\kappa|(t,E)\,\mathrm{d}t \le \|g\|_{B^\infty(J)} |||\kappa|||_{[B^\infty;L^1](J)}$;

(iv) *the function* $\kappa \star f$ *and the measure* $g \star \kappa$ *defined above satisfy* $\int_J g(s)(\kappa \star f)(s)\,\mathrm{d}s = \int_J (g \star \kappa)(\mathrm{d}s) f(s)$.

The proof is essentially the same as the proof of Theorem 2.2 (instead of using Borel measurability on the range of κ one uses Lebesgue measurability).

By (ii) above, kernels of type $[B^\infty; L^1]$ map B^∞ into L^1. For technical reasons, we need another class of kernels that map M (hence L^1) into B^∞.

3.3 Definition

Let $J \subset \mathbf{R}$ *be an interval. We say that* k *is a Volterra kernel of type* $[M; B^\infty; J; \mathbf{C}^{n \times n}]$ *(or of type* $[M; B^\infty]$ *on* J*), if* $k : J^2 \to \mathbf{C}^{n \times n}$ *is Borel measurable and bounded, and satisfies* $k(t,s) = 0$ *for* $t < s$. *The class of Volterra kernels of type* $[M; B^\infty; J; \mathbf{C}^{n \times n}]$ *is denoted by* $\mathcal{V}(M; B^\infty; J; \mathbf{C}^{n \times n})$.

In other words, a Volterra kernel of type $[M; B^\infty]$ is simply a function $k \in B^\infty(J^2; \mathbf{C}^{n \times n})$ which satisfies $k(t,s) = 0$ for $t < s$. We define the norm in $\mathcal{V}(M; B^\infty; J; \mathbf{C}^{n \times n})$ as

$$|||k|||_{[M;B^\infty](J)} = \sup_{t,s \in J} |k(t,s)|.$$

The following easy theorem details some properties of the class of Volterra kernels of type $[M; B^\infty; J]$; see also Exercise 16.

3.4 Theorem

Let $k \in \mathcal{V}(M; B^\infty; J)$ *and* μ, $\nu \in M(J)$. *Then the following claims are true:*

(i) *the kernel* $|k|(t,s)$ *belongs to* $\mathcal{V}(M; B^\infty; J; \mathbf{R})$;

(ii) *the function* $(\mu \star k)(s) = \int_J \mu(\mathrm{d}t) k(t,s)$, $s \in J$, *belongs to* $B^\infty(J)$, *and* $\|\mu \star k\|_{B^\infty(J)} \le \|\mu\|_{M(J)} |||k|||_{[M;B^\infty](J)}$;

(iii) *the function* $(k \star \nu)(t) = \int_J k(t,s)\nu(\mathrm{d}s)$, $t \in J$, *belongs to* $B^\infty(J)$, *and* $\|k \star \nu\|_{B^\infty(J)} \le |||k|||_{[M;B^\infty](J)} \|\nu\|_{M(J)}$;

(iv) *the functions* $\mu \star k$ *and* $k \star \nu$ *defined above satisfy*

$$\int_J \mu(\mathrm{d}s)(k \star \nu)(s) = \int_J (\mu \star k)(s)\nu(\mathrm{d}s).$$

Proof It is a standard result that (i) holds. In order to prove (ii) and (iii) it suffices to appeal to Fubini's theorem (in (iii) one uses the Borel measure on J^2, and in (ii) one uses the product of the Borel measure on J and the Lebesgue measure on J). Claim (iv) follows directly from Fubini's theorem. □

We define $\star$-products of kernels of different types in the obvious ways, indicated by Theorems 2.2, 3.2 and 3.4, and obtain very natural results.

3.5 Theorem
Let $a \in \mathcal{V}(M; B^\infty; J)$, $\beta \in \mathcal{M}(B^\infty; L^1; J)$, $\gamma \in \mathcal{M}(B^\infty; J)$, and $f \in L^1(J)$. Then the following claims are true:

(i) *the function $(a \star \beta)(t, E)$, which for each $t \in J$ maps the Borel set E into $(a \star \beta)(t, E) = \int_J a(t, s)\beta(s, E)\,\mathrm{d}s$, is a Volterra measure kernel of type B^∞, and*

$$\|\!|a \star \beta|\!\|_{B^\infty(J)} \leq \|\!|a|\!\|_{[M;B^\infty](J)} \|\!|\beta|\!\|_{[B^\infty;L^1](J)};$$

(ii) *the function $(\beta \star \gamma)(t, E)$, which for each $t \in J$ maps the Borel set E into $(\beta \star \gamma)(t, E) = \int_J \beta(t, \mathrm{d}s)\gamma(s, E)$, is a Volterra measure kernel of type $[B^\infty; L^1]$, and*

$$\|\!|\beta \star \gamma|\!\|_{[B^\infty;L^1](J)} \leq \|\!|\beta|\!\|_{[B^\infty;L^1](J)} \|\!|\gamma|\!\|_{B^\infty(J)};$$

(iii) *the function $(\gamma \star a)(t, s) = \int_J \gamma(t, \mathrm{d}u)a(u, s)$ is a Volterra kernel of type $[M; B^\infty]$, and $\|\!|\gamma \star a|\!\|_{[M;B^\infty](J)} \leq \|\!|\gamma|\!\|_{B^\infty(J)} \|\!|a|\!\|_{[M;B^\infty](J)}$;*

(iv) *the function $(\beta \star a)(t, s) = \int_J \beta(t, \mathrm{d}u)a(u, s)$ is a Volterra kernel of type L^1, and $\|\!|\beta \star a|\!\|_{L^1(J)} \leq \|\!|\beta|\!\|_{[B^\infty;L^1](J)} \|\!|a|\!\|_{[M;B^\infty](J)}$;*

(v) *The kernels $a \star \beta$, $\beta \star \gamma$, $\gamma \star a$ defined above and the function $a \star f$ defined in Theorem 3.4(iii) satisfy*
   (1) $(a \star \beta) \star \gamma = a \star (\beta \star \gamma)$,
   (2) $(\beta \star \gamma) \star a = \beta \star (\gamma \star a)$,
   (3) $(\gamma \star a) \star \beta = \gamma \star (a \star \beta)$,
   (4) $(\gamma \star a) \star f = \gamma \star (a \star f)$.

Theorem 3.5 does not list all possible combinations, but together with the earlier lemmas it covers the cases that we need below.

Proof of Theorem 3.5 The major part (which we omit) of the proof consists of straightforward applications of Theorems 2.2, 3.2 and 3.4. The only claims that require separate arguments are the one that says that the function $(\gamma \star a)(t, s)$ in (iii) is Borel measurable on J^2, and the one that says that the function $(\beta \star a)(t, s)$ in (iv) is Lebesgue measurable on J^2. The proofs of these claims are completely analogous, so let us prove only the first claim. As usual, it suffices to prove the special case where a is the characteristic function of a Borel set $E \subset J^2$, because from this case one gets to the general case by using a uniformly convergent sequence of simple functions and Lebesgue's dominated convergence theorem. Let $\mathcal{B}$ be the collection of all subsets E of J^2 for which the function $\gamma \star \chi_E$ is Borel measurable. This collection contains the whole space J^2, it is closed under complementations, and it is closed under countable unions as well (use Lebesgue's dominated convergence theorem). In other words, $\mathcal{B}$ is a σ-algebra. It contains all rectangles $E \times F$, where E and F are Borel measurable subsets of J, because $(\gamma \star \chi)_{E\times F}(t, s) = \gamma(t, E)\chi_F(s)$. This implies that $\mathcal{B}$ contains the product of the Borel algebra on J by itself, so $\mathcal{B}$ contains the σ-algebra of Borel sets. □

Above, we have discussed kernels of the types $[B^\infty; L^1]$ and $[M; B^\infty]$ on an interval J. Because of the Volterra nature of these kernels, one can also use localized versions of these classes of kernels.

3.6 Definition

Let $J \subset \mathbf{R}$ be an interval. We say that κ is a Volterra measure kernel of type $[B^\infty_{\mathrm{loc}}; L^1_{\mathrm{loc}}; J; \mathbf{C}^{n\times n}]$ if, for every compact subinterval K of J, the restriction of κ to K^2 is a Volterra measure kernel of type $[B^\infty; L^1; K; \mathbf{C}^{n\times n}]$. We say that k is a Volterra kernel of type $[M_{\mathrm{loc}}; B^\infty_{\mathrm{loc}}; J; \mathbf{C}^{n\times n}]$ if, for every compact subinterval K of J, the restriction of k to K^2 is a Volterra kernel of type $[M; B^\infty; K; \mathbf{C}^{n\times n}]$, or, equivalently, if k is Borel measurable and locally bounded. These classes are denoted by $\mathcal{M}(B^\infty_{\mathrm{loc}}; L^1_{\mathrm{loc}}; J; \mathbf{C}^{n\times n})$ and $\mathcal{V}(M_{\mathrm{loc}}; B^\infty_{\mathrm{loc}}; J; \mathbf{C}^{n\times n})$ respectively.

To realize the importance of these localized versions, note that if, e.g., $\kappa \in \mathcal{M}(B^\infty; J)$ then, in general, we only have $\kappa \in \mathcal{M}(B^\infty_{\mathrm{loc}}; L^1_{\mathrm{loc}}; J)$ (cf. the comment after Definition 3.1).

We are now ready to discuss the integrodifferential equations (1.6) and (1.8). In both cases we want the equations to have 'fundamental solutions', or, in other words, we want the measure kernel κ in (1.6) and (1.8) to have a differential resolvent. It is fairly clear what the appropriate assumptions are in the local theory, but in the global theory it is possible to choose a large number of different hypotheses. We return to the global theory in Section 4.

When one tries to generalize the theory for (1.4), given in Chapter 3, to (1.6), one soon discovers (after some formal manipulations) that the solution x of (1.6) ought to be given by the variation of constants formula

$$x(t) = r(t,0)x_0 + \int_0^t r(t,s)f(s)\,\mathrm{d}s, \tag{3.1}$$

where the *differential resolvent* r is a function satisfying $r(t,t) = I$ for $t \in \mathbf{R}^+$, $r(t,s) = 0$ for $s > t$, and

$$\frac{\partial}{\partial t} r(t,s) + \int_{[s,t]} \kappa(t,\mathrm{d}u) r(u,s) = 0, \tag{3.2}$$

and

$$\frac{\partial}{\partial s} r(t,s) - \int_{[s,t]} r(t,u)\kappa(u,\mathrm{d}s)\,\mathrm{d}u = 0. \tag{3.3}$$

The first of these two equations makes sense pointwise a.e., but the second equation must, in general, be interpreted as a measure equation. (When doing so, one should add a negative point mass at t to the right-hand side to get the correct equation; if also (3.2) is given a measure interpretation, then there one must add a positive point mass at s.) One can turn these two equations into equations that are much easier to deal with, by integrating (3.2) with respect to t, and (3.3) with respect to s. Doing so, one gets

$$r + \Lambda \star \kappa \star r = r + r \star \kappa \star \Lambda = \Lambda, \tag{3.4}$$

where Λ is the *integration kernel*

$$\Lambda(t,s) = \begin{cases} I, & s \le t, \\ 0, & s > t. \end{cases}$$

Clearly, Λ is a Volterra kernel of type $[M_{\mathrm{loc}}; B^\infty_{\mathrm{loc}}]$, so by Theorem 3.5, if κ is a Volterra measure kernel of type $[B^\infty_{\mathrm{loc}}; L^1_{\mathrm{loc}}]$ and r is a Volterra kernel of type $[M_{\mathrm{loc}}; B^\infty_{\mathrm{loc}}]$, then (3.4) is a well defined equation in $\mathcal{V}(M_{\mathrm{loc}}; B^\infty_{\mathrm{loc}}; J; \mathbf{C}^{n\times n})$. Here $J \subset \mathbf{R}$ can be an arbitrary interval.

As the following theorem shows, (3.4) has a unique solution.

3.7 Theorem

Let $J \subset \mathbf{R}$ be an interval, and let κ be a Volterra measure kernel of type $[B^\infty_{\mathrm{loc}}; L^1_{\mathrm{loc}}; J; \mathbf{C}^{n\times n}]$. Then there exists a unique Volterra kernel $r \in B^\infty_{\mathrm{loc}}(J^2; \mathbf{C}^{n\times n})$ that satisfies

$$r + \Lambda \star \kappa \star r = r + r \star \kappa \star \Lambda = \Lambda. \tag{3.4}$$

For each fixed $s \in J$, the function $t \mapsto r(t,s)$ is locally absolutely continuous on $J \cap [s,\infty)$, and the kernel $\frac{\partial}{\partial t} r(t,s)$ is of type $[L^1_{\mathrm{loc}}; J; \mathbf{C}^{n\times n}]$. For each fixed $t \in J$, the function $s \mapsto r(t,s)$ is left-continuous and locally of bounded variation on J. Moreover, $r(t,t) = I$ for all $t \in J$.

If J is an interval of the type $J = [S,T)$, where $-\infty < S < T \le \infty$, then r is of continuous type on J.

3.8 Definition

The function r in Theorem 3.7 is called the differential resolvent of κ.

Proof of Theorem 3.7 By Theorem 3.5(i), the kernel $\alpha = \Lambda \star \kappa$ is a Volterra measure kernel of type $[B^\infty_{\mathrm{loc}}; J; \mathbf{C}^{n\times n}]$. Moreover, since we can make $|||\kappa|||_{[B^\infty; L^1](K)}$ arbitrarily small by choosing the interval K small enough, we can make $|||\alpha|||_{B^\infty(K)}$ arbitrarily small by an appropriate choice of K. This means that α satisfies the hypothesis of Theorem 2.13, and, therefore, α has a resolvent ρ of type B^∞_{loc} on J. Define $r = \Lambda - \rho \star \Lambda$. Then by Theorem 3.5(iii), r is a Volterra kernel of type $[M_{\mathrm{loc}}; B^\infty_{\mathrm{loc}}; J; \mathbf{C}^{n\times n}]$. Since ρ is the resolvent of $\alpha = \Lambda \star \kappa$, it satisfies

$$\rho + \Lambda \star \kappa \star \rho = \rho + \rho \star \Lambda \star \kappa = \Lambda \star \kappa.$$

Therefore,

$$\begin{aligned} \Lambda - r = \rho \star \Lambda &= (\Lambda \star \kappa - \Lambda \star \kappa \star \rho) \star \Lambda \\ &= \Lambda \star \kappa \star (\Lambda - \rho \star \Lambda) = \Lambda \star \kappa \star r, \end{aligned}$$

and we conclude that r satisfies $r + \Lambda \star \kappa \star r = \Lambda$. That r also satisfies the second of the two equations, namely $r + r \star \kappa \star \Lambda = \Lambda$, follows from the analogous computation

$$\begin{aligned} \Lambda - r = \rho \star \Lambda &= (\Lambda \star \kappa - \rho \star \Lambda \star \kappa) \star \Lambda \\ &= (\Lambda - \rho \star \Lambda) \star \kappa \star \Lambda = r \star \kappa \star \Lambda. \end{aligned}$$

To prove uniqueness, let $p \in \mathcal{V}(M_{\text{loc}}; B^{\infty}_{\text{loc}}; J; \mathbf{C}^{n\times n})$ satisfy $p + p \star \kappa \star \Lambda = \Lambda$, and let $q \in \mathcal{V}(M_{\text{loc}}; B^{\infty}_{\text{loc}}; J; \mathbf{C}^{n\times n})$ satisfy $q + \Lambda \star \kappa \star q = \Lambda$. Then

$$\begin{aligned}\Lambda - p &= p \star \kappa \star (q + \Lambda \star \kappa \star q)\\ &= (p + p \star \kappa \star \Lambda) \star \kappa \star q = \Lambda \star \kappa \star q = \Lambda - q.\end{aligned}$$

Therefore, $p = q$.

To show that r is locally absolutely continuous with respect to t, one uses the identity $r + \Lambda \star \kappa \star r = \Lambda$ (which is part of (3.4)). That $\frac{\partial}{\partial t} r(t,s)$ is a kernel of type L^1 follows from (3.2) and Theorem 3.5(iv). For the verification of the claimed behaviour with respect to s, use the identity $r + r \star \kappa \star \Lambda = \Lambda$. The fact that the kernels are of Volterra type, together with (3.4), yields $r(t,t) = I$ for $t \in J$, and $r(t,s) = 0$ for $s > t$.

That r is of continuous type when J is an interval of the type $J = [S,T)$ follows from the local boundedness of r, the continuity with respect to t, and Lebesgue's dominated convergence theorem. □

The existence of the resolvent enables us to solve (1.4), and more generally, to solve the equation

$$x'(t) + \int_{[S,t]} \kappa(t, \mathrm{d}s) x(s) = f(t), \quad t \in [S,T); \quad x(S) = x_S. \tag{3.5}$$

Here $-\infty < S < T \le \infty$.

3.9 Theorem

Let J be an interval of the type $J = [S,T)$, where $-\infty < S < T \le \infty$. Let κ be a Volterra measure kernel of type $[B^{\infty}_{\text{loc}}; L^1_{\text{loc}}; J; \mathbf{C}^{n\times n}]$, let $f \in L^1_{\text{loc}}(J; \mathbf{C}^n)$, and let $x_S \in \mathbf{C}^n$. Then (3.5) has a unique locally absolutely continuous solution x on J. This solution is given by the variation of constants formula

$$x(t) = r(t,S)x_S + \int_S^t r(t,s) f(s)\,\mathrm{d}s, \quad t \in J, \tag{3.6}$$

where r is the differential resolvent of κ.

A variation of constants formula for equation (1.8) is given in Section 4 (see Theorem 4.6).

Proof of Theorem 3.9 If $x(t) = r(t,S)x_S + (r \star f)(t)$, then by (3.4), for $t \ge S$,

$$\begin{aligned}x(t) &= x_S + (\Lambda \star f)(t) - (\Lambda \star \kappa \star r)(t,S)x_S - (\Lambda \star \kappa \star r \star f)(t)\\ &= x_S + (\Lambda \star f)(t) - (\Lambda \star \kappa \star x)(t).\end{aligned}$$

This implies that x is locally absolutely continuous and satisfies (3.5) a.e.

Conversely, let x be an arbitrary locally absolutely continuous solution of (3.5). Then, integrating this equation, we get

$$\begin{aligned}
x(t) &= x_S + (\Lambda \star f)(t) - (\Lambda \star \kappa \star x)(t) \\
&= x_S + (\Lambda \star f)(t) - ((r + r \star \kappa \star \Lambda) \star \kappa \star x)(t) \\
&= x_S + (\Lambda \star f)(t) - (r \star \kappa \star (x + \Lambda \star \kappa \star x))(t) \\
&= x_S + (\Lambda \star f)(t) - (r \star \kappa \star (x_S + \Lambda \star f))(t) \\
&= (\Lambda(t,S) - (r \star \kappa \star \Lambda)(t,S))x_S + ((\Lambda - r \star \kappa \star \Lambda) \star f)(t) \\
&= r(t,S)x_S + (r \star f)(t).
\end{aligned}$$

Therefore x satisfies (3.6). □

We conclude this section with the following simple comparison result (cf. Lemma 2.15).

3.10 A Generalized Gronwall Lemma

Let J be an interval of the type $[S,T)$, with $-\infty < S < T \le \infty$, let $a \in L^1_{\text{loc}}(J;\mathbf{R})$, and let κ be a nonnegative Volterra measure kernel of type $[B^\infty_{\text{loc}}; L^1_{\text{loc}}; J; \mathbf{R}]$. Suppose that $x \in AC_{\text{loc}}(J;\mathbf{R})$ satisfies

$$x'(t) + a(t)x(t) \le (\kappa \star x)(t) + f(t), \quad t \in J.$$

Then $x(t) \le y(t)$ for $t \in J$, where y is the solution of the comparison equation

$$y'(t) + a(t)y(t) = (\kappa \star y)(t) + f(t), \quad t \in J; \quad y(S) = x(S).$$

Proof Define $c(t,s) = \exp\bigl(-\int_s^t a(u)\,du\bigr)$. It follows from the variation of constants formula for ordinary differential equations, and from the given inequality on x, that

$$x(t) \le (c \star k \star x)(t) + c(t,S)x(S) + (c \star f)(t).$$

Observe that $c \star k$ is nonnegative, and use Lemma 2.15 to get $x(t) \le y(t)$, where y is the solution of the equation

$$y(t) = (c \star k \star y)(t) + c(t,S)y(S) + (c \star f)(t).$$

However, by the variation of constants formula, one may also characterize y by the fact that it is the (unique) solution of the comparison equation given above. □

For a vector-valued version of Lemma 3.10, see Exercise 20.

4. Nonconvolution Integrodifferential Equations: Global Theory

As we already mentioned above, it is not completely obvious what class of kernels one should work with to get a global theory for (3.5) or (1.8). For this reason, we use several different classes of kernels. Primarily, our results concern (3.5), and we leave the major part of the discussion of (1.8) to the reader (see the exercises). Admittedly, the existing theory on the global behaviour of linear nonconvolution integrodifferential equations is rather meagre. The results presented here are of perturbation type. Some additional results are given in Chapter 14.

Our perturbation results are based on the following variation of constants formula for differential resolvents.

4.1 Theorem

Let $J \subset \mathbf{R}$, and let $\kappa = \kappa_1 + \kappa_2$, where κ_1 and κ_2 are Volterra measure kernels of type $[B^\infty_{\text{loc}}; L^1_{\text{loc}}; J; \mathbf{C}^{n\times n}]$. Let r_1 be the differential resolvent of κ_1, and let r be the differential resolvent of κ. Then r_1 and r satisfy

$$r + r_1 \star \kappa_2 \star r = r + r \star \kappa_2 \star r_1 = r_1. \tag{4.1}$$

For the proof, observe that $\frac{\partial}{\partial t} r + \kappa_1 \star r = -\kappa_2 \star r$ for $t \geq s$. Fix s and solve for $t \geq s$ using formula (3.6) and the fact that $r(s,s) = I$. This gives $r = r_1 - r_1 \star \kappa_2 \star r$. The second part of (4.1) is left as an exercise.

Note that (3.4) is a special case of (4.1), because Λ is the differential resolvent of the kernel which is identically zero.

As an application of Theorem 4.1 we have the following result.

4.2 Theorem

Let $J = [S, \infty)$, where $-\infty < S < \infty$, and let $\kappa = \kappa_1 + \kappa_2 + \kappa_3$ be a Volterra measure kernel of type $[B^\infty_{\text{loc}}; L^1_{\text{loc}}; J; \mathbf{C}^{n\times n}]$. Assume that

(i) *for each Borel set $E \subset J$ and each $t \in J$, κ_1 is defined by $\kappa_1(t, E) = \int_{\mathbf{R}^+} \mu(\mathrm{d}s)\chi_E(t-s)$, where $\mu \in M(\mathbf{R}^+; \mathbf{C}^{n\times n})$ satisfies $\det[zI + \hat{\mu}(z)] \neq 0$ for $\Re z \geq 0$,*

(ii) *$\kappa_2 \in \mathcal{M}(B^\infty; L^1; J; \mathbf{C}^{n\times n})$,*

(iii) *$\kappa_3 \in \mathcal{M}(B^\infty; J; \mathbf{C}^{n\times n})$, and $\lim_{T\to\infty} |||\kappa_3|||_{B^\infty[T,\infty)} = 0$.*

Then the differential resolvent r of κ is of bounded uniformly continuous type on J, and r is bounded on J^2 (i.e., r is a kernel of type $[M; B^\infty]$ on J).

(Kernels of bounded uniformly continuous type were defined in Definition **9**.5.2.)

Proof of Theorem 4.2 By Theorem **3**.3.5, we know that κ_1 has a differential resolvent r_1 of the form $r_1(t,s) = q_1(t-s)$ for $t \geq s$, with q_1, $q_1' \in L^1(J; \mathbf{C}^n)$ (and $r_1(t,s) = 0$ for $t < s$). In particular, r_1 is of bounded uniformly continuous type and of types $[M; B^\infty]$ and B^∞ on J. It follows from Theorem 3.5(i) that $r_1 \star \kappa_2$ is a measure kernel of type B^∞ on J, and, because $\mathcal{M}(B^\infty; J; \mathbf{C}^{n\times n})$ is an algebra, one has $(r_1 \star \kappa_3) \in \mathcal{M}(B^\infty; J; \mathbf{C}^{n\times n})$. It is quite easy (this is where we need $\lim_{T\to\infty} |||\kappa_3|||_{B^\infty[T,\infty)} = 0$) to check that $r_1 \star (\kappa_2 + \kappa_3)$ satisfies the assumptions of Corollary 2.9. Therefore, it has a measure resolvent ρ of type B^∞ on J. We can use ρ to solve the equation $r + r_1 \star (\kappa_2 + \kappa_3) \star r = r_1$

(contained in (4.1), with κ_2 replaced by $\kappa_2 + \kappa_3$) to get $r = r_1 - \rho \star r_1$. This, together with Theorem 3.5(iii), implies that r is both of type B^∞ and of type $[M; B^\infty]$.

It remains to show that r is of bounded uniformly continuous type on J. This assertion follows from the same equation $r + r_1 \star (\kappa_2 + \kappa_3) \star r = r_1$, because, by Theorem 3.5(ii), $\kappa_2 \star r$ is a kernel of type $[B^\infty; L^1]$, $\kappa_3 \star r$ is a kernel of type B^∞, and $r_1 = q_1$ with q_1 of convolution type and q_1, $q_1' \in L^1(J; \mathbf{C}^{n\times n})$. We leave the details to the reader. □

An analogous result is true in L^1.

4.3 Theorem
Let $J = [S, \infty)$ where $\infty < S < \infty$, and let $\kappa = \kappa_1 + \kappa_2$ be a Volterra measure kernel of type $[B^\infty_{\text{loc}}; L^1_{\text{loc}}; J; \mathbf{C}^{n\times n}]$. Assume that

(i) for each Borel set $E \subset J$ and each $t \in J$, κ_1 is defined by $\kappa_1(t, E) = \int_{\mathbf{R}^+} \mu(\mathrm{d}s)\chi_E(t-s)$, where $\mu \in M(J; \mathbf{C}^{n\times n})$ satisfies $\det[zI+\hat{\mu}(z)] \neq 0$ for $\Re z \geq 0$,

*(ii) for each bounded Borel set $E \subset J$ and each $t \in J$, $\kappa_2(t, E) = \int_E k(t, s)\,\mathrm{d}s$, where k is a kernel of type $[L^1; J; \mathbf{C}^{n\times n}]$ (see Proposition **9**.2.7(ii)), and $\lim_{T\to\infty} |||k|||_{L^1(T,\infty)} = 0$.*

Then the differential resolvent r of κ is of type L^1 on J, r is bounded on J^2, $r(t, s) \to 0$ as $t \to \infty$ for every $s \in J$, and $\frac{\partial}{\partial t} r(t, s)$ is of type L^1 on J. In particular, the mapping $f \mapsto r \star f$ is continuous from $L^1(J; \mathbf{C}^n)$ into $L^1(J; \mathbf{C}^n) \cap BC_0(J; \mathbf{C}^n)$, and the mapping $f \mapsto (r \star f)'$ is continuous from $L^1(J; \mathbf{C}^n)$ into itself.

Proof It follows from Theorem **3**.3.5 that κ_1 has a differential resolvent r_1 of convolution type, i.e., $r_1(t, s) = q_1(t - s)$, $S \leq s \leq t < \infty$, with $q_1, q_1' \in L^1(J; \mathbf{C}^n)$. Obviously, r_1 is of type L^1 on J, hence $r_1 \star \kappa_2$ is of type L^1 on J. It is not difficult to check that J may be divided into finitely many subintervals J_i such that $|||r_1 \star \kappa_2|||_{L^1(J_i)} < 1$. Therefore (see Corollary **9**.3.14), $r_1 \star \kappa_2$ has a resolvent r_2 of type L^1 on J. Solving the equation $r + r_1 \star \kappa_2 \star r = r_1$, we get $r = r_1 - r_2 \star r_1$. Thus r is of type L^1 on J. That r is bounded, i.e., that r is of type $[M; B^\infty]$, follows from the equation $r = r_1 - r_1 \star \kappa_2 \star r$ and Theorem 3.5 (since r_1 is of type $[M; B^\infty]$).

That the mapping $f \mapsto r \star f$ maps $L^1(J; \mathbf{C}^n)$ into itself is a direct consequence of the fact that r is of type L^1 on J. Since $r \star f$ is the solution of (3.5) with zero initial condition, $r \star f$ is locally absolutely continuous, and $(r \star f)' = f - \kappa \star r \star f$. Thus, $(r \star f)' \in L^1(J; \mathbf{C}^n)$, as claimed. In particular, $r \star f \in BC_0(J; \mathbf{C}^n)$. To see that $r(t, s) \to 0$ as $t \to \infty$, it suffices to observe that the functions $t \mapsto r(t, s)$ and $t \mapsto \frac{\partial}{\partial t} r(t, s)$ belong to $L^1(J; \mathbf{C}^{n\times n})$ (because r is of type L^1, and the derivative of r satisfies (3.2)).

To get the final claim that $q(t, s) \stackrel{\text{def}}{=} \frac{\partial}{\partial t} r(t, s)$ is of type L^1 on J, one observes that $(r \star f)' = f - q \star f$, hence $q \star f \in L^1(J; \mathbf{C}^n)$ whenever

$f \in L^1(J; \mathbf{C}^n)$. The conclusion now follows from Lemma **9**.10.4. □

(One can get an alternative proof of the fact that $(r \star f)(t) \to 0$ as $t \to \infty$ for every $f \in L^1(J; \mathbf{C}^n)$, by combining Lebesgue's dominated convergence theorem with the fact that r is bounded on J^2, and $r(t,s) \to 0$ as $t \to \infty$.)

Instead of looking at perturbations of convolution equations, one can equally well look at perturbations of the time-dependent ordinary differential equation

$$x'(t) + A(t)x(t) = f(t), \quad t \geq S; \quad x(S) = x_S, \tag{4.2}$$

i.e., at equations of the form

$$x'(t) + A(t)x(t) + \int_{[S,t]} \kappa_2(t, \mathrm{d}s)x(s) = f(t), \quad t \geq S; \quad x(S) = x_S. \tag{4.3}$$

Clearly, (4.3) can be written in the form

$$x'(t) + \int_{[S,t]} \kappa(t, \mathrm{d}s)x(s) = f(t), \quad t \geq S; \quad x(S) = x_S, \tag{3.5}$$

where $\kappa = \kappa_1 + \kappa_2$, with $\kappa_1(t, E) = A(t)\chi_E(t)$. We assume throughout that the differential resolvent r_1 of the unperturbed kernel κ_1 is well behaved. As is well known from the theory of ordinary differential equations, this resolvent can be expressed in terms of the n by n matrix solution of the equation

$$X'(t) + A(t)X(t) = 0, \quad t \in \mathbf{R}^+; \quad X(0) = I, \tag{4.4}$$

i.e., the so-called fundamental solution. If X is the solution of (4.4), then

$$r_1(t,s) = \begin{cases} X(t)X^{-1}(s), & s \leq t, \\ 0, & s > t. \end{cases} \tag{4.5}$$

4.4 Theorem

Let $J = [S, \infty)$, where $-\infty < S < \infty$, and let $\kappa = \kappa_1 + \kappa_2$ be a Volterra measure kernel of type $[B^\infty_{\mathrm{loc}}; L^1_{\mathrm{loc}}; J; \mathbf{C}^{n\times n}]$. Assume that

(i) *for each Borel set $E \subset J$ and each $t \in J$, $\kappa_1(t, E) = A(t)\chi_E(t)$, where $A \in L^1_{\mathrm{loc}}(J; \mathbf{C}^{n\times n})$, and for every $x_S \in \mathbf{C}^n$ and every $f \in L^\infty(J; \mathbf{C}^n)$, the solution of (4.2) belongs to $BC(J; \mathbf{C}^n)$,*

(ii) *κ_2 is a measure kernel of type $[B^\infty; J; \mathbf{C}^{n\times n}]$ that satisfies $\lim_{T\to\infty} |||\kappa_2|||_{B^\infty[T,\infty)} = 0$.*

Then the differential resolvent r of κ is of bounded continuous type on J.

If, in addition, it is true that the solution of (4.2) belongs to $BC(J; \mathbf{C}^n)$ whenever $f \in L^1(J; \mathbf{C}^n)$, then r is bounded on J^2.

If $\kappa_2(t, \mathrm{d}s) = a_2(t,s)\,\mathrm{d}s$, with a_2 a function, then the assumptions on κ_2 are satisfied provided that $\sup_{t\in J} \int_J |a_2(t,s)|\,\mathrm{d}s < \infty$ and $\lim_{T\to\infty} \sup_{T\leq t<\infty} \int_{[T,\infty)} |a_2(t,s)|\,\mathrm{d}s = 0$.

Note that the conclusion of the first part implies that, if f is bounded on J, then a solution x of (4.3) is bounded on J.

Proof of Theorem 4.4 By Theorem 3.7, κ_1 has a Volterra differential resolvent $r_1 \in B^\infty_{\text{loc}}(J^2; \mathbf{C}^{n\times n})$, which is of continuous type on J. By assumption, $r_1 \star f \in BC(J; \mathbf{C}^n)$ for every $f \in L^\infty(J; \mathbf{C}^n)$. Consequently it follows from Lemma **9**.10.2 and from the fact that r_1 is of continuous type that r_1 is of bounded continuous type on J.

The kernel $\kappa_2 \star r_1$ is of bounded type on J, see Exercise 2. By the asymptotic condition on κ_2, there exists $T < \infty$ such that $|||\kappa_2 \star r_1|||_{B^\infty(T,\infty)} < 1$. Consequently (see Lemma **9**.3.8), $\kappa_2 \star r_1$ has a resolvent q of bounded type on $[T, \infty)$. Using the fact that r_1 is of continuous type, one can argue as in the proof Theorem **9**.5.5 to show that the interval $[S, T]$ may be divided into finitely many subintervals J_m such that $|||\kappa_2 \star r_1|||_{L^\infty(J_m)} < 1$ for all m. Hence (see Lemma **9**.3.8, Theorem **9**.3.14, and Theorem **9**.5.4), $\kappa_2 \star r_1$ has a resolvent q of bounded type on J. Therefore, the equation $r + r \star \kappa_2 \star r_1 = r_1$ has the solution $r = r_1 - r_1 \star q$. Now use Theorem **9**.5.3(i) to conclude that r is of bounded continuous type on J.

Suppose, finally, that the solution of (4.2) belongs to $BC(J; \mathbf{C}^n)$ whenever $f \in L^1(J; \mathbf{C}^n)$. Then r_1 maps $L^1(J; \mathbf{C}^n)$ into $BC(J; \mathbf{C}^n)$. Now use Lemma **9**.11.3 to conclude that r_1 is of type $[L^1; L^\infty]$ on J, and then use Exercise **9**.3(iii) to see that r_1 is essentially bounded on J^2. Since $r_1(t, s)$ is continuous when $s \neq t$, this implies that r_1 is bounded, i.e., r_1 is of type $[M; B^\infty]$. Now the proof can be completed in the standard way (use the equation $r = r_1 - r_1 \star \kappa_2 \star r$). □

A similar result is true in L^1.

4.5 Theorem

Let $J = [S, \infty)$, where $-\infty < S < \infty$, and let $\kappa = \kappa_1 + \kappa_2$ be a Volterra measure kernel of type $[B^\infty_{\text{loc}}; L^1_{\text{loc}}; J; \mathbf{C}^{n\times n}]$. Let the following assumptions hold:

(i) *for each Borel set $E \subset J$ and each $t \in J$, $\kappa_1(t, E) = A(t)\chi_E(t)$, where $A \in L^1_{\text{loc}}(J; \mathbf{C}^{n\times n})$, and for every $x_0 \in \mathbf{C}^n$ and every $f \in L^1(J; \mathbf{C}^n)$ the solution of (4.2) belongs to $L^1(J; \mathbf{C}^n)$;*

(ii) *for each bounded Borel set $E \subset J$ and for each $t \in J$, $\kappa_2(t, E) = \int_E k(t, s)\, ds$, where k is a kernel of type $[L^1; J; \mathbf{C}^{n\times n}]$, and $\lim_{T\to\infty} |||k|||_{L^1(T,\infty)} = 0$.*

Then the differential resolvent r of κ is of type L^1 on J.

If, in addition, it is true that the solution of (4.2) belongs to $BC(J; \mathbf{C}^n)$ whenever $f \in L^1(J; \mathbf{C}^n)$, then r is bounded on J^2.

The first conclusion of this theorem implies that if $f \in L^1(J; \mathbf{C}^n)$, then the solution x of (3.5) satisfies $x \in L^1(J; \mathbf{C}^n)$.

Observe that the size assumption on k in (ii) can be written in the form $\sup_{s\in J} \int_J |k(t, s)|\, dt < \infty$, and $\lim_{T\to\infty} \sup_{T\le s<\infty} \int_T^\infty |k(t, s)|\, dt = 0$.

Proof of Theorem 4.5 Define r_1 as in (4.5). By assumption, $r_1 \star f \in L^1(J; \mathbf{C}^n)$ for every $f \in L^1(J; \mathbf{C}^n)$. According to Lemma **9**.10.4, this implies that r_1 is of type L^1 on J.

From the hypothesis that k is of type L^1 on J, one concludes that $r_1 \star \kappa_2$ is of type L^1 on J. The asymptotic condition on κ_2 again allows us to deduce that there exists $T < \infty$ such that $|||r_1 \star \kappa_2|||_{L^1(T,\infty)} < 1$. The continuity of r_1 permits us to split the interval $[S,T]$ into finitely many parts J_n such that $|||r_1 \star \kappa_2|||_{L^1(J_n)} < 1$ for all n. Therefore, by Corollary **9**.3.10 and Theorem **9**.3.13, we may conclude that $r_1 \star \kappa_2$ has a resolvent q of type L^1 on J. Solving the equation $r + r_1 \star \kappa_2 \star r = r_1$ with the variation of constants formula, we get $r = r_1 - q \star r_1$, and this shows that r is of type L^1 on J.

The proof of the boundedness of r is the same as in the proof of Theorem 4.4. □

We conclude this section with a short discussion of (1.8).

All the perturbation results that we have proved in this section can be extended to the case $J = \mathbf{R}$; see Exercises 25–27, 29, and 30. This fact can be used to establish results concerning the solutions of the equation

$$x'(t) + \int_{(-\infty,t]} \kappa(t, \mathrm{d}s)x(s) = f(t), \quad t \in \mathbf{R}. \tag{1.8}$$

Formally, it ought to be true that the solution of this equation is given by

$$x(t) = \int_{-\infty}^{t} r(t,s)f(s)\,\mathrm{d}s, \quad t \in \mathbf{R},$$

where r is the differential resolvent of κ. Indeed, if, for example, κ is of type B^∞ on $\mathbf{R}$, and r is of bounded continuous type on $\mathbf{R}$, then this is true.

4.6 Theorem

Let κ be a Volterra measure kernel of type $[B^\infty; \mathbf{R}; \mathbf{C}^{n\times n}]$, and suppose that the differential resolvent of κ is of bounded continuous type. Let $f \in L^\infty(\mathbf{R}; \mathbf{C}^n)$. Then (1.8) has a unique bounded and locally absolutely continuous solution x on $\mathbf{R}$. This solution is given by the variation of constants formula

$$x(t) = \int_{-\infty}^{t} r(t,s)f(s)\,\mathrm{d}s, \quad t \in \mathbf{R}. \tag{4.6}$$

Proof Write (1.8) in the form

$$x'(t) + x(t) - x(t) + \int_{(-\infty,t]} \kappa(t, \mathrm{d}s)x(s) = f(t), \quad t \in \mathbf{R}.$$

In other words, write $\kappa = \kappa_1 + \kappa_2$, where $\kappa_1(t,E) = I\chi_E(t)$, and $\kappa_2 = \kappa - \kappa_1$. The differential resolvent of κ_1 is the function

$$\mathbf{e}(t,s) = \begin{cases} \mathrm{e}^{-(t-s)}I, & s \le t, \\ 0, & s > t. \end{cases}$$

Recall that (4.1) holds, or equivalently that

$$r + (\mathbf{e} \star \kappa - \mathbf{e}) \star r = r + r \star (\kappa \star \mathbf{e} - \mathbf{e}) = \mathbf{e}. \tag{4.7}$$

If $x = r \star f$, then by (4.7),

$$x = \mathbf{e} \star f - (\mathbf{e} \star \kappa - \mathbf{e}) \star r \star f = \mathbf{e} \star f - (\mathbf{e} \star \kappa - \mathbf{e}) \star x.$$

This implies that x is locally absolutely continuous, and a differentiation shows that x satisfies (1.8) a.e.

Conversely, let x be an arbitrary locally absolutely continuous solution of (1.8). Multiply (1.8) by $\mathbf{e}(t, s)$, integrate over $\mathbf{R}$, and finally integrate by parts, to get

$$\begin{aligned}
x &= \mathbf{e} \star f - \mathbf{e} \star (\kappa \star x - x) \\
&= \mathbf{e} \star f - (r + r \star (\kappa \star \mathbf{e} - \mathbf{e})) \star (\kappa \star x - x) \\
&= \mathbf{e} \star f + (r - r \star \kappa) \star (x + \mathbf{e} \star (\kappa \star x - x)) \\
&= \mathbf{e} \star f - (r - r \star \kappa) \star \mathbf{e} \star f \\
&= (\mathbf{e} - r \star (\kappa \star \mathbf{e} - \mathbf{e})) \star f \\
&= r \star f.
\end{aligned}$$

Therefore x satisfies (4.6). □

The same proof can be used to get a variation of constants formula that can be applied in the situations discussed in Exercises 25–27, 29, and 30 (in Exercises 29 and 30 one should assume in addition that, e.g., $\limsup_{S \to -\infty} \int_S^{S+1} |A(t)|\, dt < \infty$).

5. Exercises

1. Show that the equation

$$x(t) + a(t)x\big(t - g(t)\big) = f(t), \quad t \geq 0,$$

where $0 \leq g(t) \leq t$, can be put into the form (2.1). Do the same for the equation

$$x(t) + \sum_{i=1}^{\infty} a_i(t)x\big(t - g_i(t)\big) = f(t), \quad t \geq 0,$$

where $0 \leq g_i(t) \leq t$ for $t \geq 0$. What conditions are needed for the resulting kernels to be of type B^{∞}_{loc} on $\mathbf{R}^+$?

2. Show that $\mathcal{V}(B^\infty; J; \mathbf{C}^{n\times n})$ is a right ideal in $\mathcal{M}(B^\infty; J; \mathbf{C}^{n\times n})$, i.e., show that $\alpha \star b \in \mathcal{V}(B^\infty; J; \mathbf{C}^{n\times n})$ whenever $\alpha \in \mathcal{M}(B^\infty; J; \mathbf{C}^{n\times n})$ and $b \in \mathcal{V}(B^\infty; J; \mathbf{C}^{n\times n})$ (cf. Theorem 2.3).
3. Let $\tau \in \mathbf{R}$, and let $a \in C([\tau, \infty); \mathbf{C}^{n\times n})$. For each Borel set $E \subset \mathbf{R}$ and each $t \in \mathbf{R}$, let $\kappa(t, E) = 0$ if $t < \tau$, and $\kappa(t, E) = a(t)\chi_E(\tau)$ if $t \geq \tau$. Compute the measure resolvent of k. What additional

condition was needed? Show by checking the properties of Definition 2.1 that the measure resolvent is of type B^{∞}_{loc} on $\mathbf{R}$.

4. Let $J \subset \mathbf{R}$ be an interval, and let $a \in B^{\infty}(J; \mathbf{C}^{n \times n})$ and $[I + a]^{-1} \in B^{\infty}(J; \mathbf{C}^{n \times n})$. Suppose that $\kappa(t, E) = a(t)\chi_E(t) + \lambda(t, E)$, where λ is a Volterra measure kernel of type B^{∞} on J, and that the kernel $[I + a(t)]^{-1}\lambda(t, E)$ has a measure resolvent ν on J. Show that κ has a measure resolvent ρ on J, given by

$$\rho(t, E) = a(t)[I + a(t)]^{-1}\chi_E(t) + \int_E \nu(t, \mathrm{d}s)[I + a(s)]^{-1}.$$

5. Is the following statement true or false? Let $\mu \in M(\mathbf{R}^+; \mathbf{C}^{n \times n})$ have no singular part and satisfy $\inf_{\Re z \geq 0} |\det[I + \hat{\mu}(z)]| > 0$. Suppose that $h \in BC(\mathbf{R}^+; \mathbf{C}^{n \times n})$ and that $h(t)$ converges to the unit matrix as $t \to \infty$. Let $f \in B^{\infty}_0(\mathbf{R}^+; \mathbf{C}^n)$ and let x be the solution of

$$x(t) + \int_{[0,t]} \mu(\mathrm{d}s)h(t - s)x(t - s) = f(t), \quad t \geq 0.$$

Then $x(t) \to 0$ as $t \to \infty$.

6. Consider the real scalar equation

$$x(t) + \int_{[0,t]} \mu(\mathrm{d}s)a(t - s)x(s) = f(t), \quad t \in \mathbf{R}^+,$$

where $\mu \in M(\mathbf{R}^+)$ is a positive measure. Suppose that a is continuous and $\ln(a)$ is convex on $\mathbf{R}^+$, and let $f \in BC(\mathbf{R}^+)$. Show that x is bounded on $\mathbf{R}^+$ and that $\rho(t, \cdot)$, for each $t \in \mathbf{R}^+$, is a positive measure where ρ is the measure resolvent of $\kappa(t, ds) = \mu(ds)a(t - s)$. Hint: See Sections **9**.8 and **20**.3.

7. Let $J \subset \mathbf{R}$ be an interval. It makes sense to call κ a measure kernel of continuous type on J (bounded continuous type on J) [bounded uniformly continuous type on J] if for each $t \in J$, $\kappa(t, \cdot)$ is a $\mathbf{C}^{n \times n}$-valued measure on J, and if the function $t \mapsto \kappa(t, \cdot)$ is continuous (bounded and continuous) [bounded and uniformly continuous] from J into $M(J; \mathbf{C}^{n \times n})$. We denote the classes of Volterra measure kernels of continuous, bounded continuous, and bounded uniformly continuous types on J by
   (i) $\mathcal{M}(C; J; \mathbf{C}^{n \times n})$,
   (ii) $\mathcal{M}(BC; J; \mathbf{C}^{n \times n})$, and
   (iii) $\mathcal{M}(BUC; J; \mathbf{C}^{n \times n})$,

   respectively. Extend Theorems **9**.5.3–**9**.5.5 and Corollary **9**.5.6 to these classes of kernels.

8. Give an alternative proof of Theorem 4.4 which is based on the measure resolvent of $r_1 \star \kappa_2$ instead of on the resolvent of $\kappa_2 \star r_1$. Hint: See Exercise 7.

9. Let $J \subset \mathbf{R}$ be an interval whose right end-point is $+\infty$. Show how the following classes of Volterra measure kernels are to be defined,

and extend the results of Section **9**.6 to these classes of kernels (cf. Definition 2.1):

(i) $\mathcal{M}(B_0^\infty; J; \mathbf{C}^{n\times n})$;
(ii) $\mathcal{M}(BC_0; J; \mathbf{C}^{n\times n})$;
(iii) $\mathcal{M}(B_\ell^\infty; J; \mathbf{C}^{n\times n})$;
(iv) $\mathcal{M}(BC_\ell; J; \mathbf{C}^{n\times n})$;
(v) $\mathcal{M}(B^\infty; B_0^\infty; J; \mathbf{C}^{n\times n})$;
(vi) $\mathcal{M}(B^\infty; BC_0; J; \mathbf{C}^{n\times n})$;
(vii) $\mathcal{M}(B^\infty(\mathcal{T}_T); \mathbf{C}^{n\times n})$;
(viii) $\mathcal{M}(C(\mathcal{T}_T); \mathbf{C}^{n\times n})$;
(ix) $\mathcal{M}(B^\infty(\mathcal{T}_T) \oplus B_0^\infty(J); \mathbf{C}^{n\times n})$;
(x) $\mathcal{M}(C(\mathcal{T}_T) \oplus BC_0(J); \mathbf{C}^{n\times n})$.

10. Prove the analogue of Theorem 2.14 for the following classes of kernels:
(i) $\mathcal{M}(BC; J; \mathbf{C}^{n\times n})$;
(ii) $\mathcal{M}(BUC; J; \mathbf{C}^{n\times n})$;
(iii) $\mathcal{M}(B_0^\infty; J; \mathbf{C}^{n\times n})$;
(iv) $\mathcal{M}(BC_0; J; \mathbf{C}^{n\times n})$;
(v) $\mathcal{M}(B_\ell^\infty; J; \mathbf{C}^{n\times n})$;
(vi) $\mathcal{M}(BC_\ell; J; \mathbf{C}^{n\times n})$;
(vii) $\mathcal{M}(B^\infty(\mathcal{T}_T) \oplus B_0^\infty(J); \mathbf{C}^{n\times n})$;
(viii) $\mathcal{M}(C(\mathcal{T}_T) \oplus BC_0(J); \mathbf{C}^{n\times n})$.
11. In Theorem 2.14, replace J by $\mathbf{R}$, and assume, in addition, that $\lim_{S\to-\infty} |||\kappa_2|||_{B^\infty(-\infty,S)} = 0$. Prove that k has a measure resolvent of type B^∞ on $\mathbf{R}$.
12. Improve the result in Exercise 11 by allowing κ to have one representation $\kappa = \kappa_1 + \kappa_2$ for $-\infty < s \le t \le 0$, and a different representation for $0 < s \le t < \infty$.
13. Redo Exercise 10 with J replaced by $\mathbf{R}$, and Theorem 2.14 replaced by Exercise 12.
14. Formulate and prove a version of Lemma 2.15 with $J = \mathbf{R}$ (and κ and ρ of type B^∞ instead of type B_{loc}^∞.)
15. Let $\tau \in \mathbf{R}$, and let $a \in L_{\mathrm{loc}}^1([\tau,\infty); \mathbf{C}^{n\times n})$. For each Borel set $E \subset \mathbf{R}$ and each $t \in \mathbf{R}$, let $\kappa(t,E) = 0$ if $t < \tau$, and $\kappa(t,E) = a(t)\chi_E(\tau)$ if $t \ge \tau$. Determine the differential resolvent of κ on $\mathbf{R}$ and show, by checking Definition 3.3, that it is of type $[M_{\mathrm{loc}}; B_{\mathrm{loc}}^\infty]$ on $\mathbf{R}$.
16. Disprove the following claim:
(i) *if r is Borel measurable on J^2, $r(t,s) = 0$ for $s > t$, and the mapping $f \mapsto r \star f$ is continuous from $L^1(J; \mathbf{C}^n)$ to $B^\infty(J; \mathbf{C}^n)$, then r is of type $[M; B^\infty]$ on J.*

Prove the following claim:
(ii) *if r is Borel measurable on J^2, $r(t,s) = 0$ for $s > t$, and the mapping $\mu \mapsto \int_J r(t,s)\mu(\mathrm{d}s)$ is continuous from $M(J; \mathbf{C}^n)$ to $B^\infty(J; \mathbf{C}^n)$, then r is of type $[M; B^\infty]$ on J.*

Hint: Sets of measure zero sometimes play a role.

17. Can you characterize the class of measure kernels that map $BC(J;\mathbf{C}^n)$ continuously into itself? Is this class a Banach algebra?
18. Let $J = [S,T)$, with $-\infty < S < T \le \infty$, let κ be a Volterra measure kernel of type $[B^\infty_{\rm loc}; L^1_{\rm loc}; J; \mathbf{C}^{n\times n}]$, let $g \in B^\infty_{\rm loc}(J;\mathbf{C}^n)$, $f \in L^1_{\rm loc}(J;\mathbf{C}^n)$, and let $y_S \in \mathbf{C}^n$. Show that the equation
$$(x+g)'(t) + (\kappa \star x)(t) = f(t), \quad t \in [S,T); \quad (x+g)(S) = y_S,$$
has a unique solution x in the sense that there exists a unique function $x \in L^1_{\rm loc}(J;\mathbf{C}^n)$ such that $x+g$ is locally absolutely continuous, $(x+g)(S) = y_S$, and the equation above holds a.e. This solution is given by the variation of constants formula
$$x(t) = r(t,S)y_S - g(t) + (r \star f)(t) + (r \star \kappa \star g)(t), \quad t \in J,$$
where r is the differential resolvent of κ. Hint: See Exercise **3**.24.
19. Let $J = [S,T)$, with $-\infty < S < T \le \infty$, let μ be a Volterra measure kernel of type $[B^\infty_{\rm loc}; J; \mathbf{C}^{n\times n}]$ that has a resolvent of the same type, let ν be a Volterra measure kernel of type $[B^\infty_{\rm loc}; L^1_{\rm loc}; J; \mathbf{C}^{n\times n}]$, let $g \in B^\infty_{\rm loc}(J;\mathbf{C}^n)$, let $f \in L^1_{\rm loc}(J;\mathbf{C}^n)$, and let $y_S \in \mathbf{C}^n$. Show that the equation
$$\begin{aligned} x(t) + (\mu \star x)(t) + g(t) &= y(t), \quad t \in J,\\ y'(t) + (\nu \star x)(t) &= f(t), \quad t \in J,\\ y(S) &= y_S, \end{aligned}$$
has a unique pair of solutions (x,y) in the sense that there exist unique functions $x \in B^\infty_{\rm loc}(J;\mathbf{C}^n)$ and $y \in AC_{\rm loc}(J;\mathbf{C}^n)$ such that the first equation holds for all $t \in J$, and the second equation holds for almost all $t \in J$. Hint: See Exercises 18 and **4**.17; this equation is of so-called *neutral* type.
20. Prove the following vector version of Lemma 3.10. Let J be an interval of the type $[S,T)$, with $-\infty < S < T \le \infty$, let a, $b \in L^1_{\rm loc}(J;\mathbf{R})$, with $a(t) \ge b(t)$ a.e., let κ be a Volterra measure kernel of type $[B^\infty_{\rm loc}; L^1_{\rm loc}; J; \mathbf{C}^{n\times n}]$, and let $f \in L^1_{\rm loc}(J;\mathbf{C}^n)$. Suppose that $x \in AC_{\rm loc}(J;\mathbf{C}^n)$ satisfies
$$|x'(t) + a(t)x(t)| \le (|\kappa| \star |x|)(t) + |f(t)|, \quad t \in J.$$
Then $|x(t)| \le y(t)$ for $t \in J$, where y is the solution of
$$y'(t) + b(t)y(t) = (|\kappa| \star y)(t) + |f(t)|, \quad t \in J; \quad y(S) = |x(S)|.$$
Hint: See Lemma **14**.4.3.
21. Can you prove a version of Lemma 3.10 or Exercise 20 where $J = \mathbf{R}$?
22. Complete the proof of Theorem 4.1.
23. Consider the equation
$$x'(t) + \int_{[0,t]} \mu(\mathrm{d}s)h(t-s)x(t-s) = f(t), \quad t \in \mathbf{R}^+; \quad x(0) = x_0,$$

where $h \in C(\mathbf{R}^+; \mathbf{C}^{n\times n})$, and $\lim_{t\to\infty} h(t) = h(\infty)$ exists. Suppose μ satisfies $\det[zI + \hat{\mu}(z)h(\infty)] \neq 0$ for $\Re z \geq 0$. Show that if $f \in B_0^\infty(\mathbf{R}^+; \mathbf{C}^n)$, then $x \in BC_0(\mathbf{R}^+; \mathbf{C}^n)$. Let r be the differential resolvent of the kernel in the equation above and let q be the kernel $\frac{\partial}{\partial s} r(t,s)$. Show that the mapping $f \mapsto q \star f$ maps the set of bounded measurable functions vanishing at infinity into itself. How does $f \mapsto q \star f$ map bounded functions tending to a nonzero limit? Also show that $r(t,s) \to 0$ as $t \to \infty$ for all $s \in \mathbf{R}^+$.

24. Try to find additional conditions under which the differential resolvent in Theorem 4.2 is of type BC_0 on J.
25. In Theorem 4.2, replace J by $\mathbf{R}$, and assume, in addition, that $\lim_{S\to-\infty} |||\kappa_3|||_{B^\infty(-\infty,S)} = 0$. Prove that the differential resolvent r of κ is of bounded uniformly continuous type on $\mathbf{R}$, and (pointwise) bounded on $\mathbf{R}^2$.
26. Improve the result in Exercise 25 by allowing κ to have one representation $\kappa = \kappa_1 + \kappa_2 + \kappa_3$ for $-\infty < s \leq t \leq 0$, and a different representation for $0 < s \leq t < \infty$.
27. Modify Exercises 25 and 26, replacing Theorem 4.2 by Theorem 4.3.
28. Try to find additional conditions under which the differential resolvent in Theorem 4.4 is of type BC_0 on J.
29. In Theorem 4.4, replace J by $\mathbf{R}$, assume that the equation $x'(t) + A(t)x(t) = f(t)$, $t \in \mathbf{R}$, has a unique bounded solution for every $f \in L^\infty(\mathbf{R}; \mathbf{C}^n)$, and that $\lim_{S\to-\infty} |||\kappa_2|||_{B^\infty(-\infty,S)} = 0$. Prove that the differential resolvent of κ is of bounded continuous type on $\mathbf{R}$. Can you give additional assumptions under which the solutions of (1.7) tend to zero at infinity whenever f tends to zero at infinity?
30. Prove the analogue of Exercise 29 for Theorem 4.5.
31. Let $a \in L^1_{\mathrm{loc}}(\mathbf{R}^+; \mathbf{R}^+)$ and suppose $c(t,s)$ satisfies $\int_s^\infty |c(t,s)|\,dt \leq a(s) - \epsilon$ for $0 \leq s \leq t < \infty$ and some $\epsilon > 0$. Show that the solution x of the equation
$$x'(t) + a(t)x(t) = \int_0^t c(t,s)x(s)\,ds + f(t), \quad t \geq 0; \quad x(0) = x_0, \quad (5.1)$$
satisfies $\|x\|_{L^1(\mathbf{R}^+)} \leq \epsilon^{-1}(\|f\|_{L^1(\mathbf{R}^+)} + |x_0|)$. Write (5.1) in the form of (1.6) and show that the differential resolvent r of the resulting kernel is of type L^1 on $\mathbf{R}^+$. Hint: First use Lemma 3.10, and then integrate the comparison equation. (Staffans [27], Theorem 3.1).
32. Assume that the scalar Volterra kernel a is of type L^p_{loc} on $\mathbf{R}^+$ and satisfies $a(t,s) \geq \epsilon > 0$ for $0 \leq s \leq t$. Moreover, let a be nonincreasing in t for any fixed s. Let $r(t,s)$ be the solution of $r + a \star r = \Lambda$, where $\Lambda(t,s) = 1$, if $0 \leq s \leq t < \infty$ and zero otherwise. Show that $r \geq 0$ and that r is of type L^1 on $\mathbf{R}^+$. Hint: Use Theorem **9**.8.7.
33. Let a, r, Λ be as in Exercise 32. Assume, in addition, that $a(t,s) = b(s)\Lambda(t,s) + c(t,s)$, where $b(s) \geq \epsilon > 0$, $s \geq 0$; $c(t,s) \geq 0$, $0 \leq s \leq t < \infty$; c is of bounded type on $\mathbf{R}^+$ and c is nonincreasing in t (for

any fixed s) and nondecreasing in s (for any fixed t). Show that r is of bounded type on $\mathbf{R}^+$. Hint: Fix s. Define $x(t) = (r \star \Lambda)(t,s)$, $t \geq s$; and $v(t) = x(t) + (c \star x)(t)$. Check that $0 \leq x(t) \leq v(t)$, $v'(t) + b(t)x(t) = 1$, and that $(c \star v')(t) + \epsilon(c \star x)(t) \leq (c \star \Lambda)(t)$ for $t \geq s$. Show that if t_0 is such that $v(t_0) = \sup_{s \leq u \leq t_0} v(u)$, then $x(t_0) \leq \epsilon^{-1}$ and $(c \star x)(t_0) \leq \epsilon^{-1}(c \star \Lambda)(t_0)$. (See Burton, Huang, and Mahfoud [1], pp. 263–265.)

34. Prove a perturbation result for

$$x'(t) + A(t)x(t) + \int_{[0,t]} \kappa_1(t, \mathrm{d}s)x(s) + \tfrac{\mathrm{d}}{\mathrm{d}t} \int_{[0,t]} \kappa_2(t, \mathrm{d}s)x(s) = f(t),$$

where $t \geq 0$, $x(0) = x_0$, and where κ_1 and κ_2 are small compared to $A(t)$. Hint: See Exercise 19.

6. Comments

Section 2:
Nonconvolution Volterra equations with measure kernels have been considered by several authors. For an approach based on Perron–Stieltjes integrals and on functions of bounded variation, see Schwabik, Tvrdý, and Vejvoda [1], p. 40. In this reference, it is assumed that $|\kappa|(t, J) < \infty$ for each t and, in addition, a condition on the variation of $\kappa(t, \mathrm{d}s)$ with respect to both variables is imposed. The existence of a measure resolvent and the variation of constants formula are established. These results are quoted in Fenyő and Stolle [1], part 3, p. 329. For a different approach to Volterra–Stieltjes equations, see Hinton [1] and Bitzer [1].

A Fubini theorem for measure kernels is given in Cameron and Martin [1].

Section 3:
Nonconvolution Volterra integrodifferential equations were considered in Grossman and Miller [1]. The kernel is assumed to be of the form $\kappa(t, E) = A(t)\chi_E(t) + \int_{E \cap [0,t]} B(t,s)\,\mathrm{d}s$ for $t \geq 0$ and $E \subset \mathbf{R}^+$, where $A \in L^1_{\mathrm{loc}}(\mathbf{R}^+)$ and $B \in L^1_{\mathrm{loc}}(\mathbf{R}^+ \times \mathbf{R}^+)$. With this kernel, equation (1.6) becomes

$$x'(t) + A(t)x(t) + \int_0^t B(t,s)x(s)\,\mathrm{d}s = f(t), \quad t \in \mathbf{R}^+; \quad x(0) = x_0. \quad (6.1)$$

The differential resolvent is introduced and the variation of constants formula is given.

Equation (1.6) is, in the language of functional differential equations, linear, retarded, nonautonomous and nonhomogenous. See Hale [5], Ch. 6, Sect. 2, p. 143, for the resolvent (= the fundamental matrix) and for the variation of constants formula in the context of functional differential equations. Note that in Hale [5] the existence of the fundamental matrix is not obtained by Banach algebra arguments; instead, the fact that L^∞ is the dual of L^1 is used. Some results on the adjoint equation are given, and they are applied to boundary value problems.

An L^p-theory for nonconvolution Volterra integrodifferential equations is developed in Colonius, Manitius, and Salamon [1]. It is shown that in the L^p-case, $1 \le p < \infty$, the adjoint equation is a Volterra integrodifferential equation as well.

Section 4:
There is a large literature on asymptotic properties of solutions of (6.1) when B is sufficiently small; i.e., (6.1) is considered as a perturbation of the equation $x'(t) + A(t)x(t) = f(t)$, $t \in \mathbf{R}^+$; $x(0) = x_0$; see Burton [2] for references. Many of these results are based on Lyapunov type methods, see Staffans [27] and Chapter 14. The case $A(t) \equiv A$ is analysed in Grimmer and Seifert [1] where conditions are given under which certain subspaces of $C(\mathbf{R}^+)$ are invariant under the mapping $f \mapsto r \star f$.

References

C. W. Bitzer
1. Stieltjes–Volterra integral equations, *Illinois J. Math.* **14** (1970), pp. 434–451.

T. A. Burton
2. *Volterra Integral and Differential Equations*, Academic Press, New York, 1983.

T. A. Burton, Q. Huang, and W. E. Mahfoud
1. Liapunov functionals of convolution type, *J. Math. Anal. Appl.* **106** (1985), pp. 249–272.

R. H. Cameron and W. T. Martin
1. An unsymmetric Fubini theorem, *Bull. Amer. Math. Soc.* **47** (1941), pp. 121–125.

F. Colonius, A. Manitius, and D. Salamon
1. Structure theory and duality for time varying retarded functional differential equations, *J. Differential Equations* **78** (1989), pp. 320–353.

I. Fenyő and H. W. Stolle
1. *Theorie und Praxis der linearen Integralgleichungen, 1–4*, Birkhäuser, Basel, 1982–1984.

R. Grimmer and G. Seifert
1. Stability properties of Volterra integrodifferential equations, *J. Differential Equations* **19** (1975), pp. 142–166.

S. I. Grossman and R. K. Miller
1. Perturbation theory for Volterra integrodifferential systems, *J. Differential Equations* **8** (1970), pp. 457–474.

J. K. Hale

5. *Theory of Functional Differential Equations*, Springer-Verlag, Berlin, 1977.

D. B. Hinton

1. A Stieltjes–Volterra integral equation theory, *Canad. J. Math.* **18** (1966), pp. 314–331.

Š. Schwabik, M. Tvrdý, and O. Vejvoda

1. *Differential and Integral Equations. Boundary Value Problems and Adjoints*, Reidel, Dordrecht, 1979.

O. J. Staffans

27. A direct Lyapunov approach to Volterra integrodifferential equations, *SIAM J. Math. Anal.* **19** (1988), pp. 879–901.

11

Perturbed Linear Equations

We study the existence and asymptotic behaviour of solutions of weakly nonlinear Volterra equations. We formulate two fixed point theorems that are used to derive various perturbation results. Both integral and integrodifferential equations are considered.

1. Introduction

In this chapter we take a first step away from the linear theory, developed in Chapters 2–10. We investigate nonlinear equations in which the linear part in some sense dominates the nonlinear part, at least for appropriately restricted initial data and nonhomogeneous terms. Since our techniques rely on fixed point theorems where we take the underlying spaces to consist of functions defined on $\mathbf{R}^+$ or on $\mathbf{R}$, our results yield both global existence of solutions and information about their behaviour at infinity. In Chapter 12 we shall consider the question of local existence in a more general setting.

Specifically, in this chapter, we discuss equations of the form

$$x(t) + \int_0^t k(t,s)x(s)\,\mathrm{d}s = G(x)(t), \quad t \in \mathbf{R}^+, \tag{1.1}$$

and of the form

$$x'(t) + \int_{[0,t]} \kappa(t,\mathrm{d}s)x(s) = G(x)(t), \quad t \in \mathbf{R}^+; \quad x(0) = x_0. \tag{1.2}$$

(In addition, we consider the first equation on $\mathbf{R}$.) Here k (a function) and κ (a measure) are kernels of the types considered in Chapters 9 and 10, and G is a sum of a small term and a term independent of x. To avoid undue complications, we do not consider measure kernels in (1.1). For our techniques to apply, we must have enough information on how the solutions of the linearized equation (i.e., the equation one gets when one omits the x-dependent part of G) behave.

Since we know much more about linear convolution equations than about linear nonconvolution ones, the equations that have a linear part of convolution type are especially important. These equations are of the type

$$x(t) + \int_0^t k(t-s)x(s)\,\mathrm{d}s = G(x)(t), \quad t \in \mathbf{R}^+, \tag{1.3}$$

and of the type

$$x'(t) + \int_{[0,t]} \mu(\mathrm{d}s)x(t-s) = G(x)(t), \quad t \in \mathbf{R}^+; \quad x(0) = x_0. \tag{1.4}$$

Let r denote the resolvent of the kernel k in (1.1) (see Definition **9**.3.1), and use the variation of constants formula (**9**.3.4) to rewrite (1.1) in the form

$$x(t) = G(x)(t) - \int_0^t r(t,s)G(x)(s)\,\mathrm{d}s, \quad t \in \mathbf{R}^+. \tag{1.5}$$

Similarly, if we instead let r stand for the differential resolvent of the kernel κ in (1.2) (see Definition **10**.3.8), and use the variation of constants formula (**10**.3.6), then (1.2) can be rewritten as

$$x(t) = r(t,0)x_0 + \int_0^t r(t,s)G(x)(s)\,\mathrm{d}s, \quad t \in \mathbf{R}^+. \tag{1.6}$$

Both (1.5) and (1.6) are examples of the more general equation

$$x(t) = F(x)(t) + \int_0^t r(t,s)G(x)(s)\,\mathrm{d}s, \quad t \in \mathbf{R}^+, \tag{1.7}$$

where F and G are operators that map some function space into itself, or into some other function space.

Above we discussed equations defined on $\mathbf{R}^+$. Analogous arguments applied to equations on $\mathbf{R}$ lead to

$$x(t) = F(x)(t) + \int_{-\infty}^{\infty} r(t,s)G(x)(s)\,\mathrm{d}s, \quad t \in \mathbf{R}. \tag{1.8}$$

Both (1.7) and (1.8) are particular cases of the equation

$$x(t) = F(x)(t) + \int_J r(t,s)G(x)(s)\,\mathrm{d}s, \quad t \in J, \tag{1.9}$$

where J is a subinterval of $\mathbf{R}$.

One can prove a large number of results for equations of one of the types (1.1)–(1.4) by first formulating some general fixed point theorems for (1.9), and then varying the function spaces in which one works. We carry out the first task in Section 2 and the second in Sections 3–6. The results we obtain certainly do not exhaust all possibilities. Rather, they are intended to show some relevant ideas that one can use in specific cases. In particular, we concentrate on the case where the unperturbed equations are of convolution type and leave the nonconvolution cases to the exercises; see Exercises 4, 5 and 10. Throughout this chapter we work with real scalars. This is no essential limitation, because one can interpret $\mathbf{C}^n$ as a special case of $\mathbf{R}^{2n}$.

2. Two General Perturbation Theorems

We begin by stating two fixed point theorems on which our subsequent perturbation results will be based. Both of these theorems concern (1.9). Our first theorem is a simple consequence of the contraction mapping principle. The second theorem is a straightforward corollary to Theorem **12**.1.3. We leave both proofs to the reader.

The $\star$-operation is defined as in Definition **9**.2.3 (although the two theorems are formulated so as to allow $r\star$ to be any mapping satisfying the required assumptions).

2.1 Theorem
Assume the following:

(i) *K is a closed subset of a Banach space $\mathcal{B}$ of functions taking an interval $J \subset \mathbf{R}$ into $\mathbf{R}^n$;*

(ii) *F and G are operators mapping K into $\mathcal{B}$, satisfying*

$$\|F(x) - F(y)\| \le L_F\|x - y\| \quad \text{and} \quad \|G(x) - G(y)\| \le L_G\|x - y\|,$$

for some constants L_F and L_G and for all x and y in K;

(iii) *the mapping $x \mapsto r \star x$ is a linear and continuous operator from $\mathcal{B}$ into itself, with norm ϱ;*

(iv) *the function $x \mapsto F(x) + r \star G(x)$ takes K into itself;*

(v) *the constants L_F, L_G and ϱ satisfy $L_F + L_G\varrho < 1$.*

Then (1.9) has a unique solution in K.

Our second theorem combines a contraction argument with Schauder's theorem.

2.2 Theorem
Assume the following:

(i) *K is a closed, bounded, and convex subset of a Banach space $\mathcal{B}$ of functions taking an interval $J \subset \mathbf{R}$ into $\mathbf{R}^n$;*

(ii) *F is an operator mapping K into $\mathcal{B}$, and, either*

$$\|F(x) - F(y)\| \le L_F\|x - y\|$$

for some constant $L_F < 1$ and all x and $y \in K$, or F is continuous and compact on K;

(iii) *the mapping $x \mapsto r \star G(x)$ is continuous and compact from K into $\mathcal{B}$;*

(iv) *the mapping $x \mapsto F(x) + r \star G(x)$ takes K into itself.*

Then (1.9) has at least one solution in K.

Throughout this chapter, when we apply these two theorems, we take either $J = \mathbf{R}^+$ or $\mathbf{R}$.

Let us make a few comments on hypothesis (iii) in Theorem 2.2. This condition holds, for example, when the operator G is continuous and bounded, and the mapping $x \mapsto r \star x$ is continuous and compact. If J is finite, then the second of these two requirements is satisfied when

r is a kernel of bounded uniformly continuous type, and $\mathcal{B}$ is either $L^\infty(J;\mathbf{R}^n)$ or $BC(J;\mathbf{R}^n)$, see Theorem **9**.5.3(iii). Compactness, in the case $\mathcal{B} = L^p(J;\mathbf{R}^n)$, with $p \in [1,\infty)$ and J an arbitrary interval, is analysed in Section **9**.7. However, our main interest lies in the case where $\mathcal{B} = BC(J;\mathbf{R}^n)$, with J either $\mathbf{R}^+$ or $\mathbf{R}$. In this case the mapping $x \mapsto r\star x$ is usually not compact, although it may map bounded sets into bounded equicontinuous sets (again see Theorem **9**.5.3(iii)). One can get around this lack of compactness in two different ways. One possibility is to introduce a Fréchet space in which uniform convergence on J has been replaced by uniform convergence on compact subsets of J, and then to reformulate Theorem 2.2 for this case. The other possibility is to work in a Banach space with a weighted norm. Below, we will describe this second approach.

We let η be a weight function on $\mathbf{R}$, as defined in Chapter 4 (that is, η is positive, Borel measurable, locally bounded and locally bounded away from zero). In addition, we require η to satisfy $\lim_{|t|\to\infty} \eta(t) = 0$. As in Chapter 4, we let $BC_0(\mathbf{R}^+;\eta;\mathbf{R}^n)$ be the set of continuous $\mathbf{R}^n$-valued functions ϕ on $\mathbf{R}^+$ that satisfy $\eta(t)|\phi(t)| \to 0$ as $t \to \infty$, with norm $\sup_{t\in\mathbf{R}^+} \eta(t)|\phi(t)|$. In an analogous fashion, one defines $BC_0(\mathbf{R};\eta;\mathbf{R}^n)$. Clearly, $BC(\mathbf{R}^+;\mathbf{R}^n) \subset BC_0(\mathbf{R}^+;\eta;\mathbf{R}^n)$.

Some basic relationships between $BC_0(J;\eta;\mathbf{R}^n)$ and $BC(J;\mathbf{R}^n)$ are described in the following proposition.

2.3 Proposition

Let J be either $\mathbf{R}^+$ or $\mathbf{R}$. Let K be a bounded subset of $BC(J;\mathbf{R}^n)$ and let η be a weight function on $\mathbf{R}$ satisfying $\lim_{|t|\to\infty} \eta(t) = 0$. Then the following claims are true.

(i) *A sequence of functions $\phi_k \in K$ converges (with respect to the norm in $BC_0(J;\eta;\mathbf{R}^n)$) to a limit $\phi \in BC_0(J;\eta;\mathbf{R}^n)$ if and only if ϕ_k tends to ϕ uniformly on compact subsets of J.*

(ii) *K is closed in $BC_0(J;\eta;\mathbf{R}^n)$ if and only if K is closed with respect to the topology of uniform convergence on compact subsets of J.*

(iii) *The closure of K in $BC_0(J;\eta;\mathbf{R}^n)$ is compact if and only if K is equicontinuous.*

Proof The proof of (i) is straightforward, and is left to the reader. Claim (ii) is an immediate consequence of (i). Claim (iii) follows directly from (i), and from the Arzelà–Ascoli Theorem **2**.7.5. □

The following proposition gives examples of kernels r having the compactness property that we need in order to apply Theorem 2.2. Recall that kernels of continuous type and of bounded continuous type were introduced in Definition **9**.5.2.

2.4 Proposition
Let J be either $\mathbf{R}^+$ or $\mathbf{R}$. Let K be a bounded subset of $BC(J;\mathbf{R}^n)$, and let η be a weight function on $\mathbf{R}$ satisfying $\lim_{|t|\to\infty}\eta(t) = 0$. Moreover, suppose that either (i) *or* (ii) *below holds.*

(i) *$J = \mathbf{R}^+$, r is a kernel of continuous type on $\mathbf{R}^+$, and $r \star K$ is a bounded subset of $BC(\mathbf{R}^+;\mathbf{R}^n)$.*

(ii) *$J = \mathbf{R}$ and r is a kernel of bounded continuous type on $\mathbf{R}$.*

Then the mapping $\phi \mapsto r\star\phi$ is continuous and compact from K (with the topology induced by $BC_0(J;\eta;\mathbf{R}^n)$) into $BC_0(J;\eta;\mathbf{R}^n)$.

Proof That the set $\{\, r\star\phi \mid \phi\in K \,\}$ is equicontinuous (and, in case (ii), bounded in $BC(\mathbf{R};\mathbf{R}^n)$) follows from Theorem **9**.5.3(iii). Hence, by Proposition 2.3(iii), this set has compact closure. Thus, to prove the proposition it suffices to show that $r\star\phi_k \to r\star\phi$ uniformly on compact sets whenever $\phi_k\to\phi$ uniformly on compact sets and $\{\phi_k\}\subset K$. This is not difficult: it follows from Lebesgue's dominated convergence theorem that for each fixed t one has $(r\star\phi_k)(t)\to(r\star\phi)(t)$, and this, together with equicontinuity, gives uniform convergence on compact sets. □

3. Nonlinear Convolution Integral Equations

As a first example on how to apply the fixed point theorems given in the preceding section, let us look at the two nonlinear integral equations

$$x(t) + \int_0^t g\big(t-s, x(s)\big)\,\mathrm{d}s = f(t), \quad t\in\mathbf{R}^+, \tag{3.1}$$

and

$$x(t) + \int_{-\infty}^{\infty} g\big(t-s, x(s)\big)\,\mathrm{d}s = f(t), \quad t\in\mathbf{R}. \tag{3.2}$$

Here x and f take their values in $\mathbf{R}^n$, and g maps $\mathbf{R}^+\times\mathbf{R}^n$ and $\mathbf{R}\times\mathbf{R}^n$, respectively, into $\mathbf{R}^n$. Note that (3.2) is autonomous in the following sense: if x is a solution of (3.2), then every translate $\tau_t x$ of x (defined by $(\tau_t x)(s) = x(t+s)$) is a solution of the same equation, with f replaced by the translate $\tau_t f$.

The results that we prove for (3.1) and (3.2) are very similar to each other. Let us, for the moment, concentrate on (3.1), and later return to (3.2). Prior to a formal statement, some preliminary comments are in order.

We want to study "small" solutions of (3.1), i.e., the L^∞-norms of x and f are small. In this situation it is natural to assume that $g(t,0) = 0$ for all $t\in\mathbf{R}^+$ (otherwise f may be large even if x is small). If, moreover, $\frac{\partial}{\partial x}g(t,x)|_{x=0}$ exists for all $t\geq 0$, then we can split the integral term in

(3.1) into two terms, one that is linear with respect to x, and one that is a higher order nonlinear term. Explicitly, this means that g is of the form

$$g(t,x) = k(t)x + h(t,x), \quad (t,x) \in \mathbf{R}^+ \times \mathbf{R}^n, \tag{3.3}$$

where, for each t, $k(t) = \frac{\partial}{\partial x} g(t,x)|_{x=0}$ is an n by n matrix, and the remainder $h(t,x) = o(|x|)$ as $x \to 0$. Substituting (3.3) into (3.1), we get

$$x(t) + \int_0^t k(t-s)x(s)\,\mathrm{d}s = f(t) - \int_0^t h\big(t-s, x(s)\big)\,\mathrm{d}s, \quad t \in \mathbf{R}^+. \tag{3.4}$$

Since h is a higher order term, the integral term on the right-hand side should be small compared to the other terms, if x is small. For now, we suppose that it belongs to $L^1_{\mathrm{loc}}(\mathbf{R}^+;\mathbf{R}^n)$ and, in addition, we assume that k and f are locally integrable on $\mathbf{R}^+$. Then we can apply the variation of constants formula (**2**.3.2) to obtain

$$\begin{aligned} x(t) = f(t) - \int_0^t r(t-s)f(s)\,\mathrm{d}s - \int_0^t h\big(t-s, x(s)\big)\,\mathrm{d}s \\ + \int_0^t r(t-s)\int_0^s h\big(s-v, x(v)\big)\,\mathrm{d}v\,\mathrm{d}s, \quad t \in \mathbf{R}^+, \end{aligned} \tag{3.5}$$

where r is the resolvent of k. Of course, this equation is a special case of (1.7).

Our purpose is to apply Theorem 2.1 to the preceding equation. In particular, to satisfy (iii) of that theorem, we want r to be integrable on $\mathbf{R}^+$. If k is integrable on $\mathbf{R}^+$, then, by Theorem **2**.4.1, a necessary and sufficient condition for the integrability of r is that k satisfy the half line Paley–Wiener condition $\det[I + \hat{k}(z)] \neq 0$ for $\Re z \geq 0$.

After these preliminary comments, we are ready to formulate a perturbation result for (3.1).

3.1 Theorem

Assume that

(i) $k \in L^1(\mathbf{R}^+;\mathbf{R}^{n\times n})$ *satisfies* $\det[I + \hat{k}(z)] \neq 0$ *for* $\Re z \geq 0$,

(ii) $g(t,y) = k(t)y + h(t,y)$ *for* $(t,y) \in \mathbf{R}^+ \times \mathbf{R}^n$, *where for each* y *the function* $t \mapsto h(t,y)$ *is Borel measurable on* $\mathbf{R}^+$, *and satisfies* $h(t,0) = 0$ *for* $t \in \mathbf{R}^+$,

(iii) *for all sufficiently small* $\delta > 0$, *the function* b_δ *defined by*

$$b_\delta(t) = \sup_{\substack{|u|,|v| \leq \delta \\ u \neq v}} \frac{|h(t,u) - h(t,v)|}{|u-v|}, \quad t \in \mathbf{R}^+,$$

belongs to $L^1(\mathbf{R}^+;\mathbf{R})$, *and satisfies* $\|b_\delta\|_{L^1(\mathbf{R}^+)} \to 0$ *as* $\delta \downarrow 0$.

Then there exist positive constants ϵ *and* δ *such that, for every* $f \in U \stackrel{\text{def}}{=} \{ f \in L^\infty(\mathbf{R}^+;\mathbf{R}^n) \mid \|f\|_{L^\infty(\mathbf{R}^+)} \leq \epsilon \}$, *there is a unique solution* $x(f) \in L^\infty(\mathbf{R}^+;\mathbf{R}^n)$ *of (3.1) with* $\|x(f)\|_{L^\infty(\mathbf{R}^+)} \leq \delta$. *This solution depends Lipschitz-continuously on* f, *i.e., there is a constant* M *such that,* *for all f_1 and f_2 in U,*

$$\|x(f_1) - x(f_2)\|_{L^\infty(\mathbf{R}^+)} \le M\|f_1 - f_2\|_{L^\infty(\mathbf{R}^+)}.$$

If, in addition, $f \in V$, where V is one of the spaces

(1) $L^p(\mathbf{R}^+;\mathbf{R}^n)$, $p \in [1,\infty)$, $L^\infty_\ell(\mathbf{R}^+;\mathbf{R}^n)$ *or* $L^\infty_0(\mathbf{R}^+;\mathbf{R}^n)$,
(2) $B^\infty(\mathbf{R}^+;\mathbf{R}^n)$, $B^\infty_\ell(\mathbf{R}^+;\mathbf{R}^n)$ *or* $B^\infty_0(\mathbf{R}^+;\mathbf{R}^n)$,
(3) $BC(\mathbf{R}^+;\mathbf{R}^n)$, $BUC(\mathbf{R}^+;\mathbf{R}^n)$, $BC_\ell(\mathbf{R}^+;\mathbf{R}^n)$ *or* $BC_0(\mathbf{R}^+;\mathbf{R}^n)$,
(4) $BL^p(\mathbf{R}^+;\mathbf{R}^n)$, *or* $BL^p_0(\mathbf{R}^+;\mathbf{R}^n)$, $p \in [1,\infty]$,
(5) $BV(\mathbf{R}^+;\mathbf{R}^n)$,
(6) $L^p(\mathcal{T}_S;\mathbf{R}^n) \oplus BL^p_0(\mathbf{R}^+;\mathbf{R}^n)$, $p \in [1,\infty]$, $B^\infty(\mathcal{T}_S;\mathbf{R}^n) \oplus B^\infty_0(\mathbf{R}^+;\mathbf{R}^n)$, *or* $C(\mathcal{T}_S;\mathbf{R}^n) \oplus BC_0(\mathbf{R}^+;\mathbf{R}^n)$, *where* $S > 0$,
(7) $AP(\mathbf{R}^+;\mathbf{R}^n) \oplus BC_0(\mathbf{R}^+;\mathbf{R}^n)$,

then $x(f) \in V$, and $x(f)$ depends Lipschitz-continuously on f in the sense that, for all f_1 and f_2 in $U \cap V$,

$$\|x(f_1) - x(f_2)\|_V \le M\|f_1 - f_2\|_V.$$

There is a certain redundancy in the hypothesis, since some of the spaces V contain $L^\infty(\mathbf{R}^+;\mathbf{R}^n)$; in these cases the hypothesis $f \in V$ and the conclusion $x(f) \in V$ are trivially true. However, in all the cases (1)–(7), the claims about the Lipschitz-continuity are nontrivial.

Observe that for small x the function g satisfies the so-called Carathéodory conditions, i.e., for each x the function $t \mapsto g(t,x)$ is measurable, and for almost all t and for all sufficiently small x the function $x \mapsto g(t,x)$ is continuous. This guarantees the local existence of a solution, cf. Theorem **12**.2.7.

A result similar to Theorem 3.1 is true for the whole line equation (3.2). Again, if $g(t,0) = 0$ and $\frac{\partial}{\partial x}g(t,x)|_{x=0}$ exists for all $t \ge 0$, then (3.2) can be written in the form

$$x(t) + \int_{-\infty}^{\infty} k(t-s)x(s)\,\mathrm{d}s = f(t) - \int_{-\infty}^{\infty} h\big(t-s, x(s)\big)\,\mathrm{d}s, \quad t \in \mathbf{R}.$$

If we assume that $k \in L^1(\mathbf{R};\mathbf{R}^{n\times n})$ and that k satisfies the whole line Paley–Wiener condition $\det[I + \hat{k}(i\omega)] \ne 0$ for $\omega \in \mathbf{R}$, then by Theorem **2**.4.3 k has a resolvent $r \in L^1(\mathbf{R};\mathbf{R}^{n\times n})$. Formally, we can then use the variation of constants formula on $\mathbf{R}$ (see Theorem **2**.4.6) to write the equation in the form

$$\begin{aligned} x(t) = f(t) - \int_{-\infty}^{\infty} r(t-s)f(s)\,\mathrm{d}s - \int_{-\infty}^{\infty} h\big(t-s, x(s)\big)\,\mathrm{d}s \\ + \int_{-\infty}^{\infty} r(t-s)\int_{-\infty}^{\infty} h\big(s-v, x(v)\big)\,\mathrm{d}v\,\mathrm{d}s, \quad t \in \mathbf{R}. \end{aligned} \tag{3.6}$$

3.2 Theorem

Theorem 3.1 holds with $\mathbf{R}^+$ *replaced by* $\mathbf{R}$, *(3.1) replaced by (3.2), and the condition* $\det[I+\hat{k}(z)] \neq 0$ *for* $\Re z \geq 0$ *replaced by* $\det[I+\hat{k}(\mathrm{i}\omega)] \neq 0$ *for* $\omega \in \mathbf{R}$. *In addition, it holds when* V *is one of the spaces*

(8) $L^p(\mathcal{T}_S;\mathbf{R}^n)$, $p \in [1,\infty]$, $B^\infty(\mathcal{T}_S;\mathbf{R}^n)$, $C(\mathcal{T}_S;\mathbf{R}^n)$, *or* $BV(\mathcal{T}_S;\mathbf{R}^n)$, *where* $S>0$,

(9) $AP(\mathbf{R};\mathbf{R}^n)$.

The proofs of these two theorems are completely similar, so we prove only Theorem 3.2, and leave the proof of Theorem 3.1 to the reader.

Proof of Theorem 3.2 By the transform condition, $r \in L^1(\mathbf{R};\mathbf{R}^{n\times n})$. Take $\delta > 0$ so small that

$$\bigl(1+\|r\|_{L^1(\mathbf{R})}\bigr)\|b_\delta\|_{L^1(\mathbf{R})} \leq \tfrac{1}{2}, \tag{3.7}$$

and define

$$K = \bigl\{\, \phi \in L^\infty(\mathbf{R};\mathbf{R}^n) \;\big|\; \|\phi\|_{L^\infty(\mathbf{R})} \leq \delta \,\bigr\}. \tag{3.8}$$

Next, let us show that for $\phi \in K$ and all t the function $s \mapsto h(t-s,\phi(s))$ is integrable, and the function

$$H(\phi)(t) \stackrel{\text{def}}{=} \int_{\mathbf{R}} h\bigl(t-s,\phi(s)\bigr)\,\mathrm{d}s,$$

belongs to K.

For each $t \in \mathbf{R}$ and $\phi \in K$ the function $s \mapsto h(t-s,\phi(s))$ is measurable (see the discussion on Carathéodory conditions in Section **12**.2). To see that it is integrable, it suffices to compute, using (ii) and (iii),

$$\begin{aligned} |H(\phi)(t)| &\leq \int_{\mathbf{R}} \bigl|h\bigl(t-s,\phi(s)\bigr)\bigr|\,\mathrm{d}s \\ &\leq \int_{\mathbf{R}} b_\delta(t-s)|\phi(s)|\,\mathrm{d}s \leq \delta \int_{\mathbf{R}} b_\delta(s)\,\mathrm{d}s < \infty. \end{aligned} \tag{3.9}$$

Thus we see that the function $H(\phi)$ is bounded on $\mathbf{R}$. In addition, it is measurable (see Section **12**.2). Hence $H(\phi)$ belongs to $L^\infty(\mathbf{R};\mathbf{R}^n)$ for every $\phi \in K$. Clearly, by (3.7) and (3.9),

$$\|H(\phi)\|_{L^\infty(\mathbf{R})} \leq \frac{\delta}{2\bigl(1+\|r\|_{L^1(\mathbf{R})}\bigr)}, \tag{3.10}$$

i.e., $H(\phi) \in K$, for all $\phi \in K$. Moreover, a simple computation, analogous to (3.9), shows that, for all ϕ and ψ in K,

$$|H(\phi)(t)-H(\psi)(t)| \leq \int_{\mathbf{R}} b_\delta(t-s)|\phi(s)-\psi(s)|\,\mathrm{d}s = (b_\delta * |\phi-\psi|)(t). \tag{3.11}$$

Thus, H is a Lipschitz-continuous operator from K into $L^\infty(\mathbf{R};\mathbf{R}^n)$, with Lipschitz constant $\|b_\delta\|_{L^1(\mathbf{R})}$.

We pause to observe that, since $H(\phi) \in L^\infty(\mathbf{R};\mathbf{R}^n)$ for all $\phi \in K$, the argument outlined before the proof of Theorem 3.2 about the equivalence of (3.2) and (3.6) is valid, whenever $x \in K$. Thus, we may replace the original equation (3.2) by (3.6).

Clearly, (3.6) is an equation of the form (1.8) with

$$F(x)(t) = f(t) - H(x)(t), \qquad G(x)(t) = -f(t) + H(x)(t).$$

Let us try to apply Theorem 2.1, choosing $J = \mathbf{R}$, $\mathcal{B} = L^\infty(\mathbf{R}; \mathbf{R}^n)$ and K as in (3.8). Obviously, condition (i) holds. The constants L_F, L_G and ϱ in Theorem 2.1 can be chosen to be $L_F = L_G = \|b_\delta\|_{L^1(\mathbf{R})}$ and $\varrho = \|r\|_{L^1(\mathbf{R})}$. Therefore, by (3.7), $L_F + L_G\varrho \le \frac{1}{2}$ and so conditions (ii), (iii) and (v) are satisfied. To check (iv) of Theorem 2.1, note that it follows from (3.10) that for $x \in K$

$$\|F(x) + r * G(x)\|_{L^\infty(\mathbf{R})} \le (1 + \|r\|_{L^1(\mathbf{R})})\|f\|_{L^\infty(\mathbf{R})} + \delta/2.$$

Thus, if we choose ϵ to be

$$\epsilon = \frac{\delta}{2(1 + \|r\|_{L^1(\mathbf{R})})},$$

then, for all $f \in L^\infty(\mathbf{R}; \mathbf{R}^n)$ with $\|f\|_{L^\infty(\mathbf{R})} \le \epsilon$, the function $x \mapsto F(x) + r * G(x)$ takes K into itself. Hence, Theorem 2.1 can be applied to yield the existence of a solution $x(f) \in K$ of (3.6) for every $f \in L^\infty(\mathbf{R}; \mathbf{R}^n)$ with $\|f\|_{L^\infty(\mathbf{R})} \le \epsilon$. This solution is unique in K.

To prove that $x(f)$ depends Lipschitz-continuously on f, we observe that for two functions f_1 and f_2, both with norm at most ϵ, we have

$$x(f_i) = f_i - r * f_i - H\big(x(f_i)\big) + r * H\big(x(f_i)\big), \quad i = 1, 2.$$

Subtracting these two equations from each other, and using (3.7) and (3.11), we get

$$\begin{aligned} \|x(f_1) - x(f_2)\|_{L^\infty(\mathbf{R})} &\le \big(1 + \|r\|_{L^1(\mathbf{R})}\big)\|f_1 - f_2\|_{L^\infty(\mathbf{R})} \\ &\quad + \big(1 + \|r\|_{L^1(\mathbf{R})}\big)\|b_\delta\|_{L^1(\mathbf{R})}\|x(f_1) - x(f_2)\|_{L^\infty(\mathbf{R})} \\ &\le \big(1 + \|r\|_{L^1(\mathbf{R})}\big)\|f_1 - f_2\|_{L^\infty(\mathbf{R})} + \tfrac{1}{2}\|x(f_1) - x(f_2)\|_{L^\infty(\mathbf{R})}. \end{aligned} \tag{3.12}$$

This gives us the desired Lipschitz-continuity, with Lipschitz constant $M = 2(1 + \|r\|_{L^1(\mathbf{R})})$.

Our next purpose is to show that if, in addition, $f \in V$, where V is one of the spaces listed in (1)–(9), then $x(f) \in V$, and $x(f)$ depends Lipschitz-continuously on f (with the same Lipschitz constant M as above). The proofs are to a large extent repetitions of the proof given above, but with the space L^∞ replaced by $L^\infty \cap V$.

Consider first the case where $f \in L^p(\mathbf{R}; \mathbf{R}^n)$ for some $p \in [1, \infty)$. Fix any N satisfying

$$N \ge 2\big(1 + \|r\|_{L^1(\mathbf{R})}\big)\|f\|_{L^p(\mathbf{R})},$$

and define

$$K_1 = K \cap \big\{\, \phi \in L^p(\mathbf{R}) \mid \|\phi\|_{L^p(\mathbf{R})} \le N \,\big\}.$$

By (3.7), (3.9) and Theorem **2**.2.2(i), if $\phi \in K_1$, then $H(\phi) \in L^p(\mathbf{R}; \mathbf{R}^n)$, and

$$\|H(\phi)\|_{L^p(\mathbf{R})} \le \frac{N}{2(1 + \|r\|_{L^1(\mathbf{R})})}.$$

Hence, for $x \in K_1$,

$$\begin{aligned}\|F(x) + r * G(x)\|_{L^p(\mathbf{R})} &\le (1 + \|r\|_{L^1(\mathbf{R})})\big(\|f\|_{L^p(\mathbf{R})} + \|H(x)\|_{L^p(\mathbf{R})}\big)\\ &\le (1 + \|r\|_{L^1(\mathbf{R})})\|f\|_{L^p(\mathbf{R})} + \tfrac{1}{2}N.\end{aligned}$$

Consequently, from the choice of N it follows that $x \mapsto F(x) + r * G(x)$ maps K_1 into itself.

By (3.11) and Theorem **2**.2.2, for x and y in K_1,

$$\|H(x) - H(y)\|_{L^p(\mathbf{R})} \le \|b_\delta\|_{L^1(\mathbf{R})}\|x - y\|_{L^p(\mathbf{R})}.$$

This combined with our earlier Lipschitz estimate in L^∞ shows that, with respect to the norm $\|\cdot\|_{L^\infty(\mathbf{R})} + \|\cdot\|_{L^p(\mathbf{R})}$, H is a Lipschitz-continuous operator from K_1 into $L^\infty(\mathbf{R};\mathbf{R}^n) \cap L^p(\mathbf{R};\mathbf{R}^n)$, with Lipschitz constant $\|b_\delta\|_{L^1(\mathbf{R})}$.

The (easy) verification of the fact that Theorem 2.1 can be applied, with $J = \mathbf{R}$, $\mathcal{B} = L^\infty(\mathbf{R};\mathbf{R}^n) \cap L^p(\mathbf{R};\mathbf{R}^n)$, and with K replaced by K_1, is left to the reader. According to Theorem 2.1, our equation has a solution x in K_1. By the uniqueness of our fixed point $x(f)$, this new solution must be the same as the one we found earlier.

To see that $x(f)$ depends Lipschitz-continuously on f, it suffices to use (3.7), (3.11) and Theorem **2**.2.2(i) to show that (3.12) is valid when all L^∞-norms are replaced by L^p-norms. (The proofs of the remaining claims about the Lipschitz-continuous dependence of $x(f)$ on f in the different spaces V are all almost identical to this proof, and they are left to the reader.)

To show that $x(f) \in L^\infty_\ell(\mathbf{R};\mathbf{R}^n)$ whenever $f \in L^\infty_\ell(\mathbf{R};\mathbf{R}^n)$, it suffices to observe that $L^\infty_\ell(\mathbf{R};\mathbf{R}^n)$ is closed in $L^\infty(\mathbf{R};\mathbf{R}^n)$, and to prove that the map $x \mapsto F(x)+r*G(x)$ takes $L^\infty_\ell(\mathbf{R};\mathbf{R}^n)$ into itself. By Theorem **2**.2.2(i) and since $r \in L^1(\mathbf{R}^+)$, this will follow provided H maps $L^\infty_\ell(\mathbf{R};\mathbf{R}^n)$ into itself. To see that this last fact is true, take some $x \in L^\infty_\ell(\mathbf{R};\mathbf{R}^n)$, define $y(t) = x(\infty)$ for $t \in \mathbf{R}$, note that $H(y)(t)$ is a constant, and compute

$$|H(x)(t) - H(y)(t)| \le (b_\delta * |x - y|)(t).$$

Here the right-hand side tends to zero as $t \to \infty$.

Above we have worked in L^∞, but the same arguments can be repeated with L^∞ replaced by B^∞ (the function $x \mapsto F(x)+r*G(x)$ preserves Borel measurability, and it takes bounded functions into bounded functions). Therefore, if f in (3.2) belongs to one of the spaces listed in (2), we may require the equation to hold everywhere (instead of almost everywhere), and we get a solution in B^∞.

To prove that $x(f)$ is continuous whenever f is, it suffices to show that

H preserves continuity. This is the case because

$$\begin{aligned}|H(x)(t+h) - H(x)(t)| \\ &\leq \int_{\mathbf{R}} \big|h\big(t-s, x(s+h)\big) - h\big(t-s, x(s)\big)\big| \,\mathrm{d}s \\ &\leq \int_{\mathbf{R}} b_\delta(t-s)|x(s+h) - x(s)| \,\mathrm{d}s, \qquad (3.13)\end{aligned}$$

and if $x \in BC(\mathbf{R}; \mathbf{R}^n)$, then by Lebesgue's dominated convergence theorem the right-hand side tends to zero as $h \to 0$. Moreover, if x is uniformly continuous, then so is $H(x)$.

The proofs of the cases $V = BL^p(\mathbf{R}; \mathbf{R}^n)$ and $V = BL^p_0(\mathbf{R}; \mathbf{R}^n)$ are very similar to the proof of the case $V = L^p(\mathbf{R}; \mathbf{R}^n)$, and they are left to the reader.

The next space V on the list is $BV(\mathbf{R}; \mathbf{R}^n)$. It follows from (3.13) that $H(x) \in BV(\mathbf{R}; \mathbf{R}^n)$ whenever $x \in BV(\mathbf{R}; \mathbf{R}^n) \cap K$ (cf. the proof of Theorem **3**.6.1(v)). One can proceed as in the proof of the case where $V = L^p$, throughout replacing L^p by BV, and $\|\cdot\|_{L^p(\mathbf{R})}$ by $\|\cdot\|_{BV(\mathbf{R})}$ (all the constants remain the same).

In the case (9), where $V = AP(\mathbf{R}; \mathbf{R}^n)$, the crucial question is whether H preserves almost-periodicity. However, this follows directly from (3.13) and the standard definition of almost-periodicity (cf. the proof of Theorem **2**.2.2(v) given in Section **2**.7).

The (easy) proofs of the remaining cases (6), (7) and (8) are omitted. □

In our next result we make the *a priori* assumption that the solution x of (3.1) tends to zero at infinity. Employing a submultiplicative weight function φ (see Definitions **4**.2.1 and **4**.3.2), satisfying $\varphi(t) \to \infty$ as $t \to \infty$, we obtain an explicit rate of convergence to zero, i.e., we conclude that $x(t)\varphi(t)$ remains bounded as $t \to \infty$.

As earlier, we assume that (3.1) can be written in the form

$$x(t) + \int_0^t k(t-s)x(s)\,\mathrm{d}s + \int_0^t h\big(t-s, x(s)\big)\,\mathrm{d}s = f(t), \quad t \in \mathbf{R}^+, \qquad (3.4)$$

where $k(t) = \frac{\partial}{\partial x} g(t,x)|_{x=0}$, and $h(t,x) = g(t,x) - k(t)x$.

Recall that $\omega_\varphi = -\lim_{t\to\infty} \frac{1}{t}\ln(\varphi(t))$ (see Lemma **4**.4.1).

3.3 Theorem

Let φ be a submultiplicative weight function on $\mathbf{R}^+$, which satisfies $\lim_{t\to\infty}\varphi(t) = \infty$. In addition, assume that

(i) *$k \in L^1(\mathbf{R}^+; \varphi; \mathbf{R}^{n\times n})$ satisfies $\det[I + \hat{k}(z)] \neq 0$ for $\Re z \geq \omega_\varphi$,*

(ii) *for almost every $t \in \mathbf{R}^+$ the function $y \mapsto h(t,y)$ is continuous, and for every $y \in \mathbf{R}^n$ the function $t \mapsto h(t,y)$ is measurable,*

(iii) *for every $\delta > 0$, the function b_δ defined by*

$$b_\delta(t) = \sup_{|v|\leq\delta} \frac{|h(t,v)|}{|v|}, \quad t \in \mathbf{R}^+,$$

belongs to $L^1(\mathbf{R}^+;\varphi;\mathbf{R})$ *and satisfies* $\|b_\delta\|_{L^1(\mathbf{R}^+;\varphi)} \to 0$ *as* $\delta \downarrow 0$,

(iv) $f \in BC(\mathbf{R}^+;\varphi;\mathbf{R}^n)$,

(v) $x \in BC_0(\mathbf{R}^+;\mathbf{R}^n)$ *is a solution of (3.4).*

Then $x \in BC(\mathbf{R}^+;\varphi;\mathbf{R}^n)$.

Proof It follows from Corollary **4**.4.7 and assumption (i) that the resolvent r of k satisfies $r \in L^1(\mathbf{R}^+;\varphi;\mathbf{R}^{n\times n})$. First choose δ so small that

$$\big(1 + \|r\|_{L^1(\mathbf{R}^+;\varphi)}\big)\|b_\delta\|_{L^1(\mathbf{R}^+;\varphi)} \le \tfrac{1}{2},$$

and then choose $T \ge 0$ in such a way that $|x(t)| \le \delta$ for $t \ge T$. By (3.5), we may split x as

$$x(t) = x_1(t) + x_2(t), \quad t \in \mathbf{R}^+,$$

where $x_1(t) = x(t)$ (and $x_2(t) = 0$) for $t < T$, and

$$\begin{aligned} x_1(t) &= f(t) - \int_0^t r(t-s)f(s)\,\mathrm{d}s - \int_0^T h\big(t-s, x(s)\big)\,\mathrm{d}s \\ &\quad + \int_0^t r(t-s)\int_0^T h\big(s-v, x(v)\big)\,\mathrm{d}v\,\mathrm{d}s, \quad t \ge T, \\ x_2(t) &= -\int_T^t h\big(t-s, x(s)\big)\,\mathrm{d}s \\ &\quad + \int_T^t r(t-s)\int_T^s h\big(s-v, x(v)\big)\,\mathrm{d}v\,\mathrm{d}s, \quad t \ge T. \end{aligned}$$

Define $M = \sup_{t\in\mathbf{R}^+}|x(t)|$. Then $|h(t,x(s))| \le b_M(t)|x(s)|$ for all s and t. Using this fact, assumption (iv), and Theorem **4**.1.2(ii), one easily finds that $x_1 \in BC(\mathbf{R}^+;\varphi;\mathbf{R}^n)$. Analogously one gets, for every $t \ge T$,

$$\|x_2\|_{L^\infty([T,t];\varphi)} \le \big(1 + \|r\|_{L^1(\mathbf{R}^+;\varphi)}\big)\|b_\delta\|_{L^1(\mathbf{R}^+;\varphi)}\|x\|_{L^\infty([T,t];\varphi)}.$$

This, together with our choice of δ, implies that, for all $t \ge T$,

$$\|x\|_{L^\infty([T,t];\varphi)} \le 2\|x_1\|_{L^\infty([T,t];\varphi)},$$

and the proof is complete. □

Results similar to those above are valid for the nonautonomous equations

$$x(t) + \int_0^t g\big(t,s,x(s)\big)\,\mathrm{d}s = f(t), \quad t \in \mathbf{R}^+, \tag{3.14}$$

and

$$x(t) + \int_{-\infty}^{\infty} g\big(t,s,x(s)\big)\,\mathrm{d}s = f(t), \quad t \in \mathbf{R}. \tag{3.15}$$

Let us formulate one such result for (3.14). The proof is left to the reader.

3.4 Theorem
Assume that

(i) *k is an $\mathbf{R}^{n\times n}$-valued Volterra kernel of type L^∞ on $\mathbf{R}^+$, which has a resolvent of the same type on $\mathbf{R}^+$ (cf. Definitions **9**.2.2 and **9**.3.1),*

(ii) *$g(t,s,y) = k(t,s)y + h(t,s,y)$ for $(t,s,y) \in \mathbf{R}^+ \times \mathbf{R}^+ \times \mathbf{R}^n$, where $h(t,s,y) = 0$ for $s > t$, $h(t,s,0) = 0$ for $(s,t) \in \mathbf{R}^+ \times \mathbf{R}^+$, and for each y the function $(s,t) \mapsto h(t,s,y)$ is Borel measurable on $\mathbf{R}^+ \times \mathbf{R}^+$,*

(iii) *for all sufficiently small $\delta > 0$, the function b_δ defined by*

$$b_\delta(t,s) = \sup_{\substack{|u|,|v|\le\delta \\ u\ne v}} \frac{|h(t,s,u) - h(t,s,v)|}{|u-v|}, \quad (s,t) \in \mathbf{R}^+ \times \mathbf{R}^+,$$

is a kernel of type L^∞ on $\mathbf{R}^+$, and satisfies $|||b_\delta|||_{L^\infty(\mathbf{R}^+)} \to 0$ as $\delta \downarrow 0$.

Then there exist positive constants ϵ and δ such that for every $f \in U \stackrel{\text{def}}{=} \{ f \in L^\infty(\mathbf{R}^+;\mathbf{R}^n) \mid \|f\|_{L^\infty(\mathbf{R}^+)} \le \epsilon \}$ there is a unique solution $x(f) \in L^\infty(\mathbf{R}^+;\mathbf{R}^n)$ of (3.14) with $\|x(f)\|_{L^\infty(\mathbf{R}^+)} \le \delta$. This solution depends Lipschitz-continuously on f, i.e., there is a constant M such that, for all f_1 and f_2 in U,

$$\|x(f_1) - x(f_2)\|_{L^\infty(\mathbf{R}^+)} \le M\|f_1 - f_2\|_{L^\infty(\mathbf{R}^+)}.$$

4. Perturbed Linear Integrodifferential Equations

Let us next study the perturbed linear integrodifferential equation

$$x'(t) + \int_{[0,t]} \mu(\mathrm{d}s)x(t-s) = G(x)(t), \quad t \in \mathbf{R}^+; \quad x(0) = x_0, \tag{4.1}$$

where G is some function taking $BC(\mathbf{R}^+;\mathbf{R}^n)$ into $L^1_{\text{loc}}(\mathbf{R}^+;\mathbf{R}^n)$. Assuming that a locally absolutely continuous solution of (4.1) exists, we can use the differential resolvent r of μ (see Theorem **3**.3.3) to write (4.1) in the form

$$x(t) = r(t)x_0 + (r * G(x))(t), \quad t \in \mathbf{R}^+, \tag{4.2}$$

which is an equation of type (1.7).

If we want to apply either Theorem 2.1 or Theorem 2.2 to (4.2), then the operator $x \mapsto r * x$ must be bounded. Hence it is natural to require r to be integrable. One way to assure this is to require that $\mu \in M(\mathbf{R}^+;\mathbf{R}^{n\times n})$, and that the appropriate Paley–Wiener condition holds, i.e., that $\det[zI + \hat{\mu}(z)] \ne 0$ for $\Re z \ge 0$ (see Theorem **3**.3.5).

It is not difficult to formulate a result for (4.1) that is similar to the result given in Theorem 3.1. We refer this task to the Exercises (see Exercise 9) and give, instead, a result based on Theorem 2.2.

4.1 Theorem
Assume that

(i) $\mu \in M(\mathbf{R}^+;\mathbf{R}^{n\times n})$ *satisfies* $\det[zI+\hat{\mu}(z)] \neq 0$ *for* $\Re z \geq 0$,

(ii) $G : BC(\mathbf{R}^+;\mathbf{R}^n) \to L^1_{\mathrm{loc}}(\mathbf{R}^+;\mathbf{R}^n)$, *and there exist* $\delta > 0$ *and a nondecreasing continuous function* $\eta : [0,1] \to \mathbf{R}^+$, *with* $\eta(0) = 0$, *such that*

$$\phi \in BC(\mathbf{R}^+;\mathbf{R}^n) \text{ and } \sup_{t\in\mathbf{R}^+} |\phi(t)| \leq \delta \text{ imply}$$

$$\sup_{t\in\mathbf{R}^+} \left| \int_t^{t+\tau} G(\phi)(s)\,\mathrm{d}s \right| \leq \eta(\tau),$$

for $0 < \tau \leq 1$,

(iii) *for every sequence* $\phi_k \in BC(\mathbf{R}^+;\mathbf{R}^n)$ *satisfying* $\sup_{t\in\mathbf{R}^+}|\phi_k(t)| \leq \delta$ *(*δ *as in* (ii)*), and converging uniformly on compact intervals to the limit* ϕ*, one has*

$$\int_0^t G(\phi_k)(s)\,\mathrm{d}s \to \int_0^t G(\phi)(s)\,\mathrm{d}s$$

for each $t > 0$.

Then there exists, provided $\eta(1)$ *and* $|x_0|$ *are sufficiently small, a (locally absolutely continuous) solution* x *of (4.1) satisfying* $\sup_{t\in\mathbf{R}^+}|x(t)| \leq \delta$.

The conditions involving the auxiliary function η can be reformulated as follows. The value $\eta(1)$, which is required to be sufficiently small, can be defined as

$$\sup_{t,\tau,\phi} \left| \int_t^{t+\tau} G(\phi)(s)\,\mathrm{d}s \right|,$$

where the supremum is taken over all $t \in \mathbf{R}^+$, $\tau \in [0,1]$ and $\phi \in BC(\mathbf{R}^+;\mathbf{R}^n)$ with $\sup_{t\in\mathbf{R}^+}|\phi(t)| \leq \delta$. To have the appropriate τ-dependent bound, described in (ii), we must assume that

$$\lim_{\tau\to 0} \sup_{t,\phi} \left| \int_t^{t+\tau} G(\phi)(s)\,\mathrm{d}s \right| = 0,$$

where the supremum is taken over the same t and ϕ. The explicit condition on the size of $\eta(1)$ and $|x_0|$ is given by the second inequality in (4.3).

A somewhat different assumption on G is given in Corollary 4.2.

Proof of Theorem 4.1 As we observed earlier, we can replace (4.1) by (4.2), where the differential resolvent r of μ satisfies $r \in L^1(\mathbf{R}^+;\mathbf{R}^{n\times n})$. Moreover, by Theorem **3**.3.5, $r' \in L^1(\mathbf{R}^+;\mathbf{R}^{n\times n})$.

Define two operators G_1 and G_2 on $BC(\mathbf{R}^+;\mathbf{R}^n)$, by

$$G_1(\phi)(t) = \int_{|t-1|_+}^t G(\phi)(s)\,\mathrm{d}s,$$
$$G_2(\phi)(t) = G(\phi)(t) - G_1(\phi)(t), \quad t \in \mathbf{R}^+.$$

It follows from (ii) that, if ϕ is continuous and satisfies $\sup_{t\in\mathbf{R}^+}|\phi(t)| \le \delta$, then

$$\sup_{t\in\mathbf{R}^+} |G_1(\phi)(t)| \le \eta(1).$$

Moreover, one can change the order of integration to show that, for $t \ge 0$,

$$\int_0^t G_2(\phi)(s)\,\mathrm{d}s = \int_0^1 \int_{|t+s-1|_+}^t G(\phi)(u)\,\mathrm{d}u\,\mathrm{d}s.$$

This, together with (ii), implies that, for the same class of functions ϕ,

$$\sup_{t\in\mathbf{R}^+} \left| \int_0^t G_2(\phi)(s)\,\mathrm{d}s \right| \le \eta(1).$$

Replacing $G(x)$ in (4.2) by $G_1(x)+G_2(x)$, and integrating by parts, we get

$$\begin{aligned} x(t) =& H(x)(t) \stackrel{\text{def}}{=} r(t)x_0 + \int_0^t r(t-s)G_1(x)(s)\,\mathrm{d}s \\ &+ \int_0^t G_2(x)(s)\,\mathrm{d}s + \int_0^t r'(t-s)\int_0^s G_2(x)(u)\,\mathrm{d}u\,\mathrm{d}s, \quad t \in \mathbf{R}^+. \end{aligned}$$

From the bounds on G_1 and G_2, it follows that if $\sup_{t\in\mathbf{R}^+}|\phi(t)| \le \delta$, and if $|x_0|$ and $\eta(1)$ are sufficiently small, then

$$\begin{aligned} \sup_{t\in\mathbf{R}^+} |H(\phi)(t)| \le& \sup_{t\in\mathbf{R}^+} |r(t)||x_0| \\ &+ \left(\int_{\mathbf{R}^+} \big(|r(s)| + |r'(s)|\big)\,\mathrm{d}s + 1 \right)\eta(1) \le \delta. \end{aligned} \tag{4.3}$$

Thus H maps the set

$$K \stackrel{\text{def}}{=} \{\, \phi \in BC(\mathbf{R}^+;\mathbf{R}^n) \mid \sup_{t\in\mathbf{R}^+} |\phi(t)| \le \delta \,\}$$

into itself. It follows from (ii) (more specifically, from the fact that $\eta(0) = 0$ and from the continuity of η at zero), that the set $\{\, H(\phi) \mid \phi \in K \,\}$ is equicontinuous. Pointwise convergence, together with equicontinuity, implies uniform convergence on compact sets, and therefore condition (iii), together with Proposition 2.3, guarantees that H is continuous and compact on K if K is given the topology inherited from $BC_0(\mathbf{R}^+;\eta;\mathbf{R}^n)$, where, e.g., $\eta(t) = \mathrm{e}^{-|t|}$. Thus, we can apply Theorem 2.2, where we take $F = H$ and $G = 0$, to get the desired conclusion. □

In the preceding proof, we observed that the assumptions that we had on G implied that G could be written as a sum

$$G(\phi)(t) = G_1(\phi)(t) + \tfrac{\mathrm{d}}{\mathrm{d}t} G_3(\phi)(t), \quad t \in \mathbf{R}^+,$$

where both G_1 and G_3 map $BC(\mathbf{R}^+;\mathbf{R}^n)$ into $BC(\mathbf{R}^+;\mathbf{R}^n)$ (and where, in addition, $G_3(\phi)(t) = \int_0^t G_2(\phi)(s)\,\mathrm{d}s$ is locally absolutely continuous and satisfies $G_3(\phi)(0) = 0$ for all ϕ). If we substitute this for G in (4.1), then we get the equation

$$\tfrac{\mathrm{d}}{\mathrm{d}t}\{x(t) - G_3(x)(t)\} + \int_{[0,t]} \mu(\mathrm{d}s)x(t-s) = G_1(x)(t),$$
$$t \in \mathbf{R}^+; \quad x(0) = x_0. \qquad (4.4)$$

Theorem 4.1 may be reformulated for this equation, yielding the following result.

4.2 Corollary

Assume the following:

(i) $\mu \in M(\mathbf{R}^+; \mathbf{R}^{n\times n})$ *satisfies* $\det[zI + \hat{\mu}(z)] \neq 0$ *for* $\Re z \geq 0$;

(ii) G_1 *and* G_3 *map* $BC(\mathbf{R}^+; \mathbf{R}^n)$ *into itself and* $G_3(\phi)$ *is, in addition, locally absolutely continuous on* $\mathbf{R}^+$ *with* $G_3(\phi)(0) = 0$ *for* $\phi \in BC(\mathbf{R}^+; \mathbf{R}^n)$;

(iii) *there exist constants* $\delta > 0$ *and* $\gamma > 0$ *such that for every* $\phi \in BC(\mathbf{R}^+; \mathbf{R}^n)$ *satisfying* $\sup_{t\in\mathbf{R}^+}|\phi(t)| \leq \delta$ *one has* $\|G_i(\phi)\|_{\sup(\mathbf{R}^+)} \leq \gamma$ *for* $i = 1, 3$;

(iv) *if the functions* ϕ_k *are continuous, satisfy* $\sup_{t\in\mathbf{R}^+}|\phi_k(t)| \leq \delta$ *and converge uniformly on compact intervals to the limit* ϕ, *then* $G_i(\phi_k)(t) \to G_i(\phi)(t)$ *for* $i = 1, 3$ *and all* $t \in \mathbf{R}^+$;

(v) *the set* $\{\, G_3(\phi) \mid \phi \in BC(\mathbf{R}^+; \mathbf{R}^n)$ *and* $\|\phi\|_{\sup(\mathbf{R}^+)} \leq \delta \,\}$ *is equicontinuous.*

Then there exists, provided γ *and* $|x_0|$ *are sufficiently small, a (locally absolutely continuous) solution* x *of (4.4) satisfying* $\sup_{t\in\mathbf{R}^+}|x(t)| \leq \delta$.

As a second application of the fixed point theorems to integrodifferential equations, we study the dependence of the solution on the forcing function f. Suppose we know that the solution of the equation

$$x'(t) + \int_{[0,t]} \mu(\mathrm{d}s)g\big(x(t-s)\big) = f(t), \quad t \in \mathbf{R}^+; \quad x(0) = x_0, \qquad (4.5)$$

is bounded and tends to a limit at infinity. Can we say something similar about the solution of the equation

$$y'(t) + \int_{[0,t]} \mu(\mathrm{d}s)g\big(y(t-s)\big) = h(t), \quad t \in \mathbf{R}^+; \quad y(0) = y_0, \qquad (4.6)$$

assuming that y_0 is close to x_0 and that h is close to f in some suitable sense? One answer to this question is the following.

4.3 Theorem

Let $\mu \in M(\mathbf{R}^+; \mathbf{R}^{n\times n})$, $g \in C^1(\mathbf{R}^n; \mathbf{R}^n)$, $f \in L^1_{\mathrm{loc}}(\mathbf{R}^+; \mathbf{R}^n)$ *and* $x_0 \in \mathbf{R}^n$. *Let* x *be the solution of (4.5) and suppose that* $x(\infty) = \lim_{t\to\infty} x(t)$ *exists. In addition, assume that*

$$\det[zI + \hat{\mu}(z)g'(x(\infty))] \neq 0, \qquad \Re z \geq 0. \qquad (4.7)$$

Then there exist positive constants ϵ and C such that, whenever $y_0 \in \mathbf{R}^n$ and $h \in L^1_{\mathrm{loc}}(\mathbf{R}^+;\mathbf{R}^n)$ satisfy

$$\gamma \stackrel{\mathrm{def}}{=} |y_0 - x_0| + \sup_{\substack{t\in\mathbf{R}^+\\ \tau\in[0,1]}} \left| \int_t^{t+\tau} \big(h(s) - f(s)\big)\,\mathrm{d}s \right| \le \epsilon, \tag{4.8}$$

then (4.6) has a unique solution y satisfying $\sup_{t\in\mathbf{R}^+}|y(t) - x(t)| \le C\gamma$. If, in addition, for $0 < \tau \le 1$,

$$\lim_{t\to\infty} \left| \int_t^{t+\tau} \big(h(s) - f(s)\big)\,\mathrm{d}s \right| = 0,$$

then $y(t) \to x(\infty)$ as $t \to \infty$.

Proof For $t \ge 0$, define $v(t) = y(t) - x(t)$ and

$$F(v)(t) = g\big(x(t) + v(t)\big) - g\big(x(t)\big) - g'\big(x(t)\big)v(t).$$

Subtracting (4.5) from (4.6) we find that v is a solution of the integrodifferential equation

$$\begin{aligned} v'(t) &+ \int_{[0,t]} \mu(\mathrm{d}s) g'\big(x(t-s)\big)v(t-s) \\ &= h(t) - f(t) - \big(\mu * F(v)\big)(t), \quad t \in \mathbf{R}^+; \quad v(0) = y_0 - x_0. \end{aligned}$$

Define the measure kernel κ by

$$\kappa(t,E) = \int_{t-E} \mu(\mathrm{d}s) g'\big(x(t-s)\big) = \int_{[0,t]} \mu(\mathrm{d}s) g'\big(x(t-s)\big)\chi_E(t-s),$$

and write κ as the sum of two measures, $\kappa = \kappa_1 + \kappa_3$, where for each Borel set $E \subset \mathbf{R}^+$ and each $t \ge 0$,

$$\kappa_1(t,E) = \int_{[0,t]} \mu(\mathrm{d}s) g'(x(\infty))\chi_E(t-s),$$

$$\kappa_3(t,E) = \int_{[0,t]} \mu(\mathrm{d}s) \Big(g'\big(x(t-s)\big) - g'\big(x(\infty)\big)\Big)\chi_E(t-s).$$

Our purpose is to apply Theorem **10**.4.2 with $S = 0$. To this end, note that (4.7) implies that (i) of that theorem holds. Since we take $\kappa_2 = 0$, (ii) is trivially true. Condition (iii) follows from our size condition on μ, and from the fact that $g'(x(t)) \to g'(x(\infty))$ as $t \to \infty$. We conclude that the differential resolvent r of κ is of bounded continuous type on $\mathbf{R}^+$, as well as bounded on $\mathbf{R}^+ \times \mathbf{R}^+$. The kernel $r \star \kappa$ (which according to the general theory in Section **10**.3 is a measure kernel locally of bounded type) can in this case be identified with the function

$$(r \star \kappa)(t,s) = \int_{[0,t-s]} r(t, v+s)\mu(\mathrm{d}s) g'(x(s)),$$

i.e., it is the $\star$-product of r and the convolution kernel μ, multiplied by the bounded continuous function $g'(x(s))$. In particular, $r \star \kappa$ is of bounded continuous type on $\mathbf{R}^+$, as well as bounded on $\mathbf{R}^+ \times \mathbf{R}^+$. By (**10**.3.4), for each $t > 0$ the function $s \mapsto r(t,s)$ is locally absolutely continuous on $[0,t]$, and $\frac{\partial r}{\partial s} = r \star \kappa$.

Use the variation of constants formula (**10**.3.1) to write v in the form

$$v(t) = r(t,0)(y_0 - x_0) + \big(r\star(h-f)\big)(t) + \Big(r\star\big(\mu * F(v)\big)\Big)(t), \quad t \in \mathbf{R}^+. \tag{4.9}$$

We would like to apply Theorem 2.1 to this equation, with $J = \mathbf{R}^+$, $\mathcal{B} = BC(\mathbf{R}^+;\mathbf{R}^n)$, and

$$K = \Big\{ \phi \in BC(\mathbf{R}^+;\mathbf{R}^n) \;\Big|\; \sup_{t\in\mathbf{R}^+} |\phi(t)| \le \delta \Big\}$$

for some appropriately chosen δ.

Clearly the right-hand side of (4.9) maps $C(\mathbf{R}^+;\mathbf{R}^n)$ into itself. Let $f_1(t) = \int_{|t-1|_+}^t (h(s) - f(s))\,\mathrm{d}s$, $f_2(t) = h(t) - f(t) - f_1(t)$. Then

$$\big(r\star(h-f)\big)(t) = (r\star f_1)(t) + \int_0^t f_2(s)\,\mathrm{d}s - \int_0^t \frac{\partial}{\partial s} r(t,s) \int_0^s f_2(u)\,\mathrm{d}u\,\mathrm{d}s,$$

and, by an argument similar to the one used in the proof of Theorem 4.1, we find that

$$|r(t,0)(y_0 - x_0) + \big(r\star(h-f)\big)(t)| \le M\gamma, \quad t \in \mathbf{R}^+, \tag{4.10}$$

where γ is defined in (4.8), and

$$M = \sup_{t\in\mathbf{R}^+} |r(t,0)| + |||r|||_{B^\infty(\mathbf{R}^+)} + \Big|\Big|\Big|\frac{\partial}{\partial s} r\Big|\Big|\Big|_{B^\infty(\mathbf{R}^+)} + 1.$$

Obviously,

$$\sup_{t\in\mathbf{R}^+} \Big|\Big(r\star\big(\mu * F(\phi)\big)\Big)(t)\Big| \le M\|\mu\|_{M(\mathbf{R}^+)}\|F(\phi)\|_{\sup(\mathbf{R}^+)}$$

for $\phi \in K$. Observe that $\|F(\phi_1) - F(\phi_2)\|_{\sup(\mathbf{R}^+)} = o(\|\phi_1 - \phi_2\|_{\sup(\mathbf{R}^+)})$, and so, by choosing δ small enough, we can guarantee that

$$M\|\mu\|_{M(\mathbf{R}^+)}\|F(\phi_1) - F(\phi_2)\|_{\sup(\mathbf{R}^+)} \le \tfrac{1}{2}\|\phi_1 - \phi_2\|_{\sup(\mathbf{R}^+)}$$

for ϕ_1 and $\phi_2 \in K$. Let

$$\epsilon = \frac{\delta}{2M}.$$

Then, as soon as $\gamma \le \epsilon$, the right-hand side of (4.9) maps K into itself, and it follows from Theorem 2.1, where we take $G = 0$, that (4.9) has a unique solution $v \in K$. Moreover, this solution satisfies $\sup_{t\in\mathbf{R}^+}|v(t)| \le 2M\gamma$. This proves the first of the two claims in Theorem 4.3.

To prove that $y(t) \to x(\infty)$ as $t \to \infty$, it suffices to show that the right-hand side of (4.9) maps $BC_0(\mathbf{R}^+;\mathbf{R}^{n\times n})$ into itself. This is straightforward, once one demonstrates that r and $\frac{\partial}{\partial s} r$ are of type BC_0 on $\mathbf{R}^+$ (cf. Definition **9**.6.1), and that $r(t,0) \to 0$ as $t \to \infty$. Moreover, one should observe that the additional condition on $h - f$ and Lemma **15**.9.2 imply that

$$\lim_{t\to\infty} \sup_{\tau\in[0,1]} \Big| \int_t^{t+\tau} \big(h(s) - f(s)\big)\,\mathrm{d}s \Big| = 0.$$

We leave the completion of the proof to the reader. □

5. Perturbed Ordinary Differential Equations

If we let μ in (4.1) consist of only a point mass of size A at zero, then the equation reduces to the perturbed ordinary differential equation

$$x'(t) + Ax(t) = G(x)(t), \quad t \in \mathbf{R}^+; \quad x(0) = x_0.$$

In this case, condition (i) on μ in Theorem 4.1 is equivalent to the requirement that the fundamental solution of the unperturbed equation $x'(t) + Ax(t) = 0$ tends to zero exponentially as $t \to \infty$.

One can obtain results for the perturbed nonautonomous ordinary differential equation

$$x'(t) + A(t)x(t) = G(x)(t), \quad t \in \mathbf{R}^+; \quad x(0) = x_0, \tag{5.1}$$

that are quite analogous to those given in Section 4. We write (5.1) as (cf. the paragraph prior to Theorem **10**.4.4),

$$x'(t) + \int_{[0,t]} \kappa(t, \mathrm{d}s)x(s) = G(x(t)), \quad t \in \mathbf{R}^+; \quad x(0) = x_0, \tag{5.2}$$

where $\kappa(t, E) = A(t)\chi_E(t)$. Obviously, we now need estimates on the differential resolvent $r(t, s)$ of $\kappa(t, \mathrm{d}s)$ to replace the earlier estimates we had on the differential resolvent of μ in Section 4. Recall that if X is the solution of the matrix equation

$$X'(t) + A(t)X(t) = 0, \quad t \in \mathbf{R}^+; \quad X(0) = I,$$

then $r(t, s) = X(t)X^{-1}(s)$, $0 \le s \le t < \infty$.

Using the differential resolvent, two of the standard concepts in the theory of ordinary differential equations (cf. Hale [3], Theorem 2.1, p. 84) may be formulated, e.g., as follows. The equation

$$x'(t) + A(t)x(t) = 0, \quad t \in \mathbf{R}^+, \tag{5.3}$$

is said to be *uniformly stable* on $\mathbf{R}^+$ iff

$$\sup_{0 \le s \le t < \infty} |r(t, s)| < \infty,$$

and is said to be *uniformly asymptotically stable* on $\mathbf{R}^+$ iff there is a constant $\epsilon > 0$ such that

$$\sup_{0 \le s \le t < \infty} \mathrm{e}^{\epsilon(t-s)}|r(t, s)| < \infty.$$

Slight changes in the proof of Theorem 4.1 yield the following result.

5.1 Theorem

Assume that

(i) $A \in BL^1(\mathbf{R}^+; \mathbf{R}^{n \times n})$,

(ii) *equation (5.3) is uniformly asymptotically stable on* $\mathbf{R}^+$,

(iii) $G : BC(\mathbf{R}^+; \mathbf{R}^n) \to L^1_{\mathrm{loc}}(\mathbf{R}^+; \mathbf{R}^n)$, *and there exist* $\delta > 0$ *and a nondecreasing continuous function* $\eta : [0, 1] \to \mathbf{R}^+$, *with* $\eta(0) = 0$, *such that*

$$\phi \in BC(\mathbf{R}^+;\mathbf{R}^n) \text{ and } \sup_{t\in\mathbf{R}^+} |\phi(t)| \le \delta \text{ imply}$$

$$\sup_{t\in\mathbf{R}^+} \left| \int_t^{t+\tau} G(\phi)(s)\,\mathrm{d}s \right| \le \eta(\tau)$$

for $0 < \tau \le 1$,

(iv) *for every sequence* $\{\phi_k\} \subset BC(\mathbf{R}^+;\mathbf{R}^n)$ *satisfying* $\sup_{t\in\mathbf{R}^+}|\phi_k(t)| \le \delta$ (δ *as in* (iii)), *and converging uniformly on compact intervals to the limit* ϕ, *one has*

$$\int_0^t G(\phi_k)(s)\,\mathrm{d}s \to \int_0^t G(\phi)(s)\,\mathrm{d}s$$

for each $t > 0$.

Then there exists, provided $\eta(1)$ *and* $|x_0|$ *are sufficiently small, a (locally absolutely continuous) solution* x *of (5.1) satisfying* $\sup_{t\in\mathbf{R}^+}|x(t)| \le \delta$.

Analogously to Corollary 4.2, observe that this result could have been formulated as a result for the perturbed equation

$$\frac{\mathrm{d}}{\mathrm{d}t}\big(x(t) - G_3(x)(t)\big) + A(t)x(t) = G_1(x)(t), \quad t \in \mathbf{R}^+; \quad x(0) = x_0. \tag{5.4}$$

6. L^2-Perturbations of Convolution Equations

The results in Sections 3 and 4 say that, if the nonlinear perturbation of a linear equation is small enough, then—under certain hypotheses—the nonlinear equation has a solution which behaves roughly in the same way as the solution of the linear equation. From the proofs it is not difficult to see, in each particular case, what the word 'small' means. However, the explicit conditions one gets are of limited use because they involve the L^1-norm or the sup-norm on $\mathbf{R}^+$ of the resolvent and the differential resolvent. In general, it is not easy to get good estimates on these norms. For example, the Paley–Wiener theorem gives us conditions under which the resolvent is integrable, but it gives us only crude (and complicated) estimates on the L^1-norm of the resolvent; cf. Theorem **2**.6.1.

An easy way to get a sharp estimate on the operator norm of the resolvent and the differential resolvent is to use L^2-theory. Here the key result is the following lemma; the proof is essentially contained in the proof of Theorem **2**.6.2.

6.1 Lemma

(i) *Let $f \in L^1_{\mathrm{loc}}(\mathbf{R}^+;\mathbf{C}^n)$ satisfy $\int_{\mathbf{R}^+} \mathrm{e}^{-\sigma t}|f(t)|\,\mathrm{d}t < \infty$ for every $\sigma > 0$. Then f belongs to $L^2(\mathbf{R}^+;\mathbf{C}^n)$ iff*

$$\sup_{\sigma>0} \int_{\mathbf{R}} |\hat{f}(\sigma + i\tau)|^2 \,\mathrm{d}\tau < \infty.$$

Moreover, if the norm in $\mathbf{C}^n$ is induced by an inner product, then

$$\|f\|^2_{L^2(\mathbf{R}^+)} = \frac{1}{2\pi} \sup_{\sigma>0} \int_{\mathbf{R}} |\hat{f}(\sigma + i\tau)|^2 \,\mathrm{d}\tau.$$

(ii) *Let $\mu \in M_{\mathrm{loc}}(\mathbf{R}^+;\mathbf{C}^{n\times n})$ satisfy $\int_{\mathbf{R}^+} \mathrm{e}^{-\sigma t}|\mu|(\mathrm{d}t) < \infty$ for every $\sigma > 0$, and suppose that $\hat{\mu}(z)$ is bounded in $\Re z > 0$. Then the convolution operator $\phi \mapsto \mu * \phi$ maps $L^2(\mathbf{R}^+;\mathbf{C}^n)$ continuously into itself. Moreover, if one uses an inner product norm in $\mathbf{C}^n$, then the norm of this operator is $\sup_{\Re z>0}|\hat{\mu}(z)|$.*

In order to make an efficient use of Lemma 6.1, we have to rephrase the theorems in Sections 3 and 4 so that no use is made of L^∞-estimates. In particular, this means that we have to assume global Lipschitz-continuity in, e.g., Theorem 3.1, instead of local Lipschitz-continuity.

We begin with a modification of Theorem 3.1, considering the slightly more general equation

$$x(t) + \int_0^t k(t-s)x(s)\,\mathrm{d}s = f(t) - \int_0^t h\big(t,s,x(s)\big)\,\mathrm{d}s, \quad t \in \mathbf{R}^+. \tag{6.1}$$

6.2 Theorem

Assume that the norm in $\mathbf{R}^n$ is induced by an inner product, and that

(i) *$k \in L^1_{\mathrm{loc}}(\mathbf{R}^+;\mathbf{R}^{n\times n})$ satisfies $\int_0^\infty \mathrm{e}^{-\sigma t}|k(t)|\,\mathrm{d}t < \infty$ for $\sigma > 0$, $\det[I+\hat{k}(z)] \neq 0$ for $\Re z > 0$, and*

$$\varrho \overset{\mathrm{def}}{=} \sup_{\Re z>0} \big|\hat{k}(z)[I+\hat{k}(z)]^{-1}\big| < \infty,$$

(ii) *for each $x \in \mathbf{R}^n$, the function $(t,s) \mapsto h(t,s,x)$ is Borel measurable on $\mathbf{R}^+ \times \mathbf{R}^+$, and satisfies $h(t,s,0) = 0$ for $0 \le s \le t < \infty$. Moreover, the function b defined by*

$$b(t) = \sup_{\substack{x\neq y\\ s\in\mathbf{R}^+}} \frac{|h(t+s,s,x) - h(t+s,s,y)|}{|x-y|}, \quad t \in \mathbf{R}^+,$$

belongs to $L^1(\mathbf{R}^+;\mathbf{R})$,

(iii) *$(1+\varrho)\|b\|_{L^1(\mathbf{R}^+)} < 1$.*

Then, for every $f \in L^2(\mathbf{R}^+;\mathbf{R}^n)$, there is a unique solution $x(f) \in L^2(\mathbf{R}^+;\mathbf{R}^n)$ of (6.1). This solution depends Lipschitz-continuously on f, i.e., for all $f_1, f_2 \in L^2(\mathbf{R}^+;\mathbf{R}^n)$,

$$\|x(f_1) - x(f_2)\|_{L^2(\mathbf{R}^+)} \le M\|f_1 - f_2\|_{L^2(\mathbf{R}^+)},$$

where

$$M = \frac{1+\varrho}{1-(1+\varrho)\|b\|_{L^1(\mathbf{R}^+)}}.$$

The proof follows closely that of Theorem 3.2. In particular, observe that the required analogue of (3.9) is

$$\int_0^t \big|h(t,s,x(s))\big|\,\mathrm{d}s \le \int_0^t b(t-s)|x(s)|\,\mathrm{d}s,$$

and that (by Theorem **2**.2.2(i)) the function

$$x \mapsto H(x)(t) = \int_0^t h\big(t,s,x(s)\big)\,\mathrm{d}s, \quad t \in \mathbf{R}^+,$$

maps $L^2(\mathbf{R}^+;\mathbf{R}^n)$ Lipschitz-continuously into itself, with Lipschitz constant $\|b\|_{L^1(\mathbf{R}^+)}$.

In the special case where the equation is of type

$$x(t) + \int_0^t k(t-s)x(s)\,\mathrm{d}s$$
$$= f(t) - \int_0^t k(t-s)h\big(t,s,x(s)\big)\,\mathrm{d}s, \quad t \in \mathbf{R}^+, \quad (6.2)$$

one can get a slightly different smallness condition.

6.3 Theorem

Assume that the norm in $\mathbf{R}^n$ is induced by an inner product, and that

(i) $k \in L^1_{\mathrm{loc}}(\mathbf{R}^+;\mathbf{R}^{n\times n})$ *satisfies* $\int_0^\infty \mathrm{e}^{-\sigma t}|k(t)|\,\mathrm{d}t < \infty$ *for* $\sigma > 0$, $\det[I+\hat{k}(z)] \ne 0$ *for* $\Re z > 0$, *and*

$$\varrho \stackrel{\mathrm{def}}{=} \sup_{\Re z>0} \big|\hat{k}(z)[I+\hat{k}(z)]^{-1}\big| < \infty,$$

(ii) *for each $x \in \mathbf{R}^n$, the function $(t,s) \mapsto h(t,s,x)$ is Borel measurable on $\mathbf{R}^+ \times \mathbf{R}^+$, and satisfies $h(t,s,0) = 0$ for $0 \le s \le t < \infty$. Moreover, there is a constant $\gamma > 0$ such that*

$$|h(t,s,x) - h(t,s,y)| \le \gamma|x-y|, \quad x,y \in \mathbf{R}^n, \quad 0 \le s \le t < \infty,$$

(iii) $\varrho\gamma < 1$.

Then, for every $f \in L^2(\mathbf{R}^+;\mathbf{R}^n)$, there is a unique solution $x(f) \in L^2(\mathbf{R}^+;\mathbf{R}^n)$ of (6.2). This solution depends Lipschitz-continuously on f, i.e., for all $f_1, f_2 \in L^2(\mathbf{R}^+;\mathbf{R}^n)$,

$$\|x(f_1) - x(f_2)\|_{L^2(\mathbf{R}^+)} \le M\|f_1 - f_2\|_{L^2(\mathbf{R}^+)},$$

where

$$M = \frac{1+\varrho}{1-\varrho\gamma}.$$

To prove this, one applies the variation of constants formula to get

$$x(t) = f(t) - \int_0^t r(t-s)f(s)\,\mathrm{d}s - \int_0^t r(t-s)h(t,s,x(s))\,\mathrm{d}s, \quad t \in \mathbf{R}^+.$$

A similar result can be proved for the integrodifferential equation (4.4).

6.4 Theorem

Assume that the norm in $\mathbf{R}^n$ is induced by an inner product, and that

(i) $\mu \in M_{\mathrm{loc}}(\mathbf{R}^+;\mathbf{R}^{n\times n})$ *satisfies* $\int_{\mathbf{R}^+} \mathrm{e}^{-\sigma t}|\mu|(\mathrm{d}t) < \infty$ *for* $\sigma > 0$, $\det[zI + \hat{\mu}(z)] \neq 0$ *for* $\Re z \geq 0$, $\varrho_1 \overset{\mathrm{def}}{=} \sup_{\Re z>0}\big|[zI+\hat{\mu}(z)]^{-1}\big| < \infty$, *and* $\varrho_2 \overset{\mathrm{def}}{=} \sup_{\Re z>0}\big|z[zI+\hat{\mu}(z)]^{-1}\big| < \infty$,

(ii) $G_1 : L^2(\mathbf{R}^+;\mathbf{R}^n) \to L^2(\mathbf{R}^+;\mathbf{R}^n)$, $G_3 : L^2(\mathbf{R}^+;\mathbf{R}^n) \to L^2(\mathbf{R}^+;\mathbf{R}^n) \cap AC_{\mathrm{loc}}(\mathbf{R}^+;\mathbf{R}^n)$, *and* $G_3(\phi)(0) = 0$ *for all* $\phi \in L^2(\mathbf{R}^+;\mathbf{R}^n)$,

(iii) *there exist constants* γ_1 *and* γ_3 *such that* $\|G_i(\phi) - G_i(\psi)\|_{L^2(\mathbf{R}^+)} \leq \gamma_i\|\phi-\psi\|_{L^2(\mathbf{R}^+)}$ *for all* ϕ, $\psi \in L^2(\mathbf{R}^+;\mathbf{R}^n)$ *and* $i = 1,3$,

(iv) $\varrho_1\gamma_1 + \varrho_2\gamma_3 + \gamma_3 < 1$.

Then there is a unique solution x of (4.4) that is locally absolutely continuous and belongs to $L^2(\mathbf{R}^+;\mathbf{R}^n)$.

Proof Let r be the differential resolvent of μ. Then it follows from (i) that

$$\sup_{\Re z>0}\big|(1+z)\hat{r}(z)\big| < \infty.$$

In particular, by Lemma 6.1(i), this implies that $r \in L^2(\mathbf{R}^+;\mathbf{R}^{n\times n})$. Use the variation of constants formula to write (4.4) in the form

$$x(t) = r(t)x_0 + \int_0^t r(t-s)G_1(x)(s)\,\mathrm{d}s + \tfrac{\mathrm{d}}{\mathrm{d}t}\int_0^t r(t-s)G_3(x)(s)\,\mathrm{d}s.$$

Now apply Lemma 6.1(ii) and Theorem 2.1. □

7. Exercises

1. Prove Theorem 3.1.
2. Prove Theorem 3.4, and formulate and prove a similar result for equation (3.15).
3. Can you extend Theorem 3.4 to include statements about solutions in some of the spaces V listed in (1)–(7) in Theorem 3.1? Hint: To do this you need various classes of kernels studied in Chapter 9, and

you may need additional conditions to ensure that the integral term in (3.14) belongs to V when $x \in V \cap L^\infty(\mathbf{R}^+;\mathbf{R}^n)$.

4. Prove the following result.

7.1 Theorem
Assume that

(i) *for some $\epsilon \geq 0$, the function $(t,s) \mapsto \mathrm{e}^{\epsilon(t-s)}k(t,s)$ is a $\mathbf{R}^{n\times n}$-valued kernel of bounded continuous type (cf. Definition* **9**.5.2*),*

(ii) *for each $x \in \mathbf{R}^n$, the function $s \mapsto h(s,x)$ is measurable on $\mathbf{R}^+$ and satisfies $h(s,0) = 0$ for $s \in \mathbf{R}^+$. Moreover, for some $\delta > 0$ and some $\gamma > 0$*

$$|h(s,y)| \leq \gamma|y|, \quad |y| \leq \delta,$$

(iii) *$f \in C(\mathbf{R}^+;\mathbf{R}^n)$, and $\sup_{t\in\mathbf{R}^+} \mathrm{e}^{\epsilon t}|f(t)| \leq \lambda$ for some $\lambda > 0$,*

(iv) *$\lambda + \gamma \sup_{t\geq 0} \int_0^t \mathrm{e}^{\epsilon(t-s)}|k(t,s)|\,\mathrm{d}s \leq \delta$.*

Then the equation

$$x(t) + \int_0^t k(t,s)h\big(s,x(s)\big)\,\mathrm{d}s = f(t), \quad t \in \mathbf{R}^+,$$

has a continuous solution x satisfying $|x(t)| \leq \delta\mathrm{e}^{-\epsilon t}$ for $t \in \mathbf{R}^+$.

5. Prove the following result.

7.2 Theorem
Assume that $\delta > 0$ and that the following hold.

(i) *For each $y \in \mathbf{R}^n$, the function $(t,s) \mapsto h(t,s,y)$ is Borel measurable on $\mathbf{R}^+ \times \mathbf{R}^+$, and for each $(t,s) \in \mathbf{R}^+ \times \mathbf{R}^+$ the function $y \mapsto h(t,s,y)$ is continuous on $\mathbf{R}^n$. The function b defined by*

$$b(t,s) = \sup_{|y|\leq\delta} |h(t,s,y)|, \quad 0 \leq s \leq t < \infty,$$

satisfies $\sup_{t\in\mathbf{R}^+} \int_0^t |b(t,s)|\,\mathrm{d}s < \infty$ and for each $t > 0$,

$$\lim_{\tau\downarrow 0}\left(\int_t^{t+\tau} b(t,s)\,\mathrm{d}s + \int_0^t \sup_{|y|\leq\delta} \big|h(t+\tau,s,y) - h(t,s,y)\big|\,\mathrm{d}s\right) = 0.$$

(ii) *$f \in BC(\mathbf{R}^+;\mathbf{R}^n)$.*

(iii) *$\|f\|_{\sup(\mathbf{R}^+)} + \sup_{t\geq 0} \int_0^t |b(t,s)|\,\mathrm{d}s \leq \delta$.*

Then the equation

$$x(t) + \int_0^t h\big(t,s,x(s)\big)\,\mathrm{d}s = f(t), \quad t \in \mathbf{R}^+,$$

has a continuous solution x satisfying $\sup_{t\in\mathbf{R}^+} |x(t)| \leq \delta$.

6. Prove the following result.

7.3 Theorem
Let $k \in L^1(\mathbf{R}^+;\mathbf{R}^{n\times n})$ satisfy $\det[I+\hat{k}(z)] \neq 0$ for $\Re z \geq 0$, and let $f \in C(\mathbf{R}^+;\mathbf{R}^n)$. Suppose that $g \in C(\mathbf{R}^+\times\mathbf{R}^n;\mathbf{R}^n)$ satisfies $|g(s,v)| \leq \lambda(s)(1+|v|)$ for some function $\lambda \in BC_0(\mathbf{R}^+;\mathbf{R})$ and for all $v \in \mathbf{R}^n$. Assume that the solution y of the equation

$$y(t) + \int_0^t k(t-s)y(s)\,\mathrm{d}s = f(t), \quad t \in \mathbf{R}^+,$$

is bounded on $\mathbf{R}^+$, and let x be a solution of the equation

$$x(t) + \int_0^t k(t-s)\Big\{x(s) + g\big(s,x(s)\big)\Big\}\,\mathrm{d}s = f(t), \quad t \in \mathbf{R}^+.$$

Then $\lim_{t\to\infty}|x(t)-y(t)| = 0$.

7. Show that the claims in Theorem 3.2 concerning the existence of T-periodic solutions of (3.2) remain true if one weakens the assumption $\det[I+\hat{k}(\mathrm{i}\omega)] \neq 0$ for $\omega \in \mathbf{R}$ to the assumption $\det[I+\hat{k}(\mathrm{i}\omega)] \neq 0$ for $\omega = 2\pi m/T$, $m \in \mathbf{Z}$.
8. Let $\mu \in M(\mathbf{R}^+;\mathbf{R}^{n\times n})$ and let $f \in BC_\ell(\mathbf{R}^+;\mathbf{R}^{n\times n})$. Define the measure kernel κ by $\kappa(t,E) = \int_{[0,t]} \mu(\mathrm{d}s) f(t-s)\chi_E(t-s)$, and assume that $\det[zI + \hat{\mu}(z)f(\infty)] \neq 0$ for $\Re z \geq 0$. Show that the differential resolvent r of κ satisfies $r(t,0) \to 0$ as $t \to \infty$, and that r and $\frac{\partial}{\partial s}r$ are kernels of type BC_0 on $\mathbf{R}^+$. Hint: Use (**10**.3.3).
9. Use Theorem 2.1 to prove the following modified version of Theorem 4.1.

7.4 Theorem
Assume that
(i) $\mu \in M(\mathbf{R}^+;\mathbf{R}^{n\times n})$ satisfies $\det[zI+\hat{\mu}(z)] \neq 0$ for $\Re z \geq 0$,
(ii) $G : BC(\mathbf{R}^+;\mathbf{R}^n) \to L^1_{\mathrm{loc}}(\mathbf{R}^+;\mathbf{R}^n)$, and

$$\gamma \stackrel{\text{def}}{=} \sup_{\substack{t\in\mathbf{R}^+\\ \tau\in[0,1]}} \left| \int_t^{t+\tau} G(0)(s)\,\mathrm{d}s \right| < \infty.$$

For each $\delta > 0$, define

$$\gamma_\delta \stackrel{\text{def}}{=} \sup_{t,\tau,\phi,\psi} \frac{\left|\int_t^{t+\tau}\big(G(\phi)(s) - G(\psi)(s)\big)\,\mathrm{d}s\right|}{\|\phi-\psi\|_{\sup(\mathbf{R}^+)}},$$

where the supremum is taken over all $t \in \mathbf{R}^+$, $\tau \in [0,1]$ and ϕ, $\psi \in BC(\mathbf{R}^+;\mathbf{R}^n)$ with $\phi \neq \psi$, $\|\phi\|_{\sup(\mathbf{R}^+)} \leq \delta$ and $\|\psi\|_{\sup(\mathbf{R}^+)} \leq \delta$, and suppose that
(iii) $\gamma_\delta \to 0$ as $\delta \downarrow 0$.
Then there exists, provided γ and $|x_0|$ are sufficiently small, a unique solution $x \in BC(\mathbf{R}^+;\mathbf{R}^n) \cap AC_{\mathrm{loc}}(\mathbf{R}^+;\mathbf{R}^n)$ of (4.1).

10. Formulate and prove a nonconvolution version of Theorem 4.1.
11. Prove the following result.

7.5 Theorem
Assume that $T > 0$, and that

(i) $\mu \in M(\mathbf{R};\mathbf{R}^{n\times n})$ *satisfies* $\det[zI + \hat{\mu}(z)] \neq 0$ *for* $z = \mathrm{i}\,2\pi m/T$, $m \in \mathbf{Z}$,

(ii) G *maps* $C(\mathcal{T}_T;\mathbf{R}^n)$ *into* $L^1(\mathcal{T}_T;\mathbf{R}^n)$, *and there exist* $\delta > 0$ *and a nondecreasing continuous function* $\eta : [0,T] \to \mathbf{R}^+$, *with* $\eta(0) = 0$, *such that* $\sup_{t\in\mathbf{R}}\left|\int_t^{t+\tau} G(\phi)(s)\,\mathrm{d}s\right| \leq \eta(\tau)$ *whenever* $\phi \in C(\mathcal{T}_T;\mathbf{R}^n)$ *satisfies* $\sup_{t\in[0,T]}|\phi(t)| < \delta$,

(iii) *if the functions* $\phi_k \in C(\mathcal{T}_T;\mathbf{R}^n)$ *satisfy* $\sup_{t\in\mathbf{R}}|\phi_k(t)| \leq \delta$ *and converge uniformly to the limit* ϕ *as* $k \to \infty$, *then* $\int_0^t G(\phi_k)(s)\,\mathrm{d}s \to \int_0^t G(\phi)(s)\,\mathrm{d}s$ *for each* $t \in [0,T]$.

Then there exists, provided $\eta(1)$ *is sufficiently small, a locally absolutely continuous* T*-periodic solution* x *of the equation*

$$x'(t) + (\mu * x)(t) = G(x)(t), \quad t \in \mathbf{R},$$

satisfying $\sup_{t\in\mathbf{R}}|x(t)| \leq \delta$.

12. Consider the equation

$$x(t) + \int_{-\infty}^{t} k(t-s)g\big(x(s)\big)\,\mathrm{d}s = f(t), \quad t \in \mathbf{R}, \tag{7.1}$$

where $f \in L^2(\mathcal{T}_T;\mathbf{R})$, $k \in L^1(\mathbf{R}^+;\mathbf{R})$ satisfies $1 + \frac{1}{2}(\alpha+\beta)\hat{k}(\mathrm{i}\omega) \neq 0$ for $\omega = \frac{2\pi m}{T}$, $m \in \mathbf{Z}$, and $g \in C(\mathbf{R};\mathbf{R})$ satisfies

$$\alpha \leq \frac{g(x) - g(y)}{x - y} \leq \beta,$$

for some constants $0 \leq \alpha < \beta$, and all x, $y \in \mathbf{R}$. Define

$$\gamma = \sup_{m\in\mathbf{Z}} \frac{(\alpha+\beta)\left|\hat{k}\left(\frac{2\pi i m}{T}\right)\right|}{2\left|1 + \frac{1}{2}(\alpha+\beta)\hat{k}\left(\frac{2\pi i m}{T}\right)\right|}.$$

Show that if $\gamma(\beta-\alpha) < 2$, then (7.1) has a unique solution in $L^2(\mathcal{T}_T;\mathbf{R})$.

13. Let k, f and g satisfy the assumptions in Exercise 12, except that the condition $1 + \frac{1}{2}(\alpha+\beta)\hat{k}(\mathrm{i}\omega) \neq 0$ for $\omega = \frac{2\pi m}{T}$, $m \in \mathbf{Z}$, is strengthened to $1 + \frac{1}{2}(\alpha+\beta)\hat{k}(z) \neq 0$ for $\Re z \geq 0$, γ is redefined to be

$$\gamma = \sup_{\Re z\geq 0} \frac{(\alpha+\beta)\left|\hat{k}(z)\right|}{2\left|1 + \frac{1}{2}(\alpha+\beta)\hat{k}(z)\right|},$$

and k satisfies $\int_0^\infty (1+s)|k(s)|\,\mathrm{d}s < \infty$. Let $\gamma(\beta-\alpha) < 2$, and let $h \in L^2(\mathbf{R}^+;\mathbf{R})$. Show that the equation

$$x(t) + \int_0^t k(t-s)g(x(s))\,\mathrm{d}s = f(t) + h(t), \quad t \in \mathbf{R}^+,$$

has a unique solution in $L^2_{\mathrm{loc}}(\mathbf{R}^+;\mathbf{R})$, and that this solution is of the form $x = x_1 + x_2$, where $x_1 \in L^2(\mathcal{T}_T;\mathbf{R})$, and $x_2 \in L^2(\mathbf{R}^+;\mathbf{R})$.

14. Prove Theorem 5.1.

8. Comments

Sections 1–2:

There is a large literature on perturbed Volterra equations of the form (1.1) or (1.2), with the nonlinear function $G(x)$ small in some sense. The vast majority of the papers take the approach outlined in Section 1, i.e., using the resolvent and the differential resolvent, the equations (1.1) and (1.2) are converted to (1.5) and (1.6), respectively. The latter equations are then subjected to an analysis by fixed point methods. For this technique to apply, one must, of course, have some information on the size of the resolvent r, e.g., $r \in L^1(\mathbf{R}^+)$. Frequently this size condition is formulated as an admissibility assumption. (The space B is said to be admissible with respect to the map $r\star$ if for each $f \in B$, one has $r \star f \in B$.)

Miller [10], Ch. V, Secs. 1–4, contains a general discussion of different settings for perturbation results.

Sections 3–4:

The early papers Miller [5] and [6] analyse nonlinear perturbations of, respectively, convolution and nonconvolution Volterra equations. In Miller, Nohel, and Wong [1], the equation

$$x(t) + \int_0^t a(t,s)\Big\{x(s) + g\big(s, x(s)\big)\Big\}\,\mathrm{d}s = f(t), \quad t \in \mathbf{R}^+,$$

is compared with the unperturbed equation

$$y(t) + \int_0^t a(t,s)y(s)\,\mathrm{d}s = f(t), \quad t \in \mathbf{R}^+.$$

In particular, conditions are obtained under which the perturbed equation has asymptotically periodic solutions. A key assumption is that the resolvent r of a is of type B_0^∞ on $\mathbf{R}^+$ (see Definition **9**.6.1). In Nohel [1] the asymptotic equivalence of the two equations is studied. In Strauss [1] it is shown (without the use of fixed point results) that if r is of type B_0^∞ on $\mathbf{R}^+$, and $|g(t,x)| \le \lambda(t)(1+|x|)$ for $t \in \mathbf{R}^+$, $|x| < \infty$, where $\lambda(t) \to 0$ as $t \to \infty$, and if y is bounded on $\mathbf{R}^+$, then $|x(t) - y(t)| \to 0$ for $t \to \infty$. See also Kaplan [1].

In Grossman and Miller [1] perturbed integrodifferential equations are considered using contraction arguments. Further perturbation results on nonconvolution integrodifferential equations can be found in Grimmer and Seifert [1]. For a careful overview of perturbations of Volterra equations, see Levin [11].

Cushing [1] gives conditions under which the set of bounded solutions of the perturbed equation

$$x'(t) + A(t)x(t) + \int_0^t B(t,s)x(s)\,\mathrm{d}s = G(x)(t), \quad t \in \mathbf{R}^+,$$

is locally homeomorphic to the set of bounded solutions of

$$y'(t) + A(t)y(t) + \int_0^t B(t,s)y(s)\,\mathrm{d}s = 0, \quad t \in \mathbf{R}^+.$$

See also Cushing [2].

In Miller and Nohel [1] a nonlinear extension of (part of) Theorem **7**.2.3 is given.

Section 5:
For results on (5.1), in the case where A dominates G, see Burton [2].

Section 6:
Early L^2-results are given in Grossman and Miller [1] and Miller [10], Ch. V, Sec. 6.

References

T. A. Burton
2. *Volterra Integral and Differential Equations*, Academic Press, New York, 1983.

J. M. Cushing
1. An operator equation and bounded solutions of integro-differential systems, *SIAM J. Math. Anal.* **6** (1975), pp. 433–445.
2. Bounded solutions of perturbed Volterra integrodifferential systems, *J. Differential Equations* **20** (1976), pp. 61–70.

R. Grimmer and G. Seifert
1. Stability properties of Volterra integrodifferential equations, *J. Differential Equations* **19** (1975), pp. 142–166.

S. I. Grossman and R. K. Miller
1. Perturbation theory for Volterra integrodifferential systems, *J. Differential Equations* **8** (1970), pp. 457–474.

J. K. Hale
3. *Ordinary Differential Equations*, J. Wiley, New York, 1969.

J. L. Kaplan
1. On the asymptotic behavior of Volterra integral equations, *SIAM J. Math. Anal.* **3** (1972), pp. 148–156.

J. J. Levin
11. Nonlinearly perturbed Volterra equations, *Tôhoku Math. J. (2)* **32** (1980), pp. 317–335.

R. K. Miller
5. On the linearization of Volterra integral equations, *J. Math. Anal. Appl.* **23** (1968), pp. 198–208.
6. Admissibility and nonlinear Volterra integral equations, *Proc. Amer. Math. Soc.* **25** (1970), pp. 65–71.
10. *Nonlinear Volterra Integral Equations*, W. A. Benjamin, Menlo Park, Calif., 1971.

R. K. Miller and J. A. Nohel

1. A stable manifold theorem for a system of Volterra integro-differential equations, *SIAM J. Math. Anal.* **6** (1975), pp. 506–522.

R. K. Miller, J. A. Nohel, and J. S. W. Wong

1. Perturbations of Volterra integral equations, *J. Math. Anal. Appl.* **25** (1969), pp. 676–691.

J. A. Nohel

1. Asymptotic relationships between systems of Volterra equations, *Ann. Mat. Pura Appl. (IV)* **90** (1971), pp. 149–166.

A. Strauss

1. On a perturbed Volterra integral equation, *J. Math. Anal. Appl.* **30** (1970), pp. 564–575.

12

Existence of Solutions of Nonlinear Equations

We give a number of existence and comparison results for the integral equation $x(t) = f(t) + \int_0^t g(t,s,x(s))\,\mathrm{d}s$, the functional equation $x(t) = G(x)(t)$, and the functional differential equation $x'(t) = H(x)(t)$.

1. Introduction

In Chapters 2–10 we usually obtained explicit expressions (in terms of variation of constants formulas) for the solutions to our equations. This, of course, automatically answered the questions about existence and uniqueness of solutions. In Chapter 11, global existence of solutions was obtained through the use of fixed point methods. In the remaining chapters the situation will be different. We shall analyse nonlinear equations and we shall not, in general, arrive at any definite expressions for the solutions. Our usual approach in Chapters 17–20 will, in fact, be to assume the existence of a solution, and then to prove some asymptotic facts about this solution.

An analysis of the existence of solutions of nonlinear Volterra equations is therefore well motivated before we proceed to asymptotics. We devote the current chapter to this existence analysis. But we wish to emphasize that the reader who is willing to assume that solutions exist may skip this chapter entirely (and perhaps Chapter 13 as well) or, at most, pay some attention to a few elementary results like Theorems 1.1 and 1.2, and Corollary 3.2.

We begin with a simple example, namely with the equation

$$x(t) = f(t) + \int_0^t a(t-s)g(x(s))\,\mathrm{d}s, \quad t \geq 0, \tag{1.1}$$

where a is locally integrable on $\mathbf{R}^+$ with values in $\mathbf{R}^{n\times n}$, g is a continuous function from $\mathbf{R}^n$ into itself, and f is continuous on $\mathbf{R}^+$. First, we prove that this equation has a solution on an interval $[0,T]$, where T is a sufficiently small positive number. After this, the solution is continued to a larger interval.

To obtain existence on $[0,T]$, we show that the operator defined by the right-hand side of (1.1) has a fixed point. That is, we define

$$Q(\phi)(t) = f(t) + \int_0^t a(t-s)g(\phi(s))\,\mathrm{d}s, \quad t \in [0,T],$$

and try to find a function x such that $Q(x)(t) = x(t)$ for all $t \in [0,T]$. Such a function is, of course, a solution of the original equation. We take the domain of Q to be

$$K \stackrel{\mathrm{def}}{=} \{\, \phi \in C([0,T];\mathbf{R}^n) \mid |\phi(t)| \le M \text{ for } t \in [0,T] \,\}, \tag{1.2}$$

where M is some positive real number. The numbers M and T should be chosen in such a way that Q maps K into itself, since this is a requirement common to all fixed point theorems.

It is quite clear that the function $t \mapsto Q(\phi)(t)$ is continuous (in a moment we shall establish a much stronger result), so in order to prove that Q maps K into itself it remains only to analyse the size of $|Q(\phi)(t)|$. If $\phi \in K$, then obviously we have for all $t \in [0,T]$ (see, for example, Theorem **2**.2.2(i))

$$|Q(\phi)(t)| \le \sup_{t\in[0,T]} |f(t)| + \left\{ \int_0^T |a(t)|\,\mathrm{d}t \right\} \sup_{|v|\le M} |g(v)|.$$

Hence, if we take $M \stackrel{\mathrm{def}}{=} 2\sup_{t\in[0,1]}|f(t)|$, and choose $T \le 1$ to be so small that

$$\left\{ \int_0^T |a(t)|\,\mathrm{d}t \right\} \sup_{|v|\le M} |g(v)| \le \frac{M}{2},$$

then $|Q(\phi)(t)| \le M$ for $0 \le t \le T$, and Q maps K into itself.

The set K is a bounded, closed, and convex subset of the Banach space $C([0,T];\mathbf{R}^n)$. To apply Schauder's theorem (see Theorem 1.4) it suffices, therefore, to show that Q is a compact map of K into itself. By the Arzelà–Ascoli Theorem **2**.7.5, and by what we have already proved, this amounts to showing that the set $\{\, Q(\phi) \mid \phi \in K \,\}$ is equicontinuous. The following simple estimate accomplishes the task. Let $0 \le \tau \le t$. Then

$$\begin{aligned} |Q(\phi)(t) - Q(\phi)(\tau)| &\le |f(t) - f(\tau)| + \left| \int_\tau^t a(t-s)g(\phi(s))\,\mathrm{d}s \right| \\ &\qquad + \left| \int_0^\tau \big(a(t-s) - a(\tau-s)\big)g(\phi(s))\,\mathrm{d}s \right| \\ &\le |f(t) - f(\tau)| \\ &\qquad + \sup_{|v|\le M} |g(v)| \left\{ \int_0^{t-\tau} |a(s)|\,\mathrm{d}s + \int_0^\tau |a(t-\tau+s) - a(s)|\,\mathrm{d}s \right\}. \end{aligned}$$

Since the function a is required to be integrable, and since the function f is continuous, we conclude that the set $\{ Q(\phi) \mid \phi \in K \}$ is equicontinuous on $[0, T]$.

An application of the Schauder Theorem 1.4 now tells us that there exists a fixed point x of Q, that is, there exists a solution of the original equation on $[0, T]$.

After showing that there exists a solution on some interval $[0, T]$, we may extend this solution as follows. Let $h(t) = f(t+T)+\int_0^T a(t+T-s)g(x(s))\,\mathrm{d}s$ for $t \geq 0$. It is clear that h is a continuous function of t. By using the same arguments as above, we can, therefore, find a continuous solution of the equation

$$y(t) = h(t) + \int_0^t a(t-s)g(y(s))\,\mathrm{d}s,$$

on some interval $[0, S]$. If we let $x(t) = y(t-T)$, for $t \in (T, S+T]$, then it follows from the equations satisfied by x and y that x is continuous at the point T as well, and furthermore that x satisfies (1.1) on the interval $[0, T+S]$. In this way we can extend the solution for as long as it remains bounded.

If $0 < T_1 < T_2$, if x_1 is a solution of some equation on $[0, T_1]$, and if x_2 is a solution of the same equation on $[0, T_2]$, then we say that x_2 is a *continuation* of x_1, provided that $x_1(t) = x_2(t)$ when $t \in [0, T_1]$. It follows from the Hausdorff maximality principle that every solution on an interval $[0, T]$ can be continued to a *noncontinuable solution.*

Before proceeding to more general equations, let us formulate a simple theorem that contains the results achieved above. Its proof is essentially the same as we gave above, but it may also be regarded as a corollary to Theorem 2.7, or as a corollary to Theorem 2.8.

1.1 Theorem

Let $0 < T_\infty \leq \infty$, $f \in C([0, T_\infty); \mathbf{R}^n)$, $h \in C([0, T_\infty) \times \mathbf{R}^n; \mathbf{R}^n)$, and let a be a $\mathbf{R}^{n\times n}$-valued Volterra kernel of continuous type on $[0, T_\infty)$. Then there exists a number $T \in (0, T_\infty)$ such that the equation

$$x(t) = f(t) + \int_0^t a(t,s)h\big(s, x(s)\big)\,\mathrm{d}s, \quad t \geq 0, \tag{1.3}$$

has a continuous solution on $[0, T]$.

Moreover, if x is a noncontinuable solution of (1.3), then the (maximal) interval of existence, $[0, T_{\max})$, is open to the right, and if $T_{\max} < T_\infty$ then $\limsup_{t\uparrow T_{\max}} |x(t)| = \infty$.

We recall that kernels of continuous type were defined in Definition **9**.5.2. In particular, note that if $a(t,s) = k(t-s)$, $0 \leq s \leq t$, where $k \in L^1_{\mathrm{loc}}([0, T); \mathbf{R}^{n\times n})$, then a is of continuous type on $[0, T)$.

An integral operator of the type $(K\varphi)(t) = \int_J k(t,s)h(s, \varphi(s))\,\mathrm{d}s$ is sometimes called a *Hammerstein operator.*

There is another class of results that can be proved with very small additional effort. In the discussion of (1.1) above, it is sometimes possible to restrict oneself to a smaller class of functions, i.e., to decrease the set K in (1.2). This allows us to prove a number of *comparison results.* For example, suppose that we want to compare the solution of (1.1) to the solution of the related scalar equation

$$y(t) = F(t) + \int_0^t A(t-s)G(y(s))\,\mathrm{d}s, \quad t \geq 0. \tag{1.4}$$

Here we assume that, for all $t \in \mathbf{R}^+$ and $y \in \mathbf{R}^n$,

$$|f(t)| \leq F(t), \quad |a(t)| \leq A(t), \quad |g(y)| \leq G(|y|),$$

and that G is nondecreasing in y for $y > 0$ (for example, one can take $F(t) = |f(t)|$, $A(t) = |a(t)|$, and $G(y) = \sup_{|z|\leq|y|}|g(z)|$).

We claim that the operator Q that was used in the fixed point argument above maps the set

$$L \stackrel{\text{def}}{=} \{\, \phi \in C([0,T];\mathbf{R}^n) \mid |\phi(t)| \leq y(t) \text{ for } t \in [0,T] \,\}$$

back into itself. Here y is a solution of (1.4). To see that this is true, take some $\phi \in L$. Then, for $t \in [0,T]$,

$$\begin{aligned}
|Q(\phi)(t)| &= \left| f(t) + \int_0^t a(t-s)g(\phi(s))\,\mathrm{d}s \right| \\
&\leq F(t) + \int_0^t A(t-s)G(|\phi(s)|)\,\mathrm{d}s \\
&\leq F(t) + \int_0^t A(t-s)G(y(s))\,\mathrm{d}s \\
&= y(t).
\end{aligned}$$

Thus, Q does indeed map L into itself. Hence Q maps $K \cap L$ into itself, where K is the set defined in (1.2). The same fixed point argument that we used above, but with K replaced by $K \cap L$, proves that, in this case, (1.1) has at least one solution x that (exists and) satisfies $|x(t)| \leq y(t)$ for as long as $y(t)$ exists. In particular, if y is defined for all $t \geq 0$, then (1.1) has a solution that exists for all $t \in \mathbf{R}^+$.

The preceding argument essentially proves the following theorem (which is a corollary to Theorem 2.4, and also a corollary to Theorem 2.9).

1.2 Theorem

Let the assumption of Theorem 1.1 hold. Define

$$H(s,y) = \sup_{|z|\leq|y|} |h(s,z)|, \quad s \in [0,T_\infty), \quad y \in \mathbf{R},$$

and let y be a noncontinuable solution of the equation

$$y(t) = |f(t)| + \int_0^t |a(t,s)| H(s,y(s))\,\mathrm{d}s, \quad t \geq 0, \tag{1.5}$$

defined on an interval $[0, S_{\max})$, *where* $S_{\max} \leq T_\infty$. *Then there exists a number* $T \in (0, S_{\max})$ *such that equation (1.3) has a continuous solution on* $[0,T]$ *satisfying* $|x(t)| \leq y(t)$ *for* $t \in [0,T]$.

Moreover, if x *is a solution of (1.3) on* $[0,T]$ *satisfying* $|x(t)| \leq y(t)$ *for* $t \in [0,T]$, *then* x *can be extended to a noncontinuable solution of (1.3) on a (maximal) interval* $[0, T_{\max})$, *with* $T_{\max} \geq S_{\max}$, *satisfying* $|x(t)| \leq y(t)$ *for* $t \in [0, S_{\max})$. *In particular, if* $S_{\max} = T_\infty$, *then (1.3) has a solution that is defined on* $[0, T_\infty)$.

The structure of the equations considered above is quite restrictive. Therefore, in the next section, we shall study equations of a more general kind. To this end, we make some preliminary comments, and formulate Fixed Point Theorem 1.3.

Any Volterra equation on $\mathbf{R}^+$ can be written as

$$x(t) = G(x)(t), \quad t \geq 0, \tag{1.6}$$

where, for each fixed t, the operator $x \mapsto G(x)(t)$ is allowed to depend only on the restriction of x to $[0,t]$, and not on $x(s)$ for $s > t$ (i.e., G is assumed to be causal). In Chapters 2–10 we analysed equations where G was linear and in Chapter 11 we considered perturbed linear equations. We shall now take G to be nonlinear, and shall give a number of results on the existence of solutions of equations of the type (1.6). The equation

$$x(t) = f(t) + \int_0^t g(t,s,x(s))\,\mathrm{d}s, \quad t \geq 0, \tag{1.7}$$

is an important special case of (1.6), and some of our existence results will employ the specific structure of (1.7). Another, equally important, special case of (1.6) is the functional differential equation

$$\tfrac{\mathrm{d}}{\mathrm{d}t}x(t) = H(x)(t), \quad t \geq 0; \quad x(0) = \xi, \tag{1.8}$$

where H is an operator of the same type as G in (1.6). Our existence result for (1.8) is a simple corollary to the corresponding result for (1.6).

An integral operator of the type $(K\varphi)(t) = \int_J k(t,s,\varphi(s))\,\mathrm{d}s$ is frequently called an *Uryson operator*.

In (1.6) and in (1.7) we require the equation to hold for all t if x is continuous, or if x is locally bounded and Borel measurable, and to hold for almost all t if x is locally in L^p for some $p \in [1,\infty]$. Note that if in (1.7) we take $t = 0$ then we get $x(0) = f(0)$. For continuous solutions we must, therefore, have $\lim_{t \downarrow 0} \int_0^t g(t,s,x(s))\,\mathrm{d}s = 0$ if f is continuous. A solution x of (1.8) is required to be locally absolutely continuous and to satisfy (1.8) almost everywhere.

One should note that (1.6) and (1.8) contain, as particular cases, the delay equation

$$x(t) = F(x^t), \quad t \geq 0; \quad x(t) = \varphi(t), \quad t < 0,$$

and the functional differential equation

$$x'(t) = F(x^t), \quad t \geq 0; \quad x(t) = \varphi(t), \quad t < 0,$$

where $x^t(\theta) = x(t+\theta)$, $\theta \le 0$, and where F is defined on some set of functions with domain contained in $(-\infty, 0]$. This follows from the fact that if F and φ are given, then we take

$$G(x)(t) = F(\psi^t),$$

where

$$\psi(t) = \begin{cases} x(t), & t \ge 0, \\ \varphi(t), & t < 0. \end{cases}$$

If F is defined on a set consisting of continuous functions, then we must require that $x(0) = \varphi(0-)$.

As concerns (1.6) and (1.8), we shall try to be reasonably general. Therefore, an application of the results requires some (perhaps extensive) calculations before the conditions are checked. On the other hand, for (1.7) we try to be as concrete as possible. The result of this policy is that we shall not obtain the most general theorems that can be derived for (1.7).

As corollaries to our existence theorems we get some uniqueness and some comparison results. The question of continuous dependence on data will be deferred to Chapter 13. There we give some additional uniqueness and comparison results, as well.

Our proofs are based on a fixed point theorem, that is general enough to contain both the Schauder fixed point theorem and the contraction mapping principle. This theorem employs the (Kuratowski) measure α of noncompactness, which is defined as follows: if Ω is a subset of a Banach space $\mathcal{B}$, then the *diameter* $\operatorname{diam}\{\Omega\}$ of Ω is defined by $\operatorname{diam}\{\Omega\} = \sup\{\, \|x-y\| \mid x, y \in \Omega \,\}$ and the *measure of noncompactness* $\alpha\{\Omega\}$ of Ω by

$$\alpha\{\Omega\} = \inf\{\, \epsilon > 0 \mid \Omega \text{ is covered by a finite number of sets each with diameter no larger than } \epsilon \,\}.$$

A set is said to be totally bounded iff its measure of noncompactness is zero.

See Deimling [2], p. 41, for a more detailed discussion on the properties of $\operatorname{diam}\{\Omega\}$ and $\alpha\{\Omega\}$.

1.3 Fixed Point Theorem

Let K be a closed, bounded and convex subset of some Banach space $\mathcal{B}$, and suppose that F is a continuous function from K into K with the property that $\alpha\{F(\Omega)\} < \alpha\{\Omega\}$ for each $\Omega \subset K$ satisfying $\alpha\{\Omega\} > 0$. Then the set $\{\, z \in K \mid F(z) = z \,\}$ of fixed points of F in K is nonempty and compact. If, in addition, $\|F(x) - F(y)\| < \|x - y\|$ for $x, y \in K$ and $x \ne y$, then F has only one fixed point in K.

This is Martin [1], Theorem 3.3, Ch. 4, except for the uniqueness claim, which is trivial. To get one version of the contraction mapping principle from this theorem, observe that if F is Lipschitz-continuous with Lipschitz constant γ then $\alpha\{F(\Omega)\} \le \gamma\alpha\{\Omega\}$.

As a corollary we get the Schauder Theorem 1.4, since one immediately recognizes that $\alpha\{F(\Omega)\} = 0$ if F maps bounded sets into totally bounded sets.

1.4 Schauder's Theorem
Let K be a closed, bounded and convex subset of a Banach space $\mathcal{B}$. Let F be a continuous function from K into K and suppose that the closure of $F(K)$ is compact. Then F has a fixed point in K.

2. Continuous Solutions

Our first result, Theorem 2.1 below, gives a set of sufficient conditions for the existence of a continuous solution x of the functional equation

$$x(t) = G(x)(t), \quad t \geq 0, \tag{1.6}$$

on an interval $[0, T)$. The solution curve $\{\,(t, x(t)) \mid t \in [0, T)\,\}$ is required to start at the point $(0, \xi)$, where ξ is given, and to remain in a given set $D \subset \mathbf{R}^+ \times \mathbf{R}^n$ containing $(0, \xi)$. In other words, we impose the auxiliary condition

$$x(0) = \xi; \quad (t, x(t)) \in D, \quad t \geq 0. \tag{2.1}$$

We assume that D is open in $\mathbf{R}^+ \times \mathbf{R}^n$, i.e., that D can be written as $D = D_1 \cap (\mathbf{R}^+ \times \mathbf{R}^n)$, where D_1 is open in $\mathbf{R}^{n+1}$. Clearly, there is no loss of generality if we take D to be connected.

The reason for fixing the value at zero of the functions we consider has to do with the requirement that the range of $G(x)$ consist of continuous functions. Consider, for example, the simple equation

$$x(t) = \begin{cases} g(x(t-1)), & t \geq 0; \\ \varphi(t), & t \in [-1, 0), \end{cases}$$

where g and φ are continuous functions. In this case, one defines the mapping G by

$$G(x)(t) = \begin{cases} g(\varphi(t-1)), & t \in [0, 1), \\ g(x(t-1)), & t \geq 1, \end{cases}$$

and the point is that $G(x)(t)$ will not be a continuous function of t for $t \geq 0$, unless $g(x(0)) = g(\varphi(0-))$. The easiest way to guarantee that this will be the case is to restrict the domain of G to continuous functions x satisfying $x(0) = \varphi(0-)$. Since the equation is required to hold for $t = 0$ (see (iii) of Theorem 2.1 for the general case) one gets $\varphi(0-) = g(\varphi(-1))$ as a condition for the solution to be continuous.

We shall interpret the right-hand side of (1.6) in slightly different ways, depending on the class of functions to which we require the solution to belong. In this section we take G to be a mapping that takes each function that is continuous, and satisfies the auxiliary condition (2.1) on some interval $[0, T]$, into a continuous function on the same interval.

For each interval $J \subset \mathbf{R}^+$ containing zero and each set $M \subset \mathbf{R}^+ \times \mathbf{R}^n$, we define

$$C_{\xi,M}(J; \mathbf{R}^n) \\ = \{ \phi \in C(J; \mathbf{R}^n) \mid \phi(0) = \xi \text{ and } (t, \phi(t)) \in M \text{ for } t \in J \}. \quad (2.2)$$

We define T_∞ by

$$T_\infty = \sup\{ T \in \mathbf{R}^+ \mid C_{\xi,D}([0,T]; \mathbf{R}^n) \neq \emptyset \}. \quad (2.3)$$

Note that T_∞ depends on both D and ξ and that it gives an upper bound for the interval of existence of solutions of (1.6).

We assume that, for each fixed $T \in [0, T_\infty)$, the operator G maps the set $C_{\xi,D}([0,T]; \mathbf{R}^n)$ into $C([0,T]; \mathbf{R}^n)$. This mapping is required to be *causal* or *nonanticipative*, which means that the value of $G(\phi)(t)$ is allowed to depend only on the values of ϕ on the interval $[0, t]$, and not on the values of $\phi(s)$ when $s \in (t, T]$.

We say that x is a *solution* of (1.6) and (2.1) on an interval $J = [0, T]$ or $J = [0, T)$, if $x \in C_{\xi,D}(J; \mathbf{R}^n)$, and if $x(t) = G(x)(t)$ for every $t \in J$. In particular, we require $x(0) = G(x)(0) = \xi$. Because of the causality of G, the value of $G(x)(0)$ depends only on $x(0) = \xi$ and not on $x(t)$ for $t > 0$, so it makes sense to write this condition in the form $G(\xi)(0) = \xi$.

Recall that if $0 < T_1 < T_2$, if $x_1 \in C_{\xi,D}([0,T_1]; \mathbf{R}^n)$ is a solution of (1.6) and (2.1) on $[0, T_1]$, and if $x_2 \in C_{\xi,D}([0,T_2]; \mathbf{R}^n)$ is a solution of (1.6) and (2.1) on $[0, T_2]$, then we say that x_2 is a continuation of x_1, provided that $x_1(t) = x_2(t)$ for $t \in [0, T_1]$. Moreover, every solution can be extended to a noncontinuable one on a maximal interval of existence. This interval can be either open or closed to the right, but in all the cases that we will encounter below the interval will be open to the right.

We remind the reader of the notation $|a|_+$ for $\max\{a, 0\}$.

2.1 Theorem

Let D be an open connected subset of $\mathbf{R}^+ \times \mathbf{R}^n$ containing $(0, \xi)$, define T_∞ by (2.3), and suppose that conditions (i)–(iv) *below are satisfied.*

(i) *For each $T \in [0, T_\infty)$, G maps $C_{\xi,D}([0,T]; \mathbf{R}^n)$ into $C([0,T]; \mathbf{R}^n)$, and for each $t \in [0, T]$ the mapping $\phi \mapsto G(\phi)(t)$ is continuous from $C_{\xi,D}([0,T]; \mathbf{R}^n)$ (equipped with the sup-norm) into $\mathbf{R}^n$.*

(ii) *G is causal, i.e., if $0 \le \tau < T < T_\infty$, and if ψ is the restriction of $\phi \in C_{\xi,D}([0,T]; \mathbf{R}^n)$ to $[0, \tau]$, then $G(\psi)(t) = G(\phi)(t)$ for all $t \in [0, \tau]$.*

(iii) *$G(\xi)(0) = \xi$.*

(iv) *For each compact connected subset M of D containing $(0, \xi)$ there exist a constant $\eta > 0$ and a continuous function $\delta : (0, \infty) \to (0, \infty)$, such that the following implication holds: if for some $\epsilon > 0$, $T \in (0, T_\infty)$, and for some functions $\phi, \psi \in C_{\xi,M}([0,T]; \mathbf{R}^n)$,* we have

  (1) *$\psi(s) = \phi(s)$ for $s \in [0, |T - \eta|_+]$,*

(2) $|\psi(s) - \phi(s)| \le \epsilon$ *for* $s \in [|T-\eta|_+, T]$,

then

(3) $|G(\psi)(s) - G(\phi)(s) - G(\psi)(t) + G(\phi)(t)| \le \epsilon - \delta(\epsilon)$ *for all* s *and* $t \in [|T-\eta|_+, T]$, *satisfying* $|t-s| \le \delta(\epsilon)$.

Then there exists a $T > 0$ *such that (1.6) has a continuous solution* x *on* $[0,T]$ *satisfying the auxiliary condition (2.1).*

Moreover, if x *is a noncontinuable solution of (1.6) subject to (2.1), then the (maximal) interval of existence of* x *is open to the right, i.e., it is of the form* $[0, T_{\max})$ *for some* $T_{\max} > 0$, *and* $(t, x(t))$ *leaves every compact subset of* D *as* $t \uparrow T_{\max}$.

Before proving Theorem 2.1, let us comment on the key hypothesis (iv). If G satisfies a Lipschitz-type condition with Lipschitz constant $\gamma < \frac{1}{2}$, then the implication in (iv) holds. Specifically, in this case it follows from (2) that

$$|G(\psi)(s) - G(\phi)(s)| \le \gamma\epsilon$$

for $s \in [|T-\eta|_+, T]$, and then (3) is satisfied with $\delta(\epsilon) = \epsilon - 2\gamma\epsilon$ (cf. Corollary 2.2). If G maps bounded sets into equicontinuous ones, then one can show that (iv) holds by using the inequality

$$\begin{aligned}&\big|G(\psi)(s) - G(\phi)(s) - G(\psi)(t) + G(\phi)(t)\big|\\ &\qquad \le \big|G(\psi)(s) - G(\psi)(t)\big| + \big|G(\phi)(s) - G(\phi)(t)\big|.\end{aligned}$$

See Corollary 2.3 below. Various combinations of compact and Lipschitz-continuous operators are possible as well.

We shall see from the proof that the implication in (iv) is only used in the case where one assumes, in addition to (1) and (2), that either $|G(\psi)(s) - G(\psi)(t)| < \delta(\epsilon)$ or $|G(\psi)(t) - G(\phi)(t)| < \delta(\epsilon)$. Therefore, the theorem could be somewhat improved by including this as a condition in (iv).

A slightly different assumption is given in Exercise 5.

The full strength of (iv) is needed only for the proof of the statement that $(t, x(t))$ leaves every compact subset of D as $t \uparrow T_{\max}$. If one drops this conclusion, then it would suffice to assume that (1) and (2) imply that

(3*) $|G(\psi)(s) - G(\phi)(s)| \le \epsilon - \delta(\epsilon) + |G(\psi)(t) - G(\phi)(t)|$ for all s, $t \in [|T-\eta|_+, T]$ satisfying $t \le s \le t + \delta(\epsilon)$.

(It is clear that (3*) follows from (3).)

Theorem 2.1 may seem rather intricate, but most of the results on existence found in the literature can be derived from it—a task that is not necessarily easy.

Proof of Theorem 2.1 The proof uses the classical 'method of steps', i.e., we solve the equation on successive small intervals, using Fixed Point Theorem 1.3, and show that this procedure can be continued until we reach the boundary of D.

Let $x \in C_{\xi,D}([0,\tau];\mathbf{R}^n)$ be a solution on an interval $[0,\tau]$, where $\tau \geq 0$. If $\tau = 0$, then $x = \xi$ is such a solution. Since D is open, there exist constants $T > \tau$ and $\rho > 0$ such that the set M defined by

$$\begin{aligned} M = &\{\, (s, x(s)) \mid s \in [0,\tau) \,\} \\ &\bigcup \{\, (s,y) \mid s \in [\tau,T], |y - x(\tau)| \leq \rho \,\}, \end{aligned} \tag{2.4}$$

is contained in D. Clearly $(0,\xi) \in M$ and M is compact and connected. Let η and δ be the constant and the function in (iv) corresponding to this set. We want to apply Theorem 1.3 to show that G has a fixed point in $C_{\xi,M}([0,T];\mathbf{R}^n)$.

First, let us check that G maps the closed, bounded, and convex set $C_{\xi,M}([0,T];\mathbf{R}^n)$ into itself, or rather that we can compel G to map $C_{\xi,M}([0,T];\mathbf{R}^n)$ into itself by restricting T in (2.4). (It is clearly possible to decrease the value of T, thus decreasing M, without changing η and δ, but of course $T > \tau$.) Define the function ψ by $\psi(s) = x(s)$ for $0 \leq s \leq \tau$, and by $\psi(s) = x(\tau)$ for $\tau \leq s \leq T$. Then ψ belongs to $C_{\xi,M}([0,T];\mathbf{R}^n)$; hence, by the continuity of $G(\psi)(s)$ as a function of s, and by the fact that $G(\psi)(s) = x(s)$ for $s \in [0,\tau]$ (in particular $G(\psi)(\tau) = x(\tau)$), we can guarantee that for $s \in [\tau,T]$ we have

$$|G(\psi)(s) - x(\tau)| = |G(\psi)(s) - G(\psi)(\tau)| \leq \delta(\rho),$$

provided T is chosen close enough to τ. Let $\phi \in C_{\xi,M}([0,T];\mathbf{R}^n)$. Then, by (ii), and by the fact that x is a solution of (1.6) on $[0,\tau]$, we have $G(\phi)(\tau) = x(\tau) = G(\psi)(\tau)$. Moreover, it follows from the definitions of M and ψ that $|\phi(s) - \psi(s)| = |\phi(s) - x(\tau)| \leq \rho$ for $s \in [\tau,T]$. Let us require that $T - \tau \leq \min[\eta, \delta(\rho)]$. Then $T - \eta \leq \tau$, and (1) of (iv) holds. Moreover, condition (2) of (iv) is satisfied with $\epsilon = \rho$. Then, by (iv) where we take $t = \tau$, and by the inequality above, we have for all $s \in [\tau,T]$ (here we use $T \leq \tau + \delta(\rho)$),

$$\begin{aligned} |G(\phi)(s) - x(\tau)| &\leq |G(\phi)(s) - G(\psi)(s)| + |G(\psi)(s) - x(\tau)| \\ &\leq \rho - \delta(\rho) + \delta(\rho) = \rho. \end{aligned}$$

This means that under the additional restrictions on T mentioned above, G maps the set $C_{\xi,M}([0,T];\mathbf{R}^n)$ into itself.

Next, we must show that G is continuous. Given $\epsilon > 0$, we can divide the interval $[\tau,T]$ into subintervals $[t_i, t_{i+1}]$, $1 \leq i \leq n(\epsilon)$, each of length at most $\delta(\epsilon)$. Let $\phi \in C_{\xi,M}([0,T];\mathbf{R}^n)$. Then, by (i), for each t_i, $1 \leq i \leq n(\epsilon)$, we can find a constant $0 < \delta_i \leq \epsilon$ such that, if $\psi \in C_{\xi,M}([0,t_i];\mathbf{R}^n)$ and $|\psi(s) - \phi(s)| \leq \delta_i$ for $\tau \leq s \leq t_i$, then $|G(\psi)(t_i) - G(\phi)(t_i)| \leq \delta(\epsilon)$. Hence, if $\psi \in C_{\xi,M}([0,T];\mathbf{R}^n)$, and $|\psi(s) - \phi(s)| \leq \min\{\, \delta_i \mid 1 \leq i \leq n(\epsilon) \,\} \leq \epsilon$ for $\tau \leq s \leq T$, then ϕ and ψ satisfy (1) and (2) of (iv) and, moreover, $|G(\psi)(t_i) - G(\phi)(t_i)| \leq \delta(\epsilon)$, $1 \leq i \leq n(\epsilon)$. Therefore, taking $t = t_1, t_2, \ldots, t_{n(\epsilon)}$ in (3*) (which follows from (3) of (iv)) we obtain, each time, the inequality

$$\big|G(\psi)(s) - G(\phi)(s)\big| \leq \epsilon - \delta(\epsilon) + \big|G(\psi)(t_i) - G(\phi)(t_i)\big| \leq \epsilon,$$

for s in the interval $[t_i, t_i + \delta(\epsilon)]$. This proves that G is continuous from $C_{\xi,M}([0,T];\mathbf{R}^n)$ into itself.

It remains to check the key assumption of Theorem 1.3, i.e., the assumption on the behaviour of G with respect to the measure α of noncompactness. Let $\epsilon > 0$, and let $S \subset C_{\xi,M}([0,T];\mathbf{R}^n)$ satisfy $\operatorname{diam}\{S\} \le \epsilon$. Again divide $[\tau, T]$ into subintervals $[t_i, t_{i+1}]$, $1 \le i \le n(\epsilon)$, each of length at most $\delta(\epsilon)$. The set $\{\, y \in \mathbf{R}^n \mid |y - x(\tau)| \le \rho \,\}$ is compact in $\mathbf{R}^n$; hence we can find a finite covering $\{U_j\}_{j=1}^{m(\epsilon)}$ of this set, where $\operatorname{diam}\{U_j\} \le \delta(\epsilon)/2$ for each j, $1 \le j \le m(\epsilon)$. We split S into a total of $m(\epsilon)^{n(\epsilon)}$ sets $S_{k_1,k_2,\ldots,k_{n(\epsilon)}}$ (some of which may be empty), defined by

$$S_{k_1,k_2,\ldots,k_{n(\epsilon)}} = \{\, \phi \in S \mid G(\phi)(t_i) \in U_{k_i}, \quad 1 \le i \le n(\epsilon) \,\},$$

where $1 \le k_i \le m(\epsilon)$. If ϕ_a, ϕ_b are in the same set $S_{k_1,k_2,\ldots k_{n(\epsilon)}}$, then $|G(\phi_a)(t_i) - G(\phi_b)(t_i)| \le \delta(\epsilon)/2$ for all t_i. It follows from (iv), where we take $t = t_1, t_2, \ldots, t_{n(\epsilon)}$ and recall that $\operatorname{diam}\{S\} \le \epsilon$, that for each multi-index $k_1, k_2, \ldots, k_{n(\epsilon)}$, $\operatorname{diam}\{G(S_{k_1,k_2,\ldots,k_{n(\epsilon)}})\} \le \epsilon - \delta(\epsilon)/2$; hence $\alpha\{G(S)\} \le \max\{\alpha\{G(S_{k_1,k_2,\ldots,k_{n(\epsilon)}})\}\} \le \epsilon - \delta(\epsilon)/2$. If Ω is an arbitrary subset of $C_{\xi,M}([0,T];\mathbf{R}^n)$ and $\alpha\{\Omega\} < \epsilon$, then Ω can be written as a finite union of sets, each with diameter no larger than ϵ. The preceding argument shows that $\alpha\{G(\Omega)\} \le \epsilon - \delta(\epsilon)/2$. Letting $\epsilon \downarrow \alpha\{\Omega\}$ we get $\alpha\{G(\Omega)\} \le \alpha\{\Omega\} - \delta(\alpha\{\Omega\})/2 < \alpha\{\Omega\}$ provided that $\alpha\{\Omega\} > 0$, as required by Theorem 1.3.

We have shown that Theorem 1.3 can be applied to G, considering G as a map of $C_{\xi,M}([0,T];\mathbf{R}^n)$ into itself. It follows that the given solution $x \in C_{\xi,M}([0,\tau];\mathbf{R}^n)$ of (1.6) on $[0,\tau]$ can be extended to a solution $x \in C_{\xi,M}([0,T];\mathbf{R}^n)$ of (1.6) on $[0,T]$. Using this fact, and the Hausdorff maximality principle, one finds that the solution x can be extended to a noncontinuable solution x on a maximal interval $[0, T_{\max})$.

To conclude the proof, we show that $(t, x(t))$ leaves every compact subset of D as $t \uparrow T_{\max}$. Suppose that this is not the case, i.e., suppose that $(t, x(t))$ stays in some compact subset K of D. If $\lim_{t\uparrow T_{\max}} x(t)$ exists, then necessarily $(T_{\max}, \lim_{t\uparrow T_{\max}} x(t)) \in K \subset D$, and it follows from the preceding argument that x can be continued. This contradiction shows that $\lim_{t\uparrow T_{\max}} x(t)$ cannot exist. Define

$$\beta = \limsup_{s,t\uparrow T_{\max}} |x(s) - x(t)|;$$

then $0 < \beta < \infty$. Without loss of generality, we may assume that, for some $\rho > 0$, $(s, x(t)) \in K$ for $t \in [0, T_{\max})$ and $s \in [t, t+\rho]$ (replace K by the set $\{\, (s,y) \mid \operatorname{dist}((s,y),K) \le \rho \,\}$). Let η and δ denote, respectively, the constant and the function corresponding to K given by (iv). Choose a number ϵ such that

$$\beta < \epsilon < \beta + \delta(\epsilon)/2.$$

Next, we pick a number $\tau > 0$ such that $0 < T_{\max} - \tau \le \min\{\eta, \rho\}$ and

$$|x(s) - x(\tau)| \le \epsilon, \quad s \in [\tau, T_{\max}).$$

Define the function ψ by

$$\psi(s) = \begin{cases} x(s), & s \in [0,\tau], \\ x(\tau), & s > \tau. \end{cases}$$

Then $|\psi(s) - x(s)| \le \epsilon$ for $s \in [|T_{\max} - \eta|_+, T_{\max})$. Observe that $(s, \psi(s)) \in K$ for all $s \in [0, T_{\max}]$. It follows that $G(\psi)$ is a continuous function on $[0, T_{\max}]$. Therefore, there exists a number $\tau_1 \in [\tau, T_{\max})$ such that $T_{\max} - \tau_1 < \delta(\epsilon)$ and $|G(\psi)(t) - G(\psi)(s)| \le \delta(\epsilon)/2$ for all s, $t \in [\tau_1, T_{\max})$. If we now apply (iv), then we conclude that

$$\begin{aligned} |x(t) - x(s)| &\le |G(x)(t) - G(x)(s) - G(\psi)(t) + G(\psi)(s)| \\ &\quad + |G(\psi)(t) - G(\psi)(s)| \le \epsilon - \delta(\epsilon) + \delta(\epsilon)/2 = \epsilon - \delta(\epsilon)/2, \end{aligned}$$

for all s, $t \in [\tau_1, T_{\max})$. But owing to our choice of ϵ and the definition of β we have a contradiction.

The proof of Theorem 2.1 is thus complete. □

In general, we cannot claim that the solution is unique. However, if the function G satisfies a suitable Lipschitz condition, then uniqueness does follow. We formulate this fact in Corollary 2.2 below. More generality is possible, but we prefer to give a reasonably simple statement.

2.2 Corollary

Let D be an open connected subset of $\mathbf{R}^+ \times \mathbf{R}^n$ containing $(0,\xi)$, define T_∞ by (2.3), and suppose that conditions (i)–(iii) *of Theorem 2.1 are satisfied. In addition, assume the following.*

(iv) *For each compact connected subset M of D containing $(0,\xi)$, there exist constants $\eta > 0$ and $\gamma \in (0,1)$ such that the following implication holds: if for some $T \in (0, T_\infty)$, and for some functions $\phi, \psi \in C_{\xi,M}([0,T];\mathbf{R}^n)$, we have*
  (1) $\psi(s) = \phi(s)$ *for* $s \in [0, |T-\eta|_+]$,
  (2) $|\psi(s) - \phi(s)| \le \eta$ *for* $s \in [|T-\eta|_+, T]$,
  then
  (3) $\displaystyle\sup_{s\in[|T-\eta|_+,T]} |G(\psi)(s) - G(\phi)(s)| \le \gamma \sup_{s\in[|T-\eta|_+,T]} |\psi(s) - \phi(s)|.$

Then there exists a unique noncontinuable solution x of (1.6) subject to (2.1) and the (maximal) interval of existence of x is open to the right, i.e., it is of the form $[0, T_{\max})$ for some $T_{\max} > 0$. If, moreover, $\gamma \le \frac{1}{2}$, and if the implication in (iv) *is true without the assumption* (2), *then $(t, x(t))$ leaves every compact subset of D as $t \uparrow T_{\max}$.*

The first part of this result follows from the proof of Theorem 2.1 since, as we noted above, it suffices to use the conclusion (3*) in the proof of the existence of a noncontinuable solution (in (3*), take $\delta(\epsilon) = (1-\gamma)\epsilon$ and note that (1) and (2) of (iv) in Theorem 2.1 together with (3) of (iv) in Corollary 2.2 imply (3*)). If $\gamma < \frac{1}{2}$, then the proof of the fact that a noncontinuable solution leaves every compact subset of D follows directly from Theorem 2.1 (with $\delta(\epsilon) = (1-2\gamma)\epsilon$), but if $\gamma = \frac{1}{2}$, then the proof must be reworked (we leave this as an exercise). The point is that the function ψ must go to a value that has a distance less than β from all $x(t)$; cf. Exercise 3.

Before advancing further the reader should study Exercises 1 and 2.

As the following corollary points out, if G is locally equicontinuous, then (iv) of Theorem 2.1 is satisfied.

2.3 Corollary

Let D be an open connected subset of $\mathbf{R}^+ \times \mathbf{R}^n$ containing $(0, \xi)$, define T_∞ by (2.3), and suppose that conditions (i)–(iii) of Theorem 2.1 are satisfied. In addition, assume that

(iv) *for each $T \in (0, T_\infty)$ and each compact connected subset M of D containing $(0, \xi)$, the set $\{ G(\phi) \mid \phi \in C_{\xi,M}([0,T]; \mathbf{R}^n) \}$ is equicontinuous on $[0, T]$.*

Then there exists $T > 0$ such that (1.6) has a continuous solution x on $[0, T]$ satisfying the auxiliary condition (2.1).

Moreover, if x is a noncontinuable solution of (1.6) and (2.1), then the (maximal) interval of existence of x is open to the right, i.e., it is of the form $[0, T_{\max})$ for some $T_{\max} > 0$, and $(t, x(t))$ leaves every compact subset of D as $t \uparrow T_{\max}$.

The proof is left as an exercise.

By suitably modifying the proof of Theorem 2.1 one can obtain a comparison result, in which one requires the solution curve $\{ (t, x(t)) \mid t \in \mathbf{R}^+ \}$ to belong to an *a priori* given set $K \subset \mathbf{R}^+ \times \mathbf{R}^n$. In other words, we replace (2.1) by the stronger condition

$$x(0) = \xi; \quad (t, x(t)) \in D \cap K, \quad t \geq 0. \tag{2.5}$$

In contrast to D, the set K will be closed. Moreover, note that we do not allow $(t, x(t))$ to leave K even as $t \uparrow T_{\max}$. For technical reasons we shall have to assume that each cross section $K_t = \{ y \in \mathbf{R}^n \mid (t, y) \in K \}$ is convex, and that, roughly speaking, these cross sections do not shrink in a discontinuous manner as t increases. Thus, e.g., $K = J \times K_0$, where $J = [0, T]$ or $J = [0, T)$ for some $T < T_\infty$ and $K_0 \subset \mathbf{R}^n$, is excluded. The exact result is the following.

2.4 Theorem

Let the assumptions of Theorem 2.1 hold, and let K be a closed subset of $\mathbf{R}^+ \times \mathbf{R}^n$. Define the cross sections K_t of K by $K_t = \{ y \in \mathbf{R}^n \mid (t, y) \in K \}$, and suppose that for all $T \in [0, T_\infty)$, the following conditions are satisfied:

(v) *K_T is convex, and $\xi \in K_0$;*

(vi) *every $\phi \in C_{\xi, D\cap K}([0,T]; \mathbf{R}^n)$ has, for some $\tau > T$, an extension $\psi \in C_{\xi,K}([0, \tau]; \mathbf{R}^n)$;*

(vii) *G maps $C_{\xi, D\cap K}([0,T]; \mathbf{R}^n)$ into $C_{\xi,K}([0,T]; \mathbf{R}^n)$.*

Then there exists $T > 0$ such that (1.6) has a continuous solution x on $[0, T]$ satisfying $x(0) = \xi$ and $(t, x(t)) \in D \cap K$ for $t \in [0, T]$.

Moreover, if x is a solution of (1.6) on $[0,T]$, satisfying $x(0) = \xi$, and $(t, x(t)) \in D \cap K$ for $t \in [0,T]$, then x can be extended to a solution on an interval $[0, T_{\max})$, satisfying $(t, x(t)) \in D \cap K$ for $t \in [0, T_{\max})$. This interval is maximal in the sense that $(t, x(t))$ leaves every compact subset of D as $t \uparrow T_{\max}$.

Recall that Theorem 1.2 is a consequence of Theorem 2.4. (Take $K_t = \{ x \mid |x| \le y(t) \}$ for $0 \le t < S_{\max}$ and, if $T_{\max} > S_{\max}$, $K_t = \mathbf{R}^n$ for $t \in [S_{\max}, T_{\max})$.)

Observe that Theorem 2.4 does not claim that every solution of (1.6) subject to (2.1) that is noncontinuable in the sense of Theorem 2.1 satisfies $(t, x(t)) \in K$ for every $t \in (0, T_{\max})$. It only claims that every solution has a noncontinuable extension of this type. Of course, if the solution is unique, then there is a unique noncontinuable solution, and this solution satisfies the additional conclusion.

Theorem 2.4 will be used, e.g., in the proof of existence of maximal and minimal solutions; see Section **13**.4.

Proof of Theorem 2.4 The existence of $T > 0$, and of a solution $x \in C_{\xi, D\cap K}([0,T]; \mathbf{R}^n)$ of (1.6) on $[0,T]$ is proved exactly in the same way as in the proof of Theorem 2.1 (intersect $C_{\xi,M}([0,T]; \mathbf{R}^n)$ with $C_{\xi,K}([0,T]; \mathbf{R}^n)$, and observe that this intersection is nonempty if we choose T to be small enough).

Applying the Hausdorff maximality principle to functions that satisfy (1.6), (2.5) on an interval $[0,T]$, one finds that every such solution can be extended to a noncontinuable solution, that is either defined on a closed interval $[0, T_{\max}]$, and satisfies (2.5) on $[0,T]$, or is defined on a half-open interval $[0, T_{\max})$. In the former case the solution can be continued. The fact that the latter case cannot occur, and that $(t, x(t))$ must leave every compact subset of D as $t \uparrow T_{\max}$, is proved in the same way as in the proof of Theorem 2.1. Here we need the fact that since K is closed it follows that if $\phi \in C_{\xi,D}([0,T]; \mathbf{R}^n)$ satisfies $\phi(t) \in K_t$ for all $t \in [0,T)$ then $\phi(T) \in K_T$. □

In the above we required G to map the space $C_{\xi, D\cap K}([0,T]; \mathbf{R}^n)$ into $C_{\xi,K}([0,T]; \mathbf{R}^n)$. The same type of result is true if for some properly chosen set $V([0,T])$ of continuous functions on $[0,T]$, G maps $V([0,T]) \cap C_{\xi,D}([0,T]; \mathbf{R}^n)$ into $V([0,T])$.

2.5 Theorem

Let the assumptions of Theorem 2.1 hold. For each $T \in [0, T_\infty)$, let $V([0,T])$ be a subset of $C([0,T]; \mathbf{R}^n)$ with the following properties:

(v) *$V([0,T])$ is closed and convex in $C([0,T]; \mathbf{R}^n)$, and $\xi \in V([0,0])$;*

(vi) *every $\phi \in V([0,T]) \cap C_{\xi,D}([0,T]; \mathbf{R}^n)$ has an extension $\psi \in V([0,\tau])$ for some $\tau > T$;*

(vii) *if the restriction of a function $\phi \in C_{\xi,D}([0,T]; \mathbf{R}^n)$ to $[0,\tau]$ belongs to $V([0,\tau])$ for all $\tau \in [0,T)$, then $\phi \in V([0,T])$;*

(viii) *G maps $V([0,T]) \cap C_{\xi,D}([0,T];\mathbf{R}^n)$ into $V([0,T])$.*

Then there exists $T > 0$ such that (1.6) has a continuous solution $x \in V([0,T])$ on $[0,T]$ satisfying (2.1).

Moreover, if $x \in V([0,T])$ is a solution of (1.6) on $[0,T]$ satisfying (2.1), then x can be extended to a noncontinuable solution on an interval $[0,T_{\max})$ in such a way that the restriction of x to $[0,T]$ belongs to $V([0,T])$ for all $T \in [0,T_{\max})$. The interval $[0,T_{\max})$ is maximal in the sense that $(t,x(t))$ leaves every compact subset of D as $t \uparrow T_{\max}$.

The proof is the same as the proof of Theorem 2.4.

In the scalar case one may, e.g., take $V([0,T])$ to be the set of all nonincreasing functions on $[0,T]$, or the set of all convex functions on $[0,T]$ (provided G maps these classes of functions into themselves).

Our next three theorems discuss the existence of solutions of the equation

$$x(t) = f(t) + \int_0^t g\bigl(t,s,x(s)\bigr)\,\mathrm{d}s, \quad t \geq 0. \tag{1.7}$$

They are more concrete than Theorem 2.1. The first two are specialized versions of Corollaries 2.2 and 2.3. The third theorem is quite different, and is based on a compactness result in L^1.

Let J be an interval, and suppose that g maps $J^2 \times \mathbf{R}^n$ into $\mathbf{R}^n$. We say that g satisfies *the Carathéodory conditions on $J^2 \times \mathbf{R}^n$* if

> *for each $x \in \mathbf{R}^n$, the function $(t,s) \mapsto g(t,s,x)$ is measurable on J^2, and for almost all $(t,s) \in J^2$ the function $x \mapsto g(t,s,x)$ is continuous on $\mathbf{R}^n$.*

This implies that, if x is a simple function (i.e., a function with a finite range) from J into $\mathbf{R}^n$, then $(t,s) \mapsto g(t,s,x(s))$ is measurable from J^2 into $\mathbf{R}^n$. If x is an arbitrary measurable function from J into $\mathbf{R}^n$, then there exists a sequence of simple functions x_n converging a.e. to x. The Carathéodory conditions imply that $g(t,s,x_n(s))$ converges to $g(t,s,x(s))$ for almost all $(t,s) \in J^2$; hence $(t,s) \mapsto g(t,s,x(s))$ is measurable.

(From the Carathéodory conditions one has, in addition, that, if x_n is a sequence of measurable functions from J into $\mathbf{R}^n$ that tends to a function x in measure, then $(t,s) \mapsto g(t,s,x_n(s))$ tends to $(t,s) \mapsto g(t,s,x(s))$ in measure on J^2.)

As earlier, we let D be an open connected subset of $\mathbf{R}^+ \times \mathbf{R}^n$ containing $(0,f(0))$, and require that a solution $x(t)$ of (1.7) should satisfy the auxiliary condition

$$\bigl(t,x(t)\bigr) \in D, \tag{2.6}$$

for as long as it exists (observe that, by (1.7), $x(t) - f(t) \to 0$ as $t \downarrow 0$). The function g is supposed to be given on $[0,T_\infty) \times D$, where T_∞ is defined by (2.3). We say that a function g of this type satisfies the *Carathéodory conditions* on $[0, T_\infty) \times D$, if

> *for each $y \in \mathbf{R}^n$, the function $(t,s) \mapsto g(t,s,y)$ is measurable on its domain $[0,T_\infty) \times \{\, s \in [0,T_\infty) \mid (s,y) \in D \,\}$, and for almost all $(t,s) \in [0,T_\infty)^2$ the function $y \mapsto g(t,s,y)$ is continuous on its domain $\{\, y \in \mathbf{R}^n \mid (s,y) \in D \,\}$.*

(We consider these two conditions to be vacuously satisfied when the domains of the functions above are empty.)

By specializing Corollaries 2.2 and 2.3 to equation (1.7), we get the following two results.

2.6 Theorem

Let D be an open connected subset of $\mathbf{R}^+ \times \mathbf{R}^n$ containing $(0, f(0))$, define T_∞ by (2.3), and suppose that conditions (i)–(v) *below are satisfied:*

(i) *$f \in C([0,T_\infty); \mathbf{R}^n)$;*

(ii) *g maps $[0,T_\infty) \times D$ into $\mathbf{R}^n$, $g(t,s,y) = 0$ whenever $s > t$, and g satisfies the Carathéodory conditions on $[0,T_\infty) \times D$;*

(iii) *for each $T \in (0,T_\infty)$ and each compact connected subset M of D, there is a function $m \in L^1([0,T]; \mathbf{R})$ such that $|g(T,s,y)| \le m(s)$ for almost all $s \in [0,T]$ and all $y \in M_s$ (where M_s is the cross section of M at s);*

(iv) *for each $T \in (0,T_\infty)$ and each $\phi \in C_{f(0),D}([0,T]; \mathbf{R}^n)$, the function $t \mapsto \int_0^t g(t,s,\phi(s))\,\mathrm{d}s$ is continuous on $[0,T]$ (in particular, it tends to 0 as $t \downarrow 0$);*

(v) *for each $T \in (0,T_\infty)$ and each compact connected subset M of D containing $(0, f(0))$, the function*

$$k(t,s) \overset{\text{def}}{=} \sup_{\substack{x,y \in M_s \\ x \ne y}} \frac{|g(t,s,x) - g(t,s,y)|}{|x-y|},$$

satisfies

$$\lim_{\eta \downarrow 0} \int_{|t-\eta|_+}^{t} k(t,s)\,\mathrm{d}s = 0,$$

uniformly for $t \in [0,T]$.

Then there exists a unique noncontinuable solution x of (1.7) satisfying (2.6), and defined on an interval $[0,T_{\max})$. Moreover, $(t,x(t))$ leaves every compact subset of D as $t \uparrow T_{\max}$.

2.7 Theorem

Let D be an open connected subset of $\mathbf{R}^+ \times \mathbf{R}^n$ containing $(0, f(0))$, define T_∞ by (2.3), and suppose that conditions (i)–(iv) *below are satisfied:*

(i) *$f \in C([0,T_\infty); \mathbf{R}^n)$;*

(ii) *g maps $[0,T_\infty) \times D$ into $\mathbf{R}^n$, $g(t,s,y) = 0$ whenever $s > t$, and g satisfies the Carathéodory conditions on $[0,T_\infty) \times D$;*

(iii) *for each* $T \in (0, T_\infty)$ *and each compact connected subset* M *of* D, *there is a function* $m \in L^1([0,T];\mathbf{R})$ *such that* $|g(T,s,y)| \le m(s)$ *for almost all* $s \in [0,T]$ *and all* $y \in M_s$ *(where* M_s *is the cross section of* M *at* s*);*

(iv) *for each* $T \in (0, T_\infty)$ *and each compact connected subset* M *of* D *containing* $(0, f(0))$, *the set of functions*

$$\left\{ t \mapsto \int_0^t g\big(t, s, \psi(s)\big)\,\mathrm{d}s \;\middle|\; \psi \in C_{f(0),M}([0,T];\mathbf{R}^n) \right\},$$

is equicontinuous on $[0,T]$ *(in particular, they tend to* 0 *uniformly as* $t \downarrow 0$*).*

Then there exists $T > 0$ *such that (1.7) has a continuous solution* x *on* $[0,T]$ *satisfying (2.6).*

Moreover, if x *is a noncontinuable solution of (1.7) satisfying (2.6), then the (maximal) interval of existence of* x *is open to the right, i.e., it is of the form* $[0, T_{\max})$ *for some* $T_{\max} > 0$, *and* $(t, x(t))$ *leaves every compact subset of* D *as* $t \uparrow T_{\max}$.

The proofs of Theorems 2.6 and 2.7 are fairly direct applications of Corollaries 2.2 and 2.3, respectively, and they are left to the reader as exercises.

In the proof of our third existence result for (1.7) we use a completely different technique, namely an L^p-technique. To formulate the theorem we need the space of functions $L^\infty_M(J;\mathbf{R}^n)$, which is defined as follows: for every subset M of $\mathbf{R}^+ \times \mathbf{R}^n$ and every interval $J \subset \mathbf{R}^+$, we define

$$L^\infty_M(J;\mathbf{R}^n) = \big\{ \phi \in L^\infty(J;\mathbf{R}^n) \;\big|\; (t, \phi(t)) \in M \text{ for almost all } t \in J \big\}. \tag{2.7}$$

2.8 Theorem

Let D *be an open connected subset of* $\mathbf{R}^+ \times \mathbf{R}^n$ *containing* $(0, f(0))$, *define* T_∞ *by (2.3), and suppose that conditions (i)–(iv) below are satisfied:*

(i) $f \in C([0,T_\infty);\mathbf{R}^n)$;

(ii) g *maps* $[0,T_\infty) \times D$ *into* $\mathbf{R}^n$, $g(t,s,y) = 0$ *whenever* $s > t$, *and* g *satisfies the Carathéodory conditions on* $[0,T_\infty) \times D$;

(iii) *for each* $(\tau, y) \in D$ *there exist constants* $\rho > 0$ *and* $T > \tau$ *(for which the cylinder* $[\tau, T] \times \{ z \in \mathbf{R}^n \mid |z - y| \le \rho \}$ *is contained in* D*) such that the function* $(s,t) \mapsto \sup_{|z-y|\le\rho} |g(t,s,z)|$ *belongs to* $L^1([\tau,T]^2;\mathbf{R})$;

(iv) *if* $T \in (0,T_\infty)$, M *is a compact subset of* D, *and if* $\psi \in L^\infty_M([0,T];\mathbf{R}^n)$, *then for each* $t \in [0,T]$ *the function* $s \mapsto g(t,s,\psi(s))$ *is integrable on* $[0,t]$, *and the function* $t \mapsto \int_0^t g(t,s,\psi(s))\,\mathrm{d}s$ *is continuous from* $[0,T]$ *into* $\mathbf{R}^n$.

Then there exists $T > 0$ *such that (1.7) has a continuous solution* x *on* $[0,T]$ *satisfying (2.6).*

Moreover, if x *is a noncontinuable solution of (1.7) satisfying (2.6), then the (maximal) interval of existence of* x *is open to the right, i.e., it is of the form* $[0, T_{\max})$ *for some* $T_{\max} > 0$, *and* $(t, x(t))$ *leaves every compact subset of* D *as* $t \uparrow T_{\max}$.

The reader should check that Theorem 1.1 follows from this result (as we observed earlier, it follows from Theorem 2.7 as well).

We shall return to the proof of this theorem (and to the proof of Theorem 2.9 as well) at the end of Section 4, after establishing some results on L^p-solutions of (1.6).

In Section 6 we give some examples that show that one can find some (albeit rather peculiar) situations where Theorem 2.8 is applicable to (1.7), but Theorem 2.1 is not, and, conversely, examples where the assumptions of Theorem 2.1 hold, but not those of Theorem 2.8. In the latter case it is (as one would expect) condition (iv) that fails to hold. Also note that in the results above the solution can be continued until it approaches the boundary of the domain of definition (or goes to infinity). As seen from Example 6.2, this need not be the case if the assumptions are slightly weakened, although remaining quite reasonable.

Once again we have a result that says that if the right-hand side of (1.7) maps a convex set into itself, then one can find a solution that, for as long as it exists, belongs to this set. Observe that Theorem 2.8 follows from Theorem 2.9 by taking $K = \mathbf{R}^+ \times \mathbf{R}^n$.

2.9 Theorem

Let the assumptions of Theorem 2.8 hold, and let K *be a closed subset of* $\mathbf{R}^+ \times \mathbf{R}^n$. *Define the cross sections* K_t *of* K *by* $K_t = \{ y \in \mathbf{R}^n \mid (t, y) \in K \}$, *and suppose that, for all* $T \in [0, T_\infty)$, *the following three conditions are satisfied:*

(v) K_T *is convex, and* $f(0) \in K_0$;

(vi) *every* $\phi \in C_{f(0), D\cap K}([0, T]; \mathbf{R}^n)$ *has, for some* $\tau > T$, *an extension* $\psi \in C_{f(0), K}([0, \tau]; \mathbf{R}^n)$;

(vii) *if* M *is a compact subset of* D, *and if* $\psi \in L^\infty_{M\cap K}([0, T]; \mathbf{R}^n)$, *then* $f(t) + \int_0^t g(t, s, \psi(s))\, ds \in K_t$ *for* $t \in [0, T]$.

Then there exists $T > 0$ *such that (1.7) has a continuous solution* x *on* $[0, T]$ *satisfying* $(t, x(t)) \in D \cap K$ *for* $t \in [0, T]$.

Moreover, if x *is a solution of (1.7) on* $[0, T]$, *satisfying* $(t, x(t)) \in D \cap K$ *for* $t \in [0, T]$, *then* x *can be extended to a solution on an interval* $[0, T_{\max})$, *satisfying* $(t, x(t)) \in D \cap K$ *for* $t \in [0, T_{\max})$. *This interval is maximal in the sense that* $(t, x(t))$ *leaves every compact subset of* D *as* $t \uparrow T_{\max}$.

This theorem can be applied to derive the following comparison result. Consider equation (1.7), without any restriction of the type (2.6), and assume that for all $t > 0$, all $s \in [0, t]$, and all $y \in \mathbf{R}^n$, it is true that $|f(t)| \le F(t)$ and $|g(t, s, y)| \le G(t, s, |y|)$, and that for each (s, t) the function $y \mapsto G(t, s, y)$ is nondecreasing. Let y be a nonnegative, noncontinuable solution of the comparison equation

$$y(t) = F(t) + \int_0^t G\big(t, s, y(s)\big)\,\mathrm{d}s, \quad t \geq 0, \tag{2.8}$$

defined on an interval $[0, S_{\max})$. Take the set K in Theorem 2.9 to be

$$K = \{\, (t, z) \in [0, S_{\max}) \times \mathbf{R}^n \mid |z| \leq y(t) \,\} \cup \{[S_{\max}, \infty) \times \mathbf{R}^n\}.$$

Then, provided f and g satisfy the conditions of Theorem 2.8 with $D = \mathbf{R}^+ \times \mathbf{R}^n$; it follows that (1.7) will have a solution x that exists at least on $[0, S_{\max})$, and satisfies $|x(t)| \leq y(t)$. A similar argument shows that, if we let x be any solution of (1.7), then it is possible, under the same assumptions, to find at least one solution y of (2.8) satisfying $y(t) \geq |x(t)|$ on its domain.

Recall that Theorem 1.2 follows from Theorem 2.9.

In Chapter 13 we shall prove a modified version of this result which says that every solution (as opposed to some solution) $x(t)$ of (1.7) satisfies $|x(t)| \leq y_{\max}(t)$, where $y_{\max}$ is the maximal solution of (2.8) (if the solution of (2.8) is unique, then $y_{\max}$ coincides with the solution y above).

3. Functional Differential Equations

In this section we study the equation

$$\tfrac{\mathrm{d}}{\mathrm{d}t}x(t) = H(x)(t), \quad t \geq 0, \tag{3.1}$$

under the auxiliary condition

$$x(0) = \xi; \quad \big(t, x(t)\big) \in D, \quad t \geq 0. \tag{2.1}$$

As we mentioned earlier, we require a solution of (3.1) to be locally absolutely continuous, and to satisfy (3.1) almost everywhere on the interval in question. In particular, this means that the right-hand side of (3.1) need not be defined for all t; it suffices that for each locally absolutely continuous function x the function $H(x)$ is locally in L^1. For simplicity, we shall use a slightly stronger condition, and suppose that for each continuous function x that satisfies (2.1) the function $H(x)$ is locally in L^1. Of course, H must be causal, which in the L^1-case means that if $0 \leq \tau < T$, and if $\psi \in C_{\xi,D}([0,\tau]; \mathbf{R}^n)$ is the restriction of $\phi \in C_{\xi,D}([0,T]; \mathbf{R}^n)$ to $[0,\tau]$, then $H(\psi)(t) = H(\phi)(t)$ for almost all $t \in [0,\tau]$.

We shall give only one theorem for the existence of solutions of (3.1) and (2.1). This result should be sufficient for most cases encountered in practice (although it does not cover, e.g., the existence results given in Theorem **11**.4.1 and Corollary **11**.4.2).

3.1 Theorem
Let D be an open connected subset of $\mathbf{R}^+ \times \mathbf{R}^n$ containing $(0, \xi)$, define T_∞ by (2.3), and suppose that conditions (i)–(iii) below are satisfied:
(i) *for each $T \in [0, T_\infty)$, the function H maps $C_{\xi,D}([0,T];\mathbf{R}^n)$ into $L^1([0,T];\mathbf{R}^n)$, and for each fixed $t \in [0,T]$, the mapping $\phi \mapsto \int_0^t H(\phi)(s)\,ds$ is continuous from $C_{\xi,D}([0,T];\mathbf{R}^n)$ into $\mathbf{R}^n$;*
(ii) *H is causal, i.e., if $0 \le \tau < T < T_\infty$, $\phi \in C_{\xi,D}([0,T];\mathbf{R}^n)$, and if $\psi \in C_{\xi,D}([0,\tau];\mathbf{R}^n)$ is the restriction of ϕ to $[0,\tau]$, then $H(\psi)(t) = H(\phi)(t)$ for almost all $t \in [0,\tau]$;*
(iii) *for each compact connected subset M of D containing $(0,\xi)$ there exists a function $a_M \in L^1_{\text{loc}}([0,T_\infty);\mathbf{R})$ such that if $T \in (0,T_\infty)$ and $\phi \in C_{\xi,M}([0,T];\mathbf{R}^n)$ then $|H(\phi)(s)| \le a_M(s)$ for almost all s in $[0,T]$.*

Then there exists $T > 0$ such that (3.1) has a locally absolutely continuous solution x on $[0,T]$ that satisfies (2.1).

Moreover, if x is a noncontinuable solution of (3.1) satisfying (2.1), then the (maximal) interval of existence of x is open to the right, i.e., it is of the form $[0, T_{\max})$ for some $T_{\max} > 0$, and $(t, x(t))$ leaves every compact subset of D as $t \uparrow T_{\max}$.

Proof Integrate (3.1), and define $G(\phi)(t) = \xi + \int_0^t H(\phi)(s)\,ds$, to obtain an equation of the form (1.6). Clearly this definition, together with assumptions (i) and (ii), implies that assumptions (i), (ii) and (iii) of Theorem 2.1 hold.

We want to apply Corollary 2.3. To do this, we have to verify condition (iv) in this corollary. However, the required equicontinuity follows immediately from (iii), and therefore Corollary 2.3 applies.

By Corollary 2.3, the equation $x(t) = \xi + \int_0^t H(x)(s)\,ds$ has a continuous solution x on an interval $[0,T]$. The right-hand side of this equation is locally absolutely continuous. Therefore, the continuous solution x is locally absolutely continuous, and satisfies (3.1) a.e. on $[0,T]$.

The final part of the statement follows from the corresponding part of Corollary 2.3. □

We conclude this section with a simple but useful corollary.

3.2 Corollary
Let $a \in L^1_{\text{loc}}(\mathbf{R}^+ \times \mathbf{R}^+; \mathbf{R}^{n\times n})$ be of Volterra type, let $g \in \mathbf{C}(\mathbf{R}^+ \times \mathbf{R}^n; \mathbf{R}^n)$, and suppose that $f \in L^1_{\text{loc}}(\mathbf{R}^+;\mathbf{R}^n)$. Then, for every $\xi \in \mathbf{R}^n$, there exists a locally absolutely continuous solution x of

$$x'(t) = f(t) + \int_0^t a(t,s)g(s,x(s))\,ds, \quad t \ge 0; \quad x(0) = \xi,$$

on an interval $[0,T]$, where $T > 0$.

Moreover, every solution of this equation, defined on some interval $[0,T]$, can be continued to a noncontinuable solution on $[0, T_{\max})$, where $T_{\max} > T$, and, if $T_{\max} < \infty$, then $\limsup_{t\uparrow T_{\max}} |x(t)| = \infty$.

There is no direct analogue of Theorem 2.4 for equation (3.1). Instead one can sometimes use Lyapunov-type arguments to get comparable results; see Chapter 14.

4. L^p- and B^∞-Solutions

Let us proceed to the case where a solution of the functional equation

$$x(t) = G(x)(t), \quad t \geq 0, \tag{1.6}$$

or a solution of the integral equation

$$x(t) = f(t) + \int_0^t g\big(t, s, x(s)\big)\,\mathrm{d}s, \quad t \geq 0, \tag{1.7}$$

is to be found in some L^p-space, with $1 \leq p < \infty$, or in B^∞.

When one is working in an L^p-space, it is not obvious how one can impose a condition of the type (2.1) on a solution of (1.6) or (1.7). For one thing, one is not allowed to evaluate the functions at zero; hence the condition $x(0) = \xi$ must be abandoned. The requirement that $(t, x(t)) \in D$ for almost all $t \geq 0$ does makes sense, but it is difficult to deal with when D is an open subset of $\mathbf{R}^+ \times \mathbf{R}^n$.

To avoid the difficulties we use the same technique as in Theorems 2.4 and 2.9, i.e., we replace the open set D by a set K, whose cross sections K_t are closed and convex, and we require the right-hand sides of (1.6) and (1.7) to map the spaces $L^p_K([0,T];\mathbf{R}^n)$ back into themselves. Here we define $L^p_K(J;\mathbf{R}^n)$ analogously to $L^\infty_K(J;\mathbf{R}^n)$ in (2.7), i.e., for each subset K of $\mathbf{R}^+ \times \mathbf{R}^n$, each $p \in [1,\infty)$, and each interval $J \subset \mathbf{R}^+$, we define

$$L^p_K(J;\mathbf{R}^n) = \big\{\, \phi \in L^p(J;\mathbf{R}^n) \;\big|\; (t, \phi(t)) \in K \text{ for almost all } t \in J \,\big\}. \tag{4.1}$$

In the B^∞-case we define for each interval J containing zero,

$$\begin{aligned} B^\infty_{\xi,K}(J;\mathbf{R}^n) = &\big\{\phi \in B^\infty(J;\mathbf{R}^n) \;\big|\; \phi(0) = \xi \\ &\qquad \text{and } (t, \phi(t)) \in K \text{ for } t \in J\big\}, \end{aligned} \tag{4.2}$$

and require, if x is to be a solution on $[0,T]$, that $x \in B^\infty_{\xi,K}([0,T];\mathbf{R}^n)$ and that x satisfy the equation in question everywhere on $[0,T]$.

We begin with an existence result for L^p-solutions of (1.6) in the case where G is Lipschitz-continuous.

4.1 Theorem

Let $0 < T_\infty \leq \infty$, let $p \in [1,\infty)$, let $K \subset \mathbf{R}^+ \times \mathbf{R}^n$, and suppose that conditions (i)–(iv) *below are satisfied:*

(i) *for each $T \in (0, T_\infty)$, the cross section $K_T = \{\, y \in \mathbf{R}^n \mid (T, y) \in K \,\}$ is closed and convex, and $L^p_K([0,T];\mathbf{R}^n) \neq \emptyset$;*

(ii) *for all $T \in (0, T_\infty)$, G maps $L^p_K([0,T];\mathbf{R}^n)$ into itself;*

(iii) *G is causal, i.e., if $T \in (0, T_\infty)$, $\tau \in (0, T)$, and if $\psi \in L^p_K([0,\tau];\mathbf{R}^n)$ almost everywhere equals the restriction of $\phi \in L^p_K([0,T];\mathbf{R}^n)$ to $[0,\tau]$, then $G(\psi)(t) = G(\phi)(t)$ for almost all $t \in [0,\tau]$;*

(iv) *for each $\tau \in [0, T_\infty)$, and each $\zeta \in L^p_K([0,\tau];\mathbf{R}^n)$, there exist constants $\eta > 0$ and $\gamma < 1$ such that the conditions*

  (1) *$T \in (\tau, \tau+\eta] \cap [0, T_\infty)$,*

  (2) *$\phi, \psi \in \{ \varphi \in L^p_K([0,T];\mathbf{R}^n) \mid \varphi(t) = \zeta(t)$ for almost all $t \in [0,\tau]$, and $\|\varphi\|_{L^p(\tau,T)} \le \eta \}$,*

  imply

  (3) *$\|G(\psi) - G(\phi)\|_{L^p(\tau,T)} \le \gamma \|\psi - \phi\|_{L^p(\tau,T)}$.*

Then there exists $T_{\max} \in (0, T_\infty]$, such that (1.6) has a unique noncontinuable solution $x \in L^p_{\mathrm{loc}}([0,T_{\max});\mathbf{R}^n)$ satisfying $x(t) \in K_t$ for almost all $t \in [0, T_{\max})$. If $T_{\max} < T_\infty$, then this solution satisfies $\|x\|_{L^p(0,T_{\max})} = \infty$.

Observe in particular that, if for each $T < T_\infty$ there exists a constant $M(T)$ such that $\|x\|_{L^p(0,T)} \le M(T)$ whenever $x \in L^p_K([0,T];\mathbf{R}^n)$, then necessarily $T_{\max} = T_\infty$.

Proof of Theorem 4.1 We proceed as in the proof of Theorem 2.1, i.e., we solve the equation on contiguous small intervals.

Let $0 \le \tau < T_\infty$, and suppose that we have a solution $x \in L^p_K([0,\tau];\mathbf{R}^n)$ of (1.6) on $[0,\tau]$ (if $\tau = 0$, then this assumption is trivially satisfied). Let η and γ be the constants in (iv) corresponding to τ and to $\zeta = x$, let T satisfy (1) in condition (iv), choose some $\psi \in L^p_K([0,T];\mathbf{R}^n)$ satisfying $\psi(t) = x(t)$ for almost all $t \in [0,\tau]$, and define the closed, bounded, and convex subset S of $L^p([0,T];\mathbf{R}^n)$ by

$$S = \{ \phi \in L^p_K([0,T];\mathbf{R}^n) \mid \phi(t) = x(t) \text{ for almost all } t \in [0,\tau], \text{ and } \|\phi - \psi\|_{L^p(\tau,T)} \le \tfrac{1}{2}\eta \}. \tag{4.3}$$

If necessary, decrease T until

$$\|\psi\|_{L^p(\tau,T)} \le \tfrac{1}{2}\eta,$$

and

$$\|G(\psi) - \psi\|_{L^p(\tau,T)} \le \tfrac{1}{2}(1-\gamma)\eta. \tag{4.4}$$

Observe that then every $\phi \in S$ satisfies $\|\phi\|_{L^p(\tau,T)} \le \eta$.

We want to apply Theorem 1.3 to show that G has a fixed point in S.

Let us first check that we can make G map S into S. By (ii), G maps S into $L^p_K([0,T];\mathbf{R}^n)$. Let $\phi \in S$. Then, by (iii), and by the fact that x is a solution of (1.6) on $[0,\tau]$, we have $G(\phi)(s) = G(x)(s) = x(s)$ for almost all $s \in [0,\tau]$. From (iv), (4.3), and (4.4), it follows that

$$\begin{aligned}\|G(\phi) - \psi\|_{L^p(\tau,T)} &\le \|G(\phi) - G(\psi)\|_{L^p(\tau,T)} + \|G(\psi) - \psi\|_{L^p(\tau,T)} \\ &\le \gamma\|\phi - \psi\|_{L^p(\tau,T)} + \tfrac{1}{2}(1-\gamma)\eta \\ &\le \tfrac{1}{2}\eta.\end{aligned}$$

This shows that G maps S into S.

If ψ, $\phi \in S$ then, by (iv),

$$\|G(\psi) - G(\phi)\|_{L^p(\tau,T)} \leq \gamma \|\psi - \phi\|_{L^p(\tau,T)}.$$

This implies that G is continuous on S, and that $\alpha\{G(\Omega)\} \leq \gamma\alpha\{\Omega\} < \alpha\{\Omega\}$ for each subset Ω of S with $\alpha\{\Omega\} > 0$.

We have shown that Theorem 1.3 can be applied to G mapping S into S. Consequently, G has a unique fixed point in S. In other words, the solution x, which we knew to exist on $[0, \tau]$, can be extended to a solution on a larger interval $[0, T]$. As earlier, this implies that x can be extended to a solution on a maximal interval $[0, T_{\max})$. It is readily apparent that we must have either $T_{\max} = T_\infty$, or $\|x\|_{L^p(0,T_{\max})} = \infty$ (otherwise x could be further extended).

It remains to prove that the solution of (1.6) that we obtained is unique (and not only unique in S). If this were not the case, then there would exist two distinct solutions x and y. Define

$$\tau = \sup\{\, t \in \mathbf{R}^+ \mid y(s) = x(s) \text{ for almost all } s \in [0, t] \,\}.$$

Obviously both x and y must be defined on an interval $[0, T]$ with $T > \tau$ (if not, then one of them is an extension of the other). Take T close enough to τ so that both $x \in S$ and $y \in S$, where S is defined as in (4.3). Then the preceding proof shows that (1.6) has only one solution in S, and we get a contradiction. Thus we have uniqueness. □

For an application of Theorem 4.1 to the integral equation

$$x(t) = f(t) + \int_0^t a(t,s) g(x(s))\,\mathrm{d}s, \quad t \in \mathbf{R}^+,$$

see Exercise 9.

Similar, although slightly weaker, results are true for B^∞ and L^∞. For simplicity, we give only a theorem on the existence of solutions in B^∞, and leave the formulation of the analogous result for L^∞ to the reader.

As we mentioned above, in the B^∞-case we require the solution x of (1.6) to satisfy $x(0) = \xi$ and $(t, x(t)) \in K$ for $t \geq 0$, where $\xi \in \mathbf{R}^n$ and $K \subset \mathbf{R}^+ \times \mathbf{R}^n$ are given.

4.2 Theorem

Let $0 < T_\infty \leq \infty$, let $K \subset \mathbf{R}^+ \times \mathbf{R}^n$, and suppose that conditions (i)–(iv) *below are satisfied:*

(i) *for all $T \in [0, T_\infty)$, the cross section $K_T = \{\, y \in \mathbf{R}^n \mid (T, y) \in K \,\}$ is closed and convex, and $B^\infty_{\xi,K}([0,T]; \mathbf{R}^n) \neq \emptyset$;*

(ii) *for each fixed $T \in [0, T_\infty)$, G maps $B^\infty_{\xi,K}([0,T]; \mathbf{R}^n)$ into itself;*

(iii) *G is causal, i.e., if $T \in (0, T_\infty)$, $\tau \in [0, T)$, and if $\psi \in B^\infty_{\xi,K}([0,\tau]; \mathbf{R}^n)$ is the restriction of $\phi \in B^\infty_{\xi,K}([0,T]; \mathbf{R}^n)$ to $[0, \tau]$, then $G(\psi)(t) = G(\phi)(t)$ for all $t \in [0, \tau]$;*

(iv) *for each triple (τ, ζ, r) satisfying $\tau \in [0, T_\infty)$, $\zeta \in B^\infty_{\xi,K}([0,\tau]; \mathbf{R}^n)$, $r > 0$, there exist constants $\eta > 0$, $\rho > r$ and $\gamma < 1$, such that the conditions*

(1) $T \in (\tau, \tau+\eta] \cap [0, T_\infty)$,
(2) $\phi, \psi \in \{\varphi \in B^\infty_{\xi,K}([0,T];\mathbf{R}^n) \mid \varphi(t) = \zeta(t)$ *for* $t \in [0,\tau]$, *and* $\|\varphi\|_{B^\infty[\tau,T]} \le \rho\}$,

imply

(3) $\|G(\phi)\|_{B^\infty[\tau,T]} \le \rho$,

and

(4) $\|G(\psi) - G(\phi)\|_{B^\infty[\tau,T]} \le \gamma \|\psi - \phi\|_{B^\infty[\tau,T]}$.

Then there exists $T_{\max} \in (0, T_\infty]$, *such that (1.6) has a unique noncontinuable solution* $x \in B^\infty_{\text{loc}}([0,T_{\max});\mathbf{R}^n)$ *with* $x(0) = \xi$ *and* $(t, x(t)) \in K$ *for* $t \in [0, T_{\max})$. *If* $T_{\max} < T_\infty$, *then this solution satisfies* $\|x\|_{B^\infty[0,T_{\max})} = \infty$.

Proof The proof of the fact that (1.6) has a solution $x \in B^\infty([0,T];\mathbf{R}^n)$ for some $T > 0$, and that this solution has a noncontinuable extension, is a simplified version of the proof of the corresponding part of Theorem 4.1 (define

$$S = \{\phi \in B^\infty_{\xi,K}([0,T];\mathbf{R}^n) \mid \phi(t) = x(t) \text{ for } t \in [0,\tau], \quad \text{and } \|\phi\|_{B^\infty[\tau,T]} \le \rho\},$$

use r to force ρ to be large enough to guarantee that S is nonempty, and observe that (ii) and (3) in (iv) imply that G maps S into itself).

It remains to show that the solution of (1.6) that we obtained is unique. Assume the contrary, i.e., let there exist two distinct solutions x and y. Define

$$\tau = \sup\{\, t \in \mathbf{R}^+ \mid y(s) = x(s) \text{ for all } s \in [0,t] \,\}.$$

Then both x and y must be defined on an interval $[0,T]$ with $T > \tau$ (otherwise one of them is an extension of the other). Take $r = \max\{\|x\|_{B^\infty[\tau,T]}, \|y\|_{B^\infty[\tau,T]}\}$. Let η, ρ, γ, be the constants in (iv) corresponding to τ, $\zeta = x$ and r. If necessary, decrease T to get $T \in (\tau, \tau+\eta]$. From (iv) we now have

$$\|x - y\|_{B^\infty[\tau,T]} = \|G(x) - G(y)\|_{B^\infty[\tau,T]} \le \gamma \|x - y\|_{B^\infty[\tau,T]}.$$

Since $\gamma < 1$, this cannot hold unless $x(t) \equiv y(t)$. □

The following simple corollary illustrates Theorem 4.2 (take $T_\infty = \infty$, $K = \mathbf{R}^+ \times \mathbf{R}^n$).

4.3 Corollary

Let $g \in \mathbf{C}(\mathbf{R}^+ \times \mathbf{R}^n;\mathbf{R}^n)$ *be locally Lipschitzian with respect to its second argument, let* $f \in B^\infty_{\text{loc}}(\mathbf{R}^+;\mathbf{R}^n)$, *let* a *be a kernel locally of bounded type on* $\mathbf{R}^+$ *(see Definition* **9**.5.1*), and suppose that, for each* $\tau \ge 0$, $\limsup_{t\downarrow\tau} \int_\tau^t |a(t,s)|\,ds = 0$. *Then there exists a unique noncontinuable solution of the equation*

$$x(t) = f(t) + \int_0^t a(t,s) g(s, x(s))\,ds, \quad t \ge 0,$$

defined on an interval $[0, T_{\max})$. *If* $T_{\max} < T_\infty$, *then* $\|x\|_{B^\infty[0,T_{\max})} = \infty$.

We proceed to establish an existence result in the L^p-case, $1 \le p < \infty$, for the equation

$$x(t) = f(t) + \int_0^t g\big(t, s, x(s)\big)\,\mathrm{d}s, \quad t \ge 0. \tag{1.7}$$

Our result is based on a compactness argument in L^1. The use of L^1, rather than L^p, as the space in which the operator is compact, makes it possible to use minimal assumptions on the function g in (1.7).

For simplicity, in the following theorem we do not impose any restriction of the type $(t, x(t)) \in K \subset \mathbf{R}^+ \times \mathbf{R}^n$ on a solution of (1.7). In particular, this means that the function $g(t, s, y)$ must be defined for all $y \in \mathbf{R}^n$, at least if t is sufficiently small. The restriction $(t, x(t)) \in K$ is discussed in Theorem 4.7.

4.4 Theorem

Let $J = [0, T_\infty)$, where $0 < T_\infty \le \infty$, let $p \in [1, \infty)$, and suppose that conditions (i)–(v) *below are satisfied.*

(i) $f \in L^p_{\mathrm{loc}}(J; \mathbf{R}^n)$.

(ii) *g maps $J^2 \times \mathbf{R}^n$ into $\mathbf{R}^n$, $g(t, s, y) = 0$ whenever $s > t$, and g satisfies the Carathéodory conditions on $J^2 \times \mathbf{R}^n$. Moreover, for each $T \in J$ and $\phi \in L^p([0, T]; \mathbf{R}^n)$, the function $s \mapsto g(t, s, \phi(s))$ is integrable for almost all $t \in [0, T]$.*

(iii) *For each $\tau \in J$ there exist constants $0 < r < \rho$ and $T \in (\tau, T_\infty)$, such that, if $\phi \in L^p([0, T]; \mathbf{R}^n)$ and $\|\phi\|_{L^p(\tau,T)} \le \rho$, then both $t \mapsto \int_0^\tau g(t, s, \phi(s))\,\mathrm{d}s$ and $t \mapsto \int_\tau^t g(t, s, \phi(s))\,\mathrm{d}s$ belong to $L^p([\tau, T]; \mathbf{R}^n)$, and the $L^p(\tau, T)$-norm of the latter function is at most r.*

(iv) *For each $k > 0$, the function $a_k(t, s) = \sup_{|y| \le k} |g(t, s, y)|$ belongs to $L^1_{\mathrm{loc}}(J^2; \mathbf{R})$.*

(v) *For each $T \in J$, $\phi \in L^p([0, T]; \mathbf{R}^n)$, and for each measurable subset D of $[0, T]$, the function $t \mapsto \int_D g(t, s, \phi(s))\,\mathrm{d}s$ belongs to $L^1([0, T]; \mathbf{R}^n)$. Moreover, for each $\rho > 0$ and $\epsilon > 0$ there exists $\delta > 0$, such that, whenever the Lebesgue measure of the set $D \subset [0, T]$ is at most δ, and $\|\phi\|_{L^p(0,T)} \le \rho$, then $\int_0^T |\int_D g(t, s, \phi(s))\,\mathrm{d}s|\,\mathrm{d}t \le \epsilon$.*

Then there exists $T > 0$ such that (1.7) has a solution $x \in L^p([0, T]; \mathbf{R}^n)$ on $[0, T]$. Moreover, if x is a noncontinuable solution of (1.7) on the (maximal) interval of existence $[0, T_{\max})$, and if $T_{\max} < T_\infty$, then $\|x\|_{L^p(0,T_{\max})} = \infty$.

In the extreme case $p = 1$ there is some redundancy in the hypothesis of Theorem 4.4, since (v) then implies (iii).

Proof of Theorem 4.4 Let x be a solution on some interval $[0, \tau]$. (Trivially, for $\tau = 0$ we have such a solution.) Let r, ρ and T be the constants in (iii), and define S to be the ball

$$S = \big\{\, \phi \in L^p([\tau, T]; \mathbf{R}^n) \mid \|\phi\|_{L^p(\tau,T)} \le \rho \,\big\}.$$

Let G be the operator that maps $\phi \in S$ into the function

$$G(\phi)(t) = f(t) + \int_0^\tau g\big(t,s,x(s)\big)\,\mathrm{d}s + \int_\tau^t g\big(t,s,\phi(s)\big)\,\mathrm{d}s.$$

It follows from (i) and (iii) that this function belongs to $L^p([\tau,T];\mathbf{R}^n)$. In addition, from (iii) it is clear that, if we choose T so small that the $L^p(\tau,T)$-norm of the function $t \mapsto f(t) + \int_0^\tau g(t,s,x(s))\,\mathrm{d}s$ is at most $\rho - r$, then G maps S into itself.

Next we observe that, since T is finite, the set S is a bounded and convex set in $L^1([\tau,T];\mathbf{R}^n)$. It follows from Fatou's lemma that it is closed as well, because if $\phi_n \in S$ tends to ϕ in $L^1([\tau,T];\mathbf{R}^n)$ then some subsequence ϕ_{n_k} tends to ϕ a.e., and

$$\|\phi\|_{L^p(\tau,T)} \le \liminf_{k\to\infty} \|\phi_{n_k}\|_{L^p(\tau,T)}.$$

We claim that G is continuous and compact in the L^1-topology on S. The proof of this claim is based on the fact that G can be uniformly approximated on S by operators G_k defined as

$$G_k(\phi)(t) = f(t) + \int_0^\tau g\big(t,s,x(s)\big)\,\mathrm{d}s + \int_\tau^t g_k\big(t,s,\phi(s)\big)\,\mathrm{d}s,$$

where

$$g_k(t,s,y) = \begin{cases} g(t,s,y), & |y| \le k, \\ g\left(t,s,k\frac{y}{|y|}\right), & |y| > k. \end{cases}$$

To see that this uniform approximation can be done, observe the following. Take $\phi \in S$, and define D_k and ψ by

$$D_k = \{\, s \in [\tau,T] \mid |\phi(s)| > k \,\},$$

$$\psi(t) = \phi(t), \quad t \notin D_k; \qquad \psi(t) = k\frac{\phi(t)}{|\phi(t)|}, \quad t \in D_k.$$

Then the Lebesgue measure $m(D_k)$ of D_k is at most $(\rho/k)^p$ (in particular, it tends to zero as $k \to \infty$), $\psi \in S$, and

$$\begin{aligned} \|G(\phi) - G_k(\phi)\|_{L^1(\tau,T)} &= \int_\tau^T \left| \int_{D_k} \Big\{ g\big(t,s,\phi(s)\big) - g\big(t,s,\psi(s)\big) \Big\}\,\mathrm{d}s \right| \mathrm{d}t \\ &\le \int_\tau^T \left| \int_{D_k} g\big(t,s,\phi(s)\big)\,\mathrm{d}s \right| \mathrm{d}t + \int_\tau^T \left| \int_{D_k} g\big(t,s,\psi(s)\big)\,\mathrm{d}s \right| \mathrm{d}t. \end{aligned}$$

By (v), we can make the right-hand side arbitrarily small, uniformly on S, by choosing k sufficiently large.

It is an easily proved fact that uniform convergence preserves both continuity and compactness, so it suffices to show that each operator G_k is continuous and compact on S in the L^1-topology.

Let us first show that each G_k is continuous. Let $\phi_m \in S$ be a sequence that tends to ϕ in $L^1([\tau,T];\mathbf{R}^n)$ as $m \to \infty$. Then there is a subsequence ϕ_{m_j} that tends to ϕ a.e. By (iv), for each j and almost all t, the function $s \mapsto g_k(t,s,\phi_{m_j}(s))$ is dominated by the L^1-function

$s \mapsto a_k(t,s)$, so, by the Lebesgue dominated convergence theorem, for these t, $\int_\tau^t g_k(t,s,\phi_{m_j}(s))\,\mathrm{d}s \to \int_\tau^t g_k(t,s,\phi(s))\,\mathrm{d}s$. Moreover, the function $t \mapsto \left|\int_\tau^t g_k(t,s,\phi_{m_j}(s))\,\mathrm{d}s\right|$ is dominated by the L^1-function $t \mapsto \int_\tau^t a_k(t,s)\,\mathrm{d}s$. Consequently, $G_k(\phi_{m_j})$ tends to $G_k(\phi)$ in $L^1([\tau,T];\mathbf{R}^n)$. If we apply the same argument to arbitrary subsequences of the original sequence, then we conclude that each subsequence of $G_k(\phi_m)$ contains a subsequence that converges to $G_k(\phi)$ in $L^1([\tau,T];\mathbf{R}^n)$. This implies that the sequence $G_k(\phi_m)$ itself converges to $G_k(\phi)$ in $L^1([\tau,T];\mathbf{R}^n)$, and proves that G_k is continuous in the L^1-topology on S.

It remains to show that each G_k is compact. To do this, it suffices to prove that the image of S under G_k is totally bounded in the L^1-topology. However, this follows directly from Lemma 4.5 below (because the image of S under G_k is contained in the image of the ball with radius k in $L^\infty([\tau,T];\mathbf{R}^n)$ under G_k, and that image is compact).

We have demonstrated that G is a continuous compact operator in the L^1-topology, and that G maps the bounded, closed and convex set S into itself. Therefore, by Theorem 1.4, G has a fixed point in S. This means that the original solution x of (1.7) on $[0,\tau]$ can be extended to a solution on the larger interval $[0,T]$. Clearly this implies the conclusion of Theorem 4.4. □

4.5 Lemma

Let J be a bounded interval, let U be a closed ball in $\mathbf{R}^n$, let g satisfy the Carathéodory conditions on $J^2 \times U$, and suppose that the function $a(t,s) = \sup_{y\in U}|g(t,s,y)|$ belongs to $L^1(J^2;\mathbf{R})$. Then the mapping

$$\phi \mapsto \left(t \mapsto \int_J g(t,s,\phi(s))\,\mathrm{d}s\right), \tag{4.5}$$

is continuous and compact from $L^\infty(J;U)$ into $L^1(J;\mathbf{R}^n)$.

This is a special case of Krasnosel′skiĭ *et al.* [1], Theorem 19.2, p. 401. In this case the regularity assumption on the kernel made in that reference is equivalent to the assumption $a \in L^1(J^2;\mathbf{R})$ made above; cf. Krasnosel′skiĭ *et al.* [1], lines (19.6) and (19.7), pp. 401–402, or the comment immediately following the proof of Krasnosel′skiĭ *et al.* [1], Theorem 19.2.

4.6 Remark

Assume that there exists a number $q \in [1,p)$ such that $u \in L^q([0,T];\mathbf{R}^n)$ implies that $\int_0^T\left(\int_0^t|g(t,s,u(s))|\,\mathrm{d}s\right)\mathrm{d}t < \infty$. Then it follows that condition (v) of Theorem 4.4 holds. To see this, first note that if $p > q$ then a bounded set S in $L^p([0,T];\mathbf{R}^n)$ has equi-absolutely continuous norms in $L^q([0,T];\mathbf{R}^n)$, i.e., given $\epsilon > 0$ there exists $\delta > 0$ such that $m(D) < \delta$ and $x \in S$ imply $\int_D|x(s)|^q\,\mathrm{d}s < \epsilon$. Then use Krasnosel′skiĭ *et al.* [1], Theorem 18.4, p. 383. In reference to (iii) of Theorem 4.4, note that a necessary (but not sufficient) condition for the integral operator

$\phi \mapsto \left(t \mapsto \int_0^t g(t,s,\phi(s))\,\mathrm{d}s\right)$ to be compact from $L^p([0,T];\mathbf{R}^n)$ into itself is that for every constant c the set of functions

$$\left\{ \int_0^t g(t,s,\phi(s))\,\mathrm{d}s \;\Big|\; \|\phi\|_{L^p(0,T)} \le c \right\}$$

have equi-absolutely continuous norms in $L^p([0,T];\mathbf{R}^n)$.

The corresponding result where one requires the solution curve to belong to a given set is the following.

4.7 Theorem

Let the assumptions of Theorem 4.4 hold, let $K \subset \mathbf{R}^+ \times \mathbf{R}^n$, and suppose that, for all $T \in (0,T_\infty)$, the following two conditions are satisfied:

(vi) *the cross section $K_T = \{\, y \in \mathbf{R}^n \mid (T,y) \in K \,\}$ is closed and convex, and $L^p_K([0,T];\mathbf{R}^n) \ne \emptyset$;*

(vii) *the function $t \mapsto f(t) + \int_0^t g(t,s,\psi(s))\,\mathrm{d}s$ belongs to $L^p_K([0,T];\mathbf{R}^n)$ whenever $\psi \in L^p_K([0,T];\mathbf{R}^n)$.*

Then there exists $T > 0$ such that (1.7) has a solution $x \in L^p([0,T];\mathbf{R}^n)$ on $[0,T]$ satisfying $(t,x(t)) \in K$ for almost all $t \in [0,T]$.

Moreover, if x is a solution of (1.7) on $[0,T]$, satisfying $(t,x(t)) \in K$ for almost all $t \in [0,T]$, then x can be extended to a solution on a maximal interval $[0,T_{\max})$, satisfying $(t,x(t)) \in K$ for almost all $t \in [0,T_{\max})$. If $T_{\max} < T_\infty$, then $\|x\|_{L^p(0,T_{\max})} = \infty$.

The proof is essentially the same as the proof of Theorem 4.4 (intersect S with $L^p_K([0,T];\mathbf{R}^n)$).

Theorem 4.4 has extensions to L^∞ and to B^∞. The B^∞-result, given below, is simple and easy to use.

Recall that if the function $t \mapsto \int_0^t |a(t,s)|\,\mathrm{d}s$ belongs to $B^\infty_{\mathrm{loc}}(\mathbf{R}^+;\mathbf{R})$ then a is said to be of locally bounded type.

4.8 Theorem

Let $J = [0,T_\infty)$, where $0 < T_\infty \le \infty$, and suppose that conditions (i)–(iii) *below are satisfied:*

(i) *$f \in B^\infty_{\mathrm{loc}}(J;\mathbf{R}^n)$;*

(ii) *g maps $J^2 \times \mathbf{R}^n$ into $\mathbf{R}^n$, $g(t,s,y) = 0$ whenever $s > t$; for each $(t,s) \in J \times J$ the function $y \mapsto g(t,s,y)$ is continuous; and for each $y \in \mathbf{R}^n$ the function $(t,s) \mapsto g(t,s,y)$ is Borel-measurable on $J \times J$;*

(iii) *for each $k > 0$ the function $a_k(t,s) = \sup_{|y|\le k}|g(t,s,y)|$ is a kernel of locally bounded type and satisfies $\lim_{t\downarrow\tau} \int_\tau^t a_k(t,s)\,\mathrm{d}s = 0$ for every $\tau \in J$.*

Then there exists $T > 0$ such that (1.7) has a solution $x \in B^\infty([0,T];\mathbf{R}^n)$ on $[0,T]$. Moreover, if x is a noncontinuable solution of (1.7) on the (maximal) interval of existence $[0,T_{\max})$, and if $T_{\max} < T_\infty$, then $\|x\|_{B^\infty[0,T_{\max})} = \infty$.

Proof Let x be a solution on the interval $[0,\tau]$. If $\tau = 0$, then $x(0) = f(0)$ is such a solution. Define $k = \|x\|_{B^\infty[0,\tau]}$, fix some $T \in (\tau, T_\infty)$, and let

$$r \stackrel{\text{def}}{=} \left\| f(t) + \int_0^\tau a_k(t,s)\,\mathrm{d}s \right\|_{B^\infty[\tau,T]}.$$

Let $\rho = r + 1$ and, if necessary, decrease T so that $\int_\tau^t a_\rho(t,s)\,\mathrm{d}s \le 1$, for $\tau \le t \le T$. Let S be the ball $\{\, \phi \in B^\infty([\tau,T];\mathbf{R}^n) \mid \|\phi\|_{B^\infty[\tau,T]} \le \rho \,\}$, and let G be the operator that maps $\phi \in S$ into the function

$$G(\phi)(t) = f(t) + \int_0^\tau g\big(t,s,x(s)\big)\,\mathrm{d}s + \int_\tau^t g\big(t,s,\phi(s)\big)\,\mathrm{d}s, \quad \tau \le t \le T.$$

Clearly, by the choice of T, $G(\phi) \in S$.

As in the proof of Theorem 4.4, we find that S is a bounded, convex, and closed subset of $L^1([\tau,T];\mathbf{R}^n)$, and that G is continuous and compact in the L^1-topology when viewed as a map of S into S. (Of course, now the approximation by G_k is not needed.) Therefore there exists a fixed point in S, i.e., the solution x on $[0,\tau]$ can be extended to a solution on $[0,T]$. From this fact the conclusion of Theorem 4.8 follows in the standard way. □

For completeness, we state the corresponding result for L^∞-solutions.

4.9 Theorem

Let $J = [0, T_\infty)$, where $0 < T_\infty \le \infty$, and suppose that conditions (i)–(iii) *below are satisfied:*

(i) *$f \in L^\infty_{\mathrm{loc}}(J;\mathbf{R}^n)$;*

(ii) *g maps $J^2 \times \mathbf{R}^n$ into $\mathbf{R}^n$, $g(t,s,y) = 0$ whenever $s > t$, and g satisfies the Carathéodory conditions on $J^2 \times \mathbf{R}^n$;*

(iii) *for each $k > 0$ the function $a_k(t,s) = \sup_{|y|\le k}|g(t,s,y)|$ is a kernel of type L^∞_{loc} on J (i.e., the function $t \mapsto \int_0^t a_k(t,s)\,\mathrm{d}s$ belongs to $L^\infty_{\mathrm{loc}}(J;\mathbf{R})$) and satisfies $\lim_{t\downarrow\tau} \int_\tau^t a_k(t,s)\,\mathrm{d}s = 0$ for every $\tau \in J$.*

Then there exists $T > 0$ such that (1.7) has a solution $x \in L^\infty([0,T];\mathbf{R}^n)$ on $[0,T]$. Moreover, if x is a noncontinuable solution of (1.7) on the (maximal) interval of existence $[0,T_{\max})$, and if $T_{\max} < T_\infty$, then $\|x\|_{L^\infty(0,T_{\max})} = \infty$.

We leave the proof of this theorem as an exercise.

In Section 2 we stated Theorems 2.8 and 2.9, but we gave no proofs. At this point it is possible to give simple proofs based on Theorems 4.4 and 4.7.

Proof of Theorems 2.8 and 2.9 Since Theorem 2.8 is a special case of Theorem 2.9 (take $K = \mathbf{R}^+ \times \mathbf{R}^n$), it suffices to prove Theorem 2.9.

Let x be a continuous solution of (1.7) on $[0,\tau]$ satisfying $(t,x(t)) \in D \cap K$ for $t \in [0,\tau]$ (when $\tau = 0$, we may take $x(0) = f(0)$).

We want to apply Theorem 4.7 with $p = 1$. Since this requires g to map $J^2 \times \mathbf{R}^n$ (and not only $[0,T_\infty) \times D$) into $\mathbf{R}^n$, we have to introduce an auxiliary nonlinear projection operator P, and a corresponding auxiliary

Volterra equation, which is identical to the original equation on $[0, \tau]$. We show that Theorem 4.7 can be applied to this auxiliary equation; hence it has a solution x on $[0, T]$, with $T > \tau$. Finally, we observe that, if T is sufficiently close to τ, then $Px = x$, and x is therefore a solution of the original equation.

Let ρ and T be the constants in (iii) corresponding to τ and to $y = x(\tau)$, and let U be the ball

$$U = \{ z \in \mathbf{R}^n \mid |z - x(\tau)| \le \rho \}.$$

By (vi), the function x has an extension ψ, that (if necessary, decrease the value of T) belongs to $C_{f(0),D\cap K}([0,T]; \mathbf{R}^n)$, and satisfies $|\psi(s) - x(\tau)| \le \frac{\rho}{2}$ for $s \in [\tau, T]$. For $s \in [\tau, T]$, define the nonlinear projection operators $z \mapsto P(s, z)$ in $\mathbf{R}^n$ by

$$P(s,z) = \begin{cases} z, & |z - \psi(s)| \le \frac{\rho}{2}, \\ \psi(s) + \frac{\rho}{2}(z - \psi(s))/|z - \psi(s)|, & |z - \psi(s)| > \frac{\rho}{2}. \end{cases}$$

Then, for each s, $P(s, \cdot)$ maps $\mathbf{R}^n$ into U (because $|P(s,z) - x(\tau)| \le |P(s,z) - \psi(s)| + |\psi(s) - x(\tau)| \le \frac{\rho}{2} + \frac{\rho}{2} = \rho$). Moreover, since the sets K_s are convex and $\psi(s) \in K_s$, and since $|P(s,z) - \psi(s)| \le |z - \psi(s)|$, it follows that $P(s, \cdot)$ maps K_s into itself. In particular, this means that the operator that takes the function $\phi \in L^1([\tau, T]; \mathbf{R}^n)$ into the function $s \mapsto P(s, \phi(s))$, maps $L^1([\tau, T]; \mathbf{R}^n)$ into $L^\infty_U([\tau, T]; \mathbf{R}^n)$, and $L^1_K([\tau, T]; \mathbf{R}^n)$ into $L^\infty_{U\cap K}([\tau, T]; \mathbf{R}^n)$.

Consider the equation

$$v(t) = f(t) + \int_0^\tau g(t, s, x(s))\,\mathrm{d}s + \int_\tau^t g(t, s, P(s, v(s)))\,\mathrm{d}s, \quad t \ge \tau. \quad (4.6)$$

We want to apply Theorem 4.7 to show that this equation has a solution in $L^1_K([\tau, T]; \mathbf{R}^n)$. In Theorem 4.7 it is assumed that the left end-point of the interval on which we look for a solution is zero. However, a simple change of variable changes this interval to $[\tau, \tau + T_\infty)$.

Let us check that the function $g(t, s, P(s, x))$ satisfies the assumptions imposed on g in (ii)–(v) of Theorem 4.4, where we take $J = [\tau, T)$ and $p = 1$. First note that, since $p = 1$, condition (iii) of Theorem 4.4 is implied by (v) of the same theorem. From (ii) and (iv) in Theorem 2.8 we have that (ii) of Theorem 4.4 holds. For each $k > 0$, define

$$a_k(t,s) = \sup_{|z| \le k} \big| g(t, s, P(s,z)) \big| \le \sup_{z \in U} |g(t,s,z)|.$$

Therefore, by (iii) in Theorem 2.8, condition (iv) of Theorem 4.4 is satisfied. In addition, we find that

$$\int_\tau^T \int_\tau^T \sup_{z \in \mathbf{R}^n} \big| g(t, s, P(s,z)) \big| \,\mathrm{d}s\,\mathrm{d}t < \infty,$$

from which we obtain (v) of Theorem 4.4.

To verify (i) of Theorem 4.4, we remark that, from our hypotheses (i) and (iv) in Theorem 2.8, it follows that the function

$$t \mapsto f(t) + \int_0^{\tau} g(t, s, x(s))\,\mathrm{d}s + \int_{\tau}^{t} g(t, s, x(\tau))\,\mathrm{d}s$$

is continuous on $[\tau, T]$. By (iii), the latter integral belongs to $L^1([\tau, T]; \mathbf{R}^n)$. Hence the same is true for $f(t) + \int_0^{\tau} g(t, s, x(s))\,\mathrm{d}s$. In other words, all the hypotheses of Theorem 4.4, with $p = 1$, are satisfied on the interval $[\tau, T]$.

The two final assumptions of Theorem 4.7, i.e. (vi) and (vii), are satisfied as well, because (vi) in Theorem 4.7 (with $(0, T_\infty)$ replaced by (τ, T)) follows from (v) in Theorem 2.9 and from the way in which we picked T, and (vii) is a consequence of (vii) in Theorem 2.9 and of the fact that the operator that maps ϕ into the function $s \mapsto P(s, \phi(s))$ takes $L^1_K([\tau, T]; \mathbf{R}^n)$ into $L^\infty_{K \cap U}([\tau, T]; \mathbf{R}^n)$.

Applying Theorem 4.7, with $p = 1$ and the interval $[0, T_\infty)$ replaced by $[\tau, T)$, we get a solution $v \in L^1_K([\tau, T_1]; \mathbf{R}^n)$ of (4.6) on an interval $[\tau, T_1]$, for some $T_1 < T$. If we define x by $x(t) = f(t) + \int_0^{\tau} g(t, s, x(s))\,\mathrm{d}s + \int_{\tau}^{t} g(t, s, P(s, v(s)))\,\mathrm{d}s$ for $t \in (\tau, T_1]$, then, by (i) and (iv), x is continuous on $[0, T_1]$, $x = v$ a.e. on $[\tau, T_1]$, and x satisfies (4.6) on $[\tau, T_1]$. By the continuity of the functions x and ψ (ψ is the function that was used in the definition of $P(s, z)$), there exists some T_2, $\tau < T_2 \le T_1$, such that $|x(s) - \psi(s)| \le \frac{\rho}{2}$ for $s \in [\tau, T_2]$. For s in this interval, $P(s, x(s)) = x(s)$, and we have found a continuous solution to our original equation on $[0, T_2]$. The conclusion of Theorem 2.9 now follows in the standard way. (Note that in this case there are no problems with solutions that possibly do not tend to a limit at the end-point, but that do remain in a compact set. Cf. the proof of Theorem 2.1.) □

5. Discontinuous Nonlinearities

In all the existence results on the equation

$$x(t) = f(t) + \int_0^t g\big(t, s, x(s)\big)\,\mathrm{d}s, \quad t \ge 0, \tag{1.7}$$

that we have presented so far, the function $y \mapsto g(t, s, y)$ has been continuous for fixed (t, s). For the ordinary differential equation

$$\tfrac{\mathrm{d}}{\mathrm{d}t} x(t) = f\big(t, x(t)\big),$$

one usually assumes, similarly, that f is continuous in its second argument. For a substantial number of results, however, f need only be measurable in t and locally bounded (in a suitable sense) in x. It is to be expected that the techniques employed to obtain these results may be extended to cover (1.7), with g discontinuous in y. This is, in fact, possible and we consider a special case of this situation below, where we look for solutions in L^∞. In view of the fact that our proof uses compactness in L^1 (see in particular the lines preceding (5.5)), it is more natural to work in L^∞ than in B^∞.

The concept of a solution must now be enlarged, because if we take, e.g.,

$$g(t,s,y) = \begin{cases} 1, & \text{if } y < 0, \\ -1, & \text{if } y \geq 0, \end{cases} \qquad f(t) = \gamma t, \quad \gamma \in (-1,1),$$

then (1.7) has no solution in the classical sense, as can easily be seen. However, if we extend $g(t,s,y)$ to a set-valued mapping at the points of discontinuity, allowing all the values between $\liminf_{u\to y} g(t,s,u)$ and $\limsup_{u\to y} g(t,s,u)$, and ask only that $x(t) = f(t) + \int_0^t w(t,s)\,ds$, for some $w(t,s) \in g(t,s,x(s))$, then we find that $x(t) \equiv 0$, $w(t,s) = -\gamma$ is a solution of this generalized equation.

The way in which we extend g to be a set-valued mapping is very natural. Let $\overline{\text{conv}}\{K\}$ be the closed convex hull of a set K in $\mathbf{R}^n$. For each $\epsilon > 0$ let $g_\epsilon^*(t,s,y)$ be the set

$$g_\epsilon^*(t,s,y) = \overline{\text{conv}}\{g(t,s,v) \mid |v-y| \leq \epsilon\},$$

and define $g^*(t,s,y)$ by

$$g^*(t,s,y) = \bigcap_{\epsilon>0} g_\epsilon^*(t,s,y).$$

Then $g^*(t,s,y)$ is nonempty and if $g(t,s,y)$ is continuous at y then $g^*(t,s,y) = \{g(t,s,y)\}$.

5.1 Theorem

Let $J = [0,T_\infty)$, where $0 < T_\infty \leq \infty$, and suppose that conditions (i)–(iv) *below are satisfied:*

(i) $f \in L^\infty_{\text{loc}}(J;\mathbf{R}^n)$;

(ii) *g is a measurable function from $J^2 \times \mathbf{R}^n$ into $\mathbf{R}^n$, $g(t,s,y) = 0$ whenever $s > t$, and for all $(t,s) \in J^2$ the function $y \mapsto g(t,s,y)$ is locally bounded;*

(iii) *for each $k > 0$ the function $a_k(t,s) = \sup_{|y|\leq k}|g(t,s,y)|$ is a kernel of type L^∞_{loc} and satisfies $\lim_{t\downarrow\tau}\int_\tau^t a_k(t,s)\,ds = 0$ for every $\tau \in J$;*

(iv) *$\lim_{h\to 0}\int_0^T\int_0^T \sup_{|y|\leq k}|g(t+h,s,y) - g(t,s,y)|\,ds\,dt = 0$, for all $k > 0$ and $T \in J$.*

Then there exist a constant $T > 0$, a function $x \in L^\infty([0,T];\mathbf{R}^n)$, and a measurable function w, defined on $[0,T]^2$, satisfying $\int_0^t |w(t,s)|\,ds \in L^\infty([0,T];\mathbf{R})$,

$$w(t,s) \in g^*(t,s,x(s)), \tag{5.1}$$

for almost all $(t,s) \in [0,T]^2$, and such that

$$x(t) = f(t) + \int_0^t w(t,s)\,ds, \tag{5.2}$$

for almost all $t \in [0,T]$.

Moreover, every x that is a solution of (1.7) on an interval $[0,T]$, in the sense described above, can be extended to a noncontinuable solution on a (maximal) interval $[0,T_{\max})$. If $T_{\max} < T_\infty$, then $\|x\|_{L^\infty(0,T_{\max})} = \infty$.

Proof Let p be some continuous nonnegative function on $\mathbf{R}^n$ with support in the unit ball, and such that $\int_{|v|\le 1} p(v)\,\mathrm{d}v = 1$. Define

$$\begin{aligned} p_m(v) &= m^n p(mv), \\ g_m(t,s,y) &= \int_{\mathbf{R}^n} p_m(y-v)g(t,s,v)\,\mathrm{d}v, \quad m \ge 1. \end{aligned} \tag{5.3}$$

It follows from (ii) that g_m is well defined. In addition, one has that $g_m(t,s,y) \in g^*_{1/m}(t,s,y)$.

Our intention is to apply Theorem 4.9 to the equations

$$x_m(t) = f(t) + \int_0^t g_m\big(t,s,x_m(s)\big)\,\mathrm{d}s, \quad t \ge 0, \tag{5.4}$$

in order to obtain the existence of a solution x_m for each m. We then let $m \to \infty$, and show that $x_m \to x$, where x is a solution of (1.7) in the enlarged sense.

It is straightforward to check that assumptions (ii) and (iii), together with definition (5.3), imply that for every m the function g_m satisfies the assumptions imposed on g in Theorem 4.9. Thus, for each m, we have a solution $x_m \in L^\infty([0,T_m];\mathbf{R}^n)$ of (5.4) for some $T_m > 0$. Furthermore (we leave this as an exercise), we can choose T_m so that $\inf_{m\ge1} T_m > 0$, and so that $\sup_{m\ge1}\|x_m\|_{L^\infty(0,T_m)} < \infty$. Define $T = \inf_{m\ge1} T_m$ and $r = \sup_{m\ge1}\|x_m\|_{L^\infty(0,T_m)}$, and let S be the set

$$S = \{\,\phi \in L^\infty([0,T];\mathbf{R}^n) \mid \|\phi\|_{L^\infty(0,T)} \le r\,\}.$$

It follows from (iii) that the set of functions

$$V = \left\{ t \mapsto \int_0^t g_m\big(t,s,\phi(s)\big)\,\mathrm{d}s \;\middle|\; m \ge 1, \quad \phi \in S \right\},$$

is a bounded subset of $L^\infty([0,T];\mathbf{R}^n)$. We claim that V is a compact subset of $L^1([0,T];\mathbf{R}^n)$. That this is true follows from (iv), because (iv) implies

$$\lim_{h\to 0}\int_0^T |\psi(t+h) - \psi(t)|\,\mathrm{d}t = 0,$$

uniformly for $\psi \in V$, and this is the Riesz criterion for total boundedness in $L^1([0,T];\mathbf{R}^n)$ (cf. Dunford and Schwartz [1], Theorem IV.8.20, p. 298).

Each function in the set

$$W = \{\,(s,t) \mapsto g_m(t,s,\phi(s)) \mid m \ge 1, \quad \phi \in S\,\}$$

is dominated by the L^1-function a_{r+1}. Consequently (cf. Dunford and Schwartz [1], Corollary IV.8.11, p. 294), W is weakly sequentially compact in $L^1([0,T]^2;\mathbf{R}^n)$.

Owing to the compactness of the sets V and W, we can find a subsequence m_j, and functions $x \in L^1([0,T];\mathbf{R}^n)$ and $w \in L^1([0,T]^2;\mathbf{R}^n)$, such that $x_{m_j} \to x$ in $L^1([0,T];\mathbf{R}^n)$ and $g_{m_j}(t,s,x_{m_j}(s)) \to w(t,s)$ weakly in $L^1([0,T]^2;\mathbf{R}^n)$ as $j \to \infty$. By passing to a subsequence we can assume, without loss of generality, that $x_{m_j}(t) \to x(t)$ for almost all $t \in [0,T]$. The function w must be dominated by a_{r+1}, and, therefore, the function $t \mapsto \int_0^t w(t,s)\,\mathrm{d}s$ is essentially bounded on $[0,T]$. Integrating (5.4) over $[0,\tau]$, where $\tau \in [0,T]$, we get, for all τ,

$$\begin{aligned}\int_0^\tau x(t)\,\mathrm{d}t &= \lim_{j\to\infty}\int_0^\tau x_{m_j}(t)\,\mathrm{d}t \\ &= \lim_{j\to\infty}\int_0^\tau \left\{ f(t) + \int_0^t g_{m_j}(t,s,x_{m_j}(s))\,\mathrm{d}s \right\} \mathrm{d}t \\ &= \int_0^\tau \left\{ f(t) + \int_0^t w(t,s)\,\mathrm{d}s \right\} \mathrm{d}t. \end{aligned} \tag{5.5}$$

This implies that (5.2) holds a.e. on $[0,T]$.

It remains to check (5.1). Since $x_{m_j}(s) \to x(s)$ for almost all s, it follows that

$$\epsilon_j(s) \stackrel{\text{def}}{=} \sup_{i\geq j}|x_{m_i}(s) - x(s)| \to 0,$$

a.e. on $[0,T]$, as $j \to \infty$. Now

$$g_{m_i}(t,s,x_{m_i}(s)) \in g^*_{1/m_i}(t,s,x_{m_i}(s)),$$

and, for $i \geq j$,

$$g^*_{1/m_i}(t,s,x_{m_i}(s)) \subset g^*_{\epsilon_j(s)+1/m_j}(t,s,x(s)).$$

Therefore, if we define M_j by

$$M_j = \Big\{ w \in L^1([0,T]^2;\mathbf{R}^n) \,\Big|\, w(t,s) \in g^*_{\epsilon_j(s)+1/m_j}(t,s,x(s)) \\ \text{for almost all } (t,s) \in [0,T]^2 \Big\},$$

then $g_{m_i}(t,s,x_{m_i}(s)) \in M_j$, for all $i \geq j$. The sets M_j are closed and convex in $L^1([0,T]^2;\mathbf{R}^n)$, so they are weakly closed. Therefore, the limit function w must belong to M_j for all $j \geq 1$. Taking the intersection of all the sets M_j, we get (5.1).

To continue the solution we write the equation in the form

$$x(t) = f(t) + \int_0^T w(t,s)\,\mathrm{d}s + \int_T^t g(t,s,x(s))\,\mathrm{d}s.$$

A translation reveals that we have essentially the same equation as before. Consequently, we may extend the solution as long as it remains bounded, and we obtain the desired result. □

A uniqueness result that complements Theorem 5.1 is given in Theorem **13**.5.7.

6. Examples

Next, we study some examples. The first shows that the assumptions of Theorem 2.8 do not imply those of Theorem 2.1.

6.1 Example

Take $n = 1$, let $D = \mathbf{R}^+ \times \mathbf{R}$, (then $T_\infty = \infty$), and define

$$g(t,s,y) = \begin{cases} \sin\big(s\pi/(t-1)\big)y^{1/3}, & \text{if } \ 0 \le s \le 2-t,\ 1 < t \le 2,\ y \in \mathbf{R}, \\ 0, & \text{otherwise.} \end{cases}$$

It is straightforward to check that g satisfies assumptions (ii) and (iii) of Theorem 2.8. The integrability requirement of (iv) is also easy to verify. To conclude that the continuity requirement of (iv) holds, we write $\phi(s) = \psi(s)^{1/3}$, and observe that, for $1 < t < 2$,

$$\begin{aligned} 2\int_0^t g\big(t,s,\psi(s)\big)\,\mathrm{d}s &= \int_0^{2-t} \sin\Big(\frac{s\pi}{t-1}\Big)\phi(s)\,\mathrm{d}s \\ &\quad - \int_0^{2-t} \sin\Big(\frac{s\pi}{t-1}+\pi\Big)\phi(s)\,\mathrm{d}s \\ &= \int_0^1 \sin\Big(\frac{s\pi}{t-1}\Big)\Big\{\phi(s) - \phi\big(s-(t-1)\big)\Big\}\,\mathrm{d}s \\ &\quad - \int_{2-t}^1 \sin\Big(\frac{s\pi}{t-1}\Big)\phi(s)\,\mathrm{d}s \\ &\quad + \int_0^{t-1} \sin\Big(\frac{s\pi}{t-1}\Big)\phi\big(s-(t-1)\big)\,\mathrm{d}s. \end{aligned}$$

From this relation it is easy to see that, since $\psi \in L^\infty_{\mathrm{loc}}(\mathbf{R}^+;\mathbf{R})$, it follows that $\int_0^t g(t,s,\psi(s))\,\mathrm{d}s$ is continuous at $t = 1$. Elsewhere there are no problems. Thus g, as defined above, satisfies conditions (ii), (iii) and (iv) of Theorem 2.8.

Define

$$G(\phi)(t) = \int_0^t g(t,s,\phi(s))\,\mathrm{d}s, \quad \phi \in C(\mathbf{R}^+;\mathbf{R}).$$

We claim that condition (iv) of Theorem 2.1 does not hold. Suppose it does, and take $M = [0,2] \times [-1,1]$, $\xi = 0$. Let, respectively, $\eta > 0$ and δ be the constant and the function given by (iv). Choose two positive integers p and q, such that $p > q$ and such that $\frac{1}{q} \le \frac{\eta}{2}$. Let $\epsilon \in (0,1)$ satisfy $\epsilon^{\frac{1}{3}}(\frac{1}{q} - \frac{1}{p})\frac{1}{2} > \epsilon$. Define

$$\phi_{m,\epsilon}(s) = \chi_{[1-q^{-1},1-p^{-1}]}\sin^3\Big(\frac{s\pi}{\tau_m}\Big)\epsilon,$$

where $\tau_m = (mpq)^{-1}$, and where the positive integer m is taken to be so large that $\tau_m <\min(\delta(\epsilon), \frac{1}{2}\eta)$. Then $\phi_{m,\epsilon} \in C_{\xi,M}([0,2];\mathbf{R})$ and obviously

$|\phi_{m,\epsilon}(s)| \le \epsilon$, for $s \in [0,2]$. Straightforward calculations show that

$$\begin{aligned} G(\phi_{m,\epsilon})(1+\tau_m) &= \int_0^{1+\tau_m} g(1+\tau_m, s, \phi_{m,\epsilon}(s))\,ds \\ &= \epsilon^{\frac{1}{3}} \int_{1-q^{-1}}^{1-p^{-1}} \sin^2(spmq\pi)\,ds \\ &= \epsilon^{\frac{1}{3}} \frac{1}{mpq} \int_{pqm-pm}^{pqm-qm} \sin^2(u\pi)\,du \\ &= \epsilon^{\frac{1}{3}} \frac{p-q}{pq} \int_0^1 \sin^2(u\pi)\,du = \epsilon^{\frac{1}{3}} \Big[\frac{1}{q} - \frac{1}{p}\Big]\frac{1}{2} > \epsilon. \end{aligned}$$

Now let $T = 1 - \frac{1}{q} + \eta \ge 1 + \frac{\eta}{2}$, $\phi = \phi_{m,\epsilon}$, $\psi \equiv 0$, and $s = 1$. Then, by (iv) (recall that $g(t,s,y) \equiv 0$ if $t \le 1$),

$$G(\phi_{m,\epsilon})(t) \le \epsilon - \delta(\epsilon), \quad 1 \le t \le \min\Big\{1 + \delta(\epsilon), 1 + \frac{\eta}{2}\Big\}.$$

But this conclusion is contradicted by the calculation above, and by the choice of τ_m. Therefore (iv) of Theorem 2.1 cannot hold. □

The second example shows that one cannot have $\gamma > \frac{1}{2}$ in (iv)(3) of Corollary 2.2 and still get the final conclusion of this corollary. In addition, it furnishes an example of a solution that cannot be continued, although the solution does stay in a compact subset of the domain of definition.

6.2 Example

Define the function a by

$$a(t) = \begin{cases} 0, & t \in [0, \frac{1}{2}], \\ 2 - \frac{1}{t}, & t \ge \frac{1}{2}, \end{cases}$$

and the function y by

$$y(t) = \begin{pmatrix} \cos(\frac{2\pi}{1-t}) \\ \sin(\frac{2\pi}{1-t}) \end{pmatrix}, \quad t \in [0,1).$$

Then $|y| = 1$, $t \in [0,1)$. Take $\xi = y(0)$. Let $\gamma \in (\frac{1}{2}, 1)$ be arbitrary, and define the set D to be

$$D = \mathbf{R}^+ \times \Big\{ v \in \mathbf{R}^2 \;\Big|\; \big||v| - 1\big| < 2 - \frac{1}{\gamma} \Big\}.$$

Note that D is not convex. For $T \ge 0$, and $\phi \in C_{\xi,D}([0,T];\mathbf{R}^2)$, we let $G(\phi)$ be defined by

$$G(\phi)(t) = \begin{cases} y(t)\big|1 - \gamma \sup_{s \in [a(t),t]} |\phi(s) - y(s)|\big|_+, & t \in [0,1) \cap [0,T], \\ 0, & t \in [1,\infty) \cap [0,T]. \end{cases}$$

It is obvious that $y \in C_{\xi,D}([0,1);\mathbf{R}^2)$ is a noncontinuable solution on $[0,1)$ of the equation $\phi = G(\phi)$, and, furthermore, that the solution curve remains in a compact set. Of course, $\lim_{t\uparrow 1} y(t)$ does not exist. To realize that, in fact, no solution $\phi \in C_{\xi,D}([0,1);\mathbf{R}^2)$ of $\phi = G(\phi)$ can have a limit as $t \uparrow 1$, one argues as follows. For $t \ge \frac{1}{2}$ we have $\frac{2\pi}{1-a(t)} = \frac{2\pi}{1-t} - 2\pi$ and so

$\{ y(s) \mid a(t) \le s \le t \} = \{ (\cos(\tau) \quad \sin(\tau))^T \mid 0 \le \tau \le 2\pi \}$. Therefore, if $\lim_{t\uparrow 1} \phi(t)$ exists, then (recall that ϕ takes values in D, that $|y| = 1$, and that $\gamma > \frac{1}{2}$) $\limsup_{t\uparrow 1} |\phi(t) - y(t)| \ge \gamma^{-1}$, and so $G(\phi)(t) \to 0$, as $t \uparrow 1$. By the definition of D, $\lim_{t\uparrow 1} \phi(t) \ne 0$.

To grasp why Theorem 2.1 cannot be applied, note first that conditions (i)–(iii) do hold. However, concerning (iv) we observe that if ϕ and $\psi \in C_{\xi,D}([0,1);\mathbf{R}^2)$, and if $|\phi(s) - \psi(s)| \le \epsilon$ for $s \in [0,t]$ and some $t \in (0,1)$, then in general (take, e.g., ψ constant with $|\psi| = 1$ and $\phi = (1-\epsilon)\psi$) near t) the best we can hope for is

$$\left| \sup_{s\in[a(t),t]} |\phi(s) - y(s)| - \sup_{s\in[a(t),t]} |\psi(s) - y(s)| \right| \le \epsilon.$$

But then we can only expect to get

$$|G(\psi)(s) - G(\phi)(s) - G(\psi)(t) + G(\phi)(t)| \le 2\gamma\epsilon,$$

and for the implication in (iv) of Theorem 2.1 we must have $\gamma < \frac{1}{2}$. A similar observation rules out an application of the last part of Corollary 2.2.

It is not difficult to construct an example where the assumptions of Theorem 2.1 are satisfied, but not those of Theorem 2.8.

6.3 Example
Let

$$G(\phi)(t) = \int_0^t g(t,s,\phi(s))\,\mathrm{d}s = \int_0^t k(t,s)\phi(s)\,\mathrm{d}s,$$

with $k(t,s) = \frac{1}{4}t^{-1}$, for $0 < s \le t$, and $k(t,s) = 0$ otherwise. Then the assumptions of Theorem 2.8 are not satisfied, because the function $G(\phi)$ is not continuous at zero if $\lim_{t\downarrow 0} \phi(t)$ is nonzero, but clearly the assumptions of Theorem 2.1 are satisfied if $\xi = 0$.

7. Exercises

1. Consider the real, scalar equation

$$\begin{aligned} x(t) &= f(t, x(t), x(t-1)), & t &\ge 0; \\ x(t) &= \phi(t), & -1 &\le t < 0. \end{aligned}$$

where f maps $\mathbf{R}^+ \times \mathbf{R} \times \mathbf{R}$ into $\mathbf{R}$, and where the initial function $\phi \in C([-1,0];\mathbf{R})$ is given. Formulate a statement, based on Theorem 2.1

and on Corollary 2.2, concerning existence and uniqueness of solutions of this equation.

2. Extend the results of Exercise 1 to the equation
$$x(t) = f\Big(t, x(t), \int_{[0,1]} \mu(\mathrm{d}s)x(t-s)\Big), \qquad t \ge 0;$$
$$x(t) = \phi(t), \qquad -1 \le t < 0,$$
where $\mu \in M([0,1];\mathbf{R})$.
3. Let Ω be a bounded set in $\mathbf{R}^n$ containing at least two points, and let $\rho > 0$. Show that there exists a point $y \in \mathbf{R}^n$, such that $\mathrm{dist}(y,\Omega) < \rho$, and such that $\sup_{w\in\Omega}|w-y| < \mathrm{diam}\{\Omega\}$. (This can be used in Exercises 4 and 5.)
4. Prove Corollaries 2.2 and 2.3. Hint: To show that $(t,x(t))$ leaves every compact subset of D as $t \uparrow T_{\max}$, argue as follows. If $(t,x(t))$ stays in some compact set K one defines β in the same way as in the proof of Theorem 2.1 and notes that $\beta > 0$. To get a contradiction, it suffices to construct a function $\psi \in C_{\xi,K}([0,T_{\max}];\mathbf{R}^n)$ such that $\psi(s) = x(s)$ on $[0,\tau]$ and $|\psi(s)-x(s)| \le \beta-\epsilon$ on $[\tau,T_{\max})$, where $\epsilon > 0$ and $\tau \ge \max\{0, T_{\max}-\eta\}$. To find a suitable value for $\psi(T_{\max})$, apply Exercise 3 to the set $\{\, w \in \mathbf{R}^n \mid x(t_j) \to w \text{ for some } t_j \uparrow T_{\max}\,\}$.
5. (Hard) Prove Theorem 2.1 in the case where assumption (iv) has been replaced by the following condition:
   (iv) *for each compact connected subset M of D containing $(0,\xi)$ there exist a constant $\eta > 0$, and a function $\delta : (0,\infty) \to (0,\infty)$, such that the following implication holds: if for some $\epsilon > 0$, $T \in (0,T_\infty)$, $t \in [|T-\eta|_+, T]$, and for some functions $\phi,\psi \in C_{\xi,M}([0,T];\mathbf{R}^n)$, we have*
   (1) $\psi(s) = \phi(s)$ *for* $s \in [0,|T-\eta|_+]$,
   (2) $|\psi(s)-\phi(s)| \le \epsilon$ *for* $s \in [|T-\eta|_+,T]$,
   (3) $|G(\psi)(t) - G(\phi)(t)| \le \delta(\epsilon)$,
   then
   (4) $|G(\psi)(s) - G(\phi)(s)| \le \frac{\epsilon}{2}$ *for* $s \in [t, t+\delta(\epsilon)] \cap [0,T]$.
6. Give an example where the assumptions of Theorem 2.4 are satisfied, but where there exists a solution of (1.6) on $[0,T]$ that does not belong to $C_{\xi,K}([0,T];\mathbf{R}^n)$.
7. Show that Theorems 2.6 and 2.7 follow from Corollaries 2.2 and 2.3, respectively.
8. Consider the scalar functional differential equation
$$x'(t) = f\Big(t, x(t), \int_{[0,1]} \mu(\mathrm{d}s)x(t-s)\Big), \qquad t \ge 0;$$
$$x(t) = \phi(t), \qquad -1 \le t \le 0.$$
where f maps $\mathbf{R}^+ \times \mathbf{R} \times \mathbf{R}$ into $\mathbf{R}$, $\mu \in M([0,1];\mathbf{R})$, and where the initial function $\phi \in C([-1,0];\mathbf{R})$ is given. Formulate a statement, based

on Theorem 3.1, concerning existence of solutions of this equation.

9. Show that the following result is a consequence of Theorem 4.1. Let $p^{-1}+q^{-1}=1$, with $1<q\le p<\infty$, and assume that $a\in L^2_{\mathrm{loc}}(\mathbf{R}^+\times\mathbf{R}^+;\mathbf{R}^{n\times n})$ vanishes for $s>t$, and satisfies
$$\lim_{T\downarrow\tau}\int_\tau^T\left\{\int_\tau^t|a(t,s)|^p\,\mathrm{d}s\right\}^{\frac{2}{p}}\,\mathrm{d}t=0,$$
for every $\tau\ge 0$. Suppose that $g\in C(\mathbf{R}^n;\mathbf{R}^n)$ satisfies the inequality $|g(x)-g(y)|^q\le K(|x-y|^q+|x-y|^2)$ for $x,y\in\mathbf{R}^n$ and some constant K. Finally, let $f\in L^2_{\mathrm{loc}}(\mathbf{R}^+;\mathbf{R}^n)$. Then the equation
$$x(t)=f(t)+\int_0^t a(t,s)g(x(s))\,\mathrm{d}s,\quad t\ge 0,$$
has a unique solution $x\in L^2_{\mathrm{loc}}([0,T_{\mathrm{max}});\mathbf{R}^n)$ on the maximal interval $[0,T_{\mathrm{max}})$.
Is the result true with $q=1$, $p=\infty$?
10. Consider the equation of Exercise 9. Let $f\in L^p_{\mathrm{loc}}(\mathbf{R}^+;\mathbf{R}^n)$, $p\in[1,\infty)$, $a\in L^\infty_{\mathrm{loc}}(\mathbf{R}^+\times\mathbf{R}^+;\mathbf{R}^{n\times n})$ and $g\in C(\mathbf{R}^n;\mathbf{R}^n)$. What growth conditions have to be imposed on g for Theorem 4.4 to apply?
11. Prove Theorem 4.9.
12. Formulate and prove versions of Theorems 4.8 and 4.9, analogous to Theorem 4.7, where $(t,x(t))$ is required to belong to a preassigned set K.
13. Show that in the proof of Theorem 5.1 we have $\inf_{m\ge1}T_m>0$ and $\sup_{m\ge1}\|x_m\|_{L^\infty(0,T_m)}<\infty$.
14. Let Ω be a bounded subset of $C([0,T];\mathbf{R}^n)$. Can you establish any relations between $\alpha\{\Omega\}$ (the measure of noncompactness of Ω) and
$$\limsup_{\substack{|t-s|\to0\\ s,t\in[0,1]}}\sup_{\phi\in\Omega}|\phi(t)-\phi(s)|?$$

8. Comments

Section 2:

For early work on existence of continuous solutions of (1.3) and (1.7), see Miller and Sell [1] and [3] and Miller [10], Ch. II, Theorem 1.1. For a review of existence results and references to earlier work, see Miller [10], Ch. II, Bibliography, and [16].

The operator G in equation (1.6) is sometimes called an *abstract Volterra operator* whenever it is causal.

The results of Miller [10], Ch. II, Theorem 1.1, were improved in Artstein [1], Sec. 2, p. 447. His result is our Theorem 2.7 (with $D=\mathbf{R}^+\times\mathbf{R}^n$).

In the case of the ordinary differential equation $x'(t) = g(t, x(t))$, one knows that, if x is a noncontinuable solution, then $(t, x(t))$ leaves every compact subset of the domain of g. Furthermore, one can prove that $(t, x(t))$ remains outside each compact subset once t is sufficiently close to $T_{\max}$. In general, this is not true for integral equations. Sufficient conditions for this to happen are given in Herdman [1], for the case where the domain of g is $\mathbf{R}^+ \times \mathbf{R}^n$.

A number of rather complicated conditions implying the existence of solutions of (1.7) can be found in Chen [1]. The proofs are based on Schauder's fixed point theorem.

Some simple comparison results for (1.7) are presented in Miller [10], Ch. II, Section 6. Some of these results are contained in Theorem 2.9; others are related to results discussed in Section **13**.4. In Theorems 2.4 and 2.9 the solution is required to remain in a closed set. However, in some cases it may be more interesting to know whether the solution remains in an open set, for example to know whether the solution is strictly positive for all time.

For existence results on nonlinear Volterra integral equations using the Perron–Stieltjes integral, see Schwabik [1].

Section 3:

Some basic existence results concerning functional differential equations of the form

$$x'(t) = f(t, x^t),$$

where $x^t(\theta) = x(t+\theta)$, are presented in Hale [5], Ch. 2, Secs. 2.2–2.3. It is assumed that f is continuous on an open subset of $\mathbf{R} \times C([-r, 0]; \mathbf{R}^n)$. This is a different setting from the one considered in Theorem 3.1. Note that our approach makes no use of delay-equation arguments; hence the question of whether the delay is finite or not plays no role whatsoever. Usually this is an important issue in connection with delay equations. The case where f satisfies only the Carathéodory conditions, is considered in Hale [5], Ch. 2, Sec. 2.6.

Section 4:

In Kikuchi and Nakagiri [1] and [2] compactness in L^1 is used to obtain existence results for (1.3). This is the same idea as is used in Theorem 2.8. This theorem is analogous to their result.

Section 5:

For an extensive treatment of differential equations with discontinuous right-hand sides, see Filippov [1].

Results on scalar Volterra equations of type (1.1) with discontinuous nonlinearities can be found in Maeda [1], and in Kiffe [3]. (In addition, the latter contains some extensions to the nonconvolution case.) In Kiffe [5], the equation

$$x(t) + \int_0^t a(t,s)g(x(s))\,\mathrm{d}s = f(t), \quad t \geq 0,$$

with discontinuous nonlinearity g is considered in $\mathbf{R}^n$. Our assumptions are somewhat weaker than those in Kiffe [5], and our proof is slightly different. In particular, note that we approach the function $w(t,s)$ weakly in L^1, and not weak* in L^∞.

References

Z. Artstein

1. Continuous dependence of solutions of Volterra integral equations, *SIAM J. Math. Anal.* **6** (1975), pp. 446–456.

H.-Y. Chen

1. Solutions for certain nonlinear Volterra integral equations, *J. Math. Anal. Appl.* **69** (1979), pp. 475–488.

K. Deimling

2. *Nonlinear Functional Analysis*, Springer-Verlag, Berlin, 1985.

N. Dunford and J. T. Schwartz

1. *Linear Operators, Part I: General Theory*, J. Wiley, New York, 1957.

A. F. Filippov

1. *Differential Equations with Discontinuous Righthand Sides*, Kluwer Academic Publishers, Dordrecht, 1988.

J. K. Hale

5. *Theory of Functional Differential Equations*, Springer-Verlag, Berlin, 1977.

T. L. Herdman

1. Behavior of maximally defined solutions of a nonlinear Volterra equation, *Proc. Amer. Math. Soc.* **67** (1977), pp. 297–302.

T. Kiffe

3. A discontinuous Volterra integral equation, *J. Integral Equations* **1** (1979), pp. 193–200.
5. Systems of Volterra equations with nonmonotone discontinuous nonlinearities, *J. Integral Equations* **5** (1983), pp. 341–352.

N. Kikuchi and S. Nakagiri

1. An existence theorem of solutions of non-linear integral equations, *Funkcial. Ekvac.* **15** (1972), pp. 131–138.
2. Kneser's property of solutions of non-linear integral equations, *Funkcial. Ekvac.* **17** (1974), pp. 57–66.

M. A. Krasnosel′skiĭ, P. P. Zabreĭko, E. I. Pustyl′nik, and P. E. Sobolevskiĭ (Krasnosel′skiĭ *et al.*)

1. *Integral Operators in Spaces of Summable Functions*, Noordhoff, Leyden, 1976.

H. Maeda

1. Stability considerations for a Volterra integral equation with discontinuous nonlinearity, *SIAM J. Control* **11** (1973), pp. 202–214.

R. H. Martin

1. *Nonlinear Operators and Differential Equations in Banach Spaces*, J. Wiley, New York, 1976.

R. K. Miller

10. *Nonlinear Volterra Integral Equations*, W. A. Benjamin, Menlo Park, Calif., 1971.
16. Some fundamental theory of Volterra integral equations, in *International Conference on Differential Equations, Los Angeles 1974*, H. A. Antosiewicz, ed., Academic Press, New York, 1975, pp. 568–579.

R. K. Miller and G. R. Sell

1. Existence, uniqueness and continuity of solutions of integral equations, *Ann. Mat. Pura Appl. (IV)* **80** (1968), pp. 135–152.
3. Existence, uniqueness and continuity of solutions of integral equations. An addendum, *Ann. Mat. Pura Appl. (IV)* **87** (1970), pp. 281–286.

Š. Schwabik

1. Generalized Volterra integral equations, *Czechoslovak Math. J.* **32** (1982), pp. 245–270.

13

Continuous Dependence, Differentiability, and Uniqueness

We discuss some results on the continuous dependence of solutions upon data, differentiability with respect to a parameter, and uniqueness of solutions. In addition, some comparison results are given.

1. Introduction

In the previous chapter we proved a number of results on the existence of solutions of the functional equation

$$x(t) = G(x)(t), \quad t \geq 0, \tag{1.1}$$

and of the integral equation

$$x(t) = f(t) + \int_0^t g\big(t, s, x(s)\big)\,\mathrm{d}s, \quad t \geq 0. \tag{1.2}$$

Moreover, we considered briefly the functional differential equation

$$\tfrac{\mathrm{d}}{\mathrm{d}t}x(t) = H(x)(t), \quad t \geq 0. \tag{1.3}$$

Throughout, the solutions of (1.1) and (1.3) were required to satisfy the auxiliary condition

$$x(0) = \xi; \quad (t, x(t)) \in D, \quad t \geq 0, \tag{1.4}$$

and the solutions of (1.2) were required to satisfy

$$(t, x(t)) \in D, \quad t \geq 0, \tag{1.5}$$

where D was a given subset of $\mathbf{R}^+ \times \mathbf{R}^n$.

In this chapter we give results concerning the continuous dependence of solutions on parameters and their differentiability with respect to a parameter. In addition, we consider the existence of maximal and minimal solutions, and give some comparison results. Finally, the uniqueness question is investigated in more detail. For simplicity, we concentrate on (1.1), and leave part of the task of applying our results to (1.2) and (1.3) to the reader.

The chapter begins with a discussion of continuous dependence on data in Section 2. It is quite clear that one can give several different results on this subject for (1.1)–(1.4), depending on the setting in which the existence of solutions is considered. If one assumes that the solution is unique, then the proofs become slightly simpler. However, even without this uniqueness assumption (which will be discussed further in Section 5), it is possible, without undue trouble, to prove a certain type of continuous dependence. Let us formulate one such result.

Consider the equation

$$x(t) = f(t) + \int_0^t a(t,s)h\bigl(s,x(s)\bigr)\,\mathrm{d}s, \quad t \ge 0, \tag{1.6}$$

which was studied in Section **12**.1, and its parameter-dependent version

$$x(\lambda,t) = f(\lambda,t) + \int_0^t a(\lambda,t,s)h\bigl(\lambda,s,x(\lambda,s)\bigr)\,\mathrm{d}s, \quad t \ge 0. \qquad (1.6)_\lambda$$

Without uniqueness of solutions, the best one may hope for is that, if λ_j, $f(\lambda_j,\cdot)$, $a(\lambda_j,\cdot,\cdot)$, and $h(\lambda_j,\cdot,\cdot)$ tend to λ, $f(\lambda,\cdot)$, $a(\lambda,\cdot,\cdot)$, and $h(\lambda,\cdot,\cdot)$ in an appropriate sense, then there is a subsequence of solutions $x(\lambda_j,\cdot)$ of $(1.6)_{\lambda_j}$ that converges to a solution $x(\lambda,\cdot)$ of $(1.6)_\lambda$. This is, in fact, what the next theorem claims. We assume that for each fixed $\lambda \in \mathbf{R}$ the functions f, a, and h satisfy the conditions of Theorem **12**.1.1, and that these functions depend continuously on λ. As a corollary to Theorem 2.2, we then have the following result.

1.1 Theorem

Let $0 < T_\infty \le \infty$, let $f \in C(\mathbf{R}\times[0,T_\infty);\mathbf{R}^n)$, and let $h \in C(\mathbf{R}\times[0,T_\infty)\times\mathbf{R}^n;\mathbf{R}^n)$. Assume that, for each $\lambda \in \mathbf{R}$, the function $a(\lambda,\cdot,\cdot)$ is a Volterra kernel of continuous type on $[0,T_\infty)$, and that a is continuous with respect to λ in the sense that, for each $\lambda \in \mathbf{R}$ and $T \in [0,T_\infty)$,

$$\sup_{t\in[0,T]} \int_0^t \bigl|a(\mu,t,s) - a(\lambda,t,s)\bigr|\,\mathrm{d}s \to 0 \quad \text{as } \mu \to \lambda.$$

Then the following claims are true.

(i) *If $\lambda_j \to \lambda$, and if, for each j, $x(\lambda_j,\cdot)$ is a noncontinuable solution of $(1.6)_{\lambda_j}$ (by Theorem **12**.1.1, such a solution exists), then there is a subsequence λ_{j_k} such that the corresponding solutions $x(\lambda_{j_k},\cdot)$ converge to a noncontinuable solution $x(\lambda,\cdot)$ of $(1.6)_\lambda$, uniformly on compact subsets of the maximal interval of existence $[0,T_{\max}(\lambda))$ of $x(\lambda,\cdot)$. In particular, $T_{\max}(\lambda) \le \liminf_{\lambda_{j_k}\to\lambda} T_{\max}(\lambda_{j_k})$.*

(ii) *If, in addition, for each $\lambda \in \mathbf{R}$, the solution $x(\lambda,\cdot)$ of $(1.6)_\lambda$ is unique, then $x(\lambda,\cdot)$ depends continuously on λ, that is, $x(\lambda,t)$ is continuous* *on the set* $\{ (\lambda, t) \mid \lambda \in \mathbf{R}, t \in [0, T_{\max}(\lambda)) \}$.

The task of showing that Theorem 1.1 follows from Theorem 2.2 is left to the reader (see Exercise 1).

In Section 3 we go one step further, and discuss differentiability of solutions with respect to a parameter. Again, let us formulate a simple version of the main result, i.e., of Theorem 3.1. A formal differentiation of $(1.6)_\lambda$ leads to the equation (subscripts λ and x denote partial derivatives)

$$\begin{aligned} x_\lambda(\lambda, t) = f_\lambda(\lambda, t) + \int_0^t \Big\{ a_\lambda(\lambda, t, s) h\big(\lambda, s, x(\lambda, s)\big) \\ + a(\lambda, t, s) h_\lambda\big(\lambda, s, x(\lambda, s)\big) \Big\} \, \mathrm{d}s \\ + \int_0^t a(\lambda, t, s) h_x\big(\lambda, s, x(\lambda, s)\big) x_\lambda(\lambda, s) \, \mathrm{d}s, \quad t \geq 0. \end{aligned} \tag{1.7}$$

As a corollary to Theorem 3.1 we have the following result.

1.2 Theorem
Let $0 < T_\infty \leq \infty$, *let* $f \in C(\mathbf{R} \times [0, T); \mathbf{R}^n)$ *be continuously differentiable with respect to its first variable, and let* $h \in C(\mathbf{R} \times [0, T) \times \mathbf{R}^n; \mathbf{R}^n)$ *be continuously differentiable with respect to its first and third variables. Furthermore, assume that, for each* $\lambda \in \mathbf{R}$, *the function* $a(\lambda, \cdot, \cdot)$ *is a Volterra kernel of continuous type on* $[0, T)$ *that is differentiable with respect to* λ *in the sense that there exists a Volterra kernel* $a_\lambda(\lambda, \cdot, \cdot)$ *of continuous type on* $[0, T_\infty)$ *satisfying, for each* $T \in [0, T_\infty)$,

$$\sup_{t \in [0, T]} \int_0^t \big| a(\lambda + \epsilon, t, s) - a(\lambda, t, s) - \epsilon a_\lambda(\lambda, t, s) \big| \, \mathrm{d}s = o(\epsilon) \text{ as } \epsilon \to 0.$$

Then, for each $\lambda \in \mathbf{R}$, *there is a unique solution* $x(\lambda, \cdot)$ *of equation* $(1.6)_\lambda$, *defined on the maximal interval of existence* $[0, T_{\max}(\lambda))$. *Moreover,* $x(\lambda, t)$ *is continuously differentiable with respect to* λ *on the set* $\{ (\lambda, t) \mid \lambda \in \mathbf{R}, t \in [0, T_{\max}(\lambda)) \}$, *and the derivative* x_λ *satisfies (1.7) on* $[0, T_{\max}(\lambda))$.

As one might expect, it does in fact suffice that the assumptions hold for all λ in some open set.

In Section 4, we give some results on the existence of maximal and minimal solutions, and some comparison results. These results are based on, and they extend, the comparison Theorem **12**.2.4.

Finally, in Section 5, we discuss uniqueness of solutions of (1.2).

2. Continuous Dependence

We proceed to the study of continuous dependence on λ of solutions $x(\lambda, \cdot)$

of the equation

$$x(\lambda, t) = G\big(\lambda, x(\lambda, \cdot)\big)(t), \quad t \geq 0, \tag{1.1$_\lambda$}$$

subject to the auxiliary condition

$$x(\lambda, 0) = \xi(\lambda); \quad \big(t, x(\lambda, t)\big) \in D(\lambda), \quad t \geq 0. \tag{1.4$_\lambda$}$$

The hypothesis that we use for each fixed λ is the same as in the existence Theorem **12**.2.1. For simplicity, we assume that the parameter λ is non-negative, and that (1.1) corresponds to the case $\lambda = 0$ in $(1.1)_\lambda$. Recall that T_∞ is defined by

$$T_\infty = \sup\{\, T \in \mathbf{R}^+ \mid C_{\xi,D}([0,T); \mathbf{R}^n) \neq \emptyset \,\}. \tag{2.1}$$

Note that T_∞ depends on ξ and D. Occasionally we need the corresponding λ-dependent quantity

$$T_\infty(\lambda) = \sup\{\, T \in \mathbf{R}^+ \mid C_{\xi(\lambda),D(\lambda)}([0,T); \mathbf{R}^n) \neq \emptyset \,\}. \tag{2.2}$$

2.1 Theorem

Let the hypothesis of Theorem **12**.*2.1 hold, both in its original form, and with G, ξ and D replaced, respectively, by $G(\lambda, \cdot)$, $\xi(\lambda)$ and $D(\lambda)$, with $\lambda > 0$. As $\lambda \downarrow 0$, let $G(\lambda, \cdot)$, $\xi(\lambda)$ and $D(\lambda)$ tend to G, ξ and D in the following sense:*

(v) *for each compact connected set $M \subset D$ there is a $\mu > 0$ such that $M \subset D(\lambda)$ for $\lambda \in (0, \mu)$;*

(vi) $\lim_{\lambda \downarrow 0} \xi(\lambda) = \xi$;

(vii) *if $T \in [0, T_\infty)$, if $\lambda_j \downarrow 0$, and if $\{\phi_j\} \subset C_{\xi(\lambda_j),D(\lambda_j)}([0,T]; \mathbf{R}^n)$ is a sequence that tends to a function $\phi \in C_{\xi,D}([0,T]; \mathbf{R}^n)$ uniformly on $[0,T]$, then $G(\lambda_j, \phi_j)(t) \to G(\phi)(t)$ uniformly on $[0,T]$.*

In addition, suppose that condition (iv) *in Theorem* **12**.*2.1 holds uniformly for all sufficiently small λ in the following sense.*

$(\mathrm{iv})_\lambda$ *For each compact connected subset M of D containing $(0, \xi)$ there exist constants $\mu > 0$, $\eta > 0$, and a continuous function $\delta : (0, \infty) \to (0, \infty)$, such that the following implication holds: if for some $\epsilon > 0$, $T \in (0, T_\infty)$, $\lambda \in (0, \mu)$, and for some functions ϕ and $\psi \in C_{\xi(\lambda),M}([0,T]; \mathbf{R}^n)$, we have*

(1) $\psi(s) = \phi(s)$ *for* $s \in [0, |T - \eta|_+]$,

(2) $|\psi(s) - \phi(s)| \leq \epsilon$ *for* $s \in [|T - \eta|_+, T]$,

then

(3) $|G(\lambda, \psi)(s) - G(\lambda, \phi)(s) - G(\lambda, \psi)(t) + G(\lambda, \phi)(t)| \leq \epsilon - \delta(\epsilon)$ *for all s, $t \in [|T - \eta|_+, T]$ satisfying $|t - s| \leq \delta(\epsilon)$.*

For each $\lambda > 0$, let $x(\lambda, \cdot)$ be a noncontinuable continuous solution of $(1.1)_\lambda$ and $(1.4)_\lambda$, and denote the corresponding maximal interval of existence by $[0, T_{\max}(\lambda))$ (by Theorem **12**.*2.1, such a family of solutions exists). Then the following claims are true.*

(a) *Every sequence $\lambda_j \downarrow 0$ contains a subsequence λ_{j_k} such that the corresponding solutions $x(\lambda_{j_k}, \cdot)$ converge to a noncontinuable solution x of (1.1) and (1.4), uniformly on compact subsets of the* *maximal interval of existence* $[0, T_{\max})$ *of* x. *In particular,* $T_{\max} \le \liminf_{\lambda_{j_k} \downarrow 0} T_{\max}(\lambda_{j_k})$.

(b) *If, in addition, the solution of (1.1) and (1.4) is unique, then* $x(\lambda, \cdot)$ *tends to the noncontinuable solution* x *of (1.1) and (1.4) as* $\lambda \downarrow 0$, *uniformly on compact subsets of the maximal interval of existence* $[0, T_{\max})$ *of* x. *In particular,* $T_{\max} \le \liminf_{\lambda \downarrow 0} T_{\max}(\lambda)$.

There is some redundancy in the hypothesis, since $(\text{iv})_\lambda$, together with (v)–(vii), implies that (iv) in Theorem **12**.2.1 holds, both in its original form, and with G, ξ and D replaced by $G(\lambda, \cdot)$, $\xi(\lambda)$ and $D(\lambda)$ (at least for small λ). Also observe that $(\text{iv})_\lambda$ only imposes a uniformity condition on the functions $G(\lambda, \cdot)$; no intrinsically new assumptions are made.

Note that in the nonunique case we do not claim that all solutions of (1.1) and (1.4) are obtained as limits of solutions of $(1.1)_\lambda$ and $(1.4)_\lambda$. We claim only that there is at least one solution of this type.

Although λ is taken as a continuous parameter, Theorem 2.1 requires continuity with respect to λ only at $\lambda = 0$. This means that one can immediately read off the corresponding result, where the continuous parameter $\lambda \in \mathbf{R}^+$ has been replaced by a sequence $\lambda_k \downarrow 0$. By using this result for sequences, one can easily generalize the theorem to the case where λ belongs to an arbitrary metric space.

Proof of Theorem 2.1 Let us prove the following two auxiliary claims (once these claims have been established, the remainder of the proof is very easy).

If $\lambda_j \downarrow 0$, *then there exists a* $T > 0$ *such that some subsequence of* $x(\lambda_j, \cdot)$ *converges uniformly on* $[0, T]$ *to a solution* x *of (1.1), subject to (1.4).* (2.3)

If $\tau \in (0, T_\infty)$, *if* $\lambda_j \downarrow 0$, *and if* $x(\lambda_j, \cdot)$ *converges uniformly on compact subsets of* $[0, \tau)$ *to a function* $x \in C_{\xi,D}([0, \tau]; \mathbf{R}^n)$, *then* x *is a solution of (1.1) on* $[0, \tau]$. *Moreover, there exists a* $T > \tau$ *such that* x *can be continued to a solution on* $[0, T]$, *and such that some subsequence of* $x(\lambda_j, \cdot)$ *converges uniformly to* x *on* $[0, T]$. (2.4)

First we observe that, because of (vii) and $(1.1)_\lambda$, the limit function x in (2.3) or in (2.4) must be a solution of (1.1) on $[0, T]$. Thus, as regards (2.4), it is the proof of the second part that necessitates the analysis to follow.

The proofs of (2.3) and of the second part of (2.4) are very similar to each other. The only difference is that certain constants and functions are defined in slightly different ways. We prove both claims concurrently.

In the proof of (2.4), take some $T > \tau$, and define

$$\psi(s) = \begin{cases} x(s), & s \in [0, \tau], \\ x(\tau), & s \in [\tau, T]. \end{cases} \tag{2.5}$$

By choosing T sufficiently close to τ we can arrange things so that $(\psi(s), s) \in D$ for $s \in [\tau, T]$. Then $\psi \in C_{\xi,D}([0,T]; \mathbf{R}^n)$. Next choose some $\tilde{\rho} > 0$ such that the compact set

$$K = \{\, (s, y) \mid s \in [0, T], \text{ and } |y - \psi(s)| \le \tilde{\rho} \,\}$$

is a subset of D. Let μ and η be the constants, and let δ be the function in (iv)$_\lambda$ corresponding to K. Choose some numbers ρ and ϵ such that $0 < \rho < \tilde{\rho}$ and

$$\rho < \epsilon < \min\{2\rho, \rho + (\delta(\epsilon)/4)\}, \tag{2.6}$$

and, if necessary, decrease T (but keep it greater than τ) until

$$\sup_{t\in[\tau,T]} \bigl|G(\psi)(t) - \psi(t)\bigr| < \delta(\epsilon)/4, \tag{2.7}$$

and

$$T - \tau < \min\{\delta(\epsilon), \eta\}. \tag{2.8}$$

If it happens to be true that $x(\lambda_j, \cdot)$ converges to x uniformly on $[0, \tau]$, then we define

$$M = \{\, (s, y) \mid s \in [0, T], \quad |y - \psi(s)| \le \rho \,\}, \tag{2.9}$$

and the setup for the proof of (2.4) is complete. If not, then we are forced to adjust τ, i.e., to decrease τ (by a very small amount) to get uniform convergence on $[0, \tau]$. Clearly, this forces the function ψ to change, since $\psi(s) = x(\tau)$ for $s \in [\tau, T]$, and so τ, hence $x(\tau)$, change. However, by continuity (see (i)), if the change is small enough, then neither (2.7) nor (2.8) is violated, and the (modified) set M in (2.9) is a subset of the original set K. (In the sequel we shall no longer need the set K, and the strict inequalities in (2.7) and (2.8).)

In the proof of (2.3), take $\tau = 0$, and proceed as above (with no final adjustment of τ).

As a result of the preceding construction, in both cases, we may assume that ψ is defined as in (2.5), that $x(\lambda_j, \cdot)$ converges uniformly to x on $[0, \tau]$, that (2.6)–(2.8) hold, that μ, η and δ are related to each other as in (iv)$_\lambda$ (since $M \subset K$), and that the constant T that we have chosen is strictly larger than the originally given τ.

We assert that (in both (2.3) and (2.4)) there is a number $m \ge 1$ such that for $j \ge m$ the functions $x(\lambda_j, \cdot)$ are defined on $[0, T]$, and satisfy $(s, x(\lambda_j, s)) \in M$ for $s \in [0, T]$.

To prove this assertion, let us first observe that, because of the uniform convergence, we have $(s, x(\lambda_j, s)) \in M$ for $s \in [0, \tau]$ and for all sufficiently large j.

Next, we note that, in order to prove the assertion, it suffices to show that, for j large enough,

$$(s, x(\lambda_j, s)) \in M \text{ for } s \in [\tau, T] \cap [0, T_{\max}(\lambda_j)). \tag{2.10}$$

To see that this suffices, one argues as follows. For j large enough, the point $(\tau, x(\lambda_j, \tau))$ belongs to M, and M is a compact subset of $D(\lambda_j)$. Thus, since

$x(\lambda_j, \cdot)$ is noncontinuable, the point $(s, x(\lambda_j, s))$ must eventually leave M. Condition (2.10) prevents $(s, x(\lambda_j, s))$ from leaving M in any other way than by s becoming larger than T. Clearly, this implies that $T_{\max}(\lambda_j) > T$.

Let us prove (2.10). Define the auxiliary functions ψ_j by

$$\psi_j(s) = \begin{cases} x(\lambda_j, s), & \text{for } 0 \le s \le \tau, \\ x(\lambda_j, \tau), & \text{for } \tau < s \le T. \end{cases}$$

Then $\psi_j \to \psi$ uniformly on $[0, T]$. It follows that if we choose m large enough, and $j \ge m$, then (by (vii))

$$|\psi_j(s) - \psi(s)| \le \epsilon - \rho \text{ and } |G(\lambda_j, \psi_j)(s) - G(\psi)(s)| \le \delta(\epsilon)/2 \text{ for } s \in [\tau, T]. \tag{2.11}$$

(In particular, $(s, \psi_j(s)) \in M$ for $s \in [\tau, T]$.) Fix some $j \ge m$, and define

$$t_0 = \sup\{\, s \in [\tau, T] \cap [\tau, T_{\max}(\lambda_j)) \mid |x(\lambda_j, s) - \psi(s)| < \rho \,\}. \tag{2.12}$$

Then, by the argument given immediately after (2.10), $t_0 < T_{\max}(\lambda_j)$. Moreover, $(s, x(\lambda_j, s)) \in M$ and $|x(\lambda_j, s) - \psi_j(s)| \le \epsilon$ for $s \in [0, t_0]$. Thus, by $(1.1)_{\lambda_j}$, (2.8), (2.11), and $(\text{iv})_\lambda$ (in (3), take $s = t_0$ and $t = \tau$),

$$\begin{aligned} \big|x(\lambda_j, t_0) - G(\psi)(t_0)\big| &= \big|G(\lambda_j, x(\lambda_j, \cdot))(t_0) - G(\psi)(t_0)\big| \\ &\le \big|G(\lambda_j, x(\lambda_j, \cdot))(t_0) - G(\lambda_j, \psi_j)(t_0)\big| + \delta(\epsilon)/2 \\ &\le \epsilon - \delta(\epsilon) + \delta(\epsilon)/2 \\ &= \epsilon - \delta(\epsilon)/2. \end{aligned}$$

Since, by (2.7), $|G(\psi)(t_0) - \psi(t_0)| \le \delta(\epsilon)/4$ we get by (2.6), $|x(\lambda_j, t_0) - \psi(t_0)| < \rho$. However, if $t_0 < T$, then, by (2.12), $|x(\lambda_j, t_0) - \psi(t_0)| = \rho$. We conclude that $t_0 = T$, and that (2.10) holds.

To repeat, we have shown that for $j \ge m$, where m is some sufficiently large number, the functions $x(\lambda_j, \cdot)$ are defined on $[0, T]$, and satisfy $(s, x(\lambda_j, s)) \in M$ for $s \in [\tau, T]$.

Next, we want to show that the set $S \overset{\text{def}}{=} \{x(\lambda_j, \cdot)\}_{j=m}^{\infty}$ is totally bounded in $C([\tau, T])$ (it is clearly totally bounded in $C([0, \tau])$). According to the Arzelà–Ascoli Theorem **2**.7.5, it suffices to show that the number β defined by

$$\beta = \limsup_{\substack{|t-s|\to 0 \\ s,t \in [\tau, T]}} \Big(\sup_{j \ge m} |x(\lambda_j, t) - x(\lambda_j, s)| \Big)$$

is zero. Assume that this is not the case, i.e., $\beta > 0$. It is clear that $\beta \le 2\rho$ $(= \operatorname{diam}\{\, y \mid (s, y) \in M \,\})$.

The number ϵ chosen to satisfy (2.6) is no longer needed and hence we can choose a new ϵ such that

$$\beta < \epsilon \le \beta + (\delta(\epsilon)/4). \tag{2.13}$$

By the definition of β there exists a number $\gamma > 0$ such that

$$\sup_{j \ge m} |x(\lambda_j, t) - x(\lambda_j, s)| < \frac{\epsilon + \beta}{2} \quad \text{for all } s,\ t \text{ in } [\tau, T] \text{ with } |t - s| \le \gamma. \tag{2.14}$$

Divide $[\tau, T]$ into subintervals $[t_{i-1}, t_i]$, $1 \le i \le N$, each of length at most γ. The set $L \stackrel{\text{def}}{=} \{ y \in \mathbf{R}^n \mid |y - x(\tau)| \le \rho \}$ is compact in $\mathbf{R}^n$; hence we can find a finite number p of balls B_j with radius $\frac{\epsilon-\beta}{2}$, whose centres belong to L, and whose union covers L. Without loss of generality (increase m, if necessary), suppose that $|x(\lambda_j, \tau) - x(\tau)| \le \frac{\epsilon-\beta}{2}$ for $j \ge m$. Let us split S into a total of p^N sets $S_{k_1,k_2,\dots,k_N}$ (some of which may be empty), defined by

$$S_{k_1,k_2,\dots,k_N} = \big\{ x(\lambda_j, \cdot) \in S \bigm| x(\lambda_j, t_i) \in B_{k_i}, \quad 1 \le i \le N \big\},$$

where $1 \le k_i \le N$. In the sequel, let us concentrate on one of these sets, calling it S_*, and let us restrict ourselves to the set of indices j for which $x(\lambda_j, \cdot) \in S_*$. Let ϕ be the function that satisfies $\phi(s) = x(s)$ for $s \in [0, \tau]$, is linear in each interval $[t_{i-1}, t_i]$, and whose value at each point t_i, $1 \le i \le N$, is equal to the centre of the corresponding ball B_{k_i}. Define

$$\phi_j(s) = \begin{cases} x(\lambda_j, s), & s \in [0, \tau], \\ \frac{t_1 - s}{t_1 - \tau} x(\lambda_j, \tau) + \frac{s - \tau}{t_1 - \tau} \phi(t_1), & s \in (\tau, t_1], \\ \phi(s), & s \in (t_1, T]. \end{cases}$$

Clearly $(s, \phi_j(s)) \in M$ for $s \in [0, T]$, and $\phi_j \to \phi$ in $C_{\xi,M}([0,T]; \mathbf{R}^n)$ as $j \to \infty$. It follows from (vii) that the functions $\{G(\lambda_j, \phi_j)\}$ are equicontinuous. Thus we conclude that, if s, $t \in [\tau, T]$ are such that $|t - s|$ is sufficiently small, then

$$|G(\lambda_j, \phi_j)(t) - G(\lambda_j, \phi_j)(s)| \le \delta(\epsilon)/2. \tag{2.15}$$

We claim that

$$|x(\lambda_j, s) - \phi_j(s)| \le \epsilon, \quad s \in [\tau, T]. \tag{2.16}$$

To see that this is true, observe that, by (2.14), for all s in each interval $[t_{i-1}, t_i]$, one has $|x(\lambda_j, s) - x(\lambda_j, t_{i-1})| \le \frac{\epsilon+\beta}{2}$, and that we have constructed ϕ in such a way that $|x(\lambda_j, t_{i-1}) - \phi(t_{i-1})| \le \frac{\epsilon-\beta}{2}$. Thus, $|x(\lambda_j, s) - \phi(t_{i-1})| \le \epsilon$ for $s \in [t_{i-1}, t_i]$. By the same argument, $|x(\lambda_j, s) - \phi(t_i)| \le \epsilon$ for $s \in [t_{i-1}, t_i]$. This, together with the triangle inequality and the fact that ϕ is linear in $[t_{i-1}, t_i]$, implies (2.16).

First apply (iv)$_\lambda$ with ψ replaced by $x(\lambda_j, \cdot)$ and ϕ replaced by ϕ_j, and then use (2.15) and (2.13) to get

$$\begin{aligned} |x(\lambda_j, t) - x(\lambda_j, s)| &= \big| G\big(\lambda_j, x(\lambda_j, \cdot)\big)(t) - G\big(\lambda_j, x(\lambda_j, \cdot)\big)(s) \big| \\ &\le \big| G\big(\lambda_j, \phi_j(t)\big) - G\big(\lambda_j, \phi_j(s)\big) \big| + \epsilon - \delta(\epsilon) \\ &\le \epsilon - \delta(\epsilon)/2 \le \beta - \delta(\epsilon)/4, \end{aligned}$$

if $|t - s|$ is sufficiently small. Above we have considered only one set S_*, but the same argument can be repeated for each set $S_{k_1,k_2,\dots,k_N}$, with the same conclusion. This contradicts the definition of β and completes the proof of the total boundedness.

We conclude that the claims (2.3) and (2.4) are true.

To complete the proof of claim (a), we must show, first, that one can construct a subsequence of the solutions $\{x(\lambda_j, \cdot)\}$, such that, for some number $T_{\max}$, this subsequence converges uniformly on every compact subset of

$[0, T_{\max})$ to a solution $x_{\max}$ of (1.1), and, secondly, that either $T_{\max} = T_\infty$, or $x \notin C_{\xi,D}([0, T_{\max}]; \mathbf{R}^n)$ (this guarantees that x is noncontinuable). By (2.4), to accomplish the second task, it suffices either to construct the sequence λ_j in such a way that $T_{\max} = T_\infty$, or to show that no subsequence of $\{x(\lambda_j, \cdot)\}$ converges uniformly on $[0, T_{\max}]$. (If $x \in C_{\xi,D}([0, T_{\max}]; \mathbf{R}^n)$, then, by (2.4), some subsequence would converge uniformly on $[0, T_{\max}]$.) We proceed as follows. First, choose $T_1 > 0$ such that some subsequence $x(\lambda_{1,j}, \cdot)$ of $x(\lambda_j, \cdot)$ converges uniformly on $[0, T_1]$ (by (2.3), this is possible). Let $m \geq 1$, and suppose that $T_m > 0$ has been chosen in such a way that some subsequence $x(\lambda_{m,j}, \cdot)$ of $x(\lambda_j, \cdot)$ converges uniformly on $[0, T_m]$ (to a solution of (1.1)). Choose T_{m+1} such that

$$T_{m+1} - T_m = \min\{1, \tfrac{1}{2} \sup T - \tfrac{1}{2} T_m\},$$

where the supremum is taken over all those $T \in (0, T_\infty)$ for which some subsequence of $x(\lambda_{m,j}, \cdot)$ converges uniformly on $[0, T]$. Then pick some subsequence $\lambda_{m+1,j}$ of $\lambda_{m,j}$ for which $x(\lambda_{m+1,j}, \cdot)$ converges uniformly on $[0, T_{m+1}]$. Define $T_{\max} = \lim_{m\to\infty} T_m$. Clearly, the diagonal sequence $x(\lambda_{j,j}, \cdot)$ converges uniformly on compact subsets of $[0, T_{\max})$, and, if $T_{\max} < T_\infty$, it cannot converge uniformly on $[0, T_{\max}]$. This completes the proof of claim (a).

That (b) holds follows directly from (a). □

As a corollary to Theorem 2.1 we get the following continuous dependence result for the parametrized version

$$x(\lambda, t) = f(\lambda, t) + \int_0^t g(\lambda, t, s, x(\lambda, s))\, \mathrm{d}s, \quad t \geq 0, \qquad (1.2)_\lambda$$

of (1.2), with the auxiliary condition

$$\big(t, x(\lambda, t)\big) \in D(\lambda), \quad t \geq 0. \qquad (1.5)_\lambda$$

2.2 Theorem

Let the hypothesis of Theorem **12**.*2.7 hold, both in its original form, and with f, g and D replaced, respectively, by $f(\lambda, \cdot)$, $g(\lambda, \cdot, \cdot, \cdot)$ and $D(\lambda)$, with $\lambda > 0$. As $\lambda \downarrow 0$, let $f(\lambda, \cdot)$, $g(\lambda, \cdot, \cdot, \cdot)$ and $D(\lambda)$ tend to f, g and D in the following sense:*

(v) *for each compact connected set $M \subset D$ there is a $\mu > 0$ such that $M \subset D(\lambda)$ for $\lambda \in (0, \mu)$;*

(vi) *$f(\lambda, t) \to f(t)$ as $\lambda \downarrow 0$, uniformly for t in compact subintervals of $[0, T_\infty)$;*

(vii) *if $T \in (0, T_\infty)$, if $\lambda_j \downarrow 0$, and if $\{\phi_j\} \subset C_{f(\lambda_j,0),D(\lambda_j)}([0, T]; \mathbf{R}^n)$ is a sequence that tends to a function $\phi \in C_{f(0),D}([0, T]; \mathbf{R}^n)$ uniformly on $[0, T]$, then $\int_0^T g(\lambda_j, T, s, \phi_j(s))\, \mathrm{d}s \to \int_0^T g(T, s, \phi(s))\, \mathrm{d}s$.*

In addition, suppose that

(iv)$_\lambda$ *for each $T \in (0, T_\infty)$ and each compact connected subset M of D, there exists a constant $\mu > 0$ such that the set of functions*

$$\left\{ t \mapsto \int_0^t g\big(\lambda, t, s, \psi(s)\big)\,\mathrm{d}s \;\middle|\; \psi \in C_{f(\lambda,0),M}([0,T]; \mathbf{R}^n),\ \lambda \in (0,\mu) \right\},$$

is equicontinuous on $[0,T]$ (in particular, they tend uniformly to 0 as $t \downarrow 0$).

For each $\lambda > 0$, let $x(\lambda, \cdot)$ be a noncontinuable continuous solution of (1.2)$_\lambda$ satisfying (1.5)$_\lambda$, and denote the corresponding maximal interval of existence by $[0, T_{\max}(\lambda))$ (by Theorem **12**.*2.7, such a family of solutions exists). Then the following claims are true.*

(a) *Every sequence $\lambda_j \downarrow 0$ contains a subsequence λ_{j_k} such that the corresponding solutions $x(\lambda_{j_k}, \cdot)$ converge to a noncontinuable solution x of (1.2) satisfying (1.5), uniformly on compact subsets of the maximal interval of existence $[0, T_{\max})$ of x. In particular, $T_{\max} \le \liminf_{\lambda_{j_k} \downarrow 0} T_{\max}(\lambda_{j_k})$.*

(b) *If, in addition, the solution of (1.2) satisfying (1.5) is unique, then $x(\lambda, \cdot)$ tends to the noncontinuable solution x of (1.2), subject to (1.5), as $\lambda \downarrow 0$, uniformly on compact subsets of the maximal interval of existence $[0, T_{\max})$ of x. In particular, $T_{\max} \le \liminf_{\lambda \downarrow 0} T_{\max}(\lambda)$.*

The proof of Theorem 2.2 is left to the reader. (Observe the following. By (i) of Theorem **12**.2.7, $f(\lambda, \cdot) \in C([0, T_\infty(\lambda)); \mathbf{R}^n)$. It follows from the assumption $f(\lambda, 0) \to f(0)$ as $\lambda \downarrow 0$ and from (v) that $T_\infty \le \liminf_{\lambda \downarrow 0} T_\infty(\lambda)$. Thus, for any $T < T_\infty$, the function $f(\lambda, t)$ is defined for $t \in [0, T]$ and λ sufficiently small. An analogous observation applies to $g(\lambda, \cdot, \cdot, \cdot)$ in (vii).)

Theorem 2.1 may be applied also to the parametrized version

$$\tfrac{\mathrm{d}}{\mathrm{d}t} x(\lambda, t) = H\big(\lambda, x(\lambda, \cdot)\big)(t), \quad t \ge 0, \tag{1.3$_\lambda$}$$

of (1.3), with the usual constraint

$$x(\lambda, 0) = \xi(\lambda); \quad \big(t, x(\lambda, t)\big) \in D(\lambda), \quad t \ge 0. \tag{1.4$_\lambda$}$$

This application gives us the following result. The proof is left as an exercise.

2.3 Theorem

Let the hypothesis of Theorem **12**.*3.1 hold, both in its original form, and with H, ξ and D replaced, respectively, by $H(\lambda, \cdot)$, $\xi(\lambda)$ and $D(\lambda)$, with $\lambda > 0$. As $\lambda \downarrow 0$, let $H(\lambda, \cdot)$, $\xi(\lambda)$ and $D(\lambda)$ tend to H, ξ and D in the following sense:*

(iv) *for each compact connected set $M \subset D$ there is a $\mu > 0$ such that $M \subset D(\lambda)$ for $\lambda \in (0, \mu)$;*

(v) $\lim_{\lambda \downarrow 0} \xi(\lambda) = \xi$;

(vi) *if $T \in (0, T_\infty)$, if $\lambda_j \downarrow 0$, and if $\{\phi_j\} \subset C_{\xi(\lambda_j), D(\lambda_j)}([0,T]; \mathbf{R}^n)$ is a sequence that tends to a function $\phi \in C_{\xi, D}([0,T]; \mathbf{R}^n)$ uniformly on $[0, T]$, then $H(\lambda_j, \phi_j) \to H(\phi)$ in $L^1([0,T]; \mathbf{R}^n)$.*

In addition, suppose that condition (iii) *in Theorem* **12**.*3.1 holds uniformly for all sufficiently small* λ *in the sense that*

(iii)$_\lambda$ *for each compact connected subset* M *of* D *containing* $(0,\xi)$, *there exist a constant* $\mu > 0$, *and a function* $a_M \in L^1_{\text{loc}}([0,T_\infty);\mathbf{R})$, *such that if* $T \in (0,T_\infty)$, $\lambda \in (0,\mu)$, *and* $\phi \in C_{\xi(\lambda),M}([0,T];\mathbf{R}^n)$, *then* $|H(\lambda,\phi)(s)| \le a_M(s)$ *for almost all* s *in* $[0,T]$.

For each $\lambda > 0$, *let* $x(\lambda,\cdot)$ *be a noncontinuable continuous solution of* $(1.3)_\lambda$ *satisfying* $(1.4)_\lambda$, *and denote the corresponding maximal interval of existence by* $[0,T_{\max}(\lambda))$ *(by Theorem* **12**.*3.1, such a family of solutions exists). Then the following claims are true.*

(a) *Every sequence* $\lambda_j \downarrow 0$ *contains a subsequence* λ_{j_k} *such that the corresponding solutions* $x(\lambda_{j_k},\cdot)$ *converge to a noncontinuable solution* x *of (1.3) satisfying (1.4), uniformly on compact subsets of the maximal interval of existence* $[0,T_{\max})$ *of* x. *In particular,* $T_{\max} \le \liminf_{\lambda_{j_k}\downarrow 0} T_{\max}(\lambda_{j_k})$.

(b) *If, in addition, the solution of (1.3), subject to (1.4), is unique, then* $x(\lambda,\cdot)$ *tends to the noncontinuable solution* x *of (1.3) satisfying (1.4) as* $\lambda \downarrow 0$, *uniformly on compact subsets of the maximal interval of existence* $[0,T_{\max})$ *of* x. *In particular,* $T_{\max} \le \liminf_{\lambda\downarrow 0} T_{\max}(\lambda)$.

As our last topic in this section, we consider the continuous dependence of the solutions of $(1.1)_\lambda$ in the case where we are dealing with nonlinearities of Lipschitz-type and with L^p-solutions in the spirit of Theorem **12**.4.1. For simplicity, we suppose that the set K, to which the solution curve is required to belong, does not depend on the parameter.

2.4 Theorem

Let the hypothesis of Theorem **12**.*4.1 hold, both in its original form, and with* G *and* T_∞ *replaced by* $G(\lambda,\cdot)$ *and* $T_\infty(\lambda)$, *with* $\lambda > 0$. *In addition, assume that*

(v) $T_\infty \le \liminf_{\lambda\downarrow 0} T_\infty(\lambda)$,

(vi) *if* $T \in (0,T_\infty)$, *if* $\lambda_j \downarrow 0$, *and if some sequence* $\phi_j \in L^p_K([0,T];\mathbf{R}^n)$ *tends to a function* $\phi \in L^p_K([0,T];\mathbf{R}^n)$, *then* $G(\lambda_j,\phi_j) \to G(\phi)$ *in* $L^p([0,T];\mathbf{R}^n)$,

and that condition (iv) *in Theorem* **12**.*4.1 holds uniformly for small* λ *in the sense that*

(iv)$_\lambda$ *for each* $S \in [0,T_\infty)$ *and each* $\zeta \in L^p_K([0,S];\mathbf{R}^n)$, *there exist constants* $\mu > 0$, $\eta > 0$ *and* $\gamma < 1$ *such that, for each* $\lambda \in (0,\mu)$, *the conditions*

(1) $\tau \in [0,S]$, *and* $T \in (\tau,\tau+\eta] \cap [0,S]$,

(2) $\phi,\psi \in \{\varphi \in L^p_K([0,T];\mathbf{R}^n) \mid \|\varphi - \zeta\|_{L^p(0,T)} \le \eta\}$, *and* $\psi(s) = \phi(s)$ *for almost all* $s \in [0,\tau]$,

imply

(3) $\|G(\lambda,\psi) - G(\lambda,\phi)\|_{L^p(\tau,T)} \le \gamma\|\psi - \phi\|_{L^p(\tau,T)}$.

Let x *and* $x(\lambda,\cdot)$ *be the noncontinuable solutions of* (1.1) *and* $(1.1)_\lambda$, *respectively, given by Theorem* **12**.*4.1, with maximal intervals of existence*

$[0, T_{\max})$ and $[0, T_{\max}(\lambda))$. Then $x(\lambda, \cdot)$ converges in $L^p_{\text{loc}}([0, T_{\max}); \mathbf{R}^n)$ to x. In particular, $T_{\max} \le \liminf_{\lambda \downarrow 0} T_{\max}(\lambda)$.

Proof One can proceed in much the same manner as in the proofs of Theorem 2.1 and Theorem **12**.4.1. The main effort goes into the proof of the following analogues of (2.3) and (2.4).

If $\lambda_j \downarrow 0$, then there exists a $T > 0$ such that $x(\lambda_j, \cdot)$ converges in $L^p([0, T]; \mathbf{R}^n)$ to x. (2.17)

If $\tau \in (0, T_{\max})$, if $\lambda_j \downarrow 0$, and if the sequence $x(\lambda_j, \cdot)$ converges in $L^p_{\text{loc}}([0, \tau); \mathbf{R}^n)$ to x, then there exists a $T > \tau$ such that $x(\lambda_j, \cdot)$ converges to x in $L^p([0, T]; \mathbf{R}^n)$. (2.18)

In the proof of (2.18), let $\zeta \in L^p_K([0, S]; \mathbf{R}^n)$ be some (arbitrary) extension of x, i.e., $S \in (\tau, T_\infty)$, and $\zeta(s) = x(s)$ for almost all $s \in [0, \tau]$, and let μ, η and γ be the constants in (iv)$_\lambda$ corresponding to S and ζ. Choose some $T \in (\tau, \tau + \frac{1}{2}\eta] \cap [0, S]$, and, if necessary, decrease T (but keep it $> \tau$) until

$$\|G(\zeta)(t) - \zeta(t)\|_{L^p(\tau, T)} \le \tfrac{1}{4}(1 - \gamma)\eta. \tag{2.19}$$

Finally, decrease τ (by a small amount $\le \frac{1}{2}\eta$) to get convergence of $x(\lambda_j, \cdot)$ to x in $L^p([0, \tau]; \mathbf{R}^n)$. This does not affect (2.19) (because ζ is a solution of (1.1) on $[0, \tau]$).

In the proof of (2.17) we take $\tau = 0$, and proceed as above (with no final adjustment of τ).

For j large enough, we have $T_\infty(\lambda_j) \ge \tau$, and for these j we can define the closed, bounded and convex sets

$$M_j = \{\, \phi \in L^p_K([0, T]; \mathbf{R}^n) \mid \phi(t) = x(\lambda_j, t) \text{ for almost all } t \in [0, \tau], \text{ and } \|\phi - \zeta\|_{L^p(\tau, T)} \le \tfrac{1}{2}\eta \,\}.$$

We assert that, for large j, $G(\lambda_j, \cdot)$ maps M_j into itself. To show this, first define the auxiliary function $\psi_j(t) = x(\lambda_j, t)$, $0 \le t \le \tau$, $\psi_j(t) = \zeta(t)$, $\tau < t \le T$. Then recall that $x(\lambda_j, \cdot) \to \zeta$ in $L^p([0, \tau]; \mathbf{R}^n)$, and use (vi) to conclude that for sufficiently large j one has $\|G(\lambda_j, \psi_j) - G(\zeta)\|_{L^p(\tau, T)} \le \frac{1}{4}(1 - \gamma)\eta$. Therefore it follows from (iv)$_\lambda$ and (2.19) that (for $\phi \in M_j$ and j large)

$$\begin{aligned}
&\|G(\lambda_j, \phi) - \zeta\|_{L^p(\tau, T)} \\
&\quad \le \|G(\lambda_j, \phi) - G(\lambda_j, \psi_j)\|_{L^p(\tau, T)} + \|G(\lambda_j, \psi_j) - G(\zeta)\|_{L^p(\tau, T)} \\
&\qquad + \|G(\zeta) - \zeta\|_{L^p(\tau, T)} \\
&\quad \le \tfrac{1}{2}\gamma\eta + \tfrac{1}{4}(1 - \gamma)\eta + \tfrac{1}{4}(1 - \gamma)\eta = \tfrac{1}{2}\eta.
\end{aligned}$$

Clearly, by (iv)$_\lambda$, $G(\lambda_j, \cdot)$ is a contraction on M_j. Therefore, $G(\lambda_j, \cdot)$ has a fixed point in M_j. But the fixed point of $G(\lambda_j, \cdot)$ is $x(\lambda_j, \cdot)$, so this means that $x(\lambda_j, \cdot)$ belongs to M_j. In particular, $T_{\max}(\lambda_j) > T$.

By continuity, condition (iv)$_\lambda$ implies that one can use the same constants η and γ in condition (iv) of Theorem **12**.4.1 as in (iv)$_\lambda$. Therefore,

$\|x-\zeta\|_{L^p(\tau,T)} \le \eta$. In addition, we may repeat the argument above to show that $T_{\max} > T$.

We claim that $\|x(\lambda_j,\cdot)-x\|_{L^p(\tau,T)} \to 0$ as $j\to\infty$. To see that this is true, define

$$\psi_j(s) = \begin{cases} x(\lambda_j, s), & \text{for } 0 \le s \le \tau, \\ x(s), & \text{for } \tau < s \le T. \end{cases}$$

Then $\psi_j \to x$ in $L^p([0,T];\mathbf{R}^n)$ as $j\to\infty$. By $(1.1)_\lambda$ and $(\mathrm{iv})_\lambda$, for j large enough,

$$\begin{aligned} \|x(\lambda_j,\cdot)-x\|_{L^p(\tau,T)} &= \|G(\lambda_j, x(\lambda_j,\cdot)) - G(x)\|_{L^p(\tau,T)} \\ &\le \|G(\lambda_j, x(\lambda_j,\cdot)) - G(\lambda_j,\psi_j)\|_{L^p(\tau,T)} \\ &\quad + \|G(\lambda_j,\psi_j) - G(x)\|_{L^p(\tau,T)} \\ &\le \gamma\|x(\lambda_j,\cdot)-\psi_j\|_{L^p(\tau,T)} + \|G(\lambda_j,\psi_j)-G(x)\|_{L^p(\tau,T)} \\ &= \gamma\|x(\lambda_j,\cdot)-x\|_{L^p(\tau,T)} + \|G(\lambda_j,\psi_j)-G(x)\|_{L^p(\tau,T)}. \end{aligned}$$

Therefore,

$$\|x(\lambda_j,\cdot)-x\|_{L^p(\tau,T)} \le (1-\gamma)^{-1}\|G(\lambda_j,\psi_j)-G(x)\|_{L^p(\tau,T)}$$

and this, together with (vi), implies that $\|x(\lambda_j,\cdot)-x\|_{L^p(\tau,T)} \to 0$ as $j\to\infty$, as claimed.

We conclude that (2.17) and (2.18) hold.

The completion of the proof is similar to the final part of the proof of Theorem 2.1, only simpler, and is left to the reader. □

3. Differentiability with Respect to a Parameter

In this section we study the differentiability of the solutions $x(\lambda,\cdot)$ of $(1.1)_\lambda$, subject to $(1.4)_\lambda$, with respect to the parameter λ (Theorems 3.1 and 3.2), and the differentiability of the solution x of (1.1) satisfying (1.4) with respect to the time variable t (Theorem 3.3).

The result that we obtain for $(1.1)_\lambda$, subject to $(1.4)_\lambda$, requires that $G(\cdot,\cdot)$ should be differentiable with respect to both its variables (in a rather weak sense) along the solutions of (1.1) satisfying (1.4), and the conclusion is that the solution is differentiable with respect to λ. The exact sense in which G is required to be differentiable is spelled out in (ii) of Theorem 3.1. Moreover, the derivative $x_\lambda(\lambda,\cdot)$ satisfies the equation one obtains by differentiating $(1.1)_\lambda$. Some additional comments on the differentiability of G are given in the two paragraphs following Theorem 3.1.

Below we consider only nonnegative values of the parameter λ, and we study the existence of the right-derivative at $\lambda = 0$. However, in the same way as in Section 2, this result can immediately be extended to cover the case where, e.g., λ belongs to some finite-dimensional space, or to some Banach space.

For the definition of measure kernels of type $B^{\infty}_{\mathrm{loc}}$, see Definition **10**.2.11.

3.1 Theorem

Let the assumptions of Theorem 2.1 hold. In addition, suppose that G is differentiable along the solutions of (1.1), subject to (1.4), in the sense that, for each noncontinuable continuous solution x on $[0, T_{\max})$ of (1.1) satisfying (1.4), there exist a function $h \in B^{\infty}_{\mathrm{loc}}([0, T_{\max}); \mathbf{R}^n)$, and a measure kernel κ of type $B^{\infty}_{\mathrm{loc}}$ on $[0, T_{\max})$, with the following properties:

(i) *$-\kappa$ has a measure resolvent of type $B^{\infty}_{\mathrm{loc}}$ on $[0, T_{\max})$;*

(ii) *for each $T \in [0, T_{\max})$,*

$$\sup_{t\in[0,\tau]} \Big| G(\lambda, x(\lambda, \cdot))(t) - G(x)(t) - \lambda h(t) - \int_{[0,t]} \kappa(t, \mathrm{d}s)\big(x(\lambda, s) - x(s)\big)\Big|$$
$$= o\Big(\lambda + \sup_{t\in[0,\tau]} |x(\lambda, t) - x(t)|\Big)$$
$$\text{as } \Big(\lambda + \sup_{t\in[0,\tau]} |x(\lambda, t) - x(t)|\Big) \downarrow 0,$$

uniformly for $\tau \in [0, T]$, where $x(\lambda, \cdot)$ represents an arbitrary solution of $(1.1)_\lambda$ satisfying $(1.4)_\lambda$, defined at least on $[0, T]$.

Then (1.1) has a unique noncontinuable continuous solution satisfying (1.4). Moreover, if we define $x(0, \cdot) = x$, then $x(\lambda, \cdot)$ has a right-derivative $x_\lambda \in B^{\infty}_{\mathrm{loc}}([0, T_{\max}); \mathbf{R}^n)$ with respect to λ at zero, in the sense that

$$\sup_{t\in[0,\tau]} \big|x(\lambda, t) - x(t) - \lambda x_\lambda(t)\big| = o(\lambda) \text{ as } \lambda \downarrow 0, \tag{3.1}$$

uniformly for τ in compact subsets of $[0, T_{\max})$. Furthermore, x_λ is the unique solution of the equation

$$x_\lambda(t) - \int_{[0,t]} \kappa(t, \mathrm{d}s) x_\lambda(s) = h(t), \quad t \in [0, T_{\max}). \tag{3.2}$$

Assumption (i) is very natural in the sense that one needs the resolvent of $-\kappa$ to be able to solve (3.2) for x_λ. We can reformulate assumption (ii) as follows:

(ii) *for each $\epsilon > 0$ and each $T \in [0, T_{\max})$ there exists a constant $\delta > 0$, such that if $\tau \in [0, T]$, if $x(\lambda, \cdot)$ is a noncontinuable continuous solution of $(1.1)_\lambda$ and $(1.4)_\lambda$, defined at least on $[0, T]$, and if*

$$\lambda + \sup_{t\in[0,\tau]} |x(\lambda, t) - x(t)| < \delta,$$

then

$$\sup_{t\in[0,\tau]}\Big|G\big(\lambda,x(\lambda,\cdot)\big)(t)-G(x)(t)-\lambda h(t)-\int_{[0,t]}\kappa(t,\mathrm{d}s)\big(x(\lambda,s)-x(s)\big)\Big|\le\epsilon\Big(\lambda+\sup_{t\in[0,\tau]}|x(\lambda,t)-x(t)|\Big).$$

Equation (3.2) is called a *variational equation* associated with $(1.1)_\lambda$. We warn the reader that this equation is not unique, i.e., the function h and the kernel κ are not uniquely determined by the requirements imposed on them in Theorem 3.1. To see this, let $G(\lambda,x)=G(x)=x$, and observe that, if the conclusion of Theorem 3.1 holds in the sense that (3.1) is true for some x_λ, then one may pick κ arbitrarily, and use (3.2) as a definition for h. For example, one may choose $\kappa=0$ and $h=x_\lambda$. The function h and the kernel κ are unique if G is sufficiently differentiable, and if one requires the estimate in (ii) to hold with $x(\lambda,\cdot)$ and x replaced by arbitrary continuous functions y and z, i.e., by functions that need not satisfy (1.1). In this case, h and κ are, respectively, the partial (Fréchet-type) derivatives of G with respect to its first and second arguments at $(0,x)$.

Proof of Theorem 3.1 First, we prove uniqueness.

Let x be a noncontinuable continuous solution of (1.1) and (1.4) of the type mentioned in Theorem 2.1, with maximal interval of existence $[0,T_{\max})$. Then there exist a sequence $\lambda_j\downarrow 0$, and a corresponding sequence of noncontinuable solutions $x(\lambda_j,\cdot)$, converging to x uniformly on compact subintervals of $[0,T_{\max})$. We claim that (1.1) has no other solutions (satisfying (1.4)).

If the preceding claim is false, then (1.1), subject to (1.4), has some other noncontinuable solution x_*, defined on some interval $[0,T_*)$. Let h_* and κ_* be the corresponding partial derivatives of G mentioned in the hypotheses. Define $z=x-x_*$. Since x and x_* are two different noncontinuable solutions, there exists a number $T<\min\{T_{\max},T_*\}$ such that $z(T)\neq 0$. Define

$$p(t)=z(t)-\int_{[0,t]}\kappa_*(t,\mathrm{d}s)z(s),\quad t\in[0,T].$$

Let ρ_* be the measure resolvent of $-\kappa_*$. Then

$$z(t)=p(t)-\int_{[0,t]}\rho_*(t,\mathrm{d}s)p(s)\,\mathrm{d}s,\quad t\in[0,T].\tag{3.3}$$

We claim that for each $\epsilon>0$ there is a $\delta>0$ such that for $\tau\in[0,T]$ the condition

$$\sup_{t\in[0,\tau]}|z(t)|\le\delta \text{ implies } \sup_{t\in[0,\tau]}|p(t)|\le\epsilon\sup_{t\in[0,\tau]}|z(t)|.\tag{3.4}$$

Suppose for a moment that this claim is true. Since ρ_* is a measure kernel of type B^∞ on $[0,\tau]$, it follows from (3.3) that there exists a constant K such that

$$|z(t)| \le K \sup_{s\in[0,t]} |p(s)|, \quad t \in [0,T]. \tag{3.5}$$

This, together with (3.4), implies that for each $\epsilon > 0$ there is a $\delta > 0$ such that for each $\tau \in [0,T]$ the condition $\sup_{t\in[0,\tau]}|z(t)| \le \delta$ implies $\sup_{t\in[0,\tau]}|z(t)| \le \epsilon K \sup_{t\in[0,\tau]}|z(t)|$. In particular, choosing $\epsilon < 1/K$, we find that the condition $\sup_{t\in[0,\tau]}|z(t)| \le \delta$ implies that $z(t) = 0$ for $t \in [0,\tau]$. As $z(0) = 0$, and z is continuous, this means that we must have $z(t) = 0$ for $t \in [0,T]$. Since we assumed that $z(T) \ne 0$, we get a contradiction, and this contradiction proves that the solution of (1.1) and (1.4) is unique, provided (3.4) holds.

Let us prove (3.4). Take some $\epsilon > 0$, let δ_* be the constant mentioned in the hypothesis of Theorem 3.1 corresponding to the chosen x_*, and define $\delta = \delta_*/3$. Take some $\tau \in [0,T]$, and write z in the form $z(t) = (x(t) - x(\lambda_j,t)) + (x(\lambda_j,t) - x_*(t))$ for $t \in [0,\tau]$. Observe that, for $t \in [0,\tau]$, one has $\lambda_j + |x(\lambda_j,t) - x_*(t)| \le \delta_*$ whenever $\lambda_j \le \delta$, $|x(t) - x(\lambda_j,t)| \le \delta$ and $|z(t)| \le \delta$, and use (ii) and equations (1.1) and $(1.1)_\lambda$ to get for all $t \in [0,\tau]$, $\lambda_j \le \delta$, and some positive C,

$$\begin{aligned}
|p(t)| &= \Big| z(t) - \int_{[0,t]} \kappa_*(t,\mathrm{d}s) z(s) \Big| \\
&\le \Big| x(t) - x(\lambda_j,t) - \int_{[0,t]} \kappa_*(t,\mathrm{d}s)\big(x(s) - x(\lambda_j,s)\big) \Big| \\
&\quad + \Big| G(\lambda_j, x(\lambda_j,\cdot))(t) - G(x_*)(t) \\
&\qquad - \lambda_j h_*(t) - \int_{[0,t]} \kappa_*(t,\mathrm{d}s)\big(x(\lambda_j,s) - x_*(s)\big) \Big| + \lambda_j |h_*(t)| \\
&\le C \sup_{t\in[0,\tau]} |x(t) - x(\lambda_j,t)| \\
&\quad + \epsilon\Big(\lambda_j + \sup_{t\in[0,\tau]} |x(\lambda_j,t) - x_*(t)|\Big) + \lambda_j \sup_{t\in[0,\tau]} |h_*(t)|.
\end{aligned}$$

Letting $\lambda_j \downarrow 0$ we find that (3.4) is true. This completes the proof of uniqueness.

We proceed to prove differentiability. Because of (i), it is clear that (3.2) has a unique solution x_λ, defined on $[0,T_{\max})$. Fix some $T \in [0,T_{\max})$, and let $x(\lambda,\cdot)$ be some family of noncontinuable solutions of $(1.1)_\lambda$ for $\lambda > 0$, converging uniformly to x as $\lambda \downarrow 0$ on $[0,T]$. Define $y_\lambda(t) = x(\lambda,t) - x(t) - \lambda x_\lambda(t)$. Then it follows from (1.1), $(1.1)_\lambda$, and (3.2) that

$$y_\lambda(t) - \int_{[0,t]} \kappa(t,\mathrm{d}s) y_\lambda(s) = q_\lambda(t), \quad t \in [0,T],$$

where

$$q_\lambda(t) = G\big(\lambda, x(\lambda,\cdot)\big)(t) - G(x)(t)$$
$$- \lambda h(t) - \int_{[0,t]} \kappa(t,\mathrm{d}s)\big(x(\lambda,s) - x(s)\big), \quad t \in [0,T].$$

Using the measure resolvent ρ of $-\kappa$ we get

$$y_\lambda(t) = q_\lambda(t) - \int_{[0,t]} \rho(t,\mathrm{d}s) q_\lambda(s), \quad t \in [0,T].$$

It follows from the main hypothesis (ii) that

$$\sup_{t\in[0,\tau]} |q_\lambda(t)| = o\Big(\sup_{t\in[0,\tau]} |x(\lambda,t) - x(t)|\Big) + o(\lambda),$$
$$\text{uniformly for } \tau \in [0,T]. \tag{3.6}$$

Observe that (3.5) holds with z replaced by y_λ, and p replaced by q_λ, and that

$$\sup_{t\in[0,\tau]} \big|y_\lambda(t) - \big(x(\lambda,t) - x(t)\big)\big| = O(\lambda), \text{ uniformly for } \tau \in [0,T],$$

and use these facts in (3.6) to get the desired conclusion that

$$\sup_{t\in[0,\tau]} |y_\lambda(t)| = o(\lambda) \text{ as } \lambda \downarrow 0, \text{ uniformly for } \tau \in [0,T]. \quad \square$$

In Theorem 3.1 we inferred that x_λ belongs locally to B^∞. A similar result is true if, instead, one permits x_λ to belong locally to L^p, for some $p \in [1,\infty]$.

3.2 Theorem

Let the assumptions of Theorem 2.1 hold. In addition, suppose that G is differentiable along the solutions of (1.1) in the sense that for some $p \in [1,\infty]$, and for each noncontinuable continuous solution x on $[0,T_{\max})$ of (1.1) satisfying (1.4), there exist a function $h \in L^p_{\mathrm{loc}}([0,T_{\max});\mathbf{R}^n)$ and a kernel k of type L^p_{loc} on $[0,T_{\max})$ with the following properties:

(i) *$-k$ has a resolvent of type L^p_{loc} on $[0,T_{\max})$;*

(ii) *for each $\epsilon > 0$ and each $T \in [0,T_{\max})$ there exists a constant $\delta > 0$, such that if $\tau \in [0,T]$, if $x(\lambda,\cdot)$ is a noncontinuable continuous solution of $(1.1)_\lambda$, subject to $(1.4)_\lambda$, defined at least on $[0,T]$, and if*

$$\lambda + \sup_{t\in[0,\tau]} |x(\lambda,t) - x(t)| < \delta,$$

then

$$\Big\| G\big(\lambda, x(\lambda,\cdot)\big) - G(x) - \lambda h - \int_{[0,\cdot]} k(\cdot,s)\big(x(\lambda,s) - x(s)\big)\,\mathrm{d}s \Big\|_{L^p(0,\tau)}$$
$$\le \epsilon\Big(\lambda + \|x(\lambda,t) - x(t)\|_{L^p(0,\tau)}\Big).$$

Then (1.1), subject to (1.4), has a unique noncontinuable continuous solution. Moreover, if we define $x(0,\cdot) = x$, then $x(\lambda,\cdot)$ has a right-derivative

$x_\lambda \in L^p_{\mathrm{loc}}([0,T_{\max});\mathbf{R}^n)$ with respect to λ at zero, in the sense that

$$\left\|x(\lambda,\cdot)-x-\lambda x_\lambda\right\|_{L^p(0,\tau)} = o(\lambda) \text{ as } \lambda\downarrow 0,$$

uniformly for τ in compact subsets of $[0,T_{\max})$. Moreover, x_λ is the unique solution of the equation

$$x_\lambda(t) - \int_{[0,t]} k(t,s)x_\lambda(s)\,\mathrm{d}s = h(t), \quad t\in[0,T_{\max}). \tag{3.2}$$

We leave the task of checking that the proof is an almost exact copy of the proof of Theorem 3.1 to the reader (change sup-norms to L^p-norms).

At this point the reader might want to do Exercise 10.

Now, let us leave the differentiability with respect to λ in $(1.1)_\lambda$, and instead turn to the differentiability with respect to t in (1.1).

One can get certain trivial results concerning differentiability with respect to t in (1.1), by reducing (1.1) to (1.3): if G maps continuous functions (or L^p-functions) into locally absolutely continuous functions, then it follows from (1.1) that the solutions are locally absolutely continuous, and that they satisfy (1.3) with H replaced by the t-derivative of G.

Results of a different character can be obtained from the two preceding theorems, and from Exercises 8 and 10. The idea is to take $x(\lambda,t)$ to be $x(t,\lambda)=x(t+\lambda)$, i.e., to look at the equation

$$x(t+\lambda)=G(x)(t+\lambda), \quad t\geq 0, \tag{3.7}$$

and to apply one of the results formulated above, to show that the solution is differentiable with respect to λ. As we have already observed, if G maps every continuous function into an absolutely continuous one, then the result becomes trivial, so let us assume that G does not have this property. For simplicity, we investigate (1.2) only; similar results are true for (1.1).

A differentiation of the right-hand side of (1.2) with respect to t shows that if f is locally absolutely continuous, if $t\mapsto g(t,t,x(t))$ is integrable, and if the partial derivative g_t behaves well enough, then the solution x of (1.2) must be locally absolutely continuous. This is the trivial case that we mentioned above. A different situation occurs when g_t behaves badly, but the partial derivatives h_s and h_x of the function $(t,s,x)\mapsto h(t,s,x)\stackrel{\text{def}}{=} g(t+s,s,x)$ behave well. To see this, let us recast (1.2) in the form

$$x(t)=f(t)+\int_0^t h(t-s,s,x(s))\,\mathrm{d}s, \quad t\geq 0, \tag{3.8}$$

or equivalently as

$$x(t)=f(t)+\int_0^t h(s,t-s,x(t-s))\,\mathrm{d}s, \quad t\geq 0. \tag{3.9}$$

A formal differentiation of (3.9), followed by a change of variables, yields

$$\begin{aligned} x'(t) &= f'(t)+h(t,0,x(0)) \\ &\quad + \int_0^t h_s(t-s,s,x(s))\,\mathrm{d}s + \int_0^t h_x(t-s,s,x(s))x'(s)\,\mathrm{d}s, \end{aligned} \tag{3.10}$$

which is an equation of the type (3.2). We let the reader state the exact hypotheses needed for an application of Theorem 3.2 to this case and, instead, formulate and prove the corresponding result for the simpler equation

$$x(t) = f(t) + \int_0^t k(t-s)g\big(s, x(s)\big)\,\mathrm{d}s, \quad t \geq 0. \tag{3.11}$$

Here we expect the derivative to satisfy the variational equation

$$x'(t) = f'(t) + k(t)g\big(0, f(0)\big) + \int_0^t k(t-s)g_s\big(s, x(s)\big)\,\mathrm{d}s$$
$$+ \int_0^t k(t-s)g_x\big(s, x(s)\big)x'(s)\,\mathrm{d}s, \quad t \geq 0. \tag{3.12}$$

3.3 Theorem

Let D be an open connected subset of $\mathbf{R}^+ \times \mathbf{R}^n$ containing $(0, f(0))$, define $T_\infty = \sup\{T \in \mathbf{R}^+ \mid C_{f(0),D}([0,T); \mathbf{R}^n) \neq \emptyset\}$, and suppose that the following conditions are satisfied:

(i) *$f \in AC_{\mathrm{loc}}([0, T_\infty); \mathbf{R}^n)$;*

(ii) *$k \in L^1_{\mathrm{loc}}([0, T_\infty); \mathbf{R}^{n\times n})$;*

(iii) *$g \in C(D; \mathbf{R}^n)$ is continuously differentiable with respect to its second argument, and g has a continuous L^1-derivative g_s with respect to its first variable in the sense that for each $T \in [0, T_\infty)$ the mapping $\phi \mapsto g_s(\cdot, \phi(\cdot))$ is continuous from $C_D([0,T]; \mathbf{R}^n)$ into $L^1([0,T]; \mathbf{R}^n)$, and, for each compact connected subset M of D,*

$$\frac{1}{\lambda}\int_0^T \big|g(s+\lambda, \phi(s)) - g(s, \phi(s)) - \lambda g_s(s, \phi(s))\big|\,\mathrm{d}s \to 0$$

as $\lambda \downarrow 0$, uniformly for $\phi \in C_M([0,T]; \mathbf{R}^n)$.

Then (3.11) has a unique noncontinuable continuous solution satisfying (1.5). This solution is locally absolutely continuous, and its derivative satisfies (3.12).

Proof By Theorem **12**.2.6, equation (3.11) has a unique noncontinuable continuous solution x satisfying (1.5) on its maximal interval of existence. Define

$$G(\phi)(t) = f(t) + \int_0^t k(t-s)g\big(s, \phi(s)\big)\,\mathrm{d}s,$$

$$G(\lambda, \phi)(t) = f(t+\lambda) + \int_0^\lambda k(t+\lambda-s)g\big(s, x(s)\big)\,\mathrm{d}s$$
$$+ \int_0^t k(t-s)g\big(s+\lambda, \phi(s)\big)\,\mathrm{d}s,$$

$$\xi = f(0), \qquad \xi_\lambda = x(\lambda).$$

Then x is the unique solution of (1.1), subject to (1.4), and the unique solution $x(\lambda, \cdot)$ of $(1.1)_\lambda$, subject to $(1.4)_\lambda$, is given by $x(\lambda, t) = x(t+\lambda)$.

We want to apply Theorem 3.2 with $p = 1$. To do this, we must first check that the assumptions of Theorem 2.1 are satisfied. This is straightforward. In particular, as required by Theorem 2.1, the assumption of Theorem **12**.2.1 holds, both in its original form, and with G and ξ replaced by $G(\lambda)$ and ξ_λ (in fact, with (iv) in Theorem **12**.2.1 strengthened to (iv) in Corollary **12**.2.3; cf. Theorem **12**.2.7). We leave this check to the reader.

Define

$$h(t) = f'(t) + k(t)g\big(0, f(0)\big) + \int_0^t k(t-s)g_s\big(s, x(s)\big)\,\mathrm{d}s,$$

and

$$K(t,s) = k(t-s)g_x\big(s, x(s)\big).$$

Clearly, (i) in Theorem 3.2 holds with $p = 1$ and with k replaced by K. To see that (ii) in Theorem 3.2 holds with the same replacements, write

$$\begin{aligned}
G\big(\lambda, x(\lambda,\cdot)\big)(t) - G(x)(t) &- \int_0^t K(t,s)\big(x(\lambda,s) - x(s)\big)\,\mathrm{d}s - \lambda h(t) \\
&= f(t+\lambda) - f(t) - \lambda f'(t) \\
&\quad + \int_0^\lambda k(t+\lambda-s)\Big\{g\big(s, x(s)\big) - g\big(0, x(0)\big)\Big\}\,\mathrm{d}s \\
&\quad + \Big\{\int_0^\lambda \big(k(t+\lambda-s) - k(t)\big)\,\mathrm{d}s\Big\}g\big(0, x(0)\big) \\
&\quad + \int_0^t k(t-s)\Big\{g\big(s+\lambda, x(\lambda,s)\big) - g\big(s, x(\lambda,s)\big) \\
&\quad - \lambda g_s\big(s, x(\lambda,s)\big)\Big\}\,\mathrm{d}s \\
&\quad + \lambda\int_0^t k(t-s)\Big\{g_s\big(s, x(\lambda,s)\big) - g_s\big(s, x(s)\big)\Big\}\,\mathrm{d}s \\
&\quad + \int_0^t k(t-s)\Big\{g\big(s, x(\lambda,s)\big) - g\big(s, x(s)\big) \\
&\quad - g_x\big(s, x(s)\big)\big(x(\lambda,s) - x(s)\big)\Big\}\,\mathrm{d}s.
\end{aligned}$$

It is a straightforward task to check that the $L^1(0,T)$-norms of all the terms on the right-hand side are of order $o(\lambda + \|x(\lambda,\cdot) - x\|_{L^1(0,T)})$, as required by (ii) in Theorem 3.2.

By Theorem 3.2, and by the fact that $x(\lambda, t) = x(t+\lambda)$, the solution of (3.11) is locally absolutely continuous, and the derivative x' satisfies the variational equation (3.12). □

4. Maximal and Minimal Solutions

In this section we discuss maximal and minimal solutions, i.e., solutions that are pointwise larger or smaller than all other solutions.

In the one-dimensional case the words 'larger' and 'smaller' have obvious interpretations, but for the $\mathbf{R}^n$-case we need a definition.

4.1 Definition
Let $\mathcal{K}$ be some closed, convex, strict, and solid cone in $\mathbf{R}^n$ with vertex zero (i.e., $\mathcal{K}$ is closed and convex, $a\mathcal{K} \subset \mathcal{K}$ for $a \in \mathbf{R}^+$, $\mathcal{K} \cap -\mathcal{K} = \{0\}$, and the interior of $\mathcal{K}$ is nonempty; cf. Martin [1], pp. 14–15). Let $u \in \mathbf{R}^n$ and $v \in \mathbf{R}^n$. Then $u \leq v$ (or $v \geq u$) if $v - u \in \mathcal{K}$ and $u < v$ (or $v > u$) if $v - u$ belongs to the interior of $\mathcal{K}$.

Of course, for different cones $\mathcal{K}$ one gets different relations $\leq$, $<$, $\geq$ and $>$, but below we assume that $\mathcal{K}$ has been fixed once and for all.

4.2 Definition
A solution x of one of the equations (1.1)–(1.3), subject to (1.4) or (1.5), on an interval I is called a maximal solution (minimal solution) on I, if it is larger (smaller) than all other solutions in the sense that if $\tilde{x}$ is some other solution of the same equation, defined on some interval $\tilde{I}$, then $x(t) \geq \tilde{x}(t)$ ($x(t) \leq \tilde{x}(t)$) for $t \in I \cap \tilde{I}$.

It is well known that scalar ordinary differential equations always have maximal and minimal solutions. On the contrary, scalar integral and functional equations do not, in general, have maximal and minimal solutions. Only in the case where the right-hand sides of (1.1)–(1.3) are sufficiently monotone is it possible to prove the existence of maximal and minimal solutions.

We begin by giving a result on the existence of maximal and minimal solutions of (1.1), subject to the auxiliary condition (1.4). Observe that the hypothesis of Theorem 4.3 constitutes only a slight strengthening of the basic existence result Theorem **12**.2.1. In fact, the sole addition is the monotonicity assumption (v). The proof of Theorem 4.3 relies on the continuous dependence result Theorem 2.1; in addition, the usefulness of Theorem **12**.2.4 becomes evident.

4.3 Theorem
Let D be an open connected subset of $\mathbf{R}^+ \times \mathbf{R}^n$ containing $(0, \xi)$, define T_∞ as in (2.1), and suppose that conditions (i)–(v) *below are satisfied.*

(i) *For each $T \in [0, T_\infty)$, G maps $C_{\xi,D}([0,T];\mathbf{R}^n)$ into $C([0,T];\mathbf{R}^n)$, and for each $t \in [0,T]$ the mapping $\phi \mapsto G(\phi)(t)$ is continuous from $C_{\xi,D}([0,T];\mathbf{R}^n)$ (equipped with the sup-norm) into $\mathbf{R}^n$.*

(ii) *G is causal, i.e., if $0 \leq \tau < T < T_\infty$, and if ψ is the restriction of $\phi \in C_{\xi,D}([0,T];\mathbf{R}^n)$ to $[0,\tau]$, then $G(\psi)(t) = G(\phi)(t)$ for all $t \in [0,\tau]$.*

(iii) $G(\xi)(0) = \xi$.

(iv) *For each compact connected subset* M *of* D *containing* $(0,\xi)$ *there exist a constant* $\eta > 0$, *and a continuous function* $\delta : (0,\infty) \to (0,\infty)$, *such that the following implication holds: if for some* $\epsilon > 0$, $T \in (0, T_\infty)$, *and for some functions* $\phi, \psi \in C_{\xi,M}([0,T]; \mathbf{R}^n)$ *we have*
   (1) $\psi(s) = \phi(s)$ *for* $s \in [0, |T-\eta|_+]$,
   (2) $|\psi(s) - \phi(s)| \leq \epsilon$ *for* $s \in [|T-\eta|_+, T]$,

   then
   (3) $|G(\psi)(s) - G(\phi)(s) - G(\psi)(t) + G(\phi)(t)| \leq \epsilon - \delta(\epsilon)$ *for all* $s, t \in [|T-\eta|_+, T]$, *satisfying* $|t-s| \leq \delta(\epsilon)$.

(v) *If* $\tau > 0$, *and if* $\psi, \phi \in C_{\xi,D}([0,\tau]; \mathbf{R}^n)$ *satisfy* $\psi(t) \leq \phi(t)$ *for* $t \in [0,\tau]$ *then* $G(\psi)(\tau) \leq G(\phi)(\tau)$.

Then (1.1), subject to (1.4), has a noncontinuable maximal solution x_M *and a noncontinuable minimal solution* x_m, *defined on maximal intervals* $[0, T_{M,\max})$ *and* $[0, T_{m,\max})$, *respectively.*

If, in addition, for each $t \in [0, T_{M,\max})$, $y_1 \in D_t$, *and for each* $y_2 \in D_t$, *it is true that the set* $\{ y \in \mathbf{R}^n \mid y_1 \leq y \leq y_2 \}$ *is a compact subset of* D_t *(in the scalar case this is true iff* D_t *is connected), then every noncontinuable solution of (1.1) is defined at least on the interval* $[0, T_{M,\max}) \cap [0, T_{m,\max})$.

It is evident that a maximal (or minimal) solution, if one exists, must be unique.

In the proof of Theorem 4.3 we need the following lemma.

4.4 Lemma
Under the hypothesis of Theorem **12**.*2.1 there exists some* $T > 0$ *such that every noncontinuable continuous solution of (1.1) satisfying (1.4) is defined (at least) on* $[0,T]$, *and such that the set of all solutions of (1.1) satisfying (1.4) on* $[0,T]$ *is compact in* $C([0,T]; \mathbf{R}^n)$.

Proof Checking the proof of Theorem **12**.2.1 one finds that, by choosing T and ρ properly, one can guarantee that all solutions x of (1.1) and (1.4) must exist on $[0,T]$, and satisfy $|x(t) - \xi| \leq \rho$ for $t \in [0,T]$. In the proof of Theorem **12**.2.1, we constructed the solution of (**12**.1.6) by applying the Fixed Point Theorem **12**.1.3, and although we did not need the fact at that time, the conclusion of that theorem says that the set of all fixed points is compact. □

Proof of Theorem 4.3 The existence proofs for the maximal and minimal solutions are entirely analogous, hence we consider only the former proof—leaving the latter to the reader.

The proof of the existence of a maximal solution of (1.1) is based on the construction of functions $x(\lambda, \cdot)$ that solve approximate equations of the type $(1.1)_\lambda$, are larger than any solution of (1.1), and converge to a noncontinuable maximal solution of (1.1) as $\lambda \downarrow 0$. This construction proceeds as follows.

Choose some $p \in \mathbf{R}^n$ such that $p > 0$ (i.e., p belongs to the interior of the cone $\mathcal{K}$). For each $T \in [0, T_\infty)$, $\lambda > 0$, and for each $x \in C_{\xi,D}([0,T];\mathbf{R}^n)$, define

$$G(\lambda, x)(t) = \lambda pt + G(x)(t), \quad t \in [0,T].$$

This family of functions $G(\lambda, \cdot)$ satisfies the assumption of Theorem 2.1, with $D(\lambda) = D$ and $\xi(\lambda) = \xi$. For each $\lambda > 0$ we construct a noncontinuable solution $x(\lambda, \cdot)$ of $(1.1)_\lambda$, satisfying $(1.4)_\lambda$, as follows. By Lemma 4.4, there is some $\tau > 0$ such that all noncontinuable solutions of (1.1) satisfying (1.4) are defined at least on $[0, \tau]$, and such that this set of solutions is compact in $C([0,\tau];\mathbf{R}^n)$. Denote this set of solutions of (1.1) and (1.4) on $[0,\tau]$ by V. Then V is equicontinuous. Define

$$K = \{ (t,y) \mid t \in [0,\tau], \text{ and } y \geq x(t) \text{ for all } x \in V \} \cup \{(\tau,\infty) \times \mathbf{R}^n\}.$$

Then the assumptions (v)–(vi) of Theorem **12**.2.4 are satisfied. Moreover, by (v), for each $T \in [0, T_\infty)$, the function G maps $C_{\xi,D\cap K}([0,T];\mathbf{R}^n)$ into $C_{\xi,K}([0,T];\mathbf{R}^n)$, and so do all the functions $G(\lambda, \cdot)$. Thus (vii) holds as well. By Theorem **12**.2.4, equations $(1.1)_\lambda$ and $(1.4)_\lambda$ have a noncontinuable solution $x(\lambda, \cdot)$ that is defined on some interval $[0, T_{\max}(\lambda))$ and satisfies $(t, x(\lambda,t)) \in K$ for all $t \in [0, T_{\max}(\lambda))$.

We claim that if x is any solution of (1.1), subject to (1.4), defined on some interval $[0, T_{\max})$, then, for all $\lambda > 0$, we have $x(\lambda, t) > x(t)$ for $t \in (0, T_{\max}(\lambda)) \cap (0, T_{\max})$. To prove this, we first observe that it follows from the monotonicity of G and from the way in which we defined $G(\lambda,\cdot)$ that, if $x(\lambda,t) \geq x(t)$ for all t in some interval $[0,T]$, then it is in fact true that $x(\lambda,t) \geq x(t) + \lambda pt$ on the same interval. In particular, $x(\lambda,t) > x(t)$ for $t \in (0,T]$. We constructed $x(\lambda,\cdot)$ in such a way that $x(\lambda,t) \geq x(t)$ for $t \in [0,\tau]$ (or in $[0, T_{\max}(\lambda))$ if $T_{\max}(\lambda) < \tau$), so $x(\lambda,t) \geq x(t) + \lambda pt > x(t)$ for $t \in (0,\tau]$. If it is not true that $x(\lambda,t) > x(t)$ for all $t \in (0, T_{\max}(\lambda)) \cap (0, T_{\max})$, then there is some $t > \tau$ such that $x(\lambda,t) \not> x(t)$, but $x(\lambda,s) > x(s)$ for $s \in (0,t)$. However, this means that $x(\lambda,s) \geq x(s) + \lambda ps$ for all $s \in (0,t)$ and, by continuity, $x(\lambda,t) \geq x(t) + \lambda pt > x(t)$. This contradiction shows that $x(\lambda,t) > x(t)$ for $t \in (0, T_{\max}(\lambda)) \cap (0, T_{\max})$.

By Theorem 2.1, there is some sequence $\lambda_j \downarrow 0$ such that the functions $x(\lambda_j, \cdot)$ converge to some noncontinuable solution x_M of (1.1), uniformly on compact subintervals of the maximal interval of existence $[0, T_{M,\max})$ of x_M. It is clear that this solution is maximal in the sense of Definition 4.2, because every $x(\lambda, \cdot)$ dominates all the solutions of (1.1).

(The interested reader should observe that not only is it true that $x(\lambda_j, \cdot)$ tends to x_M uniformly on compact subsets of $[0, T_{M,\max})$, but, because of the uniqueness of the limit x_M, the whole family $x(\lambda,\cdot)$ must tend to x_M as $\lambda \downarrow 0$, uniformly on compact subsets of $[0, T_{M,\max})$.)

Under the additional assumption on the cross sections D_t, the fact that every solution x satisfies $x_m(t) \leq x(t) \leq x_M(t)$ forces the curve $(t, x(t))$ to stay in a compact subset of D for t in compact subintervals of $[0, T_{M,\max}) \cap [0, T_{m,\max})$. Therefore, a noncontinuable solution must exist at least on $[0, T_{M,\max}) \cap [0, T_{m,\max})$. □

The following comparison result is a direct consequence of Theorem 4.3 combined with Theorem **12**.2.4.

4.5 Theorem
Let the assumption of Theorem 4.3 hold, and let x_M be the noncontinuable maximal solution of (1.1), subject to (1.4), defined on the maximal interval of existence $[0, T_{M,\max})$. Let $T \in [0, T_{M,\max})$, and assume that $y \in C_{\xi,D}([0,T]; \mathbf{R}^n)$ satisfies

$$y(t) \le G(y)(t), \quad t \in [0,T].$$

Then $y(t) \le x_M(t)$ for $t \in [0,T]$.

The easy proof is left to the reader. (Use Theorem **12**.2.4 to show that there is a solution x of (1.1) that satisfies $x \ge y$. Then use Theorem 4.3.)

Above we studied (1.1) under an assumption close to the one of Theorem **12**.2.1. A similar result is true for (1.2) under the assumption of Theorem **12**.2.7 (plus monotonicity).

4.6 Theorem
*Let the assumption of Theorem **12**.2.7 hold. In addition, suppose that, for all s and t, the function $x \mapsto g(t,s,x)$ is nondecreasing. Then (1.2), subject to (1.5), has a noncontinuable maximal solution x_M and a noncontinuable minimal solution x_m, defined on maximal intervals $[0, T_{M,\max})$ and $[0, T_{m,\max})$, respectively.*

If, in addition, for each $t \in [0, T_{M,\max})$, $y_1 \in D_t$, and for each $y_2 \in D_t$, it is true that the set $\{ y \in \mathbf{R}^n \mid y_1 \le y \le y_2 \}$ is a compact subset of D_t (in the scalar case this is true iff D_t is connected), then every noncontinuable solution of (1.2), subject to (1.5), is defined at least on the interval $[0, T_{M,\max}) \cap [0, T_{m,\max})$.

The assumption that g be nondecreasing should be interpreted as follows: write $D_{ts} = \{ x \mid (t,s,x) \in [0,T_\infty) \times D \}$ and require (for any t, s such that D_{ts} is nonempty) that $x, y \in D_{ts}$, $x \le y$, implies $g(t,s,x) \le g(t,s,y)$.

The proof of Theorem 4.6 is a simplified version of the proof of Theorem 4.3, and is left to the reader. (Define $f(\lambda,\cdot) = f + \lambda p$, for some $p > 0$, and $g(\lambda,\cdot,\cdot) = g$, and let $x(\lambda,\cdot)$ be an arbitrary solution of $(1.2)_\lambda$.)

The corresponding comparison result is as follows.

4.7 Theorem
Let the assumption of Theorem 4.6 hold, and let x_M be the noncontinuable maximal solution of (1.2), subject to (1.5), defined on the maximal interval of existence $[0, T_{M,\max})$. Let $T \in [0, T_{M,\max})$, and let $y \in C_D([0,T]; \mathbf{R}^n)$

satisfy

$$y(t) \leq f(t) + \int_0^t g(t,s,y(s))\,\mathrm{d}s, \quad t \in [0,T].$$

Then $y(t) \leq x_M(t)$ for all $t \in [0,T]$.

The easy proof is left to the reader.

Theorem 4.3 can, of course, be applied to (1.3) to prove that, if H in (1.1) is monotone, then under the assumptions of Theorem **12**.3.1, equation (1.3) has noncontinuable maximal and minimal solutions. However, by using a different technique one can prove a better result, which requires less monotonicity of H. This result can be described as follows. If one can split $H(x)(t)$ in two parts, one that depends only on the present value $x(t)$, and one that depends on past values $x(s)$ with $s \in [0,t)$, then only the second part is required to be monotone. This splitting is realized by requiring the order condition in (iv) to hold only for ϕ, ψ satisfying $\phi(t) > \psi(t)$, $t < T$, but $\phi(T) = \psi(T)$.

The exact formulation is slightly more complicated in the vector case than in the scalar case, so let us first give the scalar version. (We remind the reader of the notation $|a|_- = \max\{-a, 0\}$.)

4.8 Theorem

*Let the assumption of Theorem **12**.3.1 hold with $n = 1$. In addition, suppose that*

(iv) *there is a (nonnegative) function $b \in L^1_{\mathrm{loc}}([0,T_\infty);\mathbf{R})$ such that if $T \in (0,T_\infty)$, and if ϕ, $\psi \in C_{\xi,D}([0,T_\infty);\mathbf{R})$ satisfy $\phi(t) > \psi(t)$ for $t \in [0,T)$ and $\phi(T) = \psi(T)$, then*

$$\int_t^T \big|H(\phi)(s) - H(\psi)(s)\big|_-\,\mathrm{d}s = o\Big(\int_t^T b(s)\,\mathrm{d}s\Big) \text{ as } t \uparrow T.$$

Then (1.3), subject to (1.4), has a noncontinuable maximal solution x_M and a noncontinuable minimal solution x_m, defined on maximal intervals $[0,T_{M,\max})$ and $[0,T_{m,\max})$, respectively. If, in addition, the cross sections $D_t = \{\, y \in \mathbf{R} \mid (t,y) \in D \,\}$ are connected, then every noncontinuable solution of (1.1) satisfying (1.4) is defined at least on the interval $[0,T_{M,\max}) \cap [0,T_{m,\max})$.

Proof As usual, we leave the proof of the existence of a noncontinuable minimal solution to the reader.

To prove that the equation has a noncontinuable maximal solution we proceed in essentially the same way as in the proof of Theorem 4.3. Without loss of generality, suppose that $b(t) \geq 1$ for all t. For each $\lambda > 0$, define $\xi(\lambda) = \xi + \lambda$, $D(\lambda) = D$, and $H(\lambda,x)(t) = H(x)(t) + \lambda b(t)$. With these definitions, the assumption of Theorem 2.3 is satisfied (at least for λ small enough that $(0,\xi(\lambda)) \in D$).

For each sufficiently small $\lambda > 0$, let $x(\lambda, \cdot)$ be a noncontinuable solution of $(1.3)_\lambda$ satisfying $(1.4)_\lambda$. We claim that, if x is an arbitrary noncontinuable solution of (1.3) satisfying (1.4), then $x(\lambda, t) > x(t)$ on the common interval of existence of $x(\lambda, \cdot)$ and x. If not, then, since $x(\lambda, 0) > x(0)$, there exists some $T > 0$ such that $x(\lambda, t) > x(t)$ for $t \in [0, T)$, and $x(\lambda, T) = x(T)$. This, together with (iv), implies that, as $t \uparrow T$,

$$\begin{aligned} 0 &> x(t) - x(\lambda, t) \\ &= \big(x(\lambda, T) - x(\lambda, t)\big) - \big(x(T) - x(t)\big) \\ &= \lambda \int_t^T b(s)\,\mathrm{d}s + \int_t^T \Big\{H\big(x(\lambda, \cdot)\big)(s) - H(x)(s)\Big\}\,\mathrm{d}s \\ &\geq \lambda \int_t^T b(s)\,\mathrm{d}s + o\bigg(\int_t^T b(s)\,\mathrm{d}s\bigg). \end{aligned}$$

But for t sufficiently close to T the final term in this chain of inequalities is strictly positive, and we have a contradiction. This proves that, indeed, $x(\lambda, t) > x(t)$ on the common interval of existence of $x(\lambda, \cdot)$ and x.

The completion of the proof follows the same lines as the proof of Theorem 4.3. □

What makes the vector case more difficult than the scalar case is that, if $x(\lambda, t) > x(t)$ for $t \in [0, T)$ but $x(\lambda, T) \not> x(T)$, then it need not be true that $x(\lambda, T) = x(T)$. In general, it is only true that $x(\lambda, T) - x(T)$ lies on the boundary of the positive cone $\mathcal{K}$. In this case there will exist a linear functional P, that satisfies $P(x(\lambda, T) - x(T)) = 0$, and $Py > 0$ for all y in the interior of $\mathcal{K}$ (the equation $Py = 0$ defines a tangent hyperplane to $\mathcal{K}$ through the point $x(\lambda, T) - x(T)$). This means that one appropriate n-dimensional formulation of Theorem 4.8 is the following.

4.9 Theorem

Let the assumption of Theorem **12**.*3.1 hold. In addition, suppose that*

(iv) *there is a (nonnegative) function $b \in L^1_{\mathrm{loc}}([0, T_\infty); \mathbf{R})$ such that if $P : \mathbf{R}^n \to \mathbf{R}$ is a linear functional satisfying $Pu > 0$ for all $u > 0$, and $Pu = 0$ for some nonzero $u \geq 0$, if $T \in (0, T_\infty)$, and if ϕ, $\psi \in C_{\xi,D}([0, T]; \mathbf{R}^n)$ satisfy $\phi(t) > \psi(t)$ for $t \in [0, T)$ and $P\phi(T) = P\psi(T)$, then*

$$\int_t^T \big|PH(\phi)(s) - PH(\psi)(s)\big|_-\,\mathrm{d}s = o\bigg(\int_t^T b(s)\,\mathrm{d}s\bigg) \text{ as } t \uparrow T.$$

Then (1.3), subject to (1.4), has a noncontinuable maximal solution x_M and a noncontinuable minimal solution x_m, defined on maximal intervals $[0, T_{M,\max})$ and $[0, T_{m,\max})$, respectively.

If, in addition, for each $t \in [0, T_{M,\max})$, $y_1 \in D_t$, and for each $y_2 \in D_t$, it is true that the set $\{\, y \in \mathbf{R}^n \mid y_1 \leq y \leq y_2 \,\}$ is a compact subset of D_t, then every noncontinuable solution of (1.3) satisfying (1.4) is defined at least on the interval $[0, T_{M,\max}) \cap [0, T_{m,\max})$.

The proof is left to the reader.

The corresponding comparison result is the following.

4.10 Theorem

Let the assumption of Theorem 4.9 hold, and let x_M be the noncontinuable maximal solution of (1.3) satisfying (1.4), defined on the maximal interval of existence $[0, T_{M,\max})$. Let $T \in [0, T_{M,\max})$, and let $y \in C_D([0,T]; \mathbf{R}^n)$ be locally absolutely continuous and satisfy $y(0) \leq \xi$, and

$$y'(t) \leq H(y)(t), \quad t \in [0,T].$$

Then $y(t) \leq x_M(t)$ for all $t \in [0,T]$.

To prove this theorem, one repeats the argument in the proof of Theorem 4.9 to show that the approximate solutions $x(\lambda, \cdot)$ satisfy $x(\lambda, t) \geq y(t)$. Letting $\lambda \downarrow 0$ one finds that $x_M(t) \geq y(t)$ for all $t \in [0,T]$.

5. Some Uniqueness Results

In this section, we discuss some additional results on uniqueness of solutions of (1.2). Recall that we have already seen some uniqueness results for (1.1) in Corollary **12**.2.2, in Theorems **12**.4.1, **12**.4.2, 3.1 and 3.2, and some uniqueness results for various cases of (1.2) in Theorem **12**.2.6, Corollary **12**.4.3, and in Theorems 1.2, 3.3. However, the importance of uniqueness for continuous dependence motivates us to consider uniqueness problems once more.

We begin with a preliminary result that isolates a condition that implies that the solution of (1.2) is unique. Note that the existence of a solution follows from Theorem **12**.2.8.

5.1 Theorem

*Let the assumptions of Theorem **12**.2.8 hold. Suppose that for each compact subset M of D there exist a function h and a constant $\eta > 0$ satisfying the following conditions:*

(i) *h maps $[0, T_\infty)^2 \times \mathbf{R}$ into $\mathbf{R}^+$, $h(t,s,y) = 0$ whenever $s > t$ or $y \leq 0$, h satisfies the Carathéodory conditions on $[0, T_\infty)^2 \times \mathbf{R}$, and h is nondecreasing in its last argument;*

(ii) *for each $\rho > 0$ and $T \in (0, T_\infty)$ the function $\sup_{0 \leq y \leq \rho} h(t,s,y)$ belongs to $L^1([0,T]^2; \mathbf{R}^+)$;*

(iii) *if $\psi \in L^\infty_{\mathrm{loc}}([0, T_\infty); \mathbf{R})$, then, for all $t > 0$, the function $s \mapsto h(t,s,\psi(s))$ is integrable on $[0,t]$, and the function $t \mapsto \int_0^t h(t,s,\psi(s))\, ds$ is continuous on $[0, T_\infty)$;*

(iv) *the only nonnegative solution of the equation*

$$y(t) = \int_0^t h(t,s,y(s))\,\mathrm{d}s, \quad t \geq 0, \tag{5.1}$$

is the trivial solution $y \equiv 0$;

(v) *for all* (t,s,x) *and* (t,s,y) *in* $[0,T_\infty) \times M$ *with* $s \leq t \leq s+\eta$, *we have*

$$|g(t,s,x) - g(t,s,y)| \leq h(t,s,|x-y|).$$

Then (1.2), subject to (1.5), has a unique continuous solution.

Proof Suppose that $x_1 \neq x_2$ both solve (1.2) on $[0,T]$, where $T \in (0,T_\infty)$. Let M be the set $\bigcup_{i=1}^{2}\{\,(t,x_i(t)) \mid t \in [0,T]\,\}$, define $\tau = \sup\{\,t \in [0,T] \mid x_1(s) = x_2(s) \text{ for } s \in [0,t]\,\}$, and, without loss of generality, suppose that $T \leq \tau + \eta$, where η is the constant corresponding to the set M. It follows from (v) and from (1.2) that $z \stackrel{\text{def}}{=} |x_1 - x_2|$ satisfies the inequality

$$z(t) \leq \int_0^t h(t,s,z(s))\,\mathrm{d}s, \quad t \in [0,T].$$

Recall that the function $y \mapsto h(t,s,y)$ is nondecreasing and use Theorem **12**.2.9 to show that there exists a solution y of the equation

$$y(t) = \int_0^t h(t,s,y(s))\,\mathrm{d}s, \quad t \geq 0,$$

satisfying $y(t) \geq z(t)$. But, according to our assumption (iv), we must have $y \equiv 0$. This implies that $x_1 \equiv x_2$, and the proof is complete. □

Clearly, the crucial assumption in Theorem 5.1 is (iv). Functions h that have this property are sometimes called *Kamke functions.* When one wishes to verify this condition, the following simple approach is useful. First, note that the proof of Theorem 5.1 shows the validity of the following result.

5.2 Theorem
Let h satisfy conditions (i)–(iv) *in Theorem 5.1, take* $T \in (0,T_\infty)$, *and let* $x \in C([0,T];\mathbf{R})$ *satisfy*

$$x(t) \leq \int_0^t h(t,s,x(s))\,\mathrm{d}s, \quad t \in [0,T].$$

Then $x(t) \leq 0$ *for* $t \in [0,T]$.

Next, let us record the following immediate consequence of Theorem 5.2.

5.3 Corollary

Let h_1 and h_2 be two functions satisfying conditions (i)–(iii) *in Theorem 5.1. In addition, suppose that h_1 satisfies* (iv), *and that there is some constant $\eta > 0$ such that $h_2(t,s,u) \le h_1(t,s,u)$ for all $(t,s,u) \in [0,T_\infty)^2 \times [0,\eta]$ with $s \le t \le s+\eta$. Then h_2 satisfies* (iv).

Thus, if $h_2 \le h_1$, and if $x(t) = \int_0^t h_1(t,s,x(s))\,\mathrm{d}s$ has no nontrivial solution, then neither does $x(t) = \int_0^t h_2(t,s,x(s))\,\mathrm{d}s$ have. Conversely, if the latter equation has a nontrivial solution, then so does the former equation. (Provided, of course, that both h_1 and h_2 satisfy (i)–(iii).)

In the case where $h(t,s,x) = (t-s)^{\alpha-1}g(x)$, for some $\alpha > 0$, one can give a sharp, although rather specialized, result on the existence of nontrivial solutions of (5.1).

5.4 Sharp Uniqueness Theorem

Let $\alpha > 0$, and assume that the following two conditions are satisfied:

(i) *the function $g : \mathbf{R} \to \mathbf{R}$ is continuous and nondecreasing, $g(u) = 0$ for $u \le 0$, and g is continuously differentiable and (strictly) positive on $(0,\infty)$;*

(ii) *the function $u \mapsto g(u)/u$ is nonincreasing on some interval $(0,\delta]$, and, for all $p > 0$, the function $u \mapsto u\big(g(u)/u\big)^p$ is nondecreasing on some interval $[0,\delta_p]$.*

Then the equation

$$x(t) = \int_0^t (t-s)^{\alpha-1} g\big(x(s)\big)\,\mathrm{d}s, \quad t \ge 0, \tag{5.2}$$

has a nontrivial (i.e., nonzero) solution on some interval $[0,T]$ if and only if

$$\int_0^\delta \frac{\mathrm{d}u}{u\big(g(u)/u\big)^{\frac{1}{\alpha}}} < \infty. \tag{5.3}$$

The proof of Theorem 5.4 is given in Section 6. Observe that the proof makes use of the comparison result Theorem **12**.2.5.

The most restrictive condition in the hypothesis of Theorem 5.4 is (ii), which puts a very strong limitation on the growth rate of g close to zero. To see this, write (ii) in the form

(ii*) *the function $u \mapsto g(u)/u$ is nonincreasing on some interval $(0,\delta]$, but, for all $q > 0$, the function $u \mapsto u^q g(u)/u$ is nondecreasing on some interval $[0,\delta_{1/q}]$.*

One class of functions g satisfying (ii*) is

$$g(x) = x(-\ln(x))^\gamma, \quad 0 < x \le \tfrac{1}{2},$$

with $\gamma > 0$ (the values of g on $(\frac{1}{2},\infty)$ are irrelevant). Making use of Corollary 5.3, Theorem 5.4, and of this class of functions g, we get

5.5 Corollary

Let h satisfy conditions (i)–(iii) *in Theorem 5.1.*

(i) *If there exist constants $\alpha > 0$ and $\eta \in (0,1)$, such that $h(t,s,u) \le (t-s)^{\alpha-1}u(-\ln(u))^{\alpha}$ for all $(t,s,u) \in [0,T_\infty)^2 \times [0,\eta]$ with $s \le t \le s+\eta$, then (5.1) has no nontrivial solution.*

(ii) *If there exist constants $\alpha > 0$, $\gamma > \alpha$, $\eta \in (0,1)$, and $T \in [0,T_\infty)$, such that $h(t,s,u) \ge (t-s)^{\alpha-1}u(-\ln(u))^{\gamma}$ for all $(t,s,u) \in [0,T]^2 \times [0,\eta]$ with $s \le t \le s+\eta$, then (5.1) has a nontrivial solution on $[0,T]$.*

Next, we give a nonconvolution version of Theorem 5.4. This theorem is not quite as sharp as Theorem 5.4.

5.6 Theorem

*Let $T > 0$, $\alpha > 0$, $p > 1$, and let k be a nonnegative Volterra kernel of continuous type on $[0,T]$ (see Definition **9**.5.2) such that*

$$\sup_{t\in[0,T]} \int_0^t k(t,s)^p (t-s)^\beta \,\mathrm{d}s < \infty,$$

where $\beta = p(1-\alpha) - 1$. Let g satisfy

(i) *$g : \mathbf{R} \to \mathbf{R}$ is continuous and nondecreasing, $g(u) = 0$ for $u \le 0$, and g is continuously differentiable and (strictly) positive on $(0,\infty)$,*

(ii) *the function $u \mapsto g(u)/u$ is nonincreasing on some interval $(0,\delta]$, and, for all $r > 0$, the function $u \mapsto u\big(g(u)/u\big)^r$ is nondecreasing on some interval $[0,\delta_r]$,*

and suppose that

$$\int_0^\delta \frac{\mathrm{d}u}{u\big(g(u)/u\big)^{\frac{1}{\alpha}}} = \infty.$$

Let $x \in C([0,T];\mathbf{R})$ satisfy

$$x(t) \le \int_0^t k(t,s) g(x(s))\,\mathrm{d}s, \quad t \in [0,T]. \tag{5.4}$$

Then $x(t) \le 0$ on $[0,T]$.

Proof Since $g(u) = 0$ for $u \le 0$ and $g(u)$ is positive on $(0,\infty)$, we may assume that x is nonnegative on $[0,T]$. Moreover, we may without loss of generality take (since x is continuous and $x(0) = 0$) $x(t) \le \min\{1,\delta\}$ for $t \in [0,T]$. Define $q = p/(p-1)$, and use Hölder's inequality in (5.4) to get

$$\begin{aligned} x(t) \le & \left\{ \int_0^t k(t,s)^p (t-s)^\beta \,\mathrm{d}s \right\}^{\frac{1}{p}} \\ & \times \left\{ \int_0^t (t-s)^{q\alpha-1} \Big(g(x(s))\Big)^q \,\mathrm{d}s \right\}^{\frac{1}{q}}, \quad t \in [0,T]. \end{aligned} \tag{5.5}$$

Let $y = x^q$. Note that $y \le x$ since $x \le 1$ and $q > 1$, and use (ii) to conclude that $g(x(s))/x(s) \le g(y(s))/y(s)$ for $s \in [0,T]$. Thus, by (5.5)

$$y(t) \le c^{\frac{q}{p}} \int_0^t (t-s)^{q\alpha-1} \left(\frac{g(y(s))}{y(s)} \right)^q y(s)\,\mathrm{d}s, \quad t \in [0,T], \tag{5.6}$$

where $c = \sup_{t\in[0,T]} \int_0^t k(t,s)^p (t-s)^\beta\,\mathrm{d}s$. Now use Theorem 5.4 to conclude (after some computations) that if we let $h(t,s,y(s))$ denote the integrand in (5.6) then the equation $y(t) = \int_0^t h(t,s,y(s))\,\mathrm{d}s$ has no nontrivial solution. Therefore, Theorem 5.2 may be applied, and it follows that $y(t) \le 0$. Thus $y \equiv 0$. □

We conclude this section with a uniqueness result that is of a different nature from those that we have discussed so far. Here we use the monotonicity of the nonlinearity to get uniqueness, and we do not even require that the right-hand side of the equation depend continuously on x.

Below we define h^* in the same way as in Section **12**.5, i.e., we let $h^*(y) = \bigcap_{\epsilon>0} \overline{\operatorname{conv}}\{\, h(v) \mid |v-y| \le \epsilon \,\}$.

5.7 Theorem

Let the assumptions of Theorem **12**.*5.1 hold, and suppose that the function g in Theorem* **12**.*5.1 is of the form $g(t,s,x) = a(t-s)h(x)$, where $a \in L^1_{\mathrm{loc}}(\mathbf{R}^+;\mathbf{R}^n)$ is of strict positive type (see Definition* **16**.*4.4), and $-h^*$ is monotone, i.e.,*

$$\langle y_1 - y_2, h_1 - h_2 \rangle \le 0, \quad y_1, y_2 \in \mathbf{R}^n, \quad h_1 \in h^*(y_1), \quad h_2 \in h^*(y_2).$$

Then the solution of (1.2) given by Theorem **12**.*5.1 is unique.*

The proof given here depends on some of the results in Chapter 16.

Proof of Theorem 5.7 Let x_1, x_2 both satisfy (1.2) (in the enlarged sense; cf. Theorem **12**.5.1) on some interval $[0,T]$. Then there exist functions w_1 and w_2 in $L^\infty_{\mathrm{loc}}(\mathbf{R}^+;\mathbf{R}^n)$ satisfying $w_1(s) \in h^*(x_1(s))$ and $w_2(s) \in h^*(x_2(s))$ for almost all $s \in [0,T]$, such that

$$x_1(t) - x_2(t) = \int_0^t a(t-s)\big(w_1(s) - w_2(s)\big)\,\mathrm{d}s, \tag{5.7}$$

for almost all $t \in [0,T]$. Form the scalar product of $w_1 - w_2$ and (5.7), and then integrate with respect to t. By the monotonicity of $-h^*$, and since a is of positive type, it follows that

$$\int_0^T \int_0^t \Big\langle w_1(t) - w_2(t), a(t-s)\big(w_1(s) - w_2(s)\big)\,\mathrm{d}s \Big\rangle\,\mathrm{d}t = 0.$$

Now use the strict positivity of a, to conclude that $w_1(t) = w_2(t)$ for almost all $t \in [0,T]$. Consequently, $x_1(t) = x_2(t)$ for almost all $t \in [0,T]$. (If we require x_1 and x_2 to satisfy the equation for all t rather than for almost all t, then x_1 and x_2 are continuous, and $x_1(t) = x_2(t)$ for all $t \in [0,T]$.) □

6. Proof of the Sharp Uniqueness Theorem

We begin the proof of Theorem 5.4 with some simple estimates and well-known results.

6.1 Lemma
Let assumptions (i) *and* (ii) *of Theorem 5.4 hold, and let* $p > 0$ *be arbitrary. Then*

$$u\big(g(u)/u\big)^p \le \int_0^u \big(g(s)/s\big)^p \,\mathrm{d}s \le 2u\big(g(u)/u\big)^p, \quad u \in [0, \delta_{2p}], \tag{6.1}$$

and

$$\tfrac{1}{2}\big(g(u)/u\big)^p \le \tfrac{\mathrm{d}}{\mathrm{d}u}\Big(u\big(g(u)/u\big)^p\Big) \le \big(g(u)/u\big)^p, \quad u \in (0, \delta_{2p}). \tag{6.2}$$

Proof In this proof, let us denote the function $g(u)/u$ by $h(u)$.

Clearly, (6.1) follows directly from (6.2) by integration (and from the fact that (ii*) implies that, for all $q > 0$, $g(u)/u = O\big(u^{-q}\big)$ as $u \downarrow 0$; in particular, $u\big(g(u)/u\big)^p \downarrow 0$ as $u \downarrow 0$). To prove (6.2) we observe that, by assumption (ii*), for each $q > 0$ and $\epsilon > 0$, we have

$$\begin{aligned} 0 \le h(u) - h(u+\epsilon) &= (u+\epsilon)^{-q}\big((u+\epsilon)^q h(u) - (u+\epsilon)^q h(u+\epsilon)\big) \\ &\le (u+\epsilon)^{-q}\big((u+\epsilon)^q - u^q\big)h(u), \quad u \in (0, \delta_{1/q} - \epsilon), \end{aligned}$$

hence

$$0 \le -h'(u) \le \frac{q}{u} h(u), \quad u \in (0, \delta_{1/q}).$$

Taking $q = 1/(2p)$, and differentiating $uh(u)^p$ (recall that this function is nondecreasing) we get (6.2). □

The following lemma is known as Osgood's uniqueness theorem.

6.2 Lemma
Let $H \in C(\mathbf{R}; \mathbf{R})$ *satisfy* $H(y) > 0$ *for* $y > 0$ *and* $H(y) = 0$ *for* $y \le 0$, *and assume that*

$$\int_0^1 \frac{\mathrm{d}u}{H(u)} = \infty.$$

Let $x \in AC_{\mathrm{loc}}(\mathbf{R}^+; \mathbf{R})$ *satisfy the inequalities*

$$x'(t) \le H(x(t)), \quad t \ge 0; \quad x(0) \le 0$$

Then $x(t) \le x(0)$ *on* $\mathbf{R}^+$.

Proof One may, without loss of generality, assume that x is nondecreasing (if not, replace x by the function $t \mapsto \max_{0 \le s \le t} x(s)$ and use Lemma **14**.2.2). It follows that if $x(t) \not\equiv x(0)$ then there exists a point $\tau \ge 0$ such that $x(t) > 0$ when $t > \tau$ but $x(\tau) = 0$. It follows that

$$t - \tau - \epsilon = \int_{x(\tau+\epsilon)}^{x(t)} \frac{\mathrm{d}u}{H(u)},$$

and letting $\epsilon \downarrow 0$ one gets a contradiction. □

6.3 Lemma
For all $\alpha > 0$, $\beta > 0$, and $t > 0$, one has

$$\int_0^t (t-s)^{\alpha-1} s^{\beta-1}\,\mathrm{d}s = B(\alpha,\beta)t^{\alpha+\beta-1}, \tag{6.3}$$

where $B(\alpha,\beta) = \int_0^1 (1-u)^{\alpha-1}u^{\beta-1}\,\mathrm{d}u$ is the beta function evaluated at (α,β).

To prove this, simply change the integration variable. In the proofs below the exact value of $B(\alpha,\beta)$ is of no importance (although it is important that $B(\alpha,\beta) > 0$).

Proof of Theorem 5.4 Assume that (5.2) has a nontrivial solution x on some interval $[0,T]$. Without loss of generality, we may assume that x is not identically zero on any interval of the form $[0,\tau]$ (if it is, then replace x by the function $t \mapsto x(t-\tau)$, where $[0,\tau]$ is the maximal interval on which x vanishes). We claim that, in addition, we may without loss of generality assume that x is nondecreasing. To see this, let V be the set of all continuous nondecreasing functions ϕ on $\mathbf{R}^+$ satisfying $\phi(0) = 0$ and $\phi(t) \geq x(t)$ for $t \in [0,T]$, and observe that (since g is positive and nondecreasing) the operator on the right-hand side of (5.2) maps V into itself. By Theorem **12**.2.5, equation (5.2) has a solution in $V([0,T])$; i.e., (5.2) has a nondecreasing solution, which is positive on $(0,T]$. (The same argument shows that the maximal solution of (5.2), which exists according to Theorem 4.6, is nondecreasing and positive on $(0,T]$.)

In order to get a contradiction, let us assume that (5.3) does not hold, that is,

$$\int_0^\delta \frac{\mathrm{d}u}{u(g(u)/u)^{\frac{1}{\alpha}}} = \infty. \tag{6.4}$$

We have to consider several different cases depending on the value of α.

First we take the case where $\alpha = m$ is a positive integer. Equation (5.2) is then equivalent to the differential equation

$$\begin{aligned} x^{(m)}(t) &= (m-1)!\,g(x(t)), \quad t \in [0,T]; \\ x^{(i)}(0) &= 0, \quad i = 0,1,\ldots,m-1. \end{aligned} \tag{6.5}$$

For $m = 1$ it follows from Lemma 6.2 that (6.4) implies that every nonnegative solution of (6.5) must vanish identically. Let $m \geq 2$. We will show by induction that

$$x^{(m-j+1)}(t) \leq d_j\big(g(x(t))\big)^{1/j}\big(x^{(m-j)}(t)\big)^{(j-1)/j}, \quad t \in [0,T], \tag{6.6}$$

where d_j are certain constants, and $j = 1,2,\ldots,m$. By (6.5), this inequality holds for $j = 1$. Assume that (6.6) holds for a certain integer $j \leq m-1$, multiply both sides of (6.6) by $\left(x^{(m-j)}(t)\right)^{1/j}$, and integrate over $[0,t]$. This gives, since $x^{(m-j)}(0) = x^{(m-j-1)}(0) = 0$ and g is nondecreasing,

$$\begin{aligned}\tfrac{j}{j+1}\left(x^{(m-j)}(t)\right)^{(j+1)/j} &\le d_j\big(g(x(t))\big)^{1/j}\int_0^t x^{(m-j)}(s)\,\mathrm{d}s\\ &= d_j\big(g(x(t))\big)^{1/j}x^{(m-j-1)}(t), \quad t\in[0,T].\end{aligned}$$

We conclude that (6.6) holds with j replaced by $j+1$. Taking $j=m$ in (6.6) we get

$$x'(t) \le d_m\big(g(x(t))/x(t)\big)^{1/m}x(t), \quad t\in[0,T].$$

By Lemma 6.2 and (6.4) x must vanish identically. This completes the proof of the fact that (6.4) implies that (5.2) cannot have any nontrivial solution in the case where α is an integer.

Next we assume that $\alpha > 1$, and that α is not an integer. Write α in the form $\alpha = m+\beta$ where $m \ge 1$ is an integer, and $\beta \in (0,1)$. As the exponent $\alpha - 1 = m+\beta-1$ in (5.2) is positive, x is continuously differentiable (this follows from, e.g., Corollary **3**.7.4). Since both g and x are nondecreasing and g is continuously differentiable on $(0,\infty)$, we must have $g(x(t)) = \int_0^t g'(x(s))x'(s)\,\mathrm{d}s$. If we insert this expression on the right-hand side in equation (5.2) (and recall that $\alpha-1 = m+\beta-1$), then we can change the order of integration and differentiate both sides to get

$$x'(t) = \int_0^t (t-s)^{m+\beta-1}g'(x(s))x'(s)\,\mathrm{d}s, \quad t\in(0,T]. \tag{6.7}$$

This equation can be further differentiated m times, and carrying out the differentiations one gets

$$x^{(m+1)}(t) = c_1\int_0^t (t-s)^{\beta-1}g'(x(s))x'(s)\,\mathrm{d}s, \quad t\in(0,T],$$

where c_1 is a constant (the value of which, although easily computed, is irrelevant). Let

$$h(u) = g(u)/u. \tag{6.8}$$

Then, by (6.2) with $p=1$, $g'(x(t)) \le h(x(t))$. Thus, we get

$$x'(t) \le \int_0^t (t-s)^{m+\beta-1}h(x(s))x'(s)\,\mathrm{d}s, \quad t\in(0,T], \tag{6.9}$$

and

$$x^{(m+1)}(t) \le c_1\int_0^t (t-s)^{\beta-1}h(x(s))x'(s)\,\mathrm{d}s, \quad t\in(0,T]. \tag{6.10}$$

Estimate $x'(s)$ on the right-hand side of (6.10) by (6.9), interchange the order of integration, use the fact that the function $t\mapsto h(x(t))$ is nonincreasing on $[0,T]$, and finally use Lemma 6.3, to conclude that

$$\begin{aligned} x^{(m+1)}(t) &\le c_1 \int_0^t (t-s)^{\beta-1} h(x(s)) \int_0^s (s-v)^{m+\beta-1} h(x(v))x'(v)\,\mathrm{d}v\,\mathrm{d}s \\ &\le c_1 \int_0^t \int_v^t (t-s)^{\beta-1}(s-v)^{m+\beta-1}\,\mathrm{d}s \big(h(x(v))\big)^2 x'(v)\,\mathrm{d}v \\ &= c_2 \int_0^t (t-s)^{m+2\beta-1}\big(h(x(s))\big)^2 x'(s)\,\mathrm{d}s, \quad t \in (0,T], \end{aligned} \tag{6.11}$$

where $c_2 = B(\beta, m+\beta)c_1$. Thus, by Hölder's inequality, (6.1) with $p=1$, and (6.8), for all $t \in (0,T]$ we have (without loss of generality, suppose that T is so small that $x(t) \le \delta_{2(m+1)/(1-\beta)}$ on $[0,T]$)

$$\begin{aligned} x^{(m+1)}(t) &\le c_2 \left\{ \int_0^t \Big(h(x(s))\Big)^{(m+1)/(1-\beta)} x'(s)\,\mathrm{d}s \right\}^{(1-\beta)/(m+\beta)} \\ &\quad \times \left\{ \int_0^t (t-s)^{m+\beta} h(x(s))x'(s)\,\mathrm{d}s \right\}^{(m+2\beta-1)/(m+\beta)} \\ &= c_2 \left\{ \int_0^{x(t)} h(u)^{(m+1)/(1-\beta)}\,\mathrm{d}u \right\}^{(1-\beta)/(m+\beta)} \\ &\quad \times \left\{ \int_0^t (t-s)^{m+\beta} h(x(s))x'(s)\,\mathrm{d}s \right\}^{(m+2\beta-1)/(m+\beta)} \\ &\le 2c_2 \Big(h(x(t))\Big)^{(m+1)/(m+\beta)} \big(x(t)\big)^{(1-\beta)/(m+\beta)} \\ &\quad \times \left\{ \int_0^t (t-s)^{m+\beta} h(x(s))x'(s)\,\mathrm{d}s \right\}^{(m+2\beta-1)/(m+\beta)}. \end{aligned} \tag{6.12}$$

To estimate the final integral one can integrate (6.7), and use (6.2) with $p=1$ to get

$$x(t) \ge c_3 \int_0^t (t-s)^{m+\beta} h(x(s))x'(s)\,\mathrm{d}s,$$

where $c_3 = 1/(2(m+\beta))$. This substituted into (6.12) yields

$$x^{(m+1)}(t) \le c_4 \Big(h(x(t))\Big)^{(m+1)/(m+\beta)} x(t), \quad t \in (0,T],$$

where $c_4 = 2c_2/c_3$. Now we are in a position where we can repeat the induction arguments yielding (6.6) that we used in the case where $\alpha = m$ was an integer ≥ 1 (the fact that the equality in (6.5) is replaced by an inequality is of no consequence). Once we do that and use Lemma 6.2, we may conclude that (5.2) cannot have any nontrivial solution when (6.4) holds, and $\alpha > 1$ is not an integer.

Combining the two cases considered above, we find that (5.2) cannot have any nontrivial solution when (6.4) holds, and $\alpha \ge 1$.

Finally, let us consider the case where $\alpha \in (0,1)$. Assume that we have already shown that if $\alpha \geq 2^{-j}$ then (6.4) implies that (5.2) has no nontrivial solution. By the preceding argument, this is true if $j = 0$. Assume that $\alpha \geq 2^{-(j+1)}$. Since $t \mapsto h(x(t))$ is nonincreasing on $[0,T]$, it follows from (5.2) that

$$\begin{aligned} x(t) &= \int_0^t (t-s)^{\alpha-1} h(x(s)) \int_0^s (s-v)^{\alpha-1} h(x(v)) x(v) \, dv \, ds \\ &\leq \int_0^t \int_v^t (t-s)^{\alpha-1}(s-v)^{\alpha-1} \, ds \Big(h(x(v))\Big)^2 x(v) \, dv \\ &= c_5 \int_0^t (t-s)^{2\alpha-1} \Big(h(x(s))\Big)^2 x(s) \, ds, \quad t \in [0,T], \end{aligned}$$

where $c_5 = B(\alpha,\alpha)$. Since $2\alpha \geq 2^{-j}$, we can use Theorem 5.2 and the induction hypothesis to prove that $x \equiv 0$, and we conclude that the induction hypothesis is true for $\alpha \geq 2^{-(j+1)}$.

As a result, we find that for all values $\alpha > 0$ condition (6.4) implies that (5.2) cannot have any nontrivial solution.

Let us proceed to the second half of the proof, and assume that (5.3) holds. We want to show that (5.2) has a nontrivial solution. The argument we use is similar to the argument given above, but we have to revert the inequalities.

This time we begin by studying the case where α is small, more precisely, $\alpha < \frac{1}{2}$. For every $\lambda \in (0,1)$, let $x(\lambda, \cdot)$ be the maximal, noncontinuable solution of the equation

$$x(\lambda,t) = \lambda t + \int_0^t (t-s)^{\alpha-1} g(x(\lambda,s)) \, ds, \quad t \geq 0 \qquad (5.2)_\lambda$$

(see Theorem 4.6). By the same argument as we used earlier in the proof, $x(\lambda,\cdot)$ is nondecreasing and positive (for $t > 0$). Moreover, each function $x(\lambda,\cdot)$ dominates all solutions of (5.2) (see the proof of Theorem 4.3), and as $\lambda \downarrow 0$, $x(\lambda,\cdot)$ tends to the maximal solution of (5.2), uniformly on some interval $[0,T]$, with $T > 0$. In particular, for λ small enough, the functions $x(\lambda,\cdot)$ are defined on $[0,T]$.

We claim that $x(\lambda,\cdot)$ is absolutely continuous with respect to t on $[0,T]$. Since $x(\lambda,\cdot)$ and g are nondecreasing continuous functions, there exists a nonnegative, continuous measure η_λ such that $g(x(\lambda,t)) = \eta_\lambda([0,t])$. Hence, it follows from equation $(5.2)_\lambda$ that

$$x(\lambda,t) = \lambda t + \int_0^t \int_{[0,s]} (s-v)^{\alpha-1} \eta_\lambda(dv) \, ds.$$

This equation shows (see Theorem **3**.6.1(i)) that $x(\lambda,t)$ is absolutely continuous on $[0,T]$. The function $g(x(\lambda,\cdot))$ must be absolutely continuous as well, with derivative $g'(x(\lambda,\cdot))x'(\lambda,\cdot)$. We conclude that

$$x'(\lambda,t) = \lambda + \int_0^t (t-s)^{\alpha-1} g'(x(\lambda,s)) x'(\lambda,s) \, ds, \quad t \in [0,T]. \qquad (6.13)$$

By Theorem **5**.5.4, the kernel $t \mapsto t^{\alpha-1}$ has a resolvent of the first kind. It follows from Lemma 6.3 that this resolvent is the function $t \mapsto c_6 t^{-\alpha}$, where $c_6 = B(\alpha, 1-\alpha)^{-1}$ is a positive constant. Use the variation of constants formula for resolvents of the first kind (see Theorem **5**.5.3) to solve $g'(x(\lambda,t))x'(\lambda,t)$ from (6.13). After integrating the result and dropping the λ-term we have

$$g(x(\lambda,t)) \le c_6 \int_0^t (t-s)^{-\alpha} x'(\lambda,s)\,\mathrm{d}s, \quad t \in [0,T]. \tag{6.14}$$

Now use Hölder's inequality (recall that $\alpha < \frac{1}{2}$), the left-hand side of (6.2) with $p=1$, (6.8), (6.13), (6.14), and the facts that $x(\lambda,0)=0$ and that h^q is nondecreasing whenever $q<0$, to get, for $t \in (0,T]$,

$$\begin{aligned}
h(x(\lambda,t))x(\lambda,t) &\le c_6 \left\{ \int_0^t \Big(h(x(\lambda,s))\Big)^{-\alpha/(1-2\alpha)} x'(\lambda,s)\,\mathrm{d}s \right\}^{(1-2\alpha)/(1-\alpha)} \\
&\quad \times \left\{ \int_0^t (t-s)^{\alpha-1} h(x(\lambda,s))x'(\lambda,s)\,\mathrm{d}s \right\}^{\alpha/(1-\alpha)} \\
&= c_6 \left\{ \int_0^{x(\lambda,t)} h(u)^{-\alpha/(1-2\alpha)}\,\mathrm{d}u \right\}^{(1-2\alpha)/(1-\alpha)} \\
&\quad \times \left\{ \int_0^t (t-s)^{\alpha-1} h(x(\lambda,s))x'(\lambda,s)\,\mathrm{d}s \right\}^{\alpha/(1-\alpha)} \\
&\le c_6 \big(x(\lambda,t)\big)^{(1-2\alpha)/(1-\alpha)} \Big(h(x(\lambda,t))\Big)^{-\alpha/(1-\alpha)} \Big(2\big(x'(\lambda,t)-\lambda\big)\Big)^{\alpha/(1-\alpha)}.
\end{aligned}$$

Thus, we find that

$$x'(\lambda,t) \ge \lambda + c_7 \Big(h(x(\lambda,t))\Big)^{1/\alpha} x(\lambda,t), \quad t \in (0,T], \tag{6.15}$$

where $c_7 = \frac{1}{2} c_6^{\alpha/(1-\alpha)}$. Since we assume that (5.3) holds, there exists a function v that is the solution of the equation $\int_0^{v(t)} \big(h(u)^{1/\alpha} u\big)^{-1}\,\mathrm{d}u = c_7 t$. Clearly v is a nondecreasing and noncontinuable solution v of the equation

$$v'(t) = c_7 \Big(h(v(t))\Big)^{\frac{1}{\alpha}} v(t), \quad t \ge 0, \tag{6.16}$$

that is positive for $t > 0$. Since $v'(t) \to 0$ as $t \downarrow 0$, but $x(\lambda,t) > \lambda t$, it follows that we must have $x(\lambda,t) \ge v(t)$ on some initial interval $(0, T_\lambda]$, and a standard contradiction argument shows that $x(\lambda,t) > v(t)$ for as long as the two solutions exist (cf. the proof of Theorem 4.3). If we now let $\lambda \downarrow 0$, then, as we already observed above, $x(\lambda,\cdot)$ tends to the maximal solution of (5.2), and therefore (5.2) has a solution $x \ge v$. In particular, (5.2) has a nontrivial solution. This completes the proof of the fact that (5.2) has a nontrivial solution in the case where $\alpha < \frac{1}{2}$ and (5.3) holds.

To complete the proof we use an inductive argument, where for each step we double the set of admissible values of α. Suppose that we have proved the theorem for $\alpha < 2^j$. This we did above with $j = -1$. Let

$2^j \leq \alpha < 2^{j+1}$. As above, define h by (6.8) and let $c_8 = \left(B(\alpha/2, \alpha/2)\right)^{-1}$. It follows from our induction hypothesis that there exists a continuous nondecreasing solution v of the equation

$$v(t) = \sqrt{c_8} \int_0^t (t-s)^{\alpha/2-1} \left(h(v(s))\right)^{\frac{1}{2}} v(s)\,\mathrm{d}s, \quad t \geq 0,$$

that is defined on some interval $[0, T]$, and positive for $t \in (0, T]$. Because of the monotonicity of the functions v and h,

$$\begin{aligned} v(t) &= c_8 \int_0^t (t-s)^{\alpha/2-1} \left(h(v(s))\right)^{\frac{1}{2}} \int_0^s (s-u)^{\alpha/2-1} \left(h(v(u))\right)^{\frac{1}{2}} v(u)\,\mathrm{d}u\,\mathrm{d}s \\ &\leq c_8 \int_0^t (t-s)^{\alpha/2-1} \int_0^s (s-u)^{\alpha/2-1} h(v(u)) v(u)\,\mathrm{d}u\,\mathrm{d}s \\ &= \int_0^t (t-s)^{\alpha-1} g(v(s))\,\mathrm{d}s, \quad t \in [0, T], \end{aligned}$$

where the last equality follows from Lemma 6.3. As $v(t) > 0$ for $t \in (0, T]$, this contradicts the conclusion of Theorem 5.2, unless (5.2) has a nontrivial solution. This proves the induction claim, and completes the proof of Theorem 5.4. □

7. Exercises

1. Show that Theorem 1.1 follows from Theorem 2.2.
2. Show that it is possible, under the hypotheses of Theorem 2.1, to have $T_{\max} < \liminf_{\lambda_j \downarrow 0} T_{\max}(\lambda_j)$.
3. Show with an example that one cannot replace (iv)$_\lambda$ in Theorem 2.1 by the following (cf. Exercise **12**.5).

(iv)$_\lambda$* *For each compact connected subset M of D containing $(0, \xi)$ there exist constants $\mu > 0$, $\eta > 0$, and a function $\delta : (0, \infty) \to (0, \infty)$, such that the following implication holds: if for some $\epsilon > 0$, $T \in (0, T_\infty)$, $\lambda \in (0, \mu)$, $t \in [|T - \eta|_+, T]$, and for some functions $\phi, \psi \in C_{\xi(\lambda), M}([0, T]; \mathbf{R}^n)$, we have*

(1) $\psi(s) = \phi(s)$ *for* $s \in [0, |T - \eta|_+]$,

(2) $|\psi(s) - \phi(s)| \leq \epsilon$ *for* $s \in [|T - \eta|_+, T]$,

(3) $|G(\lambda, \psi)(t) - G(\lambda, \phi)(t)| \leq \delta(\epsilon)$,

then

(4) $|G(\lambda, \psi)(s) - G(\lambda, \phi)(s)| \leq \frac{\epsilon}{2}$ *for* $s \in [t, t + \delta(\epsilon)] \cap [0, T]$.

Hint: Define the functions y_λ, $\lambda > 0$ by

$$y_\lambda(t) = t^2 \sin(t/\lambda) + t, \quad t \geq 0.$$

Take $G(\lambda, \cdot)$ to be

$$G(\lambda, \varphi)(t) = y_\lambda(t) \left| 1 - 8 \sup_{0 \le \sigma \le t} \left\{ \left(1 - \tfrac{\sigma}{t}\right) \frac{|\varphi(\sigma) - y_\lambda(\sigma)|}{\sigma^2 + \lambda} \right\} \right|_+ ,$$

for $\lambda > 0$ and $t \ge 0$, and

$$G(\varphi) \equiv 0.$$

4. Formulate and prove a continuous dependence result based on Corollary **12**.2.2.
5. Prove Theorem 2.3.
6. Formulate and prove a continuous dependence result based on Corollary **12**.4.3.
7. Formulate and prove a differentiability result with respect to a parameter λ for equation (1.2).
8. Prove a version of Theorem 3.1, where (ii) is required to hold only for one specific family $x(\lambda, \cdot)$, and where the uniqueness claim is dropped.
9. Prove Theorem 3.2.
10. Formulate and prove a differentiability result similar to Theorem 3.2, but for L^p-solutions. Take the hypothesis of Theorem 2.4 with $K = \mathbf{R}^+ \times \mathbf{R}^n$ as the starting point.
11. Let the assumption of Theorem **12**.2.1 hold with $D = \mathbf{R}^+ \times \mathbf{R}^n$, and let the assumption of Theorem 4.3 hold with $D = \mathbf{R}^+ \times \mathbf{R}$, and with G and ξ replaced by F and ζ. In addition, suppose that $|\xi| \le \zeta$, and that, for all $T > 0$ and all $\phi \in C([0,T]; \mathbf{R}^n)$ it is true that $|G(\phi)(t)| \le F(|\phi|)(t)$ for $t \in [0,T]$. Let y_M be the maximal solution of the equation

$$y(t) = F(y)(t), \quad t \ge 0; \quad y(0) = \zeta, \tag{7.1}$$

with maximal interval of existence $[0, T_{M,\max})$. Prove that every noncontinuable solution x of (1.1) is defined on (at least) the interval $[0, T_{M,\max})$, and satisfies $|x(t)| \le y_M(t)$ for $t \in [0, T_{M,\max})$.
12. Prove Theorems 4.6 and 4.7.
13. Let the assumption of Theorem **12**.2.7 hold with $D = \mathbf{R}^+ \times \mathbf{R}^n$, and let the assumption of Theorem 4.6 hold with $D = \mathbf{R}^+ \times \mathbf{R}$, and with f and g replaced by F and G. In addition, suppose that, for all $t > 0$, $s \in [0,t]$ and $y \in \mathbf{R}^n$, it is true that $|f(t)| \le F(t)$, and $|g(t,s,y)| \le G(t,s,|y|)$. Let y_M be the maximal solution of the equation

$$y(t) = F(t) + \int_0^t G(t, s, y(s))\, \mathrm{d}s, \quad t \ge 0,$$

with maximal interval of existence $[0, T_{M,\max})$. Prove that every noncontinuable solution x of (1.2) is defined (at least) on the interval

$[0, T_{M,\max})$, and satisfies $|x(t)| \le y_M(t)$ for $t \in [0, T_{M,\max})$.

14. Prove Theorem 4.9.
15. Show that there exists at most one solution of the equation

$$x(t) + \int_0^t a(t-s)g(x(s))\,ds = f(t), \quad t \ge 0,$$

if $a \in L^1_{\mathrm{loc}}(\mathbf{R}^+;\mathbf{R})$ is of positive type (see Chapter 16), and $g \in C(\mathbf{R};\mathbf{R})$ is strictly increasing.

16. Prove a nonconvolution version of Exercise 15, and also a nonconvolution version of Theorem 5.7.
17. What assumptions do you need on a, k, and f, in order to apply Theorem 4.8 to the equation

$$x'(t) + a(t)x(t) + \int_0^t k\big(t,s,x(s)\big)\,ds = f(t), \quad t \ge 0; \quad x(0) = \xi.$$

Hint: Choose $b(t) = |a(t)|$.

18. Consider the scalar equation

$$x'(t) = a(t)g(x(t)) + b(t)g\big(x(t-1)\big),$$

with initial condition $x(t) = \phi(t)$, $-1 \le t \le 0$. Assume that $a \ge 0$, $a \in L^1_{\mathrm{loc}}(\mathbf{R}^+;\mathbf{R})$, $b \in L^1_{\mathrm{loc}}(\mathbf{R}^+;\mathbf{R})$ and that $g \in C(\mathbf{R};\mathbf{R})$ is nondecreasing and positive. Does this equation have a unique solution?

8. Comments

Section 2:

Miller [10], Sect. II.4, contains a number of results on continuous dependence of the solutions of (1.2) on the given functions f and g. See also Miller and Sell [1] and the correction in Miller and Sell [3]. The same problem is considered in Artstein [1] and [2], where, in addition, the question of necessary conditions for continuous dependence are given. A statement in Miller [10], p. 107 and p. 145, concerning topologies that give continuous dependence is straigthened out by Gyllenberg [1]. Note that our Theorem 2.2 (which is a corollary to Theorem 2.1) is practically the same as the sufficiency part of Artstein [1], Theorem A, p. 448. Kelley [1] studies the continuous dependence of solutions of (1.2) on the given data f and g, under assumptions analogous to those of Miller [10], Sect. II.4, and gives conditions under which the set of all solutions of (1.2) is compact and connected. For related results, see Deimling [1], Kikuchi and Nakagiri [1], Reichert [1], and Szufla [1].

Section 3:

Differentiability results given in the literature are mostly analogous to Theorem 1.2. Two reasons motivate the more general hypothesis in Theorem

3.1, including the differentiability requirement along the solutions: the assumption is relatively easy to formulate, and one can prove differentiability results under quite weak hypotheses.

Section 4:
In general, maximal and minimal solutions of ordinary differential equations are discussed in the literature under assumptions that are stronger than those that we have used in Theorem 4.9 (even in the case of an ordinary differential equation). Miller [10], Thm. II.5.1, is approximately the same as our Theorem 4.6. Some monotonicity results for (1.3) are given in Smith [1]. A counterexample showing that there does not, in general, exist a maximal solution is given by Gollwitzer and Hager [1].

Section 5:
The uniqueness result contained in Theorem 5.1 is essentially the same as Miller [10], Thm. II.6.2. Theorem 5.4 is taken from Gripenberg [15]. For some analogous results and extensions, see Bushell and Okrasiński [1].

References

Z. Artstein

1. Continuous dependence of solutions of Volterra integral equations, *SIAM J. Math. Anal.* **6** (1975), pp. 446–456.
2. Continuous dependence on parameters: on best possible results, *J. Differential Equations* **19** (1975), pp. 214–225.

P. J. Bushell and W. Okrasiński

1. Uniqueness of solutions for a class of non-linear Volterra integral equations with convolution kernel, *Math. Proc. Cambridge Philos. Soc*, to appear 1989.

K. Deimling

1. Eigenschaften der Lösungsmenge eines Systems von Volterra-Integralgleichungen, *Manuscripta Math.* **4** (1971), pp. 201–212.

H. E. Gollwitzer and R. A. Hager

1. The nonexistence of maximum solutions of Volterra integral equations, *Proc. Amer. Math. Soc.* **26** (1970), pp. 301–304.

G. Gripenberg

15. Unique solutions of some Volterra integral equations, *Math. Scand.* **48** (1981), pp. 59–67.

M. Gyllenberg

1. A note on continuous dependence of solutions of Volterra integral equations, *Proc. Amer. Math. Soc.* **81** (1981), pp. 546–548.

W. G. Kelley

1. A Kneser theorem for Volterra integral equations, *Proc. Amer. Math. Soc.* **40** (1973), pp. 183–190.

N. Kikuchi and S. Nakagiri

1. An existence theorem of solutions of non-linear integral equations, *Funkcial. Ekvac.* **15** (1972), pp. 131–138.

R. H. Martin

1. *Nonlinear Operators and Differential Equations in Banach Spaces*, J. Wiley, New York, 1976.

R. K. Miller

10. *Nonlinear Volterra Integral Equations*, W. A. Benjamin, Menlo Park, Calif., 1971.

R. K. Miller and G. R. Sell

1. Existence, uniqueness and continuity of solutions of integral equations, *Ann. Mat. Pura Appl. (IV)* **80** (1968), pp. 135–152.
3. Existence, uniqueness and continuity of solutions of integral equations. An addendum, *Ann. Mat. Pura Appl. (IV)* **87** (1970), pp. 281–286.

M. Reichert

1. Über die Fixpunktmengen einer Klasse Volterrascher Integraloperatoren in Banachräumen, *J. Reine Angew. Math.* **258** (1973), pp. 173–185.

H. Smith

1. Monotone semiflows generated by functional differential equations, *J. Differential Equations* **66** (1987), pp. 420–442.

S. Szufla

1. Solutions sets of non-linear integral equations, *Funkcial. Ekvac.* **17** (1974), pp. 67–71.

14

Lyapunov Techniques

We use methods of Lyapunov type to derive conclusions about boundedness and asymptotic behaviour of solutions of integral and functional differential equations. The emphasis is on the main ideas; we first formulate some general principles, and then illustrate these principles by applying them to a number of examples.

1. Introduction

In earlier chapters, we studied the asymptotic behaviour of linear convolution equations by means of Laplace transform methods. In addition, we gave some results for nonlinear equations based on linearization, i.e., we assumed that the equation had a small nonlinear part which was dominated by the linear part.

Here we present a different approach, a Lyapunov-type approach, which can be applied to certain nonlinear equations without a dominant linear part, as well as to certain linear equations that are not of convolution type. In particular, this approach is applicable to equations that have a dominant 'ordinary differential equation part', i.e., to equations of the type

$$x'(t) = f\big(t, x(t)\big) + F(x)(t), \quad t \in \mathbf{R}^+; \quad x(0) = \xi, \tag{1.1}$$

where the functional term $F(x)(t)$ is small in some sense compared with the term $f(t, x(t))$, and where the ordinary differential equation

$$x'(t) = f\big(t, x(t)\big), \quad t \in \mathbf{R}^+; \quad x(0) = \xi, \tag{1.2}$$

has a stable solution.

The Lyapunov approach can be illustrated by the following definitions and by Proposition 1.1.

Consider the autonomous ordinary differential equation

$$x'(t) = f\big(x(t)\big), \quad t \in \mathbf{R}^+; \quad x(0) = \xi. \tag{1.3}$$

If U is a neighbourhood of zero in $\mathbf{R}^n$, then we call a function $V : U \to \mathbf{R}$ *positive definite on* U if V is continuous, $V(0) = 0$, and $V(y) > 0$ for $y \in U$, $y \neq 0$. Let us denote the solution of (1.3) by $x(t, \xi)$ (assuming that a unique solution exists for each $\xi \in U$), and define

$$\dot{V}(\xi) = \limsup_{h \to 0} \frac{1}{h}\{V(x(h, \xi)) - V(\xi)\}.$$

Clearly, under appropriate differentiability assumptions, one has $\dot{V}(\xi) = \langle V'(\xi), f(\xi)\rangle$, where V' is the gradient of V.

The solution $x = 0$ is said to be *stable*, if for every $\epsilon > 0$ there is a $\delta > 0$ such that $|\xi| \leq \delta$ implies $|x(t, \xi)| \leq \epsilon$ for $t \geq 0$. The zero solution is called *asymptotically stable* if it is stable, and if there exists a $\gamma > 0$ such that $|\xi| \leq \gamma$ implies that $x(t, \xi) \to 0$ as $t \to \infty$.

The following result is well known; see, e.g., Hale [3], Theorem 1.1, p. 293.

1.1 Proposition

If there is a positive definite function V on U with $\dot{V} \leq 0$, then the solution $x = 0$ of (1.3) is stable. If, in addition, $-\dot{V}$ is positive definite on U, then the solution $x = 0$ is asymptotically stable.

A related result says that if $\dot{V}$ is positive definite then the zero solution is unstable. (To prove this, it suffices to observe that by Proposition 1.1 the equation is asymptotically stable in the backwards time direction. Actually, it is possible to prove instability under somewhat weaker assumptions; see Hale [3], Theorem 1.2, p. 294.)

A function V that satisfies the conditions of Proposition 1.1 is called a *Lyapunov function* for (1.3).

In the asymptotic theory for ordinary differential equations, the Lyapunov approach plays an important role. In particular, it can be proved that under quite weak assumptions there exists a Lyapunov function for (1.3) whenever the zero solution is asymptotically stable (see, e.g., Hale [3], Theorem 4.2, p. 309).

Much effort has gone into obtaining similar results for functional differential equations and for integrodifferential equations. The formal theory presents few obstacles, but the difficulties involved in the actual construction of a Lyapunov functional for, say, an arbitrary integrodifferential equation are frequently insurmountable. This fact severely restricts the applicability of the Lyapunov theory to integral and integrodifferential equations. However, as we shall see in Chapters 17–19, some of the basic ideas in the Lyapunov approach, combined with transform techniques, can be used to give a quite detailed description of the asymptotic behaviour of a special class of nonlinear convolution integral equations.

The direct analogues of Proposition 1.1 for integral and functional equations are of little practical interest. Therefore, instead of proving these analogues, we first develop some basic principles which can be used to prove that solutions of integral and functional equations are bounded. In addition, we give some examples on how these principles can be applied. Then we continue with a discussion on the corresponding problem concerning the existence of limits at infinity.

We consider real functions and equations only, but since $\mathbf{C}^n$ can be embedded in $\mathbf{R}^{2n}$ this is not a vital restriction.

2. Boundedness

Before studying a more general theory, let us look at a specific example. We let x be a solution of the scalar equation

$$\begin{aligned} x'(t) + a(t)g(x(t)) + b(t)g\big(x(t-1)\big) &= f(t), \quad t \in \mathbf{R}^+; \\ x(t) &= \varphi(t), \quad t \in [-1, 0], \end{aligned} \tag{2.1}$$

and prove the following easy boundedness result.

2.1 Proposition

Assume that

(i) a, $b \in L^1_{\mathrm{loc}}(\mathbf{R}^+; \mathbf{R})$, *and* $a(t) \geq |b(t)|$ *for almost all* $t \in \mathbf{R}^+$,

(ii) $g \in C(\mathbf{R}; \mathbf{R})$ *is odd and nondecreasing,*

(iii) $f \in L^1(\mathbf{R}^+; \mathbf{R})$,

(iv) $\varphi \in C([-1, 0]; \mathbf{R})$,

(v) $x \in AC_{\mathrm{loc}}(\mathbf{R}^+; \mathbf{R})$ *is a solution of (2.1) on* $\mathbf{R}^+$.

Then x is bounded. More precisely,

$$|x(t)| \leq \max_{-1 \leq s \leq 0} |\varphi(s)| + \int_0^t |f(s)| \, \mathrm{d}s, \quad t \in \mathbf{R}^+. \tag{2.2}$$

Clearly, formula (2.2) is equally valid in the case where f is only locally integrable.

The proof of Proposition 2.1 is based on the following lemma, which will play a crucial role later.

2.2 Lemma

Let x be a real-valued locally absolutely continuous function on $\mathbf{R}^+$, and define $z(t) = \max_{0 \leq s \leq t} x(s)$. Then z is locally absolutely continuous, $z'(t) = x'(t)$ for almost all t in the set E defined by

$$E = \{\, t \in \mathbf{R}^+ \mid x(t) = z(t) \,\},$$

and $z'(t) = 0$ for all $t \notin E$. In particular, $z(t) = x(0) + \int_0^t \chi_E(s) x'(s) \, \mathrm{d}s$ for all $t \geq 0$, and x is bounded from above if and only if $\chi_E x'$ belongs to $L^1(\mathbf{R}^+; \mathbf{R})$.

Lemma 2.2 is proved in Section 4.

Proof of Proposition 2.1 Define $y(t) = |x(t)|$ for $t \geq 0$. As x is a solution of (2.1), x is locally absolutely continuous, hence so is y (see Lemma 4.5(iii)).

If it is true for all $t \geq 0$ that $y(t) = |x(t)| \leq \max_{-1 \leq s \leq 0} |\varphi(s)|$, then (2.2) holds, and there is nothing to prove. If not, then there exists a number $\tau \geq 0$ such that $\max_{0 \leq s \leq \tau} y(s) = y(\tau) = \max_{-1 \leq s \leq 0} |\varphi(s)|$. Define z and E as in Lemma 2.2, with x replaced by y and $\mathbf{R}^+$ replaced by $[\tau, \infty)$. Then, by (2.1), (ii), Lemma 4.5(iii), and the facts that $a(t) \geq |b(t)|$ and $|g(x(t))| \geq |g(x(t-1))|$ for $t \in E$, one has, for almost all $t \in E$,

$$y'(t) = \operatorname{sign}(x(t))x'(t) \leq -a(t)|g(x(t))| + |b(t)||g(x(t-1))| + |f(t)| \leq |f(t)|.$$

The conclusion of Proposition 2.1 now follows from Lemma 2.2. □

The previous argument can be generalized in an obvious way. Above we studied the function $y = |x|$ and obtained an *a priori* bound for y, which automatically gave us an *a priori* bound for x. Here it was natural to study the function $|x|$, but in some other equations it may be more appropriate to consider some (generally nonlinear) functional $V(x)$ of x, that has properties similar to those of the function $|x|$. In other words, to a given x in, e.g., $C(\mathbf{R}^+; \mathbf{R}^n)$ or $AC_{\text{loc}}(\mathbf{R}^+; \mathbf{R}^n)$ we assign some *Lyapunov function* $V(x)(t)$. We require V to dominate x in the sense that $\sup_{t \in \mathbf{R}^+} V(x)(t) < \infty$ implies $\sup_{t \in \mathbf{R}^+} |x(t)| < \infty$. In addition, we require V to be locally absolutely continuous with respect to the t-variable. Instead of applying Lemma 2.2 to x or to $|x|$, we apply it to the function $V(x)$. In general, we shall have very little to say about how $V(x)$ depends on x (as we mentioned above, there is no universal rule for how one should choose V).

The argument outlined above can be formalized in the following principle.

2.3 First Boundedness Principle
Let $x \in C(\mathbf{R}^+; \mathbf{R}^n)$, let $V(x) \in AC_{\text{loc}}(\mathbf{R}^+; \mathbf{R})$, and assume that x is bounded whenever $V(x)$ is bounded from above. In addition, suppose that there exists a function $f \in L^1(\mathbf{R}^+; \mathbf{R})$ such that $[V(x)]'(t) \leq f(t)$ for almost all t in the set E defined by

$$E = \{ t \in \mathbf{R}^+ \mid V(x)(t) = \max_{0 \leq s \leq t} V(x)(s) \}.$$

Then x is bounded.

The converse is true too: if we define the set E as in Principle 2.3, then the function f can be chosen to be equal to $\chi_E V(x)'$, and by Lemma 2.2, this function is integrable whenever the function $V(x)$ is bounded from above.

In many cases one would like to study a whole class of functions x, e.g., all solutions of a given integral or functional equation corresponding to a given set of data. In this case, one needs a uniform bound which applies to all functions x under study. It is obvious that Principle 2.3 can be turned into a *Uniform Boundedness Principle*, suitable for this case. One simply assumes that $V(x)(0)$ is uniformly bounded in x, that a uniform upper bound on $V(x)$ gives a uniform bound on x, and that the L^1-norm of the function f can be bounded uniformly in x. See Exercise 4.

Although our boundedness Principle 2.3 and its uniform extension are fairly general, it is sometimes possible to bypass these principles by using different arguments. For example, in the proof of Theorem **9**.9.1, where we studied the equation $x(t)+\int_{-\infty}^{t} k(t,s)x(s)\,\mathrm{d}s = f(t)$, $t \in \mathbf{R}$, we did not get a bound on the derivative of $V(x) = |x|$ on the set E; instead we directly obtained a bound on $V(x)(t)$ for $t \in E$.

Observe that in the boundedness Principle 2.3 one looks exclusively at the values of $[V(x)]'(t)$ at those points where $V(x)(t)$ is larger than it has ever been before. As a result, the bound $f(t)$ that we get on $[V(x)]'(t)$ at these points must be nonnegative. Sometimes it is more appropriate to include a larger set of points, and to allow the bound on $[V(x)]'(t)$ to take negative values as well. The following easy lemma illustrates this idea.

2.4 Lemma

Let x be a real-valued locally absolutely continuous function on $\mathbf{R}^+$, let α and β be two constants satisfying $x(0) \le \alpha < \beta$, and let $f \in L^1_{\mathrm{loc}}(\mathbf{R}^+;\mathbf{R})$ be a function satisfying $\int_t^T f(s)\,\mathrm{d}s < \beta - \alpha$ whenever $0 \le t < T < \infty$. If it is true that $x'(t) \le f(t)$ for almost all t in the set E defined by

$$E = \{\, t \in \mathbf{R}^+ \mid x(t) \ge \alpha \text{ and } \max_{0\le s\le t} x(s) \le \beta \,\},$$

then $x(t) < \beta$ for all $t \ge 0$.

If the function f above is nonnegative, then this result is a simple consequence of Lemma 2.2.

Proof of Lemma 2.4 Suppose that $x(t) = \beta$ for some $t > 0$. Since $x(0) \le \alpha < \beta$, there exist numbers τ and T such that $x(s) < \beta$ for $s \in [0,\tau]$, $\alpha < x(s) < \beta$ for $s \in (\tau, T)$, $x(\tau) = \alpha$, and $x(T) = \beta$. In particular, the interval $[\tau, T]$ is contained in E. Therefore, for almost all $t \in [\tau, T]$ we have $x'(t) \le f(t)$. This implies that

$$\beta - \alpha = x(T) - x(\tau) = \int_\tau^T x'(s)\,\mathrm{d}s \le \int_\tau^T f(s)\,\mathrm{d}s < \beta - \alpha,$$

and we get a contradiction. □

Minor modifications of the proof yield the following result.

2.5 Corollary

Let x be a real-valued locally absolutely continuous function on $\mathbf{R}^+$, let α and β be two constants satisfying $x(0) \le \alpha < \beta$, and let $f \in L^1_{\mathrm{loc}}(\mathbf{R}^+;\mathbf{R})$ be a function satisfying $\sup_{0\le t<\tau<\infty} \int_t^\tau f(s)\,\mathrm{d}s < \beta - \alpha$. If it is true that $x'(t) \le f(t)$ for almost all t in the set E defined by

$$E = \{\, t \in \mathbf{R}^+ \mid x(t) \ge \alpha \text{ and } \max_{0\le s\le t} x(s) \le \beta \,\},$$

then $\sup_{t\ge 0} x(t) \le \alpha + \sup_{0\le t<\tau<\infty} \int_t^\tau f(s)\,\mathrm{d}s < \beta$.

Clearly, we get a new boundedness principle of the same type as Principle 2.3, by applying Lemma 2.4 instead of Lemma 2.2.

2.6 Second Boundedness Principle

Let $T \in \mathbf{R}$ and assume that

(i) $x \in C([T,\infty);\mathbf{R}^n)$, $V(x) \in AC_{\mathrm{loc}}([T,\infty);\mathbf{R})$, *and x is bounded whenever $V(x)$ is bounded from above,*

(ii) *α and β are two constants satisfying $V(x)(T) \le \alpha < \beta$,*

(iii) *$f \in L^1_{\mathrm{loc}}([T,\infty);\mathbf{R})$ is a function satisfying $\int_\sigma^\tau f(s)\,\mathrm{d}s < \beta - \alpha$ whenever $T \le \sigma < \tau < \infty$,*

(iv) *$[V(x)]'(t) \le f(t)$ for almost all t in the set E defined by*

$$E = \{\, t \in [T,\infty) \mid V(x)(t) \ge \alpha \text{ and } \max_{T\le s\le t} V(x)(s) \le \beta \,\}.$$

Then x is bounded.

To illustrate the use of this principle, let us prove an alternative boundedness result for equation (2.1).

2.7 Proposition

Assume that

(i) *a, $b \in L^1_{\mathrm{loc}}(\mathbf{R}^+;\mathbf{R})$, and $a(t) \ge \epsilon + (1+\epsilon)|b(t)|$ for some $\epsilon > 0$ and almost all $t \in \mathbf{R}^+$,*

(ii) *$g \in C(\mathbf{R};\mathbf{R})$ is odd and nondecreasing, $\lim_{y\to\infty} g(y+1)/g(y) = 1$, and $\lim_{y\to\infty} g(y) = \infty$,*

(iii) *$f \in L^1_{\mathrm{loc}}(\mathbf{R}^+;\mathbf{R})$, and $\sup_{t\ge 0}\int_t^{t+1}|f(s)|\,\mathrm{d}s < \infty$,*

(iv) *$\varphi \in C([-1,0];\mathbf{R})$,*

(v) *$x \in AC_{\mathrm{loc}}(\mathbf{R}^+;\mathbf{R})$ is a solution of (2.1) on $\mathbf{R}^+$.*

Then x is bounded.

Proof Define $F = \sup_{t\ge 0}\int_t^{t+1}|f(s)|\,\mathrm{d}s$. Then, for all real positive constants c, one has $\int_t^{t+c}|f(s)|\,\mathrm{d}s \le (c+1)F$ for $t \ge 0$.

Our hypothesis on g implies that we can find some $\alpha \ge \max_{-1\le s\le 0}|\varphi(s)|$ such that $g(\alpha) > F/\epsilon$. A slightly less evident fact is that by choosing α large enough we can be sure that $(1+\epsilon)g(\alpha) \ge g(\alpha+F)$. This is true because if k is an arbitrary integer, then $g(y+k)/g(y)$ can be written in the form

$$\frac{g(y+k)}{g(y)} = \frac{g(y+k)}{g(y+k-1)}\,\frac{g(y+k-1)}{g(y+k-2)}\cdots\frac{g(y+1)}{g(y)},$$

and here each factor tends to 1 as $y \to \infty$; hence $g(y+k)/g(y) \to 1$ as $y \to \infty$ as well.

Let $\beta = \alpha + F$, define $y = V(x) = |x|$, and let E be the set defined in

Principle 2.6 (take $T = 0$). Then, for almost all $t \in E$,

$$\begin{aligned} y'(t) &\leq -a(t)g(y(t)) + \big|b(t)g\big(x(t-1)\big)\big| + |f(t)| \\ &\leq -a(t)g(\alpha) + |b(t)|g(\beta) + |f(t)| \\ &\leq -\big\{\epsilon + (1+\epsilon)|b(t)|\big\}g(\alpha) + |b(t)|g(\beta) + |f(t)| \\ &\leq -\epsilon g(\alpha) + |f(t)|. \end{aligned}$$

For all $t \geq 0$ and all $c > 0$ we have

$$\int_t^{t+c} \big(-\epsilon g(\alpha) + |f(s)|\big)\,\mathrm{d}s \leq -\epsilon g(\alpha)c + (c+1)F < F.$$

Thus, Principle 2.6 applies. □

It is quite obvious from the proofs of Propositions 2.1 and 2.7 that analogous results are valid for the more general class of equations

$$\begin{aligned} x'(t) + a(t)g(x(t)) + \int_{(-\infty,t)} \kappa(t,\mathrm{d}s)g(x(s)) &= f(t), \quad t \in \mathbf{R}^+; \\ x(t) &= \varphi(t), \quad t \in (-\infty, 0], \end{aligned} \tag{2.3}$$

as well. See Exercises 5–8 below.

In Lemmas 2.2 and 2.4, we obtained boundedness results by studying the derivative $x'(t)$ at the points in the interval $[0,t]$ where $x(t)$ attains its maximum, or is close to its maximum. Sometimes it is more appropriate to investigate the behaviour of $x'(t)$ for all those t where $x(t)$ belongs to an *a priori* given set E. The following lemma is a result of this type. Here m represents the Lebesgue measure, and $|a|_\pm = \max\{\pm a, 0\}$.

2.8 Lemma

Let x and p be real-valued locally absolutely continuous functions on $\mathbf{R}^+$, let $f \in L^1_{\mathrm{loc}}(\mathbf{R}^+;\mathbf{R})$, and let $E \subset \mathbf{R}$. Define $F = \{\, t \in \mathbf{R}^+ \mid x(t) \in E \,\}$. If it is true that

$$\begin{aligned} &x'(t) + p'(t) \leq f(t) \text{ for almost all } t \in F, \\ &p'(t) \leq 0 \text{ for almost all } t \in \mathbf{R}^+ \setminus F, \end{aligned}$$

then, for all $t \geq 0$,

$$m\big(E \cap [x(0), x(t)]\big) + p(t) \leq p(0) + \int_0^t |f(s)|_+\,\mathrm{d}s, \quad \text{if} \quad x(t) \geq x(0).$$

Proof We have, for almost all $s \in \mathbf{R}^+$,

$$\begin{aligned} \chi_F(s)x'(s) + \chi_F(s)p'(s) &\leq \chi_F(s)f(s), \\ \big(1 - \chi_F(s)\big)p'(s) &\leq 0. \end{aligned}$$

Add these inequalities, and integrate over $[0,t]$ to get

$$\int_0^t \chi_F(s)x'(s)\,\mathrm{d}s + p(t) \le p(0) + \int_0^t \chi_F(s)f(s)\,\mathrm{d}s,$$
$$\le p(0) + \int_0^t |f(s)|_+\,\mathrm{d}s.$$

Since $\int_0^t \chi_F(s)x'(s)\,\mathrm{d}s = \int_{x(0)}^{x(t)} \chi_E(u)\,\mathrm{d}u$, we obtain the desired conclusion. □

This lemma may be reformulated as a boundedness principle. We leave this formulation to the reader (see Exercise 12), and proceed to give an application of the lemma to the equation

$$x'(t) + a(t)g(x(t)) + \int_{[0,t)} \kappa(t,\mathrm{d}s)g(x(s)) = f(t), \quad t \in \mathbf{R}^+;$$
$$x(0) = x_0. \tag{2.4}$$

Essentially, this is the same equation as (2.3). To get (2.4) from (2.3), one simply includes the term $\int_{(-\infty,0)} \kappa(t,\mathrm{d}s)g(\varphi(s))$ in $f(t)$.

Our basic assumptions are that a is nonnegative, that κ is nonpositive, and that a dominates κ. More precisely, if we want to apply Proposition 2.9 to equation (2.1), then we have to take $b(t) \le 0$, and $a(t) \ge -b(t+1)$ for all $t \ge 0$. In the convolution case, under the assumptions on κ that we use below, (2.4) is the differentiated version of the equation

$$x(t) + \int_0^t k(t-s)g(x(s))\,\mathrm{d}s = x_0 + \int_0^t f(s)\,\mathrm{d}s,$$

with a nonnegative and nonincreasing kernel k.

2.9 Proposition

Assume that

(i) $a \in L^1_{\mathrm{loc}}(\mathbf{R}^+;\mathbf{R})$ *is nonnegative,*

(ii) κ *is a nonpositive (real-valued) Volterra measure kernel of type* $[B^\infty_{\mathrm{loc}}; L^1_{\mathrm{loc}}]$ *on* $\mathbf{R}^+$, *and* $\left|\int_0^t \kappa(u,E)\,\mathrm{d}u\right| \le \int_0^t a(u)\chi_E(u)\,\mathrm{d}u$ *for all Borel measurable sets* $E \subset \mathbf{R}^+$ *and* $t \in \mathbf{R}^+$,

(iii) $g \in C(\mathbf{R};\mathbf{R})$,

(iv) $f \in L^1_{\mathrm{loc}}(\mathbf{R}^+;\mathbf{R})$,

(v) $x \in AC_{\mathrm{loc}}(\mathbf{R}^+;\mathbf{R})$ *is a solution of (2.4) on* $\mathbf{R}^+$.

Then x satisfies

$$m\Big(\{\, y \mid g(y) \ge 0,\ x_0 \le y \le x(t)\,\}\Big) \le \int_0^t |f(s)|_+\,\mathrm{d}s, \quad \text{if} \quad x(t) \ge x_0,$$

$$m\Big(\{\, y \mid g(y) \le 0,\ x(t) \le y \le x_0\,\}\Big) \le \int_0^t |f(s)|_-\,\mathrm{d}s, \quad \text{if} \quad x(t) \le x_0,$$

where m denotes Lebesgue measure. In particular, if $f \in L^1(\mathbf{R}^+;\mathbf{R})$, $m(\{\, y \ge 0 \mid g(y) \ge 0\,\}) = \infty$ and $m(\{\, y \le 0 \mid g(y) \le 0\,\}) = \infty$, then every solution of (2.4) is bounded.

Proof First, let $x(t) \geq x_0$.

Integrating (2.4), we get

$$x(t) + \int_{[0,t]} \beta(t, \mathrm{d}s) g(x(s)) = x_0 + \int_0^t f(u)\,\mathrm{d}u, \quad t \in \mathbf{R}^+, \tag{2.5}$$

where the measure kernel β is defined by

$$\beta(t, E) = \int_0^t \Big(a(u)\chi_E(u) + \kappa(u, E) \Big)\,\mathrm{d}u, \quad t \in \mathbf{R}^+. \tag{2.6}$$

(It follows from Theorem **10**.3.5(i) that β is a Volterra measure kernel of type B^∞_{loc}.) Note that we have assumed that β is nonnegative. Write (2.5) in the form

$$x(t) + p(t) = q(t) + x_0 + \int_0^t f(u)\,\mathrm{d}u,$$

where

$$p(t) = \int_{[0,t]} \beta(t, \mathrm{d}s) \big|g(x(s))\big|_+, \quad t \geq 0,$$

$$q(t) = \int_{[0,t]} \beta(t, \mathrm{d}s) \big|g(x(s))\big|_-, \quad t \geq 0.$$

Both p and q are nonnegative, and satisfy $p(0) = q(0) = 0$. From (2.6) one deduces that they are locally absolutely continuous, with derivatives that for almost all $t > 0$ are given by

$$p'(t) = a(t)|g(x(t))|_+ + \int_{[0,t]} \kappa(t, \mathrm{d}s) \big|g(x(s))\big|_+,$$

$$q'(t) = a(t)|g(x(t))|_- + \int_{[0,t]} \kappa(t, \mathrm{d}s) \big|g(x(s))\big|_-.$$

In particular, because κ is nonpositive, $p'(t) \leq a(t)|g(x(t))|_+$ and $q'(t) \leq a(t)|g(x(t))|_-$ for almost all $t \in \mathbf{R}^+$. Thus, if we let H be the set $H = \{\, t \in \mathbf{R}^+ \mid g(x(t)) \geq 0 \,\}$, then

$$q'(t) \leq 0 \text{ for almost all } t \in H,$$
$$p'(t) \leq 0 \text{ for almost all } t \in \mathbf{R}^+ \setminus H.$$

The first part of the conclusion is now a consequence of Lemma 2.8.

To obtain the second part of the conclusion, rewrite (2.5) as

$$y(t) + \int_{[0,t]} \beta(t, \mathrm{d}s) h(y(s)) = y_0 - \int_0^t f(u)\,\mathrm{d}u,$$

where $y(t) = -x(t)$, $h(v) = -g(-v)$, and apply the first part. □

Observe that the proof given above shows that the conclusion of Proposition 2.9 can be slightly strengthened: one can add the nonnegative function $p(t)$ to the left-hand side of the first of the two inequalities, and the nonnegative function $\int_{[0,t]} \beta(t, \mathrm{d}s)|g(-x(s))|_-$ to the left-hand side of the second inequality.

3. Existence of a Limit at Infinity

Once boundedness has been established, one may investigate the asymptotic behaviour in more detail. Of course, the simplest question to ask is whether or not the solution tends to a limit at infinity.

Let us return to the equation

$$\begin{aligned} x'(t) + a(t)g(x(t)) + b(t)g\big(x(t-1)\big) &= f(t), \quad t \in \mathbf{R}^+; \\ x(t) &= \varphi(t), \quad t \in [-1, 0], \end{aligned} \tag{2.1}$$

and to the proof of Proposition 2.1. That proof was based on a bound on the derivative of $y(t) = |x(t)|$, valid at all points t where $|x(t)|$ was larger than $|x(s)|$ for $s \in [-1, t]$. Actually, this bound is valid on a larger set of points, namely at all points where $|x(t)| \geq |x(s)|$ for $s \in [t-1, t]$. (In fact, it is valid at any point t where $|x(t)| \geq |x(t-1)|$.) This simple observation makes it possible to refine Proposition 2.1.

3.1 Proposition
Let the hypothesis of Proposition 2.1 hold. Then $\max_{t-1 \leq s \leq t} |x(s)|$ *tends to a limit as* $t \to \infty$.

The proof of this proposition is based on the following lemma.

3.2 Lemma
Let x be a real-valued locally absolutely continuous function on $\mathbf{R}^+$ that is bounded from below, and suppose that there exists a function $f \in L^1(\mathbf{R}^+; \mathbf{R})$ such that $x'(t) \leq f(t)$ for almost all those $t \in \mathbf{R}^+$ where $x'(t) > 0$. Then $x' \in L^1(\mathbf{R}^+; \mathbf{R})$. In particular, $\lim_{t \to \infty} x(t)$ exists.

We leave the proof to the reader.

Proof of Proposition 3.1 Define $z(t) = \max_{t-1 \leq s \leq t} |x(s)|$. Then z is locally absolutely continuous (see Lemma 4.6). At those points t where $z'(t)$ exists and is positive, we obviously have $z(t) = |x(t)| = \max_{t-1 \leq s \leq t} |x(s)|$, and the same argument as we used in the proof of Proposition 2.1 shows that $z'(t) = |x'(t)| \leq |f(t)|$. By Lemma 3.2, z has a limit at infinity. □

So far nothing has been said about the existence of a limit at infinity of the solution x of (2.1). In fact, the hypothesis of Proposition 2.1 does not imply that such a limit exists (see Exercise 2). However, before we carry on our discussion of (2.1), let us reformulate the argument given above as a general principle, which in spirit resembles the first boundedness Principle 2.3.

3.3 First Limit Principle
Let $x \in C(\mathbf{R}^+;\mathbf{R}^n)$, let $V(x) \in AC_{\text{loc}}(\mathbf{R}^+;\mathbf{R})$ be a function on $\mathbf{R}^+$ that is bounded from below, and suppose that x has a limit at infinity whenever $V(x)$ has a limit at infinity. In addition, suppose that there exists a function $f \in L^1(\mathbf{R}^+;\mathbf{R})$ such that $[V(x)]'(t) \le f(t)$ for almost all those $t \in \mathbf{R}^+$ where $[V(x)]'(t)$ is positive. Then x tends to a limit at infinity.

This is an obvious consequence of Lemma 3.2.

A restrictive premise in Principle 3.3 is that x is assumed to have a limit whenever $V(x)$ has a limit. Note that in Proposition 3.1 we were not able to verify this assumption; consequently, we obtained only the existence of a limit of the function $V(x)(t) = \max_{t-1\le s\le t}|x(s)|$.

In general, how does one prove that x has a limit whenever $V(x)$ has a limit? Here several different methods are available. In Proposition 3.9 below, the Lyapunov function $V(x)$ is the convolution of x by another function, and Wiener's Tauberian theorem (see Corollary **15**.6.4) can be applied. In later chapters, we will encounter examples where $V(x)$ is composed of a convolution operator and of a nonlinear mapping from $\mathbf{R}^n$ into itself. Sometimes, the mere fact that $V(x)$ is bounded implies that $x(t) \to 0$ as $t \to \infty$; this is the case, e.g., when x satisfies a Tauberian condition and boundedness of $V(x)$ implies that $x \in L^p(\mathbf{R}^+)$ for some $p \in [1,\infty)$. A classical hypothesis is that $[V(x)]'(t)$ is bounded away from zero (and is negative) whenever $x(t)$ is bounded away from zero; under appropriate additional assumptions this implies that $x(t) \to 0$ as $t \to \infty$.

3.4 Lemma
Let $x \in BUC(\mathbf{R}^+;\mathbf{R}^n)$, let $V(x) \in AC_{\text{loc}}(\mathbf{R}^+;\mathbf{R})$ be a function on $\mathbf{R}^+$ that tends to a limit at infinity, and suppose that, for almost all $t \in \mathbf{R}^+$,

$$[V(x)]'(t) \le -p(x(t)) + f(t),$$

where $p \in C(\mathbf{R}^n;\mathbf{R})$ satisfies $p(0) = 0$ and $p(y) > 0$ for $y \ne 0$, and $f \in L^1_{\text{loc}}(\mathbf{R}^+;\mathbf{R})$ satisfies $\lim_{t\to\infty}\int_t^{t+1} f(s)\,ds = 0$. Then $x(t) \to 0$ as $t \to \infty$.

Proof Clearly,

$$\begin{aligned} 0 \le \int_t^{t+1} p(x(s))\,ds &\le \int_t^{t+1} f(s)\,ds - \int_t^{t+1} [V(x)]'(s)\,ds \\ &= \int_t^{t+1} f(s)\,ds - V(x)(t+1) + V(x)(t), \end{aligned}$$

and, therefore, $\int_t^{t+1} p(x(s))\,ds \to 0$ as $t \to \infty$. Together with the nonnegativity and the uniform continuity of the function $t \mapsto p(x(t))$, this implies that $p(x(t)) \to 0$ as $t \to \infty$, or equivalently that $x(t) \to 0$ as $t \to \infty$. □

Lemma 3.4 can be used to prove the statement about asymptotic stability in Proposition 1.1.

Next, we continue our discussion of the asymptotic behaviour of solutions of (2.1). In particular, we analyse the question of the existence of a limit at infinity. Exercises 16 and 18 give two sets of conditions under which this limit exists; here we state and prove the following result.

3.5 Proposition

Let the hypothesis of Proposition 2.1 hold, and in addition suppose that (i*) $a(t) \geq (1+\epsilon)|b(t)|$ *for some* $\epsilon > 0$ *and all* $t \geq 0$.
Then, either

$$\int_t^{t+1} a(s)\,\mathrm{d}s \to 0 \text{ as } t \to \infty,$$

or

$$g(x(t)) \to 0 \text{ as } t \to \infty.$$

In particular, if $\limsup_{t\to\infty} \int_t^{t+1} a(s)\,\mathrm{d}s > 0$ *and if* $g(y) \neq 0$ *for* $y \neq 0$, *then* $x(t) \to 0$ *as* $t \to \infty$.

Proof Define $\alpha = \liminf_{t\to\infty}|x(t)|$ and $\beta = \limsup_{t\to\infty}|x(t)|$. If $g(\beta) = 0$, then g vanishes identically on the interval $[-\beta, \beta]$, and therefore $g(x(t)) \to 0$ as $t \to \infty$. Thus, in the sequel we may assume that $g(\beta) > 0$. In particular, $\beta > 0$.

Let us first treat the case where $\alpha = \beta$, i.e., the case where $|x(t)| \to \beta$ as $t \to \infty$. We claim that in this case, $\int_t^{t+1} a(s)\,\mathrm{d}s \to 0$ as $t \to \infty$.

For simplicity, let us assume that $x(t) \to \beta$ as $t \to \infty$ (the case where $x(t) \to -\beta$ is treated analogously). As in the proof of Lemma 3.4, we integrate (2.1) from t to $t+1$, to get

$$\int_t^{t+1} \Big(a(s)g(x(s)) + b(s)g\big(x(s-1)\big)\Big)\,\mathrm{d}s = -\big(x(t+1) - x(t)\big) + \int_t^{t+1} f(s)\,\mathrm{d}s.$$

As $t \to \infty$, we have $x(t+1) - x(t) \to 0$ and $\int_t^{t+1} f(s)\,\mathrm{d}s \to 0$. Therefore,

$$\lim_{t\to\infty} \int_t^{t+1} \Big(a(s)g(x(s)) + b(s)g\big(x(s-1)\big)\Big)\,\mathrm{d}s = 0. \tag{3.1}$$

Recall that $g(\beta)$ is supposed to be positive. Hence, by the continuity of g, one can find a constant $\delta > 0$ such that $(1+\epsilon)g(\beta - \delta) \geq g(\beta + \delta) + \delta$. For sufficiently large s we have $x(s) \geq \beta - \delta$ and $x(s-1) \leq \beta + \delta$ and, for almost all such s,

$$\begin{aligned} a(s)g(x(s)) + b(s)g\big(x(s-1)\big) &\geq a(s)g(\beta - \delta) - |b(s)|g(\beta + \delta) \\ &\geq \big\{(1+\epsilon)g(\beta - \delta) - g(\beta + \delta)\big\}|b(s)| \\ &\geq \delta|b(s)|. \end{aligned}$$

This together with (3.1) implies that $\lim_{t\to\infty} \int_t^{t+1} |b(s)|\,\mathrm{d}s = 0$. If we substitute this back into (3.1) and use the fact that for large s one has $g(x(s)) \geq g(\beta - \delta) > 0$, then we get the first of the two alternative conclusions, namely $\lim_{t\to\infty} \int_t^{t+1} a(s)\,\mathrm{d}s = 0$.

It remains to show that it is impossible to have $\alpha < \beta$ and $g(\beta) > 0$. Suppose that $\alpha < \beta$ and $g(\beta) > 0$. Choose $\delta > 0$ so small that $\delta \leq (\beta - \alpha)/3$

and $(1+\epsilon)g(\beta-2\delta) > g(\beta+\delta)$, and choose τ so large that $|x(s-1)| \leq \beta+\delta$ for $s \geq \tau$ and $\int_\tau^\infty |f(s)|\,\mathrm{d}s \leq \delta$. Fix some arbitrary $t \geq \tau$ such that $|x(t)| = \beta - 2\delta$, and define z and E as in Lemma 2.2, with x replaced by $|x|$ and $\mathbf{R}^+$ replaced by $[t, \infty)$. For almost all $u \in E$ we have

$$\begin{aligned} z'(u) &\leq -a(u)g(z(u)) + \Big|b(u)g\big(x(u-1)\big)\Big| + |f(u)| \\ &\leq -a(u)g(\beta - 2\delta) + |b(u)|g(\beta+\delta) + |f(u)| \\ &\leq |f(u)|. \end{aligned}$$

Thus, by Lemma 2.2, for all $v \geq t$ we have

$$|x(v)| \leq |x(t)| + \int_t^v |f(s)|\,\mathrm{d}s \leq \beta - 2\delta + \delta = \beta - \delta.$$

This contradicts the fact that $\limsup_{t\to\infty} |x(t)| = \beta$, and completes the proof. □

An examination of the proof of Proposition 3.5 reveals that the argument given there definitely does not fall within the scope of the limit principle 3.3. This is not very surprising because, although this principle can be quite useful owing to its simplicity, it is not very sharp. As we noted earlier, the limit Principle 3.3 is essentially a restatement of Lemma 3.2. Contrary to Lemma 2.2, which gives a necessary and sufficient condition for the boundedness of a locally absolutely continuous function, the condition used in Lemma 3.2 is far from necessary for the existence of a limit at infinity. For example, the function $x(t) = \sin(t)/(1+t)$ tends to zero at infinity but it is not of bounded variation; hence it cannot satisfy the assumption of Lemma 3.2.

Clearly, if we can formulate and prove a replacement for Lemma 3.2, then a limit principle different from, but of the same type as, the limit Principle 3.3 will follow. Here, one possibility is to use the same type of arguments as was used in Lemma 2.2 and in the proof of the second alternative conclusion in Proposition 3.5, i.e., to consider $x'(t)$ only when t is large and when $x(t)$ is close to $\limsup_{t\to\infty} x(t)$. This leads to the following necessary and sufficient condition for the existence of a limit at infinity.

3.6 Lemma

Let x be a real-valued locally absolutely continuous bounded function on $\mathbf{R}^+$, and suppose that for each sufficiently small $\epsilon > 0$ there exist a constant $T \geq 0$ and a function $f \in L^1_{\mathrm{loc}}(\mathbf{R}^+; \mathbf{R})$ satisfying $\sup_{T\leq t<\tau<\infty} \int_t^\tau f(s)\,\mathrm{d}s < \epsilon$ such that $x'(t) \leq f(t)$ for almost all t in the set E defined by

$$E = \big\{\, t \in [T, \infty) \;\big|\; \limsup_{s\to\infty} x(s) - \epsilon \leq x(t) \leq \limsup_{s\to\infty} x(s) \,\big\}.$$

Then x tends to a limit at infinity.

To see that this condition is necessary, observe that, if x tends to a limit $x(\infty)$ at infinity, then for each $\epsilon > 0$ one can choose T so large that $|x(t) - x(\infty)| < \epsilon/4$ for $t > T$, and then take $f(t) = x'(t)$.

Proof of Lemma 3.6 Define $\alpha = \liminf_{t\to\infty} x(t)$, $\beta = \limsup_{t\to\infty} x(t)$, and suppose that $\alpha < \beta$ (otherwise $x(t)$ tends to a limit). Define $\epsilon = (\beta-\alpha)/2$, and let T and f be the constant and the function that correspond to this ϵ. Fix some $t \geq T$ such that $x(t) = \beta - \epsilon$. It follows from Corollary 2.5, with $\mathbf{R}^+$ replaced by $[t, \infty)$, that $\sup_{v\geq t} x(v) < \beta$, contrary to our definition of β as $\beta = \limsup_{t\to\infty} x(t)$. □

As was mentioned earlier, Lemma 3.6 gives us a new limit principle.

3.7 Second Limit Principle

Assume that

(i) $x \in C(\mathbf{R}^+;\mathbf{R}^n)$, $V(x) \in AC_{\text{loc}}(\mathbf{R}^+;\mathbf{R}) \cap BC(\mathbf{R}^+;\mathbf{R})$ *and x has a limit at infinity whenever $V(x)$ has a limit at infinity,*

(ii) *For each sufficiently small $\epsilon > 0$ there exist a constant $T \geq 0$ and a function $f \in L^1_{\text{loc}}(\mathbf{R}^+;\mathbf{R})$ satisfying $\sup_{T\leq t<\tau<\infty} \int_t^\tau f(s)\,ds < \epsilon$ and such that $[V(x)]'(t) \leq f(t)$ for almost all t in the set E defined by*

$$E = \{\, t \in [T,\infty) \mid \limsup_{s\to\infty} V(x)(s) - \epsilon \leq V(x)(t) \leq \limsup_{s\to\infty} V(x)(s)\,\}.$$

Then x tends to a limit at infinity.

We conclude this section by considering some simple examples where the principles given above are applied. First, we study the functional differential equation

$$x'(t) + H(x)(t) = f(t), \quad t \in \mathbf{R}^+;$$
$$x(0) = x_0, \tag{3.2}$$

in the case where the delay is (essentially) finite. (Cf. Exercise 24.)

3.8 Proposition

Assume that

(i) *λ and ρ are positive numbers, and $0 < \lambda\rho < \frac{3}{2}$,*

(ii) *$f \in L^1_{\text{loc}}(\mathbf{R}^+;\mathbf{R})$, and $\lim_{t\to\infty} \int_0^t f(s)\,ds$ exists,*

(iii) *$H : C(\mathbf{R}^+;\mathbf{R}) \to L^1_{\text{loc}}(\mathbf{R}^+;\mathbf{R})$ satisfies*

$$-\lambda \max_{t-\rho\leq s\leq t} |\varphi(s)|_- \leq H(\varphi)(t) \leq \lambda \max_{t-\rho\leq s\leq t} |\varphi(s)|_+$$

for $t \geq \rho$ and $\varphi \in C(\mathbf{R}^+;\mathbf{R})$,

(iv) *$x \in AC_{\text{loc}}(\mathbf{R}^+;\mathbf{R})$ is a solution of (3.2) on $\mathbf{R}^+$.*

Then x is bounded and tends to a limit at infinity.

To take a simple example, let H be of the form $H(x)(t) = g(t, x(t-r(t)))$ for some functions $g \in C(\mathbf{R}^+ \times \mathbf{R};\mathbf{R})$ and $r \in C(\mathbf{R}^+;\mathbf{R})$, with $0 \leq r(t) \leq \min(t,\rho)$. The inequality in (iii) then amounts to

$$0 \leq xg(t,x) \leq \lambda x^2, \quad t \in \mathbf{R}^+, \quad x \in \mathbf{R}.$$

Proof of Proposition 3.8 We intend to apply the second limit Principle 3.7 in the form of Lemma 3.6, i.e., we choose the Lyapunov function $V(x) = x$.

Since the proof of boundedness is almost the same as the proof of the fact that the limit exists, we leave the former proof to the reader (see Exercise 23), and presume that the solution is bounded. Without loss of generality, we may take $\limsup_{t\to\infty} x(t) \geq |\liminf_{t\to\infty} x(t)|$ (otherwise replace $x(t)$ by $-x(t)$ and $H(\phi)$ by $-H(-\phi)$). Assume that $\alpha \stackrel{\text{def}}{=} \limsup_{t\to\infty} x(t) > 0$ (otherwise there is nothing to prove). Without loss of generality, suppose that $\lambda\rho \geq 1$ (increase the value of λ, if necessary). Let $\epsilon > 0$ satisfy

$$\epsilon < \frac{\left(\frac{3}{2} - \lambda\rho\right)\alpha}{\frac{5}{2} + \lambda\rho}, \tag{3.3}$$

and choose $T > 0$ to be so large that

$$|x(t)| \leq \alpha + \epsilon \text{ for } t \geq T, \tag{3.4}$$

and

$$\sup_{T\leq s\leq t<\infty} \left| \int_s^t f(u)\,du \right| < \epsilon. \tag{3.5}$$

If we are able to show that $x'(t) \leq f(t)$ for almost all those $t \geq T + 2\rho$ for which $x(t) \geq \alpha - \epsilon$, then the desired conclusion follows from Lemma 3.6. By (iii) and (3.2), this will be the case if $\min_{t-\rho\leq s\leq t} x(s) \geq 0$ for every $t \geq T+2\rho$ for which $x(t) \geq \alpha-\epsilon$. Thus, to prove the proposition, it suffices to show that x has this property.

Assume that this is not the case, i.e., assume that there exists a number $\tau \geq T + 2\rho$ such that

$$x(\tau) \geq \alpha - \epsilon \text{ and } \min_{\tau-\rho\leq s\leq\tau} x(s) < 0. \tag{3.6}$$

Define the number σ by

$$\sigma = \sup\{\, s < \tau \mid x(s) = (\alpha + \epsilon)(\lambda\rho - 1) + \epsilon \,\}. \tag{3.7}$$

Then $0 < x(\sigma) < x(\tau)$ (this follows from (3.3), and from the assumption $\lambda\rho \geq 1$), and $\tau - \rho < \sigma < \tau$. By (iii) and (3.4),

$$x'(s) \leq f(s) + \lambda(\alpha + \epsilon), \quad s \geq \sigma - \rho.$$

Integrating both sides of this inequality over (s, σ), and using (3.5), we get

$$x(\sigma) - x(s) \leq \epsilon + \lambda(\alpha + \epsilon)(\sigma - s), \quad \sigma - \rho \leq s \leq \sigma,$$

which, combined with (3.7), gives

$$x(s) \geq (\alpha + \epsilon)\bigl(\lambda\rho - 1 - \lambda(\sigma - s)\bigr), \quad \sigma - \rho \leq s \leq \sigma. \tag{3.8}$$

Let u be the point where the right-hand side vanishes, i.e.,

$$u = \sigma - \rho + \frac{1}{\lambda}.$$

Then, since $x(\sigma) > 0$, and $x(s) < 0$ for some $s \in [\tau - \rho, \sigma]$, we have $u > \tau - \rho$, or equivalently,

$$\tau - \sigma < \frac{1}{\lambda}. \tag{3.9}$$

Inequality (3.8), combined with (3.2) and (iii), implies

$$x'(t) \le f(t) + \lambda(\alpha + \epsilon)\{1 - \lambda(t - \sigma)\}, \quad \sigma \le t \le \tau.$$

We integrate both sides of this inequality over (σ, τ) to get

$$x(\tau) - x(\sigma) \le \epsilon + \lambda(\alpha + \epsilon)(\tau - \sigma) - \frac{1}{2}\lambda^2(\alpha + \epsilon)(\tau - \sigma)^2,$$

which, combined with (3.6), (3.7), and (3.9), gives

$$\alpha - \epsilon - (\alpha + \epsilon)(\lambda\rho - 1) - \epsilon < \epsilon + \frac{1}{2}(\alpha + \epsilon).$$

This contradicts (3.3) (it is the same inequality, except for the fact that the direction has been reversed) and, therefore, completes the proof. □

The next result is of a somewhat different nature. We consider the equation

$$\begin{aligned} x(t) &= k\left(P - \int_{-\infty}^{t} A(t-s)x(s)\,\mathrm{d}s\right)\int_{-\infty}^{t} a(t-s)x(s)\,\mathrm{d}s, \quad t \in \mathbf{R}^+; \\ x(t) &= \varphi(t), \quad t \in \mathbf{R}^-, \end{aligned} \tag{3.10}$$

where k and P are constants. Under reasonable assumptions (see Gripenberg [20]), it is possible to show that this equation has a bounded, strictly positive solution. Here we discuss how such a solution behaves asymptotically. Below, $*$ denotes convolution.

3.9 Proposition

Let k and P be positive constants satisfying $kP > 1$, and suppose that

(i) *$A \in L^1(\mathbf{R}^+; \mathbf{R})$ is bounded, nonnegative and nonincreasing on $\mathbf{R}^+$,*

(ii) *$a \in L^1(\mathbf{R}^+; \mathbf{R})$ is nonnegative and $\int_{\mathbf{R}^+} a(t)\,\mathrm{d}t = 1$,*

(iii) *there exists a measure $\mu \in M(\mathbf{R}^+; \mathbf{R})$, with $|\mu|(\mathbf{R}^+) < 1$, such that $A = A_0 + \mu * A_0$, where A_0 is the function $A_0(t) = \int_t^\infty a(s)\,\mathrm{d}s$, $t \in \mathbf{R}^+$,*

(iv) *$\varphi \in BUC(\mathbf{R}^-)$ is nonnegative, and the solution x of (3.10) is bounded and (strictly) positive on $\mathbf{R}^+$.*

Then

$$\lim_{t\to\infty} x(t) = \frac{kP - 1}{k\int_{\mathbf{R}^+} A(s)\,\mathrm{d}s}.$$

Proof The resolvent ρ of μ exists and belongs to $M(\mathbf{R}^+; \mathbf{R})$ since $|\mu|(\mathbf{R}^+) < 1$. Thus $A_0 = A - \rho * A$ and $A_0 \in L^1(\mathbf{R}^+; \mathbf{R})$. Extend A_0 and a as zero to $\mathbf{R}^-$, and define

$$v(t) = (A_0 * x)(t) - \frac{kP - 1}{k(1 + \mu(\mathbf{R}^+))}, \quad t \in \mathbf{R}. \tag{3.11}$$

The function v is bounded and locally absolutely continuous. From (3.10) and conditions (ii) and (iii) we get the following relation:

$$\begin{aligned}
v'(t) &= x(t) - (a*x)(t) \\
&= k\Big(P - (A*x)(t)\Big)(a*x)(t) - (a*x)(t) \\
&= (a*x)(t)\left(kP - k(A_0*x)(t) - k\int_{\mathbf{R}^+}(A_0*x)(t-s)\mu(\mathrm{d}s) - 1\right) \\
&= -k(a*x)(t)\left(v(t) + \int_{\mathbf{R}^+} v(t-s)\mu(\mathrm{d}s)\right), \quad t \in \mathbf{R}^+.
\end{aligned}$$

Since $|\mu|(\mathbf{R}^+) < 1$ and k, a, and x are nonnegative, it follows from this equation, for example with the aid of the second limit Principle 3.7 (with $f \equiv 0$), that $\lim_{t\to\infty} v(t)$ exists.

To show that $x(t)$ tends to the given limit, we apply Pitt's form of Wiener's Tauberian theorem, i.e., Corollary **15**.6.4 to (3.11). First note that we can, without loss of generality, take x continuous at 0. By (3.10) we know that $x \in BUC(\mathbf{R};\mathbf{R})$. Thus it suffices to show that $\tilde{A}_0(\omega) \neq 0$ for all $\omega \in \mathbf{R}$. Since A_0 is nonnegative and not identically zero, we have $\tilde{A}_0(0) \neq 0$. Suppose that $\tilde{A}_0(\omega_0) = 0$ for some $\omega_0 \neq 0$. Then $\tilde{A}_0(\omega_0) = (1 - \tilde{a}(\omega_0))/(\mathrm{i}\omega_0) = 0$; hence $\tilde{a}(\omega_0) = 1$. But, since a is nonnegative and $\int_0^\infty a(t)\,\mathrm{d}t = 1$, this is impossible. Hence, we conclude from Corollary **15**.6.4 that $\lim_{t\to\infty} x(t)$ exists.

It follows from (3.10) and (ii), together with Lebesgue's dominated convergence theorem, that the limit $x(\infty)$ satisfies

$$x(\infty) = k\left\{P - \left(\int_{\mathbf{R}^+} A(s)\,\mathrm{d}s\right)x(\infty)\right\}x(\infty),$$

so either $x(\infty) = 0$, or $x(\infty) = \big(kP - 1\big)/\big(k\int_{\mathbf{R}^+} A(s)\,\mathrm{d}s\big)$, as claimed. To see that the value zero can be excluded, we argue as follows. If $0 \le \tau \le t$ and $x(\tau) \ge x(s) \ge x(t)$ for $s \in [\tau, t]$, then it follows from (3.10) that

$$\begin{aligned}
x(t) \ge &\left\{kP - k\left(\int_{\mathbf{R}^+} A(s)\,\mathrm{d}s\right)x(\tau) - k\left(\int_{t-\tau}^\infty A(s)\,\mathrm{d}s\right)\sup_{t\in\mathbf{R}} x(t)\right\} \\
&\times \left(\int_0^{t-\tau} a(s)\,\mathrm{d}s\right)x(t).
\end{aligned}$$

If $x(t) \to 0$ as $t \to \infty$, then we can choose t and τ such that $t - \tau \to \infty$ as t and $\tau \to \infty$, and this gives a contradiction, since $\int_{\mathbf{R}^+} a(t)\,\mathrm{d}t = 1$, $A \in L^1(\mathbf{R}^+;\mathbf{R})$, $kP > 1$, $x(\tau) \to 0$, and $x(t) > 0$. □

4. Appendix: Local Absolute Continuity

In this appendix we first prove two auxiliary lemmas, then give a number of results describing certain operations that preserve local absolute continuity, and finally prove two lemmas that were stated earlier in this chapter.

4.1 Lemma
Let x be a function on $\mathbf{R}$ with values in $\mathbf{R}^n$, and suppose that there exist a constant $c > 0$ and a function $f \in L^1_{\mathrm{loc}}(\mathbf{R};\mathbf{R})$ such that $|x(b) - x(a)| \leq \int_a^b |f(s)|\,\mathrm{d}s$ whenever $0 < b-a < c$. Then x is locally absolutely continuous.

This follows directly from the definition of absolute continuity of a function (see, e.g., Rudin [3], Definition 7.17) and from the fact that the integral of an L^1-function is absolutely continuous.

4.2 Lemma
Let x and y belong to $AC_{\mathrm{loc}}(\mathbf{R};\mathbf{R}^n)$. If $x(t) = y(t)$ for all t in some subset E of $\mathbf{R}$, then $x'(t) = y'(t)$ for almost all $t \in E$.

Proof The set of isolated points of E has measure zero (because it is countable), and similarly the set of points $t \in E$ at which $x'(t)$ or $y'(t)$ does not exist has measure zero. If $t \in E$ belongs to neither of these two sets, then there is a sequence $t_k \to t$ such that, for each k, $t_k \in E$, $t_k \neq t$, and

$$x'(t) = \lim_{k\to\infty} \frac{x(t_k) - x(t)}{t_k - t} = \lim_{k\to\infty} \frac{y(t_k) - y(t)}{t_k - t} = y'(t).$$

This proves the lemma. □

In Lyapunov-type arguments the following consequence of Lemma 4.1 is frequently used. In this result, one can replace $|\cdot|$ by an arbitrary Lipschitz-continuous real-valued function with Lipschitz constant 1.

4.3 Lemma
Let $x \in AC_{\mathrm{loc}}(\mathbf{R};\mathbf{R}^n)$. Then $|x| \in AC_{\mathrm{loc}}(\mathbf{R};\mathbf{R})$, and $|x|'(t) \leq |x'(t)|$ for almost all $t \in \mathbf{R}$.

Exactly how $|x|'$ is related to x' depends on the norm which one uses in $\mathbf{R}^n$. In the case where $|x| = \langle x, x\rangle^{\frac{1}{2}}$, the connection is the following.

4.4 Lemma
Let $x \in AC_{\mathrm{loc}}(\mathbf{R};\mathbf{R}^n)$ and let the norm be induced by an inner product. Then $|x|'(t) = 0$ for almost all those t where $x(t) = 0$, and $|x|'(t) = \langle x(t), x'(t)\rangle / |x(t)|$ for almost all those t where $x(t) \neq 0$.

Proof That $|x|'(t) = 0$ for almost all those t where $x(t) = 0$, follows from Lemma 4.2 (take $y(t) \equiv 0$). Define $z = \langle x, x\rangle = |x|^2$. Then z is locally absolutely continuous. On the one hand, $z'(t) = 2\langle x(t), x'(t)\rangle$ for almost all $t \in \mathbf{R}$ and, on the other hand, $z'(t) = 2|x|(t)|x|'(t)$ for almost all $t \in \mathbf{R}$. This implies that $|x|'(t) = \langle x(t), x'(t)\rangle / |x(t)|$ for almost all those t where $x(t) \neq 0$. □

Our next lemma lists some simple operations on real-valued functions that preserve local absolute continuity.

4.5 Lemma
Let x and y belong to $AC_{\text{loc}}(\mathbf{R};\mathbf{R})$.

(i) *If $z = \max\{x,y\}$, then $z \in AC_{\text{loc}}(\mathbf{R};\mathbf{R})$, $z'(t) = x'(t)$ for almost all those t where $x(t) \geq y(t)$, and $z'(t) = y'(t)$ for almost all those t where $x(t) \leq y(t)$.*

(ii) *If $z = \min\{x,y\}$, then $z \in AC_{\text{loc}}(\mathbf{R};\mathbf{R})$, $z'(t) = x'(t)$ for almost all those t where $x(t) \leq y(t)$, and $z'(t) = y'(t)$ for almost all those t where $x(t) \geq y(t)$.*

(iii) *If $z = |x|$, then $z \in AC_{\text{loc}}(\mathbf{R};\mathbf{R})$ and $z'(t) = \operatorname{sign}(x(t))x'(t)$ for almost all t.*

Proof (i) If $a < b$, then

$$\begin{aligned}|z(b) - z(a)| &\leq \max\{|x(b) - x(a)|, |y(b) - y(a)|\} \\ &\leq |x(b) - x(a)| + |y(b) - y(a)| \\ &\leq \int_a^b (|x'(s)| + |y'(s)|)\,\mathrm{d}s,\end{aligned}$$

and, by Lemma 4.1, $z \in AC_{\text{loc}}(\mathbf{R};\mathbf{R})$. The two claims about $z'(t)$ follow from Lemma 4.2.

(ii) Apply (i) to $-x$ and $-y$.

(iii) This is a consequence of Lemma 4.4. □

The following lemma was needed in the proof of Proposition 3.1.

4.6 Lemma
Let $x \in AC_{\text{loc}}(\mathbf{R};\mathbf{R})$, and define $z(t) = \max_{t-1\leq s\leq t} x(s)$. Then z is locally absolutely continuous, $z'(t) = x'(t)$ for almost all those t where $z'(t) > 0$, and $z'(t) = x'(t-1)$ for almost all those t where $z'(t) < 0$.

Proof If $0 < b - a \leq 1$, and $z(b) \geq z(a)$, then

$$\begin{aligned}z(b) - z(a) &= \max_{b-1\leq s\leq b} x(s) - \max_{a-1\leq s\leq a} x(s) \\ &\leq \max_{a\leq s\leq b} x(s) - x(a) \leq \int_a^b |x'(s)|\,\mathrm{d}s.\end{aligned}$$

A similar estimate is valid when $z(b) < z(a)$, and this, together with Lemma 4.1, gives us the local absolute continuity. The claim about $z'(t)$ now follows from Lemma 4.2. □

Finally, let us give the proof of Lemma 2.2.

Proof of Lemma 2.2 If $a < b$, then

$$0 \le z(b) - z(a) = \max_{0 \le s \le b} x(s) - \max_{0 \le s \le a} x(s) \le \max_{0 \le s \le b} x(s) - x(a)$$

$$= \max_{0 \le u \le b} \int_a^u x'(s)\,\mathrm{d}s \le \int_a^b |x'(s)|\,\mathrm{d}s.$$

Thus, by Lemma 4.1, $z \in AC_{\mathrm{loc}}(\mathbf{R}^+;\mathbf{R})$. That $z'(t) = 0$ for all $t \notin E$ is obvious, and that $z'(t) = x'(t)$ for almost all $t \in E$ follows from Lemma 4.2. □

5. Exercises

1. Without recourse to Lyapunov theory, prove the following boundedness result.

   5.1 Proposition
   Assume that
   (i) $a \in L^1(\mathbf{R}^+;\mathbf{R})$ *is nonnegative,*
   (ii) $g \in C(\mathbf{R};\mathbf{R})$ *is bounded from below, and* $\limsup_{u\to-\infty} g(u) < \infty$,
   (iii) $f \in BC(\mathbf{R}^+;\mathbf{R})$,
   (iv) $x \in C(\mathbf{R}^+;\mathbf{R})$ *is a solution of the equation*

   $$x(t) + \int_0^t a(t-s)g(x(s))\,\mathrm{d}s = f(t), \quad t \in \mathbf{R}^+.$$

   Then x is bounded.

2. Show by an example that the hypothesis of Proposition 2.1 does not imply that $\lim_{t\to\infty} x(t)$ exists.
3. Show by an example that if, in Proposition 2.1, the assumption that g is odd is replaced by the condition $yg(y) \ge 0$, for all $y \in \mathbf{R}$, then the solutions need no longer be bounded.
4. Formulate a uniform boundedness principle based on the boundedness Principle 2.3.
5. Formulate and prove a result similar to Proposition 2.1 for equation (2.3).
6. Formulate and prove a result similar to Proposition 2.7 for equation (2.3).
7. Prove the following result.

5.2 Proposition
Let $a \in L^1_{\mathrm{loc}}(\mathbf{R}^+;\mathbf{R})$, let κ be a real-valued Volterra measure kernel of type $[B^\infty_{\mathrm{loc}}; L^1_{\mathrm{loc}}]$ on $\mathbf{R}^+$, and suppose that $a(t) \geq |\kappa|(t,E)$ for each $t \in \mathbf{R}^+$ and every Borel set $E \subset \mathbf{R}^+$. Then the differential resolvent of the kernel $a(s)\delta(\mathrm{d}s) + \kappa(t,\mathrm{d}s)$ (see (2.4)) is bounded on $\mathbf{R}^+ \times \mathbf{R}^+$.

Hint: See Exercise 5.

8. Prove the following result.

5.3 Proposition
Assume that

(i) *$a \in L^1_{\mathrm{loc}}(\mathbf{R}^+;\mathbf{R})$, κ is a real-valued Volterra measure kernel of type $[B^\infty; L^1]$ on $(-\infty,T]$ for every $T > 0$, $|\kappa|(\cdot,\mathbf{R}^-) \in L^\infty(\mathbf{R}^+;\mathbf{R})$, and $a(t) \geq \epsilon + |\kappa|(t,E)$ for each $t \in \mathbf{R}^+$, every Borel set $E \subset \mathbf{R}^+$, and some $\epsilon > 0$,*
(ii) *$g \in C(\mathbf{R};\mathbf{R})$ is odd and nondecreasing and satisfies $g(u) \to \infty$ as $u \to \infty$,*
(iii) *$f \in L^\infty(\mathbf{R}^+;\mathbf{R})$,*
(iv) *$\varphi \in BC((-\infty,0];\mathbf{R})$,*
(v) *$x \in AC_{\mathrm{loc}}(\mathbf{R}^+;\mathbf{R})$ is a solution of (2.3) on $\mathbf{R}^+$.*

Then x is bounded.

9. Reformulate Exercise 8 for the equation (2.1).
10. Under hypothesis (i) in Exercise 8 (with the interval $(-\infty,T]$ replaced by $[0,T)$), prove that the differential resolvent of the kernel in (2.4) is of bounded type on $\mathbf{R}^+$ (cf. Definition **9**.5.1).
11. Prove the following boundedness result.

5.4 Proposition
Let $b > 0$, and assume that

(i) *$a \in L^1(\mathbf{R}^+;\mathbf{R})$ and $\int_{\mathbf{R}^+} |a(s)|\,\mathrm{d}s < b$,*
(ii) *$g \in C(\mathbf{R};\mathbf{R})$ is nonnegative and nondecreasing, and*

$$\lim_{u\to-\infty} g(u) = 0 \text{ and } \lim_{u\to\infty} g(u) = \infty,$$

(iii) *$f \in BC_\ell(\mathbf{R}^+;\mathbf{R})$, and $\lim_{t\to\infty} f(t) > 0$,*
(iv) *$x \in AC_{\mathrm{loc}}(\mathbf{R}^+;\mathbf{R})$ is a solution of the equation*

$$x'(t) + bg(x(t)) + \int_0^t a(t-s)g(x(s))\,\mathrm{d}s = f(t), \quad t \in \mathbf{R}^+; \quad x(0) = x_0.$$

Then x is bounded.

Hint: First show that x is bounded from above, then show that x' is bounded, and finally show that there are constants T, τ and M such that $x'(t) \geq 0$ if $t \geq T$ and $x(s) \leq -M$ for $s \in [t-\tau,t]$.

12. Formulate a boundedness principle based on Lemma 2.8.
13. In Proposition 2.9, add the assumption that $yg(y) \geq 0$ for all $y \in \mathbf{R}$,

and show that then $|x(t)| \le |x_0| + \int_0^t |f(s)|\,ds$ for $t \ge 0$.

14. In Proposition 2.9, add the assumption that $yg(y) \ge 0$ for all $y \in \mathbf{R}$, drop the assumption that κ is nonpositive, assume that $\int_0^t |\kappa|(u, E)\,du \le \int_0^t a(u)\chi_E(u)\,du$ for all Borel measurable sets $E \subset \mathbf{R}^+$ and $t \in \mathbf{R}^+$, and show that then $|x(t)| \le |x_0| + \int_0^t |f(s)|\,ds$ for $t \ge 0$. Hint: What equation does the derivative of $|x(t)|$ satisfy?
15. In Exercise 14, assume in addition that $f \in L^1(\mathbf{R}^+; \mathbf{R})$ and that $\int_0^t |\kappa|(u, E)\,du + \epsilon m(E \cap [0, t]) \le \int_0^t a(u)\chi_E(u)\,du$ for all Borel measurable sets $E \subset \mathbf{R}^+$, all $t \in \mathbf{R}^+$, and some $\epsilon > 0$ (here $m(E)$ is the Lebesgue measure of E), and show that $\int_0^\infty |g(x(t))|\,dt < \infty$. Under what additional assumptions can you conclude that $x' \in L^1(\mathbf{R}^+; \mathbf{R})$?
16. Prove the following result.

5.5 Proposition
Assume that

(i) $a,\ b \in L^\infty(\mathbf{R}^+; \mathbf{R})$, *and* $|b(t + 1)| + \epsilon \le |a(t)|$ *for almost all* $t \in \mathbf{R}^+$ *and some* $\epsilon > 0$,
(ii) $g \in C(\mathbf{R}; \mathbf{R})$ *is odd and nondecreasing,*
(iii) $f \in L^1(\mathbf{R}^+; \mathbf{R})$,
(iv) $\varphi \in C([-1, 0]; \mathbf{R})$,
(v) $x \in AC_{\mathrm{loc}}(\mathbf{R}^+; \mathbf{R})$ *is a solution of (2.1) on* $\mathbf{R}^+$.

Then $x' \in L^1(\mathbf{R}^+; \mathbf{R})$. *In particular,* $\lim_{t\to\infty} x(t)$ *exists.*

Hint: See Exercise 15.

17. Prove the following result.

5.6 Proposition
Let $a \in L^1_{\mathrm{loc}}(\mathbf{R}^+; \mathbf{R})$, *let* κ *be a real-valued Volterra measure kernel of type* $[B^\infty_{\mathrm{loc}}; L^1_{\mathrm{loc}}]$ *on* $\mathbf{R}^+$, *satisfying* $\int_0^t |\kappa|(u, E)\,du + \epsilon m(E \cap [0, t]) \le \int_0^t a(u)\chi_E(u)\,du$ *for all Borel measurable sets* $E \subset \mathbf{R}^+$, *all* $t \in \mathbf{R}^+$, *and some* $\epsilon > 0$ *(here* $m(E)$ *is the Lebesgue measure of* E*). Then the differential resolvent of the kernel in (2.4) is of type* L^1 *on* $\mathbf{R}^+$.

Hint: See Exercise 15.

18. Prove the following result.

5.7 Proposition
Assume that

(i) $a,\ b \in L^1_{\mathrm{loc}}(\mathbf{R}^+; \mathbf{R})$, *and* $a(t) \ge (1 + \epsilon)|b(t)| + \epsilon$ *for almost all* $t \in \mathbf{R}^+$, *and some* $\epsilon > 0$,
(ii) $g \in C(\mathbf{R}; \mathbf{R})$ *is odd and nondecreasing and satisfies* $g(u) \to \infty$ *as* $u \to \infty$,
(iii) $f \in L^\infty_0(\mathbf{R}^+; \mathbf{R})$,
(iv) $\varphi \in C([-1, 0]; \mathbf{R})$,
(v) $x \in AC_{\mathrm{loc}}(\mathbf{R}^+; \mathbf{R})$ *is a solution of (2.1) on* $\mathbf{R}^+$.

Then $g(x(t)) \to 0$ *as* $t \to \infty$.

19. Prove that the solution in Exercise 11 satisfies

$$\left(b + \int_{\mathbf{R}^+} a(s)\,ds\right) g(x(t)) \to f(\infty) \text{ as } t \to \infty.$$

20. Prove the following result.

5.8 Proposition

Assume that $P > 0$, $b > 0$, $\varphi(0) > 0$, $a \in L^1(\mathbf{R}^+; \mathbf{R})$, and that $\int_{\mathbf{R}^+} |a(s)|\,ds < b$. Then the solution x of the equation

$$\begin{aligned} x'(t) &= x(t)\left(P - bx(t) - \int_{-\infty}^{t} a(t-s)x(s)\,ds\right), && t \in \mathbf{R}^+; \\ x(t) &= \varphi(t), && t \in \mathbf{R}^-, \end{aligned}$$

is strictly positive, and satisfies

$$x(t) \to \frac{P}{b + \int_{\mathbf{R}^+} a(s)\,ds}, \quad t \to \infty.$$

Hint: Define $y(t) = \ln\big(x(t)/\alpha\big)$, where $\alpha = P/\big(b + \int_{\mathbf{R}^+} a(s)\,ds\big)$, and apply Exercises 11 and 19.

21. Use Proposition 3.8 to show that, under appropriate assumptions on λ and $r(t)$, the resolvent of the kernel in the linear equation

$$x(t) + \lambda \int_0^t x\big(s - r(s)\big)\,ds = f(t), \quad t \in \mathbf{R}^+,$$

maps $BC_\ell(\mathbf{R}^+; \mathbf{R}^n)$ into $BC_\ell(\mathbf{R}^+; \mathbf{R}^n)$ with unchanged limit value, i.e., show that if $f \in BC_l(\mathbf{R}^+; \mathbf{R}^n)$ then $x \in BC_0(\mathbf{R}^+; \mathbf{R}^n)$. What type of conditions do you need on r if you want to write this equation in the form

$$x(t) + \int_0^t k(t,s)x(s)\,ds = f(t), \quad t \in \mathbf{R}^+,$$

for some function $k(t,s)$, which is, e.g., of continuous type?

22. Prove the following result.

5.9 Proposition

Assume that

(i) *λ and ρ are positive numbers such that $\lambda\rho < \frac{3}{2}$,*
(ii) *$g \in C(\mathbf{R}; \mathbf{R})$, and $0 \le ug(u) \le \lambda u^2$ for $u \in \mathbf{R}$,*
(iii) *$r \in C(\mathbf{R}^+; \mathbf{R})$ and $0 \le r(t) \le \rho < \infty$ for all $t \in \mathbf{R}^+$,*
(iv) *$a \in L^1_{\mathrm{loc}}(\mathbf{R}^+; \mathbf{R})$ is nonnegative, and $\int_t^{t+\rho} a(s)\,ds \le 1$, for all $t \ge 0$,*
(v) *$f \in L^1_{\mathrm{loc}}(\mathbf{R}^+; \mathbf{R})$ and $\lim_{t\to\infty} \int_0^t f(s)\,ds$ exists,*
(vi) *$\varphi \in C([-\rho, 0]; \mathbf{R})$,*
(vii) *$x \in AC_{\mathrm{loc}}(\mathbf{R}^+; \mathbf{R})$ is a solution of the equation*

$$\begin{aligned} x'(t) + a(t)g\big(x(t - r(t))\big) &= f(t), && t \ge 0; \\ x(t) &= \varphi(t), && -\rho \le t \le 0. \end{aligned}$$

Then x is bounded and tends to a limit at infinity.

23. Prove that, under the assumptions of Proposition 3.8, every solution of (3.2) on $\mathbf{R}^+$ remains bounded. Can the assumption $\lambda\rho < \frac{3}{2}$ be replaced by $\lambda\rho \leq \frac{3}{2}$ in this proof?
24. Show by an example that the functional differential equation (3.2) need not have finite delay (and need not even be a delay equation but can be an equation of advanced type), when the assumptions of Proposition 3.8 are satisfied.
25. Show that, if the requirement $\lambda\rho < \frac{3}{2}$ in Proposition 3.8 is replaced by $\lambda\rho = \frac{3}{2}$, then the conclusion is no longer valid, even if $f \in L^1(\mathbf{R}^+;\mathbf{R})$.
26. Let the assumptions (i), (ii) and (iv) of Proposition 3.9 be satisfied, and assume that k and P are positive constants such that $kP \leq 1$. Prove that $x(t) \to 0$ as $t \to \infty$.

6. Comments

Section 1:

Lyapunov methods for functional differential equations are considered extensively in Hale [5], Ch. 5. See Section 5 of this reference for a comprehensive survey of the literature. For additional comments on Lyapunov theory applied to delay equations, see Driver [1], Ch. VIII, Section 31. Lyapunov methods for integrodifferential equations are presented in Miller [10], Ch. VI, p. 337. One may also consult Burton [2].

The most common approach is to require the derivative of the (nonnegative) Lyapunov functional to be nonpositive along the solutions of the equation. Our approach resembles the so called Razumikhin approach, discussed in, e.g., Hale [5], Section 5.4. See also Seifert [1] and Haddock and Terjéki [1]. In both approaches, the main difficulty lies in finding a function, or functional, V with the desired properties.

For recent Soviet applications of Lyapunov techniques, see Kolmanovskiĭ and Nosov [1], and the references therein.

Section 2:

Proposition 2.9 is essentially the same result which is found in Levin [6] and Gripenberg [1], but the proof given here is simpler. For further boundedness results of the type discussed in Section 2, see Staffans [27].

Section 3:

A linear, constant delay version of Proposition 3.8 is established in Hale [5], pp. 132–133. A nonlinear version is found in Yorke [1], Theorem 1.1, p. 198 (this result is described in Hale [5], pp. 137–138). Observe that the constant $\frac{3}{2}$, and the strict inequality in (i), cannot be improved (cf. Exercise 25).

Equation (3.10) occurs in mathematical epidemiology. Proposition 3.9 is a special case of Gripenberg [20], Theorem, p. 318. In this reference it is shown, in addition to the result given here, that $x(t) \to 0$ when $kP \leq 1$. Moreover, the condition $A = A_0 + \mu * A_0$ is examined in more detail.

Section 5:
The result in Exercise 20 is proved in Miller [3], Theorem 1. Proposition 5.9 is closely related to the results in Yoneyama [1].

References

T. A. Burton

2. *Volterra Integral and Differential Equations*, Academic Press, New York, 1983.

R. D. Driver

1. *Ordinary and Delay Differential Equations*, Springer-Verlag, Berlin, 1977.

G. Gripenberg

1. Bounded solutions of a Volterra equation, *J. Differential Equations* **28** (1978), pp. 18–22.
20. On some epidemic models, *Quart. Appl. Math.* **39** (1981), pp. 317–327.

J. R. Haddock and J. Terjéki

1. Liapunov–Razumikhin functions and an invariance principle for functional differential equations, *J. Differential Equations* **48** (1983), pp. 95–122.

J. K. Hale

3. *Ordinary Differential Equations*, J. Wiley, New York, 1969.
5. *Theory of Functional Differential Equations*, Springer-Verlag, Berlin, 1977.

V. B. Kolmanovskiĭ and V. R. Nosov

1. *Stability of Functional Differential Equations*, Academic Press, London, 1986.

J. J. Levin

6. A bound on the solutions of a Volterra equation, *Arch. Ratl. Mech. Anal.* **52** (1973), pp. 339–349.

R. K. Miller

3. On Volterra's population equation, *J. SIAM Appl. Math.* **14** (1966), pp. 446–452.
10. *Nonlinear Volterra Integral Equations*, W. A. Benjamin, Menlo Park, Calif., 1971.

W. Rudin

3. *Real and Complex Analysis,* 3rd *ed.*, McGraw–Hill, New York, 1986.

G. Seifert

1. Liapunov–Razumikhin conditions for asymptotic stability in functional differential equations of Volterra type, *J. Differential Equations* **16** (1974), pp. 289–297.

O. J. Staffans

27. A direct Lyapunov approach to Volterra integrodifferential equations, *SIAM J. Math. Anal.* **19** (1988), pp. 879–901.

T. Yoneyama

1. On the $\frac{3}{2}$ stability theorem for one-dimensional delay-differential equations, *J. Math. Anal. Appl.* **125** (1987), pp. 161–173.

J. A. Yorke

1. Asymptotic stability for one dimensional differential-delay equations, *J. Differential Equations* **7** (1970), pp. 189–202.

15

General Asymptotics

We study limit sets, limit functions and spectra of bounded continuous functions. These concepts will be used in the remaining chapters to describe the asymptotic behaviour of solutions of integral equations.

1. Introduction

In earlier chapters we studied the asymptotic behaviour of solutions of linear convolution equations by means of Laplace transform methods. In addition, we gave some results for nonlinear equations based on linearization, i.e., we assumed that the equation had a small nonlinear part, which was dominated by the linear part. In Chapter 14, we studied the basic ideas behind Lyapunov techniques.

In Chapters 17–20 we extend our study to cover the asymptotic analysis of certain nonlinear equations to which the methods given so far do not apply. Several ingredients will enter into the theory, namely

- Lyapunov-type arguments,
- limit sets and limit equations,
- spectral analysis,
- estimates for kernels of positive type,
- monotonicity type arguments.

In this chapter we give some general results pertaining to limit sets, limit equations, and spectral analysis. The next chapter will deal with kernels of positive type.

To get an idea of how the results to be developed will be used, let us look at the nonlinear convolution equation

$$x(t) + \int_0^t a(t-s)g(x(s))\,\mathrm{d}s = f(t), \qquad t \in \mathbf{R}^+. \tag{1.1}$$

Here we can assume that a is integrable, that g and f are continuous, and that f tends to zero at infinity. Let us suppose that the solution x of (1.1) (which exists, at least locally, according to, e.g., Theorem **12**.1.1) is defined and bounded on $\mathbf{R}^+$. It follows from (1.1) that x is uniformly continuous.

One way to study the behaviour of $x(t)$ for large t is to translate x far to the left, and to see if something can be said about the translated function. That is, we define the function $\tau_t x$ to be $\tau_t x(s) = x(s+t)$ for $s \geq -t$, and study the behaviour of $\tau_t x$ as $t \to \infty$.

What can then be achieved? Owing to the uniform continuity of x, the set of functions $\{\, \tau_t x \mid t \in \mathbf{R}^+ \,\}$ is uniformly bounded and equicontinuous. Therefore, by the Arzelà–Ascoli Theorem **2**.7.6, every sequence $t_k \to \infty$ has a subsequence $t_j \to \infty$ for which $\tau_{t_j} x$ converges to some limit function y as $j \to \infty$, uniformly on compact sets. Each function $x_j = \tau_{t_j} x$ satisfies the equation

$$x_j(t) + \int_{-t_j}^{t} a(t-s) g\big(x_j(s)\big)\, \mathrm{d}s = f(t + t_j).$$

As $j \to \infty$, we have $f(t+t_j) \to 0$ and $x_j(t) \to y(t)$ for all $t \in \mathbf{R}$, and therefore Lebesgue's dominated convergence theorem implies that y must satisfy the limit equation

$$y(t) + \int_{-\infty}^{t} a(t-s) g(y(s))\, \mathrm{d}s = 0, \qquad t \in \mathbf{R}. \tag{1.2}$$

Since we need no longer worry about the function f, this equation is more lucid than the original equation (1.1). Note that y satisfies the limit equation on $\mathbf{R}$, not only on $\mathbf{R}^+$.

As we shall see later, the fact that y satisfies (1.2) implies, under appropriate assumptions on a, that y must have a specific behaviour, and this will enable us to say something about the behaviour of our original solution x of (1.1). However, before we get to that point we have to study limit sets and limit equations in some detail.

2. Limit Sets and Limit Equations

In the sequel we state and prove several results that do not depend on the particular space ($\mathbf{R}$, $\mathbf{C}$, $\mathbf{R}^n$, $\mathbf{C}^n$, $\mathbf{R}^{m\times n}$, or $\mathbf{C}^{m\times n}$) in which the values of our functions lie. In such cases, it suffices to consider the space $\mathbf{C}^n$ with arbitrary n, since, for example, $\mathbf{R}^{m\times n}$ is a subspace of $\mathbf{C}^p$ with $p = mn$. However, in some contexts it is not advisable to replace $\mathbf{R}^n$ by $\mathbf{C}^n$. One such context is the study of nonlinear functions. In these cases, we use the letter $\mathbf{E}$ to stand for either $\mathbf{R}$ or $\mathbf{C}$, so $\mathbf{E}^n$ then represents either $\mathbf{R}^n$ or $\mathbf{C}^n$, and $\mathbf{E}^{m\times n}$ either $\mathbf{R}^{m\times n}$ or $\mathbf{C}^{m\times n}$.

We begin with the fundamental notion of the limit set of a bounded uniformly continuous function (the theory can be extended to other classes of functions; see the comments in Section 11).

2.1 Definition
The limit sets $\Gamma_+(\varphi)$ and $\Gamma_-(\varphi)$ of a function $\varphi \in BUC(\mathbf{R}; \mathbf{E}^n)$ are given by

$$\Gamma_\pm(\varphi) = \{\psi : \mathbf{R} \to \mathbf{E}^n \mid \varphi(t + t_k) \to \psi(t) \text{ uniformly on compact subsets of } \mathbf{R} \text{ for some sequence } t_k \to \pm\infty\}.$$

An alternative notation for $\Gamma_+(\varphi)$ is $\Gamma(\varphi)$.

Above we have assumed that φ is defined, bounded, and uniformly continuous on all of $\mathbf{R}$. Of course, the values of φ on $\mathbf{R}^-$ are irrelevant in the definition of $\Gamma_+(\varphi)$. Therefore, if φ is defined, bounded and uniformly continuous on $\mathbf{R}^+$, we can extend φ to $\mathbf{R}$ by taking, e.g., $\varphi(t) = \varphi(0)$ for $t < 0$, and use this extended function to define $\Gamma_+(\varphi)$ as in Definition 2.1.

Most of the results given below and formulated for Γ $(= \Gamma_+)$ hold with obvious modifications when Γ is replaced by Γ_-. This fact will occasionally be used later.

2.2 Definition
For each $h \in \mathbf{R}$ and each function φ defined on $\mathbf{R}$ the function $\tau_h\varphi$ is defined by $\tau_h\varphi(s) = \varphi(s + h)$ for $s \in \mathbf{R}$.

A set of functions is said to be translation invariant if $\tau_h\varphi$ belongs to the set for all $h \in \mathbf{R}$ and all φ in this set.

The following properties of limit sets are fairly obvious.

2.3 Theorem
For each $\varphi \in BUC(\mathbf{R}; \mathbf{E}^n)$ the limit sets $\Gamma_+(\varphi)$ and $\Gamma_-(\varphi)$ are nonempty, translation invariant subsets of $BUC(\mathbf{R}; \mathbf{E}^n)$. Furthermore, they are connected and compact in the topology of uniform convergence on compact sets, and the distance (in this topology) from $\tau_t\varphi$ to $\Gamma_+(\varphi)$ (or $\Gamma_-(\varphi)$) tends to zero as $t \to \infty$, (or $t \to -\infty$).

The proof is left as an exercise.

The distance function above can be any distance function compatible with the topology of uniform convergence on compact sets. For example, one can take

$$\mathrm{d}(\psi, \varphi) = \sum_{k=1}^{\infty} 2^{-k} \frac{\sup_{|s| \le k} |\psi(s) - \varphi(s)|}{1 + \sup_{|s| \le k} |\psi(s) - \varphi(s)|}.$$

As usual, the distance from $\tau_t\varphi$ to the set $\Gamma_+(\varphi)$ is given by $\mathrm{d}(\tau_t\varphi, \Gamma_+(\varphi)) = \inf_{\psi \in \Gamma_+(\varphi)} \mathrm{d}(\tau_t\varphi, \psi)$.

2.4 Theorem
Let $\varphi \in BUC(\mathbf{R}; \mathbf{E}^n)$. If $\psi \in \Gamma(\varphi)$, then $\Gamma_\pm(\psi) \subset \Gamma(\varphi)$.

The proof is left as an exercise.

2.5 Definition
An operator F that maps a set of translation invariant functions defined on $\mathbf{R}$ into a set of functions defined on $\mathbf{R}$ is called autonomous if it commutes with translations, i.e., if $\tau_h F(\varphi) = F(\tau_h \varphi)$ for every φ in the domain of F.

An example of an autonomous operator is the mapping $\varphi \mapsto \mu * \varphi$, where $*$ denotes convolution, and where $\mu \in M(\mathbf{R})$. In fact, if $F : L^1(\mathbf{R}) \to L^1(\mathbf{R})$ is linear, continuous and translation invariant, then F must be of this form; see Stein and Weiss [1], Theorem 3.19, p. 29. Another example is given by $G(\varphi)(t) = g(\varphi(t))$, where g is some continuous function defined on $\mathbf{E}^n$.

It is important to have a suitable concept of continuity for functions mapping $BUC(\mathbf{R}; \mathbf{E}^n)$ into itself in the case where the norm topology is too strong. It turns out that in such cases it is often appropriate to assume that we have continuity with respect to so-called 'narrow convergence'. This concept is defined as follows.

2.6 Definition
A sequence of functions $\varphi_k \in BC(\mathbf{R}; \mathbf{E}^n)$ is said to converge narrowly to a limit φ in BC if $\sup_{k \geq 1} \|\varphi_k\|_{BC(\mathbf{R})} < \infty$ and $\varphi_k(t) \to \varphi(t)$ uniformly for t in compact sets.

Observe that if the functions $\varphi_k \in BUC(\mathbf{R}^n)$ converge narrowly to a function φ then $\varphi \in BC(\mathbf{R})$, but φ is not necessarily uniformly continuous. However, if $\varphi \in BUC(\mathbf{R})$ and $\psi \in \Gamma(\varphi)$, then there exists a sequence $\{t_k\}$ such that $\varphi(t + t_k)$ converges narrowly to $\psi(t)$, and in this case we know that ψ is uniformly continuous (the modulus of continuity of ψ is dominated by the modulus of continuity of φ).

2.7 Theorem
Let $\varphi \in BUC(\mathbf{R}; \mathbf{E}^n)$, and let $F : BUC(\mathbf{R}; \mathbf{E}^n) \to BUC(\mathbf{R}; \mathbf{E}^n)$ be autonomous and continuous with respect to narrow convergence. Then $\Gamma(F(\varphi)) = F(\Gamma(\varphi))$.

Proof Assume that $\psi \in \Gamma(\varphi)$. Then there exists a sequence $\{t_k\}$ such that $\tau_{t_k}\varphi \to \psi$ narrowly in $BC(\mathbf{R}; \mathbf{E}^n)$ as $t_k \to \infty$. Since F commutes with translations, i.e., $\tau_h F(\varphi) = F(\tau_h \varphi)$, we have $\tau_{t_k} F(\varphi) = F(\tau_{t_k}\varphi)$ for each k, and under the continuity assumption made on F the functions $F(\tau_{t_k}\varphi)$ will converge narrowly to $F(\psi)$. Thus $F(\psi) \in \Gamma(F(\varphi))$ and we have shown that $F(\Gamma(\varphi)) \subset \Gamma(F(\varphi))$.

Conversely, let $\phi \in \Gamma(F(\varphi))$, and take some sequence $t_k \to \infty$ such that $\tau_{t_k} F(\varphi) \to \phi$ narrowly. It follows from Theorem 2.3 that one can find a subsequence, again denoted by $\{t_k\}$, and a function $\psi \in \Gamma(\varphi)$ such that the translates $\tau_{t_k}\varphi$ converge to ψ narrowly. Since F is autonomous and continuous, one has $\phi = F(\psi)$. In particular, $\phi \in F(\Gamma(\varphi))$. We conclude that $\Gamma(F(\varphi)) \subset F(\Gamma(\varphi))$. □

Another result of the same type is the following.

2.8 Theorem
Let $\varphi \in BUC^1(\mathbf{R}; \mathbf{E}^n)$. Then every $\psi \in \Gamma(\varphi)$ belongs to $BUC^1(\mathbf{R}; \mathbf{E}^n)$, and $\psi' \in \Gamma(\varphi')$. Conversely, every $\phi \in \Gamma(\varphi')$ is of the form $\phi = \psi'$ for some $\psi \in \Gamma(\varphi)$.

Proof If $\psi \in \Gamma(\varphi)$, there exists a sequence $\{t_k\}$ tending to infinity such that $\tau_{t_k}\varphi \to \psi$ narrowly. By taking a subsequence (if necessary) we can find a function $\phi \in \Gamma(\varphi')$ such that $\tau_{t_k}\varphi' \to \phi$ narrowly. Conversely, if $\phi \in \Gamma(\varphi')$, there exists a sequence $\{t_k\}$ such that $\tau_{t_k}\varphi' \to \phi$ narrowly, and by again taking a subsequence we can find a function $\psi \in \Gamma(\varphi)$ such that $\tau_{t_k}\varphi \to \psi$ narrowly. In either case, it suffices to prove that $\psi' = \phi$.

For every $h \in \mathbf{R}$ and $t \in \mathbf{R}$, we have

$$\varphi(t + h + t_k) - \varphi(t + t_k) \to \psi(t + h) - \psi(t).$$

On the other hand, it is true that

$$\varphi(t + h + t_k) - \varphi(t + t_k) = \int_t^{t+h} \varphi'(s + t_k)\,\mathrm{d}s \to \int_t^{t+h} \phi(s)\,\mathrm{d}s.$$

This shows that $\psi(t + h) - \psi(t) = \int_t^{t+h} \phi(s)\,\mathrm{d}s$, and we immediately get both $\psi \in BUC^1(\mathbf{R}; \mathbf{E}^n)$ and $\psi' = \phi$. □

We proceed to some results concerning limit sets of solutions of functional equations.

In Chapter 12 we studied the functional equation

$$x(t) = F(x)(t), \quad t \in \mathbf{R}^+; \quad x(0) = \xi, \tag{2.1}$$

where F mapped a set of functions defined on $\mathbf{R}^+$ into itself. Clearly, this setting prevents F from being autonomous, since its domain is not translation invariant. Thus, one cannot directly apply any results concerning autonomous operators to the functional equation (2.1). The easiest way to overcome this difficulty is to modify the equation so that it fits into the new framework, and to replace (2.1) by the initial value problem

$$x(t) = \begin{cases} F(x)(t), & t \in \mathbf{R}^+; \\ \varphi(t), & t \in \mathbf{R}^-. \end{cases} \tag{2.2}$$

Here we let the initial function φ belong to $BUC(\mathbf{R}^-; \mathbf{R}^n)$, and we let the domain of the functions on which F operates be all of $\mathbf{R}$.

2.9 Theorem
Let $x \in BUC(\mathbf{R};\mathbf{R}^n)$ be a solution of (2.2) and let $F = F_1 + F_2$ be the sum of two operators F_1 and F_2 which map $BUC(\mathbf{R};\mathbf{R}^n)$ into itself and satisfy the following conditions:
(i) *F_1 is autonomous and continuous with respect to narrow convergence;*
(ii) *F_2 satisfies $F_2(\psi)(t) \to 0$ as $t \to \infty$ for every $\psi \in BUC(\mathbf{R};\mathbf{R}^n)$.*
Then every $y \in \Gamma(x)$ satisfies $y = F_1(y)$.

The proof is obvious (use Theorem 2.7). We call the equation $y = F_1(y)$ the *limit equation associated with (2.2).*

Applying Theorem 2.9 to equation (1.1) we get the following result.

2.10 Corollary
Assume that
(i) $a \in L^1(\mathbf{R}^+;\mathbf{R}^{n\times n})$,
(ii) $g \in C(\mathbf{R}^n;\mathbf{R}^n)$,
(iii) $f \in BC_0(\mathbf{R}^+;\mathbf{R}^n)$,
(iv) *$x \in BC(\mathbf{R}^+;\mathbf{R}^n)$ is a solution of (1.1) on $\mathbf{R}^+$.*
Then every $y \in \Gamma(x)$ satisfies the limit equation (1.2).

A similar result is true for the functional differential equation

$$\begin{aligned} x'(t) &= F(x)(t), \quad && t \in \mathbf{R}^+; \\ x(t) &= \varphi(t), && t \in \mathbf{R}^-. \end{aligned} \tag{2.3}$$

Here we assume that φ and x are bounded and uniformly continuous, with x being, in addition, locally absolutely continuous on $\mathbf{R}^+$. Again we assume that F can be split in two parts; one that is autonomous, and maps $BUC(\mathbf{R})$ continuously into itself with respect to narrow convergence, and one that tends to zero at infinity in a weak sense.

2.11 Theorem
Let $x \in BUC(\mathbf{R};\mathbf{R}^n) \cap AC_{\text{loc}}(\mathbf{R}^+;\mathbf{R}^n)$ be a solution of (2.3) and let $F = F_1 + F_2$ be the sum of two operators F_1 and F_2 satisfying the following conditions:
(i) *$F_1 : BUC(\mathbf{R};\mathbf{R}^n) \to BUC(\mathbf{R};\mathbf{R}^n)$ is autonomous and continuous with respect to narrow convergence;*
(ii) *$F_2 : BUC(\mathbf{R};\mathbf{R}^n) \to L^1_{\text{loc}}(\mathbf{R};\mathbf{R}^n)$, and $\int_t^{t+T} F_2(\psi)(s)\,\mathrm{d}s \to 0$ as $t \to \infty$ for every $\psi \in BUC(\mathbf{R};\mathbf{R}^n)$ and every $T > 0$.*
Then every y in $\Gamma(x)$ belongs to $BUC^1(\mathbf{R};\mathbf{R}^n)$ and satisfies $y' = F_1(y)$.

Again, we call the functional differential equation $y' = F_1(y)$ the *limit equation associated with (2.3).*

Proof of Theorem 2.11 Integrating (2.3) from t to $t+T$ we get, for all $T \in \mathbf{R}$ and all $t \geq |T|$,

$$x(t+T) - x(t) = \int_t^{t+T} \Big(F_1(x)(s) + F_2(x)(s)\Big)\,\mathrm{d}s.$$

This implies that every $y \in \Gamma(x)$ satisfies the corresponding limit equation

$$y(t+T) - y(t) = \int_t^{t+T} F_1(y)(s)\,\mathrm{d}s, \qquad t \in \mathbf{R}.$$

For each fixed t, if we divide by T and let $T \to 0$, then the right-hand side tends to $F_1(y)(t)$. Thus, y is differentiable everywhere, and y satisfies the limit equation $y' = F_1(y)$. This equation implies that $y' \in BUC(\mathbf{R};\mathbf{R}^n)$. □

2.12 Corollary

Assume that

(i) $\mu \in M(\mathbf{R}^+;\mathbf{R}^{n\times n})$,

(ii) $g \in C(\mathbf{R}^n;\mathbf{R}^n)$,

(iii) $f \in L^1_{\mathrm{loc}}(\mathbf{R}^+;\mathbf{R}^n)$ *satisfies* $\int_t^{t+T} f(s)\,\mathrm{d}s \to 0$ *as* $t \to \infty$ *for every* $T > 0$,

(iv) $x \in L^\infty(\mathbf{R}^+;\mathbf{R}^n) \cap AC_{\mathrm{loc}}(\mathbf{R}^+;\mathbf{R}^n)$ *is a solution of the equation*

$$x'(t) + \int_{[0,t]} \mu(\mathrm{d}s) g\big(x(t-s)\big) = f(t), \quad t \in \mathbf{R}^+; \quad x(0) = x_0.$$

Then every y in $\Gamma(x)$ belongs to $BUC^1(\mathbf{R};\mathbf{R}^n)$ and satisfies the limit equation

$$y'(t) + \int_{\mathbf{R}^+} \mu(\mathrm{d}s) g\big(y(t-s)\big) = 0, \quad t \in \mathbf{R}.$$

Proof For every $\psi \in BUC(\mathbf{R};\mathbf{R}^n)$ and $t \in \mathbf{R}^+$, define $F_1(\psi)(t) = -\int_{\mathbf{R}^+} \mu(\mathrm{d}s) g\big(\psi(t-s)\big)$ and $F_2(\psi)(t) = f(t) + \int_{(t,\infty)} \mu(\mathrm{d}s) g(x_0)$. Let $\varphi(t) = x_0$ for $t \in \mathbf{R}^-$. An application of Theorem 2.11 gives the assertion. □

Finally, we consider limit equations associated with periodic equations. We have the following result.

2.13 Theorem

Let $x \in BUC(\mathbf{R};\mathbf{R}^n)$ be a solution of (2.2), let $S > 0$, and let $F = F_1 + F_2$ be the sum of two operators F_1 and F_2 that map $BUC(\mathbf{R};\mathbf{R}^n)$ into itself and satisfy the following two conditions:

(i) F_1 *is continuous with respect to narrow convergence and* $F_1(\tau_S\psi) = \tau_S F_1(\psi)$ *for every* $\psi \in BUC(\mathbf{R};\mathbf{R}^n)$;

(ii) F_2 *satisfies* $F_2(\psi)(t) \to 0$ *as* $t \to \infty$ *for every* $\psi \in BUC(\mathbf{R};\mathbf{R}^n)$.

Then for every $y \in \Gamma(x)$ there exists a number $h \in [0,S)$ (which may depend on y) such that $\tau_h y = F_1(\tau_h y)$.

For the functional differential equation (2.3) an analogous result holds.

2.14 Theorem
Let $x \in BUC(\mathbf{R};\mathbf{R}^n) \cap AC_{\mathrm{loc}}(\mathbf{R}^+;\mathbf{R}^n)$ be a solution of (2.3), let $S > 0$, and let $F = F_1 + F_2$ be the sum of two operators F_1 and F_2 that satisfy the following conditions:

(i) *$F_1 : BUC(\mathbf{R};\mathbf{R}^n) \to BUC(\mathbf{R};\mathbf{R}^n)$ is continuous with respect to narrow convergence and $F_1(\tau_S\psi) = \tau_S F_1(\psi)$ for every $\psi \in BUC(\mathbf{R};\mathbf{R}^n)$;*

(ii) *$F_2 : BUC(\mathbf{R};\mathbf{R}^n) \to L^1_{\mathrm{loc}}(\mathbf{R};\mathbf{R}^n)$, and $\int_t^{t+T} F_2(\psi)(s)\,\mathrm{d}s \to 0$ as $t \to \infty$ for every $\psi \in BUC(\mathbf{R};\mathbf{R}^n)$ and every $T > 0$.*

Then every y in $\Gamma(x)$ belongs to $BUC^1(\mathbf{R};\mathbf{R}^n)$, and there exists a number $h \in [0, S)$ (which may depend on y) such that $\tau_h y' = F_1(\tau_h y)$.

The proofs are left to the reader.

3. The Structure of a Limit Set

Suppose we have some information concerning a particular limit set $\Gamma(x)$ (more than what is contained in, say, Theorem 2.3). Does this allow us to conclude anything about x? Below we pursue this question and give a few simple answers.

Our first result is obvious.

3.1 Theorem
Let $\varphi \in BUC(\mathbf{R};\mathbf{E}^n)$. Then φ tends to a limit $\varphi(\infty)$ at infinity if and only if $\Gamma(\varphi)$ contains nothing but the constant function $\psi \equiv \varphi(\infty)$.

Of course, the case where the limit set consists of exactly one function is so transparent that one does not really need the machinery of limit sets. But, as we shall see in a moment, the situation is different when the limit set contains infinitely many functions. First, we give a definition.

3.2 Definition
A function $\varphi \in BUC(\mathbf{R};\mathbf{E}^n)$ is said to be asymptotically slowly varying if $\varphi(t+T) - \varphi(t) \to 0$ as $t \to \infty$ for every $T \in \mathbf{R}$.

Note that this differs from the concept of a function being slowly varying in Karamata's sense. (See, for example, Bingham, Goldie, and Teugels [1], p. 6.)

3.3 Theorem
Let $\varphi \in BUC(\mathbf{R};\mathbf{E}^n)$. Then the following three conditions are equivalent:

(i) *φ is asymptotically slowly varying;*

(ii) *$\varphi(t+T) - \varphi(t) \to 0$ as $t \to \infty$ uniformly for T in compact sets;*

(iii) *$\Gamma(\varphi)$ contains only constant functions.*

If, in addition, $\varphi \in BUC^1(\mathbf{R}; \mathbf{E}^n)$, then these three conditions are equivalent to the condition

(iv) *$\varphi'(t) \to 0$ as $t \to \infty$.*

Proof Obviously, (ii)⇒(i)⇒(iii). The implication (iii)⇒(ii) is a consequence of the part of Theorem 2.3 which says that the distance from $\tau_t\varphi$ to $\Gamma(\varphi)$ tends to zero as $t \to \infty$.

If $\varphi \in BUC^1(\mathbf{R}; \mathbf{E}^n)$, then it follows from Theorems 2.8 and 3.1 that (iii) and (iv) are equivalent. □

For periodic functions the situation is very similar.

3.4 Definition

Let $S > 0$. A function $\varphi \in BUC(\mathbf{R}; \mathbf{E}^n)$ is said to be asymptotically drifting periodic with period S, if $\varphi(t+S) - \varphi(t) \to 0$ as $t \to \infty$.

3.5 Theorem

Let $\varphi \in BUC(\mathbf{R}; \mathbf{E}^n)$. Then the following conditions are equivalent:

(i) *φ is asymptotically drifting periodic with period S;*

(ii) *$\varphi(t+kS) - \varphi(t) \to 0$ as $t \to \infty$ uniformly for k in bounded subsets of $\mathbf{Z}$;*

(iii) *Every $y \in \Gamma(\varphi)$ is periodic with period S.*

If, in addition, $\varphi \in BUC^1(\mathbf{R}; \mathbf{E}^n)$, then these three conditions are equivalent to the condition

(iv) *φ' is asymptotically drifting periodic with period S.*

Here the equivalence of (i), (ii) and (iii) is proved in the same way as in Theorem 3.3. Since a bounded function is periodic if it has a periodic derivative, it follows from Theorem 2.8 that (iii) and (iv) are equivalent. □

If a limit set contains a nonconstant function, then it contains infinitely many different nonconstant functions, which are translates of each other, because every limit set is translation invariant. In the nontrivial (i.e., nonconstant) case, this means that if we want to count the number of functions in $\Gamma(\varphi)$ we have to ignore translates.

3.6 Definition

A translation invariant subset V of functions defined on $\mathbf{R}$ is said to be generated by a subset W of V if every $\varphi \in V$ is a translate of some function in W. We call V finitely generated, if it is generated by some finite subset.

3.7 Theorem

Every finitely generated limit set contains a periodic (possibly constant) function.

Proof We shall for simplicity prove the result for positive limit sets only. Let $\varphi \in BUC(\mathbf{R}; \mathbf{E}^n)$, and suppose that the limit set $\Gamma(\varphi)$ is generated by the functions $\{\psi_1, \psi_2, \ldots, \psi_m\}$. Here we assume that no two of the functions ψ_j are translates of each other. Define the sets E_p by

$$E_p = \bigcup_{j=1}^{m} \Big\{ \tau \;\Big|\; \tau \geq p, \quad |\varphi(t+\tau) - \psi_j(t)| \leq \frac{1}{p}, \quad |t| \leq p \Big\}.$$

Clearly, each of the sets E_p is closed and nonempty, and $E_{p+1} \subset E_p$.

Assume now that none of the functions ψ_j is periodic. Then we claim the following.

There exist positive numbers δ_p and M_p with $\delta_p \to 0$ and $M_p \to \infty$ as $p \to \infty$ such that if τ and $\sigma \in E_p$ then either $|\tau - \sigma| < \delta_p$ or $|\tau - \sigma| > M_p$. (3.1)

Suppose that this is not the case. Then there exist constants c_1 and c_2 such that $0 < c_1 < c_2$ and sequences $\{p_i\}$, $\{\tau_i\}$ and $\{\sigma_i\}$ so that $\lim_{i\to\infty} p_i = \infty$, $\tau_i, \sigma_i \in E_{p_i}$, and $c_1 < |\tau_i - \sigma_i| < c_2$ for all i. By picking a subsequence, if necessary, we may assume (see the definition of the sets E_p) that $\varphi(t+\tau_i) \to \psi_j(t)$ and $\varphi(t+\sigma_i) \to \psi_k(t)$ for some fixed numbers j and k uniformly on compact sets as $i \to \infty$. Furthermore we may assume that $\tau_i - \sigma_i \to h$ where h is some nonzero real number. But this implies that $\psi_k(t+h) = \psi_j(t)$ for all t. Since none of the functions ψ_i is periodic and since they are not translates of each other, this is impossible, and this contradiction establishes the claim (3.1).

Next, we construct two sequences $\{p_j\}$ and $\{s_i\}$ tending to ∞ so that

$$\lim_{j\to\infty} \limsup_{i\to\infty} \operatorname{dist}(s_i, E_{p_j}) = \infty.$$

Suppose for the moment that we have already found such sequences. By picking a subsequence we may assume that $\varphi(t+s_i)$ tends to some function in $\Gamma_+(\varphi)$, which by our assumption must be of the form $\psi_k(t+h)$ for some k and some $h \in \mathbf{R}$. This implies that for every p we have, for sufficiently large s_i, $|\varphi(t+s_i-h) - \psi_k(t)| \leq \frac{1}{p}$ for all t with $|t| \leq p$. But then $s_i - h \in E_p$, that is, $\operatorname{dist}(s_i, E_p) \leq h$. From this contradiction we conclude that at least one of the functions ψ_i must be periodic.

We proceed to construct the sequences $\{s_i\}$ and $\{p_j\}$. Choose p_1 to be so large that $M_{p_1} > \delta_{p_1}$ and take s_1, s_2, ... to be the midpoints of the open intervals of length at least M_{p_1} in the set $(p_1, \infty) \setminus E_{p_1}$. By (3.1) such intervals exist. (We shall modify this sequence of numbers s_i at later stages.) Next, let b_1 be the supremum of the lengths of the intervals in $(p_1, \infty) \setminus E_{p_1}$. If this number is infinite, then we are done and can choose the remainder of the sequence $\{p_j\}$ in an arbitrary fashion (recall that $E_{p+1} \subset E_p$).

If b_1 is finite, then we choose p_2 to be so large that $M_{p_2} > 3b_1 + 2\delta_{p_1}$. The idea behind this choice is that then each interval in $(p_2, \infty) \setminus E_{p_2}$ of length greater than M_{p_2} includes at least three pairwise disjoint intervals contained in $(p_1, \infty) \setminus E_{p_1}$, and of length greater than M_{p_1}. Thus we can choose a subsequence from $\{s_i\}_{i=2}^{\infty}$ (renumbering the resulting sequence) so that $\operatorname{dist}(s_i, E_{p_2}) > M_{p_1}$, i.e., we choose the points that are midpoints of 'the interval in the middle'. Next, we define b_2 to be the supremum of the lengths of the intervals in $(p_2, \infty) \setminus E_{p_2}$. If this is infinite, we are done, otherwise take p_3 so that $M_{p_3} > 3b_2 + 2\delta_{p_2}$ and proceed in the same manner. In this way, we are able to construct the desired sequences. □

4. The Spectrum of a Bounded Function

In the following two sections we shall give some characterizations of functions in $BC(\mathbf{R}; \mathbf{C}^n)$ in terms of their spectral properties. The current section presents some general results that do not depend on the dimension n of the space. In Section 5 we discuss some results pertaining to the case $n \geq 2$.

Let us begin with the definition of the spectrum of an integrable function, or more generally of a measure in $M(\mathbf{R}; \mathbf{C}^{m\times n})$.

4.1 Definition
The support of the Fourier transform of a measure $\mu \in M(\mathbf{R}; \mathbf{C}^{m\times n})$ is called the spectrum of μ, and is denoted by $\sigma(\mu)$. In other words, $\sigma(\mu)$ is the closure of the set $\{\, \omega \in \mathbf{R} \mid \tilde{\mu}(\omega) \neq 0 \,\}$.

Analogously, if $a \in L^1(\mathbf{R}; \mathbf{C}^{m\times n})$, then we define the spectrum $\sigma(a)$ of a to be the closure of the set $\{\, \omega \in \mathbf{R} \mid \tilde{a}(\omega) \neq 0 \,\}$.

The spectrum $\sigma(\mu)$ of μ can be characterized in an alternative way.

4.2 Lemma
A point $\omega_0 \in \mathbf{R}$ does not belong to the spectrum $\sigma(\mu)$ of a measure $\mu \in M(\mathbf{R}; \mathbf{C}^{m\times n})$ iff there is some function $a \in L^1(\mathbf{R}; \mathbf{C})$ satisfying $a * \mu \equiv 0$ and $\tilde{a}(\omega_0) \neq 0$. In other words, $\sigma(\mu) = \bigcap_{a\in N(\mu)} \{\, \omega \in \mathbf{R} \mid \tilde{a}(\omega) = 0 \,\}$, where $N(\mu)$ is the set $N(\mu) = \{\, a \in L^1(\mathbf{R}; \mathbf{C}) \mid a * \mu \equiv 0 \,\}$.

Proof If $a * \mu \equiv 0$, then for all $\omega \in \mathbf{R}$ we have $\tilde{a}(\omega)\tilde{\mu}(\omega) = 0$. Thus, $\tilde{a}$ must vanish on the support of $\tilde{\mu}$. This implies that $\sigma(\mu) \subset \bigcap_{a\in N(\mu)} \{\, \omega \in \mathbf{R} \mid \tilde{a}(\omega) = 0 \,\}$.

Conversely, suppose that $\omega_0 \notin \sigma(\mu)$. Then there is an interval $[\omega_0 - \epsilon, \omega_0 + \epsilon]$ on which $\tilde{\mu}$ vanishes. Choose an arbitrary function $a \in L^1(\mathbf{R}; \mathbf{C})$ such that $\tilde{a}(\omega_0) \neq 0$ and $\sigma(a) \subset [\omega_0 - \epsilon, \omega_0 + \epsilon]$. Then

$\tilde{a}(\omega)\tilde{\mu}(\omega) = 0$ for all $\omega \in \mathbf{R}$, hence $a * \mu \equiv 0$. This shows that $\omega_0 \notin \bigcap_{a \in N(\mu)} \{\, \omega \in \mathbf{R} \mid \tilde{a}(\omega) = 0 \,\}$ and so $\sigma(\mu) \supset \bigcap_{a \in N(\mu)} \{\, \omega \in \mathbf{R} \mid \tilde{a}(\omega) = 0 \,\}$. □

Above we took a to be scalar-valued. Essentially the same proof gives us the following slightly modified result, where both a and μ are matrix-valued.

4.3 Lemma

A point $\omega_0 \in \mathbf{R}$ does not belong to the spectrum $\sigma(\mu)$ of a measure $\mu \in M(\mathbf{R}; \mathbf{C}^{m \times n})$ iff there is some function $a \in L^1(\mathbf{R}; \mathbf{C}^{m \times m})$ satisfying $a * \mu \equiv 0$ and $\det[\tilde{a}(\omega_0)] \neq 0$. In other words, $\sigma(\mu) = \bigcap_{a \in N_m(\mu)} \{\, \omega \in \mathbf{R} \mid \det[\tilde{a}(\omega)] = 0 \,\}$, where $N_m(\mu)$ is the set $N_m(\mu) = \{\, a \in L^1(\mathbf{R}; \mathbf{C}^{m \times m}) \mid a * \mu \equiv 0 \,\}$.

If one wants to define the spectrum $\sigma(\varphi)$ of a function $\varphi \in BC(\mathbf{R}; \mathbf{C}^n)$, one can proceed as in Definition 4.1 and define $\sigma(\varphi)$ to be the support of the distribution Fourier transform of φ. Here we shall approach the problem in a different way, avoiding the use of distribution Fourier transforms, and turn the alternative characterization given by Lemma 4.2 into a definition.

4.4 Definition

The spectrum $\sigma(\varphi)$ of a function $\varphi \in BC(\mathbf{R}; \mathbf{C}^n)$ is given by $\sigma(\varphi) = \bigcap_{a \in N(\varphi)} \{\, \omega \in \mathbf{R} \mid \tilde{a}(\omega) = 0 \,\}$, where $N(\varphi) = \{\, a \in L^1(\mathbf{R}; \mathbf{C}) \mid a * \varphi \equiv 0 \,\}$. Equivalently, a point $\omega_0 \in \mathbf{R}$ does not belong to $\sigma(\varphi)$ iff there is some $a \in L^1(\mathbf{R}; \mathbf{C})$ satisfying $a * \varphi \equiv 0$ and $\tilde{a}(\omega_0) \neq 0$.

Obviously, if $\varphi \equiv 0$, then $\sigma(\varphi) = \emptyset$. The converse (nontrivial) statement is true as well, as we shall see in Theorem 4.7.

Let us observe that one could equally well have taken the functions a in the definition of the spectrum to be matrix-valued.

4.5 Lemma

The spectrum $\sigma(\varphi)$ of a function $\varphi \in BC(\mathbf{R}; \mathbf{C}^n)$ is given by $\sigma(\varphi) = \bigcap_{a \in N_n(\varphi)} \{\, \omega \in \mathbf{R} \mid \det[\tilde{a}(\omega)] = 0 \,\}$, where $N_n(\varphi) = \{\, a \in L^1(\mathbf{R}; \mathbf{C}^{n \times n}) \mid a * \varphi \equiv 0 \,\}$. Equivalently, a point $\omega_0 \in \mathbf{R}$ does not belong to $\sigma(\varphi)$ iff there is some $a \in L^1(\mathbf{R}; \mathbf{C}^{n \times n})$ satisfying $a * \varphi \equiv 0$ and $\det[\tilde{a}(\omega_0)] \neq 0$.

Proof Since every scalar kernel can be interpreted as a matrix kernel (multiply it by the identity matrix), we have $\sigma(\varphi) \supset \bigcap_{a \in N_n(\varphi)} \{\, \omega \in \mathbf{R} \mid \det[\tilde{a}(\omega)] = 0 \,\}$.

Conversely, suppose that $a \in L^1(\mathbf{R}; \mathbf{C}^{n \times n})$ with $\det[\tilde{a}(\omega_0)] \neq 0$, and that $a * \varphi \equiv 0$. By Lemma **6**.3.5, it is possible to find a constant $\epsilon > 0$ and a function $b \in L^1(\mathbf{R}; \mathbf{C}^{n \times n})$ such that $\tilde{b}(\omega)\tilde{a}(\omega) = I$ for $|\omega - \omega_0| \leq \epsilon$. Choose some $c \in L^1(\mathbf{R}; \mathbf{C})$ satisfying $\tilde{c}(\omega_0) \neq 0$, and $\tilde{c}(\omega) = 0$ for $|\omega - \omega_0| \geq \epsilon$.

Then $\tilde{c}(\omega)\tilde{b}(\omega)\tilde{a}(\omega) = \tilde{c}(\omega)I$ for all $\omega \in \mathbf{R}$; hence $c * b * a = cI$. Convolving $a * \varphi$ by $c * b$, we find that $c * \varphi \equiv 0$. Thus, by Definition 4.4, $\omega_0 \notin \sigma(\varphi)$. □

Next, we turn to some elementary properties of spectra.

4.6 Theorem
Let $\mu \in M(\mathbf{R}; \mathbf{C}^{m \times n})$, $\varphi \in BC(\mathbf{R}; \mathbf{C}^n)$, $\psi \in BC(\mathbf{R}; \mathbf{C}^n)$ and $\gamma \in \mathbf{R} \setminus \{0\}$. Then the following claims are true.

(i) *$\sigma(\varphi)$ is closed.*
(ii) *$\sigma(\tau_\gamma \varphi) = \sigma(\varphi)$.*
(iii) *If $\phi(t) = e^{i\gamma t}\varphi(t)$, then $\sigma(\phi) = \gamma + \sigma(\varphi)$.*
(iv) *If $\phi(t) = \varphi(\gamma t)$, then $\sigma(\phi) = \gamma\sigma(\varphi)$.*
(v) *$\sigma(\varphi + \psi) \subset \sigma(\varphi) \cup \sigma(\psi)$.*
(vi) *$\sigma(\mu * \varphi) \subset \sigma(\mu) \cap \sigma(\varphi)$.*
(vii) *If $m = n$, then $\sigma(\varphi) \subset \sigma(\mu * \varphi) \cup \{\, \omega \in \mathbf{R} \mid \det[\tilde{\mu}(\omega)] = 0 \,\}$. In particular, if $\mu * \varphi \equiv 0$, then $\sigma(\varphi) \subset \{\, \omega \in \mathbf{R} \mid \det[\tilde{\mu}(\omega)] = 0 \,\}$.*

Proof Properties (i)–(v) are fairly obvious if one uses the characterization of points that do not belong to the spectrum.

(vi) First, suppose that $\omega_0 \notin \sigma(\mu)$. Then, by Definition 4.1, there is an interval $[\omega_0 - \epsilon, \omega_0 + \epsilon]$ on which $\tilde{\mu}$ vanishes. Choose an arbitrary function $a \in L^1(\mathbf{R}; \mathbf{C})$ such that $\tilde{a}(\omega_0) \neq 0$ and $\sigma(a) \subset [\omega_0 - \epsilon, \omega_0 + \epsilon]$. Then $\tilde{a}(\omega)\tilde{\mu}(\omega) = 0$ for all $\omega \in \mathbf{R}$, hence $a * \mu \equiv 0$. Therefore, we have $a * (\mu * \varphi) = (a * \mu) * \varphi \equiv 0$, and this means that $\omega_0 \notin \sigma(\mu * \varphi)$.

Next, suppose that $\omega_0 \notin \sigma(\varphi)$. Then there exists a function $a \in L^1(\mathbf{R}; \mathbf{C})$ satisfying $a * \varphi \equiv 0$ and $\tilde{a}(\omega_0) \neq 0$. This implies that $a * (\mu * \varphi) = \mu * (a * \varphi) \equiv 0$ and therefore we have $\omega_0 \notin \sigma(\mu * \varphi)$.

(vii) Suppose that $\omega_0 \notin \sigma(\mu * \varphi)$, and that $\det[\tilde{\mu}(\omega_0)] \neq 0$. Then there exists a function $a \in L^1(\mathbf{R}; \mathbf{C})$ satisfying $\tilde{a}(\omega_0) \neq 0$ and $a * (\mu * \varphi) \equiv 0$. Define $b = a * \mu$. Then $b \in L^1(\mathbf{R}; \mathbf{C}^{n \times n})$, $b * \varphi \equiv 0$, and $\det[\tilde{b}(\omega_0)] = \tilde{a}(\omega_0)\det[\tilde{\mu}(\omega_0)] \neq 0$. Thus, by Lemma 4.5, $\omega_0 \notin \sigma(\varphi)$. The second claim in (vii) follows from the first claim together with the fact that $\mu * \varphi \equiv 0$ implies $\sigma(\mu * \varphi) = \emptyset$. □

A very important problem (in fact, the key problem of spectral synthesis, see Section 11) is the question concerning the converse of the second claim in (vii), i.e., under what additional conditions does the fact that $\sigma(\varphi) \subset \{\, \omega \in \mathbf{R} \mid \tilde{\mu}(\omega) = 0 \,\}$ imply $\mu * \varphi \equiv 0$? (This is the scalar version, the problem should be reformulated when μ is matrix-valued.) We shall return to this question later.

Our next result is one version of Wiener's Tauberian theorem. We shall encounter other versions later.

4.7 Theorem
If $\varphi \in BC(\mathbf{R}; \mathbf{C}^n)$ does not vanish identically, then $\sigma(\varphi) \neq \emptyset$.

The proof of Theorem 4.7 is based on the following modification of Theorem **6**.3.2 and Lemma **6**.3.5.

4.8 Lemma

(i) *Let ψ be a $\mathbf{C}^{m\times n}$-valued function defined on $\mathbf{R}$ that has compact support, and let $\varphi \in BC(\mathbf{R}; \mathbf{C}^n)$. Suppose that for every $\omega_0 \in \mathbf{R}$ there are a function $a_{\omega_0} \in L^1(\mathbf{R}; \mathbf{C}^{m\times n})$ and a constant $\epsilon > 0$ such that $a_{\omega_0} * \varphi \equiv 0$ and $\psi(\omega) = \tilde{a}_{\omega_0}(\omega)$ for $|\omega_0 - \omega| \le \epsilon$. Then ψ is the Fourier transform of a function $a \in L^1(\mathbf{R}; \mathbf{C}^{m\times n})$ satisfying $a * \varphi \equiv 0$.*

(ii) *Let $a \in L^1(\mathbf{R}; \mathbf{C}^{n\times n})$ and $\varphi \in BC(\mathbf{R}; \mathbf{C}^n)$, and suppose that $a * \varphi \equiv 0$. If $\det[\tilde{a}(\omega_0)] \neq 0$ for some $\omega_0 \in \mathbf{R}$, then there are a function $b \in L^1(\mathbf{R}; \mathbf{C}^{n\times n})$ and a positive number ϵ such that $\tilde{b}(\omega) = I$ if $|\omega - \omega_0| \le \epsilon$ and $b * \varphi \equiv 0$.*

The proof of (i) is essentially the same as the proof of Theorem **6**.3.2. To prove (ii), one can first apply Lemma **6**.3.5(i) to get a function $c \in L^1(\mathbf{R}; \mathbf{C}^{n\times n})$ satisfying $\tilde{c}(\omega) = [\tilde{a}(\omega)]^{-1}$ in some neighbourhood of ω_0, and then define $b = c * a$.

Proof of Theorem 4.7 Let us suppose that $\sigma(\varphi) = \emptyset$, and try to prove that then $\varphi \equiv 0$.

Define

$$\eta(t) = \frac{1}{\pi t^2}(\cos t - \cos 2t), \qquad t \in \mathbf{R} \setminus \{0\}, \tag{4.1}$$

and $\eta(0) = 3/(2\pi)$. Then

$$\tilde{\eta}(\omega) = \begin{cases} 1, & |\omega| \le 1, \\ 2 - |\omega|, & 1 \le |\omega| \le 2, \\ 0, & |\omega| > 2. \end{cases} \tag{4.2}$$

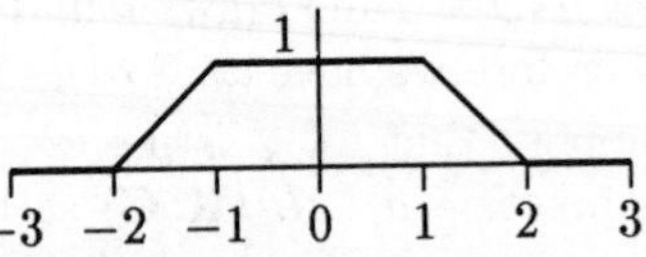

Define $\eta_k(t) = k\eta(kt)$ for $k > 0$ and $t \in \mathbf{R}$. Then $\widetilde{\eta_k}(\omega) = \tilde{\eta}(\omega/k)$, and in particular $\widetilde{\eta_k}$ has compact support. By Lemma **2**.7.4(iii), $(\eta_k * \varphi) \to \varphi$ narrowly in BC, as $k \to \infty$. Therefore, to prove that $\varphi \equiv 0$ it suffices to show that $\eta_k * \varphi \equiv 0$ for all $k > 0$.

Our assumption $\sigma(\varphi) = \emptyset$, together with Definition 4.4 and Lemma 4.8(ii), implies that for every $\omega_0 \in \mathbf{R}$ there is a function $b_{\omega_0} \in L^1(\mathbf{R}; \mathbf{C})$ satisfying $b_{\omega_0} * \varphi \equiv 0$, and $\widetilde{b_{\omega_0}}(\omega) = 1$ for all ω in some interval $[\omega_0 - \epsilon, \omega_0 + \epsilon]$. Convolving b_{ω_0} by η_k, we get a function $a_{\omega_0} \in L^1(\mathbf{R}; \mathbf{C})$ satisfying $a_{\omega_0} * \varphi \equiv 0$ and $\widetilde{a_{\omega_0}}(\omega) = \widetilde{b_{\omega_0}}(\omega)\widetilde{\eta_k}(\omega) = \widetilde{\eta_k}(\omega)$ for ω in a neighbourhood of ω_0.

By Lemma 4.8(i), $\eta_k * \varphi \equiv 0$, and the proof is complete. □

Spectra have the following continuity property.

4.9 Theorem
If $\varphi_k \to \varphi$ narrowly in $BC(\mathbf{R};\mathbf{C}^n)$, then $\sigma(\varphi) \subset \bigcap_{k\in\mathbf{N}} \overline{\bigcup_{j\geq k} \sigma(\varphi_j)}$. In particular, if $\sigma(\varphi_j) \subset \sigma(\varphi_k)$ for $j \geq k$, then $\sigma(\varphi) \subset \bigcap_{k\in\mathbf{N}} \sigma(\varphi_k)$.

Proof Let ω be an arbitrary point in $\mathbf{R} \setminus \bigcap_{k\in\mathbf{N}} \overline{\bigcup_{j\geq k} \sigma(\varphi_j)}$. Then there exist a number $\epsilon > 0$ and a sequence $j_k \to \infty$ such that $(\omega - \epsilon, \omega + \epsilon) \cap \sigma(\varphi_{j_k}) = \emptyset$. It follows from Theorem 4.6(vi) and Theorem 4.7 that if $a \in L^1(\mathbf{R};\mathbf{C})$ satisfies $\sigma(a) \subset (\omega - \epsilon, \omega + \epsilon)$ then $a * \varphi_{j_k} \equiv 0$. By the dominated convergence theorem, we get $a * \varphi \equiv 0$, and since it is possible to choose a such that $\tilde{a}(\omega) = 1$ we see that $\omega \notin \sigma(\varphi)$. □

From Theorems 4.7 and 4.9 we get a result on the spectrum of a product.

4.10 Theorem
Let $\mu \in M(\mathbf{R};\mathbf{C}^{m\times n})$, $\psi \in BC(\mathbf{R};\mathbf{C}^{m\times n})$ and $\varphi \in BC(\mathbf{R};\mathbf{C}^n)$. Then $\sigma(\mu\varphi) \subset \overline{\sigma(\mu) + \sigma(\varphi)}$ and $\sigma(\psi\varphi) \subset \overline{\sigma(\psi) + \sigma(\varphi)}$.

Here $\mu\varphi$ is the measure given by $(\mu\varphi)(E) = \int_E \mu(\mathrm{d}s)\varphi(s)$, and the spectrum of a continuous function with values in $\mathbf{C}^{m\times n}$ is defined in the same way as the spectrum of a function with values in $\mathbf{C}^n$.

Proof of Theorem 4.10 Let $\omega \in \mathbf{R} \setminus (\sigma(\mu) + \sigma(\varphi))$ be arbitrary. Define $\phi(t) = \mathrm{e}^{\mathrm{i}\omega t}\varphi(-t)$. By Theorem 4.6(iii) and (iv) we have $\sigma(\phi) = \omega - \sigma(\varphi)$. From Theorem 4.6(vi) we then get $\sigma(\mu*\phi) \subset \sigma(\mu) \cap \sigma(\phi) = \sigma(\mu) \cap (\omega - \sigma(\varphi))$, and this set is empty by our choice of ω. By Theorem 4.7, $\mu * \phi \equiv 0$. In particular, $0 = (\mu * \phi)(0) = (\widetilde{\mu\varphi})(\omega)$. Now the first claim follows from Definition 4.1.

For the second claim, we let $a_\delta(t) \stackrel{\text{def}}{=} (2\pi/3)\eta(\delta t/2)$. Then $a_\delta(0) = 1$, and $a_\delta(t) \to 1$, uniformly on compact sets as $\delta \downarrow 0$. Since $\sigma(a_\delta) = [-\delta, \delta]$ (see (4.2)), we get from the first claim proved above

$$\sigma(a_\delta\psi\varphi) \subset \overline{\sigma(a_\delta\psi) + \sigma(\varphi)} \subset \overline{\sigma(\psi) + \sigma(\varphi)} + [-\delta, \delta].$$

Since $a_\delta\psi \to \psi$ narrowly in $BC(\mathbf{R};\mathbf{C}^{m\times n})$ as $\delta \downarrow 0$, we get the desired conclusion from Theorem 4.9. □

Combining Theorem 4.6 with Theorem 4.7, we find that convolutions have the following uniqueness properties.

4.11 Theorem
Let $\mu \in M(\mathbf{R};\mathbf{C}^{m\times n})$, $\nu \in M(\mathbf{R};\mathbf{C}^{m\times n})$, and $\varphi \in BC(\mathbf{R};\mathbf{C}^n)$.

(i) *If $\sigma(\mu) \cap \sigma(\varphi) = \emptyset$, then $\mu * \varphi \equiv 0$.*
(ii) *If $\sigma(\mu - \nu) \cap \big(\sigma(\mu*\varphi) \cup \sigma(\nu*\varphi)\big) = \emptyset$, then $\mu*\varphi = \nu*\varphi$. In particular, this is true if $\sigma(\mu - \nu) \cap \sigma(\varphi) = \emptyset$.*

Proof Property (i) follows directly from Theorem 4.6(vi) and Theorem 4.7. To see that (ii) is true, it suffices to observe that, on the one hand, $\sigma((\mu - \nu) * \varphi) \subset \sigma(\mu - \nu)$ by Theorem 4.6(vi), and on the other hand,

$\sigma((\mu-\nu)*\varphi) \subset \sigma(\mu*\varphi) \cup \sigma(\nu*\varphi)$ by Theorem 4.6(v). Thus $\sigma((\mu-\nu)*\varphi) = \emptyset$, and so $(\mu - \nu) * \varphi \equiv 0$. □

The next result is fundamental to the analysis of $\sigma(\mu * \varphi)$ in the case, alluded to earlier, where $\sigma(\varphi) \subset \{\, \omega \in \mathbf{R} \mid \tilde{\mu}(\omega) = 0 \,\}$.

4.12 Theorem

Let $\mu \in M(\mathbf{R}; \mathbf{C}^{m\times n})$ and $\varphi \in BC(\mathbf{R}; \mathbf{C}^n)$, and suppose that $\tilde{\mu}(\omega_0) = 0$ for some $\omega_0 \in \mathbf{R}$. Then ω_0 is not an isolated point of $\sigma(\mu * \varphi)$.

Proof First, let us observe that we may without loss of generality take $\omega_0 = 0$ (multiply $\mu(\mathrm{d}t)$ and $\varphi(t)$ by $\mathrm{e}^{-\mathrm{i}\omega_0 t}$ and use Theorem 4.6(iii)).

Next, we show that we may as well replace the measure $\mu \in M(\mathbf{R}; \mathbf{C}^{m\times n})$ by a function $a \in L^1(\mathbf{R}; \mathbf{C}^{m\times n})$ satisfying $\tilde{a}(0) = 0$: if we define $a = \eta * \mu$, where η is the function given by (4.1), then by (vi) and (vii) of Theorem 4.6 we have $\sigma(a * \varphi) \cap [-2, 2] = \sigma(\mu * \varphi) \cap [-2, 2]$, and so the point 0 is an isolated point of $\sigma(\mu * \varphi)$ iff it is an isolated point of $\sigma(a * \varphi)$.

Thus, let us assume that $a \in L^1(\mathbf{R}; \mathbf{C}^{m\times n})$ satisfies $\tilde{a}(0) = 0$, and let us make the counterhypothesis that 0 is an isolated point of $\sigma(a * \varphi)$. For all $\gamma > 0$, define η_γ by $\eta_\gamma(t) = \gamma\eta(\gamma t)$. Thus $\sigma(\eta_\gamma) = [-2\gamma, 2\gamma]$, so there is a $\gamma_0 > 0$ such that $\sigma(\eta_\gamma) \cap \sigma(a * \varphi) = \{0\}$ for all $\gamma \le \gamma_0$. In some neighbourhood of zero we have $\widetilde{\eta_{\gamma_1}}(\omega) = \widetilde{\eta_{\gamma_2}}(\omega)$ and so, if both $\gamma_1 \le \gamma_0$ and $\gamma_2 \le \gamma_0$, it follows that $\sigma(\eta_{\gamma_1} - \eta_{\gamma_2}) \cap \sigma(a*\varphi) = \emptyset$. By Theorem 4.11(ii) we have $\eta_{\gamma_1} * (a * \varphi) = \eta_{\gamma_2} * (a * \varphi)$. Let us denote this common function by ψ. Then $\psi = \eta_\gamma * a * \varphi$ for all $\gamma \le \gamma_0$. Moreover, $\sigma(\psi) = \{0\}$, because, on the one hand, $\sigma(\psi) \subset \sigma(\eta_\gamma) \cap \sigma(a * \varphi) = \{0\}$, and, on the other hand, if $0 \notin \sigma(\psi)$ then there exists a function $b \in L^1(\mathbf{R}; \mathbf{C})$ such that $b * \psi \equiv 0$ but $\tilde{b}(0) \ne 0$. Then $(b * \eta_\gamma) * (a * \varphi) \equiv 0$ and $\widetilde{(b * \eta_\gamma)}(0) \ne 0$ so that $0 \notin \sigma(a * \varphi)$ which contradicts the counterhypothesis. Since $\sigma(\psi) \ne \emptyset$, it follows that $\psi \not\equiv 0$.

We have now shown that there exists a nonzero function ψ satisfying $\psi = \eta_\gamma * a * \varphi$ for all $\gamma \le \gamma_0$, and to complete the contradiction argument it suffices to show that necessarily $\psi \equiv 0$. (Observe that up to now we have not used the assumption $\tilde{a}(0) = 0$.) To see that this is the case one argues as follows. Fix some $\epsilon > 0$. By Lemma **2**.5.3, we can find a function $b \in L^1(\mathbf{R}; \mathbf{C}^{m\times n})$ with $\|b\|_{L^1(\mathbf{R})} \le \epsilon$, and satisfying $\tilde{b}(\omega) = \tilde{a}(\omega)$ in some interval $[-\delta, \delta]$. If we choose γ small enough that $\sigma(\eta_\gamma * \varphi) \subset \sigma(\eta_\gamma) \subset [-\delta/2, \delta/2]$, then by Theorem 4.11(ii) $\psi = a * (\eta_\gamma * \varphi) = b * (\eta_\gamma * \varphi)$. This implies that $\|\psi\|_{BC(\mathbf{R})} \le \|b\|_{L^1(\mathbf{R})}\|\eta_\gamma\|_{L^1(\mathbf{R})}\|\varphi\|_{BC(\mathbf{R})} \le 2\epsilon\|\varphi\|_{BC(\mathbf{R})}$. Here ϵ was arbitrary, and this shows that $\psi \equiv 0$. □

As an immediate consequence of Theorem 4.12, we get the following corollary.

4.13 Corollary
Let $\mu \in M(\mathbf{R}; \mathbf{C}^{m\times n})$ and $\varphi \in BC(\mathbf{R}; \mathbf{C}^n)$, and suppose that $\tilde{\mu}(\omega) = 0$ for every $\omega \in \sigma(\varphi)$. Then $\sigma(\mu * \varphi)$ is a perfect subset of the intersection of the boundary of $\sigma(\mu)$ and the boundary of $\sigma(\varphi)$ (a perfect set is a closed set with no interior and no isolated points).

It follows from Theorem 4.12 that $\sigma(\mu * \varphi)$ has no isolated points. To see that it has no interior and that it is a subset of the intersection of the boundaries of $\sigma(\mu)$ and $\sigma(\varphi)$ one employs Theorem 4.6(vi): $\sigma(\mu * \varphi) \subset \sigma(\mu) \cap \sigma(\varphi) \subset \sigma(\mu) \cap \{\, \omega \in \mathbf{R} \mid \tilde{\mu}(\omega) = 0 \,\}$, and the interior of the set $\{\, \omega \in \mathbf{R} \mid \tilde{\mu}(\omega) = 0 \,\}$ is $\mathbf{R} \setminus \sigma(\mu)$.

It is a well-known fact that every perfect set is uncountable. Therefore, the preceding corollary yields a condition under which the inclusion $\sigma(\varphi) \subset \{\, \omega \in \mathbf{R} \mid \tilde{\mu}(\omega) = 0 \,\}$ implies that $\mu * \varphi$ vanishes identically.

4.14 Corollary
Let $\mu \in M(\mathbf{R}; \mathbf{C}^{m\times n})$, let $\varphi \in BC(\mathbf{R}; \mathbf{C}^n)$, and suppose that $\tilde{\mu}(\omega) = 0$ for every $\omega \in \sigma(\varphi)$, and that the intersection of the boundaries of $\sigma(\mu)$ and $\sigma(\varphi)$ is countable (in particular, this is true if the set $\{\, \omega \in \mathbf{R} \mid \tilde{\mu}(\omega) = 0 \,\}$ is countable). Then $\mu * \varphi \equiv 0$.

Under a smallness assumption on $\tilde{\mu}$, one obtains the identical conclusion.

4.15 Theorem
Let $\mu \in M(\mathbf{R}; \mathbf{C}^{m\times n})$ and $\varphi \in BC(\mathbf{R}; \mathbf{C}^n)$, and suppose that $\tilde{\mu}(\omega) = 0$ for every $\omega \in \sigma(\varphi)$. In addition, assume that at least one of the following conditions holds:

(i) *for each $k \in \mathbf{N}$ there exists a constant c_k such that $|\tilde{\mu}(\omega_1) - \tilde{\mu}(\omega_2)| \le c_k|\omega_1 - \omega_2|^{\frac{1}{2}}$ for $|\omega_1|$, $|\omega_2| \le k$;*

(ii) *$\tilde{\mu} \in BV_{\mathrm{loc}}(\mathbf{R}; \mathbf{C}^{m\times n})$.*

Then $\mu * \varphi \equiv 0$.

Proof Since $\sigma(\tau_h\varphi) = \sigma(\varphi)$ and $(\mu * \tau_h\varphi)(0) = (\mu * \varphi)(h)$ for all $h \in \mathbf{R}$, it suffices to prove that $(\mu * \varphi)(0) = 0$.

First, we show that we may, without loss of generality, assume that $\sigma(\mu)$ and $\sigma(\varphi)$ are compact. To see this, note that, if $\eta_k(t) = k\eta(kt)$ where η is given in (4.1), then $\eta_k * \mu$ and $\eta_k * \varphi$ have compact spectra; see (4.2) and Theorem 4.6(vi), and if μ satisfies (i) or (ii), then so does $\eta_k * \mu$. Furthermore, $\widetilde{\eta_k}(\omega)\tilde{\mu}(\omega) = 0$ for $\omega \in \sigma(\eta_k * \varphi)$, and if $(\eta_k * \mu * \eta_k * \varphi)(0) = 0$ for all k, then, by Lemma **2**.7.4(iii), $(\mu * \varphi)(0) = 0$.

Let $a_\delta(t) = (2\pi/3)\eta(\delta t/2)$ and $\psi_\delta = \widetilde{a_\delta\varphi}$. Then ψ_δ is a continuous function with compact support (contained in $\sigma(\varphi) + [-\delta, \delta]$; see Theorem 4.10). Hence, by the Fourier inversion formula,

$$a_\delta(t)\varphi(t) = \frac{1}{2\pi}\int_{-\infty}^{\infty} \mathrm{e}^{\mathrm{i}\omega t}\psi_\delta(\omega)\,\mathrm{d}\omega.$$

From the dominated convergence theorem and Fubini's theorem we get (observe that $a_\delta(t) \to 1$ as $\delta \downarrow 0$)

$$(\mu * \varphi)(0) = \lim_{\delta\downarrow 0} \int_{-\infty}^{\infty} \mu(\mathrm{d}t) a_\delta(-t)\varphi(-t)$$
$$= \frac{1}{2\pi} \lim_{\delta\downarrow 0} \int_{-\infty}^{\infty} \tilde{\mu}(\omega)\psi_\delta(\omega)\,\mathrm{d}\omega. \qquad (4.3)$$

The function ψ_δ has its support in $\sigma(\varphi) + [-\delta, \delta]$, and $\tilde{\mu}$ vanishes on $\sigma(\varphi)$. Therefore we conclude that

$$\int_{-\infty}^{\infty} \tilde{\mu}(\omega)\psi_\delta(\omega)\,\mathrm{d}\omega = \int_{E(\delta)} \tilde{\mu}(\omega)\psi_\delta(\omega)\,\mathrm{d}\omega,$$

where $E(\delta) \overset{\text{def}}{=} \{\omega \in \mathbf{R} \mid \operatorname{dist}(\omega, \sigma(\varphi)) < \delta\} \setminus \sigma(\varphi)$ is bounded and open. Hence $E(\delta)$ can be written in the form $E(\delta) = \bigcup_{j=1}^{\infty} J_j(\delta)$, where each $J_j(\delta)$ is a bounded open interval and where at least one of the end-points of each interval $J_j(\delta)$ belongs to $\sigma(\varphi)$. The length of each $J_j(\delta)$ is at most 2δ, because no point of $J_j(\delta)$ belongs to $\sigma(\varphi)$, but every point is within the distance δ from a point in $\sigma(\varphi)$.

Let us assume that (i) holds. Then we get from Hölder's inequality combined with Plancherel's theorem (see Theorem **16**.8.2; without loss of generality, we may let the norm in $\mathbf{C}^m$ be Euclidean)

$$\left|\int_{E(\delta)} \tilde{\mu}(\omega)\psi_\delta(\omega)\,\mathrm{d}\omega\right|^2 \le \int_{E(\delta)} \left|\tilde{\mu}(\omega)\right|^2 \mathrm{d}\omega \int_{\mathbf{R}} \left|\psi_\delta(\omega)\right|^2 \mathrm{d}\omega$$
$$= 2\pi \int_{E(\delta)} \left|\tilde{\mu}(\omega)\right|^2 \mathrm{d}\omega \int_{\mathbf{R}} \left|a_\delta(t)\varphi(t)\right|^2 \mathrm{d}t$$
$$\le \frac{2\pi}{\delta} \int_{E(\delta)} \left|\tilde{\mu}(\omega)\right|^2 \mathrm{d}\omega \, \|a_1\|_{L^2(\mathbf{R})}^2 \|\varphi\|_{L^\infty(\mathbf{R})}^2.$$

Since $\tilde{\mu}$ vanishes on $\sigma(\varphi)$, $\tilde{\mu}$ vanishes at one (or both) of the end-points of each interval $J_j(\delta)$. Recall that $E(\delta)$ is bounded, uniformly for small δ, and that the length of $J_j(\delta)$ is at most 2δ. Therefore, it follows from (i) that $|\tilde{\mu}(\omega)| \le c\sqrt{\delta}$ on each interval $J_j(\delta)$, where c is some constant, independent of j and δ. Thus,

$$\left|\int_{E(\delta)} \tilde{\mu}(\omega)\psi_\delta(\omega)\,\mathrm{d}\omega\right|^2 \le 2\pi c^2 m(E(\delta)) \|a_1\|_{L^2(\mathbf{R})}^2 \|\varphi\|_{L^\infty(\mathbf{R})}^2,$$

where $m(E(\delta))$ is the Lebesgue measure of $E(\delta)$. But $m(E(\delta)) \downarrow 0$ as $\delta \downarrow 0$, and so the right-hand side of (4.3) vanishes and we get the desired conclusion.

Assume next that (ii) holds. Denote the distribution derivative of $\tilde{\mu}$ by α, i.e., let $\tilde{\mu}(\omega) = \alpha((-\infty, \omega])$. Since $\tilde{\mu}$ is assumed to have compact support, it follows from (ii) that $\alpha \in M(\mathbf{R}; \mathbf{C}^{m\times n})$. Recall that $\tilde{\mu}(\omega)$ vanishes at one of the end-points of each interval $J_j(\delta)$. Let Ψ_j be the integral function of ψ_δ that vanishes at the opposite end-point of $J_j(\delta)$, and integrate by parts to get

$$\left|\int_{J_j(\delta)} \tilde{\mu}(\omega)\psi_\delta(\omega)\,\mathrm{d}\omega\right| \le \int_{J_j(\delta)} |\alpha|(\mathrm{d}\omega)\big|\Psi_j(\omega)\big|.$$

Since $|\psi_\delta(\omega)| \le \frac{1}{\delta}\|a_1\|_{L^1(\mathbf{R})}\|\varphi\|_{L^\infty(\mathbf{R})}$, and since the length of $J_j(\delta)$ is at most 2δ, we have $\big|\Psi_j(\omega)\big| \le 2\|a_1\|_{L^1(\mathbf{R})}\|\varphi\|_{L^\infty(\mathbf{R})}$. Thus,

$$\begin{aligned}\left|\int_{E(\delta)} \tilde{\mu}(\omega)\psi_\delta(\omega)\,\mathrm{d}\omega\right| &\le 2\sum_{j=1}^{\infty}\int_{J_j(\delta)} |\alpha|(\mathrm{d}\omega)\,\|a_1\|_{L^1(\mathbf{R})}\|\varphi\|_{L^\infty(\mathbf{R})} \\ &= 2|\alpha|\big(E(\delta)\big)\|a_1\|_{L^1(\mathbf{R})}\|\varphi\|_{L^\infty(\mathbf{R})}.\end{aligned}$$

Since $|\alpha|\big(E(\delta)\big) \to 0$ as $\delta \downarrow 0$ (recall that $\tilde{\mu}$ is continuous and of bounded variation), we again get the desired conclusion. □

As the next result shows, the assumption that $\tilde{\mu}(\omega) = 0$ for $\omega \in \sigma(\varphi)$ does not alone (in general) imply $\mu * \varphi \equiv 0$.

4.16 Proposition
There exist a function $a \in L^1(\mathbf{R};\mathbf{C})$ and a function $\varphi \in BC(\mathbf{R};\mathbf{C})$ such that $\tilde{a}(\omega) = 0$ for $\omega \in \sigma(\varphi)$, but $a * \varphi \not\equiv 0$.

Proof Recall that an ideal A in $L^1(\mathbf{R};\mathbf{C})$ is a subspace satisfying the condition that $a * b \in A$ whenever $a \in A$ and $b \in L^1(\mathbf{R};\mathbf{C})$. It is easy to prove that every closed ideal is translation invariant (approximate the translation operator by an L^1 convolution operator). For each ideal A, let $Z(A)$ be the zero set $\bigcap_{b\in A}\{\,\omega \mid \tilde{b}(\omega) = 0\,\}$, and for each set $E \subset \mathbf{R}$ let $I(E)$ denote the closed ideal $I(E) = \{\,a \in L^1(\mathbf{R};\mathbf{C}) \mid \tilde{a}(\omega) = 0 \text{ for } \omega \in E\,\}$. By Rudin [1], Theorem 7.6.1, there exist a closed set $E \subset \mathbf{R}$ and a closed ideal A in $L^1(\mathbf{R};\mathbf{C})$, such that $E = Z(A)$ and $A \subset I(E)$, but $A \ne I(E)$. By the Hahn–Banach theorem, there exist a function $\phi \in L^\infty(\mathbf{R};\mathbf{C})$ and a function $a \in I(E) \setminus A$, such that $\int_{-\infty}^{\infty} a(-t)\phi(t)\,\mathrm{d}t \ne 0$, and $\int_{-\infty}^{\infty} b(-t)\phi(t)\,\mathrm{d}t = 0$ for every $b \in A$. Let ξ be some function in $L^1(\mathbf{R};\mathbf{C})$ with nonvanishing Fourier transform, and take $\varphi = \xi * \phi$. Then it follows that $\varphi \in BC(\mathbf{R};\mathbf{C})$ and that $a * \varphi \not\equiv 0$, but (here we use the translation invariance of the ideal A) $b * \varphi \equiv 0$ for every $b \in A$. It follows from the first part of Definition 4.4 that $\sigma(\varphi) \subset E$, and the proof is complete. □

By now we are able to identify the class of functions that have small (i.e., finite or countable) spectra.

4.17 Theorem
Let $\varphi \in BC(\mathbf{R};\mathbf{C}^n)$. Then the following claims are true.

(i) *$\sigma(\varphi) = \{0\}$ iff φ is equal to a nonzero constant.*

(ii) *ω_0 is an isolated point of $\sigma(\varphi)$ iff φ is of the form $\varphi(t) = \alpha e^{i\omega_0 t} + \phi(t)$, where α is a nonzero constant, and $\sigma(\phi) = \sigma(\varphi) \setminus \{\omega_0\}$.*

(iii) *$\sigma(\varphi) = \{\omega_1, \omega_2, \ldots, \omega_k\}$ iff φ is a trigonometric polynomial of the form $\varphi(t) = \sum_{j=1}^{k} \alpha_j e^{i\omega_j t}$, where each α_j is a nonzero constant.*

(iv) $\sigma(\varphi) \subset \{ k\omega_0 \mid k \in \mathbf{Z} \}$ *for some* $\omega_0 \neq 0$ *iff* φ *is a periodic function with period* $2\pi/\omega_0$.

(v) *If* φ *is uniformly continuous and* $\sigma(\varphi)$ *is countable, then* φ *is almost-periodic.*

Proof (i) Clearly, if φ is a nonzero constant, then $\sigma(\varphi) = \{0\}$.

Conversely, let $\sigma(\varphi) = \{0\}$. Define $\mu_h = \delta_h - \delta_0$, where δ_h and δ_0 are unit point masses at h and 0, respectively. Then $\widetilde{\mu_h}(0) = 0$, and hence, by Theorem 4.6(vi), Theorem 4.7 and Theorem 4.12, $\mu_h * \varphi \equiv 0$. This means that $\tau_h \varphi = \varphi$ for all $h \in \mathbf{R}$; hence φ is a constant.

(ii) If φ is of the given form, then clearly ω_0 is an isolated point of $\sigma(\varphi)$.

Assume that ω_0 is an isolated point of $\sigma(\varphi)$. Without loss of generality we may take $\omega_0 = 0$ (multiply $\varphi(t)$ by $\mathrm{e}^{-\mathrm{i}\omega_0 t}$). Choose some function $a \in L^1(\mathbf{R}; \mathbf{C})$, whose Fourier transform is identically 1 in some neighbourhood of zero, and vanishes on some neighbourhood of $\sigma(\varphi) \setminus \{0\}$ (for example, a rescaled version of the function η in (4.1) will do). Define $\psi = a * \varphi$. It follows from Theorem 4.6(vi) that $\sigma(\psi) = \{0\}$ and so, by (i), ψ is equal to a nonzero constant. Let $\phi = \varphi - \psi = (\delta - a) * \varphi$ (where δ is the unit point mass at zero) and so by Theorem 4.6(vi) we have $0 \notin \sigma(\phi)$. On the other hand, $\phi = \varphi - \psi$, and so (recall Theorem 4.6(v)), $\sigma(\phi) \subset \sigma(\varphi) \cup \{0\} = \sigma(\varphi)$. In particular, $\sigma(\phi) \subset \sigma(\varphi) \setminus \{0\}$. Write $\varphi = \phi + \psi$ to get $\sigma(\varphi) \subset \sigma(\phi) \cup \{0\}$. Thus $\sigma(\phi) = \sigma(\varphi) \setminus \{0\}$.

(iii) Apply (ii) k times to get (iii).

(iv) Define $\mu = \delta_{2\pi/\omega_0} - \delta_0$ (where $\delta_{2\pi/\omega_0}$ is the unit point mass at $2\pi/\omega_0$). Then φ is $2\pi/\omega_0$-periodic iff $\mu * \varphi \equiv 0$. Moreover, $\tilde{\mu}(\omega) = \mathrm{e}^{-\mathrm{i}\, 2\pi\omega/\omega_0} - 1 = 0$ iff $\omega = k\omega_0$ for some $k \in \mathbf{Z}$. On the one hand, by Theorem 4.6(vii), $\mu * \varphi \equiv 0$ implies that $\sigma(\varphi) \subset \{ \omega \in \mathbf{R} \mid \tilde{\mu}(\omega) = 0 \}$, and, on the other hand, if $\sigma(\varphi) \subset \{ \omega \in \mathbf{R} \mid \tilde{\mu}(\omega) = 0 \}$, then $\mu * \varphi \equiv 0$ by Corollary 4.14.

(v) First, we define the set of points in the spectrum where '$\tilde{\varphi}$ belongs locally to $\widetilde{AP}$' to be

$$A(\varphi) = \{ \omega \mid \textit{there exists a function } \psi \in L^1(\mathbf{R}; \mathbf{C}), \textit{ with } \tilde{\psi} \equiv 1 \textit{ in some neighbourhood of } \omega, \textit{ such that } \psi * \varphi \in AP(\mathbf{R}; \mathbf{C}^n) \}.$$

It is clear that $A(\varphi)$ may alternatively be characterized by the fact that

$$A(\varphi) = \{ \omega \mid \textit{ there exists a constant } \epsilon > 0 \textit{ such that } \psi * \varphi \in AP(\mathbf{R}; \mathbf{C}^n) \textit{ for all } \psi \in L^1(\mathbf{R}; \mathbf{C}) \textit{ with } \sigma(\psi) \subset (\omega - \epsilon, \omega + \epsilon) \}$$

(choose ϵ so small that the interval $[\omega - \epsilon, \omega + \epsilon]$ belongs to the set where the function $\tilde{\psi}$ in the first characterization of $A(\varphi)$ is identically one and recall Theorem **2**.2.2(v)). It is also clear that $A(\varphi)$ is open, and that $A(\varphi)$ contains $\mathbf{R} \setminus \sigma(\varphi)$. (These statements are true for every $\varphi \in BC(\mathbf{R}; \mathbf{C}^n)$, independently of the size of $\sigma(\varphi)$.)

By using a division of unity, one concludes that $\psi * \varphi$ is almost-periodic for every $\psi \in L^1(\mathbf{R}; \mathbf{C})$ with compact spectrum contained in $A(\varphi)$ (cf. the proof of Theorem **6**.3.2). Thus, if we can show that $A(\varphi) = \mathbf{R}$, then we can convolve φ by an approximate identity with compact spectrum (e.g., convolve φ by a sequence of functions $\eta_k \in L^1(\mathbf{R};\mathbf{C})$, where $\eta_k(t) = k\eta(kt)$, and η is given by (4.1)), and use Lemma **2**.7.4(ii) to show that φ is almost-periodic (recall that $AP(\mathbf{R})$ is closed in $BUC(\mathbf{R})$).

Since $\mathbf{R} \setminus A(\varphi)$ is contained in the countable set $\sigma(\varphi)$, we conclude that if $\mathbf{R} \setminus A(\varphi)$ is nonempty then it must contain at least one isolated point, say ω_0. We may without loss of generality assume that $\omega_0 = 0$. Choose $\epsilon > 0$ such that $(-\epsilon, \epsilon) \cap \mathbf{R} \setminus A(\varphi) = \{0\}$. Let η be the function in (4.1), and define $\eta_\delta(t) = \delta\eta(\delta t)$ for $\delta > 0$. Then $\widetilde{\eta_\delta}(\omega) = 1$ for $\omega \le \delta$, and $\sigma(\eta_\delta) = [-2\delta, 2\delta]$. Recall that the Fourier transform of η_δ' is $i\omega\widetilde{\eta_\delta}(\omega)$ (see Theorem **3**.8.4(i)), hence $\sigma(\eta_\delta') = [-2\delta, 2\delta]$. If we now choose δ_1 and δ_2 satisfying $0 < \delta_2 < \delta_1 < \epsilon/3$, then it follows that $(\eta_{\delta_1}' - \eta_{\delta_2} * \eta_{\delta_1}') * \varphi$ is almost-periodic, because the spectrum of $\eta_{\delta_1}' - \eta_{\delta_2} * \eta_{\delta_1}'$ is contained in $[-2\delta_1, -\delta_2] \cup [\delta_2, 2\delta_1]$. But $\eta_{\delta_2} * \eta_{\delta_1}' \to 0$ in $L^1(\mathbf{R};\mathbf{C})$ as $\delta_2 \to 0$ (see the proof of Lemma **2**.5.3, and note that $\widetilde{\eta_{\delta_1}'}(0) = 0$). Thus, $\eta_{\delta_2} * \eta_{\delta_1}' * \varphi \to 0$ in $BUC(\mathbf{R};\mathbf{C}^n)$ as $\delta_2 \downarrow 0$, and we conclude that $\eta_{\delta_1}' * \varphi = (\eta_{\delta_1} * \varphi)'$ is almost-periodic. By Theorem 4.19 below, a differentiable function that is bounded and has an almost-periodic derivative must be almost-periodic. We deduce that $\eta_{\delta_1} * \varphi$ is almost-periodic. Thus $0 \in A(\varphi)$, and the proof of (v) is complete. □

We leave it as an exercise to show that the converse of part (v) does not hold, i.e., if φ is almost-periodic, then it is not necessarily true that the spectrum $\sigma(\varphi)$ is countable (let φ be the Fourier transform of a discrete measure with uncountable support).

The spectra of a function and its derivative are related as follows.

4.18 Theorem

Let $\varphi \in BC^1(\mathbf{R};\mathbf{C}^n)$. Then $\sigma(\varphi') \subset \sigma(\varphi) \subset \sigma(\varphi') \cup \{0\}$.

Proof An application of Theorem 4.6(ii) and (v) shows that, if we define φ_h by $\varphi_h(t) = \frac{1}{h}(\tau_h\varphi - \varphi)$, then $\sigma(\varphi_h) \subset \sigma(\varphi)$. As $h \to 0$ we have $\varphi_h \to \varphi'$ uniformly on compact subsets. In particular, $\varphi_h \to \varphi'$ narrowly, and it follows from Theorem 4.9 that $\sigma(\varphi') \subset \sigma(\varphi)$.

Conversely, suppose that $\omega_0 \ne 0$ does not belong to $\sigma(\varphi')$. Then one can find some function $a \in L^1(\mathbf{R};\mathbf{C})$ such that $\tilde{a}(\omega_0) \ne 0$, and $a * \varphi' \equiv 0$. Without loss of generality we may assume that $a \in AC_{\mathrm{loc}}(\mathbf{R};\mathbf{C})$, and that $a' \in L^1(\mathbf{R};\mathbf{C})$ (replace a by $b * a$, where $b(t) = \chi_{\mathbf{R}^+}(t)e^{-t}$). Integrating by parts (or using Theorem **3**.7.1) one gets $a' * \varphi = a * \varphi' \equiv 0$. Defining $b = a'$, we have $\tilde{b}(\omega_0) = i\omega_0\tilde{a}(\omega_0) \ne 0$, and $b * \varphi \equiv 0$. Thus $\omega_0 \notin \sigma(\varphi)$. □

In the proof of Theorem 4.17 we needed (part of) the following result.

4.19 Theorem
Let $\varphi \in BUC^1(\mathbf{R};\mathbf{C}^n)$. Then $\varphi \in AP(\mathbf{R};\mathbf{C}^n)$ if and only if $\varphi' \in AP(\mathbf{R};\mathbf{C}^n)$.

Proof Suppose that $\varphi \in AP(\mathbf{R};\mathbf{C}^n)$. Define φ_h by $\varphi_h(t) = \frac{1}{h}(\tau_h\varphi - \varphi)$. Then $\varphi_h \in AP(\mathbf{R};\mathbf{C}^n)$. As $h \to 0$, we have $\varphi_h \to \varphi'$ in $BUC(\mathbf{R};\mathbf{C}^n)$, and we conclude that $\varphi' \in AP(\mathbf{R};\mathbf{C}^n)$.

Conversely, suppose that $\varphi' \in AP(\mathbf{R};\mathbf{C}^n)$. By studying each component of φ separately, we may, without loss of generality, assume that φ is a scalar function. For a proof of this scalar case we refer the reader to Katznelson [1], Theorem 5.21, p. 167. □

5. The Directional Spectrum in $\mathbf{C}^n$

As we saw in Theorem 4.6(vii), if $\mu \in M(\mathbf{R};\mathbf{C}^{n\times n})$ and $\varphi \in BC(\mathbf{R};\mathbf{C}^n)$, and if $\mu * \varphi \equiv 0$, then $\sigma(\varphi) \subset \{\,\omega \in \mathbf{R} \mid \det[\tilde{\mu}(\omega)] = 0\,\}$. However, in this case it is possible to say more. Let us, for simplicity, assume that $\varphi \in L^1(\mathbf{R};\mathbf{C}^n)$. Then, for any $\omega_0 \in \sigma(\varphi)$ it must be true that $\tilde{\varphi}(\omega_0)$ belongs to the null space of $\tilde{\mu}(\omega_0)$. This observation is the starting point for the results which we shall present below.

5.1 Definition
Let $\varphi \in BC(\mathbf{R};\mathbf{C}^n)$ and let $N_n(\varphi) \stackrel{\text{def}}{=} \{\,a \in L^1(\mathbf{R};\mathbf{C}^{n\times n}) \mid a * \varphi \equiv 0\,\}$. For each $\omega \in \mathbf{R}$, the set $\rho_\omega(\varphi) \stackrel{\text{def}}{=} \bigcap_{a\in N_n(\varphi)} \ker[\tilde{a}(\omega)]$ is called the spectral subspace of φ at ω. The set $\rho(\varphi) \stackrel{\text{def}}{=} \{\,(\omega, \rho_\omega(\varphi)) \mid \omega \in \mathbf{R}\,\}$ is called the directional spectrum of φ.

We want to stress the fact that $\rho_\omega(\varphi)$ is a subspace of $\mathbf{C}^n$, in particular, $\rho(\varphi)$ is a subset of $\mathbf{R} \times \mathbf{C}^n$.

If $\varphi \in BC(\mathbf{R};\mathbf{C}^n) \cap L^1(\mathbf{R};\mathbf{C}^n)$, then $\rho_\omega(\varphi) = \operatorname{span}\{\tilde{\varphi}(\omega)\}$ at any point ω where $\tilde{\varphi}(\omega) \neq 0$ (see Exercise 20).

The connection between the spectral subspaces defined above and the spectrum defined in the previous section is the following.

5.2 Theorem
Let $\varphi \in BC(\mathbf{R};\mathbf{C}^n)$. Then $\sigma(\varphi) = \{\,\omega \in \mathbf{R} \mid \rho_\omega(\varphi) \neq \{0\}\,\}$.

Proof It follows immediately from the first part of Lemma 4.5 that $\sigma(\varphi) \supset \{\,\omega \in \mathbf{R} \mid \rho_\omega(\varphi) \neq \{0\}\,\}$.

Next, let us assume that $\rho_{\omega_0}(\varphi) = \{0\}$. An equivalent way to formulate Definition 5.1 is to define $\rho_{\omega_0}(\varphi)$ to be the intersection of all the nullspaces of the operators $\tilde{b}(\omega_0) : \mathbf{C}^n \to \mathbf{C}$ where each b belongs to $L^1(\mathbf{R};\mathbf{C}^{1\times n})$ and

$b * \varphi \equiv 0$. Therefore, there must exist n functions $b_j \in L^1(\mathbf{R}; \mathbf{C}^{1\times n})$ such that $b_j * \varphi \equiv 0$ and $\cap_{j=1}^n \ker[\widetilde{b_j}(\omega_0)] = \{0\}$. This implies that the matrix B whose jth row is equal to b_j satisfies $B * \varphi \equiv 0$ and $\det[\tilde{B}(\omega_0)] \neq 0$ (if $\det[\tilde{B}(\omega_0)] = 0$, then there exists a vector $v \neq 0$ such that $\tilde{B}(\omega_0)v = 0$, and this means that $v \in \cap_{j=1}^n \ker[\widetilde{b_j}(\omega_0)] = \{0\}$). By the second part of Lemma 4.5, $\omega_0 \notin \sigma(\varphi)$. □

Reformulating Theorem 4.6 to suit the new definition we get the following result.

5.3 Theorem
Let $\mu \in M(\mathbf{R}; \mathbf{C}^{m\times n})$, $\varphi \in BC(\mathbf{R}; \mathbf{C}^n)$, $\psi \in BC(\mathbf{R}; \mathbf{C}^n)$, $\omega \in \mathbf{R}$, and $\gamma \in \mathbf{R} \setminus \{0\}$. Then the following claims are true.
(i) *$\rho(\varphi)$ is a closed subset of $\mathbf{R} \times \mathbf{C}^n$.*
(ii) *$\rho(\tau_\gamma \varphi) = \rho(\varphi)$.*
(iii) *If $\phi(t) = e^{i\gamma t}\varphi(t)$, then $\rho_\omega(\phi) = \rho_{\omega-\gamma}(\varphi)$.*
(iv) *If $\phi(t) = \varphi(\gamma t)$, then $\rho_{\gamma\omega}(\phi) = \rho_\omega(\varphi)$.*
(v) *If $\rho_\omega(\psi) = \{0\}$, then $\rho_\omega(\varphi + \psi) = \rho_\omega(\varphi)$.*
(vi) *$\tilde{\mu}(\omega)\rho_\omega(\varphi) \subset \rho_\omega(\mu * \varphi)$.*
(vii) *If $m = n$ and $\det[\tilde{\mu}(\omega)] \neq 0$, then $\tilde{\mu}(\omega)\rho_\omega(\varphi) = \rho_\omega(\mu * \varphi)$.*
(viii) *If $\mu * \varphi \equiv 0$, then $\rho_\omega(\varphi) \subset \ker[\tilde{\mu}(\omega)]$.*

Proof The proofs of claims (i)-(iv) are left as an exercise.

To prove (v), suppose that $v \notin \rho_\omega(\varphi)$. Then, by Definition 5.1, there is some $a \in L^1(\mathbf{R}; \mathbf{C}^{n\times n})$ such that $a * \varphi \equiv 0$ and $\tilde{a}(\omega)v \neq 0$. By Theorem 5.2, $\omega \notin \sigma(\psi)$. Choose some function $b \in L^1(\mathbf{R}; \mathbf{C})$ such that $\tilde{b}(\omega) \neq 0$ and such that $b * \psi \equiv 0$ (cf. Definition 4.4). Define $c = b * a$. Then $c * (\varphi + \psi) = b * (a * \varphi) + a * (b * \psi) \equiv 0$ and $\tilde{c}(\omega)v = \tilde{b}(\omega)\tilde{a}(\omega)v \neq 0$. Thus $v \notin \rho_\omega(\varphi + \psi)$. To prove the reverse inclusion, observe that $\varphi = (\varphi + \psi) + (-\psi)$ and that $\rho_\omega(-\psi) = \{0\}$, and apply the same argument with φ replaced by $\varphi + \psi$ and ψ replaced by $-\psi$.

To prove (vi), we let $v \in \rho_\omega(\varphi)$ and let $a \in L^1(\mathbf{R}; \mathbf{C}^{m\times m})$ be an arbitrary function satisfying $a * (\mu * \varphi) \equiv 0$. Then $(a * \mu) * \varphi \equiv 0$ and $\tilde{a}(\omega)\tilde{\mu}(\omega)v = 0$. Therefore, $\tilde{\mu}(\omega)v \in \rho_\omega(\mu * \varphi)$.

To establish (vii), we first note that we have, by (vi), for each $\omega_0 \in \mathbf{R}$, $\tilde{\mu}(\omega_0)\rho_{\omega_0}(\varphi) \subset \rho_{\omega_0}(\mu * \varphi)$. Choose some ω_0 so that $\det[\tilde{\mu}(\omega_0)] \neq 0$, let $w \notin \tilde{\mu}(\omega_0)\rho_{\omega_0}(\varphi)$, and $v = [\tilde{\mu}(\omega_0)]^{-1}w$. Then $v \notin \rho_{\omega_0}(\varphi)$. This means that there exists a function $a \in L^1(\mathbf{R}; \mathbf{C}^{n\times n})$ such that $a * \varphi \equiv 0$, and $\tilde{a}(\omega_0)v \neq 0$. By Theorem **6**.3.3(ii) and Lemma **6**.3.5(i), there is a function $b \in L^1(\mathbf{R}; \mathbf{C}^{n\times n})$ such that $\tilde{b}\tilde{\mu} = I$ in some neighbourhood of ω_0. Choose some $\eta \in L^1(\mathbf{R}; \mathbf{C})$ such that $\tilde{\eta}(\omega_0) \neq 0$, and such that $\tilde{\eta}$ is supported in the same neighbourhood. Then $a*\eta*b*\mu*\varphi = a*\eta*\varphi = \eta*a*\varphi \equiv 0$. Thus, if we define $c = \eta * a * b$, then $c * (\mu * \varphi) \equiv 0$. Since $\tilde{c}(\omega_0) = \tilde{\eta}(\omega_0)\tilde{a}(\omega_0)[\tilde{\mu}(\omega_0)]^{-1}$, we have $\tilde{c}(\omega_0)w = \tilde{\eta}(\omega_0)\tilde{a}(\omega_0)v \neq 0$. According to Definition 5.1, this means that $w \notin \rho_{\omega_0}(\mu * \varphi)$.

To prove (viii), we note that we may convolve μ by some function $\eta \in L^1(\mathbf{R};\mathbf{C})$ with nonvanishing Fourier transform and in this way reduce the problem to the case where μ is absolutely continuous. But then the claim is an immediate consequence of the definition of the directional spectrum. □

5.4 Remark

The converse of (vi) in Theorem 5.3 is not true, i.e., it is not true in general that $\rho_\omega(\mu * \varphi) = \tilde{\mu}(\omega)\rho_\omega(\varphi)$. To see this, one can consider the following unpretentious example. Let $\psi \in BC(\mathbf{R};\mathbf{C})$ and $a \in L^1(\mathbf{R};\mathbf{C})$ satisfy $\tilde{a}(\omega) = 0$ for all $\omega \in \sigma(\psi)$, but $a * \psi \not\equiv 0$; see Proposition 4.16. Define φ by

$$\varphi = \begin{pmatrix} a * \psi \\ \psi \end{pmatrix},$$

and take $\mu = \alpha\delta$, where δ is the unit point mass at zero and $\alpha = (1 \quad 0)$, i.e., α is the projection onto the first component. Let ω_0 be an arbitrary point in $\sigma(a * \psi)$. As $\mu * \varphi = \alpha\varphi = a * \psi$, and $\omega_0 \in \sigma(a * \psi)$, it follows that $\rho_{\omega_0}(\mu * \varphi) = \mathbf{C}$. Let η be an arbitrary function in $L^1(\mathbf{R};\mathbf{C})$ such that $\tilde{\eta}(\omega_0) = 1$, and take $b \stackrel{\text{def}}{=} (\eta \quad -\eta * a)$. Then we have $b * \varphi \equiv 0$ and $\tilde{b}(\omega_0) = (1 \quad 0)$, hence $\rho_{\omega_0}(\varphi) \subset (0 \quad \mathbf{C})^T$ (see the proof of Theorem 5.2). Therefore, $\tilde{\mu}(\omega_0)\rho_{\omega_0}(\varphi) = \{0\} \neq \rho_{\omega_0}(\mu * \varphi)$.

5.5 Remark

At first sight, one would also expect that $\rho_\omega(\phi + \varphi) \subset \operatorname{span}\{\rho_\omega(\phi) \cup \rho_\omega(\varphi)\}$ (without any restrictions of the type given in Theorem 5.3(v)), but it turns out that this is not true either. One can, for example, take a and ψ as in the previous remark and define

$$\phi = \begin{pmatrix} a * \psi \\ \psi \end{pmatrix}, \qquad \varphi = \begin{pmatrix} a * \psi \\ -\psi \end{pmatrix}.$$

We leave it to the reader (Exercise 23) to check that there exist points $\omega \in \mathbf{R}$ for which the inclusion mentioned above is not satisfied.

Theorem 4.12 has the following sharpened version.

5.6 Theorem

Let $\mu \in M(\mathbf{R};\mathbf{C}^{m\times n})$ and $\varphi \in BC(\mathbf{R};\mathbf{C}^n)$, and suppose that $\rho_{\omega_0}(\varphi) \subset \ker[\tilde{\mu}(\omega_0)]$ for some $\omega_0 \in \mathbf{R}$. Then ω_0 is not an isolated point of $\sigma(\mu * \varphi)$.

Proof In the same way as in the proof of Theorem 4.12, we may assume that $\omega_0 = 0$ and that μ has been replaced by a function $a \in L^1(\mathbf{R};\mathbf{C}^{m\times n})$. Assume that 0 is an isolated point in $\sigma(a*\varphi)$. Define η as in (4.1) and take $\eta_\delta(t) = \delta\eta(\delta t)$. Again we may choose δ to be so small that $\sigma(\eta_\delta * a * \varphi) = \{0\}$. This implies, by Theorem 4.17(i), that the function $\eta_\delta * a * \varphi$ is a nonzero constant which we may denote by v.

The set $\{\eta_\delta * \varphi\}_{\delta \le 1}$ is uniformly bounded in $BC(\mathbf{R};\mathbf{C}^n)$, and is equicontinuous because $\|(\eta_\delta * \varphi)'\|_{BC(\mathbf{R})} \le \|\eta_\delta'\|_{L^1(\mathbf{R})}\|\varphi\|_{BC(\mathbf{R})}$ and $\|\eta_\delta'\|_{L^1(\mathbf{R})} \le \delta\|\eta'\|_{L^1(\mathbf{R})}$. By the Arzelà–Ascoli Theorem **2**.7.5, some subsequence $\eta_{\delta_j} * \varphi$

converges narrowly to a limit in $BC(\mathbf{R};\mathbf{C}^n)$ as $\delta_j \to 0$. By Theorem 4.9 this limit must necessarily be a constant. Let us denote this constant by w.

We claim that $w \in \rho_0(\varphi)$. To prove this, let $b \in L^1(\mathbf{R};\mathbf{C}^{n\times n})$ satisfy $b * \varphi \equiv 0$. Then $b * \eta_{\delta_j} * \varphi \equiv 0$, and passing to the limit we get $0 = \left(\int_{\mathbf{R}} b(s)\,ds\right) w = \tilde{b}(0)w$. This proves our claim. In particular, since we assumed that $\rho_0(\varphi) \subset \ker[\tilde{a}(0)]$, this means that $\tilde{a}(0)w = 0$.

On the other hand, as $v = a * \eta_{\delta_j} * \varphi$ for all j, when we pass to the limit we get $v = \tilde{a}(0)w = 0$. This contradicts our earlier observation that v must be nonzero, and completes the proof. □

In analogy to Corollary 4.14, we have the following result.

5.7 Corollary

Let $\mu \in M(\mathbf{R};\mathbf{C}^{n\times n})$ and $\varphi \in BC(\mathbf{R};\mathbf{C}^n)$. If the set $\{\,\omega \in \mathbf{R} \mid \det[\tilde{\mu}(\omega)] = 0\,\}$ is countable, and if $\rho_\omega(\varphi) \subset \ker[\tilde{\mu}(\omega)]$ for all $\omega \in \mathbf{R}$, then $\mu * \varphi \equiv 0$.

Proof The assumptions imply that $\sigma(\varphi)$ is countable. It follows from Theorem 4.6(vi), that the same holds for $\sigma(\mu * \varphi)$. □

As the next remark shows, Theorem 4.15 has no counterpart for directional spectra in higher dimensions.

5.8 Remark

Let $a \in L^1(\mathbf{R};\mathbf{C})$ and $\psi \in BC(\mathbf{R};\mathbf{C})$ satisfy $\sigma(\psi) \subset \{\,\omega \in \mathbf{R} \mid \tilde{a}(\omega) = 0\,\}$ and $a * \psi \not\equiv 0$; see Proposition 4.16. Define the function φ to be

$$\varphi = \begin{pmatrix} a * \psi \\ \psi \end{pmatrix}.$$

As we have seen in Remark 5.4, (note that $\sigma(\varphi) \subset \sigma(a * \varphi)$)

$$\rho_\omega(\varphi) \subset \begin{pmatrix} 0 \\ \mathbf{C} \end{pmatrix}, \quad \omega \in \sigma(\varphi).$$

Let

$$A(t) = \begin{pmatrix} \xi(t) & 0 \\ 0 & 0 \end{pmatrix},$$

where ξ is some integrable function satisfying $\tilde{\xi}(\omega) \neq 0$ for all ω. Evidently, $\rho_\omega(\varphi) \subset \ker[\tilde{A}(\omega)]$ for all $\omega \in \mathbf{R}$, but nevertheless

$$A * \varphi = \begin{pmatrix} \xi * a * \psi \\ 0 \end{pmatrix} \not\equiv 0.$$

Clearly, no smoothness assumption on the function A or $\tilde{A}$, corresponding, e.g., to those given in Theorem 4.15 for the scalar case, affect this example.

6. The Asymptotic Spectrum

In this section we combine the notions of the limit set and the spectrum of a function in $BUC(\mathbf{R};\mathbf{C}^n)$, and get asymptotic results.

6.1 Definition

The closure of the set $\bigcup_{\psi\in\Gamma(\varphi)}\sigma(\psi)$, *where* φ *is a function in* $BUC(\mathbf{R};\mathbf{C}^n)$, *is called the asymptotic spectrum of* φ, *and is denoted by* $\sigma^\infty(\varphi)$. *The closed span of the set* $\bigcup_{\psi\in\Gamma(\varphi)}\rho_{\omega_0}(\psi)$ *is called the asymptotic spectral subspace of* φ *at* $\omega_0\in\mathbf{R}$ *and is denoted by* $\rho^\infty_{\omega_0}(\phi)$. *The set* $\rho^\infty(\varphi)\overset{\text{def}}{=}\{\,(\omega,\rho^\infty_\omega(\varphi))\mid\omega\in\mathbf{R}\}$ *is called the asymptotic directional spectrum of* φ.

By simply combining earlier results on limit sets with the recently presented results on spectra, we obtain several asymptotic results.

6.2 Theorem

Let $\varphi\in BUC(\mathbf{R};\mathbf{C}^n)$.

(i) $\sigma^\infty(\varphi)=\emptyset$ *iff* $\varphi(t)\to 0$ *as* $t\to\infty$.

(ii) $\sigma^\infty(\varphi)=\{0\}$ *iff* φ *is asymptotically slowly varying and does not tend to zero at infinity.*

(iii) ω_0 *is an isolated point of* $\sigma^\infty(\varphi)$ *iff* φ *is of the form* $\varphi(t)=\alpha(t)\mathrm{e}^{\mathrm{i}\omega_0 t}+\phi(t)$, *where the function* α *is asymptotically slowly varying and does not tend to zero at infinity, and* $\sigma^\infty(\phi)=\sigma^\infty(\varphi)\setminus\{\omega_0\}$.

(iv) $\sigma^\infty(\varphi)=\{\omega_1,\omega_2,\dots,\omega_k\}$ *iff* φ *is a trigonometric polynomial of the form* $\varphi(t)=\sum_{j=1}^k\alpha_j(t)\mathrm{e}^{\mathrm{i}\omega_j t}$, *where each function* α_j *is asymptotically slowly varying and does not tend to zero at infinity.*

(v) $\sigma^\infty(\varphi)\subset\{\,k\omega_0\mid k\in\mathbf{Z}\}$ *for some* $\omega_0\neq 0$ *iff* φ *is asymptotically drifting periodic with period* $2\pi/\omega_0$.

The proof of this theorem is very similar to the proof of Theorem 4.17 and is therefore omitted.

In many cases the most interesting part of Theorem 6.2 is the converse which says that if we have sufficient restrictions on $\sigma^\infty(\varphi)$, then we get asymptotic information on φ. Of course, for this to be useful we need some independent method to get such *a priori* knowledge. A typical result of this type is the following.

6.3 Theorem

Assume that $\mu\in M(\mathbf{R};\mathbf{C}^{m\times n})$, $\varphi\in BUC(\mathbf{R};\mathbf{C}^n)$, $\psi\in BUC(\mathbf{R};\mathbf{C}^n)$, $\phi\in BUC(\mathbf{R};\mathbf{C}^{m\times n})$ *and* $\omega\in\mathbf{R}$. *Then the following claims are true.*

(i) $\sigma^\infty(\varphi+\psi)\subset\sigma^\infty(\varphi)\cup\sigma^\infty(\psi)$.

(ii) $\sigma^\infty(\phi\varphi)\subset\overline{\sigma^\infty(\phi)+\sigma^\infty(\varphi)}$.

(iii) $\sigma^\infty(\mu*\varphi)\subset\sigma(\mu)\cap\sigma^\infty(\varphi)$.

(iv) *If* $m = n$, *then* $\sigma^\infty(\varphi) \subset \sigma^\infty(\mu * \varphi) \cup \{\omega \in \mathbf{R} \mid \det[\tilde{\mu}(\omega)] = 0\}$. *In particular, if* $(\mu * \varphi)(t) \to 0$ *as* $t \to \infty$, *then* $\sigma^\infty(\varphi) \subset \{\omega \in \mathbf{R} \mid \det[\tilde{\mu}(\omega)] = 0\}$.

(v) *If* $\varphi \in BUC^1(\mathbf{R}; \mathbf{C}^n)$, *then* $\sigma^\infty(\varphi') \subset \sigma^\infty(\varphi) \subset \sigma^\infty(\varphi') \cup \{0\}$.

(vi) *If* $\rho_\omega^\infty(\psi) = \{0\}$, *then* $\rho_\omega^\infty(\varphi + \psi) = \rho_\omega^\infty(\varphi)$.

(vii) $\tilde{\mu}(\omega)\rho_\omega^\infty(\varphi) \subset \rho_\omega^\infty(\mu * \varphi)$.

(viii) *If* $m = n$ *and* $\det[\tilde{\mu}(\omega)] \neq 0$, *then* $\tilde{\mu}(\omega)\rho_\omega^\infty(\varphi) = \rho_\omega^\infty(\mu * \varphi)$.

(ix) *If* $(\mu * \varphi)(t) \to 0$ *as* $t \to \infty$, *then* $\rho_\omega^\infty(\varphi) \subset \ker[\tilde{\mu}(\omega)]$.

(x) *If* $\rho_\omega^\infty(\varphi) \subset \ker[\tilde{\mu}(\omega)]$, *then* ω *is not an isolated point of* $\sigma^\infty(\mu * \varphi)$.

(xi) *If the set* $\{\omega \in \mathbf{R} \mid \det[\tilde{\mu}(\omega)] = 0\}$ *is countable, and if* $\rho_\omega^\infty(\varphi) \subset \ker[\tilde{\mu}(\omega)]$ *for every* $\omega \in \sigma^\infty(\varphi)$, *then* $(\mu * \varphi)(t) \to 0$ *as* $t \to \infty$.

The easy proof of this theorem is left to the reader.

As a corollary to Theorems 6.2(i) and 6.3(iv), we get Pitt's form of Wiener's Tauberian theorem.

6.4 Corollary

Assume that $\mu \in M(\mathbf{R}; \mathbf{C}^{n\times n})$ *satisfies* $\det[\tilde{\mu}(\omega)] \neq 0$ *for all* $\omega \in \mathbf{R}$, *let* $\varphi \in BUC(\mathbf{R}; \mathbf{C}^n)$, *let* $A \in \mathbf{C}^n$, *and suppose that* $(\mu * \varphi)(t) \to A$ *as* $t \to \infty$. *Then* $\varphi(t) \to [\tilde{\mu}(0)]^{-1}A$ *as* $t \to \infty$.

Proof By Theorem 2.9, every $\psi \in \Gamma(\varphi)$ satisfies $\mu * \psi \equiv A$. By Theorems 3.3, 6.2(i)–(ii) and 6.3(iv), ψ must be a constant, i.e., $\psi(t) \equiv B$ for some $B \in \mathbf{C}^n$. Hence $\mu * \psi \equiv \tilde{\mu}(0)B = A$, and thus $B = [\tilde{\mu}(0)]^{-1}A$. Clearly (cf. Theorem 3.1), this implies that $\varphi \to [\tilde{\mu}(0)]^{-1}A$ as $t \to \infty$. □

7. The Renewal Equation

In this section we apply some of the results developed above to the scalar renewal equation

$$x(t) - \int_{[0,t]} \beta(\mathrm{d}s)x(t-s) = f(t), \quad t \in \mathbf{R}^+. \tag{7.1}$$

Earlier, this equation was studied in Theorem **7**.4.1. Like we did in that theorem, we assume that β is a positive measure with total mass 1. In Theorem **7**.4.1 it was assumed, in addition, that β has an absolutely continuous component, and a finite first moment. Below we relax these additional assumptions, at the expense of a somewhat weaker conclusion.

As we have seen in Chapters 2–4, it suffices to study the resolvent of the kernel. We let $-\mu$ be the resolvent of $-\beta$, that is, μ solves the equations

$$\mu = \beta + \beta * \mu = \beta + \mu * \beta. \tag{7.2}$$

Since the equation is scalar, it follows from Theorem 4.1.5 that there exists a unique solution $\mu \in M_{\mathrm{loc}}(\mathbf{R}^+;\mathbf{R})$, provided $\beta(\{0\}) < 1$. Furthermore, one can easily show that μ can be represented as the infinite series

$$\mu = \sum_{j=1}^{\infty} \beta^{*j}. \tag{7.3}$$

Recall that a measure is nonarithmetic if its support is *not* contained in a set of the form $\{ j\lambda \mid j \in \mathbf{Z} \}$ for some $\lambda \in \mathbf{R}$.

Concerning the asymptotic behaviour of μ, we have the following result (where we use the convention $1/\infty = 0$).

7.1 Theorem
Let $\beta \in M(\mathbf{R}^+;\mathbf{R})$ be positive and nonarithmetic, with $\beta(\mathbf{R}^+) = 1$. Let μ be the solution of (7.2) (i.e., $-\mu$ is the resolvent of $-\beta$), and let $J \subset \mathbf{R}^+$ be an interval with finite length $m(J)$. Then μ is positive, and

$$\lim_{t\to\infty} \mu(J + t) = \frac{m(J)}{\int_{\mathbf{R}^+} t\beta(\mathrm{d}t)}. \tag{7.4}$$

That $-\beta$ has a resolvent is a consequence of the fact that, since β is positive and nonarithmetic with $\beta(\mathbf{R}^+) = 1$, we have $\beta(\{0\}) < 1$.

Proof of Theorem 7.1 By (7.3), μ is positive. Observe that, since $\beta(\mathbf{R}^+) = 1$, the function $x \equiv 1$ is the solution of the equation

$$x(t) - \int_{[0,t]} \beta(\mathrm{d}s)x(t-s) = \beta\big((t,\infty)\big), \quad t \in \mathbf{R}^+.$$

Using the variation of constants formula and the facts that the nonhomogeneous term is nonnegative and nonincreasing, and μ is positive, we get, for every $\epsilon \in \mathbf{R}^+$ and $t \geq \epsilon$,

$$\begin{aligned} x(t) = 1 = \beta\big((t,\infty)\big) + \int_{[0,t]} \mu(\mathrm{d}s)\beta\big((t-s,\infty)\big) \\ \geq \beta\big((\epsilon,\infty)\big)\mu\big([t-\epsilon,t]\big). \end{aligned} \tag{7.5}$$

Since we can choose $\epsilon > 0$ such that $\beta\big((\epsilon,\infty)\big) > 0$, we conclude that

$$\sup_{t\geq\epsilon} \mu\big([t-\epsilon,t]\big) < \infty.$$

It follows immediately from this result that

$$\sup_{t\geq 0} \mu(t+J) < \infty, \tag{7.6}$$

for every finite interval $J \subset \mathbf{R}^+$.

Let $\varphi \in C_c((0,\infty);\mathbf{R})$ be arbitrary, and define y by $y(t) = \varphi(t) + (\mu * \varphi)(t)$. Then y is a continuous solution of the equation

$$y(t) - (\beta * y)(t) = \varphi(t), \quad t \in \mathbf{R}^+.$$

Moreover, as φ is continuous and has compact support, it follows from (7.6) that y is bounded and uniformly continuous on $\mathbf{R}^+$. Observe that

$1-\tilde{\beta}(\omega)=0$ if and only if $\omega=0$, write $\varphi=(\delta-\beta)*y$ and use Theorem 6.3(iv) to conclude that $\sigma^\infty(y)\subset\{0\}$. Therefore, by Theorems 3.3 and 6.2(i)–(ii) we conclude that $\Gamma(y)=\Gamma(\mu*\varphi)$ contains only constant functions.

First, make the additional assumption that $\int_{\mathbf{R}^+} t\beta(\mathrm{d}t)<\infty$, or equivalently that the function $t\mapsto b(t)\stackrel{\text{def}}{=}\beta((t,\infty))$ is integrable. Since (cf. (7.5))

$$b(t)+(\mu*b)(t)=1,\quad t\ge 0,$$

we get

$$(\varphi*\mu*b)(t)=\int_0^t\varphi(s)\,\mathrm{d}s-(\varphi*b)(t),\quad t\in\mathbf{R}^+.$$

Now $\tilde{b}(\omega)\ne 0$ for all $\omega\in\mathbf{R}$ (because $\tilde{b}(0)=\int_{\mathbf{R}^+} t\beta(\mathrm{d}t)$, and $\tilde{b}(\omega)=(1-\tilde{\beta}(\omega))/(\mathrm{i}\omega)$ for $\omega\ne 0$), and $(\varphi*b)(t)\to 0$ as $t\to\infty$. Thus, by Corollary 6.4,

$$\lim_{t\to\infty}(\mu*\varphi)(t)=\frac{\int_{\mathbf{R}^+}\varphi(t)\,\mathrm{d}t}{\int_{\mathbf{R}^+}t\beta(\mathrm{d}t)}.$$

Let J be an arbitrary finite interval contained in $(0,\infty)$, with length $m(J)$, and let $\epsilon>0$. Then we may find two functions φ_1 and φ_2 in $C_c((0,\infty);\mathbf{R})$, such that $\varphi_1(t)\le\chi_J(t)\le\varphi_2(t)$ for $t\in\mathbf{R}^+$, and such that $\int_{\mathbf{R}^+}\varphi_1(t)\,\mathrm{d}t\ge m(J)-\epsilon$, and $\int_{\mathbf{R}^+}\varphi_2(t)\,\mathrm{d}t\le m(J)+\epsilon$. By the preceding argument, and by the fact that μ is nonnegative,

$$\begin{aligned}\frac{m(J)-\epsilon}{\int_{\mathbf{R}^+}t\beta(\mathrm{d}t)}&\le\frac{\int_{\mathbf{R}^+}\varphi_1(t)\,\mathrm{d}t}{\int_{\mathbf{R}^+}t\beta(\mathrm{d}t)}=\lim_{t\to\infty}(\mu*\varphi_1)(t)\le\liminf_{t\to\infty}\mu(J+t)\\&\le\limsup_{t\to\infty}\mu(J+t)\le\lim_{t\to\infty}(\mu*\varphi_2)(t)\\&=\frac{\int_{\mathbf{R}^+}\varphi_2(t)\,\mathrm{d}t}{\int_{\mathbf{R}^+}t\beta(\mathrm{d}t)}\le\frac{m(J)+\epsilon}{\int_{\mathbf{R}^+}t\beta(\mathrm{d}t)}.\end{aligned}$$

Clearly, this implies that $\lim_{t\to\infty}\mu(J+t)=m(J)/\int_{\mathbf{R}^+}t\beta(\mathrm{d}t)$.

In the case where b is not integrable, we choose φ to be nonnegative, and obtain for every nonnegative integrable function $b_0\le b$, the inequality

$$\limsup_{t\to\infty}\big(b_0*(\mu*\varphi)\big)(t)\le\int_{\mathbf{R}^+}\varphi(s)\,\mathrm{d}s.$$

Let $\psi\in\Gamma(\mu*\varphi)$. Then ψ is a constant; say $\psi(t)\equiv A$ for some constant A. By the preceding inequality, $A\int_{\mathbf{R}^+}b_0(s)\,\mathrm{d}s\le\int_{\mathbf{R}^+}\varphi(s)\,\mathrm{d}s$. Since $\int_0^\infty b_0(s)\,\mathrm{d}s$ can be arbitrarily large, we see that we must have $A=0$. Thus $\lim_{t\to\infty}(\mu*\varphi)(t)=0$. From this fact one easily concludes that $\mu(J+t)\to 0$ as $t\to\infty$ for every bounded interval J. □

8. A Result of Lyapunov Type

As one of the last results in this chapter, we prove a theorem which involves a certain Lyapunov argument. This result will be needed in our study of the asymptotic behaviour of solutions of Volterra equations.

8.1 Theorem

Assume that

(i) $H : BUC(\mathbf{R}; \mathbf{E}^n) \to BUC(\mathbf{R}; \mathbf{R})$ *is autonomous (i.e., commutes with translations) and is continuous with respect to narrow convergence,*

(ii) $\varphi \in BUC(\mathbf{R}; \mathbf{E}^n)$,

(iii) $\lim_{t\to-\infty} H(\psi)(t)$ *exists and equals* $\sup_{t\in\mathbf{R}} H(\psi)(t)$ *for every* $\psi \in \Gamma(\varphi)$, *and if* $\psi \in \Gamma(\varphi)$ *is nonconstant, then* $\limsup_{t\to\infty} H(\psi)(t) < \sup_{t\in\mathbf{R}} H(\psi)(t)$,

(iv) *The set* $\{ H(\psi)(0) \mid \psi \in \Gamma(\varphi) \text{ is a constant} \} \subset \mathbf{R}$ *has empty interior.*

Then $\lim_{t\to\infty} H(\varphi)(t)$ *exists, and* φ *is asymptotically slowly varying.*

Proof First, let us observe that it suffices to prove that φ is asymptotically slowly varying. If we can show this, then necessarily $\Gamma(H(\varphi))$ contains only constant functions of the type $H(\psi)$, where $\psi \in \Gamma(\varphi)$ is another constant function (see Theorem 2.7). As $\Gamma(H(\varphi))$ is connected, and the set $\{ H(\psi)(0) \mid \psi \in \Gamma(\varphi) \text{ is a constant} \}$ has no interior, this means that $\Gamma(H(\varphi))$ contains exactly one constant function. Therefore, by Theorem 3.1, $\lim_{t\to\infty} H(\varphi)(t)$ exists.

Suppose that $\Gamma(\varphi)$ contains a nonconstant function ψ. Define the real numbers α and β by $\alpha = \limsup_{t\to\infty} H(\psi)(t)$ and $\beta = \sup_{t\in\mathbf{R}} H(\psi)(t)$. Then, by the hypothesis, $\alpha < \beta$. We claim that for each $\gamma \in (\alpha, \beta)$ it is possible to find a constant function $\psi_\gamma \in \Gamma(\varphi)$ for which $H(\psi_\gamma)(0) = \gamma$. Clearly, this contradicts the assumption that the set $\{ H(\psi)(0) \mid \psi \in \Gamma(\varphi) \text{ is a constant} \}$ has empty interior. Thus, once we have established the claim, the proof will be complete.

Fix $\gamma \in (\alpha, \beta)$, and fix some sequence $\{t_n\}$ tending to $+\infty$ such that $\varphi(t+t_n) \to \psi(t)$ uniformly on compact sets. Since $\limsup_{t\to\infty} H(\psi)(t) < \gamma$ there exist positive numbers ϵ and T such that $H(\psi)(t+T) \le \gamma - \epsilon$ for all $t \ge 0$. As H is continuous with respect to narrow convergence, it follows that for every n there exists a number $p_n \ge t_n$ such that $H(\varphi)(t+p_n) < \gamma$ for all $t \in [0, n]$. On the other hand, as $\beta = \sup_{t\in\mathbf{R}} H(\psi)(t) > \gamma$, we must have $\limsup_{t\to\infty} H(\varphi)(t) > \gamma$. Define $q_n = \inf\{ s \ge p_n \mid H(\varphi)(s) = \gamma \}$. Then $q_n - p_n > n$, $H(\varphi)(t) \le \gamma$ for $p_n \le t \le q_n$, and $H(\varphi)(q_n) = \gamma$. By picking a subsequence, if necessary, we may assume that $\varphi(t+q_n) \to \eta(t)$ uniformly on compact sets. Then clearly $H(\eta)(t) \le \gamma$ for all $t \le 0$ and $H(\eta)(0) = \gamma$. But this, together with our assumptions on H, implies that $\lim_{t\to-\infty} H(\eta)(t) = \gamma$. By Theorems 2.7 and 3.1, there exists a function $\psi_\gamma \in \Gamma_-(\eta) \subset \Gamma(\varphi)$ such that $H(\psi_\gamma)(t) \equiv \gamma$ for all $t \in \mathbf{R}$. Necessarily, ψ_γ

must be a constant (because H is not allowed to map nonconstant elements of $\Gamma(\varphi)$ into constants), and the proof is complete. □

To verify assumption (iv) in Theorem 8.1, one can frequently use the following version of Sard's theorem.

8.2 Sard's Theorem
Let $E \in C^n(\mathbf{R}^n;\mathbf{R})$. Then the set $\{\, E(y) \mid y \in \mathbf{R}^n, E'(y) = 0 \,\}$ has Lebesgue measure zero. In particular, this set has no interior.

For a proof, see Sternberg [1], p. 47.

When we apply this result to the situation described in Theorem 8.1, we do it in the following form.

8.3 Corollary
Assume that $\varphi \in BUC(\mathbf{R};\mathbf{R}^n)$ and that $H : BUC(\mathbf{R};\mathbf{R}^n) \to BUC(\mathbf{R};\mathbf{R})$. Let $y \in \mathbf{R}^n$, let 1_y be the function that is identically equal to y on $\mathbf{R}$, and define the function $E : \mathbf{R}^n \to \mathbf{R}$ by $E(y) = H(1_y)(0)$. If $E \in C^n(\mathbf{R}^n;\mathbf{R})$ and $E'(y) = 0$ whenever the function 1_y belongs to $\Gamma(\varphi)$, then the set $\{\, H(\psi)(0) \mid \psi \in \Gamma(\varphi)$ is a constant $\}$ has empty interior.

9. Appendix: Two Tauberian Decomposition Results

Frequently, the functions whose limiting behaviour is studied are not assumed to be uniformly continuous but, instead, they are required to satisfy a Tauberian condition of the form $\lim_{t\to\infty,\tau\to 0}|f(t) - f(t+\tau)| = 0$. The next lemma shows that this approach gives the same results that are attainable with the tools given in this chapter.

9.1 Lemma
Let $f : \mathbf{R}^+ \to \mathbf{E}^n$. Then the following two conditions are equivalent:
(i) $\lim_{t\to\infty,\tau\to 0}|f(t) - f(t+\tau)| = 0$;
(ii) $f = f_1 + f_2$, *where f_1 is a uniformly continuous function on $\mathbf{R}^+$ and* $\lim_{t\to\infty} f_2(t) = 0$.

Proof It is obvious that (ii) ⇒(i).

Conversely, suppose that (i) holds. Choose an increasing sequence $\{t_n\} \subset \mathbf{R}^+$ such that $\lim_{n\to\infty} t_n = \infty$ and $\lim_{n\to\infty}(t_{n+1} - t_n) = 0$. Define the function f_1 by

$$f_1(t) = \begin{cases} f(t_1), & t \le t_1, \\ f(t_n) + \big(f(t_{n+1}) - f(t_n)\big)\frac{t-t_n}{t_{n+1}-t_n}, & t \in (t_n, t_{n+1}],\ n = 1, 2, \ldots. \end{cases}$$

It is easy to check that (i) implies that f_1 is uniformly continuous on $\mathbf{R}^+$, and that $\lim_{t\to\infty}\big(f(t) - f_1(t)\big) = 0$. □

The next lemma gives a useful result on asymptotic averages that will be needed in the remaining chapters.

9.2 Lemma

Let $f \in L^1_{\mathrm{loc}}(\mathbf{R}^+;\mathbf{E}^n)$, and $F \in \mathbf{E}^n$. Then the following two conditions are equivalent:

(i) $\lim_{t\to\infty} \frac{1}{T}\int_t^{t+T} f(s)\,\mathrm{d}s = F$ *for all* $T > 0$;

(ii) $f = F + f_1 + f_2$, *where* $f_1 \in BC_0(\mathbf{R}^+;\mathbf{E}^n)$, $f_2 \in L^1_{\mathrm{loc}}(\mathbf{R}^+;\mathbf{E}^n)$, *and* $\int_0^t f_2(s)\,\mathrm{d}s \to 0$ *as* $t \to \infty$.

In particular, if these conditions hold, then the function $t \mapsto \int_0^t f(s)\,\mathrm{d}s$ *is uniformly continuous.*

Proof Clearly (ii) $\Rightarrow$(i).

Conversely, suppose that (i) holds. Without loss of generality, let us take $F = 0$; (replace f by $f - F$). For each $m \in \mathbf{N}$, let

$$Q_m \stackrel{\mathrm{def}}{=} \left\{ T \in [0,1] \;\middle|\; \left| \int_t^{t+T} f(s)\,\mathrm{d}s \right| \le 1 \text{ for all } t \ge m \right\}.$$

Clearly, each Q_m is closed, hence measurable. Also, $Q_m \subset Q_{m+1}$. It follows from (i) that

$$\bigcup_{m=1}^{\infty} Q_m = [0,1].$$

In particular, for some m, the Lebesgue measure of Q_m is nonzero. Fix any such m. Then, by Lemma 9.3 below, the set $Q_m - Q_m \stackrel{\mathrm{def}}{=} \{T - S \mid T \in Q_m,\, S \in Q_m\}$ contains an interval $(-\epsilon, \epsilon)$. If both T and S belong to Q_m, then

$$\left| \int_t^{t+T-S} f(s)\,\mathrm{d}s \right| \le \left| \int_{t-S}^{t+T-S} f(s)\,\mathrm{d}s \right| + \left| \int_{t-S}^{t} f(s)\,\mathrm{d}s \right| \le 2, \quad t \ge m+1.$$

Therefore we find that, for all $T \in (-\epsilon, \epsilon)$,

$$\left| \int_t^{t+T} f(s)\,\mathrm{d}s \right| \le 2, \quad t \ge m+1,$$

hence

$$\sup_{\substack{t\in\mathbf{R}^+ \\ T\in[-1,1]}} \left| \int_t^{t+T} f(s)\,\mathrm{d}s \right| < \infty. \tag{9.1}$$

Define $f(t)$ to be zero for $t < 0$, and let $f_1(t) = \int_{t-1}^{t} f(s)\,\mathrm{d}s$ for $t \ge 0$, and $f_2 = f - f_1$. By (i), $f_1 \in BC_0(\mathbf{R}^+;\mathbf{E}^n)$. Moreover, a simple calculation shows that

$$\begin{aligned}\int_0^t f_2(s)\,\mathrm{d}s &= \int_0^t f(s)\,\mathrm{d}s - \int_0^t\int_{s-1}^s f(u)\,\mathrm{d}u\,\mathrm{d}s\\ &= \int_t^{t+1}\int_{s-1}^t f(u)\,\mathrm{d}u\,\mathrm{d}s\\ &= \int_0^1\int_{t+v-1}^t f(u)\,\mathrm{d}u\,\mathrm{d}v.\end{aligned}$$

It follows from (i), (9.1) and Lebesgue's dominated convergence theorem that the last integral tends to zero as $t \to \infty$; hence $\int_0^t f_2(s)\,\mathrm{d}s \to 0$ as $t \to \infty$. □

In the proof above, we needed the following result.

9.3 Lemma

Let $E \subset \mathbf{R}$ have positive Lebesgue measure. Then the set $E - E$ contains an open interval centered around zero.

Proof Without loss of generality, let E be a bounded set. Then $\chi_E \in L^1(\mathbf{R};\mathbf{R}) \cap L^\infty(\mathbf{R};\mathbf{R})$. We have $\chi_{E-E}(t) = 1$ if and only if $\chi_E(s)\chi_{-E}(t-s) \neq 0$ for some $s \in \mathbf{R}$. This implies that $\chi_{E-E}(t) = 1$ whenever $(\chi_E * \chi_{-E})(t) > 0$. But $\chi_E * \chi_{-E}$ is a continuous function, and $(\chi_E * \chi_{-E})(0) = m(E) > 0$. Thus $(\chi_E * \chi_{-E})(t) > 0$ in a neighbourhood of zero, i.e., $\chi_{E-E}(t) = 1$ in a neighbourhood of zero. □

10. Exercises

1. Prove Theorem 2.3.
2. Prove Theorem 2.4.
3. Give an example of a sequence of functions $\{\varphi_k\} \subset BUC(\mathbf{R};\mathbf{C})$ such that $\varphi_k \to \varphi$ narrowly as $k \to \infty$ but $\varphi \notin BUC(\mathbf{R};\mathbf{C})$.
4. Let $\varphi \in BUC(\mathbf{R};\mathbf{E}^n)$, and define

$$\alpha(t) = \frac{1}{S}\int_t^{t+S} \varphi(s)\,\mathrm{d}s.$$

   Show that φ is asymptotically drifting periodic with period S iff α is asymptotically slowly varying.
5. Let $\varphi \in BUC(\mathbf{R};\mathbf{E}^n)$ be asymptotically drifting periodic with period S. Define the Fourier coefficients α_k of φ to be the functions

$$\alpha_k(t) = \frac{1}{S}\int_t^{t+S} \mathrm{e}^{\mathrm{i}\,2\pi ks/S}\varphi(s)\,\mathrm{d}s, \qquad k \in Z.$$

Prove that each α_k is asymptotically slowly varying.

6. Let the limit set $\Gamma(\varphi)$ be finitely generated (cf. Definition 3.6) by two or more functions. Show that $\Gamma(\varphi)$ contains a nonconstant function.
7. Let $\psi_1 \equiv 0$, $\psi_2 \equiv 1$, and let ψ_3 be a nonincreasing function with $\lim_{t\to-\infty}\psi_3(t) = 1$ and $\lim_{t\to\infty}\psi_3(t) = 0$. Is it possible that the set generated (through translations) by $\{\psi_1, \psi_2, \psi_3\}$ is the limit set of some function φ?
8. Let the limit set $\Gamma(\varphi)$ be finitely generated (cf. Definition 3.6) by two or more functions, and let one of the functions in $\Gamma(\varphi)$ be a constant $\psi \equiv A$, for some $A \in \mathbf{C}^n$. Show that $\Gamma(\varphi)$ must contain a function ψ_1, different from ψ, satisfying $\psi_1(t) \to A$ as $t \to -\infty$, and a function ψ_2, different from ψ (but possibly equal to ψ_1), satisfying $\psi_2(t) \to A$ as $t \to \infty$.
9. Give an example of a limit set $\Gamma(\varphi)$ generated (through translations) by exactly m functions, where $m = 1, 2, 3, 4$. For what values of m can you allow two of the generating functions to be constants?
10. Complete the proof of Theorem 4.6.
11. Let $S \subset \mathbf{R}$ be a closed set. Give an example of an almost-periodic function φ that has $\sigma(\varphi) = S$. (This shows that the converse of Theorem 4.17(v) does not hold.)
12. Show that there exist closed sets A and $B \subset \mathbf{R}$ for which $A + B$ is not closed (cf. Theorem 4.10).
13. Show that (i) in Theorem 4.15 is satisfied if $\int_{\mathbf{R}} |t|^{1/2} |\mu|(dt) < \infty$.
14. Let $\varphi \in BC(\mathbf{R}; \mathbf{C}^n) \cap L^p(\mathbf{R}; \mathbf{C}^n)$ for some $p \in [1, \infty)$. Show that $\sigma(\varphi)$ has no isolated points.
15. Let $\varphi \in BC(\mathbf{R}; \mathbf{C}^n) \cap BV(\mathbf{R}; \mathbf{C}^n)$. Show that $\sigma(\varphi)$ has no isolated points.
16. Prove the following theorem.

10.1 Theorem
Let $\varphi \in BC(\mathbf{R}; \mathbf{C}^n)$. Then the following conditions are equivalent:
(i) *$a * \varphi \equiv 0$ whenever $a \in L^1(\mathbf{R}; \mathbf{C})$ and $\tilde{a}(\omega) = 0$ for every $\omega \in \sigma(\varphi)$;*
(ii) *φ belongs to the weak*-closure in $L^\infty(\mathbf{R}; \mathbf{C}^n)$ of the set of all linear combinations of translates of the functions $t \mapsto e^{i\omega t}$, $\omega \in \sigma(\varphi)$.*

17. Prove the following theorem.

10.2 Theorem
Let $E \subset \mathbf{R}$ be a closed set. Then the following conditions are equivalent:
(i) *$a * \varphi \equiv 0$ whenever $a \in L^1(\mathbf{R}; \mathbf{C})$, $\varphi \in BUC(\mathbf{R}; \mathbf{C})$, and $\tilde{a}(\omega) = 0$ for every $\omega \in \sigma(\varphi) \subset E$;*
(ii) *there exists exactly one closed ideal I in $L^1(\mathbf{R}; \mathbf{C})$ such that $\bigcap_{a \in I} \{\omega \in \mathbf{R} \mid \tilde{a}(\omega) = 0\} = E$.*

18. Let $a \in L^1(\mathbf{R};\mathbf{C})$ and $\varphi \in L^2(\mathbf{R};\mathbf{C}) \cap BC(\mathbf{R};\mathbf{C})$. Suppose that $\sigma(\varphi) \subset \{\, \omega \in \mathbf{R} \mid \tilde{a}(\omega) = 0 \,\}$. Show that $a * \varphi \equiv 0$. Hint: Let $\tilde{\varphi}$ be the L^2 Fourier transform of φ, and show that $\sigma(\varphi)$ is the support of $\tilde{\varphi}$.
19. Complete the proof of Theorem 5.3.
20. Let $\varphi \in BC(\mathbf{R};\mathbf{C}^n) \cap L^1(\mathbf{R};\mathbf{C}^n)$. Show that $\rho_\omega(\varphi)$ is one-dimensional at all those points where $\tilde{\varphi}(\omega) \neq 0$ (it is the span of $\tilde{\varphi}(\omega)$). Show, in addition, that if $\tilde{\varphi}(\omega) = 0$ for some $\omega \in \mathbf{R}$, and $\varphi \not\equiv 0$, then there exist points ω where $\rho_\omega(\varphi) \neq \{0\}$, but $\tilde{\varphi}(\omega) = 0$.
21. Let $\varphi \in BC(\mathbf{R};\mathbf{C}^n)$, and let ω be an isolated point of $\sigma(\varphi)$. Show that $\rho_\omega(\varphi)$ is one-dimensional.
22. Let φ, $\psi \in BC(\mathbf{R};\mathbf{C}^n)$, and let ω be an isolated point of $\sigma(\varphi) \cup \sigma(\psi)$. Show that $\rho_\omega(\varphi + \psi) \subset \operatorname{span}\{\rho_\omega(\varphi) \cup \rho_\omega(\psi)\}$, and that the inclusion is strict whenever $\{0\} \neq \rho_\omega(\varphi) \neq \rho_\omega(\psi) \neq \{0\}$.
23. Prove the claim made in Remark 5.5.
24. Let $\varphi \in C(\mathcal{T}_T;\mathbf{C}^n)$. Describe the directional spectrum of φ in terms of the Fourier coefficients of φ.
25. Let $\mu \in M(\mathbf{R};\mathbf{C}^{n\times n})$, and $\varphi \in C(\mathcal{T}_T;\mathbf{C}^n)$. Give necessary and sufficient conditions on μ in terms of Fourier transforms which imply $\mu * \varphi \equiv 0$.
26. Prove Theorem 6.2.
27. Prove Theorem 6.3.
28. Let the assumptions of Theorem 7.1 be satisfied, and let f be a continuous real-valued function on $\mathbf{R}^+$ with compact support. Show that the solution x of (7.1) tends to $\int_{\mathbf{R}^+} f(s)\,\mathrm{d}s / \int_{\mathbf{R}^+} s\beta(\mathrm{d}s)$ as $t \to \infty$.
29. Give an example of a function $E \in C^1(\mathbf{R}^2;\mathbf{R})$ such that the set $\{\, E(y) \mid y \in \mathbf{R}^2, E'(y) = 0 \,\}$ has a nonempty interior (cf. Sard's Theorem 8.2). Hint: Take $E(y_1, y_2) = g(y_1) + h(y_2)$ where g and h are integrals of functions vanishing on a Cantor set.
30. Let $\varphi \in BC(\mathbf{R})$ be almost-periodic. Show that $\sigma_{w_+}(\varphi) = \sigma(\varphi)$. (See Section 11 for the definition of σ_{w_+}.) Hint: Show that $J_{\varphi_+} \subset N(\varphi)$ by using the fact that if ϕ is almost-periodic, nonnegative, and not identically zero then $\lim_{T\to\infty} \frac{1}{2T} \int_{-T}^{T} \phi(s)\,\mathrm{d}s > 0$.

11. Comments

Section 2:

For early work on limit sets and limit equations in connection with Volterra equations, see Miller [10], Ch. III. In Miller and Sell [4] and [5], topological dynamics is applied to Volterra equations. For a treatise on dynamical (infinite-dimensional) systems and limit sets, see Hale [6].

Let $\varphi \in BUC(\mathbf{R})$. The limit set $\Gamma(\varphi)$ may—when used together with so-called Ψ-sequences—be used to synthesize φ. This is done as follows.

Let the increasing sequence $\{t_m\}$ satisfy $t_m - t_{m-1} \to \infty$ as $m \to \infty$. A Ψ-sequence associated with $\{t_m\}$ is a sequence of real-valued functions $\{\psi_m\}$, $\psi_m \in C^\infty(\mathbf{R})$, such that

$\sum_{m=1}^{\infty} \psi_m(t) = 1$ for $t \in \mathbf{R}$; $\lim_{m\to\infty} \|\psi'_m\|_{L^\infty(\mathbf{R})} = 0$; $\psi_1(t) = 1$ for $t \le t_1$, $\psi_1(t) = 0$ for $t \ge t_2$, and $\psi'_1(t) \le 0$ for $t_1 \le t \le t_2$;

$\psi_m(t) = 0$ for $t \le t_{m-1}$ and $t \ge t_{m+1}$, $\psi_m(t_m) = 1$, $\psi'_m(t) \ge 0$ for $t_{m-1} \le t \le t_m$, and $\psi'_m(t) \le 0$ for $t_m \le t \le t_{m+1}$, when $m = 2, 3, \ldots$.

The last part of Theorem 2.3 may now be used to derive the representation

$$\varphi(t) = \sum_{m=1}^{\infty} \psi_m(t) y_m(t) + \eta(t), \quad t \in \mathbf{R}^+, \tag{11.1}$$

where $\{\psi_m\}$ is a Ψ-sequence, $y_m \in \Gamma(\varphi)$ and $\eta \in BC_0(\mathbf{R}^+)$. For a thorough analysis of (11.1), of limit equations of Volterra equations and of related questions, see Levin and Shea [1]. For a topological-dynamical approach to the same problem, see Sell [1].

For $\varphi \in L^\infty(\mathbf{R}^+)$ one may define $\Gamma(\varphi)$ using the weak*-topology of $L^\infty(\mathbf{R})$, i.e., $\Gamma(\varphi)$ is the set of functions ψ such that there exists a sequence $\{t_k\}$ tending to $+\infty$ with the property that $\int_{-\infty}^{\infty} \phi(t)\varphi(t+t_k)\,dt \to \int_{-\infty}^{\infty} \phi(t)\psi(t)\,dt$ for every function $\phi \in L^1(\mathbf{R})$. See Staffans [3], p. 243.

Limit equations for functional differential equations with infinite delay and for Volterra equations have been considered in Marcus and Mizel [1].

Section 3:
The structure of limit sets and the synthesis of solutions of Volterra equations are discussed extensively in Levin [7], [10], and [12].

Section 4:
There is a large literature on spectral analysis. Easily accessible presentations can be found in, e.g., Rudin [1], Katznelson [1] and Reiter [1]. See Benedetto [1] for references up to 1975.

The spectrum $\sigma(\varphi)$ (Definition 4.4) is, of course, what is frequently called the weak*-spectrum (see Katznelson [1], p. 170). Usually the spectrum is defined for L^∞-functions rather than for BC-functions, but this is not a very significant extension. For example, if $\varphi \in L^\infty(\mathbf{R}; \mathbf{C}^n)$, then one can define $\sigma(\varphi)$ to be $\sigma(a * \varphi)$, where $a \in L^1(\mathbf{R}; \mathbf{C})$ is some function whose Fourier transform vanishes nowhere. With this definition, the theory presented here can be extended to L^∞-functions.

To keep the presentation reasonably short, we did not explicitly mention, even less analyse in any detail, the question of spectral synthesis. Let us briefly make a few comments here. The main problem of *spectral synthesis* is the following. Let $\varphi \in BC(\mathbf{R})$ with spectrum $\sigma(\varphi)$. When is it true that φ is contained in the weak*-closure of $\operatorname{span}\{e^{i\omega t}\}$, $\omega \in \sigma(\varphi)$, or—stated differently—when does $\varphi \in BC(\mathbf{R})$ admit weak* spectral synthesis? By the Hahn–Banach theorem this is equivalent to the question: do the conditions

$$a \in L^1(\mathbf{R}) \quad \text{and} \quad \tilde{a}(\sigma(\varphi)) = \{0\} \quad \text{imply} \quad a * \varphi \equiv 0? \tag{11.2}$$

By Proposition 4.16, due to Malliavin [1], the answer can be negative; see also Katznelson [1], p. 231. For a more explicit expression for the functions a and φ in Proposition 4.16, see Richards [1].

Since the implication (11.2) does not in general hold, one may examine what additional assumptions on a or on φ it takes to make it true. Corollary 4.14 shows that it does hold if $\{\, \omega \in \mathbf{R} \mid \tilde{a}(\omega) = 0 \,\}$ is a sufficiently small set, whereas Theorem 4.15(i) (due to Pollard [1]) shows that if $\tilde{\mu}$ is small in a neighbourhood of $\sigma(\varphi)$ then synthesis is possible. Theorem 4.15(ii), giving another hypothesis under which (11.2) holds, can be found in Sedig [1] and extends a result of Katznelson (see Benedetto [1], Theorem 3.2.4, p. 230).

Instead of imposing more conditions on a, we may put additional hypotheses on φ to make (11.2) hold. For example, if $\varphi = \tilde{\mu}$ for some $\mu \in M(\mathbf{R})$, then (11.2) is satisfied (see Rudin [1], p. 186). Another result is contained in Exercise 18. Atzmon [1] showed that if $\varphi \in BV((-\infty, a) \cup (a, \infty)) \cap BC(\mathbf{R})$ for some $a > 0$ then φ admits spectral synthesis. A negative synthesis result is given by Sedig [2].

An extension of Theorem 4.17(v) is given in Loomis [1], Theorem 5, p. 367.

Section 5:

Some of the results in this section can be found implicitly in Staffans [6].

Section 6:

Some results on asymptotic spectra are given in Staffans [5].

An asymptotic analysis can also be done through the use of Wiener spectra. Let $\varphi \in BC(\mathbf{R})$, and define

$$J_{\varphi+} = \Big\{\, a \in L^1(\mathbf{R}) \,\Big|\, \lim_{T\to\infty} \frac{1}{T} \int_0^T |(a * \varphi)(t)|\, dt = 0 \,\Big\},$$

$$J_{\varphi-} = \Big\{\, a \in L^1(\mathbf{R}) \,\Big|\, \lim_{T\to\infty} \frac{1}{T} \int_{-T}^0 |(a * \varphi)(t)|\, dt = 0 \,\Big\},$$

$$J_{\varphi} = \Big\{\, a \in L^1(\mathbf{R}) \,\Big|\, \lim_{T\to\infty} \frac{1}{2T} \int_{-T}^T |(a * \varphi)(t)|\, dt = 0 \,\Big\}.$$

Then the positive, negative and two-sided Wiener spectra are defined, respectively, as

$$\sigma_{w_+}(\varphi) = \bigcap_{a \in J_{\varphi_+}} \{\omega \in \mathbf{R} \mid \tilde{a}(\omega) = 0\},$$

$$\sigma_{w_-}(\varphi) = \bigcap_{a \in J_{\varphi_-}} \{\omega \in \mathbf{R} \mid \tilde{a}(\omega) = 0\},$$

and

$$\sigma_w(\varphi) = \bigcap_{a \in J_\varphi} \{\omega \in \mathbf{R} \mid \tilde{a}(\omega) = 0\}.$$

For results on Wiener spectra, see Benedetto [1], Ch. 2, Benedetto [2] and Sedig [2].

Section 7:
For results on renewal theory, see, e.g., Bingham, Goldie, and Teugels [1], p. 359, or Feller [2], Ch. XI, and references therein.

Section 8:
Sard's theorem and related questions are studied in, e.g., Sternberg [1].

References

A. Atzmon

1. Spectral synthesis of functions of bounded variation, *Proc. Amer. Math. Soc.* **47** (1975), pp. 417–422.

J. J. Benedetto

1. *Spectral Synthesis*, Teubner, Stuttgart, 1975.
2. The Wiener spectrum in spectral synthesis, *Stud. Appl. Math.* **54** (1975), pp. 91–115.

N. H. Bingham, C. M. Goldie, and J. L. Teugels

1. *Regular Variation*, Cambridge University Press, Cambridge, 1987.

W. Feller

2. *An Introduction to Probability Theory and its Applications, II,* 2nd *ed.*, J. Wiley, New York, 1971.

J. K. Hale

6. *Asymptotic Behavior of Dissipative Systems*, Amer. Math. Soc., Providence, R. I., 1988.

Y. Katznelson

1. *An Introduction to Harmonic Analysis*, Dover Publications, New York, 1976.

J. J. Levin

7. On some geometric structures for integrodifferential equations, *Adv. in Math.* **22** (1976), pp. 146–186.
10. Tauberian curves and integral equations, *J. Integral Equations* **2** (1980), pp. 57–91.
12. Synthesis of solutions of integral equations, *J. Integral Equations* **5** (1983), pp. 93–157.

J. J. Levin and D. F. Shea

1. On the asymptotic behavior of the bounded solutions of some integral equations, I, II, III, *J. Math. Anal. Appl.* **37** (1972), pp. 42–82, 288–326, 537–575.

L. H. Loomis

1. The spectral characterization of a class of almost periodic functions, *Ann. of Math.* **72** (1960), pp. 362–368.

P. Malliavin

1. Impossibilité de la synthèse spectrale sur les groupes Abéliens non compacts, *Publ. Math. Inst. Hautes Études Sci.* **2** (1959), pp. 61–68.

M. Marcus and V. Mizel

1. *Limiting Equations for Problems Involving Long Range Memory*, Mem. Amer. Math. Soc. **278**, Amer. Math. Soc., Providence, R. I., 1983.

R. K. Miller

10. *Nonlinear Volterra Integral Equations*, W. A. Benjamin, Menlo Park, Calif., 1971.

R. K. Miller and G. R. Sell

4. *Volterra Integral Equations and Topological Dynamics*, Mem. Amer. Math. Soc. **102**, Amer. Math. Soc., Providence, R. I., 1970.
5. Topological dynamics and its relation to integral equations and nonautonomous systems, in *Dynamical Systems* **1**, L. Cesari, J. K. Hale and J. LaSalle, eds., Proc. Internat. Symp., Brown Univ., Providence, R. I., 1974, Academic Press, New York, 1976, pp. 233–249.

H. Pollard

1. The harmonic analysis of bounded functions, *Duke Math. J.* **20** (1953), pp. 499–512.

H. Reiter

1. *Classical Harmonic Analysis and Locally Compact Groups*, Oxford University Press, Oxford, 1968.

I. Richards

1. On the disproof of spectral synthesis, *J. Combin. Theory* **2** (1967), pp. 61–70.

W. Rudin

1. *Fourier Analysis on Groups*, Interscience, New York, 1967.

I. Sedig

1. A note on Lipschitz functions and spectral synthesis, *Ark. Mat.* **25** (1987), pp. 141–145.
2. Some results in spectral analysis and synthesis at infinity, *Math. Scand.*, to appear 1989.

G. R. Sell
1. A Tauberian condition and skew product flows with applications to integral equations, *J. Math. Anal. Appl.* **43** (1973), pp. 388–396.

O. J. Staffans
3. Tauberian theorems for a positive definite form, with applications to a Volterra equation, *Trans. Amer. Math. Soc.* **218** (1976), pp. 239–259.
5. On the asymptotic spectra of the bounded solutions of a nonlinear Volterra equation, *J. Differential Equations* **24** (1977), pp. 365–382.
6. Systems of nonlinear Volterra equations with positive definite kernels, *Trans. Amer. Math. Soc.* **228** (1977), pp. 99–116.

E. M. Stein and G. Weiss
1. *Introduction to Fourier Analysis on Euclidean Spaces*, Princeton University Press, Princeton, 1971.

S. Sternberg
1. *Lectures on Differential Geometry*, Prentice Hall, Englewood Cliffs, N. J., 1964.

16

Convolution Kernels of Positive Type

We study the theory of convolution kernels of positive type. These results will be of central importance in Chapters 17, 18, and 19 dealing with the asymptotic behaviour of solutions of nonlinear Volterra integral equations with convolution kernels.

1. Introduction

The preceding Chapter 15 developed some general asymptotic results, in particular on limit sets, limit equations, and spectra of bounded functions. This provides us with certain tools for the analysis of the asymptotic behaviour of solutions of nonlinear convolution Volterra equations.

But a key question remains. What classes of nonlinear Volterra equations can fruitfully be analysed with these tools? When do we have bounded solutions and when can we expect solutions to vanish at infinity?

The major part of the remaining chapters addresses these questions. The present chapter develops the theory of kernels of positive type; in later chapters we apply this theory to various Volterra equations, and as a result we obtain some rather detailed asymptotic statements.

Section 2 gives the basic definitions and some important equivalences. In Section 3 we analyse some examples of functions and measures of positive type. Frequently, more specific asymptotic statements result only if the kernel μ is of strong or strict positive type, and not merely of positive type. Such kernels are discussed in Section 4. Another subclass of the set of kernels of positive type is presented in Section 5. Some technical inequalities, useful for the asymptotic analysis of solutions of Volterra equations, are obtained in Section 6. The first appendix gives certain results on semi-inner products; the second lists some elementary properties of Fourier transforms of distributions.

2. Functions and Measures of Positive Type

In what follows we work with complex scalars, i.e., we let our functions and measures take their values in $\mathbf{C}$, $\mathbf{C}^n$ or $\mathbf{C}^{m\times n}$. This simplifies the theory substantially, and is more natural in view of the fact that we make extensive use of Fourier transforms. When we apply the theory in Chapters 17, 18, and 19 we work in $\mathbf{R}$, $\mathbf{R}^n$ or $\mathbf{R}^{m\times n}$, and consequently only need real scalars.

We let $\langle\cdot,\cdot\rangle$ denote some inner product on $\mathbf{C}^n$, and we let the norm in $\mathbf{C}^n$ be the norm induced by this inner product. The norm $|A|$ of a $\mathbf{C}^{n\times n}$-valued matrix is, correspondingly, defined as $\sup_{y\in\mathbf{C}^n,|y|=1}|Ay|$.

2.1 Definition

A measure $\mu \in M_{\mathrm{loc}}(\mathbf{R}^+;\mathbf{C}^{n\times n})$ is said to be of positive type iff for every $\varphi \in L^2(\mathbf{R};\mathbf{C}^n)$ with compact support one has

$$\begin{aligned}[\![\mu,\varphi]\!] &\stackrel{\text{def}}{=} \Re\int_{\mathbf{R}} \langle \varphi(t), (\mu*\varphi)(t)\rangle\,\mathrm{d}t \\ &= \Re\int_{\mathbf{R}}\Big\langle \varphi(t), \int_{\mathbf{R}^+}\mu(\mathrm{d}s)\varphi(t-s)\Big\rangle\,\mathrm{d}t \geq 0. \qquad (2.1)\end{aligned}$$

If μ is locally absolutely continuous with respect to the Lebesgue measure, i.e., $\mu(\mathrm{d}s) = a(s)\,\mathrm{d}s$, where $a \in L^1_{\mathrm{loc}}(\mathbf{R}^+;\mathbf{C}^{n\times n})$, then the function a is said to be of positive type iff μ is of positive type. In this case we define $[\![a,\varphi]\!] = [\![\mu,\varphi]\!]$.

The set of measures of positive type is denoted by $PT(\mathbf{R}^+;\mathbf{C}^{n\times n})$.

Measures that are of positive type in the sense described above have frequently been called positive definite in the literature. See Section 10 for references.

If $\tau_t\varphi$ is an arbitrary translate of φ, then $[\![\mu,\tau_t\varphi]\!] = [\![\mu,\varphi]\!]$. This means that in Definition 2.1 one could, without loss of generality, require φ to satisfy $\varphi(t) = 0$ for, e.g., $t < 0$.

It is straightforward to check that the following result holds.

2.2 Theorem

$PT(\mathbf{R}^+;\mathbf{C}^{n\times n})$ is a closed cone in $M_{\mathrm{loc}}(\mathbf{R}^+;\mathbf{C}^{n\times n})$.

In other words, every positive linear combination of two measures of positive type is of positive type, and if the sequence $\{\mu_n\} \subset PT(\mathbf{R}^+)$ satisfies $\lim_{n\to\infty}\|\mu_n - \mu\|_{M(J)} = 0$ for some $\mu \in M_{\mathrm{loc}}(\mathbf{R}^+)$ and for each compact subset J of $\mathbf{R}^+$, then $\mu \in PT(\mathbf{R}^+)$.

As Theorem 2.4 below shows, the property of a measure μ of being of positive type is equivalent to a positivity property of the Laplace transform $\hat{\mu}$ or of the Fourier transform $\tilde{\mu}$ of μ. For the formulation of this equivalence relation, we use the concepts of a positive and a negative matrix defined in Definition 5.2.1. For the convenience of the reader we repeat the definition here. We urge the reader to recall the cautionary words and comments following this definition in Chapter 5.

2.3 Definition

Let A be an $n \times n$ matrix. We say that A is positive and write $A \succeq 0$ iff $\langle v, Av \rangle \geq 0$ for all vectors $v \in \mathbf{C}^n$, and we say that A is strictly positive and write $A \succ 0$ iff $\langle v, Av \rangle > 0$ for all nonzero $v \in \mathbf{C}^n$. If $-A \succeq 0$ or $-A \succ 0$, then we say that A is negative or strictly negative, respectively, and write $A \preceq 0$ or $A \prec 0$.

The adjoint of A is denoted by A^* and it is defined by the formula $\langle u, Av \rangle = \langle A^* u, v \rangle$ for all u and $v \in \mathbf{C}^n$. If $A = A^*$ then A is self-adjoint, and if $A = -A^*$ then A is anti-symmetric. The matrix $\Re A = \frac{1}{2}(A + A^*)$ is called the real part of A, and the matrix $\Im A = \frac{1}{2\mathrm{i}}(A - A^*)$ the imaginary part of A.

A measure $\alpha \in M(J; \mathbf{C}^{n \times n})$, where $J \subset \mathbf{R}$, is said to be positive if $\alpha(E) \succeq 0$ for every Borel set $E \subset J$.

Note that it is something quite different for a measure to be of positive type and to be positive.

Recall that the two most important properties of $\Re A$ and $\Im A$ (besides their self-adjointness) are the facts that $A = \Re A + \mathrm{i}\Im A$, and that for all $y \in \mathbf{C}^n$ we have $\Re\langle y, Ay \rangle = \langle y, \Re A y \rangle$ and $\Im \langle y, Ay \rangle = -\langle y, \Im A y \rangle$. Observe that $A = \Re A$ iff A is self-adjoint, and that $A = \mathrm{i}\Im A$ iff A is anti-symmetric.

In the sequel we shall need the fact that $\inf_{y \in \mathbf{C}^n, |y|=1} \langle y, Ay \rangle > 0$ iff $A \succ 0$. This is true owing to the continuity of the function $y \mapsto \langle y, Ay \rangle$ (every continuous function attains its minimum on a compact set). Another fact that we need is that whenever $A \succeq 0$, it is true for all $y \in \mathbf{C}^n$ that

$$|Ay|^2 = \langle y, A^2 y \rangle \leq |A| \langle y, Ay \rangle, \tag{2.2}$$

and that

$$|A| = \sup_{|y|=1} \langle y, Ay \rangle. \tag{2.3}$$

(For a proof of these relations, see Section 7).

If $\langle y, Ay \rangle$ is real for all $y \in \mathbf{C}^n$, then A is self-adjoint. This is true because then

$$\langle y, Ay \rangle = \overline{\langle y, Ay \rangle} = \langle Ay, y \rangle = \langle y, A^* y \rangle,$$

and so $\langle y, (A - A^*)y \rangle = 0$ for all y. In particular, $A - A^* \succeq 0$; hence, by (2.3),

$$|A - A^*| = \sup_{|y|=1} \langle y, (A - A^*)y \rangle = 0.$$

The Laplace and Fourier transforms of measures of positive type have positive real parts, and conversely measures whose Laplace or Fourier transforms have a positive real part are of positive type.

2.4 Theorem
Let $\mu \in M_{\mathrm{loc}}(\mathbf{R}^+;\mathbf{C}^{n\times n})$ satisfy $\int_{\mathbf{R}^+} \mathrm{e}^{-\epsilon t}|\mu|(\mathrm{d}t) < \infty$ for all $\epsilon > 0$. Then the following conditions are equivalent:

(i) $\mu \in PT(\mathbf{R}^+;\mathbf{C}^{n\times n})$;
(ii) $\langle y, \mu y\rangle \in PT(\mathbf{R}^+;\mathbf{C})$ *for all* $y \in \mathbf{C}^n$;
(iii) $\Re\hat{\mu}(z) \succeq 0$ *for* $\Re z > 0$;
(iv) $\Re\hat{\mu}(z) \succeq 0$ *in the strip* $0 < \Re z < \epsilon$ *for some* $\epsilon > 0$;
(v) $\liminf_{z\to \mathrm{i}\tau, \Re z>0} \Re\hat{\mu}(z) \succeq 0$ *for every* $\tau \in \mathbf{R}$ *and* $\liminf_{|z|\to\infty, \Re z>0} \Re\hat{\mu}(z) \succeq 0$.

In the case where $\int_{\mathbf{R}^+} |\mu|(\mathrm{d}t) < \infty$, these conditions are equivalent to

(vi) $\Re\tilde{\mu}(\tau) \geq 0$, *for* $\tau \in \mathbf{R}$.

Here we have used the notation '$\liminf_{z\to \mathrm{i}\tau, \Re z>0} \Re\hat{\mu}(z) \succeq 0$' as a shorthand for the requirement that $\liminf_{z\to \mathrm{i}\tau, \Re z>0}\langle y, \Re\hat{\mu}(z)y\rangle \geq 0$ for every $y \in \mathbf{C}^n$. Recall the notation $\tilde{\mu}(\tau) = \hat{\mu}(\mathrm{i}\tau)$.

The assumption that $\int_{\mathbf{R}^+} \mathrm{e}^{-\epsilon t}|\mu|(\mathrm{d}t) < \infty$ for $\epsilon > 0$ has been included in order to guarantee the existence of the Laplace transform $\hat{\mu}(z)$ for all z with $\Re z > 0$. Theorems 2.5 and 2.6 below show that it may actually be omitted, provided one is willing to work with Fourier transforms of distributions.

Proof of Theorem 2.4 Trivially (iii)$\Rightarrow$(v). The converse implication follows from the maximum principle for harmonic functions (the transform $\hat{\mu}(z)$ is analytic for $\Re z > 0$; see Lemma **3**.8.2). Thus (iii)$\Leftrightarrow$(v).

It is clear that (i)$\Rightarrow$(ii) and that (iii)$\Rightarrow$(iv). Thus, to prove the first part of the theorem it suffices to show that (ii)$\Rightarrow$(iii) and that (iv)$\Rightarrow$(i).

(ii)$\Rightarrow$(iii): Without loss of generality, let μ be scalar-valued (replace μ by $\langle y, \mu y\rangle$, where y is an arbitrary vector in $\mathbf{C}^n$). Fix $z = \sigma + \mathrm{i}\tau$ with $\sigma > 0$. For each $T > 0$, define

$$R(T) = [\![\mu, \chi_{[0,T]}(\cdot)\mathrm{e}^{z\cdot}]\!] = \Re \int_0^T \mathrm{e}^{zt} \int_{[0,t]} \overline{\mu}(\mathrm{d}s)\mathrm{e}^{\overline{z}(t-s)}\,\mathrm{d}t$$

$$= \Re \int_0^T \int_{[0,t]} \mathrm{e}^{2\sigma t}\overline{\mu}(\mathrm{d}s)\mathrm{e}^{-\overline{z}s}\,\mathrm{d}t = \Re \int_0^T \int_{[0,t]} \mathrm{e}^{2\sigma t}\mu(\mathrm{d}s)\mathrm{e}^{-zs}\,\mathrm{d}t.$$

Since μ is of positive type, we have $R(T) \geq 0$. Define $a(t) = 2\sigma\mathrm{e}^{-2\sigma t}$ and $b(t) = \int_{[0,t]} \mathrm{e}^{-zs}\mu(\mathrm{d}s)$ for $t \geq 0$. Then

$$\Re(a * b)(T) = 2\sigma\mathrm{e}^{-2\sigma T}R(T) \geq 0.$$

Because $a \in L^1(\mathbf{R}^+;\mathbf{R})$ with $\int_{\mathbf{R}^+} a(s)\,\mathrm{d}s = 1$, and $b(t) \to \hat{\mu}(z)$ as $t \to \infty$, we have

$$\Re\hat{\mu}(z) = \lim_{T\to\infty} \Re(a * b)(T) \geq 0.$$

Thus, (ii)$\Rightarrow$(iii).

(iv)$\Rightarrow$(i): Fix $\varphi \in L^2(\mathbf{R};\mathbf{C}^n)$, and suppose that the support of φ is contained in some compact interval J. Define μ_ϵ by $\mu_\epsilon(\mathrm{d}t) = \mathrm{e}^{-\epsilon t}\mu(\mathrm{d}t)$.

Since both $\mathrm{e}^{\epsilon t}\varphi(t)$ and $\mathrm{e}^{-\epsilon t}\varphi(t)$ tend to φ in $L^2(J;\mathbf{C}^n)$ as $\epsilon \downarrow 0$, and vanish on $\mathbf{R}\setminus J$, it follows from Theorem **3**.6.1 that

$$\lim_{\epsilon\downarrow 0}[\![\mu_\epsilon,\varphi]\!] = \lim_{\epsilon\downarrow 0}\Re\int_J\Big\langle \mathrm{e}^{-\epsilon t}\varphi(t), \int_{\mathbf{R}^+}\mu(\mathrm{d}s)\mathrm{e}^{\epsilon(t-s)}\varphi(t-s)\Big\rangle\,\mathrm{d}t = [\![\mu,\varphi]\!].$$

Thus, it suffices to prove that $\mu_\epsilon \in PT(\mathbf{R}^+;\mathbf{C}^{n\times n})$ for all $\epsilon > 0$. Obviously, $\mu_\epsilon \in M(\mathbf{R}^+;\mathbf{C}^{n\times n})$ and $\hat{\mu}_\epsilon(z) = \hat{\mu}(\epsilon + z)$ when $\Re z > 0$. Consequently, it is enough to demonstrate the implication (iv)$\Rightarrow$(i) in the special case where $\mu \in M(\mathbf{R}^+;\mathbf{C}^{n\times n})$. In this case the continuity of $\hat{\mu}(z)$ in the half plane $\Re z \geq 0$ together with (iv) implies that $\Re\hat{\mu}(\mathrm{i}\omega) \succeq 0$ for all $\omega \in \mathbf{R}$, and we can use Parseval's identity (see formula (8.2)) to get (recall that $\Re\langle y, Ay\rangle = \langle y, \Re Ay\rangle$)

$$\begin{aligned}[\![\mu,\varphi]\!] &= \Re\int_{\mathbf{R}}\langle\varphi(t),(\mu*\varphi)(t)\rangle\,\mathrm{d}t\\ &= \frac{1}{2\pi}\Re\int_{\mathbf{R}}\langle\hat{\varphi}(\mathrm{i}\omega),\hat{\mu}(\mathrm{i}\omega)\hat{\varphi}(\mathrm{i}\omega)\rangle\,\mathrm{d}\omega\\ &= \frac{1}{2\pi}\int_{\mathbf{R}}\langle\hat{\varphi}(\mathrm{i}\omega),\Re\hat{\mu}(\mathrm{i}\omega)\hat{\varphi}(\mathrm{i}\omega)\rangle\,\mathrm{d}\omega.\end{aligned}$$

Thus, the condition $\Re\hat{\mu}(\mathrm{i}\omega) \succeq 0$ implies $[\![\mu,\varphi]\!] \geq 0$.

Finally, let $\mu \in M(\mathbf{R}^+;\mathbf{C}^{n\times n})$. Then (iv) clearly implies (vi), and (vi) implies (i). □

Occasionally we use the following more general characterization of kernels of positive type. (The reader who prefers to avoid distributions should skip Theorems 2.5 and 2.6, and add the assumption $\int_{\mathbf{R}^+}\mathrm{e}^{-\epsilon t}|\mu|(\mathrm{d}t) < \infty$ whenever needed.)

2.5 Theorem

Let $\mu \in M_{\mathrm{loc}}(\mathbf{R}^+;\mathbf{C}^{n\times n})$, and extend μ to all of $\mathbf{R}$ by defining $\mu(E) = \mu(\mathbf{R}^+\cap E)$ for every Borel set $E \subset \mathbf{R}$. Then the following two conditions are equivalent:

(i) *$\mu \in PT(\mathbf{R}^+;\mathbf{C}^{n\times n})$;*

(ii) *μ induces a tempered distribution (i.e., a matrix of tempered distributions), and $\Re\tilde{\mu}$ is induced by a positive measure in $M_{\mathrm{loc}}(\mathbf{R};\mathbf{C}^{n\times n})$ (i.e., $\Re\tilde{\mu}(E) \succeq 0$ for all bounded Borel sets $E \subset \mathbf{R}$).*

Before proving Theorem 2.5 we give some definitions and some simple properties of distributions. Further properties are given in Section 8.

For ν a (scalar) distribution and φ a (scalar) test function, we let $[\nu,\varphi]$ denote the value of ν evaluated at φ. If ν is an $m\times n$ matrix of distributions ν_{ij} and φ an n-vector with components φ_j, then $[\nu,\varphi]$ is the m-vector whose components are $\sum_{j=1}^n[\nu_{ij},\varphi_j]$.

If ν is a tempered distribution ($\nu \in \mathcal{S}'$), then its Fourier transform $\tilde{\nu}$ is defined through $[\tilde{\nu},\varphi] = [\nu,\tilde{\varphi}]$, where $\tilde{\varphi} \in \mathcal{S}$ is the Fourier transform of $\varphi \in \mathcal{S}$. If $\nu \in \mathcal{S}'(\mathbf{R}^+)$ (i.e., $\nu \in \mathcal{S}'$ and ν is supported on $\mathbf{R}^+$), one defines its Laplace transform $\hat{\nu}(z)$ for $\Re z > 0$ through $[\nu, e_z]$, where e_z is any function in $\mathcal{S}$ satisfying $e_z(t) = \mathrm{e}^{-zt}$, $t \geq -1$.

We define the convolution $\nu * \varphi$ of a distribution ν and a test function φ in the standard way, i.e., as $(\nu * \varphi)(t) = [\nu, (\check{\varphi})_t]$, where $(\check{\varphi})_t(\tau) = \varphi(t-\tau)$. Recall that $(\nu * \varphi)(t)$ is a function. In particular, if $\nu \in \mathcal{S}'$ and $\varphi \in \mathcal{S}$, then $\nu * \varphi$ is a C^∞-function whose growth rate at infinity is at most polynomial. This means that we can consider $\nu * \varphi$ to be a tempered distribution.

For a $\mathbf{C}^n$-valued test function φ and an n-vector ν of distributions, we define $\langle\!\langle \varphi, \nu \rangle\!\rangle$ to be the value of the complex conjugate of ν evaluated at $\overline{\varphi}$. Thus, if $\nu = (\nu_1, \nu_2, \ldots \nu_n)$ and $\varphi = (\varphi_1, \varphi_2, \ldots \varphi_n)$, then $\langle\!\langle \varphi, \nu \rangle\!\rangle = \overline{\sum_{i=1}^n [\nu_i, \overline{\varphi}_i]}$.

For each $\nu \in \mathcal{S}'(\mathbf{R}^+; \mathbf{C}^{n\times n})$ and $\varphi \in \mathcal{S}(\mathbf{R}; \mathbf{C}^n)$, we define $[\![\nu, \varphi]\!] = \Re\langle\!\langle \varphi, \nu * \varphi \rangle\!\rangle$. Observe that if ν is induced by a measure then this definition almost reduces to Definition 2.1, the only difference being that the class of test functions is slightly different.

Proof of Theorem 2.5 The proof is very simple in the direction (ii) $\Rightarrow$(i). This follows from the fact that, if φ is a $\mathbf{C}^n$-valued test function with compact support (or more generally, if $\varphi \in \mathcal{S}$), then by standard distribution Fourier transform theory (see Rudin [2], Theorem 7.19, and use Plancherel's theorem for distributions)

$$[\![\mu, \varphi]\!] = \Re\langle\!\langle \varphi, \mu * \varphi \rangle\!\rangle = \frac{1}{2\pi}\Re\langle\!\langle \tilde{\varphi}, \tilde{\mu}\tilde{\varphi} \rangle\!\rangle = \frac{1}{2\pi}\langle\!\langle \tilde{\varphi}, \Re\tilde{\mu}\tilde{\varphi} \rangle\!\rangle \geq 0.$$

The set of test functions supported in some interval J is dense in $L^2(J; \mathbf{C}^n)$, and hence, for every $\varphi \in L^2(\mathbf{R}; \mathbf{C}^n)$ with compact support, we have $[\![\mu, \varphi]\!] \geq 0$. Thus μ is of positive type.

Conversely, assume that $\mu \in PT(\mathbf{R}^+; \mathbf{C}^{n\times n})$. Extend μ to $\mathbf{R}$ by defining $\mu(E) = \mu(E \cap \mathbf{R}^+)$ for any Borel set $E \subset \mathbf{R}$. Let $\check{\mu}$ be the measure $\check{\mu}(E) = \mu^*(-E)$, and define $\nu = \frac{1}{2}(\mu + \check{\mu})$. Then, for each $\mathbf{C}^n$-valued test function φ, we have

$$\int_{\mathbf{R}} \langle (\check{\mu} * \varphi)(t), \varphi(t) \rangle \,\mathrm{d}t = \int_{\mathbf{R}} \langle \varphi(t), (\mu * \varphi)(t) \rangle \,\mathrm{d}t;$$

hence

$$\int_{\mathbf{R}} \langle \varphi(t), (\nu * \varphi)(t) \rangle \,\mathrm{d}t = 2\int_{\mathbf{R}} \Re\langle \varphi(t), (\mu * \varphi)(t) \rangle \,\mathrm{d}t = 2[\![\mu, \varphi]\!] \geq 0.$$

Assume for the moment that the dimension n is equal to 1. Then the preceding inequality, together with Schwartz [1], Théorème XVII, p. 275, implies that ν induces a tempered distribution. From the definition of ν one sees that the same conclusion holds for the measure μ. Thus it follows that $\tilde{\nu}$ and $\tilde{\mu}$ are well defined as tempered distributions. Furthermore, it follows from Schwartz [1], Théorème XVIII, p. 276, that $\tilde{\nu}$, and consequently $\Re\tilde{\mu}$ as well, is induced by a nonnegative measure.

For the remaining case $n > 1$, we observe that, for each vector $y \in \mathbf{C}^n$, the measure $\langle y, \mu(\cdot)y\rangle$ belongs to $PT(\mathbf{R}^+; \mathbf{C})$. For every $u, v \in \mathbf{C}^n$ we have

$$4\langle u, \mu(\cdot)v\rangle = \big\langle (u+v), \mu(\cdot)(u+v)\big\rangle - \big\langle (u-v), \mu(\cdot)(u-v)\big\rangle \\ - \mathrm{i}\big\langle (u+\mathrm{i}v), \mu(\cdot)(u+\mathrm{i}v)\big\rangle + \mathrm{i}\big\langle (u-\mathrm{i}v), \mu(\cdot)(u-\mathrm{i}v)\big\rangle, \quad (2.4)$$

and this implies (choose u, v suitably so as to get only a particular component μ_{ij} on the left-hand side) that each component of μ is a combination of measures in $PT(\mathbf{R}^+; \mathbf{C})$ and, therefore, induces a tempered distribution. The same argument applied to ν shows that $\tilde{\nu}$ is induced by a measure in $M_{\mathrm{loc}}(\mathbf{R}; \mathbf{C}^{n\times n})$, hence this is true for $\Re\tilde{\mu}$ as well. The fact that $\Re\tilde{\mu}(E) \succeq 0$ for all bounded Borel sets E can be reduced to the corresponding scalar statement. □

The following result extends Theorem 2.4, by removing the assumption that $\int_{\mathbf{R}^+} \mathrm{e}^{-\epsilon t}|\mu|(\mathrm{d}t) < \infty$.

2.6 Theorem

Let $\mu \in M_{\mathrm{loc}}(\mathbf{R}^+; \mathbf{C}^{n\times n})$. Then the following conditions are equivalent:

(i) $\mu \in PT(\mathbf{R}^+; \mathbf{C}^{n\times n})$;

(ii) $\langle y, \mu y\rangle \in PT(\mathbf{R}^+; \mathbf{C})$ *for every* $y \in \mathbf{C}^n$;

(iii) $\mu \in S'(\mathbf{R}^+; \mathbf{C}^{n\times n})$ *and* $\Re\hat{\mu}(z) \succeq 0$ *for* $\Re z > 0$;

(iv) $\mu \in S'(\mathbf{R}^+; \mathbf{C}^{n\times n})$ *and, for some* $\epsilon > 0$, $\Re\hat{\mu}(z) \succeq 0$ *in the strip* $0 < \Re z < \epsilon$;

(v) $\mu \in S'(\mathbf{R}^+; \mathbf{C}^{n\times n})$, $\liminf_{z\to \mathrm{i}\tau, \Re z>0} \Re\hat{\mu}(z) \succeq 0$ *for every* $\tau \in \mathbf{R}$, *and* $\liminf_{|z|\to\infty, \Re z>0} \Re\hat{\mu}(z) \succeq 0$;

(vi) $\mu \in S'(\mathbf{R}^+; \mathbf{C}^{n\times n})$ *and* $\Re\tilde{\mu}$ *is induced by a positive measure in* $M_{\mathrm{loc}}(\mathbf{R}; \mathbf{C}^{n\times n})$.

Here the condition $\mu \in S'(\mathbf{R}^+; \mathbf{C}^{n\times n})$ means that μ induces a tempered distribution, when extended as zero to the negative half axis, and the condition $\liminf_{z\to \mathrm{i}\tau, \Re z>0} \Re\hat{\mu}(z) \succeq 0$ is given the same interpretation as in Theorem 2.4.

Proof of Theorem 2.6 Theorem 2.5 gives the equivalence of (i) and (vi). The equivalence of (i) and (ii) became evident in the last paragraph of the proof of Theorem 2.5. The equivalence of (iii) and (v) is a consequence of the maximum principle for harmonic functions. Trivially (iii)⇒(iv). Thus, to complete the proof, it suffices to show that (i)⇒(iii), and that (iv)⇒(i). Moreover, there is no loss of generality if we assume μ to be scalar-valued (replace μ by $\langle y, \mu y\rangle$ for some $y \in \mathbf{C}^n$).

(i)⇒(iii): Let σ be an arbitrary positive number, and define $\mu_\sigma(\mathrm{d}t) = \mathrm{e}^{-\sigma t}\mu(\mathrm{d}t)$ for $t \ge 0$. We claim that μ_σ is of positive type. This is true, because if we define φ_σ to be the function $\varphi_\sigma(t) = \mathrm{e}^{\sigma t}\varphi(t)$, for $\varphi \in L^2(\mathbf{R}; \mathbf{C})$ with compact support, then for each $t \in \mathbf{R}$ we have

$$Q(t) \stackrel{\mathrm{def}}{=} \Re \int_{-\infty}^{t} \varphi_\sigma(s)\overline{(\mu * \varphi_\sigma)(s)}\,\mathrm{d}s = [\![\mu, \chi_{(-\infty,t]}\varphi_\sigma]\!] \ge 0,$$

and an integration by parts gives

$$\begin{aligned}\llbracket \mu_\sigma, \varphi \rrbracket &= \Re \int_{\mathbf{R}} \varphi(t)\overline{(\mu_\sigma * \varphi)(t)}\,\mathrm{d}t = \Re \int_{\mathbf{R}} \mathrm{e}^{-2\sigma t}\varphi_\sigma(t)\overline{(\mu * \varphi_\sigma)(t)}\,\mathrm{d}t \\ &= 2\sigma \int_{\mathbf{R}} \mathrm{e}^{-2\sigma t} Q(t)\,\mathrm{d}t \geq 0.\end{aligned}$$

By Theorem 2.5, μ and μ_σ induce tempered distributions. Hence, their Fourier transforms are well defined. In the distribution sense the Fourier transform of μ_σ equals the function $\omega \mapsto \hat{\mu}(\sigma + \mathrm{i}\omega)$, and, therefore, by Theorem 2.5, the real part of this function is nonnegative in the distribution sense. But every continuous function that is nonnegative in the distribution sense is nonnegative at each point. Consequently, (i)$\Rightarrow$(iii).

(iv)$\Rightarrow$(i): Let $0 < \sigma < \epsilon$ and use the same notation as above. Then it follows from Theorem 2.5 that μ_σ is of positive type. Letting $\sigma \downarrow 0$ we conclude that μ is of positive type. □

When restricted to bounded functions, the notions of a function and a measure of positive type are straightforward extensions of Bochner's original concept of a continuous positive definite function. Bochner's theorem on the representation of a positive definite function may be given the following formulation.

2.7 Bochner's Theorem

Let $a \in L^\infty_{\mathrm{loc}}(\mathbf{R}^+; \mathbf{C}^{n\times n})$ be of positive type. Then there is a positive measure $\alpha \in M(\mathbf{R}; \mathbf{C}^{n\times n})$, such that

$$a(t) = \frac{1}{\pi}\int_{\mathbf{R}} \mathrm{e}^{\mathrm{i}\omega t}\alpha(\mathrm{d}\omega) \tag{2.5}$$

for almost all $t \in \mathbf{R}^+$. The measure α can be identified with the real part of the distribution Fourier transform of a. If a is continuous on $\mathbf{R}^+$, then $a(0) \succeq 0$.

By redefining the function a on a set of measure zero, one can make it bounded and uniformly continuous.

Recall that the conclusion $a(0) \succeq 0$ implies that $a(0)$ is self-adjoint. (The original version of Bochner's theorem requires a to be continuous, and $a(0)$ to be self-adjoint.)

Proof of Theorem 2.7 Without loss of generality we may take a to be scalar-valued (cf. Theorem 2.6 and formula (2.4)).

For the arguments below, we need a scalar nonnegative C^∞-function η with support in $[-1, 1]$, that satisfies $\int_{\mathbf{R}} \eta(t)\,\mathrm{d}t = 1$, and has a nonnegative Fourier transform. To obtain such a function, let ψ be any nonnegative C^∞-function with support in $[-\frac{1}{2}, \frac{1}{2}]$ and satisfying $\int_{\mathbf{R}} \psi(t)\,\mathrm{d}t = 1$, define $\check{\psi}(t) = \psi(-t)$ for $t \in \mathbf{R}$, and take η to be $\psi * \check{\psi}$. Then $\tilde{\eta} = |\tilde{\psi}|^2$, so $\tilde{\eta}$ is nonnegative.

Define $c(t) = a(t)$ for $t \geq 0$, and $c(t) = \overline{a(-t)}$ for $t < 0$, let η be a function of the type described in the preceding paragraph, and define η_ρ by $\eta_\rho(t) = \rho\eta(\rho t)$ for all $t \in \mathbf{R}$ and $\rho \in \mathbf{N}$. Then $\eta_\rho * c$ is continuous, and

$$|(\eta_\rho * c)(0)| \leq \|\eta_\rho\|_{L^1(\mathbf{R})}\|c\|_{L^\infty(-1/\rho,1/\rho)} = \|c\|_{L^\infty(-1/\rho,1/\rho)}.$$

In particular, this implies that

$$\limsup_{\rho\to\infty}\big|(\eta_\rho * c)(0)\big| < \infty.$$

On the other hand, we can express $(\eta_\rho * c)(0)$ in terms of Fourier transforms. If we denote the positive measure that $\Re\tilde{a}$ induces by α (cf. Theorem 2.5), then the (distribution) Fourier transform of c is the measure $\beta = 2\alpha$, and the transform of $\eta_\rho * c$ is the measure $\widetilde{\eta_\rho}(\omega)\beta(\mathrm{d}\omega)$. Because $\widetilde{\eta_\rho} \in \mathcal{S}$, and because β defines a tempered distribution, this measure is bounded. By the Fourier inversion formula,

$$(\eta_\rho * c)(0) = \frac{1}{2\pi}\int_{\mathbf{R}} \widetilde{\eta_\rho}(\omega)\beta(\mathrm{d}\omega),$$

and, therefore,

$$\limsup_{\rho\to\infty}\int_{\mathbf{R}} \widetilde{\eta_\rho}(\omega)\beta(\mathrm{d}\omega) < \infty.$$

We picked η in such a way that $\tilde{\eta} \geq 0$, and hence $\widetilde{\eta_\rho}(\omega) = \tilde{\eta}(\omega/\rho) \geq 0$ for all ρ and ω. Moreover, as $\rho \to \infty$, $\widetilde{\eta_\rho}(\omega) \to 1$ for all $\omega \in \mathbf{R}$. This means that we can apply Fatou's lemma to get $\int_{\mathbf{R}} \beta(\mathrm{d}\omega) < \infty$. The function b defined by $b(t) = \frac{1}{2\pi}\int_{\mathbf{R}} \mathrm{e}^{\mathrm{i}\omega t}\beta(\mathrm{d}\omega)$ is bounded and uniformly continuous, and by the uniqueness theorem for Fourier transforms it must be equal to c almost everywhere. In particular, it must be equal to a for almost all $t \geq 0$. □

Every continuous function of positive type has a positive limit at infinity in the following weak sense.

2.8 Theorem
Let $a \in C(\mathbf{R}^+; \mathbf{C}^{n\times n})$ be of positive type, and let α be the measure that $\Re\tilde{a}$ induces (cf. Bochner's Theorem 2.7). Then $\lim_{T\to\infty} \frac{1}{T}\int_0^T a(s)\,\mathrm{d}s$ exists and equals $\frac{1}{\pi}\alpha(\{0\})$. In particular, this limit is positive, i.e., $\alpha(\{0\}) \succeq 0$.

Proof By the Bochner Theorem 2.7, and by Fubini's theorem,

$$\frac{1}{T}\int_0^T a(s)\,\mathrm{d}s = \frac{1}{T\pi}\int_{\mathbf{R}}\int_0^T \mathrm{e}^{\mathrm{i}\omega s}\,\mathrm{d}s\,\alpha(\mathrm{d}\omega) = \frac{1}{\pi}\int_{\mathbf{R}} \theta(T\omega)\alpha(\mathrm{d}\omega),$$

where θ is the continuous function

$$\theta(\omega) = \begin{cases} 1, & \omega = 0, \\ \frac{1}{\mathrm{i}\omega}\left(\mathrm{e}^{\mathrm{i}\omega} - 1\right), & \omega \neq 0. \end{cases}$$

Letting $T \to \infty$, and using Lebesgue's dominated convergence theorem, we get the claim. □

3. Examples of Functions and Measures of Positive Type

To illustrate the previous theory, we give some examples of measures and functions of positive type. Our first example is classical.

3.1 Proposition

Let $a \in L^1_{\mathrm{loc}}(\mathbf{R}^+; \mathbf{C}^{n\times n})$, and suppose that for every $y \in \mathbf{C}^n$ the function $t \mapsto \langle y, a(t)y\rangle$ is nonnegative, nonincreasing, and convex on $(0,\infty)$. Then a is of positive type.

Most of the time when applying this proposition we take a to be a scalar function.

Proof of Proposition 3.1 It follows from Theorem 2.4 that a is of positive type if and only if the scalar function $\langle y, ay\rangle$ is of positive type for every $y \in \mathbf{C}^n$. Therefore, it suffices to prove the proposition in the scalar case.

Let us for a moment, further simplify the problem, and assume that a satisfies

$$\begin{aligned} &a \in C^2(\mathbf{R}^+; \mathbf{R}) \cap L^1(\mathbf{R}^+; \mathbf{R}) \text{ and} \\ &\lim_{t\to\infty} a(t) = \lim_{t\to\infty} a'(t) = 0. \end{aligned} \tag{3.1}$$

Then, by Theorem 2.4, to prove that a is of positive type it suffices to show that $\Re\tilde{a}(\omega) \geq 0$ for all $\omega \in \mathbf{R}$. This is easily done, because two integrations by parts give, if $\omega \neq 0$,

$$\begin{aligned} \Re\tilde{a}(\omega) &= \int_{\mathbf{R}^+} \cos(\omega t)a(t)\,\mathrm{d}t = -\frac{1}{\omega}\int_{\mathbf{R}^+} \sin(\omega t)a'(t)\,\mathrm{d}t \\ &= \frac{1}{\omega^2}\int_{\mathbf{R}^+} \big(1-\cos(\omega t)\big)a''(t)\,\mathrm{d}t \geq 0. \end{aligned}$$

By continuity, $\Re\tilde{a}(0) \geq 0$. This proves the simplified problem.

In the general case, where $a \in L^1_{\mathrm{loc}}(\mathbf{R}^+;\mathbf{R})$ is a nonnegative, nonincreasing, and convex function on $\mathbf{R}^+$, we approximate a by some sequence $\{a_m\}_{m=1}^{\infty}$ of nonnegative, nonincreasing, and convex functions on $\mathbf{R}^+$ that satisfy (3.1). This approximation should be done in such a way that $a_m \to a$ in $L^1_{\mathrm{loc}}(\mathbf{R}^+;\mathbf{R})$ as $m \to \infty$. (For example, we can take $a_m(t) = me^{-t/m}\int_t^{t+(1/m)} \eta(m(t-s))a(s)\,\mathrm{d}s$, where η is some nonnegative C^2-function on $[-1,0]$ satisfying $\eta(-1) = \eta'(-1) = \eta(0) = \eta'(0) = 0$, and $\int_{-1}^0 \eta(s)\,\mathrm{d}s = 1$.) By the preceding argument, each a_m is of positive type; hence $[\![a_m,\varphi]\!] \geq 0$ for all $\varphi \in L^2(\mathbf{R};\mathbf{C})$ with compact support. This, together with the fact that a_m tends to a in $L^1_{\mathrm{loc}}(\mathbf{R}^+;\mathbf{C})$, implies that $[\![a,\varphi]\!] \geq 0$, i.e., that a is of positive type. □

Our next result is simple but useful.

3.2 Proposition
Suppose that $\mu = A\delta + \nu$, where δ is the scalar unit point mass at zero, A is a constant matrix, and $\nu \in M(\mathbf{R}^+; \mathbf{C}^{n\times n})$. Also assume that

$$\Re A - \int_{\mathbf{R}^+} |\nu|(\mathrm{d}t) I \succeq 0.$$

Then μ is of positive type.

Proof For every $\varphi \in L^2(\mathbf{R}; \mathbf{C}^n)$ with compact support, we have

$$\begin{aligned}
[\![\mu, \varphi]\!] &= \Re \int_{\mathbf{R}} \langle \varphi(t), (\mu * \varphi)(t) \rangle \,\mathrm{d}t \\
&= \Re \int_{\mathbf{R}} \langle \varphi(t), A\varphi(t) \rangle \,\mathrm{d}t + \Re \int_{\mathbf{R}} \langle \varphi(t), (\nu * \varphi)(t) \rangle \,\mathrm{d}t \\
&\geq \Re \int_{\mathbf{R}} \langle \varphi(t), A\varphi(t) \rangle \,\mathrm{d}t - \|\nu\|_{M(\mathbf{R}^+)} \|\varphi\|^2_{L^2(\mathbf{R})} \\
&= \int_{\mathbf{R}} \big\langle \varphi(t), \big(\Re A - \|\nu\|_{M(\mathbf{R}^+)} I\big) \varphi(t) \big\rangle \,\mathrm{d}t \geq 0. \quad \square
\end{aligned}$$

Our next example is even simpler. It is useful, for example, when one discusses ordinary differential equations (i.e., equations with no delay).

3.3 Proposition
Let $\nu \in M_{\mathrm{loc}}(\mathbf{R}^+; \mathbf{C}^{n\times n})$, and let $\mu = A\delta + \nu$, where $A \in \mathbf{C}^{n\times n}$ is antisymmetric and δ is the scalar unit point mass at zero. Then μ is of positive type iff ν is of positive type.

In other words, the imaginary part of a possible point mass at zero has no effect on whether a measure is of positive type or not. This follows from the fact that for every $\varphi \in L^2(\mathbf{R}; \mathbf{C}^n)$ with compact support one has

$$[\![A\delta, \varphi]\!] = \Re \int_{\mathbf{R}} \langle \varphi(t), \mathrm{i}\Im A\varphi(t) \rangle \,\mathrm{d}t = 0.$$

In our fourth example we investigate the behaviour of trigonometric polynomials.

3.4 Proposition
Let

$$\mu(\mathrm{d}t) = \nu(\mathrm{d}t) + p\cos(\omega_0 t)\,\mathrm{d}t + q\sin(\omega_0 t)\,\mathrm{d}t,$$

where $\omega_0 \in \mathbf{R}$, p and q are constant $n\times n$ matrices, $\nu \in M_{\mathrm{loc}}(\mathbf{R}^+; \mathbf{C}^{n\times n})$ induces a tempered distribution when extended to $\mathbf{R}$ as zero on $(-\infty, 0)$, and where we suppose that the two limits $\lim_{z\to\pm\mathrm{i}\omega_0, \Re z>0} \hat{\nu}(z)$ exist. If $\omega_0 \neq 0$, then $\mu \in PT(\mathbf{R}^+; \mathbf{C}^{n\times n})$ if and only if $\nu \in PT(\mathbf{R}^+; \mathbf{C}^{n\times n})$ and $p \pm \mathrm{i}q \succeq 0$. If $\omega_0 = 0$, then $\mu \in PT(\mathbf{R}^+; \mathbf{C}^{n\times n})$ if and only if $\nu \in PT(\mathbf{R}^+; \mathbf{C}^{n\times n})$ and $p \succeq 0$.

Observe that the condition $p \pm \mathrm{i}q \succeq 0$ implies $p \succeq 0$. In particular, p is self-adjoint, so $p = \Re p$ and $\Im p = 0$. Also $\mathrm{i}q$ must be self-adjoint, so $\Re q = 0$ and $q = \mathrm{i}\Im q$.

Proof of Proposition 3.4 It follows from Theorem 2.6 that we can, without loss of generality, take $n = 1$.

For all z with $\Re z > 0$ we have

$$\Re\hat{\mu}(z) = \Re\hat{\nu}(z) + \Re\frac{zp + \omega_0 q}{z^2 + \omega_0^2}. \tag{3.2}$$

First, suppose that we know that ν is of positive type, take $\omega_0 \neq 0$, and let $p \pm \mathrm{i}q \geq 0$. Then $p \geq 0$ and q is a pure imaginary number. If we define $r = \Im q$, then $q = \mathrm{i}r$, and $p \geq |r|$. Let $z = \sigma + \mathrm{i}\tau$, where $\sigma > 0$. Then (3.2) becomes

$$\Re\hat{\mu}(z) = \Re\hat{\nu}(z) + \frac{\sigma(\sigma^2 + \tau^2 + \omega_0^2)p + 2\sigma\tau\omega_0 r}{|z^2 + \omega_0^2|^2}.$$

This implies that $\Re\hat{\mu}(z) \geq 0$, because $\Re\hat{\nu}(z) \geq 0$ (see Theorem 2.6 (iii)), $p \geq |r|$, $\tau^2 + \omega_0^2 \geq |2\tau\omega_0|$, and $\sigma > 0$. Since z was arbitrary, it follows from Theorem 2.6 that μ is of positive type. The case $\omega_0 = 0$ is treated in the same way.

Next, assume that μ is of positive type, and let $\Re z > 0$. It follows that

$$\Re\hat{\nu}(z) + \Re\frac{zp + \omega_0 q}{z^2 + \omega_0^2} \geq 0. \tag{3.3}$$

First, suppose $\omega_0 \neq 0$. If we take $z = \sigma \pm \mathrm{i}\omega_0$, where $\sigma > 0$, then

$$\Re\hat{\nu}(\sigma \pm \mathrm{i}\omega_0) + \Re\frac{(\sigma \pm \mathrm{i}\omega_0)p + \omega_0 q}{\sigma^2 \pm 2\mathrm{i}\sigma\omega_0} \geq 0. \tag{3.4}$$

Recall that $\lim_{z \to \pm\mathrm{i}\omega_0, \Re z > 0} \hat{\nu}(z)$ exists; hence

$$\lim_{z \to \pm\mathrm{i}\omega_0, \Re z > 0} (z \mp \mathrm{i}\omega_0)\hat{\nu}(z) = 0.$$

Multiply both sides of (3.4) by σ, and let $\sigma \downarrow 0$ to get

$$\frac{1}{2}\Re\{p \pm \mathrm{i}q\} \geq 0. \tag{3.5}$$

Above, we let z approach the points $\pm\mathrm{i}\omega_0$ along a line perpendicular to the imaginary axis. If instead we let z approach $\pm\mathrm{i}\omega_0$ tangentially, then we get a different result. For example, substituting $z = \pm\mathrm{i}\omega_0 + \mathrm{i}\tau + \tau^2$ into (3.3) one gets

$$\Re\hat{\nu}(z) + \Re\frac{(\pm\mathrm{i}\omega_0 + \mathrm{i}\tau + \tau^2)p + \omega_0 q}{\mp 2\omega_0\tau + \tau^2(\pm 2\mathrm{i}\omega_0 - 1 + 2\mathrm{i}\tau + \tau^2)} \geq 0.$$

If we multiply this inequality by $\tau > 0$ and let $\tau \downarrow 0$, and multiply by $\tau < 0$ and let $\tau \uparrow 0$, then we get

$$\frac{1}{2}\Re\{-\mathrm{i}p \mp q\} = \frac{1}{2}\Im\{p \mp \mathrm{i}q\} = 0. \tag{3.6}$$

From (3.5) and (3.6) we conclude that p is real and q is pure imaginary, and that $p \pm \mathrm{i}q \geq 0$. If $\omega_0 = 0$, then one can proceed in the same manner to show that $p \geq 0$.

To show that ν is of positive type, one observes that by (3.3), and by the fact that p is real and q is pure imaginary, one has $\liminf_{z\to i\tau,\Re z>0}\Re\hat{\nu}(z)\geq 0$, at least for all $\tau\neq\pm\omega_0$. But at these points one can rely on the fact that the limits $\lim_{z\to\pm i\omega_0,\Re z>0}\Re\hat{\nu}(z)$ exist, and cannot possibly be negative (in these limits we do allow tangential approach, and, therefore, $\limsup_{z\to\pm i\omega_0,\Re z\geq 0}\Re\hat{\nu}(z)\geq\limsup_{\tau\to\omega_0}\limsup_{z\to i\tau,\Re z>0}\Re\hat{\nu}(z)\geq 0$). □

Suppose that μ, $\nu\in PT(\mathbf{R}^+)$ with $\mu(\mathrm{d}t)=a(t)\,\mathrm{d}t$, where $a\in L^1_{\mathrm{loc}}(\mathbf{R}^+)$. For applications it is convenient to have conditions under which the measures $a*\mu$ and $a\mu$ are of positive type. Below we formulate three statements in this direction. In these results, it is essential that at least one of the two factors be scalar-valued. (The results are not true if both factors are allowed to take values in $\mathbf{C}^{n\times n}$. This is related to the fact that the product of two positive matrices need not be positive.)

3.5 Proposition
Let $\mu\in M_{\mathrm{loc}}(\mathbf{R}^+)$ and $a\in C(\mathbf{R}^+)$ both be of positive type, and suppose that μ or a is scalar-valued. Then the measure ν defined by $\nu(\mathrm{d}t)=a(t)\mu(\mathrm{d}t)$ is of positive type.

Proof Again we may, without loss of generality, take both a and μ scalar-valued.

Extend the function a as $a(t)=\overline{a(-t)}$ for $t\in(-\infty,0)$. It follows from the Bochner Theorem 2.7 that there exists a nonnegative measure α in $M(\mathbf{R};\mathbf{R})$, such that

$$a(t)=\frac{1}{\pi}\int_{\mathbf{R}}\mathrm{e}^{\mathrm{i}\omega t}\alpha(\mathrm{d}\omega),\quad t\in\mathbf{R}.$$

Let $\varphi\in L^2(\mathbf{R};\mathbf{C})$ have compact support. Then

$$\begin{aligned}
[\![\nu,\varphi]\!]&=\Re\int_{\mathbf{R}}\varphi(t)\overline{(\nu*\varphi)(t)}\,\mathrm{d}t\\
&=\Re\int_{\mathbf{R}}\varphi(t)\int_{\mathbf{R}^+}\overline{\mu}(\mathrm{d}s)\overline{\varphi(t-s)a(s)}\,\mathrm{d}t\\
&=\frac{1}{\pi}\Re\int_{\mathbf{R}}\varphi(t)\int_{\mathbf{R}^+}\overline{\mu}(\mathrm{d}s)\overline{\varphi(t-s)}\int_{\mathbf{R}}\mathrm{e}^{-\mathrm{i}\omega s}\alpha(\mathrm{d}\omega)\,\mathrm{d}t\\
&=\frac{1}{\pi}\Re\int_{\mathbf{R}}\left\{\int_{\mathbf{R}}\mathrm{e}^{-\mathrm{i}\omega t}\varphi(t)\int_{\mathbf{R}^+}\overline{\mu}(\mathrm{d}s)\mathrm{e}^{\mathrm{i}\omega(t-s)}\overline{\varphi(t-s)}\,\mathrm{d}t\right\}\alpha(\mathrm{d}\omega)\\
&=\frac{1}{\pi}\int_{\mathbf{R}}[\![\mu,\mathrm{e}^{-\mathrm{i}\omega\cdot}\varphi(\cdot)]\!]\alpha(\mathrm{d}\omega)\geq 0.\ \square
\end{aligned}$$

3.6 Proposition
Let $p \in [1,\infty]$ and $1/p + 1/q = 1$, let $a \in L^p_{\mathrm{loc}}(\mathbf{R}^+)$ and $b \in L^q_{\mathrm{loc}}(\mathbf{R}^+)$ both be of positive type, and suppose that a or b is scalar-valued. Then the product ab is of positive type.

Proof Again we may, without loss of generality, take both a and b scalar-valued.

Clearly, if, e.g., a is continuous, then the conclusion of the proposition follows from Proposition 3.5. In particular, by the Bochner Theorem 2.7, this is the case when $p = \infty$ or $q = \infty$.

In the remaining case where $1 < p < \infty$, and neither a nor b is continuous, we approximate a by a sequence of smooth functions, e.g. as follows. Let φ be a real nonnegative even C^∞-function with compact support, satisfying $\int_{\mathbf{R}} \varphi(s)\,ds = 1$. Define $\psi = \varphi * \varphi$ and $\psi_m(t) = m\psi(mt)$ for $t \in \mathbf{R}$. Then $\int_{\mathbf{R}} \psi_m(t)\,dt = 1$, and $\tilde{\psi}_m(\omega) \geq 0$. Define $a(t) = 0$ for $t < 0$, $c(t) = a(t)$ for $t \geq 0$, and $c(t) = \overline{a}(-t)$ for $t < 0$. Then $\tilde{c}(\omega) = 2\Re\tilde{a}(\omega) \geq 0$ for $\omega \in \mathbf{R}$. Therefore, the C^∞-functions $c_m \stackrel{\mathrm{def}}{=} c * \psi_m$ satisfy $\Re\tilde{c}_m(\omega) \geq 0$. Moreover, $c_m \to c$ in $L^p_{\mathrm{loc}}(\mathbf{R};\mathbf{C})$ as $m \to \infty$. Define a_m to be $a_m(t) = c_m(t)$ for $t \geq 0$. Then $a_m \in C(\mathbf{R}^+;\mathbf{C})$ is of positive type, and $a_m \to a$ in $L^p_{\mathrm{loc}}(\mathbf{R}^+;\mathbf{C})$. Because of the special case proved above, we know that $a_m b$ is of positive type for each m. This, combined with the fact that $a_m b \to ab$ locally in $L^1(\mathbf{R}^+;\mathbf{C})$, implies that ab must be of positive type. □

The third statement may appear somewhat technical, but it plays a crucial role in the proof of Theorem 3.8 below. Note that in the proof of the preceding result we did in fact work through a particular case of this statement.

3.7 Proposition
Let $\mu \in M_{\mathrm{loc}}(\mathbf{R}^+)$ and $a \in L^1(\mathbf{R}^+)$ both be of positive type, and suppose that the support of a is bounded, and that μ or a is scalar-valued. Define $\mu(E) = \mu(E \cap \mathbf{R}^+)$, $\check{\mu}(E) = \mu^*(-E)$ for every Borel set $E \subset \mathbf{R}$, and $a(t) = 0$ for $t < 0$, $\check{a}(t) = a^*(-t)$ for all t, and let b be the restriction to $\mathbf{R}^+$ of $(a + \check{a}) * (\mu + \check{\mu})$. Then b is of positive type.

Proof Since one of the two factors is scalar-valued, we can use Theorem 2.6 to reduce the proof of the proposition to the scalar case. Thus, let us, without loss of generality, take the dimension $n = 1$.

If a is a C^∞-function with compact support, then, in the distribution sense,

$$\Re\tilde{b}(\omega) = 4\Re\tilde{a}(\omega)\Re\tilde{\mu}(\omega), \quad \omega \in \mathbf{R},$$

and it follows from Theorem 2.5 that b is of positive type. In the general case we can approximate a (as in the proof of Proposition 3.6) by a sequence a_m of C^∞-functions, with uniformly bounded supports, that converge to a in L^1 as $m \to \infty$. If we define b_m in the same way as we defined b above, but with a replaced by a_m, then each b_m is of positive type. As $m \to \infty$, $b_m \to b$ locally in L^1, and, therefore, b must be of positive type. □

Let us finish this section with a result which relates the two conditions $a \in PT$ and $-a'' \in PT$ to each other.

3.8 Theorem

Let $a \in AC_{\mathrm{loc}}(\mathbf{R}^+; \mathbf{C}^{n\times n})$ satisfy $a' \in BV_{\mathrm{loc}}(\mathbf{R}^+; \mathbf{C}^{n\times n})$, and let α be the measure derivative of $-a'$, i.e., $\alpha([0,t]) = -a'(t)$ for almost all $t \in \mathbf{R}^+$. Then the following claims are true.

(i) *If a is of positive type, then so is α.*

(ii) *Let α be of positive type. Then a is of positive type if and only if at least one of the following two conditions is satisfied:*

(1) *$a(0)$ is self-adjoint, $A \overset{\mathrm{def}}{=} \lim_{T\to\infty} \frac{1}{T}\int_0^T a(s)\,\mathrm{d}s$ exists, and $\Re A \succeq 0$;*

(2) *$A \overset{\mathrm{def}}{=} \lim_{T\to\infty} \frac{1}{T}\int_0^T a(s)\,\mathrm{d}s$ exists, and $A \succeq 0$.*

Note that if a' is locally absolutely continuous, then $\alpha(\mathrm{d}t) = -a'(0)\delta - a''(t)\,\mathrm{d}t$. In addition, observe that it follows from Theorems 2.7 and 2.8 that both (1) and (2) are necessary for a to be positive type.

Proof of Theorem 3.8 (i) Let λ denote the C^∞-function $\lambda(\omega) = \omega$. It follows from Theorem 2.5 that $\tilde{\alpha}$ and $\tilde{a}$ are well defined as tempered distributions, and that $\tilde{\alpha} = -\mathrm{i}\lambda a(0) + \lambda^2\tilde{a}$. Since $a(0)$ is self-adjoint, $\Re\tilde{\alpha} = \lambda^2\Re\tilde{a}$. Thus, by Theorem 2.5, α must be of positive type whenever a is of positive type.

(ii) As we observed above, both (1) and (2) are necessary for a to be of positive type.

In the proof of the fact that either one of the two conditions (1) and (2), together with $\alpha \in PT$, is sufficient to imply that a is of positive type, let us first show how case (2) can be reduced to case (1), i.e., let us take for granted that (1) and $\alpha \in PT$ imply that a is of positive type, and suppose that (2) holds. Define $a_1 = a - \mathrm{i}\Im a(0)$. Then $a_1(0)$ is self-adjoint and a_1 satisfies (1); hence a_1 is of positive type. Clearly $A = \lim_{T\to\infty}\frac{1}{T}\int_0^T a(s)\,\mathrm{d}s = \mathrm{i}\Im a(0) + \lim_{T\to\infty}\frac{1}{T}\int_0^T a_1(s)\,\mathrm{d}s$. By Theorem 2.8, the second term on the right-hand side is positive; hence self-adjoint. Since A too, was supposed to be self-adjoint, this implies that $\Im a(0) = 0$. Thus $a_1 = a$ is of positive type.

It remains to show that (1) and $\alpha \in PT$ imply that a is of positive type. In the proof of this fact we may, without loss of generality, take $n = 1$ (this follows from the equivalence of (i) and (ii) in Theorem 2.6).

Define

$$b(t) = \begin{cases} a(t), & t \geq 0, \\ \overline{a(-t)}, & t < 0, \end{cases}$$

and

$$\eta(t) = \begin{cases} \frac{1}{2}, & |t| \le 1, \\ 0, & \text{otherwise}, \end{cases}$$

$$\varphi(t) = \begin{cases} -\int_t^\infty \eta(s)\,\mathrm{d}s, & t \ge 0, \\ \int_{-\infty}^t \eta(s)\,\mathrm{d}s, & t < 0, \end{cases}$$

$$\psi(t) = \int_{-\infty}^t \varphi(s)\,\mathrm{d}s, \quad t \in \mathbf{R}.$$

Then η, φ, and ψ are all supported on $[-1, 1]$. Define

$$\eta_\delta(t) = \delta\eta(\delta t), \quad \varphi_\delta(t) = \varphi(\delta t), \quad \psi_\delta(t) = \frac{1}{\delta}\psi(\delta t), \quad \delta > 0, \quad t \in \mathbf{R},$$

and define b_δ by

$$b_\delta = b - \eta_\delta * b.$$

Integrating by parts one gets

$$b_\delta = -\varphi_\delta * b' = \psi_\delta * \beta,$$

where β is the distribution derivative of $-b'$, i.e., the measure that one obtains by defining $\alpha(E) = \alpha(\mathbf{R}^+ \cap E)$ and $\breve{\alpha}(E) = \overline{\alpha(-E)}$ for every Borel set $E \subset \mathbf{R}$, and $\beta = \alpha + \breve{\alpha}$. It follows from the way in which we defined ψ_δ that

$$\widetilde{\psi_\delta}(\omega) = \tfrac{1}{\omega^2}\Big(1 - \frac{\sin(\omega/\delta)}{\omega/\delta}\Big), \quad \omega \ne 0,$$

and that $\widetilde{\psi_\delta}(0) = 1/(6\delta^2)$ (this is the number that one gets by letting $\omega \to 0$ above). In particular, as $\widetilde{\psi_\delta}(\omega) \ge 0$ for all $\omega \in \mathbf{R}$, it follows from Proposition 3.7, where $a + \breve{a}$ is now ψ_δ and $\mu + \breve{\mu}$ is now β, that the function that one gets by restricting b_δ to $\mathbf{R}^+$ is of positive type.

So far, we have not used the assumption that $A \stackrel{\text{def}}{=} \lim_{T\to\infty} \frac{1}{T}\int_0^T a(s)\,\mathrm{d}s$ exists. If this is the case, then $\lim_{\delta\to 0} b_\delta(0) = b(0) - \Re A$ exists as well, and hence, by the Bochner Theorem 2.7,

$$\lim_{\delta\to 0} \frac{1}{2\pi}\int_{\mathbf{R}} \widetilde{\psi_\delta}(\omega)\tilde{\beta}(\mathrm{d}\omega) = b(0) - \Re A.$$

We have

$$\lim_{\delta\to 0} \widetilde{\psi_\delta}(\omega) = \begin{cases} \infty, & \omega = 0, \\ 1/\omega^2, & \omega \ne 0. \end{cases}$$

Therefore, $\tilde{\beta}$ cannot have a point mass at zero, and by Fatou's lemma

$$\int_{\omega\ne 0} \frac{\tilde{\beta}(\mathrm{d}\omega)}{\omega^2} < \infty.$$

Let γ be the finite measure that has no point mass at zero and satisfies $\gamma(\mathrm{d}\omega) = \omega^{-2}\tilde{\beta}(\mathrm{d}\omega)$ for $\omega \ne 0$, and let c be the inverse Fourier transform of γ. Then the restriction of the function c to $\mathbf{R}^+$ is of positive type.

To complete the proof, it suffices to show that $b - c$ is a nonnegative constant, or, more precisely, that $b(t) = c(t) + \Re A$ for all $t \in \mathbf{R}$. Now, in the distribution sense, it is true that both $\lambda^2\tilde{b} = \tilde{\beta}$ and $\lambda^2\tilde{c} = \tilde{\beta}$, where λ is the C^∞-function $\lambda(\omega) = \omega$. Hence $\tilde{b} - \tilde{c}$ must vanish on $\mathbf{R} \setminus \{0\}$. By, e.g., Rudin [2], Theorem 6.25, this implies that $b - c$ can be expressed as a finite sum

$$b(t) = c(t) + \beta_0 + \beta_1 t + \cdots + \beta_m t^m.$$

Consequently (by Theorem 2.8 the integral of c drops out),

$$\begin{aligned} A &= \frac{1}{T} \lim_{T\to\infty} \int_0^T b(s)\,\mathrm{d}s \\ &= \lim_{T\to\infty} \left\{ \frac{1}{T} \int_0^T c(s)\,\mathrm{d}s + \sum_{j=0}^{m} \frac{1}{j+1} \beta_j T^j \right\} \\ &= \lim_{T\to\infty} \sum_{j=0}^{m} \frac{1}{j+1} \beta_j T^j. \end{aligned}$$

Therefore, we must have $\beta_0 = A$, and $\beta_j = 0$ for $j \geq 1$. Moreover, both $b(0)$ and $c(0)$ are self-adjoint, and so β_0 is self-adjoint as well. That is, $\beta_0 = \Re\beta_0 = \Re A$, as claimed. □

4. Kernels of Strong and Strict Positive Type

In applications to Volterra equations, one frequently has to impose more than mere positivity on the kernel μ, in order to obtain asymptotic results. What one needs is 'strict' or 'strong' positivity. We define these concepts below, and give some examples.

4.1 Definition

We say that a measure $\mu \in M_{\mathrm{loc}}(\mathbf{R}^+; \mathbf{C}^{n\times n})$ is of strong positive type, and write $\mu \in PT_{\mathrm{strong}}(\mathbf{R}^+; \mathbf{C}^{n\times n})$, if there exists a constant $\epsilon > 0$ for which the measure $\nu(\mathrm{d}t) = \mu(\mathrm{d}t) - \epsilon \mathrm{e}^{-t} I\,\mathrm{d}t$ is of positive type. A function $a \in L^1_{\mathrm{loc}}(\mathbf{R}^+; \mathbf{C}^{n\times n})$ is said to be of strong positive type if the measure $\mu(\mathrm{d}s) = a(s)\,\mathrm{d}s$ is of strong positive type.

If μ is of strong positive type, then, of course, μ is of positive type, and for each $\varphi \in L^2(\mathbf{R}; \mathbf{C}^n)$ with compact support one has $\epsilon[\![\mathbf{e}, \varphi]\!] \leq [\![\mu, \varphi]\!]$, where we define the function $\mathbf{e}$ by $\mathbf{e}(t) = \mathrm{e}^{-t} I$ for $t \in \mathbf{R}^+$.

The Laplace transform of $\mathbf{e}$ is $\hat{\mathbf{e}}(z) = (1+z)^{-1} I$ for $\Re z \geq 0$, and the real part of this function is

$$\Re\hat{\mathbf{e}}(\sigma + \mathrm{i}\tau) = \frac{1+\sigma}{(1+\sigma)^2 + \tau^2} I.$$

In particular, this means that μ is of strong positive type iff one of the conditions (iii)–(v) in Theorem 2.6 holds with $\Re\hat{\mu}(\sigma + \mathrm{i}\tau)$ replaced by

$\Re\hat{\mu}(\sigma + i\tau) - \epsilon(1+\sigma)/((1+\sigma)^2+\tau^2)$ for some $\epsilon > 0$, or if (vi) holds with $\Re\tilde{\mu}$ replaced by $\Re\tilde{\mu}(\omega) - \epsilon(1+\omega^2)^{-1}I$.

Some examples of measures of strong positive type are given below.

4.2 Proposition

Let $a \in L^1(\mathbf{R}^+; \mathbf{C}^{n\times n})$ be of positive type, and assume that a and a' are locally absolutely continuous, with a', $a'' \in L^1(\mathbf{R}^+; \mathbf{C}^{n\times n})$. Then a is of strong positive type if and only if $\Re\tilde{a}(\omega) \succ 0$ for all $\omega \in \mathbf{R}$, and $\Re a'(0) \prec 0$.

Proof By Definition 4.1 and Theorem 2.4, the function a is of strong positive type iff there exists some $\epsilon > 0$ such that $\Re\tilde{a}(\omega) \succeq \epsilon(1+\omega^2)^{-1}I$ for $\omega \in \mathbf{R}$. Fixing $\omega \in \mathbf{R}$, we find that we have to take

$$\epsilon \le \inf_{y\in\mathbf{C}^n, \|y\|=1} \langle y, (1+\omega^2)\Re\tilde{a}y\rangle .$$

Owing to the continuity of $\tilde{a}$, this function, which restricts ϵ from above, is continuous in ω, so on each compact subinterval J of $\mathbf{R}$ we can find some $\epsilon > 0$, for which $\Re\tilde{a}(\omega) \succeq \epsilon(1+\omega^2)^{-1}I$ if and only if $\Re\tilde{a}(\omega) \succ 0$ for all $\omega \in J$.

To complete the proof, it suffices to show that $\lim_{|\omega|\to\infty} \omega^2\Re\tilde{a}(\omega) = -\Re a'(0)$. Integrating by parts twice, one gets

$$\tilde{a}(\omega) = \frac{1}{i\omega}a(0) - \frac{1}{\omega^2}a'(0) - \frac{1}{\omega^2}\int_{\mathbf{R}^+} e^{-i\omega t}a''(t)\,dt.$$

By Theorem 2.7, $a(0) \succeq 0$. Therefore, $\Im a(0) = 0$, and by the Riemann–Lebesgue lemma

$$\lim_{|\omega|\to\infty} \int_{\mathbf{R}^+} e^{-i\omega t}a''(t)\,dt = 0.$$

Thus $\lim_{|\omega|\to\infty} \omega^2\Re\tilde{a}(\omega) = -\Re a'(0)$, and the proof is complete. □

Most convex functions are of strong positive type.

4.3 Proposition

Let $a \in L^1_{\text{loc}}(\mathbf{R}^+; \mathbf{C}^{n\times n})$, and suppose that for every $y \in \mathbf{C}^n$ the function $t \mapsto \langle y, a(t)y\rangle$ is nonnegative, nonincreasing, and convex on $(0,\infty)$. In addition, suppose that, for each nonzero $y \in \mathbf{C}^n$, it is true that $\frac{d^2}{dt^2}\langle y, a(t)y\rangle > 0$ on a set of positive measure. Then a is of strong positive type.

Here $\frac{d^2}{dt^2}\langle y, a(t)y\rangle$ represents the ordinary (i.e., not the distribution) derivative of the function $\langle y, a'(t)y\rangle$. The latter function is monotone, therefore it is locally of bounded variation, and so it has a derivative a.e.

Proof of Proposition 4.3 First, let us prove this result under the additional assumptions that a' is locally absolutely continuous on $\mathbf{R}^+$, and that a, a', $a'' \in L^1(\mathbf{R}^+; \mathbf{C}^{n\times n})$. Then $a'(0) = -\int_{\mathbf{R}^+} a''(t)\,dt \prec 0$, and hence, by Proposition 4.2, it suffices to show that $\Re\tilde{a}(\omega) \succ 0$ for all $\omega \in \mathbf{R}$. Since $a(t) \succeq 0$ for all $t \geq 0$, the function a is self-adjoint, and therefore

$$\Re\tilde{a}(\omega) = \int_{\mathbf{R}} \cos(\omega t)\Re a(t)\,\mathrm{d}t - \int_{\mathbf{R}} \sin(\omega t)\Im a(t)\,\mathrm{d}t = \int_{\mathbf{R}^+} \cos(\omega t) a(t)\,\mathrm{d}t.$$

Clearly, $\Re\tilde{a}(0) = \int_{\mathbf{R}^+} a(t)\,\mathrm{d}t \succ 0$. For $\omega \neq 0$ one can integrate by parts, in the same way as in the proof of Proposition 3.1, to show that

$$\Re\tilde{a}(\omega) = \frac{1}{\omega^2}\int_{\mathbf{R}^+} \big(1 - \cos(\omega t)\big) a''(t)\,\mathrm{d}t.$$

This, together with the facts that for each nonzero $y \in \mathbf{C}^n$ we have $\langle y, a''(t)y\rangle \geq 0$ a.e. and that $\langle y, a''(t)y\rangle > 0$ on a set of positive measure, implies that $\Re\tilde{a}(\omega) \succ 0$ for all $\omega \in \mathbf{R}$, and completes the proof under the additional assumptions.

The general case can be reduced to the special case proved above, in the following way. Denote the measure derivative of a' by α, and define $\beta(\mathrm{d}t) = \alpha(\mathrm{d}t) - \frac{\mathrm{d}^2}{\mathrm{d}t^2} a(t)\,\mathrm{d}t$. Then β contains the discrete and the singular parts of α. In particular, $\beta \succeq 0$. We can write a in the form

$$a(t) = a(\infty) - \int_t^\infty a'(s)\,\mathrm{d}s = a(\infty) + \int_t^\infty \int_{[s,\infty)} \alpha(\mathrm{d}u)\,\mathrm{d}s.$$

Define $\eta(t) = t^2\mathrm{e}^{-t}$. Then $0 < \eta(t) \leq 1$ for all $t > 0$, and it follows from Lemma **6**.2.2 that

$$\Big\langle y, \int_{\mathbf{R}^+} (1 + t^2)\eta(t)\frac{\mathrm{d}^2}{\mathrm{d}t^2} a(t) y\,\mathrm{d}t \Big\rangle < \infty.$$

Define

$$b(t) = \int_t^\infty \int_s^\infty \eta(u)\frac{\mathrm{d}^2}{\mathrm{d}u^2} a(u)\,\mathrm{d}u\,\mathrm{d}s,$$

and denote the remainder $a - b$ by c. Then

$$c(t) = a(\infty) + \int_t^\infty \int_{[s,\infty)} \Big\{\beta(\mathrm{d}u) + \big(1 - \eta(u)\big)\frac{\mathrm{d}^2}{\mathrm{d}u^2} a(u)\,\mathrm{d}u\Big\}\,\mathrm{d}s.$$

In particular, both b and c are nonnegative, nonincreasing, and convex, and b satisfies the additional smoothness and size conditions imposed on a in the proof of the special case above. By that result, b is of strong positive type, and, by Proposition 3.1, c is of positive type. This implies that $a = b + c$ is of strong positive type. □

We shall also need another, slightly weaker form of positivity, which we call 'strict' instead of 'strong'.

4.4 Definition
We say that a measure $\mu \in M_{\mathrm{loc}}(\mathbf{R}^+; \mathbf{C}^{n\times n})$ is of strict positive type, and write $\mu \in PT_{\mathrm{strict}}(\mathbf{R}^+; \mathbf{C}^{n\times n})$, if there exists a function $a \in L^1(\mathbf{R}^+; \mathbf{C}^{n\times n})$ satisfying $\Re\tilde{a}(\omega) \succ 0$ for all $\omega \in \mathbf{R}$, such that the measure ν defined by $\nu(\mathrm{d}t) = \mu(\mathrm{d}t) - a(t)\,\mathrm{d}t$ is of positive type. A function $k \in L^1_{\mathrm{loc}}(\mathbf{R}^+; \mathbf{C}^{n\times n})$ is said to be of strict positive type if the measure $\mu(\mathrm{d}s) = k(s)\,\mathrm{d}s$ is of strict positive type.

Clearly, every measure that is of strong positive type is of strict positive type, and, by Theorem 2.5, every measure of strict positive type is of positive type.

It follows from Theorem 2.6 that if a measure μ is of strict positive type then for each $\tau \in \mathbf{R}$ we must have

$$\liminf_{\substack{z\to \mathrm{i}\tau \\ \Re z>0}} \Re\hat{\mu}(z) \succ 0, \tag{4.1}$$

in the sense that for every nonzero $y \in \mathbf{C}^n$ one has

$$\liminf_{\substack{z\to \mathrm{i}\tau \\ \Re z>0}} \langle y, \Re\hat{\mu}(z)y\rangle > 0. \tag{4.2}$$

As a matter of fact, because $\inf_{y\in\mathbf{C}^n, |y|=1}\langle y, \Re\tilde{a}(\tau)y\rangle > 0$, condition (4.2) must even hold uniformly in y in the sense that

$$\inf_{\substack{y\in\mathbf{C}^n \\ |y|=1}} \liminf_{\substack{z\to \mathrm{i}\tau \\ \Re z>0}} \langle y, \Re\hat{\mu}(z)y\rangle > 0. \tag{4.3}$$

Interestingly enough, also the converse is true.

4.5 Theorem
Let $\mu \in PT(\mathbf{R}^+; \mathbf{C}^{n\times n})$. Then μ is of strict positive type iff it satisfies (4.3) for all $\tau \in \mathbf{R}$.

Proof Above we commented on the necessity of (4.3).

Conversely, suppose that (4.3) holds for each $\tau \in \mathbf{R}$, and define v by

$$v(\tau) \stackrel{\mathrm{def}}{=} \inf_{\substack{y\in\mathbf{C}^n \\ |y|=1}} \liminf_{\substack{z\to \mathrm{i}\tau \\ \Re z>0}} \langle y, \Re\hat{\mu}(z)y\rangle.$$

Then $v(\tau) > 0$ for all $\tau \in \mathbf{R}$, and, owing to the fact that we allow tangential approach in (4.3), v is lower semicontinuous. If we define

$$h(\tau) = \min\Big\{ \inf_{|\omega|\le|\tau|} v(\omega), \mathrm{e}^{-|\tau|} \Big\}, \qquad \tau \in \mathbf{R},$$

then for each $\tau \in \mathbf{R}$, $0 < h(\tau) \le v(\tau)$ (and, of course, h is even, integrable on $\mathbf{R}$ and nonincreasing on $\mathbf{R}^+$). Next, we construct an even function K that is nonincreasing and convex on $\mathbf{R}^+$, and that satisfies $0 < K(\tau) \le h(\tau)$ for all $\tau \in \mathbf{R}$. The simplest method to get such a function K is probably to define K to be the supremum of all the functions f that are even, nonincreasing, and convex on $\mathbf{R}^+$, and satisfy $0 \le f(\tau) \le h(\tau)$ for all $\tau \in \mathbf{R}$. Alternatively, one can define K explicitly to be

$$K(\omega) \stackrel{\text{def}}{=} \inf_{0\le\tau\le|\omega|\le T} \Big\{ \tfrac{T-|\omega|}{T-\tau} h(\tau) + \tfrac{|\omega|-\tau}{T-\tau} h(T) \Big\}.$$

By Proposition 3.1, this K, considered on $\mathbf{R}^+$, is of positive type. Let k be the inverse Fourier transform of K, i.e., $k(t) = \frac{1}{2\pi}\int_{\mathbf{R}} \mathrm{e}^{\mathrm{i}\omega t} K(\omega)\,\mathrm{d}\omega$. Note that k is real and continuous, in fact C^∞. From Theorem 2.7 it therefore follows that $k \in L^1(\mathbf{R};\mathbf{R})$ (and that $k \ge 0$).

If we define $a(t) = 2k(t)I$ for $t \ge 0$, then $\Re\tilde{a}(\omega) = K(\omega)I$. Because $K(\omega) \le h(\omega)$, and since $\hat{a}(z) \to 0$ as $|z| \to \infty$ in the half plane $\Re z \ge 0$ (see Lemma **2**.2.8), it follows from Theorem 2.6 (i) and (v) that $\mu(\mathrm{d}t) - a(t)\,\mathrm{d}t$ is of positive type, i.e., that μ is of strict positive type. □

The preceding proof shows, in addition, that one may, without loss of generality, take the function a in Definition 4.4 to be the identity matrix multiplied by a scalar, nonnegative C^∞-function.

Earlier, in Proposition 4.3, we observed that 'most' convex kernels are of strong positive type. However, there are some that are not of strong positive type, but are of strict positive type.

4.6 Proposition
Let $a \in L^1_{\mathrm{loc}}(\mathbf{R}^+;\mathbf{C}^{n\times n})$ and suppose that for every $y \in \mathbf{C}^n$ the function $t \mapsto \langle y, a(t)y\rangle$ is nonnegative, nonincreasing, and convex on $(0,\infty)$. Then a is of strict positive type iff there exist no $y \in \mathbf{C}^n$ and $T > 0$ such that the function $t \mapsto \langle y, a(t)y\rangle$ is linear in all the intervals $[(k-1)T, kT]$, $k \in \mathbf{N}$.

In the exceptional case, denoting the measure derivative of a' by α, we see that the support of $\langle y, \alpha y\rangle$ is contained in the set $\{\, kT \mid k \in \mathbf{N} \,\}$. A measure of this type is usually called arithmetic (cf. Section **15**.7). Thus, the conclusion of Proposition 4.6 says that a is of strict positive type iff none of the measures $\langle y, \alpha y\rangle$ is arithmetic (where α is the second derivative of a).

Proof of Proposition 4.6 First, assume that a, $a' \in L^1(\mathbf{R}^+;\mathbf{C}^{n\times n})$, and that the measure derivative α of a' satisfies $\alpha \in M(\mathbf{R}^+;\mathbf{C}^{n\times n})$. Clearly, because of the monotonicity of a,

$$\Re\tilde{a}(0) = \int_{\mathbf{R}^+} a(s)\,\mathrm{d}s \succ 0 \ \text{ iff } \ a(0) \succ 0.$$

For $\omega \ne 0$ we can perform the usual integrations by parts to get

$$\Re\tilde{a}(\omega) = \frac{1}{\omega^2}\int_{\mathbf{R}^+} \big(1 - \cos(\omega t)\big)\alpha(\mathrm{d}t).$$

Thus, $\langle y, \Re\tilde{a}(\omega)y\rangle = 0$ for some $\omega \ne 0$ iff the measure $\langle y, \alpha y\rangle$ vanishes on the intervals $(2\pi(k-1)/\omega, 2\pi k/\omega)$ for all $k \in \mathbf{N}$. Therefore, under the extra size conditions imposed above, the conclusion of the proposition is valid.

Next, let us consider the general case where a is larger at infinity and less smooth. First, assume that there exist no $y \in \mathbf{C}^n$ and $T > 0$ such that $\langle y, a(t)y\rangle$ is linear in all the intervals $[(k-1)T, kT]$, $k \in \mathbf{N}$. Then α is not arithmetic. As in the proof of Proposition 4.3, let $\eta(t) = t^2\mathrm{e}^{-t}$, $t \geq 0$, and define

$$b(t) = \int_t^\infty \int_{(s,\infty)} \eta(u)\alpha(\mathrm{d}u)\,\mathrm{d}s.$$

Then b is nonnegative, nonincreasing, and convex, and satisfies the additional assumptions imposed on a; moreover, note that the measure derivatives α and $\eta\alpha$ have identical supports. Thus, b is of strict positive type. Define

$$c(t) = a(t) - b(t) = a(\infty) + \int_t^\infty \int_{(s,\infty)} (1-\eta(u))\alpha(\mathrm{d}u)\,\mathrm{d}s.$$

Then c is of positive type. Consequently, $a = b + c$ is of positive type as well.

Finally, assume that there exist $y \in \mathbf{C}^n$ and $T > 0$ such that the function $t \mapsto \langle y, a(t)y\rangle$ is linear in all the intervals $[(k-1)T, kT]$, $k \in \mathbf{N}$. Define $b = \langle y, ay\rangle$ and $\beta = \langle y, \alpha y\rangle$. Then the support of β is contained in the set $\{\, kT \mid k \in \mathbf{N}\,\}$. Because of the piecewise linearity, we have $b(0)$, $|b'(0)| < \infty$, and thus, $b' \in L^1(\mathbf{R}^+)$ and $\beta \in M(\mathbf{R}^+)$. Therefore, if $\omega \neq 0$,

$$\liminf_{z\to i\omega, \Re z>0} \Re\hat{b}(z) = \lim_{z\to i\omega, \Re z>0} \Re\hat{b}(z) = \omega^{-2}\int_{\mathbf{R}^+} (1-\cos\omega t)\beta(\mathrm{d}t).$$

In particular, taking $\omega = \omega_0 = 2\pi/T$, we obtain $\liminf_{z\to i\omega_0, \Re z>0} \Re\hat{b}(z) = 0$. Now it follows from Theorem 4.5 that b is not of strict positive type. □

5. Anti-Coercive Measures

In the preceding sections we defined measures of positive type, and measures of strong and strict positive type. In addition, we gave some equivalence results and a few examples. In this section we pursue the study of these measures further. In particular, we show, see (5.3) and Definition 5.4, that under additional assumptions these measures satisfy a certain inverse coercivity inequality.

Our first result is closely related to Theorem 2.4.

5.1 Theorem
Let $\mu \in PT(\mathbf{R}^+;\mathbf{C}^{n\times n})$ and $\nu \in M_{\mathrm{loc}}(\mathbf{R}^+;\mathbf{C}^{m\times n})$ satisfy $\int_{\mathbf{R}^+} \mathrm{e}^{-\epsilon t}(|\mu|(\mathrm{d}t)+|\nu|(\mathrm{d}t)) < \infty$ for all $\epsilon > 0$, and let q be a positive constant. Then the following conditions are equivalent.

(i) *If $\varphi \in L^2(\mathbf{R};\mathbf{C}^n)$ has compact support, then $\nu * \varphi \in L^2(\mathbf{R};\mathbf{C}^m)$, and $\|\nu * \varphi\|^2_{L^2(\mathbf{R})} \le q[\![\mu,\varphi]\!]$.*

(ii) *$\int_0^T \left|(\nu*\varphi)(t)\right|^2 \mathrm{d}t \le q[\![\mu,\chi_{[0,T]}\varphi]\!]$ for every $\varphi \in L^2_{\mathrm{loc}}(\mathbf{R}^+;\mathbf{C}^n)$ and every $T > 0$.*

(iii) *$\hat{\nu}^*(z)\hat{\nu}(z) \preceq q\Re\hat{\mu}(z)$ for $\Re z > 0$.*

(iv) *For some $\epsilon > 0$, $\hat{\nu}^*(z)\hat{\nu}(z) \preceq q\Re\hat{\mu}(z)$ in the strip $0 < \Re z < \epsilon$.*

(v) *$\liminf_{z\to \mathrm{i}\tau,\Re z>0}\big(q\Re\hat{\mu}(z) - \hat{\nu}^*(z)\hat{\nu}(z)\big) \succeq 0$ for all $\tau \in \mathbf{R}$, and $\liminf_{|z|\to\infty,\Re z>0}\big(q\Re\hat{\mu}(z) - \hat{\nu}^*(z)\hat{\nu}(z)\big) \succeq 0$.*

In the case $n = 1$ the conditions (iii)–(v) look a little simpler, because then $\hat{\nu}^*(z)\hat{\nu}(z) = |\hat{\nu}(z)|^2$, where $|\cdot|$ is the norm in $\mathbf{C}^m$.

Proof of Theorem 5.1 As in the proof of Theorem 2.4, we note that (iii) trivially implies (v) and that the converse implication follows from the maximum principle. The implications (i)$\Rightarrow$(ii) and (iii)$\Rightarrow$(iv) are obvious. Consequently, if we show that (ii)$\Rightarrow$(iii) and (iv)$\Rightarrow$(i), then we are done.

(ii)$\Rightarrow$(iii): Without loss of generality, we may take $n = 1$. This is true, because (iii) is a statement about the behaviour of the (scalar) function $q\langle y, \Re\hat{\mu}(z)y\rangle - \left|\hat{\nu}(z)y\right|^2$, where $y \in \mathbf{C}^n$ is arbitrary, and (ii) implies

$$\int_0^T |(\nu y * \varphi)(t)|^2 \mathrm{d}t \le q[\![\langle y,\mu y\rangle, \chi_{[0,T]}\varphi]\!]$$

for arbitrary $y \in \mathbf{C}^n$. Note that $\langle y,\mu y\rangle$ is a scalar measure, whereas νy is $\mathbf{C}^m$-valued.

Fix $z = \sigma + \mathrm{i}\tau$ with $\sigma > 0$. Define $R(T)$ as in the proof of Theorem 2.4, and define $S(T)$ by

$$S(T) = \int_0^T \left| \int_{[0,t]} \nu(\mathrm{d}s)\mathrm{e}^{z(t-s)} \right|^2 \mathrm{d}t.$$

By (ii), $S(T) \le qR(T)$. Since we already know, from the proof of Theorem 2.4, that $\lim_{T\to\infty} 2\sigma\mathrm{e}^{-2\sigma T}R(T) = \Re\hat{\mu}(z)$, it suffices to show that

$$\lim_{T\to\infty} 2\sigma^{-2\sigma T}S(T) = |\hat{\nu}(z)|^2. \tag{5.1}$$

We have

$$2\sigma\mathrm{e}^{-2\sigma T}S(T) = \int_0^T 2\sigma\mathrm{e}^{-2\sigma(T-t)} \left| \int_{[0,t]} \mathrm{e}^{-zs}\nu(\mathrm{d}s) \right|^2 \mathrm{d}t = (a*b)(T),$$

where $a(t) = 2\sigma\mathrm{e}^{-2\sigma t}$ satisfies $a \in L^1(\mathbf{R}^+;\mathbf{R})$, with $\int_0^\infty a(t)\,\mathrm{d}t = 1$, and where $b(t) = \left|\int_{[0,t]} \nu(\mathrm{d}s)\mathrm{e}^{-zs}\right|^2$ tends to $|\hat{\nu}(z)|^2$ as $t \to \infty$. Thus, the claim (5.1) holds, and hence (ii) implies (iii).

(iv) ⇒(i): Analogously to the proof of Theorem 2.4, it suffices to consider the case where μ, ν are finite measures, i.e., $\mu \in M(\mathbf{R}^+; \mathbf{C}^{n\times n})$ and $\nu \in M(\mathbf{R}^+; \mathbf{C}^{m\times n})$. Then (iv) implies that

$$q\Re\hat{\mu}(\mathrm{i}\omega) \succeq \hat{\nu}^*(\mathrm{i}\omega)\hat{\nu}(\mathrm{i}\omega), \qquad \omega \in \mathbf{R}. \tag{5.2}$$

Let $\varphi \in L^2(\mathbf{R}; \mathbf{C}^n)$ have compact support. It follows from Parseval's identity and from (5.2) that

$$\begin{aligned}\int_{\mathbf{R}} |(\nu * \varphi)(t)|^2 \mathrm{d}t &= \frac{1}{2\pi}\int_{\mathbf{R}} \big|\hat{\nu}(\mathrm{i}\omega)\hat{\varphi}(\mathrm{i}\omega)\big|^2 \mathrm{d}\omega \\ &\le \frac{q}{2\pi}\int_{\mathbf{R}} \langle \hat{\varphi}(\mathrm{i}\omega), \Re\hat{\mu}(\mathrm{i}\omega)\hat{\varphi}(\mathrm{i}\omega)\rangle \,\mathrm{d}\omega = q[\![\mu, \varphi]\!],\end{aligned}$$

and so (i) holds. □

The following modification of the previous theorem parallels Theorem 2.5.

5.2 Theorem
Let $\mu \in PT(\mathbf{R}^+; \mathbf{C}^{n\times n})$ and $\nu \in M_{\mathrm{loc}}(\mathbf{R}; \mathbf{C}^{m\times n})$, and denote the positive measure $\Re\tilde{\mu}$ by α. Then (i) *of Theorem 5.1 is satisfied for some $q \ge 0$ if and only if ν induces a tempered distribution, $\tilde{\nu}$ is induced by a function $a \in L^2_{\mathrm{loc}}(\mathbf{R}; \mathbf{C}^{m\times n})$, and the measure β, defined by $\beta(\mathrm{d}\omega) = q\alpha(\mathrm{d}\omega) - a^*(\omega)a(\omega)\,\mathrm{d}\omega$, satisfies $\beta(E) \succeq 0$ for every bounded Borel set $E \subset \mathbf{R}$.*

One particular difference between this result and Theorem 5.1 is that here we do not require ν to vanish on $(-\infty, 0)$.

Proof of Theorem 5.2 Suppose that ν induces a tempered distribution with a Fourier transform a locally in L^2, and that the measure β defined above satisfies $\beta \succeq 0$. As in the proof of Theorem 2.5, one observes that in order to prove that condition (i) in Theorem 5.1 holds it suffices to consider the case where φ is a C^∞-function with compact support. Since by assumption $\mu \in PT(\mathbf{R}^+; \mathbf{C}^{n\times n})$—hence $\tilde{\mu} \in S'(\mathbf{R}; \mathbf{C}^{n\times n})$—the expression

$$\int_{\mathbf{R}} q\langle \tilde{\varphi}(\omega), \alpha(\mathrm{d}\omega)\tilde{\varphi}(\omega)\rangle - \int_{\mathbf{R}} \big|a(\omega)\tilde{\varphi}(\omega)\big|^2 \mathrm{d}\omega$$

is well defined and nonnegative. Therefore,

$$0 \le \frac{1}{2\pi}\int_{\mathbf{R}} \big|a(\omega)\tilde{\varphi}(\omega)\big|^2 \mathrm{d}\omega \le q[\![\mu, \varphi]\!],$$

and so $a\tilde{\varphi} \in L^2(\mathbf{R}; \mathbf{C}^m)$. This function, however, equals the Fourier transform of $\nu * \varphi$ and consequently, by Parseval's identity,

$$\int_{\mathbf{R}} \big|(\nu * \varphi)(t)\big|^2 \mathrm{d}t \le q[\![\mu, \varphi]\!].$$

To prove the converse statement, let us first observe that the fact that $\nu * \varphi \in L^2(\mathbf{R}; \mathbf{C}^m) \subset S'(\mathbf{R}; \mathbf{C}^m)$ for every $\mathbf{C}^m$-valued C^∞-function φ with compact support implies that ν induces a tempered distribution (see Schwartz [1], Théorème VI, p. 239). The same fact implies that the Fourier transform of ν must belong locally to L^2, because the transform of $\nu * \varphi$ is $\tilde{\nu}\tilde{\varphi}$, and this must be an L^2-function. This means that one can define β as in the statement of Theorem 5.2. Reversing the computation above, we find that, for all test functions φ, we have

$$\int_{\mathbf{R}} \langle \tilde{\varphi}(\omega), \beta(\mathrm{d}\omega)\tilde{\varphi}(\omega)\rangle \geq 0.$$

We claim that this implies $\beta \succeq 0$, i.e., that for every $y \in \mathbf{C}^n$ the scalar measure $\langle y, \beta(\mathrm{d}\omega)y\rangle$ is positive. Choose $y \in \mathbf{C}^n$, and let $\tilde{\varphi}(\omega) = \tilde{\psi}(\omega)y$, where $\psi(t)$ is an arbitrary scalar test function. Then

$$\int_{\mathbf{R}} |\tilde{\psi}(\omega)|^2 \langle y, \beta(\mathrm{d}\omega)y\rangle \geq 0.$$

However, from this it follows (see Schwartz [1], p. 276-277) that $\langle y, \beta(\mathrm{d}\omega)y\rangle$ defines a positive measure. □

Taking $\nu(\mathrm{d}t) = b(t)\,\mathrm{d}t$ and $\mu(\mathrm{d}t) = \mathrm{e}^{-t} I\,\mathrm{d}t$ in Theorem 5.2, we get the following corollary.

5.3 Corollary

Let $b \in L^2(\mathbf{R}^+; \mathbf{C}^{m\times n})$ satisfy

$$q \stackrel{\text{def}}{=} \operatorname*{ess\,sup}_{\omega\in\mathbf{R}} (1+\omega^2)\big|\tilde{b}(\omega)\big|^2 < \infty.$$

Then, for each $\varphi \in L^2(\mathbf{R}; \mathbf{C}^n)$ with compact support, the function $b * \varphi$ belongs to $L^2(\mathbf{R}; \mathbf{C}^m)$, and

$$\|b * \varphi\|^2_{L^2(\mathbf{R})} \leq q[\![\mathbf{e}, \varphi]\!],$$

where $\mathbf{e}(t) = \mathrm{e}^{-t} I$ for $t \in \mathbf{R}^+$.

We proceed to consider Theorem 5.1 in the case where $\mu = \nu$. The crucial inequality (i) then reads

$$\|\mu * \varphi\|^2_{L^2(\mathbf{R})} \leq q[\![\mu, \varphi]\!], \tag{5.3}$$

or, more explicitly,

$$\int_{\mathbf{R}} |(\mu * \varphi)(t)|^2 \mathrm{d}t \leq q\Re \int_{\mathbf{R}} \langle \varphi(t), (\mu * \varphi)(t)\rangle\,\mathrm{d}t.$$

Denote the (possibly multivalued) inverse of the convolution operator $\varphi \mapsto \mu * \varphi$ by A, and denote the scalar product in $L^2(\mathbf{R})$ by $(\cdot,\cdot)$. Then the inequality may be written $q^{-1}\|\varphi\|^2_{L^2(\mathbf{R})} \leq \Re(\varphi, A\varphi)$ for φ in the domain of A. Consequently, A is what has been called coercive; see, e.g., Deimling [2], p. 100. This motivates the use of the name 'anti-coercivity' for this property.

5.4 Definition
A measure $\mu \in M_{\text{loc}}(\mathbf{R}^+;\mathbf{C}^{n\times n})$ is said to be of anti-coercive type, with coercivity constant $q>0$, if, for every $\varphi \in L^2(\mathbf{R};\mathbf{C}^n)$ with compact support, we have $\mu * \varphi \in L^2(\mathbf{R};\mathbf{C}^n)$ and $\|\mu*\varphi\|^2_{L^2(\mathbf{R})} \le q[\![\mu,\varphi]\!]$. A function $a \in L^1_{\text{loc}}(\mathbf{R}^+;\mathbf{C}^{n\times n})$ is said to be of anti-coercive type if the measure $\mu(\mathrm{d}s) = a(s)\,\mathrm{d}s$ is of anti-coercive type.

The set of measures of anti-coercive type is denoted by $PT_{\text{aco}}(\mathbf{R}^+;\mathbf{C}^{n\times n})$.

Of course, every measure that is of anti-coercive type is of positive type, but it need not be of strong or of strict positive type.

Let us record the following immediate fact.

5.5 Lemma
Let $\mu \in M_{\text{loc}}(\mathbf{R}^+;\mathbf{C}^{n\times n})$ be of anti-coercive type, with coercivity constant q, and let $\varphi \in L^2(\mathbf{R}^+;\mathbf{C}^n)$. Then $\mu*\varphi \in L^2(\mathbf{R}^+;\mathbf{R}^n)$, and $\|\mu*\varphi\|_{L^2(\mathbf{R}^+)} \le q\|\varphi\|_{L^2(\mathbf{R}^+)}$.

Proof For all $T>0$, we have

$$\begin{aligned}\|\mu*\varphi\|^2_{L^2(0,T)} &= \|\mu*(\chi_{[0,T]}\varphi)\|_{L^2(0,T)}\\ &\le q[\![\mu,\chi_{[0,T]}\varphi]\!] \le q\|\mu*\varphi\|_{L^2(0,T)}\|\varphi\|_{L^2(0,T)}.\end{aligned}$$

Dividing by $\|\mu*\varphi\|_{L^2(0,T)}$, and letting $T\to\infty$, we get the conclusion. □

Of course, we may combine Theorems 5.1 and 5.2 with Definition 5.4, but for completeness we formulate an additional equivalence result.

5.6 Theorem
Let $\mu \in PT(\mathbf{R}^+;\mathbf{C}^{n\times n})$ and let $q>0$. Then the following conditions (i)–(v) *are equivalent:*

(i) *$\mu \in PT_{\text{aco}}(\mathbf{R}^+;\mathbf{C}^{n\times n})$ with coercivity constant q;*
(ii) *$\hat\mu^*(z)\hat\mu(z) \preceq q\Re\hat\mu(z)$ for $\Re z>0$;*
(iii) *for some $\epsilon>0$, $\hat\mu^*(z)\hat\mu(z) \preceq q\Re\hat\mu(z)$ in the strip $0<\Re z<\epsilon$;*
(iv) *$\liminf_{z\to i\tau,\Re z>0}[q\Re\hat\mu(z) - \hat\mu^*(z)\hat\mu(z)] \succeq 0$ for every $\tau\in\mathbf{R}$, and $\liminf_{|z|\to\infty,\Re z>0}[q\Re\hat\mu(z)-\hat\mu^*(z)\hat\mu(z)] \succeq 0$;*
(v) *$|\tilde\mu(\omega)| \le q$ a.e., and $\tilde\mu^*(\omega)\tilde\mu(\omega) \preceq q\Re\tilde\mu(\omega)$ for almost all $\omega\in\mathbf{R}$.*

If, in addition, $\det[qI-\mu(\{0\})]\neq 0$, so that the equation

$$q\lambda = \mu + \mu*\lambda = \mu + \lambda*\mu \tag{5.4}$$

has a solution $\lambda \in M_{\text{loc}}(\mathbf{R}^+;\mathbf{C}^{n\times n})$, then all the preceding conditions are equivalent to
(vi) *$\lambda \in PT(\mathbf{R}^+;\mathbf{C}^{n\times n})$.*

Again, in the scalar case we have $\hat\mu^*(z)\hat\mu(z) = |\hat\mu(z)|^2$, so in this case (ii)–(v) can be written in a slightly simpler form.

To interpret (vi), observe two simple facts: (a) the resolvent of a kernel of positive type (exists and) is of positive type (see Exercise 18); (b) the relation (5.4) says that $q^{-1}\mu$ is the resolvent of λ. Consequently, (vi) allows us to conclude that the resolvent of a kernel of positive type is of anti-coercive type.

To see the necessity of the assumption that λ exists, take the scalar kernel $\mu(\mathrm{d}t) = \delta(\mathrm{d}t) - \mathrm{e}^{-t}\,\mathrm{d}t$. Simple calculations (check (iii)) yield that μ is of anti-coercive type with coercivity constant $q = 1$. But (5.4) with $q = 1$ has no solution in $M_{\mathrm{loc}}(\mathbf{R}^+)$.

Recall that there is a solution λ of (5.4) if and only if $\det[qI - \mu(\{0\})] \neq 0$ (cf. Theorem **4**.1.5 and Exercise 18).

Proof of Theorem 5.6 The proof of the equivalence of (i)–(v) is straightforward and it follows the general pattern used, e.g., in the proof of Theorem 2.6. We leave the bulk of this proof to the reader, but there is one point worth mentioning. When applying Theorem 5.2, one only gets $\tilde{\mu} \in L^2_{\mathrm{loc}}(\mathbf{R}; \mathbf{C}^{n\times n})$, and not $\tilde{\mu} \in L^\infty(\mathbf{R}; \mathbf{C}^{n\times n})$, as required by the first part of (v). The pointwise estimate for $\tilde{\mu}$ can be deduced as follows. Let $y \in \mathbf{C}^n$ with $|y| = 1$. Then for almost all $\omega \in \mathbf{R}$ one has, by the second part of (v),

$$|\tilde{\mu}(\omega)y|^2 \leq q\Re\langle y, \tilde{\mu}(\omega)y\rangle \leq q|\tilde{\mu}(\omega)|.$$

Taking the supremum over all $y \in \mathbf{C}^n$ with $|y| = 1$, we get $|\tilde{\mu}(\omega)|^2 \leq q|\tilde{\mu}(\omega)|$, and this implies that $|\tilde{\mu}(\omega)| \leq q$ a.e.

For the proof of the equivalence of (i) and (vi), assume that λ exists. First, let (i) hold. Choose an arbitrary $\varphi \in L^2(\mathbf{R}; \mathbf{C}^n)$ with compact support. Let ψ be defined by

$$q\psi = \varphi + \lambda * \varphi.$$

Then $\lambda * \varphi = \mu * \psi$ and ψ is the solution of the equation

$$q\psi - \mu * \psi = \varphi. \tag{5.5}$$

Hence

$$\Re\int_{\mathbf{R}} \langle \varphi(s), (\lambda * \varphi)(s)\rangle\,\mathrm{d}s$$
$$= \Re\int_{\mathbf{R}} \big\langle \big(q\psi(s) - (\mu * \psi)(s)\big), (\mu * \psi)(s)\big\rangle\,\mathrm{d}s \geq 0. \tag{5.6}$$

Consequently, λ is of positive type.

Conversely, let $\lambda \in PT(\mathbf{R}^+; \mathbf{C}^{n\times n})$, and let $\psi \in L^2(\mathbf{R}; \mathbf{C}^n)$ be arbitrary with compact support. Define φ by (5.5). Again $\lambda * \varphi = \mu * \psi$, and (5.6) holds. Hence μ is of anti-coercive type. □

The next result is a generalization of Proposition 3.2 (see Exercise 5).

5.7 Proposition

Let $\mu \in M(\mathbf{R}^+; \mathbf{C}^{n\times n})$ be self-adjoint (i.e., $\mu(E)$ is self-adjoint for all Borel sets E), and satisfy $\mu(\{0\}) \succeq \int_{(0,\infty)} |\mu|(\mathrm{d}t)I$. Then μ is of anti-coercive type with coercivity constant $q = 2|\mu(\{0\})| + 6|\mu|((0,\infty))$.

Proof Denote the point mass $\mu(\{0\})$ by μ_0. Since μ is self-adjoint, we have

$$\Re\tilde{\mu}(\omega) = \mu_0 + \int_{(0,\infty)} \cos(\omega t)\,\mu(\mathrm{d}t)$$

and

$$\Im\tilde{\mu}(\omega) = \int_{(0,\infty)} \sin(\omega t)\,\mu(\mathrm{d}t).$$

Hölder's inequality gives

$$|\Im\tilde{\mu}(\omega)|^2 \le |\mu|\big((0,\infty)\big) \int_{(0,\infty)} \sin^2(\omega t)\,|\mu|(\mathrm{d}t).$$

Since $2|\cos(t)| + \sin^2(t) \le 2$ (this is equivalent to the inequality $\big(1 - |\cos(t)|\big)^2 \ge 0$), it follows from our assumption on μ_0 that, for every vector $y \in \mathbf{C}^n$ with $|y| = 1$, we have

$$\begin{aligned}
\int_{(0,\infty)} \sin^2(\omega t)\,|\mu|(\mathrm{d}t) &\le \int_{(0,\infty)} \big(2 - 2|\cos(\omega t)|\big)\,|\mu|(\mathrm{d}t)\\
&\le 2\int_{(0,\infty)} |\mu|(\mathrm{d}t)\,|y|^2 - 2\int_{(0,\infty)} |\cos(\omega t)|\,\langle y, \mu(\mathrm{d}t)y\rangle\\
&\le 2\langle y, \mu_0 y\rangle + 2\int_{(0,\infty)} \cos(\omega t)\,\langle y, \mu(\mathrm{d}t)y\rangle\\
&= 2\langle y, \Re\tilde{\mu}(\omega)y\rangle.
\end{aligned}$$

Thus, we conclude that

$$\big|\Im\tilde{\mu}(\omega)y\big|^2 \le \big|\Im\tilde{\mu}(\omega)\big|^2 \le 2|\mu|\big((0,\infty)\big)\langle y, \Re\tilde{\mu}(\omega)y\rangle.$$

Since $\Re\tilde{\mu}(\omega) \succeq 0$, we have (see (2.2))

$$\big|\Re\tilde{\mu}(\omega)y\big|^2 \le \big|\Re\tilde{\mu}(\omega)\big|\langle y, \Re\tilde{\mu}(\omega)y\rangle \le |\mu|(\mathbf{R}^+)\langle y, \Re\tilde{\mu}(\omega)y\rangle,$$

and, therefore,

$$\begin{aligned}
|\tilde{\mu}(\omega)y|^2 &\le 2\big|\Re\tilde{\mu}(\omega)y\big|^2 + 2\big|\Im\tilde{\mu}(\omega)y\big|^2\\
&\le \Big(2|\mu(\{0\})| + 6|\mu|\big((0,\infty)\big)\Big)\langle y, \Re\tilde{\mu}(\omega)y\rangle.
\end{aligned}$$

Now use Theorem 5.6. □

Our next result gives conditions under which monotone functions are of anti-coercive type.

5.8 Proposition

Let $a \in L^1(\mathbf{R}^+;\mathbf{C}^{n\times n})$ and assume that at least one of the conditions (i), (ii), *or* (iii) *holds:*

(i) *for each $y \in \mathbf{C}^n$ the functions $t \mapsto \langle y, a(t)y\rangle$ and $t \mapsto \langle y, -a'(t)y\rangle$ are nonnegative, nonincreasing, and convex;*

(ii) $a \in PT_{\mathrm{strong}}(\mathbf{R}^+;\mathbf{C}^{n\times n}) \cap BV(\mathbf{R}^+;\mathbf{C}^{n\times n})$;

(iii) $a = a_0 I$ *where* a_0 *is a real-valued, nonnegative, and convex function satisfying* $\limsup_{\epsilon\downarrow 0}(\int_0^\epsilon a_0(s)\,\mathrm{d}s)^2/(\int_{(0,\epsilon)} t^2\alpha(\mathrm{d}t)) < \infty$, *where* α *is the measure derivative of* a_0'.

Then a *is of anti-coercive type.*

If $\alpha(\mathrm{d}t) = a_0''(t)\,\mathrm{d}t$ with $a_0''(t)$ behaving roughly as $t^{-2}a_0(t)$ for small t, then (iii) reduces to $a_0 \in L^1(0,1)$.

Proof of Proposition 5.8 (i) This follows from Theorem **20**.2.4 and we see that the coercivity constant is $4|\int_0^\infty a(s)\,\mathrm{d}s|$.

(ii) Because a is both integrable and of strong positive type, there is an $\epsilon > 0$ for which $\Re\tilde{a}(\omega) \succeq \epsilon/(1+\omega^2)I$ for all $\omega \in \mathbf{R}$. Since a is of bounded variation, one has

$$|\tilde{a}(\omega)| = O\Big(\frac{1}{|\omega|}\Big) \text{ as } |\omega| \to \infty,$$

and this, together with the boundedness of $\tilde{a}(\omega)$, implies that $|\tilde{a}(\omega)|^2 \le M(1+\omega^2)^{-1}$ for some constant M. Therefore, a is anti-coercive with coercivity constant M/ϵ.

(iii) By Lemma **6**.2.2,

$$|\tilde{a}_0(\omega)| \le 2\int_0^{1/|\omega|} a_0(t)\,\mathrm{d}t, \qquad \omega \ne 0.$$

It is also true that (cf. Exercise 6)

$$\Re\tilde{a}_0(\omega) \ge \frac{1}{\omega^2}\int_{(0,1/|\omega|)} \big(1-\cos(\omega t)\big)\alpha(\mathrm{d}t);$$

which, together with the fact that $1-\cos t > t^2/3$ for $0 \le t \le 1$, implies

$$\liminf_{|\omega|\to\infty} \frac{\Re\tilde{a}_0(\omega)}{\int_{(0,1/|\omega|)} t^2\alpha(\mathrm{d}t)} \ge \frac{1}{3}.$$

Thus, by assumption (iii), $|\tilde{a}_0(\omega)|^2 \le q\Re\tilde{a}_0(\omega)$ for some constant q and for ω greater than some ω_0. Use Proposition 4.6 to obtain the same inequality between $\tilde{a}_0$ and $\Re\tilde{a}_0$ on the interval $|\omega| \le \omega_0$. (Note that the assumptions made in (iii) exclude the exceptional case of Proposition 4.6.) The equivalence of (i) and (v) in Theorem 5.6 now gives the desired assertion. □

6. Further Inequalities

In the previous section our analysis was based on Theorem 5.1, which only gives us bounds on the L^2-norms of various convolutions. Under appropriate conditions, it is possible to get bounds of the type

$$\Big|\int_{\mathbf{R}} \langle b(t), \varphi(t)\rangle\,\mathrm{d}t\Big|^2 \le q[\![\mu,\varphi]\!]$$

and

$$\sup_{t\in\mathbf{R}}\left|\int_{\mathbf{R}}\langle b(s),\varphi(t-s)\rangle\,\mathrm{d}s\right|^2\le q[\![\mu,\varphi]\!],$$

where b is a suitable function in L^2_{loc}. For a moment, let us argue formally, avoiding all technical details and taking the dimension n to be 1, to see what type of conditions are needed on $\tilde{b}$ for inequalities of this type to make sense. It turns out that the condition that we need is almost the same in both cases, so let us consider only, e.g., the case in which we want a bound on $\int_{\mathbf{R}} b(t)\overline{\varphi}(t)\,\mathrm{d}t$.

It follows from Parseval's identity that

$$\left|\int_{\mathbf{R}} b(t)\overline{\varphi}(t)\,\mathrm{d}t\right|=\left|\frac{1}{2\pi}\int_{\mathbf{R}}\tilde{b}(\omega)\overline{\tilde{\varphi}(\omega)}\,\mathrm{d}\omega\right|\le\frac{1}{2\pi}\int_{\mathbf{R}}\left|\tilde{b}(\omega)\overline{\tilde{\varphi}(\omega)}\right|\mathrm{d}\omega.$$

We also have (assuming $\Re\tilde{\mu}$ to be a locally absolutely continuous function)

$$[\![\mu,\varphi]\!]=\frac{1}{2\pi}\int_{\mathbf{R}}\Re\tilde{\mu}(\omega)|\tilde{\varphi}(\omega)|^2\mathrm{d}\omega.$$

If we define γ as $\gamma(\omega)=|\tilde{b}(\omega)|/\Re\tilde{\mu}(\omega)$, and use Hölder's inequality, we get

$$\left\{\int_{\mathbf{R}}\left|\tilde{b}(\omega)\overline{\tilde{\varphi}(\omega)}\right|\mathrm{d}\omega\right\}^2\le\left\{\int_{\mathbf{R}}\Re\tilde{\mu}(\omega)\gamma(\omega)^2\mathrm{d}\omega\right\}\left\{\int_{\mathbf{R}}\Re\tilde{\mu}(\omega)|\tilde{\varphi}(\omega)|^2\mathrm{d}\omega\right\}.$$

Thus, we conclude that the crucial assumption is

$$\int_{\mathbf{R}}\Re\tilde{\mu}(\omega)\gamma(\omega)^2\mathrm{d}\omega<\infty. \tag{6.1}$$

The preceding argument can be made precise, and it can be applied in the case $n>1$. However, before we do this, let us show that the two different estimates that we mentioned above are essentially equivalent.

6.1 Lemma
Let $\mu\in PT(\mathbf{R}^+;\mathbf{C}^{n\times n})$, $b\in L^2_{\mathrm{loc}}(\mathbf{R};\mathbf{C}^n)$ and $q\ge 0$. Then

$$\left|\int_{\mathbf{R}}\langle b(t),\varphi(t)\rangle\,\mathrm{d}t\right|^2\le q[\![\mu,\varphi]\!] \tag{6.2}$$

for all $\varphi\in L^2(\mathbf{R};\mathbf{C}^n)$ with compact support if and only if (μ^* is the adjoint of μ)

$$\sup_{t\in\mathbf{R}}\left|\int_{\mathbf{R}}\langle b(s),\varphi(t-s)\rangle\,\mathrm{d}s\right|^2\le q[\![\mu^*,\varphi]\!] \tag{6.3}$$

for the same class of functions φ.

Proof Fix $t\in\mathbf{R}$, and define $\psi(s)=\varphi(t-s)$. Then the inequality in (6.3) becomes

$$\left|\int_{\mathbf{R}}\langle b(s),\psi(s)\rangle\,\mathrm{d}s\right|^2\le q[\![\mu^*,\varphi]\!].$$

Thus, to prove the lemma, it suffices to show that $[\![\mu, \psi]\!] = [\![\mu^*, \varphi]\!]$. This is easily accomplished. Extend μ to $\mathbf{R}$ by defining $\mu(E) = \mu(E \cap \mathbf{R}^+)$ for every Borel set $E \subset \mathbf{R}$. Let $\check{\mu}$ be the measure $\check{\mu}(E) = \mu^*(-E)$. Then

$$\begin{aligned}
[\![\mu, \psi]\!] &= \Re \int_{\mathbf{R}} \langle \psi(v), (\mu * \psi)(v) \rangle \, \mathrm{d}v \\
&= \Re \int_{\mathbf{R}} \langle (\check{\mu} * \psi)(v), \psi(v) \rangle \, \mathrm{d}v \\
&= \Re \int_{\mathbf{R}} \langle \psi(v), (\check{\mu} * \psi)(v) \rangle \, \mathrm{d}v \\
&= \Re \int_{\mathbf{R}} \Big\langle \psi(v), \int_{\mathbf{R}^+} \mu^*(-\mathrm{d}s)\psi(v-s) \Big\rangle \, \mathrm{d}v \\
&= \Re \int_{\mathbf{R}} \Big\langle \varphi(t-v), \int_{\mathbf{R}^+} \mu^*(-\mathrm{d}s)\varphi(t-v+s) \Big\rangle \, \mathrm{d}v \\
&= \Re \int_{\mathbf{R}} \Big\langle \varphi(v), \int_{\mathbf{R}^+} \mu^*(-\mathrm{d}s)\varphi(v+s) \Big\rangle \, \mathrm{d}v \\
&= \Re \int_{\mathbf{R}} \Big\langle \varphi(v), \int_{\mathbf{R}^+} \mu^*(\mathrm{d}s)\varphi(v-s) \Big\rangle \, \mathrm{d}v \\
&= [\![\mu^*, \varphi]\!]. \quad \square
\end{aligned}$$

Let us now go back to our earlier discussion that related the two conditions mentioned in Lemma 6.1 to a Fourier transform condition. First we treat the simpler case where μ is scalar but b and φ are vectors. The expression $[\![\mu, \varphi]\!]$, with μ scalar and φ an n-vector, should be understood as $[\![\mu I, \varphi]\!]$, where I is the $n \times n$ unit matrix.

6.2 Theorem

Let $\mu \in PT(\mathbf{R}^+; \mathbf{C})$, let $b \in L^2_{\mathrm{loc}}(\mathbf{R}; \mathbf{C}^n)$, and let $q \geq 0$. Then the following conditions are equivalent:

(i) $\left|\int_{\mathbf{R}} \langle b(t), \varphi(t) \rangle \, \mathrm{d}t\right|^2 \leq q[\![\mu, \varphi]\!]$ *for all $\varphi \in L^2(\mathbf{R}; \mathbf{C}^n)$ with compact support;*

(ii) $\sup_{t\in\mathbf{R}} \left|\int_{\mathbf{R}} \langle b(s), \varphi(t-s) \rangle \, \mathrm{d}s\right|^2 \leq q[\![\overline{\mu}, \varphi]\!]$ *for all $\varphi \in L^2(\mathbf{R}; \mathbf{C}^n)$ with compact support;*

(iii) $b \in \mathcal{S}'(\mathbf{R}; \mathbf{C}^n)$, *and $\tilde{b}$ is induced by a measure β of the form $\beta(\mathrm{d}\omega) = \gamma(\omega)\alpha(\mathrm{d}\omega)$, where $\alpha = \Re\tilde{\mu}$, and γ is a Borel measurable function satisfying $\frac{1}{2\pi}\int_{\mathbf{R}} |\gamma(\omega)|^2 \alpha(\mathrm{d}\omega) \leq q$.*

Of course, if (iii) holds, then Theorem 6.2 implies that (i) and (ii) hold for functions φ having compact support and vanishing on $(-\infty, 0)$. The converse is not true, e.g., if (i) holds only for functions φ having compact support and vanishing on $(-\infty, 0)$, then (iii) need not be true. (An example is given by Corollary 6.3 below: if in that corollary $b(0) \neq 0$, and if we define b to be zero for $t < 0$, then $\tilde{b}$ does not satisfy (iii).)

Proof of Theorem 6.2 In the scalar case, $\overline{\mu} = \mu^*$ and so the equivalence of (i) and (ii) follows from Lemma 6.1 above.

If (iii) holds and $\varphi \in C^\infty(\mathbf{R}; \mathbf{C}^n)$ has compact support, then (i) is equivalent to

$$\frac{1}{4\pi}\left|\int_{\mathbf{R}} \langle \beta(\mathrm{d}\omega), \tilde{\varphi}(\omega)\rangle\right|^2 \leq \frac{q}{2\pi}\int_{\mathbf{R}} |\tilde{\varphi}(\omega)|^2 \alpha(\mathrm{d}\omega).$$

That this inequality indeed is satisfied follows from Hölder's inequality, because

$$\begin{aligned}\frac{1}{4\pi}\left|\int_{\mathbf{R}} \langle \beta(\mathrm{d}\omega), \tilde{\varphi}(\omega)\rangle\right|^2 &= \frac{1}{4\pi}\left|\int_{\mathbf{R}} \langle \gamma(\omega), \tilde{\varphi}(\omega)\rangle \alpha(\mathrm{d}\omega)\right|^2 \\ &\leq \left\{\frac{1}{2\pi}\int_{\mathbf{R}} |\gamma(\omega)|^2 \alpha(\mathrm{d}\omega)\right\}\left\{\frac{1}{2\pi}\int_{\mathbf{R}} |\tilde{\varphi}(\omega)|^2 \alpha(\mathrm{d}\omega)\right\} \\ &\leq q\frac{1}{2\pi}\int_{\mathbf{R}} |\tilde{\varphi}(\omega)|^2 \alpha(\mathrm{d}\omega).\end{aligned}$$

From this special case one can pass to the general case in the usual way (the set of C^∞-functions with compact support is dense in the set of L^2-functions with compact support).

To complete the proof, it suffices to show that (i) and (ii) together imply (iii). It follows from (ii) that $b \in \mathcal{S}'(\mathbf{R}; \mathbf{C}^n)$ (because $b * \varphi$ is continuous and bounded for all C^∞-functions with compact support; cf. Schwartz [1], Théorème VI, p. 239). Using the notation $\langle\!\langle \cdot, \cdot \rangle\!\rangle$ that we defined after Theorem 2.5, and Plancherel's theorem for distributions, we can, for each C^∞-function φ with compact support, write (i) in the form

$$\frac{1}{4\pi}\left|\langle\!\langle \tilde{\varphi}, \tilde{b} \rangle\!\rangle\right|^2 \leq \frac{q}{2\pi}\int_{\mathbf{R}} |\tilde{\varphi}(\omega)|^2 \alpha(\mathrm{d}\omega),$$

where $\alpha = \Re\tilde{\mu}$. This implies that $\tilde{b}$ is of order zero; hence $\tilde{b}$ is induced by a measure β (cf. Schwartz [1], Théorème XXXI, p. 96). Replacing $\tilde{b}$ by β above, we get

$$\frac{1}{4\pi}\left|\int_{\mathbf{R}} \langle \tilde{\varphi}(\omega), \beta(\mathrm{d}\omega)\rangle\right|^2 \leq \frac{q}{2\pi}\int_{\mathbf{R}} |\tilde{\varphi}(\omega)|^2 \alpha(\mathrm{d}\omega).$$

By continuity, the same inequality is valid if we replace $\tilde{\varphi}$ by an arbitrary continuous function ψ with compact support. The inequality that we get in this way shows that the mapping that takes $\psi \in C(\mathbf{R}; \mathbf{C}^n)$ with compact support into $\int_{\mathbf{R}} \langle \psi(\omega), \beta(\mathrm{d}\omega)\rangle$, is a continuous linear functional from (a dense subset of) the weighted L^2-space $L^2(\mathbf{R}; \alpha; \mathbf{C}^n)$ with the measure α into $\mathbf{C}$, and that the norm of this mapping is at most $(2\pi q)^{1/2}$. However, since $L^2(\mathbf{R}; \alpha; \mathbf{C}^n)$ is a Hilbert space, every linear functional on $L^2(\mathbf{R}; \alpha; \mathbf{C}^n)$ can be represented as an element in the same space, and its operator norm coincides with its Hilbert space norm. Explicitly, this means that there exists a function $\gamma \in L^2(\mathbf{R}; \alpha; \mathbf{C}^n)$ (without loss of generality, we can take it to be Borel measurable) that satisfies

$$\int_{\mathbf{R}} |\gamma(\omega)|^2 \alpha(\mathrm{d}\omega) \le 2\pi q,$$

and such that for all continuous functions ψ with compact support (or more generally, for all functions $\psi \in L^2(\mathbf{R}; \alpha; \mathbf{C}^n)$) one has

$$\int_{\mathbf{R}} \langle \psi(\omega), \beta(\mathrm{d}\omega) \rangle = \int_{\mathbf{R}} \langle \psi(\omega), \gamma(\omega) \rangle \alpha(\mathrm{d}\omega).$$

Thus, $\beta(\mathrm{d}\omega) = \gamma(\omega)\alpha(\mathrm{d}\omega)$ as claimed. □

We have the following corollary to Theorem 6.2.

6.3 Corollary
Let $b \in L^2(\mathbf{R}^+; \mathbf{C}^n)$ be locally absolutely continuous with $b' \in L^2(\mathbf{R}^+; \mathbf{C}^n)$. Define

$$q = |b(0)|^2 + \int_{\mathbf{R}^+} \Big(|b(t)|^2 + |b'(t)|^2 \Big) \, \mathrm{d}t.$$

Then, for every $\varphi \in L^2_{\mathrm{loc}}(\mathbf{R}^+; \mathbf{C}^n)$ and every $T > 0$,

$$\left| \int_0^T \big\langle b(t), \varphi(t) \big\rangle \, \mathrm{d}t \right|^2 \le q [\![\mathbf{e}, \chi_{[0,T]}\varphi]\!]$$

and

$$\left| \int_0^T \big\langle b(T-t), \varphi(t) \big\rangle \, \mathrm{d}t \right|^2 \le q [\![\mathbf{e}, \chi_{[0,T]}\varphi]\!],$$

where $\mathbf{e}(t) = \mathrm{e}^{-t}$ for $t \in \mathbf{R}^+$.

Observe that if μ is of strong positive type then one can replace $[\![\mathbf{e}, \chi_{[0,T]}\varphi]\!]$ in the above by $[\![\mu, \chi_{[0,T]}\varphi]\!]$, provided one modifies the constant q. The same remark applies to Corollary 6.4 below. In Chapters 17–19 we denote the expression $[\![\mu, \chi_{[0,T]}\varphi]\!]$ by $Q(\mu, \varphi, T)$.

Proof of Corollary 6.3 Define $b(t) = \mathrm{e}^{-|t|}b(0)$ for $t < 0$. Then, by Parseval's identity,

$$\frac{1}{2\pi} \int_{\mathbf{R}} \big|\tilde{b}(\omega)\big|^2 \mathrm{d}\omega = \int_{\mathbf{R}} |b(t)|^2 \mathrm{d}t = \tfrac{1}{2}|b(0)|^2 + \int_{\mathbf{R}^+} \big|b(t)\big|^2 \mathrm{d}t,$$

and likewise

$$\frac{1}{2\pi} \int_{\mathbf{R}} \big|\mathrm{i}\omega\tilde{b}(\omega)\big|^2 \mathrm{d}\omega = \int_{\mathbf{R}} |b'(t)|^2 \mathrm{d}t = \tfrac{1}{2}|b(0)|^2 + \int_{\mathbf{R}^+} |b'(t)|^2 \mathrm{d}t.$$

Add these two equations to get

$$\frac{1}{2\pi} \int_{\mathbf{R}} (1+\omega^2) \big|\tilde{b}(\omega)\big|^2 \mathrm{d}\omega = q.$$

Choose $\mu(\mathrm{d}t) = \mathbf{e}(t)\,\mathrm{d}t$ in Theorem 6.2. Then $\alpha(\mathrm{d}\omega) = (1+\omega^2)^{-1}\,\mathrm{d}\omega$, $\gamma(\omega) = (1+\omega^2)\tilde{b}(\omega)$ and the desired conclusions follow from Theorem 6.2. □

Above, b was defined on all of $\mathbf{R}^+$. The case where b is defined only on the interval $[0,T]$ is very similar.

6.4 Corollary
Let $T > 0$, and let $b \in L^2([0,T];\mathbf{C}^n)$ be absolutely continuous with $b' \in L^2([0,T];\mathbf{C}^n)$. Define

$$q = |b(0)|^2 + |b(T)|^2 + \int_0^T \Big(|b(t)|^2 + |b'(t)|^2\Big)\,\mathrm{d}t.$$

Then, for every $\varphi \in L^2_{\mathrm{loc}}(\mathbf{R}^+;\mathbf{C}^n)$,

$$\Big|\int_0^T \langle b(t), \varphi(t)\rangle\,\mathrm{d}t\Big|^2 \le q[\![\mathbf{e}, \chi_{[0,T]}\varphi]\!]$$

and

$$\Big|\int_0^T \langle b(T-t), \varphi(t)\rangle\,\mathrm{d}t\Big|^2 \le q[\![\mathbf{e}, \chi_{[0,T]}\varphi]\!],$$

where $\mathbf{e}(t) = \mathrm{e}^{-t}$ for $t \in \mathbf{R}^+$.

The easy proof of Corollary 6.4 is left to the reader (cf. the proof of Corollary 6.3).

A result very similar to Theorem 6.2 is true in the case where μ is matrix-valued. For simplicity, we give only a sufficient Fourier transform condition (the condition $\int_{\mathbf{R}}|\gamma(\omega)|^2|\alpha|(\mathrm{d}\omega) < \infty$ that we use below is not necessary in the matrix case).

6.5 Theorem
Let $\mu \in PT(\mathbf{R}^+;\mathbf{C}^{n\times n})$ and denote $\Re\tilde{\mu}$ by α. Let $b \in L^2_{\mathrm{loc}}(\mathbf{R};\mathbf{C}^n)$ induce a tempered distribution whose Fourier transform $\tilde{b}$ is a measure $\beta \in M_{\mathrm{loc}}(\mathbf{R};\mathbf{C}^n)$ of the form $\beta(\mathrm{d}\omega) = \alpha(\mathrm{d}\omega)\gamma(\omega)$, where γ is a Borel measurable function satisfying $\int_{\mathbf{R}}|\gamma(\omega)|^2|\alpha|(\mathrm{d}\omega) < \infty$. Define

$$q = \frac{1}{2\pi}\int_{\mathbf{R}} \langle \gamma(\omega), \alpha(\mathrm{d}\omega)\gamma(\omega)\rangle.$$

Then

$$\Big|\int_{\mathbf{R}} \langle b(t), \varphi(t)\rangle\,\mathrm{d}t\Big|^2 \le q[\![\mu, \varphi]\!]$$

and

$$\sup_{t\in\mathbf{R}}\Big|\int_{\mathbf{R}} \langle b(s), \varphi(t-s)\rangle\,\mathrm{d}s\Big|^2 \le q[\![\mu^*, \varphi]\!],$$

for every $\varphi \in L^2(\mathbf{R};\mathbf{C}^n)$ with compact support.

Proof By Lemma 6.1, it suffices to prove the first of the two claims. This proof is essentially the same as the proof of the corresponding part of Theorem 6.2, i.e., the proof that (iii) implies (i). The only differences are that now we need the polar decomposition of a matrix-valued measure in order to show that the integrals are well defined, and we also need the following version of Hölder's inequality:

$$\left|\frac{1}{2\pi}\int_{\mathbf{R}}\langle\gamma(\omega),\alpha(\mathrm{d}\omega)\tilde{\varphi}(\omega)\rangle\right|^2 \leq \left\{\frac{1}{2\pi}\int_{\mathbf{R}}\langle\gamma(\omega),\alpha(\mathrm{d}\omega)\gamma(\omega)\rangle\right\}\left\{\frac{1}{2\pi}\int_{\mathbf{R}}\langle\tilde{\varphi}(\omega),\alpha(\mathrm{d}\omega)\tilde{\varphi}(\omega)\rangle\right\}.$$

For these technical tools, see Section 7. □

We conclude this section by giving a corollary to the previous result.

6.6 Corollary
Let $a \in C(\mathbf{R}^+;\mathbf{C}^{n\times n}) \cap PT(\mathbf{R}^+;\mathbf{C}^{n\times n})$, let $\varphi \in L^2_{\mathrm{loc}}(\mathbf{R}^+;\mathbf{C}^n)$, and let $T > 0$. Then

$$\left|\int_0^T a^*(t)\varphi(t)\,\mathrm{d}t\right|^2 \leq 2|a(0)|[\![a,\chi_{[0,T]}\varphi]\!]$$

and

$$\left|\int_0^T a(T-t)\varphi(t)\,\mathrm{d}t\right|^2 \leq 2|a(0)|[\![a,\chi_{[0,T]}\varphi]\!].$$

Proof Let $y \in \mathbf{C}^n$ satisfy $|y| = 1$. Define $\mu(\mathrm{d}t) = a(t)\,\mathrm{d}t$, and define b by $b(t) = a(t)y$ for $t \geq 0$, and $b(t) = a^*(-t)y$ for $t < 0$. Then, with the notations of Theorem 6.5, we have $\alpha = \Re\tilde{a}$ and $\tilde{b} = 2\Re\tilde{a}y = 2\alpha y$. Thus, we can take $\gamma(\omega) \equiv 2y$, and Theorem 6.5 gives (with φ replaced by $\chi_{[0,T]}\varphi$)

$$\begin{aligned}\left|\left\langle y,\int_0^T a^*(t)\varphi(t)\,\mathrm{d}t\right\rangle\right|^2 &= \left|\int_0^T\langle b(t),\varphi(t)\rangle\,\mathrm{d}t\right|^2\\ &\leq \frac{4}{2\pi}\int_{\mathbf{R}}\langle y,\alpha(\mathrm{d}\omega)y\rangle[\![a,\chi_{[0,T]}\varphi]\!]\\ &= 2\langle y,a(0)y\rangle[\![a,\chi_{[0,T]}\varphi]\!]\\ &\leq 2|a(0)|[\![a,\chi_{[0,T]}\varphi]\!],\end{aligned}$$

where we also used Theorem 2.7. Letting y vary over all vectors in $\mathbf{C}^n$ with $|y| = 1$, we get the first of the two conclusions.

The second claim is proved in much the same way. □

7. Appendix: Positive Matrices and Measures

A *semi-inner product* (denoted by $[\cdot,\cdot]$) in a complex vector space $\mathcal{H}$ is a mapping from $\mathcal{H}\times\mathcal{H}$ into $\mathbf{C}$ that for arbitrary x, y, $z \in \mathcal{H}$ and $\alpha \in \mathbf{C}$ satisfies the following conditions:

(i) $[x,y]=\overline{[y,x]}$;
(ii) $[x+y,z]=[x,z]+[y,x]$;
(iii) $[\alpha x,y]=\alpha[x,y]$;
(iv) $[x,x]\geq 0$.
If, in addition,
(v) $[x,x]>0$ when $x\neq 0$,
then we call $[\cdot,\cdot]$ an *inner product* in $\mathcal{H}$.

There are several different cases in which we need this definition. One is the case where A is a positive matrix in $\mathbf{C}^{n\times n}$; see Definition 2.3. Then it is trivial to check that $[x,y]=\langle x,Ay\rangle$ is a semi-inner product in $\mathbf{C}^n$ (here $\langle\cdot,\cdot\rangle$ is the ordinary Euclidean inner product in $\mathbf{C}^n$). If $A\succ 0$, then $[\cdot,\cdot]$ is an inner product in $\mathbf{C}^n$.

Another case is where $\mu\in M_{\text{loc}}(\mathbf{R};\mathbf{C}^{n\times n})$, and $\mu\succeq 0$, i.e., $\mu(E)\succeq 0$ for all bounded Borel sets $E\subset\mathbf{R}$. In this case, we choose our space $\mathcal{H}$ to be the set $L^2(\mathbf{R};|\mu|;\mathbf{C}^n)$ of all Borel measurable, $\mathbf{C}^n$-valued functions φ that satisfy $\int_{\mathbf{R}}|\varphi(s)|^2|\mu|(\mathrm{d}s)<\infty$, and define $[\varphi,\psi]$ by

$$[\varphi,\psi]=\int_{\mathbf{R}}\langle\varphi(s),\mu(\mathrm{d}s)\psi(s)\rangle=\int_{\mathbf{R}}\langle\varphi(s),h(s)\psi(s)\rangle|\mu|(\mathrm{d}s),$$

where the second equality sign is a consequence of the polar decomposition of a matrix-valued measure (see Lemma **3**.5.9). Thus defined, $[\varphi,\psi]$ is a semi-inner product and, in particular, it satisfies (iv). This follows from the fact that the Radon–Nikodym derivative of μ with respect to $|\mu|$ (this is the matrix function $h(x)$) is positive a.e. in the sense of Definition 2.3.

In Chapter 20 we shall need similar semi-inner products where the integral is replaced by a two- or three-dimensional integral, and φ and ψ depend on two or three variables.

The fact that the two preceding constructions are semi-inner products makes it possible to apply the following easily proved result.

7.1 Lemma

Let x, $y\in\mathcal{H}$, and let $[\cdot,\cdot]$ be a semi-inner product on $\mathcal{H}$. Then $\big|[x,y]\big|^2\leq[x,x][y,y]$ for all x and y in $\mathcal{H}$.

A slight modification of the proof of Theorem 12.2 in Rudin [2] proves this lemma.

If we apply the preceding lemma to the first of the two cases mentioned above, then we get, for all $x,y\in\mathbf{C}^n$, and for all $A\succeq 0$,

$$\big|\langle x,Ay\rangle\big|^2\leq\langle x,Ax\rangle\langle y,Ay\rangle.$$

This inequality has some further consequences. For example, if we take $x=Ay$, then we get

$$\begin{aligned}|Ay|^4&\leq\langle Ay,A^2y\rangle\langle y,Ay\rangle\\&\leq|Ay||A^2y|\langle y,Ay\rangle\\&\leq|A||Ay|^2\langle y,Ay\rangle.\end{aligned}$$

In other words we get, for all $y \in \mathbf{C}^n$,

$$|Ay|^2 \le |A|\langle y, Ay\rangle,$$

which is the relation (2.2). If we let y vary over all vectors satisfying $|y| = 1$, we find that

$$|A|^2 \le |A| \sup_{|y|=1} \langle y, Ay\rangle,$$

or, equivalently,

$$|A| \le \sup_{|y|=1} \langle y, Ay\rangle.$$

On the other hand, it is trivially true that $\sup_{|y|=1}\langle y, Ay\rangle \le |A|$. Thus, we find that

$$|A| = \sup_{|y|=1} \langle y, Ay\rangle,$$

which is (2.3).

If we apply the same lemma to the second semi-inner product that we defined above we get, for all φ, $\psi \in L^2(\mathbf{R}; |\mu|; \mathbf{C}^n)$,

$$\left|\int_{\mathbf{R}} \langle \varphi(s), \mu(\mathrm{d}s)\psi(s)\rangle\right|^2 \le \left\{\int_{\mathbf{R}} \langle \varphi(s), \mu(\mathrm{d}s)\varphi(s)\rangle\right\}\left\{\int_{\mathbf{R}} \langle \psi(s), \mu(\mathrm{d}s)\psi(s)\rangle\right\}.$$

This is the matrix version of Hölder's inequality that was needed in the proof of Theorem 6.5. Similar inequalities are valid for two- and three-dimensional integrals.

8. Appendix: Fourier and Laplace Transforms of Distributions

Below we list some elementary properties of distributions and of their Fourier and Laplace transforms.

We let $\mathcal{D}$ denote the space of all C^∞-functions on $\mathbf{R}$ with compact support. The space of distributions $\mathcal{D}'$ is defined to be the set of continuous linear functionals on $\mathcal{D}$. If $\nu \in \mathcal{D}'$ and $\varphi \in \mathcal{D}$, then the value of ν evaluated at φ is denoted by $[\nu, \varphi]$. The word 'continuous' in this connection means that if φ_j is a sequence of functions in $\mathcal{D}$ supported on a common compact interval, and if the functions φ_j, together with all their derivatives, converge uniformly on the interval to a function $\varphi \in \mathcal{D}$, then $[\nu, \varphi_j] \to [\nu, \varphi]$.

To get the class of all tempered distributions, one replaces $\mathcal{D}$ above by a larger set of functions $\mathcal{S}$. We say that a C^∞-function φ belongs to $\mathcal{S}$ if $t^k\varphi^{(m)}(t) \to 0$ as $|t| \to \infty$ for all integers k, $m \ge 0$. One defines convergence in $\mathcal{S}$ by saying that $\varphi_j \to \varphi$ in $\mathcal{S}$ iff $t^k\varphi_j^{(m)}(t) \to t^k\varphi^{(m)}(t)$ uniformly in t as $j \to \infty$ for all k, $m \ge 0$. The set $\mathcal{S}'$ of tempered distributions consists of all continuous linear functionals on $\mathcal{S}$. Again, we denote the value of

$\nu \in \mathcal{S}'$ evaluated at $\varphi \in \mathcal{S}$ by $[\nu, \varphi]$. Convergence in $\mathcal{S}'$ is defined as follows: $\nu_n \to \nu$ iff $[\nu_n, \varphi] \to [\nu, \varphi]$ for all $\varphi \in \mathcal{S}$.

The main advantage of the class $\mathcal{S}$, as compared to the class $\mathcal{D}$, is that the Fourier transform is a continuous, one-to-one map of $\mathcal{S}$ onto itself. This makes it possible to define the Fourier transform $\tilde{\nu}$ of a tempered distribution ν to be the tempered distribution that maps $\varphi \in \mathcal{S}$ into $[\nu, \tilde{\varphi}]$. In other words, $\tilde{\nu}$ is defined through the relation $[\tilde{\nu}, \varphi] = [\nu, \tilde{\varphi}]$. This extended Fourier transform is a continuous, one-to-one map of $\mathcal{S}'$ onto itself.

The convolution $\nu * \varphi$ is defined by $(\nu * \varphi)(t) = [\nu, (\check{\varphi})_t]$, where $(\check{\varphi})_t(\tau) = \varphi(t - \tau)$. If $\nu \in \mathcal{S}'$ and $\varphi \in \mathcal{S}$, then $\nu * \varphi \in \mathcal{S}'$; hence the Fourier transform of $\nu * \varphi$ is well defined. It is a fairly straightforward computation to check that $\widetilde{\nu * \varphi} = \tilde{\nu}\tilde{\varphi}$, where $\tilde{\nu}\tilde{\varphi}$ is the distribution that maps $\psi \in \mathcal{S}$ into $[\tilde{\nu}, \tilde{\varphi}\psi]$ for all $\psi \in \mathcal{S}$. Note that $\tilde{\varphi} \in \mathcal{S}$ and $\psi \in \mathcal{S}$ implies $\tilde{\varphi}\psi \in \mathcal{S}$.

Every function $a \in L^1_{\mathrm{loc}}(\mathbf{R})$ that has at most polynomial growth at infinity, i.e., satisfies

$$\int_{\mathbf{R}} (1 + |s|)^{-k} |a(s)| \, \mathrm{d}s < \infty$$

for some $k > 0$, induces a tempered distribution through the formula

$$[a, \varphi] = \int_{\mathbf{R}} a(s)\varphi(s) \, \mathrm{d}s. \tag{8.1}$$

If $\eta \in \mathcal{S}$, and $\int_{\mathbf{R}} \eta(s) \, \mathrm{d}s = \tilde{\eta}(0) = 1$, then we call the sequence $\eta_k(s) = k\eta(ks)$ an *approximate convolution identity* in $\mathcal{S}$. This name is motivated by the fact that for all $\nu \in \mathcal{S}'$ and all $\varphi \in \mathcal{S}$ it is true that $\nu * \eta_k \to \nu$ in $\mathcal{S}'$, and $\varphi * \eta_k \to \varphi$ in $\mathcal{S}$, as $k \to \infty$. To prove this, one can use Fourier transforms and the easily verified fact that if instead we define η_k by $\eta_k(s) = \eta(s/k)$, where now η satisfies $\eta(0) = 1$, then $\eta_k \nu \to \nu$ in $\mathcal{S}'$ and $\eta_k \varphi \to \varphi$ in $\mathcal{S}$ as $k \to \infty$. We call a sequence η_k of the latter type an *approximate multiplicative identity.* If all of the functions η_k have some property in common (such as having compact support, or being nonnegative, or having a Fourier transform with these properties), then we say simply that the approximate identity has this property.

We say that a distribution ν vanishes on an open set Ω if $[\nu, \varphi] = 0$ for all $\varphi \in \mathcal{D}$ whose support is contained in Ω. The support of ν is defined as the complement of the largest set where ν vanishes. If the support of $\nu \in \mathcal{S}'$ is contained in $\mathbf{R}^+$, then we write $\nu \in \mathcal{S}'(\mathbf{R}^+)$.

For every distribution ν, its derivative ν', defined through the relation $[\nu', \varphi] = -[\nu, \varphi']$, is also a distribution. If ν is tempered, then ν' is tempered, and the Fourier transform of ν' is $\eta\tilde{\nu}$, where η is the C^∞-function $\eta(\omega) = \mathrm{i}\omega$.

Every $\mu \in M(\mathbf{R})$ induces a tempered distribution through the formula

$$[\mu, \varphi] = \int_{\mathbf{R}} \mu(\mathrm{d}s)\varphi(s).$$

The distribution Fourier transform of μ can be identified with the function (see Definition **3**.2.2) $\omega \mapsto \tilde{\mu}(\omega) = \int_{\mathbf{R}} \mathrm{e}^{-\mathrm{i}\omega t} \mu(\mathrm{d}t)$, i.e.,

$$[\tilde{\mu}, \varphi] = \int_{\mathbf{R}} \tilde{\mu}(\omega)\varphi(\omega)\,\mathrm{d}\omega.$$

Moreover, the Fourier transform of the distribution derivative of μ can be identified with the function $\mathrm{i}\omega\tilde{\mu}(\omega)$.

Every $a \in BV(\mathbf{R})$ induces a tempered distribution through the formula (8.1). If the distribution derivative of a is denoted by μ, then $\mu \in M(\mathbf{R})$, and $\tilde{\mu}$ is (in the distribution sense) given by $\tilde{\mu} = \psi\tilde{a}$, where ψ is the function $\psi(\omega) = \mathrm{i}\omega$. This implies that, away from the origin, $\tilde{a}$ can be identified with the function $\frac{1}{\mathrm{i}\omega}\tilde{\mu}(\omega)$. In particular, $\tilde{a}$ belongs locally to $\tilde{L}^1$ on $\overline{\mathbf{R}} \setminus 0$ (cf. Theorems **6**.3.3 and **6**.3.4).

In many proofs it is convenient to know that the set of all test functions $\varphi \in \mathcal{D}$ that vanish on $\mathbf{R}^-$ is dense in $\mathcal{S}'(\mathbf{R}^+)$, i.e., that for each $\nu \in \mathcal{S}'(\mathbf{R}^+)$ it is possible to find a sequence of functions $\varphi_k \in \mathcal{D}$, each φ_k satisfying $\varphi_k(t) = 0$ for $t < 0$, such that $\varphi_k \to \nu$ in $\mathcal{S}'$. To see that this is true, it suffices to regularize ν by 'smoothing' it (convolving it by an approximate convolution identity supported on $\mathbf{R}^+$) and 'cutting' it (multiplying it by an approximate multiplicative identity with compact support).

For every $\nu \in \mathcal{S}'(\mathbf{R}^+)$ and every $z \in \mathbf{C}$ with $\Re z > 0$ one can define the Laplace transform $\hat{\nu}(z)$ of ν at z to be $\hat{\nu}(z) = [\nu, e_z]$, where e_z is an arbitrary function in $\mathcal{S}$ that satisfies $e_z(t) = \mathrm{e}^{-zt}$ for $t \geq -1$. This definition makes sense because, as ν vanishes on $(-\infty, 0)$, the value of $[\nu, e_z]$ is independent of the values of e_z on $(-\infty, -1)$. Observe that for each $\nu \in \mathcal{S}'$ the Laplace transform $\hat{\nu}$ is a classical function, and that the Laplace transform operator is continuous in the sense that $\hat{\nu}_k(z) \to \hat{\nu}(z)$, for each $z \in \mathbf{C}$ with $\Re z > 0$, whenever $\nu_k \to \nu$ in $\mathcal{S}'(\mathbf{R}^+)$.

It is possible to define the convolution $\nu * \mu$ of two distributions in $\mathcal{S}'(\mathbf{R}^+)$. As we observed above, in our definition of $\hat{\nu}(z)$, it is possible to define $[\nu, \varphi]$ not only for functions $\varphi \in \mathcal{S}(\mathbf{R})$, but for all C^∞-functions whose restriction to $\mathbf{R}^+$ belongs to $\mathcal{S}(\mathbf{R}^+)$. This means that one can define the convolution of ν by all C^∞-functions φ whose restriction to $\mathbf{R}^-$ belongs to $\mathcal{S}(\mathbf{R}^-)$. One example of a function of this type is the convolution $\mu * \varphi$, where $\mu \in \mathcal{S}'(\mathbf{R}^+)$ and $\varphi \in \mathcal{S}$. Therefore, one can define $\nu * \mu$ to be the distribution that satisfies $(\nu * \mu) * \varphi = \nu * (\mu * \varphi)$ for all φ in $\mathcal{S}$.

Laplace transforms of distributions in $\mathcal{S}'(\mathbf{R}^+)$ have the following properties.

8.1 Lemma

Let ν, $\mu \in \mathcal{S}'(\mathbf{R}^+)$, and let $\sigma > 0$. Then the following claims are true.

(i) *The Laplace transform $\hat{\nu}(z)$ is analytic in the half plane $\Re z > 0$.*

(ii) *The Fourier transform of the distribution $\mathrm{e}^{-\sigma t}\nu$ can be identified with the function $\omega \mapsto \hat{\nu}(\sigma + \mathrm{i}\omega)$.*

(iii) $\widehat{\nu'}(z) = z\hat{\nu}(z)$ *for* $\Re z > 0$.

(iv) $(\widehat{\nu * \mu})(z) = \hat{\nu}(z)\hat{\mu}(z)$ *for* $\Re z > 0$.

Proof (i) Pick e_z in such a way that $e_z(t) = 0$ for $t \leq -2$, and define $e_{z+w}(t) = \mathrm{e}^{-wt}e_z(t)$, $t \in \mathbf{R}$. Then $\frac{1}{w}(e_{z+w} - e_z)$ tends to the function $te_z(t)$ in $\mathcal{S}$ as $w \to 0$, and, therefore,

$$\lim_{w\to 0} \frac{\hat{\nu}(z+w) - \hat{\nu}(z)}{w} = \lim_{w\to 0}[\nu, \frac{e_{z+w} - e_z}{w}] = [\nu, te_z(t)].$$

In particular, $\frac{\mathrm{d}}{\mathrm{d}z}\hat{\nu}(z)$ exists, and it is the Laplace transform of the distribution that one gets when one multiplies ν by the function t.

(ii) Let η_k be an approximate convolution identity. By definition, the Fourier transform of $e_\sigma\nu$ (where $e_\sigma(t) = \mathrm{e}^{-\sigma t}$ for $t \geq -1$) satisfies

$$[\widetilde{e_\sigma\nu}, \varphi] = [e_\sigma\nu, \tilde{\varphi}] = [\nu, e_\sigma\tilde{\varphi}], \quad \varphi \in \mathcal{S}.$$

If we fix some $\omega \in \mathbf{R}$ and choose φ to be $\varphi(t) = \overline{\eta_k}(\omega - t)$, then $\tilde{\varphi}(s) = \mathrm{e}^{\mathrm{i}\omega s}\widetilde{\overline{\eta_k}}(s)$, and we get

$$(\widetilde{e_\sigma\nu} * \overline{\eta_k})(\omega) = [\nu, \psi_{\omega,k}],$$

where $\psi_{\omega,k}(s) = \overline{\mathrm{e}^{-(\sigma+\mathrm{i}\omega)s}\widetilde{\eta_k}(s)}$ for $s \geq -1$. As $k \to \infty$, we have, on the one hand, $\widetilde{e_\sigma\nu} * \overline{\eta_k} \to \widetilde{e_\sigma\nu}$ in $\mathcal{S}'$; in particular, $\widetilde{e_\sigma\nu} * \overline{\eta_k} \to \widetilde{e_\sigma\nu}$ in $\mathcal{D}'$. On the other hand, $\psi_{\omega,k}(s) \to e_\sigma\mathrm{e}^{-\mathrm{i}\omega s}$ in $\mathcal{S}$, uniformly for ω in compact sets. This implies that $[\nu, \psi_{\omega,k}] \to \hat{\nu}(\sigma + \mathrm{i}\omega)$, uniformly for ω in compact sets. Uniform convergence on compact sets implies convergence in $\mathcal{D}'$, and, therefore, the transform $\widetilde{e_\sigma\nu}$ equals, in the distribution sense, the function $\omega \mapsto \hat{\nu}(\sigma + \mathrm{i}\omega)$.

(iii) This claim is true if we replace $\nu \in \mathcal{S}'$ by a function $\varphi \in \mathcal{S}$ vanishing on $\mathbf{R}^-$. Since this set of functions φ is dense in $\mathcal{S}'(\mathbf{R}^+)$, the claim must be true for arbitrary distributions in $\mathcal{S}'(\mathbf{R}^+)$.

(iv) This claim is true if we replace $\nu \in \mathcal{S}'(\mathbf{R}^+)$ by a function $\varphi \in \mathcal{S}$ vanishing on $\mathbf{R}^-$ (to prove this one can, e.g., reduce this claim to the corresponding claim for Fourier transforms by using (ii)). Again, the general case is a consequence of the density of functions of this type in $\mathcal{S}'(\mathbf{R}^+)$. □

A particular Fourier transform, which can be regarded as a special case of the distribution Fourier transform, is the L^2 Fourier transform. Of course, every L^2-function f induces a tempered distribution, so it has a Fourier transform in the distribution sense. It turns out that the distribution Fourier transform of f can be identified with another L^2-function. More specifically, the following result is true.

8.2 Plancherel's Theorem

The distribution Fourier transform of a function $f \in L^2(\mathbf{R}; \mathbf{C}^n)$ *can be identified with a function* $\tilde{f} \in L^2(\mathbf{R}; \mathbf{C}^n)$. *Moreover,*

$$\int_{\mathbf{R}} \langle f(t), g(t)\rangle \,\mathrm{d}t = \frac{1}{2\pi}\int_{\mathbf{R}} \langle \tilde{f}(\omega), \tilde{g}(\omega)\rangle \,\mathrm{d}\omega. \tag{8.2}$$

In particular, if one computes $\|f\|_{L^2(\mathbf{R})}$ *and* $\|\tilde{f}\|_{L^2(\mathbf{R})}$ *by using the inner product norm in* $\mathbf{C}^n$, *then* $\|f\|_{L^2(\mathbf{R})} = \frac{1}{\sqrt{2\pi}}\|\tilde{f}\|_{L^2(\mathbf{R})}$.

The relation (8.2) is usually referred to as Parseval's identity.

For a proof of this well-known theorem, see, e.g., Rudin [2], Theorem 7.9.

9. Exercises

1. Let a be the scalar function $a(t) = t^{-\alpha}\mathrm{e}^{-\beta t}\cos(\gamma t)$, $t > 0$, where α, β, γ are real constants. For what values of these constants is a of positive type? When is a of strong, strict or of anti-coercive type?
2. Suppose $a \in C(\mathbf{R}^+;\mathbf{R})$ satisfies $a(T) > a(0)$ for some $T > 0$. Can it happen that a is of positive type?
3. Let $a \in L^1_{\mathrm{loc}}(\mathbf{R}^+;\mathbf{C}^{n\times n})$ and suppose that each component a_{ij} of a satisfies $a_{ij} \in PT(\mathbf{R}^+;\mathbf{C})$. Does it follow that $a \in PT(\mathbf{R}^+;\mathbf{C}^{n\times n})$?
4. Show that Proposition 3.7 is true without the assumption that a has compact support. Hint: μ is a bounded distribution; cf. Schwartz [1], Théorème XVII, p. 275.
5. Show that if $n = 1$ or $\mu(\mathbf{R}^+) = 0$ in Proposition 5.7 then one can take the coercivity constant q to be $q = 2|\mu(\{0\})|$.
6. Let a satisfy the assumptions of Proposition 5.8(i) and let β be the measure derivative of $-a''$. Show that

$$\Re\tilde{a}(\omega) = \frac{1}{\omega^2}\int_{\mathbf{R}^+}(1-\cos(\omega t))a''(t)\,\mathrm{d}t$$
$$= \frac{1}{\omega^3}\int_{(0,\infty)}(\omega t - \sin(\omega t))\beta(\mathrm{d}t).$$

7. Let $a(t) = 1 - t/T$, $0 \le t \le T$, $a(t) = 0$, $t > T$. Is a of anti-coercive type?
8. Give an example of a measure which is of anti-coercive type but
   (i) is not of strong positive type,
   (ii) is not of strict positive type.
9. Plancherel's theorem for distributions was used in the proof of Theorems 2.5 and 6.2. Try to prove it.
10. Prove the following result.

9.1 Corollary
Let $a \in C(\mathbf{R}^+;\mathbf{C}^{n\times n}) \cap PT(\mathbf{R}^+;\mathbf{C}^{n\times n})$ *satisfy* $a(0) \succ 0$, *let* $\varphi \in L^2_{\mathrm{loc}}(\mathbf{R};\mathbf{C}^n)$, *and let* $T > 0$. *Then*

$$\left\langle \int_0^T a^*(t)\varphi(t)\,\mathrm{d}t, [a(0)]^{-1}\int_0^T a^*(t)\varphi(t)\,\mathrm{d}t\right\rangle \le 2[\![a,\chi_{[0,T]}\varphi]\!]$$

and

$$\left\langle \int_0^T a(T-t)\varphi(t)\,\mathrm{d}t, [a(0)]^{-1} \int_0^T a(T-t)\varphi(t)\,\mathrm{d}t \right\rangle \le 2[\![a, \chi_{[0,T]}\varphi]\!].$$

Hint: Apply Corollary 6.6, with a replaced by $[a(0)]^{-1/2}a(\cdot)[a(0)]^{-1/2}$, and φ replaced by $[a(0)]^{1/2}\varphi(\cdot)$.

11. Let a be a continuous scalar function of positive type on $\mathbf{R}^+$ satisfying $\lim_{t\to 0} \frac{1}{t^2}(a(0)-a(t)) = 0$. Show that a is a constant function. Hint: Use Theorem 2.7.
12. Let $\mu \in M_{\mathrm{loc}}(\mathbf{R}^+;\mathbf{R})$ and $f \in C(\mathbf{R}^+;\mathbf{R})$ with $f(0) > 0$. Suppose there exists a sequence $\{\delta_j\}$ of positive numbers tending to 0 such that for all j, $\int_{\mathbf{R}^+} |\mu|(\mathrm{d}s)|f(\delta_j s)| < \infty$ and
$$\int_{\mathbf{R}^+} \mu(\mathrm{d}s) f(\delta_j s)\cos(\omega s) \ge 0, \quad \omega \in \mathbf{R}.$$
Show that μ is of positive type.
13. Let $\mu \in M_{\mathrm{loc}}(\mathbf{R}^+;\mathbf{C}^{n\times n})$, let $f \in C(\mathbf{R}^+;\mathbf{C})$ with $f(0) > 0$ have compact support, define $f_\delta(t) = f(\delta t)$ for $\delta > 0$, and assume that $\mu f_\delta \in PT(\mathbf{R}^+;\mathbf{C}^{n\times n})$ for $\delta > 0$. Show that $\mu \in PT(\mathbf{R}^+;\mathbf{C}^{n\times n})$.
14. Let $a \in L^1_{\mathrm{loc}}(\mathbf{R};\mathbf{R})$ be real-valued and even. Suppose that, for some $T_n \to \infty$,
$$\int_0^{T_n}\int_0^{T_n} a(t-s)\mathrm{e}^{\mathrm{i}\omega(t-s)}\,\mathrm{d}t\,\mathrm{d}s \ge 0, \quad \omega \in \mathbf{R}.$$
Show that the restriction of a to $\mathbf{R}^+$ is of positive type. Hint: Use Exercise 12. See Cooper [1].
15. Prove Lemma 7.1.
16. Let $a \in C(\mathbf{R}^+;\mathbf{C})$. Show that a is of positive type if and only if $\sum_{i=1}^n \sum_{j=1}^n c_i\overline{c_j}a(t_i - t_j) \ge 0$ for all complex numbers $c_1, \ldots c_n$ and all points $t_1, \ldots t_n \in \mathbf{R}$. Here $a(-t) \stackrel{\mathrm{def}}{=} \overline{a(t)}$ for $t \ge 0$.
17. Let $\mu \in M_{\mathrm{loc}}(\mathbf{R}^+;\mathbf{R}^{n\times n})$ and assume that $\langle\cdot,\cdot\rangle$ restricted to $\mathbf{R}^n \times \mathbf{R}^n$ is real. Show that μ is of positive type if and only if
$$\int_{\mathbf{R}} \langle \varphi(t), (\mu * \varphi)(t)\rangle\,\mathrm{d}t \ge 0$$
for all $\varphi \in L^2(\mathbf{R};\mathbf{R}^n)$ with compact support.
18. Let $\mu \in PT(\mathbf{R}^+;\mathbf{C}^{n\times n})$. Show that μ has a resolvent ρ, and that this resolvent is of positive type.
19. Let $\mu \in PT(\mathbf{R}^+;\mathbf{C}^{n\times n})$. Show that the differential resolvent r of μ is of positive type, and that the distribution derivative of $-r'$ is of positive type.
20. Let $a \in L^1_{\mathrm{loc}}(\mathbf{R}^+;\mathbf{C}^{n\times n})$ be completely monotone. Show that a is of positive type.
21. Under what assumptions on the kernel is the resolvent (differential

resolvent) of strong, strict or anti-coercive type?

22. Let $\mu \in PT(\mathbf{R}^+; \mathbf{C}^{n\times n})$, let $\varphi \in C^\infty(\mathbf{R}^+; \mathbf{C})$ have compact support and satisfy $\varphi(0) = 1$. Define $\varphi_\delta(t) = \varphi(\delta t)$ for $\delta > 0$. Is it true that
$$\widehat{(\mu\varphi_\delta)}(z) \to \hat{\mu}(z), \quad \delta \to 0, \quad \Re z > 0?$$
23. Prove or disprove the claim that the differential resolvent of a nonnegative, nonincreasing, and convex kernel in $L^1_{\text{loc}}(\mathbf{R}^+; \mathbf{R})$ is always integrable.
24. Let $\mu \in PT(\mathbf{R}^+; \mathbf{C}^{n\times n})$. Show that $\mu(\{0\}) \succeq 0$.

10. Comments

Section 2:

The theory of functions of positive type can be traced back to the fundamental work in Bochner [1]. See also Mathias [1] and Bochner [4]. The functions of positive type form a subclass of the distributions of positive type; see Schwartz [1], p. 274. Observe that these references consider functions and distributions of positive type having their support on $\mathbf{R}$ and not on $\mathbf{R}^+$. The quadratic form in (2.1) is then required to be real and nonnegative. This forces μ to be Hermitian.

Systems with kernels of positive type can be physically interpreted as passive causal systems (a system is passive if it cannot emit more energy than it has absorbed). See Zemanian [1], Ch. 10.

An easily accessible discussion of functions of positive type applied to Volterra equations can be found in Nohel and Shea [2]. They also formulate an equivalence relation comparable to Theorem 2.4.

Measures of positive type on $\mathbf{R}$ were investigated in Staffans [2], where an equivalence result analogous to Theorem 2.6 can be found. Measures of positive type on locally compact Abelian groups are analysed in Berg and Forst [1].

For the original version of the representation Theorem 2.7, see Bochner [1], Satz 19. The result is extended to distributions of positive type in Schwartz [1], Théorème XVIII, p. 276. For a formulation in the context of locally compact Abelian groups, see Berg and Forst [1], p. 14. A useful result on $(C,1)$-representation is the following due to Cooper [1]. Let $a(t) \in L^1_{\text{loc}}(\mathbf{R}^+)$ be a function of positive type. Then there exists a positive measure μ that satisfies $\int_{\mathbf{R}^+} \mu(\mathrm{d}s)(1+s^2)^{-1} < \infty$ and $\lim_{s\to\infty} \frac{1}{s}\mu([0,s]) = 0$ such that
$$a(t) = \lim_{T\to\infty} \int_0^T \mu(\mathrm{d}s)\left(1 - \tfrac{s}{T}\right)\cos(st) \text{ a.e. on } (0,\infty).$$

The references mentioned above all concern scalar-valued functions or measures. Matrix-valued measures of positive type are considered in Staffans [6]. This paper contains what is essentially Theorem 2.6.

Functions of positive type are frequently employed in control theory. See, e.g., Popov [2] and Corduneanu [9].

For functions being of positive type on a compact interval, see Gripenberg [7].

Section 3:
Proposition 3.1 can be found in Titchmarsh [1], p. 170. A special case of Proposition 3.4 with $\omega_0 = 0$ is found in Staffans [6], p. 105. For different versions of Proposition 3.5, see Cooper [1], Theorem 1, and Berg and Forst [1], p. 22. For a result analogous to Proposition 3.7, see Schwartz [1], p. 278. The proof of Theorem 3.8 uses techniques similar to those of the proof of Staffans [5], Lemma 1.1.

Section 4:
Kernels of strong positive type were first used in Halanay [1]. See also Nohel and Shea [2], MacCamy and Wong [1], and MacCamy [2]. Proposition 4.3 can be found in Nohel and Shea [2], although with a different proof.

Measures of strict positive type were defined in Staffans [2], where Theorem 4.5 and Proposition 4.6 can be found. The exceptional case where a convex kernel is piecewise linear was discussed in Levin and Nohel [2] and Hannsgen [1].

Section 5:
Scalar kernels $k \in L^1(\mathbf{R}^+)$ satisfying $q\Re\tilde{k}(\omega) \geq |\tilde{k}(\omega)|^2$ were used by Barbu [1]. The implication (v)$\Rightarrow$(ii) of Theorem 5.1 (in case $\mu = \nu \in M(\mathbf{R}^+;\mathbf{R})$) is proved in Staffans [4]. This reference also contains first versions of Proposition 5.7 and Proposition 5.8(i) and (ii).

Kernels that are of strong positive type and satisfy a, a', $a'' \in L^1(\mathbf{R}^+)$ (thus anti-coercive) have been extensively used in viscoelasticity theory. Given such a kernel one can show that there exists a constant c having the property that, for all $\varphi \in C([0,T])$ and $t \in [0,T]$,

$$\int_0^t |\varphi(s)|\, ds \leq c\Big\{|\varphi(0)|^2 + [\![a, \chi_{[0,T]}\varphi]\!] + \liminf_{h\downarrow 0} [\![a, \chi_{[0,T]}\Delta_h\varphi]\!]\Big\}.$$

The difference operator Δ_h is defined by $(\Delta_h\varphi)(t) = h^{-1}(\varphi(t+h) - \varphi(t))$. For the inequality above, see Hrusa and Nohel [1]. Related estimates in the same vein can be found in Hrusa and Renardy [1] and [2].

Section 6:
Inequalities of type (6.2) were first studied in Staffans [7]. See also MacCamy and Wong [1], pp. 28-29. Theorem 6.2 is taken from Staffans [7], Sec. 8, Corollary 6.3 from Staffans [7], Proposition 4.1, and Corollary 6.6 from Staffans [2], Lemma 6.1.

An early version of the second part of Corollary 6.6 was proved by Popov in the scalar case (see Popov [2], Theorem 1, part 1°, p. 203), and by Halanay in the vector case (see Popov [2], Theorem 2, part 1°, p. 208). Popov uses the name *hyperstability* for this property.

Section 8:
For distributions and their Fourier transforms, see Donoghue [1], Rudin [2], and Schwartz [1]. Laplace transforms of distributions are considered in Zemanian [1].

Section 9:
Corollary 9.1 appears to be new.

References

V. Barbu
1. Sur une équation intégrale non-linéaire, *An. Ştiinţ. "Al. I. Cuza" Iaşi Secţ. I a Mat (N.S.)* **10** (1964), pp. 61–65.

C. Berg and G. Forst
1. *Potential Theory on Locally Compact Abelian Groups*, Springer-Verlag, Berlin, 1975.

S. Bochner
1. Monotone Funktionen, Stieltjessche Integrale und harmonische Analyse, *Math. Ann.* **108** (1933), pp. 378–410.
4. *Lectures on Fourier Integrals*, Princeton University Press, Princeton, 1959.

J. L. B. Cooper
1. Positive definite functions of a real variable, *Proc. London Math. Soc. (3)* **10** (1960), pp. 53–66.

C. Corduneanu
9. *Integral Equations and Stability of Feedback Systems*, Academic Press, New York, 1973.

K. Deimling
2. *Nonlinear Functional Analysis*, Springer-Verlag, Berlin, 1985.

W. F. Donoghue
1. *Distributions and Fourier Transforms*, Academic Press, New York, 1969.

G. Gripenberg
7. On a frequency domain condition used in the theory of Volterra equations, *SIAM J. Math. Anal.* **10** (1979), pp. 839–843.

A. Halanay
1. On the asymptotic behavior of the solutions of an integro-differential equation, *J. Math. Anal. Appl.* **10** (1965), pp. 319–324.

K. B. Hannsgen
1. Indirect Abelian theorems and a linear Volterra equation, *Trans. Amer. Math. Soc.* **142** (1969), pp. 539–555.

W. J. Hrusa and J. A. Nohel

1. The Cauchy problem in one-dimensional nonlinear viscoelasticity, *J. Differential Equations* **59** (1985), pp. 388–412.

W. J. Hrusa and M. Renardy

1. On a class of quasilinear partial integrodifferential equations with singular kernels, *J. Differential Equations* **64** (1986), pp. 195–220.
2. A model equation for viscoelasticity with a strongly singular kernel, *SIAM J. Math. Anal.* **19** (1988), pp. 257-269.

J. J. Levin and J. A. Nohel

2. On a nonlinear delay equation, *J. Math. Anal. Appl.* **8** (1964), pp. 31–44.

R. C. MacCamy

2. Remarks on frequency domain methods for Volterra integral equations, *J. Math. Anal. Appl.* **55** (1976), pp. 555–575.

R. C. MacCamy and J. S. W. Wong

1. Stability theorems for some functional equations, *Trans. Amer. Math. Soc.* **164** (1972), pp. 1–37.

M. Mathias

1. Über positive Fourier-Integrale, *Math. Zeit.* **16** (1923), pp. 103–125.

J. A. Nohel and D. F. Shea

2. Frequency domain methods for Volterra equations, *Adv. in Math.* **22** (1976), pp. 278–304.

V. M. Popov

2. *Hyperstability of Control Systems*, Springer-Verlag, Berlin, 1973.

W. Rudin

2. *Functional Analysis*, McGraw–Hill, New York, 1973.

L. Schwartz

1. *Théorie des Distributions, nouv. éd.*, Hermann, Paris, 1966.

O. J. Staffans

2. Positive definite measures with applications to a Volterra equation, *Trans. Amer. Math. Soc.* **218** (1976), pp. 219–237.
4. An inequality for positive definite Volterra kernels, *Proc. Amer. Math. Soc.* **58** (1976), pp. 205–210.
5. On the asymptotic spectra of the bounded solutions of a nonlinear Volterra equation, *J. Differential Equations* **24** (1977), pp. 365–382.
6. Systems of nonlinear Volterra equations with positive definite kernels, *Trans. Amer. Math. Soc.* **228** (1977), pp. 99–116.
7. Boundedness and asymptotic behavior of solutions of a Volterra equation, *Michigan Math. J.* **24** (1977), pp. 77–95.

E. C. Titchmarsh

1. *Introduction to the Theory of Fourier Integrals,* 2nd *ed.*, Oxford University Press, London, 1948.

A. H. Zemanian

1. *Distribution Theory and Transform Analysis*, McGraw–Hill, New York, 1965.

17

Frequency Domain Methods: Basic Results

We study the asymptotic behaviour of solutions of certain nonlinear convolution equations with kernels of positive type. The proofs are based on frequency domain methods and on arguments of Lyapunov type.

1. Introduction

In Part I we have seen that the best asymptotic results for linear convolution equations are based on resolvent theory, and on the use of Laplace or Fourier transforms. In Chapter 11 we analysed certain weakly nonlinear equations by combining these tools with perturbation theory. But the usefulness of this approach is obviously quite restricted, and only marginal deviations from linearity can be allowed.

Of course, to make statements on the asymptotic behaviour of solutions of genuinely nonlinear equations, some assumptions on the structure of the equations have to be made. Here we will consider equations where the integral term is composed of a nonlinear function $g(x)$ of the unknown solution x and a linear integral operator of convolution type. That is, we assume that the equations can be written as

$$x(t) + \int_0^t a(s)g\big(x(t-s)\big)\,\mathrm{d}s = f(t), \quad t \in \mathbf{R}^+, \tag{1.1}$$

or

$$x'(t) + \int_{[0,t]} \mu(\mathrm{d}s)g\big(x(t-s)\big) = f(t), \quad t \in \mathbf{R}^+; \quad x(0) = x_0. \tag{1.2}$$

Moreover, we take the kernels a and μ (or possibly some derivative of these kernels) to be of positive type. Under these assumptions, a reasonably satisfactory asymptotic theory exists.

Some of our results can be extended to equations where the nonlinear function g depends explicitly on t, as in the equation

$$x(t) + \int_0^t a(s) g\big(t-s, x(t-s)\big)\,\mathrm{d}s = f(t), \quad t \in \mathbf{R}^+. \tag{1.3}$$

In Section 5 we formulate several of these extensions. Equations having a more complicated structure than (1.1)–(1.3) are briefly studied in Chapter 20.

Our analysis uses the ideas developed in Chapters 14–16; in particular, we make repeated use of transform techniques and of results about kernels of positive type. However, in contrast to the linear case, we do not always treat all the terms in the equation by transform techniques; some terms may be treated in the time-domain. In general—since the proofs are based on various estimates—the conditions that we give for asymptotic conclusions are sufficient, but not necessary. We wish to emphasize that, inasmuch as a detailed theory of kernels of positive type was built up in Chapter 16, the proofs in the present chapter can be kept rather short.

As far as possible we shall state and prove our results in terms of vector-valued functions. Originally the results were proved for scalar equations only, and it is still true that the scalar equations constitute the most important applications. This hinges on the fact that the assumptions become rather restrictive in the case of systems of equations. For example, when we discuss the integrodifferential equation (1.2), we shall often assume that the nonlinear function g is the gradient of a real-valued function G. This assumption is automatically satisfied when the dimension $n = 1$ (define $G(x) = \int_0^x g(y)\,\mathrm{d}y$), but it is a significant restriction when $n > 1$.

We shall always take for granted that solutions x, for which asymptotic statements are made, exist for all $t > 0$, and are locally integrable, or even locally bounded. Note, however, that all the boundedness results below could be used to establish the global existence of such solutions (see the standard conclusion in Sections **12**.2 and **12**.3 saying that a noncontinuable solution leaves every compact set as $t \uparrow T_{\max}$).

In the asymptotic analysis, one frequently separates the proof of global boundedness from the results that describe the asymptotic behaviour in more detail. This approach is natural, since the type of conditions needed to get global boundedness are somewhat different from those used to extract more detailed asymptotic information. In particular, this is the case if the function f in (1.1)–(1.3) does not decay sufficiently rapidly at infinity; then rather little is known about global boundedness of solutions, and much more is known about the asymptotic behaviour of bounded solutions. This question will be discussed further in Chapter 19. In this chapter we concentrate on the case where f is comparatively small or well behaved at infinity; for example, $f \in L^1(\mathbf{R}^+)$ or $f \in BV(\mathbf{R}^+)$. Consequently, here we are able to prove that all solutions are bounded. Actually, our results usually give conditions under which the solution x is small at infinity, e.g.

$x \in L^2(\mathbf{R}^+)$, or $\lim_{t\to\infty} x(t) = 0$. Sometimes statements are available only on the asymptotic size of $g(x(t))$, or on a combination of $x(t)$ and $g(x(t))$. No results yielding the existence of unbounded solutions, or characterizing the behaviour of such solutions at infinity, are given.

We wish to stress that this chapter, as well as the following Chapters 18–19, does not contain any unifying main theorem. Instead, certain steps in the proofs are common to most of the results. These steps can be summarized as follows.

1. Write the given equation in the form (1.1), (1.2) or (1.3).
2. Multiply the equation (i.e., form the inner product) by one of the functions $x(t)$, $g(x(t))$, or $x'(t)$. This should be done in such a way that the terms that are quadratic in x (or in $g(x)$ or $x'(t)$) are all of the same sign (this step is usually where the Fourier transform technique enters), and so that these quadratic terms dominate the linear terms, at least for large values of $\|x\|_{L^2(0,T)}$ or $\|g\circ x\|_{L^2(0,T)}$, etc.
3. Show that all terms can be bounded independently of T.
4. Show that $x(t)$ (or $g(x(t))$, $x'(t)$) is in some sense bounded, for example that $\sup_{T\in\mathbf{R}^+}\|x\|_{L^2(0,T)} < \infty$.
5. Obtain more detailed (pointwise) asymptotic information about the unknown function.

Of course, this summary is only a rough outline, and there are distinctive features in each of the results considered below. Moreover, note that this recipe by no means eliminates the difficulties. The crucial questions, such as by what function to multiply, how to conclude that x is square integrable, etc., all remain.

In Section 2 we formulate some results on boundedness of solutions of (1.2). Sections 3 and 4 give more detailed asymptotic information on solutions of the same equation. Equations (1.1) and (1.3) are investigated in Section 5.

Throughout this and the next three chapters we let $\langle\cdot,\cdot\rangle$ denote some inner product on $\mathbf{C}^n$ that restricted to $\mathbf{R}^n$ is real-valued. Observe that the notion of the gradient depends on the inner product since the gradient g of a function G is defined by the requirement that $|G(x+h) - G(x) - \langle g(x), h\rangle|/|h| \to 0$ as $|h| \to 0$.

2. Boundedness Results for an Integrodifferential Equation

Let us first apply results from Chapter 16 to the integrodifferential equation

$$x'(t) + \int_{[0,t]} \mu(\mathrm{d}s) g\big(x(t-s)\big) = f(t), \quad t \in \mathbf{R}^+; \quad x(0) = x_0. \tag{1.2}$$

Our approach will be the one described in the introduction. We assume that g is the gradient of a real-valued function G (as we observed above, this assumption is trivially satisfied in the scalar case), let f be integrable on $\mathbf{R}^+$, and suppose that μ is of positive type.

Taking the inner product of both sides of (1.2) with $g(x(t))$, and integrating over some interval $[0,T]$, one gets the relation

$$G\big(x(T)\big) + Q(\mu, g\circ x, T) = G(x_0) + \int_0^T \big\langle g\big(x(t)\big), f(t)\big\rangle \,\mathrm{d}t, \tag{2.1}$$

where

$$Q(\mu, g\circ x, T) \stackrel{\text{def}}{=} \int_0^T \Big\langle g\big(x(t)\big), \int_{[0,t]} \mu(\mathrm{d}s) g\big(x(t-s)\big)\Big\rangle \,\mathrm{d}t. \tag{2.2}$$

Note that this double integral was denoted by $[\![\mu, \chi_{[0,T]} g(x(\cdot))]\!]$ in Chapter 16.

Since μ is of positive type, $Q(\mu, g\circ x, T)$ is nonnegative. Therefore, (2.1) gives us the inequality

$$G\big(x(T)\big) \le G(x_0) + \int_0^T \big\langle g\big(x(t)\big), f(t)\big\rangle \,\mathrm{d}t. \tag{2.3}$$

If, in some suitable way, $|g(y)|$ is dominated by $G(y)$, then it follows from this inequality that $\sup_{T>0} G(x(T)) < \infty$. Under appropriate hypotheses, this implies that x is bounded and so, by (2.1), $Q(\mu, g\circ x, T)$ is bounded, independently of T. From this fact one can obtain a more detailed description of the asymptotic behaviour of $g(x(t))$ as $t \to \infty$.

In what follows we elaborate on these comments, and begin with the following result.

2.1 Proposition

Assume that the following conditions hold:

(i) $\mu \in PT(\mathbf{R}^+; \mathbf{R}^{n\times n})$;

(ii) $g \in C(\mathbf{R}^n; \mathbf{R}^n)$ *is the gradient of a function* $G \in C^1(\mathbf{R}^n; \mathbf{R})$ *satisfying* $\lim_{|y|\to\infty} G(y) = \infty$, *and* $\limsup_{|y|\to\infty} |g(y)|/G(y) < \infty$;

(iii) $f \in L^1(\mathbf{R}^+; \mathbf{R}^n)$;

(iv) $x \in AC_{\mathrm{loc}}(\mathbf{R}^+; \mathbf{R}^n)$ *is a solution of (1.2) on* $\mathbf{R}^+$.

Then $x \in BC(\mathbf{R}^+; \mathbf{R}^n)$, *and* $\sup_{T>0} Q(\mu, g\circ x, T) < \infty$.

As the next theorem shows, the conclusion is true under a weaker hypothesis. To obtain Proposition 2.1 from Theorem 2.2, note that if (ii) of Proposition 2.1 is assumed then there exist positive constants C_1 and C_2 such that $|g(y)| \le C_1 + C_2 G(y)$ for all $y \in \mathbf{R}^n$. Obviously, this inequality can be written in the form $|g(y)| \le u(G(y))$, where u is the function $u(y) = C_1 + C_2 y$.

2.2 Theorem
Assume that

(i) $\mu \in PT(\mathbf{R}^+; \mathbf{R}^{n\times n})$,
(ii) $g \in C(\mathbf{R}^n; \mathbf{R}^n)$ *is the gradient of a function* $G \in C^1(\mathbf{R}^n; \mathbf{R})$, *and* u *is a nondecreasing function, defined on the range of* G, *and with values in* $[0, \infty]$, *such that* $|g(y)| \le u(G(y))$ *for* $y \in \mathbf{R}^n$,
(iii) $f \in L^1_{\mathrm{loc}}(\mathbf{R}^+; \mathbf{R}^n)$,
(iv) $x \in AC_{\mathrm{loc}}(\mathbf{R}^+; \mathbf{R}^n)$ *is a solution of (1.2) on* $\mathbf{R}^+$.

Then

$$\int_{G(x_0)}^{G(x(t))} \frac{\mathrm{d}s}{u(s)} \le \int_0^t |f(s)|\,\mathrm{d}s, \quad t \in \mathbf{R}^+. \tag{2.4}$$

Thus, if furthermore $\lim_{|y|\to\infty} G(y) = \infty$, *and if*

$$\int_{G(x_0)}^{\infty} \frac{\mathrm{d}s}{u(s)} > \int_{\mathbf{R}^+} |f(s)|\,\mathrm{d}s,$$

then $x \in BC(\mathbf{R}^+; \mathbf{R}^n)$, *and* $\sup_{T>0} Q(\mu, g{\circ}x, T) < \infty$.

Here we use the convention that if $b < a$ then $\int_a^b h(s)\,\mathrm{d}s = -\int_b^a h(s)\,\mathrm{d}s$. We allow the function u to take the values 0 and ∞, and employ the standard rules $1/0 = \infty$ and $1/\infty = 0$ in (2.4).

Given g and G, it is always possible to construct a function u of the type required in (ii). Define

$$u(s) \stackrel{\mathrm{def}}{=} \sup\{\, |g(y)| \mid y \in \mathbf{R}^n, G(y) \le s \,\}$$

for all s in the range of G. Clearly, this function is the smallest possible; hence it gives us the best estimate available with this method.

Proof of Theorem 2.2 In the case where G is bounded from above, the function $u(s)$ need not be defined for $s > \alpha \stackrel{\mathrm{def}}{=} \sup_{y\in\mathbf{R}^n} G(y)$. In this case, we can define $u(s)$ for $s > \alpha$ to be equal to $\lim_{s\uparrow\alpha} u(s)$. Thus, without loss of generality, we may assume that the domain of u is unbounded to the right.

First, let us prove a simplified version of the theorem, where we assume that u is strictly positive and continuous.

By (2.3) and (ii),

$$G\big(x(t)\big) \le G(x_0) + \int_0^t u\big(G\,(x(s))\big)|f(s)|\,\mathrm{d}s, \quad t \in \mathbf{R}^+. \tag{2.5}$$

Define the functions V and K by

$$V(s) = \int_{G(x_0)}^{s} \frac{1}{u(\tau)}\,\mathrm{d}\tau, \quad s \ge G(x_0);$$

$$K(t) = G(x_0) + \int_0^t u\big(G(x(s))\big)|f(s)|\,\mathrm{d}s, \quad t \ge 0.$$

Then, by (2.5), $G(x(t)) \le K(t)$, and so $u(G(x(t))) \le u\big(K(t)\big)$ for $t \in \mathbf{R}^+$. It follows that

$$\frac{u(G(x(t)))|f(t)|}{u\big(K(t)\big)} \le |f(t)|, \quad t \in \mathbf{R}^+,$$

which can be written as $\frac{\mathrm{d}}{\mathrm{d}t} V\big(K(t)\big) \le |f(t)|$. Hence,

$$\int_{K(0)}^{K(t)} \frac{1}{u(\tau)}\,\mathrm{d}\tau \le \int_0^t |f(s)|\,\mathrm{d}s. \tag{2.6}$$

If $G(x(t)) \le G(x_0)$, then (2.4) follows trivially, because the left-hand side is nonpositive and the right-hand side nonnegative. If $G(x(t)) \ge G(x_0)$, then the claim follows from (2.6), because $G(x(t)) \le K(t)$, and $K(0) = G(x_0)$. The claim concerning the boundedness of x follows immediately from the additional assumptions. Once one knows that x is bounded, the boundedness of the quadratic term is a consequence of (2.1), and of the fact that f is integrable.

Above we made the simplifying assumption that u is strictly positive and continuous. If u is bounded, but discontinuous or not strictly positive, we can replace u by the sequence of strictly positive, continuous functions $u_k(s) = 1/k + k\int_s^{s+1/k} u(y)\,\mathrm{d}y$, and apply the preceding argument with u replaced by u_k. Letting $k \to \infty$ we get (2.4).

The case where u is unbounded remains. Fix $t \in \mathbf{R}^+$ and suppose that (2.4) does not hold. Then there is some finite constant M such that $\int_{G(x_0)}^{M} (u(s))^{-1}\mathrm{d}s > \int_0^t |f(s)|\,\mathrm{d}s$. We can assume that $u(M) < \infty$, because $\int_{M_1}^{\infty} (u(s))^{-1}\mathrm{d}s = 0$, where $M_1 = \inf\{\, s \mid u(s) = \infty \,\}$. Replace u by the bounded function $\min\{u(s), u(M)\}$, and use the first part of the proof to show that

$$\int_{G(x_0)}^{G(x(\tau))} \frac{\mathrm{d}s}{u(s)} \le \int_0^{\tau} |f(s)|\,\mathrm{d}s,$$

provided $G(x(s)) \le M$ for all $s \le \tau \le t$ (the inequality $|g(y)| \le u(G(y))$ was needed only for y in the set $\{\, x(s) \mid 0 \le s \le \tau \,\}$). Thus it follows from the choice of M that $G(x(s)) < M$ for all $s \le \tau$. Hence it is possible to take $\tau = t$, and (2.4) follows. □

In the one-dimensional case, it is possible to get an explicit upper bound on the solution provided the function g happens to be nondecreasing and odd.

2.3 Corollary

Suppose that

(i) $\mu \in PT(\mathbf{R}^+; \mathbf{R})$,
(ii) $g \in C(\mathbf{R}; \mathbf{R})$ *is odd and nondecreasing on* $\mathbf{R}$,
(iii) $f \in L^1_{\mathrm{loc}}(\mathbf{R}^+; \mathbf{R})$,
(iv) $x \in AC_{\mathrm{loc}}(\mathbf{R}^+; \mathbf{R})$ *is a solution of (1.2) on* $\mathbf{R}^+$.

Then

$$|x(t)| \le |x(0)| + \int_0^t |f(s)|\,\mathrm{d}s, \quad t \in \mathbf{R}^+.$$

Proof The case where g vanishes identically is trivial, so we suppose that $\alpha \overset{\text{def}}{=} \inf\{\, y \in \mathbf{R}^+ \mid g(y) > 0 \,\} < \infty$. Define $G(y) = \int_0^y g(s)\,\mathrm{d}s$, let h denote the inverse of G on $[\alpha, \infty)$, and define $u(y) = g(h(y))$ for $y \ge 0$. Then $u(G(y)) = g(y)$ for all $y \ge 0$, and therefore, as g is odd and G is even, $|g(y)| = u(G(y))$ for all $y \in \mathbf{R}$. Because of this, we can apply Theorem 2.2 to get

$$\int_{G(x_0)}^{G(x(t))} \frac{\mathrm{d}s}{u(s)} \le \int_0^t |f(s)|\,\mathrm{d}s, \quad t \in \mathbf{R}^+.$$

Differentiating the identity $G(h(y)) = y$ (which is valid for $y \ge 0$), we get $g(h(y))h'(y) = 1$, or equivalently, $h'(y) = 1/u(y)$ for $y > 0$. Thus it follows that

$$h\big(G(x(t))\big) - h\big(G(x_0)\big) \le \int_0^t |f(s)|\,\mathrm{d}s.$$

Since $h(G(y)) \ge |y|$ with equality if $|y| \ge \alpha$, we get the desired assertion, provided $|x_0| \ge \alpha$. If this is not the case, then we define T by $T \overset{\text{def}}{=} \inf\{\, t > 0 \mid |x(t)| > \alpha \,\}$, and observe that $g(x(t)) \equiv 0$ on $[0, T]$. Therefore, by (1.2), $x'(t) = f(t)$ on that interval and so, without loss of generality, we may replace the point 0 by the point T in the analysis above. □

The reader may find it instructive to compare this corollary with Proposition **14**.2.9.

If the kernel is of strong positive type, we may also employ another estimate obtained in Chapter 16 to get global boundedness.

2.4 Theorem

Let the following hypotheses be satisfied:

(i) $\mu \in M(\mathbf{R}^+; \mathbf{R}^{n\times n})$ *is of strong positive type;*
(ii) $g \in C(\mathbf{R}^n; \mathbf{R}^n)$ *is the gradient of a function* $G \in C^1(\mathbf{R}^n; \mathbf{R})$ *satisfying* $\lim_{|y|\to\infty} G(y) = +\infty$;
(iii) $f \in L^2(\mathbf{R}^+; \mathbf{R}^n) \cap AC_{\mathrm{loc}}(\mathbf{R}^+; \mathbf{R}^n)$ *and* $f' \in L^2(\mathbf{R}^+; \mathbf{R}^n)$;
(iv) $x \in AC_{\mathrm{loc}}(\mathbf{R}^+; \mathbf{R}^n)$ *is a solution of (1.2) on* $\mathbf{R}^+$.

Then $x \in BC(\mathbf{R}^+; \mathbf{R}^n)$, *and* $\sup_{T>0} Q(\mu, g\circ x, T) < \infty$.

Proof It follows from Corollary **16**.6.3 (in particular, see the observation following this corollary), and from (2.1) and our hypothesis that

$$\begin{aligned} G\big(x(T)\big) + Q(\mu, g\circ x, T) &\le G(x(0)) + \sqrt{qQ(\mu, g\circ x, T)} \\ &\le G(x(0)) + \tfrac{1}{2}Q(\mu, g\circ x, T) + \tfrac{1}{2}q, \end{aligned}$$

where q is a constant. This inequality gives the desired assertions. □

The same proof, but with Corollary **16**.6.3 replaced by Theorem **16**.6.5, furnishes the following result.

2.5 Theorem
Assume that

(i) $\mu \in PT(\mathbf{R}^+; \mathbf{R}^{n\times n})$,

(ii) $g \in C(\mathbf{R}^n; \mathbf{R}^n)$ *is the gradient of a function* $G \in C^1(\mathbf{R}^n; \mathbf{R})$ *satisfying* $\lim_{|y|\to\infty} G(y) = +\infty$,

(iii) $f \in L^2_{\mathrm{loc}}(\mathbf{R}^+; \mathbf{R}^n)$ *induces a tempered distribution whose Fourier transform is a measure* $\beta \in M_{\mathrm{loc}}(\mathbf{R}; \mathbf{C}^n)$ *of the form* $\beta(\mathrm{d}\omega) = \alpha(\mathrm{d}\omega)\gamma(\omega)$, *where* $\alpha = \Re\tilde{\mu}$, *and* γ *is a Borel measurable function satisfying* $\int_{\mathbf{R}} |\gamma(\omega)|^2 |\alpha|(\mathrm{d}\omega) < \infty$,

(iv) $x \in AC_{\mathrm{loc}}(\mathbf{R}^+; \mathbf{R}^n)$ *is a solution of (1.2) on* $\mathbf{R}^+$.

Then $x \in BC(\mathbf{R}^+; \mathbf{R}^n)$, *and* $\sup_{T>0} Q(\mu, g\circ x, T) < \infty$.

3. Asymptotic Behaviour

In the previous section, we concentrated on boundedness, obtaining $x \in BC(\mathbf{R}^+)$ and $\sup_{T>0} Q(\mu, g\circ x, T) < \infty$. Next, we study the asymptotic behaviour of solutions that are assumed to satisfy these conclusions.

First, let us give some conditions that provide a global bound on the derivative of x. The proof is based on Corollary **16**.6.6.

3.1 Theorem
Assume that the following conditions hold:

(i) $\mu(\mathrm{d}t) = \nu(\mathrm{d}t) + a(t)\,\mathrm{d}t$, *where* $\nu \in M(\mathbf{R}^+; \mathbf{R}^{n\times n})$, $a \in C(\mathbf{R}^+; \mathbf{R}^{n\times n})$, *and both* ν *and* a *are of positive type;*

(ii) $g \in C(\mathbf{R}^n; \mathbf{R}^n)$;

(iii) $f \in L^1_{\mathrm{loc}}(\mathbf{R}^+; \mathbf{R}^n)$;

(iv) $x \in BC(\mathbf{R}^+; \mathbf{R}^n) \cap AC_{\mathrm{loc}}(\mathbf{R}^+; \mathbf{R}^n)$ *is a solution of (1.2) on* $\mathbf{R}^+$, *satisfying* $\sup_{T>0} Q(\mu, g\circ x, T) < \infty$.

Then $x' - f \in L^\infty(\mathbf{R}^+; \mathbf{R}^n)$. *In particular, if the integral of* f *is uniformly continuous, then so is* x.

Proof Since both ν and a are of positive type, it follows from (i) and (iv) that $\sup_{T>0} Q(a, g\circ x, T) < \infty$. By Corollary **16**.6.6, this implies that $a * g\circ x \in BC(\mathbf{R}^+; \mathbf{R}^n)$. Moreover, since $\nu \in M(\mathbf{R}^+; \mathbf{R}^{n\times n})$ and $g\circ x \in BC(\mathbf{R}^+; \mathbf{R}^n)$, we have $\nu * g\circ x \in B^\infty(\mathbf{R}^+)$. Together with (1.2) this gives the assertion. □

Below we will repeatedly use the fact that if $\lim_{t\to\infty} \frac{1}{T}\int_t^{t+T} f(s)\,\mathrm{d}s = F$ for all $T > 0$ then the function $t \mapsto \int_0^t f(s)\,\mathrm{d}s$ is uniformly continuous (see Lemma **15**.9.2).

If we assume that μ is of anti-coercive type, then we can say more about the behaviour of the solution.

3.2 Theorem
Let the following hypotheses hold:
(i) $\mu \in PT_{aco}(\mathbf{R}^+; \mathbf{R}^{n\times n})$;
(ii) $g \in C(\mathbf{R}^n; \mathbf{R}^n)$;
(iii) $f \in L^1_{loc}(\mathbf{R}^+; \mathbf{R}^n)$;
(iv) $x \in BC(\mathbf{R}^+; \mathbf{R}^n) \cap AC_{loc}(\mathbf{R}^+; \mathbf{R}^n)$ *is a solution of (1.2) on* $\mathbf{R}^+$, *satisfying* $\sup_{T>0} Q(\mu, g\circ x, T) < \infty$.
Then $x' - f \in L^2(\mathbf{R}^+; \mathbf{R}^n)$. *In particular, if* $\lim_{t\to\infty} \int_t^{t+T} f(s)\,\mathrm{d}s = 0$ *for all* $T > 0$, *then* x *is asymptotically slowly varying.*

Proof Since the quadratic form in $g\circ x$ is assumed to be uniformly bounded, it follows that $\mu * g\circ x$ is square integrable by the definition of anti-coercivity, and we have the first conclusion from (1.2). The second conclusion is then obvious. □

When the second conclusion of Theorem 3.2 is valid, it is easy to obtain an equation satisfied by the limit values of x.

3.3 Theorem
Assume that
(i) $\mu \in M(\mathbf{R}^+; \mathbf{R}^{n\times n})$,
(ii) $g \in C(\mathbf{R}^n; \mathbf{R}^n)$,
(iii) $f \in L^1_{loc}(\mathbf{R}^+; \mathbf{R}^n)$ *satisfies* $\lim_{t\to\infty} \frac{1}{T} \int_t^{t+T} f(s)\,\mathrm{d}s = F_0$ *for some* $F_0 \in \mathbf{R}^n$ *and all* $T > 0$,
(iv) $x \in BUC(\mathbf{R}^+; \mathbf{R}^n) \cap AC_{loc}(\mathbf{R}^+; \mathbf{R}^n)$ *is an asymptotically slowly varying solution of (1.2) on* $\mathbf{R}^+$.
Then $A_0 g(x(t)) \to F_0$ *as* $t \to \infty$, *where* $A_0 = \int_{\mathbf{R}^+} \mu(\mathrm{d}s)$. *In particular,* F_0 *belongs to the range of* A_0, *and if* A_0 *is invertible, then* $g(x(t)) \to A_0^{-1} F_0$ *as* $t \to \infty$.

Suppose that, in addition,
(i*) $A_0 = 0$ *(hence* $F_0 = 0$*), and* $\int_{\mathbf{R}^+} s|\mu|(\mathrm{d}s) < \infty$,
(iii*) $\int_0^t f(s)\,\mathrm{d}s \to F_1$ *as* $t \to \infty$.
Then $x(t) + A_1 g(x(t)) \to x_0 + F_1$ *as* $t \to \infty$, *where* $A_1 = -\int_{\mathbf{R}^+} s\mu(\mathrm{d}s)$.

Proof Since x is asymptotically slowly varying, every y in the positive limit set $\Gamma(x)$ of x is a constant. On the other hand, every $y \in \Gamma(x)$ satisfies the limit equation (cf. Theorem **15**.2.11 and its proof)

$$y'(t) + \int_{\mathbf{R}^+} \mu(\mathrm{d}s) g\big(y(t-s)\big) = F_0, \quad t \in \mathbf{R}.$$

As y is a constant, $A_0 g(y) = F_0$, where $A_0 = \int_{\mathbf{R}^+} \mu(\mathrm{d}s)$. But $\Gamma\big(A_0(g\circ x)\big) = A_0 g\big(\Gamma(x)\big)$ (see Theorem **15**.2.7), and therefore $\Gamma\big(A_0(g\circ x)\big) = \{F_0\}$. By Theorem **15**.3.1, $A_0 g(x(t)) \to F_0$ as $t \to \infty$.

When $A_0 = 0$, we integrate (1.2) to get the integral equation

$$x(t) + \int_0^t a(s)g\big(x(t-s)\big)\,\mathrm{d}s = x_0 + \int_0^t f(s)\,\mathrm{d}s,$$

where $a(t) = \mu([0,t])$ for $t > 0$. Straightforward integrations, which use the assumption $A_0 = 0$, give

$$\begin{aligned}\int_{\mathbf{R}^+} a(s)\,\mathrm{d}s &= \int_{\mathbf{R}^+}\int_{[0,t]} \mu(\mathrm{d}s)\,\mathrm{d}t \\ &= -\int_{\mathbf{R}^+}\int_{(t,\infty)} \mu(\mathrm{d}s)\,\mathrm{d}t = -\int_{\mathbf{R}^+} s\mu(\mathrm{d}s). \qquad (3.1)\end{aligned}$$

Similarly, by the second part of (i*), $a \in L^1(\mathbf{R}^+;\mathbf{R}^{n\times n})$. Since x is asymptotically slowly varying, we infer, by (iii*), that

$$x(t) + \Big(\int_{\mathbf{R}^+} a(s)\,\mathrm{d}s\Big)g\big(x(t)\big) \to x_0 + F_1 \quad \text{as} \quad t \to \infty.$$

Finally, if we replace $\int_{\mathbf{R}^+} a(s)\,\mathrm{d}s$ by $-\int_{\mathbf{R}^+} s\mu(\mathrm{d}s)$, then we get the desired conclusion. □

In Theorem 3.2, μ was supposed to be of anti-coercive type. A fairly small change in the proof permits us to prove a much more general result. The key observation is that, whenever $\mu \in M(\mathbf{R}^+)$ is a measure of positive type, it is always possible to find another measure $\nu \in M(\mathbf{R})$ that is dominated by μ in the sense of Theorem **16**.5.2, i.e., there exists a constant q such that for every $\varphi \in L^2(\mathbf{R})$ (with compact support) we have

$$\|\nu * \varphi\|^2_{L^2(\mathbf{R})} \le q[\![\mu,\varphi]\!].$$

Explicitly, take ν to be the measure whose Fourier transform $\tilde{\nu}$ is $\tilde{\nu} = 2\Re\tilde{\mu}$, that is, the measure that one gets by defining $\nu(E) = \mu(E\cap\mathbf{R}^+) + \mu(-E\cap\mathbf{R}^+)^*$. It is clear that this measure satisfies the hypothesis of Theorem **16**.5.2. We have $\tilde{\nu} = 2\Re\tilde{\mu} \succeq 0$; hence, by (**16**.2.2), for every $\omega \in \mathbf{R}$ and every $y \in \mathbf{C}^n$,

$$\begin{aligned}|\tilde{\nu}(\omega)y|^2 &\le |\tilde{\nu}(\omega)|\langle y,\tilde{\nu}(\omega)y\rangle \\ &= 4|\Re\tilde{\mu}(\omega)|\langle y,\Re\tilde{\mu}(\omega)y\rangle \\ &\le 4\|\mu\|_{M(\mathbf{R}^+)}\langle y,\Re\tilde{\mu}(\omega)y\rangle.\end{aligned}$$

In particular, we can take the constant q to be $4\|\mu\|_{M(\mathbf{R}^+)}$. (Moreover, $\tilde{\mu} \in L^2_{\mathrm{loc}}(\mathbf{R}^+)$ since $\mu \in M(\mathbf{R}^+)$.)

Let us summarize the preceding argument in the following lemma.

3.4 Lemma
Let $\varphi \in B^\infty(\mathbf{R}^+;\mathbf{C}^n)$, let $\mu \in M(\mathbf{R}^+;\mathbf{C}^{n\times n})$ be of positive type, and suppose that $K \stackrel{\mathrm{def}}{=} \sup_{T>0} Q(\mu,\varphi,T) < \infty$. Define $\nu(E) = \mu(E\cap\mathbf{R}^+) + \mu(-E\cap\mathbf{R}^+)^*$ for all Borel sets $E \subset \mathbf{R}$. Then the function ψ defined by $\psi(t) = \int_{(-\infty,t]} \nu(\mathrm{d}s)\varphi(t-s)$, $t \in \mathbf{R}$, is square integrable and $\|\psi\|^2_{L^2(\mathbf{R})} \le 4\|\mu\|_{M(\mathbf{R}^+)}K$. If, in addition, $\varphi \in BUC(\mathbf{R}^+;\mathbf{C}^n)$, then $\psi(t) \to 0$ as $t \to \infty$.

Proof Fix $T > 0$, and define φ_T to be $\varphi_T = \chi_{[0,T]}\varphi$. Applying the argument given above, with φ replaced by φ_T, we find that the function $\psi_T = \nu * \varphi_T$ satisfies

$$\|\psi_T\|^2_{L^2(\mathbf{R})} \le 4\|\mu\|_{M(\mathbf{R}^+)} Q(\mu, \varphi, T) \le 4\|\mu\|_{M(\mathbf{R}^+)} K.$$

Let $T \to \infty$. Then $\psi_T(t) \to \psi(t)$ for each fixed $t \in \mathbf{R}$, and hence, by Fatou's lemma,

$$\|\psi\|^2_{L^2(\mathbf{R})} \le \liminf_{T\to\infty} \|\psi_T\|^2_{L^2(\mathbf{R})} \le 4\|\mu\|_{M(\mathbf{R}^+)} K.$$

Next, suppose in addition that $\varphi \in BUC(\mathbf{R}^+; \mathbf{C}^n)$. Define $\varphi(t) = 0$ for $t < 0$, and write φ as the sum of two functions φ_1 and φ_2, where $\varphi_1 \in L^2(\mathbf{R}; \mathbf{C}^n) \cap B_0^\infty(\mathbf{R}; \mathbf{C}^n)$ and $\varphi_2 \in BUC(\mathbf{R}; \mathbf{C}^n)$ (e.g., define $\varphi_1(t) = \varphi(0)(1-t)$ for $t \in [0,1]$ and $\varphi_1(t) = 0$ otherwise). By Theorem **3**.6.1(i)–(iii), $\nu * \varphi_1 \in L^2(\mathbf{R}; \mathbf{C}^n)$, $(\nu * \varphi_1)(t) \to 0$ as $t \to \infty$, and $\nu * \varphi_2 \in BUC(\mathbf{R}; \mathbf{C}^n)$. Moreover, since $\nu * \varphi_2 = \psi - \nu * \varphi_1$, one has $\nu * \varphi_2 \in L^2(\mathbf{R}; \mathbf{C}^n)$. Consequently, $(\nu * \varphi_2)(t) \to 0$ as $t \to \infty$. □

This lemma enables us to prove several important results, beginning with the following.

3.5 Theorem

Let $\mu \in PT_{\mathrm{strict}}(\mathbf{R}^+; \mathbf{C}^{n\times n})$, $\varphi \in BUC(\mathbf{R}^+; \mathbf{C}^n)$, and suppose that $\sup_{T>0} Q(\mu, \varphi, T) < \infty$. Then $\varphi(t) \to 0$ as $t \to \infty$.

By choosing $\varphi = g \circ x$, we obtain a result on solutions of (1.2). The same comment applies to Theorem 3.7.

Proof of Theorem 3.5 Since μ is of strict positive type, there is a scalar function $a \in L^1(\mathbf{R}^+; \mathbf{R})$ of positive type satisfying $\Re\tilde{a}(\omega) > 0$ for $\omega \in \mathbf{R}$, and $Q(aI, \varphi, T) \le Q(\mu, \varphi, T)$ for all $T > 0$. (See the comment following the proof of Theorem **16**.4.5. Here I is the identity matrix.) In particular, this means that $\sup_{T>0} Q(aI, \varphi, T) < \infty$. Define ψ as in Lemma 3.4 with μ replaced by aI. Then, by Lemma 3.4, $\psi(t) \to 0$ as $t \to \infty$. Applying Pitt's form of Wiener's Tauberian theorem (i.e., Corollary **15**.6.4), we conclude that $\varphi(t) \to 0$ as $t \to \infty$. □

If μ is not of strict positive type, then we can still use essentially the same proof to obtain a weaker conclusion. The main difference is that one replaces the reference to Wiener's Tauberian theorem by a reference to Theorem **15**.6.3(vi). However, before we can utilise this result in the same generality as above, we need a new definition.

3.6 Definition

Let $\mu \in PT(\mathbf{R}^+; \mathbf{C}^{n\times n})$. The set

$$Z(\mu) \overset{\text{def}}{=} \left\{ \tau \in \mathbf{R} \;\middle|\; \inf_{\substack{y\in\mathbf{C}^n \\ |y|=1}} \liminf_{\substack{z\to i\tau \\ \Re z>0}} \langle y, \Re\hat{\mu}(z) y\rangle = 0 \right\}$$

is called the Fourier zero-set of μ.

If $|\mu|(\mathbf{R}^+) < \infty$, then this definition reduces to $Z(\mu) = \{\omega \in \mathbf{R} \mid \det[\Re\tilde{\mu}(\omega)] = 0\}$. To see this, it suffices to observe that if, for some ω, $\det[\Re\tilde{\mu}(\omega)] = 0$ then, obviously, there exists y such that $|y| = 1$, $\Re\tilde{\mu}(\omega)y = 0$, and, conversely, to note that if $\langle y, \Re\tilde{\mu}(\omega)y\rangle = 0$ for some $|y| = 1$, then, by (**16**.2.2), $\Re\tilde{\mu}(\omega)y = 0$, hence $\det[\Re\tilde{\mu}(\omega)] = 0$.

Note that if μ is of strict positive type then, by Theorem **16**.4.5, its Fourier zero-set is empty. Thus, the next theorem generalizes Theorem 3.5.

3.7 Theorem

Let $\mu \in PT(\mathbf{R}^+; \mathbf{C}^{n\times n})$, $\varphi \in BUC(\mathbf{R}^+; \mathbf{C}^n)$, and suppose that $\sup_{T>0} Q(\mu, \varphi, T) < \infty$. Then the asymptotic spectrum $\sigma_\infty(\varphi)$ of φ is contained in the Fourier zero-set $Z(\mu)$ of μ.

Proof Define (as in the proof of Theorem **16**.4.5),

$$v(\tau) \overset{\text{def}}{=} \inf_{\substack{y\in\mathbf{C}^n \\ |y|=1}} \liminf_{\substack{z\to i\tau \\ \Re z>0}} \langle y, \Re\hat{\mu}(z)y\rangle.$$

Then $Z(\mu) = \{\tau \in \mathbf{R} \mid v(\tau) = 0\}$. Take some $\tau \notin Z(\mu)$. Since v is lower semicontinuous and $v(\tau) > 0$, there is an $\epsilon > 0$ such that $v(\omega) > \epsilon$ for $|\omega - \tau| \le \epsilon$. Let $a \in L^1(\mathbf{R}^+; \mathbf{C})$ be the function whose Fourier transform $\tilde{a}$ satisfies $\Re\tilde{a}(\omega) = \epsilon - |\omega - \tau|$ for $|\omega - \tau| \le \epsilon$, and $\Re\tilde{a}(\omega) = 0$ otherwise. Then, by the equivalence of (i) and (v) in Theorem **16**.2.6, $\mu - aI$ is of positive type, and hence $\sup_{T>0} Q(aI, \varphi, T) < \infty$. It follows from Lemma 3.4 that if we define $b(t) = a(t)$ for $t \ge 0$ and $b(t) = \overline{a(-t)}$ for $t < 0$ then $(b * \varphi)(t) \to 0$ as $t \to \infty$. By Theorem **15**.6.3(iv), we have $\tau \notin \sigma_\infty(\varphi)$. □

This result can be further extended, provided $\mu \in M(\mathbf{R}^+)$ and $\tilde{\mu}$ vanishes whenever $\Re\tilde{\mu}$ does, or more precisely provided $\ker[\Re\tilde{\mu}(\omega)] = \ker[\tilde{\mu}(\omega)]$ for all $\omega \in Z(\mu)$. (If this equality holds for $\omega \in Z(\mu)$, then it holds for $\omega \in \mathbf{R}$.) In this case the set $\sigma^\infty(\mu * \varphi)$ is empty. We formulate the result as it applies to solutions of (1.2). In particular, we take the functions real-valued.

3.8 Theorem

Assume that

(i) *$\mu \in M(\mathbf{R}^+; \mathbf{R}^{n\times n})$ is of positive type, with*

$$\ker[\Re\tilde{\mu}(\omega)] = \ker[\tilde{\mu}(\omega)], \quad \omega \in Z(\mu),$$

and $Z(\mu)$ is countable, or $n = 1$ and $\tilde{\mu}$ is Hölder-continuous with exponent $\frac{1}{2}$, or $n = 1$ and $\tilde{\mu}$ is locally of bounded variation,

(ii) *$g \in C(\mathbf{R}^n; \mathbf{R}^n)$,*

(iii) *$f \in L^1_{\text{loc}}(\mathbf{R}^+; \mathbf{R}^n)$,*

(iv) *$x \in BUC(\mathbf{R}^+; \mathbf{R}^n) \cap AC_{\text{loc}}(\mathbf{R}^+; \mathbf{R}^n)$ is a solution of (1.2) on $\mathbf{R}^+$, satisfying $\sup_{T>0} Q(\mu, g\circ x, T) < \infty$.*

Then $\operatorname{ess\,lim}_{t\to\infty}(x'(t) - f(t)) = 0$. *In particular, if f satisfies $\lim_{t\to\infty}\int_t^{t+T} f(s)\,\mathrm{d}s = 0$ for all $T > 0$, then x is asymptotically slowly varying.*

If $n = 1$, then the condition $\ker[\Re\tilde{\mu}(\omega)] = \ker[\tilde{\mu}(\omega)]$ for $\omega \in Z(\mu)$, reduces to the assumption that $\tilde{\mu}(\omega) = 0$ whenever $\Re\tilde{\mu}(\omega) = 0$. Recall that, when x is asymptotically slowly varying, one can say more about the asymptotic behaviour of x by applying Theorem 3.3.

Proof of Theorem 3.8 It suffices to prove that $(\mu * g\circ x)(t) \to 0$ as $t \to \infty$, or equivalently, that every $y \in \Gamma(g\circ x)$ satisfies $\mu * y \equiv 0$ (cf. Theorems **15**.2.7 and **15**.3.1). It follows from Lemma 3.4 and Theorem **15**.2.7 that every such y satisfies $\nu * y \equiv 0$, where ν is the measure defined in Lemma 3.4. By Definition **15**.5.1, for each $\omega \in \mathbf{R}$ we have $\rho_\omega(y) \subset \ker[\tilde{\nu}(\omega)] = \ker[\Re\tilde{\mu}(\omega)]$, and from (i) we get that $\ker[\Re\tilde{\mu}(\omega)] = \ker[\tilde{\mu}(\omega)]$, for $\omega \in \mathbf{R}$. By applying Corollary **15**.5.7 and Theorem **15**.4.15, we get the desired conclusion that $\mu * y \equiv 0$. □

4. L^2-Estimates

In the two previous sections we found conditions for boundedness and for the existence of a limit at infinity of a solution of

$$x'(t) + \int_{[0,t]} \mu(\mathrm{d}s)g\big(x(t-s)\big) = f(t), \quad t \in \mathbf{R}^+; \quad x(0) = x_0. \tag{1.2}$$

We now turn our attention to estimates yielding square integrability.

First, let us prove the following easy extension of Theorems 2.4 and 3.2.

4.1 Theorem

Suppose that

(i) $a \in L^1(\mathbf{R}^+;\mathbf{R}^{n\times n}) \cap BV(\mathbf{R}^+;\mathbf{R}^{n\times n})$ *is of strong positive type,*

(ii) $g \in C(\mathbf{R}^n;\mathbf{R}^n)$ *is the gradient of a function* $G \in C^1(\mathbf{R}^n;\mathbf{R})$*, satisfying* $\lim_{|y|\to\infty} G(y) = +\infty$, $\langle y, g(y)\rangle > 0$ *for all nonzero* $y \in \mathbf{R}^n$*, and* $\liminf_{y\to 0}\langle y, g(y)\rangle/|y|^2 > 0$,

(iii) $f \in L^2(\mathbf{R}^+;\mathbf{R}^n) \cap AC_{\mathrm{loc}}(\mathbf{R}^+;\mathbf{R}^n)$ *and* $f' \in L^2(\mathbf{R}^+;\mathbf{R}^n)$,

(iv) $x \in AC_{\mathrm{loc}}(\mathbf{R}^+;\mathbf{R}^n)$ *is a solution of (1.2) with* $\mu(\mathrm{d}s)$ *replaced by* $a(s)\,\mathrm{d}s$.

Then x *and* $x' \in L^2(\mathbf{R}^+;\mathbf{R}^n)$.

(The assumption $a \in L^1(\mathbf{R}^+)$ above can be removed; see Theorem 4.3 below.)

Proof of Theorem 4.1 By Theorem 2.4, $x \in BC(\mathbf{R}^+;\mathbf{R}^n)$, and $K \stackrel{\text{def}}{=} \sup_{T>0} Q(a, g\circ x, T) < \infty$. According to Proposition **16**.5.8(ii), a is of anti-coercive type, and thus, by Theorem 3.2, $x' - f \in L^2(\mathbf{R}^+;\mathbf{R}^n)$, or equivalently by (iii), $x' \in L^2(\mathbf{R}^+;\mathbf{R}^n)$.

Fix $T > 0$. Then, by Corollary **16**.6.4,

$$\left|\int_0^T \langle x(t), g(x(t))\rangle\,\mathrm{d}t\right|^2 \le \left\{|x(0)|^2 + |x(T)|^2 + \int_0^T \left(|x(t)|^2 + |x'(t)|^2\right)\mathrm{d}t\right\} Q(\mathbf{e}, g\circ x, T),$$

where $\mathbf{e}(t) = \mathrm{e}^{-t}$ for $t \ge 0$ and $\mathbf{e}(t) = 0$ for $t < 0$. The hypothesis (i) implies that

$$\epsilon Q(\mathbf{e}, g\circ x, T) \le Q(a, g\circ x, T) \le K, \tag{4.1}$$

for some $\epsilon > 0$ and (ii) together with the fact that x is bounded implies that ϵ can be chosen such that, for all $t > 0$,

$$\langle x(t), g(x(t))\rangle \ge \epsilon |x(t)|^2.$$

Thus we get

$$\epsilon^3 \|x\|^4_{L^2(0,T)} \le K\left(|x(0)|^2 + |x(T)|^2 + \|x\|^2_{L^2(0,T)} + \|x'\|^2_{L^2(0,T)}\right). \tag{4.2}$$

Since $x' \in L^2(\mathbf{R}^+;\mathbf{R}^n)$ and $x \in BC(\mathbf{R}^+;\mathbf{R}^n)$, this inequality implies that $\|x\|_{L^2(0,T)}$ is bounded uniformly in T; i.e., $x \in L^2(\mathbf{R}^+;\mathbf{R}^n)$. □

If the assumption $a \in L^1(\mathbf{R}^+)$ is removed, then the proof becomes slightly more complicated. It is based on the following modification of Corollary **16**.6.4.

4.2 Lemma
Let $T > 0$, let $b \in L^2([0,T];\mathbf{C}^n)$ be absolutely continuous with $b' \in L^2([0,T];\mathbf{C}^n)$, let $\nu \in M([0,T];\mathbf{C}^{n\times n})$, and let $\varphi \in L^2_{\mathrm{loc}}(\mathbf{R}^+;\mathbf{C}^n)$. Then

$$\int_0^T \langle b(t), (\nu * \varphi)(t)\rangle\,\mathrm{d}t \le \|\nu\|_{M([0,T])}|b(T)| \sup_{t\in[0,T]} \sqrt{2Q(\mathbf{e},\varphi,t)} + \left(\|b\|_{L^2(0,T)} + \|b'\|_{L^2(0,T)}\right)\|\nu\|_{M([0,T])}\sqrt{Q(\mathbf{e},\varphi,T)}.$$

In the case where ν is the unit point mass at zero, this inequality gives an alternative bound for the integral $\int_0^T \langle b(t), \varphi(t)\rangle\,\mathrm{d}t$ studied in Corollary **16**.6.4.

Proof of Lemma 4.2 Define $b(t) = \varphi(t) = 0$ for $t < 0$ and $\nu(E) = \nu(E \cap \mathbf{R}^+)$ for every Borel set $E \subset \mathbf{R}$. Integrating by parts, one finds that

$$\int_0^T \left\langle b(t), (\nu * \varphi)(t)\right\rangle \mathrm{d}t = \left\langle b(T), (\mathbf{e} * \nu * \varphi)(T)\right\rangle + \int_0^T \left\langle b(t) - b'(t), (\mathbf{e} * \nu * \varphi)(t)\right\rangle \mathrm{d}t.$$

Since $\mathbf{e} * (\nu * \varphi) = \nu * (\mathbf{e} * \varphi)$, this implies that

$$\int_0^T \Big\langle b(t), (\nu * \varphi)(t) \Big\rangle \, dt \le |b(T)| \|\nu\|_{M([0,T])} \|\mathbf{e} * \varphi\|_{\sup([0,T])}$$
$$+ \Big(\|b\|_{L^2(0,T)} + \|b'\|_{L^2(0,T)} \Big) \|\nu\|_{M([0,T])} \|\mathbf{e} * \varphi\|_{L^2(0,T)}.$$

By Corollary **16**.6.6, $|(\mathbf{e} * \varphi)(t)|^2 \le 2Q(\mathbf{e}, \varphi, t)$ for all $t \in [0,T]$; hence $\|\mathbf{e} * \varphi\|_{\sup([0,T])} \le \sup_{t \in [0,T]} \sqrt{2Q(\mathbf{e}, \varphi, t)}$, and by Corollary **16**.5.3 $\|\mathbf{e} * \varphi\|_{L^2(0,T)} \le \sqrt{Q(\mathbf{e}, \varphi, T)}$. If we substitute these estimates above, then we get the claim. □

4.3 Theorem

Theorem 4.1 is true without the assumption $a \in L^1(\mathbf{R}^+; \mathbf{R}^{n \times n})$.

Proof Checking the proof of Theorem 4.1, we find that the argument giving $x \in BC(\mathbf{R}^+; \mathbf{R}^n)$, $\sup_{T>0} Q(a, g{\circ}x, T) < \infty$, and the inequality (4.2) is still valid. Therefore, we only need to show that $x' \in L^2(\mathbf{R}^+; \mathbf{R}^n)$, without recourse to the assumption $a \in L^1(\mathbf{R}^+; \mathbf{R}^{n \times n})$. To do this, first observe that by Theorem **16**.2.7 one may, without loss of generality, take a continuous, and then that, by Theorem 3.1, $x' \in L^\infty(\mathbf{R}^+; \mathbf{R}^n)$. Let α be the measure derivative of a, i.e., α is the measure that satisfies $\alpha([0,t]) = a(t)$ for almost all $t \in \mathbf{R}^+$. Differentiating (1.2) one gets, for almost all $t \in \mathbf{R}^+$,

$$x''(t) + (\alpha * g{\circ}x)(t) = f'(t).$$

Take the inner product of this equation and $-x(t)$, and integrate over $[0,T]$ (the first term by parts) to obtain

$$\int_0^T |x'(t)|^2 dt = \langle x(T), x'(T) \rangle - \langle x(0), x'(0) \rangle$$
$$+ \int_0^T \langle x(t), (\alpha * g{\circ}x)(t) \rangle \, dt - \int_0^T \langle x(t), f'(t) \rangle \, dt.$$

Here the first term on the right-hand side is bounded, since both x and x' are bounded. The second term is independent of T. The third term can be estimated with the aid of Lemma 4.2 and (4.1). In the last term, apply Hölder's inequality. This leads to an estimate of the type

$$\|x'\|^2_{L^2(0,T)} \le C\Big(1 + \|x\|_{L^2(0,T)} + \|x'\|_{L^2(0,T)}\Big),$$

where C is a constant independent of T. Together with (4.2), this implies that both $\|x\|_{L^2(0,T)}$ and $\|x'\|_{L^2(0,T)}$ are bounded, uniformly in T. □

5. An Integral Equation

After discussing the integrodifferential equation (1.2), we are now faced by the integral equations

$$x(t) + \int_0^t a(s)g\big(x(t-s)\big)\,\mathrm{d}s = f(t), \quad t \in \mathbf{R}^+, \tag{1.1}$$

and

$$x(t) + \int_0^t a(s)g\big(t-s, x(t-s)\big)\,\mathrm{d}s = f(t), \quad t \in \mathbf{R}^+, \tag{1.3}$$

mentioned in the introduction.

A fact that should be pointed out immediately is that (1.1) and (1.2) are closely related to each other. If a is locally of bounded variation and f is locally integrable, then x satisfies (1.1) if and only if it satisfies (1.2) with μ replaced by the distribution derivative of a, f replaced by f', and $x_0 = f(0)$. Conversely, (1.2) can always be integrated, and every solution of (1.2) satisfies (1.1) with $a(t)$ replaced by $\mu([0,t])$ and $f(t)$ replaced by $x_0 + \int_0^t f(s)\,\mathrm{d}s$. The consequence is that one may obtain results for (1.1) directly from the earlier theory developed for (1.2).

For the preceding reasoning to be valid, three conditions have to be met:

- a should be locally of bounded variation;
- f should be locally absolutely continuous;
- the measure derivative of a should be of positive type.

If any one of these three conditions fails to hold, then the results of Sections 2–4 are of no use when one studies the behaviour of solutions of (1.1). However, it may of course be the case that the kernel a itself is of positive type (instead of having a derivative of positive type). In that case one does the obvious thing, which is to take the inner product of (1.1) and $g(x(t))$, and to integrate over an interval $[0,T]$. This leads to the equation

$$\int_0^T \langle g\big(x(t)\big), x(t)\rangle\,\mathrm{d}t + \int_0^T \Big\langle g\big(x(t)\big), \int_0^t a(s)g\big(x(t-s)\big)\,\mathrm{d}s\Big\rangle\,\mathrm{d}t = \int_0^T \langle g\big(x(t)\big), f(t)\rangle\,\mathrm{d}t, \tag{5.1}$$

which immediately produces a new basic inequality, namely

$$\begin{aligned}\int_0^T \langle g\big(x(t)\big), x(t)\rangle\,\mathrm{d}t &\le \int_0^T \langle g\big(x(t)\big), x(t)\rangle\,\mathrm{d}t + Q(a, g\circ x, T) \\ &= \int_0^T \langle g\big(x(t)\big), f(t)\rangle\,\mathrm{d}t.\end{aligned} \tag{5.2}$$

The only difference, compared to the basic estimate (2.3) for the integrodifferential equation (1.2), is that the term $\int_0^T \langle g(x(t)), x'(t)\rangle\,\mathrm{d}t = G(x(t)) - G(x_0)$ has been replaced by $\int_0^T \langle g(x(t)), x(t)\rangle\,\mathrm{d}t$. In particular, there is no longer any need to assume that g is the gradient of a continuously differentiable function G. Instead, it is natural to assume that $\langle g(y), y\rangle \geq 0$ for all $y \in \mathbf{R}^n$.

In our discussion of (1.2) we could not, without an additional monotonicity assumption, let g be explicitly dependent on t (cf. Exercise 5), but in the estimate (5.2) it causes no trouble to let g depend not only on x, but also explicitly on t. That is, instead of studying (1.1), we can equally well study (1.3). The only difference is that (5.2) is replaced by

$$\int_0^T \langle g(t,x(t)), x(t)\rangle \, dt \leq \int_0^T \langle g(t,x(t)), x(t)\rangle \, dt + Q(a, g(\cdot, x(\cdot)), T) = \int_0^T \langle g(t,x(t)), f(t)\rangle \, dt. \quad (5.3)$$

In the discussion above, we have taken a to be of positive type. If this is not true, but a only satisfies the weaker condition

$$\Re \hat{a}(z) + A \succeq 0 \quad \text{for} \quad \Re z \geq 0,$$

for a matrix $A \succeq 0$ (this condition is always satisfied for some A when $a \in L^1(\mathbf{R}^+)$), then $a + A\delta$ is of positive type, and

$$Q(a, g\circ x, T) \geq -\int_0^T \langle g(x(t)), Ag(x(t))\rangle \, dt.$$

In this case, the inequality (5.2) should be replaced by

$$\int_0^T \langle g(x(t)), x(t) - Ag(x(t))\rangle \, dt \leq \int_0^T \langle g(x(t)), x(t)\rangle \, dt + Q(a, g\circ x, T) = \int_0^T \langle g(x(t)), f(t)\rangle \, dt, \quad (5.4)$$

and, similarly, (5.3) should be replaced by

$$\begin{aligned} \int_0^T \langle g(t,x(t)), x(t) - Ag(t,x(t))\rangle \, dt &\leq \int_0^T \langle g(t,x(t)), x(t)\rangle \, dt + Q(a, g(\cdot, x(\cdot)), T) \\ &= \int_0^T \langle g(t,x(t)), f(t)\rangle \, dt. \end{aligned} \quad (5.5)$$

To be able to utilise these two inequalities, we have to strengthen the assumptions $\langle g(y), y\rangle \geq 0$ and $\langle g(t,y), y\rangle \geq 0$ to the corresponding assumptions $\langle g(y), y - Ag(y)\rangle \geq 0$ and $\langle g(t,y), y - Ag(t,y)\rangle \geq 0$ for all $y \in \mathbf{R}^n$ (and almost all $t \in \mathbf{R}^+$).

Starting with one of the estimates (5.2)–(5.5), instead of with (2.3), one can develop a theory for (1.1) and (1.3) which is quite similar to the theory that we developed earlier for (1.2). The main difference is that, owing to the absence of the term $G(x(t))$ in the estimate, it is more difficult to get pointwise bounds on $x(t)$. On the other hand, the new term $\int_0^T \langle g(x(t)), x(t)\rangle \, dt$ makes it easier to get L^2-estimates for x.

In the sequel we list a few results for (1.3). The proofs are more or less obvious modifications of the corresponding proofs in Sections 2 and 3, so we either merely outline them, or leave them to the reader.

5.1 Theorem
Suppose that

(i) $a \in L^1_{\text{loc}}(\mathbf{R}^+; \mathbf{R}^{n\times n})$, *and* $a + A\delta$ *is of positive type for some constant matrix* $A \succeq 0$,

(ii) $g : \mathbf{R}^+ \times \mathbf{R}^n \to \mathbf{R}^n$ *satisfies the Carathéodory conditions and there exist a function* $h \in L^2(\mathbf{R}^+; \mathbf{R})$ *and a constant* $\epsilon > 0$ *such that, for almost all* $t \in \mathbf{R}^+$,

$$\langle g(t,y), y - Ag(t,y)\rangle \geq \epsilon |g(t,y)|^2 - h(t)|g(t,y)|, \quad y \in \mathbf{R}^n,$$

(iii) $f \in L^2(\mathbf{R}^+; \mathbf{R}^n)$,

(iv) $x \in L^2_{\text{loc}}(\mathbf{R}^+; \mathbf{R}^n)$ *is a solution of (1.3) on* $\mathbf{R}^+$.

Then $g(\cdot, x(\cdot)) \in L^2(\mathbf{R}^+; \mathbf{R}^n)$, *and* $\sup_{T>0} Q(a + A\delta, g(\cdot, x(\cdot)), T) < \infty$. *If, in addition,* $a \in L^1(\mathbf{R}^+; \mathbf{R}^{n\times n})$, *then* $x \in L^2(\mathbf{R}^+; \mathbf{R}^n)$.

Of course, if $A = 0$, then a is of positive type. To clarify (ii), recall that g satisfies the Carathéodory conditions (see Section **12**.2) if for each $y \in \mathbf{R}^n$ the function $t \mapsto g(t,y)$ is measurable on $\mathbf{R}^+$, and if for almost all $t \in \mathbf{R}^+$ the function $y \mapsto g(t,y)$ is continuous on $\mathbf{R}^n$.

Proof of Theorem 5.1 Because of the growth condition on g in (ii), and the positivity of A, we have

$$\epsilon |g(t,x(t))|^2 \leq |g(t,x(t))|\big(|x(t)| + h(t)\big),$$

hence

$$|g(t,x(t))| \leq \tfrac{1}{\epsilon}\big(|x(t)| + h(t)\big).$$

We conclude that the function $\psi(t) \stackrel{\text{def}}{=} g(t,x(t))$ belongs locally to L^2. It follows from (5.5) and (ii) that

$$\begin{aligned}\epsilon \|\psi\|^2_{L^2[0,T]} &\leq \int_0^T \big(h(t) + |f(t)|\big)|\psi(t)|\,\mathrm{d}t \\ &\leq \|\psi\|_{L^2[0,T]}\Big(\|h\|_{L^2(\mathbf{R}^+)} + \|f\|_{L^2(\mathbf{R}^+)}\Big).\end{aligned}$$

This implies that $\psi \in L^2(\mathbf{R}^+; \mathbf{R}^n)$. □

If a is not only of positive type, but actually of anti-coercive type, then the combined sign and growth condition (ii) on g can be slightly weakened.

5.2 Theorem

Let the following hypotheses hold:

(i) $a \in L^1_{\mathrm{loc}}(\mathbf{R}^+; \mathbf{R}^{n\times n})$ *is of anti-coercive type with coercivity constant* $q > 0$;

(ii) $g : \mathbf{R}^+ \times \mathbf{R}^n \to \mathbf{R}^n$ *satisfies the Carathéodory conditions, and there exist a function* $h \in L^2(\mathbf{R}^+; \mathbf{R})$ *and a constant* $\epsilon > 0$ *such that, for almost all* $t \in \mathbf{R}^+$,

$$\langle g(t,y), y\rangle \geq \epsilon |g(t,y)|^2 - \tfrac{1}{q}|y|^2 - h(t)|g(t,y)|, \quad y \in \mathbf{R}^n;$$

(iii) $f \in L^2(\mathbf{R}^+; \mathbf{R}^n)$;

(iv) $x \in L^2_{\mathrm{loc}}(\mathbf{R}^+; \mathbf{R}^n)$ *is a solution of (1.3) on* $\mathbf{R}^+$.

Then x *and* $g(\cdot, x(\cdot)) \in L^2(\mathbf{R}^+; \mathbf{R}^n)$, *and* $\sup_{T>0} Q(a, g(\cdot, x(\cdot)), T) < \infty$.

Proof Again, write $\psi(t) = g(t, x(t))$. In almost the same way as in the proof of Theorem 5.1, one shows that $\psi \in L^2_{\mathrm{loc}}(\mathbf{R}^+; \mathbf{R}^n)$. By the coercivity assumption and by (1.3),

$$\begin{aligned} qQ(a,\psi,T) &\geq \|a * \psi\|^2_{L^2(0,T)} \\ &= \|f - x\|^2_{L^2(0,T)} \\ &\geq \|x\|^2_{L^2(0,T)} + \|f\|^2_{L^2(0,T)} - 2\|x\|_{L^2(0,T)}\|f\|_{L^2(0,T)}. \end{aligned}$$

If we use this fact in (5.3), and invoke (ii), then we obtain

$$\epsilon\|\psi\|^2_{L^2(0,T)} \leq \|\psi\|_{L^2(0,T)}\Big(\|h\|_{L^2(\mathbf{R}^+)} + \|f\|_{L^2(\mathbf{R}^+)}\Big) + \frac{2}{q}\|x\|_{L^2(0,T)}\|f\|_{L^2(0,T)}. \quad (5.6)$$

By (1.3), $\|x\|_{L^2(0,T)} \leq \|a * \psi\|_{L^2(0,T)} + \|f\|_{L^2(0,T)}$. Recall from Lemma **16**.5.5 that $\|a * \psi\|_{L^2(0,T)} \leq q\|\psi\|_{L^2(0,T)}$. Thus $\|x\|_{L^2(0,T)} \leq q\|\psi\|_{L^2(0,T)} + \|f\|_{L^2(0,T)}$. Combining this inequality with (5.6), we obtain the desired conclusions. □

Above, if a is of positive type (i.e., if $A = 0$ in Theorem 5.1), then one can get a pointwise bound on $x - f$ (and hence on x if f is bounded) in the following way.

5.3 Theorem

Assume that

(i) $a = b + c$ *where* $b \in C(\mathbf{R}^+; \mathbf{R}^{n\times n})$ *and* $c \in L^2(\mathbf{R}^+; \mathbf{R}^{n\times n})$, *and both* b *and* c *are of positive type,*

(ii) $g : \mathbf{R}^+ \times \mathbf{R}^n \to \mathbf{R}^n$ *satisfies the Carathéodory conditions,*

(iii) $f \in L^2_{\mathrm{loc}}(\mathbf{R}^+; \mathbf{R}^n)$,

(iv) $x \in L^2_{\mathrm{loc}}(\mathbf{R}^+; \mathbf{R}^n)$ *is a solution of equation (1.3) on* $\mathbf{R}^+$ *such that* $g(\cdot, x(\cdot)) \in L^2(\mathbf{R}^+; \mathbf{R}^n)$ *and* $\sup_{T>0} Q(a, g(\cdot, x(\cdot)), T) < \infty$.

Then $x - f \in BC(\mathbf{R}^+; \mathbf{R}^n)$.

The short (omitted) proof is based on Corollary **16**.6.6.

Both in Theorem 5.1 and in Theorem 5.2 we require ϵ to be strictly positive. We can allow ϵ to be zero if we are willing to assume more on a and f, and to relate the size of $\tilde{f}$ to the size of $\tilde{a}$ (for simplicity we take the matrix A in Theorem 5.1 to be zero; no significant difficulties appear if $A \neq 0$).

5.4 Theorem

Let the following hypotheses hold:

(i) *$a \in L^1_{\text{loc}}(\mathbf{R}^+; \mathbf{R}^n)$ is of positive type;*

(ii) *g satisfies the Carathéodory conditions on $\mathbf{R}^+ \times \mathbf{R}^n$, and there exists a function $h \in L^1(\mathbf{R}^+; \mathbf{R})$ such that, for almost all $t \in \mathbf{R}^+$,*

$$\langle g(t,y), y\rangle \geq -h(t), \quad y \in \mathbf{R}^n;$$

(iii) *$f \in L^2_{\text{loc}}(\mathbf{R}^+; \mathbf{R}^n)$ induces a tempered distribution whose Fourier transform is a measure $\beta \in M_{\text{loc}}(\mathbf{R}; \mathbf{C}^n)$ of the form $\beta(\mathrm{d}\omega) = \alpha(\mathrm{d}\omega)\gamma(\omega)$, where $\alpha = \Re\tilde{a}$, and γ is a Borel measurable function satisfying $\int_{\mathbf{R}} |\gamma(\omega)|^2 |\alpha|(\mathrm{d}\omega) < \infty$;*

(iv) *$x \in L^2_{\text{loc}}(\mathbf{R}^+; \mathbf{R}^n)$ is a solution of (1.3) on $\mathbf{R}^+$, for which $g(\cdot, x(\cdot)) \in L^2_{\text{loc}}(\mathbf{R}^+; \mathbf{R}^n)$.*

Then $\sup_{T>0} Q(a, g(\cdot, x(\cdot)), T) < \infty$.

If, in addition to (i)*, a is of anti-coercive type, then $x - f \in L^2(\mathbf{R}^+; \mathbf{R}^n)$, and if, in addition to* (i)*, $a \in C(\mathbf{R}^+; \mathbf{R}^n)$, then $x - f \in BC(\mathbf{R}^+; \mathbf{R}^n)$.*

For the proof, it suffices to invoke Theorem **16**.6.5, Definition **16**.5.4, and Corollary **16**.6.6.

Let us finally describe one additional approach. If g is (independent of t and) continuously differentiable, then one can differentiate (1.1) to obtain

$$x'(t) + \int_0^t a(s) g'\big(x(t-s)\big) x'(t-s)\,\mathrm{d}s = f'(t) - a(t) g(x(0)).$$

Define $y(t) = x'(t)$, $g_0(t,y) = g'(x(t))y$ and $f_0(t) = f'(t) - a(t)g(x(0))$ to turn this equation into the equation

$$y(t) + \int_0^t a(s) g_0\big(t-s, y(t-s)\big)\,\mathrm{d}s = f_0(t), \quad t \in \mathbf{R}^+, \tag{5.7}$$

which is of the form (1.3). Under appropriate assumptions on a, g and f, one can derive a number of results regarding this equation. Observe, in particular, that if a is continuous then we can use Corollary **16**.6.6, to let the quadratic form $Q(a, g_0(\cdot, y(\cdot)), T)$ dominate the term

$$\int_0^T \Big\langle g_0\big(t, y(t)\big), a(t) g\big(x(0)\big)\Big\rangle\,\mathrm{d}t = \int_0^T \Big\langle a^*(t) g_0\big(t, y(t)\big), g\big(x(0)\big)\Big\rangle\,\mathrm{d}t,$$

appearing in the identity that one gets by taking the inner product of $g_0(\cdot, y(\cdot))$ and (5.7), and integrating over $[0, T]$.

6. Exercises

1. Prove an $\mathbf{R}^n$-version of Corollary 2.3 with G radially invariant (i.e., $G(x) = G(y)$ whenever $|x| = |y|$) and convex. Apply your result to the linear equation that you get by taking $g(x) = x$ and $G(x) = \frac{1}{2}|x|^2$. More generally, you can take $G(x) = \frac{1}{k}|x|^k$ for some $k > 1$.
2. By applying the results of this chapter, determine the asymptotic behaviour and size at infinity of solutions of the equation

$$x'(t) + \int_0^t \big(x(t-s)\big)^3 (1+s)^{-\alpha} \cos(s)\,\mathrm{d}s = (1+t)^{-\beta}, \quad t > 0,$$

where $x(0) = x_0$ and α, β are positive constants.
3. Prove the following result.

6.1 Theorem
Assume that
(i) $\mu = \nu + \epsilon\delta$, *where* $\nu \in PT(\mathbf{R}^+; \mathbf{R}^{n\times n})$, $\epsilon > 0$, *and* δ *is the identity point mass at zero,*
(ii) $g \in C(\mathbf{R}^n; \mathbf{R}^n)$ *is the gradient of a function* $G \in C^1(\mathbf{R}^n; \mathbf{R})$, *satisfying* $\lim_{|y|\to\infty} G(y) = +\infty$,
(iii) $f \in L^2(\mathbf{R}^+; \mathbf{R}^n)$,
(iv) $x \in AC_{\mathrm{loc}}(\mathbf{R}^+; \mathbf{R}^n)$ *is a solution of (1.2) on* $\mathbf{R}^+$.
Then $x \in BC(\mathbf{R}^+; \mathbf{R}^n)$, $\sup_{T>0} Q(\mu, g\circ x, T) < \infty$, *and* $g\circ x \in L^2(\mathbf{R}^+; \mathbf{R}^n)$.

4. Prove the following result.

6.2 Theorem
Suppose that
(i) $\mu \in M(\mathbf{R}^+; \mathbf{R}^{n\times n})$ *is of positive type and satisfies* $\Re\tilde{\mu}(\omega) \geq \epsilon\frac{\omega^2}{1+\omega^2}$ *for some* $\epsilon > 0$ *and all* $\omega \in \mathbf{R}$,
(ii) $g \in C(\mathbf{R}^n; \mathbf{R}^n)$ *is the gradient of a function* $G \in C^1(\mathbf{R}^n; \mathbf{R})$, *satisfying* $\lim_{|y|\to\infty} G(y) = +\infty$,
(iii) $f \in L^2(\mathbf{R}^+; \mathbf{R}^n)$ *satisfies* $\lim_{t\to\infty} \int_0^t f(s)\,\mathrm{d}s = F$, *and the function* $t \mapsto \int_0^t f(s)\,\mathrm{d}s - F$ *belongs to* $L^2(\mathbf{R}^+; \mathbf{R}^n)$,
(iv) $x \in AC_{\mathrm{loc}}(\mathbf{R}^+; \mathbf{R}^n)$ *is a solution of (1.2) on* $\mathbf{R}^+$.
Then $x \in BC(\mathbf{R}^+; \mathbf{R}^n)$, *and* $\sup_{T>0} Q(\mu, g\circ x, T) < \infty$.

5. Let $\mu \in PT(\mathbf{R}^+; \mathbf{R}^{n\times n})$, $f \in L^1_{\mathrm{loc}}(\mathbf{R}^+; \mathbf{R}^n)$, $G \in C^1(\mathbf{R}^+ \times \mathbf{R}^n; \mathbf{R})$, and let $g(t,x) = \frac{\partial}{\partial x}G(t,x) = G_x(t,x)$. Consider the equation

$$x'(t) + \int_{[0,t]} \mu(\mathrm{d}s) g\big(t-s, x(t-s)\big) = f(t), \quad t \in \mathbf{R}^+;$$
$$x(0) = x_0. \quad (6.1)$$

Take the inner product of both sides of this equation and $g(t, x(t))$, and integrate over some interval $[0, T]$ to obtain

$$\begin{aligned} G(T, x(T)) &- \int_0^T G_t(t, x(t))\,\mathrm{d}t \\ &+ \int_0^T \Big\langle g(t, x(t)), \int_{[0,t]} \mu(\mathrm{d}s) g\big(t-s, x(t-s)\big)\Big\rangle\,\mathrm{d}t \\ &= G(0, x_0) + \int_0^T \langle g(t, x(t)), f(t)\rangle\,\mathrm{d}t. \end{aligned} \tag{6.2}$$

Try to generalize as many of the results in Sections 2–4 as you can to equation (6.1), replacing (2.1) by (6.2). A natural assumption is that, for each $x \in \mathbf{R}^n$, the function $t \mapsto G(t, x)$ is nonincreasing and bounded from below; hence $G(\infty, x)$ exists, and $G(t, x) \geq G(\infty, x)$.

6. Let the assumptions of Theorem 3.8 hold, and assume that

$$\lim_{t\to\infty} \frac{1}{T} \int_t^{t+T} f(s)\,\mathrm{d}s = F,$$

for every $T > 0$. Show that $F = 0$.

7. Let $x \in BUC(\mathbf{R}^+; \mathbf{R}) \cap AC_{\mathrm{loc}}(\mathbf{R}^+; \mathbf{R})$ satisfy

$$x'(t) + \int_0^T \Big(1 - \frac{s}{T}\Big) g\big(x(t-s)\big)\,\mathrm{d}s = f(t), \quad t > 0,$$

where $g \in C(\mathbf{R}; \mathbf{R})$, $f \in L^1_{\mathrm{loc}}(\mathbf{R}^+; \mathbf{R})$ and $f(t) \to 0$ as $t \to \infty$. Can you say something more specific about the asymptotic behaviour of x?

8. Prove the following result.

6.3 Theorem
Assume that
(i) $\mu(\mathrm{d}t) = \nu(\mathrm{d}t) + \epsilon I\,\mathrm{d}t$, *where* $\nu \in PT(\mathbf{R}^+; \mathbf{R}^{n\times n})$, *and* $\epsilon > 0$,
(ii) $g \in C(\mathbf{R}^n; \mathbf{R}^n)$ *is the gradient of a function* $G \in C^1(\mathbf{R}^n; \mathbf{R})$, *satisfying* $\lim_{|y|\to\infty} G(y) = +\infty$,
(iii) $f \in BV(\mathbf{R}^+; \mathbf{R}^n)$,
(iv) $x \in AC_{\mathrm{loc}}(\mathbf{R}^+; \mathbf{R}^n)$ *is a solution of (1.2) on* $\mathbf{R}^+$.
Then $x \in BC(\mathbf{R}^+; \mathbf{R}^n)$, *and* $\sup_{T>0} Q(\mu, g\circ x, T) < \infty$.

9. Verify the following statement.

6.4 Theorem
Assume that
(i) $a \in L^2(\mathbf{R}^+; \mathbf{R}^{n\times n}) \cap PT(\mathbf{R}^+; \mathbf{R}^{n\times n})$,
(ii) $g \in C^1(\mathbf{R}^n, \mathbf{R}^n)$ *is Lipschitz-continuous and there exists a positive constant* ϵ *such that* $\epsilon|v|^2 \leq \langle g'(y)v, v\rangle$ *for all* v, $y \in \mathbf{R}^n$,
(iii) $f \in AC_{\mathrm{loc}}(\mathbf{R}^+; \mathbf{R}^n)$ *and* $f' \in L^2(\mathbf{R}^+; \mathbf{R}^n)$,
(iv) $x \in C(\mathbf{R}^+; \mathbf{R}^n)$ *is a solution of (1.1).*
Then $x' \in L^2(\mathbf{R}^+; \mathbf{R}^n)$.

If, furthermore, $f(\infty) \stackrel{\text{def}}{=} \lim_{t\to\infty} f(t)$ *and* $\alpha \stackrel{\text{def}}{=} \lim_{t\to\infty} \int_0^t a(s)\,\mathrm{d}s$ *exist, and* $(\int_0^t a(s)\,\mathrm{d}s - \alpha) \in L^2(\mathbf{R}^+;\mathbf{R}^{n\times n})$, *then* $\lim_{t\to\infty}(x(t) + \alpha g(x(t))) = f(\infty)$.

7. Comments

Sections 2 and 3:

Ordinary differential equations have been studied with the methods of these sections in Popov [2]. The early history is reviewed in Popov [2], pp. 239–240. For the literature up to 1973 on frequency techniques applied to Volterra equations, see Corduneanu [9], Ch. 3, and, in particular, pp. 137-142. One of the first results for (1.2) which makes use of the fact that the kernel is of positive type is given in Corduneanu [1]. That result resembles Theorem **18**.2.1.

Results that in a sense are more similar to those given above are found in Halanay [1], Londen [2], [4], and MacCamy and Wong [1]. The equations in Halanay [1] and MacCamy and Wong [1] have kernels of strong positive type. The equations in Londen [2] and Londen [4] have nonnegative, nonincreasing, and convex kernels; cf. Proposition **16**.3.1. The proofs in Londen [2] and [4] are different from our proofs, resembling some of the proofs in Section **20**.5. Earlier results in the same direction are found in Levin [1], Levin and Nohel [1], and Hannsgen [2]. For an extension to systems of equations, see Weis [1]. In Levin and Nohel [3], a weakly perturbed version of (1.2) is considered. Extensions to (1.1) are given in Levin [2], Londen [5] and [7], with a nonnegative and nonincreasing kernel; cf. Proposition **16**.3.2 and the discussion at the beginning of Section 5. A thorough discussion of (1.2) is given in Nohel and Shea [2]. For (1.1) in a somewhat more abstract setting, see MacCamy [2].

One version of Proposition 2.1 is found in Londen [2], Theorem 2. Theorem 2.2 and Corollary 2.3 are found in Staffans [15], Theorem 2.1 and Corollary 3.2. Theorem 2.4 is extracted from Staffans [7], Theorem 1.1 and Proposition 4.1, and Theorem 2.5 is hinted at in Staffans [7], Section 8.

Theorem 2.2 has been generalized in Karakostas [1]. More elaborate boundedness criteria for (1.2) in the case where the kernel is nonnegative, nonincreasing, and convex, and has compact support, have been obtained in Kuen and Rybakowski [1].

Theorem 3.1 is taken from Staffans [2], Theorem 6.3, and Theorem 3.2 from Staffans [4], Theorem 1. Lemma 3.4 and Theorem 3.5 appeared first in Staffans [1] and Staffans [2]. Theorem 3.7 is found in Staffans [3], Theorem 3.1 and Proposition 7.1 (this reference also contains further results in the same direction). A scalar version of Theorem 3.8 was given in Staffans [5],

Theorem 3.2, and the n-dimensional version of Theorem 3.8 was proved in Staffans [6], Theorem 6.1.

The idea that information about the asymptotic behaviour of a function can be extracted from the boundedness of a quadratic form is used in Popov [2], § 21, p. 210–213.

Section 4:
An early version of Theorem 4.1 is found in Staffans [7], Theorem 4.3. It was improved to Theorem 4.3 in Londen [15], Theorem 1.

Section 5:
Most of the results in this section are modelled after Staffans [9]. One version of Theorem 5.2 is found in Barbu [1].

Section 6:
Theorem 6.1 is taken from Staffans [7], Proposition 4.1, Theorem 6.2 from Staffans [7], Proposition 5.1, and Theorem 6.3 from Staffans [7], Theorem 2.1.

References

V. Barbu

1. Sur une équation intégrale non-linéaire, *An. Ştiinţ. "Al. I. Cuza" Iaşi Secţ. I a Mat (N.S.)* **10** (1964), pp. 61–65.

C. Corduneanu

1. Sur une équation intégrale de la théorie du réglage automatique, *C. R. Acad. Sci. Paris* **256** (1963), pp. 3564–3567.
9. *Integral Equations and Stability of Feedback Systems*, Academic Press, New York, 1973.

A. Halanay

1. On the asymptotic behavior of the solutions of an integro-differential equation, *J. Math. Anal. Appl.* **10** (1965), pp. 319–324.

K. B. Hannsgen

2. On a nonlinear Volterra equation, *Michigan Math. J.* **16** (1969), pp. 365–376.

G. Karakostas

1. On a Staffans type inequality for vector Volterra integro-differential equations, *Colloq. Math.* **52** (1987), pp. 313–317.

S. M. Kuen and K. P. Rybakowski

1. Boundedness of solutions of a system of integro-differential equations, *J. Math. Anal. Appl.* **112** (1985), pp. 378–390.

J. J. Levin

1. The asymptotic behavior of the solution of a Volterra equation, *Proc. Amer. Math. Soc.* **14** (1963), pp. 534–541.
2. The qualitative behavior of a nonlinear Volterra equation, *Proc. Amer. Math. Soc.* **16** (1965), pp. 711–718.

J. J. Levin and J. A. Nohel

1. Note on a nonlinear Volterra equation, *Proc. Amer. Math. Soc.* **14** (1963), pp. 924–929.
3. Perturbations of a nonlinear Volterra equation, *Michigan Math. J.* **12** (1965), pp. 431–447.

S-O. Londen

2. The qualitative behavior of the solutions of a nonlinear Volterra equation, *Michigan Math. J.* **18** (1971), pp. 321–330.
4. On some nonlinear Volterra integrodifferential equations, *J. Differential Equations* **11** (1972), pp. 169–179.
5. On the solutions of a nonlinear Volterra equation, *J. Math. Anal. Appl.* **39** (1972), pp. 564–573.
7. On a nonlinear Volterra integral equation, *J. Differential Equations* **14** (1973), pp. 106–120.
15. A Volterra equation with L^2-solutions, *SIAM J. Math. Anal.* **18** (1987), pp. 168–171.

R. C. MacCamy

2. Remarks on frequency domain methods for Volterra integral equations, *J. Math. Anal. Appl.* **55** (1976), pp. 555–575.

R. C. MacCamy and J. S. W. Wong

1. Stability theorems for some functional equations, *Trans. Amer. Math. Soc.* **164** (1972), pp. 1–37.

J. A. Nohel and D. F. Shea

2. Frequency domain methods for Volterra equations, *Adv. in Math.* **22** (1976), pp. 278–304.

V. M. Popov

2. *Hyperstability of Control Systems*, Springer-Verlag, Berlin, 1973.

O. J. Staffans

1. Nonlinear Volterra integral equations with positive definite kernels, *Proc. Amer. Math. Soc.* **51** (1975), pp. 103–108.
2. Positive definite measures with applications to a Volterra equation, *Trans. Amer. Math. Soc.* **218** (1976), pp. 219–237.
3. Tauberian theorems for a positive definite form, with applications to a Volterra equation, *Trans. Amer. Math. Soc.* **218** (1976), pp. 239–259.
4. An inequality for positive definite Volterra kernels, *Proc. Amer. Math. Soc.* **58** (1976), pp. 205–210.
5. On the asymptotic spectra of the bounded solutions of a nonlinear Volterra equation, *J. Differential Equations* **24** (1977), pp. 365–382.
6. Systems of nonlinear Volterra equations with positive definite kernels, *Trans. Amer. Math. Soc.* **228** (1977), pp. 99–116.
7. Boundedness and asymptotic behavior of solutions of a Volterra equation, *Michigan Math. J.* **24** (1977), pp. 77–95.
9. Some energy estimates for a nondifferentiated Volterra equation, *J. Differential Equations* **32** (1979), pp. 285–293.
15. A bound on the solutions of a nonlinear Volterra equation, *J. Math. Anal. Appl.* **83** (1981), pp. 127–134.

D. G. Weis

1. Note on a Volterra integro-differential equation system, *Proc. Amer. Math. Soc.* **45** (1974), pp. 214–216.

18

Frequency Domain Methods: Additional Results

We consider some alternative energy estimates for nonlinear convolution equations. First we show how to obtain asymptotic results by interpolating between the undifferentiated and the differentiated versions of the equation. After that we recast the equation in various ways to get equations where the kernels are of positive type.

1. Introduction

In the preceding chapter our results and estimates were based entirely on the fact that either the kernel in the integral equation

$$x(t) + \int_0^t a(s)g\big(x(t-s)\big)\,\mathrm{d}s = f(t), \quad t \in \mathbf{R}^+, \tag{1.1}$$

was of positive type, or the kernel in the integrodifferential equation

$$x'(t) + \int_{[0,t]} \mu(\mathrm{d}s)g\big(x(t-s)\big) = f(t), \quad t \in \mathbf{R}^+; \quad x(0) = x_0, \tag{1.2}$$

was of that type. As we observed at the beginning of Section **17**.5, this gives us two different methods to approach (1.1), one that assumes that the kernel in (1.1) is of positive type, and one that assumes that the kernel in the differentiated version of (1.1) (i.e., (1.2) with μ replaced by the distribution derivative of a, f replaced by f', and x_0 replaced by $f(0)$) is of positive type. Below we shall discuss some alternative results, where neither of the two equations is required to have a kernel of positive type, but where some combination or transformation of the two kernels is assumed to be of positive type. The method of proof is basically the following. First, the equation under study ((1.1) or (1.2)) is manipulated so as to have a convolution term with the particular combination as kernel. Next, some energy estimates are developed for this manipulated equation.

To exemplify the basic idea, suppose, e.g., that the kernel a in (1.1) is not of positive type but only satisfies $\Re\{(q+z)\hat{a}(z)\} \succeq -A$, for $\Re z > 0$, where A is some self-adjoint matrix and q is some positive constant. In this case we convert (1.1) into an equation having a convolution term with a kernel ν satisfying $\hat{\nu}(z) = (q+z)\hat{a}(z)+A$, and if possible build up appropriate energy estimates for the converted equation. (See Corollary 2.4 for this particular example.) Of course, the necessary assumptions on g and f, and the size assumption on a, will depend on the particular conversion performed. In general, one cannot avoid some sign condition of the type $\langle y, g(y)\rangle \geq 0$, $y \in \mathbf{R}^n$. Moreover, in many cases g is required to be sublinear.

The first results, presented in Section 2, are fairly straightforward. In later sections the complexity of the equations increases, and we have to make more and more stringent assumptions about the nonlinearity.

For the convenience of the reader, we give a summary of the different positivity assumptions considered in this and in the previous chapter. Below, ϵ, q and r are positive constants and $c \in L^1_{\text{loc}}(\mathbf{R}^+;\mathbf{R})$ is (at least) nonnegative and nonincreasing. Thus, e.g., if $\Re\{(1+z\hat{c}(z))\hat{a}(z)\} \succeq -A$ for $\Re z > 0$, some self-adjoint matrix A and some sufficiently monotone function c, then the results of Section 4 can be applied to (1.1).

$$\left.\begin{array}{ll} \Re\hat{a}(z) \succeq -A, & \Re z > 0 \\ \Re\hat{a}(z) \succ 0, & \Re z \geq 0 \\ \Re\hat{a}(z) \succeq q\hat{a}^*(z)\hat{a}(z), & \Re z > 0 \end{array}\right\} \quad \text{Section } \mathbf{17}.5$$

$$\left.\begin{array}{ll} \Re\{z\hat{a}(z)\} \succeq 0, & \Re z > 0 \\ \Re\{z\hat{a}(z)\} \succ 0, & \Re z \geq 0 \\ \Re\{z\hat{a}(z)\} \succeq \epsilon I, & \Re z > 0 \\ \Re\{z\hat{a}(z)\} \succeq \epsilon\Re\{\frac{1}{1+z}\}I, & \Re z > 0 \\ \Re\{z\hat{a}(z)\} \succeq q|z|^2\hat{a}^*(z)\hat{a}(z), & \Re z > 0 \end{array}\right\} \quad \left\{\begin{array}{l} \text{Sections } \mathbf{17}.2\text{–}\mathbf{17}.4 \\ \text{beginning of Section } \mathbf{17}.5 \\ \text{Section } \mathbf{17}.6 \end{array}\right.$$

$$\left.\begin{array}{ll} \Re\{(q+z)\hat{a}(z)\} \succeq -A, & \Re z > 0 \\ \Re\{(q+z)\hat{a}(z)\} \succeq -A + \epsilon\Re\{\frac{1}{1+z}\}I, & \Re z > 0 \end{array}\right\} \quad \text{Section 2}$$

$$\left.\begin{array}{ll} \Re\{\frac{1}{z}(\hat{a}(z)+A)\} \succeq \epsilon\Re\{\frac{1}{1+z}\}I, & \Re z > 0 \\ \Re\{\frac{z+q}{z+r}\hat{a}(z)\} \succeq -qA\Re\{\frac{1}{z+r}\}, & \Re z > 0 \end{array}\right\} \quad \text{Section 3}$$

$$\left.\begin{array}{ll} \Re\{(1+z\hat{c}(z))\hat{a}(z)\} \succeq -A, & \Re z > 0 \\ \Re\{\frac{z+q}{z+r}\hat{a}(z)\} \succeq -\frac{q}{r}A, & \Re z > 0 \\ \Re\{\frac{1}{1+z\hat{c}(z)}\hat{a}(z)\} \succeq 0, & \Re z > 0 \\ \Re\{\frac{z+q}{z+r}\hat{a}(z)\} \succeq 0, & \Re z > 0 \\ \Re\{\frac{q+i\omega}{r+i\omega}\tilde{a}(\omega)\} \succeq -\frac{r(q^2+\omega^2)}{q(r^2+\omega^2)}A, & \omega \in \mathbf{R} \end{array}\right\} \quad \text{Section 4}$$

We also encounter the following Laplace transform conditions on the kernel μ in (1.2).

$$\left.\begin{array}{ll} \Re\{(1+\hat{c}(z))\hat{\mu}(z)\} \succeq 0, & \Re z > 0 \\ \Re\{\frac{1}{1+z\hat{c}(z)}\hat{\mu}(z)\} \succeq 0, & \Re z > 0 \\ \Re\{\frac{z+q}{z+r}\hat{\mu}(z)\} \succeq 0, & \Re z > 0 \end{array}\right\} \quad \text{Section 5}$$

As in the previous chapter, we let $\langle\cdot,\cdot\rangle$ denote some inner product on $\mathbf{C}^n$ that restricted to $\mathbf{R}^n$ is real-valued. Recall that the definition of the gradient involves the inner product.

2. Interpolation between an Integral and an Integrodifferential Equation

Consider (1.1), as well as the differentiated version of this equation, namely

$$x'(t) + \int_{[0,t]} \mu(\mathrm{d}s) g\big(x(t-s)\big) = f'(t), \quad t \in \mathbf{R}^+; \quad x(0) = f(0), \tag{2.1}$$

where μ is the distribution derivative of a, i.e., $a(t) = \mu([0,t])$ for almost all $t \in \mathbf{R}^+$.

So far, our analysis of (1.1) has mainly been based on the identity that we obtain by taking the inner product of (1.1) and $g(x(t))$ and then integrating, that is,

$$\int_0^T \langle g(x(t)), x(t)\rangle \,\mathrm{d}t + \int_0^T \Big\langle g(x(t)), \int_0^t a(s) g\big(x(t-s)\big)\,\mathrm{d}s \Big\rangle \,\mathrm{d}t = \int_0^T \langle g(x(t)), f(t)\rangle \,\mathrm{d}t. \tag{2.2}$$

Similarly, our study of (2.1) has been based on the corresponding identity

$$G\big(x(T)\big) + \int_0^T \Big\langle g(x(t)), \int_{[0,t]} \mu(\mathrm{d}s) g\big(x(t-s)\big)\Big\rangle \,\mathrm{d}t = G\big(f(0)\big) + \int_0^T \langle g(x(t)), f'(t)\rangle \,\mathrm{d}t. \tag{2.3}$$

Sometimes, one can achieve better results by using a combination of (2.2) and (2.3). To this end, multiply (2.2) by a positive number q and add the result to (2.3), to get

$$G(x(t)) + q\int_0^T \langle g(x(t)), x(t)\rangle \,\mathrm{d}t + Q(qa+\mu, g\circ x, T) = G\big(f(0)\big) + \int_0^T \Big\langle g(x(t)), qf(t) + f'(t)\Big\rangle \,\mathrm{d}t. \tag{2.4}$$

The (albeit hybrid) notation $Q(qa+\mu,\varphi,T)$ is self-explanatory. The basic assumption in the results that will be discussed next is that q can be chosen in such a way that $qa(t)\,\mathrm{d}t + \mu(\mathrm{d}t)$ is of positive type, is of strong or strict positive type, or dominates a multiple of the Dirac measure.

We remark that to get (2.4) one can just as well first multiply (1.1) by q, add the result to (2.1), form the inner product with $g(x(t))$, and integrate over $[0,T]$. In other words, one first manipulates (1.1) to turn it into a special case of the mixed equation

$$x'(t) + qx(t) + \int_{[0,t]} \nu(\mathrm{d}s) g\big(x(t-s)\big) = F(t), \quad t \in \mathbf{R}^+; \quad x(0) = x_0,$$

where ν is of positive type, after which one takes the inner product with $g(x(t))$ and integrates.

The same approach applies equally well to the somewhat more general equation

$$x'(t) + h(x(t)) + \int_{[0,t]} \nu(\mathrm{d}s) g\big(x(t-s)\big) = F(t), \quad t \in \mathbf{R}^+;$$
$$x(0) = x_0, \quad (2.5)$$

where we have replaced the linear term $qx(t)$ by a nonlinear term $h(x(t))$. Clearly, taking the inner product of $g(x(t))$ and (2.5), and integrating, one gets

$$G(x(t)) + \int_0^T \big\langle g(x(t)), h(x(t))\big\rangle\,\mathrm{d}t + Q(\nu, g{\circ}x, T)$$
$$= G\big(x_0\big) + \int_0^T \Big\langle g(x(t)), F(t)\Big\rangle\,\mathrm{d}t. \quad (2.6)$$

First, we formulate a result that gives us square integrability of the solution.

2.1 Theorem

Assume that

(i) $\nu \in M_{\mathrm{loc}}(\mathbf{R}^+; \mathbf{R}^{n\times n})$, *and* $\nu + A\delta$ *is of positive type for some constant self-adjoint matrix* A,

(ii) $g \in C(\mathbf{R}^n; \mathbf{R}^n)$ *is the gradient of a function* $G \in C^1(\mathbf{R}^n; \mathbf{R})$ *satisfying* $\lim_{|y|\to\infty} G(y) = \infty$, $h \in C(\mathbf{R}^n; \mathbf{R}^n)$, *and* $\langle g(y), h(y) - Ag(y)\rangle \geq \epsilon |g(y)|^2$ *for some* $\epsilon > 0$ *and all* $y \in \mathbf{R}^n$,

(iii) $F \in L^2(\mathbf{R}^+; \mathbf{R}^n)$,

(iv) $x \in AC_{\mathrm{loc}}(\mathbf{R}^+; \mathbf{R}^n)$ *is a solution of (2.5) on* $\mathbf{R}^+$.

Then $x \in BC(\mathbf{R}^+; \mathbf{R}^n)$ *and* $g{\circ}x \in L^2(\mathbf{R}^+; \mathbf{R}^n)$. *Moreover, it follows that* $\sup_{T>0} Q(\nu + A\delta, g{\circ}x, T) < \infty$.

If, in addition, $|g(y)| \geq \epsilon|y|$ *for some* $\epsilon > 0$ *and all* $y \in \mathbf{R}^n$, *or* $\nu \in M(\mathbf{R}^+; \mathbf{R}^{n\times n})$ *and* $\langle x, h(x)\rangle \geq \epsilon|x|^2$ *for some* $\epsilon > 0$ *and all* $x \in \mathbf{R}^n$, *then* $x \in L^2(\mathbf{R}^+; \mathbf{R}^n)$.

The (omitted) proof follows very closely that of Theorem **17**.5.1.

Observe that if $A = 0$ and $h(y) = y$ then (ii) requires $\langle g(y), y\rangle \geq 0$, and G is automatically bounded from below, because then we have, for all

$y \in \mathbf{R}^n$,

$$G(y) = G(0) + \int_0^1 \tfrac{\mathrm{d}}{\mathrm{d}t} G(ty)\,\mathrm{d}t = G(0) + \int_0^1 \langle g(ty), y\rangle\,\mathrm{d}t$$

$$= G(0) + \int_0^1 \tfrac{1}{t}\langle g(ty), ty\rangle\,\mathrm{d}t \ge G(0).$$

In the following two theorems we have basically the same situation as above. The proofs are essentially the same as the proofs of the corresponding results for the integrodifferential equation given in Chapter 17. Recall, moreover, that if ν is of strict positive type, and if we can prove that the quadratic form $Q(\nu, g\circ x, T)$ is bounded, uniformly in T, then it follows from Theorem **17**.3.5 that $g\circ x \to 0$ as $t \to \infty$, provided x is uniformly continuous.

2.2 Theorem

Let the following hypotheses hold.

(i) $\nu \in M_{\mathrm{loc}}(\mathbf{R}^+; \mathbf{R}^{n\times n})$, *and* $\nu + A\delta$ *is of positive type for some constant self-adjoint matrix* A.

(ii) $g \in C(\mathbf{R}^n; \mathbf{R}^n)$ *is the gradient of a function* $G \in C^1(\mathbf{R}^n; \mathbf{R})$ *satisfying* $\lim_{|y|\to\infty} G(y) = \infty$, $h \in C(\mathbf{R}^n; \mathbf{R}^n)$, *and* $\langle g(y), h(y) - Ag(y)\rangle \ge 0$ *for all* $y \in \mathbf{R}^n$. *In addition, there exists some nonnegative, nondecreasing function* u, *defined on the range of* G, *such that* $|g(y)| \le u(G(y))$ *for all* $y \in \mathbf{R}^n$, *and such that*

$$\int_{G(x_0)}^{\infty} \frac{\mathrm{d}s}{u(s)} = \infty.$$

(iii) $F \in L^1(\mathbf{R}^+; \mathbf{R}^n)$.

(iv) $x \in AC_{\mathrm{loc}}(\mathbf{R}^+; \mathbf{R}^n)$ *is a solution of (2.5) on* $\mathbf{R}^+$.

Then $x \in BC(\mathbf{R}^+; \mathbf{R}^n)$, *and* $\sup_{T>0} Q(\nu + A\delta, g\circ x, T) < \infty$.

The proof is the same as the proof of Theorem **17**.2.2.

2.3 Theorem

Suppose that

(i) $\nu \in M_{\mathrm{loc}}(\mathbf{R}^+; \mathbf{R}^{n\times n})$, *and* $\nu + A\delta$ *is of strong positive type for some constant self-adjoint matrix* A,

(ii) $g \in C(\mathbf{R}^n; \mathbf{R}^n)$ *is the gradient of a function* $G \in C^1(\mathbf{R}^n; \mathbf{R})$ *satisfying* $\lim_{|y|\to\infty} G(y) = \infty$, $h \in C(\mathbf{R}^n; \mathbf{R}^n)$, *and* $\langle g(y), h(y) - Ag(y)\rangle \ge 0$ *for all* $y \in \mathbf{R}^n$,

(iii) $F \in L^2(\mathbf{R}^+; \mathbf{R}^n) \cap AC_{\mathrm{loc}}(\mathbf{R}^+; \mathbf{R}^n)$ *and* $F' \in L^2(\mathbf{R}^+; \mathbf{R}^n)$,

(iv) $x \in AC_{\mathrm{loc}}(\mathbf{R}^+; \mathbf{R}^n)$ *is a solution of (2.5) on* $\mathbf{R}^+$.

Then $x \in BC(\mathbf{R}^+; \mathbf{R}^n)$, *and* $\sup_{T>0} Q(\nu + A\delta, g\circ x, T) < \infty$.

The proof is essentially the same as the proof of Theorem **17**.2.4.

Applying the three preceding theorems to (1.1) in the way that we described at the beginning of this section, we get the following corollaries.

2.4 Corollary

Assume that the following conditions are satisfied:

(i) $q \in \mathbf{R}$, $a \in BV_{\mathrm{loc}}(\mathbf{R}^+;\mathbf{R}^{n\times n})$, μ *is the measure that satisfies* $\mu([0,t]) = a(t)$ *for almost all* $t \in \mathbf{R}^+$, A *is a self-adjoint matrix, and the measure* $qa(t)\,\mathrm{d}t + \mu(\mathrm{d}t) + A\delta(\mathrm{d}t)$ *is of positive type;*

(ii) $g \in C(\mathbf{R}^n;\mathbf{R}^n)$ *is the gradient of a function* $G \in C^1(\mathbf{R}^n;\mathbf{R})$ *satisfying* $\lim_{|y|\to\infty} G(y) = \infty$, *and* $\langle g(y), qy - Ag(y)\rangle \ge \epsilon |g(y)|^2$ *for some* $\epsilon > 0$ *and all* $y \in \mathbf{R}^n$;

(iii) $f \in L^2(\mathbf{R}^+;\mathbf{R}^n) \cap AC_{\mathrm{loc}}(\mathbf{R}^+;\mathbf{R}^n)$ *and* $f' \in L^2(\mathbf{R}^+;\mathbf{R}^n)$;

(iv) $x \in C(\mathbf{R}^+;\mathbf{R}^n)$ *is a solution of (1.1) on* $\mathbf{R}^+$.

Then $x \in AC_{\mathrm{loc}}(\mathbf{R}^+;\mathbf{R}^n) \cap BC(\mathbf{R}^+;\mathbf{R}^n)$, $g \circ x \in L^2(\mathbf{R}^+;\mathbf{R}^n)$, *and* $\sup_{T>0} Q(qa + \mu + A\delta, g\circ x, T) < \infty$.

If, in addition, $a \in L^1(\mathbf{R}^+;\mathbf{R}^{n\times n}) \cap BV(\mathbf{R}^+;\mathbf{R}^{n\times n})$, *then both* x *and* x' *belong to* $L^2(\mathbf{R}^+;\mathbf{R}^n)$. *In particular, in this case* $x(t) \to 0$ *as* $t \to \infty$.

2.5 Corollary

Suppose that the following conditions hold.

(i) $q \in \mathbf{R}$, $a \in BV_{\mathrm{loc}}(\mathbf{R}^+;\mathbf{R}^{n\times n})$, μ *is the measure that satisfies* $\mu([0,t]) = a(t)$ *for almost all* $t \in \mathbf{R}^+$, A *is a self-adjoint matrix, and the measure* $qa(t)\,\mathrm{d}t + \mu(\mathrm{d}t) + A\delta(\mathrm{d}t)$ *is of positive type.*

(ii) $g \in C(\mathbf{R}^n;\mathbf{R}^n)$ *is the gradient of a function* $G \in C^1(\mathbf{R}^n;\mathbf{R})$ *satisfying* $\lim_{|y|\to\infty} G(y) = \infty$, *and* $\langle g(y), qy - Ag(y)\rangle \ge 0$ *for all* $y \in \mathbf{R}^n$. *In addition, there exists some nonnegative, nondecreasing function* u, *defined on the range of* G, *such that* $|g(y)| \le u(G(y))$ *for all* $y \in \mathbf{R}^n$, *and such that*

$$\int_{G(x_0)}^{\infty} \frac{\mathrm{d}s}{u(s)} = \infty.$$

(iii) $f \in L^1(\mathbf{R}^+;\mathbf{R}^n) \cap AC_{\mathrm{loc}}(\mathbf{R}^+;\mathbf{R}^n)$ *and* $f' \in L^1(\mathbf{R}^+;\mathbf{R}^n)$.

(iv) $x \in C(\mathbf{R}^+;\mathbf{R}^n)$ *is a solution of (1.1) on* $\mathbf{R}^+$.

Then $x \in BC(\mathbf{R}^+;\mathbf{R}^n)$, *and* $\sup_{T>0} Q(qa + \mu + A\delta, g\circ x, T) < \infty$.

2.6 Corollary

Assume that

(i) $q \in \mathbf{R}$, $a \in BV_{\mathrm{loc}}(\mathbf{R}^+;\mathbf{R}^{n\times n})$, μ *is the measure that satisfies* $\mu([0,t]) = a(t)$ *for almost all* $t \in \mathbf{R}^+$, A *is a self-adjoint matrix, and the measure* $qa(t)\,\mathrm{d}t + \mu(\mathrm{d}t) + A\delta(\mathrm{d}t)$ *is of strong positive type,*

(ii) $g \in C(\mathbf{R}^n;\mathbf{R}^n)$ *is the gradient of a function* $G \in C^1(\mathbf{R}^n;\mathbf{R})$ *satisfying* $\lim_{|y|\to\infty} G(y) = \infty$, *and* $\langle g(y), qy - Ag(y)\rangle \ge 0$ *for all* $y \in \mathbf{R}^n$,

(iii) $f \in C^1(\mathbf{R}^+;\mathbf{R}^n)$ *has a locally absolutely continuous derivative, and* f, f', *and* f'' *all belong to* $L^2(\mathbf{R}^+;\mathbf{R}^n)$,

(iv) $x \in C(\mathbf{R}^+;\mathbf{R}^n)$ *is a solution of (1.1) on* $\mathbf{R}^+$.

Then $x \in BC(\mathbf{R}^+;\mathbf{R}^n)$, *and* $\sup_{T>0} Q(qa + \mu + A\delta, g\circ x, T) < \infty$.

Of course, the crucial positivity conditions on a above can also be expressed by means of Laplace transforms. In Corollaries 2.4 and 2.5 the kernel a is supposed to satisfy

$$\Re\{(q+z)\hat{a}(z)\} \succeq -A \text{ for } \Re z > 0,$$

and in Corollary 2.6 this inequality is strengthened to

$$\Re\{(q+z)\hat{a}(z)\} \succeq -A + \epsilon\Re\left\{\tfrac{1}{1+z}\right\} I \text{ for } \Re z > 0,$$

and some $\epsilon > 0$.

3. Integral Equations Remoulded by Partial Integration

So far we have essentially considered situations where the kernel a in (1.1) is of positive type, or has a derivative of positive type, and interpolated versions of these cases. Suppose, instead, that a has some integral function b that is of positive type. If $b(t) = A + \int_0^t a(s)\,\mathrm{d}s$, then (1.1) can be rewritten in the form

$$x(t) - Ag(x(t)) + \tfrac{\mathrm{d}}{\mathrm{d}t}\int_0^t b(s)g\big(x(t-s)\big)\,\mathrm{d}s = f(t), \quad t \in \mathbf{R}^+. \tag{3.1}$$

Clearly, for this equation to make sense it is not necessary that A be equal to $b(0)$ and from now on we will not assume that this is the case.

Let us first make some comments on the existence of a solution of (3.1). If we suppose that b is locally absolutely continuous, then (3.1) can be written in the form

$$x(t) + (b(0) - A)g(x(t)) + \int_0^t b'(s)g\big(x(t-s)\big)\,\mathrm{d}s = f(t), \quad t \in \mathbf{R}^+. \tag{3.2}$$

If, moreover, the function $x \mapsto x + (b(0) - A)g(x)$ has an inverse h, and if we define y to be the function $y(t) = x(t) + (b(0) - A)g(x(t))$, then y satisfies an equation of the type (1.1), namely

$$y(t) + \int_0^t b'(s)g\big(h\big(y(t-s)\big)\big)\,\mathrm{d}s = f(t), \quad t \in \mathbf{R}^+.$$

We can apply a standard existence theorem to this equation (see, e.g., Theorem **12**.1.1), and consequently, under the additional assumptions mentioned, (3.1) has a solution.

Whereas the form (3.2) of (3.1) is appropriate when one wants to prove that the equation has a solution, it is not the best possible form for an asymptotic analysis in the case where b is of positive type. Typically, to be able to exploit the fact that b is a kernel of positive type, we would take the inner product of the equation and an auxiliary function (which up to now has been $g \circ x$), and then integrate. If this approach is to be applied to (3.1), then the differentiation should be carried out in such a way that the possibility of capitalizing on the presumed positivity of b is not eliminated. In other words, one should assume g to be continuously differentiable, and write (3.1) in the form

$$x(t) - Ag(x(t)) + \int_0^t b(s)g'\big(x(t-s)\big)x'(t-s)\,ds$$
$$= f(t) - b(t)g(x(0)), \quad t \in \mathbf{R}^+. \tag{3.3}$$

The obvious thing to do here is to form the inner product with $g'(x(t))x'(t)$, and integrate. Doing so, and assuming A to be self-adjoint and g to be the gradient of a function G, one gets

$$H\big(x(T)\big) + Q\big(b, g'(x(\cdot))x'(\cdot), T\big)$$
$$= H\big(x(0)\big) + \int_0^T \Big\langle g'(x(t))x'(t), f(t) - b(t)g(x(0))\Big\rangle\,dt, \tag{3.4}$$

where

$$H(y) = \Big\langle g(y), y - \tfrac{1}{2}Ag(y)\Big\rangle - G(y), \quad y \in \mathbf{R}^n. \tag{3.5}$$

From this identity one can derive, e.g., the following result.

3.1 Theorem

Let the following conditions hold:

(i) $b \in C(\mathbf{R}^+; \mathbf{R}^{n\times n})$ *is of strong positive type, and A is a self-adjoint matrix;*

(ii) $g \in C^1(\mathbf{R}^n; \mathbf{R}^n)$ *is the gradient of a function* $G \in C^1(\mathbf{R}^n; \mathbf{R})$, *and*

$$H(y) \ge -(\tfrac{1}{2} - \epsilon)\Big\langle y - Ag(y), [b(0)]^{-1}\big(y - Ag(y)\big)\Big\rangle, \tag{3.6}$$

for all $y \in \mathbf{R}^n$ and some $\epsilon > 0$ (where H is given in (3.5));

(iii) $f \in L^2(\mathbf{R}^+; \mathbf{R}^n) \cap AC_{\mathrm{loc}}(\mathbf{R}^+; \mathbf{R}^n)$ *and* $f' \in L^2(\mathbf{R}^+; \mathbf{R}^n)$;

(iv) $x \in AC_{\mathrm{loc}}(\mathbf{R}^+; \mathbf{R}^n)$ *is a solution of (3.1) on* $\mathbf{R}^+$.

Then $x - Ag{\circ}x \in BC(\mathbf{R}^+; \mathbf{R}^n)$, *and* $\sup_{T>0} Q(b, g'(x(\cdot))x'(\cdot), T) < \infty$.

If, in addition, b is of anti-coercive type and $b \in L^2(\mathbf{R}^+; \mathbf{R}^{n\times n})$, *then* $x - Ag{\circ}x \in L^2(\mathbf{R}^+; \mathbf{R}^n)$.

Before proceeding to the proof, let us make some comments. Since b is of strong positive type we have $b(0) \succ 0$, and hence $[b(0)]^{-1}$ exists and is positive. Thus the inner product on the right-hand side of (3.6) is non-negative. In the linear case where $g(y) = Ny$ for some self-adjoint matrix N, we can choose G to be the function $G(y) = \frac{1}{2}\langle Ny, y\rangle = \frac{1}{2}\langle g(y), y\rangle$, and (3.6) becomes

$$\tfrac{1}{2}\langle Ny, y - ANy\rangle \ge -(\tfrac{1}{2} - \epsilon)\Big\langle y - Ny, [b(0)]^{-1}(y - Ny)\Big\rangle, \tag{3.7}$$

which (for $\epsilon < \frac{1}{2}$) is weaker than $\langle Ny, y - ANy\rangle \ge 0$. This last inequality is the linear version of the familiar condition $\langle g(y), y - Ag(y)\rangle \ge 0$ used (in slightly different forms), e.g., in Theorems 2.1–2.3.

One way to guarantee that (3.6) holds in the general nonlinear case is to require G to be convex, normalize G so that $G(0) = 0$, and assume that

$$(\tfrac{1}{2} - \epsilon)\Big\langle y - Ag(y), [b(0)]^{-1}(y - Ag(y))\Big\rangle \geq \tfrac{1}{2}\langle g(y), Ag(y)\rangle, \quad y \in \mathbf{R}^n,$$

because then (3.6) is implied by the general inequality

$$G(x) - G(y) \geq \langle g(y), x - y\rangle, \quad x, y \in \mathbf{R}^n,$$

satisfied by every convex function G with gradient g (take $x = 0$).

Proof of Theorem 3.1 Without loss of generality, suppose that $\epsilon < 1$. In (3.4), we split the term $Q(b, g'(x(\cdot))x'(\cdot), T)$ in $(1-\epsilon)Q(b, g'(x(\cdot))x'(\cdot), T)$ and $\epsilon Q(b, g'(x(\cdot))x'(\cdot), T)$, apply the second part of Corollary **16**.9.1 to the first of these two terms, and use (3.3) to get (we have denoted $x(T) - Ag(x(T))$ by $\psi(T)$, and $f(T) - b(T)g(x(0))$ by $h(T)$)

$$\begin{aligned} H\big(x(T)\big) + \tfrac{1}{2}(1-\epsilon)\Big\langle \psi(T) - h(T), [b(0)]^{-1}\big(\psi(T) - h(T)\big)\Big\rangle& \\ + \epsilon Q\big(b, g'(x(\cdot))x'(\cdot), T\big)& \\ \leq H\big(x(0)\big) + \int_0^T \Big\langle g'(x(t))x'(t), f(t) - b(t)g(x(0))\Big\rangle\, \mathrm{d}t.& \end{aligned}$$

This together with (3.5) and the positivity assumption (3.6) implies that

$$\begin{aligned} &\tfrac{1}{2}\epsilon\Big\langle \psi(T), [b(0)]^{-1}\psi(T)\Big\rangle + \epsilon Q(b, g'(x(\cdot))x'(\cdot), T) \\ &\quad \leq H\big(x(0)\big) + \tfrac{1}{2}(1-\epsilon)\Big\langle 2\psi(T) - h(T), [b(0)]^{-1}h(T)\Big\rangle \\ &\quad + \int_0^T \Big\langle g'(x(t))x'(t), f(t)\Big\rangle\, \mathrm{d}t - \int_0^T \Big\langle b^*(t)g'(x(t))x'(t), g(x(0))\Big\rangle\, \mathrm{d}t. \end{aligned}$$

Since b is continuous and of strong positive type, this implies that the left-hand side in this inequality is bounded uniformly in T (use Corollaries **16**.6.3 and **16**.6.6, respectively, for the two last terms and note that $h(T)$ is bounded, uniformly in T), and the main conclusion of the theorem follows.

The final claim is a consequence of (3.3) and Definition **16**.5.4. □

As we indicated earlier, the result that we have just proved can be applied to (1.1), provided we let $A = b(0)$. With this choice, the growth condition (3.6) simplifies into

$$\tfrac{1}{2}\langle y, A^{-1}y\rangle - G(y) \geq \epsilon\Big\langle y - Ag(y), A^{-1}(y - Ag(y))\Big\rangle, \quad y \in \mathbf{R}^n.$$

Equivalently (recall that the matrix $A^{-1} = [b(0)]^{-1}$ is of strict positive type, and decrease the value of ϵ, if necessary),

$$\tfrac{1}{2}\langle y, A^{-1}y\rangle - G(y) \geq \epsilon\big|y - Ag(y)\big|^2, \quad y \in \mathbf{R}^n. \tag{3.8}$$

3.2 Corollary

Assume that

(i) $a \in L^1_{\mathrm{loc}}(\mathbf{R}^+;\mathbf{R}^{n\times n})$, $A \succ 0$, *and the function* $b(t) = A + \int_0^t a(s)\,\mathrm{d}s$ *is of strong positive type,*

(ii) $g \in C^1(\mathbf{R}^n;\mathbf{R}^n)$ *is the gradient of a function* $G \in C^2(\mathbf{R}^n;\mathbf{R})$, *and* $\frac{1}{2}\langle y, A^{-1}y\rangle - G(y) \ge \epsilon\big|y - Ag(y)\big|^2$ *for all* $y \in \mathbf{R}^n$ *and some* $\epsilon > 0$,

(iii) $f \in L^2(\mathbf{R}^+;\mathbf{R}^n) \cap AC_{\mathrm{loc}}(\mathbf{R}^+;\mathbf{R}^n)$ *and* $f' \in L^2(\mathbf{R}^+;\mathbf{R}^n)$,

(iv) $x \in AC_{\mathrm{loc}}(\mathbf{R}^+;\mathbf{R}^n)$ *is a solution of (3.1) on* $\mathbf{R}^+$.

Then $x - Ag{\circ}x \in BC(\mathbf{R}^+;\mathbf{R}^n)$, *and* $\sup_{T>0} Q(b, g'(x(\cdot))x'(\cdot), T) < \infty$.

If, in addition, b *is of anti-coercive type, and* $b \in L^2(\mathbf{R}^+;\mathbf{R}^{n\times n})$, *then* $x - Ag{\circ}x \in L^2(\mathbf{R}^+;\mathbf{R}^n)$.

In terms of Laplace transforms, the condition on a in (i) is

$$\Re\Big\{\tfrac{1}{z}\big(\hat{a}(z) + A\big)\Big\} \succeq \epsilon\Re\Big\{\tfrac{1}{1+z}\Big\}I, \quad \Re z > 0.$$

Note that (ii) essentially requires g to be sublinear.

There are other ways of transforming (1.1) than the one used above. For example, by choosing the kernel b and the forcing term F appropriately, one can turn (1.1) into an equation of the form

$$\frac{\mathrm{d}}{\mathrm{d}t}\Big\{x(t) + \int_0^t b(s)g\big(x(t-s)\big)\,\mathrm{d}s\Big\} + q\{x(t) - Ag(x(t))\} \\ + r\int_0^t b(s)g\big(x(t-s)\big)\,\mathrm{d}s = F(t), \quad t \in \mathbf{R}^+; \quad x(0) = x_0. \tag{3.9}$$

This equation has been derived as follows. Assume that a is locally absolutely continuous, and differentiate (1.1) to get

$$x'(t) + a(0)g(x(t)) + \int_0^t a'(s)g\big(x(t-s)\big)\,\mathrm{d}s = f'(t), \quad t \in \mathbf{R}^+; \quad x(0) = f(0).$$

Add q times (1.1) to this equation, and observe that the result is identical to (3.9), provided

$$b'(t) + rb(t) = a'(t) + qa(t), \quad t \in \mathbf{R}^+; \quad b(0) = a(0) + qA,$$

and

$$F(t) = f'(t) + qf(t), \quad t \in \mathbf{R}^+; \quad x_0 = f(0). \tag{3.10}$$

Solving the equation governing b, one gets the explicit formula

$$b(t) = a(t) + (q - r)\int_0^t \mathrm{e}^{-rs}a(t-s)\,\mathrm{d}s + qA\mathrm{e}^{-rt}, \quad t \in \mathbf{R}^+. \tag{3.11}$$

(An easy way to convince oneself of the equivalence of (3.9) and (1.1) with the substitutions (3.10) and (3.11), is to use Laplace transforms. This is possible even when a and b are not differentiable, provided one uses the integrated version

$$\begin{aligned}x(t) &+ \int_0^t b(s)g\big(x(t-s)\big)\,\mathrm{d}s \\ &+ \int_0^t \Big\{qx(s) - Ag\big(x(s)\big) + r\int_0^s b(v)g\big(x(s-v)\big)\,\mathrm{d}v\Big\}\,\mathrm{d}s \\ &= x_0 + \int_0^t F(s)\,\mathrm{d}s, \quad t \in \mathbf{R}^+,\end{aligned}$$

of (3.9). The same integrated version solves the question of the existence of a solution of (3.9), because it is an equation to which the standard existence results in Chapter 12 can be applied. Moreover, if g is continuously differentiable, then we can use Theorem **13**.3.3 to show that the solution x is locally absolutely continuous, and satisfies (3.9).)

If we assume, as we have done earlier in this section, that g is continuously differentiable, and if x is locally absolutely continuous, then we can differentiate the first integral in (3.9) in the same way as we did when passing from (3.1) to (3.3). After that we take the inner product of (3.9) and $g'(x(t))x'(t) + rg(x(t))$, and integrate over $[0,T]$, to get (as usual, we let g be the gradient of a function G)

$$\begin{aligned}rG\big(x(T)\big) &+ qH\big(x(T)\big) + Q\Big(b, g'(x(\cdot))x'(\cdot) + rg(x(\cdot)), T\Big) \\ &+ \int_0^T \Big\langle g'(x(t))x'(t), x'(t)\Big\rangle\,\mathrm{d}t \\ &+ rq\int_0^T \Big\langle g(x(t)), x(t) - Ag(x(t))\Big\rangle\,\mathrm{d}t \\ &= rG(x_0) + qH(x_0) \\ &+ \int_0^T \Big\langle g'(x(t))x'(t) + rg(x(t)), F(t) - b(t)g(x_0)\Big\rangle\,\mathrm{d}t, \quad (3.12)\end{aligned}$$

where H is the same function as in (3.5).

From the identity (3.12) one can immediately deduce the following theorem.

3.3 Theorem

Assume that

(i) $b \in L^2(\mathbf{R}^+; \mathbf{R}^{n\times n})$ *is of positive type, A is a self-adjoint matrix,* $q > 0$, *and* $r > 0$,

(ii) $g \in C^1(\mathbf{R}^n; \mathbf{R}^n)$ *is the gradient of a function* $G \in C^2(\mathbf{R}^n; \mathbf{R})$,

$$\lim_{|y|\to\infty} \Big\{(r-q)G(y) + q\Big\langle g(y), y - \tfrac{1}{2}Ag(y)\Big\rangle\Big\} = \infty,$$

and there is an $\epsilon > 0$ *such that*

$$\Big\langle g(y), y - Ag(y)\Big\rangle \ge \epsilon\big|g(y)\big|^2, \quad y \in \mathbf{R}^n,$$

and

$$\langle g'(x)y, y\rangle \geq \epsilon\big|g'(x)y\big|^2, \quad x, y \in \mathbf{R}^n,$$

(iii) $F \in L^2(\mathbf{R}^+; \mathbf{R}^n)$,
(iv) $x \in AC_{\mathrm{loc}}(\mathbf{R}^+; \mathbf{R}^n)$ *is a solution of (3.9) on* $\mathbf{R}^+$.
Then $x \in BC_0(\mathbf{R}^+; \mathbf{R}^n)$, $g{\circ}x \in L^2(\mathbf{R}^+; \mathbf{R}^n)$, $g'(x(\cdot))x'(\cdot) \in L^2(\mathbf{R}^+; \mathbf{R}^n)$, *and* $\sup_{T>0} Q(b, g'(x(\cdot))x'(\cdot) + rg(x(\cdot)), T) < \infty$.

We leave the proof of this result to the reader.

Observe that the hypothesis of the theorem implies that G is convex, because (ii) requires $\langle g'(x)y, y\rangle \geq 0$ for all x and $y \in \mathbf{R}^n$ (see Exercise 2).

Applying this result to (1.1), we get the following result.

3.4 Corollary
Assume that
(i) $a \in L^2(\mathbf{R}^+; \mathbf{R}^{n\times n})$, $A \succeq 0$, $q > 0$, $r > 0$, *and*

$$\Re\left\{\tfrac{z+q}{z+r}\hat{a}(z)\right\} \succeq -qA\Re\left\{\tfrac{1}{z+r}\right\}, \quad \Re z > 0,$$

(ii) $g \in C^1(\mathbf{R}^n; \mathbf{R}^n)$ *is the gradient of a function* $G \in C^2(\mathbf{R}^n; \mathbf{R})$, *and*

$$\lim_{|y|\to\infty} \left\{(r-q)G(y) + q\Big\langle g(y), y - \tfrac{1}{2}Ag(y)\Big\rangle\right\} = \infty,$$

and there is an $\epsilon > 0$ *such that*

$$\Big\langle g(y), y - Ag(y)\Big\rangle \geq \epsilon\big|g(y)\big|^2, \quad y \in \mathbf{R}^n,$$

and

$$\langle g'(x)y, y\rangle \geq \epsilon\big|g'(x)y\big|^2, \quad x, y \in \mathbf{R}^n,$$

(iii) $f \in L^2(\mathbf{R}^+; \mathbf{R}^n) \cap AC_{\mathrm{loc}}(\mathbf{R}^+; \mathbf{R}^n)$ *and* $f' \in L^2(\mathbf{R}^+; \mathbf{R}^n)$,
(iv) $x \in AC_{\mathrm{loc}}(\mathbf{R}^+; \mathbf{R}^n)$ *is a solution of (1.1) on* $\mathbf{R}^+$.
Then $x \in BC_0(\mathbf{R}^+; \mathbf{R}^n)$, $g{\circ}x \in L^2(\mathbf{R}^+; \mathbf{R}^n)$, $g'(x(\cdot))x'(\cdot) \in L^2(\mathbf{R}^+; \mathbf{R}^n)$, *and* $\sup_{T>0} Q(b, g'(x(\cdot))x'(\cdot) + rg(x(\cdot)), T) < \infty$ *where* b *is defined by (3.11).*

4. Integral Equations Remoulded by Convolutions

In this section we consider some additional ways of transforming the equation

$$x(t) + \int_0^t a(s)g\big(x(t-s)\big)\,\mathrm{d}s = f(t), \quad t \in \mathbf{R}^+. \tag{1.1}$$

One possible transformation is to take the convolution of this equation and some function c, differentiate the result and then add what one gets to (1.1). The equation that one obtains is of the type

$$\tfrac{\mathrm{d}}{\mathrm{d}t}\int_0^t c(s)x(t-s)\,\mathrm{d}s + h(x(t)) + \int_0^t b(s)g\big(x(t-s)\big)\,\mathrm{d}s = F(t), \qquad t\in\mathbf{R}^+, \tag{4.1}$$

where

$$h(y)=y, \quad b=\tfrac{\mathrm{d}}{\mathrm{d}t}(c*a)+a, \quad F=\tfrac{\mathrm{d}}{\mathrm{d}t}(c*f)+f. \tag{4.2}$$

As will be seen from the results below, for technical reasons we have to assume that c is (scalar and) bounded, nonnegative, and nonincreasing.

Clearly, to get an energy type estimate for (4.1), it is very tempting to take the inner product of (4.1) and $g(x(t))$, and integrate. Doing so, one obtains the identity

$$\int_0^T \Big\langle g(x(t)), \tfrac{\mathrm{d}}{\mathrm{d}t}(c*x)(t)\Big\rangle\,\mathrm{d}t + \int_0^T \big\langle g(x(t)), h(x(t))\big\rangle\,\mathrm{d}t + Q(b, g\circ x, T) = \int_0^T \big\langle g(x(t)), F(t)\big\rangle\,\mathrm{d}t. \tag{4.3}$$

In this identity all the terms, except for the first, are of a familiar nature and are easy to estimate.

The first term in (4.3) can be estimated with the aid of the following lemma.

4.1 Lemma

Let $c\in BV_{\mathrm{loc}}(\mathbf{R}^+;\mathbf{R})$, $g\in C(\mathbf{R}^n;\mathbf{R}^n)$, $G\in C(\mathbf{R}^n;\mathbf{R})$, $x\in L^2_{\mathrm{loc}}(\mathbf{R}^+;\mathbf{R}^n)$, and suppose that $g\circ x\in L^2_{\mathrm{loc}}(\mathbf{R}^+;\mathbf{R}^n)$ and $G\circ x\in L^1_{\mathrm{loc}}(\mathbf{R}^+;\mathbf{R}^n)$. Define $H:\mathbf{R}^n\times\mathbf{R}^n\to\mathbf{R}$ by

$$H(x,y)=G(x)-G(y)-\langle g(y), x-y\rangle,$$

and let γ be the measure that satisfies $\gamma([0,t])=c(t)$ for almost all $t\in\mathbf{R}^+$. Then

$$\int_0^T \Big\langle g(x(t)), \tfrac{\mathrm{d}}{\mathrm{d}t}(c*x)(t)\Big\rangle\,\mathrm{d}t = \int_0^T c(t)\Big(G\big(x(T-t)\big)-G(0)\Big)\,\mathrm{d}t + \int_0^T c(t)H\big(0,x(t)\big)\,\mathrm{d}t - \int_{[0,T]}\int_s^T H\big(x(t-s),x(t)\big)\,\mathrm{d}t\,\gamma(\mathrm{d}s).$$

In particular, if c is nonnegative and nonincreasing, G is convex, $G(y)\geq G(0)$ for all $y\in\mathbf{R}^n$, and g is the gradient of G, then all the terms in the preceding identity are nonnegative.

The straightforward proof of this lemma is left to the reader. The identity above contains a total of ten terms, and they all cancel pairwise (change the order of integration in the last integral, and observe that $(c*x)'=(\gamma*x)$). The nonnegativity of the last two terms is due to the facts that $H(x,y)\geq 0$ for all x and y (because a convex function lies above its tangent plane), and that $H(x,x)=0$ for all x (this is needed in the last term to take care of the fact that γ contains a positive point mass at zero; observe that γ is negative on $(0, \infty)$).

Combining (4.3) with Lemma 4.1, one can get the following result.

4.2 Theorem
Assume that

(i) $b \in L^1_{\text{loc}}(\mathbf{R}^+; \mathbf{R}^{n\times n})$, $A \succeq 0$, $b + A\delta$ *is of positive type, and* c *is a bounded, nonnegative, and nonincreasing function on* $\mathbf{R}^+$,

(ii) $h \in C(\mathbf{R}^n; \mathbf{R}^n)$, $g \in C(\mathbf{R}^n; \mathbf{R}^n)$ *is the gradient of a convex function* $G \in C^1(\mathbf{R}^n; \mathbf{R})$ *satisfying* $G(y) \geq G(0)$ *for all* $y \in \mathbf{R}^n$, *and* $\langle g(y), h(y) - Ag(y)\rangle \geq \epsilon |g(y)|^2$ *for some* $\epsilon > 0$ *and all* $y \in \mathbf{R}^n$,

(iii) $F \in L^2(\mathbf{R}^+; \mathbf{R}^n)$,

(iv) $x \in L^2_{\text{loc}}(\mathbf{R}^+; \mathbf{R}^n)$ *is a solution of equation (4.1) on* $\mathbf{R}^+$ *such that* $h{\circ}x \in L^2_{\text{loc}}(\mathbf{R}^+; \mathbf{R}^n)$ *and* $G{\circ}x \in L^1_{\text{loc}}(\mathbf{R}^+; \mathbf{R}^n)$.

Then $g{\circ}x \in L^2(\mathbf{R}^+; \mathbf{R}^n)$, *and* $\sup_{T>0} Q(b + A\delta, g{\circ}x, T) < \infty$.

The easy proof of this theorem is left to the reader (observe that $|g(y)| \leq \frac{1}{\epsilon}|h(y)|$ for all $y \in \mathbf{R}^n$; hence $g{\circ}x \in L^2_{\text{loc}}$).

Applying the preceding theorem to equation (1.1) (with h, b and F defined as in (4.2)), we get the following result.

4.3 Corollary
Let the following conditions hold:

(i) $a \in L^1_{\text{loc}}(\mathbf{R}^+; \mathbf{R}^{n\times n})$, *and there exist a matrix* $A \succeq 0$ *and a bounded, nonnegative, and nonincreasing function* c *on* $\mathbf{R}^+$ *such that the kernel* $\frac{\mathrm{d}}{\mathrm{d}t}(c * a) + a + A\delta$ *is of positive type;*

(ii) $g \in C(\mathbf{R}^n; \mathbf{R}^n)$ *is the gradient of a convex function* $G \in C^1(\mathbf{R}^n; \mathbf{R})$ *and* $\langle g(y), y - Ag(y)\rangle \geq \epsilon|g(y)|^2$ *for some* $\epsilon > 0$ *and all* $y \in \mathbf{R}^n$;

(iii) $f \in L^2(\mathbf{R}^+; \mathbf{R}^n)$;

(iv) $x \in L^2_{\text{loc}}(\mathbf{R}^+; \mathbf{R}^n)$ *is a solution of (1.1) on* $\mathbf{R}^+$.

Then $g{\circ}x \in L^2(\mathbf{R}^+; \mathbf{R}^n)$.

If, in addition, $a \in L^1(\mathbf{R}^+; \mathbf{R}^{n\times n})$, *then* $x \in L^2(\mathbf{R}^+; \mathbf{R}^n)$, *and if* $a \in L^2(\mathbf{R}^+; \mathbf{R}^{n\times n})$, *then* $\operatorname{ess\,lim}_{t\to\infty}(x(t) - f(t)) = 0$.

(In order to see that $G{\circ}x \in L^1_{\text{loc}}(\mathbf{R}^+)$, observe that $|g(y)| \leq |y|/\epsilon$, and $G(y) \leq G(0) + |y|^2/\epsilon$ for all $y \in \mathbf{R}^n$.)

In terms of Laplace transforms, the positivity assumption in (i) means that for some c of the type mentioned above,

$$\Re\{(1 + z\hat{c}(z))\hat{a}(z)\} \succeq -A, \quad \Re z > 0.$$

In particular, if we choose $c(t)$ to be $\frac{r-q}{q}\mathrm{e}^{-rt}$, with $0 < q < r$, then we get

$$\Re\left\{\tfrac{z+q}{z+r}\hat{a}(z)\right\} \succeq -\tfrac{q}{r}A, \quad \Re z > 0. \tag{4.4}$$

This is a weaker condition than the one that we met in Corollary 3.4. Also our present assumptions on g and f are weaker than the corresponding assumptions in Corollary 3.4.

Above we replaced the original kernel a by a new kernel $b = \frac{\mathrm{d}}{\mathrm{d}t}(c*a)+a$. Another possibility is to reverse this transformation, and to let b be the solution of the equation $a = \frac{\mathrm{d}}{\mathrm{d}t}(c * b) + b$ (by an obvious modification of Theorem 4.1.7, this equation has a solution $b \in L^1_{\mathrm{loc}}(\mathbf{R}^+)$ whenever $c \in BV_{\mathrm{loc}}(\mathbf{R}^+)$ satisfies $c(0) \neq -1$, and $a \in L^1_{\mathrm{loc}}(\mathbf{R}^+)$). Then (1.1) becomes

$$x(t) + \int_0^t b(s)g\big(x(t-s)\big)\,\mathrm{d}s + \frac{\mathrm{d}}{\mathrm{d}t}\int_0^t (c*b)(s)g\big(x(t-s)\big)\,\mathrm{d}s = f(t), \quad t \in \mathbf{R}^+. \tag{4.5}$$

Here it is quite natural to form the inner product with $\frac{\mathrm{d}}{\mathrm{d}t}(c * g\circ x)(t) + g(x(t))$, and to integrate. This leads to the identity

$$\int_0^T \Big\langle \frac{\mathrm{d}}{\mathrm{d}t}(c * g\circ x)(t), x(t)\Big\rangle\,\mathrm{d}t + \int_0^T \langle g(x(t)), x(t)\rangle\,\mathrm{d}t + Q(b, (c*g\circ x)' + g\circ x, T) = \int_0^T \Big\langle \frac{\mathrm{d}}{\mathrm{d}t}(c * g\circ x)(t) + g(x(t)), f(t)\Big\rangle\,\mathrm{d}t. \tag{4.6}$$

The first term in (4.6) can be estimated with the following lemma.

4.4 Lemma

Let $c \in BV_{\mathrm{loc}}(\mathbf{R}^+;\mathbf{R})$, $g \in C(\mathbf{R}^n;\mathbf{R}^n)$, $G \in C(\mathbf{R}^n;\mathbf{R})$, $x \in L^2_{\mathrm{loc}}(\mathbf{R}^+;\mathbf{R}^n)$, and suppose that $g\circ x \in L^2_{\mathrm{loc}}(\mathbf{R}^+;\mathbf{R}^n)$ and $G\circ x \in L^1_{\mathrm{loc}}(\mathbf{R}^+;\mathbf{R}^n)$. Define $H : \mathbf{R}^n \times \mathbf{R}^n \to \mathbf{R}$ by

$$H(x,y) = G(x) - G(y) - \langle g(y), x-y\rangle,$$

and let γ be the measure that satisfies $\gamma([0,t]) = c(t)$ for almost all $t \in \mathbf{R}^+$. Then

$$\int_0^T \Big\langle \frac{\mathrm{d}}{\mathrm{d}t}(c * g\circ x)(t), x(t)\Big\rangle\,\mathrm{d}t = \int_0^T c(t)\big(G(x(t)) - G(0)\big)\,\mathrm{d}t + \int_0^T c(t)H\big(0, x(T-t)\big)\,\mathrm{d}t - \int_{[0,T]}\int_s^T H\big(x(t), x(t-s)\big)\,\mathrm{d}t\,\gamma(\mathrm{d}s).$$

In particular, if c is nonnegative and nonincreasing, G is convex, $G(y) \geq G(0)$ for all $y \in \mathbf{R}^n$, and g is the gradient of G, then all the terms in the preceding identity are nonnegative.

The proof of this lemma is very similar to the proof of Lemma 4.1.

It is quite obvious how one should estimate all the other terms in (4.6). For example, the following result is easy to prove.

4.5 Theorem
Assume that

(i) $b \in L^1_{\text{loc}}(\mathbf{R}^+;\mathbf{R}^{n\times n})$ *is of positive type and* c *is a bounded, nonnegative, and nonincreasing function on* $\mathbf{R}^+$,

(ii) $g \in C(\mathbf{R}^n;\mathbf{R}^n)$ *is the gradient of a convex function* $G \in C^1(\mathbf{R}^n;\mathbf{R})$, *and* $\langle g(y), y\rangle \ge \epsilon |g(y)|^2$ *for some* $\epsilon > 0$ *and all* $y \in \mathbf{R}^n$,

(iii) $f \in L^2(\mathbf{R}^+;\mathbf{R}^n)$,

(iv) $x \in L^2_{\text{loc}}(\mathbf{R}^+;\mathbf{R}^n)$ *is a solution of (4.5) on* $\mathbf{R}^+$.

Then $g\circ x \in L^2(\mathbf{R}^+;\mathbf{R}^n)$, *and* $\sup_{T>0} Q(b,(c * g\circ x)' + g\circ x, T) < \infty$.

If, in addition, $b \in L^1(\mathbf{R}^+;\mathbf{R}^{n\times n})$, *then* $x \in L^2(\mathbf{R}^+;\mathbf{R}^n)$, *and if* $b \in L^2(\mathbf{R}^+;\mathbf{R}^{n\times n})$, *then* $\operatorname{ess\,lim}_{t\to\infty}(x(t) - f(t)) = 0$.

The theorem above requires b itself to be of positive type. In some of our earlier results it sufficed to have $b + A\delta$ of positive type, for some positive matrix A. It is possible to include such a matrix A here too, at the expense of additional size restrictions on b and c, and a more complicated proof. The key observation is that the last term on the left-hand side in (4.6) is almost of the following type.

4.6 Lemma
Let μ, $\nu \in M(\mathbf{R}^+;\mathbf{R}^{n\times n})$ *and let* $\varphi \in L^2(\mathbf{R}^+;\mathbf{R}^n)$. *Then*

$$\int_{\mathbf{R}^+} \Big\langle (\mu * \varphi)(t), (\nu * \varphi)(t) \Big\rangle \,\mathrm{d}t = \frac{1}{2\pi}\int_{\mathbf{R}} \Big\langle \tilde\varphi(\omega), \Re\{\tilde\mu^*(\omega)\tilde\nu(\omega)\}\tilde\varphi(\omega)\Big\rangle \mathrm{d}\omega.$$

This lemma is an immediate consequence of Parseval's identity.

As $Q(b,(c * g\circ x)' + (g\circ x), T)$ is not quite of the form described in Lemma 4.6, we need additional growth assumptions on b and c compared to Theorem 4.5, namely $b \in L^1(\mathbf{R}^+) \cap L^2(\mathbf{R}^+)$ and $c \in L^1(\mathbf{R}^+)$.

4.7 Theorem
Let the following hypotheses be satisfied:

(i) $c \in L^1(\mathbf{R}^+;\mathbf{R})$ *is bounded, nonnegative, and nonincreasing,* $A \succeq 0$, *and* $b \in L^1(\mathbf{R}^+;\mathbf{R}^{n\times n}) \cap L^2(\mathbf{R}^+;\mathbf{R}^{n\times n})$ *satisfies* $|1 + \mathrm{i}\omega\tilde c(\omega)|^2 \Re\tilde b(\omega) \succeq -A$ *for all* $\omega \in \mathbf{R}$;

(ii) $g \in C(\mathbf{R}^n;\mathbf{R}^n)$ *is the gradient of a convex function* $G \in C^1(\mathbf{R}^n;\mathbf{R})$, *and* $\langle g(y), y - Ag(y)\rangle \ge \epsilon|g(y)|^2$ *for some* $\epsilon > 0$ *and all* $y \in \mathbf{R}^n$;

(iii) $f \in L^2(\mathbf{R}^+;\mathbf{R}^n) \cap BC_0(\mathbf{R}^+;\mathbf{R}^n)$;

(iv) $x \in L^2_{\text{loc}}(\mathbf{R}^+;\mathbf{R}^n)$ *is a solution of (4.5) on* $\mathbf{R}^+$.

Then $x \in BC_0(\mathbf{R}^+;\mathbf{R}^n) \cap L^2(\mathbf{R}^+;\mathbf{R}^n)$.

Proof First let us prove the theorem under the additional assumption that g is bounded. Let γ be the measure derivative of c, and observe that $\tilde\gamma(\omega) = \mathrm{i}\omega\tilde c(\omega)$ for all $\omega \in \mathbf{R}$, and that $(c * g\circ x)' + (g\circ x) = (\gamma + \delta) * g\circ x$.

We define $\varphi_T = \chi_{[0,T]} g \circ x$, and replace $Q(b, (c * g \circ x)' + (g \circ x), T)$ in (4.6) by

$$\int_{\mathbf{R}^+} \Big\langle \big((\gamma + \delta) * \varphi_T\big)(t), \big(b * (\gamma + \delta) * \varphi_T\big)(t) \Big\rangle \, \mathrm{d}t - W(T),$$

where

$$W(T) = \int_T^\infty \Big\langle \big((\gamma + \delta) * \varphi_T\big)(t), \big(b * (\gamma + \delta) * \varphi_T\big)(t) \Big\rangle \, \mathrm{d}t.$$

By Lemmas 4.4 and 4.6 and our assumption on c, b, g and f, we get from (4.6) the inequality

$$\epsilon \| g \circ x \|^2_{L^2(0,T)} - W(T) \le \big(2c(0) + 1\big) \| f \|_{L^2(0,T)} \| g \circ x \|_{L^2(0,T)}, \tag{4.7}$$

provided we estimate the right-hand side of (4.6) by using the fact that $\int_{(0,\infty)} |\gamma|(\mathrm{d}s) \le c(0)$.

Let us estimate $W(T)$. Define $K = \sup_{y \in \mathbf{R}^n} |g(y)|$ (recall that for the moment we let g be bounded). Then, for $t > T$ we have $(\delta * \varphi_T)(t) = 0$, and

$$|(\gamma * \varphi_T)(t)| \le \int_{t-T}^t \big|g\big(x(t-s)\big)\big| |\gamma|(\mathrm{d}s) \le K c(t-T),$$

hence $\| (\gamma + \delta) * \varphi_T \|_{L^1(T,\infty)} \le K \| c \|_{L^1(\mathbf{R}^+)}$. Clearly, for all $t \in \mathbf{R}^+$,

$$\Big| \big(b * (\gamma + \delta) * \varphi_T\big)(t) \Big| \le K(2c(0) + 1) \| b \|_{L^1(\mathbf{R}^+)}.$$

This means that W is a bounded function of T.

As W is bounded, it follows from (4.7) that $\| g \circ x \|_{L^2(0,T)}$ is bounded, independently of T. This means that $g \circ x \in L^2(\mathbf{R}^+; \mathbf{R}^n)$. If we use this fact in (4.5), we find that $x \in BC_0(\mathbf{R}^+; \mathbf{R}^n) \cap L^2(\mathbf{R}^+; \mathbf{R}^n)$, as claimed.

A closer analysis of the situation above will reveal that the bound that we get on the sup-norm of x is independent of the bound that we assumed on g. As a matter of fact, the bound is even independent of g in the sense that we get the same bound for all functions g that are bounded, are gradients of a convex function, and satisfy the crucial growth condition

$$\langle g(y), y - Ag(y) \rangle \ge \epsilon |g(y)|^2$$

(with the same matrix A and the same constant ϵ). This will eventually allow us to remove the additional boundedness assumption on g. To see this, recall that we have shown that x and $g \circ x \in L^2(\mathbf{R}^+; \mathbf{R}^n)$, take $T = \infty$ in (4.6), observe that all terms are finite, and use Lemma 4.6 to get (cf. (4.7))

$$\epsilon \| g \circ x \|^2_{L^2(\mathbf{R}^+)} \le \big(2c(0) + 1\big) \| f \|_{L^2(\mathbf{R}^+)} \| g \circ x \|_{L^2(\mathbf{R}^+)}.$$

Thus

$$\| g \circ x \|_{L^2(\mathbf{R}^+)} \le \frac{2c(0)+1}{\epsilon} \| f \|_{L^2(\mathbf{R}^+)}.$$

Since the bound that we get on $\| g \circ x \|_{L^2(\mathbf{R}^+)}$ is independent of g, also the bound that we found for the sup-norm of x is independent of g.

Let us denote the bound for x that we found above by α. We claim that, even in the case where g is not bounded, it is still true that $|x(t)| \le \alpha$ for

$t \in \mathbf{R}^+$. If this is not the case, then there exists some $T > 0$ such that $|x(T)| > \alpha$, but $|x(t)| \le 2\alpha$ for $t \in [0,T]$. Suppose for the moment that we can redefine $g(y)$ for $|y| > 2\alpha$ in such a way that the new function, which we denote by g_α, is bounded, is the gradient of a convex function G_α, and satisfies the same growth condition as g does. It follows from the existence result in Theorem **12**.1.1 that, since x is a solution of the equation

$$x(t) + \int_0^t b(s) g_\alpha\big(x(t-s)\big)\,\mathrm{d}s + \frac{\mathrm{d}}{\mathrm{d}t}\int_0^t (c*b)(s) g_\alpha\big(x(t-s)\big)\,\mathrm{d}s = f(t), \quad t \in \mathbf{R}^+, \qquad (4.8)$$

on the interval $[0,T]$, it can be extended to a maximally defined solution of the same equation. Let us denote this maximally defined solution of (4.8) by x_α. The result that we have just proved gives us an *a priori* bound on $|x_\alpha(t)|$ and x_α is maximally defined. Therefore, it must be true that x_α exists on all of $\mathbf{R}^+$, and that $|x_\alpha(t)| \le \alpha$ for all $t \in \mathbf{R}^+$. But this contradicts our assumption $|x(T)| > \alpha$. Thus, the only thing that remains to be checked is that we can find a function $g_\alpha(y)$ that is bounded, is the gradient of a convex function G_α, coincides with $g(y)$ for $|y| \le 2\alpha$, and satisfies the crucial growth condition. This part of the proof is left to the reader (see Exercise 12). □

Interpreting Theorem 4.5 in terms of (1.1), we get the following corollary.

4.8 Corollary

Assume that

(i) $a \in L^1_{\mathrm{loc}}(\mathbf{R}^+;\mathbf{R}^{n\times n})$, *$c$ is a bounded, nonnegative, and nonincreasing function on $\mathbf{R}^+$, and the solution b of the equation $b + \frac{\mathrm{d}}{\mathrm{d}t}(c*b) = a$ is of positive type,*

(ii) $g \in C(\mathbf{R}^n;\mathbf{R}^n)$ *is the gradient of a convex function* $G \in C^1(\mathbf{R}^n;\mathbf{R})$, *and* $\langle g(y), y\rangle \ge \epsilon |g(y)|^2$ *for some* $\epsilon > 0$ *and all* $y \in \mathbf{R}^n$,

(iii) $f \in L^2(\mathbf{R}^+;\mathbf{R}^n)$,

(iv) $x \in L^2_{\mathrm{loc}}(\mathbf{R}^+;\mathbf{R}^n)$ *is a solution of (1.1) on* $\mathbf{R}^+$.

Then $g\circ x \in L^2(\mathbf{R}^+;\mathbf{R}^n)$, *and* $\sup_{T>0} Q(b, (c*g\circ x)' + g\circ x, T) < \infty$.

If, in addition, $a \in L^1(\mathbf{R}^+;\mathbf{R}^{n\times n})$, *then* $x \in L^2(\mathbf{R}^+;\mathbf{R}^n)$, *and if* $a \in L^2(\mathbf{R}^+;\mathbf{R}^{n\times n})$, *then* $\operatorname{ess\,lim}_{t\to\infty}(x(t) - f(t)) = 0$.

To prove this, observe that the measure resolvent ρ of γ belongs to $M(\mathbf{R}^+;\mathbf{R})$ (cf. Exercise **4**.1).

In terms of Laplace transforms, the positivity assumption on a in Corollary 4.8 can be written as

$$\Re\Big\{\frac{1}{1 + z\hat{c}(z)}\hat{a}(z)\Big\} \succeq 0, \quad \Re z > 0.$$

In particular, if we choose $c(t)$ to be $\frac{q-r}{r}\mathrm{e}^{-qt}$, with $0 < r < q$, then we get

$$\Re\Big\{\tfrac{z+q}{z+r}\hat{a}(z)\Big\} \succeq 0, \quad \Re z > 0.$$

This differs from the example given in connection with Corollary 4.3 in the sense that there we had $0 < q < r$, and here we have $0 < r < q$. (In addition, the right-hand side was $-\frac{q}{r}A$, not 0.)

An analogous application of Theorem 4.7 to (1.1), gives the following result.

4.9 Corollary
Assume that

(i) $a \in L^1(\mathbf{R}^+;\mathbf{R}^{n\times n}) \cap L^2(\mathbf{R}^+;\mathbf{R}^{n\times n})$, $c \in L^1(\mathbf{R}^+;\mathbf{R})$ *is bounded, nonnegative, and nonincreasing,* $A \succeq 0$, *and* $\Re\{(1+\mathrm{i}\omega\tilde{c}(\omega))\tilde{a}(\omega)\} \succeq -A$ *for all* $\omega \in \mathbf{R}$,

(ii) $g \in C(\mathbf{R}^n;\mathbf{R}^n)$ *is the gradient of a convex function* $G \in C^1(\mathbf{R}^n;\mathbf{R})$, *and* $\langle g(y), y - Ag(y)\rangle \geq \epsilon|g(y)|^2$ *for some* $\epsilon > 0$ *and all* $y \in \mathbf{R}^n$,

(iii) $f \in L^2(\mathbf{R}^+;\mathbf{R}^n) \cap BC_0(\mathbf{R}^+;\mathbf{R}^n)$,

(iv) $x \in L^2_{\mathrm{loc}}(\mathbf{R}^+;\mathbf{R}^n)$ *is a solution of (1.1) on* $\mathbf{R}^+$.

Then $x \in BC_0(\mathbf{R}^+;\mathbf{R}^n) \cap L^2(\mathbf{R}^+;\mathbf{R}^n)$.

Again, if we choose $c(t)$ to be $\frac{q-r}{r}\mathrm{e}^{-qt}$, where $0 < r < q$, then the positivity assumption on a in Corollary 4.9, formulated in terms of Fourier transforms, requires that

$$\Re\left\{\frac{r-\mathrm{i}\omega}{q-\mathrm{i}\omega}\tilde{a}(\omega)\right\} \succeq -\frac{r}{q}A, \quad \omega \in \mathbf{R}.$$

One can multiply this inequality by $\frac{q^2+\omega^2}{r^2+\omega^2}$ to get the equivalent condition

$$\Re\left\{\frac{q+\mathrm{i}\omega}{r+\mathrm{i}\omega}\tilde{a}(\omega)\right\} \succeq -\frac{r(q^2+\omega^2)}{q(r^2+\omega^2)}A, \quad \omega \in \mathbf{R}.$$

Since $A \succeq 0$, this is a weaker condition than the one which we met in (i) of Corollary 3.4.

5. Integrodifferential Equations Remoulded by Convolutions

The two main results of the preceding section were formulated for equations (4.1) and (4.5), respectively, but, as we saw, they could be applied to (1.1). We can use an analogous approach to obtain results on the integrodifferential equation

$$x'(t) + \int_{[0,t]} \mu(\mathrm{d}s)g(x(t-s)) = f(t), \quad t \in \mathbf{R}^+; \quad x(0) = x_0. \qquad (1.2)$$

If we convolve this equation by some function c, and add the result to the original one, then we get the equation

$$x'(t) + \int_0^t c(s)x'(t-s)\,\mathrm{d}s + \int_{[0,t]} \nu(\mathrm{d}s)g\big(x(t-s)\big) = F(t), \quad t \in \mathbf{R}^+; \quad x(0) = x_0, \tag{5.1}$$

where

$$\nu = \mu + c * \mu, \quad F = f + c * f.$$

Taking the inner product of this equation and $g(x(t))$, and integrating, we get the identity (assuming that g is the gradient of G)

$$G\big(x(T)\big) + \int_0^T \Big\langle g(x(t)), (c * x')(t) \Big\rangle\,\mathrm{d}t + Q(\nu, g\circ x, T) = G(x_0) + \int_0^T \langle g(x(t)), F(t) \rangle\,\mathrm{d}t. \tag{5.2}$$

Here the second term on the left-hand side is almost the same term which we had in (4.3), and it can be estimated as follows.

5.1 Lemma

Let $c \in BV_{\mathrm{loc}}(\mathbf{R}^+;\mathbf{R})$, $g \in C(\mathbf{R}^n;\mathbf{R}^n)$, $G \in C(\mathbf{R}^n;\mathbf{R})$, and let $x \in AC_{\mathrm{loc}}(\mathbf{R}^+;\mathbf{R}^n)$. Define $H : \mathbf{R}^n \times \mathbf{R}^n \to \mathbf{R}$ by

$$H(x,y) = G(x) - G(y) - \langle g(y), x - y \rangle,$$

and let γ be the measure that satisfies $\gamma([0,t]) = c(t)$ for almost all $t \in \mathbf{R}^+$. Then

$$\begin{aligned}\int_0^T \Big\langle g(x(t)), (c * x')(t) \Big\rangle\,\mathrm{d}t &= \int_0^T c(T-t)\Big(G(x(t)) - G(x(0))\Big)\,\mathrm{d}t \\ &\quad + \int_0^T c(t)H\big(x(0), x(t)\big)\,\mathrm{d}t \\ &\quad - \int_{[0,T]} \int_s^T H\big(x(t-s), x(t)\big)\,\mathrm{d}t\,\gamma(\mathrm{d}s).\end{aligned}$$

In particular, if $c \in L^1(\mathbf{R}^+;\mathbf{R}^n)$ is nonnegative and nonincreasing, G is convex and bounded from below, and g is the gradient of G, then all the terms in the preceding identity are bounded from below.

This is the same result as Lemma 4.1, except for the fact that the term

$$\int_0^T c(t)\langle g(x(t)), x(0) \rangle\,\mathrm{d}t$$

has been subtracted from both sides.

From (5.2) and Lemma 5.1 we get the following result.

5.2 Theorem
Assume the following.

(i) $\nu \in PT(\mathbf{R}^+;\mathbf{R}^{n\times n})$ *and* $c \in L^1(\mathbf{R}^+;\mathbf{R})$ *is bounded, nonnegative, and nonincreasing on* $\mathbf{R}^+$.

(ii) $g \in C(\mathbf{R}^n;\mathbf{R}^n)$ *is the gradient of a convex function* $G \in C^1(\mathbf{R}^n;\mathbf{R})$ *satisfying* $\lim_{|y|\to\infty} G(y) = \infty$. *In addition, there exists some nonnegative, nondecreasing function* u*, defined on the range of* G*, such that* $|g(y)| \le u(G(y))$ *for all* $y \in \mathbf{R}^n$*, and such that*

$$\int_{G(x_0)}^{\infty} \frac{\mathrm{d}s}{u(s)} = \infty.$$

(iii) $F \in L^1(\mathbf{R}^+;\mathbf{R}^n)$.

(iv) $x \in C(\mathbf{R}^+;\mathbf{R}^n)$ *is a solution of (5.1) on* $\mathbf{R}^+$.

Then $x \in BC(\mathbf{R}^+;\mathbf{R}^n)$*, and* $\sup_{T>0} Q(\nu, g\circ x, T) < \infty$.

The proof is the same as that of Theorem **17**.2.2.

Let us reformulate this as a result for our original equation.

5.3 Corollary
Assume the following.

(i) $\mu \in M_{\mathrm{loc}}(\mathbf{R}^+;\mathbf{R}^{n\times n})$*, and there exists a function* $c \in L^1(\mathbf{R}^+;\mathbf{R})$ *that is bounded, nonnegative, and nonincreasing on* $\mathbf{R}^+$ *such that* $\mu + c*\mu$ *is of positive type.*

(ii) $g \in C(\mathbf{R}^n;\mathbf{R}^n)$ *is the gradient of a convex function* $G \in C^1(\mathbf{R}^n;\mathbf{R})$ *satisfying* $\lim_{|y|\to\infty} G(y) = \infty$. *In addition, there exists some nonnegative, nondecreasing function* u*, defined on the range of* G*, such that* $|g(y)| \le u(G(y))$ *for all* $y \in \mathbf{R}^n$*, and such that*

$$\int_{G(x_0)}^{\infty} \frac{\mathrm{d}s}{u(s)} = \infty.$$

(iii) $f \in L^1(\mathbf{R}^+;\mathbf{R}^n)$.

(iv) $x \in C(\mathbf{R}^+;\mathbf{R}^n)$ *is a solution of (1.2) on* $\mathbf{R}^+$.

Then $x \in BC(\mathbf{R}^+;\mathbf{R}^n)$*, and* $\sup_{T>0} Q(\mu + c*\mu, g\circ x, T) < \infty$.

If we choose $c(t) = (q-p)\mathrm{e}^{-pt}$ for some $0 < p < q$, then the positivity condition on μ requires, in terms of Laplace transforms, that

$$\Re\Big\{\tfrac{z+q}{z+p}\hat{\mu}(z)\Big\} \succeq 0, \quad \Re z > 0.$$

Extending the second of the two approaches discussed in Section 4 to (1.2), we get an integrodifferential equation analogous to (4.5), namely

$$x'(t) + \int_{[0,t]} \nu(\mathrm{d}s) g\big(x(t-s)\big) + \tfrac{\mathrm{d}}{\mathrm{d}t}\int_0^t (c*\nu)(s) g\big(x(t-s)\big)\,\mathrm{d}s$$
$$= f(t), \quad t \in \mathbf{R}^+; \quad x(0) = x_0. \quad (5.3)$$

It is clear that this equation has the same solutions as (1.2), provided ν is the solution of the equation $\mu = \frac{\mathrm{d}}{\mathrm{d}t}(c*\nu) + \nu$. If c is nonnegative, nonincreasing, and bounded, then such a solution always exists.

As in (4.5), we take the inner product of (5.3) and $g\circ x + \frac{\mathrm{d}}{\mathrm{d}t}(c * g\circ x)$, and then integrate. This gives

$$\begin{aligned} G\bigl(x(T)\bigr) + \int_0^T \Bigl\langle \tfrac{\mathrm{d}}{\mathrm{d}t}(c * g\circ x)(t), x'(t)\Bigr\rangle\,\mathrm{d}t + Q(\nu, g\circ x + (c * g\circ x)', T) \\ = G(x_0) + \int_0^T \Bigl\langle g(x(t)) + \tfrac{\mathrm{d}}{\mathrm{d}t}(c * g\circ x)(t), f(t)\Bigr\rangle\,\mathrm{d}t. \quad (5.4) \end{aligned}$$

The second term on the right-hand side is similar to the corresponding term in (4.6), and it can be estimated as follows.

5.4 Lemma

Let $c \in AC_{\mathrm{loc}}(\mathbf{R}^+;\mathbf{R})$ with $c' \in BV_{\mathrm{loc}}(\mathbf{R}^+;\mathbf{R})$, let $g \in C(\mathbf{R}^n;\mathbf{R}^n)$ be the gradient of a function $G \in C^1(\mathbf{R}^n;\mathbf{R})$, and let $x \in AC_{\mathrm{loc}}(\mathbf{R}^+;\mathbf{R}^n)$. Define $H : \mathbf{R}^n \times \mathbf{R}^n \to \mathbf{R}$ by

$$H(x,y) = G(x) - G(y) - \langle g(y), x - y\rangle,$$

and let γ be the measure that satisfies $\gamma([0,t]) = c'(t)$ for almost all $t \in \mathbf{R}^+$. Then

$$\begin{aligned} \int_0^T \Bigl\langle \tfrac{\mathrm{d}}{\mathrm{d}t}(c * g\circ x)(t), x'(t)\Bigr\rangle\,\mathrm{d}t = c(T)G\bigl(x(T)\bigr) - c(0)G\bigl(x(0)\bigr) \\ - \int_0^T c'(t)G(x(t))\,\mathrm{d}t \\ - \int_0^T c'(t)H\bigl(x(T), x(T-t)\bigr)\,\mathrm{d}t \\ + \int_{[0,T]}\int_s^T H\bigl(x(t), x(t-s)\bigr)\,\mathrm{d}t\,\gamma(\mathrm{d}s). \end{aligned}$$

In particular, if c is nonnegative, nonincreasing, and convex, and G is convex and bounded from below, then all the terms in this identity are bounded from below.

The simplest way to prove this identity is to first integrate by parts to get

$$\begin{aligned} \int_0^T \Bigl\langle \tfrac{\mathrm{d}}{\mathrm{d}t}(c * g\circ x)(t), x'(t)\Bigr\rangle\,\mathrm{d}t = c(0)\bigl(G\bigl(x(T)\bigr) - G\bigl(x(0)\bigr)\bigr) \\ + \int_0^T c'(t)\Bigl\langle g\bigl(x(T-t)\bigr), x(T)\Bigr\rangle\,\mathrm{d}t - \int_0^T \Bigl\langle \tfrac{\mathrm{d}}{\mathrm{d}t}(c' * g\circ x)(t), x(t)\Bigr\rangle\,\mathrm{d}t, \end{aligned}$$

and then to apply Lemma 4.4 (with c replaced by c') to the last term.

From (5.4) and Lemma 5.4 one can deduce, e.g., the following result.

5.5 Theorem
Assume the following.

(i) $\nu \in PT(\mathbf{R}^+;\mathbf{R}^{n\times n})$ *and c is a bounded, nonnegative, nonincreasing, and convex function on* $\mathbf{R}^+$.

(ii) $g \in C(\mathbf{R}^n;\mathbf{R}^n)$ *is the gradient of a convex function* $G \in C^1(\mathbf{R}^n;\mathbf{R})$ *satisfying* $\lim_{|y|\to\infty} G(y) = \infty$. *In addition, there exists some nonnegative, nondecreasing function u, defined on the range of G, such that* $|g(y)| \le u(G(y))$ *for all* $y \in \mathbf{R}^n$, *and such that*

$$\int_{G(x_0)}^{\infty} \frac{\mathrm{d}s}{u(s)} = \infty.$$

(iii) $f \in L^1(\mathbf{R}^+;\mathbf{R}^n)$.

(iv) $x \in AC_{\mathrm{loc}}(\mathbf{R}^+;\mathbf{R}^n)$ *is a solution of (5.3) on* $\mathbf{R}^+$.

Then $x \in BC(\mathbf{R}^+;\mathbf{R}^n)$, *and* $\sup_{T>0} Q(\nu, g\circ x + (c * g\circ x)', T) < \infty$.

Applying this result to (1.2) we get the following result.

5.6 Corollary
Assume the following.

(i) $\mu \in M_{\mathrm{loc}}(\mathbf{R}^+;\mathbf{R}^{n\times n})$ *and there exists a bounded, nonnegative, nonincreasing, and convex function c on $\mathbf{R}^+$, such that the measure ν which is the unique solution of the equation* $(1 + c(0))\nu + c' * \nu = \mu$ *is of positive type.*

(ii) $g \in C(\mathbf{R}^n;\mathbf{R}^n)$ *is the gradient of a convex function* $G \in C^1(\mathbf{R}^n;\mathbf{R})$ *satisfying* $\lim_{|y|\to\infty} G(y) = \infty$. *In addition, there exists some nonnegative, nondecreasing function u, defined on the range of G, such that* $|g(y)| \le u(G(y))$ *for all* $y \in \mathbf{R}^n$, *and such that*

$$\int_{G(x_0)}^{\infty} \frac{\mathrm{d}s}{u(s)} = \infty.$$

(iii) $f \in L^1(\mathbf{R}^+;\mathbf{R}^n)$.

(iv) $x \in AC_{\mathrm{loc}}(\mathbf{R}^+;\mathbf{R}^n)$ *is a solution of (1.2) on* $\mathbf{R}^+$.

Then $x \in BC(\mathbf{R}^+;\mathbf{R}^n)$, *and* $\sup_{T>0} Q(\nu, g\circ x + (c * g\circ x)', T) < \infty$.

If we choose c to be $c(t) = \frac{q-r}{r}\mathrm{e}^{-qt}$ for some $q > r > 0$, then the positivity condition on μ, formulated in terms of Laplace transforms, requires that

$$\Re\left\{\tfrac{z+q}{z+r}\hat{\mu}(z)\right\} \succeq 0, \quad \Re z > 0, \tag{5.5}$$

Observe that this is the same condition as we derived from Corollary 5.3, except for the fact that there $q < r$, and here $q > r$.

6. Exercises

1. In Corollary 2.6, let $g(y) = By$, for some constant self-adjoint matrix

B and compare the result with the asymptotic results obtained in Chapter 2.

2. Let $G \in C^2(\mathbf{R}^n; \mathbf{R})$, let g be the gradient of G, and suppose that, for all $x \in \mathbf{R}^n$ and $y \in \mathbf{R}^n$, $\langle g'(x)y, y \rangle \geq 0$. Show that G is convex. Hint: Show that, for all $v \in \mathbf{R}^n$ and $w \in \mathbf{R}^n$,
$$G(v+w) = G(v) + \int_0^1 \tfrac{\mathrm{d}}{\mathrm{d}\lambda} G(v+\lambda w)\,\mathrm{d}\lambda \geq G(v) + \langle g(v), w \rangle,$$
substitute $v = \lambda x + (1-\lambda)y$, and either $v+w=x$ or $v+w=y$, and combine the resulting inequalities.
3. Under the assumptions (i)–(iv) of Corollary 3.2, can you have $x \notin BC(\mathbf{R}^+; \mathbf{R}^n)$, $g(x(\cdot)) \notin BC(\mathbf{R}^+; \mathbf{R}^n)$? If the additional hypotheses on b are made, can you have $x \notin L^2(\mathbf{R}^+; \mathbf{R}^n)$?
4. Consider the scalar version of (1.1) under the assumptions of Corollary 4.3, with $a \in L^1(\mathbf{R}^+; \mathbf{R})$ and $A = 0$. Suppose that $\Re\tilde{a}(\omega) < 0$ for $\omega_1 < \omega < \omega_2$ and $\Re\tilde{a}(\omega) \geq 0$ otherwise. If $c(t) = \frac{r-q}{q}\mathrm{e}^{-rt}$, how would you pick q and r? What limitations have to be put on $\Im\tilde{a}$, for Corollary 4.3 to be applicable? How are your conclusions affected if the nonlinearity allows $A > 0$?
5. In Theorem 5.5 and Corollary 5.6 the conclusions are that both x and a particular quadratic form are bounded. Can you prove something more about the asymptotic behaviour of x and $g(x(\cdot))$? Are any additional assumptions necessary?
6. In Corollary 5.6, let $g(y) = By$ for some constant self-adjoint matrix B and compare the result with the asymptotic results obtained in Chapter 3.
7. Prove the following result.

6.1 Theorem

Assume the following.

(i) $a \in L^1(\mathbf{R}^+; \mathbf{R})$, $\tilde{a}(0) \geq 0$, $|\tilde{a}(\omega)| = O(|\omega|^{-1})$ *as* $|\omega| \to \infty$, *and* $|\tilde{a}(\omega) - \tilde{a}(0)| = O(|\omega|)$ *as* $\omega \to 0$. *Moreover, there exists a nonnegative Borel measure* μ *on* $\mathbf{R}^+$ *such that* $\mu(\mathbf{R}^+) \leq 1$, $\int_{\mathbf{R}^+} t\mu(\mathrm{d}t) < \infty$, *and, for some* $\epsilon > 0$,
$$\Re\Big\{\big(1 - \overline{\tilde{\mu}(\omega)}\big)\tilde{a}(\omega)\Big\} \geq \epsilon\omega^2, \quad \omega \in \mathbf{R}.$$

(ii) $g \in C(\mathbf{R}; \mathbf{R})$ *is nondecreasing, Lipschitz-continuous, and satisfies* $|g(y)| \leq c|y|$ *for some* $c < \infty$ *and all* $y \in \mathbf{R}$.

(iii) $f \in L^2(\mathbf{R}^+; \mathbf{R})$.

Then there exists a unique solution $x \in L^2(\mathbf{R}^+; \mathbf{R})$ *of (1.1) on* $\mathbf{R}^+$ *satisfying*
$$\|x\|_{L^2(\mathbf{R}^+)} \leq k\|f\|_{L^2(\mathbf{R}^+)},$$
where the constant k *depends on* c, ϵ, *and the kernel* a *only.*

If in addition

(i*) $a \in L^2(\mathbf{R}^+; \mathbf{R})$ *and* $\mu(\mathbf{R}^+) < 1$, *then the constant k is independent of a.*

Hint: Consult Gripenberg [13], Theorem 1.

8. Extend the result in Theorem 6.1 to the vector case.
9. Establish a result similar to Theorem 4.5 for the equation

$$x(t) + \frac{\mathrm{d}}{\mathrm{d}t}\int_0^t (c * b)(s) g\big(x(t-s)\big)\,\mathrm{d}s = f(t), \quad t \in \mathbf{R}^+.$$

10. Prove Theorem 3.3.
11. Prove Theorem 4.2.
12. Assume that $A \succeq 0$, $g \in C(\mathbf{R}^n; \mathbf{R}^n)$ is the gradient of a convex function $G \in C^1(\mathbf{R}^n; \mathbf{R})$, and $\langle g(y), y - Ag(y)\rangle \geq \epsilon |g(y)|^2$ for some $\epsilon > 0$ and all $y \in \mathbf{R}^n$. For each $\alpha > 0$, construct a function g_α such that g_α satisfies the same assumptions as g, $g_\alpha(y) = g(y)$ when $|y| \leq \alpha$ and g_α is bounded. Hint: Let $H(y) = \sup_{x \in \mathbf{R}^n}(\langle x, y\rangle - G(x))$, $\beta = \sup_{|x| \leq \alpha} |g(x)|$, and let $I_\beta(y) = 0$ if $|y| \leq \beta$ and $+\infty$ otherwise. Finally let $G_\alpha(x) = \sup_{y \in \mathbf{R}^n}(\langle y, x\rangle - H(y) - I_\beta(y))$.
13. In (4.4) and (5.5), $c(t)$ was taken to be an exponentially decreasing function. Can you conceive of a case where it would be preferable to use the function $c(t) = (1 - \frac{t}{T})$, $t \leq T$, $c(t) = 0$, $t > T$, or (in (4.4)) $c(t) = 1$, $t \leq T$, $c(t) = 0$, $t > T$?

7. Comments

Section 1:

Asymptotic results for (1.2) in the case where the nonlinearity satisfies an *a priori* one-sided bound were obtained in Levin [4].

Section 2:

Results of the type presented here were developed and extensively used by Popov, mainly for the control of ordinary differential equations. See, e.g., Popov [2], § 25, [3], and [4], Theorem 6.4.1.

An early extension of the results by Popov to (1.1) is given in Corduneanu [1]. In particular, the results in this article may be compared with Corollary 2.6 above, (take $A = 0$). Related results are found in Corduneanu [9], Chapter 3 (which contains an extensive reference list), Miller [10], Theorems 7.3–7.5, and Nohel and Shea [2], Theorem 4.

Section 3:

Corollary 3.2 has been modelled after Popov [5], Theorem on p. 574, but our proof is quite different from the proof given in Popov [5].

Section 4:

Lemmas 4.1 and 4.4 are taken, respectively, from Staffans [19], Lemmas 3.2 and 3.1, and Theorems 4.2 and 4.5 are adapted from Staffans [19], Corollary

1.2. In the case where $g(x) = x$ and $G(x) = \frac{1}{2}\langle x, x \rangle$, the identities in Lemmas 4.1 and 4.4 are the same, and this case is found already in Londen [7], formula (2.19). Some versions of Theorems 4.5 and 4.7 are found in Gripenberg [13]. (The proof of Staffans [19], Theorem 1.1 with condition (1.2), is incorrect; the term corresponding to the term $W(T)$ in our proof of Theorem 4.7 has been discarded, and it is not clear whether Staffans [19], Theorem 1.1 with condition (1.2), is true or not.)

Section 5:
The first version of Lemma 5.4 is found in Engler [1], Lemma 1.

References

C. Corduneanu

1. Sur une équation intégrale de la théorie du réglage automatique, *C. R. Acad. Sci. Paris* **256** (1963), pp. 3564–3567.
9. *Integral Equations and Stability of Feedback Systems*, Academic Press, New York, 1973.

H. Engler

1. A version of the chain rule and integrodifferential equations in Hilbert spaces, *SIAM J. Math. Anal.* **13** (1982), pp. 801–810.

G. Gripenberg

13. Stability problems for some nonlinear Volterra equations, *J. Integral Equations* **2** (1980), pp. 247–258.

J. J. Levin

4. Boundedness and oscillation of some Volterra and delay equations, *J. Differential Equations* **5** (1969), pp. 369–398.

S-O. Londen

7. On a nonlinear Volterra integral equation, *J. Differential Equations* **14** (1973), pp. 106–120.

R. K. Miller

10. *Nonlinear Volterra Integral Equations*, W. A. Benjamin, Menlo Park, Calif., 1971.

J. A. Nohel and D. F. Shea

2. Frequency domain methods for Volterra equations, *Adv. in Math.* **22** (1976), pp. 278–304.

V. M. Popov

2. *Hyperstability of Control Systems*, Springer-Verlag, Berlin, 1973.
3. Dichotomy and stability by frequency-domain methods, *Proc. IEEE* **62** (1974), pp. 548–562.
4. Stability-spaces and frequency-domain conditions, in *Calculus of Variations and Control Theory*, D. Russell, ed., Academic Press, New York, 1976, pp. 371–390.
5. Applications of the saturability technique in the problem of stability of nonlinear systems, *Nonlinear Anal.* **1** (1977), pp. 571–581.

O. J. Staffans

19. On the stability of a Volterra equation with a monotone nonlinearity, *J. Integral Equations* **7** (1984), pp. 239–248.

19

Combined Lyapunov and Frequency Domain Methods

We study the asymptotic behaviour of bounded solutions of a nonlinear integrodifferential equation of convolution type. The emphasis is on achieving results under minimal assumptions on the forcing function. In return, stronger assumptions on the kernel or on the nonlinearity are needed.

1. Introduction

In this chapter we continue our investigation of the asymptotics of the nonlinear Volterra equations initiated in Chapters 17 and 18. Primarily we will devote our attention to the integrodifferential equation

$$x'(t) + \int_{[0,t]} \mu(\mathrm{d}s) g\big(x(t-s)\big) = f(t), \quad t \in \mathbf{R}^+; \quad x(0) = x_0, \tag{1.1}$$

where we assume that μ is a measure of positive type, g is continuous, and f is locally integrable and tends to a limit in a weak sense.

Recall that when we analysed (1.1) in Chapters 17 and 18 we imposed some global size conditions on f and showed that the quadratic form

$$Q(\mu, g \circ x, T) = \int_0^T \Big\langle g(x(t)), \int_{[0,t]} \mu(\mathrm{d}s) g\big(x(t-s)\big)\Big\rangle \,\mathrm{d}t$$

was bounded uniformly in T. (Or we assumed that $Q(\mu, g \circ x, T)$ was uniformly bounded and analysed the possible implications of this assumption. In this case, the asymptotic conditions on f could be kept to a minimum.)

In the present chapter there will be no restrictions on how slowly the nonhomogeneous term f can tend to its limit at infinity. This makes it impossible for us to prove that an arbitrary solution of (1.1) is bounded; instead we have to *assume* that (1.1) has a bounded solution. Even so, we will not be able to show that the quadratic form $Q(\mu, g \circ x, T)$ is bounded; consequently our proofs have to exploit a different kind of argument.

If we want x to have a limit at infinity, then the weakest possible asymptotic assumption on f can be read off from the following partial converse to Theorem **17**.3.3.

1.1 Proposition

Assume that

(i) $\mu \in M(\mathbf{R}^+; \mathbf{R}^{n\times n})$,
(ii) $g \in C(\mathbf{R}^n; \mathbf{R}^n)$,
(iii) $f \in L^1_{\mathrm{loc}}(\mathbf{R}^+; \mathbf{R}^n)$,
(iv) $x \in AC_{\mathrm{loc}}(\mathbf{R}^+; \mathbf{R}^n)$ *satisfies (1.1), and* $x(\infty) = \lim_{t\to\infty} x(t)$ *exists.*

Let $A_0 = \mu(\mathbf{R}^+)$. *Then*

$$\lim_{t\to\infty} \frac{1}{T}\int_t^{t+T} f(s)\,\mathrm{d}s = A_0 g\,(x(\infty)), \quad T > 0. \tag{1.2}$$

If, in addition to (i),
(i*) $A_0 = 0$, *and* $\int_{\mathbf{R}^+} s|\mu|(\mathrm{d}s) < \infty$,
then

$$\lim_{t\to\infty} \int_0^t f(s)\,\mathrm{d}s = x(\infty) + A_1 g\,(x(\infty)) - x_0, \tag{1.3}$$

where $A_1 = -\int_{\mathbf{R}^+} s\mu(\mathrm{d}s)$.

The straightforward proof of Proposition 1.1 is left to the reader.

When A_0 is invertible it is natural, in view of the preceding result, to suppose that f satisfies

$$\lim_{t\to\infty} \frac{1}{T}\int_t^{t+T} f(s)\,\mathrm{d}s = F_0, \quad T > 0, \tag{1.4}$$

for some $F_0 \in \mathbf{R}^n$, and when $A_0 = 0$ it is natural to assume that

$$\lim_{t\to\infty} \int_0^t f(s)\,\mathrm{d}s = F_1, \tag{1.5}$$

for some $F_1 \in \mathbf{R}^n$. In the intermediate case where A_0 is neither invertible nor zero, we shall use a combination of these two conditions (we project (1.4) onto the range of A_0, and project (1.5) onto the nullspace of A_0; see Section 4).

It is not surprising that stronger restrictions than those in Chapter 17 will have to be imposed on g or μ as a price to be paid for allowing a larger nonhomogeneous term f. In Section 2 we impose a moment condition on μ that enables us to obtain asymptotic statements without any restrictions on the nonlinearity (except continuity). In Section 3 we show that if A_0 is invertible then the additional requirement that g be locally Lipschitz-continuous permits us to remove this moment condition on μ. In Section 4 we use a linear transformation of the nonlinear equation and make assumptions about the solution of the linear equation and about the differential resolvent of μ.

2. A Kernel with a Finite First Moment

Our first result is a new version of the result that one gets by combining Theorem **17**.3.8 with Theorem **17**.3.3. In the new result we make only the necessary (cf. Proposition 1.1) assumptions on the behaviour of f at infinity, but instead we assume that μ has a finite first moment, and that (1.1) has a bounded solution.

2.1 Theorem

Let x_0, F_0, and $F_1 \in \mathbf{R}^n$, and assume that

(i) $\mu \in PT(\mathbf{R}^+; \mathbf{R}^{n\times n})$ *satisfies* $\int_{\mathbf{R}^+} s|\mu|(\mathrm{d}s) < \infty$,

$$\ker[\Re\tilde{\mu}(\omega)] = \ker[\tilde{\mu}(\omega)], \quad \omega \in \mathbf{R},$$

and the set $\{\, \omega \in \mathbf{R} \mid \det[\Re\tilde{\mu}(\omega)] = 0 \,\}$ *is countable if* $n > 1$,

(ii) $g \in C(\mathbf{R}^n; \mathbf{R}^n)$ *is the gradient of a function* $G \in C^1(\mathbf{R}^n; \mathbf{R})$,

(iii) $f \in L^1_{\mathrm{loc}}(\mathbf{R}^+; \mathbf{R}^n)$,

(iv) $x \in BUC(\mathbf{R}^+; \mathbf{R}^n) \cap AC_{\mathrm{loc}}(\mathbf{R}^+; \mathbf{R}^n)$ *is a solution of (1.1) on* $\mathbf{R}^+$.

Let $A_0 = \int_{\mathbf{R}^+} \mu(\mathrm{d}s)$ *and* $A_1 = -\int_{\mathbf{R}^+} s\mu(\mathrm{d}s)$. *Then the following conclusions hold:*

(a) *if*

(1) A_0 *is invertible,*

(2) $\lim_{t\to\infty} \frac{1}{T}\int_t^{t+T} f(s)\,\mathrm{d}s = F_0$ *for all* $T > 0$,

(3) *the set* $\{\, G(y) - \langle A_0^{-1}F_0, y\rangle \mid g(y) = A_0^{-1}F_0 \,\}$ *has empty interior,*

then x is asymptotically slowly varying, and

$$\lim_{t\to\infty} g\left(x(t)\right) = A_0^{-1}F_0;$$

(b) *if*

(4) $A_0 = 0$,

(5) $\lim_{t\to\infty} \int_0^t f(s)\,\mathrm{d}s = F_1$,

(6) *the set* $\{\, G(y) + \frac{1}{2}\langle g(y), A_1 g(y)\rangle \mid y + A_1 g(y) = x_0 + F_1 \,\}$ *has empty interior,*

then x is asymptotically slowly varying, and

$$\lim_{t\to\infty} \left(x(t) + A_1 g\left(x(t)\right)\right) = x_0 + F_1.$$

In Theorem 4.5 below we treat the intermediate case where A_0 is neither invertible nor zero. It turns out that this can be reduced to a special case of (b) above. Actually, (a) too can be reduced to (b), but it is easier and more instructive to give a direct proof of (a).

Most of the time when one applies this theorem, the set $\{\omega \in \mathbf{R} \mid \det[\Re\tilde{\mu}(\omega)] = 0\}$ is likely to be empty, in which case $\ker[\Re\tilde{\mu}(\omega)] = \ker[\tilde{\mu}(\omega)] = \{0\}$ for all $\omega \in \mathbf{R}$.

The assumption (3) is trivially satisfied in the scalar case; see Sard's Theorem **15**.8.2. By the same theorem, it is satisfied in the multi-dimensional case if $g \in C^{n-1}(\mathbf{R}^n;\mathbf{R}^n)$. Since A_1 is invertible and self-adjoint (this is established in the proof of part (b)), the same arguments can be applied to (6). (In Sard's theorem, take $E(y) = G(y)+\frac{1}{2}\langle x_0+F_1-y, A_1^{-1}(x_0+F_1-y)\rangle$.)

As we have seen earlier, the assumption that g is the gradient of another function G holds automatically in the scalar case $n = 1$. The theorem is not true in the multi-dimensional case without this assumption, not even for linear g. To see this, one can take $n = 2$, $\mu = I\delta$,

$$g(y) = \begin{pmatrix} 0 & -1 \\ 1 & 0 \end{pmatrix} y,$$

$f(t) \equiv 0$, and $x_0 = \begin{pmatrix} 0 \\ 1 \end{pmatrix}$. Then (1.1) becomes

$$\begin{pmatrix} x_1'(t) \\ x_2'(t) \end{pmatrix} + \begin{pmatrix} 0 & -1 \\ 1 & 0 \end{pmatrix} \begin{pmatrix} x_1(t) \\ x_2(t) \end{pmatrix} = \begin{pmatrix} 0 \\ 0 \end{pmatrix}, \quad t \in \mathbf{R}^+; \quad \begin{pmatrix} x_1(0) \\ x_2(0) \end{pmatrix} = \begin{pmatrix} 0 \\ 1 \end{pmatrix}.$$

The solution

$$\begin{pmatrix} x_1(t) \\ x_2(t) \end{pmatrix} = \begin{pmatrix} \sin(t) \\ \cos(t) \end{pmatrix}$$

of this equation is not asymptotically slowly varying, and $g(x(t))$ does not tend to a limit at infinity.

The assumption in (i) on the equality of the nullspaces is essential for the conclusion of the theorem, see Exercise 8.

Proof of Part (a) of Theorem 2.1 First let us show that we may, without loss of generality, take $F_0 = 0$: define $g_0(y) = g(y) - A_0^{-1}F_0$, $G_0(y) = G(y) - \langle A_0^{-1}F_0, y\rangle$ and $f_0(t) = f(t) - \mu([0,t])A_0^{-1}F_0$; then g_0 is the gradient of G_0 and x satisfies (1.1) with g and f replaced by g_0 and f_0.

Our proof of the fact that x is asymptotically slowly varying is based on Theorem **15**.8.1. To apply that theorem we need a Lyapunov functional. It turns out that the appropriate functional is the mapping $H(\varphi) = G\circ\varphi$. It is clear that this function H commutes with translations and is continuous with respect to narrow convergence. We assumed that the set $\{\, G(y) \mid g(y) = 0 \,\}$ has empty interior. Thus, in order to be able to apply Theorem **15**.8.1 we need only establish the following two claims:

$$g(y(t)) \equiv 0 \textit{ whenever } y \in \Gamma(x) \textit{ is constant;} \tag{2.1}$$

If $y \in \Gamma(x)$ is not constant, then $\lim_{t\to-\infty} G(y(t))$ exists and $\lim_{t\to-\infty} G(y(t)) = \sup_{t\in\mathbf{R}} G(y(t)) > \limsup_{t\to\infty} G(y(t))$. (2.2)

To prove these two claims we shall use the fact that, by Corollary **15**.2.12, every $y \in \Gamma(x)$ satisfies the limit equation

$$y'(t) + \int_{\mathbf{R}^+} \mu(\mathrm{d}s) g(y(t-s)) = 0, \quad t \in \mathbf{R}. \tag{2.3}$$

The proof of (2.1) is very easy: if $y \in \Gamma(x)$ is constant, then it follows from (2.3) and the fact that A_0 is invertible that $g(y(t)) \equiv 0$.

To prove (2.2) we take the inner product of (2.3) and $g(y(t))$, and integrate over the interval $[\tau, T]$ to get

$$G(y(T)) + [\![\mu, \chi_{[\tau,T]} g\circ y]\!] = G(y(\tau)) - R(\tau, T), \tag{2.4}$$

where

$$[\![\mu, \chi_{[\tau,T]} g\circ y]\!] = \int_\tau^T \Big\langle g(y(t)), \int_{[0,t-\tau]} \mu(\mathrm{d}s) g\big(y(t-s)\big)\Big\rangle \, \mathrm{d}t,$$

and

$$R(\tau, T) = \int_\tau^T \Big\langle g(y(t)), \int_{(t-\tau,\infty)} \mu(\mathrm{d}s) g\big(y(t-s)\big)\Big\rangle \, \mathrm{d}t.$$

The function $R(\tau, T)$ is bounded uniformly in τ and T, because

$$|R(\tau, T)| \le \Big(\sup_{t\le\tau} \big|g(y(t))\big|\Big)\Big(\sup_{t\in\mathbf{R}} \big|g(y(t))\big|\Big) \int_{\mathbf{R}^+} s|\mu|(\mathrm{d}s). \tag{2.5}$$

Both $G(y(T))$ and $G(y(\tau))$ in (2.4) are bounded uniformly in τ and T, and this means that $[\![\mu, \chi_{[\tau,T]} g\circ y]\!]$ is bounded uniformly in τ and T. Define the measure ν in the same way as in Lemma **17**.3.4, i.e., let $\nu(E) = \mu(E\cap\mathbf{R}^+) + \mu(-E\cup\mathbf{R}^+)^*$ for $E \subset \mathbf{R}$. As in the proof of Lemma **17**.3.4 (let $\tau \to -\infty$), we conclude that, for each T, the function $\psi_T = \nu * \big(\chi_{(-\infty,T]}) g\circ y\big)$ belongs to $L^2(\mathbf{R}; \mathbf{R}^n)$ and satisfies

$$\|\psi_T\|_{L^2(\mathbf{R})} \le K \limsup_{\tau\to-\infty} [\![\mu, \chi_{[\tau,T]} g\circ y]\!], \tag{2.6}$$

where $K = 4\|\mu\|_{M(\mathbf{R}^+)}$. Since $g\circ y \in BUC(\mathbf{R}; \mathbf{R}^n)$, it follows that $\psi_T(t) \to 0$ as $t \to -\infty$ and so (by Theorem **15**.2.7) each $z \in \Gamma_-(\chi_{(-\infty,T]} g\circ y)$ satisfies $\nu * z \equiv 0$. By Theorem **15**.4.15(i) we conclude that $\mu * z \equiv 0$ and so (use (2.3) and the fact that A_0 is invertible) z is constant and $g(z(t)) \equiv 0$. Therefore, $g(y(t)) \to 0$ as $t \to -\infty$. This gives the first claim in (2.2), namely that $\lim_{t\to-\infty} G(y(t))$ exists (because G is a constant on every connected set where g vanishes). Moreover, since $g(y(t)) \to 0$ as $t \to -\infty$, it follows from (2.5) that $R(\tau, T) \to 0$ as $\tau \to -\infty$. This, together with (2.4) and (2.6) implies that

$$G(y(T)) + K^{-1}\|\psi_T\|_{L^2(\mathbf{R})} \le \lim_{t\to-\infty} G(y(t)). \tag{2.7}$$

In particular, this proves the second claim in (2.2), namely that $\lim_{t\to-\infty} G(y(t)) = \sup_{t\in\mathbf{R}} G(y(t))$.

Define $\psi = \nu * g\circ y$, let $T \to \infty$ in (2.7), and observe that, by Fatou's lemma,

$$\|\psi\|_{L^2(\mathbf{R})} \le \liminf_{T\to\infty} \|\psi_T\|_{L^2(\mathbf{R})}. \tag{2.8}$$

In particular, $\psi \in L^2(\mathbf{R}; \mathbf{R}^n)$. In the same way as above we conclude that $g(y(t)) \to 0$ as $t \to \infty$, and that $\lim_{t\to\infty} G(y(t))$ exists. Moreover, by (2.7) and (2.8), either $\psi \equiv 0$, or $\lim_{t\to\infty} G(y(t)) < \lim_{t\to-\infty} G(y(t))$. If $\psi = \nu * g\circ y \equiv 0$, then $\mu * g\circ y \equiv 0$ (this follows from Theorem **15**.4.15(i)); hence by (2.3), y is a constant. This completes our proof of the claim (2.2), as well as the proof of part (a) of Theorem 2.1. □

Preliminary Remarks on the Proof of Part (b) of Theorem 2.1 We begin by showing that we may without loss of generality assume that $x_0 + F_1 = 0$. To see this, define $g_1(y) = g(y + x_0 + F_1)$ and $x_1(t) = x(t) - x_0 - F_1$ for all $y \in \mathbf{R}^n$ and $t \in \mathbf{R}^+$ to get the equation

$$x_1'(t) + \int_{[0,t]} \mu(\mathrm{d}s) g_1\bigl(x_1(t-s)\bigr) = f(t), \quad t \in \mathbf{R}^+; \quad x_1(0) = -F_1.$$

This is an equation of the type (1.1) for which the constant $x_0 + F_1$ is zero. Thus we may from now on take $x_0 + F_1 = 0$.

A substantial portion of the proof of Part (a) given above remains valid in case (b). It is still true that the function ψ defined in the proof of Part (a) belongs to $L^2(\mathbf{R}; \mathbf{R}^n)$, and that $\mu * g \circ y$ tends to zero at $\pm\infty$. This together with (2.3) implies that every $y \in \Gamma(x)$ is asymptotically slowly varying at $+\infty$ and $-\infty$. However, we can no longer conclude that $g(y(t)) \to 0$ as $t \to \pm\infty$, and that $\lim_{t\to\pm\infty} G(y(t))$ exists.

On the other hand, as $A_0 = 0$, it is possible to show that every $y \in \Gamma(x)$ satisfies another limit equation, namely

$$y(t) + \int_{\mathbf{R}^+} a(s) g\bigl(y(t-s)\bigr)\,\mathrm{d}s = 0, \quad t \in \mathbf{R}, \tag{2.9}$$

where $a(t) = \mu([0,t])$ for all $t \in \mathbf{R}^+$. To prove this, it suffices to integrate (1.1), to observe that $a \in L^1(\mathbf{R}^+; \mathbf{R}^{n\times n})$, and to use Corollary **15**.2.10 (recall that we now assume that $x_0 + F_1 = 0$). This, together with the fact that y is asymptotically slowly varying at $+\infty$ and $-\infty$, implies that $x(t) + A_1 g(x(t)) \to 0$ as $t \to \pm\infty$ (observe that $\int_{\mathbf{R}^+} a(s)\,\mathrm{d}s = -\int_{\mathbf{R}^+} s\mu(\mathrm{d}s) = A_1$). Moreover, if we define E by $E(y) = G(y) + \frac{1}{2}\langle y, A_1^{-1} y\rangle$, then the gradient of E is the function $y \mapsto g(y) + A_1^{-1} y$, and therefore $\lim_{t\to\pm\infty} E(y(t))$ exists (we shall observe later in the proof that one may, without loss of generality, take $A_1 \succ 0$; in particular, A_1 is then self-adjoint and invertible).

In view of the preceding remarks, it is tempting to try to apply Theorem **15**.8.1 with $H(\varphi) = E \circ \varphi$, but there seems to be no obvious way of proving that $\lim_{t\to-\infty} E(y(t)) = \sup_{t\in\mathbf{R}} E(y(t))$ for all $y \in \Gamma(x)$. Therefore, we have to use a different function H.

One can track down an appropriate function H by manipulating the term $[\![\mu, \chi_{[\tau,T]} g \circ y]\!]$ in (2.4) in the following way.

2.2 Lemma
Let $\mu \in PT(\mathbf{R}^+; \mathbf{R}^{n\times n})$ satisfy $\int_{\mathbf{R}^+} s|\mu|\,\mathrm{d}s < \infty$ and $\int_{\mathbf{R}^+} \mu(\mathrm{d}s) = 0$. Let $\varphi \in C(\mathbf{R}; \mathbf{R}^n)$ and τ, $T \in \mathbf{R}$, $\tau < T$. Define $\nu(E) = \mu(E \cap \mathbf{R}^+) + \mu(-E \cup \mathbf{R}^+)^*$ for $E \subset \mathbf{R}$, $a(t) = \mu([0,t])$ for $t \in \mathbf{R}^+$, $a_\diamond(t) = a(t)$ for

$t \in \mathbf{R}^+$ *and* $a_\diamond(t) = -a(-t)^*$ *for* $t < 0$, *and* $A_1 = \int_{\mathbf{R}^+} a(s)\,\mathrm{d}s$. *Then*

$$\begin{aligned}\int_{\mathbf{R}} \Big| a_\diamond(t-T)\varphi(T) - a_\diamond(t-\tau)\varphi(\tau) + \int_{[t-T,t-\tau]} \nu(\mathrm{d}s)\varphi(t-s)\Big|^2 \mathrm{d}t \\ \le 4\|\mu\|_{M(\mathbf{R}^+)} \Big\{ [\![\mu, \chi_{[\tau,T]}\varphi]\!] \\ - \int_0^{T-\tau} \Big(\big\langle \varphi(\tau), a(t)\varphi(\tau+t)\big\rangle + \big\langle \varphi(T), a(t)\varphi(T-t)\big\rangle\Big)\,\mathrm{d}t \\ + \frac{1}{2}\Big(\big\langle \varphi(\tau), A_1\varphi(\tau)\big\rangle + \big\langle \varphi(T), A_1\varphi(T)\big\rangle\Big) \\ - \Big\langle \varphi(T), \Big(\int_{T-\tau}^{\infty} a(s)\,\mathrm{d}s\Big)\varphi(\tau)\Big\rangle\Big\}. \end{aligned} \tag{2.10}$$

Proof First we show that it suffices to prove Lemma 2.2 under the additional hypothesis $\varphi \in C^1(\mathbf{R}; \mathbf{R}^n)$. Suppose the lemma is true whenever $\varphi \in C^1(\mathbf{R}; \mathbf{R}^n)$. Given an arbitrary $\varphi \in C(\mathbf{R}; \mathbf{R}^n)$ one can take a sequence $\varphi_n \in C^1(\mathbf{R}; \mathbf{R}^n)$ converging uniformly on $[\tau, T]$ to φ. Then each φ_n satisfies (2.10) with φ replaced by φ_n. The right-hand side of (2.10) with φ replaced by φ_n converges toward the right-hand side of the inequality. Each term inside the absolute value sign on the left-hand side with φ replaced by φ_n converges to the appropriate limit. Hence by Fatou's lemma the desired inequality holds.

Thus, assume that $\varphi \in C^1(\mathbf{R}; \mathbf{R}^n)$. Define $b(t) = \int_t^\infty a(s)\,\mathrm{d}s$ for $t \in \mathbf{R}^+$. Then, by Theorem **16**.3.8(ii), b is of positive type. In particular, by Bochner's theorem, $A_1 = b(0) \succeq 0$, and so A_1 is self-adjoint. Using this fact and integrating by parts, one gets the identity

$$\begin{aligned}[\![\mu, \chi_{[\tau,T]}\varphi]\!] = \Re \int_\tau^T \Big\langle \varphi'(t), \int_0^{t-\tau} b(s)\varphi'(t-s)\,\mathrm{d}s\Big\rangle\,\mathrm{d}t \\ + \int_0^{T-\tau} \Big(\big\langle \varphi(\tau), a(t)\varphi(\tau+t)\big\rangle + \big\langle \varphi(T), a(t)\varphi(T-t)\big\rangle\Big)\,\mathrm{d}t \\ - \tfrac{1}{2}\Big(\big\langle \varphi(\tau), A_1\varphi(\tau)\big\rangle + \big\langle \varphi(T), A_1\varphi(T)\big\rangle\Big) \\ + \big\langle \varphi(T), b(T-\tau)\varphi(\tau)\big\rangle. \end{aligned} \tag{2.11}$$

Integrating by parts one can, furthermore, show that

$$\begin{aligned}a_\diamond(t-T)\varphi(T) - a_\diamond(t-\tau)\varphi(\tau) + \int_{[t-T,t-\tau]} \nu(\mathrm{d}s)\varphi(t-s) \\ = \int_{t-T}^{t-\tau} a_\diamond(s)\varphi'(t-s)\,\mathrm{d}s, \end{aligned} \tag{2.12}$$

for all $t \in \mathbf{R}$ for which $\nu(\{t-\tau\}) = \nu(\{t-T\}) = 0$, i.e. for almost all t. Define $\psi = \chi_{[\tau,T]}\varphi'$. If we use (2.11), (2.12), a change of variables and

Fubini's theorem, we find that the inequality (2.10) is equivalent to

$$\int_{\mathbf{R}}\left|\int_{\mathbf{R}} a_{\diamond}(t-s)\psi(s)\,\mathrm{d}s\right|^2 \mathrm{d}t \le 4\|\mu\|_{M(\mathbf{R}^+)}[\![b,\psi]\!]. \tag{2.13}$$

Therefore, by Theorem **16**.5.2, it suffices to show that the measure $\gamma(\mathrm{d}\omega) = 4\|\mu\|_{M(\mathbf{R}^+)}\beta(\mathrm{d}\omega) - \tilde{a}_{\diamond}^*(\omega)\tilde{a}_{\diamond}(\omega)\,\mathrm{d}\omega$ is positive, where β denotes the measure induced by the real part of $\tilde{b}$. Since

$$\tilde{a}_{\diamond}(\omega) = 2\mathrm{i}\Im\tilde{a}(\omega) = -\frac{2\mathrm{i}}{\omega}\Re\tilde{\mu}(\omega), \quad \omega \ne 0,$$

we have

$$\tilde{a}_{\diamond}^*(\omega)\tilde{a}_{\diamond}(\omega) = \frac{4}{\omega^2}\left[\Re\tilde{\mu}(\omega)\right]^2.$$

On the other hand, as μ is the second distribution derivative of $-b$, it is true in the distribution sense that $\tilde{\mu}(\omega) = \omega^2\tilde{b}(\omega) + \mathrm{i}\omega A_1$. This implies that, for $\omega \ne 0$, $\Re\tilde{b}(\omega)$ can be identified with the function $\omega \mapsto \omega^{-2}\Re\tilde{\mu}(\omega)$. By Theorem **16**.2.8, $\Re\tilde{b}$ does not have a point mass at zero. In other words, the measure γ above is absolutely continuous, and satisfies

$$\gamma(\mathrm{d}\omega) = \frac{4}{\omega^2}\Big(\|\mu\|_{M(\mathbf{R}^+)}\Re\tilde{\mu}(\omega) - \left[\Re\tilde{\mu}(\omega)\right]^2\Big)\mathrm{d}\omega, \quad \omega \ne 0.$$

This function is positive, because $|\Re\tilde{\mu}(\omega)| \le \|\mu\|_{M(\mathbf{R}^+)}$. This completes the proof of Lemma 2.2. □

Proof of Part (b) of Theorem 2.1 (continues) First, let us return to our earlier claim that $A_1 \succ 0$. With the notations used in Lemma 2.2, $A_1 = b(0)$. By the Bochner Theorem **16**.2.7,

$$A_1 = \frac{1}{\pi}\int_{\mathbf{R}} \Re\tilde{b}(\omega)\,\mathrm{d}\omega = \frac{1}{\pi}\int_{\mathbf{R}} \frac{1}{\omega^2}\Re\tilde{\mu}(\omega)\,\mathrm{d}\omega.$$

If $n > 1$, then our assumption that the set $\{\,\omega \in \mathbf{R} \mid \det[\Re\tilde{\mu}(\omega)] = 0\,\}$ is countable implies that $\Re\tilde{\mu}(\omega) \succ 0$ a.e.; hence $A_1 \succ 0$, as claimed. In the case $n = 1$, we have either $\mu \equiv 0$, in which case the theorem is trivially true, or $\Re\tilde{\mu}(\omega) > 0$ on a set of positive measure, in which case $A_1 > 0$.

Define, for $\tau < T$,

$$\begin{aligned} S(\tau,T) = &-\Big\langle g(y(\tau)), \int_0^{T-\tau} a(t)g(y(t+\tau))\,\mathrm{d}t\Big\rangle \\ &- \int_\tau^T \Big\langle g(y(t)), \int_{(t-\tau,\infty)} \mu(\mathrm{d}s)g\big(y(t-s)\big)\Big\rangle\,\mathrm{d}t. \end{aligned} \tag{2.14}$$

Then S can be written in the alternative form

$$S(\tau,T) = \int_0^{T-\tau} \Big\langle g\left(y(t+\tau)\right), \int_{(t,\infty)} \mu(\mathrm{d}s)\big\{g(y(\tau)) - g\big(y(t-s+\tau)\big)\big\}\Big\rangle\,\mathrm{d}t.$$

In particular (recall that μ is extended as zero on $(-\infty,0)$),

$$|S(\tau,T)| \le \Big(\sup_{t\in\mathbf{R}}|g(y(t))|\Big)\int_{\mathbf{R}}\int_{(t,\infty)} |\mu|(\mathrm{d}s)\Big|g(y(\tau)) - g\big(y(t-s+\tau)\big)\Big|\,\mathrm{d}t.$$

We already know that $g \circ y$ is asymptotically slowly varying at (plus and) minus infinity, and therefore $\lim_{\tau\to-\infty}\bigl(g(y(\tau)) - g(y(t-s+\tau))\bigr) = 0$. By the two-dimensional version of Lebesgue's dominated convergence theorem,

$$S(\tau,T) \to 0, \quad \text{as } \tau \to -\infty. \tag{2.15}$$

Combining (2.4) and (2.14) with Lemma 2.2, we get

$$\begin{aligned}
\int_{\mathbf{R}} \Bigl| a_\diamond(t-T)g(y(T)) - a_\diamond(t-\tau)g(y(\tau)) + \int_{[t-T,t-\tau]} \nu(\mathrm{d}s) g\bigl(y(t-s)\bigr)\Bigr|^2 \,\mathrm{d}t \\
\le 4\|\mu\|_{M(\mathbf{R}^+)} \Bigl\{ G\left(y(\tau)\right) - G\left(y(T)\right) + S(\tau,T) \\
- \int_0^{T-\tau} \Bigl\langle g\left(y(T)\right), a(t) g\bigl(y(T-t)\bigr)\Bigr\rangle \,\mathrm{d}t \\
+ \frac{1}{2}\Bigl(\langle g(y(\tau)), A_1 g(y(\tau))\rangle + \langle g\left(y(T)\right), A_1 g\left(y(T)\right)\rangle\Bigr) \\
- \Bigl\langle g(y(T)), \int_{T-\tau}^{\infty} a(s)\,\mathrm{d}s\, g(y(\tau))\Bigr\rangle \Bigr\}.
\end{aligned}$$

If we here let $\tau \to -\infty$, then the sum of all those terms inside the braces on the right-hand side that depend on T tends to $-H(y)(T)$, where

$$\begin{aligned}
H(y)(T) \stackrel{\text{def}}{=} G\left(y(T)\right) - \frac{1}{2}\langle g(y(T)), A_1 g(y(T))\rangle \\
+ \Bigl\langle g(y(T)), \int_{-\infty}^{T} a(T-s) g(y(s))\,\mathrm{d}s \Bigr\rangle.
\end{aligned}$$

The remaining two terms are $G(y(\tau)) + \frac{1}{2}\langle g(y(\tau)), A_1 g(y(\tau))\rangle$. Because $g \circ y$ is asymptotically slowly varying at minus infinity, it follows from Lebesgue's dominated convergence theorem that

$$\lim_{\tau\to-\infty}\Bigl(H(y)(\tau) - G\left(y(\tau)\right) - \frac{1}{2}\langle g(y(\tau)), A_1 g(y(\tau))\rangle\Bigr) = 0.$$

Therefore, Fatou's lemma implies that

$$\begin{aligned}
0 \le \int_{\mathbf{R}} \Bigl| \int_{[t-T,\infty)} \nu(\mathrm{d}s) g\left((y(t-s)\right) + g(y(T)) a_\diamond(t-T)\Bigr|^2 \,\mathrm{d}t \\
\le 4\|\mu\|_{M(\mathbf{R}^+)} \Bigl(\liminf_{\tau\to-\infty} H(y)(\tau) - H(y)(T)\Bigr).
\end{aligned} \tag{2.16}$$

This implies that $\limsup_{T\to-\infty} H(y)(T) \le \liminf_{\tau\to-\infty} H(y)(\tau)$, hence $\lim_{t\to-\infty} H(y)(t)$ exists. It is clear that $H(y)(T) \le \lim_{t\to-\infty} H(y)(t)$ for all $T \in \mathbf{R}$. Moreover, if $\limsup_{T\to\infty} H(y)(T) = H(y)(-\infty)$, then it follows from (2.16) and Fatou's lemma that $\nu * (g \circ y) \equiv 0$. As before, this implies that y must be a constant.

We are almost at the point where we can apply Theorem **15**.8.1 to complete the proof. The only thing that remains to be verified is that the set $S \stackrel{\text{def}}{=} \{\, H(y)(0) \mid y \in \Gamma(x) \text{ is a constant} \,\}$ has empty interior. If $y \in \Gamma(x)$ is constant, then $y + A_1 g(y) \equiv 0$, and $H(y)(0) = E(y)$, where E is the function that we already encountered above, namely $E(y) = G(y) + \frac{1}{2}\langle y, A_1^{-1} y\rangle$. Therefore, S can be expressed in the form $\{\, E(y) \mid y + A_1 g(y) = 0 \,\}$, or equivalently in the form $\{\, G(y) + \frac{1}{2}\langle g(y), A_1 g(y)\rangle \mid y + A_1 g(y) = 0 \,\}$. This is the set mentioned in the assumption of Part (b). □

3. Lipschitz-Continuous Nonlinearity

Below we continue our investigation of (1.1), replacing the moment condition on μ in Theorem 2.1 by the assumption that g is Lipschitz-continuous. Here we shall only discuss the simpler of the two cases listed in Theorem 2.1, namely (a) where $\int_{\mathbf{R}^+} \mu(ds)$ is invertible. (It is not known if a similar result is true when $\int_{\mathbf{R}^+} \mu(ds)$ is not invertible.)

3.1 Theorem
Assume that $F_0 \in \mathbf{R}^n$ and that

(i) *$\mu \in M(\mathbf{R}^+; \mathbf{R}^{n\times n})$ is of positive type, $A_0 \stackrel{\text{def}}{=} \int_{\mathbf{R}^+} \mu(ds)$ is invertible, and*

$$\ker[\Re\tilde{\mu}(\omega)] = \ker[\tilde{\mu}(\omega)], \quad \omega \in \mathbf{R},$$

(ii) *$g \in C(\mathbf{R}^n; \mathbf{R}^n)$ is the gradient of a function $G \in C^1(\mathbf{R}^n; \mathbf{R})$, g is locally Lipschitz-continuous, and the set*

$$\{\, G(y) - \langle A_0^{-1} F_0, y\rangle \mid g(y) = A_0^{-1} F_0 \,\},$$

has empty interior,

(iii) *$f \in L^1_{\text{loc}}(\mathbf{R}^+; \mathbf{R}^n)$ and $\lim_{t\to\infty} \frac{1}{T}\int_t^{t+T} f(s)\,ds = F_0$ for all $T > 0$,*

(iv) *$x \in BUC(\mathbf{R}^+; \mathbf{R}^n) \cap AC_{\text{loc}}(\mathbf{R}^+; \mathbf{R}^n)$ is a solution of (1.1) on $\mathbf{R}^+$.*

Then x is asymptotically slowly varying, and $\lim_{t\to\infty} g(x(t)) = A_0^{-1} F_0$.

One can, in the same way as we did in the proof of Part (a) of Theorem 2.1, show that one may take $F_0 = 0$. If we do so, then the limit equation corresponding to (1.1) is again

$$y'(t) + \int_{\mathbf{R}^+} \mu(ds) g(y(t-s)) = 0, \quad t \in \mathbf{R}. \tag{2.3}$$

It turns out that Theorem 3.1 is a more or less direct consequence of the following statement on the integrability properties of the bounded solutions of (2.3).

3.2 Theorem
Assume that

(i) $\mu \in M(\mathbf{R}^+; \mathbf{R}^{n\times n})$ *is of positive type,* $A_0 \stackrel{\text{def}}{=} \int_{\mathbf{R}^+} \mu(\mathrm{d}s)$ *is invertible, and*

$$\ker[\Re\tilde{\mu}(\omega)] = \ker[\tilde{\mu}(\omega)], \quad \omega \in \mathbf{R},$$

(ii) $g \in C(\mathbf{R}^n; \mathbf{R}^n)$ *is the gradient of a function* $G \in C^1(\mathbf{R}^n; \mathbf{R})$, *and* g *is locally Lipschitz-continuous,*

(iii) $y \in BUC^1(\mathbf{R}^+; \mathbf{R}^n)$ *is a solution of (2.3) on* $\mathbf{R}$.

Then $g\circ y \in L^2(\mathbf{R}; \mathbf{R}^n)$ *and* $y' \in L^2(\mathbf{R}; \mathbf{R}^n)$.

Let us first show how Theorem 3.1 follows from Theorem 3.2, and then prove Theorem 3.2.

Proof of Theorem 3.1 As in the proof of Theorem 2.1(a), we shall employ Theorem **15**.8.1 with $H(\varphi) = G\circ\varphi$.

As earlier, if $y \in \Gamma(x)$ is a constant, then $g(y) = 0$. If $y \in \Gamma(x)$ is not constant, then, by Theorem 3.2, $g\circ y$ and $y' \in L^2(\mathbf{R}; \mathbf{R}^n)$. Consequently both the limits $G(y(\pm\infty)) \stackrel{\text{def}}{=} \lim_{t\to\pm\infty} G(y(t))$ exist. Take the inner product of both sides of equation (2.3) and $g(y(t))$, and integrate over $(-\infty, T)$. This gives (the double integral is well defined since $\mu \in M(\mathbf{R}^+; \mathbf{R}^{n\times n})$ and $g\circ y \in L^2(\mathbf{R}; \mathbf{R}^n)$)

$$G(y(T)) - G(y(-\infty)) + [\![\mu, \chi_{(-\infty,T)} g\circ y]\!] = 0, \quad t \in \mathbf{R}. \tag{3.1}$$

Define $g_T \stackrel{\text{def}}{=} \chi_{(-\infty,T)} g\circ y$, and transform the quadratic form above to get

$$G(y(T)) - G(y(-\infty)) + \frac{1}{2\pi}\int_{\mathbf{R}} \langle \tilde{g}_T(\omega), \Re\tilde{\mu}(\omega)\tilde{g}_T(\omega)\rangle \mathrm{d}\omega = 0.$$

Observe that the function $z \mapsto \det[\hat{\mu}(z)]$ belongs to the space $H^\infty(\{ z \mid \Re z > 0 \}; \mathbf{C})$, and hence, by, e.g., Rudin [3], Theorem 17.18, $\hat{\mu}(\mathrm{i}\omega) \neq 0$ a.e. (the other possibility that μ vanishes identically is excluded by the fact that $A_0 \neq 0$). If $g_T \equiv 0$ then it follows from (2.3) that y is a constant and otherwise, i.e., if y is not a constant, then necessarily $[\![\mu, g_T]\!] > 0$, and, by (3.1)

$$G(y(-\infty)) > G(y(T)), \quad T \in \mathbf{R}. \tag{3.2}$$

Taking $T = \infty$ in (3.1) we observe that if y is not constant then $G(y(\infty)) < G(y(-\infty))$. The conclusion now follows from Theorem **15**.8.1. □

Proof of Theorem 3.2 The proof of Theorem 3.2 is based on two independent time domain estimates.

(a) The estimate that one gets by taking the inner product of (2.3) and $g(y(t))$, and integrating. If μ has a finite first moment, then we can use exactly the same estimate as in the preceding section. If μ does not have a finite first moment, then we approximate μ by a measure having a finite first moment, and use the fact that the set of all measures that have a finite first moment is dense in $M(\mathbf{R}^+; \mathbf{R}^{n\times n})$.

(b) The estimate that one gets for y' directly from (2.3), or equivalently, the estimate that one gets by taking the inner product of (2.3) and $y'(t)$ and integrating. Again, if μ has a finite first moment this estimate is very easy; otherwise we approximate μ by a measure that has a finite first moment. Observe that, because of the Lipschitz continuity of g, this gives us an estimate on $(g\circ y)'$.

After proving these time domain estimates, we move into the frequency domain to complete the proof.

Let us begin with some preliminary comments.

Let y be a solution of (2.3) on $\mathbf{R}$. For each $\tau > 0$, define m_τ by

$$m_\tau = \sup_{t\in\mathbf{R}} \int_t^{t+\tau} \left|g(y(s))\right|^2 \mathrm{d}s.$$

By Theorem **3**.6.1(iv) (it is not essential that the norm in BL^p is defined over intervals of length 1), it is true for each $\nu \in M(\mathbf{R};\mathbf{R}^{n\times n})$ that

$$\sup_{t\in\mathbf{R}} \int_t^{t+\tau} \left|(\nu * g\circ y)(s))\right|^2 \mathrm{d}s \le \|\nu\|^2_{M(\mathbf{R})} m_\tau. \tag{3.3}$$

For each $\tau > 0$, choose some translate y_τ of y ($y_\tau(t) = y(t+h(\tau))$), in such a way that

$$\int_0^\tau |g(y_\tau(s))|^2 \mathrm{d}s \ge \frac{1}{2} m_\tau. \tag{3.4}$$

Moreover, define

$$\left.\begin{aligned} b(t) &= \int_{(t,\infty)} |\mu|(\mathrm{d}s), \quad t > 0,\\ \varphi_\tau &= \chi_{[0,\tau]} g\circ y_\tau,\\ \gamma &= \sup_{t\in\mathbf{R}}\Big(\left|G(y(t))\right| + \left|g(y(t))\right|\Big). \end{aligned}\right\} \tag{3.5}$$

To prove the theorem it suffices to show that $\sup_{\tau>0}\|\varphi_\tau\|_{L^2(\mathbf{R})} < \infty$, because then $\sup_{\tau>0} m_\tau < \infty$, and hence $g\circ y \in L^2(\mathbf{R};\mathbf{R}^n)$. This together with (2.3) implies that $y' \in L^2(\mathbf{R};\mathbf{R}^n)$.

Let us work out Estimate (a). Take the inner product of $g(y(t))$ and (2.3), and integrate over $[0,\tau]$ to get

$$G\left(y(\tau)\right) + [\![\mu, \varphi_\tau]\!] = G\left(y(0)\right) - R(\tau), \tag{3.6}$$

where $R(\tau)$ is the familiar term

$$R(\tau) = \int_0^\tau \Big\langle g(y(t)), \int_{(t,\infty)} \mu(\mathrm{d}s) g\big(y(t-s)\big)\Big\rangle \,\mathrm{d}t.$$

To estimate this term (in the case where μ does not have a finite first moment) we fix some $T > 0$, and split μ in two parts: $\mu = \mu_T + \nu_T$, where $\mu_T(E) = \mu(E \cap [0,T])$ and $\nu_T = \mu - \mu_T$. Writing $\mu = \mu_T + \nu_T$ above, and using (3.3) with ν replaced by ν_T we get

$$|R(\tau)| \le \|g\circ y\|^2_{\sup(\mathbf{R})} \int_0^T \left(b(s) - b(T)\right) \mathrm{d}s + b(T) m_\tau.$$

This together with (3.4)–(3.6) leads to the inequality

$$[\![\mu, \varphi_\tau]\!] \le 2\gamma + \gamma^2 \int_0^T \big(b(s) - b(T)\big)\,\mathrm{d}s + 2b(T)\|\varphi_\tau\|^2_{L^2(\mathbf{R})}, \tag{3.7}$$

which is the first of our two main inequalities in the time domain.

Our second main estimate, which is an estimate on $\|y'_\tau\|_{L^2(0,\tau)}$, is based directly on (2.3). For $t \in [0, \tau]$ we can write (2.3) in the form

$$y'_\tau(t) = (\mu * \varphi_\tau)(t) + \int_{(t,\infty)} \mu(\mathrm{d}s) g\big(y_\tau(t-s)\big).$$

Therefore

$$\|y'_\tau\|_{L^2(0,\tau)} \le \|\mu * \varphi_\tau\|_{L^2(\mathbf{R})} + S(\tau),$$

where

$$S(\tau) = \|\mu * (g\circ y_\tau - \varphi_\tau)\|_{L^2(0,\tau)}.$$

Splitting μ in μ_T and ν_T as above, one can easily estimate $S(\tau)$ by

$$S(\tau) \le \|g\circ y\|_{\sup(\mathbf{R})} \left\{ \int_0^T \big(b(s) - b(T)\big)^2 \mathrm{d}s \right\}^{\frac{1}{2}} + b(T)\sqrt{m_T}.$$

These inequalities, together with (3.4) and (3.5), give us a preliminary version of our second main inequality, namely

$$\begin{aligned} \|y'_\tau\|_{L^2(0,\tau)} &\le \|\mu * \varphi_\tau\|_{L^2(\mathbf{R})} \\ &\quad + \gamma \left\{ \int_0^T \big(b(s) - b(T)\big)^2 \mathrm{d}s \right\}^{\frac{1}{2}} + 2b(T)\|\varphi_\tau\|_{L^2(\mathbf{R})}. \end{aligned} \tag{3.8}$$

The preceding inequality still needs to be modified, because we need an estimate on $\varphi'_\tau = (g\circ y_\tau)'$ rather than an estimate on y'_τ. Because of the Lipschitz continuity of g, such an estimate is readily obtained from (3.8). As y is continuously differentiable and g is Lipschitz-continuous, φ_τ is locally absolutely continuous, and there is a constant $\lambda > 0$ such that

$$|\varphi'_\tau(t)| \le \lambda |y'_\tau(t)|$$

for almost all $t \in [0, \tau]$. In particular, this means that

$$\|\varphi'_\tau\|_{L^2(0,\tau)} \le \lambda \|y'_\tau\|_{L^2(0,\tau)}. \tag{3.9}$$

Combining this with (3.8), we get the final form of our second main time domain estimate, namely

$$\begin{aligned} \frac{1}{\lambda}\|\varphi'_\tau\|_{L^2(0,\tau)} &\le \|\mu * \varphi_\tau\|_{L^2(\mathbf{R})} \\ &\quad + \gamma \left\{ \int_0^T \big(b(s) - b(T)\big)^2 \mathrm{d}s \right\}^{\frac{1}{2}} + 2b(T)\|\varphi_\tau\|_{L^2(\mathbf{R})}. \end{aligned} \tag{3.10}$$

To complete the proof we move into the frequency domain. The argument that follows can be slightly simplified if μ is of anti-coercive type. (If μ is not of anti-coercive type, then we have to use the fact that μ is 'almost' of anti-coercive type.)

In the sequel, to simplify the notation, let the letter C represent a generic positive constant, whose value may change from one line to the next, but does not depend on τ (nor on T). Likewise, we let $C(T)$ represent a function of T (which in general tends to infinity as $T \to \infty$), but which is independent of τ.

Transforming (3.7) into the frequency domain, we get the inequality

$$\int_{\mathbf{R}} \Big\langle \widetilde{\varphi_\tau}(\omega), \Re\tilde{\mu}(\omega)\widetilde{\varphi_\tau}(\omega) \Big\rangle \mathrm{d}\omega \leq C(T) + Cb(T) \int_{\mathbf{R}} \big|\widetilde{\varphi_\tau}(\omega)\big|^2 \mathrm{d}\omega. \tag{3.11}$$

In particular, since $\Re\tilde{\mu}(0) \succ 0$, and as we can take $b(T)$ arbitrarily close to zero by choosing T large enough, it follows that there is some $\beta > 0$, such that (for a new constant C)

$$\int_{-\beta}^{\beta} \big|\widetilde{\varphi_\tau}(\omega)\big|^2 \mathrm{d}\omega \leq C + \int_{|\omega|>\beta} \big|\widetilde{\varphi_\tau}(\omega)\big|^2 \mathrm{d}\omega. \tag{3.12}$$

The next thing to do is to transform (3.10) into the frequency domain as well, but before we do this let us observe that, taking into account the possible discontinuities of φ_τ at 0 and at τ, we have

$$\big|\omega\widetilde{\varphi_\tau}(\omega)\big| \leq 2\gamma + \big|\widetilde{v_\tau}(\omega)\big|,$$

where $v_\tau = \chi_{[0,\tau]}\varphi'_\tau$. In particular,

$$\int_{|\omega|>\beta} \sqrt{1+|\omega|}\, \big|\widetilde{\varphi_\tau}(\omega)\big|^2 \mathrm{d}\omega \leq C + C\int_{\mathbf{R}} \big|\widetilde{v_\tau}(\omega)\big|^2 \mathrm{d}\omega.$$

Using this fact and (3.12) in the transformed version of (3.10), we get

$$\begin{aligned}
&\int_{\mathbf{R}} \sqrt{1+|\omega|}\, \big|\widetilde{\varphi_\tau}(\omega)\big|^2 \mathrm{d}\omega \\
&\qquad \leq C(T) + C\int_{\mathbf{R}} \big|\tilde{\mu}(\omega)\widetilde{\varphi_\tau}(\omega)\big|^2 \mathrm{d}\omega + Cb(T)\int_{\mathbf{R}} \big|\widetilde{\varphi_\tau}(\omega)\big|^2 \mathrm{d}\omega.
\end{aligned}$$

We observe that $|\omega|^{-1/4}\big|\tilde{\mu}(\omega)\big| \to 0$ as $\omega \to \infty$, and if we choose T large enough we find that there is a constant K such that

$$\int_{\mathbf{R}} \big|\widetilde{\varphi_\tau}(\omega)\big|^2 \mathrm{d}\omega \leq C(T) + C\int_{|\omega|\leq K} \big|\tilde{\mu}(\omega)\widetilde{\varphi_\tau}(\omega)\big|^2 \mathrm{d}\omega.$$

We can now fix some $\epsilon > 0$ small enough, and define

$$\Omega_\tau = \big\{\, |\omega| \leq K \;\big|\; \big|\tilde{\mu}(\omega)\widetilde{\varphi_\tau}(\omega)\big| \geq \epsilon\big|\widetilde{\varphi_\tau}(\omega)\big| > 0 \,\big\},$$

to get

$$\int_{\mathbf{R}} \big|\widetilde{\varphi_\tau}(\omega)\big|^2 \mathrm{d}\omega \leq C(T) + C\int_{\omega\in\Omega_\tau} \big|\tilde{\mu}(\omega)\widetilde{\varphi_\tau}(\omega)\big|^2 \mathrm{d}\omega. \tag{3.13}$$

This substituted back into (3.11) gives

$$\int_{\mathbf{R}} \Big\langle \widetilde{\varphi_\tau}(\omega), \Re\tilde{\mu}(\omega)\widetilde{\varphi_\tau}(\omega) \Big\rangle \mathrm{d}\omega \leq C(T) + Cb(T)\int_{\omega\in\Omega_\tau} \big|\tilde{\mu}(\omega)\widetilde{\varphi_\tau}(\omega)\big|^2 \mathrm{d}\omega.$$

But owing to our assumption on $\tilde{\mu}$, $\langle y, \Re\tilde{\mu}(\omega)y\rangle \geq \delta\big|\tilde{\mu}(\omega)y\big|^2$ for some $\delta > 0$ independent of τ and for all $y \in \mathbf{R}^n$, $\omega \in \Omega_\tau$ (because the set

$\{ (\omega, y) \in \mathbf{R} \times \mathbf{R}^n \mid |y| = 1,\ |\omega| \le K \text{ and } |\tilde{\mu}(\omega) y| \ge \epsilon \}$ is compact, and the function $(\omega, y) \mapsto \langle y, \Re\tilde{\mu}(\omega) y\rangle / |\tilde{\mu}(\omega) y|^2$ is continuous and strictly positive on this set). This means that, if we choose T large enough, the last inequality gives a bound on $\int_{\omega\in\Omega_T} |\tilde{\mu}(\omega)\widetilde{\varphi_T}(\omega)|^2 \mathrm{d}\omega$, which substituted back into (3.13) gives the desired bound on $\|\varphi_T\|_{L^2(\mathbf{R})}$, and completes the proof. □

4. A Linear Transformation of the Nonlinear Equation

In the two previous sections we have studied (1.1) in the two cases where μ has a finite first moment and $\mu(\mathbf{R}^+)$ is either invertible or zero, and where g is Lipschitz-continuous and $\mu \in M(\mathbf{R}^+)$. In the case where μ is of positive type but not finite, i.e., $\mu \notin M(\mathbf{R}^+)$, then neither of the two results applies. Nevertheless, it is sometimes possible to say something about the asymptotic behaviour of the solutions of (1.1) by comparing the solutions of (1.1) with the solutions of the corresponding linear equation

$$v'(t) + \int_{[0,t]} \mu(\mathrm{d}s) v(t-s) = f(t), \quad t \in \mathbf{R}^+; \quad v(0) = x_0. \tag{4.1}$$

We recall from Chapter 3 that the solution v of (4.1) is given by

$$v(t) = r(t)x_0 + \int_0^t r(t-s) f(s)\,\mathrm{d}s, \tag{4.2}$$

where r is the differential resolvent of μ, i.e., the solution of the resolvent equations

$$r'(t) + (\mu * r)(t) = r'(t) + (r * \mu)(t) = 0, \quad t \in \mathbf{R}^+; \quad r(0) = I. \tag{4.3}$$

It is not surprising that in the approach that we describe below, although we do not assume that μ necessarily is finite, we do put a global size restriction on r (or more precisely, on the second derivative of r).

Although one usually associates the differential resolvent r of μ with the linear equation (4.1), it can also be used when one studies the asymptotic behaviour of the nonlinear equation (1.1). The idea is the following. Adding the term $\mu{*}x$ to both sides of (1.1), and then using the variation of constants formula (4.2) and the resolvent equation (4.3), we get

$$x(t) - \int_0^t r'(s) h\big(x(t-s)\big)\,\mathrm{d}s = v(t), \quad t \in \mathbf{R}^+,$$

where v is the solution of (4.1), and

$$h(y) = g(y) - y, \quad y \in \mathbf{R}^n. \tag{4.4}$$

Differentiating the first of these two equations, and denoting the measure derivative of $-r'$ by ρ (by Theorem **3**.3.1, $-r'$ is locally of bounded variation), we get

$$x'(t) + \int_{[0,t]} \rho(\mathrm{d}s)h\big(x(t-s)\big) = v'(t), \quad t \in \mathbf{R}^+; \quad x(0) = x_0. \tag{4.5}$$

Clearly, this is an equation of the same type as (1.1). Moreover, as the following lemma shows, the new kernel ρ is of positive type whenever the original kernel μ is.

4.1 Lemma

(i) *Let $\mu \in PT(\mathbf{R}^+;\mathbf{C}^{n\times n})$. Then both the differential resolvent r of μ and the distribution derivative ρ of $-r'$ are of positive type.*

(ii) *In addition to* (i), *suppose that, for some real $\omega_0 \neq 0$, the limit $\lim_{z\to \mathrm{i}\omega_0, \Re z>0} \hat{\mu}(z) \stackrel{\text{def}}{=} \tilde{\mu}(\omega_0)$ exists, and that this limit satisfies*

$$\ker[\Re\tilde{\mu}(\omega_0)] = \ker[\tilde{\mu}(\omega_0)].$$

Then the limit $\tilde{\rho}(\omega_0) \stackrel{\text{def}}{=} \lim_{z\to \mathrm{i}\omega_0, \Re z>0} \hat{\rho}(z)$ exists, and satisfies

$$\ker[\Re\tilde{\rho}(\omega_0)] = \ker[\tilde{\rho}(\omega_0)].$$

Moreover, $\det[\Re\tilde{\mu}(\omega_0)] \neq 0$ *iff* $\det[\Re\tilde{\rho}(\omega_0)] \neq 0$.

Proof We begin with a proof of (i). By the first part of Theorem **16**.3.8, to prove (i) it suffices to show that r is of positive type. Moreover, one can without loss of generality assume that $\mu \in M(\mathbf{R}^+;\mathbf{C}^{n\times n})$. (If not, then we approximate μ by a sequence μ_k of finite measures of positive type, and observe that, by Theorem **3**.3.4, the differential resolvent of μ_k tends to the differential resolvent of μ in $L^1_{\mathrm{loc}}(\mathbf{R}^+;\mathbf{C}^{n\times n})$ as $k\to\infty$. For example, one can define $\eta(t) = 1-t$ for $t\in[0,1]$, $\eta(t)=0$ for $t\in(1,\infty)$, $\eta_k(t) = \eta(t/k)$, and $\mu_k(\mathrm{d}t) = \eta_k(t)\mu(\mathrm{d}t)$. That μ_k is of positive type follows from Propositions **16**.3.1 and **16**.3.5.)

Thus, let us prove that r is of positive type under the additional assumption $\mu \in M(\mathbf{R}^+;\mathbf{C}^{n\times n})$. First we claim that $\int_{\mathbf{R}^+} \mathrm{e}^{-\epsilon t}|r(t)|\,\mathrm{d}t < \infty$ for all $\epsilon > 0$. The proof of this fact is based on Corollary **3**.3.6. In order to apply that result, we have to check that $\det[zI + \hat{\mu}(z)] \neq 0$ for $\Re z > 0$. However, this follows from Lemma **5**.2.2 and from the fact that $\Re\{zI + \hat{\mu}(z)\} \succeq (\Re z)I \succ 0$ for $\Re z > 0$. Thus, $\int_{\mathbf{R}^+} \mathrm{e}^{-\epsilon t}|r(t)|\,\mathrm{d}t < \infty$ for all $\epsilon > 0$. Moreover,

$$\hat{r}(z) = \big[zI + \hat{\mu}(z)\big]^{-1}, \quad \Re z > 0,$$

hence

$$\Re\hat{r}(z) = \big[\overline{z}I + \hat{\mu}^*(z)\big]^{-1}\big[(\Re z)I + \Re\hat{\mu}(z)\big]\big[zI+\hat{\mu}(z)\big]^{-1} \succeq 0, \quad \Re z > 0.$$

By Theorem **16**.2.4, r is of positive type. This completes the proof of (i).

It remains to prove (ii). In this proof we need to know that $\mathrm{i}\omega_0 I + \tilde{\mu}(\omega_0)$ is invertible. To see this, first observe that $\Re\tilde{\mu}(\omega_0) \succeq 0$. If, for some $v \in \mathbf{C}^n$, $(\mathrm{i}\omega_0 I + \tilde{\mu}(\omega_0))v = 0$, then, by Lemma **5**.2.2, $\Re\{\mathrm{i}\omega_0 I + \tilde{\mu}(\omega_0)\}v =$

$\Re\{\tilde{\mu}(\omega_0)\}v = 0$. By the assumption in (ii), this implies that $\tilde{\mu}(\omega_0)v = 0$. Thus $0 = (\mathrm{i}\omega_0 I + \tilde{\mu}(\omega_0))v = \mathrm{i}\omega_0 v$, and, as $\omega_0 \neq 0$, this forces v to be zero. We conclude that, indeed, $\mathrm{i}\omega_0 I + \tilde{\mu}(\omega_0)$ is invertible.

Let us proceed with the proof of (ii). Since $\hat{\rho}(z) = z\hat{\mu}(z)[zI + \hat{\mu}(z)]^{-1}$ for $\Re z > 0$, we have

$$\hat{\rho}(z) = \left[\overline{z}I + \hat{\mu}^*(z)\right]^{-1}\left(|z|^2\hat{\mu}(z) + z\hat{\mu}^*(z)\hat{\mu}(z)\right)\left[zI + \hat{\mu}(z)\right]^{-1} \succeq 0, \quad \Re z > 0,$$

and letting $z \to \mathrm{i}\omega_0$ we get

$$\tilde{\rho}(\omega_0) = \left[-\mathrm{i}\omega_0 I + \tilde{\mu}^*(\omega_0)\right]^{-1}\left(\omega_0^2\tilde{\mu}(\omega_0) + \mathrm{i}\omega_0\tilde{\mu}^*(\omega_0)\tilde{\mu}(\omega_0)\right)\left[\mathrm{i}\omega_0 I + \tilde{\mu}(\omega_0)\right]^{-1},$$

and

$$\Re\tilde{\rho}(\omega_0) = \left[-\mathrm{i}\omega_0 I + \tilde{\mu}^*(\omega_0)\right]^{-1}\left(\omega_0^2\Re\tilde{\mu}(\omega_0)\right)\left[\mathrm{i}\omega_0 I + \tilde{\mu}(\omega_0)\right]^{-1}.$$

The desired conclusions follow immediately from these two formulas. □

The following theorem gives quite satisfactory results for the case where we know that the first moment of ρ is finite.

4.2 Theorem

Assume the following.

(i) $\mu \in PT(\mathbf{R}^+;\mathbf{R}^{n\times n})$, *the limits* $\lim_{z\to \mathrm{i}\omega, \Re z>0}\hat{\mu}(z)$ *exist for all* $\omega \neq 0$ *(and are denoted by* $\tilde{\mu}(\omega)$*),*

$$\ker[\Re\tilde{\mu}(\omega)] = \ker[\tilde{\mu}(\omega)], \quad \omega \in \mathbf{R},$$

and the set $\{\, \omega \in \mathbf{R}\setminus\{0\} \mid \det[\Re\tilde{\mu}(\omega)] = 0 \,\}$ *is countable if* $n > 1$.

(ii) $\int_{\mathbf{R}^+} t|\rho|(\mathrm{d}t) < \infty$, *where* ρ *is the measure derivative of* $-r'$ *and* r *is the differential resolvent of* μ.

(iii) $g \in C(\mathbf{R}^n;\mathbf{R}^n)$ *is the gradient of a function* $G \in C^1(\mathbf{R}^n;\mathbf{R})$.

(iv) $f \in L^1_{\mathrm{loc}}(\mathbf{R}^+;\mathbf{R}^n)$.

(v) *The solution* v *of (4.1) tends to a limit* $v(\infty)$ *at infinity.*

(vi) *The set*

$$\left\{\, G(y) + \frac{1}{2}\langle g(y) - y, B(g(y) - y)\rangle - \frac{1}{2}|y|^2 \;\middle|\; Bg(y) + (I - B)y = v(\infty) \,\right\}$$

has empty interior, where $B \stackrel{\text{def}}{=} -\int_{\mathbf{R}^+} t\rho(\mathrm{d}t)$.

(vii) $x \in BUC(\mathbf{R}^+;\mathbf{R}^n) \cap AC_{\mathrm{loc}}(\mathbf{R}^+;\mathbf{R}^n)$ *is a solution of (1.1) on* $\mathbf{R}^+$.

Then x *is asymptotically slowly varying, and*

$$\lim_{t\to\infty}\left(Bg(x(t)) + (I - B)x(t)\right) = v(\infty).$$

Proof Since $\int_{\mathbf{R}^+} t|\rho|(\mathrm{d}t) < \infty$, we have $\rho \in M(\mathbf{R}^+;\mathbf{R}^{n\times n})$, hence $r' \in BV(\mathbf{R}^+;\mathbf{R}^{n\times n})$, and $r'(\infty) = \lim_{t\to\infty} r'(t)$ exists. This limit must be zero owing to the fact that r is bounded (every continuous function of positive type is bounded). Observe that $\rho(\mathbf{R}^+) = r'(\infty)$; hence $\rho(\mathbf{R}^+) = 0$. Now apply first Lemma 4.1, and then Theorem 2.1 with the replacements $\mu \to \rho$, $g(y) \to h(y) = g(y) - y$, $G(y) \to G(y) - \frac{1}{2}|y|^2$, $f \to v'$, and (1.1) $\to$ (4.5). □

If μ does not belong to $M(\mathbf{R}^+)$, then one cannot give any general result concerning whether or not the first moment of ρ is finite. (Some results in this direction can be found in Chapters 5 and 6.) However, it is important to note that ρ under certain assumptions inherits a moment condition from μ.

4.3 Lemma

Assume that

(i) $\mu \in PT(\mathbf{R}^+;\mathbf{R}^{n\times n})$ *satisfies* $\int_{\mathbf{R}^+} s|\mu|(\mathrm{d}s) < \infty$, *and*

$$\ker[\Re\tilde{\mu}(\omega)] = \ker[\tilde{\mu}(\omega)], \quad \omega \in \mathbf{R},$$

(ii) r *is the differential resolvent of* μ *and* ρ *is the distribution derivative of* $-r'$.

Define $A_0 = \int_{\mathbf{R}^+} \mu(\mathrm{d}s)$ *and* $A_1 = -\int_{\mathbf{R}^+} s\mu(\mathrm{d}s)$, *and let* P *be the orthogonal projection of* $\mathbf{R}^n$ *onto* $\ker[A_0]$. *Then*

(1) $\rho(\mathbf{R}^+) = 0$, $\int_{\mathbf{R}^+} t|\rho|(\mathrm{d}t) < \infty$, *and*

$$-\int_{\mathbf{R}^+} t\rho(\mathrm{d}t) = I - r(\infty) = I - [I + PA_1]^{-1}P,$$

(2) $\int_{\mathbf{R}^+} \big|r(t)(I-P)\big|\,\mathrm{d}t < \infty$, *and*

$$\int_{\mathbf{R}^+} r(t)(I-P)\,\mathrm{d}t = [I + PA_1]^{-1}[P + A_0]^{-1}(I-P).$$

Proof Clearly, $A_0 = \hat{\mu}(0)$. Thus, by (i), $\ker[A_0] = \ker[\Re A_0]$. Since $\Re A_0 = \frac{1}{2}(A_0 + A_0^*)$, we conclude that $\ker[A_0^*] = \ker[A_0]$. But $\mathrm{range}[A_0] = (\ker[A_0^*])^\perp$, and hence $\mathrm{range}[A_0] = (\ker[A_0])^\perp$. This implies that, if we denote the orthogonal projection from $\mathbf{R}^n$ onto $\ker[A_0]$ by P, then $I-P$ is the orthogonal projection onto $\mathrm{range}[A_0]$. Moreover, the restriction of A_0 to $\mathrm{range}[A_0]$ maps $\mathrm{range}[A_0]$ one-to-one onto itself. Without loss of generality, let us assume that the first p of our basis vectors belong to $\mathrm{range}[A_0]$, and the remaining $n-p$ to $\ker[A_0]$. Then A_0, $\hat{\mu}(z)$ and P are structured as

$$A_0 = \begin{pmatrix} A & 0 \\ 0 & 0 \end{pmatrix}, \quad \hat{\mu}(z) = \begin{pmatrix} A_{11}(z) & A_{12}(z) \\ A_{21}(z) & A_{22}(z) \end{pmatrix}, \quad P = \begin{pmatrix} 0 & 0 \\ 0 & I \end{pmatrix}, \tag{4.6}$$

where A is an invertible $p\times p$ matrix with $\Re A \succ 0$, $A_{11}(0) = A$, $A_{12}(0) = 0$, $A_{21}(0) = 0$ and $A_{22}(0) = 0$.

By (4.6), and by the fact that $zI + A$ is invertible for $\Re z \geq 0$ (since $\Re\{zI + A\} \succ 0$), we can write

$$[zI + A_0]^{-1}zI = \begin{pmatrix} [zI+A]^{-1}zI & 0 \\ 0 & I \end{pmatrix},$$

$$[zI + A_0]^{-1}(I-P) = \begin{pmatrix} [zI+A]^{-1} & 0 \\ 0 & 0 \end{pmatrix},$$

$$[zI + A_0]^{-1}A_0 = \begin{pmatrix} [zI+A]^{-1}A & 0 \\ 0 & 0 \end{pmatrix}, \quad \Re z \geq 0.$$

(For $z = 0$ we define the left-hand sides by the corresponding right-hand sides. The latter are well defined and continuous for $\Re z \geq 0$.) Thus, by Theorems **6**.3.2–**6**.3.4, the three functions above are Laplace transforms of finite measures, with total masses

$$\begin{pmatrix} 0 & 0 \\ 0 & I \end{pmatrix} = P, \quad \begin{pmatrix} A^{-1} & 0 \\ 0 & 0 \end{pmatrix} = [P + A_0]^{-1}(I - P), \quad \text{and}$$

$$\begin{pmatrix} I & 0 \\ 0 & 0 \end{pmatrix} = (I - P),$$

respectively. Since $\frac{1}{z}(\hat{\mu}(z) - A_0)$ is the transform of the L_1-function $t \mapsto -\int_{[t,\infty)} \mu(\mathrm{d}s)$, we conclude that

$$[zI + A_0]^{-1}\hat{\mu}(z) = [zI + A_0]^{-1}A_0 + [zI + A_0]^{-1}zI\tfrac{1}{z}(\hat{\mu}(z) - A_0)$$

is the transform of a finite measure. Thus, $[zI + A_0]^{-1}[zI + \hat{\mu}(z)]$ is the transform of a finite measure.

We claim that the last function above can be inverted for all z with $\Re z \geq 0$, and that the inverse is the transform of a finite measure. The inversion does not present any difficulties when $z \neq 0$ (cf. the proof of Lemma 4.1), but the point $z = 0$ does present a problem. To see what the function looks like at zero, let us write it as

$$[zI + A_0]^{-1}[zI + \hat{\mu}(z)] = \begin{pmatrix} [zI + A]^{-1}[zI + A_{11}(z)] & [zI + A]^{-1}A_{12}(z) \\ \frac{1}{z}A_{21}(z) & I + \frac{1}{z}A_{22}(z) \end{pmatrix}. \tag{4.7}$$

As $z \to 0$, it tends to $I + P\lim_{z\to 0}\frac{1}{z}(\hat{\mu}(z) - A_0) = I + PA_1$, so our next task is to show that this matrix is invertible.

Suppose that $I + PA_1$ is not invertible. Then there exists a vector $v \in \mathbf{R}^n$ such that $v + PA_1v = 0$. In particular, $Pv = v$, i.e., $v \in \ker[A_0]$. On the other hand, $A_1 = \lim_{\sigma\in\mathbf{R},\sigma\downarrow 0}\frac{1}{\sigma}(\hat{\mu}(\sigma) - \hat{\mu}(0)) = \lim_{\sigma\in\mathbf{R},\sigma\downarrow 0}\frac{1}{\sigma}(\hat{\mu}(\sigma) - A_0)$. Since μ is of positive type we have $\hat{\mu}(\sigma) \succeq 0$, and since $v \in \ker[A_0] = \ker[\Re A_0]$ we conclude that $\langle v, \Re A_1v\rangle \geq 0$. Recall that $v + PA_1v = 0$, hence $\langle v, v + PA_1v\rangle = 0$. But, as $Pv = v$, and P is self-adjoint, $\langle v, v + PA_1v\rangle = \langle v, v + A_1v\rangle$, hence $\Re\langle v, v + PA_1v\rangle \geq |v|^2 > 0$. This contradiction shows that $I + PA_1$ must be invertible.

We conclude that the matrix function in (4.7) is invertible for all z with $\Re z \geq 0$. By Theorems **6**.3.2–**6**.3.4, the function $[zI + \hat{\mu}(z)]^{-1}[zI + A_0]$ is the transform of a finite measure. This, combined with our earlier results for the functions $[zI + A_0]^{-1}zI$, $[zI + A_0]^{-1}(I - P)$, and $[zI + A_0]^{-1}\hat{\mu}(z)$, implies that $[zI + \hat{\mu}(z)]^{-1}zI$, $[zI + \hat{\mu}(z)]^{-1}(I - P)$, and $[z + \hat{\mu}(z)]^{-1}\hat{\mu}(z)$ are transforms of finite measures, as well.

From the resolvent equation (4.3) we get

$$\hat{\rho}(z) = [zI + \hat{\mu}(z)]^{-1}z\hat{\mu}(z),$$

and if we apply the results derived above we see that $\rho \in M(\mathbf{R}^+; \mathbf{R}^{n\times n})$ with $\hat{\rho}(0) = 0$, and that $r(I-P) \in L^1(\mathbf{R}^+; \mathbf{R}^{n\times n})$, with $\int_{\mathbf{R}^+} r(t)(I-P)\,\mathrm{d}t = [I + PA_1]^{-1}[P + A_0]^{-1}(I - P)$. Recall that $\hat{\mu}'(z)$ is the Laplace transform of the finite measure $-t\mu(\mathrm{d}t)$, and that $\hat{\rho}'(z)$ is the Laplace transform of the measure $-t\rho(\mathrm{d}t)$. Since

$$\hat{\rho}'(z) = [zI + \hat{\mu}(z)]^{-1}\big(\hat{\mu}(z) + z\hat{\mu}'(z)\big) + [zI + \hat{\mu}(z)]^{-1} zI\big(I + \hat{\mu}'(z)\big)[zI + \hat{\mu}(z)]^{-1}\hat{\mu}(z),$$

we conclude that $\int_{\mathbf{R}^+} t|\rho|(\mathrm{d}t) < \infty$. Now $\hat{\rho}(0) = 0$, and we see that $-\int_{\mathbf{R}^+} t\rho(\mathrm{d}t) = I - r(\infty)$. On the other hand we have $r(\infty) = \lim_{z\to 0} z\hat{r}(z)$, so we immediately get the desired conclusion that $r(\infty) = [I + PA_1]^{-1}P$. □

Applying Lemma 4.3 to (4.2) it is possible to derive the following result for the solution v of (4.1).

4.4 Proposition

Let μ satisfy the assumptions of Lemma 4.3, and define A_0, A_1 and P as in that lemma. Let $f \in L^1_{\mathrm{loc}}(\mathbf{R}^+;\mathbf{R}^n)$ satisfy

$$\lim_{t\to\infty} \frac{1}{T}\int_t^{t+T} (I - P)f(s)\,\mathrm{d}s = F_0, \quad T > 0, \tag{4.8}$$

and

$$\lim_{t\to\infty} \int_0^t Pf(s)\,\mathrm{d}s = F_1. \tag{4.9}$$

Then the solution v of (4.1) is bounded, and

$$\lim_{t\to\infty} v(t) = [I + PA_1]^{-1}([P + A_0]^{-1}F_0 + Px_0 + F_1).$$

Proof We can rewrite (4.2) and use an integration by parts, to get

$$v(t) = r(t)x_0 + \int_0^t r(t-s)(I-P)f(s)\,\mathrm{d}s + \int_0^t Pf(s)\,\mathrm{d}s + \int_0^t r'(t-s)\int_0^s Pf(u)\,\mathrm{d}u\,\mathrm{d}s.$$

From this decomposition combined with Lemma 4.3 the result follows immediately. □

Combining Theorem 4.2 with Lemma 4.3 and Proposition 4.4, we get the following result.

4.5 Theorem

Assume that

(i) *$\mu \in PT(\mathbf{R}^+;\mathbf{R}^{n\times n})$ satisfies $\int_{\mathbf{R}^+} s|\mu|(\mathrm{d}s) < \infty$,*

$$\ker[\Re\tilde{\mu}(\omega)] = \ker[\tilde{\mu}(\omega)], \quad \omega \in \mathbf{R},$$

and the set $\{\,\omega \in \mathbf{R} \mid \det[\Re\tilde{\mu}(\omega)] = 0\,\}$ is countable if $n > 1$,

(ii) *$g \in C(\mathbf{R}^n;\mathbf{R}^n)$ is the gradient of a function $G \in C^1(\mathbf{R}^n;\mathbf{R})$.*

Define $A_0 = \int_{\mathbf{R}^+} \mu(\mathrm{d}s)$, $A_1 = -\int_{\mathbf{R}^+} s\mu(\mathrm{d}s)$, and let P be the orthogonal projection of $\mathbf{R}^n$ onto $\ker[A_0]$, and assume in addition that

(iii) $f \in L^1_{\mathrm{loc}}(\mathbf{R}^+;\mathbf{R}^n)$ *satisfies* $\lim_{t\to\infty} \frac{1}{T}\int_t^{t+T}(I-P)f(s)\,\mathrm{d}s = F_0$ *for* $T>0$ *and* $\lim_{t\to\infty}\int_0^t Pf(s)\,\mathrm{d}s = F_1$,

(iv) *the set*

$$\Big\{ G(y) + \frac{1}{2}\langle g(y)-y, (I-[I+PA_1]^{-1}P)(g(y)-y)\rangle - \frac{1}{2}|y|^2 \;\Big|\; (I-P)g(y) + P\big(y+A_1 g(y)\big) = [P+A_0]^{-1}F_0 + Px_0 + F_1 \Big\}$$

has empty interior,

(v) $x \in BUC(\mathbf{R}^+;\mathbf{R}^n)\cap AC_{\mathrm{loc}}(\mathbf{R}^+;\mathbf{R}^n)$ *is a solution of (1.1) on* $\mathbf{R}^+$.

Then x is asymptotically slowly varying, and

$$\lim_{t\to\infty}\Big\{(I-P)g(x(t)) + P\big(x(t)+A_1 g(x(t))\big)\Big\} = [P+A_0]^{-1}F_0 + Px_0 + F_1.$$

Finally we give an easy result concerning global boundedness of solutions.

4.6 Theorem

Assume that

(i) $\mu \in PT(\mathbf{R}^+;\mathbf{R}^{n\times n})$,

(ii) $g \in C(\mathbf{R}^n;\mathbf{R}^n)$ *is the gradient of a function* $G\in C^1(\mathbf{R}^n;\mathbf{R})$, *and*

$$|g(v)| \le c(1+G(v)),\quad G(v)\ge \epsilon|v|^2 - c,\quad v\in\mathbf{R}^n,$$

for some positive constants c and ϵ,

(iii) $f = f_1+f_2$ *where* $f_1\in L^1(\mathbf{R}^+;\mathbf{R}^n)$ *and* $f_2\in BV(\mathbf{R}^+;\mathbf{R}^n)$,

(iv) r_λ *is , for each* $\lambda>0$, *the differential resolvent of* $\lambda\mu$, *and for some* $\lambda<2\epsilon$,

$$r_\lambda,\ r'_\lambda \in L^1(\mathbf{R}^+;\mathbf{R}^{n\times n}).$$

(v) $x\in AC_{\mathrm{loc}}(\mathbf{R}^+;\mathbf{R}^n)$ *is a solution of (1.1) on* $\mathbf{R}^+$.

Then $x\in BC(\mathbf{R}^+;\mathbf{R}^n)$.

Note that one can combine the results of this theorem with those in Chapter 12 to show that every local solution can be continued to a globally defined bounded one.

Proof of Theorem 4.6 A trivial modification of the argument which gave us (4.5) shows that

$$x'(t) + \int_{[0,t]}\rho_\lambda(\mathrm{d}s)h_\lambda(x(t-s)) = z'_\lambda(t),\quad t\in\mathbf{R}^+;\quad x(0)=x_0. \tag{4.10}$$

where ρ_λ is the measure derivative of $-r'_\lambda$, $h_\lambda(v) = g(v)-\lambda v$ and $z_\lambda = r_\lambda x_0 + r_\lambda * f$. Observe that it follows from our assumptions that $z'_\lambda\in L^1(\mathbf{R}^+)$. By Proposition **17**.2.1, x is bounded. □

5. Exercises

1. Prove Proposition 1.1.
2. Consider the scalar equation

$$x'(t) + \int_{\mathbf{R}} \mu(\mathrm{d}s)g\big(x(t-s)\big) = f(t), \quad t \in \mathbf{R}, \tag{5.1}$$

where we assume that

(i) $\mu = \mu_1 + \mu_2$ with $\mu_1(\mathrm{d}t) = \rho_1\delta_0(\mathrm{d}t)$, $\mu_2 \in M(\mathbf{R};\mathbf{R})$ and $\|\mu_2\|_{M(\mathbf{R})} < \rho_1$,

(ii) $g \in C(\mathbf{R};\mathbf{R})$,

(iii) $f \in L^\infty(\mathbf{R};\mathbf{R})$, $f(\infty) = \lim_{t\to\infty} f(t)$ exists,

(iv) $x \in BC(\mathbf{R};\mathbf{R}) \cap AC_{\mathrm{loc}}(\mathbf{R};\mathbf{R})$ satisfies (5.1) on $\mathbf{R}$, and the set $S \stackrel{\mathrm{def}}{=} \{\, c \mid g(c)\mu(\mathbf{R}) = f(\infty)\,\}$ is nonempty and has empty interior.

Show that $\lim_{t\to\infty} x(t) = c$, $\lim_{t\to\infty} x'(t) = 0$ for some $c \in S$. Is the assumption $\|\mu_2\|_{M(\mathbf{R})} < \rho_1$ necessary? (See Shilepsky [1]. The corresponding Volterra equation on $\mathbf{R}^+$ was analysed in Shilepsky [2].)

3. A model for the growth of a single biological population may be described by the scalar equation

$$\frac{N'(t)}{N(t)} = a - bN(t) - \int_0^t N(t-s)f(s)\,\mathrm{d}s,$$
$$t \in \mathbf{R}^+; \quad N(0) = N_0, \tag{5.2}$$

where a and b are positive constants, N is the population size and f is a hereditary term. Show that if $f \in C(\mathbf{R}^+;\mathbf{R}) \cap L^1(\mathbf{R}^+;\mathbf{R})$ and $\int_{\mathbf{R}}|f(s)|\,\mathrm{d}s < b$, then there exists a unique solution $N(t)$ of (5.2) such that $\lim_{t\to\infty} N(t)$ exists. (See Miller [3].)

4. Let $\mu \in M(\mathbf{R}^+;\mathbf{C}^{n\times n})$ and let r be the differential resolvent of μ. Suppose that both r and the distribution derivative of $-r'$ are of positive type. Does it follow that μ is of positive type?
5. Prove the following result.

5.1 Theorem
Assume that

(i) $a \in L^1(\mathbf{R}^+;\mathbf{R}^+)$ *is bounded, nonnegative, and nonincreasing on* $\mathbf{R}^+$,

(ii) $g \in C(\mathbf{R};\mathbf{R})$,

(iii) $f \in C_\ell(\mathbf{R}^+;\mathbf{R})$,

(iv) $x \in BC(\mathbf{R}^+;\mathbf{R})$ *is a solution of the equation*

$$x(t) + \int_0^t a(t-s)g(x(s))\,\mathrm{d}s = f(t), \quad t \in \mathbf{R}^+.$$

Then x is asymptotically slowly varying and

$$\lim_{t\to\infty}\left\{x(t)+g(x(t))\int_0^\infty a(s)\,\mathrm{d}s\right\}=f(\infty).$$

6. Try to prove Theorem 3.1 in the case where A_0 is not invertible (i.e., the case corresponding to Theorem 2.1(b)). (Open problem.)
7. Give an example that shows that $\lim_{t\to\infty} x(t)$ need not exist under the assumptions of Theorem 2.1. Hint: Take $g \equiv 0$.
8. Show by an example that the conclusion of Theorem 2.1 does not hold if the assumption that $\ker[\Re\tilde{\mu}(\omega)] = \ker[\tilde{\mu}(\omega)]$ for $\omega \in \mathbf{R}$ is removed.
9. Show that the following conclusions can be added to Lemma 4.3:
   (3) *the matrices $I + PA_1P$ and $I + A_1P$ are invertible, and*
   $$\begin{aligned}[I+PA_1]^{-1}P &= [I+PA_1P]^{-1}P\\ &= P[I+PA_1P]^{-1} = P[I+A_1P]^{-1};\end{aligned}$$
   (4) $\int_{\mathbf{R}^+}\big|(I-P)r(t)\big|\,\mathrm{d}t<\infty$, *and*
   $$\int_{\mathbf{R}^+}(I-P)r(t)\,\mathrm{d}t=(I-P)[P+A_0]^{-1}[I+A_1P]^{-1};$$
   (5) $\int_{\mathbf{R}^+}t\big|(I-P)r(t)(I-P)\big|\,\mathrm{d}t<\infty$ *and*
   $$\begin{aligned}\int_{\mathbf{R}^+}t(I-P)r(t)(I-P)\,\mathrm{d}t &= (I-P)[P+A_0]^{-1}\\ &\quad\times[I+A_1P]^{-1}(I+A_1)[I+PA_1]^{-1}[P+A_0]^{-1}(I-P);\end{aligned}$$
   (6) $\int_{\mathbf{R}^+}t\big|(I-P)r'(t)\big|\,\mathrm{d}t<\infty$ *and*
   $$\begin{aligned}&\int_{\mathbf{R}^+}t(I-P)r'(t)\,\mathrm{d}t\\ &\quad=(I-P)[P+A_0]^{-1}[I+A_1P]^{-1}A_1P[I+A_1P]^{-1};\end{aligned}$$
   (7) $\int_{\mathbf{R}^+}t\big|r'(t)(I-P)\big|\,\mathrm{d}t<\infty$ *and*
   $$\begin{aligned}&\int_{\mathbf{R}^+}tr'(t)(I-P)\,\mathrm{d}t\\ &\quad=[I+PA_1]^{-1}PA_1[I+PA_1]^{-1}[P+A_0]^{-1}(I-P).\end{aligned}$$

6. Comments

Section 2:
A scalar version of Theorem 2.1 was first proved in Staffans [10], Theorems 1a and 1b, p. 292. Our proof is based on that exposition. There it was assumed that the set $\{\,\omega\in\mathbf{R}\mid\det[\Re\tilde{\mu}(\omega)]=0\,\}$ is countable. This countability assumption was removed in Londen [10]. Earlier related results are given in Londen [9], Theorems 2 and 3.

Levin [7], Theorem 2.5, uses a different approach to obtain a particular case of Theorem 2.1.

Section 3:
Theorem 3.1 was established in Londen [12] with more or less the same proof as we use here.

Section 4:
The observation that (1.1) can be rewritten in the form of (4.5) was made and exploited in Gripenberg [9]. An earlier scalar version of Theorem 4.2 was obtained in Londen [13].

References

G. Gripenberg

9. On nonlinear Volterra equations with nonintegrable kernels, *SIAM J. Math. Anal.* **11** (1980), pp. 668–682.

J. J. Levin

7. On some geometric structures for integrodifferential equations, *Adv. in Math.* **22** (1976), pp. 146–186.

S-O. Londen

9. On the variation of the solutions of a nonlinear integral equation, *J. Math. Anal. Appl.* **52** (1975), pp. 430–449.
10. On a Volterra integrodifferential equation with L^∞-perturbation and noncountable zero-set of the transformed kernel, *J. Integral Equations* **1** (1979), pp. 275–280.
12. On an integral equation with L^∞-perturbation, *J. Integral Equations* **5** (1983), pp. 49–57.
13. On some integral equations with locally finite measures and L^∞-perturbations, *SIAM J. Math. Anal.* **14** (1983), pp. 1187–1203.

R. K. Miller

3. On Volterra's population equation, *J. SIAM Appl. Math.* **14** (1966), pp. 446–452.

W. Rudin

3. *Real and Complex Analysis,* 3rd *ed.*, McGraw–Hill, New York, 1986.

C. Shilepsky

1. A note on the asymptotic behavior of an integral equation, *Proc. Amer. Math. Soc.* **33** (1972), pp. 111–113.
2. The asymptotic behavior of an integral equation with an application to Volterra's population equation, *J. Math. Anal. Appl.* **48** (1974), pp. 764–779.

O. J. Staffans

10. On a nonlinear integral equation with a nonintegrable perturbation, *J. Integral Equations* **1** (1979), pp. 291–307.

20

Monotonicity Methods

Here we study results related to the monotonicity of Volterra integral operators, and extend some of the results in Chapter 16. In particular, we consider nonconvolution kernels of positive type, log-convex kernels, kernels of anti-accretive type, and kernels of totally invariant type. Some results on convolution equations with nonseparable nonlinearities are included.

1. Introduction

Recall that a locally integrable convolution kernel k is of positive type if

$$\Re \int_{\mathbf{R}^+} \langle \varphi(t), (k * \varphi)(t) \rangle \, \mathrm{d}t \geq 0$$

for all $\varphi \in L^2(\mathbf{R}^+)$ with compact support. In general, an operator A in a Hilbert space H is called *monotone* if $\Re\langle x - y, Ax - Ay \rangle \geq 0$ for all x and $y \in H$. For a linear operator A this can be simplified into $\Re\langle x, Ax \rangle \geq 0$ for all $x \in H$. Thus, for any $T > 0$, a convolution kernel of positive type defines a monotone operator on $L^2(0, T)$.

This chapter is devoted to a study of results related to various monotonicity properties of Volterra operators and kernels. We consider equations where

- the kernel is of positive, but not of convolution type,

or

- the kernel has some additional monotonicity properties that enable us to draw stronger conclusions,

or

- the structure of the equation is too complicated for the techniques of Chapters 16–19 to be applicable.

In Section 2 we consider nonconvolution kernels of positive type, and in Section 3 so-called log-convex kernels that are not necessarily of convolution type. Section 4 is devoted to the theory of kernels of anti-accretive and totally invariant types, and Section 5 gives some asymptotic results on Volterra integral and integrodifferential equations of the form

$$x(t) + \int_0^t k\big(t-s, x(s)\big)\,\mathrm{d}s = f(t), \quad t \in \mathbf{R}^+,$$

and

$$x'(t) + \int_0^t k\big(t-s, x(s)\big)\,\mathrm{d}s = f(t), \quad t \in \mathbf{R}^+.$$

As in Chapter 16, we define the central concepts for kernels with values in $\mathbf{C}^{n\times n}$. When we apply these concepts to nonlinear integral equations we throughout let the kernels take their values in $\mathbf{R}^{n\times n}$. It should be pointed out that in several cases the assumptions are quite restrictive when the dimension n is greater than 1.

Most of the results of this chapter require the norm to be induced by an inner product (that is real when restricted to $\mathbf{R}^n \times \mathbf{R}^n$). More precisely, this is the case in all the results in Sections 2 and 5, and, in addition, in several of the results in Section 4, namely Theorem 4.4 and Theorems 4.13–4.15.

2. Nonconvolution Kernels of Positive Type

We begin by extending some of the results in Chapter 16 to nonconvolution kernels. Here all our results involve convexity, and we get no analogue of the standard Laplace transform condition used throughout in Chapter 16. For simplicity, we restrict ourselves to kernels that are locally integrable functions defined on $\mathbf{R}^+ \times \mathbf{R}^+$.

The general nonconvolution definition of positivity is self-evident.

2.1 Definition

A $\mathbf{C}^{n\times n}$-valued Volterra kernel k is of positive type if it is of type L^2_{loc} on $\mathbf{R}^+$ and if for all $\varphi \in L^2(\mathbf{R}^+; \mathbf{C}^n)$ with compact support one has

$$\Re \int_{\mathbf{R}^+} \big\langle \varphi(t), (k \star \varphi)(t) \big\rangle\,\mathrm{d}t \geq 0.$$

For convolution kernels that are locally integrable functions, this definition is identical to Definition **16**.2.1.

The next result gives a simple sufficient condition for a kernel to be of positive type. It is an extension of Proposition **16**.3.1.

2.2 Theorem
Assume that k is a $\mathbf{C}^{n\times n}$-valued kernel that is of type L^2_{loc} on $\mathbf{R}^+$ and satisfies, for all $0 \le u < s < t < v$, the inequalities

$$\begin{aligned} k(t,s) &\succeq 0, \\ k(t,u) &\preceq k(t,s), \\ k(t,s) &\succeq k(v,s), \\ k(t,s) - k(t,u) &\succeq k(v,s) - k(v,u). \end{aligned}$$

(t, s) (v, s)
(t, u) (v, u)

Then k is of positive type.

One may, without loss of generality, assume that $k(t,0) \equiv 0$, that

$$k(t,s) = \lim_{v\downarrow t, u\uparrow s} k(v,u), \quad 0 \le s < t < \infty, \tag{2.1}$$

and, in case k is locally bounded, that (2.1) is valid for $0 \le s \le t < \infty$. The proofs of these facts are left to the reader as an exercise. Then it follows from the assumptions (the proof of this is another exercise) that there exists a nonnegative measure α supported on $\{\, (t,s) \mid 0 \le s < t < \infty \,\}$, such that, for all $0 \le s < t < \infty$ (or $0 \le s \le t < \infty$ if k is locally bounded), α is finite on the semi-infinite rectangle $(t,\infty) \times [0,s)$, and

$$k(t,s) = k(\infty,s) + \iint_{(t,\infty)\times[0,s)} \alpha(\mathrm{d}v\,\mathrm{d}u), \quad 0 \le s < t < \infty \tag{2.2}$$

(or $0 \le s \le t < \infty$ if k is locally bounded). Moreover this measure satisfies (see Exercise 9), for each $T > 0$,

$$\iint_{\{0\le u<v\le T\}} (v-u)^2 |\alpha|(\mathrm{d}v\,\mathrm{d}u) < \infty. \tag{2.3}$$

Proof of Theorem 2.2 By continuity, it suffices to consider the case where φ is continuous and is supported on a finite interval $[0,T]$. Moreover, we may without loss of generality suppose that $k(t,s) = 0$ for $t \ge T$ (multiply $k(t,s)$ by $\chi_{[0,T)}(t)$; this affects neither the integral in Definition 2.1, nor the conditions listed in Theorem 2.2). Then the term $k(\infty,s)$ in (2.2) is absent. We get, after a change of order of integration, justified by (2.3),

$$\begin{aligned} &\Re \int_{\mathbf{R}^+} \langle \varphi(t), (k \star \varphi)(t) \rangle \,\mathrm{d}t \\ &= \Re \int_{\mathbf{R}^+} \Big\langle \varphi(t), \int_0^t k(t,s)\varphi(s)\,\mathrm{d}s \Big\rangle \,\mathrm{d}t \\ &= \iint_{\{0\le u<v<\infty\}} \Re \int_u^v \Big\langle \varphi(t), \alpha(\mathrm{d}v\,\mathrm{d}u) \int_u^t \varphi(s)\,\mathrm{d}s \Big\rangle \,\mathrm{d}t \\ &= \tfrac{1}{2} \iint_{\{0\le u<v<\infty\}} \Big\langle \int_u^v \varphi(s)\,\mathrm{d}s, \alpha(\mathrm{d}v\,\mathrm{d}u) \int_u^v \varphi(s)\,\mathrm{d}s \Big\rangle \ge 0. \end{aligned} \tag{2.4}$$

This completes the proof. □

We shall not pursue the asymptotics of equations with nonconvolution kernels of positive type in any detail. Instead, we limit ourselves to a few estimates, Theorems 2.3 and 2.4, both helpful for prospective asymptotic work, and give a corollary to Theorem **17**.2.2.

Our first estimate extends Corollary **16**.6.6.

2.3 Theorem

Let k be a locally bounded kernel satisfying the assumptions of Theorem 2.2, and suppose that k has been normalized so that (2.1) and (2.2) hold for all $0 \le s \le t < \infty$. Then

$$\left| \int_0^T k(T,s)\varphi(s)\,\mathrm{d}s \right|^2 \le 2|k(T,T)|\,\Re \int_0^T \langle \varphi(t), (k \star \varphi)(t) \rangle\,\mathrm{d}t,$$

for every $T > 0$ and every $\varphi \in L^1([0,T]; \mathbf{C}^n)$.

Proof Let $T > 0$ be arbitrary. Without loss of generality, suppose that φ is a continuous function supported on $[0,T]$, and that $k(t,s)$ vanishes for $t > T$; see the proof of Theorem 2.2. Define the vector $\mathbf{w}$ by

$$\mathbf{w} = (k \star \varphi)(T) = \int_0^T k(T,s)\varphi(s)\,\mathrm{d}s.$$

If $\mathbf{w} = 0$, then the claim is true. If not, we may, without loss of generality, assume that $|\mathbf{w}| = 1$ (divide φ by $|\mathbf{w}|$). Then we conclude from (2.2) and the matrix version of Hölder's inequality given in Section **16**.7 that (the change of order of integration is justified by (2.3))

$$\begin{aligned}
\left| \int_0^T k(T,s)\varphi(s)\,\mathrm{d}s \right|^2 &= \left| \int_0^T \langle \mathbf{w}, k(T,s)\varphi(s) \rangle\,\mathrm{d}s \right|^2 \\
&= \left| \int_0^T \Big\langle \mathbf{w}, \iint_{(T,\infty)\times[0,s)} \alpha(\mathrm{d}v\,\mathrm{d}u)\varphi(s) \Big\rangle\,\mathrm{d}s \right|^2 \\
&= \left| \iint_{(T,\infty)\times[0,T)} \Big\langle \mathbf{w}, \alpha(\mathrm{d}v\,\mathrm{d}u) \int_u^T \varphi(s)\,\mathrm{d}s \Big\rangle \right|^2 \\
&\le \iint_{(T,\infty)\times[0,T)} \langle \mathbf{w}, \alpha(\mathrm{d}v\,\mathrm{d}u)\mathbf{w} \rangle \\
&\quad \times \iint_{(T,\infty)\times[0,T)} \Big\langle \int_u^T \varphi(s)\,\mathrm{d}s, \alpha(\mathrm{d}v\,\mathrm{d}u) \int_u^T \varphi(s)\,\mathrm{d}s \Big\rangle \\
&= \Big\langle \mathbf{w}, \iint_{(T,\infty)\times[0,T)} \alpha(\mathrm{d}v\,\mathrm{d}u)\mathbf{w} \Big\rangle
\end{aligned}$$

$$\times \iint_{(T,\infty)\times[0,T)} \left\langle \int_u^v \varphi(s)\,\mathrm{d}s, \alpha(\mathrm{d}v\,\mathrm{d}u) \int_u^v \varphi(s)\,\mathrm{d}s \right\rangle$$

$$\le |k(T,T)| \iint_{\{0\le u<v<\infty\}} \left\langle \int_u^v \varphi(s)\,\mathrm{d}s, \alpha(\mathrm{d}v\,\mathrm{d}u) \int_u^v \varphi(s)\,\mathrm{d}s \right\rangle.$$

By invoking (2.4) we get the desired conclusion. □

Next we consider a nonconvolution analogue of Proposition **16**.5.8(i).

2.4 Theorem

Let the assumptions of Theorem 2.2 hold, and assume in addition that $k(t,0) \equiv 0$, and that, for all $0 \le u < s < t < v$ and all $h > 0$,

$$\begin{aligned} &k(t,s) - k(t,u) - k(v,s) + k(v,u) \\ &\qquad \succeq k(t+h,s) - k(t+h,u) - k(v+h,s) + k(v+h,u), \end{aligned} \tag{2.5}$$

and that

$$\sup_{s>0} \int_s^\infty |k(t,s)|\,\mathrm{d}t < \infty. \tag{2.6}$$

Then

$$\|k \star \varphi\|^2_{L^2(\mathbf{R}^+)} \le 4\left(\sup_{s>0}\left|\int_s^\infty k(t,s)\,\mathrm{d}t\right|\right)\left(\Re \int_{\mathbf{R}^+} \langle \varphi(t), (k\star\varphi)(t)\rangle\,\mathrm{d}t\right),$$

for every $\varphi \in L^2(\mathbf{R}^+;\mathbf{C}^n)$ with compact support.

Observe, in particular, that by taking $u = 0$ in (2.5) one gets the inequality

$$k(t,s) - k(v,s) \succeq k(t+h,s) - k(v+h,s), \tag{2.7}$$

for $0 < s < t < v$ and $h > 0$.

Proof of Theorem 2.4 Clearly, (2.7) implies that $k(t,s)$ is differentiable with respect to t, and that $-\frac{\partial}{\partial t}k(t,s)$ has the same monotonicity properties as k has in Theorem 2.2. Of course, $\frac{\partial}{\partial t}k(t,s)$ need not be of type L^2_{loc}, but one can use that same argument as in the discussion following Theorem 2.2 to show that there is a nonnegative measure β supported on $\{(t,s) \mid 0 \le s < t < \infty\}$, such that, for all $0 \le s < t < \infty$, β is finite on the semi-infinite rectangle $(t,\infty)\times[0,s)$, and, for almost all (t,s),

$$\frac{\partial}{\partial t}k(t,s) = -\iint_{(t,\infty)\times[0,s)} \beta(\mathrm{d}v\,\mathrm{d}u). \tag{2.8}$$

Because of the integrability condition (2.6), we have $k(\infty,s) \equiv 0$; hence one can integrate the preceding equation twice to get

$$k(t,s) = \iint_{(t,\infty)\times[0,s)} (v-t)\beta(\mathrm{d}v\,\mathrm{d}u), \quad 0 \le s < t < \infty,$$

and

$$\int_t^\infty k(v,s)\,\mathrm{d}v = \tfrac{1}{2} \iint_{(t,\infty)\times[0,s)} (v-t)^2\beta(\mathrm{d}v\,\mathrm{d}u), \quad 0 \le s \le t < \infty. \tag{2.9}$$

The measure β satisfies

$$\iint_{(t,\infty)\times[0,s)} (v-t)^2|\beta|(\mathrm{d}v\,\mathrm{d}u) < \infty, \quad 0 \le s < t < \infty, \tag{2.10}$$

and, for every $T > 0$,

$$\iiint_{\{0\le u<t<v<T\}} (t-u)^2|\beta|(\mathrm{d}v\,\mathrm{d}u)\,\mathrm{d}t < \infty.$$

To prove the theorem, it suffices to consider the case where φ is a continuous function with compact support in $\mathbf{R}^+$ and where $T > 0$ is such that the support of φ is contained in $[0,T]$, and to show that

$$\|k\star\varphi\|^2_{L^2(0,T)} \le 4\Big(\sup_{s>0}\Big|\int_s^\infty k(t,s)\,\mathrm{d}t\Big|\Big)\Big(\Re\int_{\mathbf{R}^+}\langle\varphi(t),(k\star\varphi)(t)\rangle\,\mathrm{d}t\Big).$$

To do this we first observe that, by a computation similar to (2.4),

$$\Re\int_{\mathbf{R}^+}\langle\varphi(t),(k\star\varphi)(t)\rangle\,\mathrm{d}t$$

$$= \tfrac{1}{2}\iiint_{\{0\le u<t<v<\infty\}}\Big\langle\int_u^t\varphi(s)\,\mathrm{d}s, \beta(\mathrm{d}v\,\mathrm{d}u)\int_u^t\varphi(s)\,\mathrm{d}s\Big\rangle\,\mathrm{d}t. \tag{2.11}$$

Define the function w by

$$w(t) = (k\star\varphi)(t) = \int_0^t k(t,s)\varphi(s)\,\mathrm{d}s, \quad 0 < t < T,$$

and let $w(t)$ be zero for $t > T$. If $\|w\|_{L^2(\mathbf{R}^+)} = 0$, then the claim is true. If not, we may, without loss of generality, assume that $\|w\|_{L^2(\mathbf{R}^+)} = 1$ (divide φ by $\|w\|_{L^2(\mathbf{R}^+)}$). A computation similar to the one given in the proof of Theorem 2.3 shows that

$$\int_{\mathbf{R}^+}\Big\langle w(t), \int_0^t k(t,s)\varphi(s)\,\mathrm{d}s\Big\rangle\,\mathrm{d}t$$

$$= \iiint_{\{0\le u<t<v<\infty\}}\Big\langle (v-t)w(t), \beta(\mathrm{d}v\,\mathrm{d}u)\int_u^t\varphi(s)\,\mathrm{d}s\Big\rangle\,\mathrm{d}t.$$

Apply Hölder's inequality, and use (2.9)–(2.11) to get

$$\int_0^T\Big|\int_0^t k(t,s)\varphi(s)\,\mathrm{d}s\Big|^2\,\mathrm{d}t$$

$$\le \iiint_{\{0\le u<t<v<\infty\}}\Big\langle (v-t)w(t), \beta(\mathrm{d}v\,\mathrm{d}u)(v-t)w(t)\Big\rangle\,\mathrm{d}t$$

$$\times \iiint_{\{0\le u<t<v<\infty\}} \left\langle \int_u^t \varphi(s)\,ds, \beta(dv\,du) \int_u^t \varphi(s)\,ds \right\rangle dt$$

$$= 4 \int_{\mathbf{R}^+} \left\langle w(t), \left(\int_t^\infty k(v,t)\,dv \right) w(t) \right\rangle dt$$

$$\times \Re \int_{\mathbf{R}^+} \langle \varphi(t), (k \star \varphi)(t) \rangle \, dt$$

$$\le 4 \sup_{t>0} \left| \int_t^\infty k(v,t)\,dv \right| \left(\Re \int_{\mathbf{R}^+} \langle \varphi(t), (k \star \varphi)(t) \rangle \, dt \right).$$

This completes the proof. □

As an example of how one can apply the results developed above to the nonlinear equation

$$x'(t) + \int_0^t k(t,s) g\big(x(t-s)\big)\,ds = f(t), \quad t \in \mathbf{R}^+; \quad x(0) = x_0, \tag{2.12}$$

we give the following result.

2.5 Corollary

Assume that

(i) *k is a $\mathbf{R}^{n\times n}$-valued Volterra kernel of positive type on $\mathbf{R}^+$,*

(ii) *$g \in C(\mathbf{R}^n;\mathbf{R}^n)$ is the gradient of a function $G \in C^1(\mathbf{R}^n;\mathbf{R})$, and there exists some nonnegative, nondecreasing function u, defined on the range of G, such that $|g(y)| \le u(G(y))$ for all $y \in \mathbf{R}^n$,*

(iii) *$f \in L^1_{\mathrm{loc}}(\mathbf{R}^+;\mathbf{R}^n)$,*

(iv) *$x \in AC_{\mathrm{loc}}(\mathbf{R}^+;\mathbf{R}^n)$ is a solution of (2.12) on $\mathbf{R}^+$.*

Then

$$\int_{G(x_0)}^{G(x(t))} \frac{ds}{u(s)} \le \int_0^t |f(s)|\,ds, \quad t \in \mathbf{R}^+. \tag{2.13}$$

Thus, if furthermore $\lim_{|y|\to\infty} G(y) = \infty$, and

$$\int_{G(x_0)}^\infty \frac{ds}{u(s)} > \int_{\mathbf{R}^+} |f(s)|\,ds,$$

then $x \in BC(\mathbf{R}^+;\mathbf{R}^n)$ and $\sup_{T>0} \int_0^T \langle g(x(t)), (k \star (g\circ x))(t) \rangle \, dt < \infty$.

The proof is the same as the proof of Theorem **17**.2.2.

In a similar way one can get a number of extensions of other results in Chapter 17. We leave the details to the reader.

3. Log-Convex Kernels

An idea often used in the analysis of nonlinear problems is to rewrite the equation at hand as a linear one, by absorbing a part of the nonlinearity into a new unknown function that depends on the solution. In the context of integral equations, one can use this idea for example in the following form. Consider the scalar equation

$$x(t) + \int_0^t k(t,s)g\big(s,x(s)\big)\,\mathrm{d}s = f(t), \quad t \in \mathbf{R}^+. \tag{3.1}$$

If we let $b(t) = g(t,x(t))/x(t)$, then (3.1) can be written as the linear equation

$$x(t) + \int_0^t k(t,s)b(s)x(s)\,\mathrm{d}s = f(t), \quad t \in \mathbf{R}^+.$$

Of course, for this approach to be useful, we must be able to say something about the solution of the new equation without having an exact knowledge of the auxiliary function b. Note that, in the case where k is a convolution kernel, the new kernel $(t,s) \mapsto k(t-s)b(s)$ is not of convolution type, so one cannot use Laplace or Fourier transform methods. However, there is one class of kernels that is not affected by a transformation of this kind, i.e., the logarithmically convex kernels discussed in Theorem **9**.8.6. Recall that k is a kernel of this type if k is nonnegative and if $k(v,s)k(t,u) \le k(t,s)k(v,u)$ for almost all $s \le u \le v \le t$. Clearly, if b is nonnegative, and if k satisfies this inequality, then so does $k(t,s)b(s)$. Moreover, the number $M = \operatorname{ess\,sup}_{0\le s\le u\le t} k(t,s)/k(u,s)$, which bounds the L^∞-operator norm of the resolvent of the kernel (see Theorem **9**.8.5) is not affected by the multiplication by b.

Using this method, one can prove the following result (where we employ a number of simplifying assumptions).

3.1 Theorem

Assume that

(i) *k is a nonnegative, scalar kernel of continuous type on $\mathbf{R}^+$ satisfying*

$$k(v,s)k(t,u) \le k(t,s)k(v,u),$$

for almost all $s \le u \le v \le t$,

(v, u) (t, u)
(v, s) (t, s)

and

$$M \overset{\text{def}}{=} \operatorname*{ess\,sup}_{0\le s\le u\le t} \frac{k(t,s)}{k(u,s)} < \infty,$$

(ii) *$g \in C(\mathbf{R}^+ \times \mathbf{R}; \mathbf{R})$ satisfies $yg(t,y) \ge 0$ for all $t \in \mathbf{R}^+$ and all $y \in \mathbf{R}$,*

(iii) *$f \in C(\mathbf{R}^+; \mathbf{R})$,*

(iv) *$x \in C(\mathbf{R}^+; \mathbf{R})$ is a solution of (3.1).*

Then

$$|x(t)| \le (M+1) \sup_{s\in[0,t]} |f(s)|, \quad t \in \mathbf{R}^+.$$

If furthermore

(i*) *k is of type B_0^∞ on $\mathbf{R}^+$, i.e., $\lim_{t\to\infty} \int_0^T k(t,s)\,\mathrm{d}s = 0$ for every $T > 0$,*

(ii*) $\sup_{t\in\mathbf{R}^+,|y|\le K}|g(t,y)|<\infty$ *for all* $K>0$,
(iii*) $\lim_{t\to\infty} f(t)=0$,
then $x(t)\to 0$ *as* $t\to\infty$.

Proof For each number $\epsilon>0$ and each $t\in\mathbf{R}^+$, define $c_\epsilon(t)=\epsilon\,\mathrm{sign}(x(t))$, $b_\epsilon(t)=0$ if $x(t)=0$, and $b_\epsilon(t)=g(t,x(t))/(x(t)+c_\epsilon(t))$ if $x(t)\neq 0$. Then b_ϵ is nonnegative and locally bounded, $|c_\epsilon(t)|\le\epsilon$ for $t\in\mathbf{R}^+$, and

$$g\big(t,x(t)\big)=b_\epsilon(t)\big(x(t)+c_\epsilon(t)\big),\quad t\in\mathbf{R}^+.$$

Substituting this into (3.1) we get

$$x(t)+\int_0^t k_\epsilon(t,s)x(s)\,\mathrm{d}s=f(t)-\int_0^t k_\epsilon(t,s)c_\epsilon(s)\,\mathrm{d}s,\quad t\in\mathbf{R}^+,\tag{3.2}$$

where $k_\epsilon(t,s)=k(t,s)b_\epsilon(s)$. As we observed earlier, k_ϵ satisfies the same assumptions as k does, with the same number M. The solution x of (3.2) can be given in the variation of constants form

$$x(t)=f(t)-\int_0^t r_\epsilon(t,s)\big(f(s)+c_\epsilon(s)\big)\,\mathrm{d}s,\tag{3.3}$$

where r_ϵ is the resolvent of k_ϵ. It follows from Theorems **9**.8.5 and **9**.8.6 that r_ϵ is nonnegative and satisfies

$$|||r_\epsilon|||_{L^\infty(\mathbf{R}^+)}\le M.$$

Thus we have

$$|x(t)|\le(M+1)\sup_{s\in[0,t]}|f(s)|+M\epsilon,$$

and since ϵ was arbitrary we get the desired conclusion.

If $\lim_{t\to\infty}f(t)=0$, then we know from the first part of the theorem (already proved) that x is bounded. This implies that, for each fixed ϵ, the function b_ϵ is bounded. Since r_ϵ is nonnegative we have $r_\epsilon(t,s)\le k_\epsilon(t,s)$, and since b_ϵ is bounded it follows from our assumptions that r_ϵ is of type B_0^∞. Using this fact in the variation of constants formula (3.3) we get (cf. Theorem **9**.6.2)

$$\limsup_{t\to\infty}|x(t)|\le M\epsilon.$$

This shows that $x(t)\to 0$ as $t\to\infty$. □

In the convolution case, condition (i) in Theorems 3.1 and 3.2 is equivalent to log-convexity, see the discussion in Section **9**.8.

If we replace f in (3.1) by $k\star f$, then we obtain the equation

$$x(t)+\int_0^t k(t,s)g\big(s,x(s)\big)\,\mathrm{d}s=\int_0^t k(t,s)f(s)\,\mathrm{d}s,\quad t\in\mathbf{R}^+.\tag{3.4}$$

For this equation one gets a slightly modified version of Theorem 3.1.

3.2 Theorem
Assume that

(i) k *is a nonnegative, scalar kernel of continuous type on* $\mathbf{R}^+$ *satisfying*

$$k(v,s)k(t,u) \leq k(t,s)k(v,u), \qquad \text{for almost all } s \leq u \leq v \leq t,$$

and

$$M \stackrel{\text{def}}{=} \operatorname*{ess\,sup}_{0 \leq s \leq u \leq t} \frac{k(t,s)}{k(u,s)} < \infty,$$

(ii) $g \in C(\mathbf{R}^+ \times \mathbf{R}; \mathbf{R})$ *satisfies* $yg(t,y) \geq 0$ *for all* $t \in \mathbf{R}^+$ *and all* $y \in \mathbf{R}$,

(iii) $f \in C(\mathbf{R}^+; \mathbf{R})$,

(iv) $x \in C(\mathbf{R}^+; \mathbf{R})$ *is a solution of (3.4).*

Then

$$|x(t)| \leq M \sup_{s \in [0,t]} |f(s)|, \quad t \in \mathbf{R}^+.$$

If in addition f *is nonnegative, then so is* x.

The proof of this theorem, which resembles the proof of Theorem 3.1, is left to the reader.

The two theorems above can be considered as models for the results which we shall develop in Section 4. The crucial property that we used in the proof of Theorem 3.1 was the uniform boundedness of the norm of the resolvent operator. In the last claim in Theorem 3.2 it is in addition important that the resolvent be nonnegative.

4. Kernels of Anti-Accretive and Totally Invariant Types

The basic content of this section may roughly be explained as follows. In Chapter 16 and in Section 2 we considered kernels of positive type, which simply means that they are monotone in the Hilbert space L^2. Here we study two other forms of positivity, namely anti-accretivity in L^p-spaces, and complete invariance. The former property is an accretivity property of the inverse of the kernel, and the latter is a certain cone-preserving property. As will be seen below, they are related to each other. Moreover, a kernel that is anti-accretive in L^2 is of positive type, and conversely (Theorem 4.4).

We use the notation r_λ for the resolvent kernel of λk, i.e., r_λ is the solution of the equations

$$r_\lambda + \lambda(k \star r_\lambda) = r_\lambda + \lambda(r_\lambda \star k) = \lambda k.$$

Throughout this section it is important that r_λ should exist for all $\lambda > 0$, and be locally of the same type as k. Recall that the spectrum of k is the set of all complex numbers μ for which $-\frac{1}{\mu}k$ does not have a resolvent in the appropriate algebra. Hence an equivalent formulation of the assumption on the existence of resolvents of λk of the appropriate class for all $\lambda > 0$ is to say that the spectrum of k does not intersect $(-\infty, 0)$.

Let us first define what it means for a kernel to be of anti-accretive type.

4.1 Definition
A $\mathbf{C}^{n\times n}$-valued Volterra kernel k that is of type L^p_{loc} on $\mathbf{R}^+$ is said to be of L^p-anti-accretive type, $p \in [1,\infty]$, if the spectrum of k does not intersect $(-\infty, 0)$, and if for all $\lambda > 0$ and all $f \in L^p(\mathbf{R}^+;\mathbf{C}^n)$ one has

$$\|r_\lambda \star f\|_{L^p(\mathbf{R}^+)} \le \|f\|_{L^p(\mathbf{R}^+)}.$$

Note that for $1 < p < \infty$ the given condition is weaker than the requirement $|||r_\lambda|||_{L^p(\mathbf{R}^+)} \le 1$ (for this norm, see Section **9**.2), because $|||r_\lambda|||_{L^p(\mathbf{R}^+)}$ is the norm of the operator $f \mapsto |r_\lambda| \star f$ instead of the norm of $f \mapsto r_\lambda \star f$. For $p = 1$ and ∞ the given condition is, in the scalar case, equivalent to the requirement that $|||r_\lambda|||_{L^p(\mathbf{R}^+)} \le 1$; this follows from the proofs of Lemmas **9**.10.2 and **9**.10.4.

The following result is an immediate consequence of Definition 4.1 and Riesz' interpolation theorem.

4.2 Lemma
If $1 \le p \le r \le \infty$, if k is both L^p-anti-accretive and L^r-anti-accretive, then k is L^s-anti-accretive for all $s \in [p, r]$.

Anti-accretivity can be characterized in an alternative fashion.

4.3 Lemma
A $\mathbf{C}^{n\times n}$-valued Volterra kernel k that is of type L^p_{loc} on $\mathbf{R}^+$ and has a spectrum that does not intersect $(-\infty, 0)$, is of L^p-anti-accretive type, $p \in [1,\infty]$, iff for all $\lambda > 0$, all $T > 0$, and all $\varphi \in L^p([0,T];\mathbf{C}^n)$ one has

$$\|\lambda k \star \varphi\|_{L^p(0,T)} \le \|\varphi + \lambda k \star \varphi\|_{L^p(0,T)}. \tag{4.1}$$

Proof First, let us observe that since r_λ is a Volterra kernel the condition given in Definition 4.1 is equivalent to the condition that for all $\lambda > 0$, all $T > 0$ and all $f \in L^p([0,T];\mathbf{C}^n)$ one has

$$\|r_\lambda \star f\|_{L^p(0,T)} \le \|f\|_{L^p(0,T)}. \tag{4.2}$$

Let $\varphi \in L^p([0,T];\mathbf{C}^n)$, and define f by $f = \varphi + \lambda k \star \varphi$. Since λk has a resolvent, every $f \in L^p([0,T];\mathbf{C}^n)$ can be written in this way (for some $\varphi \in L^p([0,T];\mathbf{C}^n)$). Then it follows from the resolvent equation that $\lambda k \star \varphi = r_\lambda \star f$, and (4.1) implies (4.2). Therefore the kernel k is of anti-accretive type.

Conversely, let k be anti-accretive, i.e., assume that (4.2) holds. For $f \in L^p([0,T];\mathbf{C}^n)$, define $\varphi = f - r_\lambda \star f$. Again, note that every $\varphi \in L^p([0,T];\mathbf{C}^n)$ can be written in this way. Then (4.2) is converted into

(4.1) and so the latter condition is necessary for k to be of anti-accretive type. □

Recall that if $\mathcal{B}$ is an arbitrary Banach space with norm $|\cdot|$ and A is an operator, possibly multivalued, defined on a subset $D(A) \subset \mathcal{B}$, then A is called accretive if

$$u_j \in D(A) \quad v_j \in A(u_j) \quad j = 1, 2, \text{ implies that}$$
$$|u_1 - u_2| \le |u_1 - u_2 + \lambda(v_1 - v_2)| \text{ for every } \lambda > 0.$$

Clearly, the characterization of L^p-anti-accretivity given in Lemma 4.3 can be interpreted as a requirement that, for each $T > 0$, the inverse of the operator $\varphi \mapsto k \star \varphi$ should be accretive on $L^p([0,T]; \mathbf{C}^n)$. This fact motivates the use of the name 'anti-accretivity' for this property.

It is not surprising that a kernel is of L^2-anti-accretive type if and only if it is of positive type.

4.4 Theorem
Let k be a Volterra kernel that is of type L^2_{loc} on $\mathbf{R}^+$ and has a spectrum that does not intersect $(-\infty, 0)$, and suppose that the norm in $\mathbf{C}^n$ is induced by an inner product. Then k is of L^2-anti-accretive type if and only if k is of positive type.

In particular, it follows from Lemma 4.2 that, if k is of both L^1-anti-accretive type and L^∞-anti-accretive type, then k is of positive type.

Proof of Theorem 4.4 Let $T > 0$, and let $\varphi \in L^2(\mathbf{R}^+; \mathbf{C}^n)$. As the norm in $\mathbf{C}^n$ is induced by an inner product, we have

$$\|\varphi + \lambda k \star \varphi\|^2_{L^2(0,T)} - \|\lambda k \star \varphi\|^2_{L^2(0,T)} = \|\varphi\|^2_{L^2(0,T)} + 2\lambda \Re \int_0^T \langle \varphi(t), (k \star \varphi)(t) \rangle \, dt. \quad (4.3)$$

If k is of positive type, then it follows from Lemma 4.3 that k is of L^2-anti-accretive type. Conversely, if k is of L^2-anti-accretive type, then it follows from Lemma 4.3 that the left-hand side of (4.3) is nonnegative. Dividing by λ and then letting $\lambda \to \infty$ one finds that k is of positive type. □

We have the following result on scalar convolution kernels.

4.5 Lemma
Let p, $q \in [1, \infty]$ satisfy $1/p + 1/q = 1$ and let $k \in L^1_{\text{loc}}(\mathbf{R}^+; \mathbf{C})$ be a scalar convolution kernel that is of L^p-anti-accretive type. Then k is of L^q-anti-accretive type and of positive type.

The proof follows from the fact that since a convolution operator is autonomous we can apply the results in Stein [1], p. 95, and then Lemma 4.2 and Theorem 4.4.

The most important example of a kernel of L^p-anti-accretive type in the cases $p = 1$ and $p = \infty$ is the following.

4.6 Theorem

Let either $p = 1$ or $p = \infty$. Let k be a (real-valued) nonnegative Volterra kernel that is of type L^p_{loc} on $\mathbf{R}^+$ and satisfies

$$k(v,s)k(t,u) \le k(t,s)k(v,u) \text{ for almost all } s \le u \le v \le t,$$

and

$$\operatorname*{ess\,sup}_{0\le s\le u\le t} \frac{k(t,s)}{k(u,s)} \le 1 \quad \text{if} \quad p = \infty,$$

or

$$\operatorname*{ess\,sup}_{0\le s\le u\le t} \frac{k(t,s)}{k(t,u)} \le 1 \quad \text{if} \quad p = 1.$$

Then k is of L^p-anti-accretive type.

This follows directly from Theorems **9**.8.5 and **9**.8.6.

We defer a further treatment of kernels of anti-accretive type to Theorems 4.13–4.16, and now bring in the second class of kernels mentioned in the introduction, i.e., the kernels of totally invariant type. To do this, we need the concept of a cone, defined as follows.

4.7 Definition

A subset $\mathcal{C}$ of a vector space is called a cone if $\lambda_1 x_1 + \lambda_2 x_2 \in \mathcal{C}$ whenever λ_1, $\lambda_2 \in \mathbf{R}^+$ and x_1, $x_2 \in \mathcal{C}$.

In particular, every subspace is a cone.

4.8 Definition

Let $p \in [1,\infty]$ and let $\mathcal{C}$ be a cone in $L^p_{\mathrm{loc}}(\mathbf{R}^+;\mathbf{C}^n)$. A $\mathbf{C}^{n\times n}$-valued Volterra kernel k that is of type L^p_{loc} on $\mathbf{R}^+$ is said to be of $\mathcal{C}$-totally invariant type if the spectrum of k does not intersect $(-\infty,0)$ and

$$r_\lambda \star f \in \mathcal{C} \text{ for all } \lambda > 0 \text{ and all } f \in \mathcal{C}.$$

In the special case where k is a scalar kernel and $\mathcal{C}$ is the cone of (pointwise) nonnegative functions in $L^p_{\mathrm{loc}}(\mathbf{R}^+;\mathbf{C})$, it is easy to show that k is of $\mathcal{C}$-totally invariant type iff r_λ is nonnegative. In this case, we make the following definition.

4.9 Definition

Let $p \in [1,\infty]$. A scalar real-valued Volterra kernel k that is of type L^p_{loc} on $\mathbf{R}^+$, with a spectrum that does not intersect $(-\infty,0)$, is said to be of totally positive type if r_λ is nonnegative for all $\lambda > 0$.

In particular, since $-\lambda k$ is the resolvent of $-r_\lambda$, it follows from Proposition **9**.8.1 that k must be nonnegative if it is of totally positive type.

The class of log-convex kernels provides us with an example of kernels of totally positive type.

4.10 Theorem
Let $p \in [1,\infty]$. Let $\mathcal{K}$ be a closed cone in $\mathbf{C}^n$, and let $\mathcal{C}$ be the cone of functions in $L^p_{\mathrm{loc}}(\mathbf{R}^+;\mathbf{C}^n)$ taking their values in $\mathcal{K}$. Let k be a nonnegative Volterra kernel that is of type L^p_{loc} on $\mathbf{R}^+$ and satisfies

$$k(v,s)k(t,u) \le k(t,s)k(v,u) \text{ for almost all } s \le u \le v \le t.$$

Then k is of $\mathcal{C}$-totally invariant type.

This is because, by Theorem **9**.8.6, r_λ is nonnegative for all $\lambda > 0$, and nonnegative, scalar kernels map the cone $\mathcal{C}$ into itself.

The following theorem records some simple properties of kernels of $\mathcal{C}$-totally invariant type.

4.11 Theorem
Let $p \in [1,\infty]$. Let $\mathcal{C}$ be a cone in $L^p_{\mathrm{loc}}(\mathbf{R}^+;\mathbf{C}^n)$, and let k be a $\mathbf{C}^{n\times n}$-valued Volterra kernel that is (of type L^p_{loc} on $\mathbf{R}^+$ and) of $\mathcal{C}$-totally invariant type. Then the following claims are true.

(i) *The kernel r_λ is $\mathcal{C}$-totally invariant for each $\lambda > 0$.*

(ii) *If for all $T > 0$ the set $\{\, f \in L^p([0,T];\mathbf{C}^n) \mid f$ is the restriction to $[0,T]$ of some function $g \in \mathcal{C}\,\}$ is closed in $L^p([0,T];\mathbf{C}^n)$, and if $\mathcal{C}$ is completely determined by its local structure in the sense that a function g belongs to $\mathcal{C}$ iff for each $T > 0$ its restriction to $[0,T]$ equals the restriction of some function in $\mathcal{C}$, then the operator $\varphi \mapsto k \star \varphi$ maps $\mathcal{C}$ into itself.*

(iii) *For each $T > 0$ the function $y(\lambda)(t,s) \stackrel{\mathrm{def}}{=} \frac{1}{\lambda} r_\lambda(t,s)$ is infinitely many times continuously differentiable with respect to λ in the topology of $\mathcal{V}(L^p;[0,T];\mathbf{C}^{n\times n})$, and $(-1)^m \frac{\mathrm{d}^m}{\mathrm{d}\lambda^m} y(\lambda) \star f \in \mathcal{C}$ for all $f \in \mathcal{C}$, $m \ge 1$, and $\lambda > 0$.*

To see that some restrictions on the cone $\mathcal{C}$ are needed in (ii), one can for example take $k(t,s) = 1$ for $0 \le s \le t$ and $k(t,s) = 0$ otherwise, and $\mathcal{C} = BC_0(\mathbf{R}^+)$. It is easy to check that k is $\mathcal{C}$-totally invariant, and that $k \star f$ does not belong to $BC_0(\mathbf{R}^+)$ for every f in this set. In this case, it is not true that $\mathcal{C}$ is completely determined by its local structure.

Proof of Theorem 4.11 (i) This follows from the fact that the resolvent of μr_λ is $\frac{\mu}{1+\mu} r_{\lambda+\mu\lambda}$.

(iii) Clearly $y(\lambda)$ satisfies

$$y(\lambda) + \lambda y(\lambda) \star k = y(\lambda) + \lambda k \star y(\lambda) = k.$$

Using the implicit function theorem in the Banach space of kernels of type L^p on $[0,T]$, we conclude that y is differentiable with respect to λ, and that

$$y^{(1)}(\lambda) + k \star \lambda y^{(1)}(\lambda) = -k \star y(\lambda).$$

It follows from the variation of constants formula that

$$y^{(1)}(\lambda) = -k \star y(\lambda) + \lambda y(\lambda) \star k \star y(\lambda) = -y(\lambda) \star y(\lambda).$$

Repeated differentiation of this equation gives

$$y^{(m)}(\lambda) = (-1)^m\, m!(y(\lambda))^{\star(m+1)}.$$

From this formula one at once gets the desired conclusion. □

Next we give some simple examples on how the concepts developed above can be applied.

Our first example is an extension of Theorem 3.2. The equation studied in that theorem was

$$x(t) + \int_0^t k(t,s)g\big(s,x(s)\big)\,\mathrm{d}s = \int_0^t k(t,s)f(s)\,\mathrm{d}s, \quad t \in \mathbf{R}^+, \tag{3.4}$$

with $n = 1$. For this equation we now have the following result.

4.12 Theorem

Assume the following.

(i) *$\mathcal{K}$ is a closed cone in $\mathbf{R}^n$, and $\mathcal{C}$ is the cone $C(\mathbf{R}^+;\mathcal{K})$ consisting of all continuous functions that take their values in $\mathcal{K}$.*

(ii) *k is a $\mathbf{R}^{n\times n}$-valued kernel of continuous type on $\mathbf{R}^+$ which is $\mathcal{C}$-totally invariant.*

(iii) *$g \in C(\mathbf{R}^+ \times \mathbf{R}^n;\mathbf{R}^n)$, and for each $\rho > 0$ and $T > 0$,*

$$M(T,\rho) \stackrel{\text{def}}{=} \sup_{\substack{t\in[0,T]\\ x,y\in B_\rho, x\neq y}} \frac{|g(t,x)-g(t,y)|}{|x-y|} < \infty,$$

where B_ρ is the ball $\{v \in \mathbf{R}^n \mid |v| \le \rho\}$. Furthermore, if $y + \mu g(t,y) \in \mathcal{K}$ for some $t \in [0,T]$, $y \in B_\rho$, and some $\mu < 1/M(T,\rho)$, then $y \in \mathcal{K}$.

(iv) *$f \in \mathcal{C}$.*

Then there exists a unique solution x of (3.4) that satisfies $x(t) \in \mathcal{K}$ on its whole interval of existence.

Proof By Theorem **12**.1.1, there exists a solution of (3.4) on an interval $[0,T]$. By the Lipschitz condition in (iii), this solution is unique. See, for example, Corollary **12**.2.2.

Let $\rho = 2\sup_{t\in[0,T]}|x(t)|$. Take $v \in B_{3\rho/4}$, fix $t \in [0,T]$, and, for each μ satisfying $0 < \mu < (4M(T,\rho)\rho + 4\sup_{t\in[0,T]}|g(t,0)|)^{-1}$, consider the equation $y = v - \mu g(t,y)$. By the Lipschitz condition in (iii), and by our choice of μ, the right-hand side defines a contraction mapping of B_ρ into itself. Thus there exists a unique $y \in B_\rho$ such that $y + \mu g(t,y) = v$. Let $v \mapsto J_\mu(t,v)$ be the inverse of the mapping $y \mapsto y + \mu g(t,y)$ and define $g_\mu(t,v)$ by

$$g_\mu(t,v) = \tfrac{1}{\mu}\big(v - J_\mu(t,v)\big).$$

It is easy to check that the functions $v \mapsto J_\mu(t,v)$ and $v \mapsto g_\mu(t,v)$ are Lipschitz-continuous in v on the set $\{\,(t,v) \in [0,T] \times \mathbf{R}^n \mid |v| \le 3\rho/4\,\}$. Moreover, $g_\mu(t,v) \to g(t,v)$ as $\mu \downarrow 0$ for all $t \in [0,T]$ and all $v \in B_{3\rho/4}$.

Let us approximate (3.4) by the equation

$$x_\mu(t) + \int_0^t k(t,s) g_\mu\big(s, x_\mu(s)\big)\,\mathrm{d}s = \int_0^t k(t,s) f(s)\,\mathrm{d}s, \quad t \in [0,T]. \tag{4.5}$$

It is clear that this equation has a unique solution which exists at least as long as it stays in $B_{3\rho/4}$ and $t \le T$ (see Theorem **12**.1.1 and use the fact that g_μ is Lipschitz-continuous). If we use the definition of g_μ and then solve x_μ by using the resolvent of $\frac{1}{\mu}k$, we see that this equation can be written in the form

$$x_\mu(t) = \int_0^t r_{1/\mu}(t,s)\big(J_\mu(s, x_\mu(s)) + \mu f(s)\big)\,\mathrm{d}s, \quad t \in [0,T].$$

From the last part of (iii) follows that J_μ maps $B_{3\rho/4} \cap \mathcal{K}$ into $\mathcal{K}$. Therefore, we can apply Theorem **12**.2.4 to show that $x_\mu(t) \in \mathcal{K}$ as long as $t \le T$ and $x(t) \in B_{3\rho/4}$.

Let $\mu \downarrow 0$. It follows from Theorem **13**.2.1 that the functions x_μ converge to the solution x of (3.4). (In principle we could use Theorem **13**.1.1 too, but there it is assumed that the nonlinear function is defined everywhere, which is not the case here.) Hence $x(t) \in \mathcal{K}$ for $t \in [0,T]$, and the proof is complete. □

We now return to the concept of L^∞-anti-accretivity to obtain some results on the qualitative behaviour of the solutions of the equation

$$x(t) + \int_0^t k(t,s) g\big(s, x(s)\big)\,\mathrm{d}s = f(t), \quad t \in \mathbf{R}^+. \tag{3.1}$$

For simplicity, we consider only the case where we have continuous solutions, although this result could be generalized to other situations as well.

The following theorem is an extension of Theorem 3.1.

4.13 Theorem

Assume that

(i) *k is a $\mathbf{R}^{n\times n}$-valued Volterra kernel of continuous type and of L^∞-anti-accretive type on $\mathbf{R}^+$,*

(ii) *$f \in BC(\mathbf{R}^+;\mathbf{R}^n)$,*

(iii) *$g \in C(\mathbf{R}^+ \times \mathbf{R}^n;\mathbf{R}^n)$, and $\liminf_{|v|\to\infty} \inf_{t\in\mathbf{R}^+} \langle g(t,v), v - f(t)\rangle > 0$,*

(iv) *$x \in C(\mathbf{R}^+;\mathbf{R}^n)$ is a solution of (3.1).*

Then x is bounded.

If in addition

(ii*) *k is $BC_0(\mathbf{R}^+;\mathbf{R}^n)$-totally invariant,*
(iii*) $\lim_{t\to\infty} f(t) = 0$,
(iv*) $\inf_{\rho\le|v|\le 1/\rho} \liminf_{t\to\infty}\langle g(t,v),v\rangle > 0$ *for all* $\rho\in(0,1)$,
then $x(t)\to 0$ *as* $t\to\infty$.

Note that the assumption that k is BC_0-totally invariant means that for each $\lambda > 0$ the operator $f \mapsto \int_0^t r_\lambda(t,s)f(s)\,\mathrm{d}s$ maps $BC_0(\mathbf{R}^+;\mathbf{R}^n)$ into itself. This assumption was verified in the course of the proof of Theorem 3.1 for the kernel in that theorem.

Proof of Theorem 4.13 By assumption (iii) there exist positive numbers δ and ρ such that

$$\inf_{t\in\mathbf{R}^+}\big\langle g(t,v), v - f(t)\big\rangle \ge \delta \text{ if } |v|\ge\rho. \tag{4.6}$$

Let $T>0$ be arbitrary. Choose $\lambda\in(0,1)$ so small that

$$\frac{\lambda}{2}\sup_{t\in[0,T]}\big|g\big(t,x(t)\big)\big|^2 < \delta. \tag{4.7}$$

The L^∞-accretivity assumption and Lemma 4.3 imply that we have

$$\sup_{t\in[0,T]}\left|\int_0^t k(t,s)g\big(s,x(s)\big)\,\mathrm{d}s\right| \le \sup_{t\in[0,T]}\left|\int_0^t k(t,s)g\big(s,x(s)\big)\,\mathrm{d}s + \lambda g\big(t,x(t)\big)\right|. \tag{4.8}$$

Let $\tau\in[0,T]$ be a point where the supremum is achieved on the right-hand side, i.e.,

$$\left|\int_0^\tau k(\tau,s)g\big(s,x(s)\big)\,\mathrm{d}s + \lambda g\big(\tau,x(\tau)\big)\right| = \sup_{t\in[0,T]}\left|\int_0^t k(t,s)g\big(s,x(s)\big)\,\mathrm{d}s + \lambda g\big(t,x(t)\big)\right|.$$

It follows from (3.1) and (4.8) that

$$\big|f(\tau)-x(\tau)\big| \le \big|f(\tau)-x(\tau)+\lambda g\big(\tau,x(\tau)\big)\big|.$$

Square this inequality and use the fact that the norm in $\mathbf{R}^n$ is generated by the inner product. This yields

$$\Big\langle g\big(\tau,x(\tau)\big), x(\tau)-f(\tau)\Big\rangle \le \frac{\lambda}{2}\Big|g\big(\tau,x(\tau)\big)\Big|^2. \tag{4.9}$$

Then it follows from (4.6), (4.9), and our choice of λ in (4.7) that

$$|x(\tau)|\le\rho.$$

Use this fact in (3.1) to get

$$\left|\int_0^\tau k(\tau,s)g\big(s,x(s)\big)\,\mathrm{d}s+\lambda g\big(\tau,x(\tau)\big)\right| \le \rho + \sup_{t\in[0,T]}|f(t)| + \lambda\sup_{t\in[0,T]}\big|g\big(t,x(t)\big)\big|.$$

If we recall our choice of τ, and once more use the fact that x solves (3.1), then we conclude that for each $t \in [0,T]$ we have

$$|x(t)| \le \rho + 2 \sup_{t\in[0,T]} |f(t)| + 2\lambda \sup_{t\in[0,T]} \big|g\big(t, x(t)\big)\big|.$$

Since it follows from our choice of λ that $\lambda \sup_{t\in[0,T]}|g(t,x(t))| \le \sqrt{2\lambda\delta} \le \sqrt{2\delta}$, and since T was arbitrary we have obtained an upper bound for $|x(t)|$.

To establish the second claim, we can use an argument similar to the one used in the proof of Theorem 4.14 below. We leave the details to the reader. □

Next we give a result on the asymptotic equivalence of solutions of the two equations

$$x_i(t) + \int_0^t k(t,s)g\big(s, x_i(s)\big)\,\mathrm{d}s$$
$$= f_i(t) + \int_0^t k(t,s)h_i(s)\,\mathrm{d}s, \quad t \in \mathbf{R}^+, \quad i = 1,2. \quad (4.10)$$

That is, we want to find out under what conditions it is true that $x_1(t) - x_2(t) \to 0$ as $t \to \infty$ if $f_1(t) - f_2(t) \to 0$ and $h_1(t) - h_2(t) \to 0$ as $t \to \infty$, and x_1 and x_2 are solutions of (4.10) with $i = 1$ and 2, respectively.

4.14 Theorem

Assume that

(i) *k is a $\mathbf{R}^{n\times n}$-valued Volterra kernel of continuous and L^∞-anti-accretive type on $\mathbf{R}^+$, and k is $BC_0(\mathbf{R}^+;\mathbf{R}^n)$-totally invariant,*

(ii) *f_i and $h_i \in BC(\mathbf{R}^+;\mathbf{R}^n)$, $i = 1,2$, and $\lim_{t\to\infty}(f_1(t) - f_2(t)) = \lim_{t\to\infty}(h_1(t) - h_2(t)) = 0$,*

(iii) *$g \in C(\mathbf{R}^+ \times \mathbf{R}^n;\mathbf{R}^n)$ satisfies*

$$\inf_{\substack{|u|\le 1/\rho, |v|\le 1/\rho \\ |u-v|>\rho}} \liminf_{t\to\infty} \big\langle g(t,u) - g(t,v), u - v\big\rangle > 0$$

for all $\rho > 0$,

(iv) *$x_i \in BC(\mathbf{R}^+;\mathbf{R}^n)$ is a solution of (4.10) for $i = 1$ and 2.*

Then $\lim_{t\to\infty}\big(x_1(t) - x_2(t)\big) = 0$.

Proof Let $\xi(t) = x_1(t) - x_2(t)$, $\psi(t) = h_1(t) - h_2(t)$, $\gamma(t) = g(t, x_1(t)) - g(t, x_2(t)) - \psi(t)$, and $\varphi(t) = f_1(t) - f_2(t)$, for $t \in \mathbf{R}^+$. All four of these functions are bounded and continuous, and they satisfy the equation

$$\xi(t) + (k \star \gamma)(t) = \varphi(t), \quad t \in \mathbf{R}^+. \quad (4.11)$$

Moreover, $\psi(t)$ and $\varphi(t) \to 0$ as $t \to \infty$.

Let us assume that $\limsup_{t\to\infty}|\xi(t)| = \delta > 0$, and try to show that this hypothesis leads to a contradiction.

Since k is L^∞-anti-accretive and BC_0-totally invariant, it follows that for each $\lambda > 0$ and each function $\phi \in BC(\mathbf{R}^+;\mathbf{R}^n)$ we must have

$$\limsup_{t\to\infty}\big|(r_{1/\lambda}\star\phi)(t)\big| \le \limsup_{t\to\infty}|\phi(t)|.$$

As a consequence of this fact we have for each $\lambda > 0$ (cf. Lemma 4.3)

$$\limsup_{t\to\infty}|(k\star\gamma)(t)| \le \limsup_{t\to\infty}\big|(k\star\gamma)(t)+\lambda\gamma(t)\big|.$$

Combining this result with (4.11), and recalling that $\varphi(t)\to 0$ as $t\to\infty$, we conclude that

$$\limsup_{t\to\infty}\big|\xi(t)\big| \le \limsup_{t\to\infty}\big|\xi(t)-\lambda\gamma(t)\big|.$$

As the norm in $\mathbf{R}^n$ is generated by the inner product, this implies that

$$\delta^2 = \limsup_{t\to\infty}\big|\xi(t)\big|^2 \le \limsup_{t\to\infty}\Big(\big|\xi(t)\big|^2 - 2\lambda\langle\xi(t),\gamma(t)\rangle\Big) + \lambda^2 M, \tag{4.12}$$

where $M = \limsup_{t\to\infty}|\gamma(t)|^2$. It follows from (iii) and the fact that $\psi\to 0$ as $t\to\infty$ that

$$\liminf_{t\to\infty}\langle\xi(t),\gamma(t)\rangle = \liminf_{t\to\infty}\Big\langle x_1(t)-x_2(t), g\big(t,x_1(t)\big)-g\big(t,x_2(t)\big)\Big\rangle \ge 0.$$

Moreover, there is a constant $\epsilon > 0$ such that, if t is large enough and $|\xi(t)| \ge \delta/2$, then

$$\langle\xi(t),\gamma(t)\rangle \ge \epsilon.$$

If we use these facts in (4.12) we find that

$$\delta^2 \le \max\Big\{\tfrac14\delta^2, \delta^2 - 2\lambda\epsilon\Big\} + \lambda^2 M.$$

However, for sufficiently small positive values of λ the right-hand side is less than δ^2, and we have obtained the desired contradiction. □

Our final result concerning the equation

$$x(t) + \int_0^t k(t-s)g(x(s))\,\mathrm{d}s = f(t), \quad t\in\mathbf{R}^+, \tag{4.13}$$

is a statement on the existence of asymptotically periodic solutions. It is possible to prove a similar result for a somewhat more general equation, but we shall leave the generalizations to the reader.

4.15 Theorem

Let the following hypotheses hold:

(i) $k \in L^1_{\mathrm{loc}}(\mathbf{R}^+;\mathbf{R}^{n\times n})$ *is a convolution kernel of* L^∞*-anti-accretive type;*

(ii) $g\in C(\mathbf{R}^n;\mathbf{R}^n)$ *is strictly monotone, i.e.,* $\langle g(u)-g(v), u-v\rangle > 0$ *if* u, $v\in\mathbf{R}^n$ *and* $u\ne v$*, and for each* $M>0$ *we have*

$$\liminf_{|u|\to\infty}\inf_{|v|\le M}\langle g(u+v),u\rangle > 0;$$

(iii) $f\in C(\mathcal{T}_T;\mathbf{R}^n) \oplus BC_0(\mathbf{R}^+;\mathbf{R}^n)$;

(iv) $x\in BC(\mathbf{R}^+;\mathbf{R}^n)$ *is a solution of (4.13).*

Then $x\in C(\mathcal{T}_T;\mathbf{R}^n)\oplus BC_0(\mathbf{R}^+;\mathbf{R}^n)$*, that is,* x *is asymptotically periodic with period* T.

Proof Again, let r_λ denote the resolvent kernel of λk and rewrite (4.13) in the form

$$x(t) - f(t) = \int_0^t r_\lambda(t-s)\Big(x(s) - f(s) - \tfrac{1}{\lambda} g\big(x(s)\big)\Big)\,\mathrm{d}s, \quad t \in \mathbf{R}^+.$$

If one here replaces $x(t) - f(t)$ by a new unknown function z, ignores terms which vanish at infinity, and requires z to be periodic, one finds that z has to be a solution of the equation

$$z(t) = \int_{-\infty}^t r_\lambda(t-s)\Big(z(s) - \tfrac{1}{\lambda} g\big(z(s) + f_T(s)\big)\Big)\,\mathrm{d}s, \quad t \in \mathbf{R}. \tag{4.14}$$

(Here f_T is a periodic function such that $f(t) - f_T(t) \to 0$ as $t \to \infty$.) Our first task is to show that for some sufficiently large $\lambda > 0$ this equation has a periodic solution.

Let $M \overset{\text{def}}{=} \sup_{t\in\mathbf{R}}|f_T(t)|$. By (ii) we can choose N so large that

$$\langle g(u+v), u\rangle \geq \delta > 0 \text{ when } |u| \geq N/2 \text{ and } |v| \leq M.$$

Next, we choose λ so large that

$$\sup_{|u|\leq N, |v|\leq M} |g(u+v)| \leq \min\{\sqrt{\lambda\delta}, \lambda N/2\}.$$

By considering separately the case where $0 \leq |u| \leq N/2$ and the case where $N/2 < |u| \leq N$, we conclude that

$$\Big|u - \tfrac{1}{\lambda} g(u+v)\Big| \leq N \text{ if } |u| \leq N \text{ and } |v| \leq M.$$

If we use this inequality and recall the definition of L^∞-anti-accretivity, then it is straightforward to show that the operator G defined by

$$G(\varphi)(t) = \int_{-\infty}^t r_\lambda(t-s)\Big\{\varphi(s) - \tfrac{1}{\lambda} g\big(\varphi(s) + f_T(s)\big)\Big\}\,\mathrm{d}s, \quad t \in \mathbf{R},$$

maps the set

$$S \overset{\text{def}}{=} \Big\{\varphi \in C(\mathcal{T}_T; \mathbf{R}^n) \;\Big|\; \sup_{t\in\mathbf{R}}|\varphi(t)| \leq N\Big\}$$

into itself. Since r_λ is integrable (recall Lemma **9**.10.2) we conclude that the image of S under G is an equicontinuous set. As the set S consists of periodic functions, it follows that $G(S)$ must be compact. Thus we may apply Schauder's theorem (see Theorem **12**.1.4) to conclude that G has a fixed point z. This fixed point is a solution of (4.14).

To proceed, we define the functions y and v on $\mathbf{R}^+$ by

$$y(t) = z(t) + f_T(t);$$

$$v(t) = \int_{-\infty}^0 r_\lambda(t-s)\Big\{z(s) - \tfrac{1}{\lambda} g\big(y(s)\big)\Big\}\,\mathrm{d}s.$$

Then y satisfies the equation

$$y(t) + \int_0^t k(t-s) g(y(s))\,\mathrm{d}s = f_T(t) + v(t) + \lambda \int_0^t k(t-s) v(s)\,\mathrm{d}s, \quad t \in \mathbf{R}^+.$$

Obviously we have $\lim_{t\to\infty} v(t) = 0$ and hence it follows from our assumptions and Theorem 4.14 that $x(t) - y(t) \to 0$ as $T \to \infty$ (since in the convolution case a kernel of L^∞-anti-accretive type is of BC_0-invariant type as well).

Scalar convolution kernels that are of L^∞-anti-accretive type (or, equivalently, of L^1-anti-accretive type) may be characterized in the following manner.

4.16 Theorem
Let $k \in L^1_{\mathrm{loc}}(\mathbf{R}^+;\mathbf{R})$ not be identically zero. Then k is of L^∞-anti-accretive type if and only if there exist a nonnegative number γ and a function $q \in L^1_{\mathrm{loc}}(\mathbf{R}^+;\mathbf{R})$ that satisfies $q(t) \ge \|q\|_{\mathrm{var}([t,\infty))}$ for all $t > 0$, such that

$$\gamma k(t) + \int_0^t q(t-s)k(s)\,\mathrm{d}s = 1, \quad t \in \mathbf{R}^+. \tag{4.15}$$

It is easy to show that $\gamma = 0$ iff k is unbounded (this is Exercise 15). Note that if $q(t) \ge \|q\|_{\mathrm{var}([t,\infty))}$ and $\lim_{t\to\infty} q(t) = 0$ then q is nonincreasing.

Proof of Theorem 4.16 Let us first assume that k is of L^∞-anti-accretive type. This implies that for each $\lambda > 0$ we have $\int_{\mathbf{R}^+} |r_\lambda(t)|\,\mathrm{d}t \le 1$. We intend to prove that the function $t \mapsto \lambda_k\big(1 - \int_0^t r_{\lambda_k}(s)\,\mathrm{d}s\big)$ tends to a measure $\gamma\delta(\mathrm{d}t) + q(t)\,\mathrm{d}t$ in the weak*-measure sense for some sequence $\lambda_k \to \infty$.

Define

$$v_\lambda(t) = \lambda\Big(1 - \int_0^t r_\lambda(s)\,\mathrm{d}s\Big), \quad t > 0, \quad \lambda > 0.$$

Since $\int_{\mathbf{R}^+} |r_\lambda(t)|\,\mathrm{d}t \le 1$ it follows that v_λ is nonnegative and that $v_\lambda(t) \ge \|v_\lambda\|_{\mathrm{var}([t,\infty))}$ for all $t > 0$. In particular we have

$$v_\lambda(t) \le 2 \inf_{s\in(0,t)} v_\lambda(s). \tag{4.16}$$

By integrating the resolvent equation, or substituting the expression for v into the expression below, we see that

$$\tfrac{1}{\lambda} v_\lambda(t) + \int_0^t v_\lambda(t-s)k(s)\,\mathrm{d}s = 1, \quad t > 0, \quad \lambda > 0. \tag{4.17}$$

By Lemma 4.5 we know that k is of positive type. Therefore it follows from Theorem **16**.2.6 that $\hat{k}(\sigma) \ge 0$ for every $\sigma > 0$ and since $\hat{k}$ is harmonic and not identically zero we have $\hat{k}(\sigma) > 0$. Fix some $\sigma > 0$. Calculating Laplace transforms we get

$$\int_{\mathbf{R}^+} \mathrm{e}^{-\sigma t} v_\lambda(t)\,\mathrm{d}t = \hat{v}_\lambda(\sigma) = \frac{1}{\sigma}\,\frac{1}{1/\lambda + \hat{k}(\sigma)}, \quad \lambda > 0,$$

and so

$$\int_0^t \mathrm{e}^{-\sigma s} v_\lambda(s)\,\mathrm{d}s \le \frac{1}{\sigma\hat{k}(\sigma)}, \quad t > 0, \quad \lambda > 0.$$

Utilising (4.16) in this inequality we get

$$0 \le v_\lambda(t) \le \frac{2}{\sigma \hat{k}(\sigma) \int_0^t \mathrm{e}^{-\sigma s}\,\mathrm{d}s}, \quad t > 0, \quad \lambda > 0, \tag{4.18}$$

and since v_λ is nonnegative, we have, for each $T > 0$,

$$\sup_{\lambda > 0} \int_0^T \big|v_\lambda(t)\big|\,\mathrm{d}t < \infty.$$

The last conclusion implies that for some numbers $\lambda_k \to \infty$ the sequence v_{λ_k} must converge locally in the weak*-measure sense to a locally finite measure. Now (4.18) shows that v_λ is dominated by a function that is bounded on each interval of the form $[\epsilon, T]$, where $0 < \epsilon < T < \infty$, and this implies that we may, without loss of generality, assume (by passing to a subsequence) that v_{λ_k} converges weak* in L^∞ locally on $(0, \infty)$ to a locally bounded function q. Clearly, this implies that the measure to which v_{λ_k} converges is the sum of a point mass at zero, and the function q. Each of the functions v_{λ_k} satsifies $v_{\lambda_k}(t) \ge \|v_{\lambda_k}\|_{\mathrm{var}([t,\infty))}$, and therefore also the limit q must satisfy $q(t) \ge \|q\|_{\mathrm{var}([t,\infty))}$ for all $t > 0$ (and we may without loss of generality assume that the derivative of v_{λ_k} converges to the derivative of q in the weak*-measure sense). If the size of the point mass at zero is γ, then $v_{\lambda_k} * k \to \gamma k + q * k$ locally in the weak*-measure sense, and by passing to the limit in (4.17) we find that γ and q satisfy (4.15). This completes the proof of the first part of the theorem.

Conversely, suppose that there is a nonnegative constant γ and a function $q \in L^1_{\mathrm{loc}}(\mathbf{R}^+; \mathbf{R})$ that satisfies $q(t) \ge \|q\|_{\mathrm{var}([t,\infty))}$ for all $t > 0$, such that (4.15) holds. Let η be an arbitrary continuously differentiable function on $\mathbf{R}^+$, with $\eta(0) = 0$. Fix some $\lambda > 0$, and define φ and ψ by

$$\varphi = \frac{1}{\lambda} r_\lambda * \eta, \qquad \psi = \gamma \varphi' + q * \varphi'.$$

Then it follows from (4.15) that

$$k * \psi = \varphi,$$

and this together with the resolvent equation implies that

$$\eta = \lambda \varphi + \psi.$$

Let $T > 0$ be arbitrary and let $\tau > 0$ be some point where $|\varphi(\tau)| = \max_{t \in [0,T]} |\varphi(t)|$. We may without loss of generality assume that $\varphi(\tau) > 0$ unless $\varphi \equiv 0$; otherwise replace η by $-\eta$. Then

$$\varphi'(\tau) \ge 0.$$

Since q is of bounded variation on each interval of the form $[\epsilon, \infty)$ where $\epsilon > 0$ (and since we may, without loss of generality, assume that q is right-continuous), we have $q(t) = \mu((t, \infty)) + q(\infty)$ where μ is a measure that is finite on (ϵ, ∞) for each $\epsilon > 0$ and satisfies $q(t) \ge |\mu|((t, \infty))$. Since φ is continuously differentiable and $t|\mu|((t, \infty)) \to 0$ as $t \to 0$ (because q is locally integrable), and since τ was chosen so that $\varphi(\tau) = \max_{t\in[0,T]}|\varphi(t)|$, we can integrate by parts to get

$$\int_0^\tau q(\tau-s)\varphi'(s)\,\mathrm{d}s = \lim_{n\to\infty}\int_{1/n}^\tau q(s)\varphi'(\tau-s)\,\mathrm{d}s$$

$$= \lim_{n\to\infty}\Big(\mu((\tfrac{1}{n},\infty))\varphi(\tau-\tfrac{1}{n}) - \int_{(1/n,\tau)}\mu(\mathrm{d}s)\varphi(\tau-s) + q(\infty)\varphi(\tau-\tfrac{1}{n})\Big)$$

$$\geq \lim_{n\to\infty}\Big(q(\tfrac{1}{n})\varphi(\tau-\tfrac{1}{n}) - |\mu|((\tfrac{1}{n},\tau))\sup_{t\in[0,\tau]}|\varphi(t)|\Big) \geq 0.$$

Thus

$$\psi(\tau) \geq \gamma\varphi'(\tau) \geq 0.$$

We conclude that, for all $T > 0$,

$$\lambda \sup_{t\in[0,T]}|\varphi(t)| \leq \sup_{t\in[0,T]}\big|\lambda\varphi(t) + \psi(t)\big|,$$

or equivalently

$$\sup_{t\in[0,T]}\big|\big(r_\lambda * \eta\big)(t)\big| \leq \operatorname*{ess\,sup}_{t\in[0,T]}|\eta(t)|.$$

This is true for all C^1-functions η with $\eta(0) = 0$, and hence, by the fact that this set of functions is weak*-dense in L^∞, the same inequality is true for all $\eta \in L^\infty([0,T];\mathbf{R})$. We conclude that k is of L^∞-anti-accretive type, and the proof is complete. □

Scalar convolution kernels that are of totally positive type (and that have a Laplace transform) may be characterized in the following manner.

4.17 Theorem
Let $k \in L^1_{\mathrm{loc}}(\mathbf{R}^+;\mathbf{R})$ satisfy $0 < \int_0^\infty \mathrm{e}^{-\sigma t}|k(t)|\,\mathrm{d}t < \infty$ for some $\sigma > 0$. Then k is of totally positive type if and only if there exist a nonnegative number γ and a nonincreasing function $q \in L^1_{\mathrm{loc}}(\mathbf{R}^+;\mathbf{R})$ such that

$$\gamma k(t) + \int_0^t q(t-s)k(s)\,\mathrm{d}s = 1, \quad t \in \mathbf{R}^+. \tag{4.15}$$

Proof Let us first assume that k is of totally positive type. This implies that for each $\lambda > 0$ we have $r_\lambda(t) \geq 0$. In addition, we know that k is nonnegative (cf. the discussion following Definition 4.9). Again we intend to prove that the function $t \mapsto \lambda_k\big(1 - \int_0^t r_{\lambda_k}(s)\,\mathrm{d}s\big)$ tends to a measure $\gamma\delta(\mathrm{d}t) + q(t)\,\mathrm{d}t$ in the weak*-measure sense for some sequence $\lambda_k \to \infty$.

We define v_λ as in the proof of Theorem 4.16 and note that since $r_\lambda(t) \geq 0$ it follows that v_λ is nonincreasing. From (4.17) we therefore get the inequality

$$v_\lambda(t) \leq \frac{\lambda}{1 + \lambda\int_0^t k(s)\,\mathrm{d}s}, \quad t > 0, \quad \lambda > 0.$$

Since k is not identically zero and is of totally positive type, we have $\int_0^t k(s)\,\mathrm{d}s > 0$ for every $t > 0$ (see Exercise 16), and the preceding inequality gives

$$v_\lambda(t) \leq \frac{1}{\int_0^t k(s)\,\mathrm{d}s}, \quad t > 0, \quad \lambda > 0. \tag{4.19}$$

By Lemma **2**.3.4, and since $\mathrm{e}^{-\sigma t}k(t)$ is integrable for some $\sigma > 0$, $\hat{r}_\lambda(\varrho)$ is well defined for ϱ sufficiently large (possibly depending on λ) and satisfies

$$\hat{r}_\lambda(\varrho) = \frac{\lambda\hat{k}(\varrho)}{1+\lambda\hat{k}(\varrho)} < 1. \tag{4.20}$$

Thus $\int_0^\infty \mathrm{e}^{-\varrho t}|r_\lambda(t)|\,\mathrm{d}t < 1$ and therefore (4.20) holds for all $\varrho \geq \sigma$. Since $t \mapsto \mathrm{e}^{-\sigma t}r_\lambda(t)$ is the resolvent of the function $t \mapsto \lambda\mathrm{e}^{-\sigma t}k(t)$ and r_λ is nonnegative, we conclude that the function $t \mapsto \mathrm{e}^{-\sigma t}k(t)$ is of L^∞-anti-accretive type. It follows from Theorem 4.16 that there exist a nonnegative constant γ_σ and a nonnegative function $q_\sigma \in L^1_{\mathrm{loc}}(\mathbf{R}^+;\mathbf{R})$ such that

$$\gamma_\sigma\mathrm{e}^{-\sigma t}k(t) + \int_0^t q_\sigma(t-s)\mathrm{e}^{-\sigma s}k(s)\,\mathrm{d}s = 1, \quad t \in \mathbf{R}^+. \tag{4.21}$$

If $\gamma_\sigma > 0$, then it follows that k must be continuous and $k(0) = 1/\gamma_\sigma$. Hence there exists a number $\delta > 0$ such that $k(t) \geq \delta$ on $[0,\delta]$. Thus by (4.17) we have

$$\int_0^t v_\lambda(s)\,\mathrm{d}s \leq \tfrac{1}{\delta} \quad \text{if} \quad 0 \leq t \leq \min\bigl\{\delta, \sup\{\, s \geq 0 \mid v_\lambda(s) \geq 0\,\}\bigr\}.$$

Since v_λ is nonincreasing, this inequality implies that, for each $T > 0$,

$$\sup_{\lambda>0} \int_0^T |v_\lambda(s)|_+\,\mathrm{d}s < \infty. \tag{4.22}$$

If, on the other hand, $\gamma_\sigma = 0$, then it follows from (4.21) that $\int_0^t q_\sigma(t-s)k(s)\,\mathrm{d}s \geq 1$ for $t > 0$. Taking the convolution of both sides of (4.17) by q_σ we therefore get

$$\int_0^t v_\lambda(s)\,\mathrm{d}s \leq \int_0^t q_\sigma(s)\,\mathrm{d}s, \quad 0 \leq t \leq \sup\{\, s \geq 0 \mid v_\lambda(s) \geq 0\,\}.$$

But this means that (4.22) holds in this case as well.

If we combine the Laplace transform formula

$$\int_{\mathbf{R}^+} \mathrm{e}^{-\sigma t}v_\lambda(t)\,\mathrm{d}t = \hat{v}_\lambda(\sigma) = \frac{1}{\sigma}\,\frac{1}{1/\lambda + \hat{k}(\sigma)}$$

with (4.22), then we see that, for each $T > 0$,

$$\sup_{\lambda>0} \int_0^T \bigl|v_\lambda(t)\bigr|\,\mathrm{d}t < \infty.$$

Now we can complete the proof of the first part in the same way as in the proof of Theorem 4.16 since (4.19) provides the necessary pointwise bound uniformly in λ.

Conversely, suppose that there is a nonnegative constant γ and a nonincreasing function $q \in L^1_{\mathrm{loc}}(\mathbf{R}^+;\mathbf{R})$ such that (4.15) holds. Let us for the moment assume that $q(\infty) = \lim_{t\to\infty} q(t) > -\infty$. If we use the same notation and argument as in the proof of Theorem 4.16 but choose $\lambda > \max\{-q(\infty), 0\}$ and observe that now μ is a nonnegative measure, then we see that

$$\int_0^\tau q(\tau - s)\varphi'(s)\,\mathrm{d}s \geq q(\infty)\varphi(\tau).$$

Thus

$$\psi(\tau) \geq \gamma\varphi'(\tau) + q(\infty)\varphi(\tau) \geq q(\infty)\varphi(\tau).$$

We conclude that, for all $T > 0$,

$$(\lambda + q(\infty)) \sup_{t\in[0,T]} |\varphi(t)| \leq \sup_{t\in[0,T]} \big|\lambda\varphi(t) + \psi(t)\big|,$$

or equivalently, since $\lambda > -q(\infty)$,

$$\sup_{t\in[0,T]} \big|(r_\lambda * \eta)(t)\big| \leq \frac{\lambda}{\lambda + q(\infty)} \sup_{t\in[0,T]} |\eta(t)|.$$

Consequently we have

$$\|r_\lambda\|_{L^1(\mathbf{R}^+)} \leq \frac{\lambda}{\lambda + q(\infty)}.$$

Taking the convolution of both sides of (4.15) by r_λ and using the resolvent equation we get

$$\int_0^t r_\lambda(s)\,\mathrm{d}s + \tfrac{\gamma}{\lambda} r_\lambda(t) + \tfrac{1}{\lambda}(r_\lambda * q)(t) = 1, \quad t \geq 0,$$

and by taking Laplace transforms we obtain

$$\hat{r}_\lambda(\varrho) = \frac{\lambda}{\lambda + \varrho\gamma + \varrho\hat{q}(\varrho)}, \quad \varrho > 0.$$

Since $\varrho\hat{q}(\varrho) \to q(\infty)$ as $\varrho \downarrow 0$ we see that

$$\int_{\mathbf{R}^+} r_\lambda(s)\,\mathrm{d}s = \frac{\lambda}{\lambda + q(\infty)}.$$

This is the same number as we obtained as an upper bound for $\|r_\lambda\|_{L^1(\mathbf{R}^+)}$, and therefore we conclude that r_λ must be nonnegative. Since the only restriction on λ was that $\lambda > \max\{-q(\infty), 0\}$ we see from Exercise 23 that k is of totally positive type.

To complete the proof we observe that the values of k on an interval $[0, T]$ do not depend on the values of q on (T, ∞) and therefore no generality was lost when we assumed that $q(\infty) > -\infty$. □

The following result is evident from Theorem 4.16.

4.18 Corollary
Let $k \in L^1_{\text{loc}}(\mathbf{R}^+;\mathbf{R})$ of L^∞-anti-accretive type. If k is unbounded, then k has a function resolvent r of the first kind (cf. Definition **5**.*5.1), and $r(t) \geq \|r\|_{\text{var}([t,\infty))}$ for all $t > 0$.*

In the scalar case, this corollary is an extension of Theorem **5**.5.4.

Formula (4.15) can be used to write the equation

$$x(t) + \int_0^t k(t-s)g\big(s,x(s)\big)\,\mathrm{d}s = f(t), \quad t \in \mathbf{R}^+. \tag{4.23}$$

in a new 'variation of constants' form. Clearly, if we convolve this equation by q, and add a γ-multiple of the same equation, then we get

$$\gamma x(t) + \int_0^t q(t-s)x(s)\,\mathrm{d}s + \int_0^t g\big(s,x(s)\big)\,\mathrm{d}s = \gamma f(t) + \int_0^t q(t-s)f(s)\,\mathrm{d}s, \quad t \in \mathbf{R}^+. \tag{4.24}$$

For $\gamma = 0$ this is essentially a special case of the variation of constants equation of the first kind (**5**.5.2). The same formula is valid if k is locally absolutely continuous and $k(0) \neq 0$: we take $\gamma = 1/k(0)$, and let $-k(0)q(\cdot)$ be the resolvent of $k'(\cdot)/k(0)$. Of course, in this case there is no guarantee that q is nonnegative and nonincreasing.

The formula (4.24), and in particular its differentiated version, have successfully been employed in proofs of existence results for abstract equations; see Section 7.

5. Nonlinear Nonseparable Convolution Equations

In this section we shall study the behaviour of solutions of the integral equation

$$x(t) + \int_0^t k\big(t-s,x(s)\big)\,\mathrm{d}s = f(t), \quad t \geq 0, \tag{5.1}$$

and of the corresponding integrodifferential equation

$$x'(t) + \int_0^t k\big(t-s,x(s)\big)\,\mathrm{d}s = f(t), \quad t \geq 0; \quad x(0) = x_0. \tag{5.2}$$

These equations are of convolution type, but they are neither linear nor of the form studied in Chapters 17–19. Some results of perturbation nature for these equations are given in Chapter 11, but here we give a completely different analysis where the linear part of the equation is irrelevant.

In Chapters 17–19 we studied (5.1) and (5.2) under the assumption that $k(t,v) = a(t)g(v)$. Frequently, in (5.1) we assumed that $\langle v, g(v)\rangle \geq 0$, and in (5.2) we took g to be the gradient of a nonnegative function G. Here we have to combine these conditions, and additionally to assume that the functions playing the same role as G are *convex*. Moreover, we are forced to restrict the behaviour of the function $t \mapsto k(t,v)$. It is not enough for this function to be of positive type or to have a derivative of positive type. Instead, in (5.1) it must be bounded, nonnegative, and nonincreasing and, in (5.2), it is in addition required to be convex.

The results that shall be presented below are based on different versions of the following basic identity.

5.1 Lemma

Let $T > 0$, and assume that the following conditions are satisfied.

(i) $k \in C([0,T] \times \mathbf{R}^n; \mathbf{R}^n)$, $K \in C([0,T] \times \mathbf{R}^n; \mathbf{R})$ *and for each fixed* $v \in \mathbf{R}^n$ *the functions* $t \mapsto k(t,v)$ *and* $t \mapsto K(t,v)$ *are absolutely continuous. Moreover, if we denote the derivatives of these functions by* k' *and* K', *then, for each* $M > 0$,

$$\int_0^T \sup_{|v| \le M} \big(|k'(t,v)| + |K'(t,v)|\big)\,\mathrm{d}t < \infty;$$

(ii) $\varphi \in B^\infty([0,T]; \mathbf{R}^n)$.

Then the following identity is valid:

$$\begin{aligned}
&\int_0^T \Big\langle \Big(\tfrac{\mathrm{d}}{\mathrm{d}t}\int_0^t k\big(t-s,\varphi(s)\big)\,\mathrm{d}s\Big), \varphi(t)\Big\rangle\,\mathrm{d}t \\
&\quad = \int_0^T \Big(K\big(t,\varphi(t)\big) - K(t,0)\Big)\,\mathrm{d}t + \int_0^T H\big(T-s,0,\varphi(s)\big)\,\mathrm{d}s \\
&\qquad - \int_0^T \int_0^t H'\big(t-s,\varphi(t),\varphi(s)\big)\,\mathrm{d}s\,\mathrm{d}t,
\end{aligned} \tag{5.3}$$

where

$$\begin{aligned}
H(t,u,v) &= K(t,u) - K(t,v) - \langle k(t,v), u-v\rangle, \\
H'(t,u,v) &= \tfrac{\mathrm{d}}{\mathrm{d}t} H(t,u,v) = K'(t,u) - K'(t,v) - \langle k'(t,v), u-v\rangle.
\end{aligned}$$

To prove this identity, observe that it is true for $T = 0$, differentiate with respect to T, and perform some calculations.

The reader may find it instructive to compare this identity with those given in Lemmas **18**.4.1, **18**.4.4 and **18**.5.1.

The fact that makes this identity interesting is that under suitable assumptions on the functions k and K we can make all the terms on the right-hand side nonnegative. For this, one assumes that the function $v \mapsto k(t,v)$ is the gradient of the function $v \mapsto K(t,v)$, hence $k'(t,v)$ is the gradient of the function $v \mapsto K'(t,v)$. Moreover, one assumes that the functions $v \mapsto k(t,v)$ and $v \mapsto -k'(t,v)$ are monotone. (As usual, monotonicity of a function $g : \mathbf{R}^n \to \mathbf{R}^n$ means that $\langle g(u) - g(v), u - v\rangle \ge 0$ for $u, v \in \mathbf{R}^n$.) It is a well-known fact that if g is both monotone and the gradient of a function G then G is convex, and

$$G(u) - G(v) - \langle g(v), u-v\rangle \ge 0, \quad u \in \mathbf{R}^n, \quad v \in \mathbf{R}^n. \tag{5.4}$$

The second and the third term of the right-hand side of the identity in Lemma 5.1 are of this type.

It is now clear what one should do in (5.1): differentiate the equation, multiply by $x(t)$, integrate over $[0, T]$, and estimate the terms that one gets.

5.2 Theorem

(i) *Assume the following.*

(1) $k \in C(\mathbf{R}^+ \times \mathbf{R}^n; \mathbf{R}^n)$, *and for each fixed* $t \in \mathbf{R}^+$ *the function* $v \mapsto k(t, v)$ *is the gradient of a function* $v \mapsto K(t, v)$. *Moreover, for each* $v \in \mathbf{R}^n$ *the function* $t \mapsto k(t, v)$ *is locally absolutely continuous, and its derivative* $k'(t, v)$ *satisfies* $\int_{\mathbf{R}^+} \sup_{|v| \le M} |k'(t, v)| \, \mathrm{d}t < \infty$ *for every* $M < \infty$.

(2) *For almost all* $t \in \mathbf{R}^+$ *the function* $v \mapsto -k'(t, v)$ *is monotone in* v.

(3) $\langle k(t, v), v \rangle \ge K(t, v) - K(t, 0) \ge 0$ *for each* $t > 0$.

(4) $f \in AC_{\mathrm{loc}}(\mathbf{R}^+; \mathbf{R}^n)$.

(5) $x \in C(\mathbf{R}^+; \mathbf{R}^n)$ *is a solution of (5.1).*

Then x *satisfies*

$$|x(t)| \le |x(0)| + \int_0^t |f'(t)| \, \mathrm{d}t, \quad t \in \mathbf{R}^+. \tag{5.5}$$

(ii) *In addition to (1)–(5), suppose that*

(6) $\int_{\mathbf{R}^+} |f'(t)| \, \mathrm{d}t < \infty$,

(7) *the set of points* t *for which the function* $v \mapsto -K'(t, v)$ *is strictly convex has positive Lebesgue measure.*

Then x *is asymptotically slowly varying, that is,*

$$\lim_{t \to \infty} \big(x(t) - x(t - T) \big) = 0, \quad T \in \mathbf{R},$$

and

$$\lim_{t \to \infty} k\big(\infty, x(t)\big) = 0.$$

(iii) *In addition to (1)–(7) above, assume that*

(8) $\int_{\mathbf{R}^+} \sup_{|v| \le M} |k(t, v) - k(\infty, v)| \, \mathrm{d}t < \infty$ *for every* $M < \infty$.

Then

$$\lim_{t \to \infty} \left\{ x(t) + \int_0^t k\big(\infty, x(s)\big) \, \mathrm{d}s + \int_{\mathbf{R}^+} \Big(k\big(s, x(t)\big) - k\big(\infty, x(t)\big) \Big) \, \mathrm{d}s \right\} = f(\infty).$$

Observe that if the function $v \mapsto k(t, v)$ is monotone, then the inequality $\langle k(t, v), v \rangle \ge K(t, v) - K(t, 0)$ in (3) is satisfied (compare (5.4)). Also note that, under appropriate additional assumptions on the functions $v \mapsto k(\infty, v)$ and $v \mapsto \int_{\mathbf{R}^+} \big(k(s, v) - k(\infty, v) \big) \, \mathrm{d}s$, one can deduce that x tends to a limit.

Strict convexity means that the inequality (5.4) is strict for all $u \ne v$. As a reminder, the norm $|\cdot|$ is induced by the inner product $\langle \cdot, \cdot \rangle$.

Proof of Theorem 5.2 (i) Differentiate (5.1) to get

$$x'(t) + \tfrac{\mathrm{d}}{\mathrm{d}t}\int_0^t k(t-s, x(s))\,\mathrm{d}s = f'(t), \quad t \in \mathbf{R}^+. \tag{5.6}$$

Take the inner product of this equation and $x(t)$, integrate over $[0, T]$, and use (2), (3), and Lemma 5.1 to get

$$|x(T)|^2 \le |x(0)|^2 + 2\int_0^T |f'(s)||x(s)|\,\mathrm{d}s.$$

To show that this inequality implies (5.5), one argues as in the proof of Theorem **17**.2.2, with $G(x(t)) = |x(t)|^2$, $u(s) = 2\sqrt{s}$, and with f replaced by f'.

(ii) It is clear from the first part of the proof together with Lemma 5.1 that

$$\int_{\mathbf{R}^+}\int_0^t -H'\big(s, x(t), x(t-s)\big)\,\mathrm{d}s\,\mathrm{d}t < \infty, \tag{5.7}$$

where we have used the same notation as in Lemma 5.1 (observe that $-H'(t,u,v)$ is nonnegative, so the integral converges absolutely). It follows from (1), (6), (5.6), and the boundedness of x that x is uniformly continuous. This, together with (1), implies that the inner integral in (5.7) is uniformly continuous, and therefore

$$\int_0^t -H'\big(s, x(t), x(t-s)\big)\,\mathrm{d}s \to 0 \text{ as } t \to \infty.$$

If we take some y in the limit set $\Gamma(x)$ of x, then it follows from Lebesgue's dominated convergence theorem that

$$\int_{\mathbf{R}^+} -H'\big(s, y(t), y(t-s)\big)\,\mathrm{d}s = 0, \quad t \in \mathbf{R}.$$

In particular, this means that, for all $t \in \mathbf{R}$, we have $-H'(s, y(t), y(t-s)) = 0$ for almost all $s \in \mathbf{R}^+$. If we define

$$M = \big\{\, t \in \mathbf{R}^+ \mid \textit{the function } v \mapsto -K'(t,v) \textit{ is strictly convex} \,\big\},$$

then M has positive Lebesgue measure, and we conclude that, for all $t \in \mathbf{R}$, $y(t) = y(t-s)$ for almost all $s \in M$. Define

$$U = \big\{\, t \in \mathbf{R} \mid m\big(M \cap (t-\epsilon, t+\epsilon)\big) = 0 \textit{ for some } \epsilon > 0 \,\big\},$$

where m represents the Lebesgue measure. Then U is open, and $M \cap U$ is a countable union of sets of measure zero, hence $m(M \cap U) = 0$. Let $M_U = M \setminus U$. Then M_U has positive Lebesgue measure, and because of the continuity of y we have

$$y(t) = y(t-s), \quad t \in \mathbf{R}, \quad s \in M_U$$

(every open interval containing $s \in M_U$ contains a set of positive measure on which the function $v \mapsto y(t) - y(t-v)$ vanishes; hence, by continuity, $y(t-s) = y(t)$). In particular, $y(t) = y(t - s_1 + s_2)$ for all $t \in \mathbf{R}$ and all s_1 and $s_2 \in M_U$. But the set

$$M_U - M_U = \{\, s_1 - s_2 \mid s_1 \in M_U, s_2 \in M_U \,\}$$

contains some open interval containing zero (see Lemma **15**.9.3), and we conclude that $y(t) = y(t-s)$ for all s in some neighbourhood of zero. Clearly this implies that y is a constant. By Theorem **15**.3.3, x is asymptotically slowly varying.

Since x is asymptotically slowly varying, it follows from (5.6) and Theorem **15**.2.11 that every constant $y \in \Gamma(x)$ satisfies

$$k(0,y) + \int_0^\infty k'(s,y)\,\mathrm{d}s = k(\infty,y) = 0.$$

Therefore $k(\infty, x(t)) \to 0$ as $t \to \infty$ by, e.g., Theorem **15**.2.7.

(iii) The proof of (iii) is left to the reader (cf. the proof of the second claim in (ii)). □

Next we turn to the equation

$$x'(t) + \int_0^t k\big(t-s, x(s)\big)\,\mathrm{d}s = f(t), \quad t \in \mathbf{R}^+; \quad x(0) = x_0. \tag{5.2}$$

Here too, one could differentiate the equation, take the inner product of what one gets and $x(t)$, and integrate over $[0,T]$. Under the assumptions of Theorem 5.2, the resulting double integral will be bounded from below. However, the first term turns out to be

$$\int_0^T \langle x(t), x''(t)\rangle\,\mathrm{d}t,$$

and this term is difficult to estimate from below. To avoid this obstacle, take the inner product of the differentiated equation and $x'(t)$; then the first term causes no trouble. Of course, we then need a new estimate for the double integral. This estimate is given in the following lemma.

5.3 Lemma

Let $T > 0$, and assume that the following conditions are satisfied.

(i) *$k \in C([0,T] \times \mathbf{R}^n; \mathbf{R}^n)$, and for each fixed $t \in [0,T]$ the function $v \mapsto k(t,v)$ is the gradient of a function $v \mapsto K(t,v)$. Moreover, for each $v \in \mathbf{R}^n$ the function $t \mapsto k(t,v)$ is absolutely continuous, and its derivative $k'(t,v)$ satisfies $\int_0^T \sup_{|v|\le M}|k'(t,v)|\,\mathrm{d}t < \infty$ for every $M < \infty$.*

(ii) *The function $v \mapsto -k'(t,v)$ is monotone in v for all $t \in [0,T]$, and the function $v \mapsto k'(t,v) - k'(s,v)$ is monotone in v for all $0 \le s \le t \le T$.*

(iii) *$\varphi \in AC_{\mathrm{loc}}([0,T]; \mathbf{R}^n)$.*

Then

$$\int_0^T \Big\langle \frac{\mathrm{d}}{\mathrm{d}t}\Big(\int_0^t k\big(t-s,\varphi(s)\big)\,\mathrm{d}s\Big), \varphi'(t)\Big\rangle\,\mathrm{d}t$$
$$\ge K\big(T,\varphi(T)\big) - K\big(0,\varphi(0)\big) - \int_0^T K'\big(t,\varphi(t)\big)\,\mathrm{d}t, \tag{5.8}$$

where $K'(t,v)$ denotes the derivative of the function $t \mapsto K(t,v)$.

Proof If we strengthen our smoothness assumptions on k to include the local absolute continuity of k', then

$$\begin{aligned}\int_0^T \Big\langle \tfrac{\mathrm{d}}{\mathrm{d}t}\Big(\int_0^t k(t-s,\varphi(s))\,\mathrm{d}s\Big), \varphi'(t)\Big\rangle\,\mathrm{d}t \\ = K\big(T,\varphi(T)\big) - K\big(0,\varphi(0)\big) - \int_0^T K'\big(t,\varphi(t)\big)\,\mathrm{d}t \\ - \int_0^T H'\big(T-s,\varphi(T),\varphi(s)\big)\,\mathrm{d}s \\ + \int_0^T \int_0^t H''\big(t-s,\varphi(t),\varphi(s)\big)\,\mathrm{d}s\,\mathrm{d}t, \qquad (5.9)\end{aligned}$$

where

$$H'(t,u,v) = K'(t,u) - K'(t,v) - \langle k'(t,v), u-v\rangle,$$
$$H''(t,u,v) = \tfrac{\mathrm{d}}{\mathrm{d}t}H(t,u,v) = K''(t,u) - K''(t,v) - \langle k''(t,v), u-v\rangle.$$

To prove this, one again observes that it is true for $T = 0$, and then one differentiates both sides with respect to T. Because of the monotonicity assumption (ii), the last two terms are nonnegative, and therefore the given inequality is true for this class of kernels.

To prove the general case, one approximates the kernel by a smoother kernel. Clearly, for each $h > 0$, the result that we have just proved can be applied to the kernel $k_h(t,v) = \frac{1}{h}\int_t^{t+h} k(s,v)\,\mathrm{d}s$ (let $k(t,v) = k(T,v)$ for $t > T$), and hence the inequality is satisfied for the kernels k_h. Since k is continuous it follows that

$$\lim_{h\downarrow 0} \sup_{\substack{0\le t\le T\\ |v|\le M}} \big|k_h(t,v) - k(t,v)\big| = 0,$$

for every $M < \infty$. Moreover, for every $v \in \mathbf{R}^n$ and almost every $t \in [0,T]$,

$$\lim_{h\downarrow 0} k_h'(t,v) = k'(t,v).$$

A similar statement is true if we replace k and k_h by K and K_h, where K_h is defined in the analogous way. These facts permit us to pass to the limit in (5.8), and to replace the functions k_h and K_h by k and K. □

Lemma 5.3 enables us to prove the following theorem about (5.2). For this equation, our boundedness results are not wholly satisfactory.

5.4 Theorem

(i) *Assume the following.*

(1) $k \in C(\mathbf{R}^+ \times \mathbf{R}^n; \mathbf{R}^n)$, *and for each fixed* $t \in \mathbf{R}^+$ *the function* $v \mapsto k(t,v)$ *is the gradient of a function* $v \mapsto K(t,v)$. *Moreover, for each* $v \in \mathbf{R}^n$ *the function* $t \mapsto k(t,v)$ *is locally absolutely continuous, and its derivative* $k'(t,v)$ *satisfies* $\int_{\mathbf{R}^+} \sup_{|v|\le M} |k'(t,v)|\,\mathrm{d}t < \infty$ *for every* $M < \infty$.

(2) *The function* $v \mapsto -k'(t,v)$ *is monotone in* v *for almost all* $t \in \mathbf{R}^+$, *and the function* $v \mapsto k'(t,v) - k'(s,v)$ *is monotone in* v *for almost all* $0 \le s \le t < \infty$.

(3) *For each* $v \in \mathbf{R}^n$ *the function* $t \mapsto K(t,v)$ *is nonnegative and nonincreasing on* $\mathbf{R}^+$.

(4) $f \in AC_{\mathrm{loc}}(\mathbf{R}^+;\mathbf{R}^n)$.

(5) $x \in AC_{\mathrm{loc}}(\mathbf{R}^+;\mathbf{R}^n)$ *is a solution of (5.2).*

Then x *satisfies*

$$|x'(t)| \le \sqrt{|x'(0)|^2 + 2K(0,x(0))} + \int_0^t |f'(s)|\,\mathrm{d}s, \quad t \in \mathbf{R}^+.$$

If in addition

(6) $\lim_{|v|\to\infty} K(\infty,v) = +\infty$,

and

(7) $f' \in L^1(\mathbf{R}^+;\mathbf{R}^n)$,

then $x \in BC(\mathbf{R}^+;\mathbf{R}^n)$.

(ii) *In addition to* (1)–(5) *and* (7) *above, suppose that*

(8) $x \in BC(\mathbf{R}^+;\mathbf{R}^n)$,

(9) *for each* $v \in \mathbf{R}^n$ *the function* $t \mapsto k'(t,v)$ *is locally absolutely continuous on* $(0,\infty)$, *and its derivative* $k''(t,v)$ *satisfies* $\int_\epsilon^T \sup_{|v|\le M}|k''(t,v)|\,\mathrm{d}t < \infty$ *for all* $\epsilon < T < \infty$ *and every* $M < \infty$,

(10) *the set of points* $t \in \mathbf{R}^+$ *for which the function* $v \mapsto K''(t,v)$ *is strictly convex has positive Lebesgue measure.*

Then x *is asymptotically slowly varying, that is,*

$$\lim_{t\to\infty} \big(x(t) - x(t-T)\big) = 0, \quad T \in \mathbf{R},$$

and

$$\lim_{t\to\infty} k(\infty, x(t)) = 0.$$

(iii) *In addition to* (1)–(5), (7), *and* (8) *above, assume that*

(11) $\int_{\mathbf{R}^+} \sup_{|v|\le M}|k(t,v) - k(\infty,v)|\,\mathrm{d}t < \infty$ *for every* $M < \infty$.

Then

$$\lim_{t\to\infty}\left\{\int_0^t k(\infty,x(s))\,\mathrm{d}s + \int_{\mathbf{R}^+}\Big(k\big(s,x(t)\big) - k\big(\infty,x(t)\big)\Big)\,\mathrm{d}s\right\} = f(\infty).$$

Proof The proof of (i) is entirely analogous to the proof of (i) in Theorem 5.2, and is left to the reader.

The proof of (ii) is strongly reminiscent of the proof of the corresponding part of Theorem 5.2. The main difference is that one replaces (5.7) by

$$\int_{\mathbf{R}^+}\int_\epsilon^{\min\{t,T\}} H''(s,x(t),x(t-s))\,\mathrm{d}s\,\mathrm{d}t < \infty, \tag{5.10}$$

where $0 < \epsilon < T < \infty$ is chosen in such a way that the set of points $t \in [\epsilon, T]$ for which the function $v \mapsto K''(t,v)$ is strictly convex has positive

Lebesgue measure (to get (5.10) one uses the same approximations as one does in the proof of Lemma 5.3). Again, we leave the details to the reader.

The proof of (iii) is left to the reader as well. □

6. Exercises

1. Let k_j be a sequence of kernels of positive type that converges to a kernel k of type L^2_{loc} in $L^1_{\mathrm{loc}}(\mathbf{R}^+ \times \mathbf{R}^+; \mathbf{C}^{n\times n})$. Show that k is of positive type. Hint: It suffices to test the positivity of k against continuous functions φ with compact support.
2. Let k be of type L^2_{loc} on $\mathbf{R}^+$. Show that the following three conditions are equivalent:
   (i) *k is of positive type;*
   (ii) *For all $T > 0$, $\chi_{[0,T)}(t)k(t,s)$ is of positive type;*
   (iii) *For all $S \geq 0$, $\chi_{(S,\infty)}(s)k(t,s)$ is of positive type.*
3. Show that each of the kernels k below is of positive type:
   (i) *$k(t,s) = a(t-s)$ for $0 \leq s < t < \infty$, where $a \in L^1_{\mathrm{loc}}(\mathbf{R}^+; \mathbf{C}^{n\times n})$ is nonnegative, nonincreasing, and convex on $\mathbf{R}^+$, i.e., the function $t \mapsto \langle y, a(t)y\rangle$ is nonnegative, nonincreasing, and convex on $\mathbf{R}^+$ for each $y \in \mathbf{C}^n$;*
   (ii) *$k(t,s) = a(t)$ for $0 \leq s < t < \infty$, where a is nonnegative and nonincreasing on $\mathbf{R}^+$, i.e., the function $t \mapsto \langle y, a(t)y\rangle$ is nonnegative and nonincreasing on $\mathbf{R}^+$ for each $y \in \mathbf{C}^n$;*
   (iii) *$k(t,s) = a(s)$ for $0 \leq s < t < \infty$, where a is nonnegative and nondecreasing on $\mathbf{R}^+$, i.e., the function $s \mapsto \langle y, a(s)y\rangle$ is nonnegative and nondecreasing on $\mathbf{R}^+$ for each $y \in \mathbf{C}^n$.*
4. Show that each of the kernels l below is of positive type:
   (i) *$l(t,s) = a(t-s)k(t,s)$ for $0 \leq s < t < \infty$, where a is scalar-valued, continuous, nonnegative, nonincreasing, and convex on $\mathbf{R}^+$, and k satisfies the assumption of Theorem 2.2;*
   (ii) *$l(t,s) = a(t)k(t,s)$ for $0 \leq s < t < \infty$, where a is scalar-valued, nonnegative, and nondecreasing on $\mathbf{R}^+$, and k satisfies the assumption of Theorem 2.2;*
   (iii) *$l(t,s) = a(s)k(t,s)$ for $0 \leq s < t < \infty$, where a is scalar-valued, nonnegative, and nonincreasing on $\mathbf{R}^+$, and k satisfies the assumption of Theorem 2.2.*
5. Give an example of a simple function k that is of positive type.
6. Let $k(t,s) = a\big((t-s)\gamma(t)\big)$, where a satisfies the assumptions in Exercise 3(i), $\gamma \in C(\mathbf{R}^+; \mathbf{R})$ is nonnegative, nonincreasing, and the function $t \mapsto t\gamma(t)$ is nondecreasing. Show that k is of positive type.
7. Prove the claim made after Theorem 2.2 that one may, without loss

of generality, assume that $k(t,0) \equiv 0$, and that

$$k(t,s) = \lim_{v\downarrow t, u\uparrow s} k(v,u), \quad 0 \le s < t.$$

Hint: First reduce the claim to the scalar case. Then show that the set of discontinuities of k is contained in a countable union of lines parallel to the s-axis or the t-axis; hence the measure of the set where k is discontinuous has measure zero.

8. Let k satisfy the assumptions of Theorem 2.2, and suppose that k is normalized as in the preceding exercise. Show that there exists a nonnegative measure α supported on $\{\,(t,s) \mid 0 \le s < t\,\}$, such that, for all $0 \le s < t < \infty$, α is finite on the semi-infinite rectangle $(t,\infty)\times[0,s)$, and (2.2) holds. Moreover, if $k(T,T)$ is finite for some $T>0$, then α is finite on the semi-infinite rectangle $(T,\infty)\times[0,T)$, and (2.2) holds with (t,s) replaced by (T,T), provided $k(T,T) = \lim_{t\downarrow T, s\uparrow T} k(t,s)$. Hint: First reduce the claim to the scalar case. Let $0 \le u < s < t < v$, $E = \{\,(\tau,\sigma) \mid t < \tau \le v,\, u \le \sigma < s\,\}$ and define $\alpha(E) = k(t,s) - k(t,u) - k(v,s) + k(v,u)$. Show that α can be extended as a positive measure to all Borel sets $\subset \mathbf{R}^2$ so that (2.2) holds.
9. Prove that the measure in the preceding exercise satisfies, for each $T > 0$,
$$\iint_{\{0\le u<v\le T\}} (v-u)^2 |\alpha|(dv\,du) < \infty.$$
Hint: Reduce the problem to the scalar case, and let φ in (2.4) be a positive constant.
10. Prove the claims made in the beginning of the proof of Theorem 2.4 concerning the measure β, and check that all the integrals in that proof converge (absolutely).
11. Show that Theorems 3.1 and 3.2 are valid in the case where x and $f \in B^\infty_{\text{loc}}$, and k is of type B^∞.
12. Show that the resolvent of μr_λ is $\frac{\mu}{1+\mu} r_{\lambda+\mu\lambda}$.
13. Show that if k is of L^p-anti-accretive type, then so is r_λ (cf. Exercise 12).
14. Show that if $k \in L^1_{\text{loc}}(\mathbf{R}^+;\mathbf{R})$ is positive and if the function $t \to \ln(k(t))$ is nonincreasing and convex, then k is of L^∞-anti-accretive type.
15. Show that the function k in Theorem 4.16 satisfies $k(t) \le 1/\gamma$ if $\gamma > 0$, and that $\gamma = 0$ whenever k is unbounded. Also show that $q(\infty) > 0$ iff $k \in L^1(\mathbf{R}^+;\mathbf{R})$.
16. Show that if a scalar convolution kernel k is of totally positive type, and $k(t) = 0$ on some interval $[0,\epsilon]$, then $k \equiv 0$. Hint: Show that if $T = \sup\{\,t \in \mathbf{R}^+ \mid k(s) = 0 \text{ for almost all } s \le t\,\}$ then on the interval

$[2T, 3T]$ one has $r_\lambda(t) = \lambda k(t) - \lambda^2 (k * k)(t)$. Integrate over $[2T, 3T]$, and let $\lambda \to \infty$.

17. Show that the last part of Theorem 3.2 can be deduced from Theorem 4.12 in the case where g is locally Lipschitz-continuous in its second argument.
18. Complete the proof of Theorem 4.13.
19. Extend Theorem 4.13 so that it can be applied to (4.10).
20. Give an example of a scalar convolution kernel $k \in L^1_{\mathrm{loc}}(\mathbf{R}^+; \mathbf{R})$ that is totally positive and of L^∞-anti-accretive type, but is not log-convex. (Nohel [3], Section 4.2, p. 58.)
21. Give an example of a scalar convolution kernel $k \in L^1_{\mathrm{loc}}(\mathbf{R}^+; \mathbf{R})$ that is of totally positive type, but not of L^∞-anti-accretive type.
22. Give an example of a scalar convolution kernel $k \in L^1_{\mathrm{loc}}(\mathbf{R}^+; \mathbf{R})$ that is of L^∞-anti-accretive type, but not of totally positive type.
23. Let $p \in [1, \infty]$, $\lambda_0 > 0$, and let k be a real-valued Volterra kernel that is of type L^p_{loc} on $\mathbf{R}^+$ such that the spectrum of k does not intersect $(-\infty, 0)$ and $r_\lambda \geq 0$ for all $\lambda \geq \lambda_0$. Show that k is of totally positive type, i.e., show that the restriction $\lambda \geq \lambda_0$ can be removed. Hint: Use Theorem 4.11(iii).
24. Show that a scalar convolution kernel $k \in L^1_{\mathrm{loc}}(\mathbf{R}^+; \mathbf{R})$ is both of totally positive type and of L^∞-anti-accretive type if and only if $r_\lambda(t) \geq 0$ and $1 - \int_0^t r_\lambda(s)\, ds \geq 0$ for $\lambda > 0$ and $t \in \mathbf{R}^+$.
25. Assume that $k \in L^1_{\mathrm{loc}}(\mathbf{R}^+; \mathbf{R})$ is of L^∞-anti-accretive type and that k is nonnegative and $k \notin L^1(\mathbf{R}^+; \mathbf{R})$. Show that k is of totally positive type.
26. Assume that $k \in L^1_{\mathrm{loc}}(\mathbf{R}^+; \mathbf{R})$ is of totally positive type and that $\int_0^\infty e^{-\sigma t} |k(t)|\, dt < \infty$ for all $\sigma > 0$. Show that k is of L^∞-anti-accretive type.
27. Show that the conclusion of Theorem 5.2 remains true (with some obvious modifications) if one adds a term of the type $\int_0^t g(x(s))\, ds$ to the left-hand side of (5.1), where g is continuous and satisfies $\langle v, g(v) \rangle \geq 0$ for all $v \in \mathbf{R}^n$.
28. Prove the following result.

6.1 Theorem

Assume the following.

(i) $k \in C(\mathbf{R}^+ \times \mathbf{R}^n; \mathbf{R}^n)$, *and for each fixed* $t \in \mathbf{R}^+$ *the function* $v \mapsto k(t, v)$ *is the gradient of a function* $v \mapsto K(t, v)$. *Moreover, for each* $v \in \mathbf{R}^n$ *the function* $t \mapsto k(t, v)$ *is locally absolutely continuous, and its derivative* $k'(t, v)$ *satisfies* $\int_0^T \sup_{|v| \leq M} |k'(t, v)|\, dt < \infty$ *for each* $T < \infty$ *and for each* $M < \infty$.

(ii) *The function* $v \mapsto -k'(t, v)$ *is monotone in* v *for all* $t \in \mathbf{R}^+$, *and the function* $v \mapsto k'(t, v) - k'(s, v)$ *is monotone in* v *for all* $0 \leq s \leq t < \infty$.

(iii) *For each* $v \in \mathbf{R}^n$ *the function* $t \mapsto K(t,v)$ *is nonnegative and nonincreasing.*
(iv) $f \in AC_{\mathrm{loc}}(\mathbf{R}^+;\mathbf{R}^n)$ *and* $f' \in L^2(\mathbf{R}^+;\mathbf{R}^n)$.
(v) $x \in C(\mathbf{R}^+;\mathbf{R}^n)$ *is a solution of (5.1).*

Then $x \in AC_{\mathrm{loc}}(\mathbf{R}^+;\mathbf{R}^n)$ *and* $x' \in L^2(\mathbf{R}^+;\mathbf{R}^n)$.

29. Show that the conclusion of Theorem 5.4 remains true (with some obvious modifications) if one adds a term of the type $\int_0^t g(x(s))\,ds$ to the left-hand side of (5.2), where g is the gradient of a function $G \in C^1(\mathbf{R}^n;\mathbf{R})$ that satisfies $\lim_{|y|\to\infty} G(y) = \infty$.
30. Under the assumptions of Lemma 5.1 (with the same assumptions on ψ and η as on φ), show that

$$\begin{aligned}\int_0^T \langle k(0,\varphi(t)),\psi(t)\rangle\,dt &+ \int_0^T\int_0^t \langle k'(t-s,\varphi(t)),\psi(s)\rangle\,ds\,dt\\ &= \int_0^T \Big(K\big(T-s,\eta(s)\big) - K(T-s,0)\Big)\,ds\\ &+ \int_0^T \Big(\langle k(s,\varphi(s)),\psi(s)\rangle - K(s,\eta(s)) + K(s,0)\Big)\,ds\\ &- \int_0^T\int_0^t \Big(K'\big(t-s,\eta(s)\big) - K'\big(t-s,\eta(t)\big)\\ &\qquad - \langle k'(t-s,\varphi(t)),\psi(s)-\psi(t)\rangle\Big)\,ds\,dt.\end{aligned}$$

31. Let
$$k(t,x) = \sum_{i=1}^{m} a_i(t) g_i(x),$$
where each $k_i : \mathbf{R}^+ \to \mathbf{R}^+$ is bounded, nonnegative, and nonincreasing, and each $g_i \in C(\mathbf{R};\mathbf{R})$ satisfies $vg_i(v) \geq 0$ for all $v \in \mathbf{R}$. Show that the conclusion of Theorem 5.2(i) remains valid for this class of kernels. Hint: Differentiate the equation, multiply by $\mathrm{sign}(x(t))$, and integrate. If k_i is locally absolutely continuous, then you can use the identity in Exercise 30 with $\psi(t) = \eta(t) = g_i(x(t))$, $\varphi(t) = \mathrm{sign}(x(t))$, $k(t,v) = a_i(t)v$, and $K(t,v) = a_i(t)|v|$. The same identity can be extended to the case where k is less smooth, if one replaces k' by a measure. (Cf. Levin [5].)

7. Comments

Section 2:
An early scalar version of Theorem 2.2 is found in Levin [3], and an extension to the vector case in Burton [1], Theorem 3, p. 397. Exercises 3,

4 and 6 are inspired by Levin [3], p. 185. Scalar nonconvolution kernels of positive type are considered in Kiffe [1] and [4], where, in addition, some asymptotic results for (2.12) can be found. An analysis of

$$x(t) + \int_0^t a(t,\tau)g(x(\tau))\,\mathrm{d}\tau = f(t), \quad t \in \mathbf{R}^+,$$

is given in Kiffe [2]. See also Staffans [27], Section 5, and the references mentioned there.

For scalar versions of Theorems 2.3 and 2.4, see Gripenberg [16], Theorems 1 and 2, respectively.

Sections 3–4:
The theory presented in Sections 3 and 4 has emerged gradually over the years. Roberts and Mann [1] discuss a scalar equation with a log-convex convolution kernel, and prove that the solution is nondecreasing. See also Padmavally [1]. In Levinson [1] the kernel $t^{-1/2}$ is studied, and a result analogous to Theorem 4.15 is proved. Extensions and additional monotonicity results are given in Friedman [1] and [2], and in Miller [4] where it is shown that if $k \in L^1(0,1)$, k is positive, continuous, and nonincreasing on $(0,\infty)$, and $k(t)/k(t+T)$ is nonincreasing on $(0,\infty)$ for each $T > 0$, then $0 \le r(t) \le k(t)$ and $\int_0^\infty r(t)\,\mathrm{d}t \le 1$. Miller [4] also analyses the existence of asymptotically periodic solutions. Kernels of type $k(t-s)b(s)$, with $k(t)$ log-convex, are discussed in Miller, Nohel, and Wong [1], Section 2c, and in Miller [10], p. 217. Conditions implying that the (convolution) resolvent $r(t)$ is positive and nonincreasing are given in Gripenberg [2]. See also Levin [9] for comments on log-convex kernels.

It follows from results in Gripenberg [24] that, if k is positive, nonincreasing, continuous, and locally but not globally integrable on $\mathbf{R}^+$, then k is log-convex if and only if the kernel k^μ is L^∞-anti-accretive for all $\mu \in (0,1]$. If k is integrable the claim must be slightly modified.

In a series of papers focused on existence, asymptotic behaviour and positivity of solutions of abstract Volterra equations, Clément [1], Clément and Nohel [1], [2] and Nohel [3], Section 4.2, have considered kernels satisfying $r_\lambda(t) \ge 0$ and $1 - \int_0^t r_\lambda(s)\,\mathrm{d}s \ge 0$ for $\lambda > 0$, $t \in \mathbf{R}^+$. In these papers, which have provided much of the impetus for Section 4, such kernels are called completely positive. In our terminology, these kernels are both of totally positive type and of L^∞-anti-accretive type.

For a convolution version of Theorem 4.11(iii), see Clément and Nohel [1], p. 371. One version of Theorem 4.12 is given in Clément and Nohel [1], Theorem 3. A combination of Theorems 4.16 and 4.17 was proved in Clément and Nohel [2], Theorem 2.2, and in Gripenberg [24], Theorem 2. Parts of this theorem are found in Gripenberg [12]. These results plays a crucial role in Prüss [2], which looks at hyperbolic Volterra equations in Banach spaces. In Prüss [2], Propositions 1 and 2, various properties for kernels k of both totally positive and L^∞-anti-accretive type are deduced. In particular, it is shown that $1/\lambda\hat{k}(\lambda)$ and $-\hat{k}'(\lambda)/\hat{k}(\lambda)^2$ are completely monotone on $(0, \infty)$. These properties are used to study wave propagation and regularity.

The variation of constants formula (4.24) (sometimes referred to as 'MacCamy's trick') was first used in MacCamy [3] in the case where $k(0) \neq 0$ and k is locally absolutely continuous to prove existence of solutions of an abstract equation. For further applications of this formula, see, e.g., Clément, MacCamy, and Nohel [1].

Section 5:
The key identities (5.3) and (5.9) were first used in the linear scalar case (i.e., the case where $k(t,v) = a(t)v$) in Londen [2] and [7]. In this case the identity in Exercise 30 is essentially equivalent to (5.3). In the case where $k(t,v) = a(t)g(v)$, the identities in Lemma 5.1 and Exercise 30 are found in Staffans [19], and Lemma 5.3 is given in Engler [1]. Techniques related to those used in the proof of the claims (ii) in Theorems 5.2 and 5.4 are to be found in Engler and Staffans [1], in Staffans [20], and in Londen [2] and [7]. The result in Exercise 31 is proved in Engler [2].

The scalar nonconvolution equation

$$x(t) + \int_0^t g\big(t, s, x(s)\big)\,\mathrm{d}s = f(t)$$

is investigated in Engler [3]. Equation (5.2) is discussed in Engler [5], where conditions are given which imply that solutions decay exponentially.

In some cases, the smoothness assumptions in Theorems 5.2 and 5.4 on the kernel and on the function f can be weakened considerably. See Engler and Staffans [1] and Staffans [20], which consider the equations

$$x'(t) + \int_0^t h(x(s))\,\mathrm{d}s + \sum_{k=1}^{m} \int_0^t a_k(t-s) g_k(x(s))\,\mathrm{d}s = f(t), \quad t \in \mathbf{R}^+,$$

and

$$x(t) + \int_0^t h(x(s))\,\mathrm{d}s + \sum_{k=1}^{m} \int_0^t a_k(t-s) g_k(x(s))\,\mathrm{d}s = f(t), \quad t \in \mathbf{R}^+,$$

respectively.

References

T. A. Burton

1. An integrodifferential equation, *Proc. Amer. Math. Soc.* **79** (1980), pp. 393–399.

Ph. Clément

1. On abstract Volterra equations with kernels having a positive resolvent, *Israel J. Math.* **36** (1980), pp. 193–200.

Ph. Clément, R. C. MacCamy, and J. A. Nohel

1. Asymptotic properties of solutions of nonlinear abstract Volterra equations, *J. Integral Equations* **3** (1981), pp. 185–216.

Ph. Clément and J. A. Nohel

1. Abstract linear and nonlinear Volterra equations preserving positivity, *SIAM J. Math. Anal.* **10** (1979), pp. 365–388.
2. Asymptotic behavior of solutions of nonlinear Volterra equations with completely positive kernels, *SIAM J. Math. Anal.* **12** (1981), pp. 514–535.

H. Engler

1. A version of the chain rule and integrodifferential equations in Hilbert spaces, *SIAM J. Math. Anal.* **13** (1982), pp. 801–810.
2. Bounds and asymptotics for a scalar Volterra integral equation, *J. Integral Equations* **7** (1984), pp. 209–227.
3. On nonlinear scalar Volterra integral equations. I, *Trans. Amer. Math. Soc.* **291** (1985), pp. 319–336.
4. A note on scalar Volterra integral equations. II, *J. Math. Anal. Appl.* **115** (1986), pp. 363–395.
5. Asymptotic properties of solutions of nonlinear Volterra integro-differential equations, *Resultate Math.* **13** (1980), pp. 65–80.

H. Engler and O. J. Staffans

1. On a Volterra integrodifferential equation with several nonlinearities, *Houston J. Math.* **11** (1985), pp. 299–306.

A. Friedman

1. On integral equations of Volterra type, *J. Analyse Math.* **11** (1963), pp. 381–413.
2. Periodic behavior of solutions of Volterra integral equations, *J. Analyse Math.* **15** (1965), pp. 287–303.

G. Gripenberg

2. On positive, nonincreasing resolvents of Volterra equations, *J. Differential Equations* **30** (1978), pp. 380–390.
12. On Volterra equations of the first kind, *Integral Equations Operator Theory* **3/4** (1980), pp. 473–488.
16. On some positive definite forms and Volterra integral operators, *Applicable Anal.* **11** (1981), pp. 211–222.
24. Volterra integral operators and logarithmic convexity, *Math. Scand.* **50** (1982), pp. 209–220.

T. Kiffe

1. On nonlinear Volterra equations of nonconvolution type, *J. Differential Equations* **22** (1976), pp. 349–367.
2. A Volterra equation with a nonconvolution kernel, *SIAM J. Math. Anal.* **8** (1977), pp. 938–949.
4. The asymptotic behavior of bounded solutions of a nonconvolution Volterra equation, *J. Differential Equations* **31** (1979), pp. 99–108.

J. J. Levin

3. A nonlinear Volterra equation not of convolution type, *J. Differential Equations* **4** (1968), pp. 176–186.
5. On a nonlinear Volterra equation, *J. Math. Anal. Appl.* **39** (1972), pp. 458–476.
9. Resolvents and bounds for linear and nonlinear Volterra equations, *Trans. Amer. Math. Soc.* **228** (1977), pp. 207–222.

N. Levinson

1. A nonlinear Volterra equation arising in the theory of superfluidity, *J. Math. Anal. Appl.* **1** (1960), pp. 1–11.

S-O. Londen

2. The qualitative behavior of the solutions of a nonlinear Volterra equation, *Michigan Math. J.* **18** (1971), pp. 321–330.
7. On a nonlinear Volterra integral equation, *J. Differential Equations* **14** (1973), pp. 106–120.

R. C. MacCamy

3. Stability theorems for a class of functional differential equations, *SIAM J. Appl. Math.* **30** (1976), pp. 557–576.

R. K. Miller

4. On Volterra integral equations with nonnegative integrable resolvents, *J. Math. Anal. Appl.* **22** (1968), pp. 319–340.
10. *Nonlinear Volterra Integral Equations*, W. A. Benjamin, Menlo Park, Calif., 1971.

R. K. Miller, J. A. Nohel, and J. S. W. Wong

1. Perturbations of Volterra integral equations, *J. Math. Anal. Appl.* **25** (1969), pp. 676–691.

J. A. Nohel

3. Nonlinear Volterra equations for heat flow in materials with memory, in *Integral and Functional Differential Equations*, T. L. Herdman, S. M. Rankin III, and H. W. Stech, eds., Lecture Notes in Pure and Applied Mathematics **67**, Marcel Dekker, New York, 1981, pp. 3–82.

K. Padmavally

1. On a non-linear integral equation, *J. Math. Mech.* **7** (1958), pp. 533–555.

J. Prüss

2. Positivity and regularity of hyperbolic Volterra equations in Banach spaces, *Math. Ann.* **279** (1987), pp. 317–344.

J. H. Roberts and W. R. Mann

1. On a certain nonlinear integral equation of the Volterra type, *Pacific J. Math.* **1** (1951), pp. 431–445.

O. J. Staffans

19. On the stability of a Volterra equation with a monotone nonlinearity, *J. Integral Equations* **7** (1984), pp. 239–248.
20. A note on a Volterra equation with several nonlinearities, *J. Integral Equations* **7** (1984), pp. 249–252.
27. A direct Lyapunov approach to Volterra integrodifferential equations, *SIAM J. Math. Anal.* **19** (1988), pp. 879–901.

E. M. Stein
1. *Singular Integrals and Differentiability Properties of Functions*, Princeton University Press, Princeton, 1970.

BIBLIOGRAPHY

The small numbers at the end of a reference give the pages where the reference in question is cited. A number ending in * indicates that the particular reference is referred to more than once on that page.

N. H. Abel

1. Résolution d'un problème de mécanique, in *Oeuvres* **1**, Christiania, 1881, pp. 97–101. 159, 166

R. R. Akhmerov and V. G. Kurbatov

1. Exponential dichotomy and stability of neutral type equations, *J. Differential Equations* **76** (1988), pp. 1–25.

Z. Akcasu, G. S. Lellouche, and L. S. Shotkin

1. *Mathematical Methods in Nuclear Reactor Dynamics*, Academic Press, New York, 1971. 9

V. Alexiades

1. Almost periodic solutions of an integrodifferential system with infinite delay, *Nonlinear Anal.* **5** (1981), pp. 401–410.

L. Amerio and G. Prouse

1. *Almost-Periodic Functions and Functional Equations*, Van Nostrand Reinhold, New York, 1971. 73

Z. Artstein

1. Continuous dependence of solutions of Volterra integral equations, *SIAM J. Math. Anal.* **6** (1975), pp. 446–456. 379, 422*
2. Continuous dependence on parameters: on best possible results, *J. Differential Equations* **19** (1975), pp. 214–225. 422

M. Ash

1. *Nuclear Reactor Kinetics*, McGraw–Hill, New York, 1965. 9

K. B. Athreya and P. E. Ney

1. *Branching Processes*, Springer-Verlag, Berlin, 1972. 6

K. E. Atkinson

1. An existence theorem for Abel integral equations, *SIAM J. Math. Anal.* **5** (1974), pp. 729–736. 166

F. V. Atkinson and J. R. Haddock

1. Criteria for asymptotic constancy of solutions of functional differential equations, *J. Math. Anal. Appl.* **91** (1983), pp. 410–423.

F. V. Atkinson, J. R. Haddock, and O. J. Staffans

1. Integral inequalities and exponential convergence of solutions of differential equations with bounded delay, in *Ordinary and Partial Differential Equations*, W. N. Everitt and B. D. Sleeman, eds., Proc. Dundee, Scotland 1982, Lecture Notes in Mathematics **964**, Springer-Verlag, Berlin, 1982, pp. 56–68.

A. Atzmon

1. Spectral synthesis of functions of bounded variation, *Proc. Amer. Math. Soc.* **47** (1975), pp. 417–422. 487

V. Barbu

1. Sur une équation intégrale non-linéaire, *An. Ştiinţ. "Al. I. Cuza" Iaşi Secţ. I a Mat (N.S.)* **10** (1964), pp. 61–65. 534, 560
2. *Nonlinear Semigroups and Differential Equations in Banach Spaces*, Noordhoff, Leyden, 1976.

V. Barbu and S. I. Grossman

1. Asymptotic behavior of linear integrodifferential systems, *Trans. Amer. Math. Soc.* **173** (1972), pp. 277–288. 220

R. Bellman and K. L. Cooke

1. *Differential-Difference Equations*, Academic Press, New York, 1963. 105*, 107*, 108*

J. J. Benedetto

1. *Spectral Synthesis*, Teubner, Stuttgart, 1975. 486, 487, 488
2. The Wiener spectrum in spectral synthesis, *Stud. Appl. Math.* **54** (1975), pp. 91–115. 488

J. B. Bennett

1. Stability properties of a Volterra integral equation, *J. Math. Anal. Appl.* **61** (1977), pp. 475–489.
2. Total stability for a Volterra integral equation, *Nonlinear Anal.* **5** (1981), pp. 615–623.

C. Berg and G. Forst

1. *Potential Theory on Locally Compact Abelian Groups*, Springer-Verlag, Berlin, 1975. 161*, 533*, 534*

S. R. Bernfeld and J. R. Haddock

1. Liapunov–Razumikhin functions and convergence of solutions of functional differential equations, *Applicable Anal.* **9** (1979), pp. 235–245.

C. Bernier and A. Manitius

1. On semigroups in $\mathbf{R}^n \times L^p$ corresponding to differential equations with delays, *Canad. J. Math.* **30** (1978), pp. 897–914. 220

S. Bernstein

1. Sur les fonctions absolument monotones, *Acta Math.* **51** (1928), pp. 1–66. 165

A. S. Besicovitch

1. *Almost Periodic Functions*, Cambridge University Press, Cambridge, 1932. 73

I. Bihari

1. A generalization of a lemma of Bellman and its application to uniqueness problems of differential equations, *Acta Math. Acad. Sci. Hungar.* **8** (1957), pp. 81–94.

N. H. Bingham, C. M. Goldie, and J. L. Teugels

1. *Regular Variation*, Cambridge University Press, Cambridge, 1987. 204, 458, 488

C. W. Bitzer

1. Stieltjes–Volterra integral equations, *Illinois J. Math.* **14** (1970), pp. 434–451. 309

S. Bochner
1. Monotone Funktionen, Stieltjessche Integrale und harmonische Analyse, *Math. Ann.* **108** (1933), pp. 378–410. 533*
2. Completely monotone functions of the Laplace operator for torus and sphere, *Duke Math. J.* **3** (1937), pp. 488–502. 165
3. Completely monotone functions in partially ordered spaces, *Duke Math. J.* **9** (1942), pp. 519–526. 165
4. *Lectures on Fourier Integrals*, Princeton University Press, Princeton, 1959. 533

H. Bohr
1. *Almost Periodic Functions*, Chelsea, New York, 1947. 73

J. G. Borisovič and A. S. Turbabin
1. On the Cauchy problem for linear non-homogeneous differential equations with retarded argument, *Soviet Math. Doklady* **10** (1969), pp. 401–405. 220

J. M. Bownds and J. M. Cushing
1. On strong stability for linear integral equations, *Math. Systems Theory* **7** (1973), pp. 193–200.

F. Brauer
1. On a nonlinear integral equation for population growth problems, *SIAM J. Math. Anal.* **6** (1975), pp. 312–317.
2. Constant rate harvesting of populations governed by Volterra integral equations, *J. Math. Anal. Appl.* **56** (1976), pp. 18–27.
3. Perturbations of the nonlinear renewal equation, *Adv. in Math.* **22** (1976), pp. 32–51.
4. Asymptotic stability of a class of integro-differential equations, *J. Differential Equations* **28** (1978), pp. 180–188. 105

W. E. Brumley
1. On the asymptotic behavior of solutions of differential-difference equations of neutral type, *J. Differential Equations* **7** (1970), pp. 175–188.
2. Periodic solutions of linear Volterra equations, *Funkcial. Ekvac.* **27** (1984), pp. 229–253.

J. A. Burns, E. M. Cliff, and T. L. Herdman
1. A state-space model for an aeroelastic system, in *Proc. of the* 22nd *IEEE Conference on Decision and Control 1983, San Antonio, Texas*, pp. 1074–1077. 12

J. A. Burns, E. M. Cliff, T. L. Herdman, and J. Turi (Burns *et al.*)
1. On integral transforms appearing in the derivation of the equations of an aeroelastic system, in *Nonlinear Analysis and Applications*, V. Lakshmikantham, ed., Lecture Notes in Pure and Applied Mathematics **109**, Marcel Dekker, New York, 1987, pp. 89–98. 12

J. A. Burns and T. L. Herdman
1. Adjoint semigroup theory for a Volterra integrodifferential system, *Bull. Amer. Math. Soc.* **81** (1975), pp. 1099–1102. 220, 221
2. Adjoint semigroup theory for a class of functional differential equations, *SIAM J. Math. Anal.* **7** (1976), pp. 729–745. 219, 220

J. A. Burns, T. L. Herdman, and H. W. Stech

1. Linear functional differential equations as semigroups on product spaces, *SIAM J. Math. Anal.* **14** (1983), pp. 98–116.

J. A. Burns, T. L. Herdman, and J. Turi

1. Well-posedness of functional differential equations with nonatomic D operators, in *Trends in the Theory and Practice of Non-Linear Analysis*, V. Lakshmikantham, ed., Proc. 6th Int. Conf., Arlington, Math. Stud. **110**, Elsevier, 1985, pp. 71–77. 12
2. Neutral functional integro-differential equations with weakly singular kernels, *J. Math. Anal. Appl.*, to appear 1989. 12

T. A. Burton

1. An integrodifferential equation, *Proc. Amer. Math. Soc.* **79** (1980), pp. 393–399. 648
2. *Volterra Integral and Differential Equations*, Academic Press, New York, 1983. 310, 339, 448
3. *Stability and Periodic Solutions of Ordinary and Functional Differential Equations*, Academic Press, New York, 1985.

T. A. Burton, Q. Huang, and W. E. Mahfoud

1. Liapunov functionals of convolution type, *J. Math. Anal. Appl.* **106** (1985), pp. 249–272. 309

P. J. Bushell and W. Okrasiński

1. Uniqueness of solutions for a class of non-linear Volterra integral equations with convolution kernel, *Math. Proc. Cambridge Philos. Soc*, to appear 1989. 423

R. H. Cameron and W. T. Martin

1. An unsymmetric Fubini theorem, *Bull. Amer. Math. Soc.* **47** (1941), pp. 121–125. 309

V. Capasso, E. Grosso, and S. L. Paveri-Fontana

1. *Mathematics in Biology and Medicine*, V. Capasso, E. Grosso and S. L. Paveri-Fontana, eds., Proceedings, Bari 1983, Lecture Notes in Biomathematics **57**, Springer-Verlag, Berlin, 1985. 7

R. W. Carr and K. B. Hannsgen

1. A nonhomogeneous integrodifferential equation in Hilbert space, *SIAM J. Math. Anal.* **10** (1979), pp. 961–984. 11, 186, 187
2. Resolvent formulas for a Volterra equation in Hilbert space, *SIAM J. Math. Anal.* **13** (1982), pp. 459–483. 11, 188

H.-Y. Chen

1. Solutions for certain nonlinear Volterra integral equations, *J. Math. Anal. Appl.* **69** (1979), pp. 475–488. 380

J. Chover and P. Ney

1. The non-linear renewal equation, *J. Analyse Math.* **21** (1968), pp. 381–413.

Ph. Clément

1. On abstract Volterra equations with kernels having a positive resolvent, *Israel J. Math.* **36** (1980), pp. 193–200. 649

Ph. Clément, O. Diekmann, M. Gyllenberg, H. J. A. M. Heijmans, and H. R. Thieme (Clément, Diekmann, *et al.*)

1. Perturbation theory for dual semigroups I. The sun-reflexive case, *Math. Ann.* **277** (1987), pp. 709–725. 221
2. Perturbation theory for dual semigroups II. Time-dependent perturbations in the sun-reflexive case, *Proc. Roy. Soc. Edinburgh Ser A.* **109** (1988), pp. 145–172. 221
3. Perturbation theory for dual semigroups III. Nonlinear Lipschitz continuous perturbations in the sun-reflexive case, in *Volterra Integrodifferential Equations in Banach Spaces and Applications*, G. da Prato and M. Iannelli, eds., Proc., Trento 1987, Research Notes in Mathematics, Pitman, London, to appear 1989. 221
4. Perturbation theory for dual semigroups IV. The intertwining formula and the canonical pairing, in *Trends in Semigroup Theory and Applications*, Ph. Clément, S. Invernizzi, E. Mitidieri, and I. I. Vrabie, eds., Proc., Trieste, 1987, Marcel Dekker, New York, 1989, pp. 95–116. 221
5. A Hille-Yosida theorem for a class of weakly* continuous semigroups, *Semigroup Forum* **38** (1989), pp. 157–178. 221

Ph. Clément, H. J. A. M. Heijmans, S. Angenent, C. J. van Duijn, and B. de Pagter (Clément, Heijmans, *et al.*)

1. *One-Parameter Semigroups*, North-Holland, Amsterdam, 1987. 221

Ph. Clément, R. C. MacCamy, and J. A. Nohel

1. Asymptotic properties of solutions of nonlinear abstract Volterra equations, *J. Integral Equations* **3** (1981), pp. 185–216. 650

Ph. Clément and J. A. Nohel

1. Abstract linear and nonlinear Volterra equations preserving positivity, *SIAM J. Math. Anal.* **10** (1979), pp. 365–388. 649*
2. Asymptotic behavior of solutions of nonlinear Volterra equations with completely positive kernels, *SIAM J. Math. Anal.* **12** (1981), pp. 514–535. 649*

B. D. Coleman and V. J. Mizel

1. Norms and semi-groups in the theory of fading memory, *Arch. Ratl. Mech. Anal.* **23** (1966), pp. 87–123. 137, 221
2. On the general theory of fading memory, *Arch. Ratl. Mech. Anal.* **29** (1968), pp. 18–31. 137, 221
3. On the stability of solutions of functional-differential equations, *Arch. Ratl. Mech. Anal.* **30** (1968), pp. 173–196. 221

F. Colonius, A. Manitius, and D. Salamon

1. Structure theory and duality for time varying retarded functional differential equations, *J. Differential Equations* **78** (1989), pp. 320–353. 310

J. L. B. Cooper

1. Positive definite functions of a real variable, *Proc. London Math. Soc. (3)* **10** (1960), pp. 53–66. 532, 533, 534

C. Corduneanu

1. Sur une équation intégrale de la théorie du réglage automatique, *C. R. Acad. Sci. Paris* **256** (1963), pp. 3564–3567. 559, 586
2. Sur une équation intégrale non-linéaire, *An. Ştiinţ. "Al. I. Cuza" Iaşi Secţ. I a Mat (N.S.)* **9** (1963), pp. 369–375.
3. Problémes globaux dans la théorie des équations intégrales de Volterra, *Ann. Mat. Pura Appl. (IV)* **67** (1965), pp. 349–363.
4. Sur certaines équations fonctionnelles de Volterra, *Funkcial. Ekvac.* **9** (1966), pp. 119–127.
5. *Almost Periodic Functions*, Interscience, New York, 1968. 73
6. Admissibility with respect to an integral operator and applications, *SIAM Stud. Appl. Math.* **5** (1969), pp. 55–63.
7. Stability of some linear time-varying systems, *Math. Systems Theory* **3** (1969), pp. 151–155.
8. Some differential equations with delay, in *Proceedings of Equadiff III, 1972*, M. Ráb and J. Vosmanský, eds., Czechoslovak Conference on Differential Equations and their Applications, J. E. Purkyně University, Brno, 1973, pp. 105–114. 108
9. *Integral Equations and Stability of Feedback Systems*, Academic Press, New York, 1973. 278, 534, 559, 586
10. *Integral Equations and Applications*, Cambridge University Press, Cambridge, to appear 1990. 14

C. Corduneanu and V. Lakshmikantham

1. Equations with unbounded delay: a survey, *Nonlinear Anal.* **4** (1980), pp. 831–877.

C. Corduneanu and N. Luca

1. The stability of some feedback systems with delay, *J. Math. Anal. Appl.* **51** (1975), pp. 377–393.

T. L. Cromer

1. Asymptotically periodic solutions to Volterra integral equations in epidemic models, *J. Math. Anal. Appl.* **110** (1985), pp. 483–494.

K. S. Crump

1. On systems of renewal equations, *J. Math. Anal. Appl.* **30** (1970), pp. 425–434.

J. M. Cushing

1. An operator equation and bounded solutions of integro-differential systems, *SIAM J. Math. Anal.* **6** (1975), pp. 433–445. 338
2. Bounded solutions of perturbed Volterra integrodifferential systems, *J. Differential Equations* **20** (1976), pp. 61–70. 339
3. Strong stability of perturbed systems of Volterra integral equations, *Math. Systems Theory* **7** (1973), pp. 360–366.
4. *Integrodifferential equations and delay models in population dynamics*, Springer Lect. Notes in Biomathematics 20, Berlin, 1977.

R. Datko

1. Representation of solutions and stability of linear differential-difference equations in a Banach space, *J. Differential Equations* **29** (1978), pp. 105–166.

A. E. Degance and L. E. Johns

1. The stability of vector renewal equations pertaining to heterogeneous chemical reaction systems, *Quart. Appl. Math.* **34** (1976), pp. 69–83.

K. Deimling

1. Eigenschaften der Lösungsmenge eines Systems von Volterra-Integralgleichungen, *Manuscripta Math.* **4** (1971), pp. 201–212. 422
2. *Nonlinear Functional Analysis*, Springer-Verlag, Berlin, 1985. 346, 515

M. C. Delfour

1. The largest class of hereditary systems defining a C_0 semigroup on the product space, *Canad. J. Math.* **32** (1980), pp. 969–978. 220
2. Status of the state space theory of linear hereditary differential systems with delays in state and control variables, in *Analysis and Optimization of Systems*, A. Bensoussan and J. L. Lions, eds., Proc. of the 4th Internat. Conf. on Analysis and Optimization of Systems, Lecture Notes in Control and Information Sciences **28**, Springer-Verlag, Berlin, 1980, pp. 83–96. 220

M. C. Delfour and A. Manitius

1. The structural operator F and its role in the theory of retarded systems I–II, *J. Math. Anal. Appl.* **73–74** (1980), pp. 466–490,359–381. 220

O. Diekmann

1. Limiting behaviour in an epidemic model, *Nonlinear Anal.* **1** (1977), pp. 459–470. 7
2. Thresholds and travelling waves for the geographical spread of infection, *J. Math. Biol.* **6** (1978), pp. 109–130. 6, 7
3. *Volterra integral equations and semigroups of operators*, Report TW 197, Stichting Mathematisch Centrum, Amsterdam, 1980. 220*

O. Diekmann and S. A. van Gils

1. Invariant manifolds of Volterra integral equations of convolution type, *J. Differential Equations* **54** (1984), pp. 139–180. 220

O. Diekmann and H. G. Kaper

1. On the bounded solutions of a nonlinear convolution equation, *Nonlinear Anal.* **2** (1978), pp. 721–737. 7

G. Doetsch

1. *Handbuch der Laplace-Transformation. I–III*, Birkhäuser, Basel, 1950–1956. 72, 108, 193

W. F. Donoghue

1. *Distributions and Fourier Transforms*, Academic Press, New York, 1969. 535

R. Doss

1. On the almost periodic solutions of a class of integro-differential-difference equations, *Ann. of Math. (2)* **81** (1965), pp. 117–123.

R. D. Driver

1. *Ordinary and Delay Differential Equations*, Springer-Verlag, Berlin, 1977. 448

N. Dunford and J. T. Schwartz

1. *Linear Operators, Part I: General Theory*, J. Wiley, New York, 1957. 277, 373*

H. Engler

1. A version of the chain rule and integrodifferential equations in Hilbert spaces, *SIAM J. Math. Anal.* **13** (1982), pp. 801–810. 587, 650
2. Bounds and asymptotics for a scalar Volterra integral equation, *J. Integral Equations* **7** (1984), pp. 209–227. 650
3. On nonlinear scalar Volterra integral equations. I, *Trans. Amer. Math. Soc.* **291** (1985), pp. 319–336. 650
4. A note on scalar Volterra integral equations. II, *J. Math. Anal. Appl.* **115** (1986), pp. 363–395. 650
5. Asymptotic properties of solutions of nonlinear Volterra integro-differential equations, *Resultate Math.* **13** (1980), pp. 65–80. 650

H. Engler and O. J. Staffans

1. On a Volterra integrodifferential equation with several nonlinearities, *Houston J. Math.* **11** (1985), pp. 299–306. 650*

W. Feller

1. On the integral equation of renewal theory, *Ann. Math. Statist.* **12** (1941), pp. 243–267. 6
2. *An Introduction to Probability Theory and its Applications, II,* 2nd *ed.*, J. Wiley, New York, 1971. 204, 488

I. Fenyő and H. W. Stolle

1. *Theorie und Praxis der linearen Integralgleichungen, 1–4*, Birkhäuser, Basel, 1982–1984. 73, 277, 278, 279, 309

A. F. Filippov

1. *Differential Equations with Discontinuous Righthand Sides*, Kluwer Academic Publishers, Dordrecht, 1988. 380

A. M. Fink

1. *Almost Periodic Differential Equations*, Springer-Verlag, Berlin, 1974. 73

A. M. Fink and W. R. Madych

1. On certain bounded solutions of $g * \mu = f$, *Proc. Amer. Math. Soc.* **75** (1979), pp. 235–242.

A. Friedman

1. On integral equations of Volterra type, *J. Analyse Math.* **11** (1963), pp. 381–413. 165*, 649*
2. Periodic behavior of solutions of Volterra integral equations, *J. Analyse Math.* **15** (1965), pp. 287–303. 649
3. Monotonicity of solutions of Volterra integral equations in Banach space, *Trans. Amer. Math. Soc.* **138** (1969), pp. 129–148.

A. Friedman and M. Shinbrot

1. Volterra integral equations in Banach space, *Trans. Amer. Math. Soc.* **126** (1967), pp. 131–179.

I. M. Gel′fand

1. Über absolut konvergente trigonometrische Reihen und Integrale, *Mat. Sb.* **9** (1941), pp. 51–66. 136, 137*

I. M. Gel′fand, D. A. Raikov, and G. E. Shilov

1. *Commutative Normed Rings*, Chelsea, New York, 1964. 126, 136, 137, 199, 200

S. A. van Gils

1. *Linear Volterra convolution equations: semigroups, small solutions and convergence of projection operators*, Report TW 248, Stichting Mathematisch Centrum, Amsterdam 1983. 220*

I. Gohberg, P. Lancaster, and L. Rodman

1. *Matrix Polynomials*, Academic Press, New York, 1982.

È. I. Gol′dengershel′

1. Spectrum of a Volterra operator on the half-line and Tauberian theorems of the Paley–Wiener type, *Sibirsk. Math. J.* **20** (1980), pp. 364–370.

H. E. Gollwitzer

1. Admissibility and integral operators, *Math. Systems Theory* **7** (1973), pp. 219–231. 278

H. E. Gollwitzer and R. A. Hager

1. The nonexistence of maximum solutions of Volterra integral equations, *Proc. Amer. Math. Soc.* **26** (1970), pp. 301–304. 423

R. Gorenflo and S. Vessella

1. *Basic theory and some applications of Abel integral equations*, Preprint 237/1986, Fachbereich Mathematik, Freie Universität, Berlin. 166

S. R. Grace and B. S. Lalli

1. Asymptotic behaviour of certain second order integro-differential equations, *J. Math. Anal. Appl.* **76** (1980), pp. 84–90.

R. Grimmer

1. Existence of periodic solutions of functional differential equations, *J. Math. Anal. Appl.* **72** (1979), pp. 666–673.
2. Resolvent operators for integral equations in a Banach space, *Trans. Amer. Math. Soc.* **273** (1982), pp. 333–349.

R. Grimmer and G. Seifert

1. Stability properties of Volterra integrodifferential equations, *J. Differential Equations* **19** (1975), pp. 142–166. 310, 338

G. Gripenberg

1. Bounded solutions of a Volterra equation, *J. Differential Equations* **28** (1978), pp. 18–22. 448
2. On positive, nonincreasing resolvents of Volterra equations, *J. Differential Equations* **30** (1978), pp. 380–390. 165*, 649*
3. *On Volterra equations with nonconvolution kernels*, Report-HTKK-MAT-A118, Helsinki University of Technology, 1978. 279*
4. A Volterra equation with nonintegrable resolvent, *Proc. Amer. Math. Soc.* **73** (1979), pp. 57–60. 187
5. On the boundedness of solutions of Volterra equations, *Indiana Univ. Math. J.* **28** (1979), pp. 279–290.
6. On rapidly decaying resolvents of Volterra equations, *J. Integral Equations* **1** (1979), pp. 241–247. 138
7. On a frequency domain condition used in the theory of Volterra equations, *SIAM J. Math. Anal.* **10** (1979), pp. 839–843. 534
8. On the asymptotic behavior of resolvents of Volterra equations, *SIAM J. Math. Anal.* **11** (1980), pp. 654–662. 187
9. On nonlinear Volterra equations with nonintegrable kernels, *SIAM J. Math. Anal.* **11** (1980), pp. 668–682. 612

10. On the resolvents of nonconvolution Volterra kernels, *Funkcial. Ekvac.* **23** (1980), pp. 83–95. 278*, 279*
11. On the resolvents of Volterra equations with nonincreasing kernels, *J. Math. Anal. Appl.* **76** (1980), pp. 134–145. 73, 187
12. On Volterra equations of the first kind, *Integral Equations Operator Theory* **3/4** (1980), pp. 473–488. 165*, 166*, 649*
13. Stability problems for some nonlinear Volterra equations, *J. Integral Equations* **2** (1980), pp. 247–258. 586, 587
14. On the behavior of solutions of Volterra equations with nonconvolution kernels, *J. Integral Equations* **3** (1981), pp. 83–92.
15. Unique solutions of some Volterra integral equations, *Math. Scand.* **48** (1981), pp. 59–67. 423
16. On some positive definite forms and Volterra integral operators, *Applicable Anal.* **11** (1981), pp. 211–222. 649
17. Integrability of resolvents of systems of Volterra equations, *SIAM J. Math. Anal.* **12** (1981), pp. 585–594. 186, 187
18. Asymptotic solutions of some nonlinear Volterra integral equations, *SIAM J. Math. Anal.* **12** (1981), pp. 595–602.
19. A Tauberian problem for a Volterra integral operator, *Proc. Amer. Math. Soc.* **82** (1981), pp. 576–582.
20. On some epidemic models, *Quart. Appl. Math.* **39** (1981), pp. 317–327. 7, 440, 449
21. On the convergence of solutions of Volterra equations to almost-periodic functions, *Quart. Appl. Math.* **39** (1981), pp. 363–373. 10
22. Decay estimates for resolvents of Volterra equations, *J. Math. Anal. Appl.* **85** (1982), pp. 473–487.
23. Stability of periodic solutions of some integral equations, *J. Reine Angew. Math.* **331** (1982), pp. 16–31.
24. Volterra integral operators and logarithmic convexity, *Math. Scand.* **50** (1982), pp. 209–220. 649*
25. Asymptotic estimates for resolvents of Volterra equations, *J. Differential Equations* **46** (1982), pp. 230–243. 279
26. Two Tauberian theorems for nonconvolution Volterra integral operators, *Proc. Amer. Math. Soc.* **89** (1983), pp. 219–225.
27. An estimate for the solution of a Volterra equation describing an epidemic, *Nonlinear Anal.* **7** (1983), pp. 161–165. 7
28. An eigenvalue problem for a Volterra integral operator with infinite delay, *SIAM J. Math. Anal.* **16** (1985), pp. 541–547.

S. I. Grossman and R. K. Miller

1. Perturbation theory for Volterra integrodifferential systems, *J. Differential Equations* **8** (1970), pp. 457–474. 107, 309, 338, 339
2. Nonlinear Volterra integrodifferential systems with L^1-kernels, *J. Differential Equations* **13** (1973), pp. 551–566. 108

M. E. Gurtin and R. C. MacCamy

1. Nonlinear age-dependent population dynamics, *Arch. Ratl. Mech. Anal.* **54** (1974), pp. 281–300. 6

M. Gyllenberg

1. A note on continuous dependence of solutions of Volterra integral equations, *Proc. Amer. Math. Soc.* **81** (1981), pp. 546–548. 422

J. R. Haddock

1. Some new results on stability and convergence of solutions of ordinary and functional differential equations, *Funkcial. Ekvac.* **19** (1976), pp. 247–269.

J. R. Haddock and T. Krisztin

1. Estimates regarding the decay of solutions of functional differential equations, *Nonlinear Anal.* **8** (1984), pp. 1395–1408.
2. On the rate of decay of solutions of functional differential equations with infinite delay, *Nonlinear Anal.* **10** (1986), pp. 727–742.

J. R. Haddock and J. Terjéki

1. Liapunov–Razumikhin functions and an invariance principle for functional differential equations, *J. Differential Equations* **48** (1983), pp. 95–122. 448

A. Halanay

1. On the asymptotic behavior of the solutions of an integro-differential equation, *J. Math. Anal. Appl.* **10** (1965), pp. 319–324. 534, 559*
2. *Differential Equations, Stability, Oscillations, Time Lags*, Academic Press, New York, 1966. 108

J. K. Hale

1. Linear functional differential equations with constant coefficients, in *Contributions to Differential Equations* **2**, a serial issued under the auspices of RIAS and the University of Maryland, J. Wiley, New York, 1963, pp. 291–319. 220
2. Dynamical systems and stability, *J. Math. Anal. Appl.* **26** (1969), pp. 39–59. 137
3. *Ordinary Differential Equations*, J. Wiley, New York, 1969. 203, 330, 426*
4. Functional differential equations with infinite delays, *J. Math. Anal. Appl.* **48** (1974), pp. 276–283. 220
5. *Theory of Functional Differential Equations*, Springer-Verlag, Berlin, 1977. 105, 107*, 108*, 137*, 203*, 220*, 221*, 309*, 380*, 448*
6. *Asymptotic Behavior of Dissipative Systems*, Amer. Math. Soc., Providence, R. I., 1988. 486

J. K. Hale, E. F. Infante, and F–S. P. Tsen

1. Stability in linear delay equations, *J. Math. Anal. Appl.* **105** (1985), pp. 533–555.

J. K. Hale and J. Kato

1. Phase space for retarded equations with infinite delay, *Funkcial. Ekvac.* **21** (1978), pp. 11–41. 137, 220

J. K. Hale and K. R. Meyer

1. *A Class of Functional Equations of Neutral Type*, Mem. Amer. Math. Soc. **76**, Amer. Math. Soc., Providence, R. I., 1967. 137

K. B. Hannsgen

1. Indirect Abelian theorems and a linear Volterra equation, *Trans. Amer. Math. Soc.* **142** (1969), pp. 539–555. 106, 150, 184, 534
2. On a nonlinear Volterra equation, *Michigan Math. J.* **16** (1969), pp. 365–376. 184, 559
3. A Volterra equation with completely monotonic convolution kernel, *J. Math. Anal. Appl.* **31** (1970), pp. 459–471. 166
4. A Volterra equation with parameter, *SIAM J. Math. Anal.* **4** (1973), pp. 22–30. 184, 188
5. Note on a family of Volterra equations, *Proc. Amer. Math. Soc.* **46** (1974), pp. 239–243. 184, 187
6. A Volterra equation in Hilbert space, *SIAM J. Math. Anal.* **5** (1974), pp. 412–416. 188
7. A linear Volterra equation in Hilbert space, *SIAM J. Math. Anal.* **5** (1974), pp. 927–940. 188
8. Uniform boundedness in a class of Volterra equations, *SIAM J. Math. Anal.* **6** (1975), pp. 689–697. 188
9. Continuous parameter dependence in a class of Volterra equations, *SIAM J. Math. Anal.* **7** (1976), pp. 45–58. 188
10. The resolvent kernel of an integrodifferential equation in Hilbert space, *SIAM J. Math. Anal.* **7** (1976), pp. 481–490. 188
11. Uniform L^1 behavior for an integrodifferential equation with parameter, *SIAM J. Math. Anal.* **8** (1977), pp. 626–639. 5, 188
12. An L^1 remainder theorem for an integrodifferential equation with asymptotically periodic solution, *Proc. Amer. Math. Soc.* **73** (1979), pp. 331–337. 188
13. An integrodifferential equation asymptotically of convolution type, *Proc. Amer. Math. Soc.* **74** (1979), pp. 71–78.
14. A Wiener–Lévy theorem for quotients with applications to Volterra equations, *Indiana Univ. Math. J.* **29** (1980), pp. 103–120.
15. A uniform approximation for an integrodifferential equation with parameter, *J. Integral Equations* **2** (1980), pp. 117–131. 5, 188

K. B. Hannsgen and C. Shilepsky

1. A boundedness theorem for Volterra equations, *J. Differential Equations* **10** (1971), pp. 378–387.

K. B. Hannsgen and R. L. Wheeler

1. Complete monotonicity and resolvents of Volterra integrodifferential equations, *SIAM J. Math. Anal.* **13** (1982), pp. 962–969. 165
2. Behavior of the solution of a Volterra equation as a parameter tends to infinity, *J. Integral Equations* **7** (1984), pp. 229–237. 188
3. A singular limit problem for an integrodifferential equation, *J. Integral Equations* **5** (1983), pp. 199–209. 165, 188*
4. Uniform L^1-behavior in classes of integrodifferential equations with completely monotone kernels, *SIAM J. Math. Anal.* **15** (1984), pp. 579–594. 165
5. Time delays and boundary feedback stabilization in one-dimensional viscoelasticity, in *Proc. Conf. Control of Distributed Parameter Systems*, Vorau, 1986. 11

F. B. Hanson, A. Klimas, G. V. Ramanathan, and G. Sandri (Hanson *et al.*)

1. Uniformly valid asymptotic solution to a Volterra equation on an infinite interval, *J. Math. Phys.* **14** (1973), pp. 1592–1600. 4, 5
2. Analysis of a model for transport of charged particles in a random magnetic field, *J. Math. Anal. Appl.* **44** (1973), pp. 786–798. 4, 5*

G. H. Hardy and J. E. Littlewood

1. Some new properties of Fourier constants, *Math. Ann.* **97** (1926), pp. 159–209. 188

N. D. Hayes

1. Roots of the transcendental equation associated with a certain difference-differential equation, *J. London Math. Soc.* **25** (1950), pp. 226–232. 105

D. Henry

1. The adjoint of a linear functional differential equation and boundary value problems, *J. Differential Equations* **9** (1971), pp. 55–66. 221
2. Linear autonomous neutral functional differential equations, *J. Differential Equations* **15** (1974), pp. 106–128. 220

T. L. Herdman

1. Behavior of maximally defined solutions of a nonlinear Volterra equation, *Proc. Amer. Math. Soc.* **67** (1977), pp. 297–302. 380

H. W. Hethcote, M. A. Lewis, and P. Van den Driessche

1. An epidemiological model with a delay and a nonlinear incidence rate, *J. Math. Biol.* **27** (1989), pp. 49–64. 7

H. W. Hethcote, H. W. Stech, and P. Van den Driessche

1. Periodicity and stability in epidemic models: a survey, in *Differential equations and applications in ecology, epidemics, and population problems*, K. L. Cooke, ed., Proc. Conf., Claremont/Calif. 1981, Academic Press, New York, 1981, pp. 65–82. 7

E. Hewitt and K. A. Ross

1. *Abstract Harmonic Analysis, 1–2*, Springer-Verlag, Berlin, 1963–1970. 108, 137

E. Hewitt and K. Stromberg

1. *Real and Abstract Analysis*, Springer-Verlag, Berlin, 1965. xiii, 64, 73, 90, 91, 92, 108*

E. Hille and R. S. Phillips

1. *Functional Analysis and Semi-Groups*, Amer. Math. Soc., Providence, R. I., 1957. xiii, 124, 137*, 211*

E. Hille and J. D. Tamarkin

1. On the absolute integrability of Fourier transforms, *Fund. Math.* **25** (1935), pp. 329–352. 183, 188

D. B. Hinton

1. A Stieltjes–Volterra integral equation theory, *Canad. J. Math.* **18** (1966), pp. 314–331. 309

H. Hochstadt

1. *Integral Equations*, J. Wiley, New York, 1973.

F. Holland

1. Harmonic analysis on amalgams of L^p and l^q, *J. London Math. Soc.(2)* **10** (1975), pp. 295–305. 74

W. J. Hrusa and J. A. Nohel

1. The Cauchy problem in one-dimensional nonlinear viscoelasticity, *J. Differential Equations* **59** (1985), pp. 388–412. 534

W. J. Hrusa, J. A. Nohel, and M. Renardy

1. Initial value problems in viscoelasticity, *Appl. Mech. Rev.* **41** (1988), pp. 371–378. 10

W. J. Hrusa and M. Renardy

1. On a class of quasilinear partial integrodifferential equations with singular kernels, *J. Differential Equations* **64** (1986), pp. 195–220. 534
2. A model equation for viscoelasticity with a strongly singular kernel, *SIAM J. Math. Anal.* **19** (1988), pp. 257-269. 534

I. A. Ibragimov

1. A remark on the ergodic theorem for Markov chains, *Theor. Probability Appl.* **20** (1975), pp. 174–176. 187

M. N. Islam

1. Periodic solutions of nonlinear integral equations, *Ann. Mat. Pura Appl. (IV)* **150** (1988), pp. 129–139.

G. S. Jordan

1. Asymptotic stability of a class of integrodifferential systems, *J. Differential Equations* **31** (1979), pp. 359–365. 108

G. S. Jordan, O. J. Staffans, and R. L. Wheeler

1. Local analyticity in weighted L^1-spaces and applications to stability problems for Volterra equations, *Trans. Amer. Math. Soc.* **274** (1982), pp. 749–782. 138, 185, 187, 199, 200*, 204*
2. Convolution operators in a fading memory space: the critical case, *SIAM J. Math. Anal.* **18** (1987), pp. 366–386. 204*
3. Subspaces of stable and unstable solutions of a functional differential equation in a fading memory space: the critical case, *SIAM J. Math. Anal.* **18** (1987), pp. 1323–1340. 204*

G. S. Jordan and R. L. Wheeler

1. On the asymptotic behavior of perturbed Volterra integral equations, *SIAM J. Math. Anal.* **5** (1974), pp. 273–277. 279
2. Linear integral equations with asymptotically almost periodic solutions, *J. Math. Anal. Appl.* **52** (1975), pp. 454–464.
3. A generalization of the Wiener–Lévy theorem applicable to some Volterra equations, *Proc. Amer. Math. Soc.* **57** (1976), pp. 109–114. 187
4. Asymptotic behavior of unbounded solutions of linear Volterra integral equations, *J. Math. Anal. Appl.* **55** (1976), pp. 596–615. 203
5. Structure of resolvents of Volterra integral and integrodifferential systems, *SIAM J. Math. Anal.* **11** (1980), pp. 119–132. 203
6. Rates of decay of resolvents of Volterra equations with certain nonintegrable kernels, *J. Integral Equations* **2** (1980), pp. 103–110. 5, 138
7. Weighted L^1-remainder theorems for resolvents of Volterra equations, *SIAM J. Math. Anal.* **11** (1980), pp. 885–900. 203

K. Jörgens

1. *Linear Integral Operators*, Pitman, London, 1982. 277, 279

R. P. Kanwal

1. *Linear Integral Equations*, Academic Press, New York, 1971.

J. L. Kaplan

1. On the asymptotic behavior of Volterra integral equations, *SIAM J. Math. Anal.* **3** (1972), pp. 148–156. 279*, 338*

F. Kappel

1. *Laplace-transform methods and linear autonomous functional-differential equations*, Berichte der Mathematisch-statistischen Sektion im Forschungszentrum Graz, Report nr. 64, 1976. 203
2. *Linear autonomous functional differential equations in the state space C*, Technical Report 34, Technische Universität Graz, Universität Graz, 1984. 203, 220

F. Kappel and K. Kunish

1. Invariance results for delay and Volterra equations in fractional order Sobolev spaces, *Trans. Amer. Math. Soc.* **304** (1987), pp. 1–51. 73

F. Kappel and W. Schappacher

1. Autonomous nonlinear functional differential equations and averaging approximations, *Nonlinear Anal.* **2** (1978), pp. 391–422.
2. Some considerations to the fundamental theory of infinite delay equations, *J. Differential Equations* **37** (1980), pp. 141–183. 137, 220

F. Kappel and H. K. Wimmer

1. An elementary divisor theory for autonomous linear functional differential equations, *J. Differential Equations* **21** (1976), pp. 134–147. 203

F. Kappel and K. P. Zhang

1. A neutral functional differential equations with nonatomic difference operator, *J. Math. Anal. Appl.* **113** (1986), pp. 311–343.
2. Equivalence of functional-differential equations of neutral type and abstract Cauchy problems, *Monatsh. Math* **101** (1986), pp. 115–133.

G. Karakostas

1. On a Staffans type inequality for vector Volterra integro-differential equations, *Colloq. Math.* **52** (1987), pp. 313–317. 559

Y. Katznelson

1. *An Introduction to Harmonic Analysis*, Dover Publications, New York, 1976. 72, 102, 108, 472, 486*, 487*

W. G. Kelley

1. A Kneser theorem for Volterra integral equations, *Proc. Amer. Math. Soc.* **40** (1973), pp. 183–190. 422

W. O. Kermack and A. G. McKendrick

1. A contribution to the mathematical theory of epidemics, *Proc. Roy. Soc. London Ser. A* **115** (1927), pp. 700–721. 7

T. Kiffe

1. On nonlinear Volterra equations of nonconvolution type, *J. Differential Equations* **22** (1976), pp. 349–367. 649
2. A Volterra equation with a nonconvolution kernel, *SIAM J. Math. Anal.* **8** (1977), pp. 938–949. 649
3. A discontinuous Volterra integral equation, *J. Integral Equations* **1** (1979), pp. 193–200. 380
4. The asymptotic behavior of bounded solutions of a nonconvolution Volterra equation, *J. Differential Equations* **31** (1979), pp. 99–108. 649
5. Systems of Volterra equations with nonmonotone discontinuous nonlinearities, *J. Integral Equations* **5** (1983), pp. 341–352. 380, 381

N. Kikuchi and S. Nakagiri

1. An existence theorem of solutions of non-linear integral equations, *Funkcial. Ekvac.* **15** (1972), pp. 131–138. 380, 422
2. Kneser's property of solutions of non-linear integral equations, *Funkcial. Ekvac.* **17** (1974), pp. 57–66. 380

A. J. Klimas and G. Sandri

1. Foundation of the theory of cosmic-ray transport in random magnetic fields, *Astrophys. J.* **169** (1971), pp. 41–56. 4, 5

V. B. Kolmanovskiĭ and V. R. Nosov

1. *Stability of Functional Differential Equations*, Academic Press, London, 1986. 448

P. Koosis

1. *Introduction to H_p Spaces*, Cambridge University Press, Lecture Note Series 40, Cambridge, 1980. 182

M. A. Krasnosel′skiĭ, P. P. Zabreĭko, E. I. Pustyl′nik, and P. E. Sobolevskiĭ (Krasnosel′skiĭ *et al.*)

1. *Integral Operators in Spaces of Summable Functions*, Noordhoff, Leyden, 1976. 254*, 255*, 277*, 279*, 367*

N. N. Krasovskiĭ

1. *Stability of Motion*, Stanford University Press, Stanford, California, 1963 [translation of *Nekotorye zadachi teoriĭ ustoĭchivosti dvizheniya*, Gos. Fiz. Mat. Lit., Moscow, 1959.] 203, 220

T. Krisztin

1. On the convergence of solutions of functional differential equations with infinite delay, *J. Math. Anal. Appl.* **109** (1985), pp. 509–521.
2. On the rate of convergence of solutions of functional differential equations, *Funkcial. Ekvac.* **29** (1986), pp. 1–10.

T. Krisztin and J. Terjéki

1. On the rate of convergence of solutions of linear Volterra equations, *Boll. Un. Mat. Ital. B (7)* **2** (1988), pp. 427–444. 203

S. M. Kuen and K. P. Rybakowski

1. Boundedness of solutions of a system of integro-differential equations, *J. Math. Anal. Appl.* **112** (1985), pp. 378–390. 559

V. G. Kurbatov

1. Stability of functional-differential equations, *Differencial'nye Uravenija* **17** (1981), pp. 963–972.

G. S. Ladde, V. Lakshmikantham, and B. G. Zhang

1. *Oscillation Theory of Differential Equations with Deviating Arguments*, Marcel Dekker, New York, 1987.

V. Lakshmikantham and S. Leela

1. *Differential and Integral Inequalities, I, II*, Academic Press, New York, 1969.

C. E. Langenhop

1. Periodic and almost periodic solutions of Volterra integral differential equations with infinite memory, *J. Differential Equations* **58** (1985), pp. 391–403. 108

M. J. Leitman and V. J. Mizel

1. On fading memory spaces and hereditary integral equations, *Arch. Ratl. Mech. Anal.* **55** (1974), pp. 18–51. 137
2. Asymptotic stability and the periodic solutions of $x(t) + \int_{-\infty}^{t} a(t - s)g(s, x(s))ds = f(t)$, *J. Math. Anal. Appl.* **66** (1978), pp. 606–625.

J. J. Levin

1. The asymptotic behavior of the solution of a Volterra equation, *Proc. Amer. Math. Soc.* **14** (1963), pp. 534–541. 559
2. The qualitative behavior of a nonlinear Volterra equation, *Proc. Amer. Math. Soc.* **16** (1965), pp. 711–718. 559
3. A nonlinear Volterra equation not of convolution type, *J. Differential Equations* **4** (1968), pp. 176–186. 648, 649
4. Boundedness and oscillation of some Volterra and delay equations, *J. Differential Equations* **5** (1969), pp. 369–398. 586
5. On a nonlinear Volterra equation, *J. Math. Anal. Appl.* **39** (1972), pp. 458–476. 648
6. A bound on the solutions of a Volterra equation, *Arch. Ratl. Mech. Anal.* **52** (1973), pp. 339–349. 448
7. On some geometric structures for integrodifferential equations, *Adv. in Math.* **22** (1976), pp. 146–186. 486, 612
8. Some a priori bounds for nonlinear Volterra equations, *SIAM J. Math. Anal.* **7** (1976), pp. 872–897.
9. Resolvents and bounds for linear and nonlinear Volterra equations, *Trans. Amer. Math. Soc.* **228** (1977), pp. 207–222. 649
10. Tauberian curves and integral equations, *J. Integral Equations* **2** (1980), pp. 57–91. 486
11. Nonlinearly perturbed Volterra equations, *Tôhoku Math. J. (2)* **32** (1980), pp. 317–335. 338
12. Synthesis of solutions of integral equations, *J. Integral Equations* **5** (1983), pp. 93–157. 486

J. J. Levin and J. A. Nohel

1. Note on a nonlinear Volterra equation, *Proc. Amer. Math. Soc.* **14** (1963), pp. 924–929. 559
2. On a nonlinear delay equation, *J. Math. Anal. Appl.* **8** (1964), pp. 31–44. 534
3. Perturbations of a nonlinear Volterra equation, *Michigan Math. J.* **12** (1965), pp. 431–447. 559
4. A system of nonlinear integrodifferential equations, *Michigan Math. J.* **13** (1966), pp. 257–270.

5. The integrodifferential equations of a class of nuclear reactors with delayed neutrons, *Arch. Ratl. Mech. Anal.* **31** (1968), pp. 151–172.

J. J. Levin and D. F. Shea

1. On the asymptotic behavior of the bounded solutions of some integral equations, I, II, III, *J. Math. Anal. Appl.* **37** (1972), pp. 42–82, 288–326, 537–575. 108, 486

B. W. Levinger

1. A folk theorem in functional differential equations, *J. Differential Equations* **4** (1968), pp. 612–619. 203

N. Levinson

1. A nonlinear Volterra equation arising in the theory of superfluidity, *J. Math. Anal. Appl.* **1** (1960), pp. 1–11. 10, 649

D. F. Lima

1. *Hopf Bifurcation in Equations with Infinite Delay*, Ph. D. Thesis, Brown University, Providence, R. I., 1977. 137

R. Ling

1. Asymptotic behavior of integral and integrodifferential equations, *J. Math. Phys.* **18** (1977), pp. 1574–1576. 5

S-O. Londen

1. On the asymptotic behavior of the solution of a nonlinear integrodifferential equation, *SIAM J. Math. Anal.* **2** (1971), pp. 356–367. 9
2. The qualitative behavior of the solutions of a nonlinear Volterra equation, *Michigan Math. J.* **18** (1971), pp. 321–330. 559*, 650*
3. Stability analysis of nonlinear point reactor kinetics, in *Adv. Nuclear Sci. Tech.* **6**, 1972, pp. 45–63. 9
4. On some nonlinear Volterra integrodifferential equations, *J. Differential Equations* **11** (1972), pp. 169–179. 9, 559*
5. On the solutions of a nonlinear Volterra equation, *J. Math. Anal. Appl.* **39** (1972), pp. 564–573. 559
6. On the asymptotic behavior of the solutions of a nonlinear Volterra equation, *Ann. Mat. Pura Appl. (IV)* **93** (1972), pp. 263–269.
7. On a nonlinear Volterra integral equation, *J. Differential Equations* **14** (1973), pp. 106–120. 559, 587, 650*
8. On the asymptotic behavior of the bounded solutions of a nonlinear Volterra equation, *SIAM J. Math. Anal.* **5** (1974), pp. 849–875.
9. On the variation of the solutions of a nonlinear integral equation, *J. Math. Anal. Appl.* **52** (1975), pp. 430–449. 611
10. On a Volterra integrodifferential equation with L^∞-perturbation and noncountable zero-set of the transformed kernel, *J. Integral Equations* **1** (1979), pp. 275–280. 611
11. L^2-solutions of some integral equations with positive definite kernels, *J. Math. Anal. Appl.* **78** (1980), pp. 455–465.
12. On an integral equation with L^∞-perturbation, *J. Integral Equations* **5** (1983), pp. 49–57. 612
13. On some integral equations with locally finite measures and L^∞-perturbations, *SIAM J. Math. Anal.* **14** (1983), pp. 1187–1203. 612
14. On some nonintegrable Volterra kernels with integrable resolvents including some applications to Riesz potentials, *J. Integral Equations* **10** (1985), pp. 241–289. 187

15. A Volterra equation with L^2-solutions, *SIAM J. Math. Anal.* **18** (1987), pp. 168–171. 560

S-O. Londen and O. J. Staffans

1. *Volterra Equations*, S-O. Londen and O. J. Staffans, eds., Proceedings, Otaniemi, Finland, 1978, Lecture Notes in Mathematics **737**, Springer-Verlag, Berlin, 1979.

L. H. Loomis

1. The spectral characterization of a class of almost periodic functions, *Ann. of Math.* **72** (1960), pp. 362–368. 487

S. M. V. Lunel

1. A sharp version of Henry's theorem on small solutions, *J. Differential Equations* **62** (1986), pp. 266–274.
2. *Exponential Type Calculus for Linear Delay Equations*, Ph. D. Thesis, Centrum voor Wiskunde en Informatica, Amsterdam, 1988. 203*

R. C. MacCamy

1. Nonlinear Volterra equations on a Hilbert space, *J. Differential Equations* **16** (1974), pp. 373–393.
2. Remarks on frequency domain methods for Volterra integral equations, *J. Math. Anal. Appl.* **55** (1976), pp. 555–575. 534, 559
3. Stability theorems for a class of functional differential equations, *SIAM J. Appl. Math.* **30** (1976), pp. 557–576. 650

R. C. MacCamy and J. S. W. Wong

1. Stability theorems for some functional equations, *Trans. Amer. Math. Soc.* **164** (1972), pp. 1–37. 534*, 559*
2. Exponential stability for a nonlinear functional differential equation, *J. Math. Anal. Appl.* **39** (1972), pp. 699–705.

A. G. J. MacFarlane

1. *Frequency-Response Methods in Control Systems*, A. G. J. MacFarlane, ed., Selected Reprint Series, IEEE Press, New York, 1979. 8, 74

H. Maeda

1. Stability considerations for a Volterra integral equation with discontinuous nonlinearity, *SIAM J. Control* **11** (1973), pp. 202–214. 380

P. Malliavin

1. Impossibilité de la synthèse spectrale sur les groupes Abéliens non compacts, *Publ. Math. Inst. Hautes Études Sci.* **2** (1959), pp. 61–68. 487

A. Manitius

1. Completeness and F-completeness of eigenfunctions associated with retarded functional differential equations, *J. Differential Equations* **35** (1980), pp. 1–29. 220

M. Marcus

1. *Introduction to Modern Algebra*, Marcel Dekker, New York, 1978. 228

M. Marcus and V. Mizel

1. *Limiting Equations for Problems Involving Long Range Memory*, Mem. Amer. Math. Soc. **278**, Amer. Math. Soc., Providence, R. I., 1983. 486

P. Marocco

1. A study of asymptotic behaviour and stability of the solutions of Volterra equations using topological degree, *J. Differential Equations* **43** (1982), pp. 235–248.

R. H. Martin

1. *Nonlinear Operators and Differential Equations in Banach Spaces*, J. Wiley, New York, 1976. 346, 403

J. L. Massera and J. J. Schäffer

1. *Linear Differential Equations and Function Spaces*, Academic Press, New York, 1966. 73, 277

M. Mathias

1. Über positive Fourier-Integrale, *Math. Zeit.* **16** (1923), pp. 103–125. 533

J. A. J. Metz and O. Diekmann

1. *The Dynamics of Physiologically Structured Populations*, J. A. J. Metz and O. Diekmann, eds., Lecture Notes in Biomathematics **68**, Springer-Verlag, Berlin, 1986. 6

R. K. Miller

1. Asymptotic behavior of nonlinear delay-differential equations, *J. Differential Equations* **1** (1965), pp. 293–305.
2. Asymptotic behavior of solutions of nonlinear Volterra equations, *Bull. Amer. Math. Soc.* **72** (1966), pp. 153–156.
3. On Volterra's population equation, *J. SIAM Appl. Math.* **14** (1966), pp. 446–452. 449, 610
4. On Volterra integral equations with nonnegative integrable resolvents, *J. Math. Anal. Appl.* **22** (1968), pp. 319–340. 165*, 186*, 649*
5. On the linearization of Volterra integral equations, *J. Math. Anal. Appl.* **23** (1968), pp. 198–208. 338
6. Admissibility and nonlinear Volterra integral equations, *Proc. Amer. Math. Soc.* **25** (1970), pp. 65–71. 338
7. Almost-periodic behavior of solutions of a nonlinear Volterra system, *Quart. Appl. Math.* **28** (1971), pp. 553–570. 10
8. Asymptotically almost periodic solutions of a nonlinear Volterra system, *SIAM J. Math. Anal.* **2** (1971), pp. 435–444. 10
9. Asymptotic stability properties of linear Volterra integrodifferential equations, *J. Differential Equations* **10** (1971), pp. 485–506. 108
10. *Nonlinear Volterra Integral Equations*, W. A. Benjamin, Menlo Park, Calif., 1971. 73, 278*, 338*, 339*, 379*, 380*, 422*, 423*, 448*, 485*, 586*, 649*
11. Asymptotic stability and perturbations for linear Volterra integrodifferential systems, in *Delay and Functional Differential Equations and their Applications*, Academic Press, New York, 1972, pp. 257–268. 108
12. A system of Volterra integral equations arising in the theory of superfluidity, *An. Ştiinţ. "Al. I. Cuza" Iaşi Secţ. I a Mat (N.S.)* **19** (1973), pp. 349–364. 10
13. Linear Volterra integrodifferential equations as semigroups, *Funkcial. Ekvac.* **17** (1974), pp. 39–55. 220
14. Structure of solutions of unstable linear Volterra integrodifferential equations, *J. Differential Equations* **15** (1974), pp. 129–157. 203
15. A system of renewal equations, *SIAM J. Appl. Math.* **29** (1975), pp. 20–34. 204

16. Some fundamental theory of Volterra integral equations, in *International Conference on Differential Equations, Los Angeles 1974*, H. A. Antosiewicz, ed., Academic Press, New York, 1975, pp. 568–579. 379

R. K. Miller and J. A. Nohel

1. A stable manifold theorem for a system of Volterra integro-differential equations, *SIAM J. Math. Anal.* **6** (1975), pp. 506–522. 203, 204, 339

R. K. Miller, J. A. Nohel, and J. S. W. Wong

1. Perturbations of Volterra integral equations, *J. Math. Anal. Appl.* **25** (1969), pp. 676–691. 278*, 279*, 338*, 649*
2. A stability theorem for nonlinear mixed integral equations, *J. Math. Anal. Appl.* **25** (1969), pp. 446–449.

R. K. Miller and G. R. Sell

1. Existence, uniqueness and continuity of solutions of integral equations, *Ann. Mat. Pura Appl. (IV)* **80** (1968), pp. 135–152. 379, 422
2. A note on Volterra integral equations and topological dynamics, *Bull. Amer. Math. Soc.* **74** (1968), pp. 904–908.
3. Existence, uniqueness and continuity of solutions of integral equations. An addendum, *Ann. Mat. Pura Appl. (IV)* **87** (1970), pp. 281–286. 379, 422
4. *Volterra Integral Equations and Topological Dynamics*, Mem. Amer. Math. Soc. **102**, Amer. Math. Soc., Providence, R. I., 1970. 220, 485
5. Topological dynamics and its relation to integral equations and nonautonomous systems, in *Dynamical Systems* **1**, L. Cesari, J. K. Hale and J. LaSalle, eds., Proc. Internat. Symp., Brown Univ., Providence, R. I., 1974, Academic Press, New York, 1976, pp. 233–249. 485

M. S. Mousa, R. K. Miller, and A. N. Michel

1. Stability analysis of hybrid composite dynamical systems: descriptions involving operators and differential equations, *IEEE Trans. Automat. Control* **31** (1986), pp. 216–226. 8

T. Naito

1. Adjoint equations of autonomous linear functional differential equations with infinite retardations, *Tôhoku Math. J. (2)* **28** (1976), pp. 135–143. 220
2. On autonomous linear functional differential equations with infinite retardations, *J. Differential Equations* **21** (1976), pp. 297–315. 137, 220
3. On linear autonomous retarded equations with an abstract phase space for infinite delay, *J. Differential Equations* **33** (1979), pp. 74–91. 137, 220

P. Ney

1. The asymptotic behavior of a Volterra-renewal equation, *Trans. Amer. Math. Soc.* **228** (1977), pp. 147–155. 204

J. A. Nohel

1. Asymptotic relationships between systems of Volterra equations, *Ann. Mat. Pura Appl. (IV)* **90** (1971), pp. 149–166. 338
2. Asymptotic equivalence of Volterra equations, *Ann. Mat. Pura Appl. (IV)* **96** (1973), pp. 339–347.
3. Nonlinear Volterra equations for heat flow in materials with memory, in *Integral and Functional Differential Equations*, T. L. Herdman, S. M. Rankin III, and H. W. Stech, eds., Lecture Notes in Pure and Applied Mathematics **67**, Marcel Dekker, New York, 1981, pp. 3–82. 13*, 647*, 649*

J. A. Nohel, R. C. Rogers, and A. E. Tzavaras

1. Weak solutions for a nonlinear system in viscoelasticity, *Comm. Partial Differential Equations* **13** (1988), pp. 97–127.

J. A. Nohel and D. F. Shea

1. Stability of a nonlinear Volterra equation, *Boll. Un. Mat. Ital. (4)* **11** (1975), pp. 498–510.
2. Frequency domain methods for Volterra equations, *Adv. in Math.* **22** (1976), pp. 278–304. 188, 533, 534*, 559*, 586*

R. Noren

1. A singular limit problem for a linear Volterra equation, *Quart. Appl. Math.* **46** (1988), pp. 169–179. 188
2. A singular limit problem for a Volterra equation, *SIAM J. Math. Anal.* **19** (1988), pp. 1103–1107. 188
3. Uniform L^1 behavior for the solution of a Volterra equation with a parameter, *SIAM J. Math. Anal.* **19** (1988), pp. 270–286. 188

H. Nyquist

1. Regeneration theory, *Bell System Tech. J.* **11** (1932), pp. 126–147. 74*

W. Okrasiński

1. On the existence and uniqueness of non-negative solutions of a certain non-linear convolution equation, *Ann. Polon. Math.* **36** (1979), pp. 61–72.
2. On a non-linear convolution equation occurring in the theory of water percolation, *Ann. Polon. Math.* **37** (1980), pp. 223–229.
3. Non-negative solutions of some nonlinear integral equations, *Ann. Polon. Math.* **44** (1984), pp. 209–218.

W. E. Olmstead and R. A. Handelsman

1. Asymptotic solution to a class of nonlinear Volterra integral equations. II, *SIAM J. Appl. Math.* **30** (1976), pp. 180–189.

K. Padmavally

1. On a non-linear integral equation, *J. Math. Mech.* **7** (1958), pp. 533–555. 649

R. E. A. C. Paley and N. Wiener

1. *Fourier Transforms in the Complex Domain*, Amer. Math. Soc., Providence, R. I., 1934. 73

A. Pazy

1. *Semigroups of Linear Operators and Applications to Partial Differential Equations*, Springer-Verlag, Berlin, 1983. 211, 213*

M. Z. Podowski

1. Nonlinear stability analysis for a class of differential-integral systems arising from nuclear reactor dynamics, *IEEE Trans. Automat. Control* **31** (1986), pp. 98–107. 9
2. A study of nuclear reactor models with nonlinear reactivity feedbacks: stability criteria and power overshoot evaluation, *IEEE Trans. Automat. Control* **31** (1986), pp. 108–115. 9

H. Pollard

1. The harmonic analysis of bounded functions, *Duke Math. J.* **20** (1953), pp. 499–512. 487

V. M. Popov

1. Sur certaines inégalités intégrales concernant la théorie du réglage automatique, *C. R. Acad. Sci. Paris* **256** (1963), pp. 3568–3570.
2. *Hyperstability of Control Systems*, Springer-Verlag, Berlin, 1973. 534, 535*, 559*, 560*, 586*
3. Dichotomy and stability by frequency-domain methods, *Proc. IEEE* **62** (1974), pp. 548–562. 586
4. Stability-spaces and frequency-domain conditions, in *Calculus of Variations and Control Theory*, D. Russell, ed., Academic Press, New York, 1976, pp. 371–390. 586
5. Applications of the saturability technique in the problem of stability of nonlinear systems, *Nonlinear Anal.* **1** (1977), pp. 571–581. 586*
6. Monotonicity and mutability, *J. Differential Equations* **31** (1979), pp. 337–358.

J. Prüss

1. On linear Volterra equations of parabolic type in Banach spaces, *Trans. Amer. Math. Soc.* **301** (1987), pp. 691–721.
2. Positivity and regularity of hyperbolic Volterra equations in Banach spaces, *Math. Ann.* **279** (1987), pp. 317–344. 649*
3. Bounded solutions of Volterra equations, *SIAM J. Math. Anal.* **19** (1988), pp. 133–149.

M. Reichert

1. Über die Fixpunktmengen einer Klasse Volterrascher Integraloperatoren in Banachräumen, *J. Reine Angew. Math.* **258** (1973), pp. 173–185. 422

H. Reiter

1. *Classical Harmonic Analysis and Locally Compact Groups*, Oxford University Press, Oxford, 1968. 486

M. Renardy, W. J. Hrusa, and J. A. Nohel

1. *Mathematical Problems in Viscoelasticity*, Longman Scientific & Technical and J. Wiley, Essex and New York, 1987. 10

G. E. H. Reuter

1. Über eine Volterrasche Integralgleichung mit totalmonotonem Kern, *Arch. Math.* **7** (1956), pp. 59–66. 165, 186

D. W. Reynolds

1. On linear singular Volterra integral equations of the second kind, *J. Math. Anal. Appl.* **103** (1984), pp. 230–262. 166, 278

I. Richards

1. On the disproof of spectral synthesis, *J. Combin. Theory* **2** (1967), pp. 61–70. 487

J. H. Roberts and W. R. Mann

1. On a certain nonlinear integral equation of the Volterra type, *Pacific J. Math.* **1** (1951), pp. 431–445. 649

W. Rudin

1. *Fourier Analysis on Groups*, Interscience, New York, 1967. 469, 486, 487
2. *Functional Analysis*, McGraw–Hill, New York, 1973. xiii, 54, 68*, 121*, 137*, 162*, 182*, 183*, 238*, 239*, 256*, 496*, 507*, 526*, 531*, 535*
3. *Real and Complex Analysis,* 3rd *ed.*, McGraw–Hill, New York, 1986. xiii, 49, 61, 64*, 66*, 73*, 90*, 91*, 92*, 93*, 95*, 96*, 108*, 137*, 145*, 151*, 442*, 599*

D. L. Russell

1. A Floquet decomposition for Volterra equations with periodic kernel and a transform approach to linear recursion equations, *J. Differential Equations* **68** (1987), pp. 41–71. 203, 279

D. Salamon

1. *Control and Observation of Neutral Systems*, Pitman, London, 1984. 203, 220

Š. Schwabik

1. Generalized Volterra integral equations, *Czechoslovak Math. J.* **32** (1982), pp. 245–270. 380

Š. Schwabik, M. Tvrdý, and O. Vejvoda

1. *Differential and Integral Equations. Boundary Value Problems and Adjoints*, Reidel, Dordrecht, 1979. 309

L. Schwartz

1. *Théorie des Distributions, nouv. éd.*, Hermann, Paris, 1966. 108, 496*, 514*, 515*, 522*, 531*, 533*, 534*, 535*

I. Sedig

1. A note on Lipschitz functions and spectral synthesis, *Ark. Mat.* **25** (1987), pp. 141–145. 487
2. Some results in spectral analysis and synthesis at infinity, *Math. Scand.*, to appear 1989. 487, 488

G. Seifert

1. Liapunov–Razumikhin conditions for asymptotic stability in functional differential equations of Volterra type, *J. Differential Equations* **16** (1974), pp. 289–297. 448
2. Almost periodic solutions for delay-differential equations with infinite delay, *J. Differential Equations* **41** (1981), pp. 416–425.

G. R. Sell

1. A Tauberian condition and skew product flows with applications to integral equations, *J. Math. Anal. Appl.* **43** (1973), pp. 388–396. 486

D. F. Shea and S. Wainger

1. Variants of the Wiener–Lévy theorem, with applications to stability problems for some Volterra integral equations, *American J. Math.* **97** (1975), pp. 312–343. 138, 169, 184, 186, 187

C. Shilepsky

1. A note on the asymptotic behavior of an integral equation, *Proc. Amer. Math. Soc.* **33** (1972), pp. 111–113. 610
2. The asymptotic behavior of an integral equation with an application to Volterra's population equation, *J. Math. Anal. Appl.* **48** (1974), pp. 764–779. 610

J. A. Shohat and J. D. Tamarkin

1. *The Problem of Moments*, Amer. Math. Soc., Providence, R. I., 1943. 165

H. Smith

1. Monotone semiflows generated by functional differential equations, *J. Differential Equations* **66** (1987), pp. 420–442. 423

M. C. Smith

1. On a nonlinear Volterra equation of nonconvolution type, *J. Differential Equations* **32** (1979), pp. 294–309.

F. Smithies

1. *Integral Equations*, Cambridge University Press, Cambridge, 1958. 277

O. J. Staffans

1. Nonlinear Volterra integral equations with positive definite kernels, *Proc. Amer. Math. Soc.* **51** (1975), pp. 103–108. 559
2. Positive definite measures with applications to a Volterra equation, *Trans. Amer. Math. Soc.* **218** (1976), pp. 219–237. 533, 534*, 559*
3. Tauberian theorems for a positive definite form, with applications to a Volterra equation, *Trans. Amer. Math. Soc.* **218** (1976), pp. 239–259. 486, 559
4. An inequality for positive definite Volterra kernels, *Proc. Amer. Math. Soc.* **58** (1976), pp. 205–210. 534, 559
5. On the asymptotic spectra of the bounded solutions of a nonlinear Volterra equation, *J. Differential Equations* **24** (1977), pp. 365–382. 487, 534, 559
6. Systems of nonlinear Volterra equations with positive definite kernels, *Trans. Amer. Math. Soc.* **228** (1977), pp. 99–116. 487, 534*, 560*
7. Boundedness and asymptotic behavior of solutions of a Volterra equation, *Michigan Math. J.* **24** (1977), pp. 77–95. 534*, 559*, 560*
8. On the holomorphic properties of the nonlinearity in a Volterra equation, *J. Math. Anal. Appl.* **64** (1978), pp. 48–60.
9. Some energy estimates for a nondifferentiated Volterra equation, *J. Differential Equations* **32** (1979), pp. 285–293. 560
10. On a nonlinear integral equation with a nonintegrable perturbation, *J. Integral Equations* **1** (1979), pp. 291–307. 611
11. A nonlinear Volterra equation with rapidly decaying solutions, *Trans. Amer. Math. Soc.* **258** (1980), pp. 523–530.
12. A Volterra equation with square integrable solution, *Proc. Amer. Math. Soc.* **78** (1980), pp. 213–217.
13. On asymptotically almost periodic solutions of a convolution equation, *Trans. Amer. Math. Soc.* **266** (1981), pp. 603–616.
14. A priori bounds for a discontinuous Volterra equation, *J. Integral Equations* **3** (1981), pp. 231–243.
15. A bound on the solutions of a nonlinear Volterra equation, *J. Math. Anal. Appl.* **83** (1981), pp. 127–134. 559

16. On a neutral functional differential equation in a fading memory space, *J. Differential Equations* **50** (1983), pp. 183–217. 137, 204
17. A neutral FDE with stable D-operator is retarded, *J. Differential Equations* **49** (1983), pp. 208–217.
18. The null space and the range of a convolution operator in a fading memory space, *Trans. Amer. Math. Soc.* **281** (1984), pp. 361–388. 137, 204
19. On the stability of a Volterra equation with a monotone nonlinearity, *J. Integral Equations* **7** (1984), pp. 239–248. 586*, 587*, 650*
20. A note on a Volterra equation with several nonlinearities, *J. Integral Equations* **7** (1984), pp. 249–252. 650*
21. Semigroups generated by a convolution equation, in *Infinite-Dimensional Systems.*, F. Kappel and W. Schappacher, eds., Proc. of the Conference on Operator Semigroups and Applications held in Retzhof, Austria, June 5–11, 1983, Lecture Notes in Mathematics **1076**, Springer-Verlag, Berlin, 1984, pp. 209–226. 221
22. Some well-posed functional equations which generate semigroups, *J. Differential Equations* **58** (1985), pp. 157–191. 221
23. On a nonconvolution Volterra resolvent, *J. Math. Anal. Appl.* **108** (1985), pp. 15–30. 278*
24. On the almost periodicity of solutions of an integrodifferential equation, *J. Integral Equations* **8** (1985), pp. 249–260.
25. Extended initial and forcing function semigroups generated by a functional equation, *SIAM J. Math. Anal.* **16** (1985), pp. 1034–1048. 137*, 216*, 221*
26. Semigroups generated by a neutral functional differential equation, *SIAM J. Math. Anal.* **17** (1986), pp. 46–57. 221
27. A direct Lyapunov approach to Volterra integrodifferential equations, *SIAM J. Math. Anal.* **19** (1988), pp. 879–901. 308, 310, 448, 649
28. On the stable and unstable subspaces of a critical functional differential equation, *J. Integral Equations Appl.*, to appear 1989. 204

H. W. Stech

1. On the adjoint theory for autonomous linear functional differential equations with unbounded delays, *J. Differential Equation* **27** (1978), pp. 421–443. 220*

H. Stech and M. Williams

1. Stability in a class of cyclic epidemic models with delay, *J. Math. Biol.* **11** (1981), pp. 95–103. 7

E. M. Stein

1. *Singular Integrals and Differentiability Properties of Functions*, Princeton University Press, Princeton, 1970. 277, 624

E. M. Stein and G. Weiss

1. *Introduction to Fourier Analysis on Euclidean Spaces*, Princeton University Press, Princeton, 1971. 68*, 72*, 231*, 454*

S. Sternberg

1. *Lectures on Differential Geometry*, Prentice Hall, Englewood Cliffs, N. J., 1964. 481, 488

A. Strauss

1. On a perturbed Volterra integral equation, *J. Math. Anal. Appl.* **30** (1970), pp. 564–575. 278, 279, 338

S. Szufla

1. Solutions sets of non-linear integral equations, *Funkcial. Ekvac.* **17** (1974), pp. 67–71. 422

E. C. Titchmarsh

1. *Introduction to the Theory of Fourier Integrals,* 2nd *ed.*, Oxford University Press, London, 1948. 89*, 534*

Z. B. Tsalyuk

1. Asymptotic properties of solutions of the regeneration equation, *Differential Equations* **6** (1970), pp. 852–854. 204

S. Vessella

1. Stability results for Abel equations, *J. Integral Equations* **9** (1985), pp. 125–134. 166

R. B. Vinter

1. On the evolution of the state of linear differential delay equations in M^2: properties of the generator, *J. Inst. Math. Appl.* **21** (1978), pp. 13–23. 220

V. Volterra

1. Sulla inversione degli integrali definiti, *R. C. Accad. Lincei (5)* **5** (1896), pp. 177–185. 72*
2. Sulla inversione degli integrali multipli, *R. C. Accad. Lincei (5)* **5** (1896), pp. 289–300. 72
3. Sull' inversione degli integrali definiti, *Atti Accad. Torino* **31** (1896), pp. 311–323, 400–408, 537–567, 693–708. 72
4. *Theory of Functionals and of Integral and Integro-Differential Equations*, Dover Publications, New York, 1959. 3, 72
5. *Leçons sur la Théorie Mathématique de la Lutte pour la Vie*, Gauthier–Villars, Paris, 1931. 5

E. Wagner

1. Über die Asymptotik der Lösungen linearer Volterrascher Integralgleichungen 2. Art vom Faltungstyp, *Beiträge Anal.* **11** (1978), pp. 165–183.
2. Zur Asymptotik der Lösungen linearer Volterrascher Integralgleichungssysteme zweiter Art vom Faltungstyp mit nichtnegativen Kernen, *Math. Nachr.* **90** (1979), pp. 173–187. 204

J. A. Walker

1. *Dynamical Systems and Evolution Equations*, Plenum Press, New York, 1980.

G. F. Webb

1. *Theory of Nonlinear Age-Dependent Population Dynamics*, Marcel Dekker, New York, 1985. 6

D. G. Weis

1. Note on a Volterra integro-differential equation system, *Proc. Amer. Math. Soc.* **45** (1974), pp. 214–216. 559

R. L. Wheeler

1. A note on systems of linear integrodifferential equations, *Proc. Amer. Math. Soc.* **35** (1972), pp. 477–482.

D. V. Widder

1. Necessary and sufficient conditions for the representation of a function as a Laplace integral, *Trans. Amer. Math. Soc.* **33** (1931), pp. 851–892. 165
2. *The Laplace Transform*, Princeton University Press, Princeton, 1946. 72, 108, 143, 144, 145, 150, 165

N. Wiener

1. On the representation of functions by trigonometrical integrals, *Math. Z.* **24** (1925), pp. 575–616. 73

N. Wiener and H. R. Pitt

1. On absolutely convergent Fourier–Stieltjes transforms, *Duke Math. J.* **4** (1938), pp. 420–436. 137

J. S. W. Wong and R. Wong

1. Asymptotic solutions of linear Volterra integral equations with singular kernels, *Trans. Amer. Math. Soc.* **189** (1974), pp. 185–200. 5, 188

A. Yanagiya

1. On some integrodifferential equations, *Math. Japon.* **29** (1984), pp. 295–308.

T. Yoneyama

1. On the $\frac{3}{2}$ stability theorem for one-dimensional delay-differential equations, *J. Math. Anal. Appl.* **125** (1987), pp. 161–173. 449

T. Yoneyama and J. Sugie

1. On the stability region of scalar delay-differential equations, *J. Math. Anal. Appl.* **134** (1988), pp. 408–425.

J. A. Yorke

1. Asymptotic stability for one dimensional differential-delay equations, *J. Differential Equations* **7** (1970), pp. 189–202. 448

A. C. Zaanen

1. *Linear Analysis*, North-Holland, Amsterdam, 1964. 277*

A. H. Zemanian

1. *Distribution Theory and Transform Analysis*, McGraw–Hill, New York, 1965. 533, 535

A. Zygmund

1. *Trigonometric Series, I–II,* 2nd *ed.*, Cambridge University Press, Cambridge, 1979. 57*

INDEX

Numbers following entries are page numbers; numbers in brackets refer to the theorem, lemma, etc. where the index word appears. Boldface type indicates that the entry is defined or explained.

Observe that a large number of the entries are grouped under certain key words, such as *differential resolvent*, *kernel*, *resolvent*, *solution of a (non)linear equation*, and *spectrum*.